2nd EDITION

GRAPHICS FOR ENGINEERS
VISUALIZATION, COMMUNICATION, AND DESIGN

JERRY S. DOBROVOLNY
Professor and Head of the Department of General Engineering

DAVID C. O'BRYANT
Associate Professor and Assistant Head of the Department of General Engineering

UNIVERSITY OF ILLINOIS AT URBANA-CHAMPAIGN

JOHN WILEY & SONS
NEW YORK   CHICHESTER   BRISBANE   TORONTO   SINGAPORE

Copyright © 1984, by John Wiley & Sons, Inc.

All rights reserved. Published simultaneously in Canada.

Reproduction or translation of any part of
this work beyond that permitted by Sections
107 and 108 of the 1976 United States Copyright
Act without the permission of the copyright
owner is unlawful. Requests for permission
or further information should be addressed to
the Permissions Department, John Wiley & Sons.

Library of Congress Cataloging in Publication Data:

Dobrovolny, Jerry S.
  Graphics for engineers.

    Rev. ed. of: Graphics for engineers/by Randolph P. Hoelscher.
    Includes index.
    1. Engineering graphics.   I. O'Bryant, David C.
II. Hoelscher, Randolph P. (Randolph Philip), 1890–
Graphics for engineers.   III. Title.
T353.D63 1984   604.2   84-3547
ISBN 0-471-87124-9

Printed in the United States of America

10 9 8 7 6 5 4 3 2 1

PREFACE

Engineers, designers, drafters, and technicians must work together to build the many products and projects, both large and small, that are necessary for our modern life. Teamwork and an intercommunication of ideas are essential in this task.

The language used in engineering drawing, together with supporting graphical systems for computation and representation, must be thoroughly understood by all who carry on the work of engineering. This book has been designed so as to enable the student who plans to enter the field of engineering or production can learn this language and use it proficiently. It contains the latest standards that have been nationally and internationally accepted by the various national standardization bodies. Therefore, it serves as an excellent reference book in engineering offices.

This is the latest of a series of books based on the successful organization and teaching methods of a department that has been in existence for over 80 years. During this time, it has had consecutive overlapping leadership of experienced individuals, all of them graduate engineers. It represents a philosophy of teaching in engineering drawing and graphic sciences that is in harmony with the current trend in engineering education practice. The same sound and valid objectives that guided the preparation of first edition of *Graphics for Engineers* by Hoelscher, Springer, and Dobrovolny have been followed in this edition.

The current edition has been somewhat shortened to bring it more into line with subject matter being taught in college engineering graphics courses. The material in the chapters on dimensioning has been revised to include the use of the metric system. Since almost all schools use some type of problem workbooks in teaching graphics, the problem sets after each chapter have been omitted. The major objectives of the book are as follows:

1. To provide a textbook so clear in its verbal discussion and pictorial illustration that it can be readily understood by the student. To accomplish this, more step-by-step illustrations and explanations have been given.
2. To make a textbook that presents the best in drafting practice but emphasizes the development of the reasoning process in its theoretical discussion rather than manual skill.
3. To stress the fundamentals of descriptive geometry in covering the work in a basic graphics course.
4. To present the material in such a way that it will stimulate creative imagination, develop visual perception in three dimensions, and promote original thinking, useful in engineering design.
5. To eliminate all material that is not essential for engineering students or useful in a reference work on drawing for practicing engineers. Thus, it serves as both a learning and reference material.
6. To adhere rigorously to third-quadrant projection, not only for the three principal views but also for auxiliary views, which is the basis for the unification of drawing practice among the United States, Great Britain, and Canada. Thus, the plane of projection is always between the object and the viewer. Consequently, the reference line, if used, is always between the adjacent views.

7. To facilitate understanding of the material by a strategic use of color. In some chapters color is used not only for distinguishing planes of projection but also for the more functional purpose of aiding the student in differentiating between auxiliary planes, cutting planes, and other surfaces. In the chapter "Auxiliary Views," reference lines between the various views and the projecting lines to them are also drawn in color. This makes the construction for each view stand out clearly.

The chapter "Auxiliary Views" has been expanded to include all of the basic topics in descriptive geometry. This enables a rigorous treatment of the theory of projection in solving spatial visualization problems.

The chapters "Sectional Views," "Fasteners," and "Shop Terms and Processes" have been placed ahead of "Basic Dimensioning," since the information in those chapters is necessary for the shape description on any engineering part before size dimensions can be applied. Although this order seems logical, each chapter is self-contained, and the instructor can change the assignment schedule to suit any objectives.

On the subject of dimensioning, two chapters have been presented. This is the area in which the greatest advancement has been made in recent years. The first chapter, "Basic Dimensioning," covers all of the material needed by students in such fields as civil engineering and architecture and in general engineering practice. The second chapter, "Production Dimensioning," is quite essential for students who will go into mass-production industries, such as the automotive industries, aeronautics, and space vehicles.

In the chapters on axonometric and oblique projections, conventional methods of construction have been placed first. The exact theory of projection has been presented at the end of each chapter for those who wish to give a more rigorous course. Considerable proficiency in making these two types of pictorial drawings can be attained by the conventional methods. Many schools will not have time for more thorough study.

To meet the problem situation we have provided 15 workbooks, with the aid of staff members at the University of Illinois, to cover a variety of courses. These are available from the Stipes Publishing Company, 10 Chester Street, Champaign, Illinois 61820. At present there are

1. Five workbooks for straight drawing courses, designated Series A, B, C, D, and E.
2. Two workbooks for straight descriptive geometry courses, designated Series 1 and 2.
3. Seven workbooks for combination courses, designated Series 12, 13, 15, 16, 31, 32, and 33 (Series 31, 32, and 33 are metric).
4. One workbook for advanced drawing and geometry courses with application to design, designated Series 22.

Full-scale solution files are available to instructors for workbooks, Series A, B, C, D, and E; Series 1 and 2; and Series 12, 13, 15, 16, 22, 31, 32, and 33. Workbooks save a great deal of the student's time since the layout for the solution is printed and the student can forego the labor of making it.

ACKNOWLEDGMENTS

We are indebted to several of our colleagues for valuable assistance in the development of this book. Chapter 20, "Computer Graphics and CAD/CAM," was prepared by Professor Michael H. Pleck. We wish to thank Tadeusz Kaczor for his assistance in preparing illustrations and Mrs. Marilyn Butler for typing the manuscript. We express our appreciation to each of these persons for their contribution to the usefulness of this book. Our thanks go also to Professors H. H. Jordan, R. P. Hoelscher, and C. H. Springer, the originators of a long series of books of which this one is the culmination. We also express our appreciation for the many valuable suggestions and criticisms received from members of the staff of the University of Illinois, both at Urbana and at the Chicago campus. We would also like to thank the following reviewers and questionnaire respondents: Professor David Carlson, Michigan Technological University; Professor Norman Powers, University of Wisconsin; Professor Lowell K. Dirksen, Washburn University; Professor Moustafa R. Mustafa, Old Dominion University; and Dr. Michael Wozny, Rensselaer Polytechnic University. To these individuals and many others at other schools we express our gratitude. Many industrial concerns have contributed drawings for illustration purposes, for which credit is given in the appropriate places.

Urbana, Illinois *Jerry S. Dobrovolny*
January 1984 *David C. O'Bryant*

CONTENTS

1 AN INTRODUCTION TO ENGINEERING GRAPHICS   1
2 LETTERING   13
3 USE AND CARE OF INSTRUMENTS   31
4 GEOMETRIC CONSTRUCTION   51
5 SKETCHING   85
6 ORTHOGRAPHIC PROJECTION   101
7 AUXILIARY VIEWS   135
8 SECTIONAL VIEWS   193
9 FASTENERS   213
10 SHOP TERMS AND PROCESSES   245
11 BASIC DIMENSIONING   277
12 PRODUCTION DIMENSIONING   303
13 AXONOMETRIC PROJECTION   351
14 OBLIQUE PROJECTION   377
15 PERSPECTIVE   393
16 CHARTS AND DIAGRAMS   423
17 GRAPHIC VECTOR ANALYSIS   439
18 INTERSECTIONS AND DEVELOPMENTS   457
19 MAP DRAWING   499
20 COMPUTER GRAPHICS AND CAD/CAM   519
APPENDIX   557
INDEX   607

CHAPTER 1
AN INTRODUCTION TO ENGINEERING GRAPHICS

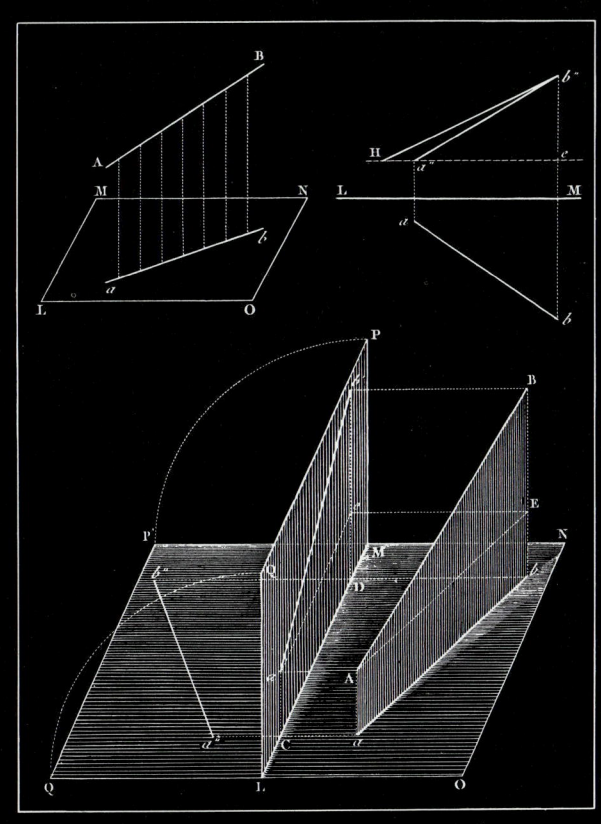

CHAPTER 1

1.1 ORIGIN OF PROJECTION DRAWING

The representation of three-dimensional objects on two-dimensional surfaces by means of geometric drawings, such as plans and elevations, has involved a gradual change. Through the centuries drawings developed from crude pictorial of prehistoric man, such as that in Fig. 1.1, through a period of highly artistic drawings to the present well-developed types of industrial drawing. In Mesopotamia, maps were made on clay tablets such as that shown in Fig. 1.2. In the early Middle Ages, most of the construction work was concerned with buildings. During the Roman period, drawings of building plans were made before construction was undertaken. Frequently construction problems were worked out by the mason or builder from general specifications as the work progressed. However, very few examples of these drawings have been preserved.

One early Egyptian drawing made on papyrus that shows two views of a shrine without dimensions has been found. Pictorial drawings seemed to have been quite common during the Middle Ages. See Fig. 1.3.

Two elevations are in existence of the west front of the Cathedral of Orvieto, supposed to have been made by Lorenzo del Maitano of Siena soon after 1310. These are not true front views, since each is in slight perspective. By the end of the fifteenth century, there were draftsmen who could make true elevations. One of the earliest examples of the use of plan and elevation is

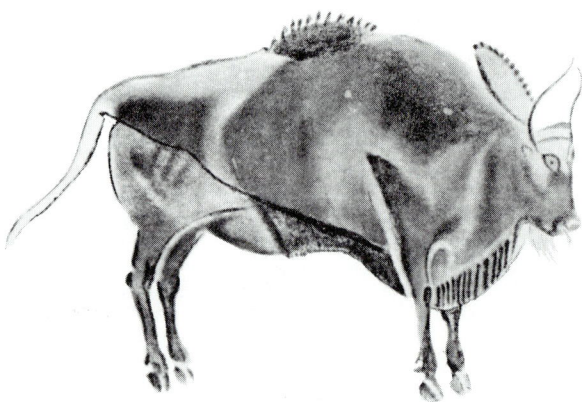

FIG. 1.1. Late Palaeolithic representation of a bison (from Singer, Holmyard, and Hall, *A History of Technology,* Vol. 1, Oxford University Press, London/New York, 1954).

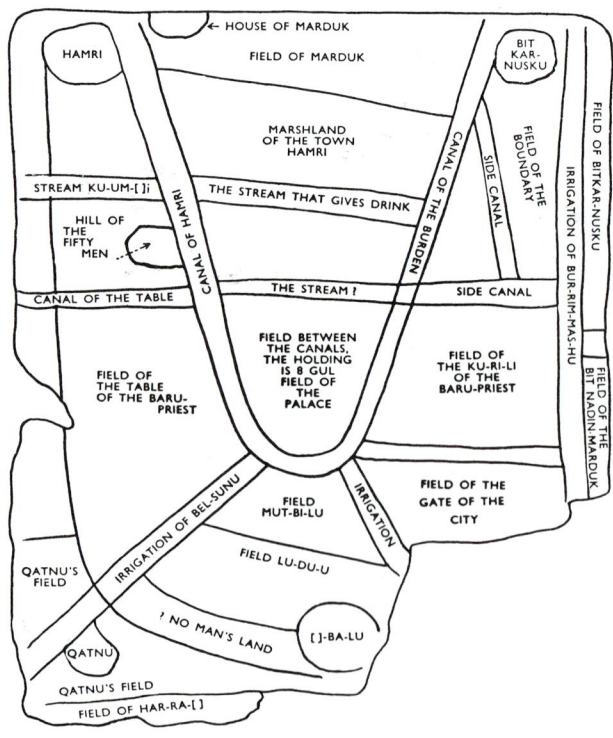

FIG. 1.2. Map of fields and canals near Nippur (from Singer, Holmyard, and Hall, *A History of Technology,* Vol. 1, Oxford University Press, London/New York, 1954).

included in an album of drawings in the Vatican Library, drawn by Giuliano da Sangallo. The date on the title page is 1465, but the book was not actually completed until 1490.

The drawings of early architects and engineers contained the basic idea of the theory that was to be developed into our modern forms of geometric projection. The system of right-angle projection on planes set up perpendicular to each other was first completely worked out by Gaspard Monge in the eighteenth century and was used to solve geometric problems. See Fig. 1.4. The work of Monge is the basis of descriptive geometry, which is the fundamental theory underlying all modern industrial drawing.

1.2 FUNCTION OR PURPOSE OF ENGINEERING DRAWING

In order that the student may understand the reason for studying engineering graphics, this chapter gives a brief overall view of the scope of the subject and its place in engineering practice.

The ideas for all the works of man are conceived in some person's mind. These ideas may require extensive computations. But they do not end there. Inevitably what is sometimes referred to as *hardware* must be produced. This hardware can rarely be produced without drawings. Even though computers are being designed to make drawings, some kind of preliminary drawing or layout is necessary to program the computer.

Some of the more common uses of drawings are stated in the following paragraphs.

1.2.1 Design of Machines. In designing machines and other structures, the engineer or technician must first form a clear mental picture of the thing to be made. He/she must then convey this idea to others. This cannot be done by the written word alone, but it can be done by drawings combined with verbal instructions or specifications.

1.2.2 Fabrication Details. Most machines require a number of separate parts. These parts

FIG. 1.3. First stages in the excavation of a mine (from Agricola, *De Re Metallica*, 1556).

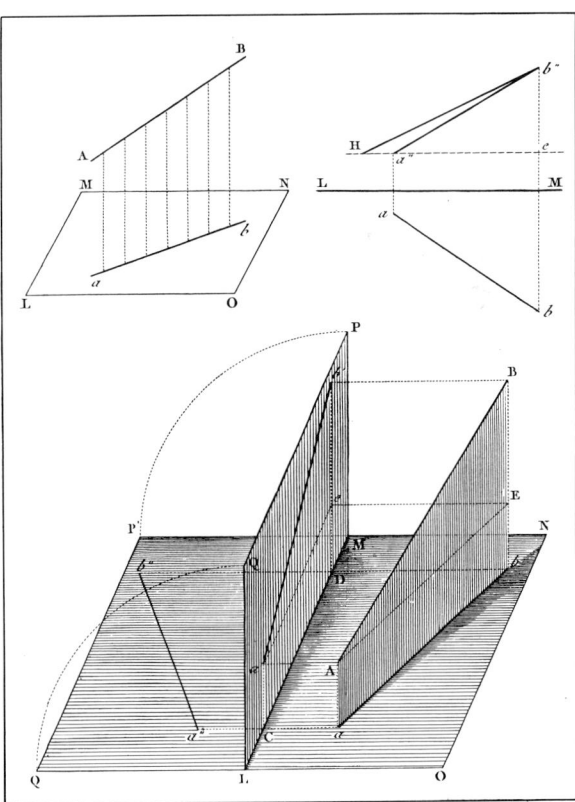

FIG. 1.4. Typical illustration from *Descriptive Geometrie* by Gaspard Monge.

are then assembled, either permanently or in a definite position, as moving parts. In either event, the parts must fit together exactly. This requires that the size and location of parts be held to very close tolerances. See Chapter 12.

1.2.3 Design of Buildings. The basic drawings for buildings consist of floor plans for each story, elevations for each side, and numerous sectional views and large-scale detail drawings of various parts. Even a relatively simple home will require a considerable amount of such drawings.

1.2.4 Detail Drawings. Larger structures will require separate sets of steel or reinforced concrete framing plans, plumbing plans, heating and ventilating plans, as well as drawings, for elevators, electrical wiring, and any special equipment that may be involved.

1.2.5 Cost Estimates. Before any contracts for construction or manufacturing can be made, cost estimates must be prepared. Such estimates are made from drawings and specifications. The drawing must delineate the exact size and shape of all parts. The kind and quality of materials to be used must be clearly stated.

1.2.6 Highway and Railroad Construction Plans. In order to construct highways or railroads, maps must be made showing the exact location of the center line of the road, the "right of way" lines, and the boundary lines of adjacent property.

In addition to the plan, a profile showing the elevation of the ground along the center line must be made. This drawing also shows the grade line for the finished road. At frequent intervals along the road, cross-sectional views must be made showing the "cut" or "fill" required. See Chapter 19.

Other details such as drainage, culverts, bridges, and viaducts must also be supplied before any contracts for construction can be entered into.

1.2.7 Department of Defense. Drawings for military equipment must be made to the specification and standards required by the Department of Defense. Thousands of such drawings are made annually. Thus, a battleship or submarine will require 40,000 to 50,000 drawings.

1.2.8 Aerospace Applications. A very large amount of complex mathematics is necessary to place a satellite in orbit. Many sciences are also involved, but nothing happens until "hardware" is designed and produced. Satellites, missiles, and space platforms require thousands of drawings. Some of these parts involve close and very accurate tolerances in order to function properly.

1.2.9 Sales Presentations. In many projects, companies will submit *proposals* as a part of the sales promotion. These are very frequently printed booklets describing in detail what the company proposes to furnish or supply. The proposals are illustrated with many kinds of drawings including those of a pictorial character such as isometric, oblique, or perspective.

1.3 MEANING OF TERMS

In the area of graphics, a number of terms have been used, such as *mechanical drawing, engineering drawing, technical drawing, engineering graphics,* and *graphic science.* Many of these terms are used loosely to mean the same thing. Some are misnomers. To clarify the meanings of terms as used in this book, the following paragraphs explain the terms and the contents of the field of drawing as used in industry and science.

1.3.1 Instrumental and Freehand Drawing. The older term *mechanical drawing* is really a misnomer since it seems to imply the drawing of mechanical things. It really means drawings made with instruments or instrumental drawings, as contrasted with freehand drawing, which is done without the aid of instruments. Hence, with the advent of computer graphics we have three comprehensive types of drawings: instrumental, freehand, and computer generated. Drawings of any kind may be made by either of the three methods, although it should be quite clear that where very accurate line work is required, instruments or a computer must be used.

1.3.2 Engineering Graphics. This is the most inclusive term now applied to drawings of all varieties made with pen or pencil, except those that may be classified as pure art. It ranges from the simple three-view drawing of a machine part to the most complex graphical layout of nomographs or graphical calculus computations.

This extensive group of drawings divides itself quite naturally into two major groups: *projection drawings,* or drawings based on a geometrical theory of projection, and *nonprojection drawings,* having a fundamental basis of algebraic mathematics. An outline of these two major categories is given in the paragraphs that follow.

1.4 PROJECTION DRAWING

This type includes drawings based on a fundamental geometric theory of projection. It includes the major portion of drawings made in industry. Some kinds of projection drawing, particularly in the pictorial area, may be made by "rule of thumb," but the underlying theoretical construction on which all rules are based is orthographic projection. *This term simply means that the projecting lines from the object to the plane of projection are at right angles to that plane.*

1.4.1 Orthographic Multiview Drawing. Drawings of this variety involve one, two, three, or more views based on right-angle projection. The planes on which the views are made are also at right angles to each other. This is the type most commonly used in industry. Fig. 1.5 shows such a three-view drawing. For a complete discussion see Chapter 6.

1.4.2 Pictorial Projections. These drawings show three faces of an object in one view. Since they look like photographs, they are called *pictorial drawings*. Many types of pictorial drawings can be made by rule-of-thumb methods. In all cases these methods are based on orthographic projection theory. The fact that the underlying basis for all forms of projection is orthographic projection should become clear to the student as he/she proceeds through the text.

1.4.2.1 Axonometric Projection. All axonometric projections are purely orthographic. Three types of axonometric drawings are possible, depending on the position of the object relative to the plane of projection.

a. ***Isometric projection*** results when the plane of projection makes equal angles with the three principal faces of the object or when the axes of the object make equal angles with the plane of projection. See Fig. 1.6a. For a complete discussion see Chapter 13.

b. ***Dimetric projection*** results when two axes of the object make equal angles with the plane of projection and the third axis has a different value. This type of drawing gives a more pleasing appearance, as shown in Fig. 1.6b. See also Chapter 13.

c. ***Trimetric projection*** is produced when the three axes of the object each make different angles with the plane of projection. These should always be made in true projection because rule-of-thumb methods are too slow. See Fig. 1.6c and Chapter 13.

1.4.2.2 Oblique Projection. An oblique projection results when parallel projecting lines make an angle other than 90° with the plane of projection. Obviously, many types of oblique projection can be made. A few of these have been given special names.

a. ***Cavalier Projection.*** When the projecting lines make an angle of 45° with the plane of projection, the drawing is called a *Cavalier projection* or drawing. An example is shown in Fig. 1.7a. For further details see Chapter 14.

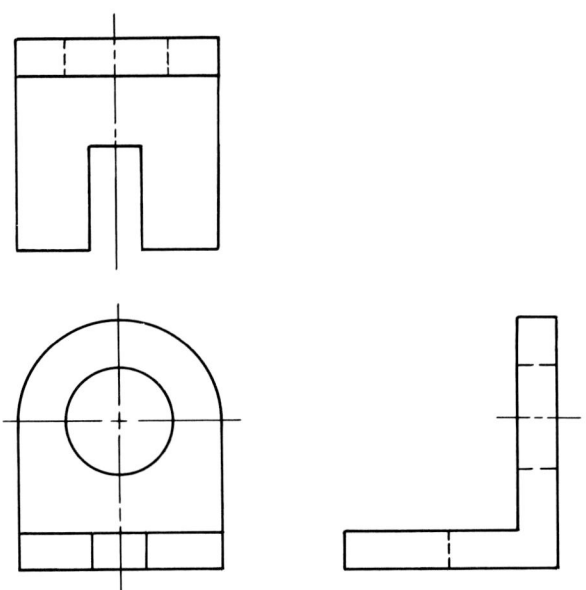

FIG. 1.5. Three-view orthographic projection.

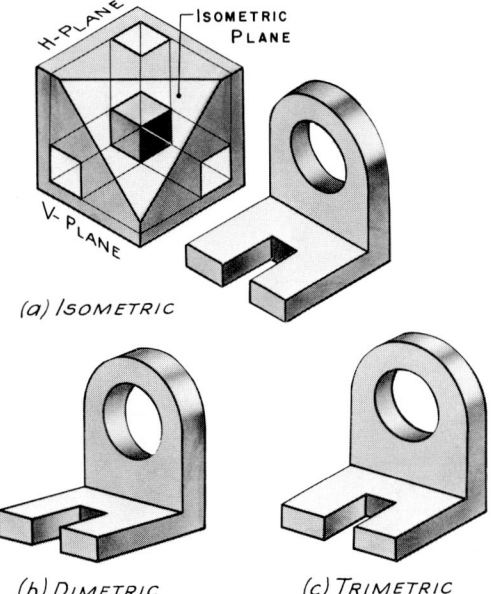

FIG. 1.6. Axonometric projection.

6 AN INTRODUCTION TO ENGINEERING GRAPHICS

b. **Cabinet Projection.** When the angle that the projecting lines make with the plane of projection is such that the scale on the receding axis in the drawing is just one half as long as the other two axis, the result is called a *Cabinet drawing.* See Fig. 1.7b and also Chapter 14.

c. **Clinographic Projection.** In Cavalier and Cabinet projection, the principal face of the object is made parallel to the plane of projection. For some purposes, however, it may be desirable to turn the object at an angle with the plane of projection. In the fields of minerology and crystallography such a system is used. The angles are shown in Fig. 1.8. For further discussion see Chapter 14.

1.4.2.3 Perspective. When the projecting lines converge to a point the drawing is called a *perspective.* Three kinds of perspectives may be made to serve different purposes. If the principal face of the object is parallel to the plane of projection and there is only one vanishing point, the drawing is called a *parallel* or *one-point perspective.* See Fig. 1.9a.

If two faces of the object are at an angle with the plane of projection, while the third face is perpendicular to it, two principal vanishing points occur and the drawing is called an *angular* or *two-point perspective.* See Fig. 1.9b.

When all three principal faces of the object are inclined to the plane of projection, the drawing is called an *oblique* or *three-point perspective.* See Fig. 1.9c. For further discussion see Chapter 15.

1.4.3 Descriptive Geometry. This is the science of orthographic projection theory underlying all types of projection. It may be noted here that the geometric construction for the kinds of projection mentioned in the preceding paragraphs is based on orthographic projection. The study of descriptive geometry provides a method for solving the problems relation to points, lines, planes, and other surfaces in space. Figure 1.10 illustrates a problem of this kind. See Chapter 7.

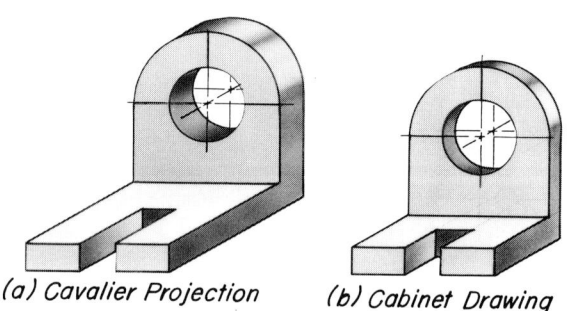

(a) Cavalier Projection   (b) Cabinet Drawing

FIG. 1.7. Oblique projection.

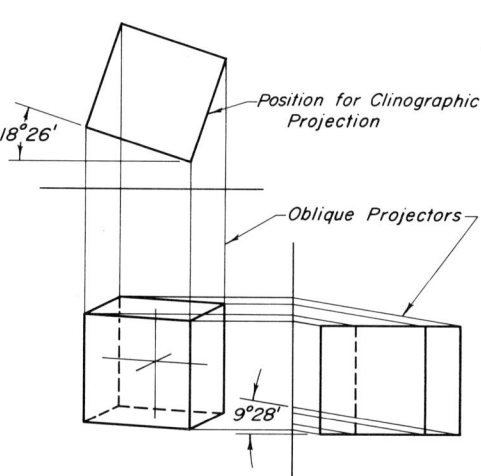

FIG. 1.8. Clinographic projection of a cube.

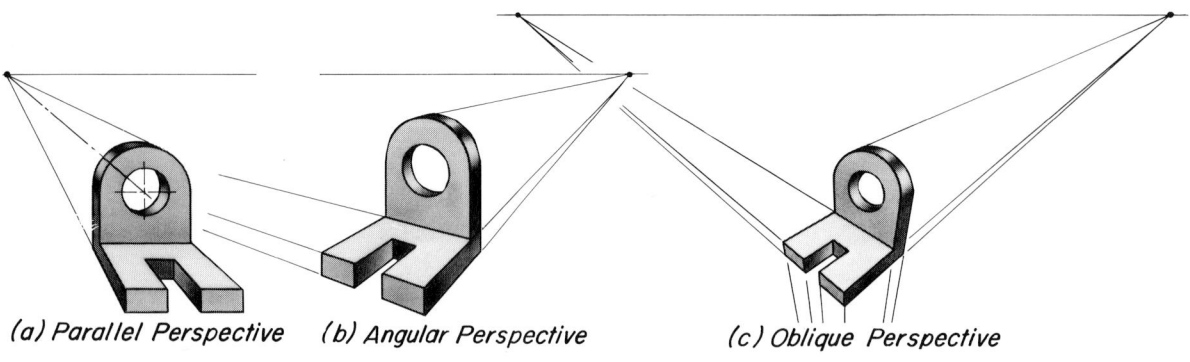

(a) Parallel Perspective   (b) Angular Perspective   (c) Oblique Perspective

FIG. 1.9. Types of perspective.

1.5 SUMMARY   7

1.4.3.1 Space Relationships. Descriptive geometry can be used to solve the relationship between parts of an object, such as the distance from a point to a plane or the true shape of the face of an object. Clearance problems such as the relationship between moving parts or the access to bolts and screws can best be solved by descriptive geometry.

1.4.3.2 Problems of Motion. By means of kinematics, a graphical method closely related to descriptive geometry, the problems of motion can be readily solved on the drawing board. For example, the need to change the rotary motion of a motor to reciprocating linear motion at various speeds and accelerations is a common one.

1.5 SUMMARY

The kinds of projection discussed in this chapter, the relationships of the geometric elements entering into each type of projection, and the classification and name of each special form of projection are summarized in Table 1.1. Each type is discussed in detail in later chapters.

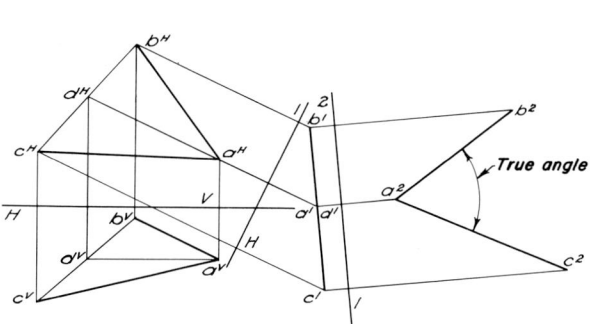

FIG. 1.10. Descriptive geometry.

TABLE 1.1 TYPES OF PROJECTIONS

Major Classifications	Subdivision	Number of Planes of Projection	Relation of Lines of Sight to Plane of Projection	Relation of Lines of Sight to Each Other	Location of Point of Sight in Relation to Plane of Projection	Position of Enclosing Cube with Relation to Plane of Projection
Orthographic	Multiple view	As many as necessary	Perpendicular	Parallel	Infinite distance	Faces parallel to planes
	Isometric	One	Perpendicular	Parallel	Infinite distance	Faces equally inclined to plane
	Dimetric	One	Perpendicular	Parallel	Infinite distance	Two faces equally inclined to plane
	Trimetric	One	Perpendicular	Parallel	Infinite distance	Three faces at different angles to plane
Oblique	Cavalier	One	45°	Parallel	Infinite distance	Principal face parallel to plane
	Cabinet	One	63°26′	Parallel	Infinite distance	Principal face parallel to plane
	Clinographic	One	80°32′	Parallel	Infinite distance	Principal face at angle of 18°26′
	General	One	Any angle except those above	Parallel	Infinite distance	Principal face parallel to plane
Perspective	One-point	One	Various angles	Converge at a point	Finite distance	Principal face parallel to plane
	Two-point	One	Various angles	Converge at a point	Finite distance	Two faces inclined and one face perpendicular to plane
	Three-point	One	Various angles	Converge at a point	Finite distance	Three faces inclined to plane

1.6 SHOP AND CONSTRUCTION METHODS

Before one can proceed very far in their profession, the young engineer or technician must become familiar with methods of manufacture in the shop or construction in the field. It is clearly a waste of time and money to make a drawing, however beautiful and accurate, of a part that cannot be made economically in the shop. Since shop courses have almost disappeared from engineering education, the technical students as a part of their training in graphic expression should learn the simpler fundamentals of shop and construction methods as presented in several chapters of this book.

1.7 DIMENSIONING

A shape description of a part shown by one or more projected views is totally inadequate unless dimensions are given showing the exact size of each part and the location of parts in relation to each other. It is this phase of drawing that requires the greatest care and study. Engineers are giving a great deal of attention to the standardization of dimensioning so that there may be no ambiguity. The dimensioning practice given in this book represents the most recently adopted American standards, as set forth in the latest ANSI Y14-5M standard. The letters ANSI mean American National Standards Institute.

1.8 PROFESSIONAL ASPECTS OF DRAWING

The young person who plans to enter a technical industry will frequently find his/her work in one of the specialized fields, such as civil engineering, mechanical engineering, or architecture. In the larger industries and in research institutions the work tends to cross over professional lines. In any event, the fundamental principles of drawings as explained in this book will be the same in all these areas. After training in basic drawing, the student may wish to go further in specialized fields. The difference between these areas of special drawing practices lies in the symbols, conventional practices, standards, and dimensioning methods used in each area.

1.9 STANDARDIZATION OF DRAFTING

In order to establish drafting practice throughout the country on a firm basis so that drawings made anywhere will have clear and unmistakable meaning everywhere, engineers have established standards in drafting. Those standards were set up first in industrial companies, later in engineering societies, and finally in a collective

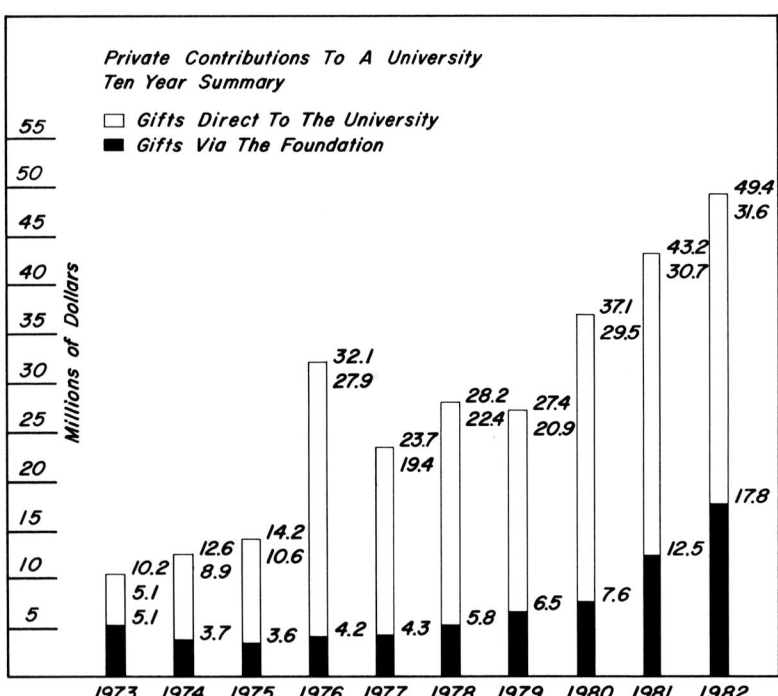

FIG. 1.11. Bar diagram.

effort through the American Standards Association. The drafting procedures shown and recommended in this book reflect the latest revision of these American standards, which are in general agreement with those of Great Britian and Canada.

The student should become familiar with the work of ANSI and that of the professional engineering society with which he is most concerned, for it is only through this standardization that modern mass production is possible.

1.10 NONPROJECTION DRAWING

The following phases of engineering graphics are algebraic in character, rather than geometric, except for certain parts of the first class. The theory of projection is not involved in any of the others.

1.10.1 Charts and Diagrams. These are drawings used to show graphically the relationship between facts. Two variables are usually involved, as shown in Fig. 1.11. A chart involving three variables in pictorial form is shown in Fig. 1.12. Three-dimensional charts of this kind follow the rules for axonometric projection. A three-variable chart may also be made as a plane figure instead of being pictorial in form. In this situation the third variable is shown by a series of curves.

1.10.2 Vector Diagrams. Problems involving the relationship between quantities that have direction as well as size may be solved graphically as shown in Fig. 1.13. This is a stress diagram for a roof truss. The graphical method is rapid, sufficiently accurate for all practical purposes, and self-checking.

1.11 LEGAL ASPECTS

The drawings for buildings, bridges, dams, and other major construction projects have always been the basis for legal contracts in which the builder agrees to erect the structure in accordance with the plans prepared by the engineer for the owner. Such drawings must always be clear and unmistakable in meaning. The drawing, in fact, becomes a legal document. If it is subject to more than one interpretation, litigation may arise causing unnecessary delay and expense.

In the machine industry, modern mass production has brought with it the letting of contracts for the manufacture of machine parts in large quantity. Machine parts, for interchangeable assembly, must be finished very accurately to the dimensions specified. The engineer can seldom go into his/her own shop and tell the supervisor what he/she wants since the part may

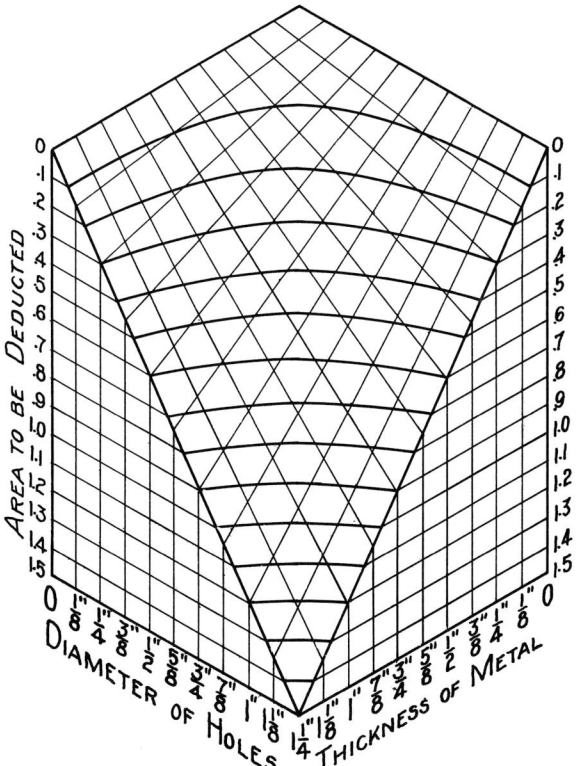

FIG. 1.12. Three-variable chart in pictorial form.

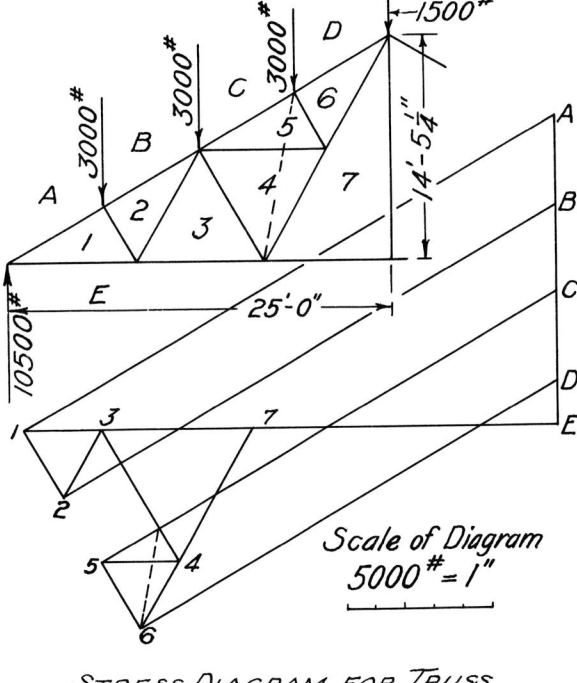

FIG. 1.13. Vector diagram.

be produced in a plant hundreds of miles away. The drawing itself must tell the whole story. It must be made not merely so that it can be understood, but it must be so clear in its meaning that it cannot be misunderstood or misinterpreted by either accident or intention.

This places on the technical person who makes the drawing or directs the work of producing it a heavy responsibility to understand thoroughly both the fundamental theory and the conventional practice of drafting that one studies in engineering graphics courses.

1.12 ENGINEERING DESIGN

Engineering design is the process of creating something new. It may be a new product to be manufactured and sold. It may be a new component that is a part of an old product that will make that product perform more efficiently. Or it may be a new system of some kind, such as the complete water and sanitary system of a large building, that involves new ideas. Essentially, engineering design is a problem-solving process. This may involve mathematical and graphical methods. Ingenuity and imagination are essential in the development of an engineering design. It also involves a familiarity with materials of all kinds, as well as new processes in computation and production.

1.13 ENGINEERING METHOD

The term *engineering method* refers to the manner in which engineers solve the problems presented to them. There is no single standard set of steps that the engineer must follow in solving all problems, but certain broad principles are usually employed in a manner and sequence that are chosen to fit the problem to be solved. A list of these general principles has been outlined in the following paragraphs.

1.13.1 Statement of the Problem. The problem may be clearly stated in the contract that an engineer signs or in the instructions that are given by an employer. If such a statement is not made, the engineer must formulate the problem mentally with a clear vision of what the end product of the work is to be. This may or may not be "spelled out" verbally.

1.13.2 Scientific Principles. After the problem has been clearly stated, the next step is to identify the laws of physical science that might apply to the problem. This is done by a careful analysis, breaking the problem down into its component parts and setting up the necessary mathematical equations that apply to the problem.

1.13.3 Proposed Solution. Most engineering problems have more than one solution. In the solution stage of the engineering method, it is necessary to make some assumptions, determine the limiting conditions of the problem, and make the calculations that will produce the answer. Such items as factors of safety, building code requirements, and the economics of the situation have to be taken into consideration.

1.13.4 Evaluation. When more than one solution is feasible, the various solutions must be compared and a selection made of the solution that best fits all of the conditions of the problem. For larger engineering projects the scope is often too wide to be considered here.

1.13.5 Checking. After a solution has been chosen, the entire problem must be checked from beginning to end:

a. To see that all limiting conditions have been satisfied.
b. To see that any public laws that may apply have been duly met.
c. To see that all computations are correct.
d. To see that all drawings involved are correct and cannot be misinterpreted.
e. To see that the latest development in new materials, especially suited for the project, have been employed.
f. To see that the best methods of production can be used.

1.13.6 Report. Finally, a report of some kind must be presented. This may require written specifications and detailed drawings, or it may be an informal discussion, with drawings submitted, by a younger person to their superior officer in a company.

1.14 ENGINEERING GRAPHICS AND DESIGN

The student should realize that engineering graphics is an integral part of the whole process of engineering design from the point where computation begins to the end result of the project represented by drawings. Many computations can be made graphically and the end product is almost always described by one or more drawings.

For the beginner, drawing is an excellent tool

for testing and developing the imagination. Many problems in engineering graphics can have more than one solution. These are sometimes referred to as *open-ended problems*.

1.15 COMPUTER AIDED DESIGN

Today the computer is used widely to solve engineering problems and to generate engineering drawings. The use of the computer in manufacturing is widely accepted. The so-called CAD/CAM (Computer Aided Design/Computer Aided Manufacturing) interface will be more important in the future than it is today.

The form of the information now supplied on drawings will continue to change. One thing that will remain the same will be the need for the engineer to be able to visualize in a three-dimensional way the projects being worked on. Therefore, it becomes even more important for the engineer to understand the fundamental theories of projection as are developed in the chapters that follow in this book.

CHAPTER 2

LETTERING

CHAPTER 2

2.1 INTRODUCTION

Handwritten lettering is used on engineering drawings because it is more uniform and more easily read than script. In much of industry the trend is toward typing of all notes and dimensions on drawings. Naturally on computer generated drawings the lettering is not accomplished by hand. Even though machines will generate much of our lettering for us, it is still necessary for engineers to know the proper form for lettering and to use it in their work. Having correct dimensions on a drawing is extremely important if mating parts are to fit and function together.

2.2 STYLE IN LETTERING

To designate the style of lettering desired it is necessary to prescribe several characteristics.
   a. Original or basic form (Gothic, Roman, Text such as Old English).
   b. Character of stroke (single, double, or filled in).
   c. Slant or vertical.
   d. Capitals or small letters (upper- or lowercase).

2.3 GOTHIC STYLE

When making drawings of machine parts, the style of lettering used by the majority of engineers is single-stroke, vertical, Gothic capitals. Some engineers use a combination of capitals and lowercase or small letters. The capital alphabets and numbers in vertical and slant style are illustrated in Figs. 2.1 and 2.2, respectively.

2.3.1 Microfont Alphabet. The National Micrographic Association has adopted the Gothic-style microfont alphabet and is intended for general usage. See Fig. 2.3. This alphabet is recommended on drawings for microfilm reproduction.

2.4 USES OF OTHER STYLES

The civil engineer is more likely to use the slant, single-stroke, Gothic, upper- and lowercase lettering. On maps he/she sometimes uses the more decorative style known as Modern Roman. This style is illustrated in Fig. 2.26. The use of an ellipse guide in lettering is illustrated in Fig. 2.27.

The architect uses a variation of the single-stroke Gothic that is distinguished by the addi-

ABCDEFGHIJKLMNOP
QRSTUVWXYZ&
1234567890

FIG. 2.1. The single-stroke vertical alphabet.

ABCDEFGHIJKLMNO
PQRSTUVWXYZ
1234567890

FIG. 2.3. The microfont alphabet.

ABCDEFGHIJKLMNOP
QRSTUVWXYZ&
1234567890

FIG. 2.2. The single-stroke slant alphabet.

tion of curves and seraphs. This is shown in Fig. 2.4.

For signs, inscriptions, diplomas, and the like, the double-stroke or filled-in letters are frequently used. The double-stroke Gothic is shown in Fig. 2.28. The most decorative and the most difficult to make is the Old English, shown in Fig. 2.29. These alphabets are shown for reference so that an engineer can learn to make them if the occasion should arise. It is not intended that the young engineer learn anything but the single-stroke Gothic.

2.5 SIZE OF LETTERS

When the size of any lettering is given, it means the height of the capital letters. When used in engineering lettering, the small letters have their bodies two thirds the height of the capitals. Such small letters as b, d, and h have stems called *ascenders*, which are made the same height as the capitals. Such letters as g, j, and p have *descenders* that extend one third the height of the capital letters below the lower guideline.

Numerals have the same height as the capital letters. Fractions are at least one and one half times the height of the capital letters. The numbers in the numerator and denominator are each two thirds the height of the capital letters.

On a drawing that is to be microfilmed, the size of the lettering must be large enough to stand the great reduction that is necessary. No letters should be less than $\frac{1}{8}$ in. and even larger letters are frequently preferred by most companies.

2.6 GUIDELINES

Lettering should never be done without guidelines. These guidelines consist of three parallel lines, one for the baseline of the lettering, one for the top of the capital letters, and one for the top of the small letters. See Fig. 2.5.

The Braddock lettering triangle, illustrated in Fig. 2.6, gives a convenient method of drawing these lines. There are several columns of holes in the triangle by means of which guidelines for various sizes of letters may be made. The height of letter for each column is given below the column in thirty seconds of an inch. Thus, num-

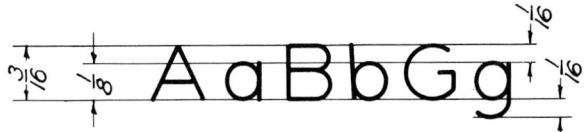

FIG. 2.5. Guidelines and height of letters.

FIG. 2.4. Styles used in architectural drawing.

ber 4 will give guidelines for letters $\frac{4}{32}$ or $\frac{1}{8}$ in. high. The pencil point is inserted in each hole successively, and a line is drawn by sliding the triangle along a T-square, as shown in Fig. 2.6.

The Ames lettering guide Fig. 2.7 can also be used to draw guidelines, using only one set of holes. The size is changed by rotating the center disk to the desired index number. The sizes are also specified in thirty seconds of an inch. Uniformly spaced lines for cross-hatching may also be done with this instrument.

All guidelines should be made lightly so that they may be erased and so that they will not show on a print. If done very lightly, they need not be erased as they will not reproduce in most reproductive processes. A sharp 4H pencil should be used for drawing guidelines.

2.7 SLOPE OF LETTERS

Whether using vertical or slant lettering, it is important that the stems of all letters have the same slope. To accomplish this, light parallel lines may be drawn across the guidelines as shown in Fig. 2.8. For slant letters those lines may vary from 65 to 75° with the horizontal. For vertical lettering it is important that the letters be as nearly vertical as possible, because the eye can easily pick up errors in the slope of vertical lines.

2.8 ELEMENTS OF LETTERS

In engineering lettering, each letter is composed of one or more of each of two elements. These elements are the stem (straight line) and the oval.

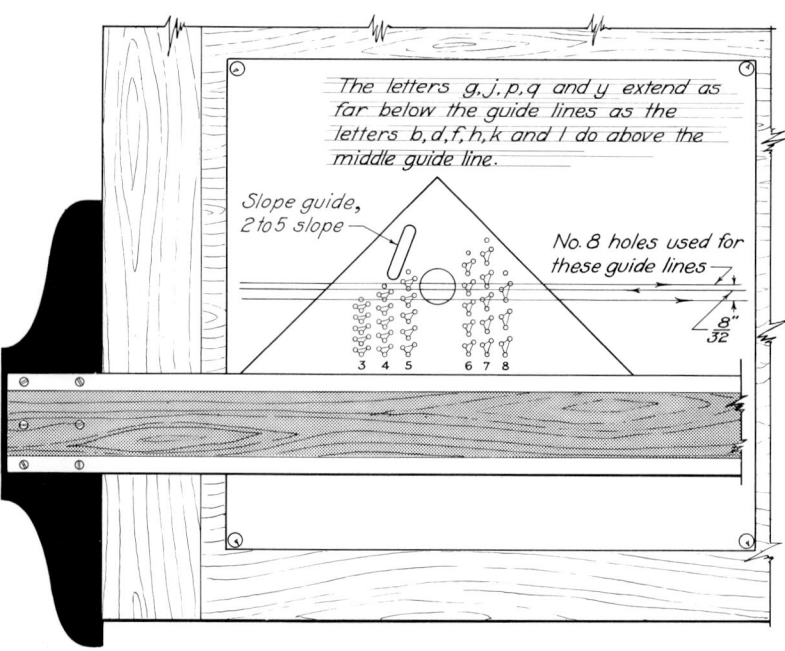

FIG. 2.6. The Braddock lettering triangle.

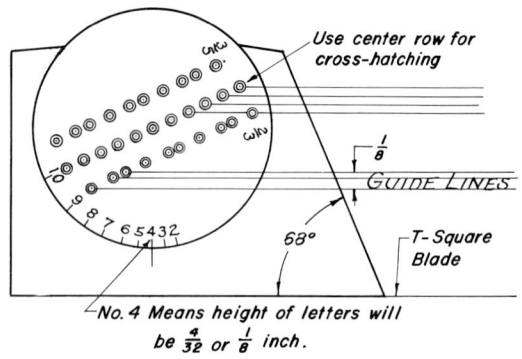

FIG. 2.7. The Ames lettering instrument.

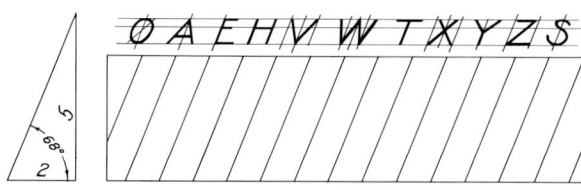

FIG. 2.8. Slope guidelines.

2.8.1 Stems. The stems are the straight lines, vertical or slant, that form a part of more than four fifths of the letters of the alphabet. The horizontal lines in such letters as E, F, and H are not classified as stems but must be made as carefully and as neatly as the stems.

The following rules are applicable in forming all stems.

a. The stems must be uniform in weight, thickness, and height.
b. The stems must be perfectly straight without hooks or curls at either end.
c. The slopes of stems must be uniform. The stems must be parallel to each other throughout any piece of lettering.

Examples of stems and ovals are shown in Fig. 2.9.

2.8.2 Ovals. The oval forms the curved parts of the letters. Although the ovals should be five sixths as wide as they are high, it is better to make them too broad than too narrow. The oval is formed inside a parallelogram and is tangent at the center point at each side. Parallelogram and oval are shown in Fig. 2.10. The standard width is indicated on the drawing. The ovals are usually made with two strokes beginning at the top and meeting at the bottom without a perceptible joint. However, for small letters when using the pencil the ovals are often made with a single stroke.

In vertical lettering the slope line coincides with the major axis of the ellipse. In slant lettering, the slope line will be about 68°, while the major axis of the ellipse will be nearer 45° with the horizontal. See Fig. 2.10.

2.9 COMBINATION OF STEM AND OVAL

In letters in which the stem and oval are combined, the stem is always tangent to the oval. This is illustrated in Fig. 2.11. The weight of the stroke for the stem and the oval must be the same. There must be no enlargement at the point of tangency. In letters such as lowercase a, b, and d, the oval must be a complete ellipse. The weight of the stroke will vary directly with the size of the lettering.

2.10 TECHNIQUE OF LETTERING

To attain the proper technique for lettering, it is necessary to consider the following.

a. The selection of the proper pencil or pen.
b. The position of the hand for lettering.
c. The proper direction of the strokes.
d. Practice of lettering technique is necessary to attain proficiency.

2.10.1 Selection and Sharpening of the Pencil. The most important technique in lettering is the making of a good solid black line. This depends on selecting a pencil that has the proper hardness and on keeping it well sharpened. To make a good black line, a soft pencil should be selected. This should be 2H or softer depending on the purpose of the drawing and the method of reproduction. If a blacker line is desired, it is better to select a softer pencil than to use more pressure on the pencil.

To get a good clean-cut line requires that the pencil be sharpened properly and frequently. For best results a conical point slightly rounded should be used.

FIG. 2.9. Stems and ovals.

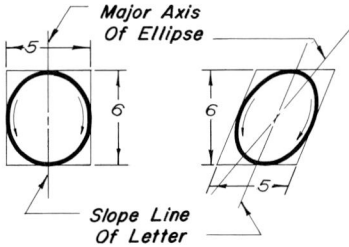

FIG. 2.10. Direction of strokes in making the oval.

FIG. 2.11. Joining stems and ovals.

2.10.2 Selection and Use of a Pen. In inking lettering the thickness of the line depends on the pen used rather than on the pressure. Figure 2.12 shows the thickness or weight of line that can be obtained from each pen point. The ball-point pen makes it easier to maintain a uniform weight of line. Tank-type pens shown in Fig. 2.13 are used extensively at the present time. Whenever lettering or line work is discontinued, the pen must be cleaned thoroughly in order that it will be ready to use the next time it is needed.

2.10.3 Position of the Hand. For good lettering the hand must hold the pencil or pen firmly, but there should be no tension caused by gripping too tightly. The best way to hold the pencil is shown in Fig. 2.14.

The following rules, if carefully followed, will give a good foundation for improving the ability to letter.

a. The forearm should rest on the desk.
b. The index finger should be kept as straight as possible along the pencil. This will help to avoid tension in the hand.
c. The strokes should be make with as little bending of the finger as possible.
d. The forearm should be at an angle of 75 to 80° with the line of lettering. The paper may be located in any convenient position.
e. All strokes should be made from top to bottom and left to right.

With the exception of the last two, these rules apply to either right-handed or left-handed lettering. Some left-handed people prefer to get the hand completely above the lettering, whereas others work with the arm at an angle of about 45° with the lettering.

2.11 THE ART OF LETTERING

The art of good lettering depends on obtaining uniformity in six items.

a. Uniformity of shape.
b. Uniformity of style.
c. Uniformity of size.
d. Uniformity of weight.
e. Uniformity of slope.
f. Uniformity of spacing.

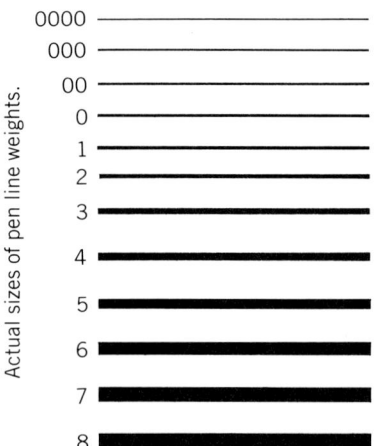

FIG. 2.12. Pen line width.

FIG. 2.13. Pen set (courtesy of Teledyne-Post).

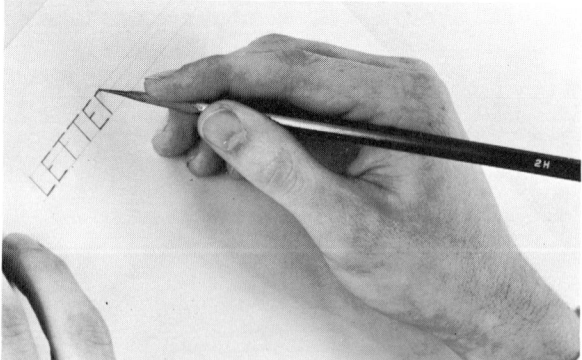

FIG. 2.14. Position of hand for lettering.

Without uniformity in these six fundamentals, the lettering will not look well and may even be difficult to read.

2.11.1 Uniformity of Shape. This requires thorough familiarity with the proper shape of the letters and that each letter be made the same every time.

2.11.2 Uniformity of Style. Gothic, Roman, and Old English letters should not be mixed in the same piece of lettering. Also, uppercase and lowercase letters should not be mixed except for purposes of capitalization of proper nouns. This is the most common error and should be carefully avoided. Such an error is not likely to occur when all capitals are being used.

2.11.3 Uniformity of Size. Uniformity of size requires that guidelines be drawn for all lettering and that they be faithfully followed. A slight overrunning or failure to touch guidelines is very noticeable when the guidelines have been removed. The width of the letters must also be uniform. It is better to have the ovals too wide than too narrow.

2.11.4 Uniformity of Weight. All straight strokes and ovals must have the same weight and must maintain that weight throughout their entire length.

2.11.5 Uniformity of Slope. The stems and the axes of all letters must be parallel, whether they are vertical or inclined. Light slope guidelines drawn across the horizontal guidelines at the proper angle will help to keep the letters at the proper slope.

2.11.6 Uniformity of Spacing. This is the basis of good composition and requires that the letters in the words and the words in the sentences be properly spaced so that they can be easily read. Figure 2.15 shows the effect of poor spacing. *As a general rule it is best to make the letters broad and the space between them small.* The spacing between words should be sufficient for the letter N to be lettered there. Because of the great variety in the shape of the letters, no actual spacing can be recommended. *If the white spaces between the letters appear equal and the white spaces between the words are made to appear equal, the lettering will read well.* This results in making letters with straight sides farther apart and letters with included space closer together. See Fig. 2.16.

When the space available for lettering is smaller than normal or larger than normal, the letters may be compressed or expanded, as illustrated in Fig. 2.17. In each case the change is made in the letters themselves rather than in the spacing between letters.

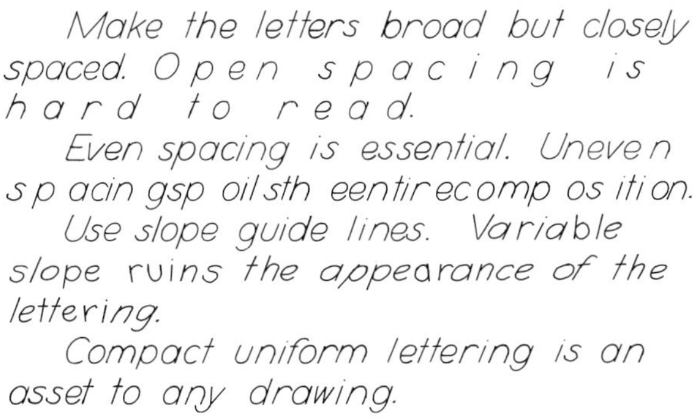

Make the letters broad but closely spaced. Open spacing is hard to read.
Even spacing is essential. Uneven spacingspoilstheentirecomposition.
Use slope guide lines. Variable slope ruins the appearance of the lettering.
Compact uniform lettering is an asset to any drawing.

FIG. 2.15. Study of composition of lettering.

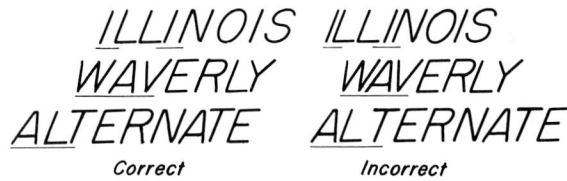

Correct   Incorrect

FIG. 2.16. Spacing of letters.

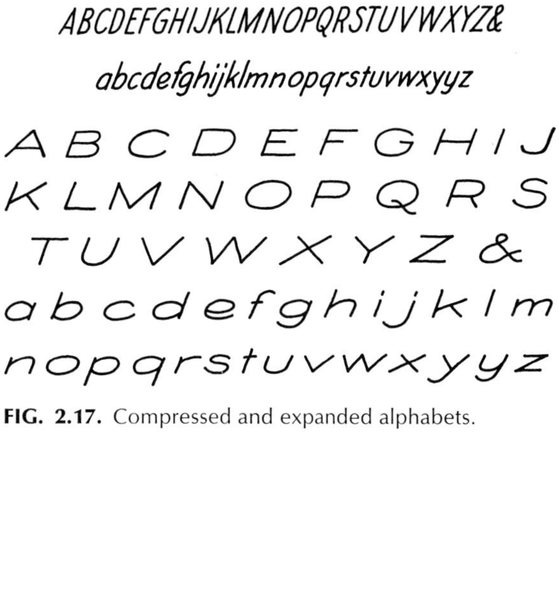

ABCDEFGHIJKLMNOPQRSTUVWXYZ&
abcdefghijklmnopqrstuvwxyyz

A B C D E F G H I J
K L M N O P Q R S
T U V W X Y Z &
a b c d e f g h i j k l m
n o p q r s t u v w x y y z

FIG. 2.17. Compressed and expanded alphabets.

2.12 THE MECHANICS OF LETTERING

The engineer should be thoroughly familiar with the shape and characteristics of every letter of the alphabet. Figure 2.18 illustrates these letters in both vertical and slant form. It gives the approved width of each letter and the number of strokes and direction of each stroke for the easiest and most rapid construction. *These strokes*

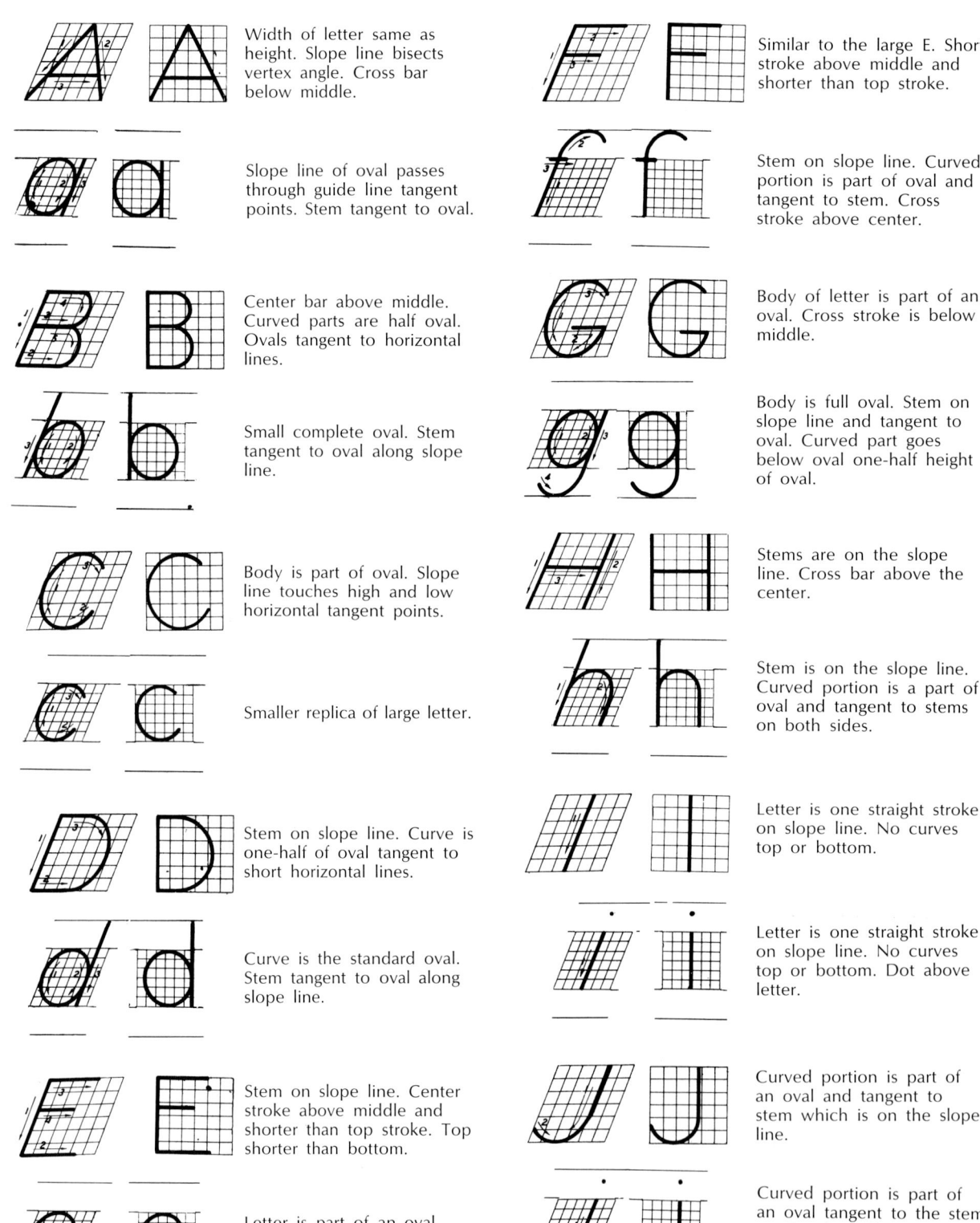

A	Width of letter same as height. Slope line bisects vertex angle. Cross bar below middle.
a	Slope line of oval passes through guide line tangent points. Stem tangent to oval.
B	Center bar above middle. Curved parts are half oval. Ovals tangent to horizontal lines.
b	Small complete oval. Stem tangent to oval along slope line.
C	Body is part of oval. Slope line touches high and low horizontal tangent points.
c	Smaller replica of large letter.
D	Stem on slope line. Curve is one-half of oval tangent to short horizontal lines.
d	Curve is the standard oval. Stem tangent to oval along slope line.
E	Stem on slope line. Center stroke above middle and shorter than top stroke. Top shorter than bottom.
e	Letter is part of an oval. Center stroke slightly below the middle and horizontal.
F	Similar to the large E. Short stroke above middle and shorter than top stroke.
f	Stem on slope line. Curved portion is part of oval and tangent to stem. Cross stroke above center.
G	Body of letter is part of an oval. Cross stroke is below middle.
g	Body is full oval. Stem on slope line and tangent to oval. Curved part goes below oval one-half height of oval.
H	Stems are on the slope line. Cross bar above the center.
h	Stem is on the slope line. Curved portion is a part of oval and tangent to stems on both sides.
I	Letter is one straight stroke on slope line. No curves top or bottom.
i	Letter is one straight stroke on slope line. No curves top or bottom. Dot above letter.
J	Curved portion is part of an oval and tangent to stem which is on the slope line.
j	Curved portion is part of an oval tangent to the stem on the slope line. Dot above letter. Total height equal to capital letter.

FIG. 2.18. Details of making Gothic letters.

2.12 THE MECHANICS OF LETTERING

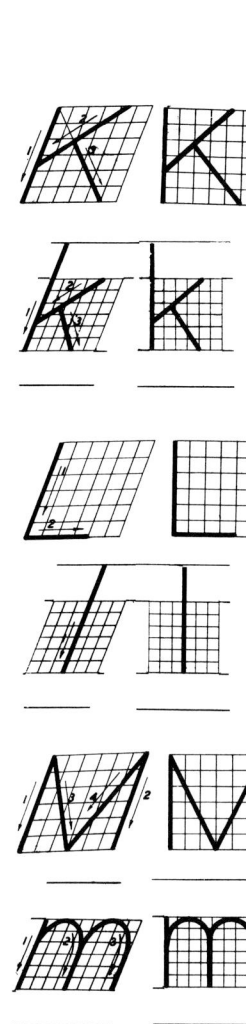

Stroke 1 on slope line. Stroke 2 touches stem below center. Stroke 3 would go through top of stroke 1 if extended. Letter wider at bottom than top.

Make stem of letter on slope line. Lower part of letter same as capital.

Stroke one is on the slope line. Stroke two is horizontal.

Small letter is one straight stroke on slope line. No curves top or bottom.

Outside strokes are on the slope line. Inside strokes form letter V with slope line as bisector.

Stems are on the slope line. Curved portions are part of the oval tangent to stem above the center.

Outside lines are on the slope line. Inside line joins top and bottom of outside lines.

Stems are on the slope line. Curved portions are part of oval, tangent to the stems above the center.

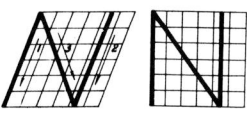

Letter is a complete oval. Tangent points with horizontal at top and bottom on slope line.

Small letter also a complete oval just like capital letter except for size.

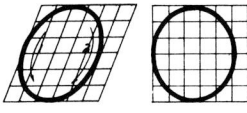

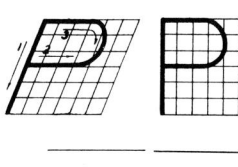

Stem is on slope line. Curved portion is part of an oval and tangent to the short horizontal lines.

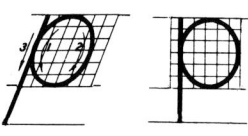

Body is a complete oval tangent to the stem which is on the slope line. Height of stem same as capital letters.

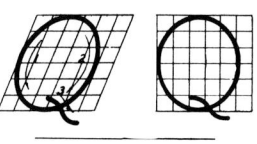

Body is complete oval. Cross line may be straight or with a slight reverse curve.

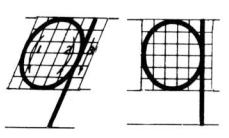

Body is a complete oval tangent to the stem on the slope line. Total height same as capitals.

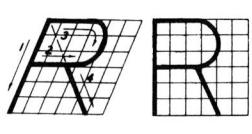

Main stem is on slope line. Curved portion is part of oval and tangent to short horizontal lines. Letter wider at bottom than top.

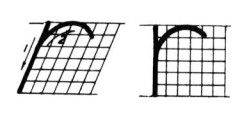

Stem is on slope line. Curved portion is part of an oval tangent to the stem.

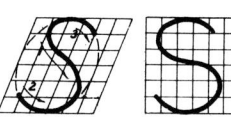

Letter fits inside a standard oval. Top part is smaller than bottom.

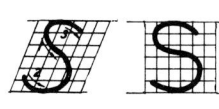

Letter is the same as the capital except only two-thirds the height of the capital.

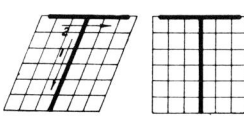

Stem which is on the slope line bisects the horizontal top line.

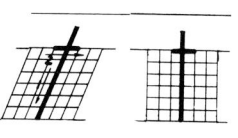

Stem on slope line, not quite as tall at the capital. Cross bar on guide line.

22 LETTERING

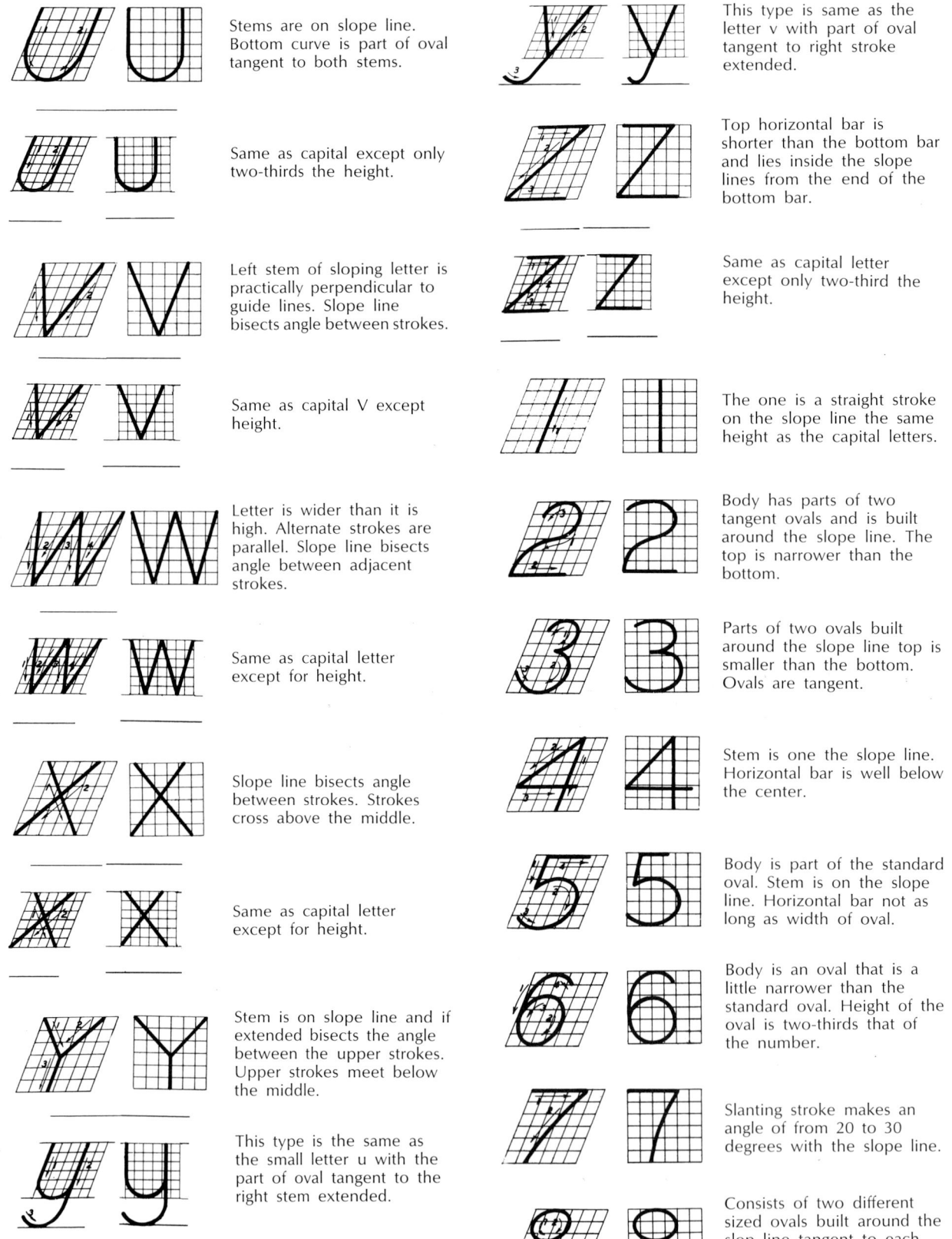

U	Stems are on slope line. Bottom curve is part of oval tangent to both stems.
u	Same as capital except only two-thirds the height.
V	Left stem of sloping letter is practically perpendicular to guide lines. Slope line bisects angle between strokes.
v	Same as capital V except height.
W	Letter is wider than it is high. Alternate strokes are parallel. Slope line bisects angle between adjacent strokes.
w	Same as capital letter except for height.
X	Slope line bisects angle between strokes. Strokes cross above the middle.
x	Same as capital letter except for height.
Y	Stem is on slope line and if extended bisects the angle between the upper strokes. Upper strokes meet below the middle.
y	This type is the same as the small letter u with the part of oval tangent to the right stem extended.
y (alt)	This type is same as the letter v with part of oval tangent to right stroke extended.
Z	Top horizontal bar is shorter than the bottom bar and lies inside the slope lines from the end of the bottom bar.
z	Same as capital letter except only two-third the height.
1	The one is a straight stroke on the slope line the same height as the capital letters.
2	Body has parts of two tangent ovals and is built around the slope line. The top is narrower than the bottom.
3	Parts of two ovals built around the slope line top is smaller than the bottom. Ovals are tangent.
4	Stem is one the slope line. Horizontal bar is well below the center.
5	Body is part of the standard oval. Stem is on the slope line. Horizontal bar not as long as width of oval.
6	Body is an oval that is a little narrower than the standard oval. Height of the oval is two-thirds that of the number.
7	Slanting stroke makes an angle of from 20 to 30 degrees with the slope line.
8	Consists of two different sized ovals built around the slop line tangent to each other. Top oval smaller than lower one.

should always be made with one continuous motion rather than by using an overlapping sketch stroke. The strokes must not be hurried, but with practice they can be made rapidly and with confidence to give the best-appearing letters. The unique characteristics of each letter, together with hints for forming them, are given in Fig. 2.18.

The letters and numerals divide themselves naturally into three groups.

a. Those made entirely with straight lines.
b. Those made entirely with ovals.
c. Those made by using a combination of straight lines and ovals.

2.12.1 Capital Letters Made Entirely with Straight Lines. Letters such as H, F, E, I, L, N, and T, and the numeral 1 form this group. In each of these letters the stems lie on the slope line. Figure 2.18 indicates the width of each letter and the desirable position of the crosslines.

Such letters as A, V, M, W, K, Y, X, and Z, and the numerals 4 and 7 are also made with straight lines, but each has some characteristic that must be carefully observed to obtain the correct letters. For many of them the principal characteristic is the fact that the slope line bisects the angle between the two strokes. In others, such as K, special hints for the construction are given.

2.12.2 Letters Made Entirely with Curves. Included in this group are such letters as O, Q, C, G, and S, and the numerals 3, 6, 8, 9, and 0. In making these letters it is usually best to follow the direction and order of the strokes given. However, for lowercase letters being made with pencil, many engineers prefer to use one complete stroke for an oval.

2.12.3 Letters Made with a Combination of Curves and Straight Lines. Such letters as B, D, J, P, R, and U, and the numerals 2 and 5 are included in this group. The direction and number of strokes indicated in Fig. 2.18 should be followed rigorously. The straight lines must be tangent to the ellipse in every case.

2.12.4 The Rule of Stability. The stability of several letters indicates another classification of capital letters. Letters such as A, B, E, H, R, X, and Y require special consideration because of the placing of the crossbar. *In each case the crossing or the crossbar is arranged to make the white spaces within the letter appear equal.* Instruction is given in Fig. 2.18, showing the best method of doing this in each case.

2.12.5 Lowercase or Small Letters. These may be divided into the same groups as the capitals. Complete instructions for forming them are listed and illustrated in Fig. 2.18. One of the major differences between capitals and small letters lies in the fact that some of the small letters have ascenders and some have descenders. The body of the lowercase letter is two thirds the height of the capital letters. The ascenders go up to the full height of the capital letters, except that of t, which may be made a little lower. The descenders go one third of the height of the capitals below the lower guideline. Special instructions are given in Fig. 2.18 when needed.

2.12.6 Fractions. Fractions frequently occur when dimensioning a drawing. *They should be made large.* The most common error is making them too small to be easily read. The fractions should extend above and below the capitals or integers. They should be at least one and one half the height of the numeral or capital letter. *The numerator and denominator must each be at least two thirds as high as the capital letter.* The line between them is horizontal and neither the numerator nor the denominator may touch this line.

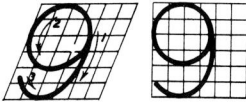
Body of figure is an oval similar to the bottom of the 6. The lower curve is part of a similar oval.

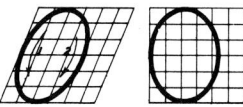
The complete oval for the zero is narrower than the oval for the letters. All other ovals used in numerals are based on this.

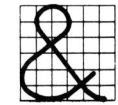
Symbols are based on ovals and straight lines balanced around the slope line.

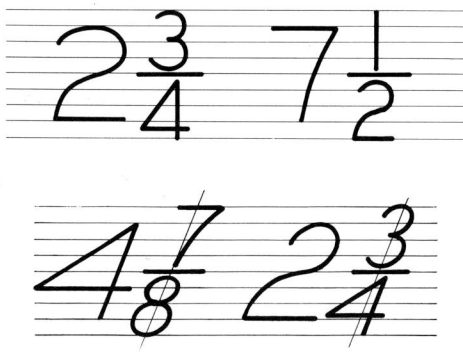

FIG. 2.19. Integers and fractions.

24 LETTERING

To obtain the proper slope of a fraction, a slope guideline should be placed through the center of the fraction, and both numerator and denominator must be built around the slope line. See Fig. 2.19.

2.13 LARGE AND SMALL CAPITALS
In a sentence or a paragraph of lettering, large and small capital letters may be combined, just as capitals and lowercase letters are used together. The large and small capitals are usually easier to read than capital and lowercase letters. See Fig. 2.20.

2.14 TITLES
In most industrial drafting rooms, printed title blocks are used. Therefore, the drafter does not have to design the title but merely letter in a few words in the appropriate place.

Figure 2.21 shows the dimensions of a typical title block for a large drawing. It is placed in the lower right-hand corner of the drawing. The information to be placed in each block is listed as follows.

   a. Signatures or initials of persons preparing, checking, and approving the drawing. The initials should be accompanied by a date.
   b. Drawing title.
   c. Name and address of design activity.
   d. Approval by design activity.
   e. Approval by other than design activity.
   f. Code identification number.
   g. Drawing size.
   h. Drawing number.
   i. Scale.
   j. Weight.
   k. Sheet number and number of sheets; thus, sheet number 2 of 5.

LARGE AND SMALL CAPITALS ARE FREQUENTLY USED INSTEAD OF UPPER AND LOWERCASE LETTERS.

EITHER VERTICAL OR SLANT LETTERS MAY BE USED.

FIG. 2.20. Composition with capital letters.

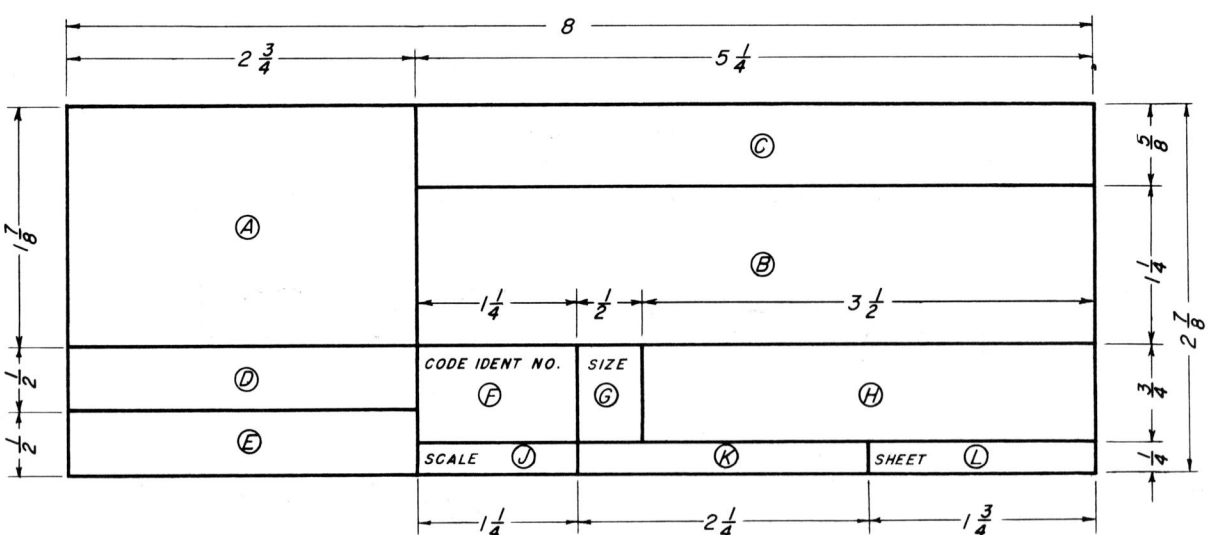

FIG. 2.21. Title-block layout.

The information necessary in a parts list is shown in Fig. 2.22. This is frequently placed above the title. The revision block showing all changes may be placed in the upper right-hand corner of the drawing. Figure 2.23 shows a sample revision block. Capital letters should be used throughout in all of these forms.

2.15 MECHANICAL LETTERING

When perfection in lettering is desired and time is available, several kinds of lettering devices are available. These have a template and a tube pen by means of which the letters can be copied. One of the first to be used was the Wrico, which consisted of a template in which holes had been cut in the shape of the letters. Two settings were necessary for most of the letters. See Fig. 2.24.

The one most commonly used today is the Leroy guide. This guide has the letters engraved in the template. While one point follows the letters on the template, the tube pen inks the letters on the drawing. See Fig. 2.25.

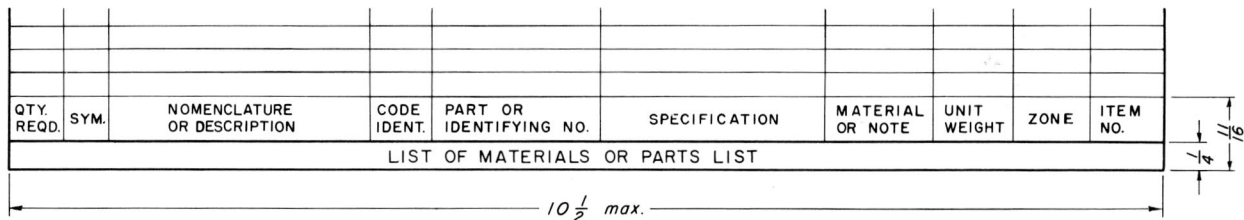

FIG. 2.22. Parts list.

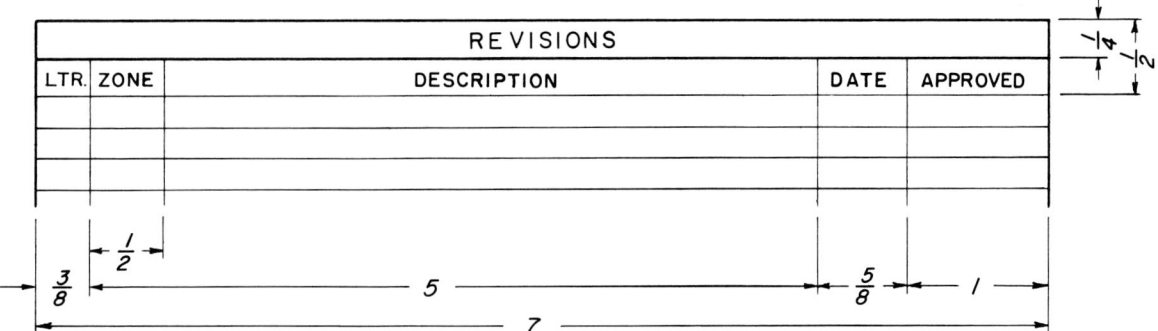

FIG. 2.23. Format for revisions.

FIG. 2.24. Wrico lettering instruments (courtesy of Wood-Regan Instrument Co.).

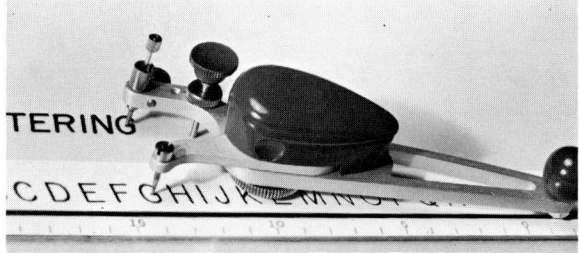

FIG. 2.25. Leroy lettering guides (courtesy Keuffel and Esser Co.).

FIG. 2.26. Modern Roman slant letters.

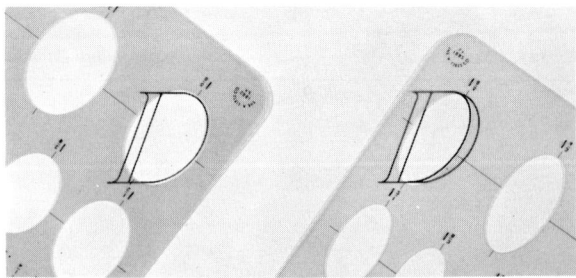

FIG. 2.27. Use of ellipse guides in making Roman lettering.

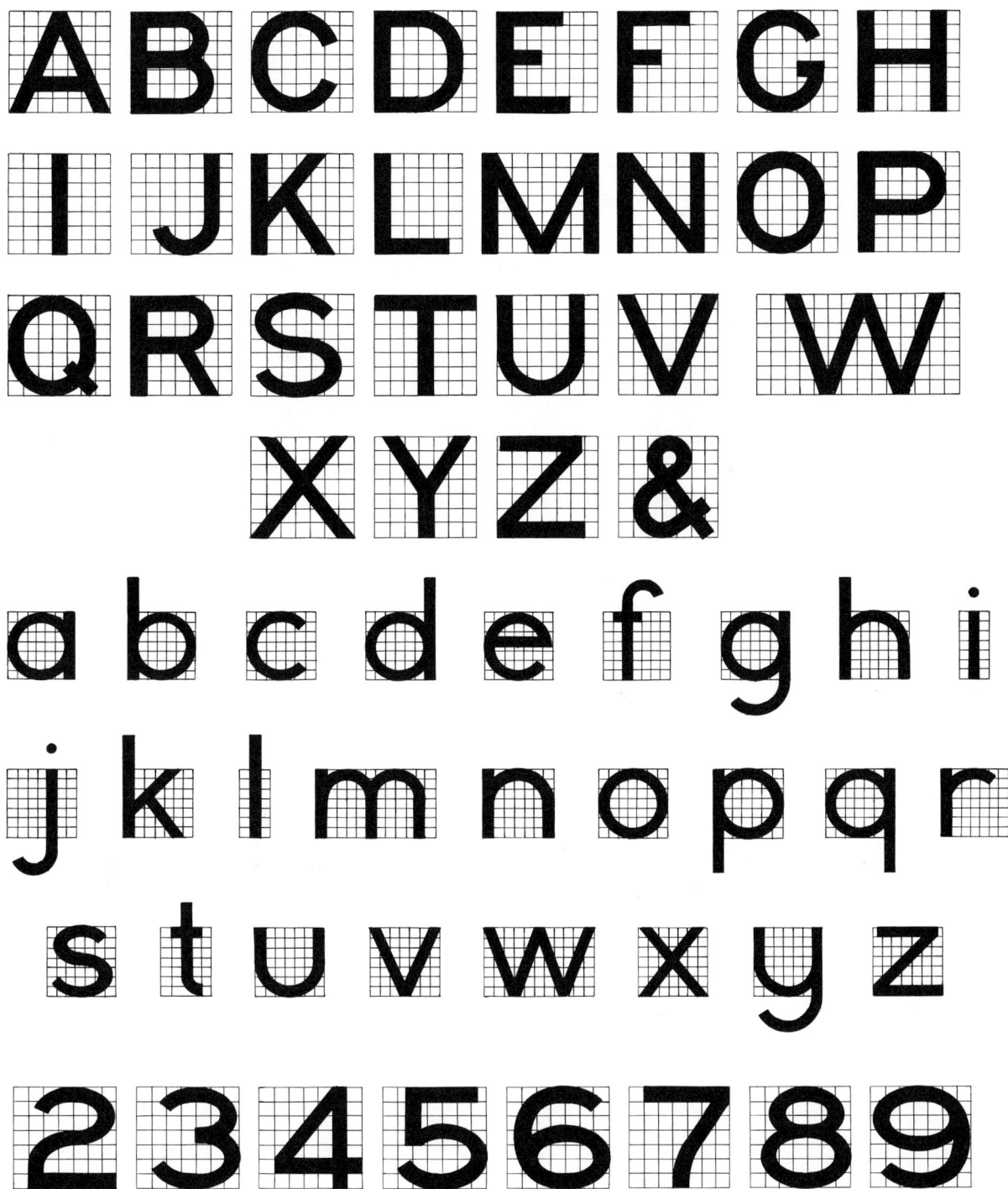

FIG. 2.28. Gothic letters.

FIG. 2.29. Old English Text alphabet (courtesy Ross F. George).

SELF-STUDY QUESTIONS

Before trying to answer these questions, read the chapter carefully. Then, without reference to the text, answer as many questions as possible. For those that cannot be answered, the number in parentheses following the question number gives the article in which the answer can be found. Look it up and write down the answer. Check the answers that you did give to see that they are correct.

2.1 (**2.4**) The style of lettering most frequently used for engineering drawings is _____.

2.2 (**2.5**) The height of the _____ letters is accepted as the height of a line of lettering.

2.3 (**2.5**) The body of the lowercase letters is _____ the height of the capital letters.

2.4 (**2.5**) The height of the ascenders in lowercase lettering is _____ the height of the capital letters.

2.5 (**2.5**) The length of the descenders in lowercase lettering is _____ the height of the capital letters.

2.6 (**2.7**) The accepted slope for slant lettering is approximately _____ degrees.

2.7 (**2.6**) To maintain uniformity of size _____ should always be used.

2.8 (**2.8**) The two elements that are used to form all engineering letters are the _____ and the _____.

2.9 (**2.8.2**) In the two parallelograms shown below, sketch an inclined and a vertical letter O. Show the major axes of the ellipses by a center line (long, short, long dashes).

2.10 (**2.8.2**) In the guidelines shown below, sketch an inclined and a vertical letter O. Show the slope line of each by a center line.

2.11 (**2.12.4**) The rule of stability states that, for certain letters having two parts, the white spaces within the letter should _____ equal.

2.12 (**2.12.1**) In the guidelines shown below, sketch the slant letters M, A, N, L, Y, Z, and X. Show direction of the slope line of each letter by a center line.

2.13 (**2.11.6**) Is it true that the spacing between letters should always be uniform? _____

2.14 (**2.11.6**) Words should be spaced so that the white spaces between words _____ uniform.

2.15 (**2.11.6**) When the space available for lettering is small, the lettering can be compressed by reducing the width of the _____.

2.16 (**2.11**) In good lettering there should be uniformity of style, _____, shape, size, and weight of letters.

2.17 (**2.15**) The most commonly used aid for mechanical lettering is the _____ lettering guide.

2.18 (**2.5**) The smallest size of lettering that should be used for microfilming is _____ inch.

2.19 (**2.14**) Important items of information that should be included in a title are drawing title, _____, date, signature of designer, and drawing number.

2.20 (**2.14**) The revision block is used to show _____ that have been made on the drawing.

CHAPTER 3

USE AND CARE OF INSTRUMENTS

CHAPTER 3

3.1 INTRODUCTION

There are two fundamental purposes for an engineering graphics course. The first of these is to give the student an understanding of the principles on which engineering drawing is based, and to be able to read and interpret drawings. The second purpose is to give the student instruction and practice in the use of instruments so that he/she may acquire a workerlike facility in their manipulation. The engineer's objective is not to produce a drawing that will just "get by" but rather to secure the best results. In doing this, one must keep in mind correctness, accuracy, and appearance while producing the drawing in the least possible time.

To learn the correct form of handling instruments, one should study the proper use of them as explained in this chapter, so that awkward and useless movements may be avoided. No attempt is made here to illustrate the hundreds of pieces of special drafting tools on the market. The engineer who learns to use the regular equipment discussed in this chapter can use the special tools without further instruction.

3.2 REGULAR EQUIPMENT

The following list of equipment constitutes what is called a *set* or *kit of instruments*. Several cases of instruments are illustrated in Fig. 3.1.

Case of Instruments
1 Large compass
1 Beam compass with attachments
1 Hairspring divider
1 Box of leads
1 Screwdriver

Other Equipment
1 T-square
1 Architect's scale
1 Engineer's scale
1 Metric scale
1 12 in. 30–60° triangle
1 8 in. 30–60° triangle
1 6 in. 45° Braddock lettering triangle

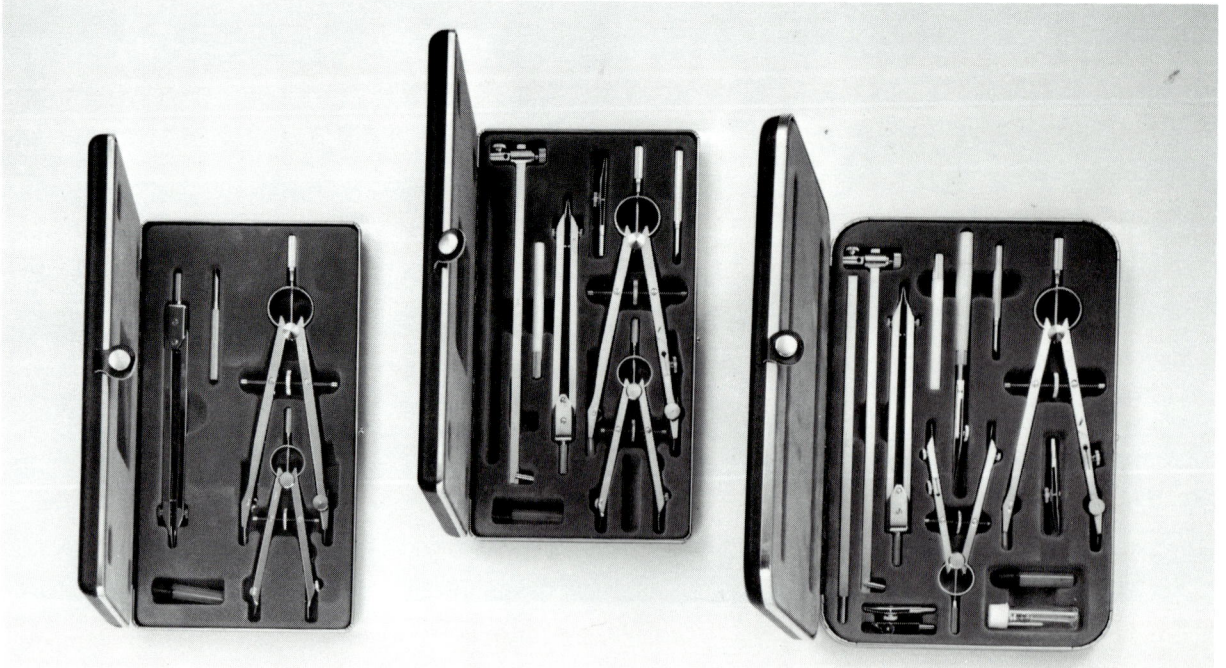

FIG. 3.1. Instrument cases (courtesy of Teledyne Post).

2 Irregular curves
1 6 in. protractor
1 Erasing shield
1 Eraser
1 Pen attachment
Assorted pencils or lead holders

Some manufacturers also provide instrument and equipment sets that hold all the equipment in one case with the possible exception of the T-square. See Fig. 3.2.

3.3 PENCILS

Special pencils are used for drawing. They are of uniform size, hexagonal in shape, with varying

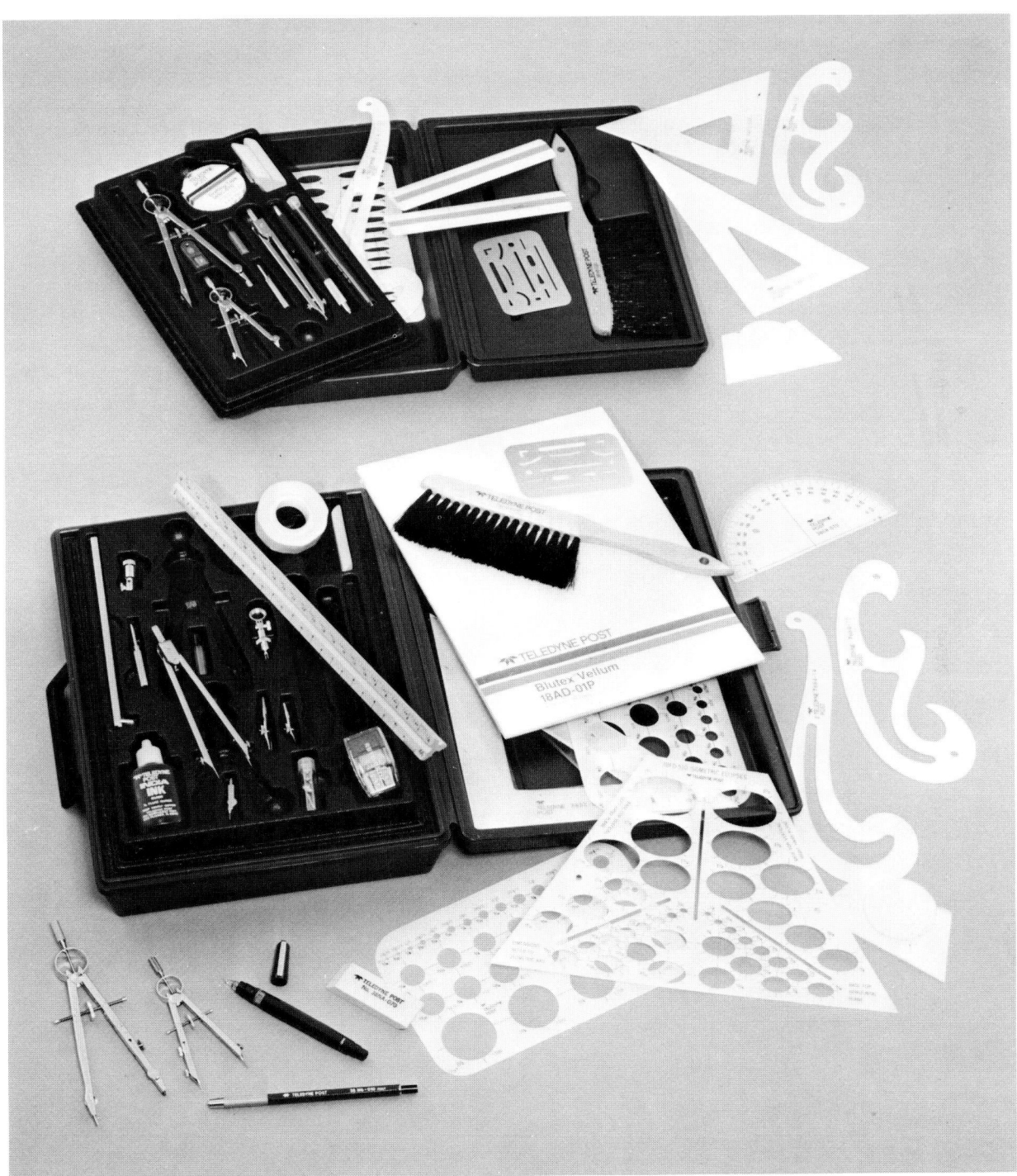

FIG. 3.2 Instrument and equipment sets (courtesy of Teledyne Post).

size of lead as shown in Fig. 3.3. Eighteen degrees of hardness are supplied by the manufacturer, ranging from 7B, the softest and blackest, to 9H, which is the hardest.

Although all manufacturers use the same system of marking their pencils, it will be found that the pencils from different manufacturers having the same hardness number do not actually have the same hardness. It is for this reason that most drafters usually prefer to use only one brand of pencil. Refill pencils, shown in Fig. 3.4, are also on the market together with refill leads of the varying degrees of hardness for the entire range. The 0.5-mm pencil is gaining wide acceptance and is currently the most widely used lead holder.

For general layout work in drawing, the 4H and 5H pencils are the most useful. The harder varieties are used in graphic statics and other graphical computation methods where fine lines and extreme accuracy are required. For making a finished pencil drawing, the H and 2H are more desirable since they give a sharp black line. For sketching and artwork, the softer grades are used. The drafter should learn to choose the quality of pencil appropriate to the work in hand.

More important than the quality of the pencil is the condition in which it is kept. The proper shape for a pencil point is shown in Fig. 3.5. The tapered wood portion should be about $\frac{7}{8}$ in. long, and $\frac{3}{8}$ in. of lead should be exposed. The lead should be brought to a point by means of a file or sandpaper. For a conical point as in Fig. 3.5a, the pencil should be rotated slowly while it is rubbed back and forth. The pencil should be inclined to the direction of motion, as shown in Fig. 3.6. To produce the wedge point, as in Fig. 3.5b, the opposite sides must be filed down. The bevel point, as in Fig. 3.5c, is made by filing on one side. In refill pencils, the lead should be filed down to a good point just as frequently as in the regular pencil. The 0.5-mm lead does not, in general,

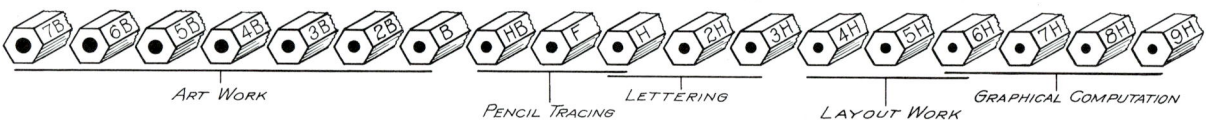

FIG. 3.3. Grades of drawing pencils.

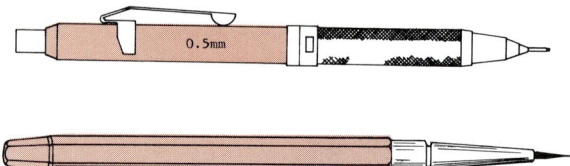

FIG. 3.4. Refill drawing pencils.

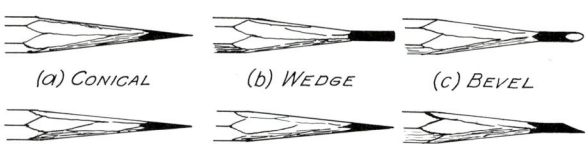

FIG. 3.5. Correct pencil point shapes.

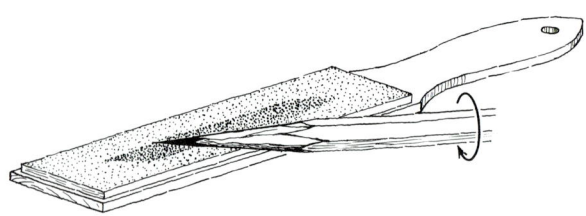

FIG. 3.6. Sharpening the drawing pencil.

require filing except in special circumstances where an extremely sharp point is required.

The wedge point is limited in use to drawing long straight lines since it does not wear down so rapidly. The conical point may be used for all general drafting purposes and always should be used for lettering. The bevel point is recommended for use in the compass, as it has the same advantages there as the wedge point has for straight lines, as well as making it possible to draw very small circles. It is also used with the softer pencils for purposes of shading.

In drawing a straight line, the pencil should be held in a plane perpendicular to the paper along the edge of the T-square or triangle and should be inclined in the direction of motion at an angle of 60 to 75° with the paper. See Fig. 3.7. Note that the pencil is held in a somewhat different manner from that used in lettering. The pencil should not be allowed to rock back and forth transversely to the direction of motion. For preliminary work, just enough pressure should be applied to the pencil to make a firm light line that can be readily seen, but not enough pressure to make a groove in the paper. Graphite in a groove cannot be easily erased and spoils the appearance of a drawing. The importance of clean-cut pencil work cannot be overemphasized; it tends toward both speed and accuracy. The proper direction for drawing lines in various positions is shown in Figs. 3.8 and 3.9. The general criterion is to draw away from the body.

3.4 ERASERS AND ERASING TOOLS

Only a pencil eraser should be used to remove lines made in error or to change lines because of an alteration in design. This applies to both pencil and ink lines. A grit eraser, a razor blade, or another sharp instrument will destroy the surface of the paper and ruin the drawing.

When there are many lines close together and only one line needs to be removed or changed,

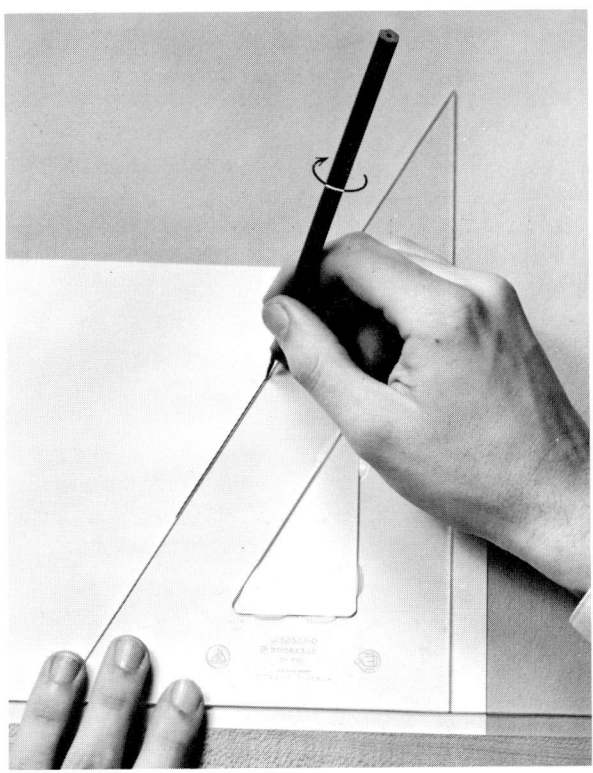

FIG. 3.7. Drawing with a pencil.

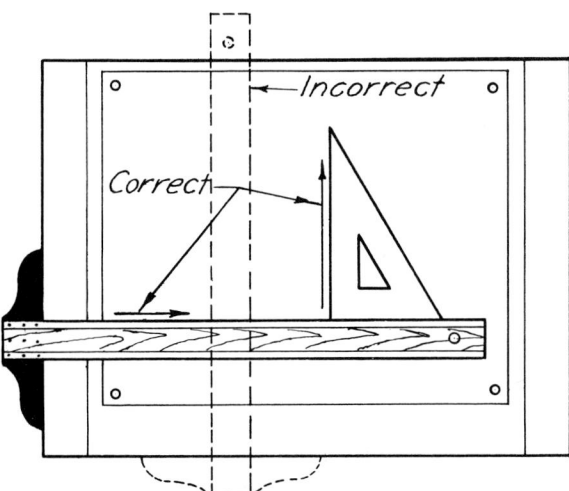

FIG. 3.8. Direction of ruling with T-square and triangle.

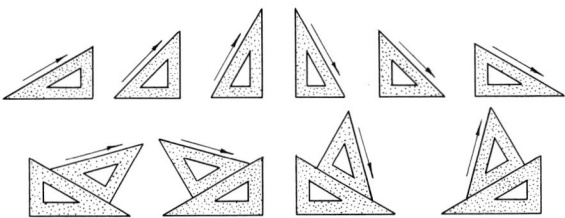

FIG. 3.9. Direction of drawing inclined lines.

the others may be protected by using an erasing shield, as shown in Fig. 3.10. One of the openings is placed over the line to be removed in such a way that other lines do not show, and then the erasure is made through this opening.

For large erasures a motor-driven eraser saves time and energy. It must be used carefully with a light touch or the surface may be damaged. See Fig. 3.10.

When extensive changes are required, a new tracing may be made by reproduction methods, masking out any desired portions of the original drawing. Photographic drawings are used in some instances when changes are to be made in an existing plant facility.

Art gum, which is a softer variety of rubber, should be used to clean the drawing. When cleaning is necessary, it should be done before the drawing is inked. Pencil drawings, finished with a soft pencil that makes a deep black line, cannot be successfully cleaned with art gum since the pencil lines smudge very easily. It is far better to keep the drawing clean than to try to scrub it with art gum after it has been soiled. The following suggestions, if observed, will help to keep the drawing clean.

a. In moving the T-square, bear down on the head so that the blade is raised slightly from the paper so as not to smear the lines.
b. The hands are always somewhat oily—keep them off the paper.
c. Use a hard pencil for layout work; such as, 6H – 9H.
d. Pick up the triangles rather than sliding them.
e. When finishing a drawing with a soft pencil, cover all views, except the one you are working on, with a clean sheet of paper.
f. Blow graphite particles, which flake off the soft pencil, from the sheet.
g. Use a brush or soft cloth to brush gently erasing crumbs off the sheet rather than the flat of the hand.
h. Use a hard, smooth-surfaced paper, if it is suitable for the type of drawing being made.
i. Use of a finely ground cleansing material on the drawing during work will keep both the drawing and the instruments clean. Several varieties of this material may be purchased.

3.5 LETTERING PENS

Several manufacturers have pen sets on the market that are equipped with different-size points for different line widths. See Fig. 3.11. The different line widths attainable are shown in Fig. 3.12.

3.6 DRAWING BOARDS

Drawing boards vary greatly in size, from small ones 12 × 15 in. to large vertical boards 6 or 7 ft high × 10 or 12 ft long. Regardless of size, the surface should be free from cracks, and it should be a plane. Soft white pine or bass wood is a most suitable material. At least one edge should be straight as a base for the T-square.

FIG. 3.10. Use of the erasing shield and motor eraser.

FIG. 3.11. Pen set—Stadler Mars (courtesy of Teledyne Post).

3.7 T-SQUARE

The common T-square consists of two parts, the blade and the head, which should be rigidly fastened together. A variety of kinds and sizes may be obtained from the vendors. Various lengths may be obtained, ranging from 18 to 72 in.

To keep the T-square in good condition, careful handling is required. It should not be used as a hammer nor be allowed to drop to the floor. The upper or drawing edge should not be used as a guide for the knife in cutting paper. If the head should become loose, the screws must be removed and the two parts glued together. The screws may then be replaced, and the T-square is ready for use again.

3.8 PARALLEL RULES

On large drawing boards, of both the horizontal and the vertical type, a parallel rule permanently attached to the table is used in place of a T-square. It is a large straight edge that operates by means of a wire-cable arrangement. It always remains parallel to its previous position as it is moved up and down the board, as shown in Fig. 3.13. On the vertical boards, there is a ledge at the bottom of the board to hold drafting tools.

3.9 DRAFTING MACHINE

In many offices a drafting machine similar to the one shown in Fig. 3.14 is used in place of the T-square, triangles, and scales. In operation, this device keeps the two blades always parallel to their original position no matter where they are moved on the sheet. The two blades, which may be simple straight edges or straight edges with scales along them, are accurately set at right angles to each other. The blades are removable, hence a variety of scales may be used. The adjusting head has a protractor with vernier attachment so that the blades may be set and clamped at any desired angle. Considerable saving of time results from the use of this machine

3.10 TRIANGLES

Triangles are the instruments used by the drafter in connection with the T-square to draw lines at various angles with the horizontal. The two most common varieties that every drafter should possess are the 30–60° and the 45° triangles. With these two triangles, angles of 15, 30, 45, 60, 75, and 90° with the horizontal can be drawn as shown in Fig. 3.15. The pencil or pen should be moved in the direction indicated by the arrows.

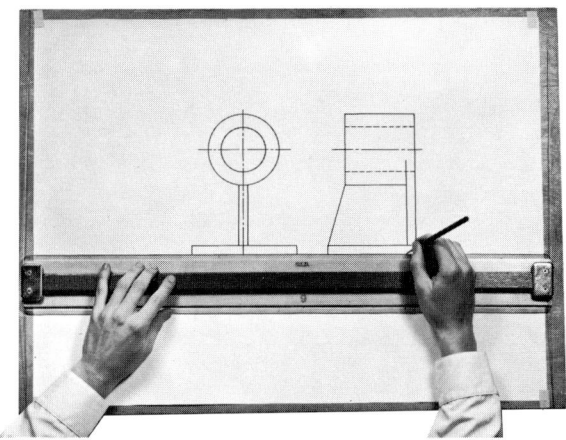

FIG. 3.13. Parallel rule.

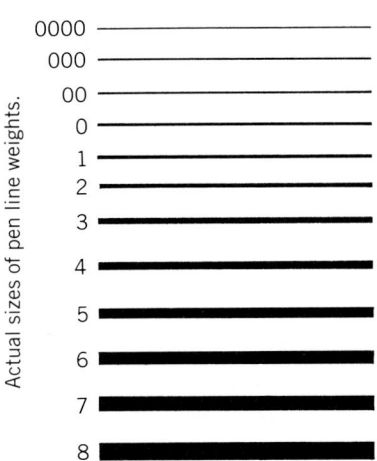

FIG. 3.12. Line widths.

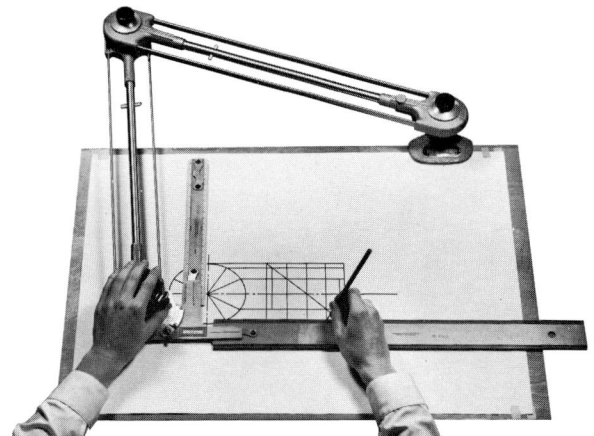

FIG. 3.14. Drafting machine.

Triangles range in size from 4 to 18 in. The length of the long leg of the triangle determines the size. A thickness of about 0.08 of an inch is recommended.

3.11 USE OF TRIANGLES

Besides drawing lines of various angles as indicated in Fig. 3.15, the triangles may be used in pairs or singly with a T-square to draw one line parallel to another as shown in Fig. 3.16 (also see Section 4.2.4).

A good method of drawing a line perpendicular to another line is the following.

a. Place the hypotenuse of one triangle along the original line as shown in the first position in Fig. 3.17a.
b. Support this triangle with another triangle as in Fig. 3.17a or with the blade of the T-square.
c. Rotate the first triangle through 90° into the second position as shown in Fig. 3.17a.
d. Draw the line along the rotated position of the hypotenuse to obtain a line perpendicular to the original line.

Another method is illustrated in Fig. 3.17b, using the following steps (also see Section 4.2.6).

a. Align one leg of a triangle with the original line.
b. Support the hypotenuse of the triangle with the blade of the T-square or another triangle.
c. Slide the triangle along the hypotenuse to the second position.
d. Draw a line along the other leg of the triangle, thereby making a line perpendicular to the original line. Method (b) in Fig. 3.17 is better and quicker than method (a) in Fig. 3.17, since the triangle is slid along the T-square and does not have to be picked up and rotated.

In drawing a line in any direction, the general rule may be laid down that the direction of motion of the pen or pencil should be away from the body of the drafter and not toward it. To draw a straight line between any two points (not horizontal or vertical), place the pencil on one point, bring the triangle up to it, and then rotate the triangle about the pencil point as a pivot until the edge touches the other point. Then draw the line.

3.12 STANDARD DRAWING SHEET SIZES

Two series of drawing sheet sizes are recommended by the American National Standards Institute (ANSI Y-14.1-1980) as listed in Table 3.1.

3.12.1 Metric Comparison. The international (metric) drawing size is given in Table 3.2. While the sizes are not exactly the same they are similar enough so as not to cause undo difficulty.

3.13 DRAWING MEDIA

The type of drawing media used in industry varies from company to company and with the use that will be made of the drawings. Some companies use a buff-colored detailing paper for their original layouts. A 25% rag stock paper is normally used when good erasing quality is required.

Many industrial drawings are made directly on tracing paper. The drawing is finished with a medium pencil (H or F). These drawings can then be used for making "prints" directly from the drawings without any intermediate operations such as tracing or microfilming. Tracing paper may be obtained in several weights and with corresponding transparency. A paper that has been oil-treated is called *vellum*.

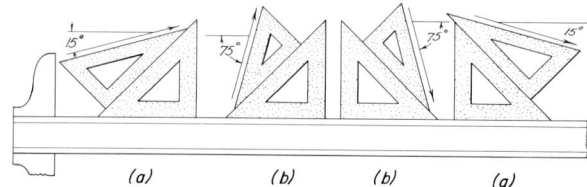

FIG. 3.15. Drawing angles of 15 and 75°.

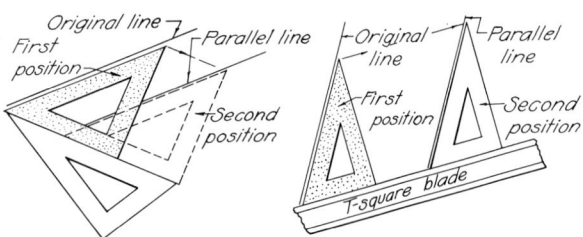

FIG. 3.16. Drawing parallel lines.

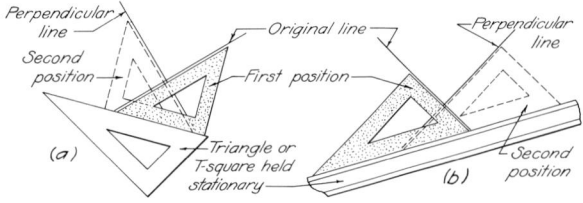

FIG. 3.17. Drawing perpendicular lines.

TABLE 3.1 STANDARD DRAWING SHEET SIZES

Flat Sizes

Size Designation	Width (Vertical)	Length (Horizontal)	Margin Horizontal	Margin Vertical
A (horizontal)	8.5	11.0	0.38	0.25
A (vertical)	11.0	8.5	0.25	0.38
B	11.0	17.0	0.38	0.62
C	17.0	22.0	0.75	0.50
D	22.0	34.0	0.50	1.00
E	34.0	44.0	1.00	0.50
F	28.0	40.0	0.50	0.50

Roll Sizes

Size Designation	Width (Vertical)	Length (Horizontal) Min	Length (Horizontal) Max	Margin Horizontal	Margin Vertical
G	11.0	22.5	90.0	0.38	0.50
H	28.0	44.0	143.0	0.50	0.50
J	34.0	55.0	176.0	0.50	0.50
K	40.0	55.0	143.0	0.50	0.50

Note: All dimensions are in inches. 1 in. = 25.4 mm.

TABLE 3.2 INTERNATIONAL AND U.S. DRAWING SIZE COMPARISON

International Designation	Width mm	Width in.	Length mm	Length in.	Nearest U.S. Size Letter	Nearest U.S. Size in.
A0	841	33.11	1189	46.81	E	34.0 × 44.0
A1	594	23.39	841	33.11	D	22.0 × 34.0
A2	420	16.54	594	23.39	C	17.0 × 22.0
A3	297	11.69	420	16.54	B	11.0 × 17.0
A4	210	8.27	297	11.69	A	8.5 × 11.0

3.13.1 Tracing Cloth. There are two types of tracing cloth that can be used when greater permanence is required. One is called a *pencil tracing cloth,* which has been treated to make the surface white so that the pencil lines show more clearly.

Regular tracing cloth has a bluish color and is quite transparent. It should be used exclusively for ink drawing since pencil lines do not show up well on this cloth. However, some companies do use pencil on ink tracing cloth. The cloth has a dull side, and the other side is extremely glossy. The glossy side does not take ink well, and it is recommended that the dull side be used for inking. Erasures of ink or pencil lines should be made only with a pencil eraser.

When the making of a tracing is to extend over several days, it is recommended that one view at a time be fully completed rather than working over the entire area. The cloth is quite responsive to changes in the moisture content of the air and will expand or shrink a great deal from one day to the next.

3.13.2 Glass Cloth. In recent years a great variety of plastic materials have been introduced for use as drawing media. Some of these are Mylar, Kronaflex, glass cloth, and drafting film. These materials have the advantage of not shrinking or expanding because of temperature and humidity. They hold their size and are dimensionally stable. Both pencil and ink can be used to draw on them.

A specially coated film is used for undimensioned drawings. The drawings are scribed onto the film with a metal scriber. The pointed metal scriber removes the opaque coating on the film. The aircraft industry uses this method of producing many of its drawings. This process is also used in making electronic circuitry layouts. These drawings are so accurate that they can be used as a template directly.

Photographic reproductions of drawings on a film base are used widely when making revisions on drawings. The parts of the drawing to be revised can be removed and the new revision can be drawn onto the same film master, thereby creating a new drawing. This saves a considerable amount of time by not having to redraw the parts of the drawing that do not require any changes.

3.14 PAPER FASTENERS

The drawing paper may be held on the board by means of fine-wire staples. These are driven into the board by a special stapler. The staples are quite fine and do not damage the board as much as thumbtacks, and they offer little obstruction to the T-square and triangles.

Masking tape is used extensively in fastening papers to the board by sticking small pieces of it across the corners of the paper.

3.15 SCALES

Drawings of large objects must be made smaller than the object because of the limited paper size. Some objects can be drawn full-size. Thus, if 1 in. on the drawing represents 2 in. on the object, the drawing is said to be half-size or half-scale and should be marked: Scale $\frac{1}{2}$ in. = 1 in. If 1 in. on the drawing represents $\frac{1}{2}$ in. on the object, it is said to be double-size and should be marked: Scale 2 in. = 1 in.

There are five basic scales in common use. They are so calibrated that they can be used directly without making any arithmetic computation for reduction. The five scales are: architect's, civil engineer's, mechanical engineer's, metric, and decimal. They can be obtained in either triangular or in several flat shapes as shown in Fig. 3.18. The reductions available are listed in Table 3.3.

3.15.1 The Architect's Scale. The basic unit at the end of the scale represents one foot and is subdivided into 12 parts to represent inches. See Fig. 3.19. On the larger scales, such as 3 in. = 1 ft. 0 in., the inch division is further subdivided so that the smallest subdivision may represent $\frac{1}{8}$ in. The smaller scales, on the other hand, do not have the basic unit divided into so many divi-

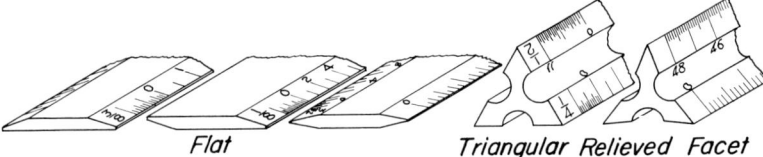

FIG. 3.18. Types of scales.

TABLE 3.2

Architect's Scales (size)		Civil Engineer's Scales	Mechanical Engineer's Scales	Metric Scales	Decimal Scale
Full	1″ = 0′1″	1″ = 10.0′ [a]	2″ = 1″	1:100	1.00″ = 1.00″
Quarter	3″ = 1′0″	1″ = 20.0′	1½″ = 1″	1:200	0.50″ = 1.00″
One eighth	1½″ = 1′0″	1″ = 30.0′	1″ = 1″	1:250	0.375″ = 1.00″
One twelfth	1″ = 1′0″	1″ = 40.0′	$\frac{3}{4}$″ = 1″	1:300	0.25″ = 1.00″
	$\frac{3}{4}$″ = 1′0″	1″ = 50.0′	$\frac{1}{2}$″ = 1″	1:400	
	$\frac{1}{2}$″ = 1′0″	1″ = 60.0′	$\frac{1}{4}$″ = 1″	1:500	
	$\frac{3}{8}$″ = 1′0″	1″ = 80.0′	$\frac{1}{8}$″ = 1″		
	$\frac{1}{4}$″ = 1′0″	1″ = 100.0′			
	$\frac{3}{16}$″ = 1′0″				
	$\frac{1}{8}$″ = 1′0″				
	$\frac{3}{32}$″ = 1′0″				

[a] Or some integral power of 10 ft as, for example, 100 or 1000 ft. The civil engineer's scales are convenient for scaling forces in a vector diagram.

sions. The smallest subdivision on the $\frac{1}{8}$ in. = 1 ft. 0 in. scale represents 2 in.

3.15.2 The Civil Engineer's Scale. The civil engineer's scale marked 10 has each inch subdivided into tenths. Figure 3.20 shows a 20 and 40 scale. These scales are fully divided and are used on map drawings as well as various other works in civil engineering. The scales available have the inch divided into 10, 20, 30, 40, 50, and 60 parts. These scales are also useful in graphic statics and nomography where the units may represent feet, pounds, stress, or pounds per square inch.

3.15.3 The Mechanical Engineer's Scale. On these scales the major end unit represents 1 in. and the subdivisions represent the commonly used fractions of an inch, $\frac{1}{2}, \frac{1}{4}, \frac{1}{8}, \frac{1}{16}$, and so forth, as shown in Fig. 3.21. These scales may be obtained either open- or full-divided. Fractions used in industry are always multiples of $\frac{1}{2}$, such as $\frac{1}{4}, \frac{1}{8}$, and $\frac{1}{16}$. Such fractions as $\frac{1}{3}, \frac{1}{5}, \frac{1}{6}, \frac{1}{7}$, or other odd numbers are never used.

3.15.4 The Metric Scale. The metric scale, which is being more widely used in this country, is set up on a decimal basis. The scales are 1 to 100, 200, 250, 300, 400, and 500. Thus, the 1:100 scale is 1 cm (10 mm) to 1 m and the 1:500 scale is 2 cm (20 mm) to 10 m. Using different powers of 10 and the six different scales on the metric scale it is possible to accommodate the needs of the drafter. See Fig. 3.22.

3.15.5 The Decimal Scale. Throughout industry in general, following the metric trend, there is a strong tendency toward decimal dimensioning rather than fractional. When parts are dimensioned in decimals, two-place decimals are used. These are made in even numbers, that is, fiftieths of an inch, so that when halved (diameters to radii) a two-place decimal results. A scale with graduations in fiftieths of an inch is shown in Fig. 3.23. Half-size and quarter-size decimal scales may be used in the drafting room. For additional information about these decimal scales, see ANSI Y-14.5-1973.

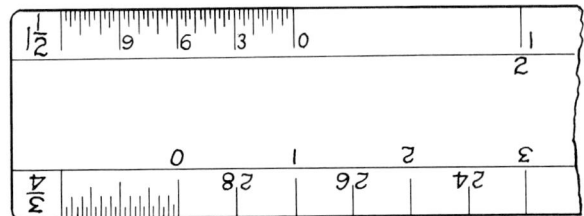

FIG. 3.19. Architect's scale.

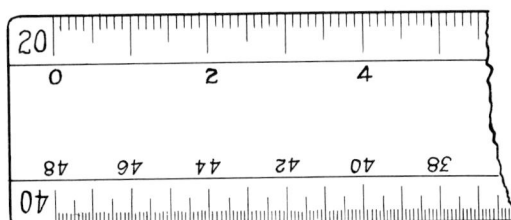

FIG. 3.20. Civil engineer's scale.

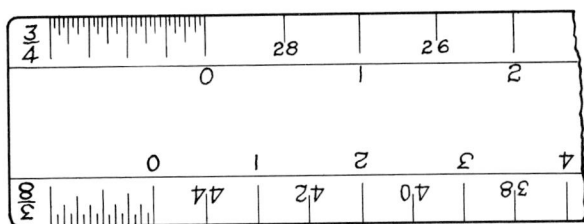

FIG. 3.21. Mechanical engineer's scale.

FIG. 3.22. Metric scale.

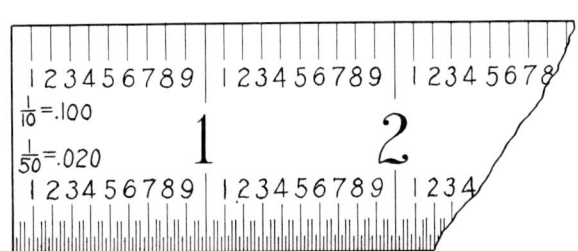

FIG. 3.23. Decimal scale.

3.16 USE OF SCALES

In Fig. 3.24, two illustrations show the use of scales. Note that in all cases the inner end of the first subdivided unit is marked 0 and the numbering goes from that point toward the other end of the scale. There are two lines of numbers, one of which applies to the scale at one end and the other to the scale at the opposite end. The larger of these scales is always just twice the smaller. To mark of 4 ft $3\frac{1}{2}$ in., set the scale with the division numbered 4 at one end and count off the $3\frac{1}{2}$ in. in the subdivided unit at the other end, as shown in Fig. 3.24a. Or, again, to mark off $4\frac{7}{16}$ in., set the mark numbered 4 at one end and count off the $\frac{7}{16}$-in. division in the opposite direction from the 0 as shown in Fig. 3.24b. To lay off a number of equal spaces, a scale rather than a divider should be used, since a minute error in setting a divider may result in a large cumulative error.

When a number of unequal distances are to be set off, the total of which lies within the length of the scale, it is best to add these distances arithmetically and then mark off the distances with the scale always at the same beginning or reference point. This avoids cumulative errors of setting.

To use the scale properly, one should place the scale on the paper with the working edge farthest from him and then, looking down over it, mark off the required dimensions with a very sharp, conical-pointed pencil or a needle point, as shown in Fig. 3.25. Note that a blunt pencil is of little value for this purpose, since the drafter cannot see where the point actually touches the paper. The pencil or needle point should be held in a vertical position.

3.17 CASE OF INSTRUMENTS

Instruments may be purchased singly or in groups, in cases designed to hold them. An adequate set for general drafting purposes should contain approximately the items illustrated in Fig. 3.1.

The following paragraphs make clear the proper methods for handling the various pieces of the set, what to look for in selecting them, and how to care for them and keep them in proper working order. The use to which they are subjected and the care they receive, when not in use, determine to a large extent how long they will last.

3.18 COMPASS

The large compass is one of the most important instruments in the drafter's kit. It should be adjusted before it is used for the first time and then maintained in that condition. One leg is arranged to hold either a lead or a pen. The pen should be put in the compass first and then the needle point

FIG. 3.24. Use of scales.

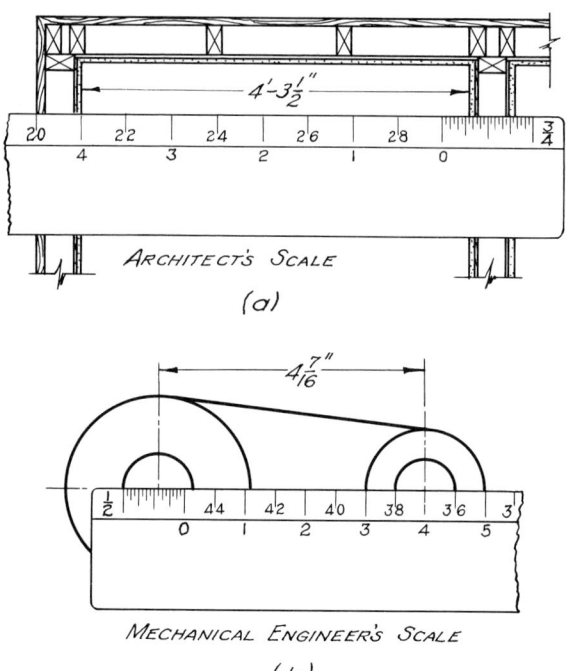

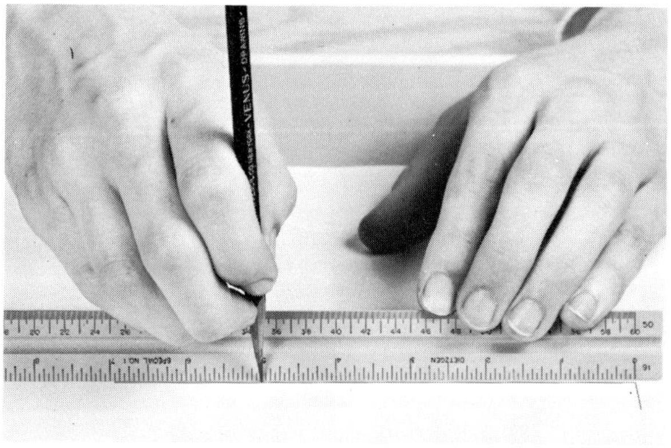

FIG. 3.25. Laying off dimensions with a scale.

adjusted so that it is about $\frac{1}{32}$ in. longer than the pen, as shown in Fig. 3.26. Thus when the needle point sinks into the paper the pen will be perpendicular to the paper and just touch it when the compass is held in a vertical position. The needle point, once adjusted, should not be changed. The length of the lead should be adjusted to match the needle point.

The lead for the compass should be of the same hardness as that used in pencil work. The lead should be sharpened to a bevel point on the outside as shown in Fig. 3.27, since it is easier to sharpen in this manner and permits the drawing of smaller circles.

In drawing small circles, the legs of the compass may be kept straight; but for the larger ones, one or both legs should be bent as in Fig. 3.28 in order that the pen may be perpendicular to and have both nibs touching the paper. When drawing circles in ink, it is best to move the pen clockwise and to go around the circle only once. However, to secure good black pencil lines it may be necessary to go over the circle several times. When a number of concentric circles are to be drawn, a horn center is very convenient. It prevents wearing a large hole in the paper.

For circles larger than the compass will accommodate, the beam compass should be used. This requires more skill since both hands must be employed, one to hold the needle point at the center and the other to move the pencil or pen point as shown in Fig. 3.29.

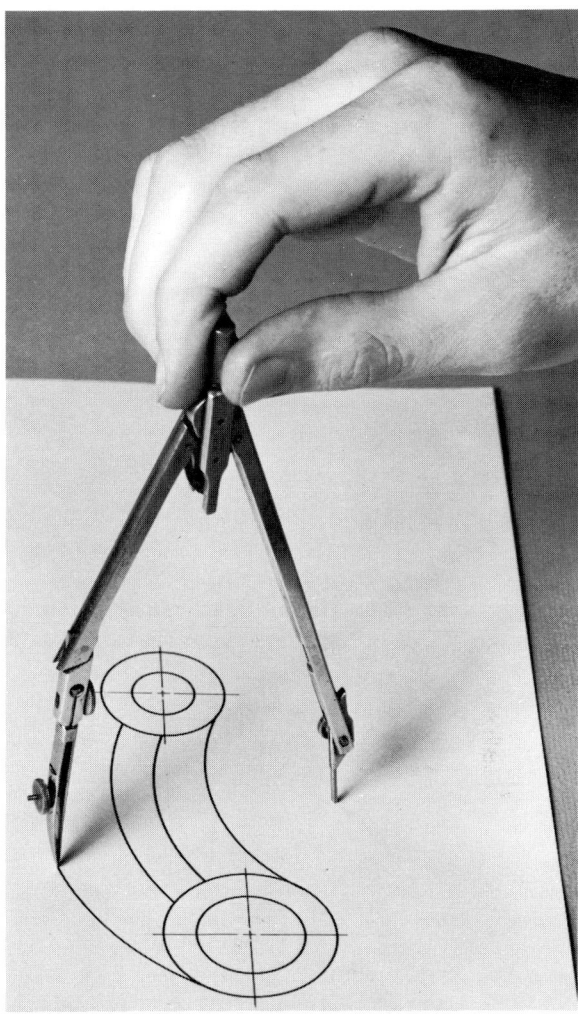

FIG. 3.28. Drawing with a compass.

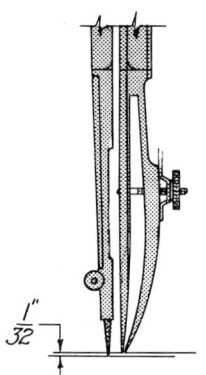

FIG. 3.26. Adjustment of a compass point.

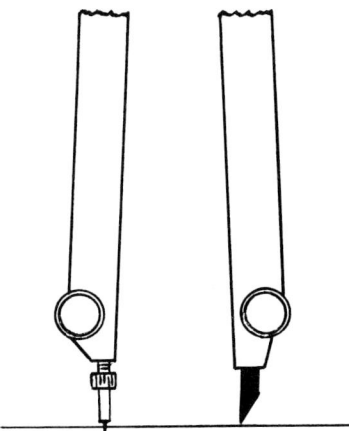

FIG. 3.27. Sharpening a compass lead.

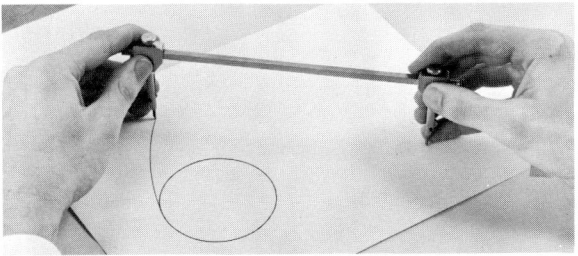

FIG. 3.29. Drawing with a beam compass.

3.19 DIVIDERS

Dividers are used chiefly for transferring distances and occasionally for dividing spaces into equal parts. To set the divider it should be held as shown in Fig. 3.30. If the distance to be set permits, the second and third fingers are placed inside to help control the movement of the points. If the space to be set is small, the divider must first be opened wider than the space and then closed down to the proper measurement by pressure from the thumb and fingers; or the hairspring adjustment may be used.

To step off distances, grasp the knurled top of the divider between thumb and first finger and rotate first in one direction and then in the other. This avoids taking a new hold on the instrument, which would be necessary if it were always turned in the same direction. The points of the instrument should not be pushed through the paper, but instead only dents need be made. These may be identified by immediately touching them with a sharp, soft pencil to make a dot or by lightly encircling them, after which they readily can be found.

3.20 BOW INSTRUMENTS

Bow instruments are small instruments that are used for drawing small arcs and circles and they can be obtained in a number of styles. Those with the adjusting screw on the outside are called *side-adjusting,* whereas those with the screw nut in the center are called *center-adjusting*. Which type to use is a matter of preference among drafters.

The most common fault with these instruments is the wearing and stripping of the threads on the adjusting screw. If the spring pressure is too great, they will wear rapidly. One of their chief advantages is the accuracy with which they may be set and the fact that they can be laid aside temporarily without danger of losing the setting.

3.21 IRREGULAR CURVES

There is such a wide variety of irregular or French curves produced that it is impossible to give more than a sampling of the different kinds within the limits of this book. In Fig. 3.31 are shown a few curves of common usefulness that are simply designated as irregular curves.

The irregular curve is one of the most difficult instruments to use skillfully, especially when doing ink work. The skill required lies not only in the handling of the pen but also, to a considerable extent, in the placing of the curve. The irregular curve should be made to fit as many of the points as possible at one time. In no case should the line be drawn when the curve does not fit at least three points and have the proper curvature with regard to the next points on both sides. Then when the curve is moved forward, the last two points should be rematched and a portion of the curve that was previously drawn

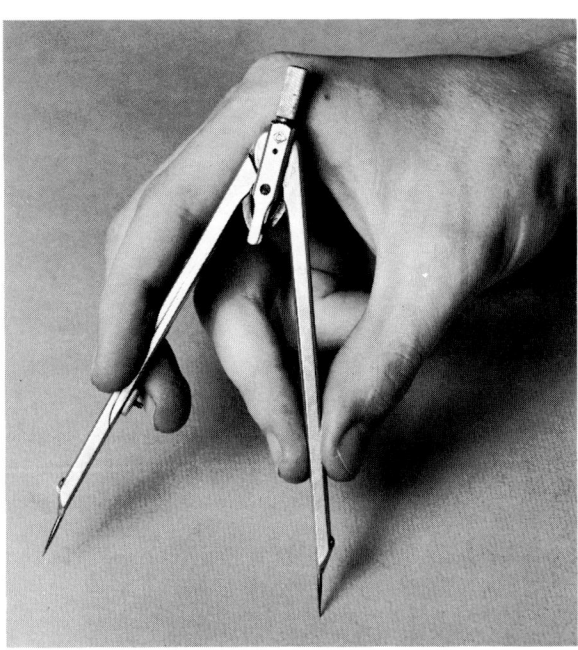

FIG. 3.30. Setting the divider.

FIG. 3.31. Irregular curves.

can be retraced to avoid humps in the final curve.

Figure 3.32 shows how the curve may be set several times to complete one curved line. In this figure two curves are used. Curve No. 1 fits the central part of the sine curve up to the point b. Curve No. 2 fits the upper part down to the point a^1. An overlap occurs between the points a and b, and each curve should be used to the center of the space between these points. Note that points a^1 and b^1 on curve No. 2 if reversed would coincide with a and b and the upper portion of the curved line under curve No. 1. Figure 3.32 shows that curve No. 2 comes tangent to the horizontal construction line at the top. Points of tangency should always be used as guides in setting the curve.

The pen should be held perpendicular to the paper in drawing an irregular curve.

When inking it may be advisable to tape three coins under the triangle and irregular curve. This raises the triangle or irregular curve above the paper and prevents ink from running under the triangle or curve and smearing.

3.22 INKING

In inking straight lines and curves that join each other, the curve should be inked first and then care taken to stop the arc exactly at the tangent point as shown in Fig. 3.33. The ruling pen should then be lined up with the end of the arc by looking down squarely over the pen, as shown in Fig. 3.33, before actually touching it to the paper. In order that the arc and straight line may have exactly the same width, which is absolutely necessary, a few sample arcs are made on scratch paper, which should be of the same kind as that on which the drawing is made. Then the width of the ruling pen is adjusted by drawing lines to these arcs and changing until a perfect match is secured. In all cases the ink line should straddle the pencil line over which it is drawn.

For convenience, a line gage may be made up, with the weights of lines in proper proportion drawn on a piece of heavy paper. This can be used as a standard to maintain uniformity in the weight of lines from one drawing to the next. For larger drawing a heavier set of lines should be used.

One of the common faults of beginners is that of running the lines beyond their proper stopping points. It should be remembered that the ink flows just a little ahead of the pen; therefore, when the line being drawn is to end upon another line, the pen must be stopped at the near

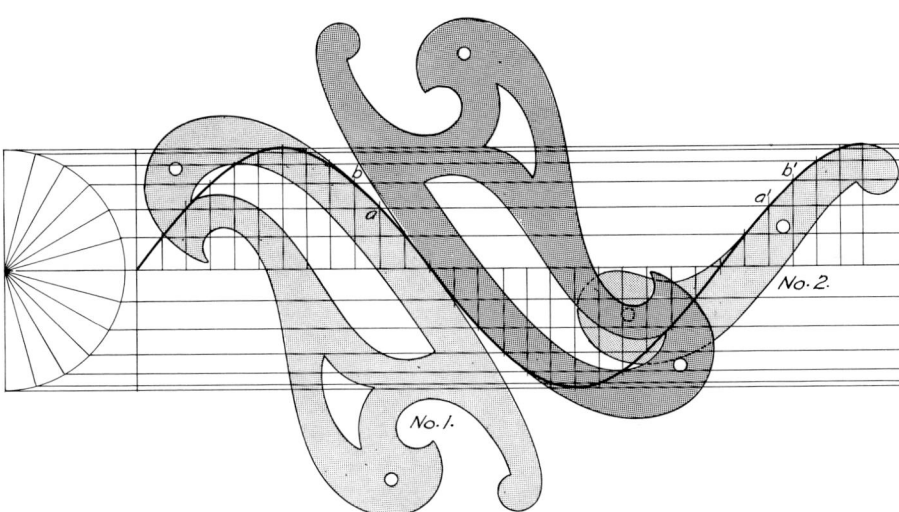

FIG. 3.32. Use of irregular curves.

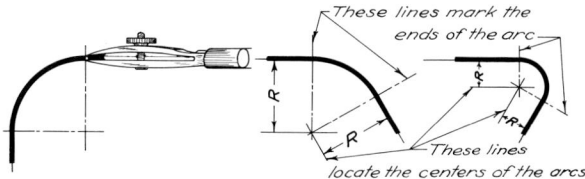

FIG. 3.33. Joining arcs and straight lines.

46 USE AND CARE OF INSTRUMENTS

side of the line. A little practice will enable the drafter to determine just when to stop the pen to avoid overrunning.

3.23 DRAWING INK

One cannot discuss the use of the ruling pen without giving consideration to drawing inks because the ink has a decided influence on the results produced. Drawing inks may be obtained in black and a wide variety of colors. The black ink is a combination of carbon and a solvent. The carbon, however, is in suspension and not in solution; consequently it is a very thick ink and the bottle will need to be shaken periodically if not in constant use. Make sure the top is securely fastened. The solvent contains alcohol and evaporates very readily. The ink bottle should therefore always be kept tightly closed, except when filling the pen. This practice also avoids upsetting an open ink bottle with the consequent spoiling of drawings and equipment. Most of the colored inks are true solutions and are much thinner than the black India ink. One must be particularly careful not to fill the pen too full with colored inks because the ink will run out of the pen much more readily when it touches the paper.

Since India ink dries very rapidly, it forms little cakes of carbon on the inside of the nibs; therefore, the pen must be cleaned frequently while in use. It must be always cleaned before it is put away. The pen is cleaned by inserting a piece of chamois or soft cloth between the nibs while the ink is still moist and then pulling the cloth out. A clogged pen is one of the most common causes of poor lines. It should be noted also that due to evaporation the ink gradually thickens in the bottle. It then dries out very rapidly in the pen and is difficult to use. Ammonia may be used to thin the ink. Normally, it is best to purchase a fresh bottle. Another cause of poor lines is the presence of lint, dust, and dirt on the paper. The ruling pen will pick these up and cause a sudden widening of the line.

3.24 QUALITY OF INK LINE

In order to give a drawing "life and vigor" there must be a variation in the weight of the different lines employed. The outline of the object should stand out sharply, with the hidden lines somewhat less prominent. Dimension lines, auxiliary lines, center lines, and cross-hatching should be still lighter. The weight and character of lines shown in Fig. 3.34 are those recommended by the ANSI. These weights of lines may be varied somewhat in accordance with the size and nature of the drawing, but in any event three distinct weights should be maintained.

In drawing hidden lines or any other line of an interrupted character, the pen should be brought to a full stop before it is lifted from the paper. This will produce a square-ended line as shown in Fig. 3.35, whereas lifting the pen while it is in motion causes the line to fade out with a ragged end. The secret of making good-looking hidden lines lies in

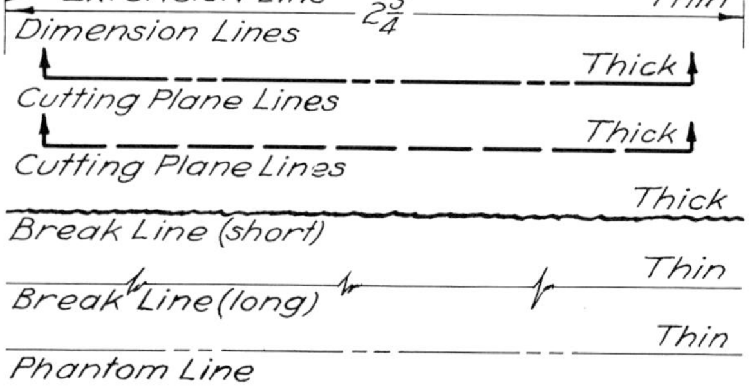

FIG. 3.34. Proper weight of lines (courtesy ANSI).

FIG. 3.35. Hidden line technique.

making the dashes of uniform length and the spaces between them also uniform but very small, say about $\frac{1}{32}$ in.

When a number of lines converge to a point, the best practice is to run only the two outside lines to the point and stop all intermediate lines along the arc of a circle just large enough to prevent the lines from touching. If the converging lines are very numerous as in charts, two arcs may be used and alternate lines may be stopped on the inner and outer arcs as illustrated in Fig. 3.36.

3.25 ORDER OF INKING

In inking a drawing or making a tracing, a certain order or procedure should be followed to give the best results in the least time. A good drawing must have uniformity and contrast. That is, there must be uniformity among lines of any one kind and contrast between lines of different kinds. To obtain this result, the lines should be inked in the following order.

   a. Visible outlines, all of the same weight.
      1. Circles and arcs of circles and other curved lines.
      2. Horizontal lines beginning at the top.
      3. Vertical lines beginning at the left.
      4. Inclined lines.
   b. Invisible outlines in the same order as in (a) above.
   c. Center lines in the same order as in (a) above. Some drafters prefer to ink center lines first.
   d. Cross-hatching light lines, evenly spaced.
   e. Dimension lines in the same order as in (a) above.
   f. Dimensions, arrowheads, and other lettering.
   g. Borderlines, title box, trim lines.

The importance of drawing arcs and circles first should not be overlooked for one can always make a straight line tangent to one or two arcs, but it is extremely difficult to make an arc or curve tangent to two straight lines, particularly if one of them should be just slightly out of place.

3.26 TECHNIQUE OF DRAFTING

The primary essentials of good drafting techniques are speed, accuracy, and neatness. A few basic rules to help achieve these qualities are as follows.

   a. Do not sharpen pencil over the drawing board.
   b. Knock excess graphite off file or sandpaper pad on the leg of table or chair at a point near the floor. Do this immediately after sharpening pencil.
   c. Always keep pencil sharp.
   d. Do not fill pen or compass over the drawing board.
   e. Time will be saved by using a scale guard on the triangular scales to indicate the scale in use. Otherwise, a good deal of time is lost hunting for the correct scale.

3.27 PENCIL TECHNIQUE

In making a pencil drawing it is important to maintain a good, sharp point on the pencil. Generally a harder pencil such as a 4H or 5H is used to lay out the problem. Then a 3H or softer pencil is used to draw in the detail of the drawing. Some of the things to consider in making a drawing look "sharp" are as follows.

   a. Do not overrun corners.
   b. Do not press hard on the pencil to cause grooves.
   c. Maintain proper weight distinction between different types of lines.
   d. Remove all construction lines.
   e. Do not make center lines or projection lines too long.

3.28 MICROFILM TECHNIQUE

Many companies are using microfilming methods to produce shop prints of their drawings. Often the drawings are reduced as much as 30 to 1

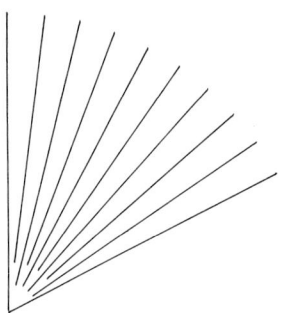

FIG. 3.36. Drawing converging lines.

when they are photographed onto microfilm. Therefore, the techniques discussed in Articles 3.25 and 3.27 must be adapted to take into consideration the allowances for reductions.

The line work must be sharp and uniform in weight. It is better to make the lines heavier than normal to ensure their being reproduced. In the case of cross-hatching, the lines must not be too close together.

3.29 SECTION LINERS

For small areas the drafter spaces the cross-hatching by eye. When the area is large or when the work is very important and is to be reproduced, the cross-hatching can be spaced with section liners.

The drafter may construct devices of his/her own that will aid in the even spacing of cross-hatching lines. By scratching a line on the triangle, say $\frac{1}{16}$ in. from the edge and parallel to it, this line can be used as a guide in doing the pencil work. The first cross-hatching line is drawn and then the triangle is moved up so that the line on the triangle is over the one on the drawing and a second line is drawn. Two or more lines can be made on the triangle to give different spacings or different spacings can be made on the various edges as illustrated in Fig. 3.37.

A small piece of heavy cardboard or thin veneer can be cut to the shape of the hole in the triangle but slightly smaller. By shifting the triangle and block alternately, accurate spacing may be obtained. For small areas the middle line of holes in the Ames lettering instrument may be used.

3.30 SPECIAL CELLULOID TOOLS

A variety of special celluloid forms are on the market whose purpose and usefulness are self-evident. A number of these are shown in Fig. 3.38. Ellipse guides, which may be obtained in a wide variety of sizes, are particularly useful in pictorial drawing. Circle guides are valuable time-savers.

3.31 PROTRACTORS

Protractors may be obtained in several sizes, made of plastic, or in the more expensive varieties of polished steel having vernier attachments for accurate setting of the ruling edge. Several types are shown in Fig. 3.39. They are used to measure angles and are particularly useful in map work. The usual divisions are in degrees and fractions thereof, though other divisions such as percentage parts of a circle may be obtained.

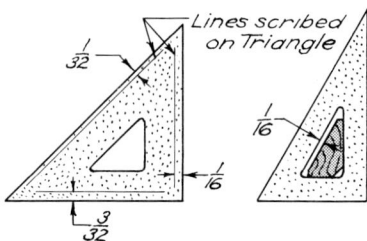

FIG. 3.37. Aids in section lining.

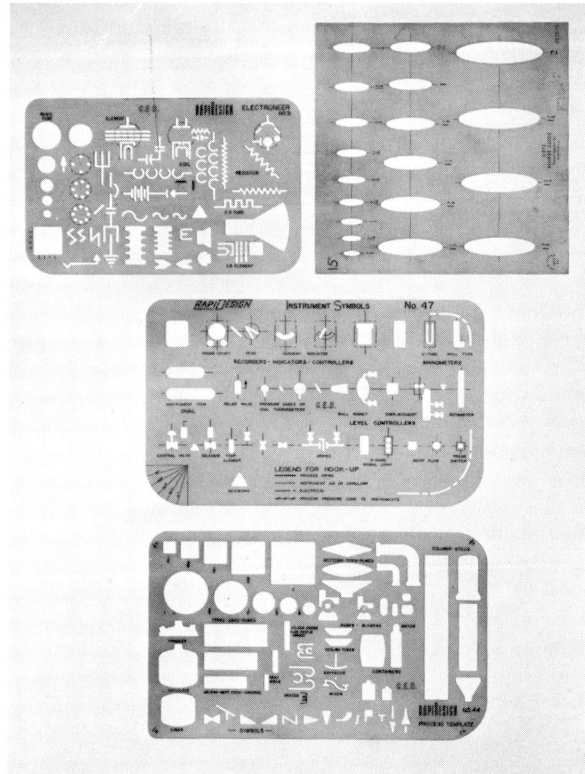

FIG. 3.38. Special drafting templates.

3.31 SPECIAL CELLULOID TOOLS   49

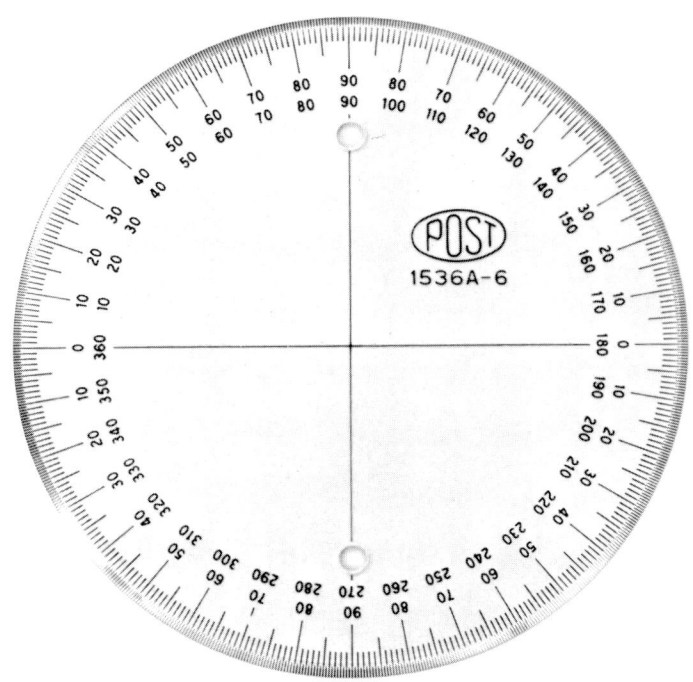
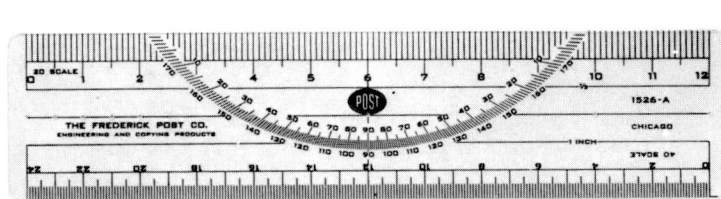
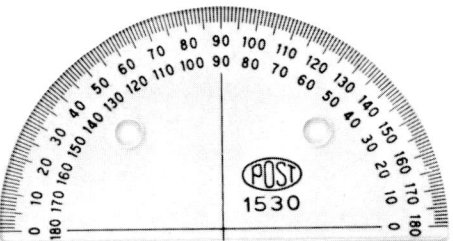

FIG. 3.39. Protractors (courtesy of Teledyne Post).

SELF-STUDY QUESTIONS

Before trying to answer these questions, read the chapter carefully. Then, without reference to the text, answer as many questions as possible. For those that cannot be answered, the number in parentheses following the question number gives the article in which the answer can be found. Look it up and write down the answer. Check the answers that you did give to see that they are correct.

3.1 (**3.1**) To make an accurate drawing _____ are used.

3.2 (**3.3**) The softest drawing pencil is _____.

3.3 (**3.3**) The hardest drawing pencil is _____.

3.4 (**3.3**) For layout work, a good pencil to use would be _____.

3.5 (**3.3**) For sketching the _____, pencils are used.

3.6 (**3.3**) For finishing the drawing, a _____ pencil could be used.

3.7 (**3.3**) For general purpose work, the _____ point is used.

3.8 (**3.3**) In drawing a line the pencil should be _____ in the direction of motion.

3.9 (**3.3**) The _____ point is used for drawing long lines.

3.10 (**3.3**) The bevel point is used in the _____.

3.11 (**3.4**) To remove lines, the _____ should be used.

3.12 (**3.4**) To protect the surrounding parts of the drawing, an _____ should be used.

3.13 (**3.16**) To avoid cumulative errors, a _____ should be used in laying off a series of equal distances along a line.

3.14 (**3.17**) A _____ is used for inking straight lines.

3.15 (**3.17**) The _____ of a ruling pen must be properly sharpened.

3.16 (**3.22**) The _____ should be inked first and then the tangent lines.

3.17 (**3.18**) The _____ is used to draw arcs and circles.

3.18 (**3.18**) The _____ _____ is used for drawing very large circles.

3.19 (**3.19**) The _____ are used to transfer distances.

3.20 (**3.27**) To keep the drawing neat, do not _____ corners.

3.21 (**3.27**) In finishing a drawing remove all _____.

3.22 (**3.25**) In inking a drawing the _____ lines should be inked first.

3.23 (**3.30**) To measure angles on a drawing, the _____ is used.

3.24 (**3.7**) The two parts of the T-square are the _____ and the _____.

3.25 (**3.9**) The drafting machine takes the place of the _____, _____, and _____.

3.26 (**3.10**) The two most common triangles in use are the _____ and the _____.

3.27 (**3.11**) Two triangles can be used to draw one line _____ to another line.

CHAPTER 4

GEOMETRIC CONSTRUCTION

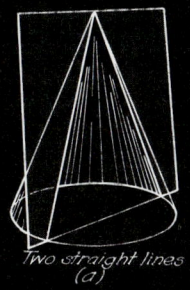
Two straight lines
(a)

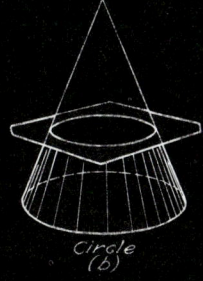
Circle
(b)

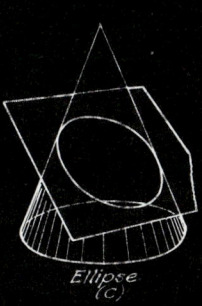
Ellipse
(c)

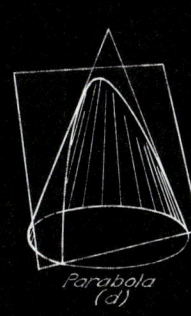
Parabola
(d)

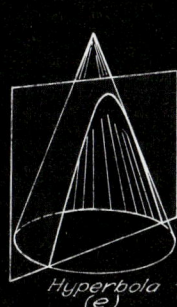
Hyperbola
(e)

CHAPTER 4

4.1 INTRODUCTION

To be successful in drafting or engineering design and in engineering practice in general, engineers must have at their command many of the simpler geometrical constructions shown in this chapter. A great variety of some of the more common as well as some of the more difficult constructions are presented. These constructions make this book an excellent reference for use in subsequent courses as well as in engineering practice. The drafter, technician, and engineer have a wide variety of drafting equipment available to aid them in making various geometric constructions. By using combinations of these drafting tools, many shortcuts are possible to make work simpler. Where possible, a practical application of the particular geometric construction is shown in the same figure in which the construction procedure is outlined.

4.2 LINES

It is often necessary to divide lines or spaces into a given number of parts and to draw lines parallel or perpendicular to each other. The following paragraphs deal with a number of simple geometrical constructions that can be used by the drafter and engineer in their work.

4.2.1 Bisect a Line—Compass Method.
A line can be bisected by using a compass and straight edge as shown in Fig. 4.1. The construction detail is as follows.

a. Set the compass using a radius r larger than half the length of the line segment AB to be bisected.

b. Using points A and B, respectively, draw two arcs with radius r from each point as indicated in Fig. 4.1b. It is important to have a good sharp point on the lead in the compass and to be certain that the steel point is placed at the exact end points of the line.

c. At the intersections C and D of the construction arcs, draw a line between them using a straight edge or triangle as is shown in Fig. 4.1c. Line CD is the perpendicular bisector of the line AB and divides the line into two equal parts.

4.2.2 Bisect a Line—Triangle Method.
A line can be bisected by using a triangle and the T-square as shown in Fig. 4.2. The construction details are outlined below.

a. Through the two end points of line AB, draw two lines through each of the points as shown in Fig. 4.2b. Care must be taken to have the line pass through points A and B and to have the construction lines each make the same angle with given line AB.

b. Points C and D determined by the intersections of the four construction lines are used to draw line CD as shown in Fig. 4.2c. This line is the perpendicular bisector of the line segment AB.

4.2.3 Line Parallel to Another Line through a Given Point—Compass Method.
By definition, parallel lines meet at infinity. This means that they are constantly a uniform distance apart. In the case of drawing a line through a point parallel to a given line, the uniform distance is the

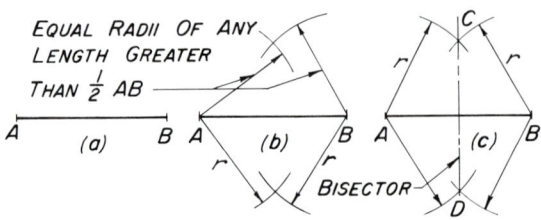

FIG. 4.1. Bisecting a line using a compass.

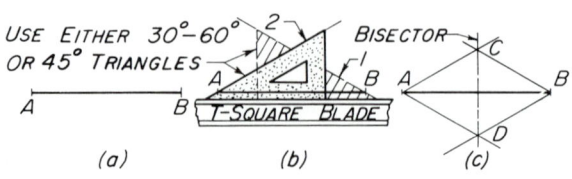

FIG. 4.2. Bisecting a line with a triangle.

shortest distance from the point to the line. This is the perpendicular distance from the point to the line. Figure 4.3 shows the steps in drawing a line through point A parallel to line AB. The steps are as follows.

a. Set the compass with radius R which is the shortest distance from point A to line CD as shown in Fig. 4.3b. The shortest distance R is the perpendicular distance from A to CD.
b. Using the radius R, strike two or more arcs with centers on the given line BC as shown in Fig. 4.3c.
c. Through point A draw a line tangent to the construction arcs. This line is parallel to line BC.

4.2.4 Line Parallel to Another Line through a Given Point—Triangle Method. Figure 4.4 shows the procedure that can be used to construct a line parallel to another line a given distance apart by using two triangles or a triangle and T-square. The construction is performed as follows.

a. Place a triangle in position 1 along line BC as shown in Fig. 4.4b.
b. Place a second triangle or the T-square in contact with the first triangle as shown in Fig. 4.4b.
c. Move the first triangle along the second triangle from position 1 to position 2, making sure that in position 2 it lines up with point A as shown in Fig. 4.4c.
d. Through point A draw a line along the edge of the first triangle in position 2. This line is parallel to the given line BC and passes through point A.

4.2.5 Line Parallel to a Curved Line. Figure 4.5 shows the compass method applied to drawing a curved line parallel to another curved line a given distance r apart. The compass is set at radius r as in Fig. 4.5. A series of arcs are drawn from points along the given curved line. A sufficient number of arcs must be drawn close enough to each other so that a smooth curve can be drawn tangent to these arcs as shown in Fig. 4.5. This method breaks down when the curve is too sharply concave.

4.2.6 Line Perpendicular to Another Line through a Given Point—Triangle Method. In Articles 4.2.1 and 4.2.2 two construction methods are discussed by which a perpendicular bisector of a line segment is obtained. Many times it is necessary merely to draw a line through a point perpendicular or at 90° to a line. The most convenient method of doing this is shown in Fig. 4.6 by using a triangle and a T-square or two triangles. The construction is as follows.

a. Place triangle in position 1 aligning one edge along line AB as shown in Fig. 4.6b.
b. Place a T-square or another triangle in contact with the first triangle as shown in Fig. 4.6b.
c. Keeping the T-square in place, rotate the triangle 90° into position 2 passing through point C as shown in Fig. 4.6c.
d. Draw a line through point C along the edge of the triangle to line AB.

This construction not only provides a method for constructing a perpendicular line through a point to a line but also enables the determination of the shortest distance from a point to a line.

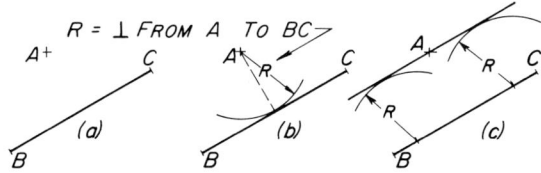

FIG. 4.3. Drawing a line through a point parallel to a line.

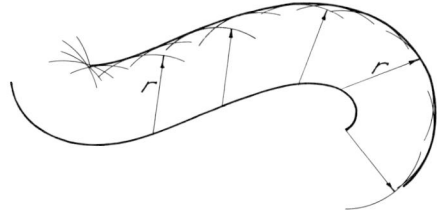

FIG. 4.5. Drawing a curved line parallel to another.

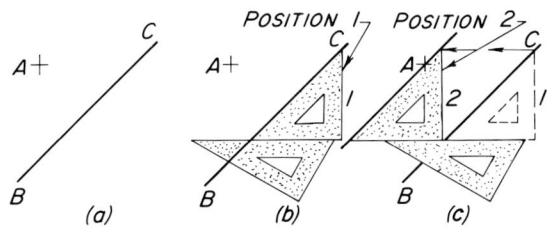

FIG. 4.4. Drawing a line through a point parallel to a line.

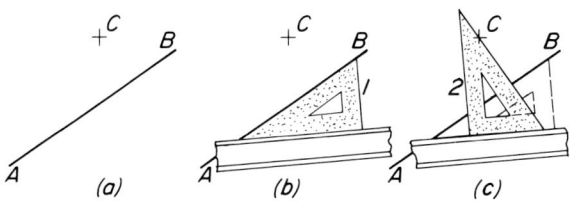

FIG. 4.6. Drawing one line perpendicular to another.

54 GEOMETRIC CONSTRUCTION

4.2.7 Divide a Line or Space into a Given Number of Equal Parts—Even Parts and Even Lengths. When a line must be divided into an equal number of parts, a number of methods are available. When the line is an even dimension so that it can be readily divided into easily measurable units on a scale, the points are merely marked off on the line along the scale.

As an example, if a 2-in. line is to be divided into four equal parts, lay off $\frac{1}{2}$-in. increments along the line with a scale and a good sharp-pointed pencil. A construction method for dividing a line in halves, fourths, eighths, and so forth is described in Article 4.2.11 and shown in Fig. 4.10. This is based on the principle discussed in Article 4.2.2.

4.2.8 Divide any Line or Space into a Given Number of Equal Parts or Odd Number of Parts. To divide a line into an odd number of equal parts, a construction similar to that shown in Fig. 4.7 is used. Thus in Fig. 4.7a the 2-in. line AB is to be divided into five equal parts. Proceed as follows.

a. Draw a line AC through point A making a convenient angle with line AB as in Fig. 4.7a.
b. Lay off five equal spaces along line AC as shown in Fig. 4.7b.
c. Draw a construction line from point 5 to point B.
d. Draw lines parallel to line 5-B through points 4, 3, 2, and 1, respectively, as shown in Fig. 4.7c. The points at which these parallel lines intersect the given line AB determine the five equal segments on line AB.

Figure 4.7d illustrates the application of this principle to practical problems.

4.2.9 Divide a Line or Space into Proportional Parts—Parallel Line Principle. Figure 4.8 shows how the principle of Fig. 4.7 can be applied to divide a line into any type of proportional parts. To construct a single scale, as in Fig. 4.8, proceed as follows.

a. From either end of line AB draw a line like AC at any convenient angle.
b. Mark off on this line the desired units.
c. For a scale of squares mark the significant points of the squares desired as in Fig. 4.8, for example, 1, 4, 9, 16, and 25.
d. Connect the end point 25 for squares to B.
e. From the other points 1, 4, 9, and 16, draw lines parallel to 25-B.
f. Mark the points on line AB as 1, 2, 3, 4, and 5 thus making a scale of the squares of these numbers.

For the logarithmic scale use as many points on the original scale as desired. To construct a logarithmic scale proceed in the same manner, marking off logarithms of numbers on the line AC.

4.2.10 Construction of Logarithmic Scales—Triangle Principle. It is often desirable to have available a series of scales of different lengths for some function such as logarithms. This can be accomplished as shown in Fig. 4.9.

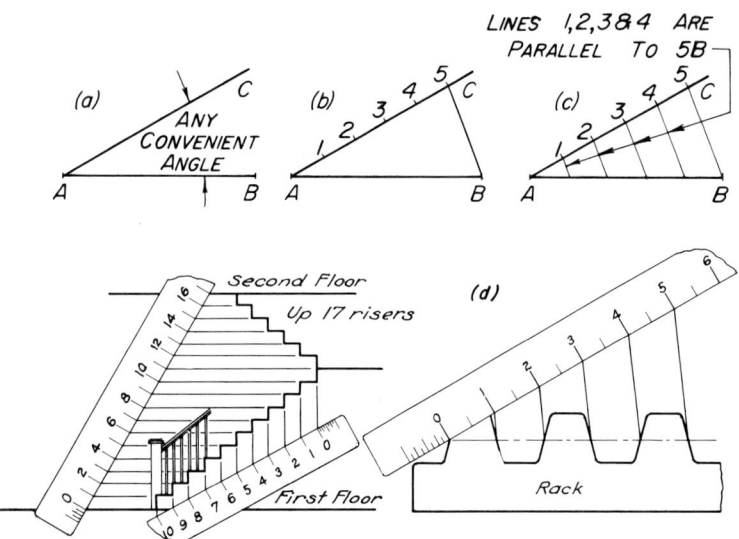

FIG. 4.7. Dividing a line into any number of equal spaces.

a. Construct the original scale or use one already made and fasten it to the drawing paper.
b. Select a point A approximately opposite the center of the original scale and at a perpendicular distance equal to the length of the scale. This distance is for convenience only. Thus, if the original scale is 5 in. long and point A is 5 in. from the scale, as shown in Fig. 4.9, a line 4 in. from A will give a 4-in. scale.
c. Draw lines from the significant points on the original scale to point A.
d. For a 3-in. scale, fold the drawing paper along the 3-in. line.

4.2.11 Divide a Line or Space into Fractional Parts. The construction for dividing a line into fractional parts is shown in Fig. 4.10. Assume that line AB in Fig. 4.10a is to be divided. The procedure is as follows.

a. Construct a rectangle of any convenient size on line AB.
b. Draw both diagonals.

c. A vertical line from intersection E divides AB into two equal parts at M.
d. For a one-third part of AB draw a line from D to M.
e. Where line DM crosses the original diagonal at F, draw a vertical line to AB, thus locating G the one-third point.
f. Further subdivision can be made in a similar manner, as shown in Fig. 4.10b.

4.2.12 Line through a Point P Making a Common Intersection With Two Other Lines AB and CD when their Intersection is Inaccessible. The steps in the construction of a line passing through a given point intersecting two other lines when their intersection is inaccessible are shown in Fig. 4.11 and are as follows.

a. With P as one vertex, draw any convenient triangle 1 as shown in Fig. 4.11b, having the other vertices on the two given lines.
b. At any other convenient place draw a triangle 2 whose sides are parallel to triangle 1 as shown.
c. The line through P and the corresponding

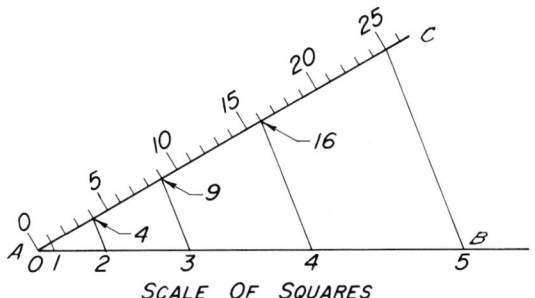

FIG. 4.8. Proportional functional scales.

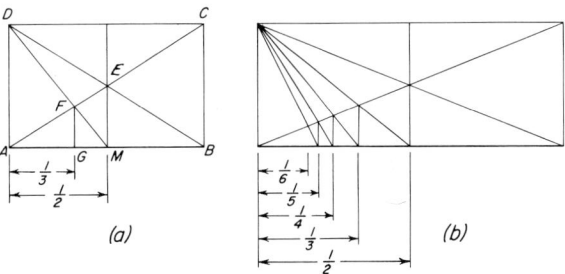

FIG. 4.10. Dividing a line or space into fractional parts.

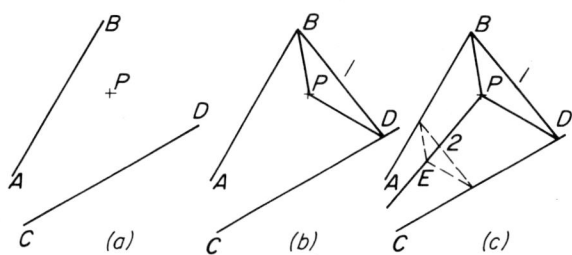

FIG. 4.11. Intersecting lines when intersection is inaccessible.

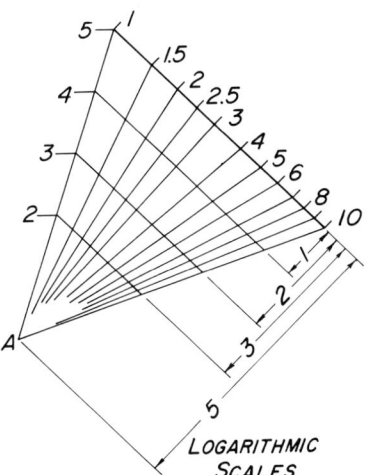

FIG. 4.9. Functional scale of different lengths.

point E will meet the given lines in a common intersection.

4.3 ANGLES, DEFINITION OF TERMS

The meanings of terms used in discussing angles are shown and defined in Fig. 4.12. The construction of angles of various sizes are commonly required in an engineering office.

4.3.1 Bisect an Angle. Figure 4.13 illustrates the method of bisecting any given angle.

a. With any convenient radius R and the center at A, draw an arc 1 across angle BAC intersecting AC at N and AB at M, as shown in Fig. 4.13a.

b. With any radius greater than one half the arc MN and centers at M and N, draw arcs 2 and 3 intersecting at point P as in Fig. 4.13b.

c. A line from A through this intersection P divides angle BAC into two equal parts, as shown in Fig. 4.13c.

4.3.2 Layout an Angle Equal to a Given Angle. One angle can be constructed equal to another as shown in Fig. 4.14.

a. With A as a center, draw an arc of any convenient radius R across the angle, thus locating m and n (Fig. 4.14b).

b. With m as a center, set the compass to radius r, so that the arc will pass through n.

c. At the new location on A'B' reconstruct the angle using the same radius R to draw arc 1, then with radius r and center at m' draw arc

2. The intersection of arcs 1 and 2 locates n'. Draw A'n' to make the required angle, as in Fig. 4.14c.

4.3.3 Line Perpendicular (90°) to Another Line at the End of the Line. One line may be drawn at right angles (90°) to another line as shown in Fig. 4.15. The construction is as follows.

a. Bisect the given line AB as shown in Fig. 4.15a.

b. At any point C on the bisector and a radius CB, draw an arc through B greater than a semicircle (Fig. 4.15b).

c. Draw the diameter BCD.

d. Draw line DA, which is at right angles to line AB (Fig. 4.15c).

The geometric principle used in this construction is that lines drawn from the ends of a diameter of a circle to any point on the circumference of it form a right angle with each other.

4.3.4 Line Perpendicular (90°) to Another Line by the 3, 4, 5 Method. For constructing a 3–4–5 right triangle, the steps are as follows.

a. On the given line AD in Fig. 4.16a lay off three equal spaces from B to C of any convenient length.

b. At B draw a line in a convenient direction and on it lay off four spaces from B to E each equal to those between B and C (Fig. 4.16b).

c. At C draw another line and on it lay off five

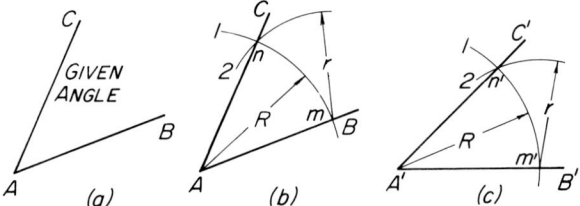

FIG. 4.14. Constructing an angle equal to a given angle.

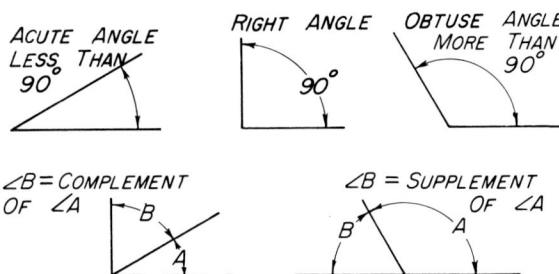

FIG. 4.12. Definition of terms for angles.

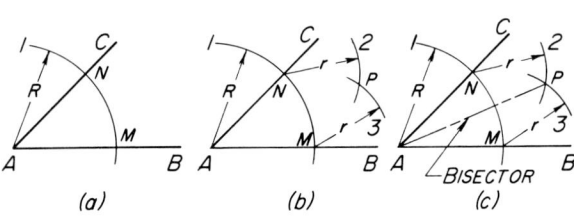

FIG. 4.13. Bisecting an angle.

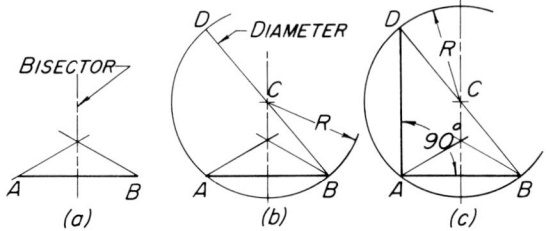

FIG. 4.15. Perpendicular to a line at the end of the line.

equal spaces from C to F of the same length as the others, as in Fig. 4.16c.

d. With B as a center and BE as a radius, draw an arc to the right.
e. With C as a center and CF as a radius, draw an arc intersecting the first one at G (Fig. 4.16c).
f. Draw the line GB, which will be at right angles to line AD.

This method is used in the field to lay out right angles in construction. It was used in ancient times by the Egyptians.

4.3.5 Lay Out Any Given Angle with a Protractor. Assume that any angle is to be constructed at B on line AC, as shown in Fig. 4.17.

a. Place the protractor along line AC with its center at B.
b. Mark a point D at the required angle (59.5° clockwise from A in this case).
c. Remove the protractor and draw line BD.

4.3.6 Lay Out an Angle by the Tangent Method. Assume that the angle (35°) is to be laid out at A on line AB as shown in Fig. 4.18.

a. On the given line AB measure off 10 spaces to x, at any convenient scale.
b. Look up the tangent of the required angle in a table of tangents or use a pocket calculator (35° in this case = 0.700) and multiply it by 10. The tangent of an angle is the ratio of the opposite side over the adjacent side in a right-angle triangle.
c. Measure this distance on a line perpendicular to AB, from x thus locating y as shown in Fig. 4.18b.
d. Connect A with point y, thus making angle BAC equal to 35°.

4.3.7 Lay Out an Angle by the Sine Method. For laying out any angle by the sine method such as 35°, the construction is shown in Fig. 4.19 and the steps are as follows.

a. Lay out 10 equal spaces, to any convenient scale, from A to x on line AB as shown in Fig. 4.19a.
b. Multiply the sine of the angle by 10. The sine of 35° = 0.574. Therefore, 0.574 × 10 = 5.74.
c. With x as a center and 5.74 as a radius, draw an arc.
d. Draw a line from A tangent to the arc as shown in Fig. 4.19c, thus constructing the required angle. This construction is based on the definition of the sine of an angle, namely, the opposite side over the hypotenuse in a right triangle.

4.4 CONSTRUCTION OF POLYGONS

Any geometric figure enclosed entirely by straight lines may be called a polygon. Regular

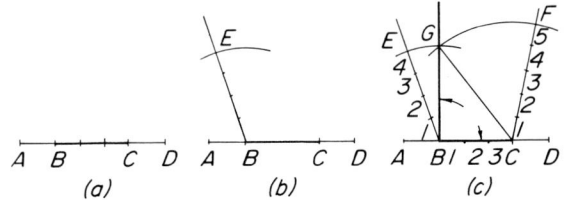

FIG. 4.16. Perpendicular to a line by the 3, 4, 5 method.

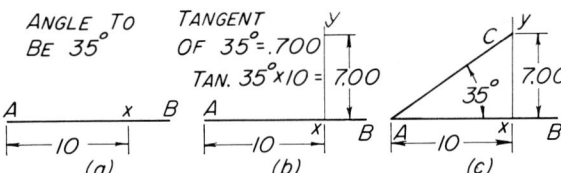

FIG. 4.18. Plotting angles by the tangent method.

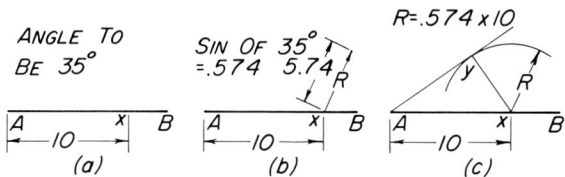

FIG. 4.19. Plotting angles by the sine method.

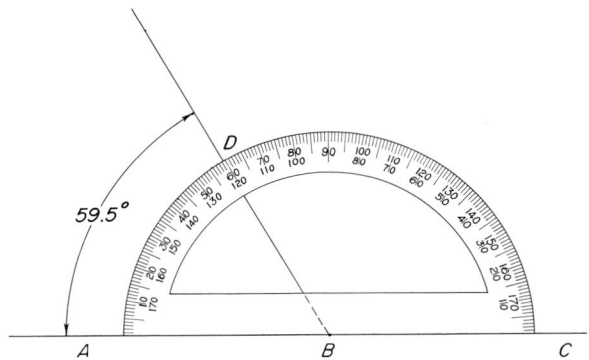

FIG. 4.17. Plotting angles by the protractor method.

58 GEOMETRIC CONSTRUCTION

polygons have their sides and interior angles equal. The more common polygons are shown in Fig. 4.20.

4.5 OTHER TYPES OF POLYGONS
Other plane figures, with their proper names, are illustrated in Fig. 4.21. In the case of the last two figures the term *trapezium* is sometimes applied to either as a generic term for all irregular four-sided figures.

4.6 CONSTRUCTION OF TRIANGLES
To construct a triangle, three parts must be known; for example,

a. Three sides.
b. Two sides and included angle.
c. Two angles and the side between them.

4.6.1 Construct a Triangle with Three Sides Given. The construction of a triangle with three sides given is shown in Fig. 4.22 and is as follows.

a. Lay out the longest *side a* as a base.
b. With the ends *A* and *B* of this line as centers, draw two arcs having radii equal to the lengths of the other two *sides b* and *c* as shown in Fig. 4.22c.
c. From *A* and *B* draw lines to *x*, thus completing the triangle.

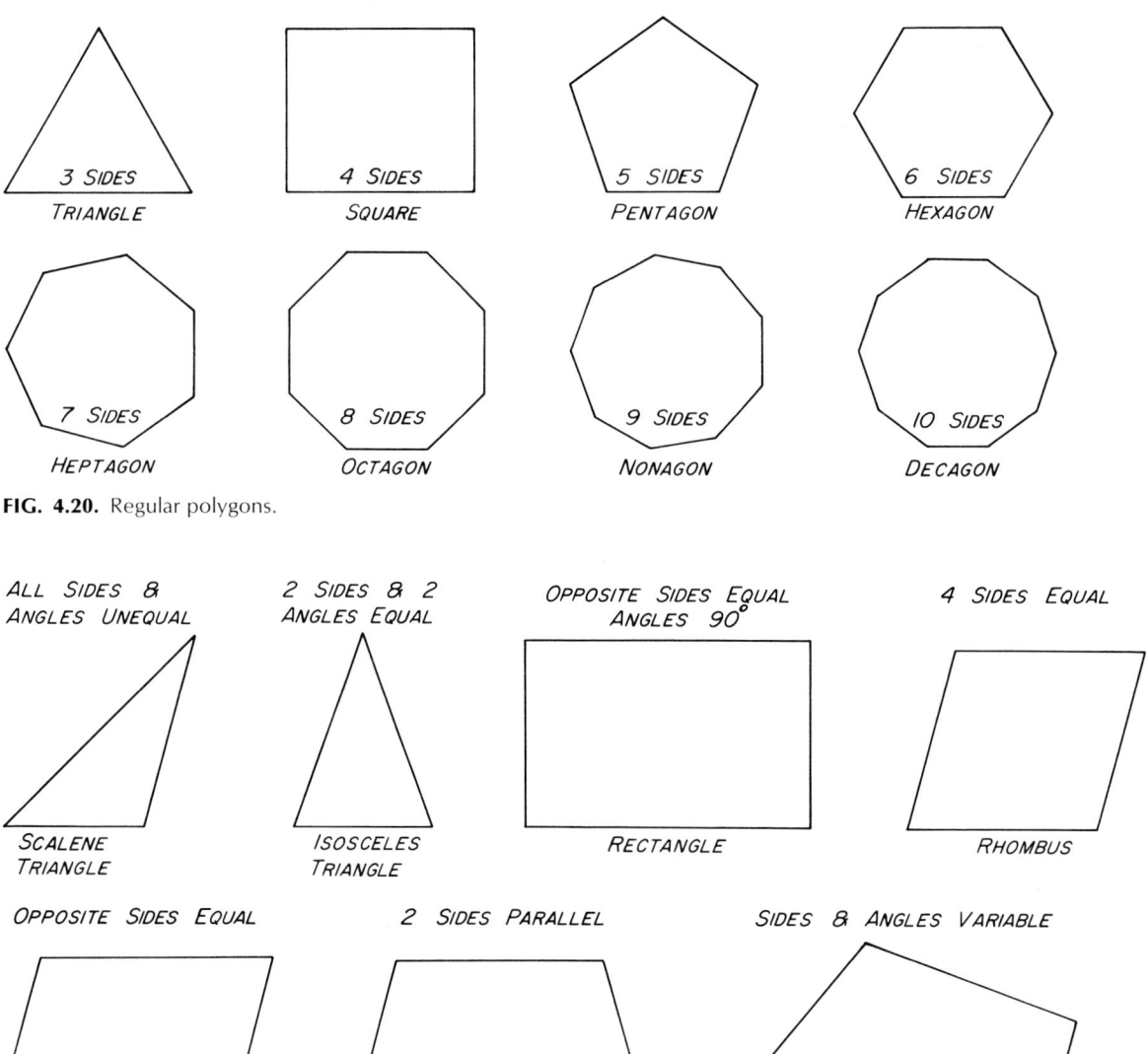

FIG. 4.20. Regular polygons.

FIG. 4.21. Other common geometric plane figures.

4.6.2 Construct a Triangle with Two Sides and the Included Angle Given.
To construct a triangle with two sides and an included angle given, proceed as shown in Fig. 4.23.

a. Lay out longest side AB equal to a as a base.
b. At the end A construct the given angle (by any method).
c. With center A and a radius equal to side b, draw an arc cutting the second side of the angle and locating the end x of the side.
d. Draw a line from B to x, thus completing the triangle (Fig. 4.23c).

4.6.3 Lay Out a Triangle with Two Angles and the Included Side Given.
To lay out a triangle with one side and two angles given, proceed as shown in Fig. 4.24a.

a. Lay out the given side a as a base.
b. At one end A lay out one of the given angles—45° in this case. See Fig. 4.24b.
c. At the other end of the base lay out the second given angle—30° in this case.
d. Extend the sides of the two angles until they intersect at C as in Fig. 4.24c. ABC is the required triangle.

4.7 CONSTRUCT A SQUARE
A square can be designated by having the length of one of its sides given or by having the length of its diagonal specified, as indicated in Fig. 4.25. With this basic information given, the square can be constructed by several methods that are discussed in the following paragraphs.

4.7.1 Square with Length of Side Given—First Method.
Figure 4.25a shows the construction of the square ABCD when side AB is given. The steps are as follows.

a. Erect a perpendicular to AB at A as shown in Fig. 4.25a.
b. With A as a center, draw an arc having a radius AB until it crosses the perpendicular at C.
c. With B and C as centers and with the same radius, draw two arcs intersecting at D.
d. Draw CD and BD, thus completing the square.

4.7.2 Square with Length of Side Given—Second Method.
Figure 4.25b shows another construction of a square, with the length of one of the sides given as AB. The steps are as follows.

a. Bisect line AB to determine the midpoint of the line at x.
b. With x as the center and Ax as a radius, draw a circle with AB as diameter.
c. Draw the four sides of the square by drawing a line tangent to the circle at points A, B, and the two points at which the bisector intersects the circle.

4.7.3 Square with Length of Diagonal Given.
Figure 4.25c shows the construction of a square with the diagonal AB given. The steps are as follows.

a. Bisect the diagonal AB as shown in Fig. 4.25c.

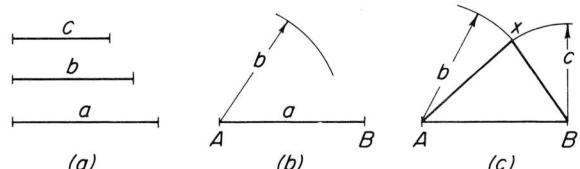

FIG. 4.22. Constructing a triangle. Three sides given.

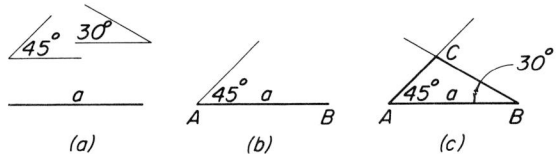

FIG. 4.24. Constructing a triangle. Two angles and side between given.

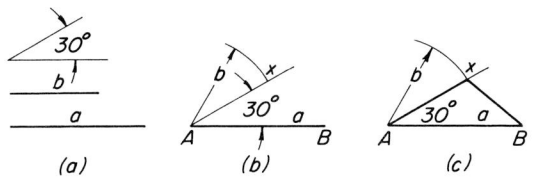

FIG. 4.23. Constructing a triangle. Two sides and included angle given.

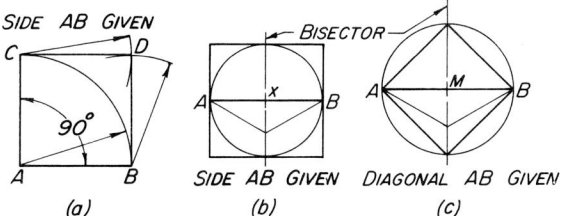

FIG. 4.25. Constructing a square.

b. At the center M draw a circle with AB as a diameter.

c. With AB as one diagonal and the bisector as the other, draw the inscribed square using a T-square and 45° triangle.

4.8 CONSTRUCT A PENTAGON INSCRIBED IN A GIVEN CIRCLE

The construction of a pentagon in a given circle with AB as the diameter is shown in Fig. 4.26. The steps are as follows.

a. Bisect the radius CB and locate point D.
b. Locate point E on the circle directly above point C as in Fig. 4.26b.
c. Strike an arc with D as the center and DE as the radius. The intersection where the arc strikes the diameter AB locates point F.
d. With E as a center and a radius EF, swing an arc intersecting the circle at G.
e. EG is one side of the pentagon.
f. Step this side off five times around the circle. See Fig. 4.26c.

For other methods of construction see Articles 4.10.1 and 4.10.2.

4.9 CONSTRUCT A HEXAGON

A hexagon may be specified and constructed in three different ways, as shown in Fig. 4.27.

4.9.1 Hexagon with Length of Side AB Given.
The steps in constructing a hexagon with one side given are as follows.

a. Draw a circle with the length of a side as a radius, as in Fig. 4.27a.
b. Step off the radius six times around the circle.
c. Connect the six points to form the hexagon.

4.9.2 Hexagon with Distance across Corners Given.
To draw a hexagon with the distance across corners given, the following steps are used.

a. Draw a circle with the distance across corners as a diameter.
b. With the ends of the diameter as centers, draw two arcs as shown in Fig. 4.27b, thus locating six points including the ends of the diameter.
c. Connect the six points to form the hexagon.

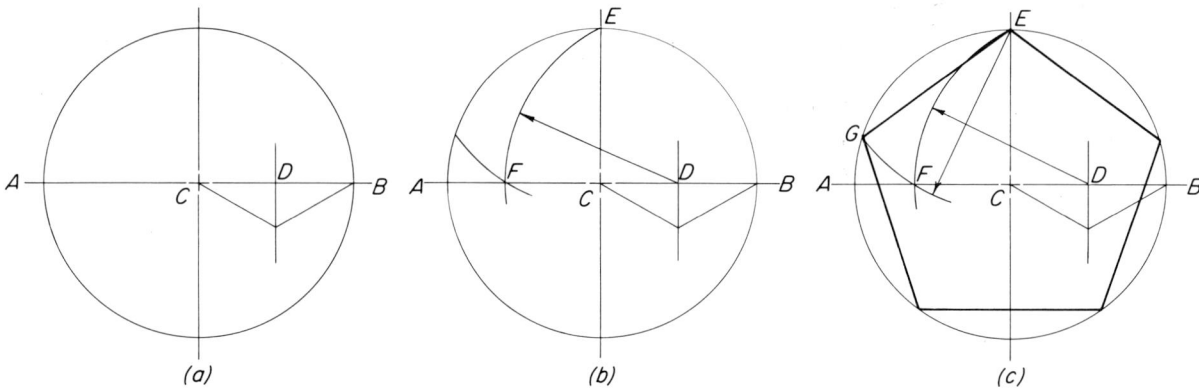

FIG. 4.26. Constructing a pentagon within a given circle.

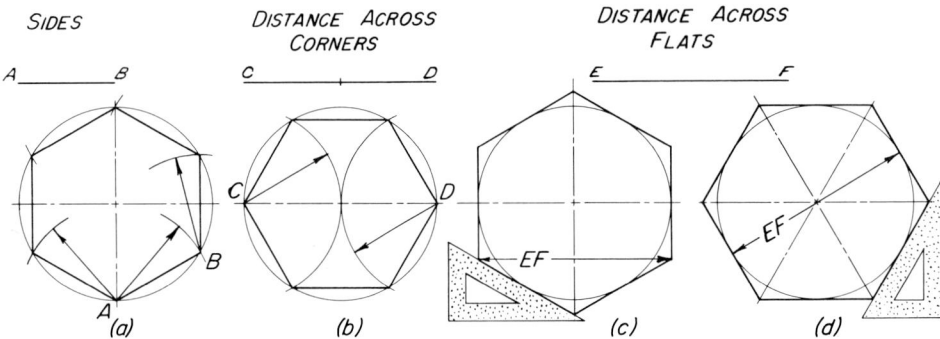

FIG. 4.27. Constructing a hexagon.

4.9.3 Hexagon with Distance across Flats Given.
Figures 4.27c and d illustrate two ways of drawing a hexagon when the distance across flats is given.

a. Draw a circle with the distance across flats, EF, as a diameter.
b. Circumscribe a hexagon tangent to the circle in either the position of Fig. 4.27c or d using a T-square and 30–60° triangles as shown.

4.10 CONSTRUCT A REGULAR POLYGON HAVING ANY GIVEN NUMBER OF SIDES

Regular polygons having any specified number of equal sides may be specified and constructed in several ways, as illustrated in Figs. 4.28 and 4.29.

4.10.1 Inscribe a Polygon within a Given Circle.
The steps in constructing a polygon within a given circle are as follows:

a. Draw the circle and divide its diameter into the specified number of equal parts, for example, five parts as shown in Fig. 4.28a.
b. With ends of the diameter A and B as centers and a radius equal to the diameter, draw two arcs intersecting at C, as shown in Fig. 4.28b.
c. From point C draw a line through the second division point of the diameter until it crosses the circle at D, as in Fig. 4.28c.
d. The chord AD is one side of the polygon.
e. Step off the distance AD the proper number of times around the circle to complete the polygon.

This method is empirical. It gives exact results for the square and hexagon and is only very slightly in error for other polygons.

4.10.2 Polygon with Length of Side Given.
In Fig. 4.29 a second method of constructing regular polygons is shown.

a. Lay out AB equal to the given side.
b. With A as a center, draw a semicircle with AB as a radius.
c. Divide the semicircle into the required number of equal parts by trial.

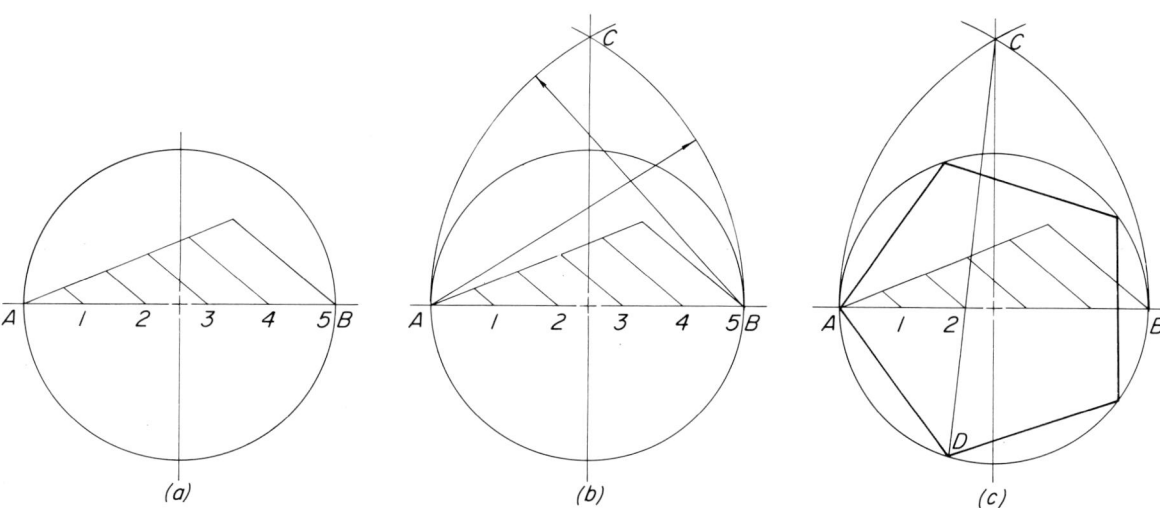

FIG. 4.28. Constructing any regular polygon within a circle.

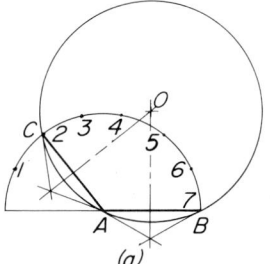 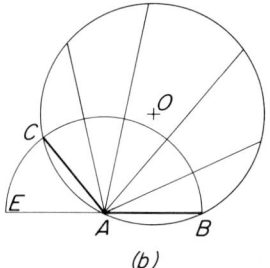 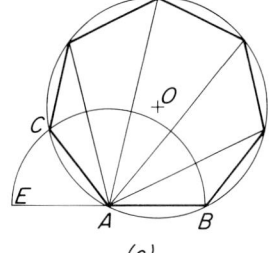

FIG. 4.29. Constructing any regular polygon with length of side given.

62 GEOMETRIC CONSTRUCTION

d. Through point A and the second division point C draw a line AC forming the second side of the polygon.

e. Bisect AB and AC and find the intersection O of these bisectors (see Article 4.2.2).

f. With O as a center, draw a circle through A, B, and C. This circle contains all the corners of the polygon.

g. From A draw lines through the other division points of the semicircle until they intersect the larger circle, as shown in Fig. 4.29b.

h. These points are corners of the polygon. See Fig. 4.29c.

4.11 CONSTRUCT A REGULAR OCTAGON

A regular octagon may be specified in two ways, as shown in Fig. 4.30. Either the distance across flats or the distance across corners may be given.

4.11.1 Octagon with Distance across Flats Given.
To construct an octagon with the distance across flats given, see Fig. 4.30.

a. Draw a circle having the distance across flats as a diameter, as shown in Fig. 4.30a.

b. With a T-square and 45° triangle, draw lines tangent to the circle, first horizontally, then vertically, and finally at 45° in each direction, as shown in Figs. 4.30a and b.

4.11.2 Octagon with Distance across Corners Given.
To construct an octagon with the distance across corners given, as shown in Fig. 4.30, by being inscribed in a circle, take the following steps.

a. Draw a circle having a diameter equal to the distance across corners.

b. Draw the horizontal and vertical center lines as in Fig. 4.30c.

c. Draw 45° center lines in each direction.

d. Connect all points where the center lines cross the circle as in Fig. 4.30d.

4.12 REPRODUCE A GIVEN PLANE FIGURE
Plane figures may be composed of either straight lines, curved lines, or combinations thereof, as shown in Figs. 4.31 and 4.32.

4.12.1 Plane Figure with Straight Sides Only.
The construction for duplicating a plane figure with straight sides is shown in Fig. 4.31.

a. Select one corner, for example, A in Fig. 4.31a, and draw straight lines from A through all other corners.

b. With A as a center, draw an arc across the lines radiating from A and have the arc completely outside the figure, as shown in Fig. 4.31b.

c. Draw a line A'B' for the new figure in the position desired.

d. Draw a new arc with radius R' equal to R.

e. Draw new arcs 2', 3', and 4' to locate radial lines.

f. With A as a center and arcs successively equal to AC, AD, and AE, reconstruct the plane figure as shown in Fig. 4.31c.

4.12.2 Plane Figure with Straight Lines and Irregular Curves.
The construction for making figures with straight lines and irregular curves is as follows.

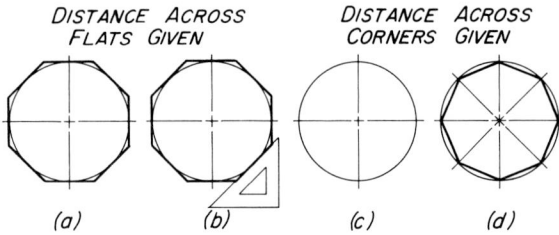

FIG. 4.30. Constructing an octagon.

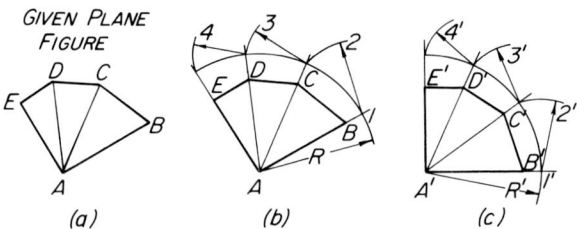

FIG. 4.31. Reproducing a given plane figure.

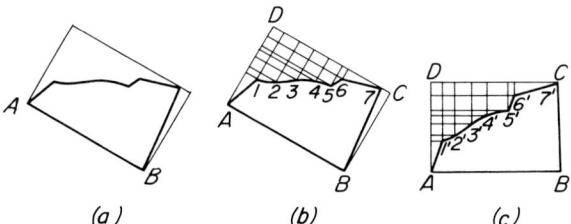

FIG. 4.32. Reproducing a given plane figure having irregular curves.

a. With the longest straight side AB as a base, draw a rectangle ABCD just enclosing the figure, as in Fig. 4.32a.
b. Carefully select points on the contour of the figure and draw coordinates through them as shown in Fig. 4.32b.
c. Reproduce the entire set of coordinates in the required new position and draw the figure through the relocated points as in Fig. 4.32c.

4.13 CIRCLES AND LINES TANGENT TO THEM

In order to think correctly about circles and lines tangent to them, one must remember the definition of a circle. It is the locus of points, lying in a plane, at a given distance from a point.

4.13.1 Construct a Circle through Three Points.
To construct a circle through three points, the following steps are used.

a. Connect points A, B, and C with two intersecting lines as in Fig. 4.33a.
b. Find the perpendicular bisector of each line.
c. The intersection of the bisectors is the center of the circle.

4.13.2 Construct a Circle within a Square.
A circle may be constructed within a square by the following steps, as shown in Fig. 4.33b.

a. Divide either center line of the square (in this case the vertical one) into any number of equal parts.
b. Divide the side that is perpendicular to the center line used in step a into the same number of equal parts.
c. Draw line B1 to intersect line Aa, line B2 to intersect Ab, and so on.
d. These intersections are points on the circle. Draw a smooth curve through them.

4.13.3 Draw a Circle of Radius r through a Point and Tangent to a Straight Line.
To draw a circle through a point and tangent to a straight line, the following steps are used.

a. Draw arc 1 with radius r with point A as the center, as in Fig. 4.34a.
b. Draw line 2 parallel to the given line BC at a distance r from it (see Articles 4.2.3 and 4.2.4).
c. The intersection of arc 1 and line 2 is the center of the tangent arc. The center of the arc is on a perpendicular to the line at the point of tangency (point where the arc ends).

An application to a drawing problem is shown in Fig. 4.34b.

4.13.4 Draw a Circle with a Given Radius r through a Point and Tangent to a Circle.
The steps in drawing a circle through a given point and tangent to a given circle as shown in Fig. 4.35a are as follows.

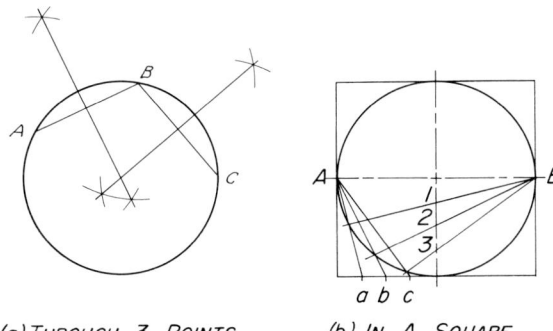

(a) THROUGH 3 POINTS   (b) IN A SQUARE

FIG. 4.33. Drawing a circle (a) through three points and (b) in a square.

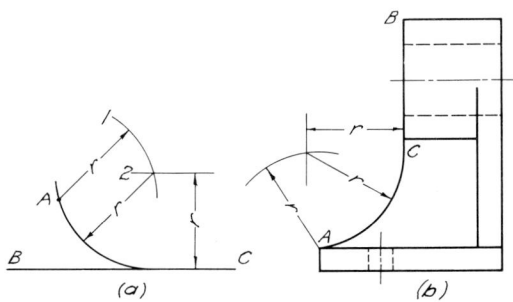

FIG. 4.34. Constructing a circle through a point and tangent to a line.

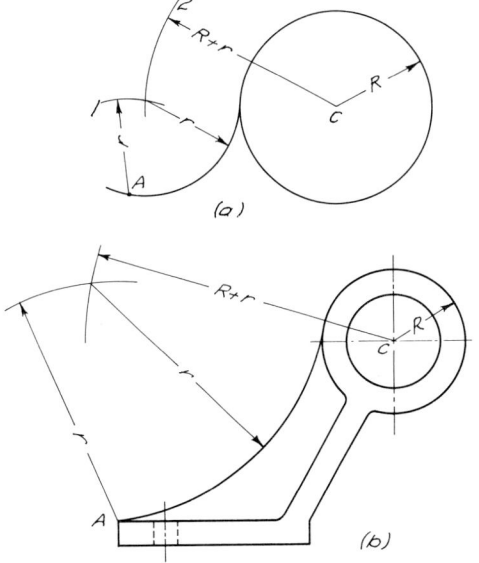

FIG. 4.35. Constructing a circle through a point and tangent to a circle.

64 GEOMETRIC CONSTRUCTION

a. Draw arc 1 with radius r and the center at A.
b. Draw arc 2 with the center of the circle C as a center and a radius $R + r$.
c. The intersection of these arcs is the center of the tangent circle or arc. The point where the line connecting the center of the arc and the center of the circle crosses the circle is the point of tangency (point where one arc ends and the other begins).

A practical application is shown in Fig. 4.35b.

4.13.5 Draw a Circle of Radius r Tangent to Two Straight Lines. The construction for drawing a circle tangent to two intersecting straight lines is shown in Fig. 4.36.

a. Draw two lines parallel to the given lines at the distance r as in Figs. 4.36a and b. The intersection of these construction lines determine the center of the arc at O.
b. Determine the tangent points of the arc by constructing perpendicular lines through point O to the given lines as in Figs. 4.36a–c.
c. Set the compass with the radius r and O as the center and draw the arc, being careful to stop the arc at the predetermined tangent points.

4.13.6 Draw a Circle of Radius r Tangent to a Straight Line and Another Circle. To draw a circle tangent to a line and a circle, the following steps are used.

a. Draw a line parallel to the given line a distance r from it as shown in Fig. 4.37a.
b. Draw an arc of radius $R + r$ concentric with the given circle.
c. The intersection of the line and arc is the required center. A practical application is shown in Fig. 4.37b.

4.13.7 Circle Tangent to Two Other Circles —External Tangent r Radius. To draw a circle tangent externally to two other circles, perform the following steps.

a. Using the center of the small circle and the radius $R + r$, draw arc 1 as in Fig. 4.38.
b. Using the center of the large circle and the radius $R_1 + r$, draw arc 2 as in Fig. 4.38. The intersection of arc 1 and arc 2 determines the center of the tangent circle to be drawn.
c. Set the compass at the radius r and, using the center determined in (b) above, draw the required circle. The tangent points are on a line joining the center of the arc with the center of each circle.

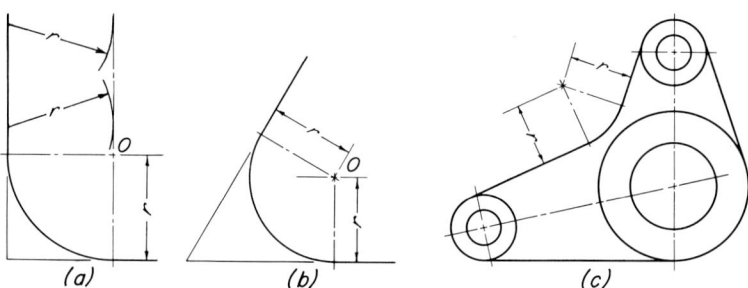

FIG. 4.36. Constructing a circle tangent to two intersecting lines.

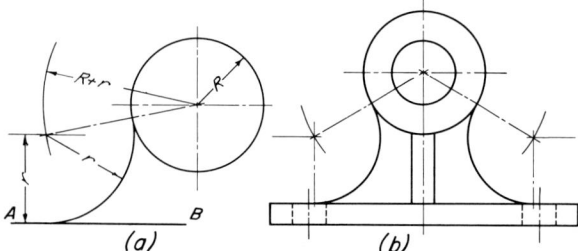

FIG. 4.37. Constructing a circle tangent to a straight line and a circle.

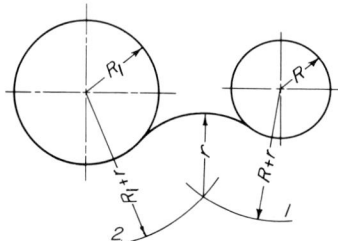

FIG. 4.38. Constructing a circle tangent to two circles.

4.13.8 Circle Tangent to Two Other Circles—Internal Tangent and Radius r.
In drawing a circle tangent to two other circles, both given circles will be inside of the tangent arc or circle. See Fig. 4.39. The radius of the tangent circle must be greater than that of either of the given circles.

a. Draw arc 1 with the radius $r - R$, using the center of the circle with radius R as in Fig. 4.39a.

b. Draw arc 2 with the radius $r - R_1$, using the center of the circle with radius R_1 as in Fig. 4.39a. The intersection of arc 1 and arc 2 determines the center of the tangency circle.

c. Through the point of the intersection of arcs 1 and 2, draw the tangent circle with radius r.

d. The tangent points are on the line joining the centers of the circle and arc.

A practical application of this construction is shown in Fig. 4.39b.

4.13.9 Circle Tangent Internally to One Given Circle and Tangent Externally to Another Given Circle.
To draw an arc or circle tangent internally to one circle and tangent externally to another, the construction shown in Fig. 4.40 is used. The tangent circle has a radius of r, is tangent externally to the circle with radius R_1, and tangent internally to the circle with radius R.

a. Draw arc 1 with the radius $r - R$, using the center of the circle with radius R.

b. Draw arc 2 with the radius $r + R_1$, using the center of the circle with radius R_1. The intersection of arc 1 and arc 2 determines the center of the tangent circle.

c. Through the point of intersection of arcs 1 and 2 draw the tangent circle using the radius R.

4.13.10 Connecting Parallel Lines with Reverse Curves Using Equal Arcs.
The problems in this paragraph and in the two that follow arise in the location of highways and railroads. In Fig. 4.41 lines AF and BE are to be connected at A and B with reverse curves.

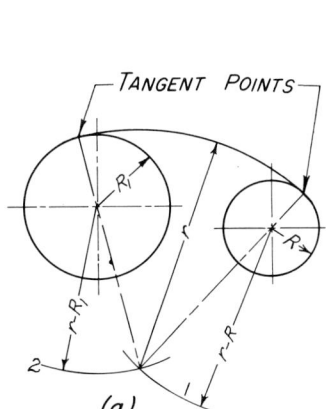
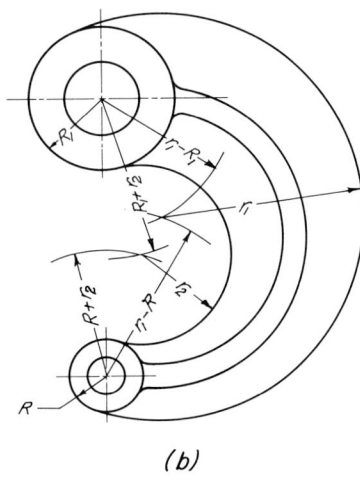

FIG. 4.39. Constructing a circular arc tangent to two other circles.

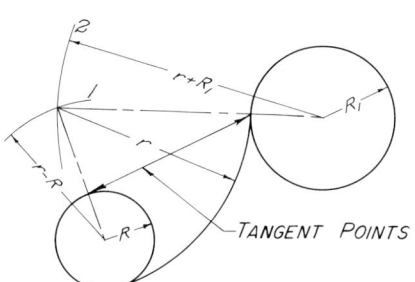

FIG. 4.40. Constructing a circular arc tangent to two circles.

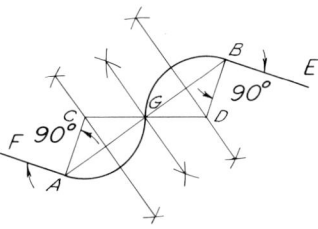

FIG. 4.41. Connecting parallel lines by equal arcs.

a. Draw a straight line between A and B.
b. Draw perpendicular lines to the given lines at A and B as shown in Fig. 4.41 (see Article 4.3.3).
c. Divide line AB into two equal parts at G.
d. Bisect AG and GB.
e. The intersection of the bisectors with the perpendiculars from A and B locates the required centers at C and D. Draw the arcs with radii CA and BD. These arcs are tangent to each other at G.

4.13.11 Connecting Parallel Lines with Reverse Curves or Unequal Radii. In Fig. 4.42 lines AF and BG are to be connected at A and B by a reverse curve made up of unequal arcs.

a. Draw a straight line between A and B.
b. Construct a perpendicular to AF at A and another to BG at B.
c. Locate point D on the perpendicular to AF so that AD is equal to the given radius r of one of the arcs.
d. Using the center D and the radius r, draw the arc from A until it intersects line AB at point C.
e. Extend line DC until it intersects the perpendicular to BG from B to locate point E. Point E is the center for the second arc.
f. Draw the second arc using E as the center and EB as the radius from point B to C.

4.13.12 Connecting Two Nonparallel Lines by a Reverse Curve. In Fig. 4.43 lines AL and BM are to be connected by a reverse curve.

a. On a perpendicular to BM at B step off the chosen radius BC.
b. With C as a center, draw an arc tangent to the line at B as shown in Fig. 4.43.
c. On a perpendicular from A draw AD equal to BC and on the same side of the two lines as BC.
d. Join D and C with a straight line.
e. Bisect line CD.
f. Extend the bisector until it intersects the perpendicular from A at F. This is the center of an arc that is tangent to the line at A and to the other arc. Note that in all cases the tangent points are on the line joining the centers.

4.13.13 Rectifying an Arc. To obtain the length of an arc of a circle, two common geometrical construction procedures can be used. The first method, shown in Fig. 4.44, uses the following steps.

a. Draw a line BD tangent to arc BC at B as in Fig. 4.44.
b. Divide arc BC into an equal number of chord lengths with a set of dividers.
c. Lay the same number of chord lengths along line BD. The greater the number of chords used, the greater will be the accuracy.

An alternate method to rectify arc AB is illustrated in Fig. 4.45.

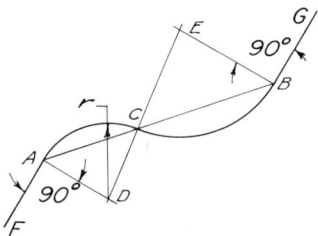

FIG. 4.42. Connecting parallel lines by unequal arcs.

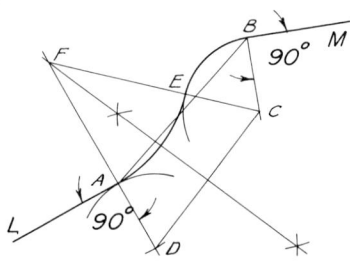

FIG. 4.43. Connecting nonparallel lines by a reverse curve.

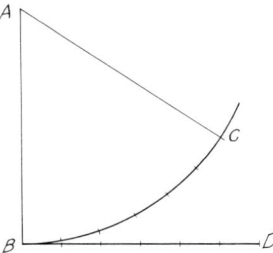

FIG. 4.44. Rectifying an arc.

a₁. Draw the chord AB and extend it beyond A to point E so that AE equals ½AB as shown in Fig. 4.45.

b₁. Draw a tangent line to the arc AB at A by making line AD perpendicular to AC.

c₁. Draw an arc with E as the center and EB as the radius until it intersects the tangent line to determine point D. AD is approximately equal in length to arc AB. For angles up to 60° the error is less than 1 in 1000.

This method of rectifying an arc is used to show the path of travel of the end of a spring for clearances. See Fig. 4.46.

4.13.14 To Make a Circular Arc Equal in Length to a Given Straight Line. In Fig. 4.47 it is required to lay off the length of line AB along the given arc. One method is to take the dividers to divide the line segment AB into a number of equal parts and then lay the same number of equal lengths along the arc as chord lengths.

Another method is illustrated in Fig. 4.47 and consists of the following steps.

a. Lay off the given length of line AB on a line tangent to the arc.

b. Divide this length into four equal parts.

c. With the first point D adjacent to the tangent point as a center draw an arc DB butting the circle at C.

d. Arc AC is equal in length to line AB with an error of less than 6 parts in 1000 for angles less than 90°. This principle is used to determine the rebound limits of a spring as shown in Fig. 4.48.

4.14 CONIC SECTIONS

The ellipse, parabola, and hyperbola are curves that may be cut from a right circular cone by a plane. Since they are cut from a cone, they are called *conic sections*. See Figs. 4.49c–e. Straight lines and circles may also be cut from a cone. See Figs. 4.49a and b. They are not commonly referred to as conic sections.

a. **The ellipse.** The *ellipse* may be defined as a section cut from a right circular cone by a plane making an angle with the axis greater than that made by the elements of the cone. See Fig. 4.49c. The angle must be less than a right angle. A 90° angle produces the circle.

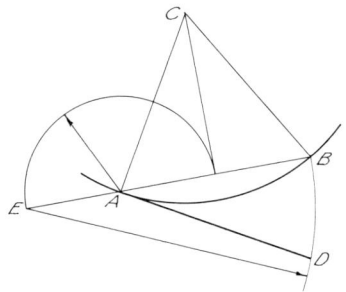

FIG. 4.45. Alternate method of rectifying an arc.

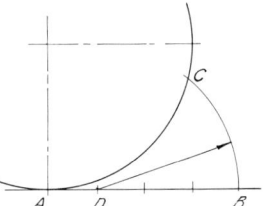

FIG. 4.47. Constructing an arc equal to a straight line.

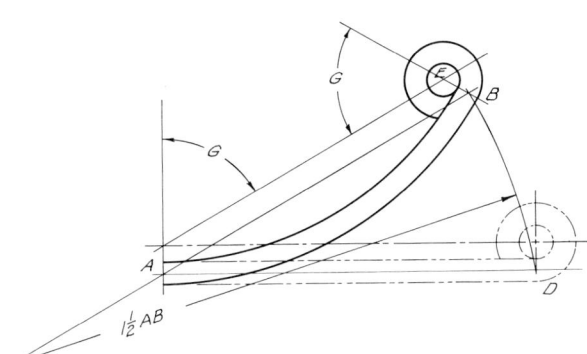

FIG. 4.46. Application of rectifying an arc (courtesy C. S. Mobley, Automotive Industries).

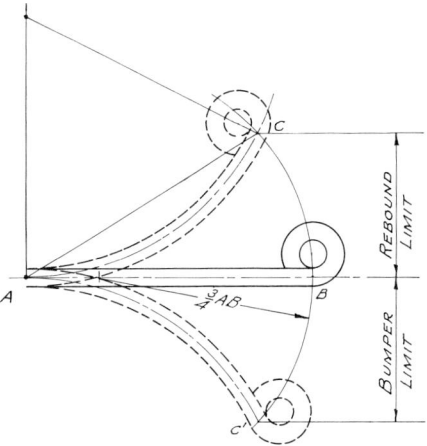

FIG. 4.48. Application of Fig. 4.47 (courtesy C. S. Mobley, Automotive Industries).

The ellipse may also be defined in mathematical form by the equation

$$\frac{x^2}{a^2} + \frac{y^2}{b^2} = 1.$$

This standard form of the equation places the major axis of the ellipse on the x-axis and the center of the ellipse at the origin of the coordinates. In the above equation the sum of the distances from the foci to any point P on the curve is 2a.

b. **Parabola.** The parabola may be defined as a section cut from a cone by a plane making an angle with the axis, equal to the angle made by the elements with the axis. The cutting plane is therefore parallel to one of the elements. The parabola may be defined mathematically by the equation $y^2 = 2px$, where the x-axis is also the axis of the curve and the vertex of the curve is on the y-axis and $p/2$ is the distance from the vertex to the focus and also to the directrix.

c. **Hyperbola.** The hyperbola may be defined as a section cut from a cone by a plane inclined to the axis at an angle less than the angle made by the elements. See Fig. 4.49e. The section is usually taken parallel to the axis. This produces a hyperbola with two symmetric branches.

The standard form of the equation for the hyperbola is

$$\frac{x^2}{a^2} - \frac{y^2}{b^2} = 1.$$

The axis of this hyperbola is on the x-axis and it is symmetrical about the y-axis.

4.14.1 Ellipse as a Section of a Cone. The construction of an ellipse as a section of a cone is shown in Fig. 4.50. The steps are as follows.

a. Pass a cutting plane AB showing as an edge view in the front view as shown in Fig. 4.50b.

b. Draw 12 equally spaced elements of the cone as shown in Fig. 4.50c. Space them equally in the top view and then project them to the front view.

c. The points at which the respective ele-

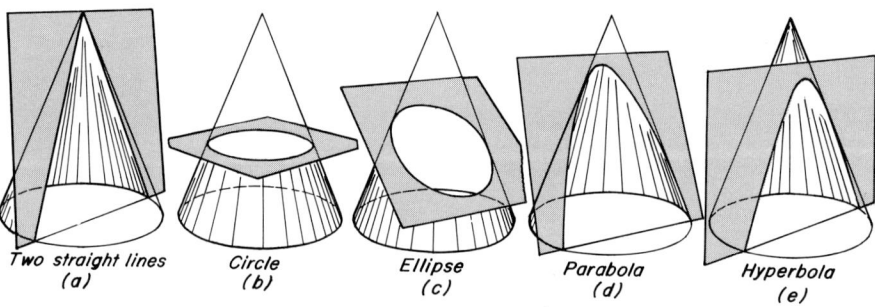

FIG. 4.49. The conic sections.

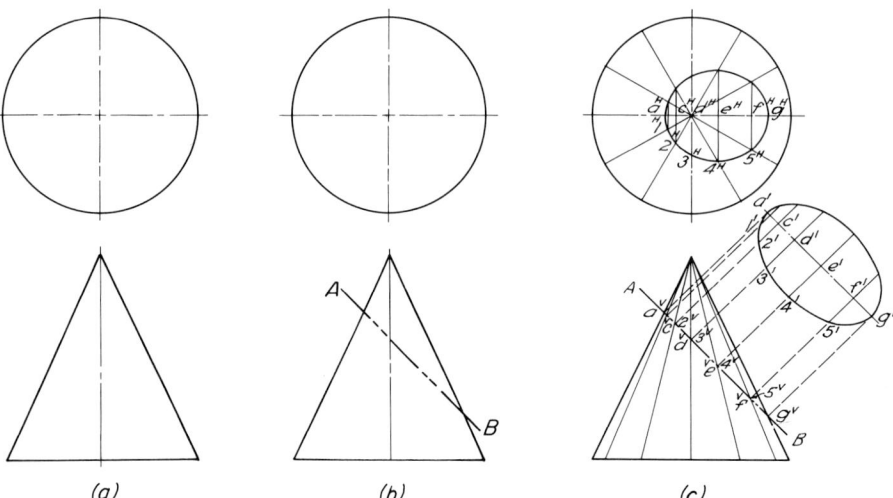

FIG. 4.50. The ellipse as a section of a cone.

ments pierce the plane can be seen by inspection at 1^V, 2^V, 3^V, and so forth.

d. Project these to the top view and draw the curve. This is an ellipse, but it is not the true shape of the one cut by plane AB.

e. To obtain the true shape of this ellipse, draw a center line a^1g^1 parallel to AB at a convenient place. See Fig. 4.50c.

f. Draw lines from 1^V, 2^V, 3^V, and so forth, in the front view, perpendicular to this center line.

g. Set off on these lines the distances f^H5^H, e^H4^H, and so forth, in the top view as f^15^1, e^14^1, and so forth, in the auxiliary view. Draw a smooth curve through these points to obtain the true shape of the curve.

4.14.2 Draw an Ellipse by Using the Foci. Assume that the major and minor axes of an ellipse are given as in Fig. 4.51a.

a. To locate the foci, set the compass to a length AC, equal to one half the major axis.

b. With point D, one end of the minor axis, as a center draw an arc cutting the major axis AB at F and F_1, thus locating the foci.

c. Divide the space from F to the center into any convenient number of parts.

d. As an example, set the compass to the distance A-4 and, with F and F_1 as centers, describe four arcs as in Fig. 4.51c.

e. Then set the compass to the distance B-4 and with the same centers describe four arcs intersecting the first four. These intersections are points on the ellipse.

f. Repeat the process, using the other points on the axis giving distances A-1, B-1, A-2, B-2, and so on. Draw a smooth curve through the points. This construction method is based on the principle that $r_a + r_b = AB$ (the major axis).

4.14.3 Draw an Ellipse by the Two-Circle Method. Assume the major and minor axes given as in Fig. 4.52a.

a. Draw two concentric circles, one with the major axis as a diameter and the other with the minor axis as a diameter, as shown in Fig. 4.52b.

b. Draw radial lines through the center intersecting both circles as in Fig. 4.52c.

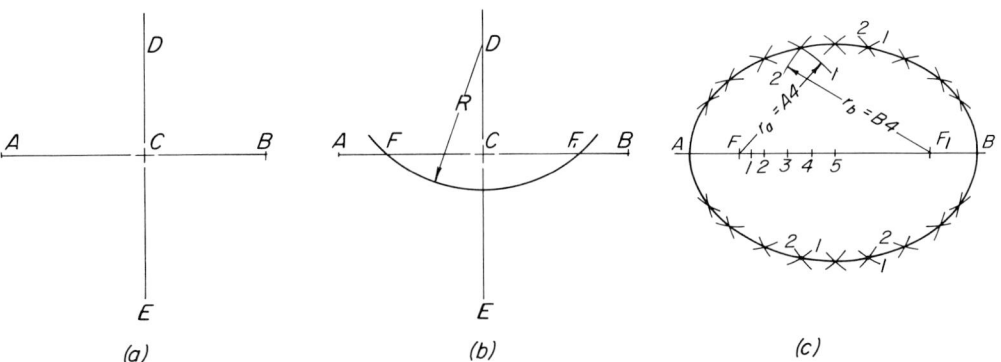

FIG. 4.51. The ellipse constructed from foci.

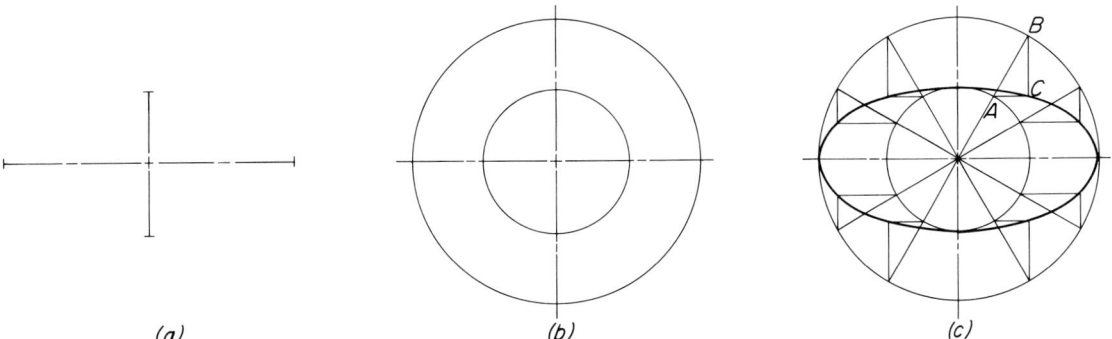

FIG. 4.52. Two-circle method for constructing an ellipse.

70 GEOMETRIC CONSTRUCTION

c. From the points on the inner circle draw horizontal lines like AC.

d. From the points on the outer circle draw vertical lines like BC, C is a point on the ellipse.

e. Repeat the process for all points on the two circles and then draw the curve.

4.14.4 Construct an Ellipse by the Intersection Method, Having the Major and Minor Axes Given. This construction depends on the construction of a circle as shown in Fig. 4.33b. This construction is geometrically correct.

a. Draw the rectangle that will enclose the ellipse, using the major and minor axes as center lines.

b. Divide the short center line into any number of equal parts as shown in Fig. 4.53.

c. Divide the long side of the rectangle into the same number of equal parts.

d. Draw B-1 to intersect A-a, B-2 to intersect A-b, and so on. These intersections are points on the ellipse.

e. The other half of the ellipse may be done in the same manner.

4.14.5 Draw an Ellipse Having the Conjugate Axes Given. One of the simplest methods of making this construction is shown in Fig. 4.54. This is the same method as explained in Article 4.14.4.

4.14.6 Ellipse by the Four-Center Approximate Method with Major and Minor Axes Given. To construct an ellipse by this method, follow the steps given below and shown in Fig. 4.55.

a. Draw a rectangle on the major and minor axes and construct arc AE with the radius equal to the semimajor axis, as shown in Fig. 4.55a.

b. Draw the diagonal BC and swing the arc with radius CE, thus locating point F in Fig. 4.55b.

c. Bisect line FB in the usual manner and extend the bisector to cross both axes at M and N as in Fig. 4.55c.

d. Locate P and S with a compass on the basis of symmetry with M and N. Make MS and NP equal to NM.

e. Draw arcs with the four centers at M, N, P, and S tangent to side of rectangle and to each other. These arcs will be tangent to each other at r, t, u, and v, as in Fig. 4.55d.

4.14.7 Ellipse by the Trammel Method. The trammel method is a very convenient method for plotting any number of points on an ellipse, since it leaves the drawing free of all construction lines.

a. *First method.* On the edge of a strip of paper, lay off the distance cd equal to the semiminor axis of the ellipse as shown in Fig. 4.56a. Also lay off ca on the same side of c as the point d, equal to the semimajor axis. By moving the strip of paper so that d is on the major axis and a on the minor axis, c will always be on the ellipse. Any number of positions for c may be located, and a smooth curve through them will give the ellipse.

b. *Second method.* This is similar to the preceding method except that a and d are laid off on opposite sides of c, as shown in Fig. 4.56b. In all other respects the procedure is the same. This scheme is a little more accurate when the difference between the major and minor axes is small.

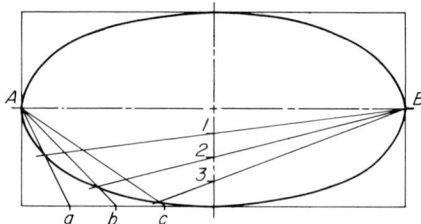

FIG. 4.53. Ellipse by intersection of lines.

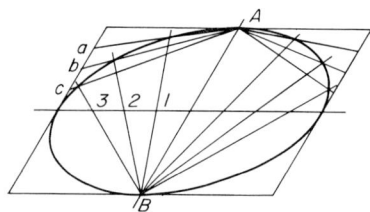

FIG. 4.54. Ellipse constructed on conjugate axes.

4.14 CONIC SECTIONS 71

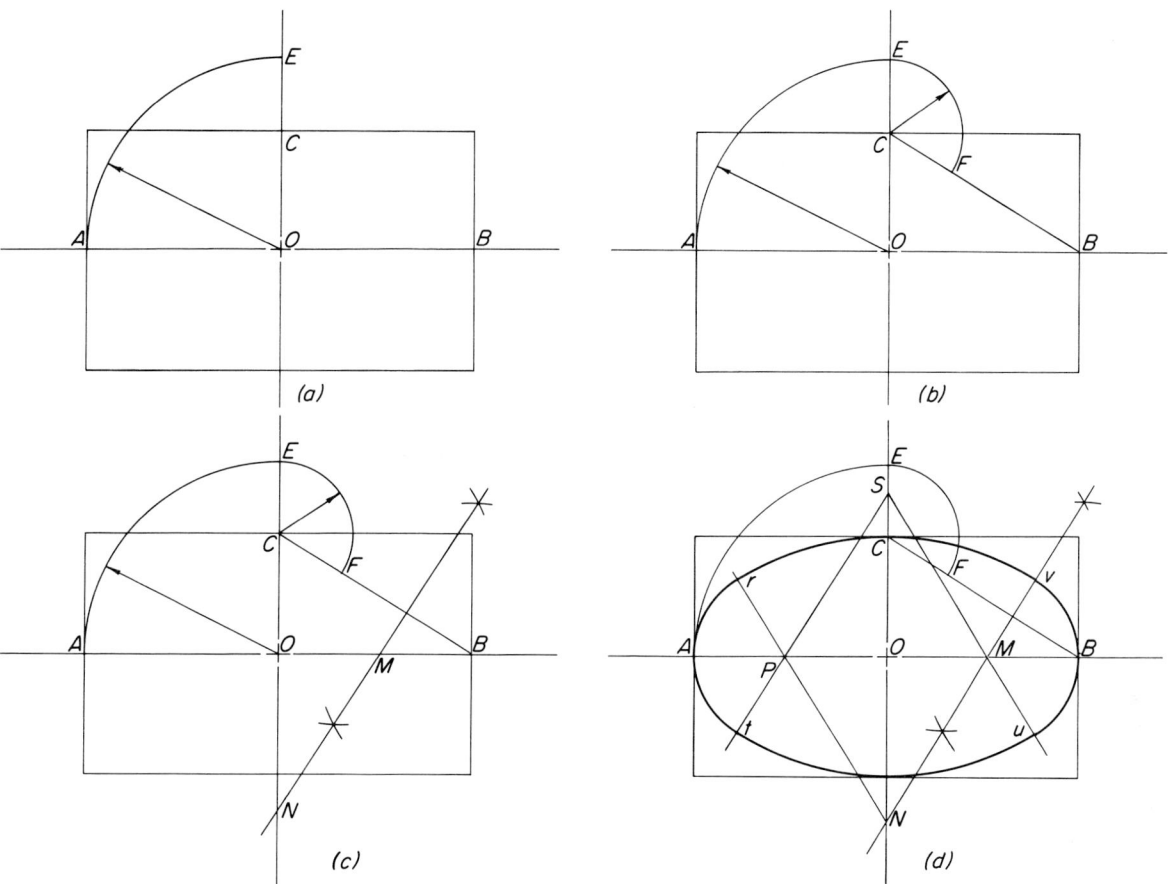

FIG. 4.55. Four-center approximate ellipse.

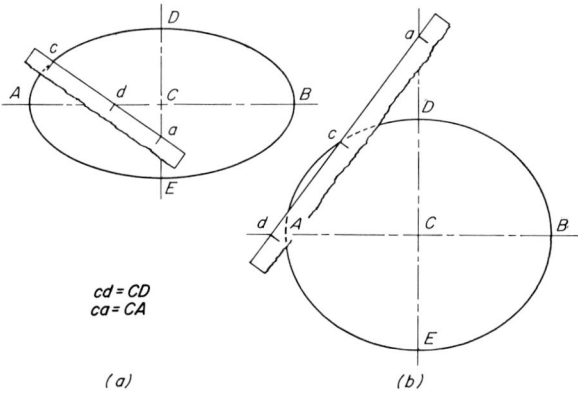

cd = CD
ca = CA

(a)　　　　(b)

FIG. 4.56. Ellipse by the trammel method.

72 GEOMETRIC CONSTRUCTION

4.14.8 Ellipse by the Diagonal Method. An ellipse based on the oblique or perspective view of a circle can be drawn in a variety of positions by the diagonal method. See Fig. 4.57.

a. *Ellipse in a rectangle.* An ellipse can be constructed in a rectangle by the following steps.
1. On one side of the rectangle draw a semicircle and divide it into six equal parts with a 30–60° triangle as shown in Fig. 4.57a.
2. Draw horizontal lines across the rectangle at the division points.
3. Draw a diagonal across the rectangle as shown in Fig. 4.57b.
4. Number the horizontal lines from the center outward in both directions as shown in Fig. 4.57c.
5. Number the vertical lines inward from both ends.
6. Make a dot where lines having the same number cross. These are points in the ellipse.
7. Draw a smooth curve through the points.

b. *Ellipse in isometric.* This is the same problem as drawing an ellipse with the conjugate axes given. See Fig. 4.57d. This ellipse can be constructed with the same steps as in (a) above.

c. *Ellipse in oblique.* For the oblique parallelogram the same principle using the diagonal can again be applied as in (a). See Fig. 4.57e.

d. *Ellipse in perspective.* Having the perspective rectangle (actually a trapezoid) laid out, proceed in the same manner as in (a). See Fig. 4.57f.

4.14.9 Four-Center Approximate Ellipse in a Rhombus. When a circle is tangent to two intersecting straight lines as AB and AC in Fig. 4.58a, the distance Ae must be equal to Ad. Therefore, in the case of the rhombus the arcs must be tangent at the midpoint of the sides. Hence, the center of each arc must lie on a perpendicular to the side at its midpoint.

a. In both Figs. 4.58b and c, erect perpendiculars to the midpoints of the sides and extend these perpendiculars until they intersect.
b. With these intersection points as centers, draw four arcs tangent to the sides of the rhombus and, of course, to each other.

These ellipses are not quite accurate, but they are satisfactory and very useful in pictorial drawing.

4.14.10 The Parabola. The parabola is one of the most useful curves not only in engineering computations but also in actual construction. Reflectors for light and sound are made in parabolic form, as are vertical curves on highways and railroads, just to mention two common applications.

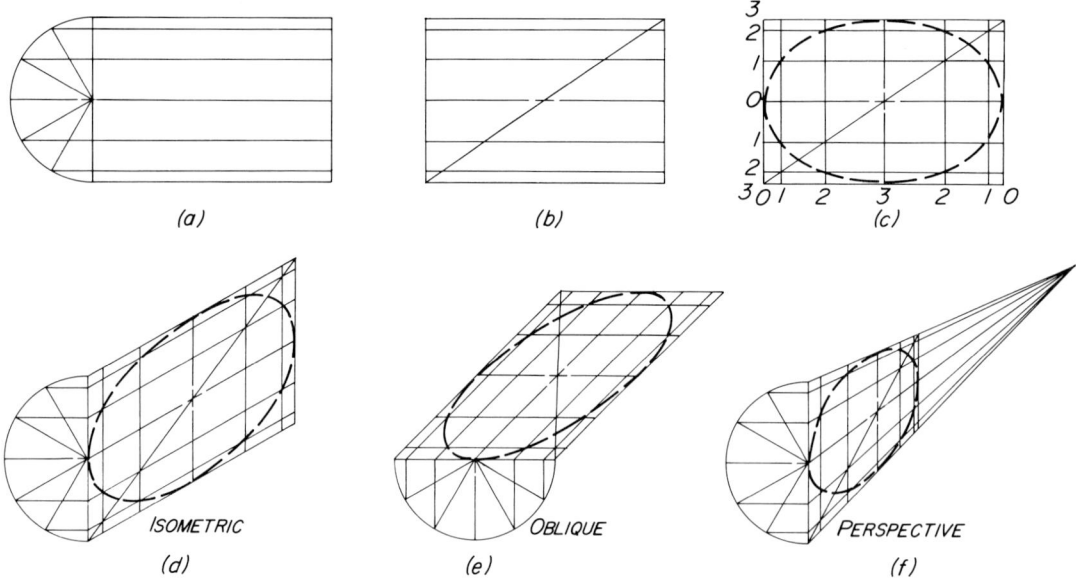

FIG. 4.57. Ellipse by the diagonal method.

4.14.11 Draw a Parabola as a Section of a Cone.
To draw a parabola as a section of a cone, the following steps are necessary, as shown in Fig. 4.59.

a. Draw the cutting plane A-A parallel to the outside element $a^v b^v$ in the front view. The cutting plane appears edgewise. See Fig. 4.59a.

b. Draw a convenient number of equally spaced elements in the top view of the cone and project through to the front view as in Fig. 4.59b.

c. The piercing points of these elements with plane A-A can be seen by inspection in the front view at 1^v, 2^v, 3^v, and so forth.

d. Project these points to the top view to obtain a view of the parabola. This does not show the true shape.

e. To obtain the true shape, draw a center line parallel to plane A-A at any convenient place.

f. Project the points 1^v, 2^v, and so forth to it as in Fig. 4.59b.

g. Measure from this center line the corresponding distances in the top view as indicated for distances a and b. A smooth curve through these points will show the true shape of the parabola.

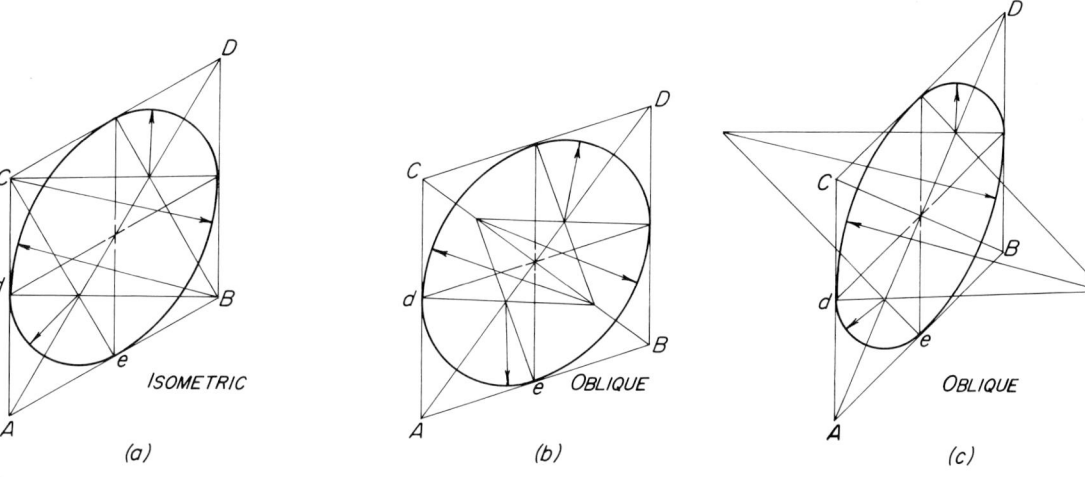

FIG. 4.58. Four-center approximate ellipse in a rhombus.

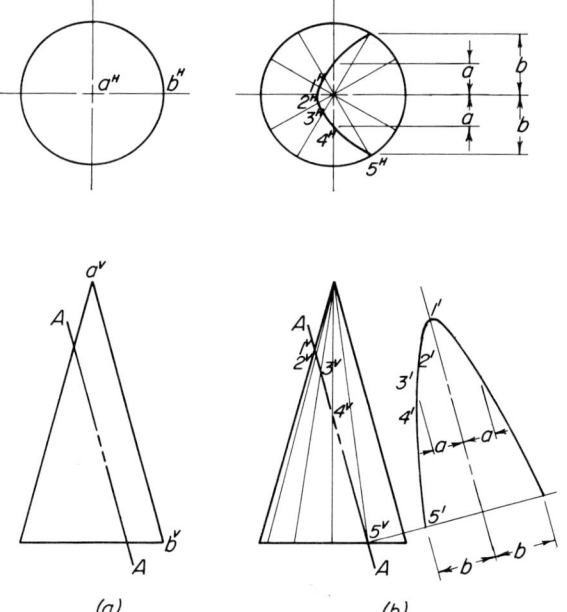

FIG. 4.59. Parabola as a section of a cone.

4.14.12 Draw a Parabola with the Focus and Directrix Given. The parabola may be described as the locus of a point, where the distance from the focus always equals the perpendicular distance from the directrix. With AB the directrix and F the focus given as in Fig. 4.60, a series of points may be located by fulfilling the condition specified above. Thus, for any given distance r, draw an arc with center at F and radius r. Likewise draw a line parallel to AB at r distance from it. The intersections of the arc and line locate two points on the curve. Locate a sufficient number of points to determine accurately the curve.

4.14.13 Draw a Parabola with One Point A at the Vertex, and Two Symmetrically Placed Points B and C Given. The construction is as follows.

a. Divide BC in Fig. 4.61 into any even number of parts and the sides perpendicular to it into half as many parts.

b. Through the points on BC, draw lines parallel to the axis.

c. From A the vertex, draw lines to the points on the sides.

d. The intersections of the lines in pairs locate points on the parabola.

4.14.14 Draw a Parabola as an Element of a Hyperbolic Paraboloid. The hyperbolic paraboloid is a warped surface composed of straight line elements as shown in the top, front, and side views of Fig. 4.62.

The curve tangent to the elements in the front view is a parabola. A section cut perpendicular to the axis, like plane A-B, gives a hyperbola as shown in the side view.

4.14.15 Draw a Parabola by Offsets from a Line. In Fig. 4.63 the construction of a parabola with its vertex at B and passing through C and D is shown. The steps are as follows.

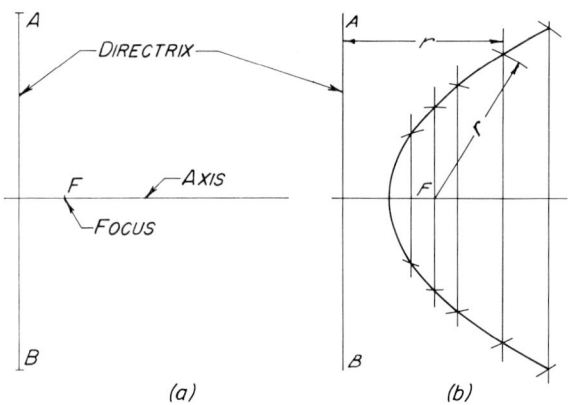

FIG. 4.60. Parabola, focus, and directrix given.

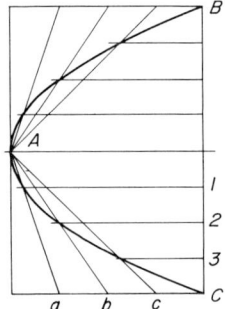

FIG. 4.61. Parabola by intersection method.

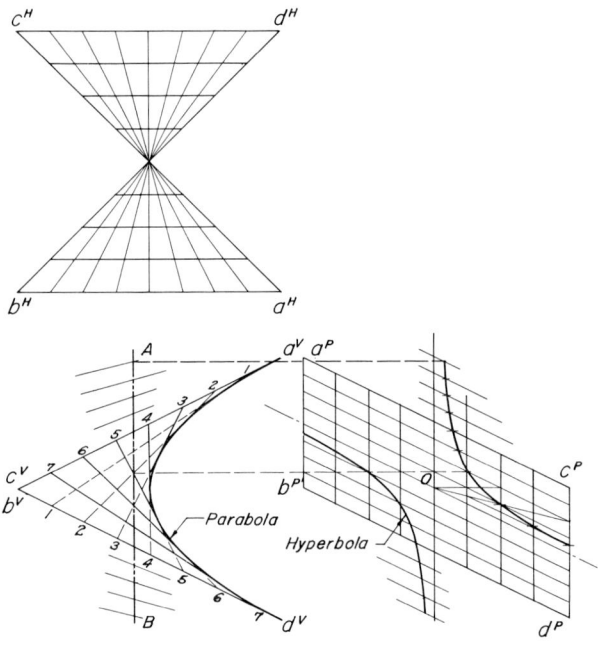

FIG. 4.62. Parabola as an element of a hyperbolic paraboloid.

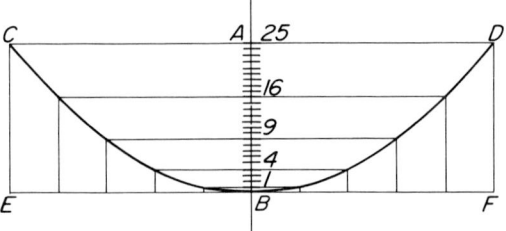

FIG. 4.63. Parabola by offsets from a tangent line.

a. Divide *BE* and *BF* each into a number of equal spaces, in this case five, as shown in Fig. 4.63. Erect perpendicular lines to *EF* at each point.
b. The offset from a straight-line tangent to a parabolic curve is proportional to the square of the distance above the line from the tangent point. Therefore, divide the distance *AB* into 25 equal parts — proportional to the square of numbers from 1 to 5.
c. Mark the points 1, 4, 9, 16, and 25 on *AB*, and draw horizontal lines through them.
d. The proper intersection of horizontal and vertical lines as shown in Fig. 4.63 locates points on the parabola.

4.14.16 Draw a Parabola by Offsets from Two Intersecting Lines. This is the situation as it occurs in highway and railroad construction. See Fig. 4.64.

The two intersecting lines *EB* and *BD* in Fig. 4.64 may represent the grade lines on a highway. Each line makes the same angle with the horizontal. Assume that the curve is to be tangent to the grade line 250 ft horizontally from *B* on each line at *E* and *D*. This is the usual practice in engineering. To draw the curve proceed as follows.

a. Connect *E* and *D* and draw a line perpendicular to it from *B*. This line bisects *DE* at *A*.
b. Find the midpoint of *AB* at *C*. Point *C* is the vertex of the parabola.
c. Divide the tangent *BE* and *BD* into five equal parts as shown in Fig. 4.64.
d. Divide *BC* into 25 equal parts.
e. From points 1, 4, 9, and 16 on this scale, draw lines parallel to the tangents until they intersect lines drawn through the one-fifth division points parallel to the axis *BA*.

These points lie on the parabola.

4.14.17 The Hyperbola. The hyperbola can be plotted by a number of methods. Three of the more useful methods are discussed in the following paragraphs.

4.14.18 Draw a Hyperbola as a Section of a Cone. In Fig. 4.65 the construction of a hyperbola as a section of a cone is shown. Proceed as follows.

a. Draw three views of a right circular cone as shown in Fig. 4.65.
b. Draw a cutting plane *A-B* parallel to the axis of the cone so that it shows edgewise in the front view.
c. Draw 12 or more elements of the cone and find their piercing points with the plane *A-B* by inspection at a^v, 1^v, 2^v, 3^v, and so forth.
d. Project these points to the corresponding elements of the cone in the side view.
e. Draw a smooth curve through the points.

4.14.19 Draw a Hyperbola with the Foci and Vertices Given. With the axis and the foci *F* and F_1 given, locate the vertices *a* and *b* as in Fig. 4.66. The construction is as follows.

a. Mark off a series of points 1, 2, 3, 4, and so forth to the right of F_1.
b. With *F* and F_1 as centers and a radius *a*-3, draw four arcs. Only one is shown in the figure.

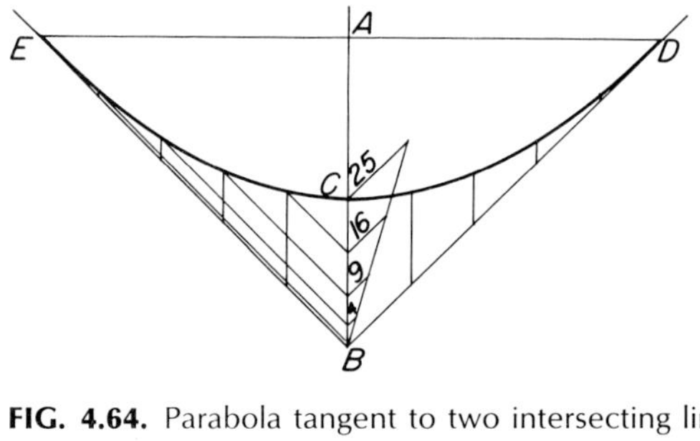

FIG. 4.64. Parabola tangent to two intersecting line by offsets.

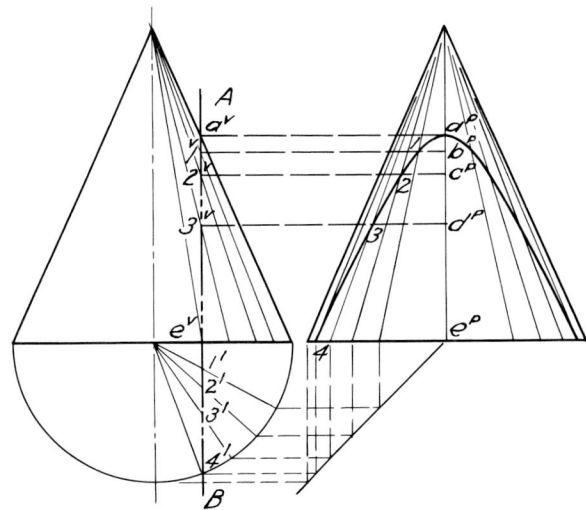

FIG. 4.65. The hyperbola on a section of a cone.

76 GEOMETRIC CONSTRUCTION

c. With F and F_1 again as centers and a radius b-3, draw four more arcs intersecting the first four.
d. The intersections are points on the curve.
e. Repeat the process with the other points 1, 2, 4, and so forth as far as needed.
f. Draw a smooth curve through the points.

Note that the constant difference between the radii is the distance ab. The asymptotes of the curve can be located by drawing a circle having FF_1, the distance between foci, as its diameter. Lines drawn through the vertices a and b perpendicular to the axis intersect the circle at points on the asymptotes.

4.14.20 Draw a Hyperbola with the Asymptotes and One Point on the Curve Given. The construction that can be made with the following steps is illustrated in Fig. 4.67.

a. Through the given point P draw two lines PM and PN, respectively, parallel to the asymptotes.
b. From the origin O draw a series of radial lines intersecting PN in points 1, 2, 3, and so forth, and PM in the corresponding points 1', 2', 3', and so forth.
c. The radial lines should be distributed on both sides of point P.
d. Draw a line through 1 and 1' parallel to the corresponding asymptotes. The intersections of these lines are points on the hyperbola.

When the asymptotes are at right angles to each other, the hyperbola is called a *rectangular* or *equilateral hyperbola*. This curve occurs in thermodynamics in the study of the expansion of gases.

4.14.21 Draw a Conic through Five Given Points. The solution of this problem, which occurs in joining surfaces as in aircraft work, is an application of Pascal's theorem which states, "Opposite pairs of sides of a hexagonal figure inscribed in a conic intersect in three points which lie in a straight line." Thus, in Fig. 4.68 the

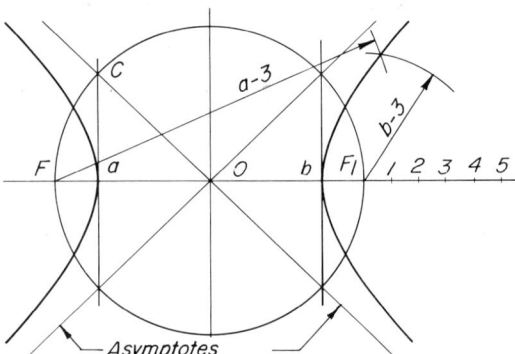

FIG. 4.66. To draw a hyperbola with the foci given.

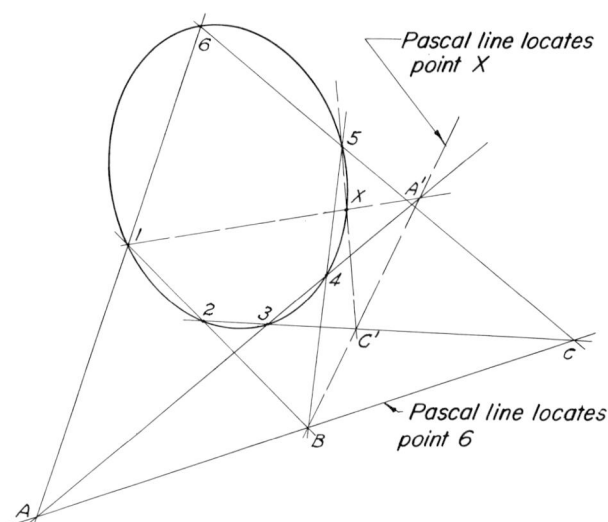

FIG. 4.68. Conic through five points.

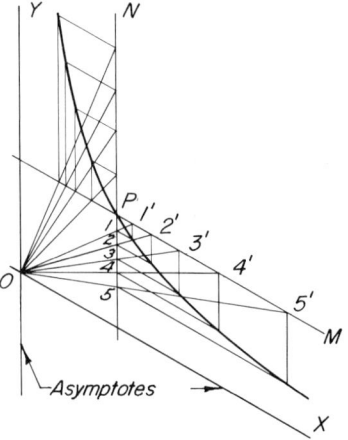

FIG. 4.67. Hyperbola through a point with asymptotes given.

sides 1-6 and 3-4 intersect at A, 1-2 and 4-5 intersect at B, and 2-3 and 5-6 intersect at C, and line ABC is referred to as *Pascal's line*.

If only points 1 to 5 are known, any sixth point X may be found on the conic as follows.

a. Extend the opposite sides 1-2 and 4-5 until they intersect at B. This point must be on every Pascal line of the conic.

b. Draw another Pascal line through B, intersecting the lines 2-3 and 3-4 as at C' and A'.

c. Draw the remaining sides of the hexagon from 1 to A' and from 5 to C', thus locating point X, the sixth corner, at their intersection.

As many points as desired may be located on the conic by drawing successive Pascal lines through B and proceeding as outlined above.

4.15 TANGENT PROBLEMS

It is a common problem for a drafter to locate a tangent to some geometric curve with considerable accuracy. The following situations are among those usually encountered.

4.15.1 Tangent to a Circle through a Point on the Circle. This problem is usually solved in the drafting room by the method shown in Fig. 4.69.

a. Bring a triangle resting on another straight edge up to the figure so that one of the edges of the right angle of the triangle coincides with line BC joining the center and the point of tangency. See dotted-line triangle.

b. Then slide the triangle along the straight edge until the other leg of the right angle passes through B. This edge is tangent to the circle at B.

Another way, using geometric methods, to construct such a tangent is shown in Fig. 4.70.

a_1. With P as a center, draw an arc through C until it cuts the circle at A. See Fig. 4.70a.

b_1. With A as a center and radius CA draw a semicircle that will pass through P. See Fig. 4.70b.

c_1. Extend line CA until it crosses the semicircle at D.

d_1. Line DP is the required tangent since it is perpendicular to the radius CP.

4.15.2 Draw a Line Tangent to a Circle through a Point Outside. A tangent of this type may be constructed as shown in Fig. 4.71. The steps are as follows.

a. Draw a line from P to C, the center of the circle as in Fig. 4.71a.

b. Bisect line PC at A.

c. With A as a center and radius AP, draw a semicircle cutting the original circle at T. See Fig. 4.71b.

d. T is the tangent point. Draw PT.

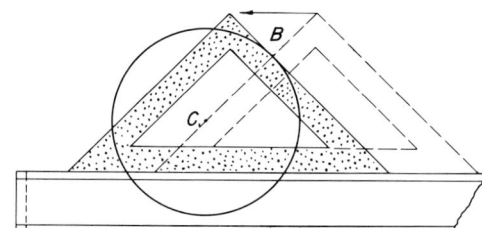

FIG. 4.69. Drafter's method of drawing a tangent to a circle.

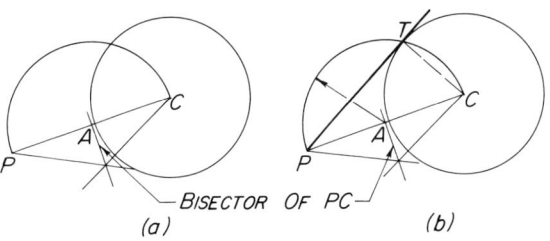

FIG. 4.71. Line tangent to a circle through a point outside.

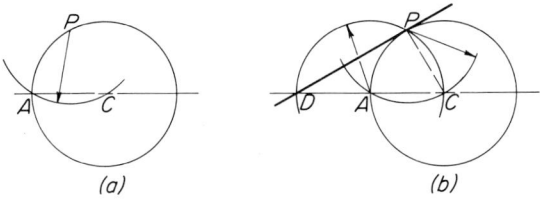

FIG. 4.70. Line tangent to a circle at a point on the circle.

78 GEOMETRIC CONSTRUCTION

4.15.3 Draw a Line Tangent to Two Circles on the Same Side. To draw a line tangent to two given circles with radii R and r as in Fig. 4.72, take the following steps.

a. At the center B of the larger circle, draw one concentric with it, having a radius $R - r$.
b. From the center A of the smaller circle, draw a tangent AC to the inner circle whose center is at B.
c. Draw the tangent to the original two circles PT parallel to AC.

4.15.4 Draw a Line Tangent to Two Circles Crossing between Them. The construction for a tangent in this position is shown in Fig. 4.73. The Procedure is as Follows.

a. With B the center of the larger circle, draw another circle concentric with it having a radius $R + r$ as in Fig. 4.73.
b. Draw a line AP from A, the center of the smaller circle tangent to the outer circle, whose center is at B.
c. The tangent to both circles, MN, is parallel to AP.

4.15.5 Draw a Tangent to an Ellipse from a Point P on It.

Method A. One method for making this construction is shown in Fig. 4.74. Proceed as follows.

a. From P draw lines to both foci F and F_1 as in Fig. 4.74. If the foci are not given, the construction for finding them is shown in Fig. 4.51.
b. Extend FP to any convenient point A.
c. Bisect the angle F_1PA. The bisector is tangent to the ellipse at P.

Method B. Another method for drawing a tangent is based on the fact that an ellipse may be constructed as an oblique view of a circle, as shown in Fig. 4.75.

a_1. Having the ellipse given, describe a circle about it having the major axis as a diameter.
b_1. Project point A from the ellipse to the circle parallel to the minor axis.
c_1. Draw a tangent to the circle at A_1 and extend it until it cuts the major axis at X.

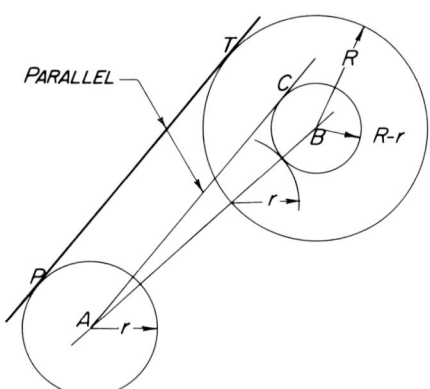

FIG. 4.72. Line tangent to two circles on same side.

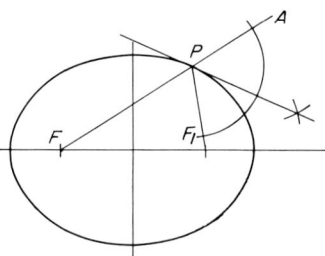

FIG. 4.74. Use of foci in drawing tangents to an ellipse.

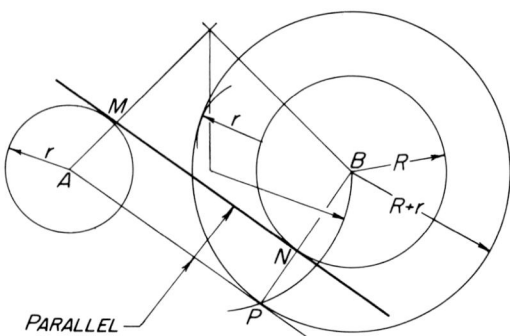

FIG. 4.73. Tangent to two circles on opposite sides.

d_1. A line from X to point A on the ellipse is the required tangent.

4.15.6 Draw a Line Tangent to an Ellipse from a Point Y Outside. The construction in Fig. 4.75 for drawing a line tangent to an ellipse from an external point is as follows.

a. Project Y to G on the revolved position of the circle.
b. Rotate G to G_1 on the edgewise view of the circle.
c. From G_1 project horizontally across to Y_1 directly under Y.
d. Draw the tangent to the circle from Y_1 at F_1.
e. From F_1 project perpendicular to the axis to F, which will be the tangent point on the ellipse.
f. Draw the tangent FY.

4.15.7 Find the Axis of a Parabolic Curve. This axis may be found in the following manner, as shown in Fig. 4.76.

a. Draw any two parallel chords across the curve as in Fig. 4.76.
b. Bisect the chords at E and F. The line EF is parallel to the axis.
c. Draw a third chord GH perpendicular to EF.
d. Bisect this chord at M. A line through M parallel to EF is the axis of the parabola.

4.15.8 Draw a Line Tangent to a Parabola from a Point P on the Curve with the Axis Given. Such a tangent may be found in the following manner, as shown in Fig. 4.77.

a. From P draw a perpendicular to the axis to locate point O as in Fig. 4.77. With a divider step off VA equal to VO.
b. Line AP is the tangent.

4.15.9 Find the Focus of a Parabola Having the Axis Given. The focus of a parabola may be found by the following steps, as shown in Fig. 4.77.

a. Find a tangent at some point on the curve as in Article 4.15.8, Fig. 4.77.
b. Bisect the tangent.
c. Extend the bisector to cut the axis at F, which is the focus.

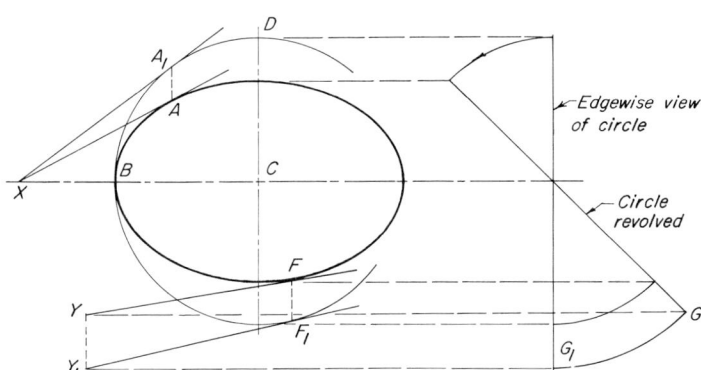

FIG. 4.75. Alternate method: line tangent to an ellipse.

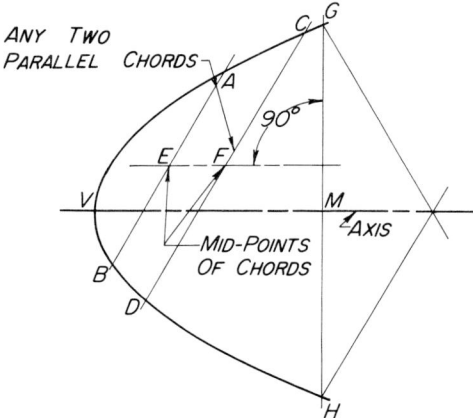

FIG. 4.76. Finding axis of given parabola.

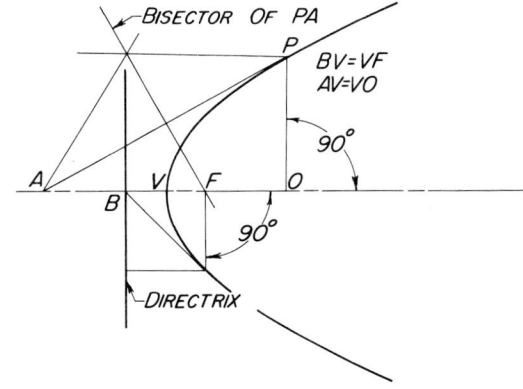

FIG. 4.77. Finding focus and directrix with parabola and axis given.

4.15.10 Find the Directrix of a Parabola Having the Axis, Focus, and Vertex Given.
With these given conditions, the directrix may be found as follows and as shown in Fig. 4.77.

a. Set the divider to the distance FV. See Fig. 4.77.

b. Step off the distance VB equal to FV. Point B is on the directrix.

c. Draw a line perpendicular to the axis through point B. This is the directrix.

4.15.11 Draw a Line Tangent to a Parabola at a Point P on It.

Method A. With the focus and directrix given as in Fig. 4.78, proceed as follows.

a. From point P draw a line perpendicular to the directrix.

b. Draw a line from P to the focus F.

c. The bisector of the angle APF between these lines is the tangent to the curve.

Method B. Another method for drawing a tangent to a parabola at a point P on it when only the curve and axis are given is as follows. See Fig. 4.79.

a_1. Through point P draw a line PQ parallel to the axis as in Fig. 4.79.

b_1. Draw two lines parallel to PQ and at any equal distances from it.

c_1. Extend these lines to cut the parabola at A and C and draw the line AC.

d_1. A line through P parallel to AC is tangent to the parabola.

4.15.12 Draw a Line Tangent To a Hyperbola at a Point P on the Curve. Axis and foci given.
Under the conditions given in Fig. 4.80 the construction for the tangent is as follows.

a. From point P draw lines to the foci F and F_1 as in Fig. 4.80.

b. The bisector of the angle FPF_1 between these lines is the tangent at P.

4.15.13 Find the Directrix and Asymptotes of a Hyperbola with Axis and Foci Given.
Under the conditions specified and shown in Fig. 4.80, the construction is as follows.

a. From one focus F_1 draw a perpendicular to the axis until it crosses the curve at A as in Fig. 4.80.

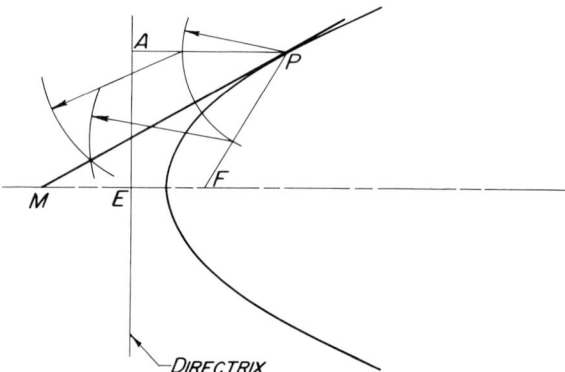

FIG. 4.78. Tangent to a parabola, focus and directrix given.

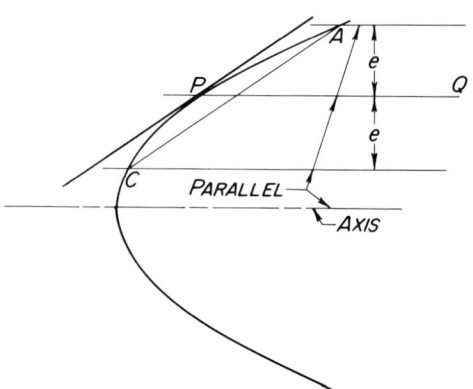

FIG. 4.79. Tangent through point on parabola, axis given.

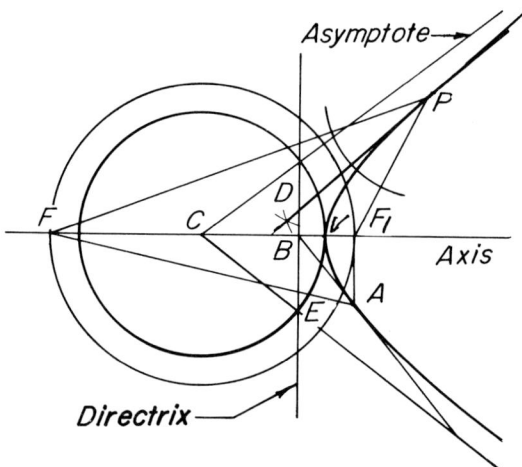

FIG. 4.80. Line tangent to a hyperbola, axis and foci given.

b. From A draw a line to the other focus F.
c. Bisect the angle F_1AF.
d. Where the bisector crosses the axis at B is a point on one directrix.
e. Through B draw a line perpendicular to the axis. This line is the directrix.
f. With the midpoint between F and F_1 labeled C as a center, draw a circle with radius CV. Point V is the vertex of the given hyperbola.
g. Where this circle crosses the directrix, locate two points D and E on the asymptotes.
h. Lines from C through D and E locate the asymptotes.

4.16 CONSTRUCTION OF OTHER CURVES COMMON IN PRACTICAL WORK

Although the conic sections have a prominent place in engineering work, a number of other mathematical curves are widely used. Among these are the involute, the helix, the cycloids, and the spiral of Archimedes.

4.16.1 The Involute. The *involute* may be defined as a curve that would be described if a string were unwound from some geometrical surface with the string kept taut and the end point describing the curve. Three involutes are shown in Fig. 4.81, one each for a triangle, a square, and a circle. The involute need not begin on the surface, although it may as in the triangle and the circle.

a. In the equilateral triangle, the curve consists of a series of 120° arcs with the radius increasing each time by the length of a side of the triangle. The corners of the triangle are the centers. This is also true for the square or any other figure with straight sides.
b. In the square, Fig. 4.81b, the beginning point of the involute has been made at point E, which is three times the length AB from A, arc BF is four times the length AB, CG is five times AB, and so on.
c. In the circle the circumference is divided into any convenient number of equal parts. Beginning at point 1, a tangent is drawn, and on it the length of the arc 0-1 is stepped off to locate A. At point 2 another tangent is drawn and a length equal to arc 0-1-2 is stepped off, and so on for as many points as may be desired. A smooth curve is then drawn through these plotted points.

The involute is the basic curve for gear teeth and for the impeller of centrifugal pumps.

4.16.2 The Cycloid. The cycloids form a group of curves generated by the path of a fixed point on the circumference of a rolling circle. When the circle rolls on a straight line, the path of the point is called simply a *cycloid* as in Fig. 4.82. If the circle rolls on the outside of another circle, the path of the point is called an *epicycloid*, whereas if the circle rolls on the inside of another circle, the path of the point is called a *hypocycloid*. These curves have a practical application in the design of gear teeth. To draw the cycloid in Fig. 4.82, divide the rolling circle into 12 or more equal parts and lay out on the straight line 12 divisions equal to the arcs of the circle. As the

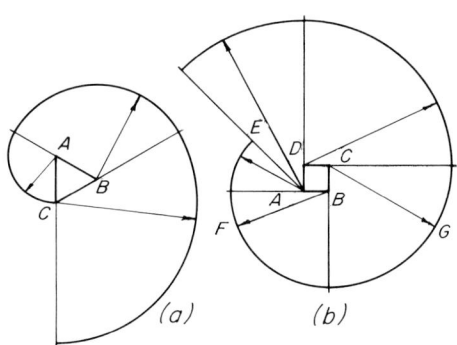

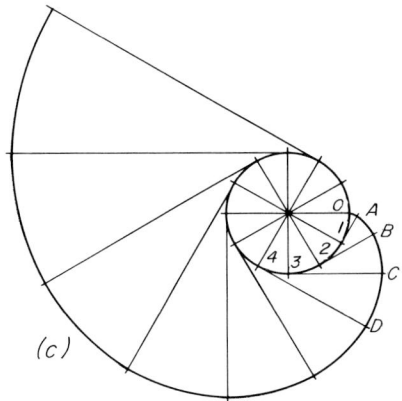

FIG. 4.81. Involutes.

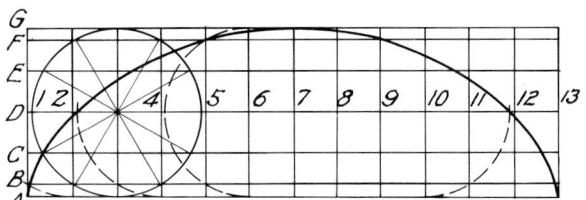

FIG. 4.82. Cycloid.

82 GEOMETRIC CONSTRUCTION

circle rolls along, the center will occupy successively the positions 1, 2, 3, and so forth, while the point on the circumference will rise to the elevation on the horizontal lines A, B, C, D, and so forth. The intersections of the circle in its successive positions with the horizontal lines will give the required points.

4.16.3 The Epicycloid and the Hypocycloid. The constructions of the epicycloid and hypocycloid are quite similar to each other and can be readily grasped from Fig. 4.83. Note that in both cases there is a separate line of centers for the rolling circles, and, in place of the horizontal lines A, B, C, and so forth, we have the arcs A, B, C, and so forth.

4.16.4 The Spiral of Archimedes. In the spiral of Archimedes, the radius of curvature increases directly as the angle through which it rotates. We may assume an arbitrary amount by which the radius shall increase in passing through a certain angle. See Fig. 4.84. With any convenient point as a center, draw a circle and divide it into 12 equal parts. Divide the radius into the same number of equal parts. One of the divisions of the radius is then the increment by which the radius of curvature increases in passing through an angle of 30°. Then beginning at the center and intersecting radius 1' by arc 1, radius 2' by arc 2, and so on, 12 points on the curve can be found. The curve thus generated is commonly used in cam design.

4.16.5 The Helix. The helix is a space or three-dimensional curve. It is described by the path of a point that moves around a cylinder at a uniform angular rate while also moving parallel to the axis at a uniform linear rate. To represent the helix, draw the circle and divide it into 12 equal parts as shown in Fig. 4.85. On the front view of the cylinder draw 12 equal divisions of the desired length. Project the points on the circumference to the corresponding horizontal line as shown in the figure. Draw a smooth curve through the points.

A conical helix may be drawn as shown in Fig. 4.85b. The curve is drawn first in the top view. Beginning at 1, one twelfth of the distance between the two circles is stepped off from the outer circle at point 2. Additional twelfths are stepped off at each succeeding radial line. The curve is then projected to the front view.

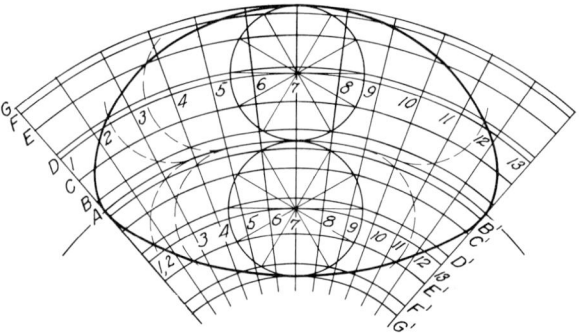

FIG. 4.83. Epicycloid and hypocycloid.

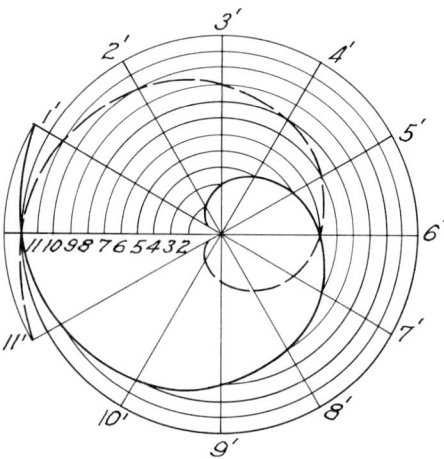

FIG. 4.84. Spiral of Archimedes.

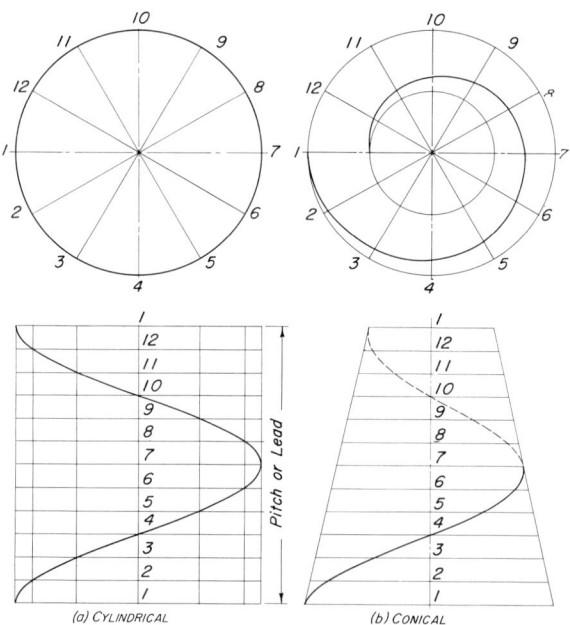

FIG. 4.85. Helix.

SELF-STUDY QUESTIONS

Before trying to answer these questions, read the chapter carefully. Then, without reference to the text, answer as many questions as possible. For those that cannot be answered, the number in parentheses following the question number gives the article in which the answer can be found. Look it up and write down the answer. Check the answers that you did give to see that they are correct.

4.1 (**4.2.1**) The bisector of a line or an angle is a line that divides a line into _____ parts.

4.2 (**4.2.3**) Parallel lines are the _____ distance apart throughout their length.

4.3 (**4.2.3**) Parallel lines are said to _____ when they are extended to infinity.

4.4 (**4.2.6; 4.3.4**) If two lines are perpendicular to each other, the angle between them is _____ degrees.

4.5 (**4.3.4**) Sketch the layout of a right triangle whose sides have the ratio of $3:4:5$.

4.6 (**4.5**) An isosceles triangle has only _____ equal sides.

4.7 (**4.5**) A scalene triangle has three _____ sides.

4.8 (**4.3**) To find the complement of an acute angle, subtract the angle from _____ degrees. Make a sketch and label the angles.

4.9 (**4.3**) To find the supplement of an obtuse angle, subtract the angle from _____ degrees. Make a sketch and label the angles.

4.10 (**4.4**) A pentagon has _____ sides.

4.11 (**4.5**) A parallelogram having four equal sides is called a _____.

4.12 (**4.5**) A trapezoid is a four-sided figure having two sides _____.

4.13 (**4.7.2**) If a circle is inscribed in a square, the circle is _____ to the sides of the square.

4.14 (**4.13**) A circle is the locus of points _____ from the center.

4.15 (**4.14**) Name the three principal conic sections.

4.16 (**4.14**) If a plane cuts a right circular cone parallel to its base, the shape of the section is _____.

4.17 (**4.14**) To cut straight lines from a cone, the cutting plane must pass through the _____.

4.18 (**4.14**) Which conic section is a single closed curve? _____

4.19 (**4.14**) Which conic section is a single open-ended curve? _____

4.20 (**4.14.10**) Name one common use of a parabola. _____

4.21 (**4.16.1**) What is a practical use of the involute curve? _____

4.22 (**4.14.2**) An ellipse has _____ foci.

4.23 (**4.14.2**) The sum of the distances from the foci to a point on an ellipse is equal to the _____ axis.

4.24 (**4.13.4**) If two circles are tangent to each other, the point of tangency lies on a line connecting _____.

4.25 (**4.13.3**) If a straight line is tangent to a circle, the radial line to the point of tangency is _____ to the straight line.

4.26 (**4.16.4**) The spiral of Archimedes is sometimes used in the design of _____.

CHAPTER 5

SKETCHING

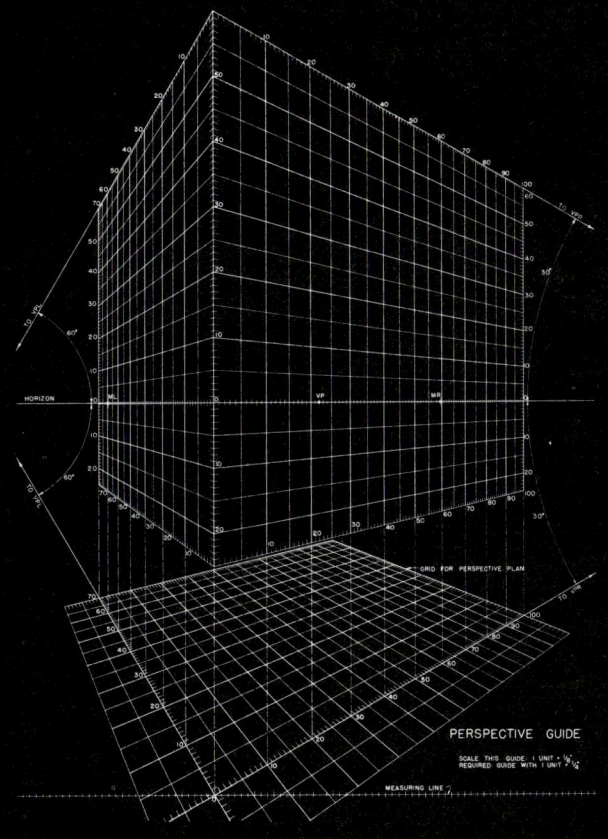

CHAPTER 5

5.1 INTRODUCTION

Many times the scientist or engineer finds him- or herself in a position in which he/she cannot express ideas mathematically or by words alone. In situations of this nature, the ideas may be expressed quite adequately by sketches. These sketches can then be used as a basis for further analysis and design.

It is sometimes assumed that skill in sketching can be acquired more rapidly than proficiency with instruments. This is not true. The same knowledge of the principles of projection is required in both cases, but it takes much more practice and effort to draw two parallel lines freehand than with instruments. It is simple to draw a circle with a compass but quite another matter to do it successfully freehand. This chapter is concerned chiefly with acquiring these sketching skills. How to make particular types of sketches is discussed in the latter part of this chapter where the basic arrangement of views is explained. The theory of various types of projection is explained in later chapters.

5.2 USES OF FREEHAND SKETCHES

Technical sketches are used for a wide variety of purposes, among which the following are the most important.

a. To transmit information, obtained in the field or shop, to the engineering office. This occurs when repairs have to be made or when changes in an existing structure are being considered.
b. To convey the ideas of the designer to the drafter.
c. To make studies of the layout of the views required in an instrumental drawing. See Fig. 5.1.
d. As a means of making preliminary studies of a design to show how it functions.
e. To compute stresses in a design. See Fig. 5.2.
f. To provide a basis for communicating between engineers, technicians, and craftspersons.

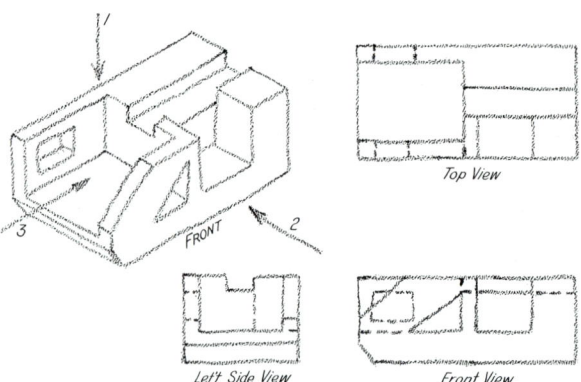

FIG. 5.1. Three-view sketch study for drawing layout.

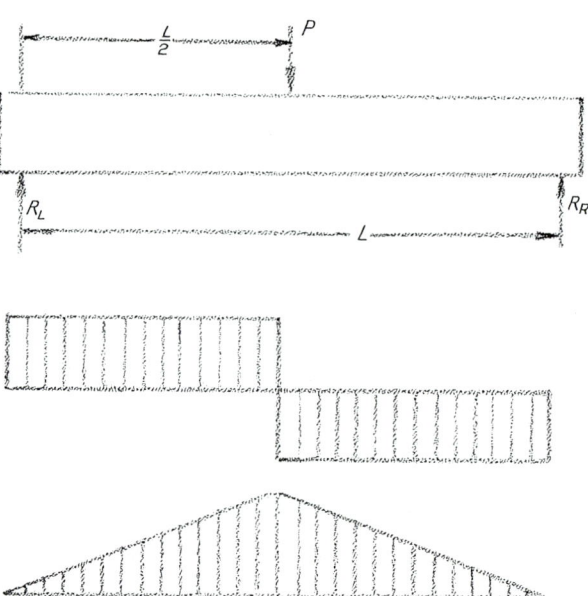

FIG. 5.2. Sketch for stress computation.

g. To furnish a three-dimensional picture of an object that will help to interpret the orthographic views.

h. To be used as shop drawings for manufacturing. These are usually made on coordinate tracing paper with grid lines that will not show on a blueprint. See Fig. 5.3.

i. To serve as a teaching aid when discussing problems in the classroom.

5.3 MATERIALS FOR SKETCHING

One of the advantages of sketching is that only a minimum of equipment is required. This normally includes pencil, paper, and eraser.

5.3.1 Pencils and Erasers. Most drafters prefer a soft pencil for sketching in the range from F to 2H. The H grade pencil is a good all-around tool. It should have a conical point and be frequently resharpened. A good eraser that will not smudge is essential.

5.3.2 Papers. Sketches have become so extensively used in industry that several different types of commercial paper are available. These are of great assistance to the engineer and technician in making industrial sketches.

a. *Blank paper.* For the experienced person blank paper of bond or ledger quality is satisfactory.

b. *Rectangular coordinate paper.* For multiview drawing, rectangular coordinate paper is an aid to the beginner. In many offices it is regularly used to produce dimensioned working drawings for the shop. Such coordinate papers can be obtained in a wide variety of divisions, both decimal and fractional. For ordinary purposes, subdivisions of $\frac{1}{8}$ or $\frac{1}{4}$ in. are satisfactory, as shown in Fig. 5.3.

Rectangular ruled paper is also useful in making oblique pictorial sketches, since the front face of an object will be the same as in multiview drawings. See Fig. 5.4.

c. *Isometric ruled paper.* In the case of pictorial sketches, isometric paper provides a guide to the correct position of the axes. See Fig. 5.5. It also provides units of measurement in proportioning.

d. *Perspective grids.* Perspective grids can be obtained in several sizes and with varying coordinate rulings. These are used most frequently for architectural projects. See Fig. 5.6.

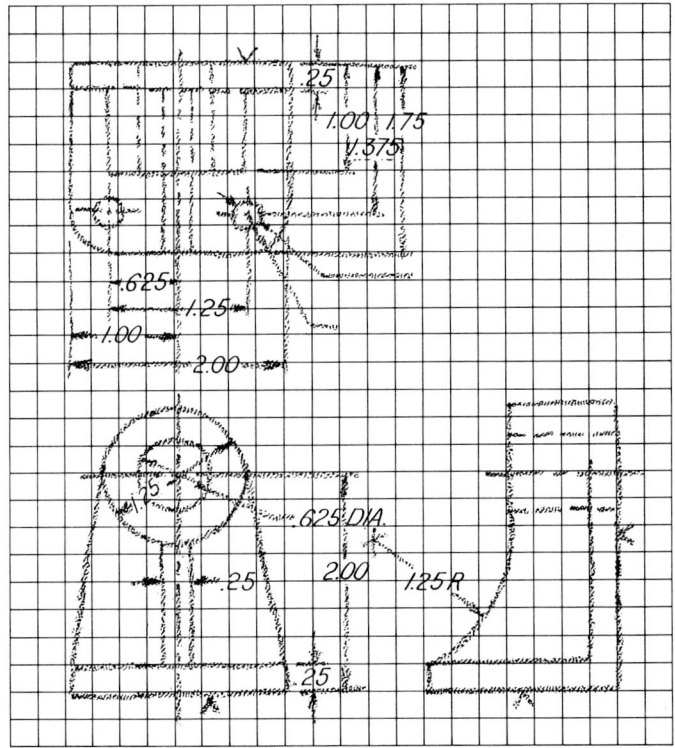

FIG. 5.3. Shop sketch for manufacturing.

88 SKETCHING

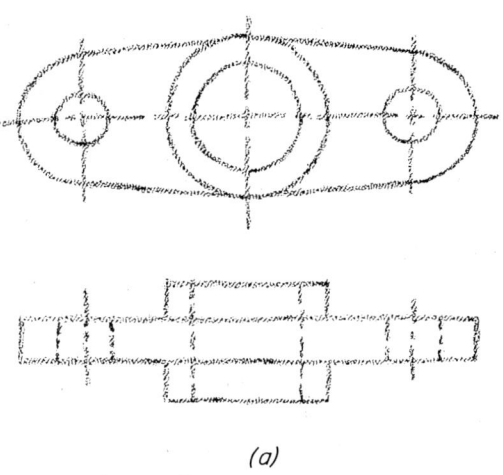

(a)

FIG. 5.4. Oblique sketch on coordinate paper.

(b)

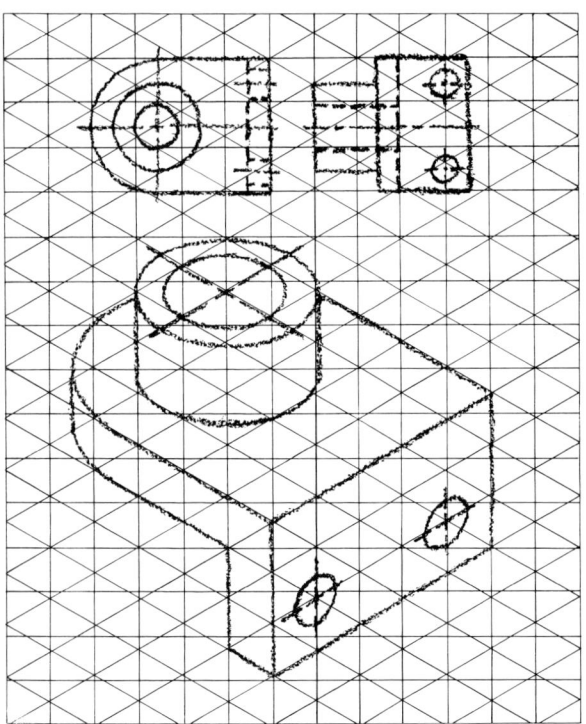

FIG. 5.5. Isometric sketch on coordinate paper.

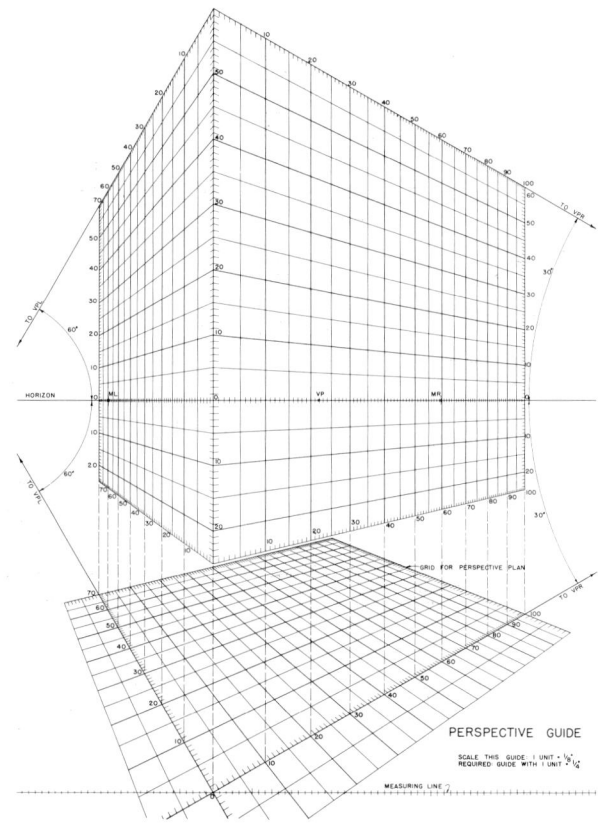

FIG. 5.6. Perspective grid (courtesy of Grace Wilson).

5.4 TOOLS FOR MEASUREMENT

Although freehand sketches are not made to scale, it is often necessary to make measurements of the object to be sketched. In an emergency, measurements must be made when a broken part is to be replaced or when work is to be done in the field on large projects.

a. *Field work.* For measuring large projects, for example, a change in the layout and location of equipment in a shop, a 6-ft rule or even a 50-ft steel tape may be needed. For a plot plan, surveying instruments may be used.

b. *Shop work.* Sketches may be made of manufactured parts that are machined to close tolerances. Measurements of such objects may require the use of steel scales with divisions of one hundredth of an inch. Ordinary calipers or micrometer calipers may be needed. It may be necessary to use surface gages, depth gages, and thread gages.

c. *Use of measuring tools.* For either type of work the person making the sketches must know how to use the tool he/she needs to obtain the dimensions required. Some of these tools are shown in Fig. 5.7.

5.5 SKETCHING STRAIGHT LINES

It is important to be able to sketch lines rapidly whether they are horizontal, vertical, or inclined. The same general principles apply in all cases.

a. Determine and mark the end points of the line to be sketched.
b. Hold the pencil in a normal writing position as in Figs. 5.8 and 5.9.
c. Place the pencil point at one of the end points.
d. Sketch a light line with one stroke, or short overlapping strokes, by keeping the eye fixed on the point toward which the line is being drawn.

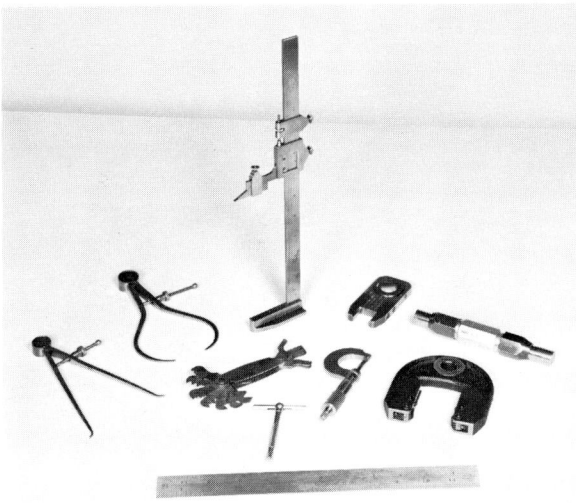

FIG. 5.7. Measuring tools.

FIG. 5.9. Sketching a vertical line.

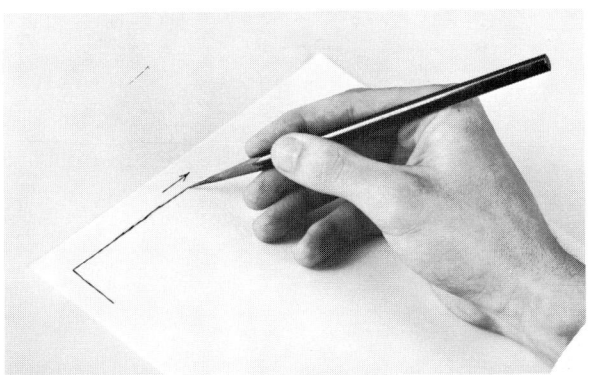

FIG. 5.8. Sketching a horizontal line.

e. Go over the line to remove waviness or roughness and make it of the required weight and character. The final sketch stroke has a freedom and character that are entirely different from the clean-cut precision of the mechanically ruled line. The difference between ruled lines and sketched lines is illustrated in Fig. 5.10.

5.5.1 Sketching Horizontal Lines. For drawing horizontal lines, the forearm should be approximately at right angles to the line being sketched, as shown in Fig. 5.8. The strokes should be made from left to right according to the principles outlined in Article 5.5.

5.5.2 Vertical Lines. The same type of stroke should be used for vertical lines as described above. When using the finger and wrist movement, the forearm should be approximately parallel to the line or at an angle of not more than 45°. Draw the stroke from top to bottom as shown in Fig. 5.9.

5.5.3 Inclined Lines. Inclined lines, whether on plain paper or on coordinate paper, are usually drawn between two given points or at some specified angle. On coordinate paper, angles of 15, 30, 45, 60, and 75° can be estimated with sufficient accuracy as shown in Fig. 5.11. If difficulty is experienced in drawing inclined lines, the paper may be rotated until the lines are about horizontal. See Fig. 5.12.

5.5.4 Parallel Lines. It is frequently necessary to draw lines parallel to each other quite accurately. Figure 5.13 shows one method that can be used. This consists of holding the pencil with the fingers as far from the point as convenient and then moving the hand so that the little finger slides along the original line.

If this is not possible, the method shown in Fig. 5.14 will produce good results. The pencil is held as shown, directly above the line, and then moved to a parallel position. The pencil is moved lengthwise to draw the line, always keeping the pencil parallel to the original line.

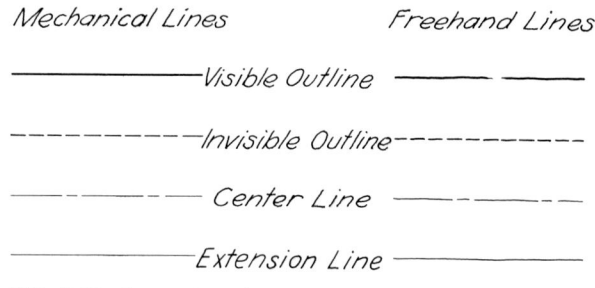

FIG. 5.10. Comparison between ruled lines and sketched lines.

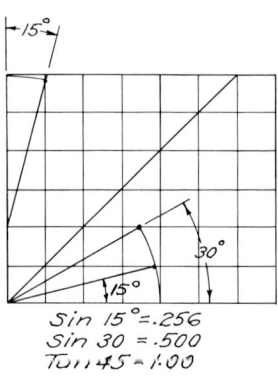

FIG. 5.11. Plotting angles.

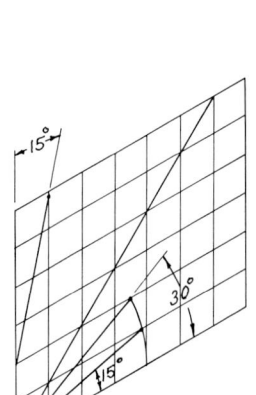

FIG. 5.12. Drawing inclined lines by turning paper.

FIG. 5.13. Drawing parallel lines.

5.5.5 Border Lines. Border lines and some horizontal and vertical lines may be made easily by holding the edge of the paper parallel to the edge of the board. Then draw the line with the third and fourth fingers of the hand sliding along the edge of the board as a guide. See Fig. 5.15.

5.6 SKETCHING CIRCLES

Circles occur frequently in engineering drawings. It is therefore necessary for the engineering designer to sketch them easily.

5.6.1 Trammel Method. On the edge of a piece of paper mark two points at a distance equal to the radius of the circle. With one point at the center, mark off as many points on the circumference as desired, as shown in Fig. 5.16. Sketch the circle through these points.

5.6.2 Enclosing-Square Method. On coordinate paper, a circle may be sketched in its enclosing square as shown in Fig. 5.17. On plain paper, the square can be sketched very lightly, the center lines drawn, and then the circle sketched, making the arcs tangent to the sides of the square at the midpoints.

5.6.3 Semimechanical Method. A semimechanical method for drawing circles is to hold two pencils intersecting each other and having the distance between the points equal to the desired radius. By holding one pencil perpendicular to the paper with its point at the center of the circle, the paper may be revolved about the point while the other pencil describes the circle. This method is illustrated in Fig. 5.18.

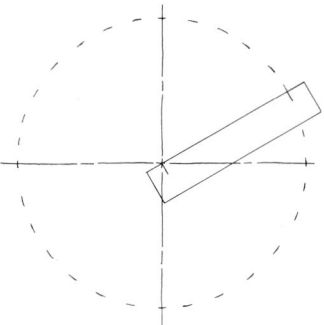

FIG. 5.16. Sketching circles by trammel.

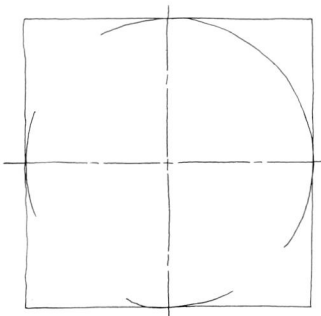

FIG. 5.17. Sketching a circle in a square.

FIG. 5.14. Drawing parallel lines.

FIG. 5.15. Sketching border line along edge of drawing board.

FIG. 5.18. Method of drawing circle with two pencils.

5.7 SKETCHING AN ELLIPSE

There are several methods in common use for sketching an ellipse. Four methods are described below.

5.7.1 Trammel Method. For the ellipse, mark off three points, *a*, *b*, and *c*, as shown in Fig. 5.19, with *bc* equal to the semiminor axis and *ac* equal to the semimajor axis. Keeping points *a* and *b* always on the center line of the ellipse, plot as many positions of point *c* as desired and sketch the ellipse (also see Article 4.14.8).

5.7.2 Rectangle Method. An ellipse may be sketched in a rectangle by drawing the center lines in the rectangle and making the arcs tangent at the midpoints of the sides in the same manner as for a circle in a square. Additional points may also be obtained as shown in Fig. 5.20 (also see Article 4.14.4).

5.7.3 Circles in Pictorial Sketches. Except for circles in the front face of an oblique sketch, all circles show as elipses in pictorial drawings. These may be drawn by the enclosing pictorial square or parallelogram method as shown in Fig. 5.21. Arcs are always tangent at the midpoint of the sides of the parallelogram.

5.7.4 Free Arm Movement. Another method of drawing ellipses is with a free arm movement as shown in Fig. 5.22. This method may be used to sketch a circle or ellipse so that it will go through certain points or be tangent to one or more straight lines or circles. For ellipses the forearm should be held approximately perpendicular to the major axis of the ellipse with the fourth and fifth fingers riding lightly on the paper. The procedure is as follows:

a. Move the arm freely so that the point of the pencil describes an ellipse just above the desired position on the paper.

b. After two or three complete circuits, allow the pencil to touch the paper lightly and draw several ellipses. Some of these will be inside the desired position and some outside.

c. With these trial lines as a basis, a sketch stroke ellipse may be drawn to satisfy the desired conditions of position or tangency.

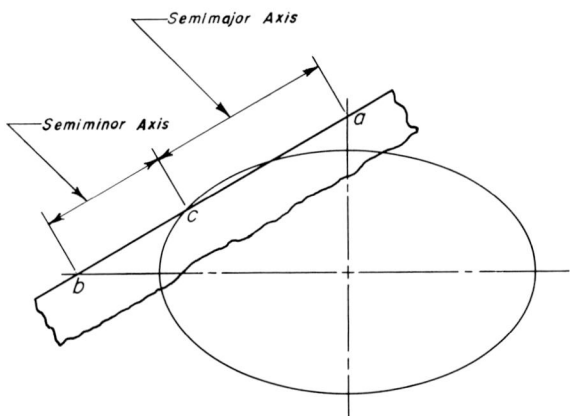

FIG. 5.19. Ellipse by the trammel method.

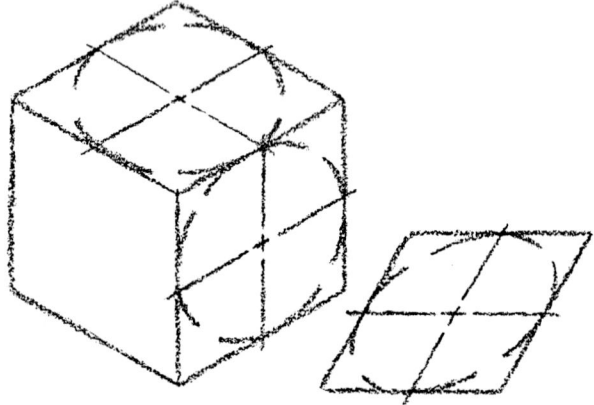

FIG. 5.21. Sketching an ellipse in a parallelogram.

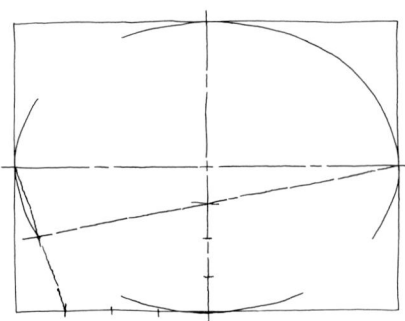

FIG. 5.20. Sketching an ellipse in a rectangle.

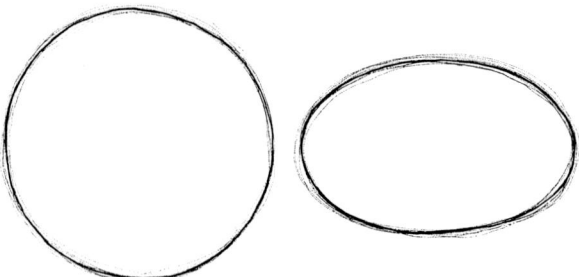

FIG. 5.22. Sketching circle or ellipse by use of the finding line.

d. Erase the trial lines and sketch stroke to the point where the ellipse is faintly visible.

e. Draw a smooth curve as shown in Fig. 5.22.

5.8 PROPORTIONING

It is important to maintain the relative proportions between the overall length, width, and height while making a sketch of an object. With practice, one can develop the ability to divide a line in half by eye and then these halves can be again divided to get fourths. To aid the beginner, several methods can be used to develop an eye for proportioning.

5.8.1 Counting Squares. When coordinate paper is used, the proportioning can be done by "counting squares." This amounts to selecting a scale whereby the sketcher assumes a given number of squares to equal a unit of measure such as an inch or a foot or a fraction of either.

5.8.2 Scrap Paper. Scales are assumed not to be available in sketching; however, a piece of scrap paper with a straight edge can be marked off to represent a given unit of measure. To get halves and fourths, the paper can be folded.

5.8.3 Geometric Constructions. Figure 5.23a shows the steps in sketching a rectangle three times as long as it is wide. With AB given in a vertical position, erect perpendiculars to AB at A and B. Bisect the right angles at A and B with lines AC and BD to form the square ABCD. Through the intersection of AC and BD, sketch a line MN parallel to AD and BC. Sketch BF through N and AE through N to locate points E and F on BC and AD extended. CDFE forms a square equal to ABCD. Repeat the process to obtain the third square.

A more accurate method is shown in Fig. 5.23b. Here the long axis is assumed as AB. A square ABCD is sketched and the diagonals drawn to locate point E, the midpoint of BC. AE is sketched in to locate point F on diagonal BD. Line GK is drawn parallel to AB through F, thus making AG and KB equal to one third the length of AB. Figure 5.23c shows how this method can be used to obtain various other subdivisions of a line segment such as $\frac{1}{2}$, $\frac{1}{3}$, $\frac{1}{4}$, and $\frac{1}{5}$ (also see Article 4.2.11).

5.8.4 Proportioning Large Objects. One convenient method for obtaining relative proportions when drawing large objects is illustrated in Fig. 5.24. The sketch board is held almost perpendicular to the line of sight. The arm is extended full length with the pencil held in the fingers. By holding the pencil between the eye

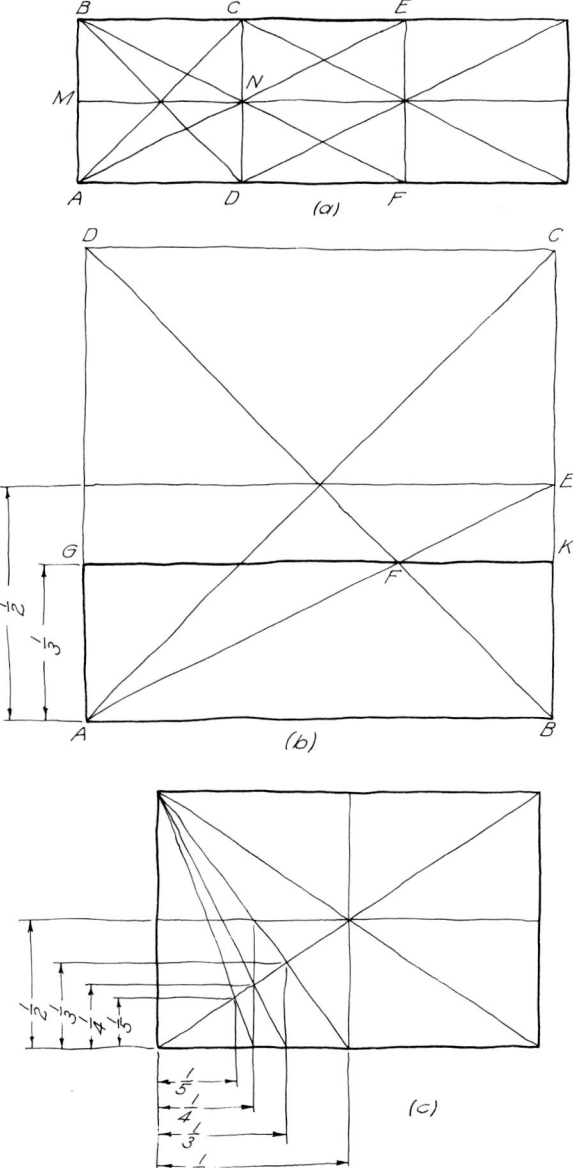

FIG. 5.23. Method or proportioning or subdividing a rectangle.

and the object, one end of the pencil can be made to coincide with one end of a line and the thumb moved along the pencil until it coincides with the other end of the line. This length may then be used to obtain the relative proportions of the object. It is also possible to get the position of a line by holding the pencil parallel to any line of the object and then moving it to a parallel position on the drawing. Sketching in this manner results in a perspective drawing.

5.9 TYPES OF SKETCHES

The engineer makes several kinds of sketches. Multiview orthographic sketches are used when discussing a problem with the staff. The theory of multiview sketching is presented in Chapter 6.

For less experienced persons a three-dimensional sketch that looks like a picture may be used. There are three types of three-dimensional sketches, namely, *axonometric, oblique,* and *perspective.*

The theory of axonometric sketching with its three subdivisions—isometric, dimetric, and trimetric—is explained in Chapter 13.

Oblique projection theory, with two subdivisions—Cavalier and Cabinet drawing—is explained in Chapter 14.

Perspective sketching is explained in Chapter 15.

5.10 MULTIVIEW SKETCHES

Multiview sketches may consist of one-view illustrations as in Fig. 5.25. These can be used only for very simple objects where the third dimension is a thickness given by a note. Gaskets, shims, and washers are in this class. For two- and three-view sketches a standard arrangement of views has been established, which has been adopted in the United States, Great Britain, and Canada, as shown in Fig. 5.26. While all six views of an object have been shown in this illustration, no more views should be made than are necessary to describe the shape of the object.

5.11 SKETCHING PRACTICE

Good technique in sketching can only be developed by practice. The student may use anything for a model to practice sketching. Practicing of sketching of lines in various positions will help develop this necessary skill.

5.12 SKETCHING MULTIVIEW DRAWINGS

In developing the theory and practice of multiview drawing, it is convenient to employ pictorial drawings that are used as a reference for sketching top, front, and side views of an object. In engineering design it is also convenient to sketch the various views from pictorial representations. This process is illustrated in Fig. 5.27. As a basis for proportioning, the object is assumed to be enclosed in a box to determine the relative dimensions. In Fig. 5.27a the bracket is in the proportion $2 \times 2 \times 3$. The steps in sketching the three views are presented in the following articles.

5.12.1 Sketching the Top View. To sketch the top view the enclosing box of the bracket is viewed from the top as indicated by arrow 1 in Fig. 5.27a. The proportions of the box are to be 2×3. The method of Fig. 5.23a is used to obtain these proportions.

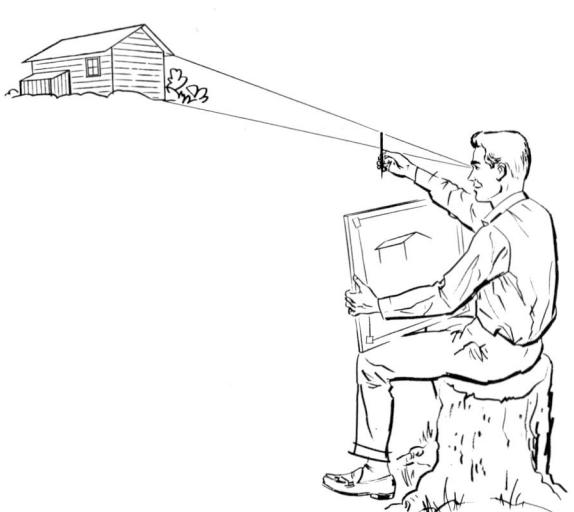

FIG. 5.24. Method of obtaining proportions by means of a pencil.

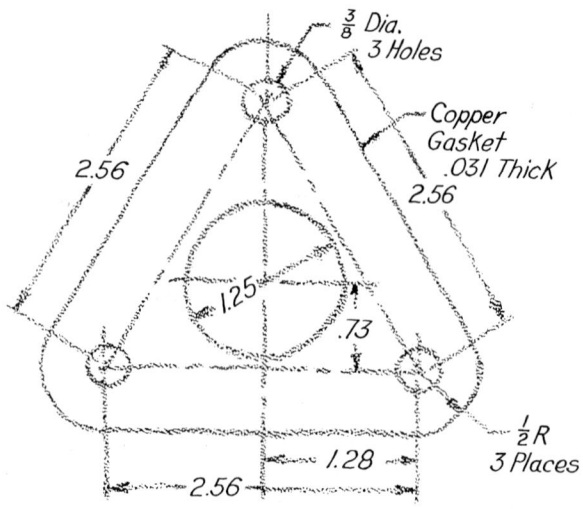

FIG. 5.25. One-view sketch.

5.12 SKETCHING MULTIVIEW DRAWINGS   95

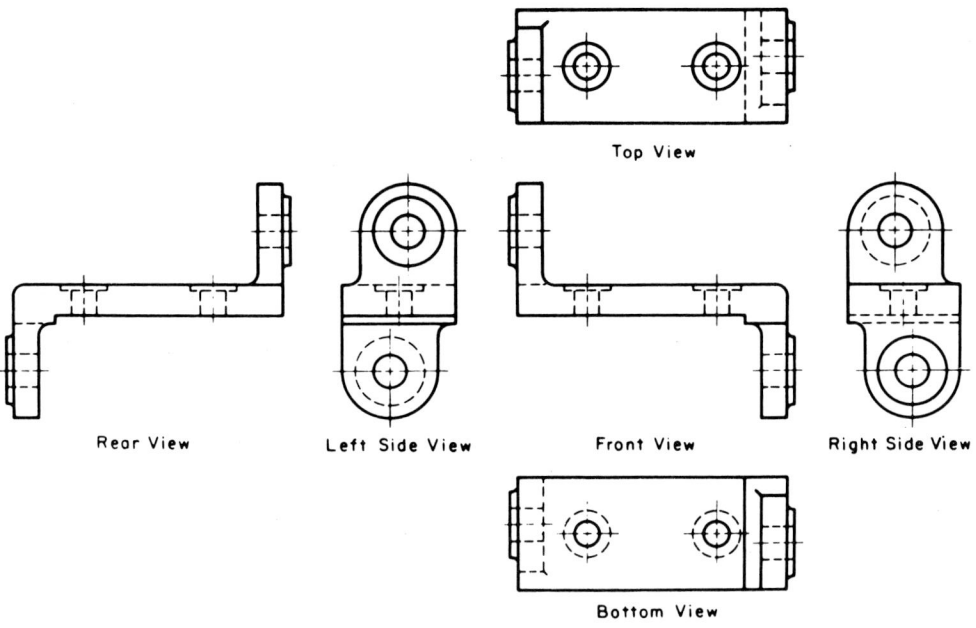

THE SIX POSSIBLE PRINCIPAL VIEWS OF AN OBJECT

FIG. 5.26. Standard arrangement of six views of an object (courtesy ANSI).

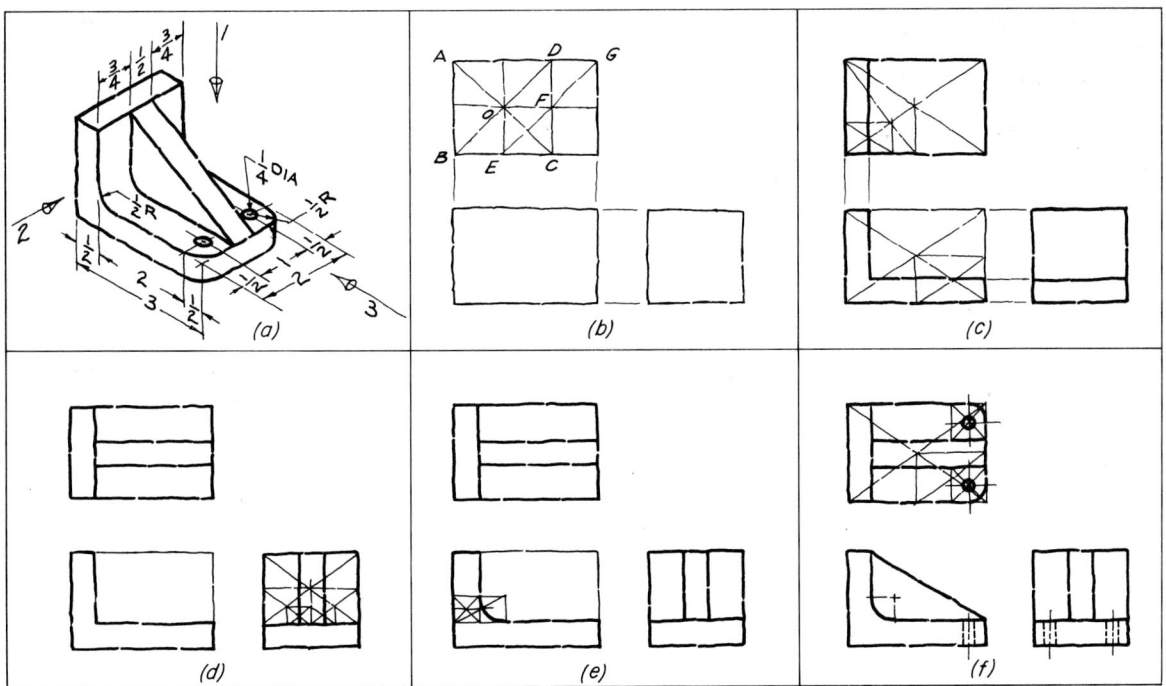

FIG. 5.27. Sketching three views of an object.

a. Make line *AB*, representing two units, any length desired to make a sketch of suitable size.
b. Sketch the square *ABCD* and draw the diagonals.
c. Draw center lines through *O*.
d. Draw line *EF* to *G* on the upper line.
e. Complete the rectangle which is now in the proportion 2 × 3.

5.12.2 Sketching the Front View. To sketch the front view of the enclosing box the bracket is viewed from the front as indicated by arrow 2 in Fig. 5.27*a*. The box will be the same size as the top view in this case.

a. The length can be sketched downward from the ends of the top view as in Fig. 5.27*b*.
b. The height can be transferred from the top view with a piece of scratch paper. This makes the front view also in the proportion 2 × 3, as it should be.

5.12.3 Sketching the Side View. The side view is obtained by viewing the bracket from the side as indicated by arrow 3 in Fig. 5.27*a*. The enclosing box has a proportion of 2 × 2. This can be obtained as follows.

a. The height can be obtained by sketching across from the front view as in Figs. 5.27*b* and *c*.
b. The depth can be transferred by a piece of paper from the top view. The proportion for the end view is therefore 2 × 2.

5.12.4 Sketching in the Details. To sketch in the details for each of the views, the proportion of the parts can be obtained by subdividing the views with diagonals or by eye.

a. In the top view note that the thickness of the vertical back is one sixth the length of the whole object. This thickness can be obtained by dividing the existing left-hand one third in two parts as in Fig. 5.27*c*.
b. The thickness of the base is one fourth the total height. This proportion can be obtained by diagonals in the front view as in Fig. 5.27*c*.
c. From the dimensioned pictorial it can be seen that the sloping web has a thickness equal to one fourth of the total depth.
d. This can be done by a series of diagonals in the side view as in Fig. 5.27*d*.
e. The thickness of the base can be sketched across from the front view to the end view.
f. Mark tangent points in front view and sketch arcs. See Figs. 5.27*e* and *f*.
g. Locate center of small holes and the concentric arcs by eye. Mark tangent points and sketch arcs.
h. Erase construction lines and make outlines firm.

5.13 PICTORIAL SKETCHING

The pictorial sketch is used for the following purposes.

a. To clarify or explain two or three views on the orthographic drawings.
b. To explain ideas to other persons.
c. As an aid in computation and design. Pictorial sketches may be made in isometric, oblique, or perspective. Of these three, isometric and oblique are the easiest to make. See Figs. 5.4 and 5.5.

5.14 ISOMETRIC SKETCHING

The first step in making an isometric sketch is to draw a box that will just enclose the object. In Fig. 5.28 assume that the ratio of the three edges of the box are 2 × 3 × 4. The steps are as follows.

a. Draw a vertical line *AB* two units long as in Fig. 5.28*a*.
b. On this line construct a square and bisect the sides with the aid of the diagonals as in Fig. 5.28*b*.
c. Using point *O* as the center, sketch an arc with a radius of one unit as shown in Fig. 5.28*b*. Through point *B* sketch a line tangent to the arc. This line will therefore recede at an angle of 30° with the horizontal. Repeat on the right side to obtain the other 30° receding line. This makes the three axes at 120° with each other. See Fig. 5.28*e*.
d. Through point *A* sketch a line parallel to line *BD* as in Fig. 5.28*c*.
e. Locate point *E* on the receding line from *A* by extending line *BC* until it intersects the line at *E*. Line *BE* becomes one of the diagonals of the square in isometric (actually a rhombus). See Fig. 5.28*c*.
f. Draw a line *EH* parallel to *AB* and draw the diagonal *AH*. Locate the horizontal center line and points *F* and *G* as in Fig. 5.28*c*.
g. From *A* draw a line through *F* to *D* on the upper line, thus doubling the square.
h. Divide the second square in half by diagonals and the vertical line *MN*, thus making *BM* three units long.

i. In Fig. 5.28d — repeat the construction by laying out two equal squares, thus making this side four units long. The box is then completed by drawing the two remaining sides of the top as in Fig. 5.28e.

5.14.1 Object with Isometric Lines Only.
The object shown in Fig. 5.29a is composed entirely of isometric lines that are parallel to the three axes. To sketch such an object, proceed as follows.

a. Construct an isometric box using the proper proportions for the object in Fig. 5.29a. See Fig. 5.29b.
b. Lay out the principal details, for example, the groove running across the top as shown in Figs. 5.29c and d.
c. Next sketch in the remainder of the details in proper proportion as shown in Figs. 5.29e and f.
d. Remove construction lines and finish the sketch as in Fig. 5.29f.

5.14.2 Object with Nonisometric Lines.
The general procedure for sketching an object with nonisometric lines is the same as that explained in the preceding paragraph, with the following exceptions. All nonisometric lines must be determined by locating their end points by means of isometric lines as shown for the top of the object in Fig. 5.30c.

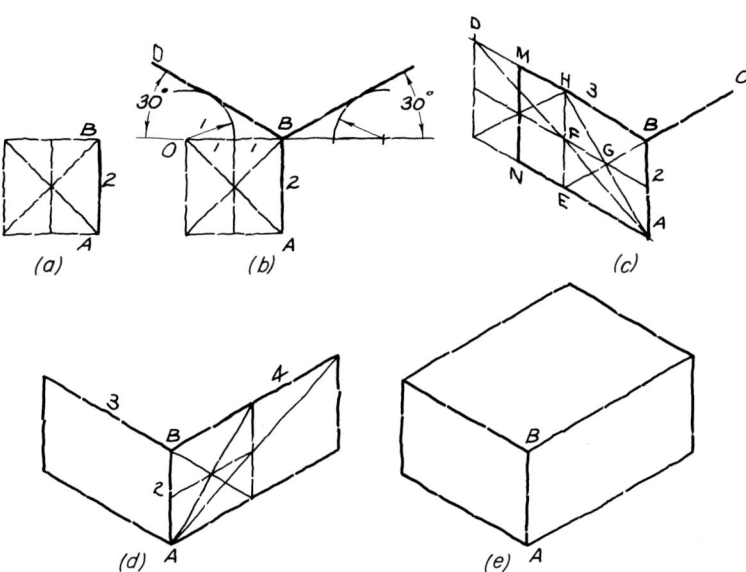

FIG. 5.28. Laying out an isometric box.

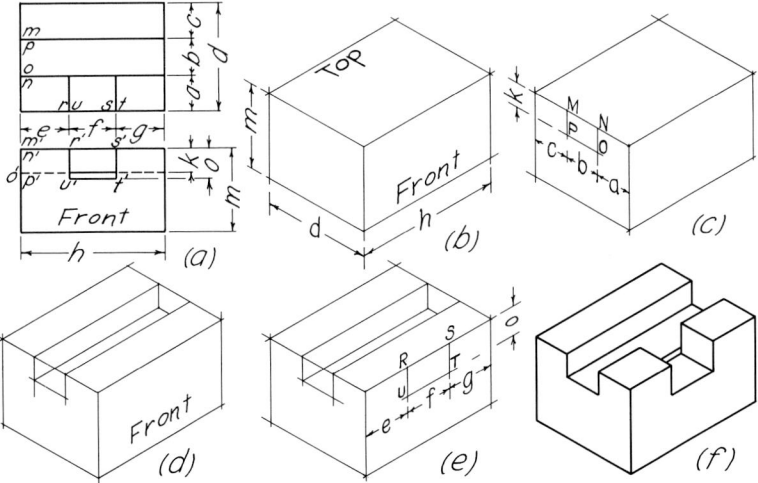

FIG. 5.29. Construction of an isometric drawing having isometric lines only.

98 SKETCHING

Next sketch in the lower step by estimating the upward distance equal to distance n, then along with width for the front and back-edge distances h and m, and then the depth f and g as shown in Fig. 5.30d. Having these eight points located, draw the sloping or nonisometric lines as in Fig. 5.30e. Finish the sketch in the usual manner as in Fig. 5.30f.

5.14.3 Cylindrical Objects. Assume an object given as in Fig. 5.31a. Instead of one large enclosing box, draw several boxes in proper proportion and location that will just enclose the separate parts as in Figs. 5.31b and c. These boxes must have their geometric centers lying on a common isometric axis, which slopes 30° up to the right in this case. Sketch the ellipses in the proper parallelograms. Draw the isometric tangent lines between them. Finish the sketch by erasing all construction and making the outlines firm. Hidden lines are not shown in pictorial drawings unless absolutely necessary for an understanding of the sketch.

5.15 OBLIQUE SKETCHING

Oblique sketching differs from isometric mainly in the position of the three principal axes. In isometric, the axes were at 120° with each other. In oblique, two axes are at right angles to each other; the third one may be at any desired angle with the horizontal. It is usually between 30 and 45°. See Fig. 5.32.

The sketch is again based on an enclosing box. This time, however, the front of the box is either a square or a rectangle, as the case may be. The front face, therefore, is just like an orthographic view and circles show as true circles. The ellipses in the top and side may be constructed by the four-center approximate method as shown in Fig. 5.32.

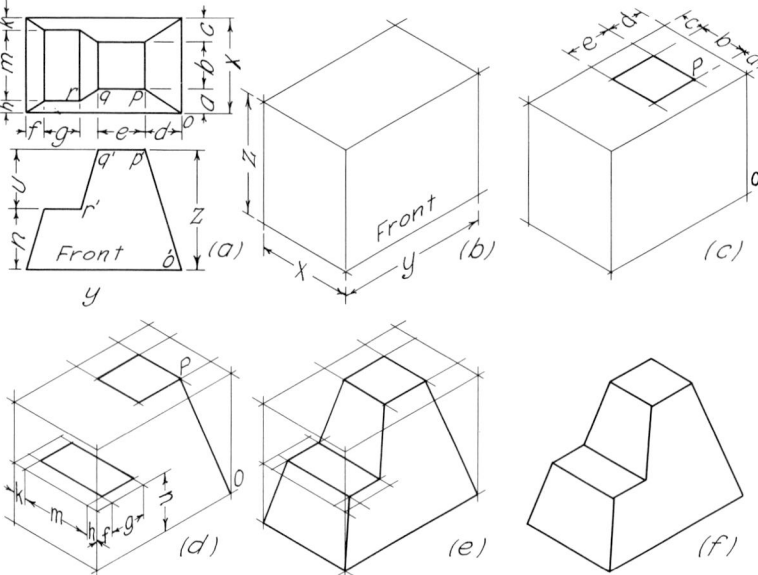

FIG. 5.30. Construction of an isometric drawing having nonisometric lines.

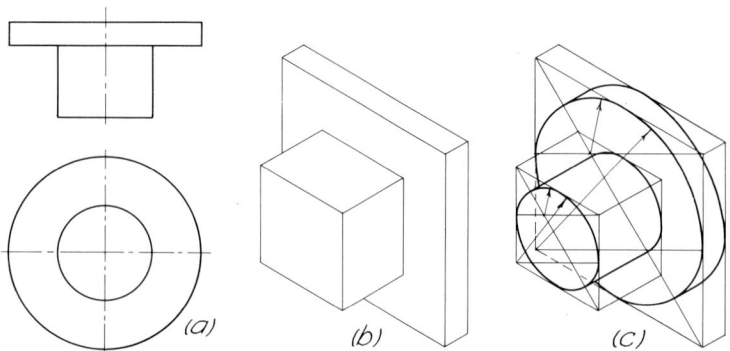

FIG. 5.31. Steps in making an isometric drawing of a cylindrical object.

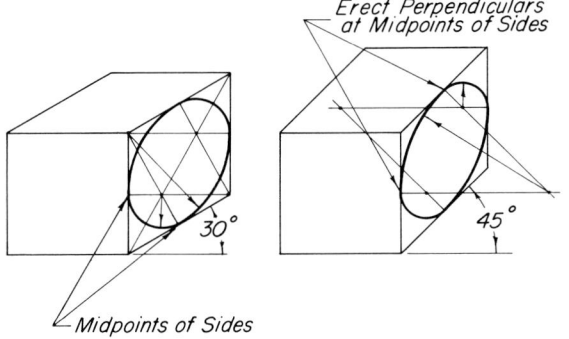

FIG. 5.32. Four-center method of constructing a circle in cavalier projection.

5.15.1 Box Method. By proper selection of the inclined axis either the side or the top face may be emphasized. The steps in construction are similar to those for isometric, as shown in Fig. 5.33.

5.15.2 Center-Line Construction. For objects that are largely cylindrical in shape, as the object in Fig. 5.34a, a centerline method of construction can be used advantageously.

a. The center lines are laid out first in their proper proportion and position as in Figs. 5.34b and c.

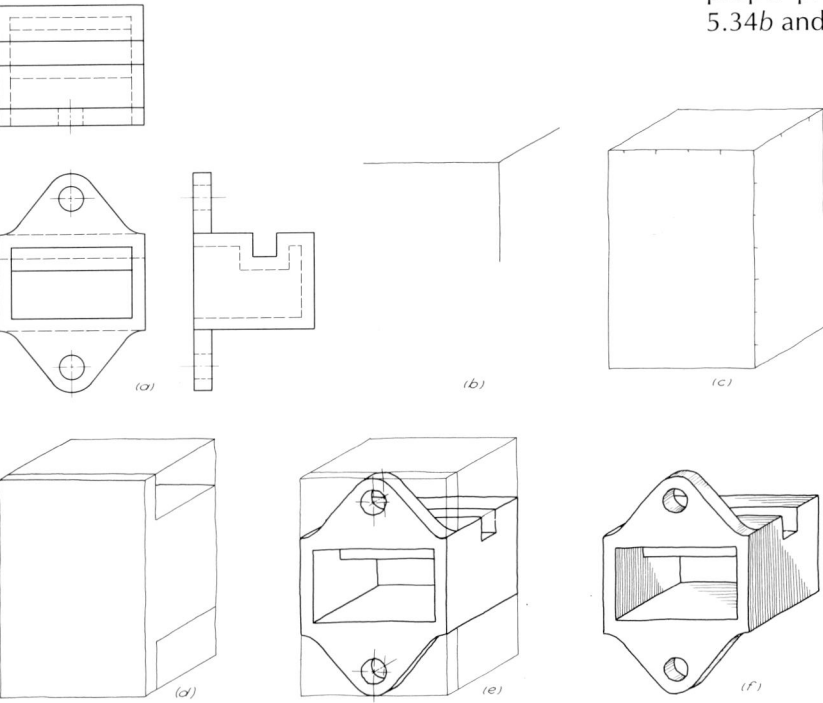

FIG. 5.33. Steps in making an oblique sketch.

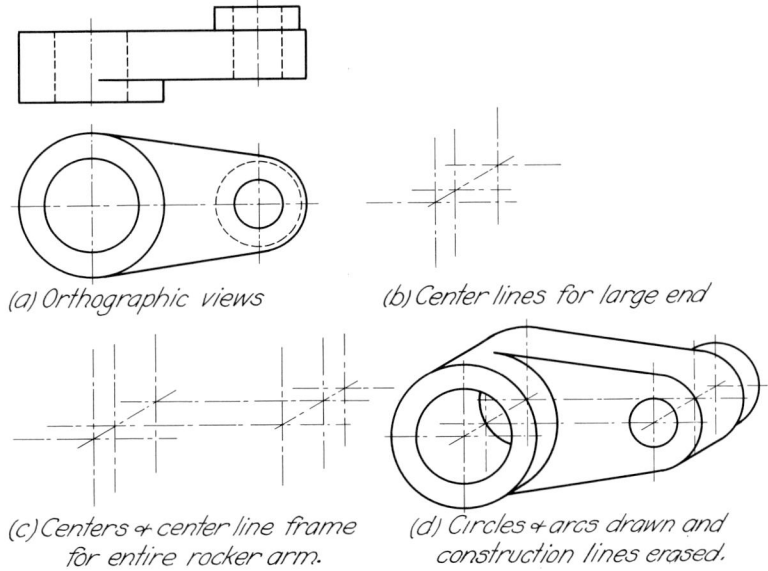

(a) Orthographic views (b) Center lines for large end

(c) Centers & center line frame for entire rocker arm. (d) Circles & arcs drawn and construction lines erased.

FIG. 5.34. Center-line method of construction.

b. Draw the circles and partial circles by the trammel method, using a scrap of paper.

c. Draw the tangent lines between parallel circles as in Fig. 5.34d.

5.16 PERSPECTIVE SKETCHING

A perspective of an object is approximately the view as seen by the observer if he were to look at it with one eye. The most important difference between perspective and other forms of pictorial is that receding lines converge instead of remaining parallel. In perspective, lines that are parallel to the picture plane will show parallel on the drawing, but all parallel receding lines are drawn to converge at a point. For large objects proportioning may be done with a pencil as shown in Fig. 5.24. The complete theory of perspective is given in Chapter 15.

SELF-STUDY QUESTIONS

In the questions that follow fill in the blank spaces or underline the correct word where a choice is given. If you cannot answer, look up the paragraph number immediately following the question number.

5.1 (**5.2**) Name three uses of freehand sketches. _____ _____ _____

5.2 (**5.3.1**) What equipment, besides paper and measuring instruments, is needed for sketching? _____

5.3 (**5.3.2**) Name three kinds of commercially available coordinate papers. _____ _____ _____

5.4 (**5.5.3**) Show by a sketch how to approximate angles of 30 and 45° on rectangular coordinate paper.

5.5 (**5.6.1**) Show by a well-labeled and noted sketch how to draw a circle, using a piece of paper that has a straight edge.

5.6 (**5.8.2**) How may a given space be divided accurately into halves and quarters using only a strip of paper? _____

5.7 (**5.9**) Name four different types of sketches used by engineers. _____ _____ _____ _____

5.8 (**5.10**) What type of objects can be shown by one (nonpictorial) view? _____ _____

5.9 (**5.12.1**) In making a top view, how does the engineer imagine him- or herself to be looking at the object? _____

5.10 (**5.12.2**) For making a front view the drafter imagines him- or herself to be looking from the _____.

5.11 (**5.12.1**) For the standard arrangement of views the drafter places the top view _____ _____ _____ view.

5.12 (**5.12.3**) The side view is placed to the _____ _____ _____ of the front view.

5.13 (**5.13**) Name two purposes for which pictorial sketches may be used. _____ _____

5.14 (**513c**) Name three types of pictorial sketches. _____ _____ _____

5.15 (**5.14**) The first step in making an isometric sketch is to lay out a _____ which will just enclose the _____.

5.16 (**5.14.1**) Isometric lines are those that are _____ to the three _____ axes.

5.17 (**5.14.2**) In sketching nonisometric lines, the _____ _____ must be located first.

5.18 (**5.15**) In oblique sketching two of the axes are at _____ _____ to each other and the third may be _____ _____ _____ with the first two.

5.19 (**5.15.2**) In oblique sketching, the layout may be made around _____ _____ instead of using the _____ method.

CHAPTER 6
ORTHOGRAPHIC PROJECTION

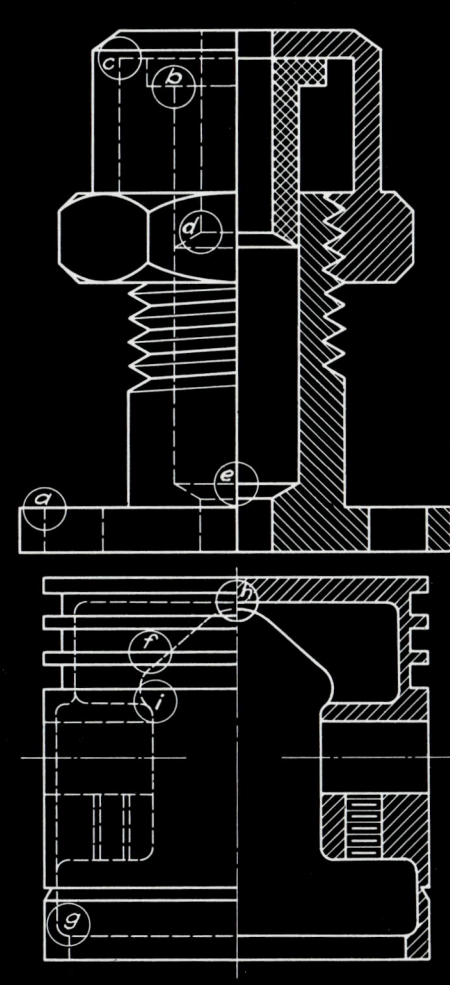

CHAPTER 6

6.1 DEFINITION

A projection is a drawing that represents or outlines a three-dimensional object on a two-dimensional surface. It is drawn on a plane of projection as it would be seen by looking through the plane of projection from a certain point of sight. See Fig. 6.1. In engineering the most commonly used type of projection is called *orthographic projection.*

Orthographic projection is the representation of an object on a plane of projection when the lines of sight from the eye to the object are perpendicular to the plane of projection. See Fig. 6.2.

6.2 ELEMENTS OF PROJECTION

In making any projection, there are four factors to be considered. These factors are called the *elements of projection*. They are

   a. The point of sight.
   b. The lines of sight.
   c. The plane of projection.
   d. The object.

They are all illustrated in Fig. 6.1.

6.3 AXIOMS OF ORTHOGRAPHIC PROJECTION

In orthographic projection, the relationships among the four elements of projection are definite and remain constant. These relationships identify orthographic projection and are listed in the following paragraphs.

6.3.1 Point of Sight. The point of sight is the real or imaginary position of the eye of the observer when viewing the object. The eye of the observer may be placed at any position desired for this purpose.

For orthographic projection the point of sight is considered to be at an infinite distance from the object.

6.3.2 Lines of Sight. All lines joining the eye or point of sight with points on the object are called *lines of sight. In orthographic projection, the lines of sight for any view are parallel to one another and perpendicular to the plane of projection.* See Fig. 6.2.

6.3.3 Plane of Projection. The plane upon which the object is projected is called the *plane*

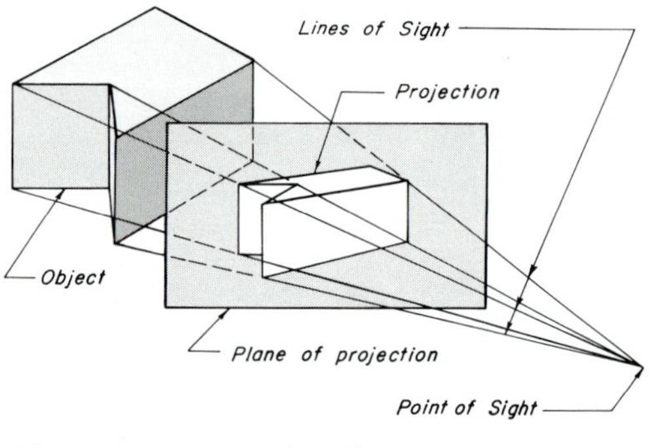

FIG. 6.1. Projection of an object.

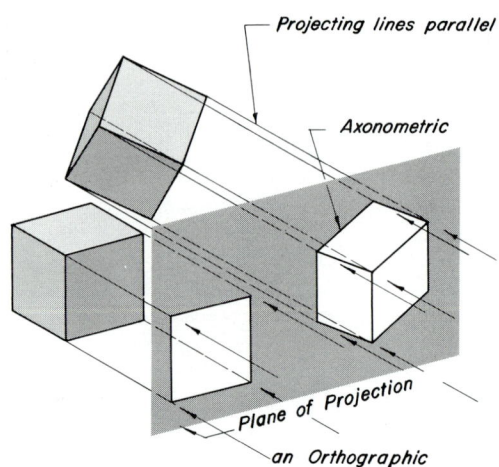

FIG. 6.2. Orthographic projection.

of projection. Since the lines of sight must be perpendicular to the planes in orthographic projection, it is necessary to have a different plane for each point of sight. Thus, for a one-view drawing there would be one plane and one point of sight and for a three-view drawing there would be three planes of projection and three points of sight.

The planes most commonly used are a *horizontal plane*, a *vertical plane*, which is perpendicular to the horizontal plane, and a *profile plane*, which is perpendicular to both the horizontal plane and the vertical plane. These are illustrated in Fig. 6.3. As a group they are called the *principal planes of projection*.

6.3.4 Object. Any object, either real or imaginary, may be drawn in orthographic projection. Theoretically, it can have any relation to the plane of projection. However, the usual purpose of making an orthographic projection is to give a complete and accurate description of the shape of the object. To accomplish this, certain rules for setting up the object have been established.

a. Place the object in its natural position or in the position in which it is to be used.
b. Place the object so that its faces will be parallel to the principal planes of projection.
c. Turn the object so that its most important or most descriptive face is parallel to the vertical plane.
d. Select the views that will show the most visible lines.

6.4 THE PROJECTION

The projection is obtained by finding the points at which the lines of sight pierce the plane of projection and connecting them in the proper order. *Every line on the object must show as a point or a line on the projection*. If the line can be seen from the point of sight, it is represented by a solid line known as a visible outline. See Fig. 6.4. If the line is hidden behind some other part of the object, it is represented by a dashed line known as an invisible outline. See Fig. 6.4. All of the various kinds of lines that may be found on a drawing or projection are shown in Fig. 6.4.

6.5 QUADRANTS OR ANGLES

When the vertical plane, or V-plane, has been set up perpendicular to the horizontal plane, or H-plane, these two planes divide all space into four quadrants or angles, as shown in Fig. 6.3. In discussing these quadrants, the observer is considered to be in front of the V-plane and facing it. Thus, the part of an object closest to the observer is said to be the front and the part farthest from the observer, the rear. Right and left are also taken from this position of the observer.

As indicated in Fig. 6.3, that portion of space in

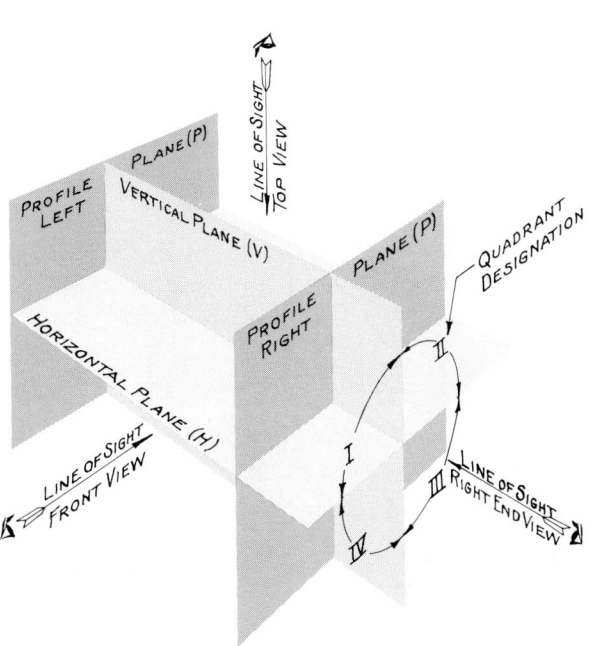

FIG. 6.3. The principal coordinate planes and quadrants.

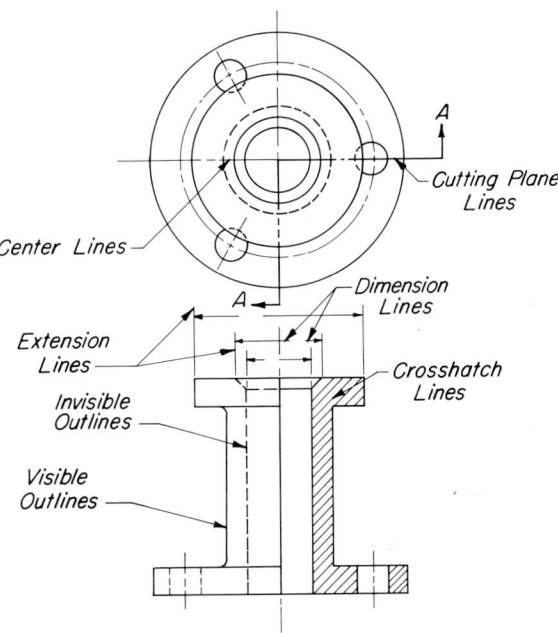

FIG. 6.4. Use of types of lines.

104 ORTHOGRAPHIC PROJECTION

front of the V-plane and above the H-plane is called *first quadrant*. That portion of space behind the V-plane and above the H-plane is called the *second quadrant*. The *third quadrant* is behind the V-plane and below the H-plane, and the *fourth quadrant* is in front of the V-plane and below the H-plane.

It is possible to place an object in any of the quadrants for purposes of orthographic projection but *in the United States almost all drawings are made with the object in the third quadrant.* This has come to be known as third-angle projection. In Europe most of the drawings are made in first-angle projection. Figure 6.5 shows an object in each quadrant together with its projections.

The position of the profile plane, or P-plane, has no effect on the quadrant. It may be placed either to the right or to the left as desired. *In third-quadrant projection the plane of projection always lies between the point of sight and the object.*

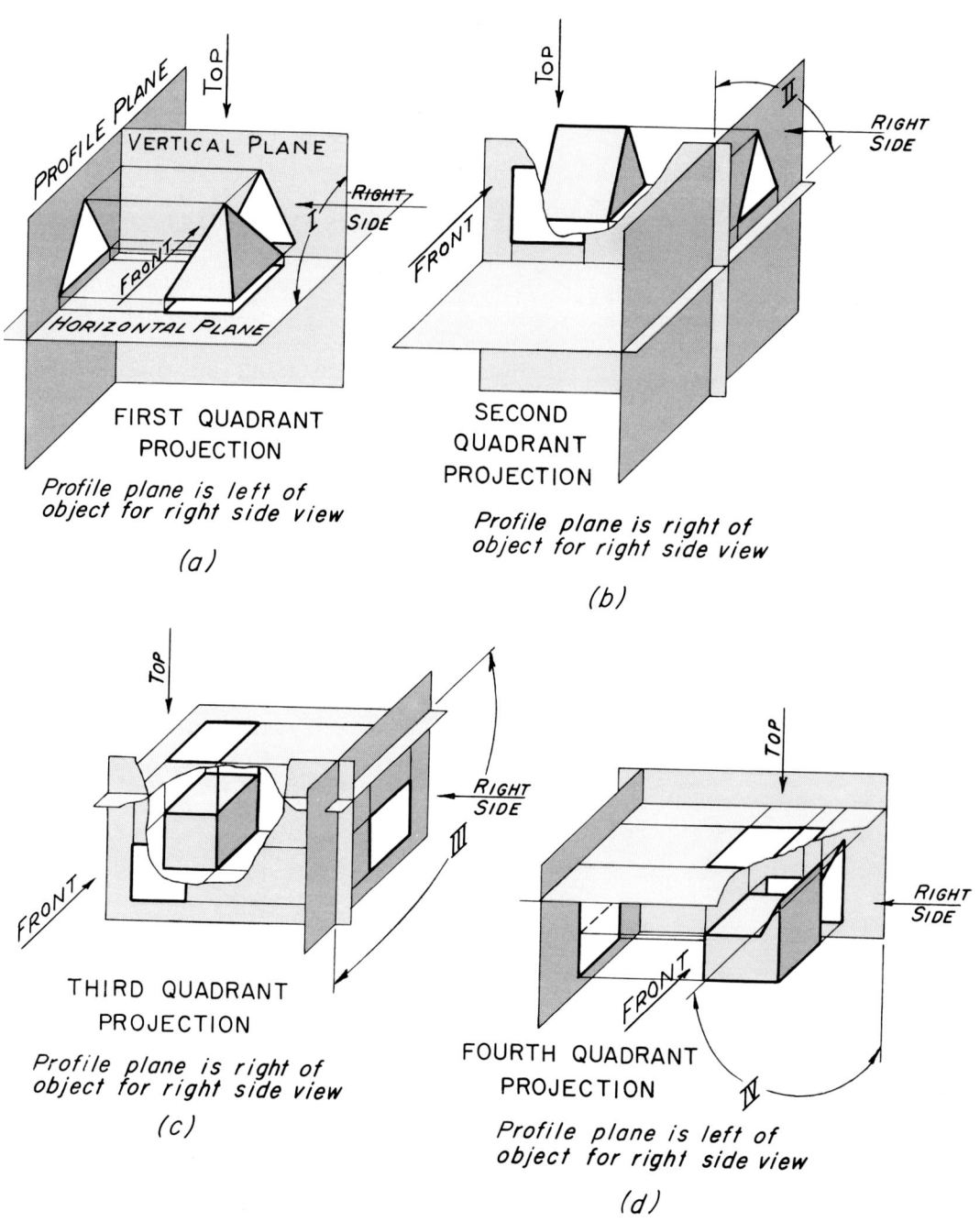

FIG. 6.5. Space location of objects in the four quadrants.

6.6 ROTATION OF PRINCIPAL PLANES

When projections, such as those shown in Fig. 6.5, have been made on the various planes, they are still of little use until they can be brought together on a single sheet of paper. To do this, the planes of projection are considered to be the sides of a transparent box surrounding the object, as shown in Fig. 6.6. The sides of the box are then unfolded as shown in Fig. 6.6a until they lie in the vertical plane. The direction of rotation of the planes must be such that the lines of sight, if rotated with the planes, would point to the front of the sheet of paper as shown in Fig. 6.6b. *In third and first quadrant this means that the plane of projection always revolves away from the object.* This is indicated by the curved arrows in Fig. 6.6a.

There are six principal views that may be drawn, as shown in Fig. 6.7b. The front, top, right-side, and left-side views are the views most used. The rear view and the bottom view are

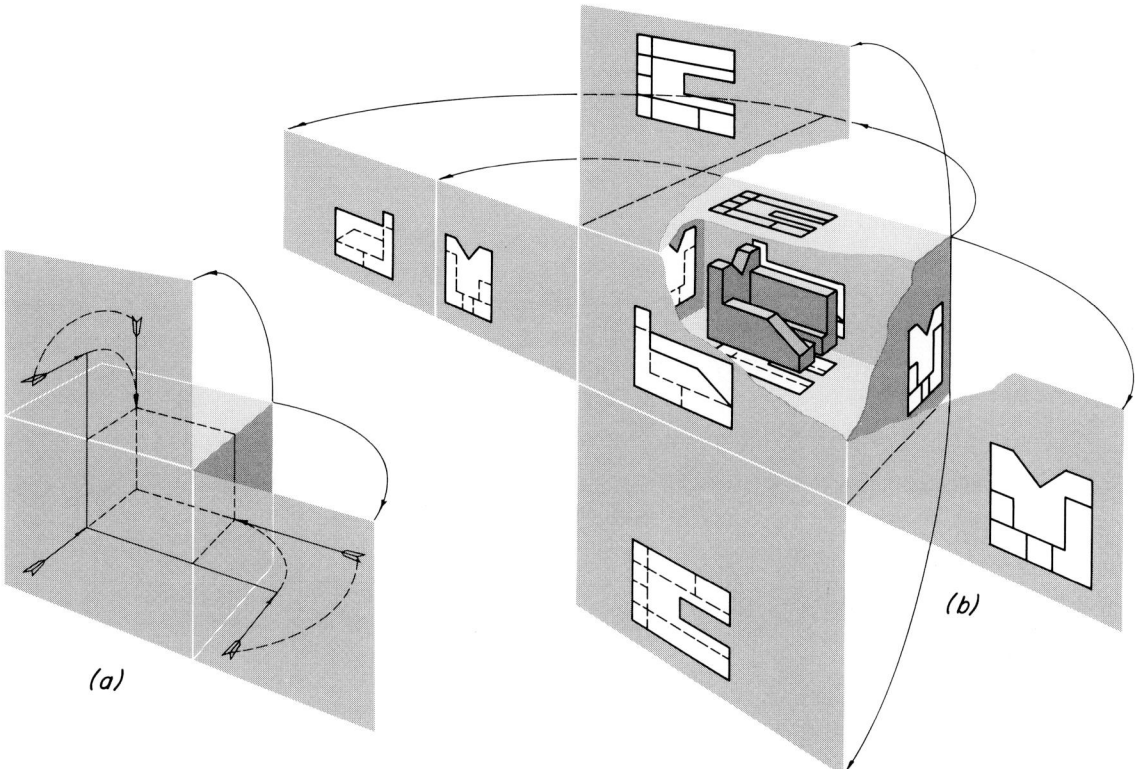

FIG. 6.6. Revolution of coordinate planes.

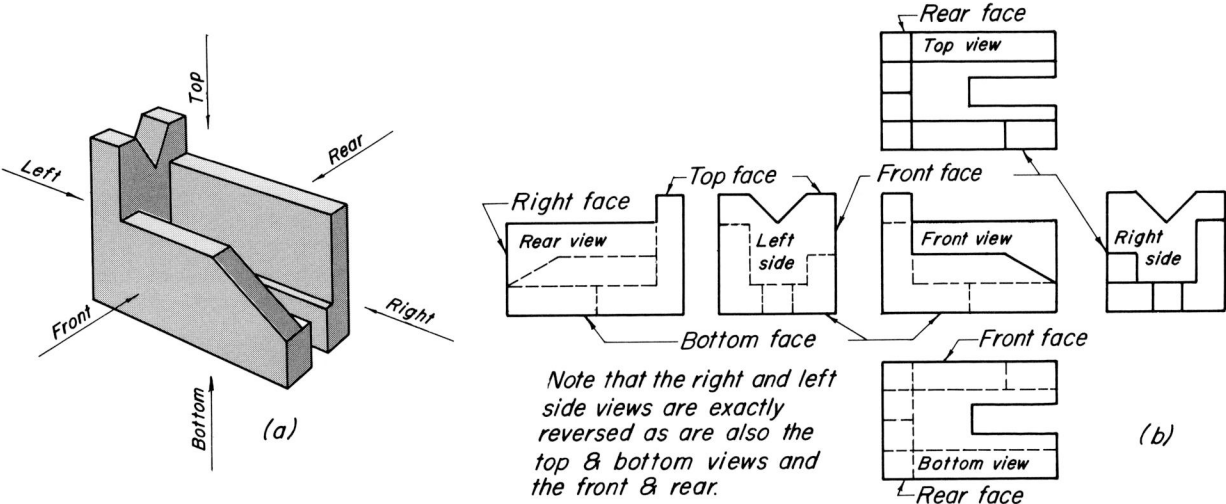

FIG. 6.7. Arrangements of views.

used occasionally. The direction from which the object is viewed in each of these views is shown in Fig. 6.7a.

There is a definite arrangement in which these views must be placed in third-angle projection as shown in Fig. 6.7b.

a. The top view must be directly above the front view in vertical alignment so that points may be projected vertically from one view to the other.

b. The right-side view is directly to the right of the front view in horizontal alignment so that points may be projected horizontally from one view to the other.

c. The left-side view is directly to the left of the front view.

d. The rear view may be placed to the left of the left-side view or to the right of the right-side view.

e. The bottom view must be directly below the front view in vertical alignment so that points may be projected vertically from one view to the other.

These relationships are referred to as having the views in *projection*.

6.7 REFERENCE LINES
The intersection of any two planes of projection form a line of intersection. In orthographic projection this line is called a *reference line*. Therefore, the intersection between the H- and V-planes forms the H-V reference line as shown in Fig. 6.11a. When viewing the H-projection the H-V reference line represents the edge view of the V-plane. When viewing the V-projection the H-V reference line represents the edge view of the H-plane.

6.8 PROJECTION LINES
Projection lines are used to transfer distances and maintain alignment between views. To maintain the left and right relationship between the top and the front view, projection lines are drawn vertically between them. When drawing the side view, the above and below relationships are obtained by drawing projecting lines horizontally between the front view and the side view. To obtain the front and back relationships in the side view, these distances are obtained from the top view. The distance can be transferred by using any of the methods shown in Fig. 6.8. The method shown in Fig. 6.8c, using the 45° mitering line, is the best since it is easier to maintain accuracy when a single angle must be made. Figure 6.9 demonstrates how these distances can be transferred by using dividers.

6.9 PROJECTION OF POINTS
Every corner of an object can be considered as an abstract point and can be projected accordingly. Theoretically a point has no dimensions but does have a location. It can be located in space by measuring in three directions from the established planes of projection or from any other planes that appear edgewise in any view. A point also may be located in space by measuring from an established point in three directions, each

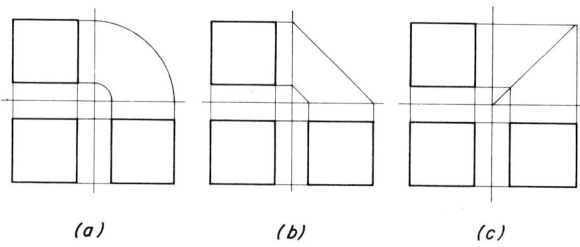

FIG. 6.8. Methods of transferring distances from top to side views.

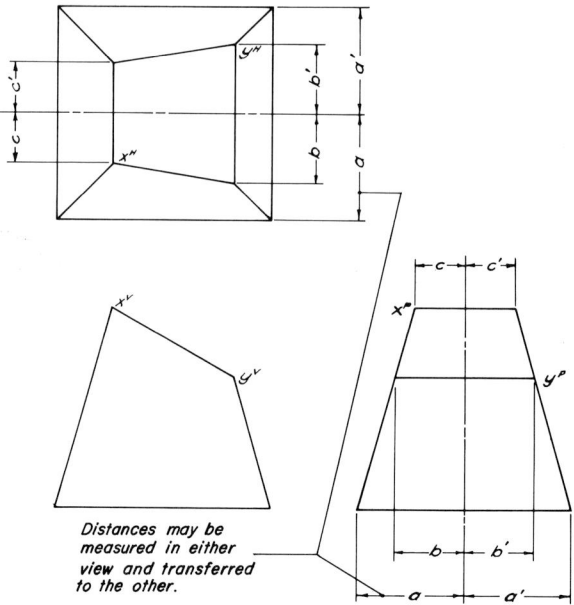

FIG. 6.9. Transferring distances by measurement.

perpendicular to one of the three principal planes of projection.

A point in space is located on a drawing by two or more projections, which are identified by an accepted system of lettering. See Fig. 6.10.

6.10 NOMENCLATURE

Each point in space is identified by a capital letter such as *A*. See Fig. 6.10. The capital letter will refer only to the point in space and will not appear on the drawing except for occasional use on a pictorial view. Projections of point *A* are designated by the lowercase letter *a*. To differentiate the projections of the point, the horizontal projection or top view of point *A* will be marked a^H. The vertical projection or front view will be marked a^V, and the profile projection or side view will be marked a^P. In each case the superscript refers to the plane on which the point has been projected. Thus the projection of point *A* on an auxiliary plane 1 will be marked a^1, and on a second auxiliary plane 2 it will be marked a^2.

6.11 LOCATION OF A POINT

The principal views or projections of a point may be located when its distances from the three principal planes are known. These distances must be measured in the view in which the plane shows edgewise. Thus in Fig. 6.11a the reference line marked *H-V* represents the edgewise view of the *H*-plane when looking at the vertical and the profile planes of projection. The same reference line, *H-V*, becomes the edgewise view of the *V*-plane when seen from the top. The vertical reference line marked *V-P* is the edgewise view of the *P*-plane when seen from the top or front. It also represents the edgewise view of the *V*-plane when viewed from the profile point of sight. *Any two views that lie on the same perpendicular to a reference line are called adjacent views.*

In Fig. 6.11a, the vertical projection or front view, a^V, of the point *A* has been located by measuring *p* distance to the left of the *P*-plane and *h* distance below the *H*-plane. The horizontal projection, a^H, of the point *A* has been located

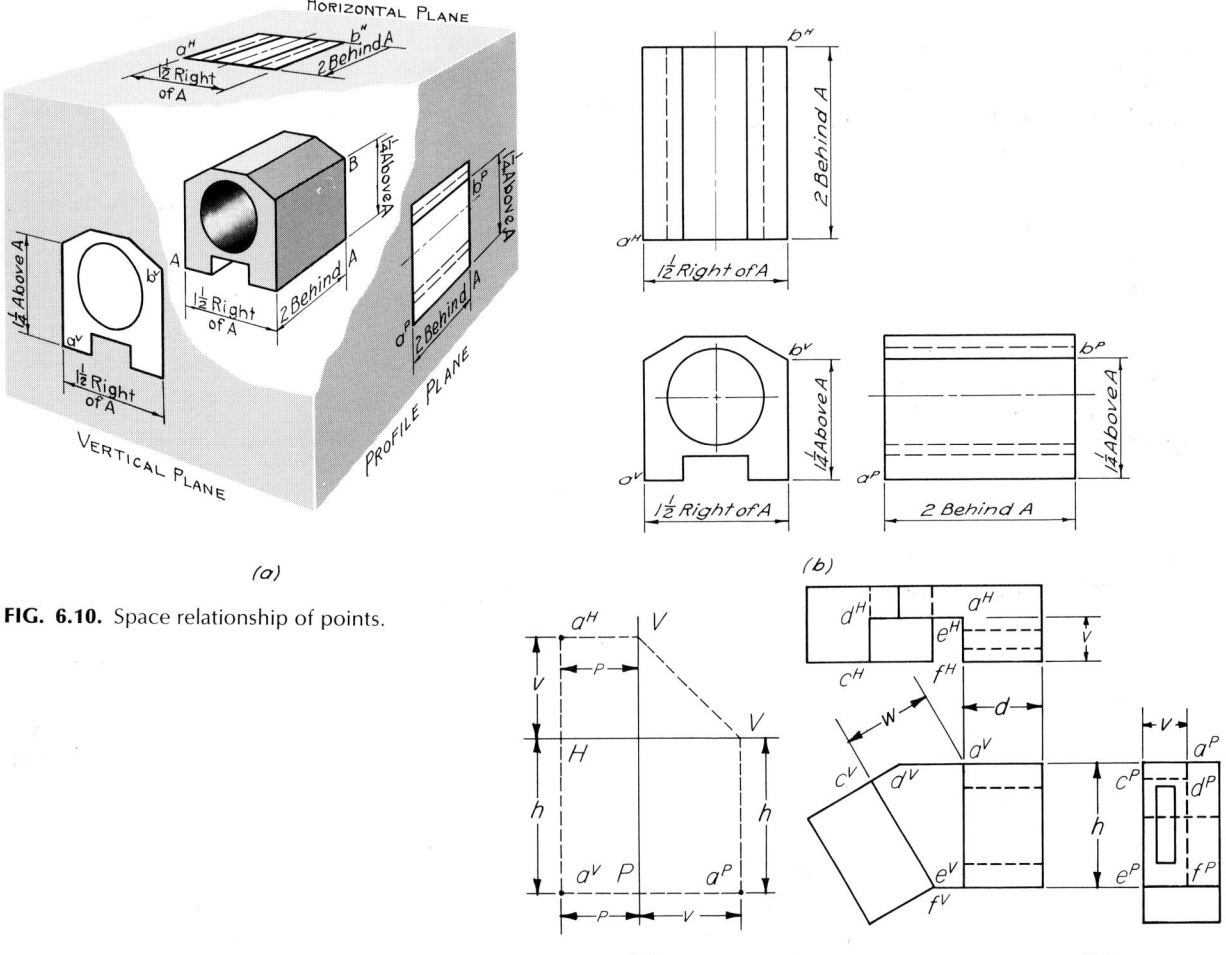

FIG. 6.10. Space relationship of points.

FIG. 6.11. Distances from points to planes.

by measuring the same p distance left of the P-plane and v distance behind the V-plane. The profile projection, a^P, may be located by drawing projection lines, as in Fig. 6.8, or by measuring the h distance below the H-plane and the v distance behind the V-plane. In every such problem the projection, a^H, must be in vertical alignment above a^V, and a^P must be in horizontal alignment with a^V. The projection a^P must be to the right of a^V if a right-side view is taken and to the left if a left-side view is taken.

The distance from a point to any plane may be measured in the view in which the plane shows edgewise. Thus, in Fig. 6.11b the distance w from point A to the plane $CDEF$ will show in the front view.

A point may also be located from some other point as in Fig. 6.10. It can be seen that distances right or left and above or below may be measured in the front view. In the top view, distances right or left and infront or behind can be measured. The profile view will show distances above or below and in front or behind as indicated in Fig. 6.10.

6.12 LINES

A line may be defined as the *path of a moving point*. Lines therefore have location, direction, and length. They may be either straight or curved.

On the drawing board or in the field, a straight line may be determined by the following.

  a. Two points.
  b. One point and a direction.

6.12.1 Position of Lines. Lines may be grouped into three distinct classifications with respect to their relationship to the three principal planes of projection. These classifications are.

  a. Perpendicular to one of the principal planes and parallel to the other two.
  b. Parallel to one of the principal planes and inclined to the other two.
  c. Oblique to all three principal planes.

The angle that a line or plane makes with the H-plane is called θ, with the V-plane ϕ, and with the P-plane π.

It is easy to remember these notations if we think of the angle θ, which is the angle that a line or plane makes with the H-plane, as a circle with a *horizontal* line through it. Likewise, the angle ϕ, which is the angle that a line or plane makes with the V-plane, is a circle with a *vertical* line through it. The angle π, which a line or plane makes with the P-plane, is easy to remember.

6.12.2 Point Projection of a Line. When a line is perpendicular to a plane of projection, it will project as a point on that plane. The line will be parallel to any other plane of projection that is also perpendicular to the original plane of projection.

6.12.3 True Length Projection of a Line. When a line is parallel to a plane of projection in orthographic projection, it will project in true length upon the plane to which it is parallel. Therefore, in the case where a straight line is inclined to all three principal planes of projection, it *will not* project in true length on any of the three principal planes of projection (H, V, or P).

6.13 LINES PERPENDICULAR TO ONE OF THE PRINCIPAL PLANES

When a line is perpendicular to a principal plane, it projects on that plane as a point. This is illustrated in the following paragraphs.

6.13.1 Line Perpendicular to the Horizontal Plane. Figure 6.12a shows the line AB, which is perpendicular to the H-plane. The horizontal projection of the line, $a^H b^H$, projects as a point. This proves that the line is perpendicular to the H-plane. In that position the line must be parallel to the V- and P-planes. Therefore the vertical projection of the line, $a^V b^V$, and the profile projection of the line, $a^P b^P$, both show the true length of the line. The three orthographic views of this line are shown in Fig. 6.12b. Figure 6.12c shows a pictorial view of an object having 12 such lines. One of these is lettered AB and shown in the orthographic projections in Fig. 6.12d. When reading the orthographic drawing in Fig. 6.12d, the student should be able to recognize and visualize the three projections of all 12 of these lines and letter them if desired.

6.13.2 Line Perpendicular to the Vertical Plane. Figure 6.13 shows the pictorial positions and projections of a line that is perpendicular to the V-plane. As in the previous figure, one projection, $a^V b^V$, is a point and the other two projections are true-length views of the line. As a reading exercise the student should be able to recognize and letter the three projections of all such lines on the object. There are 12 of these lines in Fig. 6.13c.

6.13.3 Line Perpendicular to the Profile Plane. Figure 6.14 shows the pictorial view and projections of a line that is perpendicular to the P-plane. Here the profile projection is a point and the horizontal and vertical projections are in true length. There are 12 similar lines on the object. The student should be able to recognize all of these lines and letter them if necessary.

6.14 LINES PARALLEL TO ONE PRINCIPAL PLANE AND INCLINED TO THE OTHER TWO

When a line is parallel to a principal plane, the projection on that plane will be true length. In the true-length view the angle that the line makes with two of the principal planes can be measured. The angle with the H-plane is called θ, with the V-plane ϕ, and with the P-plane π.

If a line is parallel to the horizontal plane, it is called an *H-parallel*. If it is parallel to the V-plane, it is called a *V-parallel,* and if it is parallel to the P-plane, it is called a *P-parallel*.

6.14.1 Line Parallel to the Horizontal Plane.
In Figs 6.15a and b, line AB is parallel to the horizontal plane and inclined to the vertical and profile planes. Therefore, both the vertical and profile projections appear parallel to the horizontal reference line that represents the edgewise view of the horizontal plane in those projections.

In this position the horizontal projection shows the true length of the line. It also shows the true size of the angles ϕ, and π. Note that $\theta = 0°$ for all horizontal lines.

When measuring the angle between a line and

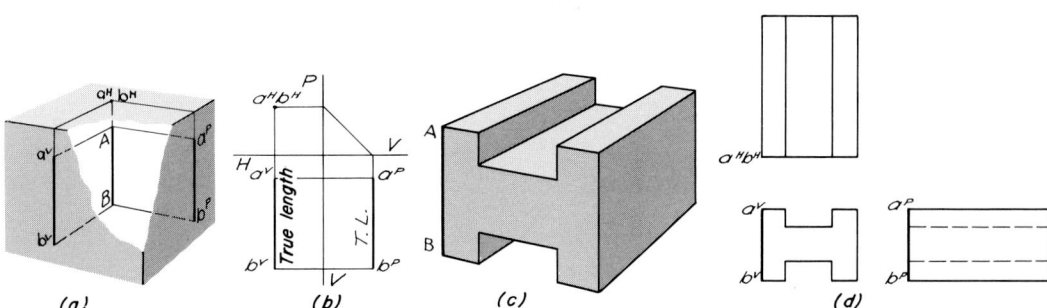

FIG. 6.12. Line perpendicular to the horizontal plane.

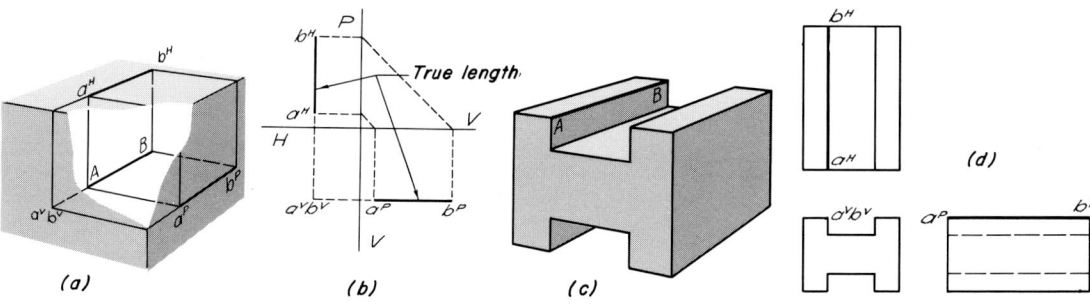

FIG. 6.13. Line perpendicular to the vertical plane.

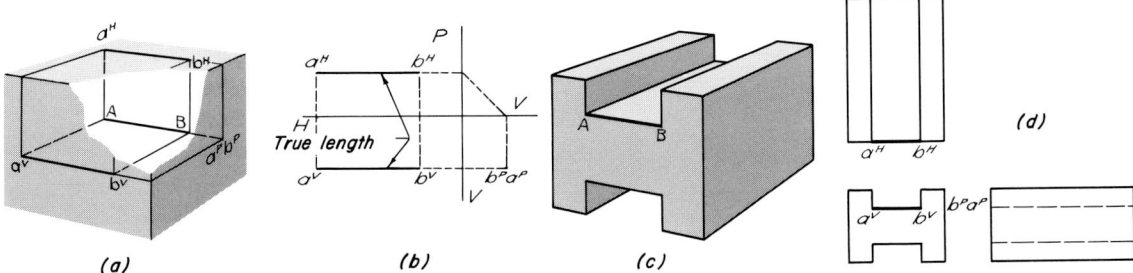

FIG. 6.14. Line perpendicular to the profile plane.

a plane, it is necessary to have the true length of the line and the edgewise view of the plane in the same view.

Figures 6.15c and d show a line of this type on an object. In learning to read the drawing, the student should recognize and locate four such lines on the object.

6.14.2 Line Parallel to the Vertical Plane. In like manner, a line parallel to the V-plane and inclined to the H-plane and the P-plane is shown in Figs. 6.16a and b. In this case the horizontal and profile projections are parallel to the reference line that represents the edgewise view of the V-plane. The vertical projection, $a^V b^V$, is the true length of the line and the angles θ and π also

are true size. Such a line is shown on an object in Figs. 6.16c and d. There are four such lines on the object, which should be identified by the student. Note that all lines parallel to the V-plane have $\phi = 0$.

6.14.3 Line Parallel to the Profile Plane. Figure 6.17 shows a line parallel to the P-plane and inclined to the H-plane and V-plane. In this figure the profile projection, $a^P b^P$, shows the true length of the line and the true size of the angles θ and ϕ. Note that all lines parallel to the P-plane have $\pi = 0$. The other two projections are parallel to the vertical reference line that represents the edgewise view of the P-plane. There are four such lines on this object. See Figs. 6.17c and d.

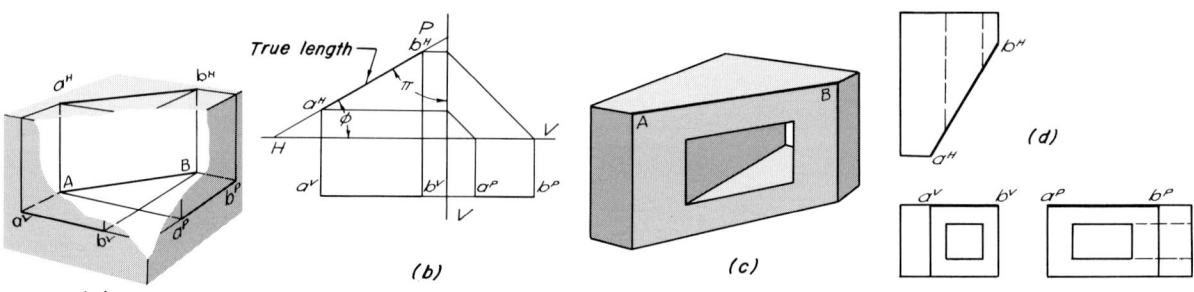

FIG. 6.15. Line parallel to the horizontal plane.

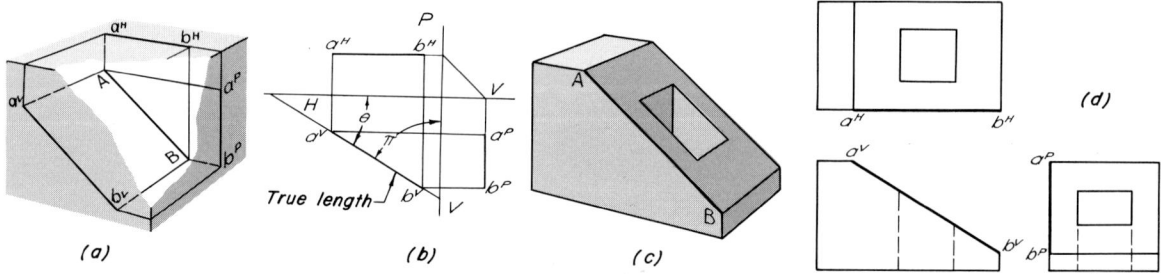

FIG. 6.16. Line parallel to the vertical plane.

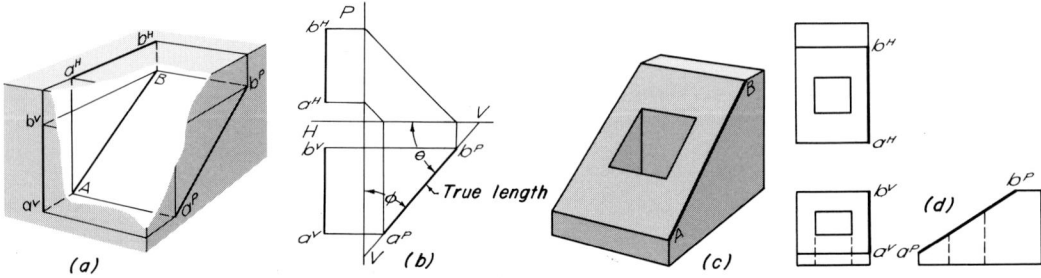

FIG. 6.17. Line parallel to the profile plane.

6.15 LINES OBLIQUE TO ALL THREE PRINCIPAL PLANES

A line that is oblique to all three principal planes has all of its projections inclined to the reference lines representing the principal planes of projection. *Its true length does not show in any of these views nor do the true angles that the line forms with the principal planes* (Table 6.1). Such a line is shown in Fig. 6.18.

TABLE 6.1 POSSIBLE POSITIONS OF A *LINE IN SPACE* RELATIVE TO *H, V,* AND *P* PLANES

H	V	P
⊥	∥	∥
∥	⊥	∥
∥	∥	⊥
∥	/	/
/	∥	/
/	/	∥
/	/	/

6.16 PLANES

A plane is a flat surface that may be located or determined by the following.

a. A point and a line.
b. Three points.
c. Two intersecting lines.
d. Two parallel lines.
e. A line and an angle with some reference plane.

A plane surface has length and breadth but no thickness. A plane may be extended as far as desired in any direction but cannot be bent or broken. All planes may be divided into three groups.

a. Parallel to one of the principal planes and perpendicular to the other two. These are sometimes called *normal planes.*
b. Perpendicular to one of the principal planes and inclined to the other two. These are referred to as *inclined planes.*
c. Oblique to all three principal planes. These are called simply *oblique planes.*

For the purpose of reading and understanding orthographic projections, the ability to recognize and visualize planes in their various positions is very important. Recognizing the general shape of a plane is valuable. A plane will retain the same general shape in every view. If a plane is shaped like the letter U, as in Figs. 6.20c and d, it will retain that same general shape whenever it appears as an area. In Figs. 6.24c and d, the shaded plane is a trapezoid and therefore whenever it appears as an area it will still be a trapezoid.

6.17 PLANES PARALLEL TO ONE PRINCIPAL PLANE OF PROJECTION AND PERPENDICULAR TO THE OTHER TWO

These planes have two principal views that show edgewise. These views are always parallel to the reference lines that represent the edgewise view of the adjacent plane. This adjacent view in each case shows the true shape of the plane surface.

6.17.1 Parallel to the Horizontal. In Fig. 6.19a, the plane *ABCD* is parallel to the *H*-plane. In Fig. 6.19b, the plane shows edgewise in the vertical and profile views, and these views are parallel to the horizontal reference line. The horizontal reference line in these views represents the edgewise view of the *H*-plane. Planes parallel to the *H*-plane are called *H-parallels.* The hori-

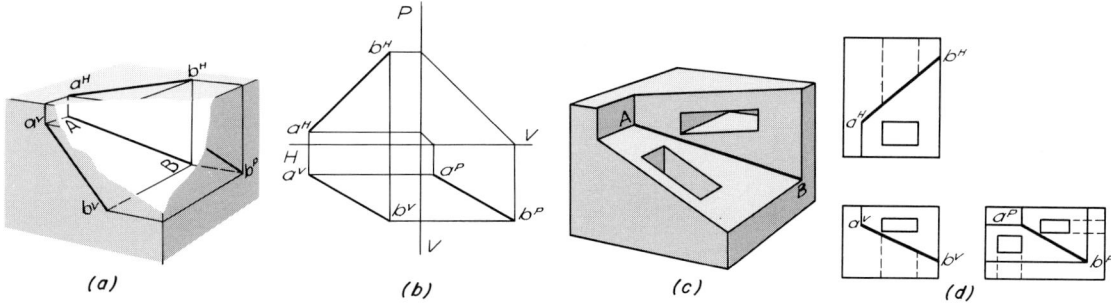

FIG. 6.18. Line oblique to all principal coordinate planes.

zontal projection, $a^H b^H c^H d^H$, gives the true shape of the plane figure.

Figures 6.19c and d show an object on which such a plane surface has been shaded. The vertical and profile views are indicated by a heavy line.

6.17.2 Parallel to the Vertical. In Fig. 6.20a, the plane ABCD is parallel to the vertical plane. In Fig. 6.20b, the plane shows edgewise in the horizontal and profile views. These views are parallel to the reference line that represents the edgewise view of the vertical plane in each case. The vertical projection $a^V b^V c^V d^V$ gives the true shape of the plane figure. Such a plane on an object is shown as a shaded area in Figs. 6.20c and d. In Fig. 6.20d the horizontal and profile views of the plane are shown as heavy lines. Planes parallel to the V-plane are called V-parallels. There is one other V-parallel plane on the object. The student should recognize and identify it and label the projections.

6.17.3 Parallel to the Profile. In Fig. 6.21a the plane ABCD is parallel to the profile plane. In Fig. 6.21b the plane projects edgewise in both the horizontal and vertical projections. These views in each case are parallel to the reference line that represents the edgewise view of the profile plane. The profile projection $a^P b^P c^P d^P$ gives the true shape of the plane figure. Planes parallel to the P-plane are called P-parallels.

In Fig. 6.21c the shaded area identifies a plane on an object in this position. In Fig. 6.21d the horizontal and vertical projections are shown as heavy lines and the profile projection of the area is shaded.

There are three other P-parallel planes on the object. The student should identify each and label the projections.

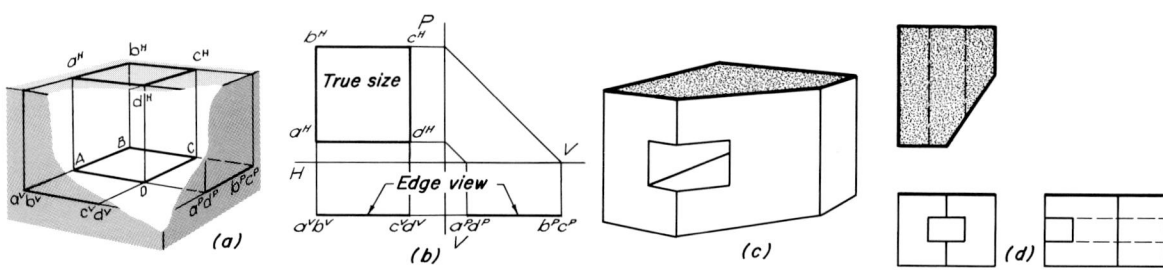

FIG. 6.19. Plane parallel to the horizontal plane.

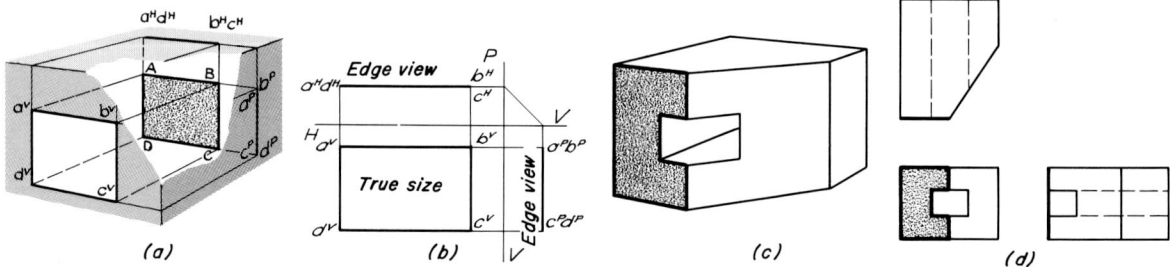

FIG. 6.20. Plane parallel to the vertical plane.

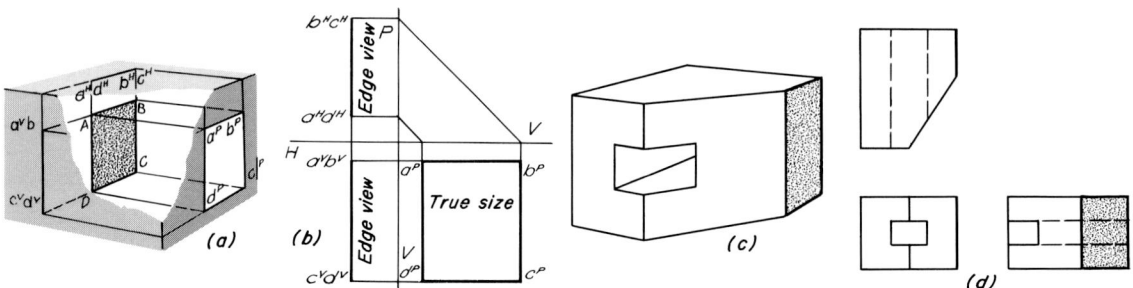

FIG. 6.21. Plane parallel to the profile plane.

6.18 PLANES PERPENDICULAR TO ONE PRINCIPAL PLANE OF PROJECTION AND INCLINED TO THE OTHER TWO

These planes have one principal view that shows edgewise. This view is inclined to both of the principal reference lines. The other two views show as areas, but they are not of true size.

The angle between two planes can be measured when both planes show edgewise in the same view. Therefore it is possible to measure the angles between one of these planes and two of the principal planes in the view that shows the given plane edgewise.

6.18.1 Planes Perpendicular to the Horizontal Plane.
Figure 6.22a shows a pictorial view of a plane that is perpendicular to the horizontal plane and inclined to the V- and P-planes. In this case, as illustrated in Fig. 6.22b, the horizontal projection or top view of the plane must be edgewise and inclined to the reference lines that represent the edgewise views of the V-plane and P-plane. The angle ϕ between the edgewise view of the plane ($a^H b^H c^H d^H$) and the horizontal reference line that represents the edgewise view of the V-plane is the angle that the plane makes with the V-plane. The angle π between the edgewise view of the plane ($a^H b^H c^H d^H$) and the vertical reference line that represents the edgewise view of the P-plane is the angle between the plane ABCD and the P-plane.

The other two views of the plane show as areas that are not of true shape. This plane is frequently called an *H-projecting plane.*

Figure 6.22c shows a pictorial view of this

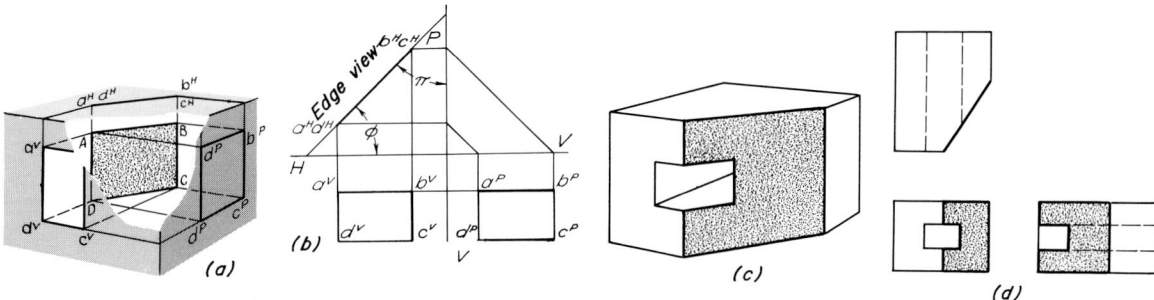

FIG. 6.22. Plane perpendicular to the horizontal plane.

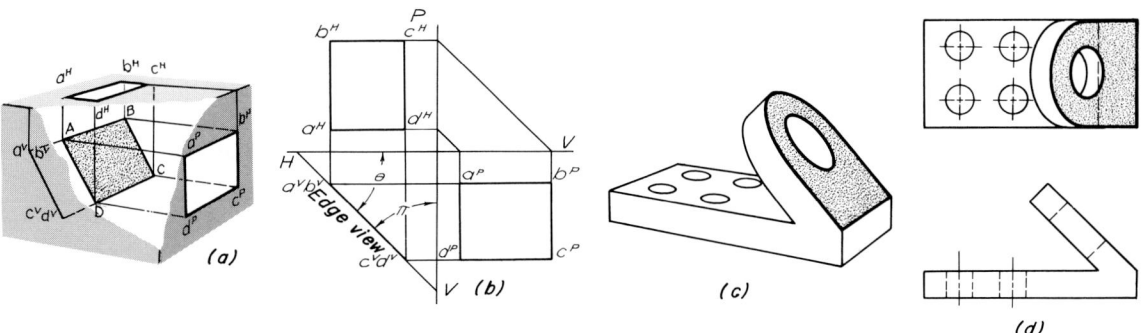

FIG. 6.23. Plane perpendicular to the vertical plane.

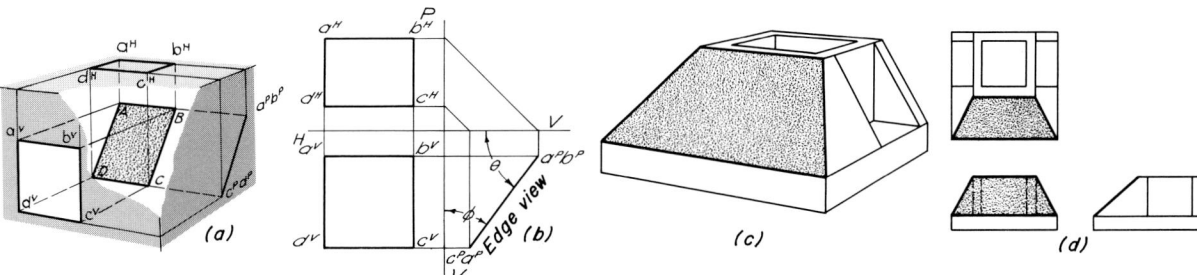

FIG. 6.24. Plane perpendicular to the profile plane.

plane on an object, and Fig. 6.22d shows the projections of the object with the edgewise view made heavy and the other two shaded. Examine Fig. 6.22d to see if any other similar planes can be found.

6.18.2 Planes Perpendicular to the Vertical Plane. Figure 6.23a shows a plane in the pictorial form that is perpendicular to the V-plane and inclined to the H-plane and P-plane. Figure 6.23b shows the three principal projections of the plane with the edgewise view of the plane ABCD and the angle θ and π indicated in the vertical projection. The horizontal and profile projections of plane ABCD show as areas that are not true in shape. This plane is commonly called a V-projecting plane.

Figures 6.23c and d show a plane perpendicular to the V-plane and inclined to the H-plane and P-plane, in a pictorial and in a two-view drawing. Identify any other planes on the object that are also perpendicular to the V-plane and inclined to the H-plane and P-plane.

6.18.3 Planes Perpendicular to the Profile Plane. A plane perpendicular to the profile and inclined to the H-plane and V-plane is shown pictorially in Fig. 6.24a. The three views of this plane are shown in Fig. 6.24b, with the edgewise view and the angles θ and ϕ indicated in the profile projection. The two views of the plane ABCD which show as areas are not of true shape. Figures 6.24c and d show such a plane on an object with the areas shaded and the edgewise view made heavy.

Examine Fig. 6.24d to see if any other planes of this kind can be found on the object.

6.19 PLANES OBLIQUE TO ALL THREE PRINCIPAL PLANES OF PROJECTION

A plane of this kind does not have an edgewise view in any of the three principal views and consequently none of the angles with the principal planes can be measured. All three principal projections will show as areas that are not of true shape. Figures 6.25a and b show a plane of this kind in pictorial form and in orthographic projection. Figures 6.25c and d show an object in pictorial form and in projection with a plane oblique to all three principal planes.

At the present time it is possible to describe these planes only as oblique to H, V, and P (Table 6.2). In the following chapters, methods will be shown to determine the true shape of the plane and the true size of the angles θ, ϕ, and π. Examine Fig. 6.25d to see if any similar planes can be found.

TABLE 6.2 POSSIBLE POSITIONS OF A PLANE IN SPACE RELATIVE TO H, V, AND P PLANES

H	V	P
∥	⊥	⊥
⊥	∥	⊥
⊥	⊥	∥
⊥	/	/
/	⊥	/
/	/	⊥
/	/	/

6.20 SOLIDS

Points, lines, and planes are the building blocks from which three-dimensional objects, known as solids, are constructed. The purpose of orthographic projection is to enable one person to make an engineering drawing of a solid that will so completely describe the shape of an object that another person can visualize it without further information.

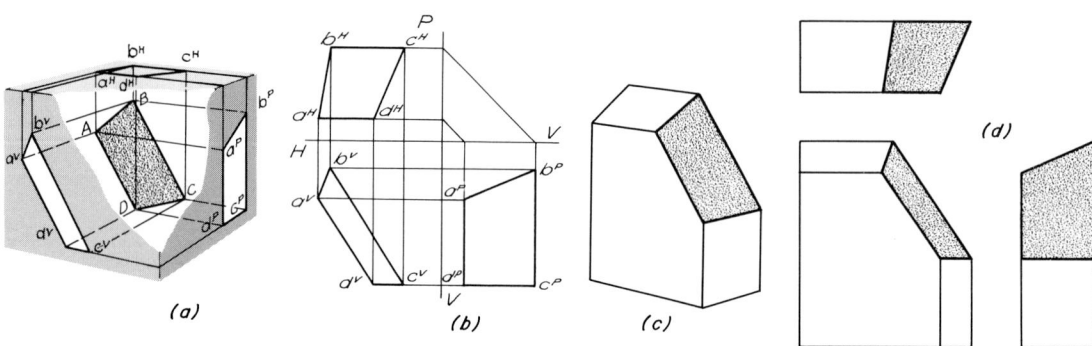

FIG. 6.25. Plane oblique to the H-, V-, and P-planes.

6.20.1 Solids Bounded by Plane Surfaces. There are many solids bounded by plane surfaces, but the most common ones are illustrated in Fig. 6.26. The cube and the rectangular solid are used more than any other in engineering work. The prism and pyramid also appear frequently. For complete description three views are almost always necessary. However, common usage has come to accept two views by assuming that the simplest form is the one meant. Thus, in Fig. 6.27, if only the top and the front views were given, the object would be accepted as a cube. The additional side views show possible shapes having the same top and front views. If any of these were desired, the side view would have to be shown.

6.20.2 Solids Bounded by Plane and Single-Curved Surfaces. Objects such as the cylinder, cone, and frustum are shown in Fig. 6.28. These solids are bounded by plane surfaces on the ends and single-curved surfaces on the sides. For these objects, two views are sufficient even though three views are shown.

Cylindrical holes in a part form a good illustration of this shape. They are shown by two dotted lines that form the boundary of the surface in one view and by a circle in the other view. See Fig. 6.10.

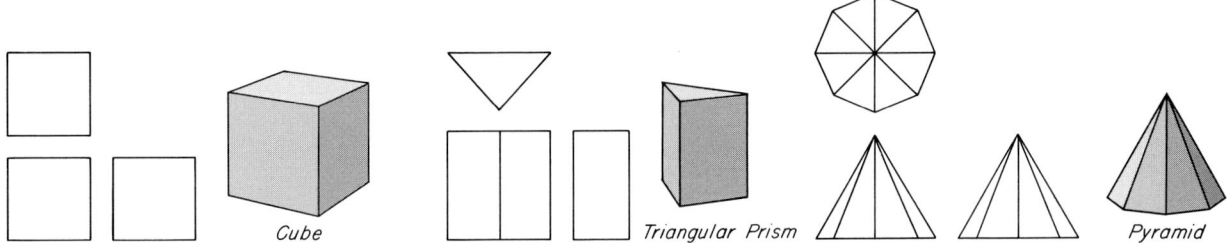

FIG. 6.26. Solids bounded by plane surfaces.

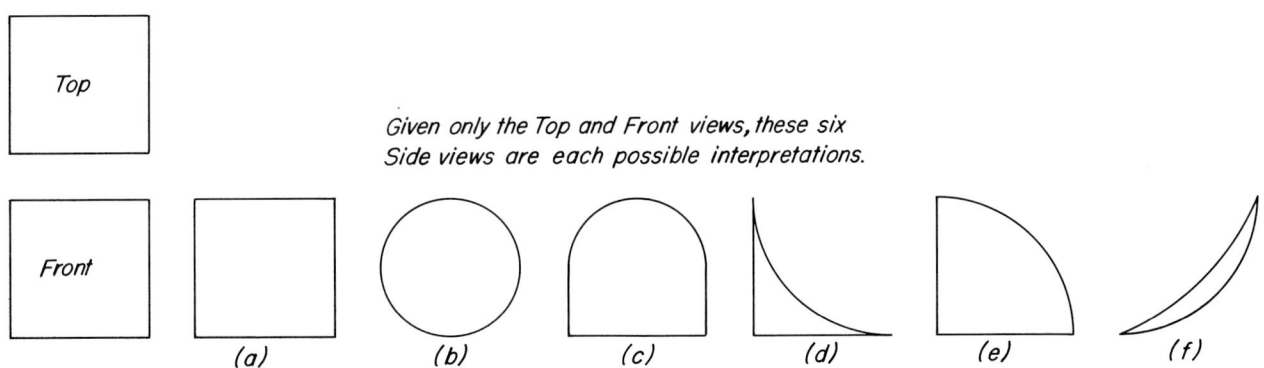

FIG. 6.27. Interpretation of views.

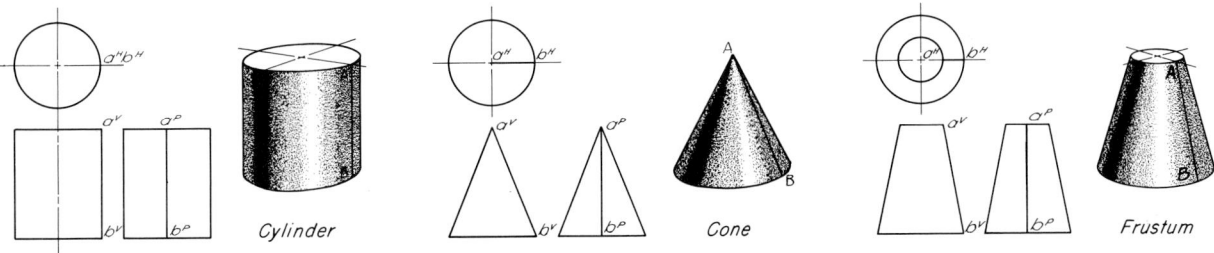

FIG. 6.28. Solids bounded by plane and single-curved surfaces.

6.20.3 Solids Bounded by Double-Curve Surfaces.
The sphere, spheroid, and torus are examples of objects bounded by double-curved surfaces. They are illustrated in Fig. 6.29. Two views are sufficient.

6.20.4 Surfaces Involving Warped Surfaces.
There is a wide variety of warped surfaces that may become a bounding surface of solid objects. A few common objects involving helicoids are shown in Fig. 6.30.

6.21 READING A DRAWING
The interpretation or reading of a drawing requires an analysis of the various views of the object. It is necessary to find the various projections of each individual point and the relation of each point to the others. The location, length, and direction of each line must be visualized. The planes formed by these points and lines as well as their position and shape must be evident. And, finally, the simple solids of which the part is composed must be separated mentally or pictorially before the object as a whole can be completely understood.

6.21.1 Interpretation of Lines.
A line may represent one or more of three things in any sketch or drawing.

a. The intersection of two plane faces.
b. The edgewise view of a plane surface.
c. The outside line of a curved or cylindrical object.

The engineer or technician must always have these interpretations in mind when reading or making a drawing.

In Fig. 6.31, lines AB, BC, and AC represent the intersection of two plane surfaces. The H- and V-projections of BC also represent the edgewise views of plane CBE. The H-projection of AC represents the edgewise view of plane ACF. Similar meanings can be pointed out for many lines on this figure.

In Fig. 6.32 the long line, AB, in the top view of the cylinder or the inclined line AB in the front view of the cone represents the outside line or edge of a curved surface.

6.21.2 Interpretation of Planes or Areas.
Areas in any view of a drawing may have a number of possible meanings. The adjacent view must always be examined and sometimes all views must be considered. An area may represent one of the following things in a drawing.

a. An area in one view that shows as a straight line in another view is a plane surface. See AFC in Fig. 6.31.

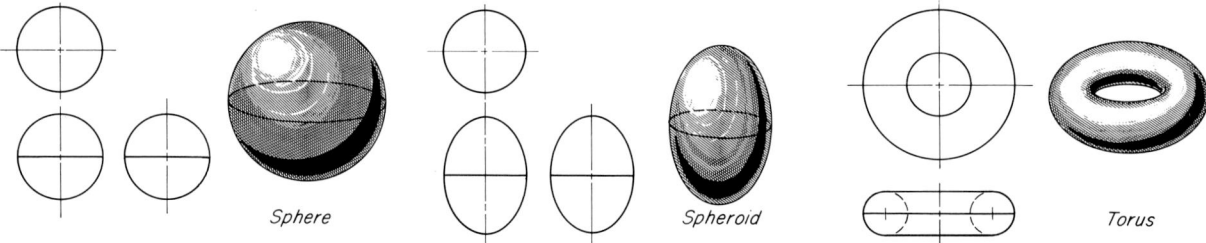

FIG. 6.29. Solids bounded by double-curved surfaces.

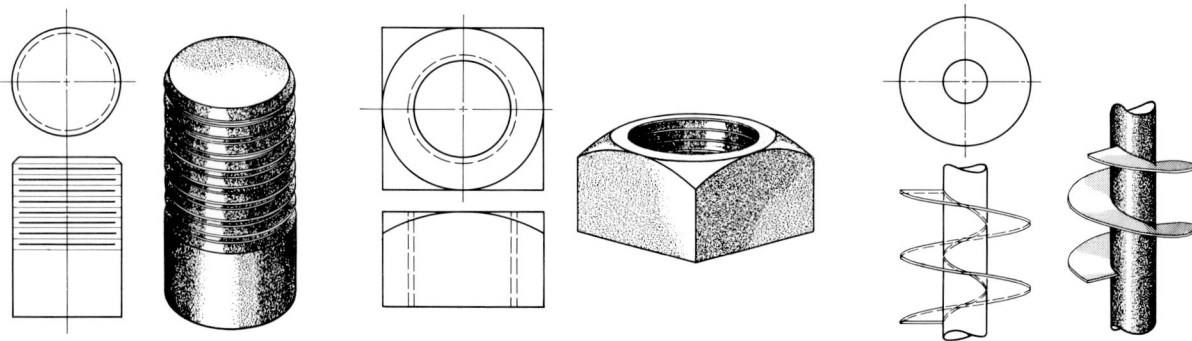

FIG. 6.30. Solids involving warped surfaces.

b. A surface that shows as an area in two views and as a straight line in the third view is a plane surface. See the shaded areas in Fig. 6.24d.
c. A surface that shows as an area in three views bounded by straight lines is a plane surface or a warped surface. For a plane surface see the shaded area in Fig. 6.25d. For a warped surface see Figs. 7.91 through 7.96.
d. A surface that is curved in one direction only is called a *single-curved surface*. See Figs. 6.32 and 6.33. They usually show as an area in one view and as a curved line in an adjacent view if the simplest views are shown.
e. A surface that is curved in two directions is called a *double-curved surface*. It is illustrated in Fig. 6.29 and Fig. 6.34.

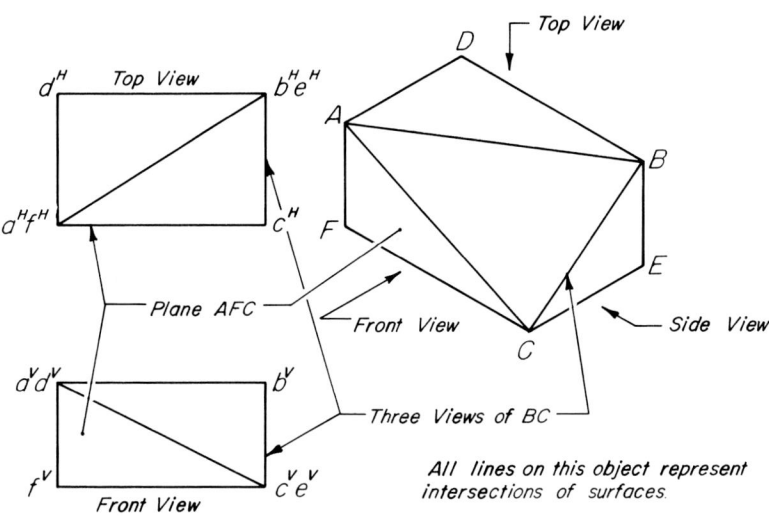

FIG. 6.31. Lines representing intersection of planes.

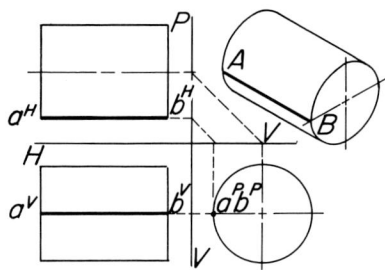

FIG. 6.32. Interpretation of lines on a cylinder.

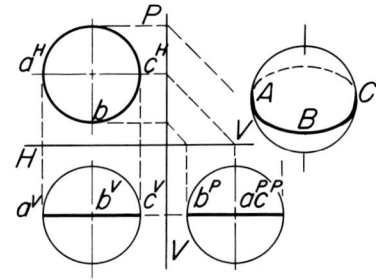

FIG. 6.34. Interpretation of lines on a sphere.

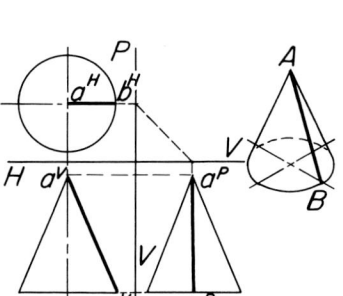

FIG. 6.33. Interpretation of lines on a cone.

6.21.3 Interpretation of Solids. One method of interpreting solids is to break the object into its basic elementary solids such as prisms, pyramids, and cylinders. In Fig. 6.35 the rectangular solids, the prisms, and the cylinders have been separated and drawn as individual units. The holes have even been drawn as cylinders. Sketching the small units is a good method of interpreting a drawing, but if these parts can be visualized in the mind, much time can be saved. Separating these units is an essential process when the volume or weight is to be computed, as is often the case for a new product.

For the sake of clarity, this breakdown was made from the pictorial view, but the same thing can be done from the orthographic projections.

Figures 6.13 to 6.25 show the pictorial view of an object from which to visualize the position of the line or plane. In learning to read a drawing this process is reversed, and it is frequently worthwhile to sketch the pictorial form of all or part of the object from the orthographic projections.

There are three approaches to the problem of reading a drawing.

a. Analyze the location, position, length, and shape of the various points, lines, planes, and solids.
b. Make a pictorial sketch of all or part of the object.
c. Break the object down into its individual units, from which the shape can be visualized and the weight computed. Some even suggest carving these parts out of soap or modeling clay.

Various exercises have been developed to give experience in reading drawings. In making the third view of an object, it is sometimes found that more than one solution is possible. Figure 6.27 emphasizes the necessity for a third view in such a case.

6.22 SELECTION OF VIEWS IN MULTIVIEW DRAWINGS

The three rules previously stated governing the placement of the object will automatically control the views to be drawn.

a. The views should be drawn with the object in its natural position. Thus, in Fig. 6.36a,

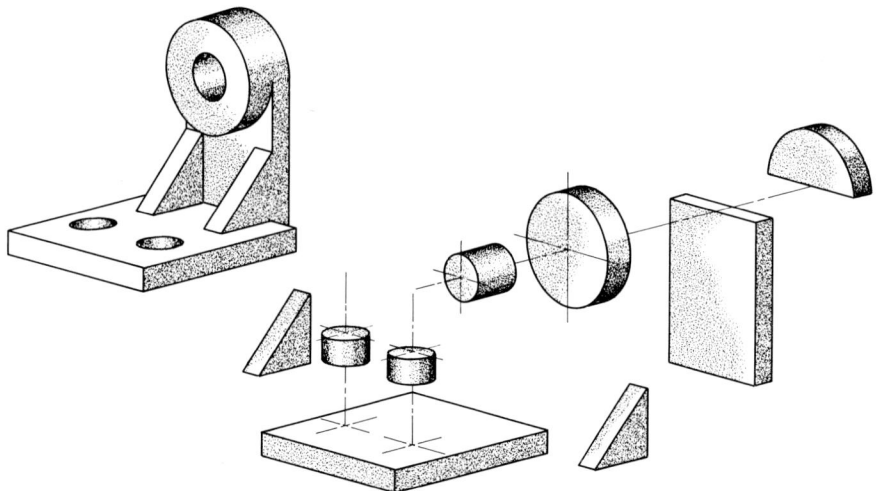

FIG. 6.35. Breakdown of object into component parts.

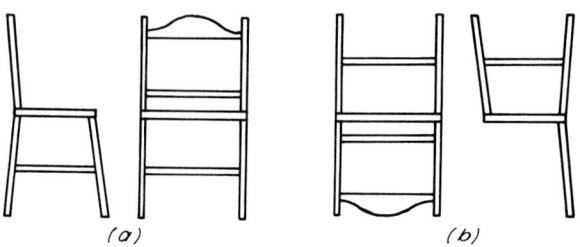

FIG. 6.36. Natural position of object.

the two views shown can be easily recognized as a chair. However, the same views inverted, as in Fig. 6.36b, have little resemblance to a chair.

b. The front view should should show the most distinctive outline or contour. Figure 6.37 shows the pictorial drawings of several objects and the proper selection of front view in each case.

c. Having the object in its natural position and the front view having been selected, the top view is automatically chosen.

d. In selecting the proper side view, when the two are similar, the one having the fewest invisible lines should be chosen. Figure 6.38 illustrates this point. The right- and left-side views have the same lines, but the right-side view is preferable because more of the lines are visible from the right side.

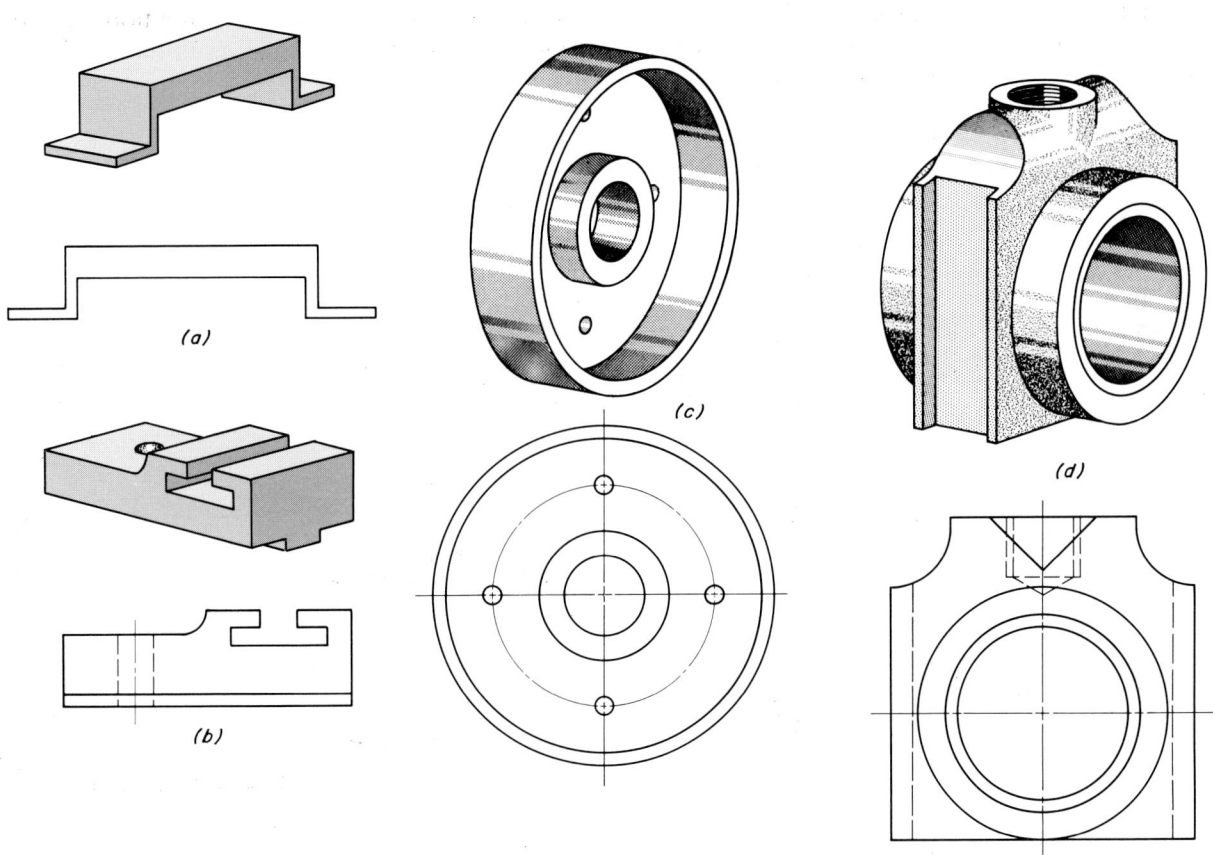

FIG. 6.37. Principal contour of object in front view.

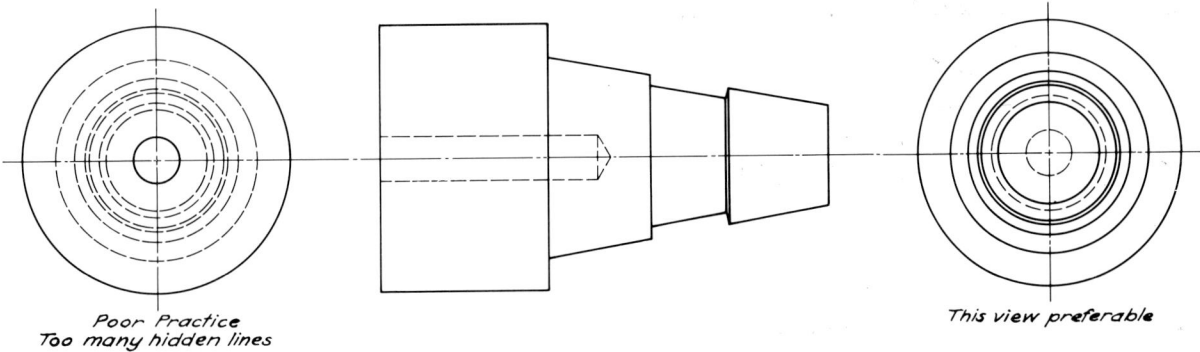

FIG. 6.38. Selection of views to avoid hidden lines.

6.23 NUMBER OF VIEWS TO BE DRAWN

After the object is placed in the proper position and the desired front view is selected, the next step is to decide on the number of views to be drawn. This requires a careful study of the object because *it is essential that all of the necessary information be given*. However, for the sake of economy, *it is very important that no more views be drawn than the minimum number required for clear understanding*.

In selecting the proper views to be shown, the final dimensioning or specifying sizes must be considered. The views chosen must be such that as many dimensions as possible can be shown in true length and in a convenient position.

6.24 ONE-VIEW DRAWINGS

A single orthographic view is often adequate to represent thin, flat objects having uniform thickness if the thickness is specified by a note. Simple cylindrical parts may require only one view, provided that the diameter is given. Such objects as shims, gaskets, steel springs, and many parts cut from thin plate may be represented in a single view, as in Fig. 6.39. The top view of the shim in Fig. 6.39 shows its true length and width together with the cuttings that must be made. The thickness and material are shown in the note. No additional information is necessary for its manufacture. Such a one-view drawing, when executed to full scale, is sometimes called a *pattern* or *template*. Similarly, the one-view drawing of the cylindrical part shown in Fig. 6.40 is sufficient for all purposes of manufacture. The fact that the part is cylindrical is shown by the two dimensions with the notation "Dia" or "ϕ" added to indicate diameter measurements. Except for extremely simple objects of the kind just mentioned, one-view orthographic drawings are rarely used.

6.25 TWO-VIEW DRAWINGS

The most common two-view drawing is obtained by projecting the object on the *V*- and *H*-planes. The method of obtaining the projection of an object on these two planes, the rotation of the horizontal plane, and the resultant arrangement of views are shown in Fig. 6.41. The two views are called the vertical and horizontal projections or, more commonly, the *front* and *top* views. From the pictorial view in Fig. 6.41*b* it can be seen that

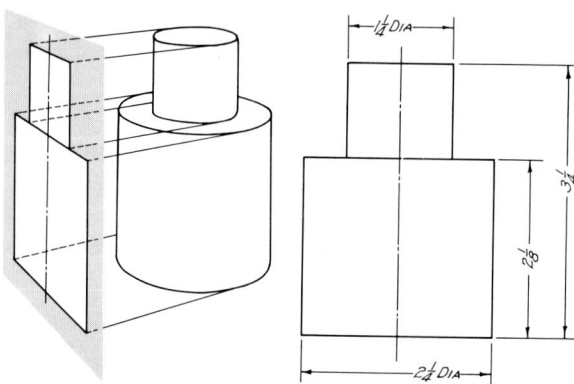

FIG. 6.40. One-view drawing of cylindrical part.

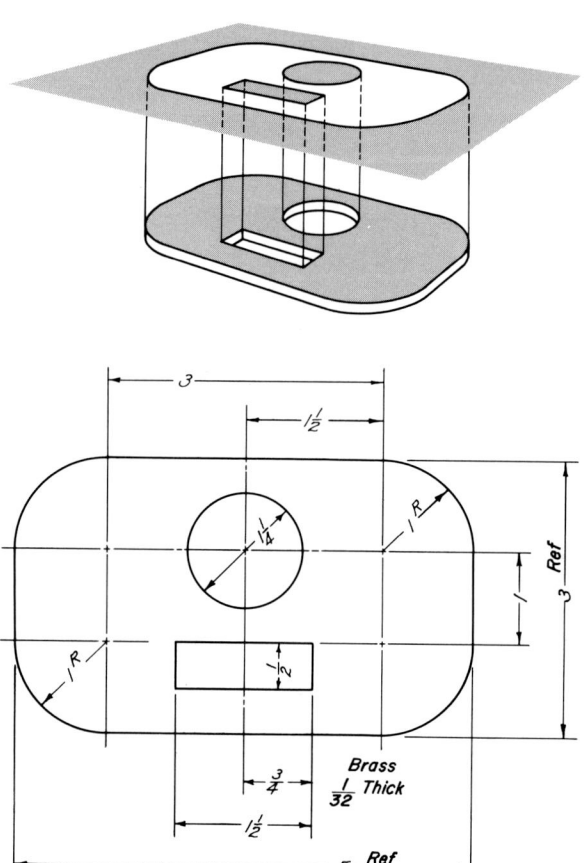

FIG. 6.39. One-view drawing of flat part.

the vertical distances and those right or left are shown in the front view. Distances right or left and in front or behind are shown in the top view. These two views enable all three major dimensions to be specified.

Another common arrangement of a two-view drawing is shown in Fig. 6.42. In this drawing, projecting the object on the vertical and profile planes shows the front and side views, respectively. The front view shows the vertical distances and those right or left. The side view shows the vertical distances and those in front or behind as indicated in Fig. 6.42.

When top and front views are used, the top view must be above the front view in exact vertical alignment. When the front and right-side views are used, the right-side view is to the right of the front view in exact horizontal alignment. This relationship is often referred to as the views being *in projection*.

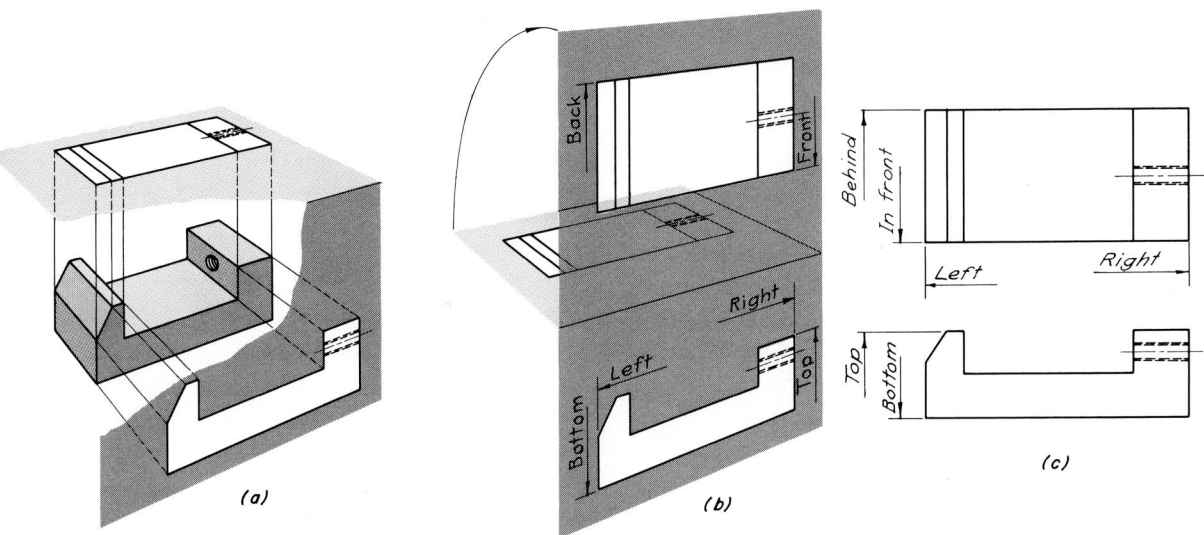

FIG. 6.41. Third quadrant arrangement of two-view drawing.

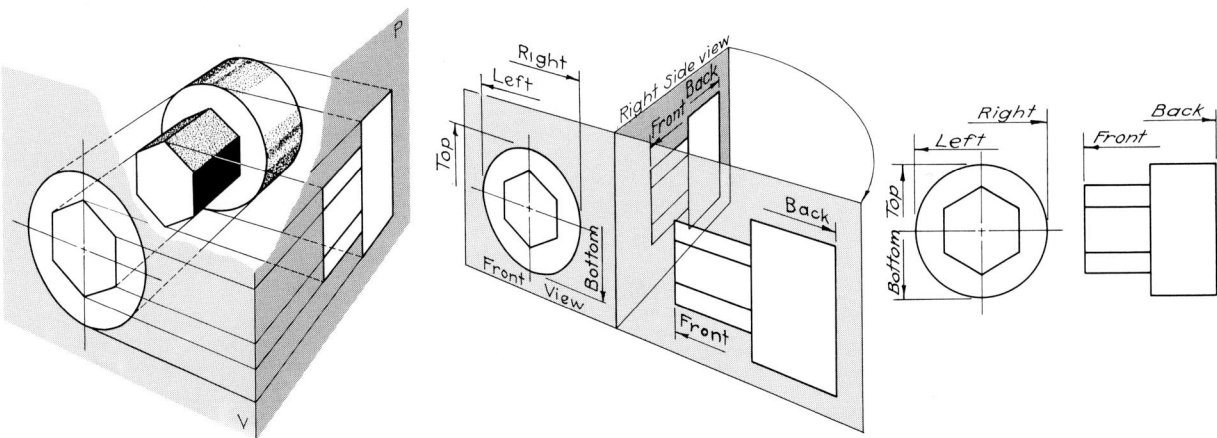

FIG. 6.42. Arrangement of front and side views.

6.26 THREE-VIEW DRAWINGS

Many objects cannot be satisfactorily described with two-view drawings, even with the addition of notes and symbols. In such cases three principal views—horizontal, vertical, and profile—are used, as shown in Fig. 6.43a. The method of revolving the planes is shown in Fig. 6.43b. The subsequent arrangement of views is shown in Fig. 6.43c.

If the right- and left-side views are equally effective in showing the shape of the object, the right-side view is usually drawn.

In a three-view drawing, each of the major distances is shown in two places. The vertical distances (heights) are shown in the front and side views. The right or left distances (widths) are shown in the front and top views. The front or back distances (depths) are shown in the top and side views. See Fig. 6.43c.

6.27 ALTERNATE ARRANGEMENT OF VIEWS

Occasionally the standard arrangement of views as shown in Fig. 6.43 is not satisfactory because of the necessity to clear a certain portion of the sheet for a title, bill of materials, or some other purpose. In that event, the planes may be revolved and the views arranged as shown in Fig. 6.44. This is not considered a desirable arrangement and should not be used unless absolutely necessary.

In this arrangement the front view is always below the top view an in exact vertical alignment with the top view. The right-side view is to the right of the top view and in exact horizontal alignment with the top view.

6.28 DETERMINATION OF VISIBILITY

When drawing a projection of an object, it is important to remember that *every line on the*

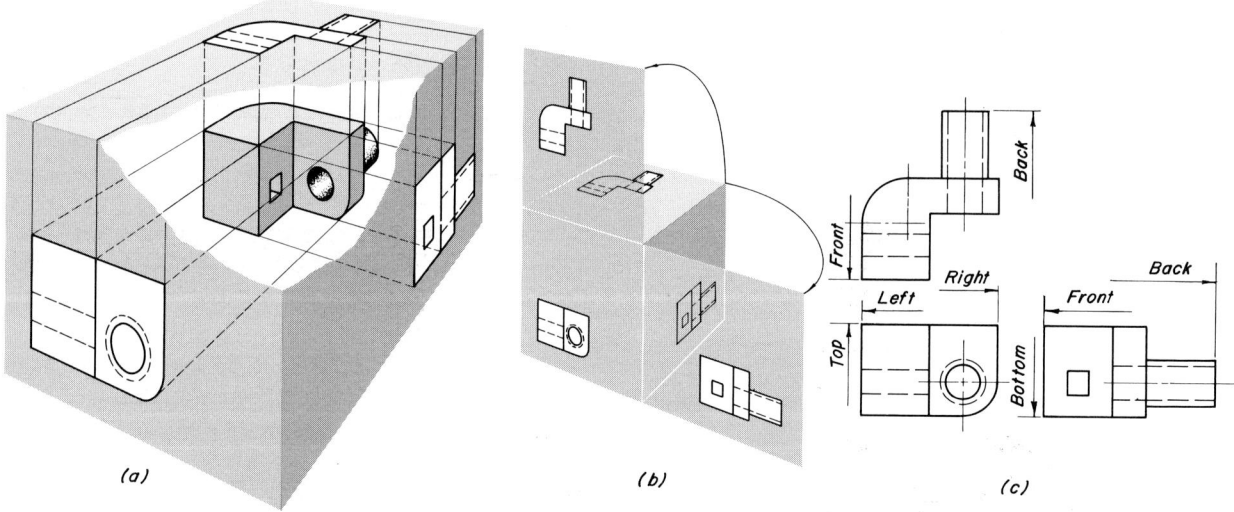

FIG. 6.43. Third quadrant arrangement of three views.

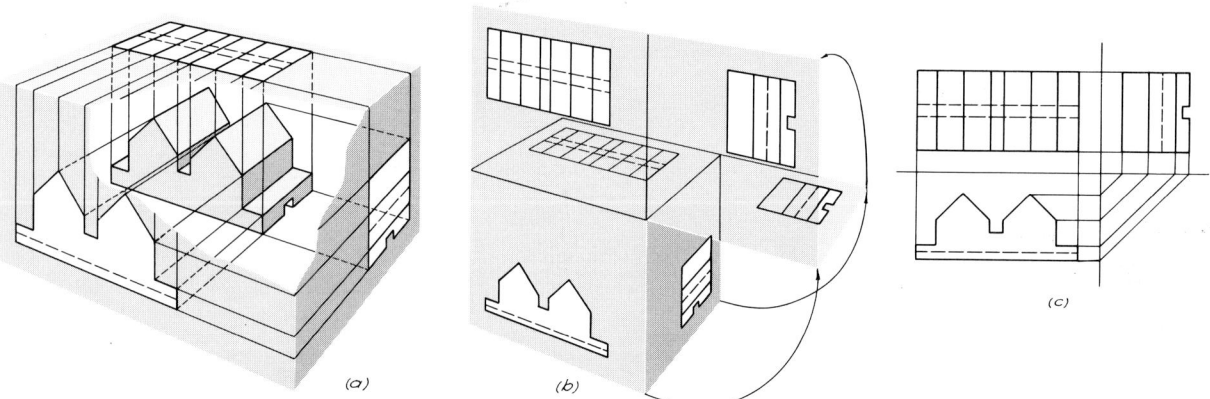

FIG. 6.44. Alternate arrangement of views.

object, such as corners and intersections, must be shown in some manner in every view. Since some lines on an object are invisible from a particular point of sight, there must be some method of distinguishing the visible line from the invisible. Dashed lines are used to represent the invisible line. See Figs. 6.43 and 6.44. These lines must be very carefully made, with the dashes approximately $\frac{3}{16}$ in. long and not more than $\frac{1}{16}$ in. apart.

Figure 6.45 shows the pictorial sketch of an object with the correct projections. In the top view, face ABCD will be edgewise, so that only AB will be visible. Lines CD, KN, and LM, which are invisible from the top, will not show in dashed form because they are below AB. The visible lines take precedence over the invisible lines. The edges of the holes and the edge of the face EFGH will show as dashed lines in the top view. In the side view, the top edge of the large hole, although invisible, will not show as dashed because it is behind the visible line of the notch. The lower edge of the hole will show as dashed because there is no other line between it and the profile point of sight.

A dashed line takes precedence over a center line when they happen to coincide.

The rules governing visibility in any view are the following.

a. In the top view, the highest parts will be visible. These are determined from the front view.

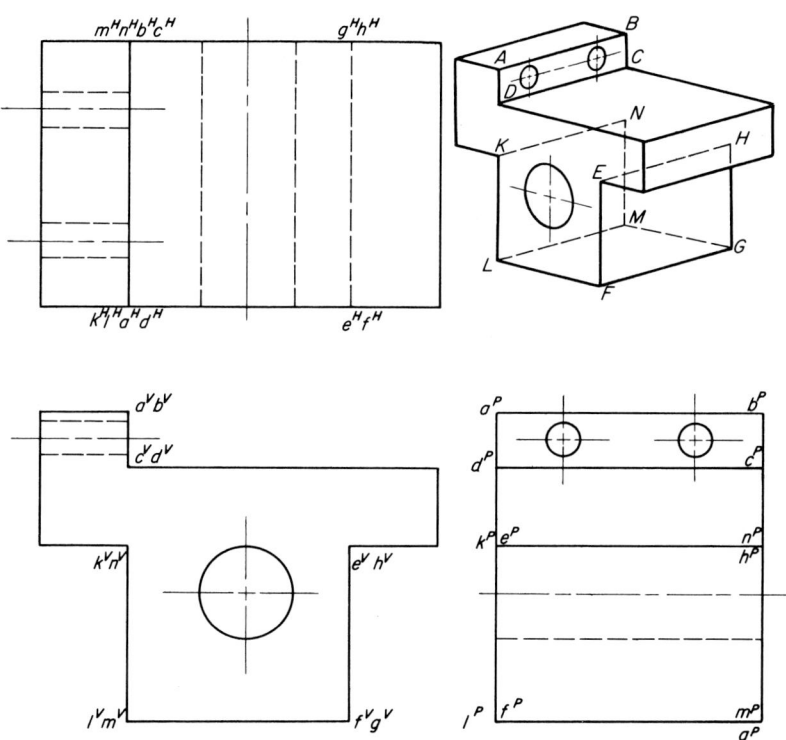

FIG. 6.45. Three-view drawing with visibility of lines.

124 ORTHOGRAPHIC PROJECTION

b. In the front view, the parts nearest the observer (front parts) will be visible. These are determined from the top or side view.

c. In the right-side view, the parts on the right will be visible. They are determined from the front or top views.

Figure 6.46 shows the visiblity of a more complicated object.

6.29 ELIMINATION OF INVISIBLE LINES

Sometimes there is an advantage in choosing a combination of views such as shown in Fig. 6.47 to reduce the number of invisible lines to be represented. This figure shows a front and two partial end views. Each end view contains only the features on one side of the central flange. This simplifies both the making and the reading of the drawing.

6.30 TECHNIQUE OF DRAWING INVISIBLE OUTLINES

Various conditions arise in the representation of invisible outlines, and all nationally recognized standards have agreed on the proper method of showing them.

These conditions are illustrated on the drawing in Fig. 6.48 and are lettered to correspond to the list shown in the figure.

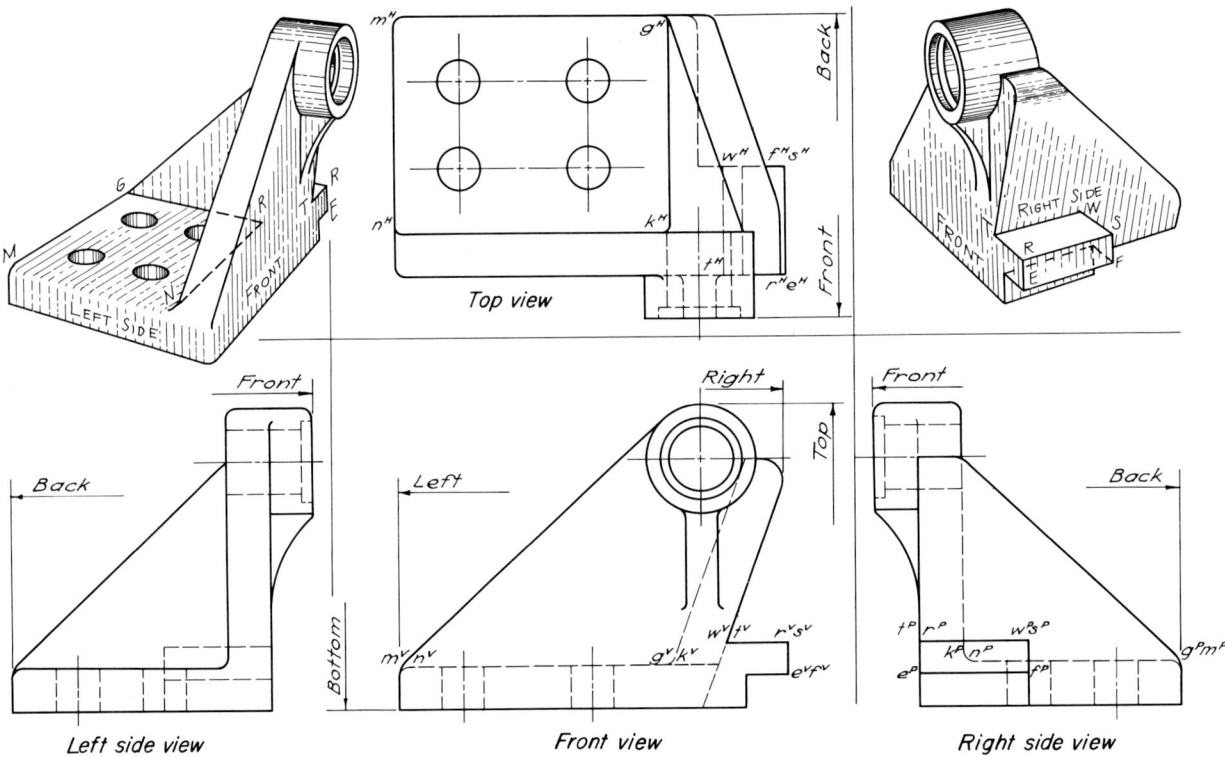

FIG. 6.46. Determining visibility of lines.

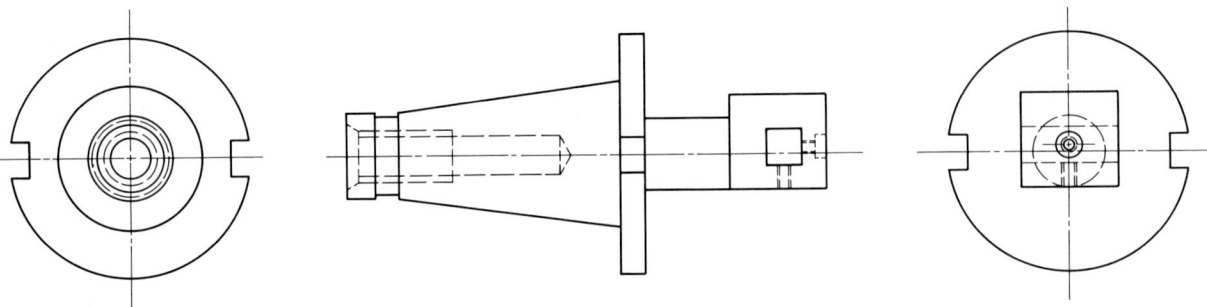

FIG. 6.47. Use of two end views to avoid hidden lines.

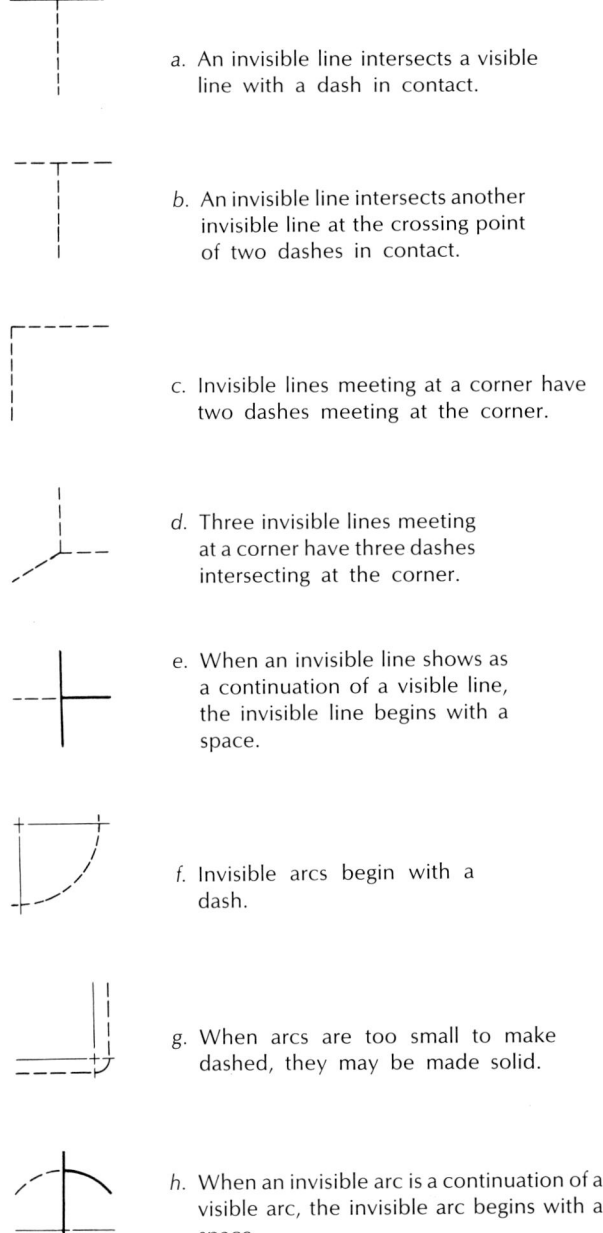
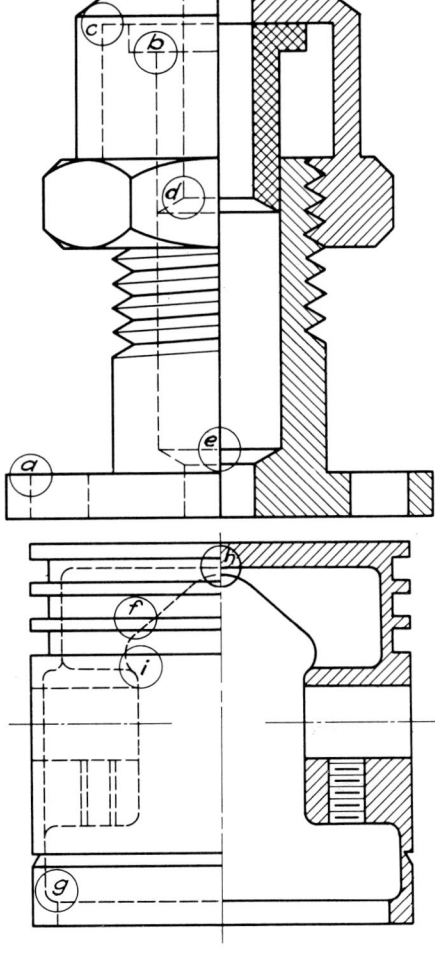

a. An invisible line intersects a visible line with a dash in contact.

b. An invisible line intersects another invisible line at the crossing point of two dashes in contact.

c. Invisible lines meeting at a corner have two dashes meeting at the corner.

d. Three invisible lines meeting at a corner have three dashes intersecting at the corner.

e. When an invisible line shows as a continuation of a visible line, the invisible line begins with a space.

f. Invisible arcs begin with a dash.

g. When arcs are too small to make dashed, they may be made solid.

h. When an invisible arc is a continuation of a visible arc, the invisible arc begins with a space.

i. When two invisible arcs meet, the intersection at the point of tangency is located with a dash on each arc.

FIG. 6.48. Invisible-line technique.

6.31 PROJECTION OF CURVED LINES

Curved lines on a drawing may consist of circles, arcs of circles, parts of ellipses, hyperbolas, parabolas, other geometric curves, or just irregular curves. For noncircular curves a sufficient number of points on the curve must be located so that a smooth curve can be drawn through them. These points must be close together when the curvature is sharp but may be spread out when the curvature is flat. Figure 6.49 shows an object with a curved line in the top view. This curve, which is a portion of an ellipse, has been determined by projecting points from the other two views. The proper way to use an irregular curve to connect the points is discussed in Article 3.21.

6.32 PROJECTION OF CURVED SURFACES

Such surfaces as cones, cylinders, and spheres often form the exterior or interior surfaces of objects to be drawn. A curved surface is usually represented by drawing the outlines of the object as seen from the appropriate point of sight. Thus, a cylinder would be a circle in one view and a rectangle in the adjacent view. A sphere would be a circle in every view.

When curved surfaces occur in combination with each other or with plane surfaces, it is frequently necessary to decide whether to show the intersections or the tangencies. In Fig. 6.50a the tangency should be shown in the top view because both curved surfaces are tangent to a vertical plane. This is not true in Fig. 6.50b where the line may be omitted in the top view.

Frequently the theoretical intersections are projected as lines when it is necessary to give meaning to the drawing. This is shown in Figs. 6.51 and 6.52.

The intersections of two unfinished surfaces

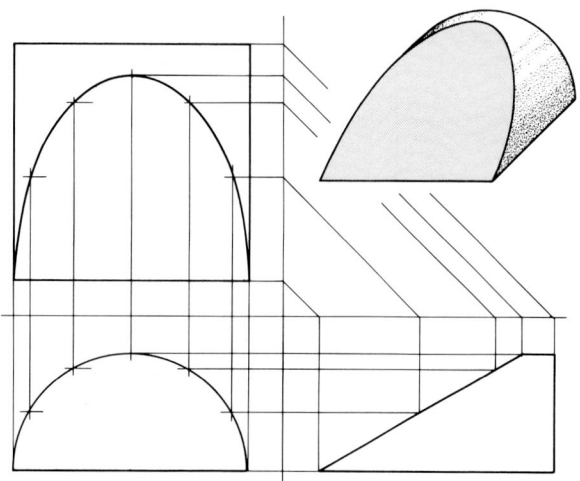

FIG. 6.49. Projection of a curved line.

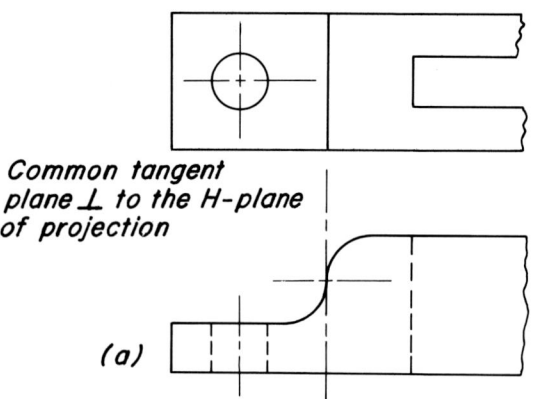
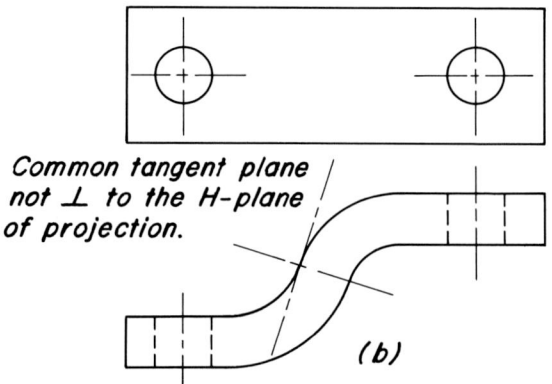

FIG. 6.50. Representation of curved contours.

6.32 PROJECTION OF CURVED SURFACES

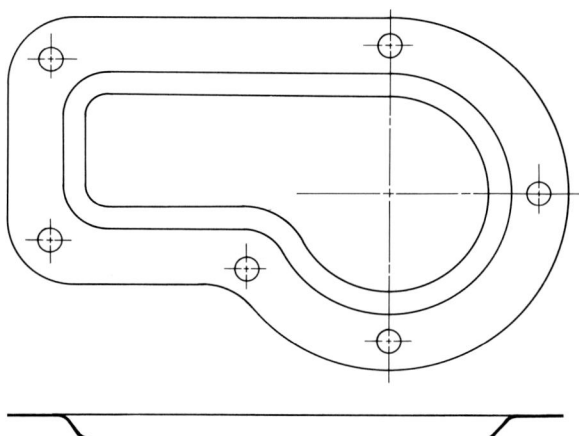

FIG. 6.51. Representation of curved contours.

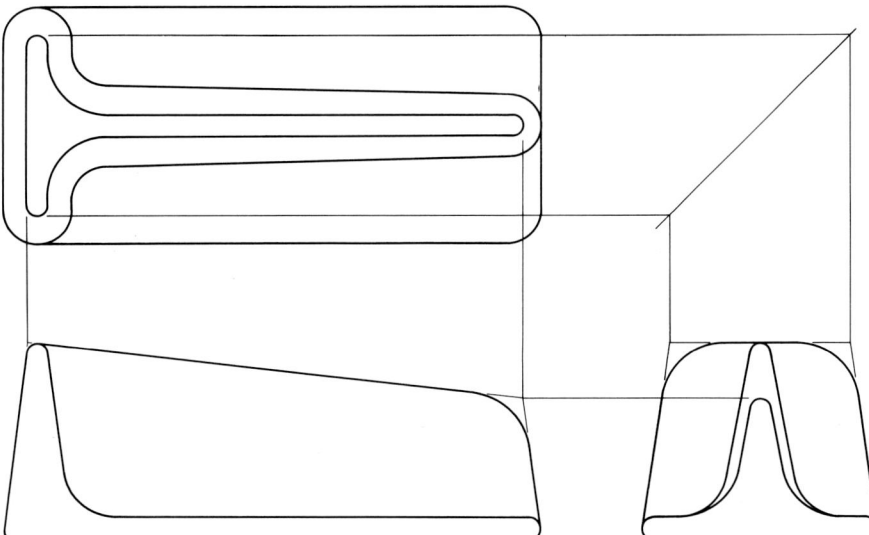

FIG. 6.52. Representing streamlined object.

on a casting will always have the corners rounded or the angles filleted as in Fig. 6.53. These rounds and fillets are always shown with a line in the adjacent view just as though the intersection were sharp. This is seen in Fig. 6.54a. When surfaces are streamlined, as in Fig. 6.54c, all intersections are shown to make it possible to interpret the drawing.

6.33 PARTIAL VIEWS

If the space is limited on the drawing sheet and the object is symmetrical about a central cutting

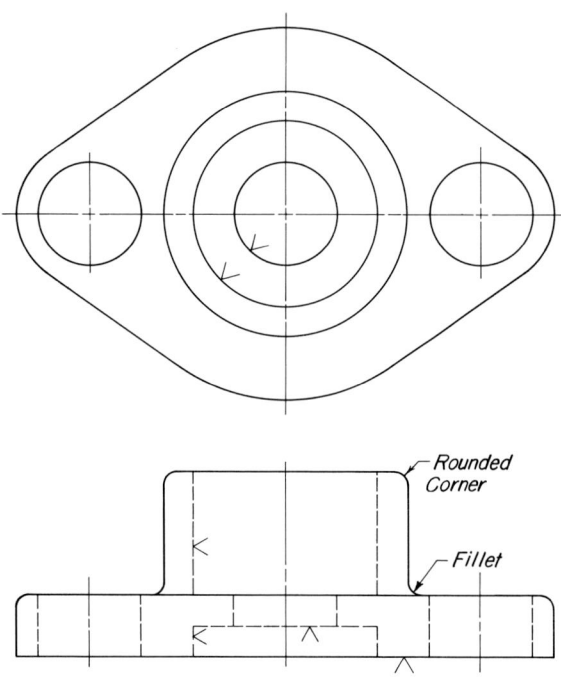

FIG. 6.53. Rounded and filleted corners.

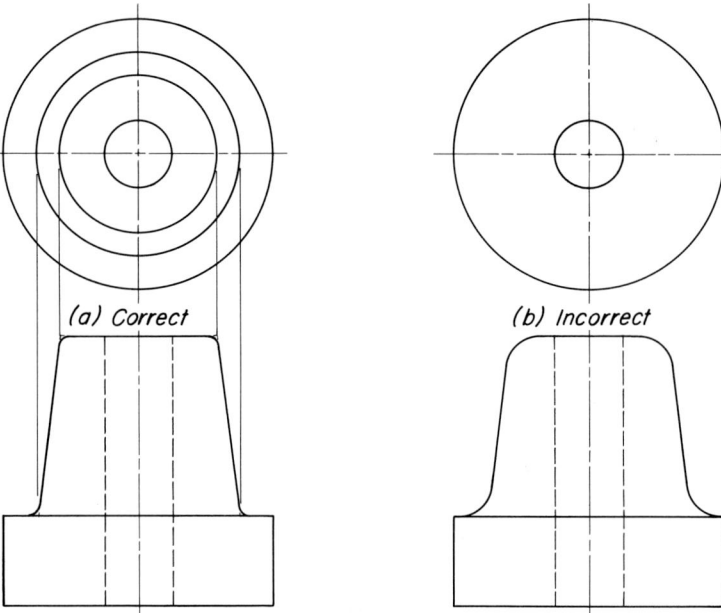 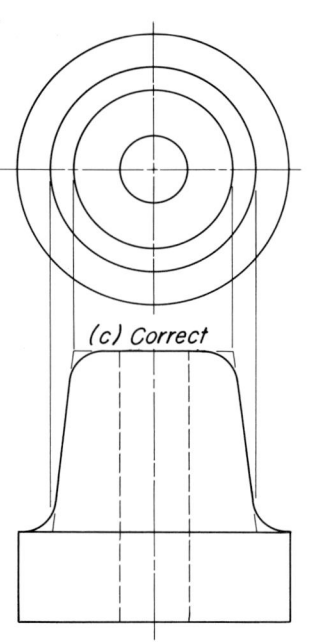

FIG. 6.54. Representing rounded corners.

plane, one half of a view may be omitted, as has been done in Fig. 6.55. If the front view is a full or a half section, it is customary to draw the rear half of the top view, as in Figs. 6.55a and b. If the front view is not sectioned and only half of the top view is shown, it should be the front half of the object, as in Fig. 6.55c. If the partial view is drawn exactly to the center line, it should be marked "symmetrical about the center line."

6.34 RUNOUT LINES

When the intersection between two surfaces is filleted, standard practice is to use a conventional representation, as illustrated in Fig. 6.56. These curved extensions of true projections are commonly referred to as *runout lines* for rounded or filleted corners. They have no specified radius of curvature but are done to suggest the general shape of the theoretical line of intersection.

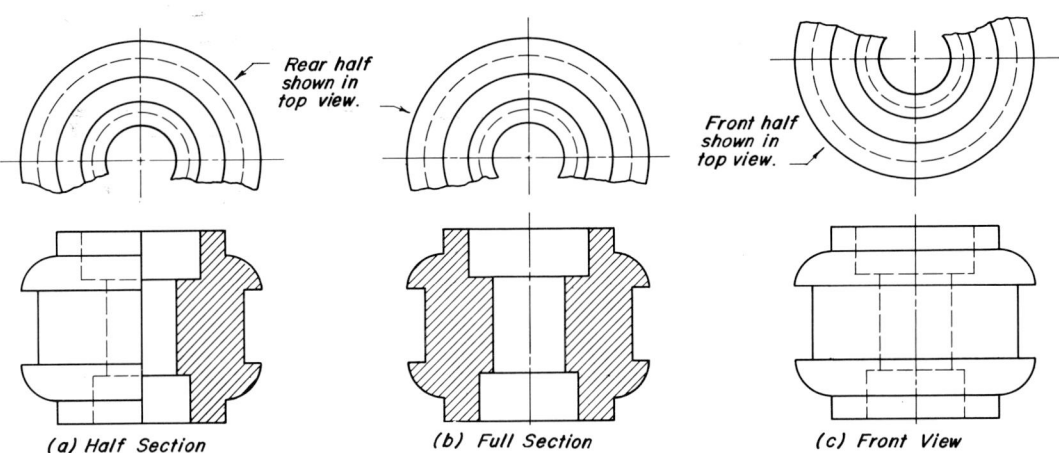

FIG. 6.55. Arrangement of partial top views.

FIG. 6.56. Standard runouts (courtesy ANSI).

130 ORTHOGRAPHIC PROJECTION

6.35 CONVENTIONAL PRACTICES

Certain violations of the true projections are sometimes desirable. Some of these violations are discussed below. This may be done for any one of a number of reasons.

    a. Clarity.
    b. Ease of making or reading a drawing.
    c. To provide information needed in the shop.

6.35.1 Odd Number of Axes.

When an object has three or any other odd number of axes, the projections may become confused by the excess lines needed to show the correct projections. In such a case the part that does not show on the center line may be shown in the front view in a revolved position. This is illustrated in Fig. 6.57 where the lug on the right side has been shown in the front view as though it were actually

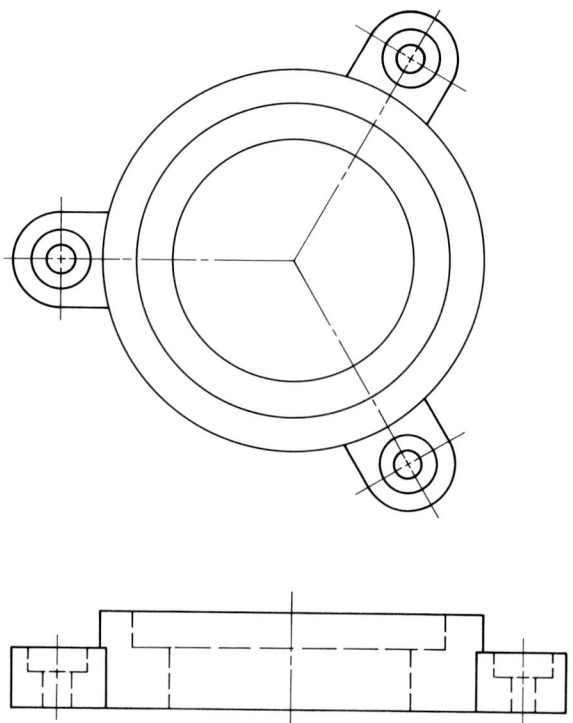

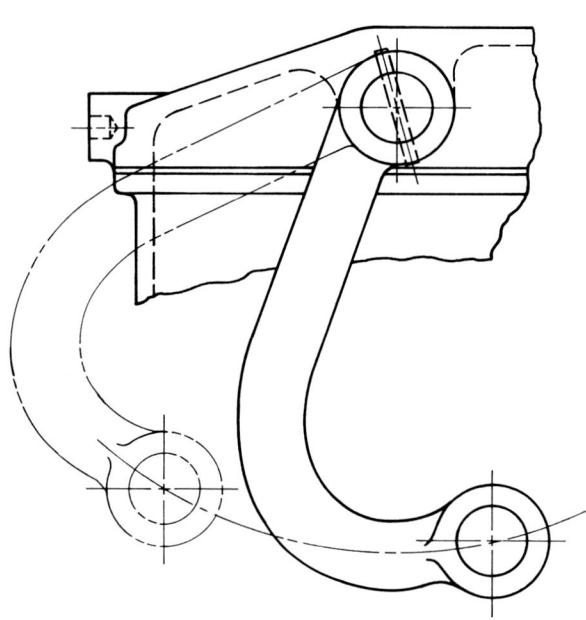

FIG. 6.58. Alternate position shown by phantom lines.

FIG. 6.57. Rotation for odd number of axes.

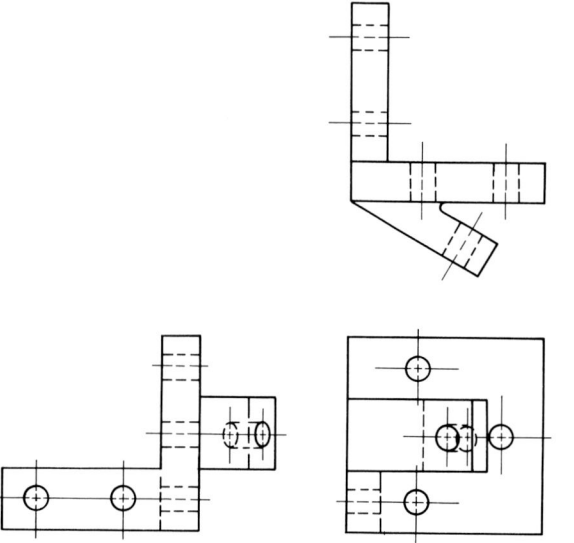

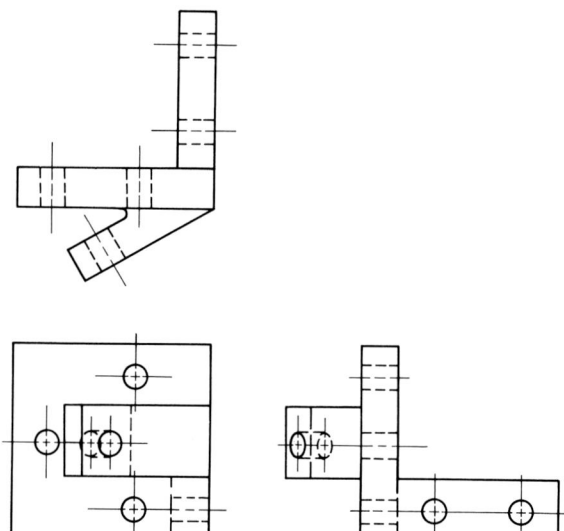

FIG. 6.59. Right- and left-hand parts—only one to be drawn.

6.35 CONVENTIONAL PRACTICES 131

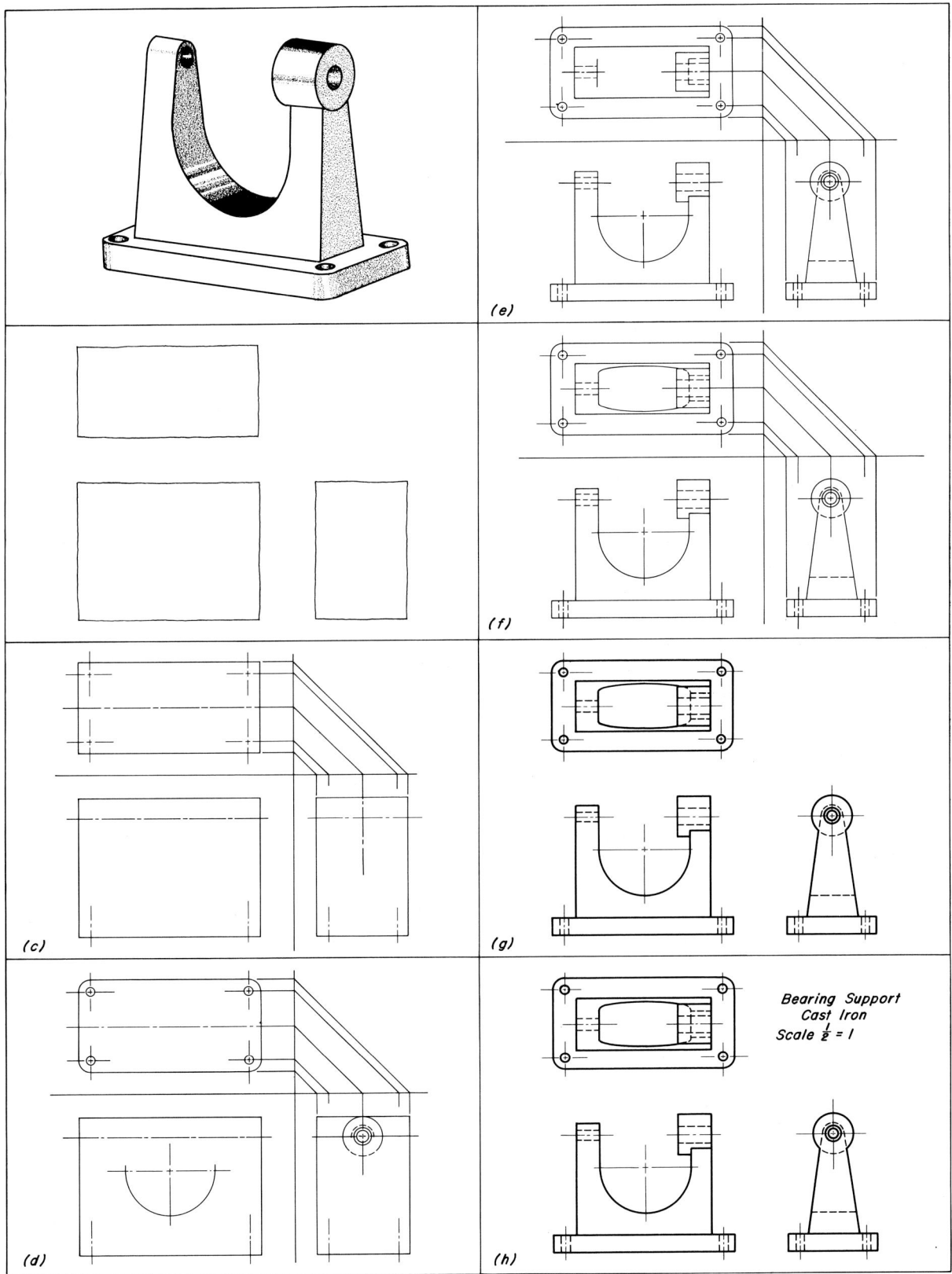

FIG. 6.60. Development of an orthographic projection.

on the center line. The top view shows the true position of the lug. This can be done only with cylindrical objects where the revolution does not change the true relation with the center line.

6.35.2 Alternate Positions. When a part is designed to operate in two positions, it is frequently drawn in one position and the other position is indicated by phantom lines. See Fig. 6.58.

6.35.3 Right- and Left-Hand Parts. In actual practice, parts frequently work in pairs, one of which is called *right-handed* and the other *left-handed*. It is not necessary to draw both parts. They may be listed in the bill of materials as one right-handed and one left-handed of the same drawing. The left-hand part is the mirror image of the right-hand part. Figure 6.59 shows a drawing of both parts.

6.36 LAYOUT OF A THREE-VIEW DRAWING

In making a three-view drawing, it is necessary to develop a method of procedure that will enable the drafter to proceed rapidly. The following paragraphs suggest a method of procedure for the beginner, which he/she may vary for convenience and desires as experience is gained.

The complete layout of a three-view drawing may be subdivided into steps that are illustrated in Fig. 6.60.

a. Choose the number and arrangement of views.
b. Make a freehand layout of the areas for each view. From this determine the scale of the drawing or the size of the paper and the spacing of the views. See Fig. 6.60*b*.
c. Make an accurate mechanical layout of the outlines of the views, using reference lines. Then add the main center lines and the center lines of the details (Fig. 6.60c).
d. Draw the circular parts (Fig. 6.60*d*).
e. Draw the straight lines (Fig. 6.60e).
f. Locate intersections and irregular curves (Fig. 6.60*f*).
g. Clean up the drawing with art gum or a soft eraser. Take out excess lines, construction lines, and reference lines (Fig. 6.60g).
h. Go over all lines to make them heavier and to improve the technique (Fig. 6.60*h*).
i. Put in title and scale. *Every drawing, whether dimensioned or not, must have a title and must have the scale specified.*

6.37 FIRST-QUADRANT DRAWING

In Europe and in most of the world, an object to be drawn is placed in the first quadrant (first angle) instead of the third, as in the United States and Canada.

The rules of projection are the same, except that in the first quadrant the object is between the plane and the point of sight. As in third quadrant, the plane must be revolved away from the object, to bring the point of sight to the front of the paper. This is shown in Fig. 6.61.

The actual difference between third- and first-quadrant drawings lies in the arrangement of the views. In first-quadrant drawings the top view is below the front view, the right-side view is to the left of the front view, and the left-side view is to the right of the front view. This gives a rather awkward arrangement, but with a little practice it soon becomes possible to read the first-quadrant drawings.

With so many corporations conducting business in a number of countries, drawings drawn in

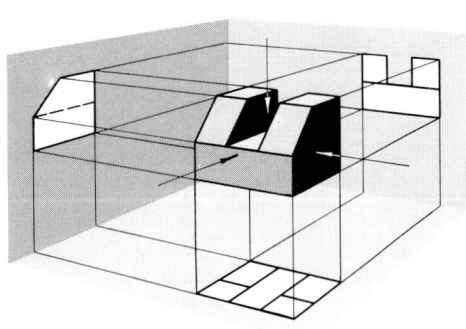 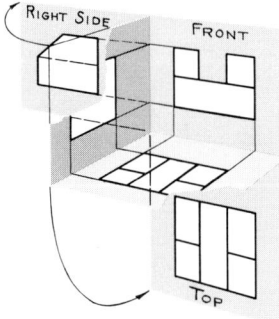 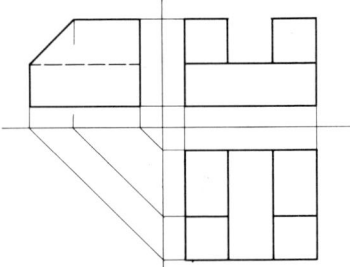

FIG. 6.61. First-angle projection.

third-angle projection may cause confusion in Europe and drawings drawn in Europe, confusion in the United States. Thus, drawings that will be utilized out of the country of origin should contain the appropriate symbol. Figure 6.62, third-angle projection, should be placed in a prominent place on drawings drawn in this country. Drawings initiated by overseas divisions of domestic companies should use Fig. 6.63, first-angle projection.

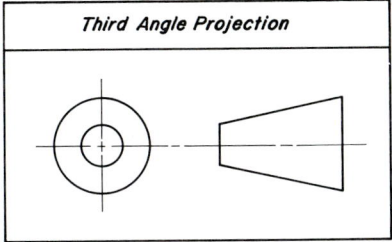

FIG. 6.62. Third-angle projection symbol.

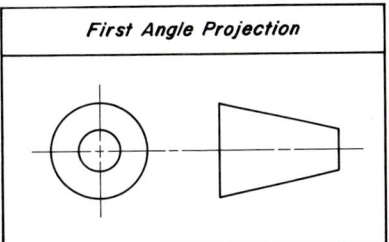

FIG. 6.63. First-angle projection symbol.

SELF-STUDY QUESTIONS

Before trying to answer these questions, read the chapter carefully. Then, without reference to the text, answer as many questions as possible. For those that cannot be answered, the number in parentheses following the question number gives the article in which the answer can be found. Look it up and write down the answer. Check the answers that you did give to see that they are correct.

6.1 (**6.1**) In orthographic projection, the lines of sight are _____ to the plane of projection.

6.2 (**6.5**) In orthographic projection in the United States, the plane of projection is between the object and the _____.

6.3 (**6.3.1**) In orthographic projection, the point of sight is at an _____ distance from the object.

6.4 (**6.3.3**) In orthographic projection there is _____ plane for each view.

6.5 (**6.3.3**) The three principal planes of projection are _____, _____, and _____.

6.6 (**6.3.4**) To make the most useful orthographic projection, there are four rules for placing the object.
   1. Place the object in its _____ position.
   2. Place the object so that its faces will be _____ to the principal planes of projection.
   3. Place the object so that its most descriptive contour will become the _____ view.
   4. Select the views that will show the fewest _____ lines.

6.7 (**6.4**) Every line on an object must show as a _____ or a _____ on the projection.

6.8 (**6.5**) The planes of projection divide all space into _____ quadrants.

6.9 (**6.5**) In the United States, the _____ quadrant is used for almost all drawings.

6.10 (**6.5**) The space described by the third quadrant is _____ the horizontal plane and _____ the vertical plane.

6.11 (**6.6**) In multiview drawing, the _____ view is always above the _____ view.

6.12 (**6.6**) In multiview drawing, the right-side view is to the _____ of the _____ view.

6.13 (**6.9**) A point in space may be definitely located by _____ projections.

6.14 (**6.10**) The projection of point A on the horizontal plane is lettered _____.

6.15 (**6.10**) The projection of point A on the third auxiliary plane is lettered _____.

6.16 (**6.11**) If point B is behind point A, the distance may be measured in the _____ or _____ projections.

6.17 (**6.11**) Vertical distances can be measured in the _____ or _____ views.

6.18 (**6.11**) Right and left distances can be measured in the _____ or _____ views.

6.19 (**6.14.1**) If the vertical projection of a line is parallel to the horizontal reference line, the true length of the line will show in the _____ projection.

6.20 (**6.14.3**) If the horizontal projection of a line is parallel to the vertical reference line, the true length of the line will show in the _____ projection.

6.21 (**6.14.1**) In measuring the angle between a line and a plane, it is necessary to have the _____ _____ of the line and the _____ _____ of the plane in the same view.

6.22 (**6.14.2**) If a line is parallel to the vertical plane, the angle π between the line and the profile plane will show in the _____ projection.

6.23 (**6.14.1**) If a line is parallel to the horizontal plane, the angle ϕ between the line and the vertical plane will show in the _____ projection.

6.24 (**6.15**) If a line is oblique to all three coordinate planes, its true length will show in _____ principal views.

6.25 (**6.17.2**) A plane that is parallel to the vertical plane will show edgewise in the _____ and _____ projections.

6.26 (**6.18.3**) When a plane is perpendicular to the profile plane and inclined to the vertical, the angle θ will show as the angle between the _____ projection of the plane and the horizontal reference line.

6.27 (**6.18**) To measure the angle between two planes, it is necessary to have both planes showing _____ in the same view.

6.28 (**6.19**) When a plane is oblique to all three principal coordinate planes, all principal projections will show as areas that _____ _____ true size.

6.29 (**6.24**) To make a one-view drawing complete, a _____ must be added.

6.30 (**6.21.3**) A _____ drawing is often useful in reading a drawing.

6.31 (**6.26**) If two side views give the same information, the _____ side view is usually drawn.

6.32 (**6.28**) The proper visibility of lines in the top view may be determined by a careful study of the _____ view.

6.33 (**6.28**) The proper visibility of lines in the front view may be determined by a careful study of the _____ view.

6.34 (**6.28**) When a visible and an invisible line coincide, the _____ line takes preference.

6.35 (**6.28**) When a center line and an invisible line coincide, the _____ line takes preference.

6.36 (**6.48(e)**) When an invisible line is a continuation of a visible line, the invisible line begins with a _____.

CHAPTER 7

AUXILIARY VIEWS

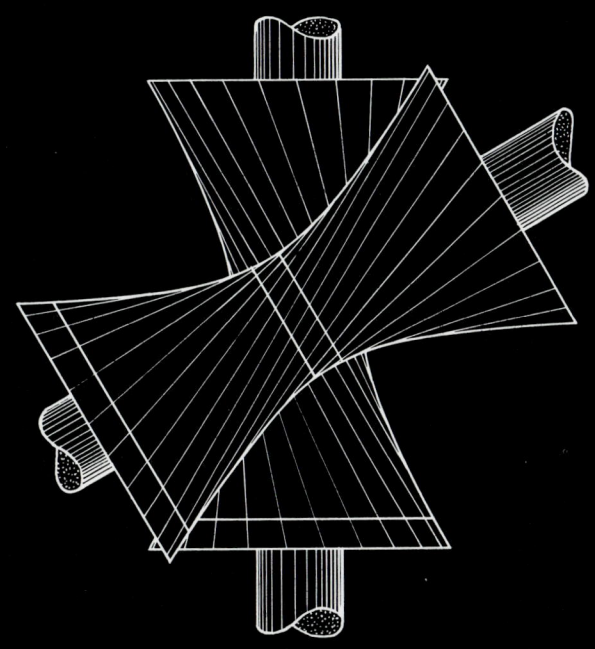

CHAPTER 7

7.1 NEED FOR AUXILIARY VIEWS

On many objects there are lines that do not show in true length and faces that do not show in true size in any of the three principal views. In manufacturing and construction it is also necessary in many cases to know the true value of the angle between lines and between plane faces. When these values do not appear in the three-principal views, it is necessary to set up additional planes of projection and make views that will give the necessary information. These additional planes are called *auxiliary planes*.

7.2 PLACING THE AUXILIARY PLANE OF PROJECTION

To serve the purposes described above, the auxiliary plane must be placed to satisfy the following conditions.

a. It must be perpendicular to one of the planes of projection already in use.

b. It must be parallel or perpendicular to some line or plane face of the object.

c. For the first auxiliary plane, the choice of which of the three principal planes to use is

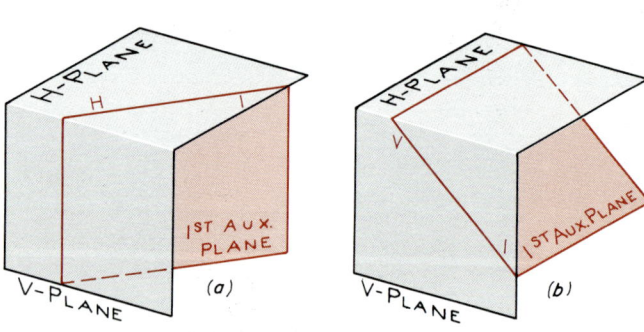
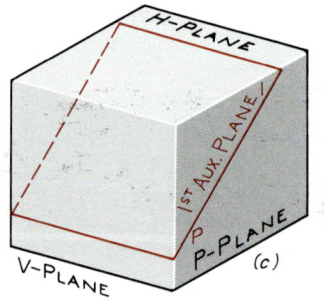

FIG. 7.1. Method of placing auxiliary planes.

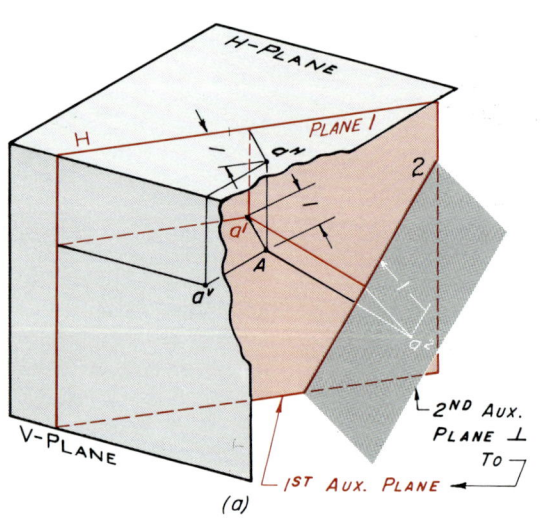
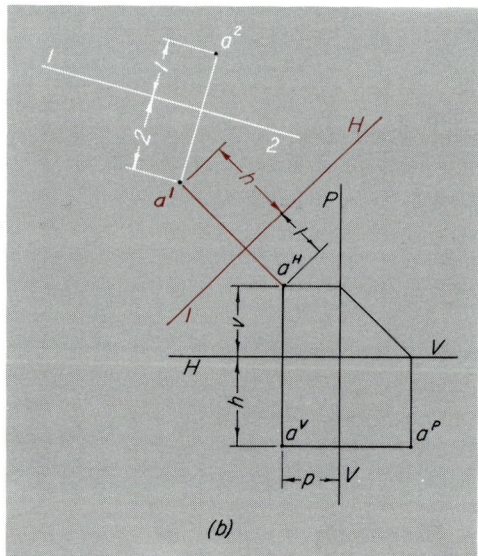

FIG. 7.2. Nomenclature and theory of auxiliary projection.

determined by the shape and position of the object and the problem to be solved.

7.3 NOTATION OF AUXILIARY PLANES AND VIEWS

A method of marking the H-, V-, and P-planes and their projections was established in Article 6.10. A similar system is used for the auxiliary views. The first auxiliary plane is numbered 1, the second 2, and so on.

When the first auxiliary plane is perpendicular to the H-plane, the line of intersection of the two planes, called a *reference line,* is labeled H-1 as in Fig. 7.1a. This means that the auxiliary plane will be revolved about this line until it coincides with the H-plane. When the auxiliary plane is perpendicular to the V-plane, the reference line is marked V-1 as in Fig. 7.1b. When it is perpendicular to the P-plane, it is called P-1 as in Fig. 7.1c.

In Fig. 7.2 the projections of point A on the H- and V-planes are shown as well as its projections on auxiliary planes 1 and 2. The auxiliary projections of the point are labeled with the lowercase a with an appropriate subscript. Thus a^1 is the projection on auxiliary plane 1, and a^2 is the projection on plane 2. The reference line between the H-projection and the 1-projection is labeled H-1. The reference line between the 1-projection and the 2-projection is labeled 1-2. *The adjacent projections in every case lie on a perpendicular to the reference line between them.*

The distance from the point in space to any plane is marked with the letter or number of the plane. Thus the distance from the point to the H-plane is marked h, from the V-plane, v; from the P-plane, p; from the 1-plane, 1; and from the 2-plane, 2. These distances in each case are indicated in Fig. 7.2.

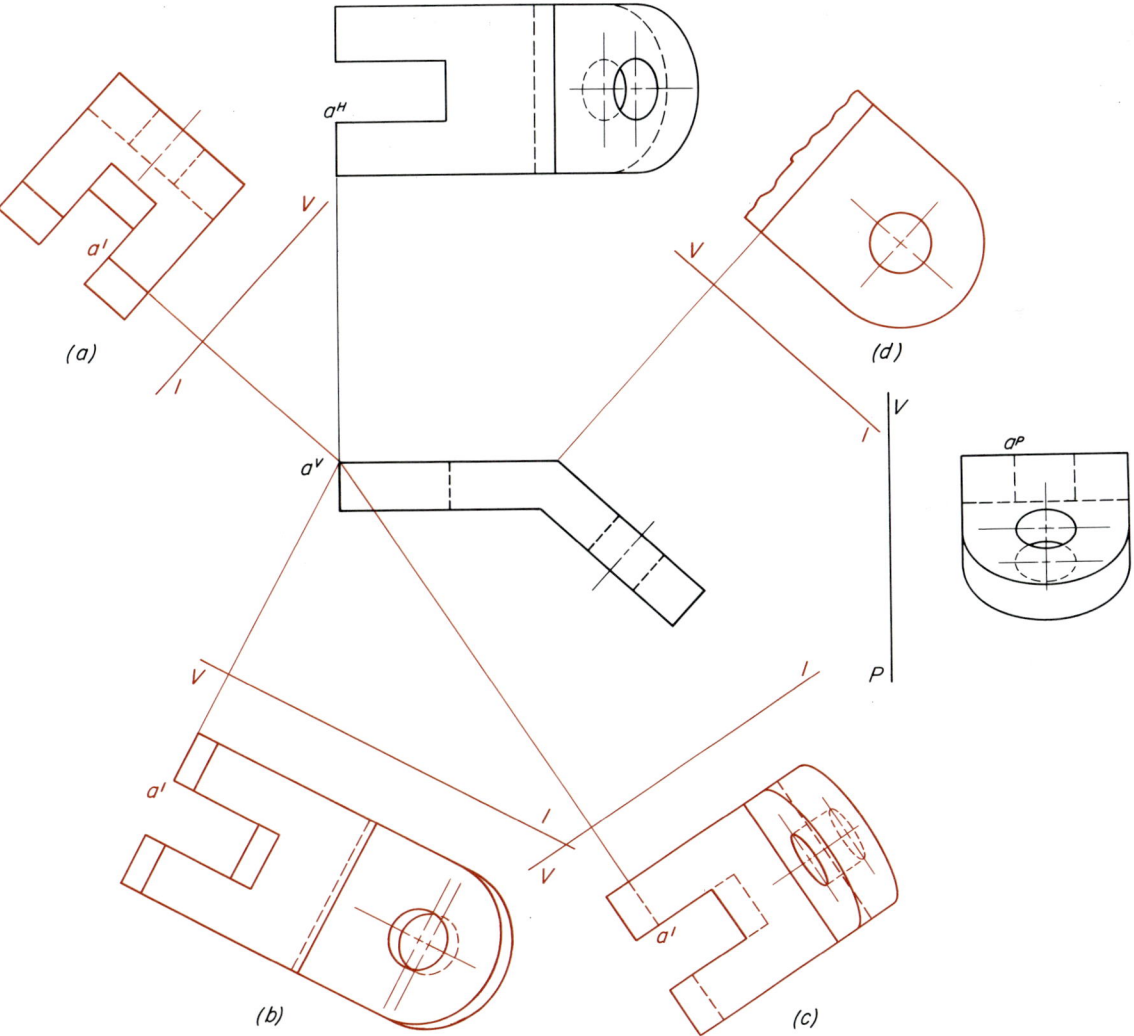

FIG. 7.3. Relationship of the auxiliary plane to an object.

7.4 RELATIONSHIP OF THE AUXILIARY PLANE TO THE OBJECT

It is possible to place auxiliary planes in any relation to the object, as shown in Fig. 7.3. The object can be projected on any one of these planes, thus giving a view in any desired direction. In Fig. 7.3 the projections marked a, b, and c are correct projections and each could possibly be useful for some purpose. However, projection d is the one that is most useful because it gives the true size and shape of the inclined face. Another feature of this view is that only the inclined face is shown. This is customary in most auxiliary projections.

It is also noticeable that point A has been projected to each view and in each case it has been labeled a^1. From this it can be seen that it is possible to have several first auxiliary projections in a single problem. Every first auxiliary projection is labeled a^1, although there is no relation between them.

7.5 THEORY OF MAKING AUXILIARY PROJECTIONS

The orthographic theory of projection, as discussed in Chapter 6, is used when making auxiliary projections. In Fig. 7.2b point A has been projected upon all three principal planes of projection, as well as into the first and second auxiliary planes.

The principle of the projections of a point on two adjacent planes being "in projection" is used to obtain the auxiliary views. Therefore, a^1 and a^H lie on a line perpendicular to the H-1 reference line; and a^1 and a^2 lie on a line perpendicular to the 1-2 reference line. The distance from the respective planes is indicated on Fig. 7.2b.

7.6 USES OF AUXILIARY VIEWS

There are four fundamental operations for which auxiliary planes are used. The ability to solve these four problems is the basis for the successful solution of many problems in drawing and engineering geometry. They are

a. Find the true length of an oblique line.
b. Find the point projection of a line.
c. Find the edgewise view of a plane.
d. Find the true shape of a plane surface.

7.7 TRUE LENGTH OF AN OBLIQUE LINE

When a line is parallel to one of the principal planes of projection, its projection upon that plane will be the true length of the line. Its projection on the adjacent planes will be parallel to the reference line.

When a line is oblique to all three principal planes of projection, as line AB in Fig. 7.4, none of its principal projections show in true length and none are parallel to any of the reference lines. To obtain the true length of the line, an auxiliary plane is used that is passed parallel to line AB. The line projected upon this plane will show the true length of the line.

In Fig. 7.4 the position of the auxiliary plane is shown by the reference line V-1. When viewed in the vertical projection, the V-1 line represents the edge view of the first auxiliary plane. When viewed in the first auxiliary plane, the V-1 line represents the edge view of the V-plane. The steps in projecting line AB onto the auxiliary plane are as follows.

a. At any convenient place draw the V-1 reference line parallel to the vertical projection $a^V b^V$ of the line.
b. Through points a^V and b^V draw perpendiculars to reference line V-1. These lines determine one locus of projection a^1 and b^1.
c. To locate a^1 and b^1, the distance they each lie from the V-plane is required. This distance v is obtained in the horizontal projection and is measured from the H-V reference line.
d. To locate a^1, lay off the distance v from the horizontal projection along the perpendicular to V-1 through a^V. Locate b^1 in the same manner.
e. Draw a line connecting a^1 and b^1, thus obtaining the true length of line AB.

7.7.1 Slope of a Line. In engineering practice it is often required to obtain the slope and bearing of a line. *The slope of a line is defined as the angle the line makes with the horizontal plane, the angle θ.* To obtain the angle θ, the line must project upon a plane of projection in true length along with the edgewise view of the H-plane. In

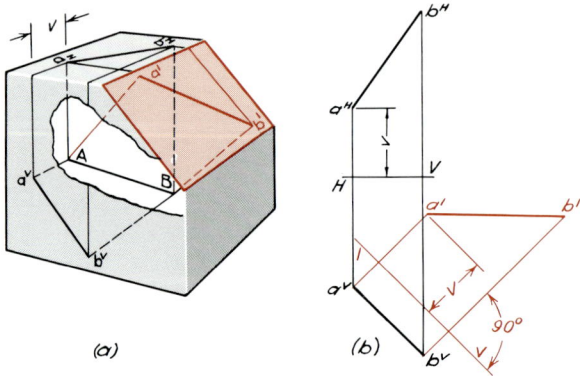

FIG. 7.4. True length of an inclined line.

the case of a line that is parallel to the V-plane, the angle θ will project in its true value in the V-projection as in Figs. 6.16b and 7.5a. When the line is parallel to the P-plane, the angle θ will project in its true value in the P-projection as in Figs. 6.17b and 7.5b.

When the line is oblique to all three principal planes of projection, an auxiliary plane of projection must be used. This plane must be passed parallel to the line and perpendicular to the H-plane; therefore, it must be taken off the top view. The line is then projected upon the auxiliary plane as shown in Fig. 7.5c. The angle θ is the slope of the line.

The slope of a line must also be expressed as a plus or minus to show direction with respect to the H-plane. Thus, in Fig. 7.5c the slope from A to B is negative (downhill) and from B to A it is positive (uphill).

7.7.2 Other Methods of Expressing Slope. As a result of varying practice in engineering areas, different methods of specifying the slope of a line have developed.

a. On **railroads and highways** the slope of the grade line (center line) is expressed in percentage as shown in Fig. 7.6a.

b. In **structural work** the slope of a member is expressed in terms of run and rise as shown in Fig. 7.6b. The larger dimension is always 12 in. whether it is horizontal or vertical.

c. In **roof construction** the slope of the roof is called the *pitch*. The pitch is the ratio of the rise to the span. See Fig. 7.6c.

d. When **concrete or stone walls** have a very steep slope, this is called the *batter*. The batter is specified as the ratio of the horizontal distance to the vertical distance, as shown in Fig. 7.6d.

7.7.3 Bearing of a Line. *The bearing of a line is defined as the smaller of the two angles that the horizontal projection of the line makes with a north and south line.* It may be measured from the north or south.

On a map the north and south line is frequently a vertical line with north at the top. Thus, east will be to the right and west to the left. The bearing should be taken from the horizontal projection and marked with the angle and direction as shown in Fig. 7.7.

It should be noted that the bearing of a line is simply a directed line segment. Thus, the bearing of BA in Fig. 7.7b is just 180° from that of AB in Fig. 7.7a.

7.7.4 Azimuth of a Line. The term *azimuth* is sometimes used to define the direction of a line. *This means the angle that the horizontal projection of the line makes with a north and south line, measured clockwise from the north*, as illustrated in Fig. 7.8. For further discussion see Chapter 19, Fig. 19.18.

7.7.5 To Determine the Projections of a Line of Given Length, Bearing, and Slope. This may best be illustrated with an example problem. Let it be required to draw a line AB from a known point A in Fig. 7.9, 200 ft long, having a bearing North 45° East and a slope of 30° downward from A to B at a scale of 100 ft = 1 in.

The student must analyze the three known facts about this line to determine the order in which the facts will be used to solve the problem.

a. Clearly the H-projection of a line having the required bearing can be drawn, but since it is a sloping line, its true length cannot be shown immediately. The H-projection $a^H c^H$ of random length will therefore be drawn having the correct bearing as shown in Fig. 7.9a.

b. An auxiliary reference line may be placed parallel to this projection and A projected

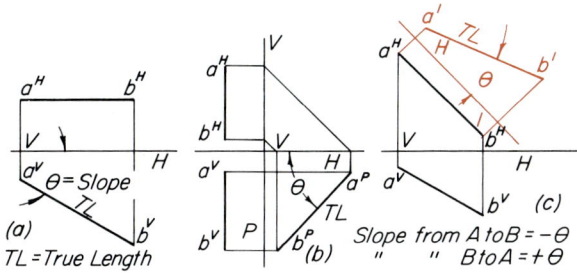

FIG. 7.5. Slope of a line.

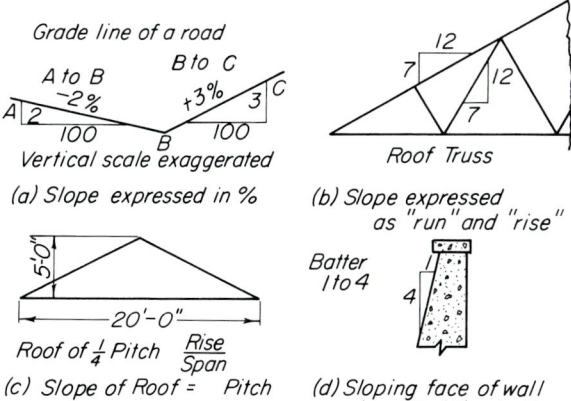

FIG. 7.6. Engineering methods of giving a slope.

to this plane. The slope can be drawn as shown in Fig. 7.9b.

c. This auxiliary view also gives the true length of the line AC, hence a length of 2 in. (200 ft) can be scaled off from a^1 and the end marked b^1.

d. The projection of B can be returned from b^1 to b^H on the horizontal projection of AC, as in Fig. 7.9c. From these two views the vertical projection can be constructed, thus completing the problem as in Fig. 7.9c.

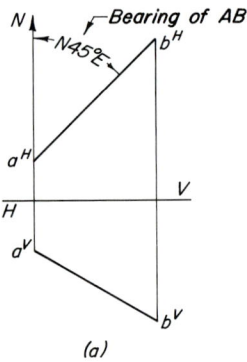
(a)

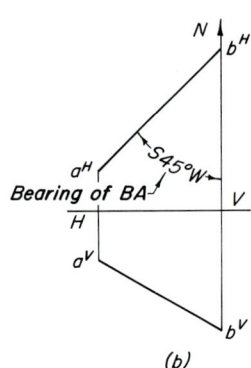
(b)

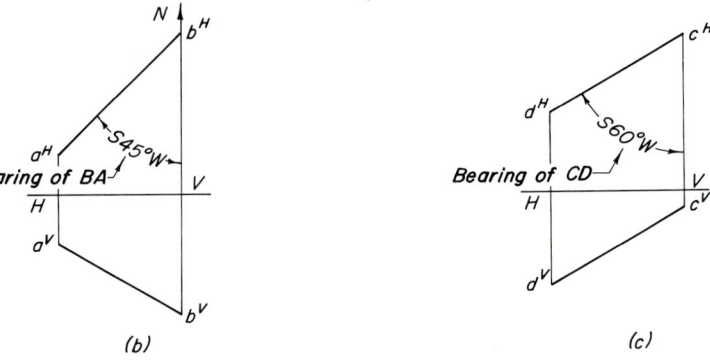
(c)

FIG. 7.7. Bearing of a line.

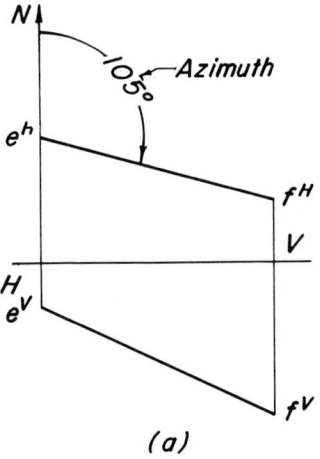
(a)

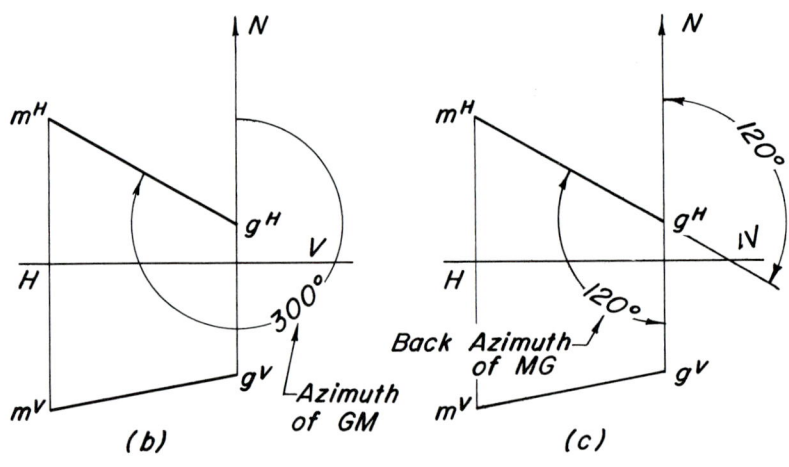
(b)   (c)

FIG. 7.8. Azimuth of a line.

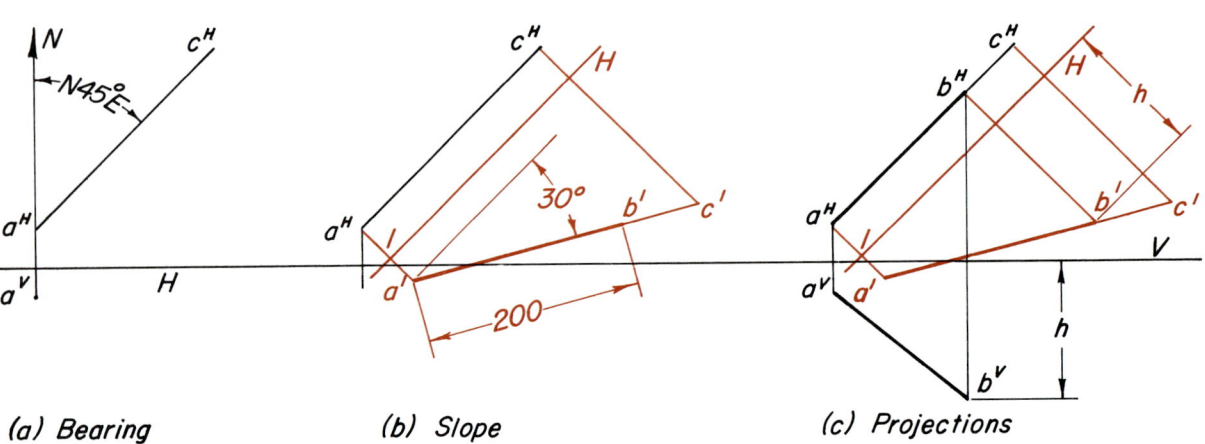

(a) Bearing   (b) Slope   (c) Projections

FIG. 7.9. Constructing a line having a given bearing, slope, and length.

7.7.6 True Length of an Oblique Line—Construction Cone or Rotation Method.
If the line is considered to be an element of a right circular cone whose axis is perpendicular to one of the principal planes, the cone may be revolved about its axis until the line becomes parallel to one of the principal planes.

In Fig. 7.10, four different cones have been set up, any one of which may be used to determine the true length of the line *AB*. In actual practice only that portion of the cone between the projection and the revolved position need be constructed.

Each cone gives not only the true length of the line but also the angle that the line makes with the coordinate plane to which the axis of the cone is perpendicular. Thus, Figs. 7.10a–c show the angles θ, ϕ, and π, respectively, and Fig. 7.10d shows the angle α that the line makes with the auxiliary plane 1.

7.8 POINT PROJECTION OF A LINE

There are many problems that require the point projection of a line to determine the solution. These include the clearance between two members, the angle between two planes, the distance between parallel lines, the right section of a cylinder or prism, and many others.

To find the point projection of a line, the auxiliary plane must be set up perpendicular to the true-length projection of the line.

7.8.1 Line Parallel to One Principal Plane and Inclined to the Others.
When a line is parallel to the vertical plane as shown for line *AB* in Fig. 7.11a, any plane that is perpendicular to line *AB* must be perpendicular to the V-plane since the true length of the line appears in the front view. For this reason only one auxiliary plane is needed. This plane is shown in Fig. 7.11a with the reference line marked V-1. In Fig. 7.11b the orthographic solution is shown. The following steps were necessary.

a. Set up the V-1 reference line perpendicular to $a^V b^V$.

b. Construct a projection line through a^V and b^V perpendicular to V-1.

c. Measure the distance v from each point to the vertical plane in the H-projection.

d. Lay out this distance on the perpendicular through a^V and b^V from the V-1 reference line to obtain $a^1 b^1$. This is the point projection of line *AB*.

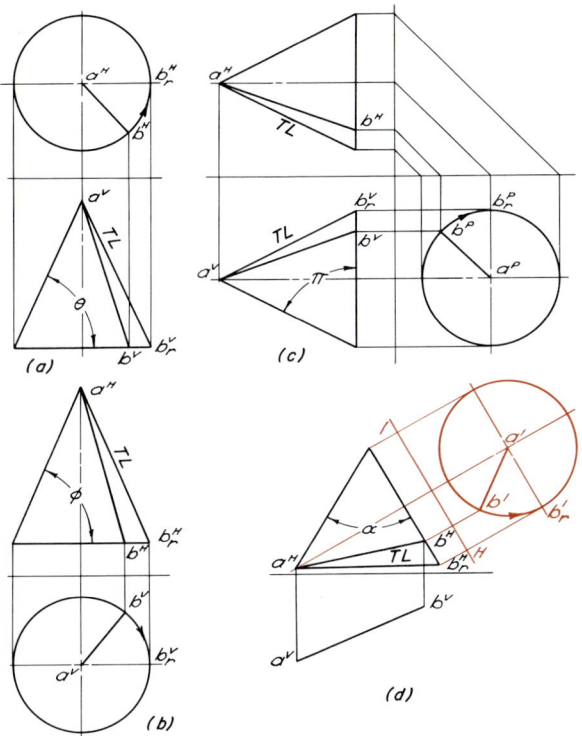

FIG. 7.10. True length of a line by rotation.

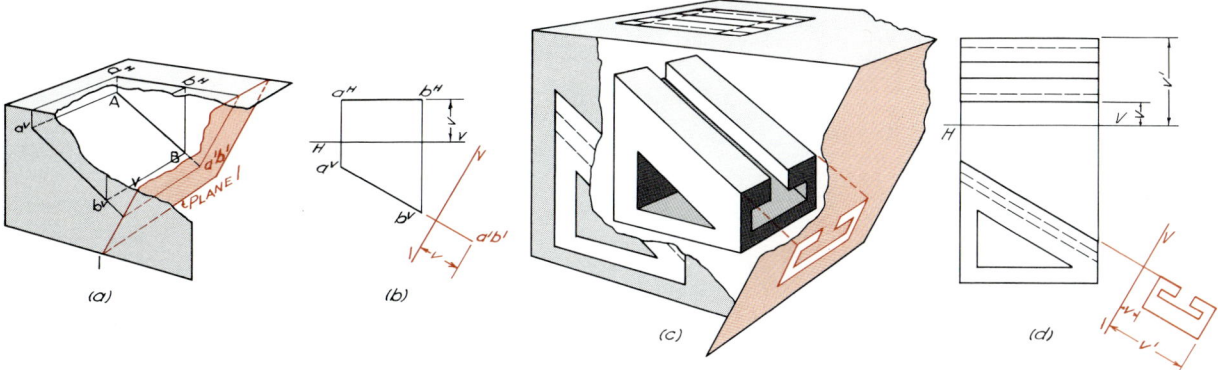

FIG. 7.11. True shape of T-slot. Auxiliary plane perpendicular to V-plane.

Figure 7.11c shows an object with a T-slot in which all lines are parallel to line AB in Figs. 7.11a and b. If the size of the T-slot is known, it may be necessary to have an end view in order to complete the other projections or to design the cutting tool. Figure 7.11c shows a pictorial of the object with the H-, V-, and auxiliary planes. Figure 7.11d shows the orthographic projections giving the top, front, and auxiliary views. The steps necessary in making these views are as follows.

e. Construct the outlines of the top and front views.
f. Construct the V-1 reference line perpendicular to the V-projection of the inclined line.
g. Draw lines from the V-projection perpendicular to the V-1 reference line.
h. Measure distances v and v^1 in the H-projection.
i. Lay out the distances v and v^1 on the perpendicular drawn in step g, by measuring from the V-1 reference line.
j. Construct the cross section of the T-slot from the known dimensions.
k. Project the lines back to the H- and V-projections.

7.8.2 Line Oblique to All of the Principal Planes.
To obtain the point projection of a line inclined to the three principal planes of projection, the true length of the line must first be obtained. The steps for the solution of this problem are illustrated in Fig. 7.12a and are as follows.

a. Set up the first auxiliary plane parallel to line AB by making reference line H-1 parallel to $a^H b^H$.
b. Construct lines through a^H and b^H perpendicular to the H-1 reference line.
c. Measure the distance h from each point to the H-plane as shown for point A in the V-projection.
d. Lay off these distances on the perpendiculars through a^H and b^H by measuring from the H-1 reference line. The true length of AB is shown at $a^1 b^1$, and it is now possible to set up a plane perpendicular to line AB and to the first auxiliary plane.
e. Set plane 2 perpendicular to plane 1 and the line AB by making the 1-2 reference line perpendicular to $a^1 b^1$.

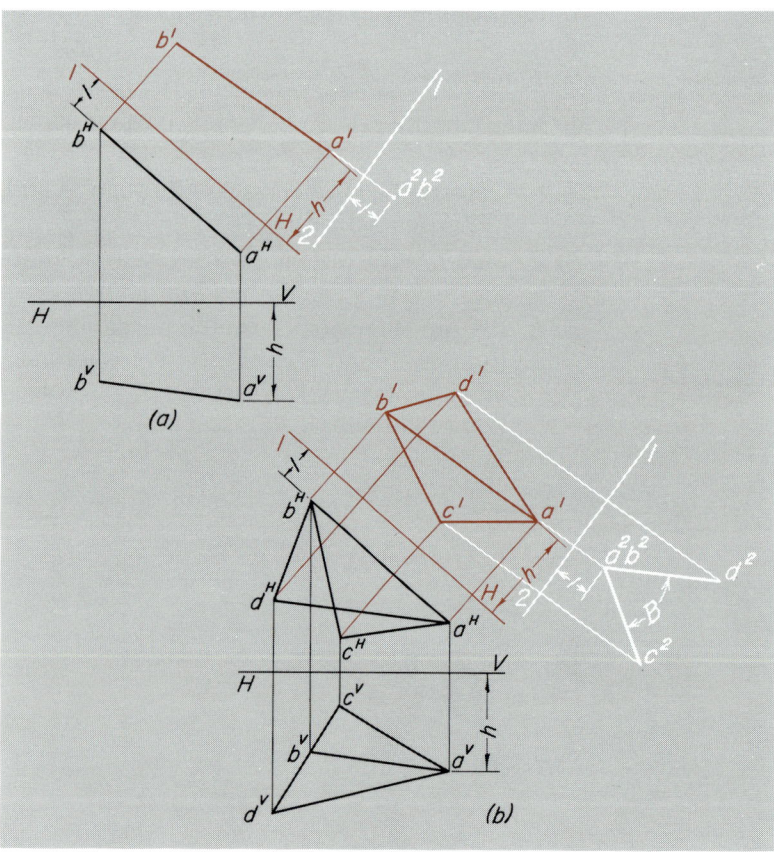

FIG. 7.12. Endwise view of a line and angle between planes. Double auxiliary.

f. Through a^1 and b^1 construct a line perpendicular to the 1-2 reference line.

g. Measure the distance from a^H and b^H to the H-1 reference. This gives the distance from the points to the first auxiliary plane.

h. Lay out this distance on the perpendicular through a^1 and b^1. The distance is laid out from the 1-2 reference line to locate a^2b^2 as shown. This is the point projection of the line.

In Fig. 7.12b line AB is the line of intersection between planes ABC and ABD and is identical to line AB in Fig. 7.12a. To obtain its point projection, the construction in Fig. 7.12b is exactly the same as described in Fig. 7.12a. By following these steps carefully and projecting points C and D also, the second auxiliary view is obtained. It can be seen in the second auxiliary view that both planes show edgewise so that the angle between them must show in true size.

7.9 LINE IN A PLANE

Before proceeding to the next step in auxiliary plane projection it is necessary to explain the method of constructing a line in a plane. *A straight line will lie in a plane when any two points on the line lie in the plane.* In each of the illustrations of Fig. 7.13 the plane surface is represented as a triangle. In this case each of the bounding lines is a line in the plane. In Fig. 7.13a point M was chosen as a point on AB and N was taken as a point on BC. Line MN must therefore lie in the plane because it connects two points that lie in the plane.

A horizontal line, called an H-parallel, may be constructed in the plane by making the V-projection, e^Vf^V, parallel to the H-V reference line as in Fig. 7.13b. Point E is made to lie on line AB by placing e^H on a^Hb^H. Point F is made to lie on line BC by placing f^H on b^Hc^H. Then since E and F lie in the plane, the line joining them must lie in the plane.

The same reasoning will prove that line EF in Fig. 7.13c may be made a V-parallel in the plane ABC by making e^Hf^H parallel to H-V. It also follows that the line EF in Fig. 7.13d is a P-parallel in plane ABC because the H- and V-projections were made parallel to the V-P reference line and points E and F were made to lie in the plane.

7.10 EDGEWISE VIEW OF A PLANE

A plane figure shows edgewise when projected on an auxiliary plane that is perpendicular to any

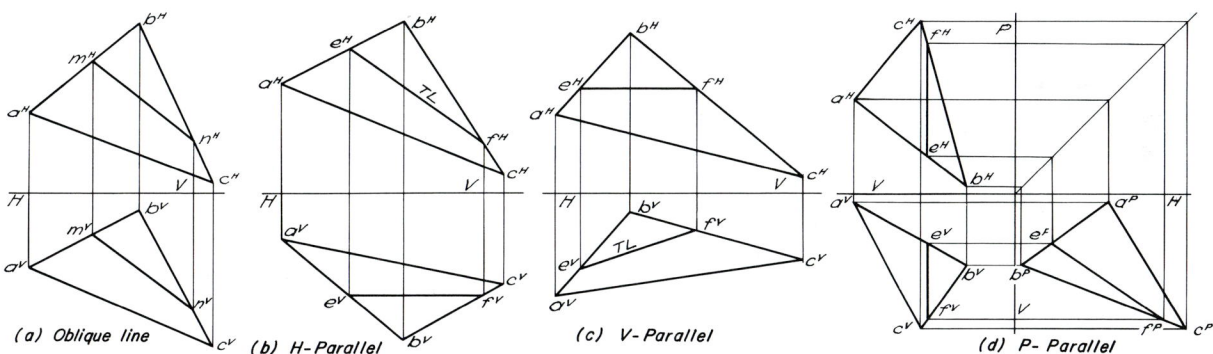

FIG. 7.13. Line in a plane.

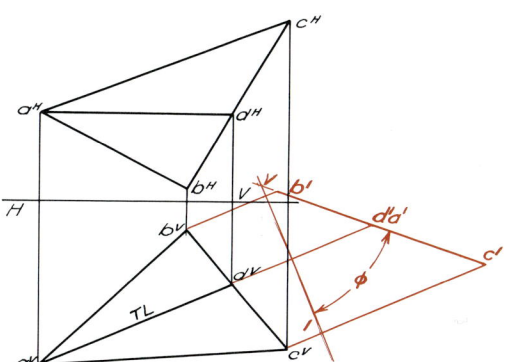

FIG. 7.14. Edgewise view of a plane.

144 AUXILIARY VIEWS

line in the plane. If the line chosen is an oblique line, it will be necessary to use two auxiliary planes as illustrated for line *AB* in Fig. 7.12. However, if an *H*- or *V*-parallel is chosen lying in the plane as shown in Fig. 7.14, only one auxiliary plane is required to obtain the edge view. The edgewise view of the plane is obtained in the following steps.

a. Construct a *V*-parallel in the plane as *AD* in Fig. 7.14.
b. Set up an auxiliary plane perpendicular to *AD* by drawing *V*-1 perpendicular to $a^v d^v$ since $a^v d^v$ is the true length of *AD*.
c. Project points *A*, *B*, and *C* on the first auxiliary plane.
d. The projection $a^1 b^1 c^1$ will be a straight line, which proves that this is the edgewise view.

7.11 TRUE SIZE OF A PLANE FIGURE

To find the true size and shape of a plane figure it is necessary to set up an auxiliary plane parallel to the given plane. In actual practice the views of an object are frequently so arranged that one of the principal views of a face is an edgewise view. In such a situation only one auxiliary plane is needed. When an edgewise view is not given, two auxiliary views are necessary.

7.11.1 Edgewise View Given. In the pyramid shown in Fig. 7.15 each of the inclined planes shows edgewise in either the front view or the side view. In this problem it is desired to find the true size and shape of the face *ACD*. Figure 7.15a shows the pictorial of the object and the planes of projection. Since the face *ACD* shows edgewise in the front view, it is necessary to set up an auxiliary plane perpendicular to the vertical plane and parallel to the face.

This is done in Fig. 7.15b by making the *V*-1 reference line parallel to the edgewise view of the plane. The auxiliary plane has been revolved about the *V*-1 reference line until it coincides with the vertical plane. In these two views, it can be seen that the distance *v* from point *C* to the vertical plane shows in the top view as the distance from c^H to the *H*-*V* reference line, and in the auxiliary view as the distance from c^1 to the

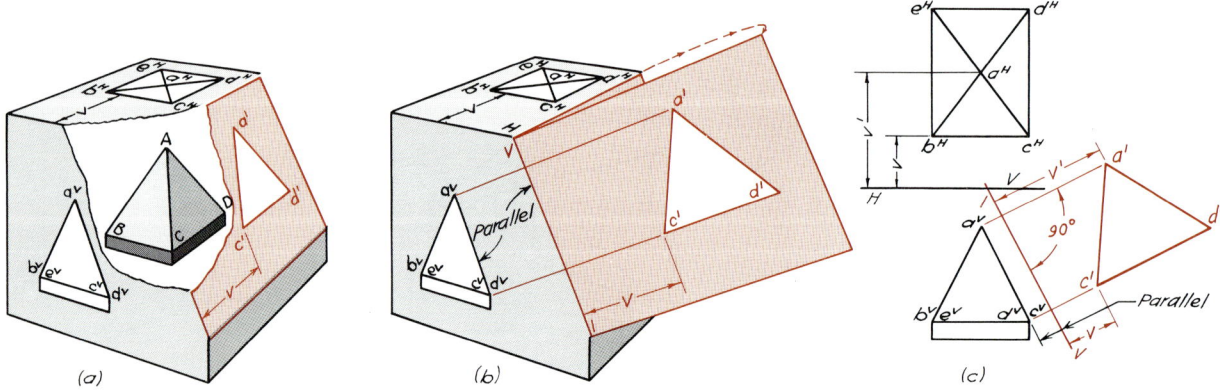

FIG. 7.15. True shape of an inclined face. Auxiliary plane perpendicular to *V*-plane.

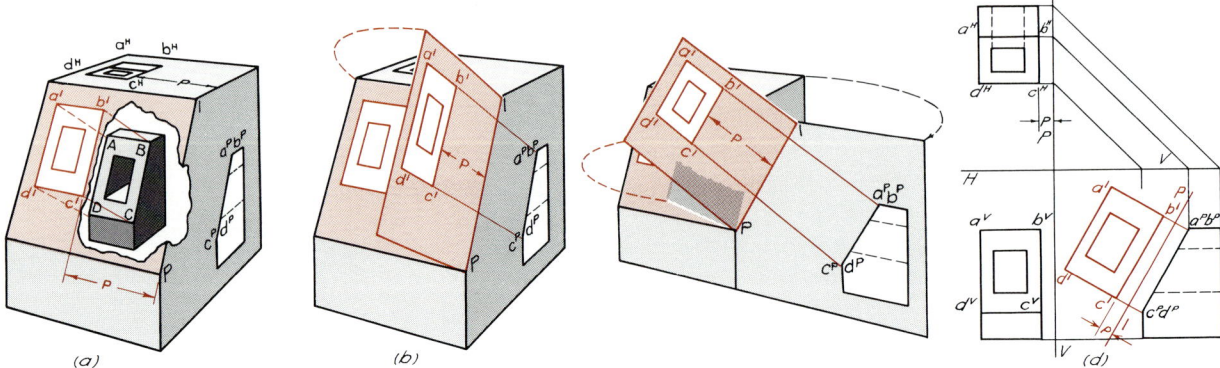

FIG. 7.16. True shape of inclined face. Auxiliary plane perpendicular to *P*-plane.

V-1 reference line. The orthographic projections of this problem are shown in Fig. 7.9c. The steps to be followed to obtain the true size are

a. Place the V-1 reference line parallel to the edgewise view $a^V d^V c^V$.
b. Construct lines through a^V, d^V, and c^V perpendicular to the V-1 reference line.
c. Measure the distance from each of these points to the vertical plane in the horizontal projection. They are marked v and v^1.
d. Lay out these distances on the perpendiculars through a^V, d^V, and c^V. Measurements are laid out from the V-1 reference line to locate a^1, b^1, and c^1. The area $a^1 b^1 c^1$ is the true size and shape of the given triangle.

Another example is shown in Fig. 7.16. In this case the edgewise view is in the profile projection. The auxiliary plane is therefore set up perpendicular to the profile plane and the distances measured are the distances from the points to the profile plane.

7.11.2 Oblique Plane.
When a plane does not show edgewise in any of the principal views, it is necessary to set up the first auxiliary plane so that an edgewise view will be obtained. The second auxiliary plane will then be made parallel to the plane, as explained in the previous article.

Figure 7.17 shows the method of finding the true size of an oblique plane. The following steps are necessary.

a. Construct an H- or V-parallel in the plane. In Fig. 7.17a, a V-parallel CD is used.
b. Set up an auxiliary plane perpendicular to the plane ABC by making the V-1 reference line perpendicular to $c^V d^V$.
c. Project the plane ABC onto plane 1 to obtain the edgewise view $a^1 b^1 c^1$.
d. Set up the second auxiliary plane 2 parallel to plane ABC. This is done by making the 1-2 reference line parallel to $a^1 b^1 c^1$. See Fig. 7.17b.
e. Construct lines through a^1, b^1, and c^1 perpendicular to the 1-2 reference line.
f. Measure the distance that each point is from the first auxiliary plane in the V-projection. This is illustrated for point A in Fig. 7.17b.
g. Lay out these distances on the perpendiculars from a^1, b^1, and c^1. Measurements are made from the 1-2 reference line to determine the projections a^2, b^2, and c^2. This determines the true size of the triangle ABC.

7.12 ELIMINATION OF REFERENCE LINES BETWEEN VIEWS

In all of the foregoing discussion, the planes of projection, represented by reference lines, were used as the reference planes. All measurements were made from these reference planes when constructing the various views. The necessity of having a reference line between adjacent views sometimes causes the views to be spaced too far apart for efficient use of the drawing paper.

In order to avoid this difficulty, a plane of symmetry on the object or some face of the

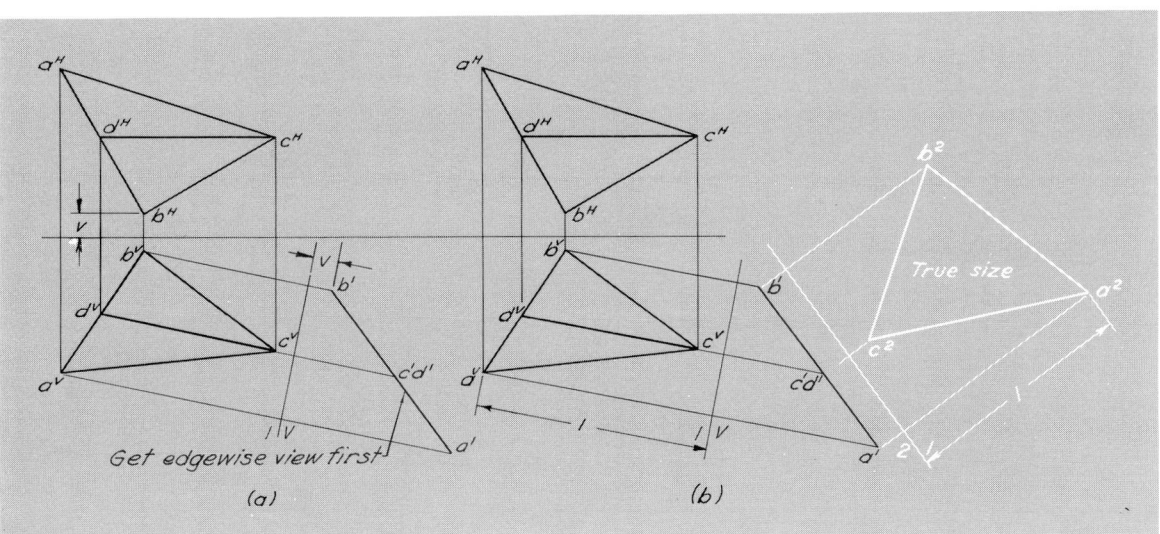

FIG. 7.17. True size of a plane figure by auxiliary projection.

146 AUXILIARY VIEWS

object may be used as a reference. The only condition to be met is that these surfaces be parallel to the plane of projection that would have been used if reference lines had been drawn. These reference planes may be indicated on the drawing if necessary, but frequently it is not considered necessary.

Figure 7.18a shows a drawing having an auxiliary view laid out in the usual manner by the use of reference lines. Projecting lines are perpendicular to the reference lines and the proper distances have been laid out on these projecting lines.

7.12.1 Face of Object Used as a Reference Plane.
In Fig. 7.18b one face of the object has been used as a reference plane. The procedure given below is then followed.

a. Select a face of the object that is parallel to the principal plane on which the inclined face shows edgewise. Use this face as the reference plane. Thus, if the inclined face shows edgewise in the vertical projection, the face used as a reference plane must be parallel to the vertical plane.

b. Construct projecting lines perpendicular to the edgewise view of the inclined face that appears in the front view in this case.

c. Place the line that represents the edgewise view of the reference plane at a convenient distance from the edgewise view. See Fig. 7.18b.

d. Measure distances such as m from the reference plane in the top view.

e. Lay out these distances from the reference plane in the auxiliary view, as shown for the distance m in Fig. 7.18b.

7.12.2 Center Line Used as a Reference Plane.
It is frequently more convenient to use a plane through the center line of a symmetrical object as the reference plane. This has been done in Fig. 7.18c. The auxiliary projection lines are perpendicular to the edgewise view of the inclined face, and all distances are measured and laid out from that center line as indicated for the m distances. The use of the center line is particularly helpful when circles are to be projected, because several points can be laid out with one setting of the dividers.

7.13 USE OF AUXILIARY VIEWS FOR THE CONSTRUCTION OF PRINCIPAL VIEWS

The auxiliary view may be used to advantage when the shape of the face is known and when it may be drawn in true shape in the auxiliary view. This is illustrated in Fig. 7.19. The true shape of the flange may be drawn in the auxiliary view from known dimensions. From the auxiliary view and the adjacent side view, the front view can be constructed. To do this a series of points, similar to A and B can be assumed on the side and auxiliary views, from which they can be projected back to the front view. The following procedures may be used.

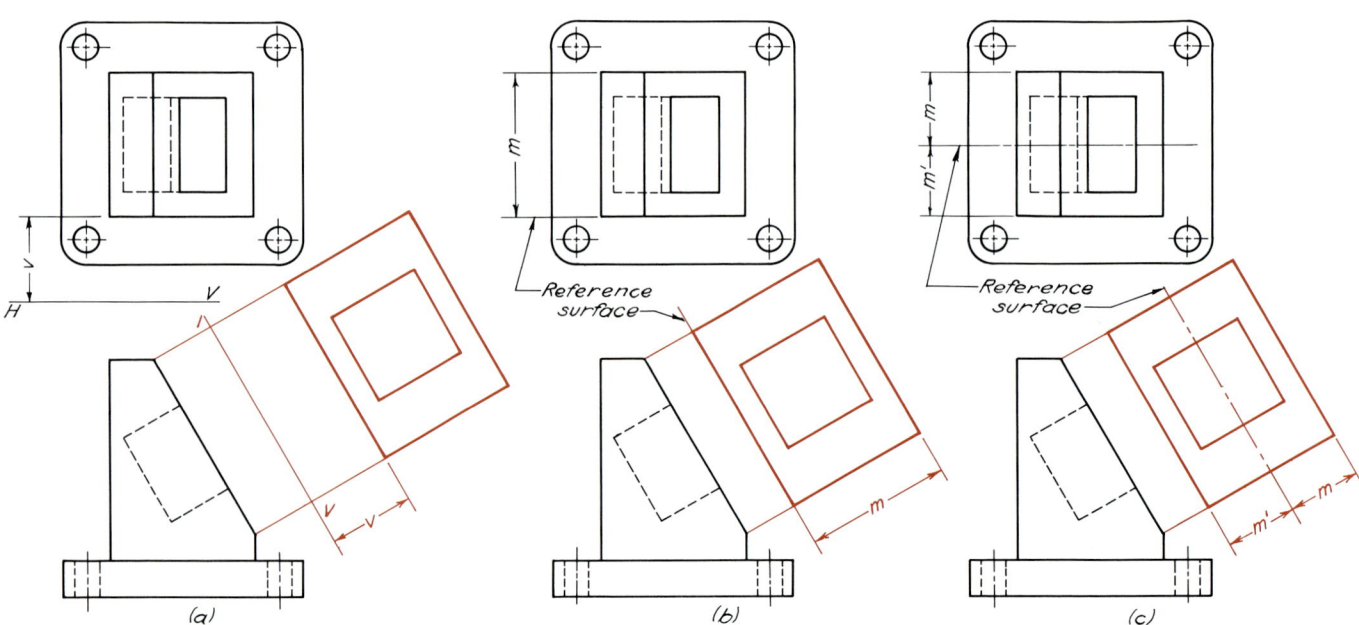

FIG. 7.18. Use of reference planes on an object.

a. Choose any points, one on the upper and one on the lower circles in the auxiliary view, whose projections are a^1 and b^1.

b. From a^1b^1 draw projecting lines perpendicular to the edgewise view of the inclined face in the side view.

c. Mark the positions of a^p and b^p at the place where the perpendiculars cross the edgewise views of the top and bottom faces of the lug.

d. From the positions of a^p and b^p draw horizontal projecting lines to the front view.

e. Measure the distance x from the center line to a^1b^1 in the auxiliary projection.

f. Lay off the distance x on each of the horizontal projecting lines (drawn in step d) on both sides of the center line in the front view. This locates the various positions of a^v and b^v.

g. Repeat the process for as many points on the circle as desired.

h. Connect the points in the front view to form the ellipses.

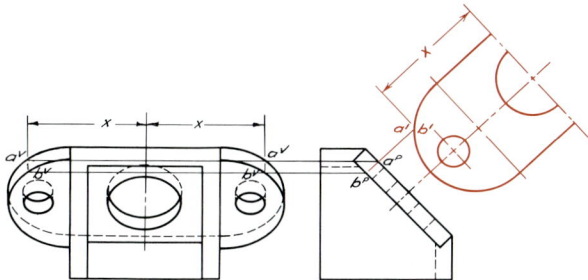

FIG. 7.19. Front view constructed from known auxiliary view.

7.14 ELIMINATION OF MEASUREMENTS IN AUXILIARY PROJECTION

When reference lines are used, it is sometimes convenient to project the points back from the auxiliary view to the principal views without the necessity of measuring any distances. This method is illustrated in Fig. 7.20. The construction may be completed in the following steps.

a. From the point of intersection of the H-V reference line and V-1 reference line, erect a new line to bisect the angle between the two reference lines.

b. Select projections of points on the auxiliary view, such as a^1 and b^1. Construct through them a line parallel to the V-1 reference line.

c. From the point where this line intersects the bisector, construct a line parallel to the H-V reference line.

d. From the auxiliary projections a^1 and b^1, construct projection lines perpendicular to the V-1 reference line. The projections a^v and b^v will be found at the place where this projection line crosses the edgewise views of the top and bottom faces of the angle.

e. From projections a^v and b^v, construct vertical projection lines to locate projections a^H and b^H on the horizontal lines drawn in step c.

f. Repeat the process to obtain as many points as desired. Connect them with a smooth curve to complete the horizontal projection.

7.15 PARTIAL VIEWS

Objects such as that shown in Fig. 7.21 would be rather difficult to draw if the complete projections were to be made. For this reason it is common practice to show one complete view and two partial views. One of these views may be an auxiliary. This simplifies the drawing and makes it easier to read. Since this object is symmetrical,

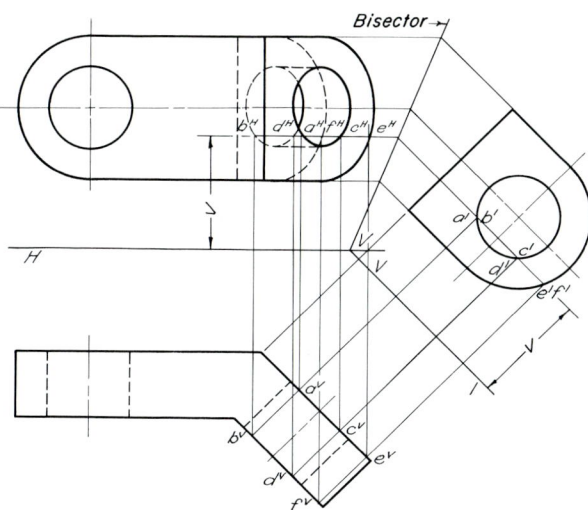

FIG. 7.20. Top view constructed from known shape of auxiliary view.

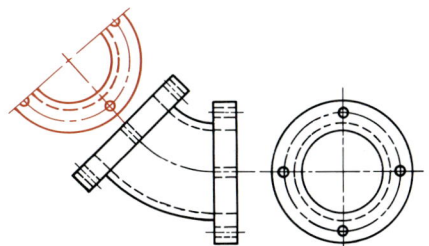

FIG. 7.21. Partial auxiliary view.

only one half of the auxiliary view is drawn. If space is limited, the side view could have been made as a half-view also.

7.16 PROCEDURE IN LAYING OUT A DRAWING HAVING ONE AUXILIARY VIEW

For speed and efficiency in his work, the drafter should develop a regular procedure for laying out a drawing. Figure 7.22 gives a step-by-step layout that may be varied to suit conditions. The object to be drawn is shown pictorially in Fig. 7.22a. The various steps and the order in which they should be taken are listed below.

7.16.1 Draw Front View and Locate Top View. In Fig. 7.22b the following steps are taken.

a. Draw the front view.
b. Draw the auxiliary projection lines perpendicular to the edgewise view of the inclined plane.
c. Locate and outline the top view so that the desired clearance between the top and auxiliary views will be maintained.

7.16.2 Locate Center Line. Figure 7.22c reveals the next two steps.

a. Draw the horizontal center line in the top view.
b. Locate the same center line in the auxiliary view so that the desired spacing between front and auxiliary views will be maintained.

7.16.3 Outline Auxiliary View. Figure 7.22d shows additional construction as follows.

a. Transfer distance a with dividers from the top view to the auxiliary view.
b. Outline the auxiliary view of the inclined face.

7.16.4 Complete Details of Auxiliary View. Figure 7.22e shows the auxiliary view complete in three steps.

a. Locate the center of the circle in the auxiliary view.
b. Draw the circle.
c. Divide the circle into eight (or more) equal parts.

7.16.5 Construct Projection Lines. In Fig. 7.22f the first steps for the completion of front and top views are made.

a. Project the 3-7 center line back to the front view and from there to the top view.
b. Project points 1 and 5, which are on the original center line, back to the front view and from there to the top view.

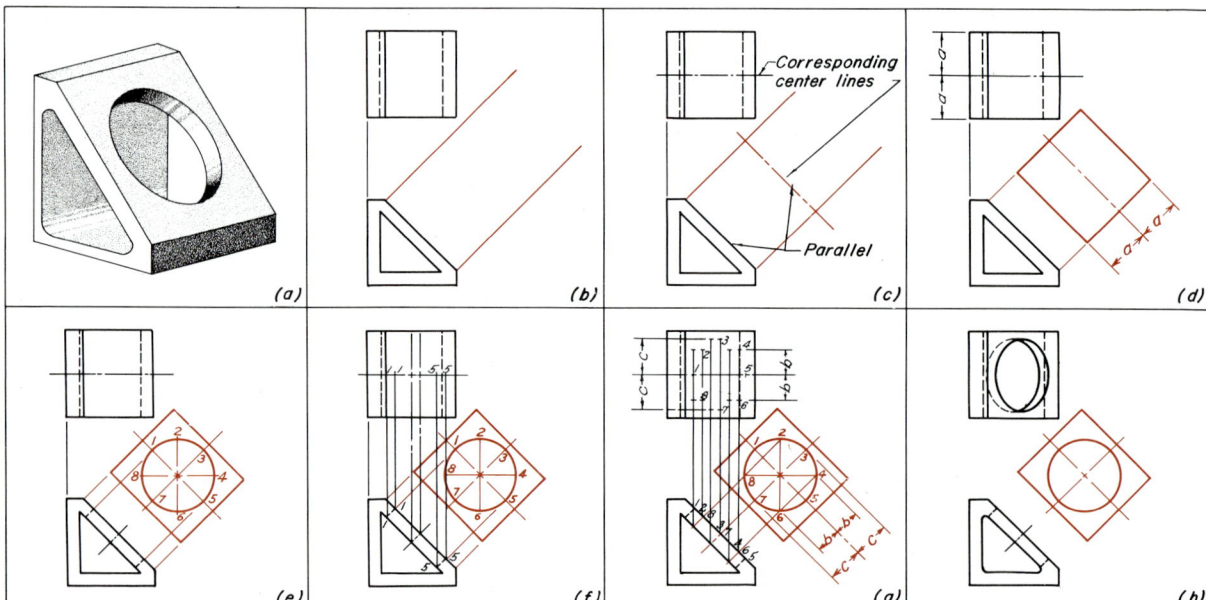

FIG. 7.22. Constructing the views of an object with the aid of an auxiliary view.

7.16.6 Complete All Projections. In Fig. 7.22g the work of projection is completed.

a. Project all points from the circle to the edgewise view of the inclined faces in the front view.
b. Project these points from the front view to the top view.
c. Transfer distances b and c from the auxiliary view to the top view on the proper projection line.

7.16.7 Finish Drawing. Figure 7.22h shows the necessary steps for finishing the drawing.

a. Connect the points with a smooth curve, showing the proper visibility.
b. Clean up the drawing.
c. Work over the lines to give them the proper weight and character.

7.17 PROCEDURE IN LAYING OUT A DRAWING HAVING TWO AUXILIARY VIEWS

In theory the procedure is an extension of that used for a single auxiliary plane. In practice a few difficulties may be encountered that can be cleared up with a step-by-step analysis. The object chosen for this illustration is shown pictorially in Fig. 7.23. Note that the first auxiliary plane, numbered 1, is perpendicular to the horizontal plane and parallel to the cylinder. The second auxiliary plane, numbered 2, is perpendicular to plane 1 and to the axis of the cylinder.

The following step-by-step analysis, illustrated in Fig. 7.24, gives a procedure that may be followed in laying out a drawing of this kind. The steps taken in each part of the figure have been listed in proper order following the figure number.

7.17.1 Locate the Layout of the Front and Top Views. In Fig. 7.24a the following steps are taken.

a. Make top and front views of the main body of the casting.
b. Lay out the center line and edge lines of the supporting web in the top view at the proper angle.

7.17.2 Lay Out First Auxiliary View. Figure 7.24b reveals the next four steps.

a. Place auxiliary reference line H-1 parallel to the top view of the web.
b. Draw auxiliary projection of the nearest face of the bracket.
c. Draw center line of the cylinder at the proper angle and at the proper height according to design specifications.

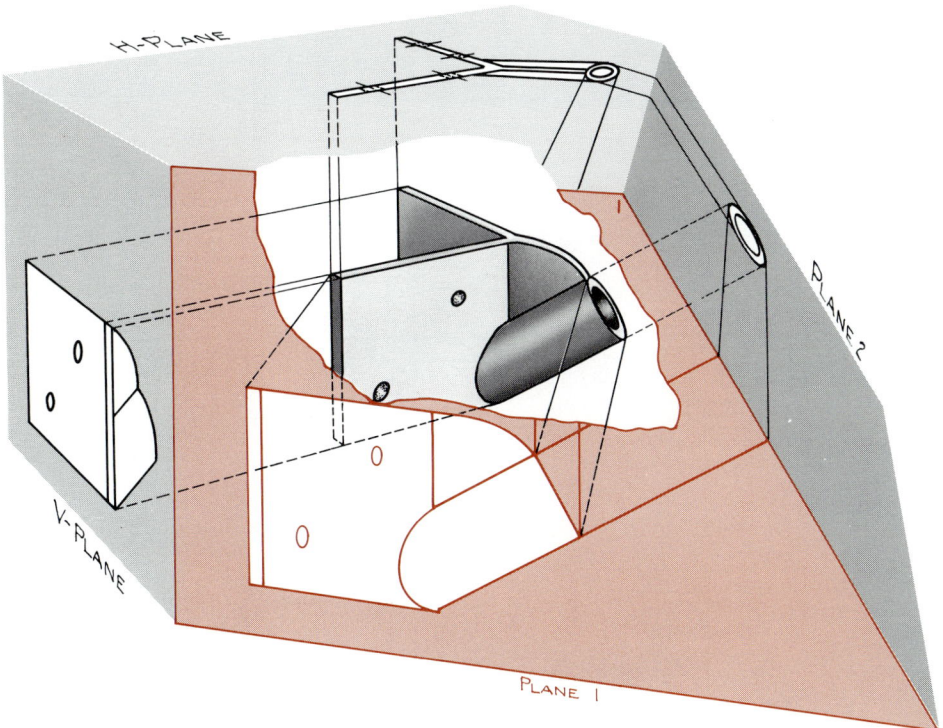

FIG. 7.23. Pictorial drawing of auxiliary planes.

d. Complete the first auxiliary projection of the web by drawing the arc tangent to the surface of the bracket and to the end face of the cylinder.

7.17.3 Draw Second Auxiliary View.
Figure 7.24c shows further auxiliary view construction.

a. Locate the center line marked x^2 parallel to the edgewise view of the end of the cylinder. This is perpendicular to the axis of the cylinder. Place this line at a convenient distance from the cylinder. (Note that this is the same center line as the one marked x^H in the top view.)

b. Extend the center line of the cylinder to form the other center line for the second auxiliary projection.

c. Draw the circles representing the end view of the cylinder.

7.17.4 Begin Completion of Top View.
Figure 7.18d illustrates the transfer of points from the second auxiliary view to the top view.

a. Divide the circles into eight (or more) equal parts.
b. Project these points to the edge view of the end of the cylinder in the first auxiliary projection.
c. Project the points from the first auxiliary view perpendicular to the H-1 reference line.
d. Measure, with dividers, the distances marked a and b in the second auxiliary view and lay them out as indicated in the top view.
e. Connect the points with a smooth curve.

7.17.5 Complete First Auxiliary View.
Figure 7.24e completes the intersection of the cylindrical part in the first auxiliary view.

a. Lay out a portion of the web in the second auxiliary projection.
b. Project the intersection of the web and cylinder from the second auxiliary view to the first auxiliary projection.
c. Draw elements of the cylinder such as 6, 7, and 8 in the top view.
d. Find the points 6^H, 7^H, and 8^H where these elements cut the face of the bracket in the top view.
e. Project these points back to the first auxiliary projection to obtain 6^1, 7^1, 8^1, and so forth.

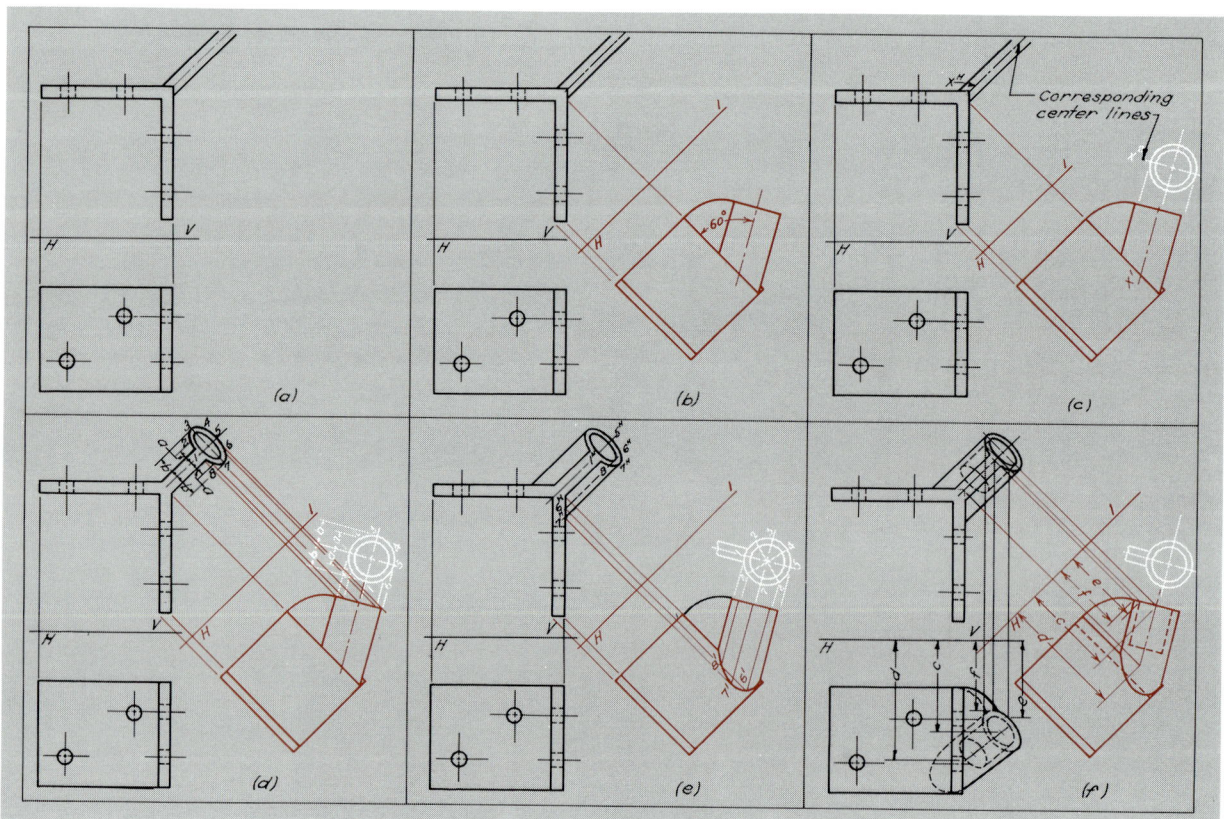

FIG. 7.24. Steps in making complete views of object with the aid of two auxiliary views.

f. Connect these points to form the intersection of the cylinder with the side face of the bracket.

7.17.6 Finish Front View. Figure 7.24f shows steps for completing the cylindrical part of the front view.

  a. Project points from the top view of the end of the cylinder vertically to the front view.
  b. Measure distances such as e and f in the auxiliary view and lay them out as indicated in the front view.
  c. Connect these points to form the ellipses representing the end of the cylinder.
  d. Locate points on the front view of the web in a similar manner.
  e. Locate the intersection of the cylinder with the back face of the bracket and any other necessary lines.
  f. Determine the visibility of all lines. Finally, clear up the drawing and erase all construction lines. Work over all lines, giving them the proper weight and character.

7.18 RELATIONSHIPS BETWEEN POINTS, LINES, AND PLANES

The solution of many design problems requires the understanding of the relationship between points, lines, and planes. The various combinations possible require the use of principal planes of projections as well as auxiliary planes of projections. Some of the more common useful relationships are as follows.

  a. The shortest distance between two points is a straight line.
  b. Two intersecting lines determine a plane.
  c. Two parallel lines determine a plane.
  d. A point and a line determine a plane.
  e. Two intersecting planes determine a line.

7.19 RELATIONSHIP BETWEEN LINES

There are three possible relationships between two lines. Two lines can be parallel to one another, they can intersect, or they can be nonparallel and nonintersecting. In the latter case they are referred to as *skew lines*. Two intersecting lines can be either inclined or perpendicular to each other.

7.19.1 Identifying Parallel Lines. By definition, lines are said to be parallel when they meet at infinity. In projection drawing, two lines are parallel when their respective projections on any and all planes are parallel.

In Fig. 7.25 this principle has been illustrated for four projections of the lines. By examining Figs. 7.26 and 7.27 it will be seen that when two projections of a set of lines are both parallel to the same reference line, the lines may or may not be

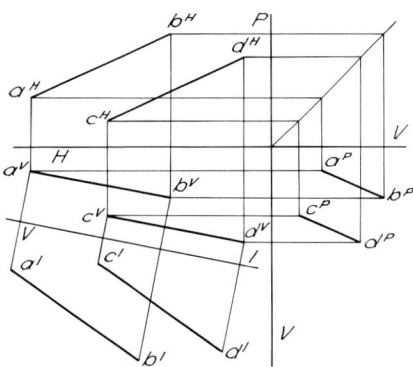

Corresponding projections parallel

FIG. 7.25. Parallel lines.

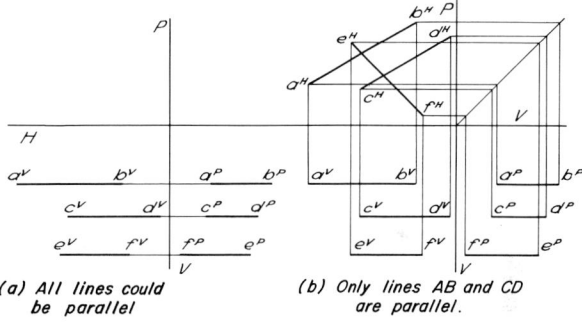

(a) All lines could be parallel

(b) Only lines AB and CD are parallel.

FIG. 7.26. Identifying parallel lines when two views are inadequate.

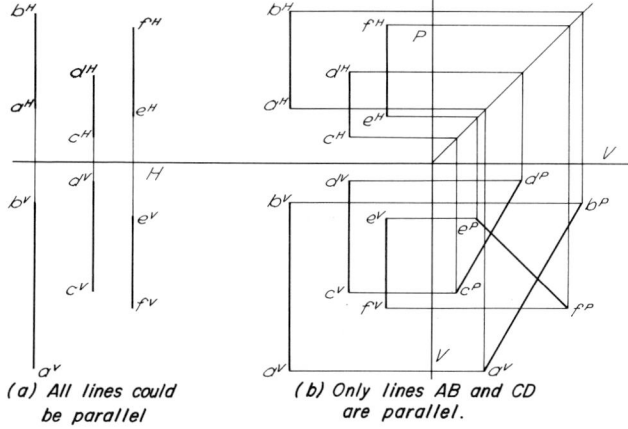

(a) All lines could be parallel

(b) Only lines AB and CD are parallel.

FIG. 7.27. Identifying parallel lines. Two views inadequate.

parallel. In these cases a third view is necessary, as shown in Figs. 7.26b and 7.27b. In both illustrations it can be seen that the line EF is not parallel to the other two lines.

7.19.2 Drawing a Line through a Point Parallel to Another Line. This problem occurs regularly in solving stress problems by means of vectors. Thus to draw a line through C in Fig. 7.28a it is necessary to draw the H-projection $c^H d^H$ parallel to $a^H b^H$ and the V-projection $c^V d^V$ parallel to $a^V b^V$.

As can be seen from Figs. 7.26 and 7.27, two projections will suffice except where the adjacent projections are both perpendicular to the reference line between them. In this situation three views will be necessary.

7.19.3 Intersecting Lines. Two intersecting lines intersect at a point. They also form a plane. The point of intersection must be in projection in the various adjacent views, as point O in Fig. 7.29a. In Fig. 7.29b the crossing point of lines AB and CD in the horizontal projection appear to intersect at point EF. However, when E and F are projected upon the respective vertical projections of lines AB and CD, they do not coincide. Therefore, lines AB and CD are not intersecting lines.

To obtain the true size of the angles between two intersecting lines, they must be projected upon a plane of projection that is parallel to them. When the intersecting lines are parallel to the H, V, or P planes, the true size of the angle will appear on the H, V, or P projections, respectively. When the two lines lie in a plane perpendicular to H, V, and P, the angle can be obtained as discussed in Article 7.11.

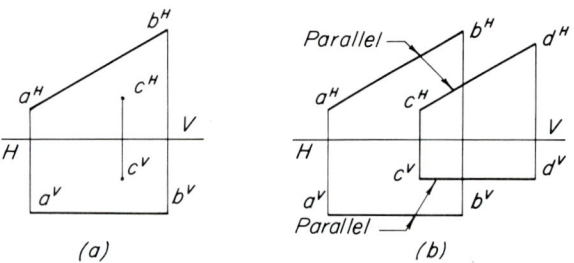

FIG. 7.28. Line through a point parallel to a given line.

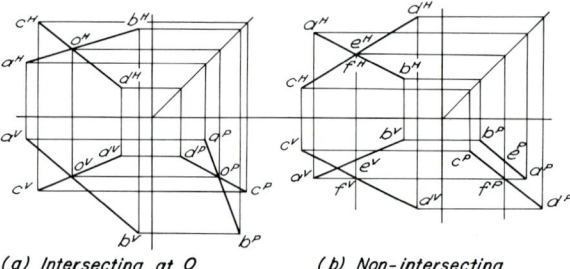

FIG. 7.29. Intersecting lines.

7.19.4 Line through a Point Making a Given Angle with a Given Line. A line and a point determine a plane. Therefore, to perform the necessary construction, the true size of the plane must be obtained by passing a plane parallel to a plane containing the line and point. In Fig. 7.30, line CD and point A are given. The easiest way to obtain the edgeview of the plane containing line CD and point A is to draw a V-parallel from a^H intersecting the line CD at point e^H and project point E to the vertical projections at e^V. AE and CD now form a plane and since AE appears in true length in the V-projection, the first auxiliary plane

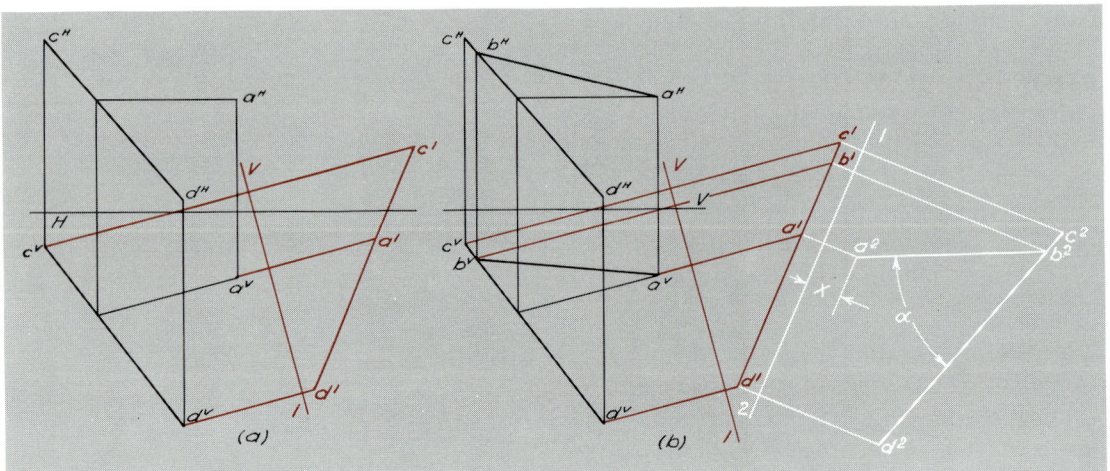

FIG. 7.30. Line making specified angle with another line.

is passed perpendicular to a^Ve^V. The point and line are then projected as an edgewise $a^1e^1c^1d^1$ as shown in Fig. 7.30a.

A second auxiliary plane is passed parallel to $a^1e^1c^1d^1$ and the point and line are projected upon it as in Fig. 7.30b to obtain $a^2c^2d^2$. Through a^2 the required α is constructed intersecting line c^2d^2 at point b^2. Point B is then projected back to b^1, b^V, and b^H, respectively, as is shown in Fig. 7.30.

7.19.5 Distance between Two Parallel Lines by the Point Projection Method. *The distance between the point projections of the two lines will be the shortest distance between the lines.*

Project both lines on an auxiliary plane perpendicular to the lines and measure the distance between the point projections, as illustrated in Fig. 7.31. Proceed as follows:

a. Set up the first auxiliary plane with reference line V-1 parallel to the front view of the lines.

b. Obtain the first auxiliary view of the two lines at a^1b^1 and c^1d^1 in Fig. 7.31a. These are true-length views.

c. Set up the second auxiliary plane with reference line 1-2 perpendicular to the true-length projections of the lines as in Fig. 7.31b and project them onto the second auxiliary plane to obtain a^2b^2 and c^2d^2. The distance between these points is the distance between the lines.

7.19.6 Distance between Two Parallel Lines by the Plane Method. *The true distance between the lines will show in that projection, which gives the true size of the plane of the two lines.* The procedure is as follows.

a. Draw a V-parallel AE between the two lines as in Fig. 7.32a and obtain the edge view of the plane of the lines.

b. Draw the second auxiliary plane with reference line 1-2 parallel to the first auxiliary view of the two lines and obtain the true shape of the plane.

c. The true distance can be measured anywhere between c^2d^2 and a^2b^2 perpendicular to them, as in Fig. 7.32b.

7.19.7 Distance from a Point to a Line by the Point Projection Method. *The shortest distance from a point to a line will show on a view that gives the point projection of the line.*

a. *Line Parallel to a Plane of Projection.* For a line like AB in Fig. 7.33 parallel to H-plane, the true length appears in the top view and an auxiliary plane reference line H-1 can be set up perpendicular to the line at once. The distance between c^1 and a^1b^1 is the true distance of the point from the line. Since c^1d^1 is perpendicular to a^1b^1 and shows in true length, the projection c^Hd^H must be parallel to the reference line H-1.

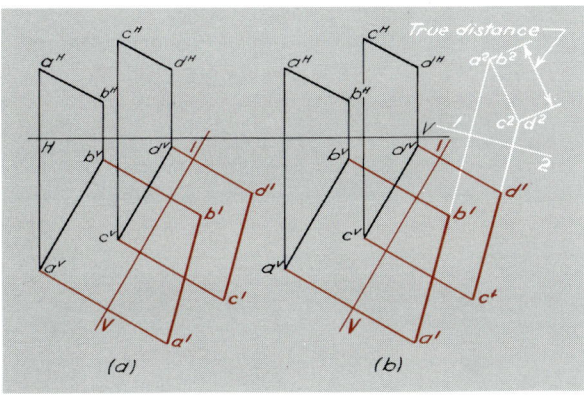

FIG. 7.31. Distance between two parallel lines: point projection method.

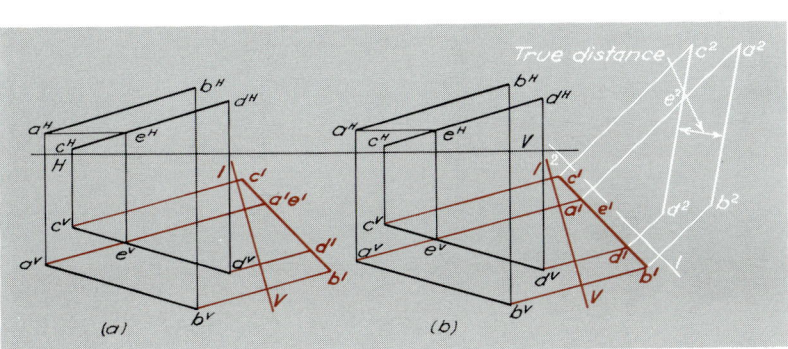

FIG. 7.32. Distance between two parallel lines: plane method.

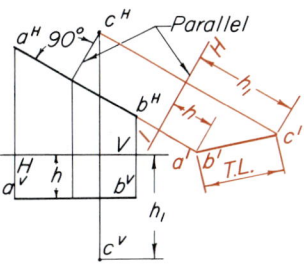

FIG. 7.33. Distance from a point to a line.

b. *Oblique Line.* For an oblique line like CD in Fig. 7.34 the first step is to find a true-length view of the line. This is obtained on a first auxiliary plane parallel to $c^V d^V$, represented by reference line V-1. Carry the point A along in the auxiliary view.

The second step is accomplished by obtaining a point projection of the line on the second auxiliary plane at $c^2 d^2$ in Fig. 7.34. The distance from a^2 to $c^2 d^2$ is the required distance.

Since the shortest distance is the perpendicular from A to CD, $a^1 b^1$ must be at right angles to $c^1 d^1$ or parallel to reference line 1-2. The H- and V-projections of B are carried back by direct projection to b^V and b^H.

7.19.8 Distance from a Point to a Line by the Plane Method. *The true distance between the point and the line will show on a view that gives the true size of the plane of the point and line.*

Project the point and line on an auxiliary plane parallel to the plane of the point and the line. The true length of the perpendicular may be measured on this view and projected back directly. See Fig. 7.35.

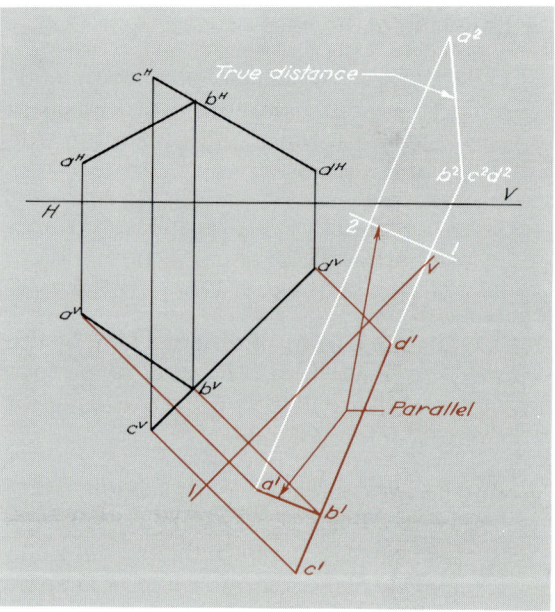

FIG. 7.34. Distance from a point to a line: point projection method.

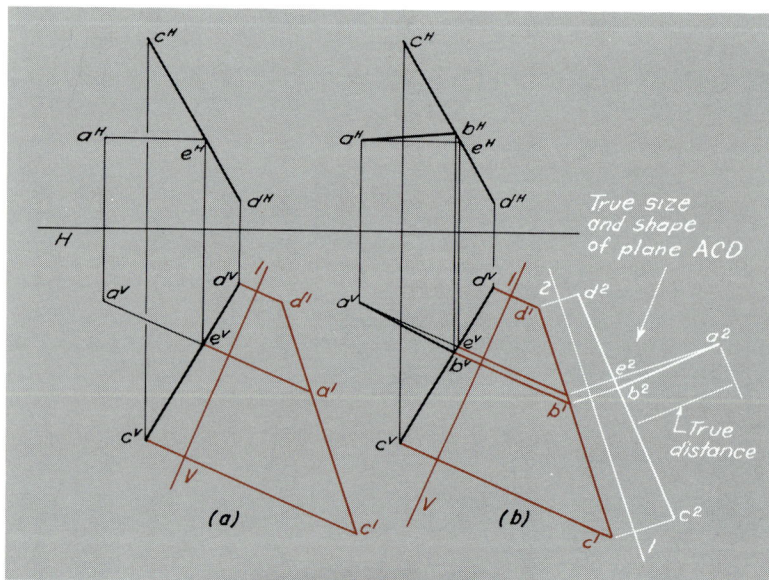

FIG. 7.35. Distance from a point to a line: plane method.

7.19.9 Revolve a Point Around an H- or V-Parallel. When a point is revolved about a line, it revolves in a circle that lies in a plane perpendicular to the line and the center of the circle is a point on the line. It is sometimes very useful to revolve a point around a line into some specified position. In Fig. 7.36 line AB is an H-parallel and point C is to be revolved about AB until AB and C are in a plane parallel to the H-plane. The following steps will solve the problem.

a. Set up an auxiliary plane perpendicular to AB, by making the reference line perpendicular to $a^H b^H$.
b. Find $a^1 b^1$ and c^1 in the auxiliary view.
c. With $a^1 b^1$ as a center revolve c^1 until it is in the plane through $a^1 b^1$ parallel to H at C_r^1. See Fig. 7.36a.
d. In the top view draw the edge view of a plane through C^H perpendicular to $a^H b^H$, the axis of rotation.
e. Project C_r^1 back to C_r^H, and then to C_r^V as shown in Fig. 7.36a.

7.19.10 Revolve a Point Around a Line into the H- and V-Planes. In the situation shown in Fig. 7.36a the rotation of point C into the H- and V-planes can be accomplished by continuing the rotation into the edge view of the H-plane as shown in Fig. 7.36b. The revolved point may fall at either of two places as shown. (C_{rh}^1)

In Fig. 7.36c the intersection of the plane of rotation with the V-plane, which is a vertical line, has been shown in the first auxiliary view, and the circle of rotation is simply continued until C^1 strikes this line in the V-plane at two places (Crv). Projection of the points can be returned to H- and V-projections in the usual manner.

9.19.11 Perpendicular Lines. One line is perpendicular to another line when it lies in a plane perpendicular to the other line. A line constructed perpendicular to another line may have an infinite number of positions. Thus, in Fig. 7.37a, each spoke of the wheels lies in a plane perpendicular to the axle joining them. When the axle is horizontal, the wheels will project edgewise on the horizontal plane. This edgewise view is perpendicular to the horizontal projection of the axle. It follows therefore that *if one of two perpendicular lines is parallel to a principal plane, the projections of both lines on that plane will be at 90° to each other.* Thus, in Fig. 7.37b, the axle AB is parallel to the horizontal plane, and every spoke of the wheel projects in the line $c^H a^H d^H$. Since the wheel may be revolved into any position, the vertical projection of a spoke may have any direction, for example, $a^V e^V$ in Figs. 7.37b and c. If the length of the perpendicular is to be determined or constructed, this may be done by a single auxiliary projection as shown in Fig. 7.38.

Although the preceding discussion has been limited to the horizontal plane, the same principles apply to any principal plane or any auxiliary plane. See Figs. 7.39 and 7.40.

If the given line AB in Fig. 7.40, to which another is to be made perpendicular, is not parallel to a principal plane, an auxiliary plane may be

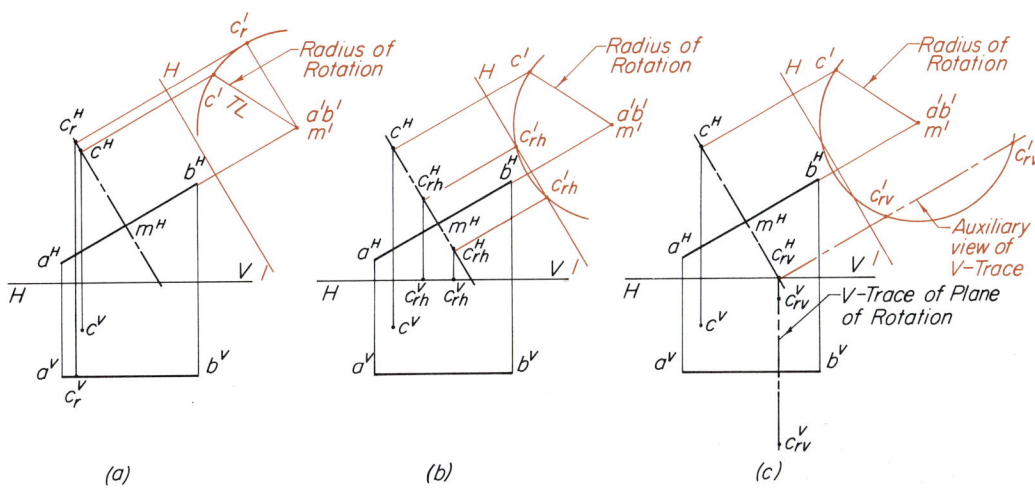

FIG. 7.36. Rotation of a point around a line into the H- and V-planes.

156 AUXILIARY VIEWS

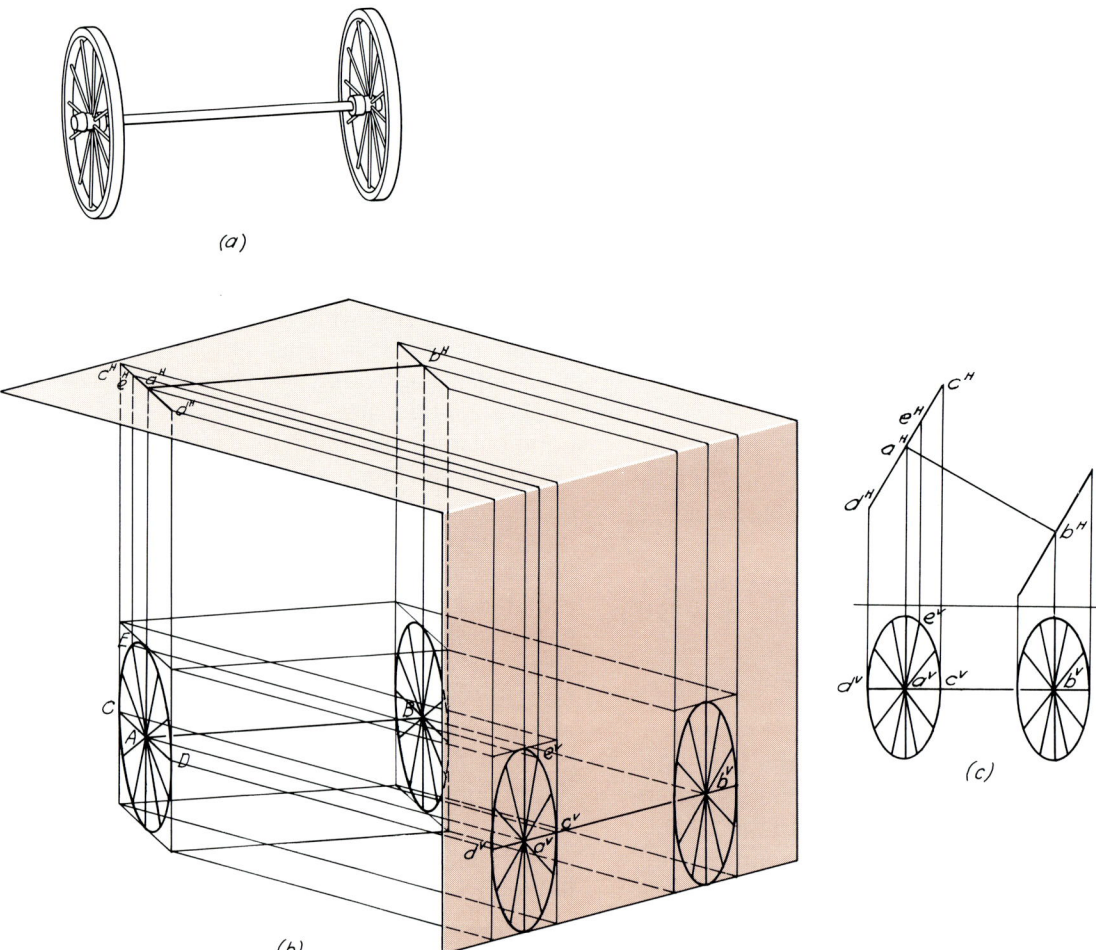

FIG. 7.37. Perpendicular lines.

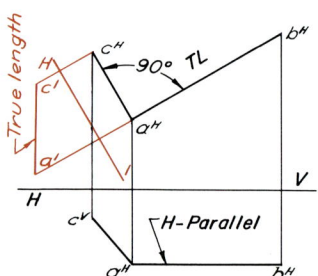

FIG. 7.38. Perpendicular lines, one an H-parallel.

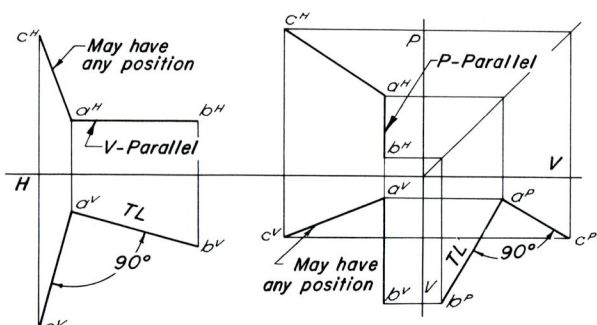

FIG. 7.39. Perpendicular lines, V- and P-parallels.

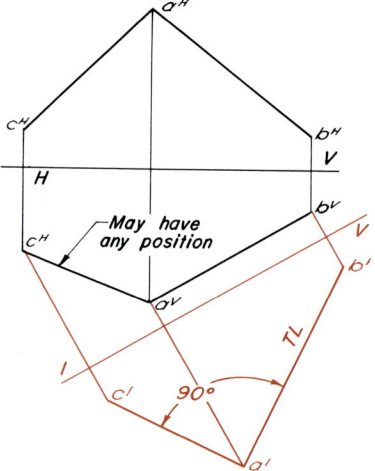

FIG. 7.40. Perpendicular lines.

set up parallel to the line. The auxiliary view will then show the right angle in true size, as shown in Fig. 7.40. After the auxiliary projections have been drawn at right angles to each other, the vertical projection of the perpendicular may be drawn in any direction. The horizontal projection is then obtained in the usual manner.

There are three situations in which a line may be drawn at right angles to an oblique line directly. This construction is useful in many problems. Thus, if we consider AC in Fig. 7.41a to be any spoke of the wheel in Fig. 7.37, the axle may be drawn by making $a^H b^H$ perpendicular to $a^H c^H$ and $a^V b^V$ parallel to the reference line. Lines BA and AC are then at right angles to each other. Figure 7.41b shows another solution in which the axle AB is drawn parallel to the vertical plane, and Fig. 7.41c shows a third solution with the axle parallel to the profile plane.

7.20 VECTOR DIAGRAMS

A vector is a line used to represent any quantity that has both magnitude and direction, for example, a force. The direction of the line is the same as the direction of the force, and the length of the line is equal to the magnitude of the force to some convenient scale. Vector diagrams can be used to solve graphically problems involving vector quantities. The geometry involved is an application of the principal of parallel lines. See Chapter 17 for a more complete treatment of vectors.

7.21 CONCURRENT COPLANAR FORCE SYSTEMS

If there are several forces acting on a body, the vector sum of these forces, a single force that would produce the same effect as the combined forces, is called the *resultant* of the forces. A force of equal magnitude and in the opposite direction is called the *equilibrant*; in other words, it is the force that will hold the whole system in equilibrium.

7.21.1 Forces in a Plane Parallel to the V-Plane. In Fig. 7.42 let it be required to construct the vector diagram to determine the load in the two ropes supporting the 100-lb weight.

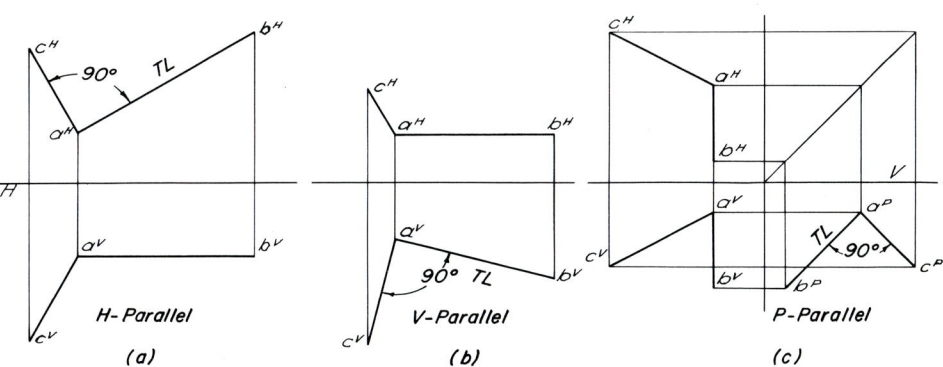

FIG. 7.41. Perpendicular lines.

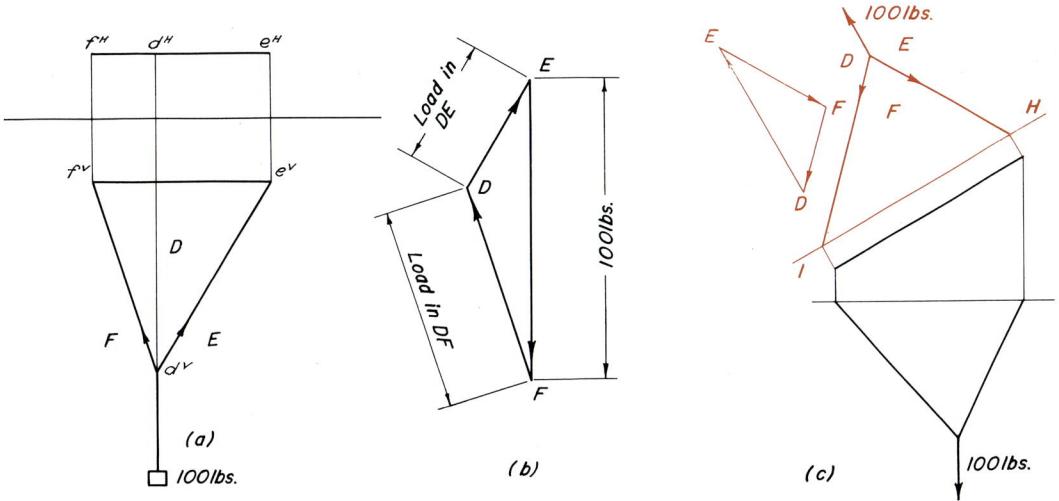

FIG. 7.42. Vector diagram.

Since the ropes are parallel to the vertical plane, the vertical projection will show all lines in true length. Hence the horizontal projection that is parallel to the reference line need not be considered. In vector diagrams it is customary to letter the spaces between the forces as D, E, and F, shown in Fig. 7.42a. The lines of the vector diagram are then designated by the two letters adjacent to them as DE, EF, and FD. This is known as *Bow's notation*.

To solve the problem proceed as follows.

a. In Fig. 7.42b draw a line EF parallel to the vertical rope and make the length equal to 100 lb to some convenient scale, for example, 1 in. = 50 lb.
b. From end E draw a line parallel to $e^v d^v$ of indefinite length.
c. From F draw another line parallel to $d^v f^v$.
d. These two lines will intersect at D, and the true length of lines ED and FD will give the load in the inclined ropes at the same scale used in drawing the first line.

7.21.2 Forces in Oblique Positions. When the ropes are not parallel to one of the principal planes, as in Fig. 7.42c, the same construction may be used by setting up an auxiliary plane parallel to the two ropes, as shown in Fig. 7.42c. The auxiliary view will show all lines and angles in the true size and length. Using the auxiliary view the steps are exactly the same as in the preceding problem.

7.21.3 Three Concurrent Noncoplanar Forces. Frequently it is necessary to find the resultant of several forces that act through one point but do not all lie in the same plane. These are called *concurrent noncoplanar forces*. The resultant is the single force that can be used to replace the entire system of forces. This problem can be solved by constructing a diagram similar to the vector diagram of the previous paragraph, with the exception that it will not form a closed polygon and that it *must be constructed simultaneously in two views*. The line necessary to close the polygon will be the resultant. The equilibrant is a force of the same magnitude but with the direction reversed. Thus, in Fig. 7.43a, a group of three forces whose directions are indicated by lines AB, AC, and AD and whose magnitudes are specified in pounds are acting on point A. The solution of this problem is shown in Fig. 7.43. The spaces between the forces should first be lettered in one view as O, M, and N in Figs. 7.43a and b. The forces are laid out in order by proceeding in a counterclockwise direction around point A.

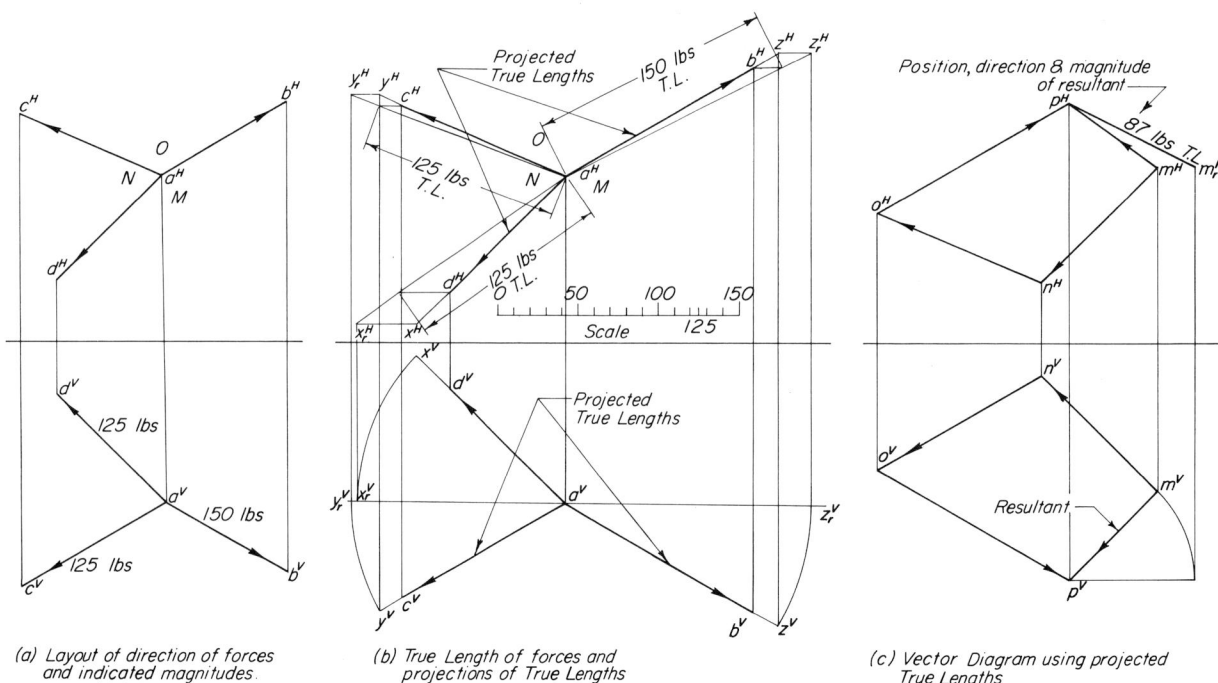

(a) Layout of direction of forces and indicated magnitudes. Not to scale

(b) True Length of forces and projections of True Lengths

(c) Vector Diagram using projected True Lengths

FIG. 7.43. Three-dimensional vector diagram.

a. *Find the True-Length Projections of the Forces.* The simplest method of procedure is to lay out the forces in proper direction and then to obtain the true-length projections as in Fig. 7.43b. Proceed as follows.
  1. Lay out lines AX, AY, and AZ in the proper direction of random length, a little longer than the true lengths, as shown in Fig. 7.43b.
  2. Find the true length of each line by rotation at $a^H x_r^H$, $a^H y_r^H$, and $a^H z_r^H$.
  3. Measure off the required true length for each force as shown in the top view of Fig. 7.43b.
  4. Return this true length to the original projections.
  5. These projected true lengths may then be used directly in the vector diagram of Fig. 7.43c.
b. *Draw the Vector Diagram.* The vector diagram may be drawn by taking the forces in consecutive order either clockwise or counterclockwise. In Fig. 7.43c they have been taken counterclockwise. The forces will be designated by the capital letters between which they lie. Proceed as follows (see top view in Fig. 7.43c).
  1. Begin with force line AD, which will be marked MN since it lies between these letters. Select the projections of points m^H and m^V so that there will be ample room for the diagram. Then draw $m^H n^H$ and $m^V n^V$ of the same length as $a^H d^H$ and $a^V d^V$, the projected true length of this force, from Fig. 7.43b.
  2. Draw NO in the same manner.
  3. Draw from O the force AB parallel and equal to the true-length projections in Fig. 7.43b and letter the end point P.
  4. Line MP with the arrow as indicated is the resultant of this force system.
  5. Find the true length of this line, which scales 87 lb.

7.22 RELATIONSHIP OF PLANES

The relationship of inclined and oblique planes to the principal planes has been presented in Chapter 6 and earlier in this chapter. Further relationships of planes in space such as parallel, perpendicular, and those at specified angles with each other are presented in the following paragraphs.

7.22.1 True Shape of a Plane Figure by Rotation. *If a horizontal line is drawn in a plane, the plane may be revolved about this line as an axis until the plane is parallel to the H-plane. In this position it may be projected on the H-plane in true shape.*

The steps are as follows and as shown in Fig. 7.44.

  a. In the plane ABCDE draw an H-parallel such as AF.
  b. Place an auxiliary plane with reference line H-1 perpendicular to $a^H f^H$ and get the edge view of the plane upon it at $b^1 c^1 a^1 d^1 e^1$.
  c. In the auxiliary view revolve the corners of the plane until they are in a plane parallel to H, as at e_r^1 to b_r^1.
  d. Project the revolved position back to the H-projection and connect the points e_r^H, d_r^H, and so forth, to form the true shape of the figure.

See Article 7.19.9 for method of revolving a point about a line.

It should be noted that a V-parallel could also be used as the axis of rotation, thus bringing the true-shape figure in the vertical plane.

7.22.2 Construction of a Plane Figure in a Specified Plane. A plane is determined in a number of ways. The more common methods are those by two intersecting lines or two parallel lines, to which all the other methods such as that by three points can be reduced. In the plane determined by lines AB and AC in Fig. 7.45, let it be required to construct a square of a given size with one side parallel to AC and $\frac{1}{2}$ in. from it. The

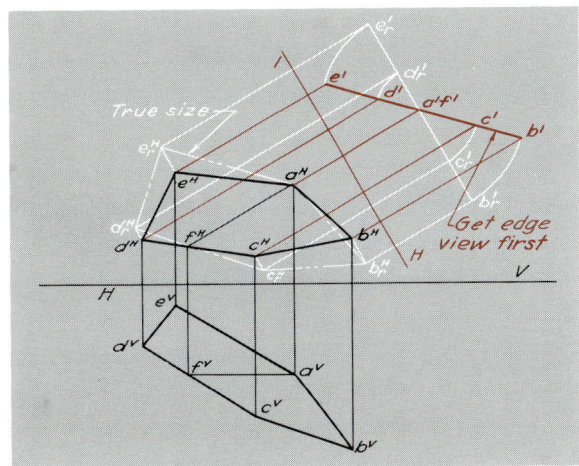

FIG. 7.44. True size of a plane figure by rotation.

first step is to find the true value of the angle between the lines *AB* and *AC* by means of two auxiliary views, that is, the true size and shape of the plane *ABC*, as shown in Fig. 7.45*b*. In the second auxiliary view the square may be drawn in its proper size and position. The projections are then carried back to the horizontal and vertical projections in the usual manner, as shown in Fig. 7.45*c*.

This problem could also be solved by the rotation method.

7.22.3 Angle That a Plane Makes with the H-, V-, and P-Planes. *If the edge view of both planes is obtained in any single view, the angle between the planes shows in true value in this view.*

Two situations arise for the problem. In the first situation the planes are inclined to two of the planes of projection and are perpendicular to the third and the second situation, when the planes are oblique to all three planes of projection.

7.22.4 Inclined Planes. This situation is discussed in Articles 6.18.1, 6.18.2, and 6.18.3 of Chapter 6.

7.22.5 Oblique Planes. The angles θ, ϕ, and π for oblique planes may be found as follows.

a. *Angle with the H-plane.* In Fig. 7.46 it is required to find the angle θ that the plane *ABC* makes with the *H*-plane. This can be done in the following manner.
  1. Draw an *H*-parallel *BD* in the plane.
  2. Obtain an end view of line *BD* and also the edge view of the plane on the auxiliary plane 1.
  3. The true value of θ shows between the reference line *H*-1 and the edge view of the plane.

b. *Angle with the V-plane.* For the angle ϕ between a given plane *ABC* and the *V*-plane proceed in the same manner as in (a) above, using a *V*-parallel as shown in Fig. 7.47.

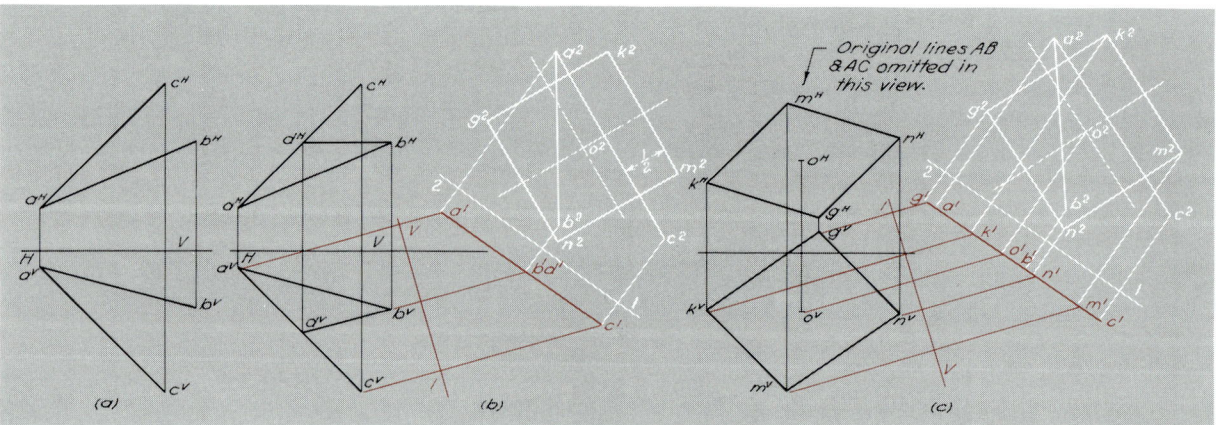

FIG. 7.45. Construction of a plane figure in a specified position.

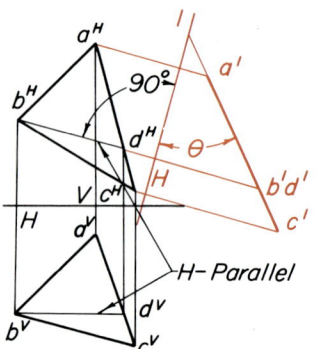

FIG. 7.46. Angle θ of an oblique plane with the *H*-plane.

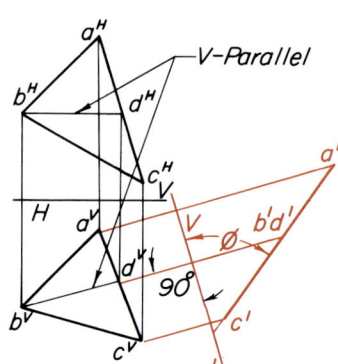

FIG. 7.47. Angle ϕ of an oblique plane with the *V*-plane.

c. *Angle with the P-plane.* For the angle π between a given plane *ABC* and the *P*-plane proceed in the same manner as in (a) above, using a *P*-parallel and as shown in Fig. 7.48.

7.22.6 Strike of a Plane. *The strike of a plane is defined as the bearing of a horizontal line in the plane.* Hence, to obtain the strike of any plane, it is only necessary to draw an *H*-parallel in the plane and measure the angle that its *H*-projection makes with a north and south line. This is illustrated in Fig. 7.49. The strike of a plane together with the *dip* is used in geology and mining to describe the position of a vein of ore, coal, or rock that may occur in the form of planar beds.

7.22.7 Dip of a Plane. *Since the slope of a plane is defined as the angle that the plane makes with the horizontal plane, the true value of this angle will show in a plane perpendicular to any horizontal line in the plane.*

In geology and mining problems the slope, in a downward direction, is referred to as the *dip*. Dip is expressed as an angle and a general direction, such as 46° NW. When the plane appears edgewise in either the front or side views, the slope or dip may be measured directly as the angle between the plane and the horizontal reference line, as shown in Figs. 7.50a and b.

For an oblique plane, the dip may be obtained by drawing an *H*-parallel in the plane as in Fig. 7.50c and then finding the edge view on an auxiliary plane perpendicular to the *H*-parallel as in Fig. 7.50c. The angle between the edge view and the auxiliary reference line is the dip. The method of showing the dip on a map is illustrated in Fig. 7.50c and consists of an arrow drawn in the horizontal projection perpendicular to the strike line, pointing down the slope. The angle of the dip is placed near the arrow. The complete symbol for strike and dip is shown in Fig. 7.51.

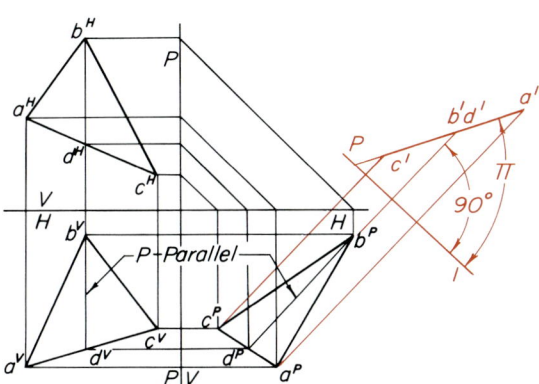

FIG. 7.48. Angle π of an oblique plane with the *P*-plane.

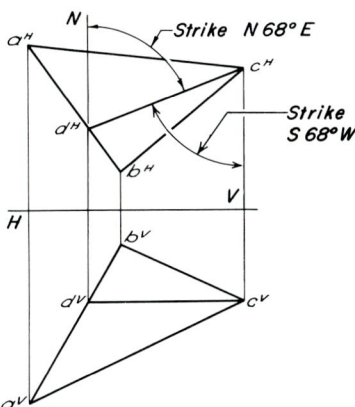

FIG. 7.49. Strike of a plane.

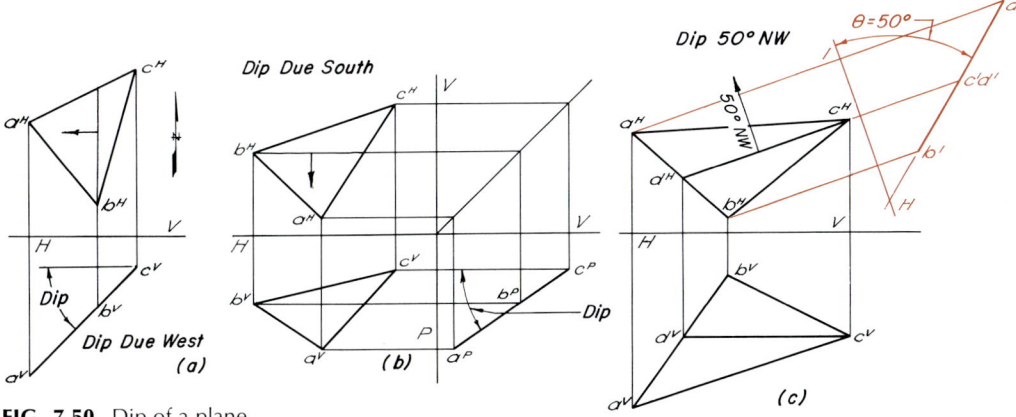

FIG. 7.50. Dip of a plane.

7.22.8 Plane through a Point with Specified Strike and Dip. Let it be required to represent a plane through point A in Fig. 7.51 having a strike of N 60° W and a dip of 30° NE.

a. Through point A draw a horizontal line having a strike of N 60° W as shown in Fig. 7.51b.

b. Draw an auxiliary plane 1 perpendicular to $a^H b^H$, and find the point projection of AB at $a^1 b^1$ as shown in Fig. 7.51c.

c. Through $a^1 b^1$ draw the edge view of the plane making an angle of 30° downward from the horizontal plane in a northeasterly direction. This determines the plane that is identified by the notes along the strike and dip lines in the top view.

d. To give greater usefulness to the top and front views, a point c^1 may be located in the edge view of the plane and then returned to the top and front views.

e. The plane could also be further identified by extending the edge view to the H-plane and drawing the outcrop line, that is, the line of intersection with the H-plane. This line has not been drawn.

7.22.9 Elevations of Points in the Plane. Elevations in the plane ABC in Fig. 7.51d can be obtained by measuring perpendicular to the 1-H reference line as shown in Fig. 7.51d. These elevations will be relative to that of the H-plane.

7.22.10 Distance between Two Parallel Planes. *The distance between two parallel planes can be measured directly between the edge view of the two planes.*

The edge view of a plane can be obtained on an auxiliary plane placed perpendicular to any line in the plane. This could be cumbersome and may require two auxiliary views. If a V-parallel is chosen as the line, one auxiliary view will solve the problem as in Fig. 7.52. An H-parallel would be equally good.

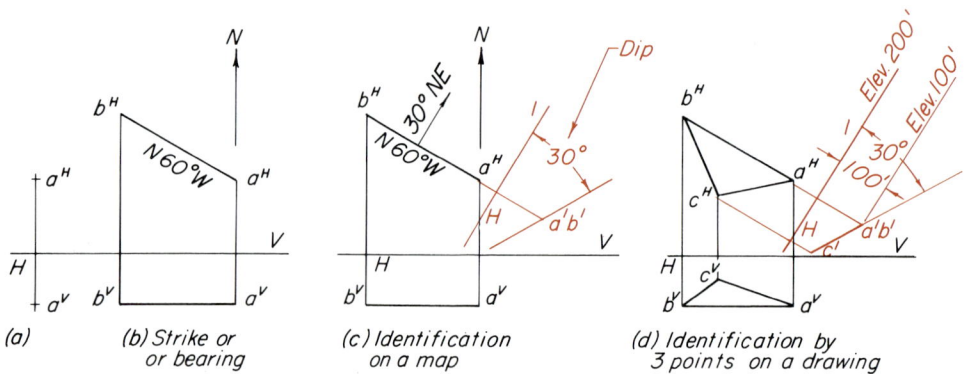

FIG. 7.51. Plane through a point having a specified strike and dip.

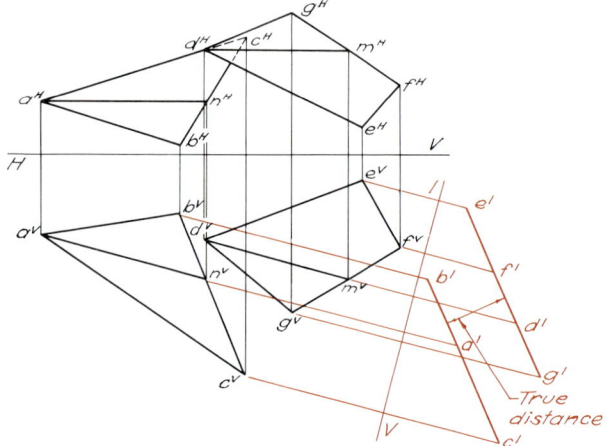

FIG. 7.52. Distance between two parallel planes.

7.22.11 Draw a Plane through a Point Parallel to a Given Plane. *One plane is parallel to a second plane when two intersecting lines in the first are respectively parallel to two lines in the second or when the edge view of one is made parallel to the other.*

a. Thus, in Fig. 7.53a, the hexagon *ABCDEF* is parallel to the triangle *MNO* because *AB* is parallel to *MN* and *CD* is parallel to *NO*. Thus, by the simple principle of parallel lines it is possible to lay out one plane parallel to another.

b. When two planes are parallel, they will appear parallel in any principal or auxiliary view in which they appear edgewise. This method may therefore be used to determine the parallel relationship or to construct one plane parallel to another, as shown in Fig. 7.53b. If the edgewise views are parallel, the other projections may have any shape as long as they remain in projection.

7.23 RELATIONSHIPS OF LINES AND PLANES

Some problems concerning planes alone cannot all be solved without considering some aspects of the relationships between lines and planes given in the following articles.

7.23.1 Point in Which a Line Pierces a Plane—Edgewise View Method. *Whenever a plane appears as an edgewise view, the piercing point of a line with that plane can be seen by inspection to be at the crossing point of the proper projection of the line with the edgewise view of the plane.*

a. *Edge view of plane given.* Figure 7.54a shows the edgewise view of plane *ABC* in the vertical projection and therefore the piercing point of the line *MN* with the plane

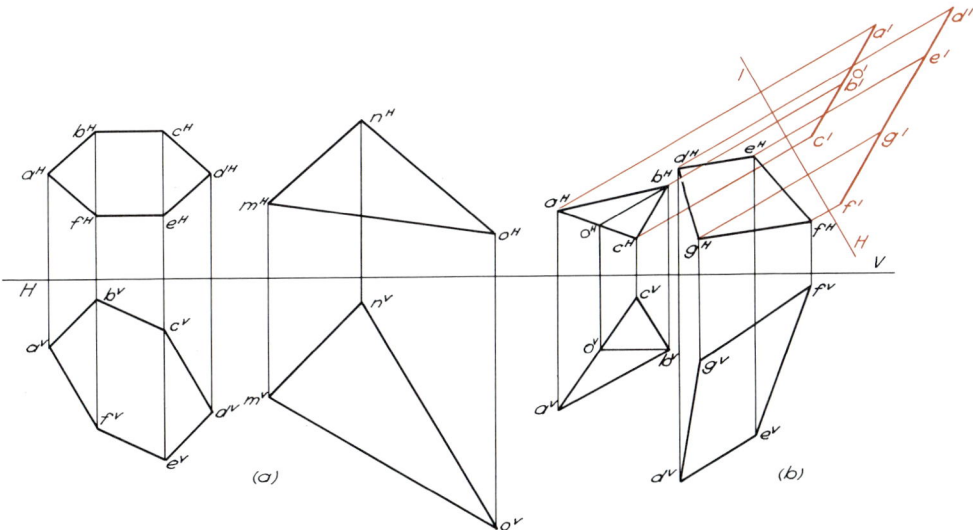

FIG. 7.53. Constructing one plane parallel to another.

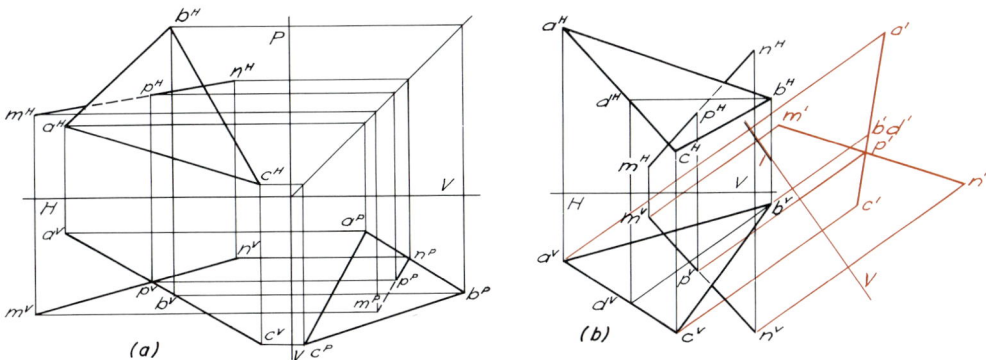

FIG. 7.54. Point through which a line pierces a plane: edgewise view method.

164 AUXILIARY VIEWS

can be located in that view at p^V by inspection. The remaining projections of the piercing point are carried to the other projections of the line in the usual manner.

b. *Oblique plane.* When the plane does not appear edgewise in any principal view, an auxiliary view can be made that will give an edgewise view of the plane by using an H- or V-parallel of the plane. The point where the auxiliary projection of the line crosses the edgewise view of the plane will be the auxiliary projection of the point where the line pierces the plane. See Fig. 7.54b. The other views are found by projecting back from the auxiliary view.

7.23.2 Piercing Point of a Line with the Principal Planes.

a. *Oblique lines.* The piercing point of a line with the principal planes can be found in the same manner since the reference line is always an edgewise view of one of the principal planes. Thus, in Fig. 7.55a, when one is looking at the vertical projection, $a^V b^V$, of a line, the horizontal reference line is the edgewise view of the H-plane. Hence extend $a^V b^V$ until it pierces the H-plane at h^V. The horizontal projection is then at h^H on $a^H b^H$ extended, and h^P on $a^P b^P$ extended.

In the same manner, when one is looking at $a^H b^H$, the top view of the line in Fig. 7.55b, the horizontal reference line represents the edgewise view of the V-plane. Hence, extend $a^H b^H$ until it pierces the V-plane at v^H. The vertical projection v^V is directly below v^H on $a^V b^V$ extended.

The profile plane appears edgewise in both the front and the top views, and the piercing point may be determined simply by extending both projections of AB until they cross the profile reference line at p^H and p^V. See Fig. 7.55c. It will be noted in the above discussion that the piercing point with the principal planes is always given the letter of the plane. Thus, h^H, h^V, and h^P for

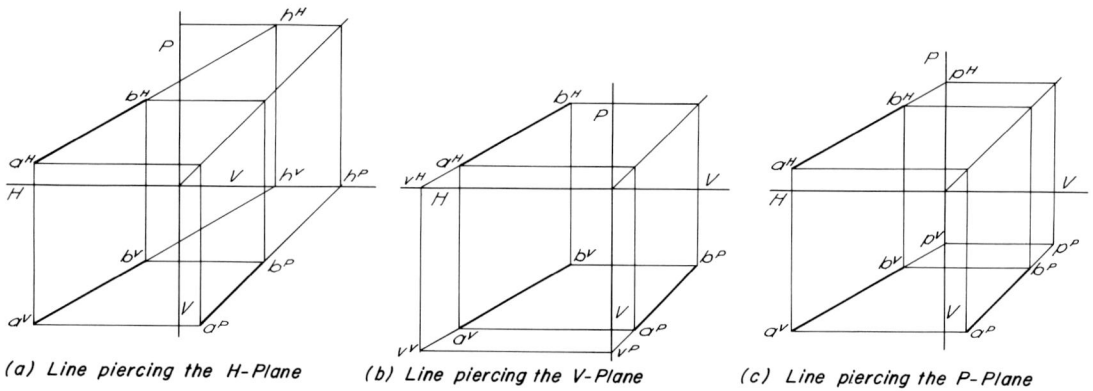

(a) Line piercing the H-Plane   (b) Line piercing the V-Plane   (c) Line piercing the P-Plane

FIG. 7.55. Points through which a line pierces the coordinate planes.

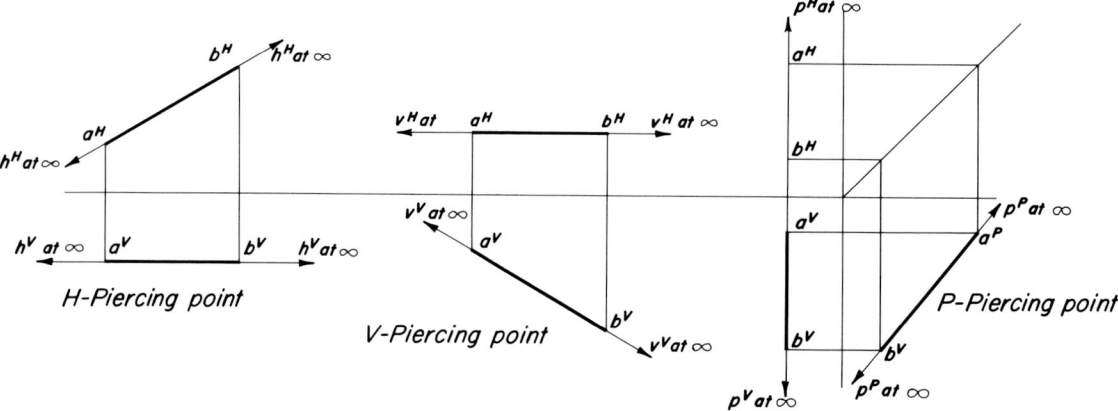

H-Piercing point

V-Piercing point

P-Piercing point

FIG. 7.56. Piercing points at infinity.

the H-plane and v^H, v^V, and v^P for the V-plane, and so on.

b. *Lines parallel to one or two principal planes.* If a line is parallel to one of the principal planes, it is said to pierce it at infinity. Obviously, this cannot be shown on a drawing. It may, however, be indicated by an arrow with the symbol for infinity at the end of the line, as illustrated in Fig. 7.56.

When a line is parallel to the profile plane, its H- and V-piercing points are obtained by making the profile view. Since both H- and V-planes appear edgewise in this view, the piercing points can be obtained by inspection, as shown in Fig. 7.57. The other views can then be obtained by projection.

The piercing point of a line with the principal planes has many applications, for example, in oblique projection (Chapter 14) and perspective (Chapter 15).

In other problems, such as outcrop problems in mining (see Fig. 19.30), highway, or railroad cuts and fills (see Figs. 19.31 and 19.32), the principle of piercing points of lines with other surfaces must be used.

7.23.3 Point in Which a Line Pierces a Plane—Cutting Plane Method.
If any plane is passed through the line and the line of intersection of this plane with the given plane is found, the piercing point will be the crossing of the given line with the line of intersection. This method, as illustrated in Fig. 7.58, is often more convenient than the auxiliary plane method. It is used extensively for finding the intersection of any two surfaces.

The procedure may be stated briefly as follows.

a. Pass a projecting plane through the line such as X-X in Fig. 7.58a. A projecting plane always shows edgewise in the view in which it is used.
b. Find the intersection MN of the given plane with the projecting plane (Fig. 7.58b).
c. Find the point of intersection P of the given line with this line of intersection MN (Fig. 7.58c).

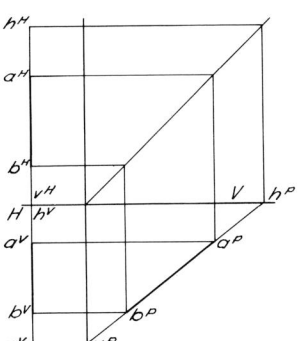

FIG. 7.57. Piercing points of a P-parallel.

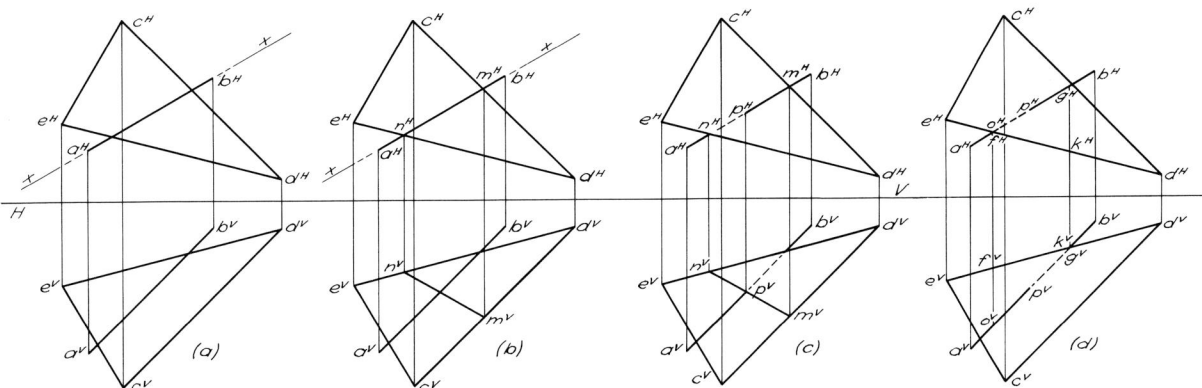

FIG. 7.58. Point in which a line pierces a plane: cutting plane method.

d. Since lines AB and MN both lie in the plane X-X, they must intersect or be parallel. In this case they intersect at point P, in Fig. 7.58c, which is the piercing point of AB with plane CDE. If AB were parallel to MN, line AB would be parallel to the plane.

7.23.4 Visibility. In order to complete the drawing in Fig. 7.58, it is necessary to show which portion of line AB is visible. In the top view of Fig. 7.58d, it is necessary to determine which of lines AB and DE is above the other by examining the apparent crossing point $o^H f^H$ of their projections. By drawing a vertical projecting line from $o^H f^H$ it can be seen from the front view that f^V on line DE is above o^V on line AB. At this point, therefore, line AB goes below DE and the plane and is therefore invisible in the top view from this point until it emerges above the plane at the piercing point.

The visibility in the front view can be determined in the same manner by observing the apparent crossing point of the vertical projection of lines AB and DE at $g^V k^V$. By erecting a vertical projecting line from $g^V k^V$ to the top view, it can be seen that k^H on DE is in front of g^H on AB. Hence AB goes behind DE and the plane, and, at this point, becomes invisible until it emerges at the piercing point.

7.23.5 Line Parallel to a Plane. A line is parallel to a plane when it is parallel to any line in that plane.

a. *Line method.* Thus, in Fig. 7.59, let it be required to construct a line through O parallel to plane CDE. One may draw any line AB in plane CDE and then draw through O the projections of OM parallel to the corresponding projections of AB. Any edge line of triangle CDE could also have been used; for example, ON is parallel to CD.

b. *Edge view method.* A line may also be drawn parallel to a plane by making one projection of the line parallel to the corresponding edgewise view of the plane, as shown in Fig. 7.60. The other projections may have any convenient position so long as they are in correct alignment with each

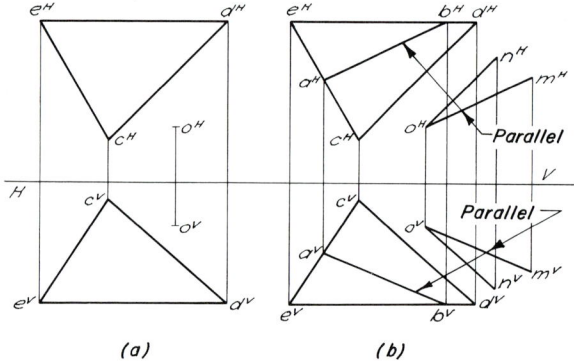

FIG. 7.59. Lines parallel to a plane.

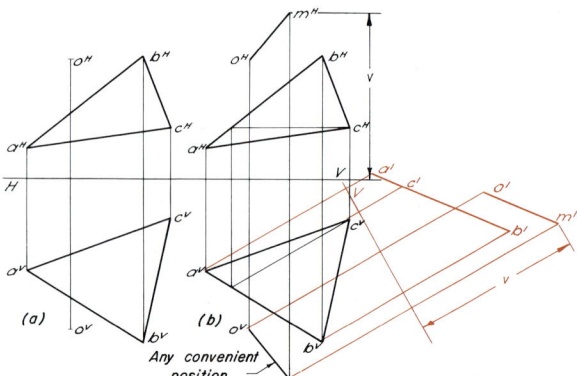

FIG. 7.60. Line parallel to a plane.

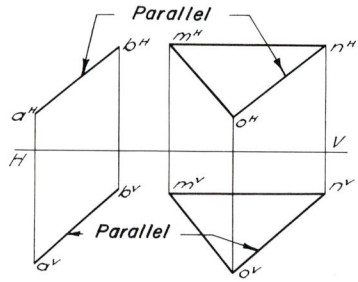

FIG. 7.61. Plane parallel to one line.

other. For example, let it be required to draw a line through O in Fig. 7.60a parallel to plane ABC. The edgewise view of the plane with the corresponding projection of O may be obtained by auxiliary projection as shown in Fig. 7.60b. Through o^1 draw o^1m^1 parallel to $a^1b^1c^1$. The adjacent projection of $o^V m^V$ may be made in any direction. The projection $o^H m^H$ is found in the usual way.

7.23.6 Plane Parallel to One or Two Lines. *If any plane contains a line parallel to a second line, that plane is parallel to this second line.*

a. *Parallel to one line.* When only one line is involved, there are an infinite number of solutions, but if two nonparallel lines are involved, only one solution is possible.

To construct a plane through a point parallel to a given line it is only necessary to draw a line ON through the point parallel to the given line AB, as shown in Fig. 7.61. Any plane such as OMN containing this line satisfies the condition.

b. *Parallel to two lines.* If a plane is to be constructed through a point parallel to two given lines, two lines must be drawn through the point respectively parallel to the two given lines, as shown in Fig. 7.62. The plane of the two lines is the required plane. Line ON is parallel to CD, and OM is parallel to AB, thus satisfying the conditions.

7.23.7 Angle between a Line and a Plane. *The true angle between a line and a plane will show on an auxiliary plane that is simultaneously parallel to the line and perpendicular to the plane.* The solution of this problem may require three auxiliary views as illustrated in Fig. 7.63. First, secure an edge view of the plane. Second, make a true-shape view of the plane as in Fig. 7.63a. Third, construct another edge view of the plane on an auxiliary plane parallel to the line. See Fig. 7.63b. The view gives the required angle. The angle could also be obtained by revolving the line in the second auxiliary view until it is parallel to the reference line 1-2.

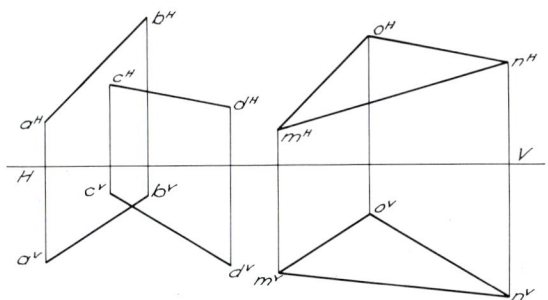

FIG. 7.62. Plane parallel to two lines.

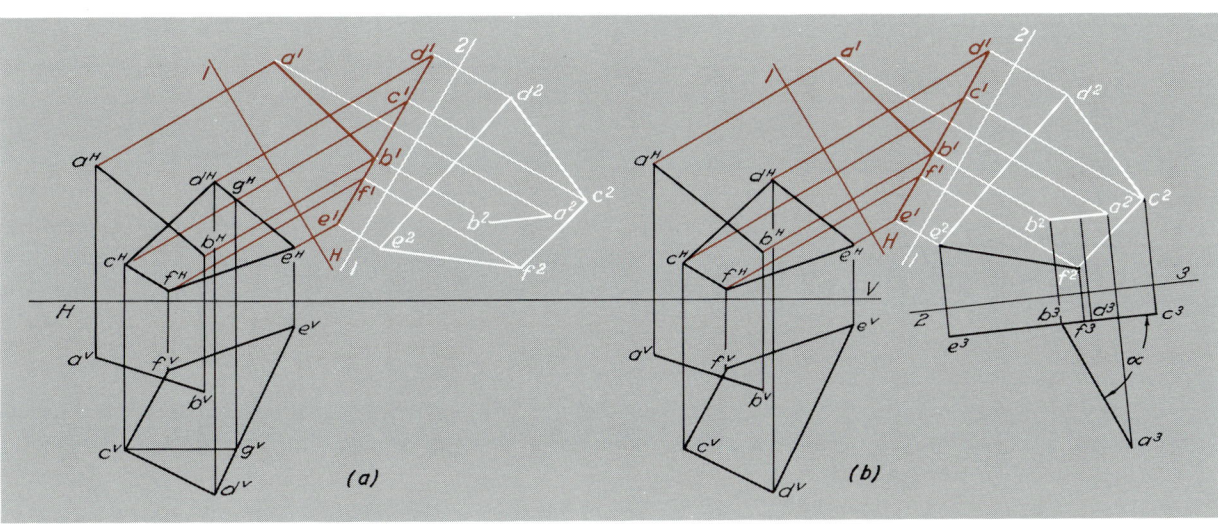

FIG. 7.63. Angle between a line and a plane.

7.23.8 Line Perpendicular to a Plane. *A line is perpendicular to a plane when it is perpendicular to any two nonparallel lines in the plane.*

a. *Line method.* Theoretically any two intersecting lines may be used but to make the construction as easy as possible, it is best to use an *H*-parallel and a *V*-parallel. Figure 7.64a shows a plane *ABC* with the *H*-parallel *AD* and the *V*-parallel *AE* already constructed. To make *AG* perpendicular to *ABC*, $a^H g^H$ is drawn perpendicular to $a^H d^H$ and $a^V g^V$ is drawn perpendicular to $a^V e^V$, as in Fig. 7.64b, thus making *AG* at right angles to both lines *AD* and *AE*, as explained in Article 7.19.11. Any line parallel to *AG* will also be perpendicular to the plane.

b. *Edge view method.* The edgewise view of a plane may also be used to construct a perpendicular. In Fig. 7.65, the plane *ABC* shows edgewise in the front view. Any line perpendicular to this plane must also be parallel to the *V*-plane. Line *DE* is therefore drawn at right angles to plane *ABC* by making $d^V e^V$ perpendicular to $a^V b^V c^V$ and $d^H e^H$ parallel to the reference line.

This method may be used for any plane by obtaining first an edge view by auxiliary projection, as shown in Fig. 7.66. In the auxiliary view, draw $d^1 e^1$ perpendicular to $b^1 a^1 c^1$ and the adjacent view $d^H e^H$ parallel to the auxiliary reference line since $d' e'$ is true length. The remaining projection is obtained in the usual way.

7.23.9 Projection of a Point on a Plane. *The piercing point of a line through the point drawn perpendicular to the plane is the projection (orthographic) of the point on the plane.*

This is an application of the preceding paragraph. For this purpose the edge view method is best, as shown in Fig. 7.67.

Given point *O* and plane *ABCD*, proceed as follows.

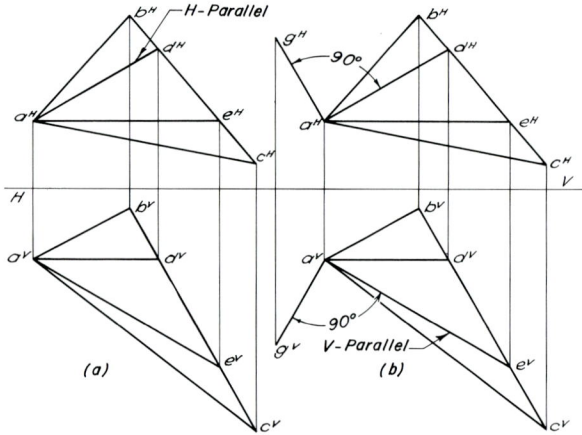

FIG. 7.64. Line perpendicular to a plane.

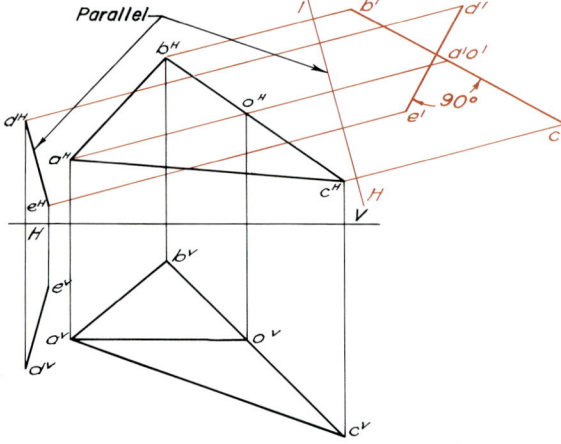

FIG. 7.66. Line perpendicular to an oblique plane: edgewise view method.

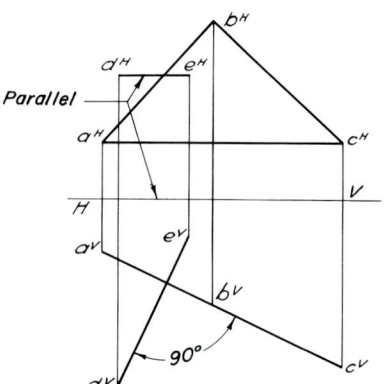

FIG. 7.65. Line perpendicular to an inclined plane: edgewise view method.

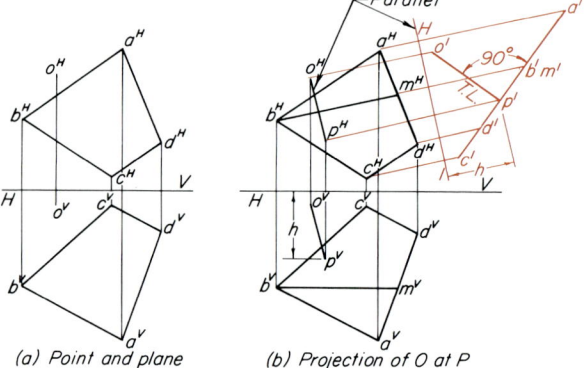

FIG. 7.67. Projection of a point on an oblique plane.

a. Using an *H*-parallel of the plane obtain an edge view. Obtain also the auxiliary projection of the point at o^1.
b. From o^1 draw a line perpendicular to the plane and label the point where it pierces the plane p^1.
c. The horizontal projection of the line *OP* is drawn perpendicular to *BM* or parallel to the 1-*H* reference line.
d. The vertical projection can be obtained from the preceding projections in the usual way.

7.23.10 Plane Perpendicular to a Line. This is simply the reverse of the preceding problem and may be solved by either procedure discussed in that paragraph.

a. *Line method.* The first solution, shown in Fig. 7.68a, consists of drawing two lines *BC* and *BD* each perpendicular to line *AB*, as explained in Article 7.19.11 and illustrated in Fig. 7.41. The plane of these two lines is the required plane.

b. *Edge view method.* In the second method shown in Figs. 7.68b and c, the edgewise view of the plane is drawn perpendicular to the true-length view of the line. The adjacent view may be drawn in any position so long as the points are in proper alignment. In Fig. 7.68b, the true-length view of the line is the top view or horizontal projection, but the same principles would apply in any view.

For an oblique line as shown in Fig. 7.68c, it is first necessary to obtain a true-length view of the line by auxiliary projection. Beginning in the auxiliary view the procedure is the same as outlined above.

7.23.11 Construction of a Solid in a Specified Position. As an illustration, let it be required to construct a right pentagonal pyramid with line *AB* of Fig. 7.69a as an axis and point *C* as one

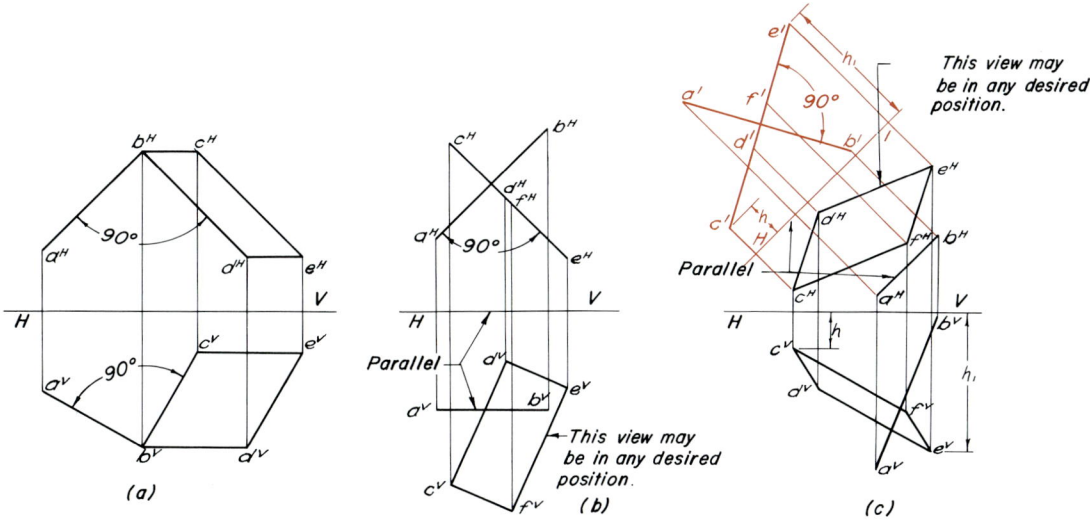

FIG. 7.68. Plane perpendicular to a line.

corner of the base. To solve this problem, the following steps are necessary.

a. Project the point and line on an auxiliary plane perpendicular to the line. See Fig. 7.69a.
b. Construct the true size of the base in the second auxiliary projection and letter the corners. See Fig. 7.69b.
c. Construct the elevation of the pyramid in the first auxiliary projection. See Fig. 7.69b.
d. Project the pyramid back to the vertical projection as shown in Fig. 7.69c.
e. Project the pyramid back to the horizontal projection as shown in Fig. 7.69d.
f. Determine the visibility in each view.

7.23.12 Plane Perpendicular to a Plane. *If a line is perpendicular to a given plane, any plane containing the line is perpendicular to the given plane.*

a. *Line method.* Thus, in Fig. 7.70a, line AB has been constructed perpendicular to plane DEFG by the method discussed in Article 7.23.8. Since plane ABC contains line AB, the plane is perpendicular to DEFG.
b. *Edge view method.* If an edge view of one plane is found as $a^1b^1c^1$ in Fig. 7.70b, another plane such as DEFG may be set up perpendicular to ABC, or at any other angle, by making the edge view $d^1e^1f^1g^1$ at the prescribed angle with $a^1b^1c^1$. The adjacent projection of DEFG may be placed in any convenient position so long as it is in proper alignment with $d^1e^1f^1g^1$.

7.23.13 Intersection of Two Planes. *The intersection of two planes is a straight line that is determined by two points each common to both planes.*

a. *Cutting plane method.* The problem, therefore, resolves itself into finding two

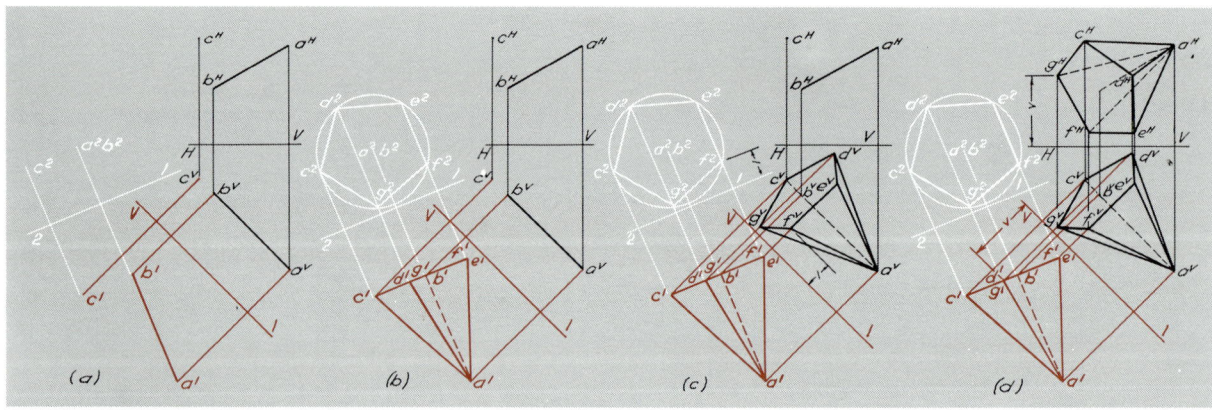

FIG. 7.69. Construction of a solid in a specified position.

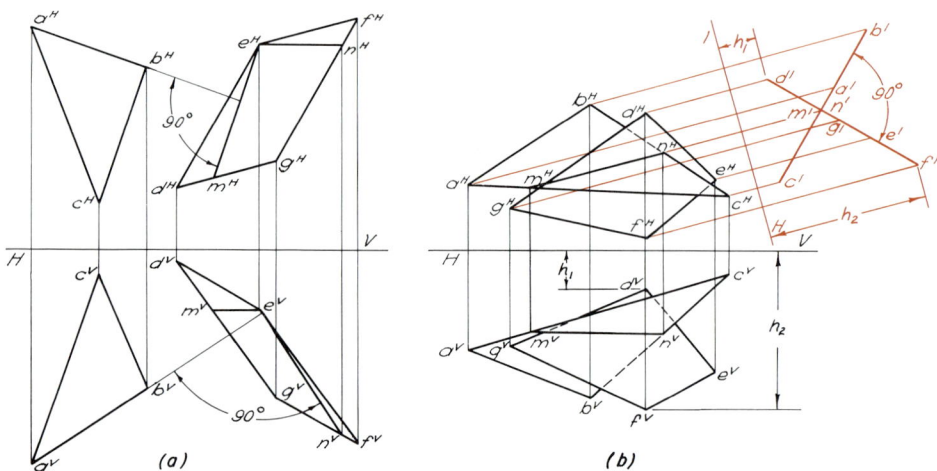

FIG. 7.70. Plane perpendicular to another plane.

points common to both planes. Using the method described in Article 7.23.3, the point O in Fig. 7.71a, where line BC pierces plane DEFG, is found. Point P in Fig. 7.71b in which DG pierces plane ABC locates the second point of line OP. Line OP shown in Fig. 7.71c is the required line of intersection. Visibility may be determined as described in Article 7.23.4 and shown in Figs. 7.71c and d.

b. *Strike and dip method.* Another method for determining the line of intersection of two planes is useful for geologists when the planes are determined by strike and dip. In Fig. 7.72, two planes through point A are determined by their strikes and dips as they would appear on a geologic map. By placing an auxiliary plane perpendicular to AB, the edgewise view of one of the planes can be obtained by laying off the dip as indicated in Fig. 7.72b.

Another auxiliary plane perpendicular to AC will make it possible to construct the edgewise view of the second plane by laying off the dip as shown in Fig. 7.72b. If a cutting plane T-T is passed parallel to the horizontal, as in Fig. 7.72c the edgewise views of this plane may be shown in both auxiliary views. The line of intersection of this cutting plane with the plane through AB is DE and with the plane through AC it is KG. Point F where these two lines intersect is a point on both planes and therefore one point on their line of intersection. Line AF is therefore the line of intersection of the two given planes.

7.23.14 Outcrop of a Plane. *The line in which any plane such as a vein of coal or a bed of stone intersects the surface of the earth is called an outcrop.* The outcrop on a horizontal surface is a strike line and may be found by obtaining the points in which any two lines in the plane of the

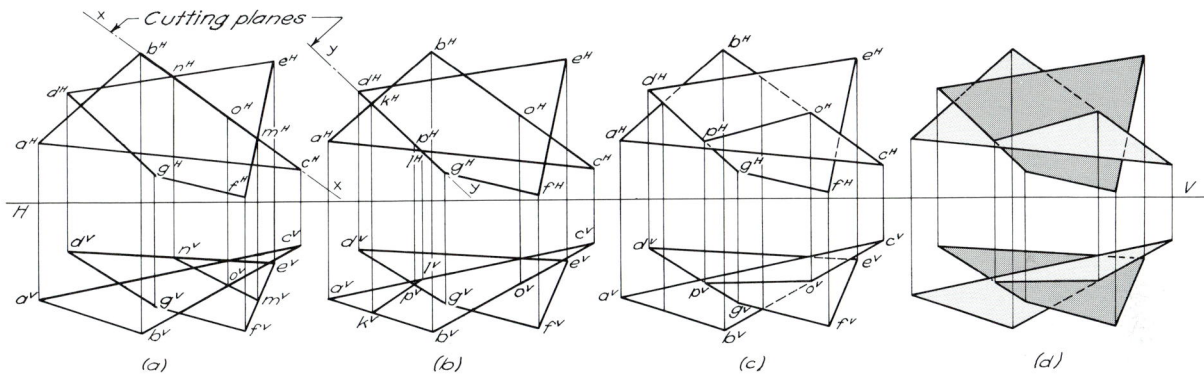

FIG. 7.71. Intersection of two planes.

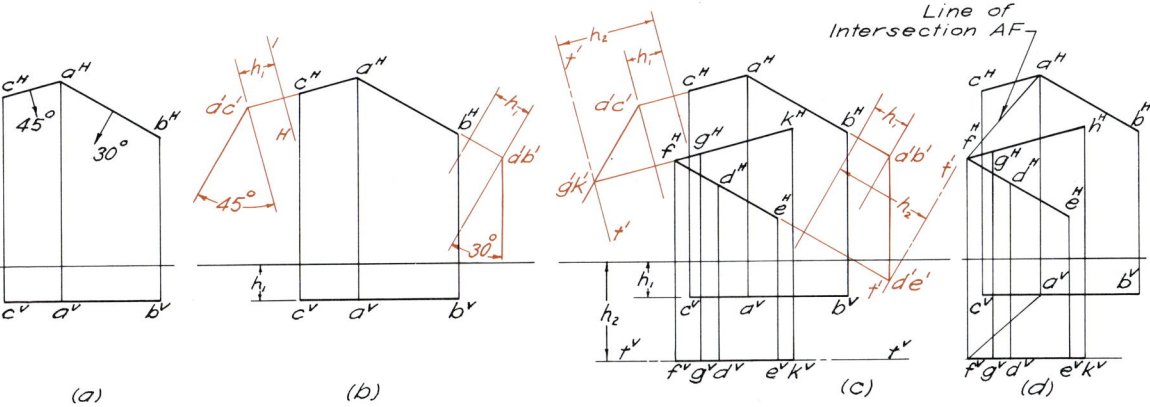

FIG. 7.72. Intersection of two planes when strike and dip are given.

vein pierce the horizontal surface. Joining these two points with a straight line, as shown in Fig. 7.73, gives the outcrop line. This outcrop can also be found by setting up an auxiliary plane perpendicular to any strike line in the plane to obtain the edgewise view of the plane.

The point where the auxiliary reference line and the edgewise view of the plane intersect is the point projection, e^1f^1, of a line that lies in the horizontal plane and in the plane of the vein. Line EF is the outcrop that can be located in the horizontal view at e^Hf^H by projecting from e^1f^1 perpendicular to the 1-H reference line. This line is indefinite in length, so the projections e^H and f^H may be placed anywhere along the outcrop as shown in Fig. 7.73. The outcrop on a vertical plane such as the side of an excavation can be found by obtaining the piercing points of any two lines in the vein with the given vertical plane and joining them together. Usually the surface of the earth is defined by contours; the method of finding the outcrop in such cases is explained in Article 19.26.

7.23.15 Angle between Two Planes. *When two planes show edgewise on the same plane of projection, the true angle between the planes will show between the edge views.*

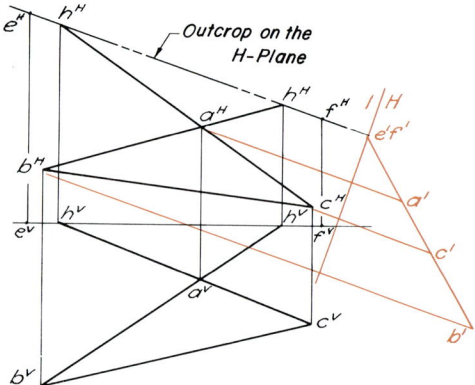

FIG. 7.73. Outcrop of a plane.

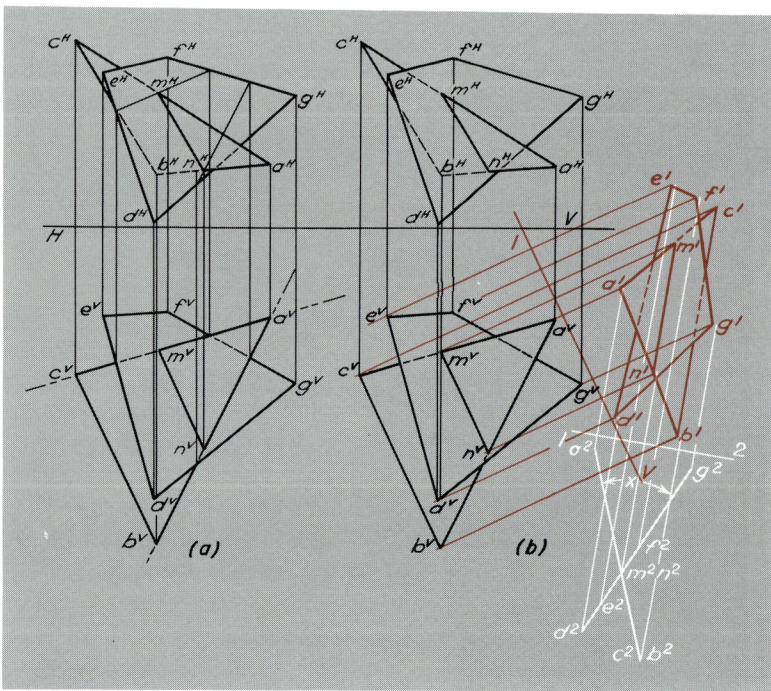

FIG. 7.74. Angle between two planes.

When the angle between two given planes is required, the edge views of both planes can be found on an auxiliary plane perpendicular to their line of intersection. In Fig. 7.74a, it is therefore necessary to find the line of intersection MN of the two planes by the usual methods. In Fig. 7.74b, the first auxiliary plane is set up parallel to MN and a second auxiliary plane perpendicular to MN, thus giving the edge view of both planes and the angle between them.

An application of this problem is found very frequently in structural engineering. Thus, in Fig. 7.75 a bent gusset plate is shown attached to a steel channel. It is necessary for fabrication to find the angle between the two portions of this plate and also the developed flat pattern.

Angle X between the faces ABCD and ABEF is found as described in the preceding paragraph and illustration.

Having the true angle X between the faces in the second auxiliary view as shown in Fig. 7.75, find the true shape of one face on an auxiliary plane parallel to it. Then revolve the other face about the line of intersection, until it is also parallel to the same auxiliary plane, as explained in Article 7.22.1. This layout will give the true size of the plate before bending.

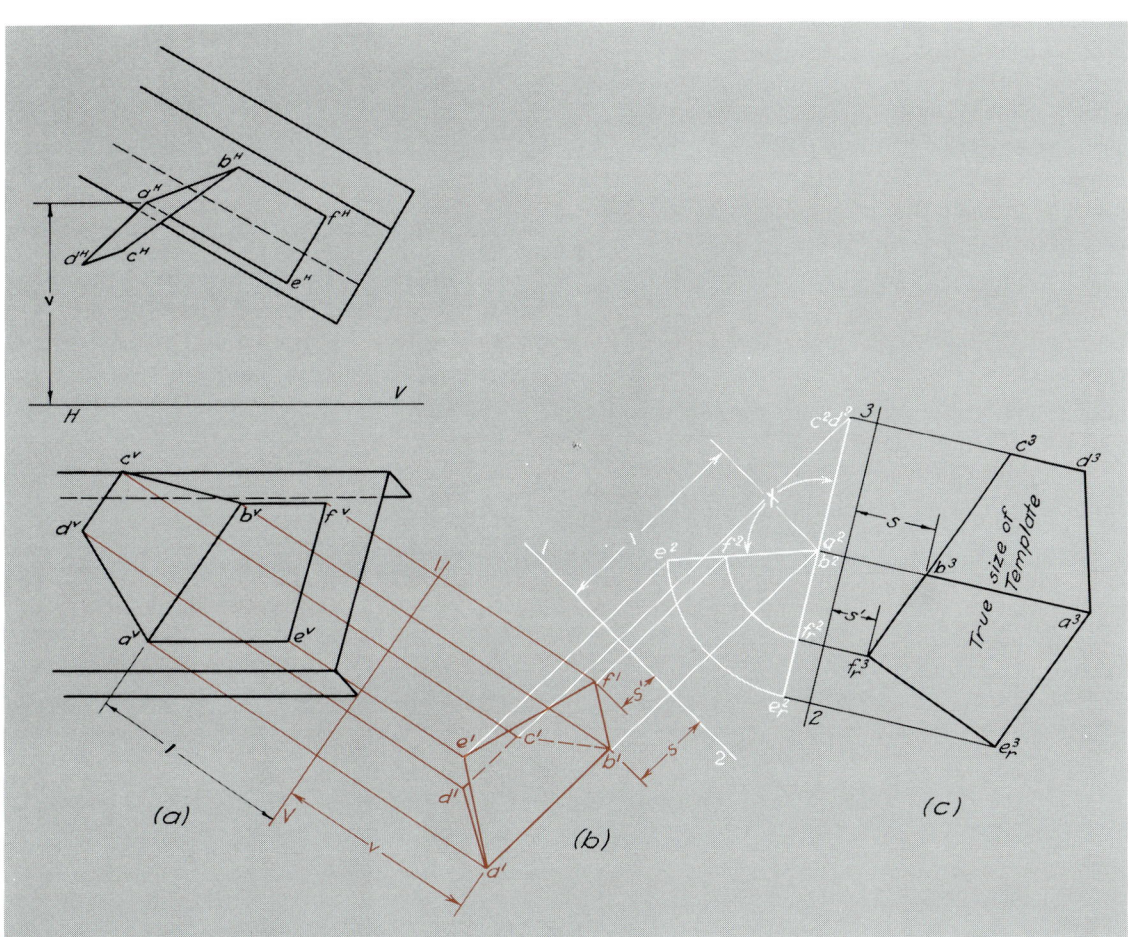

FIG. 7.75. True size of a gusset plate.

7.24 SKEW LINES

Lines that are nonparallel and nonintersecting are called skew lines. Skew lines can occur in various engineering situations, such as in mining and piping.

7.24.1 Shortest Distance between Two Skew Lines. *The shortest distance between two nonparallel, nonintersecting lines is the true length of their common perpendicular.*

a. *Point projection method.* This problem can be solved by projecting both lines on a plane perpendicular to one of them as in Fig. 7.76a by means of auxiliary projection. Having this projection, draw a line from a^2b^2 perpendicular to c^2d^2 as in Fig. 7.76b and letter it m^2n^2. Point n^1 may be found on c^1d^1 by direct projection. Construct m^1n^1 parallel to reference line 1-2. Since it is true length in the second auxiliary and since MN is perpendicular to AB, it is parallel to the second auxilary plane. Return the projections of M and N to the original views by the usual methods.

b. *Parallel plane method.* The shortest perpendicular between two skew lines will project as a point on a plane parallel to both skew lines.

1. In Fig. 7.77 determine a plane parallel to AB and CD by drawing a line AE through A parallel to CD. Any plane parallel to AB and AE is parallel to both lines. See Fig. 7.77.
2. By means of an H-parallel, EF, in plane ABE obtain an edge view of the plane. The two skew lines will show parallel in this auxiliary view.
3. On a second auxiliary plane obtain a true-shape view of the plane ABE as in Fig. 7.77a. The two skew lines appear in true length in this view.
4. The crossing point m^2n^2 of the lines in the true-length view locates the end view of their common perpendicular. Return this line to the original view.

7.24.2 Shortest Horizontal Line between Two Skew Lines. This problem is best solved by the parallel plane method. The first auxiliary plane must be set up perpendicular to the H-plane. See Article 7.24.1(b) and Fig. 7.77b.

a. In Fig. 7.77 follow steps 1 and 2 in Article 7.24.1(b).
b. Set up the second auxiliary plane perpendicular to the first auxiliary reference line H-1, since this reference line represents the H-plane in the first auxiliary view. Project the two lines on this plane.
c. The crossing point r^3s^3 of the two lines locates the shortest horizontal line between the two skew lines.

7.24.3 Shortest Line of Specified Slope or Grade between Two Skew Lines. This problem also is best solved by the parallel plane method. Again, the first auxiliary plane must be perpendicular to the H-plane. See Article 7.24.1(b) and Fig. 7.77c.

a. In Fig. 7.77c follow steps 1 and 2 of Article 7.24.1(b).
b. With the first auxiliary reference line as a datum, at some convenient place in the first auxiliary view draw a line of the specified slope or grade. In Fig. 7.77c the grade line is +15% from W to T.

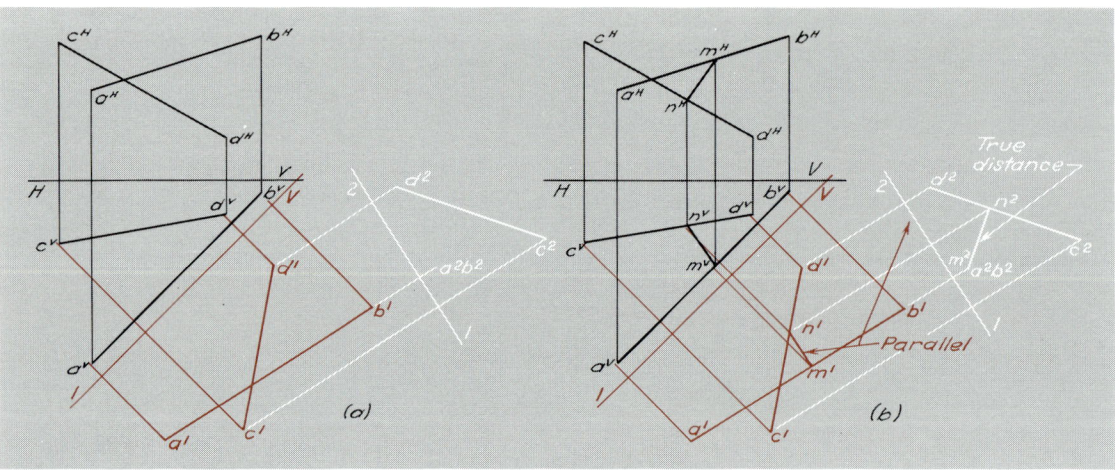

FIG. 7.76. Shortest distance between two nonparallel, nonintersecting lines.

c. Set up a second auxiliary plane perpendicular to the grade line and project both skew lines on this plane. The crossing point x^4y^4 of the lines locates the shortest 15% grade line between them. The projections of this line can be returned to the original views in the usual manner.

7.24.4 Grade Line Having a Given Bearing and Slope between Two Skew Lines. In the preceding problem (Article 7.24.3) the grade line was assumed to be parallel to the first auxiliary plane represented by the reference line H-1 in Fig. 7.77c.

If the grade line is to have some other bearing and slope, the problem may be solved as follows.

a. Lay out the grade line with the proper bearing and slope.
b. Find its vertical projection.
c. Then proceed in a manner similar to that for line EF in the following article.

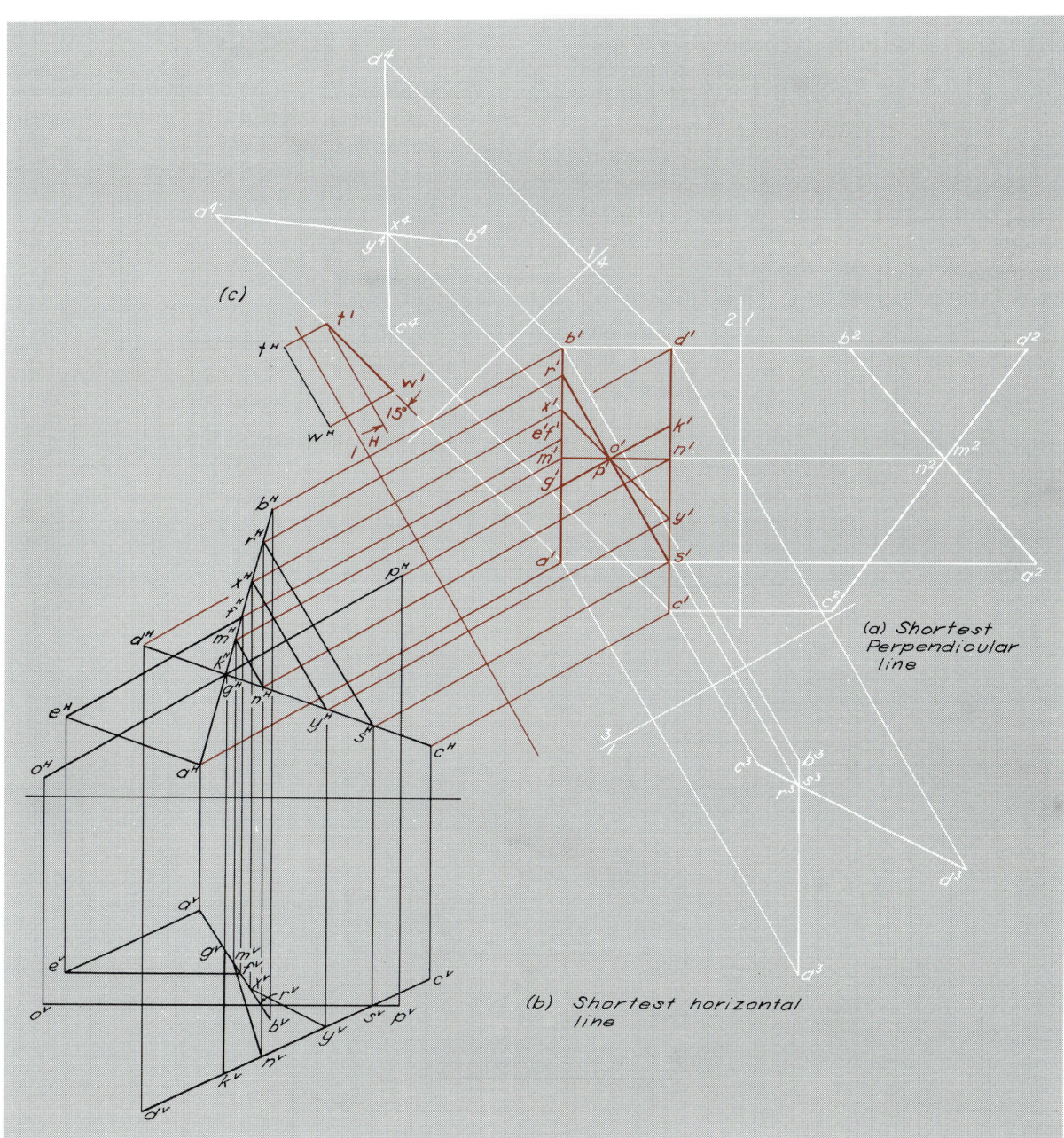

FIG. 7.77. Shortest line of a given slope intersecting two skew lines.

7.24.5 Line Parallel to a Given Line and Intersecting Two Nonparallel, Nonintersecting Lines.
Let it be required to draw a line parallel to EF and intersecting AB and CD as shown in Fig. 7.78.

If the point projection of one of the lines is found, the required line can be drawn in that projection through the point and parallel to the given line until it intersects the other of the two skew lines.

The solution for this particular problem is found by projecting all three lines on a plane perpendicular to AB, as shown in Fig. 7.78b. Since parallel lines always have their projections parallel, draw a line from a^2b^2 parallel to e^2f^2 until it intersects c^2d^2 at n^2, as in Fig. 7.78c. In the first auxiliary view, draw a line from n^1 parallel to e^1f^1 until it intersects a^1b^1 at m^1, thus determining a line between AB and CD parallel to EF. Return line MN to the original views in the usual manner.

The solution could also have been obtained by placing an auxiliary plane perpendicular to line EF, in which case the required line MN would have shown as a point in the second auxiliary view. Article 17.8 and Fig. 17.7 give a practical use of this method.

7.25 CONSTRUCTION CONES
The right circular cone provides a very useful tool in the solution of certain problems involving angles between lines and planes.

a. First, it should be noted that the elements of a cone all make the same angle with the plane of the base.

b. Second, these elements also make a constant angle with the axis of the cone. The problems in the following paragraphs illustrate the use of the cone. Review Article 7.7.6. See Fig. 7.10.

7.25.1 Line through a Point Making a Given Angle with H-, V-, or P-Plane.
A right circular cone having the specified base angle may be set up with its vertex at the given point and its axis perpendicular to the required plane. Any element of this cone will satisfy the conditions of the problem.

As an illustration, let it be required to draw a line through point A in Fig. 7.79, making an angle of 45° with the H-plane. The solution is as follows.

a. Through A draw a line AB of any convenient length perpendicular to the H-plane.

b. With AB as an axis, draw a right circular cone whose elements make 45° with the base.

c. Any element such as AC will satisfy the conditions of the problem.

A similar construction for the vertical and profile planes can be made by drawing the axis of the cone perpendicular to either of these planes, as shown in Figs. 7.80 and 7.81.

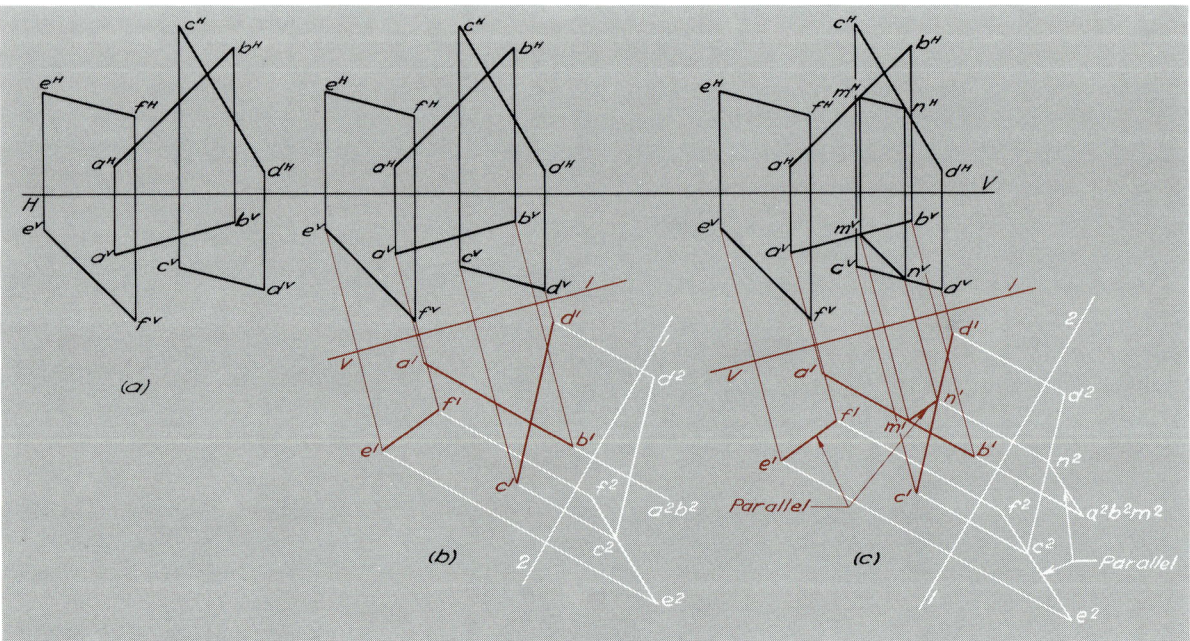

FIG. 7.78. Line parallel to a given line and intersecting two skew lines.

7.25.2 Line Through a Point Making Given Angles with Both *H*- and *V*-Planes.
If two cones are drawn from the same point as a vertex with one having its axis perpendicular to the H-plane and the other perpendicular to the V-plane and each having the proper base angle and elements of the same length, the intersection of these cones will satisfy the conditions of the problem.

For simplicity of solution the bases of both cones should be on the surface of a sphere whose center is at the given point through which the line is to be constructed. The outer circles in Fig. 7.82 represent the projections of the sphere. Every element of both cones is therefore a radius of the sphere and consequently the elements of both cones are of equal length. If the solution is possible, the cones will intersect, and since the elements are all the same length and the bases lie on the surface of the sphere, the bases will intersect each other at one or two points. These points, when connected with the apex of the cones, will be the required lines since they are elements of both cones.

Let it be required to construct a line through point A in Fig. 7.82, making an angle of 30° with the *H*-plane and 45° with the *V*-plane. The construction is as follows.

a. With point A as a center, draw the *H*- and *V*-projections of a sphere whose radius will be equal to the desired length of elements for the cones.

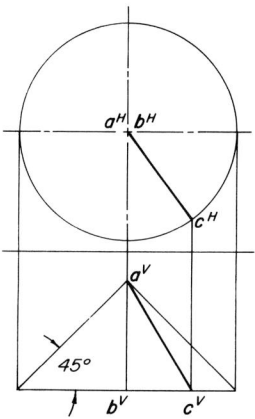

FIG. 7.79. Line making a specified angle with the *H*-plane: cone method.

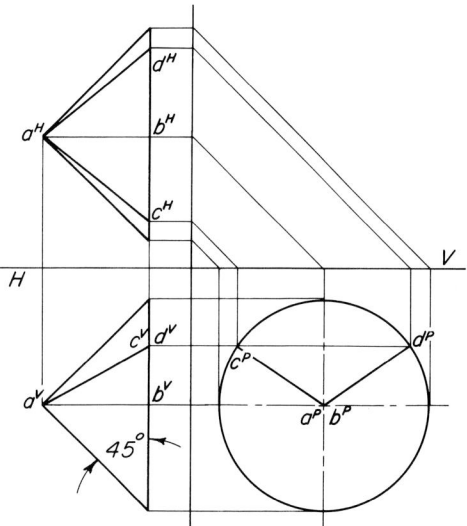

FIG. 7.81. Line making a specified angle with the *P*-plane.

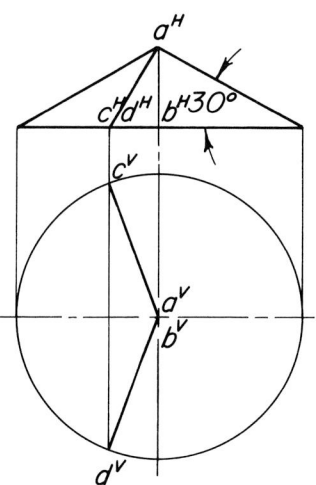

FIG. 7.80. Line making a specified angle with the *V*-plane.

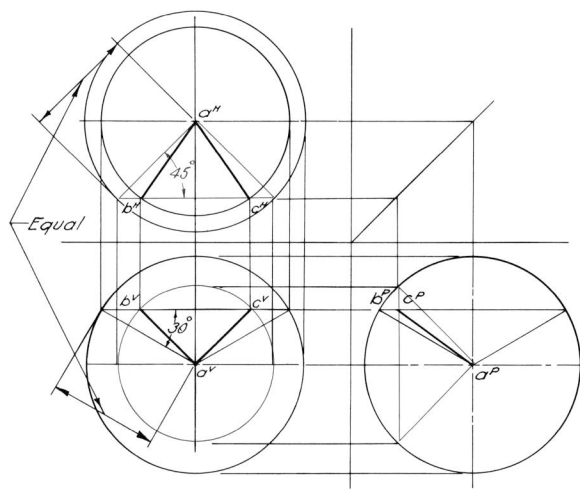

FIG. 7.82. Line making specified angles with both *H*- and *V*-planes.

b. With *A* as the apex, construct a right circular cone with its axis perpendicular to the *H*-plane, its elements equal to the radius of the sphere, and making an angle of 30° with the *H*-plane.

c. With the same apex, construct another cone with its axis perpendicular to the *V*-plane, its elements making 45° with the *V*-plane, and equal in length to the radius of the sphere.

d. The bases will now intersect each other at points *B* and *C* in Fig. 7.82.

e. Joining these points to the apex gives lines *AB* and *AC*, which satisfy the conditions of the problem.

In this problem the profile view is useful since both axes are parallel to profile and consequently both cones show as triangles.

7.25.3 Line through a Point Making a Given Angle with Any Two Planes. The solution of this problem illustrated in Fig. 7.83 is the same as the preceding one but represents a more general case. Here the line of intersection of the two planes has been made perpendicular to the *V*-plane. The solution involves the construction of two cones with axes perpendicular respectively to the two planes, as shown in Fig. 7.83.

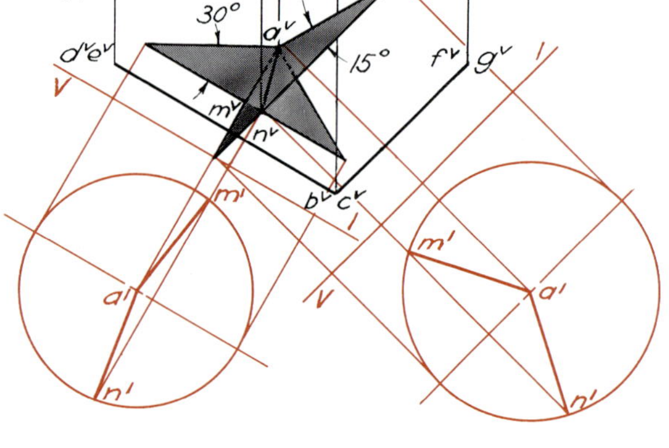

If the line of intersection of the two planes is not perpendicular to one of the principal planes, the first step in the solution will be to find the edge view of both planes on a plane perpendicular to their line of intersection. This auxiliary plane will be parallel to the axes of both cones and therefore will show both cones as triangles. The remainder of the solution is similar to that shown in Fig. 7.83.

7.25.4 Line Making Specified Angles with Two Intersecting Lines. *If a right circular cone having the specified vertex angle is constructed around each line, with the vertex of both cones at the point of intersection of the two lines, the intersection of these cones will give the specified lines.* Let it be required to draw lines making 75° with both *AB* and *BC*.

The solution may be found in the following manner.

a. Draw an *H*-parallel *BD* in the plane of the two lines as in Fig. 7.84.

b. Find the edge view $a^1b^1c^1$ of the plane of the lines in the first auxiliary view.

c. Find the true-shape view (angle between the lines $a^2b^2c^2$) in the second auxiliary view.

FIG. 7.83. Line making specified angles with any two planes.

d. At the intersection of the two lines draw a sphere of convenient diameter and construct the cones around each line with their bases on the sphere. Note that the specified angle of 75° must be between the element and the axis of the cones.

e. The crossing points of the bases of the cones locates two lines which fulfill the requirement of the problem.

f. In order to return the projections of the lines to the original views, make a third auxiliary view of the base of one of the cones and locate the two points on the base. This gives the necessary measurements for the return of the two required lines.

A second illustration of this problem is shown in Fig. 7.85. In this case the specified angles are 60° with AB and 45° with BC. It will be noted in this figure that the two cones are tangent to each other. Therefore, only one solution, namely, line BM is possible. From the figure it can be seen that in this situation a third auxiliary view is not necessary.

7.25.5 Line Intersecting Two Skew Lines and Making Specified Angles with Them. *Since the elements of a right circular cone make a constant angle with the axis, construction cones may be used to draw a line making a given angle with another line or two separate lines. Lines parallel to these constructed lines may then be drawn intersecting the two given skew lines.*

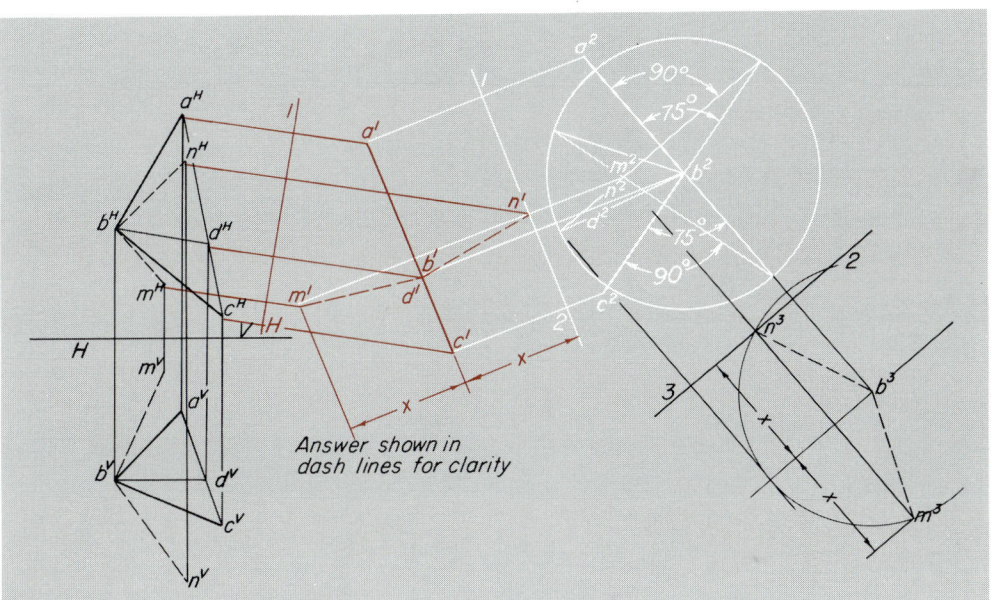

FIG. 7.84. Line making specified angles with two intersecting lines.

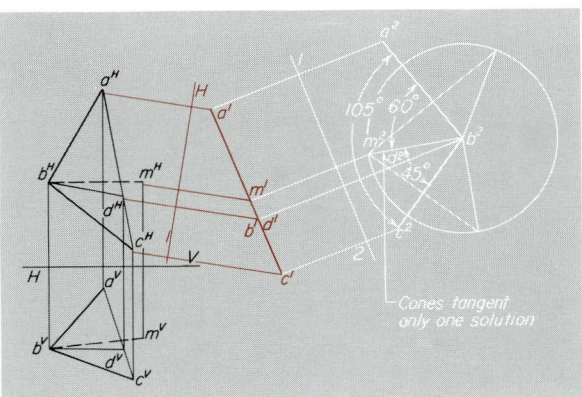

FIG. 7.85. Line making specified angles with two intersecting lines. One solution only.

180 AUXILIARY VIEWS

Let it be required to construct a line intersecting AB and CD in Fig. 7.86 making 60° with AB and 45° with CD. The solution is as follows.

a. Through any point O construct two lines, OE and OF, respectively, parallel to AB and CD. Figure 7.86a.

b. Project both lines on plane 1 to get the edgewise view at $o^1e^1f^1$ and on plane 2 to get the true size as $e^2o^2f^2$ in Fig. 7.86a.

c. In the second auxiliary view construct one view of the desired sphere, and using o^2e^2 and o^2f^2 as axes draw the cones using the specified angles. Figure 7.86b.

d. Set up a third auxiliary plane parallel to either one of the bases of the cones and draw the circle representing that base (Fig. 7.86b).

e. In the second auxiliary projection the bases of the cones intersect at m^2n^2. Line OM and ON are lines through point O parallel to the required line, and they may be projected back to the H- and V-projections (Fig. 7.86b).

f. In Fig. 7.86c draw the H- and V-projections of MO, one of the lines just found. This is shown at the far right in Fig. 7.86c.

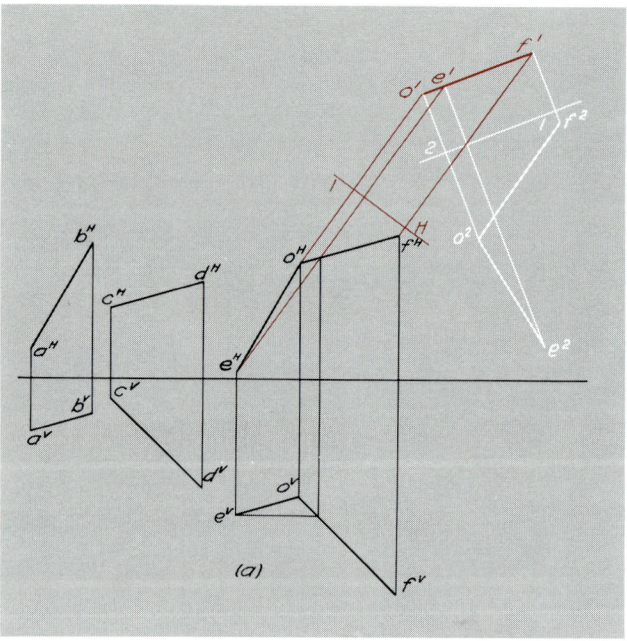

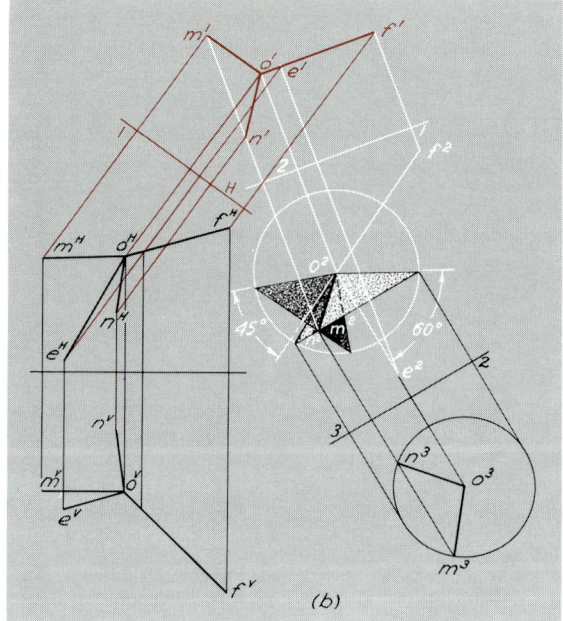

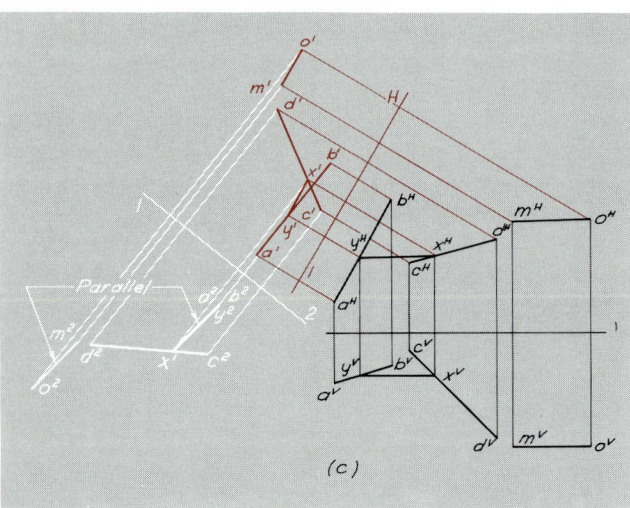

FIG. 7.86. Line intersecting two skew lines and making specified angles with them.

g. With AB and CD in Fig. 7.86c parallel to their original positions, draw an auxiliary plane 1 parallel to AB (a^Hb^H) and find all three lines AB, CD, and MO on this plane.

h. Set up an auxiliary plane 2 perpendicular to a^1b^1 and get the projections of all three lines AB, CD, and OM on this plane.

i. From a^2b^2 draw a line parallel to m^2o^2 until it intersects c^2d^2 at x^2.

j. Project x^2 back to x^1 on c^1d^1 and from x^1 draw x^1y^1 parallel to m^1o^1.

k. From this view project x and y back to the H- and V-projections in the usual manner.

7.25.6 Plane through a Given Line Making a Specified Angle with a Given Plane.

The plane may be made tangent to a construction cone, which has been set up so that the elements make the required angle with the given plane.

a. *Given plane horizontal.* As an example, let it be required to construct a plane containing line AB in Fig. 7.87 and making 60° with the horizontal plane.

1. Choose any point on AB and make that point the apex of the cone. In this case, the apex is at A and a cone whose elements make 60° with the H-plane has been constructed.

2. Next find the point P, where AB pierces the plane of the base.

3. From this point, construct a line PC tangent to the base of the cone. Lines AP and PC determine the required plane. Line PC can be drawn tangent to the circle on either side.

b. *Given plane oblique.* Let line AB be the given line and CDEF the given oblique plane as shown in Fig. 7.88.

1. The procedure is the same as for the problem in Fig. 7.87, except that an edge view of the plane along with the line must be first obtained, as shown in the first auxiliary view of Fig. 7.88.

2. Second, find a true-shape view of the plane CDEF in which the base of the cone lies. Then the piercing point of the line with the plane must be obtained, in a second auxiliary view.

3. On this view, second auxiliary view, the tangent line p^2r^2 can be drawn from the piercing point to the circle of the base.

4. The two lines AP and PR determine the plane.

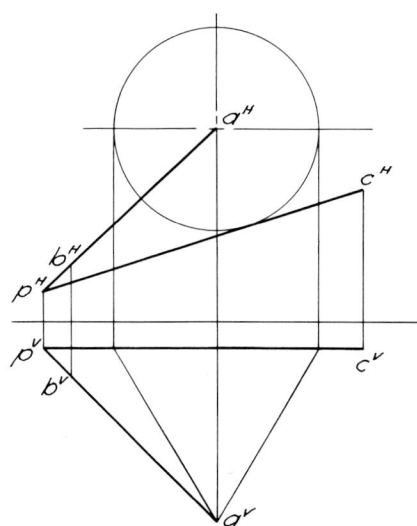

FIG. 7.87. Plane containing a given line and making a specified angle with the H-plane.

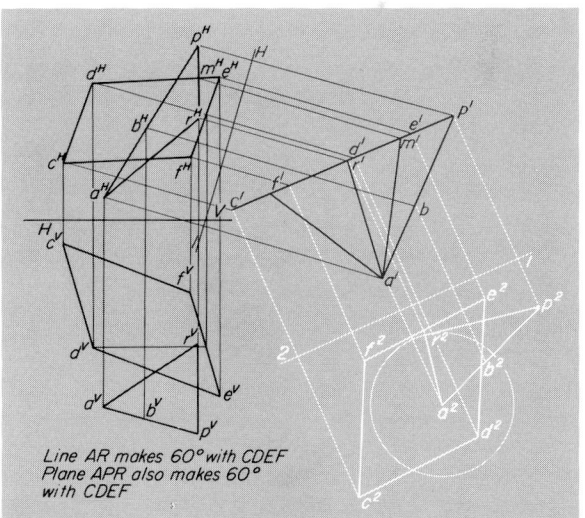

Line AR makes 60° with CDEF
Plane APR also makes 60° with CDEF

FIG. 7.88. Plane through a given line making a specified angle with an oblique plane.

182 AUXILIARY VIEWS

7.25.7 Planes Making Specified Angles with Two Given Planes. *If two lines are constructed making angles with two planes that are complements of the required angles, the required plane may be constructed perpendicular to either of these lines.*

In Fig. 7.89 the planes ABCD and CDEF have been placed so that the line of intersection of the planes is perpendicular to the H-plane and therefore both planes appear edgewise. In a more general case it might be necessary to make one or two auxiliary projections to get the edge view of the planes. Let it be required to draw a plane that makes 60° with ABCD and 45° with CDEF.

The construction is as follows.

a. Determine the complements of the specified angles: 30° with ABCD and 45° with CDEF.

b. By means of construction cones making the above angles with each plane find the lines OM and ON that meet the required condition as shown in Fig. 7.89b.

c. Construct planes perpendicular to OM and ON by drawing two lines perpendicular to OM as shown in Fig. 7.89c and likewise for ON as shown in Fig. 7.89d. See Article 7.19.11 for drawing lines perpendicular to a given line. Plane ORS is one solution and OXY is the other.

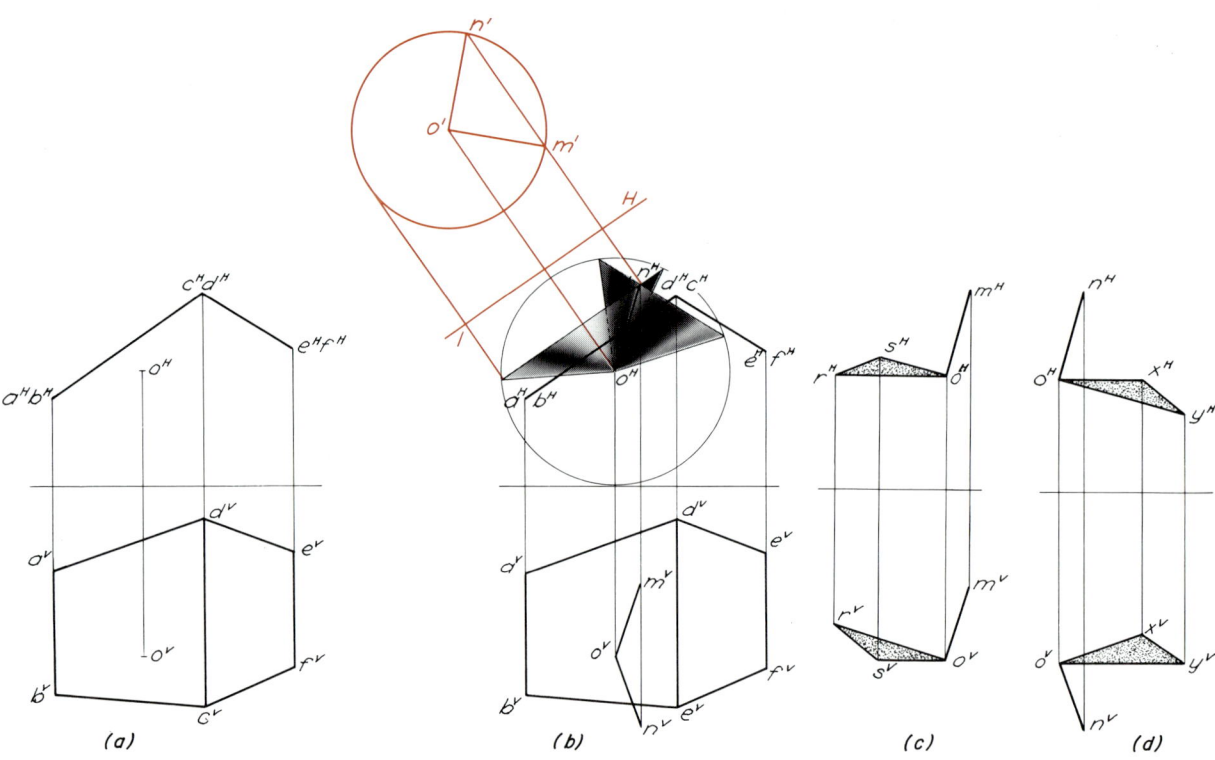

FIG. 7.89. Plane making specified angles with two planes.

7.25.8 Line Making a Specified Angle with a Given Line and a Given Plane.
Let it be required to draw a line making 60° with line AB and 30° with plane MNO. If a construction cone with vertex at any point on the line AB is made having the elements at 60° with the line AB as an axis and then a second cone with the same apex and having elements that makes 30° with plane MNO is drawn, the intersection of these cones will give the required line.

a. In Fig. 7.90 obtain first an edge view of the plane and in the same view a true-length projection of the line. In Fig. 7.90 this requires three auxiliary views. In some problems two auxiliary views might be enough.

b. Draw the standard sphere with center at point E on the line AB extended. This extension of AB was made to prevent too much overlapping of the views.

c. Draw the appropriate construction cones as shown in Fig. 7.90.

d. The intersection of the cones gives the required lines, EC and ED.

e. Return them to the original views in the usual manner. In Fig. 7.90c lines $b^2c_1^2$ and $b^2d_1^2$ have been drawn through B parallel to e^2c^2 and e^2d^2. Line BD only has been returned to the front and top views.

7.26 LOCUS PROBLEMS

The concept of loci appears in some descriptive geometry problems. The solution of such problems usually requires that the general locus be determined first and then further limiting conditions be applied. For example, let it be required to find the locus of points equidistant from two given points and in a given plane. The general locus of points equidistant from two given points is the plane-perpendicular bisector of the line joining the points. This plane must be determined first. The second condition limits the locus to points lying in a particular given plane. Hence, the answer is the line of intersection of the two planes. The loci most commonly used are listed below.

a. The locus of points equidistant from two given points is the plane-perpendicular bisector of the line joining the points.

b. The locus of points equidistant from three given points is the line of intersection of the two plane-perpendicular bisectors of the two lines joining the three points. This will be perpendicular to the plane of the three points.

c. The locus of points equidistant from four given points (not lying in a plane) is a single point, which is the center of a sphere having the four points on its surface. This problem can be solved by finding the locus of points equidistant from three points as in (b) above and then finding the piercing point of this line with the plane-perpendicular bisector of the line joining the fourth point to any one of the first three.

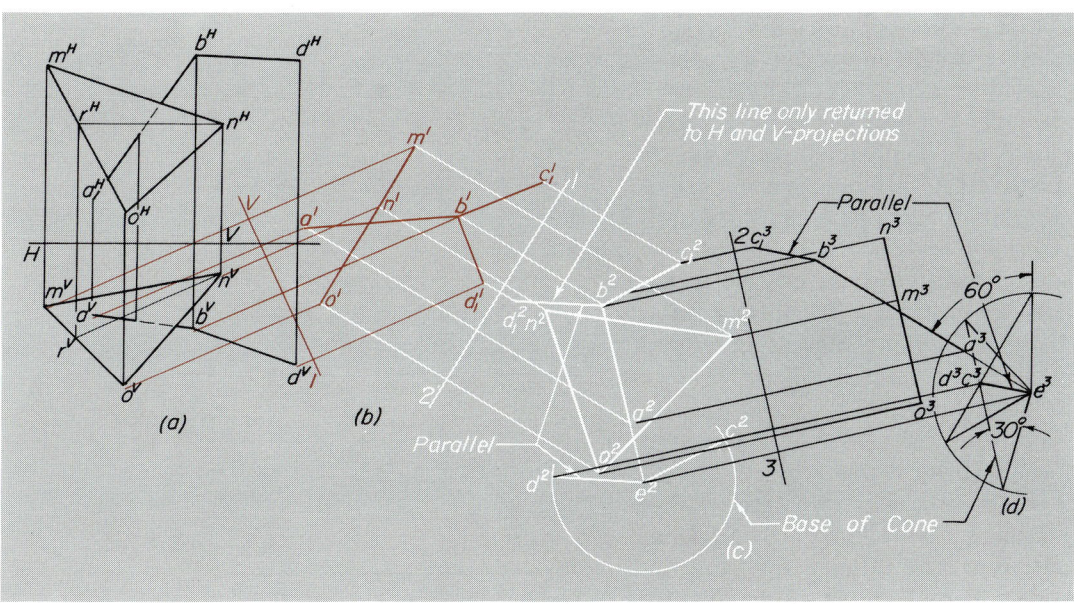

FIG. 7.90. Line making a specified angle with a given line and an oblique plane.

d. The locus of points at a given distance from a plane is two parallel planes, one on each side of the given plane at the specified distances from it.
e. The locus of points at a given distance from a line is the surface of a right circular cylinder having the line as an axis and a radius equal to the specified distance.
f. The locus of points at a given distance from a given point is the surface of a sphere with the given point as its center and a radius equal to the given distance.
g. The locus of lines making a specified angle with a given line is the surface of a right circular cone having the given line as an axis and whose elements make the specified angle with the line.
h. The locus of lines making a specified angle with a given plane is the surface of a right circular cone whose axis is perpendicular to the plane and whose elements make the specified angle with the plane.

7.27 WARPED SURFACES

A *warped surface* is one that is generated by a straight line moving according to certain specifications that vary with the different surfaces. No two positions of the generating line will lie in the same plane. These surfaces are nondevelopable; that is, they cannot be formed from metal without stretching or warping the metal. Some of these surfaces are illustrated in Fig. 18.1 in Chapter 18 and will be discussed briefly in the following paragraphs.

7.27.1 Hyperbolic Paraboloid. The *hyperbolic paraboloid* or warped quadrilateral is a surface generated by a straight line, called a *generatrix*, moving so that it always touches two nonparallel, nonintersecting lines, called *linear directrices*, and remains parallel to a plane director. It is a doubly ruled surface because it has two sets of linear directrices, two plane directors, and two sets of generating lines.

If the surface is defined by giving four bounding lines such as ABCD in Fig. 7.91, it is called a *warped quadrilateral*. Elements may be drawn in the surface by dividing either set of linear directrices (opposite sides) into the same number of equal spaces and drawing lines connecting the division points. If a plane is passed through AB parallel to CD, this plane will be one of the plane directrices and one set of elements will all be parallel to that plane. The other plane may be found by passing a plane through AD parallel to BC.

The hyperbolic paraboloid is sometimes used as the basis or framework of some practical structure and as such should be recognized by the student. The hyperbolic paraboloid is also used in the design of the bow of a boat, or in any transition surface connecting planes of different slope.

Figure 7.92 shows the projections of a hyperbolic paraboloid having two lines AB and CD as linear directrices and the horizontal plane as the plane director. Since the horizontal reference line represents the edgewise view of the horizontal plane in the vertical projection, that projection of the elements may be drawn parallel to the reference line. Therefore, to construct one view of any hyperbolic paraboloid, it is only necessary to find the edgewise view of the plane director by means of one or more auxiliary planes and draw

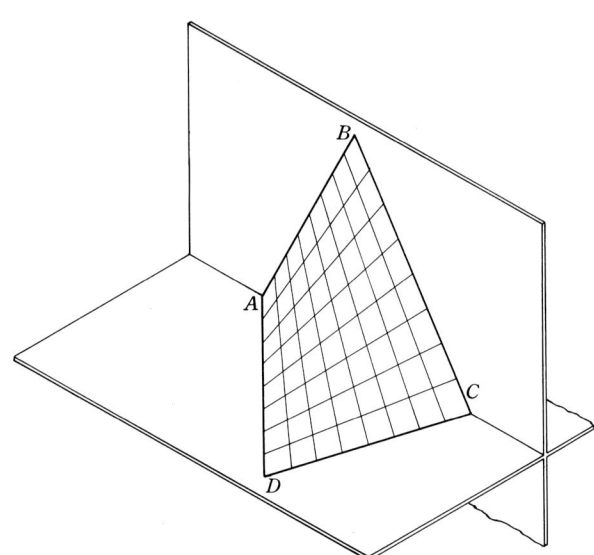

FIG. 7.91. Warped quadrilateral.

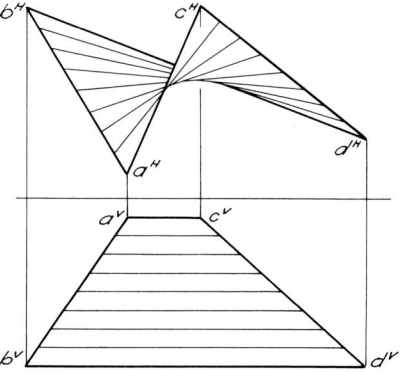

FIG. 7.92. Hyperbolic paraboloid with the H-plane as a plane director.

the elements in that view. They may then be projected to the other views by means of their intersections with the linear directrices as was done in Figure 7.92.

It is interesting to note that the true-length line of various slopes in Fig. 7.77 are elements of a hyperbolic paraboloid that are perpendicular to and intersect the line *OP*.

7.27.2 Hyperboloid of Revolution of One Sheet. This surface is generated by a right line that revolves about another nonparallel, nonintersecting right line as an axis. It may also be generated by a line touching three circles whose planes are perpendicular to a common axis through their centers. When the radius of the middle, or gorge, circle becomes zero, the surface is a cone, and when this radius becomes the same as the radius of the other two circles, the surface is a cylinder. Thus the cone and cylinder become the limits of the hyperboloid of revolution. Since this is a surface of revolution, a plane passed perpendicular to the axis of revolution cuts a circle from the surface.

The surface is doubly ruled, since two different lines may be revolved about the axis to give the same surface. These lines make equal angles with the base but slope in opposite directions. Figure 7.93 gives the projections of a hyperboloid of revolution, showing both sets of generatrices, *AB* and *CD*. The other elements shown are of the *AB* generation, no other positions of *CD* being shown.

The hyperboloid of revolution may be generated also by revolving a hyperbola about its conjugate axis. The surface is sometimes represented by showing its contour lines as illustrated in Fig. 7.94, these contour lines being the opposite branches of the hyperbola.

a. *Construction 1.* Given three curvilinear directrices, as in Fig. 7.93, divide the base

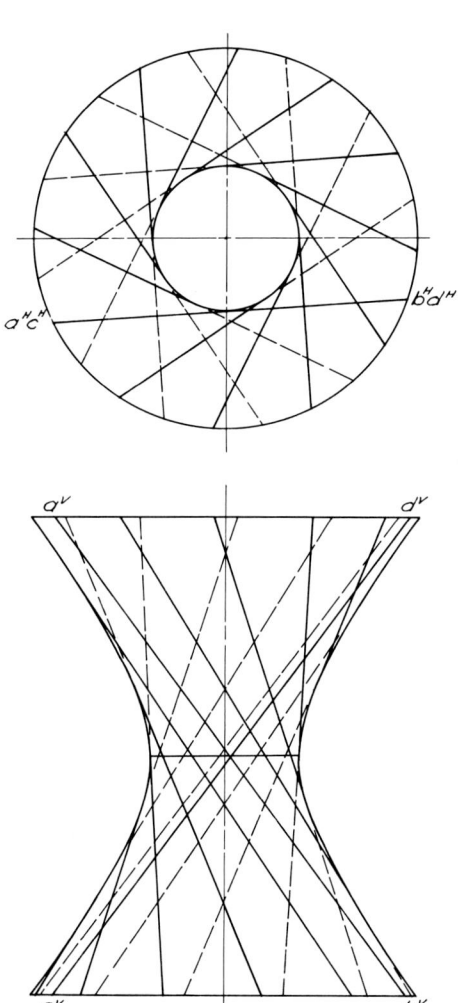

FIG. 7.93. Hyperboloid of revolution.

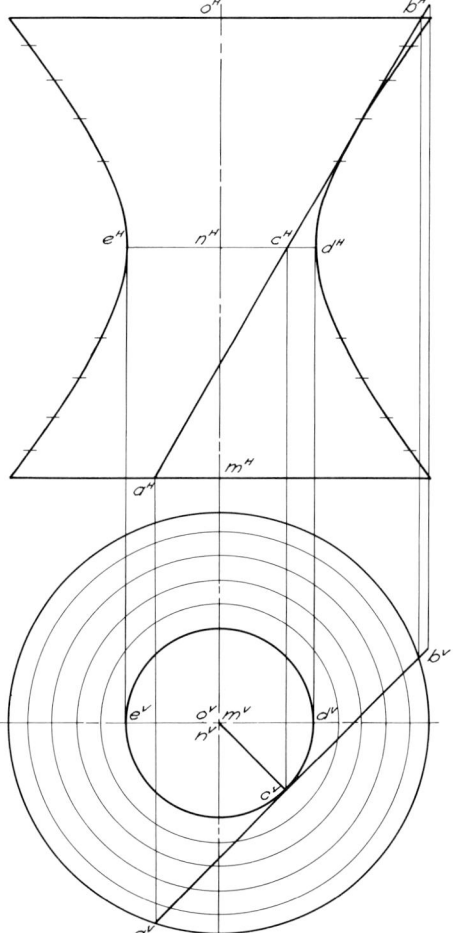

FIG. 7.94. Hyperboloid of revolution.

circle into a number of equal parts in the plan view and draw the horizontal projection of the elements through these points tangent to the gorge circle. Project the ends of the elements to the elevation and draw in the elements.

b. *Construction 2.* Given the axis *OM* and the generatrix *AB*, as in Fig. 7.94, draw the gorge circle with a radius *CN* equal to the perpendicular distance from *OM* to *AB*. Project c^V up to c^H on $a^H b^H$, and through c^H draw the vertical projection of the gorge circle perpendicular to the axis. Then locate the limiting points *D* and *E* on the gorge circle, the vertices of the contour hyperbola. By drawing other circles in the horizontal projection this process is repeated and other points on the hyperbola obtained.

The hyperboloid of revolution has practical applications in mechanism, the most important being found in the pitch surfaces of skew gears. If any two right lines not in the same plane are taken as axes, a third line may be taken in a plane parallel to the other two lines so that the hyperboloids formed about the two axes by using the third line as a generatrix will be tangent to each other. They will operate together as the pitch surface of gears, since all elements will in turn assume positions common to both surfaces. Figure 7.95 illustrates this use of the hyperboloid of revolution or rolling hyperboloid.

7.27.3 The Helicoid. A surface generated by a right line moving so that it always touches a helix and its axis, and making a constant angle with its axis, is called a *helicoid* and is the warped surface most frequently encountered in drafting. It occurs in the surface of screw threads, screw propellers, conveyors, circular staircases, and chutes. If the generatrix is perpendicular to the axis of the helix, the surface is a *right helicoid* as illustrated by the surface of a square thread. If the generatrix is inclined to the axis, it is an *oblique helicoid* such as the surface of a V-thread. The helicoid is actually a limitless surface but is usually considered only as the portion contained within a cylinder concentric with the helix.

a. *Construction of a right helicoid.* The first step is to construct the helix with a given pitch on the surface of a given cylinder. To do this the pitch is divided into equal parts, and the circle representing the cylinder is divided into the same number of equal parts (24 parts is very convenient), as shown in Fig. 7.96. The points on the circle are then projected up to the horizontal line drawn through the corresponding point on the axis. From the points on the helix, lines are drawn intersecting the axis and perpendicular to it. These lines form elements of the right helicoid.

b. *Construction of an oblique helicoid.* Given the limiting cylinder, the pitch of the helix, and the angle that the elements make with

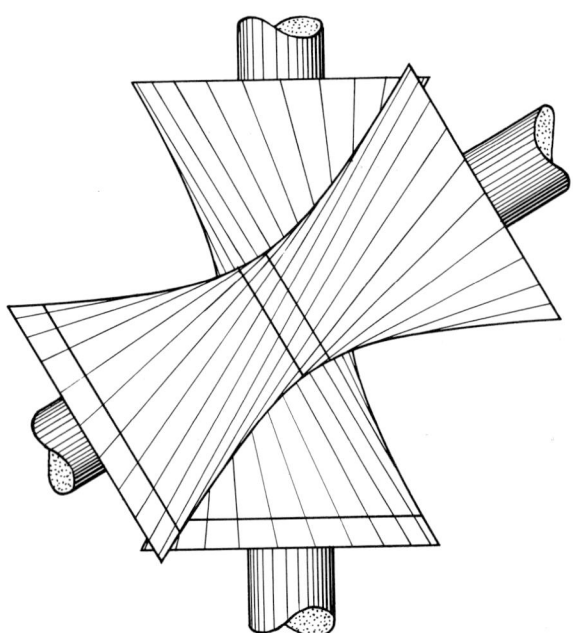

FIG. 7.95. Rolling hyperboloids.

the axis, first construct the helix in the same manner as explained for Fig. 7.96. Then draw in the element AB (Fig. 7.97), making it parallel to the horizontal plane. In this position the true angle that the element makes with the axis shows in the horizontal projection, and the horizontal projection $a^H b^H$ can be drawn. Because each element must move the same proportionate distance along the axis as it does along the curve, the other elements are drawn by laying out horizontal distances starting at b^H equal to the horizontal distances between the points on the helix and joining these points on the axis to the corresponding points on the helix.

7.27.4 Conoids. A *conoid* is a warped surface having a plane director and two linear directrices, one of which is a straight line and the other a curve. The curve may have any form, but closed plane curves are most commonly thought of.

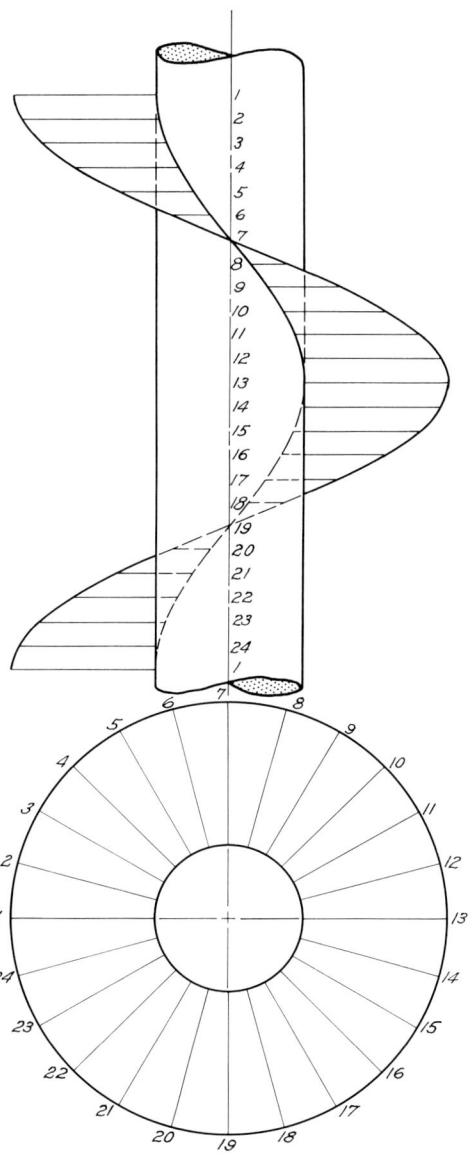

FIG. 7.96. Right helicoid.

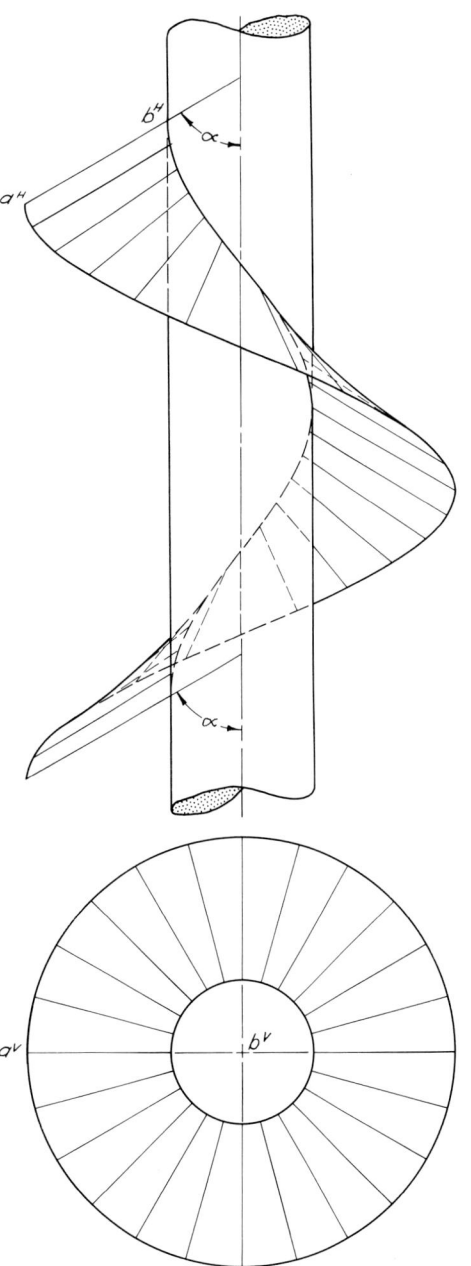

FIG. 7.97. Oblique helicoid.

If the straight-line directrix is parallel to the plane of the curved directrix and also perpendicular to the plane director, the surface is called a *right conoid*, as shown in Fig. 7.98a. A common application of this form is shown in Fig. 7.98b, where a roof must change from a curved to a flat section.

When the linear directrix, the curved directrix, and the plane director are oblique to each other, the surface is called an *oblique conoid*. Figure 7.99 shows the projections of an oblique conoid.

7.27.5 Cylindroid. A surface generated by a straight line moving so that it always remains parallel to a plane director and at the same time touches two plane curves, not lying in the same plane, is called a *cylindroid*. These curves are usually parts of circles or ellipses. See Fig. 7.100.

Construction. Since this problem is more frequently encountered in arch construction where the plane director is likely to be the horizontal plane, it is so illustrated here. However, it should be clear that any plane director and any plane curves, not lying in the same plane, can be used. Elements of the surface can be found by obtaining the edgewise view of the plane director and drawing elements in that view to intersect the two curvilinear directrices. See Fig. 7.100.

In Fig. 7.100, the two curvilinear directors *ABC* and *DEF* are to be joined by a cylindroid whose plane director is the *H*-plane. This means that elements can be drawn in the vertical projection by making them parallel to the reference line. They may then be carried to the *H*-projection to complete the required views.

7.27.6 General Procedure in Drawing Warped Surfaces. There are many more varieties of warped surfaces and other methods of generation, but the drawing of the surfaces usually consists of drawing elements touching certain curved or right-line directrices or parallel to certain directors. With a general knowledge of engineering geometry the student should be able to follow any of the necessary constructions. Accurate development of warped surfaces is not possible, but it is frequently necessary to get an approximate development. This is done by dividing the surface into small triangles and laying out these triangles side by side in true size. This is called the method of *triangulation*.

Intersections of warped surfaces with other surfaces may be found by following the general procedure given in Chapter 18 for all intersections. Usually the best method is to find the points where elements of the warped surface pierce the other surface.

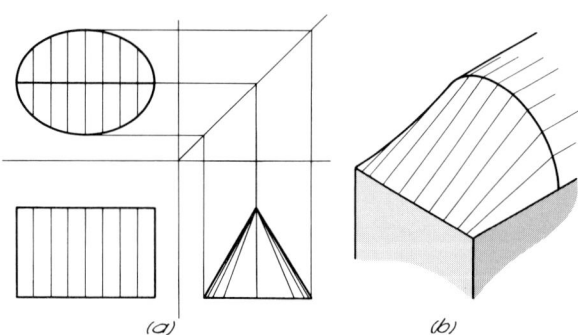

FIG. 7.98. Right conoid.

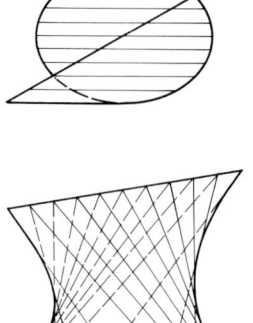

FIG. 7.99. Oblique conoid.

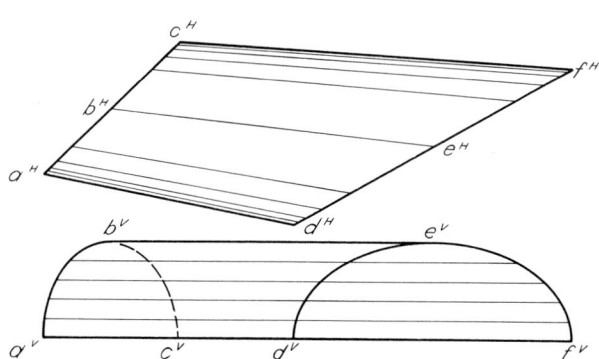

FIG. 7.100. Cylindroid.

SELF-STUDY QUESTIONS

Before trying to answer these questions, read the chapter carefully. Then, without reference to the text, answer as many questions as possible. For those that cannot be answered, the number in parentheses following the question number gives the article in which the answer can be found. Look it up and write down the answer. Check the answers that you did give to see that they are correct.

7.1 (**7.7**) To find the true length of an oblique line, an auxiliary plane must be set up _____ to the line.

7.2 (**7.2**) The first auxiliary plane is always set up _____ to one of the principal planes of projection.

7.3 (**7.3**) The projection of point A on auxiliary plane 3 is labeled _____.

7.4 (**7.8.2**) The distance from a^H to the H-1 reference line is the distance from point A to the _____ plane.

7.5 (**7.4**) It is possible to have _____ projections marked a^1 in a single problem.

7.6 (**7.2**) To be useful, an auxiliary plane should be set up _____ or perpendicular to a face or line of an object.

7.7 (**7.8.2**) The second auxiliary plane is set up _____ to the first auxiliary plane.

7.8 (**7.6**) The four fundamental operations for which auxiliary planes are used are
   1.
   2.
   3.
   4.

7.9 (**7.5**) The distance from a^1 to the 1-2 reference line is the distance from A to the _____ plane.

7.10 (**7.7**) In the first auxiliary view, the V-1 reference line represents the edge view of the _____ plane.

7.11 (**7.8**) To find the point projection of a line, the auxiliary plane must be set up _____ to the line.

7.12 (**7.8.2**) To find the point projection of a line, it is necessary to have the _____ of the line first.

7.13 (**7.8.1**) When the true length of the line shows in the vertical projection, the auxiliary plane must be set up _____ to the vertical plane and to the line, to find the point projection.

7.14 (**7.7**) When a line is oblique to all the principal planes, it requires _____ plane(s) to find the true length.

7.15 (**7.8.2**) When a line is oblique to all the principal planes, it requires _____ auxiliary planes to find the point projection.

7.16 (**7.10**) To find the edgewise view of a plane, the auxiliary plane should be set up _____ to some line in the plane.

7.17 (**7.10**) To find the edgewise view of a plane, it is best to work with a line _____ to a principal plane of projection.

7.18 (**7.11**) To find the true shape of a plane, an auxiliary plane should be set up _____ to the plane.

7.19 (**7.11.2**) To find the true shape of a plane, the first step is to find the _____ of the plane.

7.20 (**7.11.1**) To find the true shape of a plane, the reference line should be set up _____ to the edge view of the plane.

7.21 (**7.8**) To find the true size of the right section of a cylinder, the auxiliary plane should be set up _____ to the elements of the cylinder.

7.22 (**7.12**) In place of a reference plane, some _____ of the object or some axis of symmetry of the object may be used.

7.23 (**7.15**) To save space on a drawing, it is possible to show _____ views of symmetrical parts.

7.24 (**7.8.2**) Show by a sketch how to find the point projection of an oblique line.

7.25 (**7.19**) There are _____ possible relationships between two lines.

7.26 (**7.19.3**) Two intersecting lines intersect at a _____.

7.27 (**7.19.4**) A line and a point determine a _____.

7.28 (**7.7.2b**) In structural work the slope of a line is given by stating or showing the _____ and rise.

7.29 (**7.7.3**) The bearing of a line is the smaller of the two angles that the _____ projection of the line makes with a _____ and south line.

7.30 (**7.7.4**) The azimuth of a line is the angle that the _____ projection makes with a north and south line measured clockwise from the north.

7.31 (**7.19.5**) The shortest distance between two parallel lines is the distance between their point projections on a plane _____ to them.

7.32 (**7.19.6**) The true distance between two parallel lines will show on the _____ shape view of the plane of the two lines.

7.33 (**7.20**) A vector is a line that is used to represent any quantity that has both _____ and direction.

7.34 (**7.22.3**) The angle that an inclined or oblique plane makes with the H-plane shows in any view in which the _____ planes appear _____ wise.

7.35 (**7.22.6**) The strike of a plane is the direction or bearing of a _____ line in the plane.

7.36 (**7.22.7**) In geology and mining the slope of a plane in a downward direction is called the _____ of the plane.

7.37 (**7.22.10**) The distance between two parallel planes can be measured in a view where _____ planes appear _____ wise.

7.38 (**7.23.1**) The piercing point of a line with a plane can be located by inspection in a view that shows the plane _____ along with the corresponding projection of the line.

7.39 (**7.23.6**) A plane can be drawn parallel to a line by making it pass through a second line that is _____ to the first line.

7.40 (**7.23.5**) A line is parallel to a plane when it is _____ to any line in the plane.

7.41 (**7.23.10**) A plane can be drawn perpendicular to a line by passing it through two lines each of which is _____ to the given line.

7.42 (**7.23.10**) A plane can be passed perpendicular to a line by making the _____ of the plane perpendicular to the true-length view of the line.

7.43 (**7.23.13**) The intersection of two planes is a straight line passing through two points each _____ to both planes.

7.44 (**7.23.15**) When two planes show edgewise in the same view, the _____ between them shows in this view in true size.

7.45 (**7.24**) Two lines that are nonparallel and nonintersecting are called _____ lines.

7.46 (**7.26a**) The locus of points equidistant from two given points is the plane perpendicular _____ of the line joining the points.

7.47 (**7.26d**) The locus of points at a given distance from a plane is two parallel _____, one on each side of the given plane.

7.48 (**7.26e**) The locus of points at a given distance from a straight line is the surface of a _____ circular _____ having the line as an axis and the given distance as a radius.

7.49 (**7.27.3**) A surface generated by a right line moving so that it always touches a helix and the axis of the helix and makes a constant angle with the axis is called a _____.

7.50 (**7.27.3**) The _____ is the surface found on screw threads, screw propellers, and conveyors.

7.51 (**7.27.4**) A conoid is a warped surface having a plane director and two linear directrices, one of which is a _____ line and the other a curve.

CHAPTER 8
SECTIONAL VIEWS

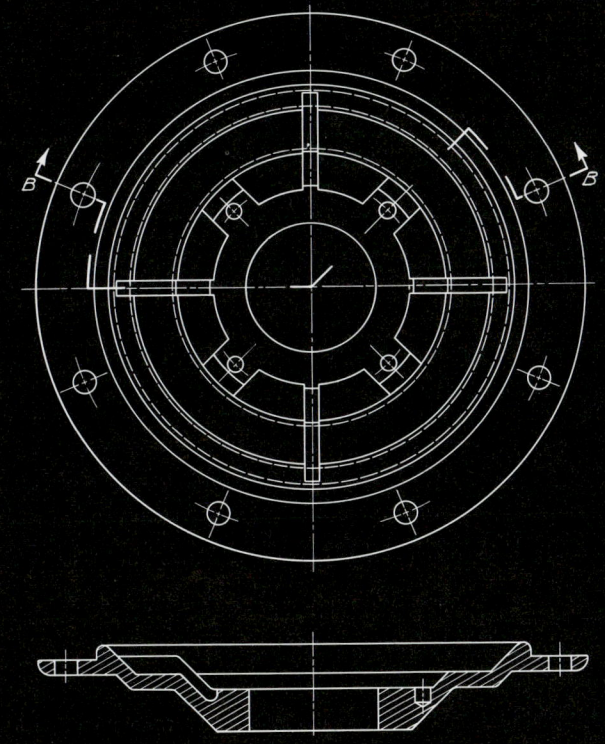

CHAPTER 8

8.1 PURPOSE OF SECTIONAL VIEWS

With the principles of projection thus far discussed, the interior parts of an object that are hidden from view can only be represented by dash lines. When these dash lines become numerous, the drawing is difficult to interpret. To overcome this difficulty the sectional view is used.

A *sectional view* is any view seen when a portion of the object nearest the observer has been imagined removed by cutting planes, thus revealing the interior construction. See Fig. 8.1. The term *cutting plane* comes from the fact that a section view looks as if the object has been "cut" with a saw. The inclined lines on the cut portion of the object can be thought of as saw tooth marks left on the material. See Section 8.5. Figure 8.2 shows an object with a right-side view and the left-side view in section. Using the section with the front view, one can interpret the drawing more easily than by using the right-side view with the front view.

A sectional view, or section as it is called, is also used to show the exact shape of exterior parts that are so rounded or curved or change shape so rapidly that the usual two- or three-view

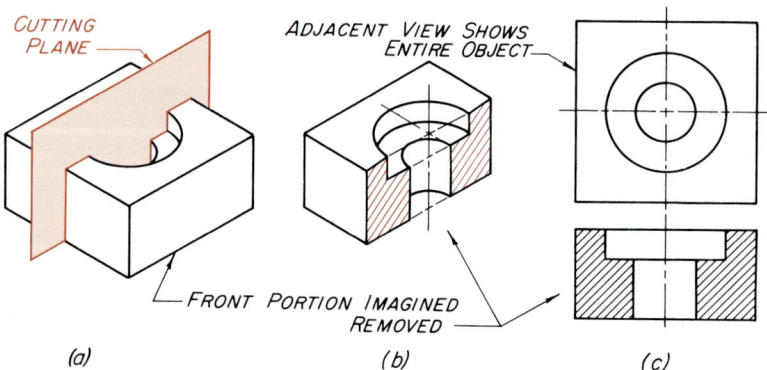

FIG. 8.1. Theory of sectional views.

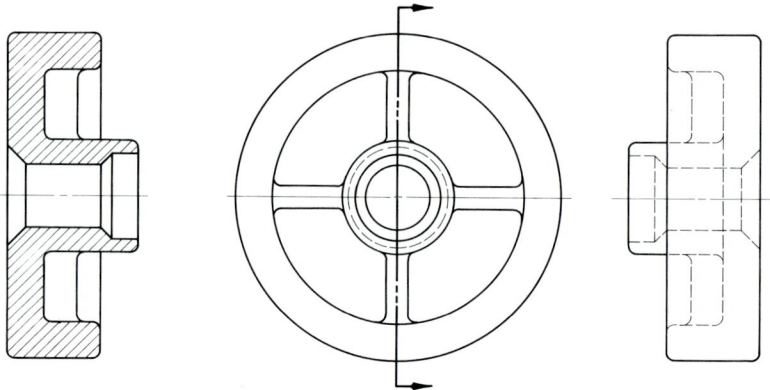

FIG. 8.2. Sectional view.

drawings do not reveal their true form. The fenders on cars, housings for revolving parts, ship propellers, and airplane wings are of this type.

8.2 CUTTING PLANE SYMBOLS

Since the location of cutting planes must frequently be shown, a standard type of line is used for this purpose. See Fig. 8.3. This distinguishes it from other lines used on the drawing. It should also be noted that for certain types of sections a center line is sometimes used to show the location of the cutting plane.

8.3 LOCATION OF CUTTING OR SECTION PLANES

To show the interior construction of an object, the main cutting plane is passed through an axis of symmetry parallel to one of the principal planes of projection, as shown in Fig. 8.4. Other cutting planes, if necessary, are passed parallel or perpendicular to the main cutting plane, as shown in Fig. 8.5.

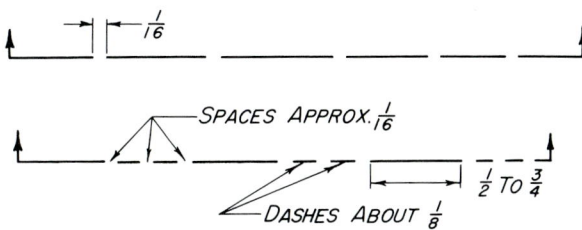

FIG. 8.3. Symbol for cutting plane line.

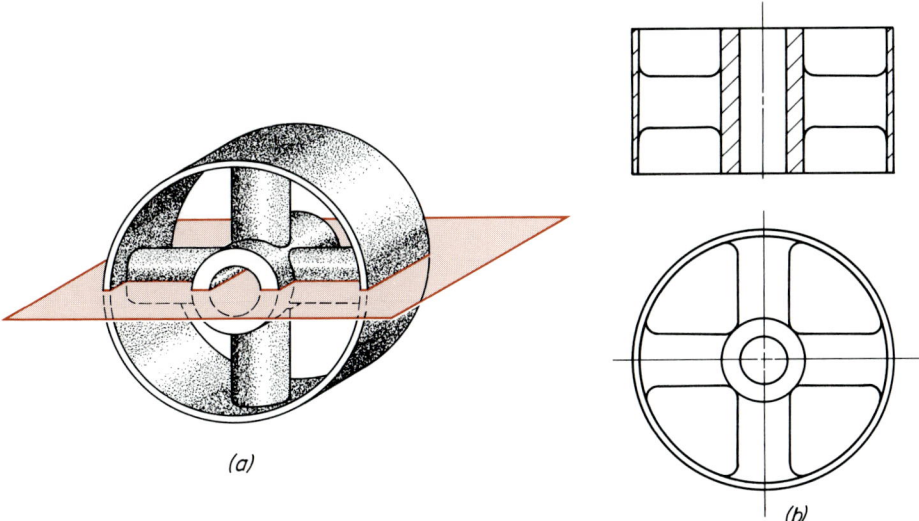

FIG. 8.4. Cutting plane parallel to horizontal plane.

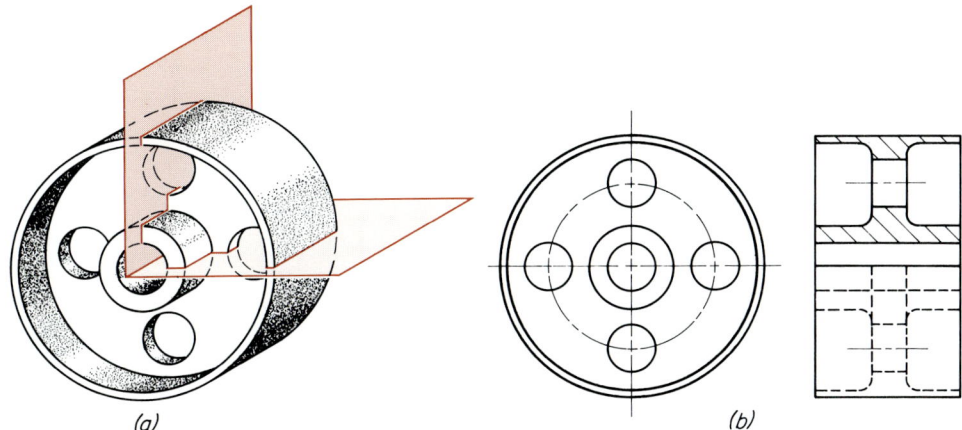

FIG. 8.5. Principal cutting plane parallel to profile plane.

196 SECTIONAL VIEWS

The cutting plane need not be continuous but may be offset as shown in Fig. 8.6. It may also be turned through an angle, as shown in Fig. 8.17. This gives rise to various types of sectional views, as discussed later.

To show the true shape of a part such as a spoke of a wheel or a connecting arm, the plane may be passed perpendicular to the axis, as shown in Fig. 8.7a and then revolved about its central axis, in place, as in Fig. 8.7b.

When a cutting plane is passed on an axis of symmetry, as shown in Figs. 8.14 and 8.15, it is not necessary to indicate the location of the plane. When, however, the cutting plane is offset, as in Figs. 8.6 and 8.16, its location must be shown in the view where it appears edgewise. In this chapter cutting plane lines are used in several drawings, whereas in actual practice they would not be. That is, in full and half-sections. They are placed in these drawings in order to make it clear to the young engineer the actual position of the cutting plane, for example, Figs. 8.2 and 8.13.

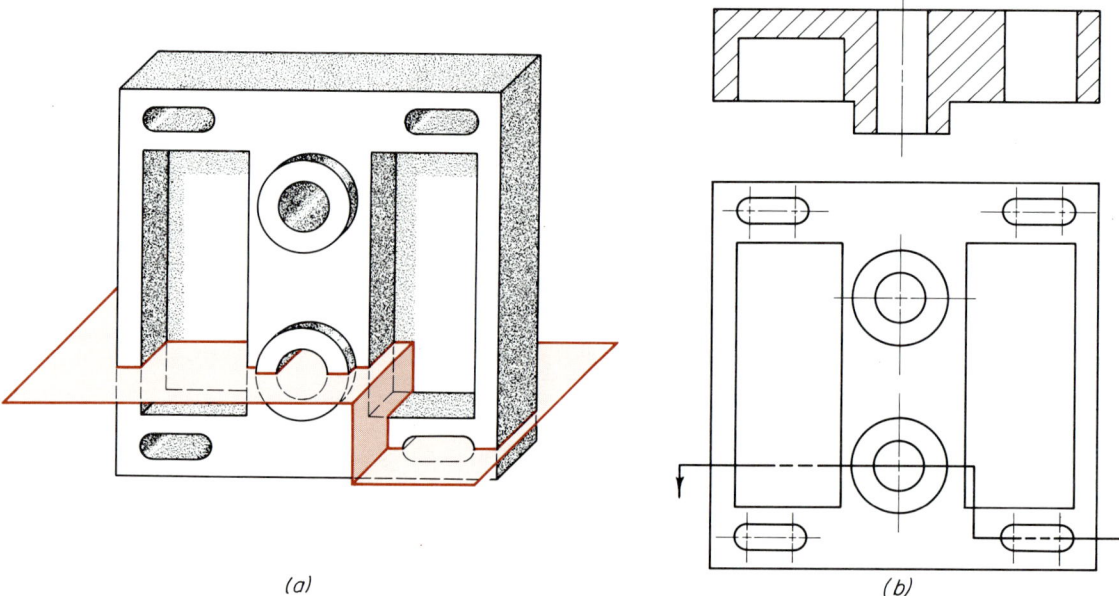

FIG. 8.6. Cutting plane parallel to horizontal but offset.

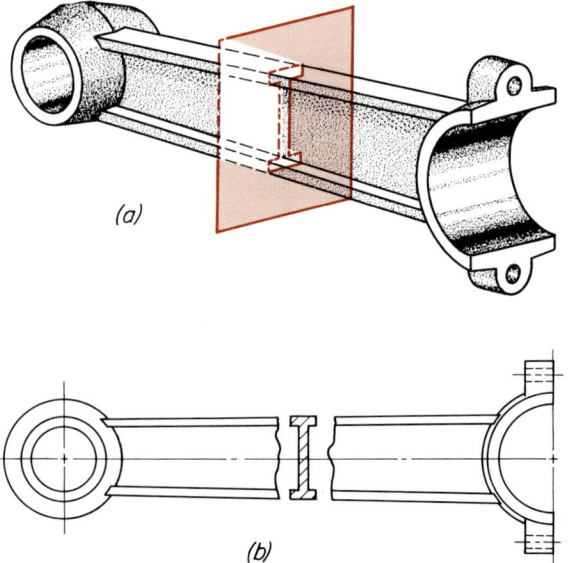

FIG. 8.7. Position of cutting plane for a revolved section.

8.4 DRAWING THE SECTIONED VIEWS

The view that is to be made in section is represented as though the portion of the object nearest the observer were actually cut away by the section planes and removed. It should be noted that the other views are not affected in any way and always represent the entire object. See Figs. 8.14 to 8.16.

8.5 SECTION LINING

In order that those parts of an object that have been cut by the section plane may stand out on the drawing, a conventional scheme called *cross-hatching,* or section lining, is employed. This is done by drawing light, inclined lines on those parts of the view of the object where the plane actually cuts the material of which the object is made.

8.5.1 Direction of Section Lines.
The section lines may be drawn at any angle to the horizontal, but 45° lines are usually employed. The direction should be chosen so that the cross-hatch lines will not be parallel or perpendicular to any of the bounding lines of the area being shaded. See Fig. 8.8.

8.5.2 Spacing of Section Lines.
Section lines should be spaced uniformly and aproximately $\frac{3}{32}$ to $\frac{1}{8}$ in. apart. See Fig. 8.9d. For very large areas the spacing may be $\frac{3}{16}$ in.

8.5.3 Weight of Section Lines.
The weight and thickness of section lines should be the same as those for dimension lines. In all cases they should come completely up to the outlines of the object. Figure 8.9 shows examples of good and faulty section lining.

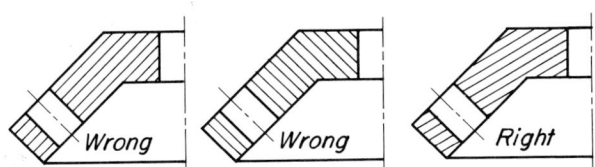

FIG. 8.8. Correct and incorrect section lining.

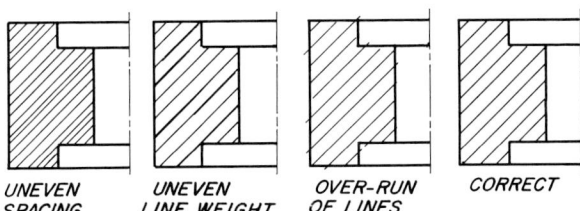

FIG. 8.9. Common errors in section lining.

8.5.4 Cross-hatching Adjacent Parts.
When several adjacent parts of a machine are of the same material and are cut by a section plane, each part should be distinguished from its neighbor by a change in slope of the section lining. The more nearly the crosshatch lines approximate 90° to each other, the better the effect will be. See Fig. 8.10.

If the section plane cuts the same part at different places in the object, the section lining should be given the same slope in the corresponding places on the drawing, as shown in Figs. 8.13 and 8.29.

When three adjacent parts of an assembly are cut, two of the parts can usually be crosshatched at 45°, but a different angle such as 30° or 60° must be used for the third part. See Fig. 8.10.

8.5.5 Partial or Outline Section Lining.
In some instances, the appearance of large areas that must be sectioned can be improved by carrying the cross-hatching only a short distance from the outlines and leaving the central portions blank, as in Fig. 8.11. Not only is time saved by this practice, but also the appearance of the drawing is improved by eliminating large shaded areas.

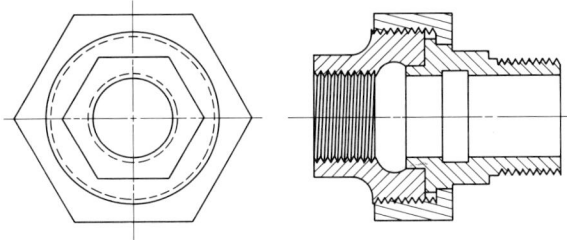

FIG. 8.10. Sectioning adjacent parts of the same material.

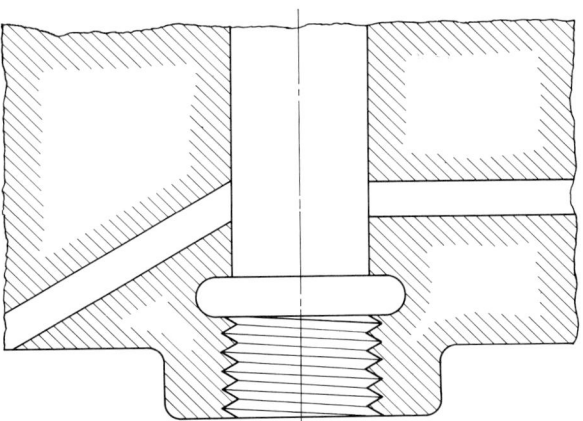

FIG. 8.11. Partial section lining.

8.6 HIDDEN LINES IN SECTIONED VIEWS

It is standard practice to omit all hidden lines in the sectioned portion of a view. In making a detail of a single part, the invisible lines may be shown on the unsectioned half of the view if they are needed for dimensioning.

In making a half-sectioned assembly, the invisible outlines are usually omitted from the unsectioned half as well as from the sectioned half so that one side shows internal construction and the other half the external appearance. See Fig. 8.12.

In several of the drawings in this chapter (namely, 8.5, 8.15, 8.25, and 8.26) in the nonsectioned portion of the drawing, all hidden lines have been included. This was done in order that the young engineer might more easily interpret the drawings.

8.7 VISIBLE LINES BEHIND THE SECTION PLANE

Visible outlines appearing behind the cutting plane must always be shown in the sectioned view. Beginners frequently omit them. Correct and incorrect practice is shown in Fig. 8.13.

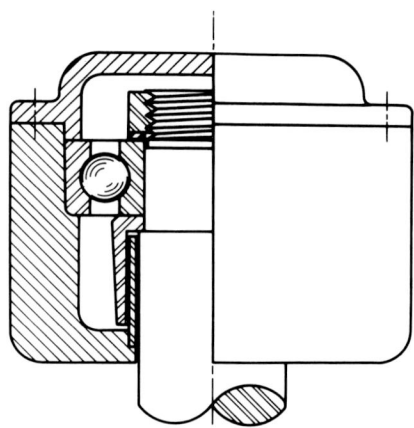

FIG. 8.12. Hidden lines omitted in a sectional drawing.

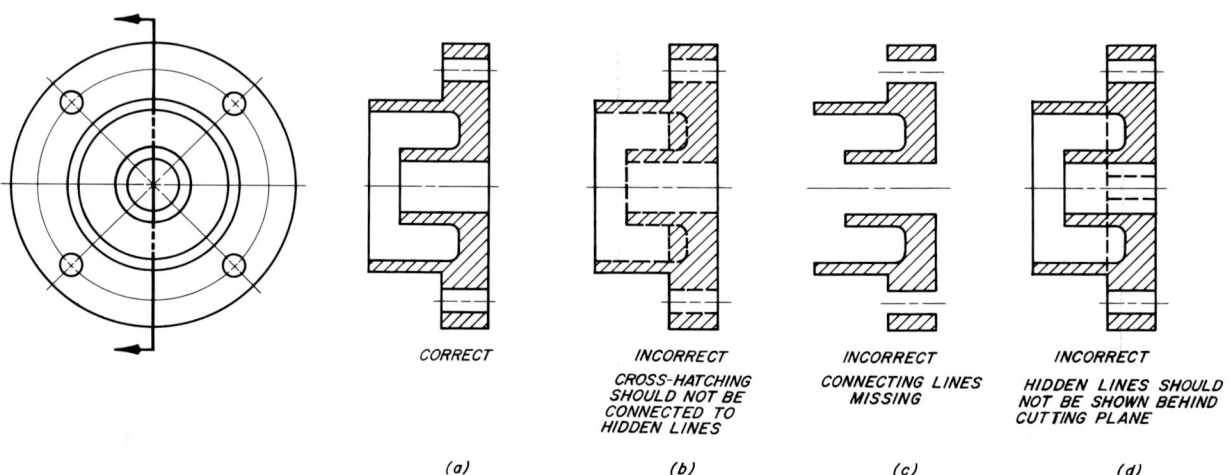

(a) CORRECT

(b) INCORRECT — CROSS-HATCHING SHOULD NOT BE CONNECTED TO HIDDEN LINES

(c) INCORRECT — CONNECTING LINES MISSING

(d) INCORRECT — HIDDEN LINES SHOULD NOT BE SHOWN BEHIND CUTTING PLANE

FIG. 8.13. Visible lines behind cutting plane shown.

8.8 FULL SECTIONS

When the section plane passes completely through the object on the plane of symmetry, as shown in Fig. 8.14a, one half of the object is imagined to be removed in drawing the section view. Figure 8.14b shows the object as it is imagined to appear in making the section view. From this it can be seen that the entire front view will be in section, which gives it the name *full section*. The correct orthographic projections of the object with a full section in the front view are shown in Fig. 8.14c. The following points should be observed in making a drawing with one view in full section.

a. In making the sectioned view, one half of the object is imagined removed.
b. The adjacent view shows the entire object, unless a part is shown broken away beyond the center line to save space as in Fig. 6.55.
c. Invisible lines behind the section are omitted.
d. Visible lines behind the section are shown.
e. Only the parts actually cut by the section plane are crosshatched.
f. The position of the cutting plane is not shown on the final drawings, for full sections.

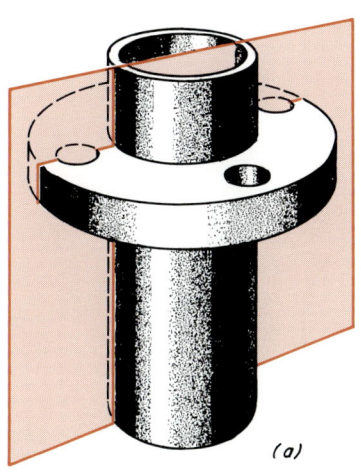
(a)

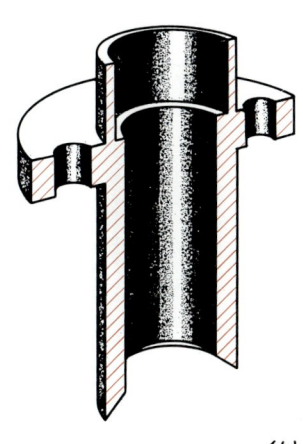
(b)

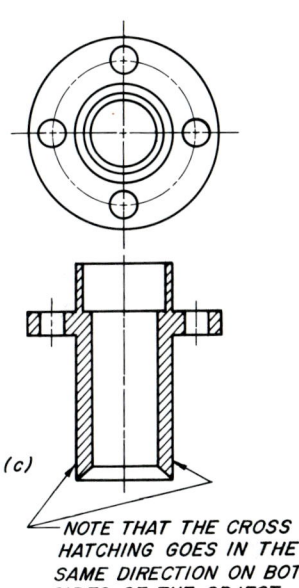
(c)

NOTE THAT THE CROSS HATCHING GOES IN THE SAME DIRECTION ON BOTH SIDES OF THE OBJECT

FIG. 8.14. A full-sectioned view.

200 SECTIONAL VIEWS

8.9 HALF SECTIONS

When two perpendicular cutting planes are passed partway through the object on planes of symmetry, as shown in Fig. 8.15a, one quarter of the object is imagined to be removed in drawing the sectioned view. Figure 8.15b shows the casting as it would appear in making the sectioned view. From this it can be seen that only one half of the front view will show in section and the other half will be an external view. From this the name *half section* is derived for such a view.

The following rules should be observed in making a half section.

a. One half of the sectioned view will be in section, while the other half is an external view.
b. Hidden lines will normally be omitted from both sides of the view unless necessary for clearness of interpretation.
c. Hidden lines may be shown on the external half, if needed for dimensioning.
d. The line separating the sectioned half from the external half should preferably be a solid line but the use of a center line is also approved by the ANSI. See Figs. 8.15c and d.

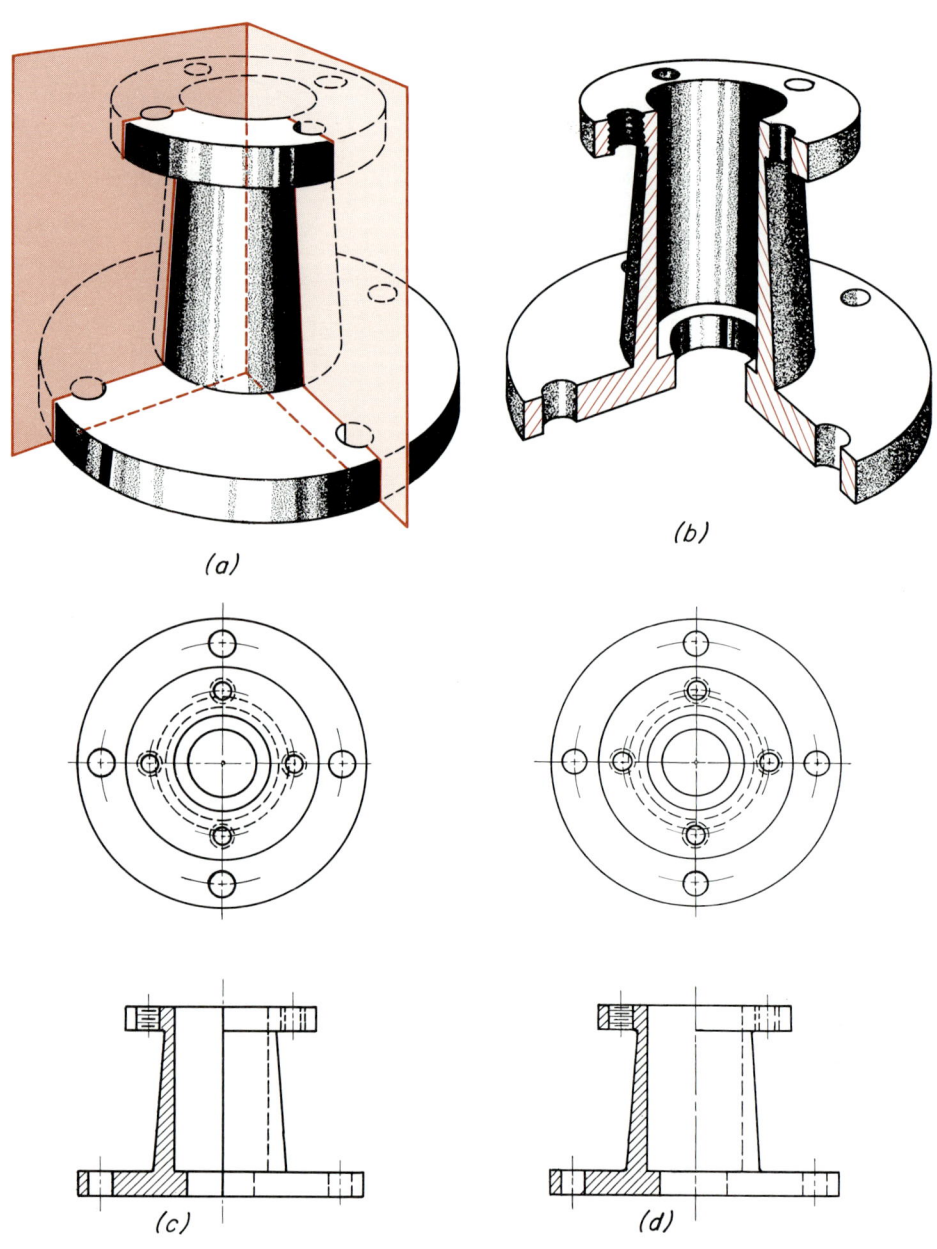

FIG. 8.15. A half-sectioned view.

e. The location of the cutting plane is not shown in the view adjacent to the sectioned view. See top view in Figs. 8.15c and d.

8.10 OFFSET SECTIONS

The section cutting planes are almost always passed through axes of symmetry of the object. If several such axes of symmetry occur, not coinciding with a single central axis of the object, the cutting plane is offset, as in Fig. 8.16, to include two or more of these axes of symmetry. No particular difficulty is experienced in reading a drawing when this is done.

The following rules should be observed for offset sections.

a. The location of the cutting plane must be shown by the proper symbolic line in the adjacent view. See Figs. 8.6 and 8.16.
b. Arrows should be placed on the ends of the cutting plane line, always pointing away from the sectioned view. This shows the direction in which the view is made.
c. No indication of the offset or break in the cutting plane is shown in the sectioned view.

8.11 ALIGNED SECTION

In sectioning certain objects, more information can be given if the section plane is bent or turned at an angle, as shown in Fig. 8.17. The various

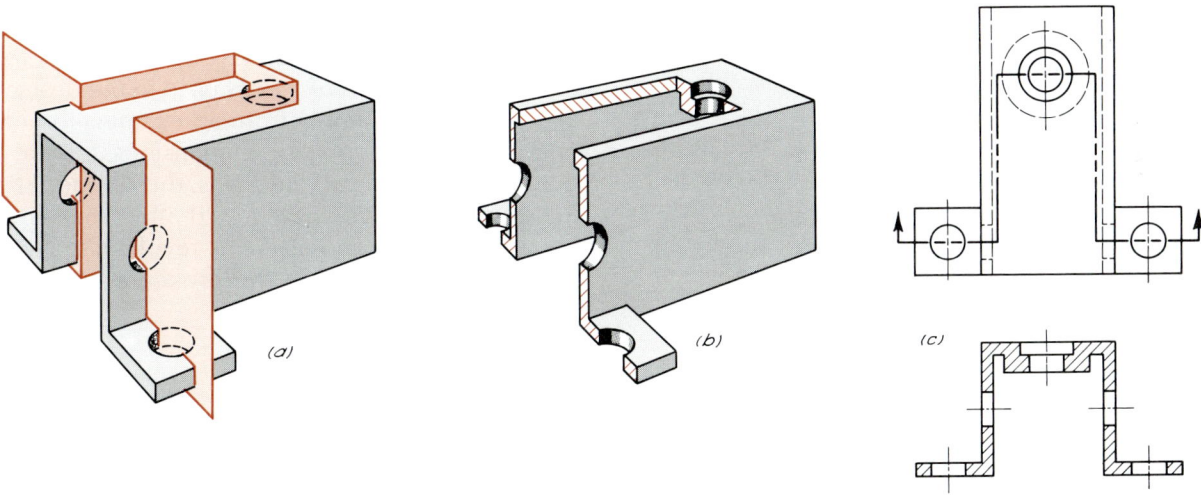

FIG. 8.16. An offset section.

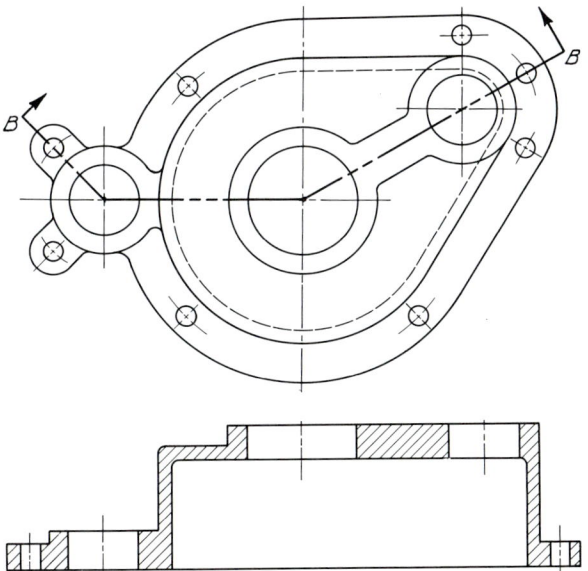

FIG. 8.17. Section plane turned at an angle: aligned section.

segments of the cutting plane are imagined to be revolved until they are parallel to the principal plane of projection. This shows all cut portions of the object in true size in the sectioned view. This is known as an *aligned section*.

In this type of sectional view the following rules should be observed.

a. No indication of the bends in the cutting plane is shown in the sectioned view.
b. The exact location of the cutting plane must be shown in the adjacent view by the proper symbolic line. See Figs. 8.17 and 8.18.
c. Arrows must be placed at the ends of the cutting plane line, showing the direction in which the object is viewed in the sectioned view.
d. The arrows always point away from the section.
e. The sectioned view is made as though the parts at an angle were rotated to be parallel to the plane of projection.
f. The sectioned view, therefore, shows the true shape of the entire section.

8.12 REVOLVED SECTIONS

In many cases the cross section of some part of a structure or machine is necessary for the purpose of giving the shape or size description. Frequently the easiest way to obtain this without drawing an extra view is by means of a revolved section. The revolved section is obtained by passing a cutting plane through the member perpendicular to one of the principal planes of projection and then revolving the cross section thus obtained about its own axis of symmetry until it is parallel to the plane of projection. In this position the section will show the true shape directly on that view. This method is particularly useful for structural shapes, spokes of wheels, arms, handles, and the like. Several examples of revolved sections are shown in Figs. 8.19 and 8.20.

The advantages of these cross sections are very great, inasmuch as they convey instantly to the mind the shapes of pieces used in the design of any structure, simply from an examination of one view of the object. Dimensions are frequently placed on such sections, thereby adding to their effectiveness. Many drafters make a break in the piece and place the revolved section in the opening between the broken ends. This

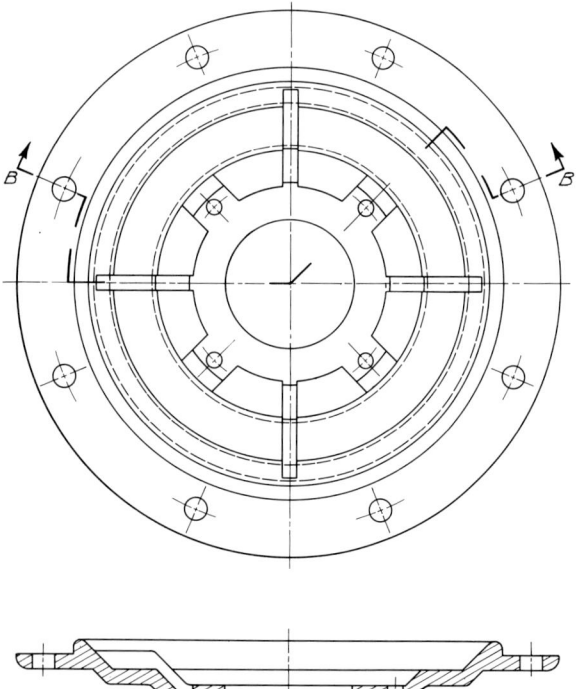

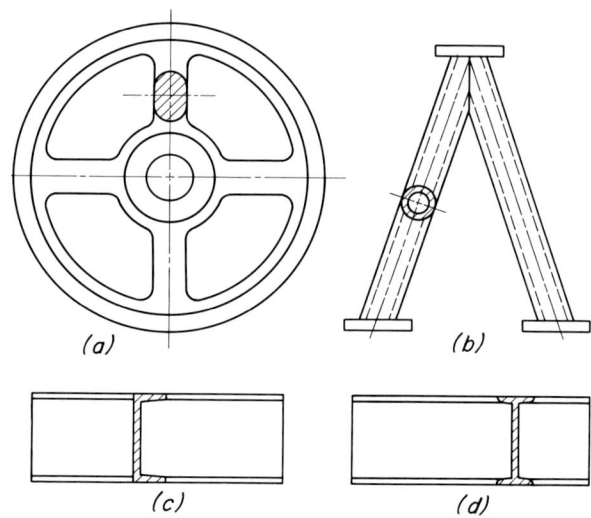

FIG. 8.19. Revolved section.

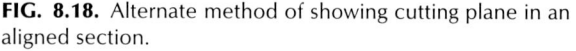

FIG. 8.18. Alternate method of showing cutting plane in an aligned section.

8.13 REMOVED SECTIONS

convention may be used for any revolved section but is almost a necessity for sections such as the one in the front view of Fig. 8.20. It should be observed in this connection that the revolved section always shows the true shape of the cross section of the piece regardless of the direction of the axis or boundary lines of the part sectioned. Streamlined objects, such as car fenders and propeller blades, are sometimes shown by taking sections at regular intervals.

8.13 REMOVED SECTIONS

In complicated drawings it frequently happens that it is necessary to clarify the construction of certain parts of a machine when it is not desirable to take a full or half section. This may be done by drawing what is known as a *removed* or *detail* section. The removed section is similar to the revolved section except that it is not revolved in place but removed to another area on the drawing. The location and extent of the cutting plane are indicated on the principal views by means of the usual symbol, with arrows indicating the direction of viewing the section. The cutting-plane line is marked with a capital letter at each end of the cutting plane. See Fig. 8.21.

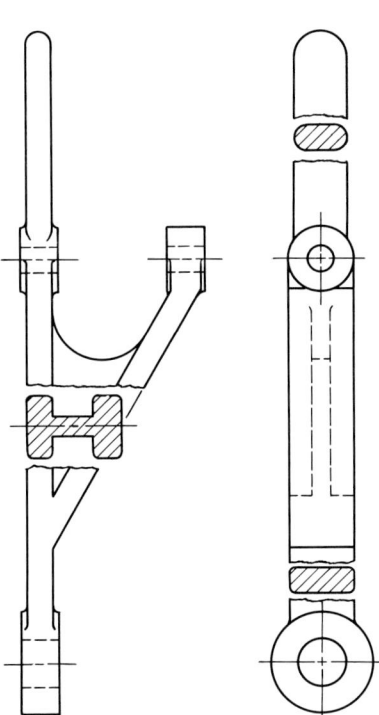

FIG. 8.20. Alternate method of showing a revolved section.

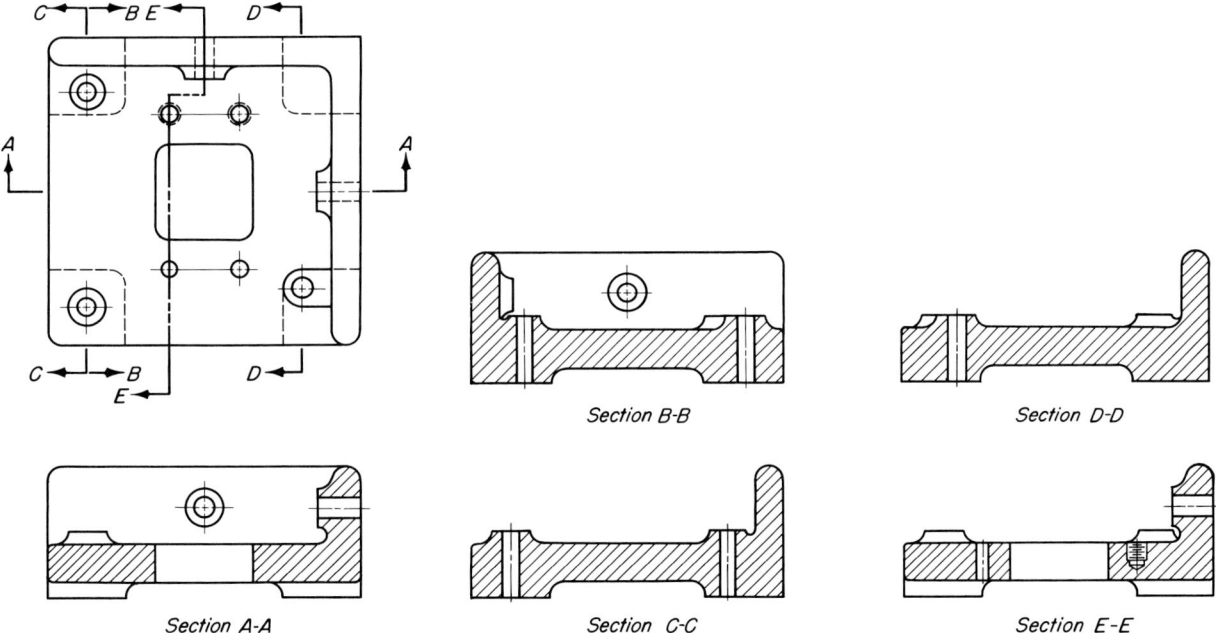

FIG. 8.21. Marking removed sections.

8.13.1 Drawing the Removed Section. The following points should be observed for removed sections.

a. The removed section is drawn at some clear place on the drawing sheet to give it a well-balanced appearance.
b. It should be drawn in the same position that it would have in the original view of the object, that is, it should not be upside down nor at an angle.
c. The location of the cutting plane must be clearly labeled, as in Fig. 8.21
d. If a scale larger than that of the original view is used, it must be specified under the title.
e. Removed sections may also be placed on a center line extended, when the center line serves as a cutting-plane line, as in Fig. 8.22.

8.13.2 The Removed Section on a Separate Sheet. In some cases the sheet on which a removed section would normally be drawn is so full that there is no room for the removed section. When this occurs, observe the following rules.

a. The removed section may be drawn on another sheet. This sheet may have other details on it.
b. The location of the cutting plane must be marked by letters and the number of the drawing sheet on which it occurs.
c. The removed section must be labeled, the scale given, and the sheet number given which shows the location of the section on the principal views. In other words, the section and its location must be cross-referenced.

8.13.3 Removed Section on Zoned Drawings. For ease of reference, very large drawings are frequently zoned by marking off equal spaces horizontally and vertically along the border lines. The horizontal spaces are lettered A, B, C, and so forth, and the vertical spaces are numbered as shown in Fig. 8.23. For example, the area in vertical band C and horizontal row 3 is designated zone C-3. Thus any removed section can easily be located when the zone is specified.

If the removed section is not on the same zoned sheet with its cutting-plane location on a principal view, both drawings must be cross-referenced by sheet and zone numbers, as well as by letters such as Section A-A. The cutting-plane location would likewise have to be indicated in a similar manner.

8.13.4 Advantages of Removed Sections. The removed section gives considerable freedom of choice.

a. The first advantage of the removed section lies in the fact that the sectioned views may be drawn to a much larger scale than the

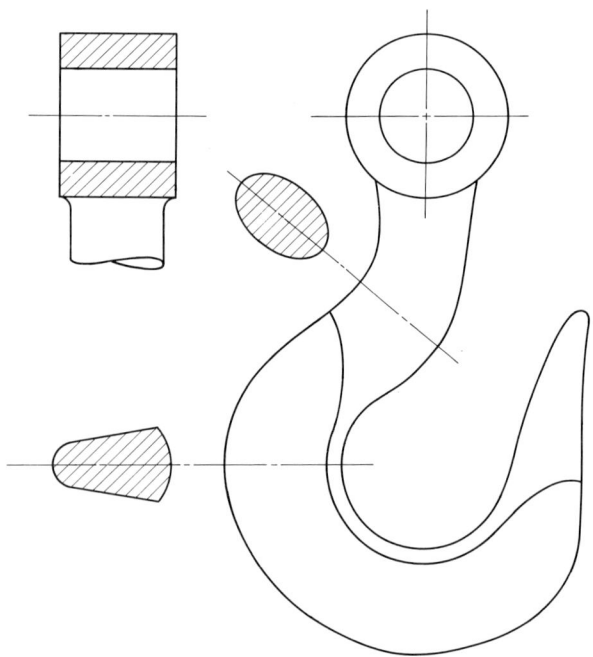

FIG. 8.22. Removed section on extended center line.

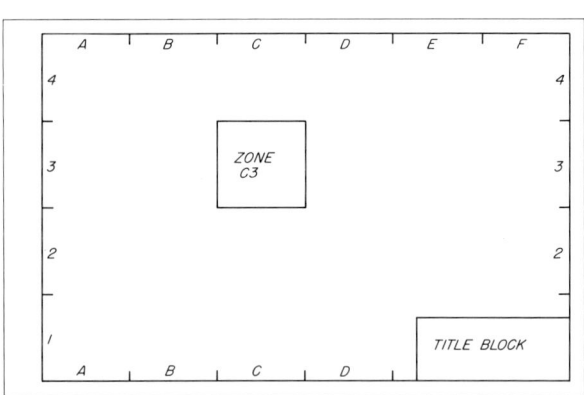

FIG. 8.23. Zoned drawing sheet.

main or principal views, thus showing the details more clearly.

b. A second advantage is that the main views showing the general outlines of the object are not confused by a large number of broken lines or cross-section lines.

c. A third advantage may be found in the better balanced drawing sheets, since the removed section may be placed in any open space on the sheet.

8.14 AUXILIARY SECTIONS

Another type of removed section is the auxiliary section. The section is easily understood since it is made in projection with one of the principal views, as shown in Fig. 8.24. The auxiliary plane is set parallel to the cutting plane and the auxiliary section drawn as though the object had been actually cut on the section plane. When the position of the section plane is shown, the arrows should point away from the sectional view.

8.15 BROKEN-OUT SECTIONS

In some instances a single interior detail needs to be made clearer than any of the principal views are able to make it and yet it does not justify a full or half section. Or it may be possible that a full or half section would eliminate some exterior detail that must be preserved. In a situation of this kind it is possible to break out a small portion of the outer part covering just the area desired. The broken part is outlined with an irregular freehand line to mark the break. See Fig. 8.25. This is called a *broken-out section*.

Another type of section that may be classified as a broken-out section is coming into use. This type of section, shown in Fig. 8.26, functions as a half section and serves its purpose better. Drawings that have been sectioned exactly to the center line have sometimes been manufactured exactly as shown, even though noted "*symmetrical about ℄*." Thus three fourths of a part may be made and shipped. This error is not likely to occur with the type of section shown in Fig. 8.26.

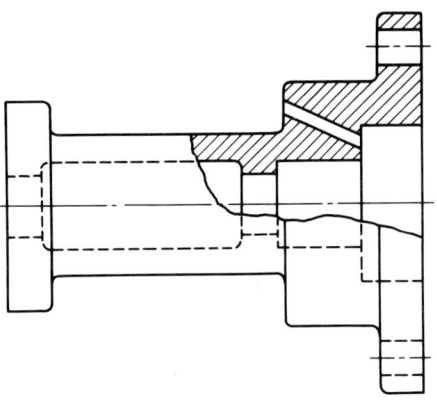

FIG. 8.25. Broken-out section.

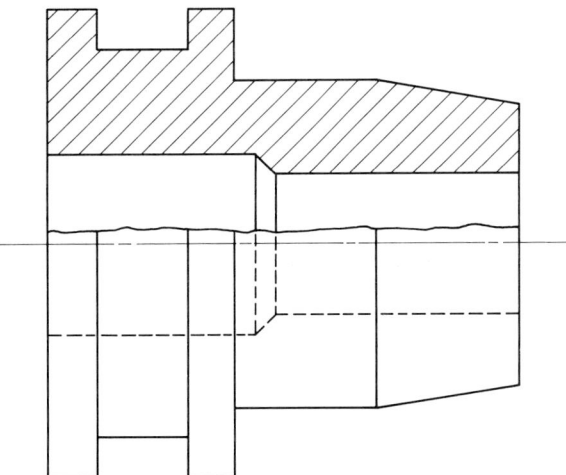

FIG. 8.26. Broken-out section.

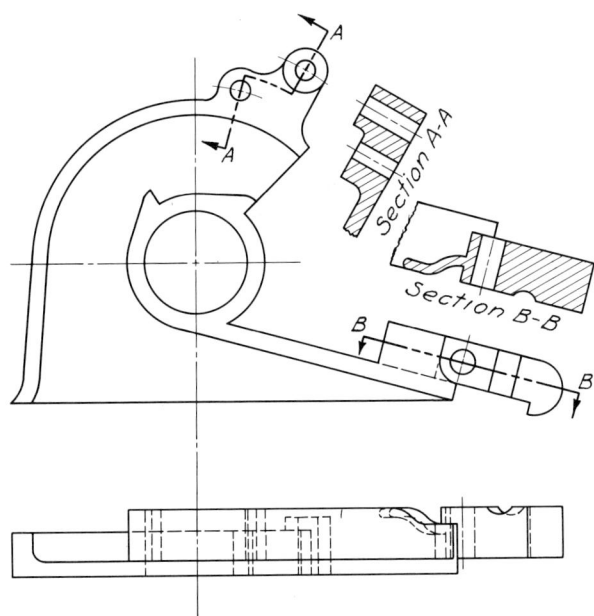

FIG. 8.24. Auxiliary sections.

8.16 PHANTOM SECTION

One type of section that is seldom used is the invisible or phantom section illustrated in Fig. 8.27. This section is marked by dotted crosshatching and is used to emphasize some inner part while at the same time retaining all the exterior construction. It is also used occasionally to show the method of attaching or the manner in which an adjacent part joins the part being drawn.

8.17 THIN SECTIONS

When thin members such as gaskets, plates, channels, and angles are shown in section, they may be made solid black if the scale of the drawing is small. Adjacent pieces should then be separated by a space, as shown in Fig. 8.28.

8.18 CONVENTIONAL PRACTICE IN SECTIONING

In ordinary two- and three-view drawings the rules and principles of orthographic projection are usually strictly followed. When sectional views are introduced, however, situations arise in which the drawing can be more readily understood and more easily made if some principles of projection are violated. Some of these situations are discussed in the following paragraphs.

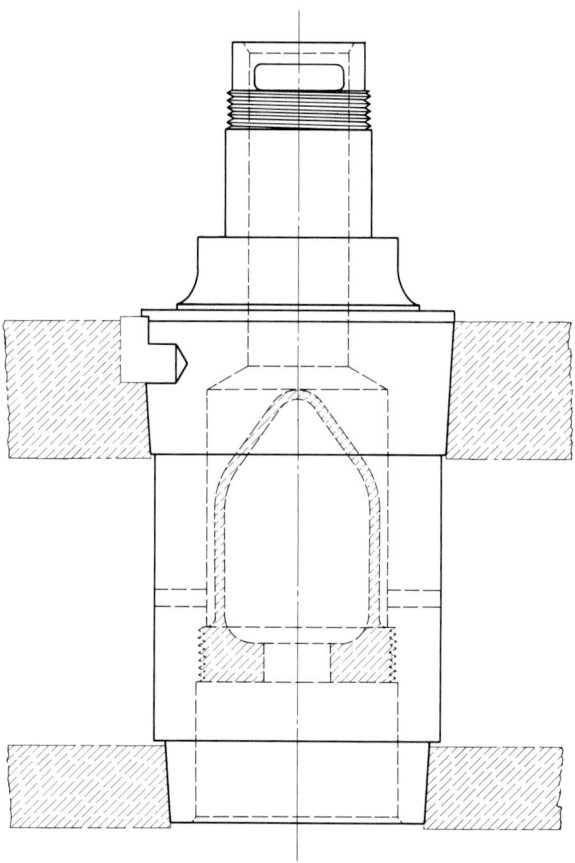

FIG. 8.27. A phantom section.

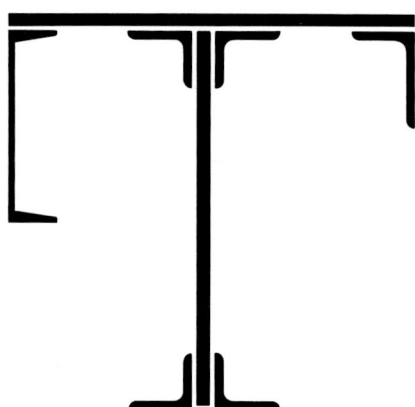

FIG. 8.28. Thin members in section.

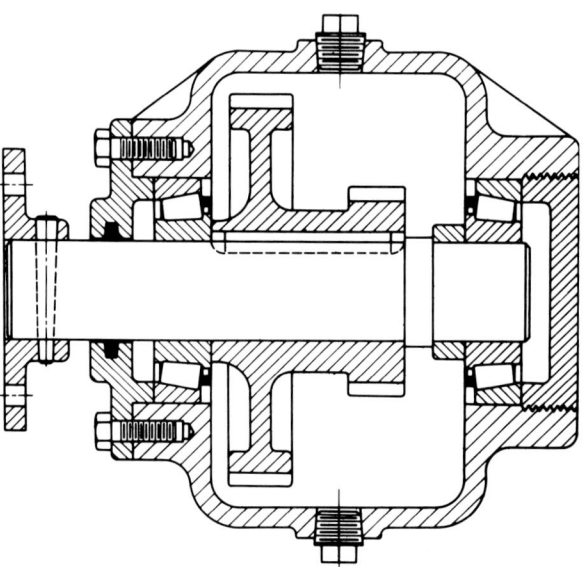

FIG. 8.29. Solid shafts, pins, screws, and keys are not sectioned.

8.18.1 Solid Shafts and Bars, Bolts, Pins, Keys, and Screws in Sections. When the cutting plane passes through the longitudinal axes of solid cylinders such as shafts, bolts, and screws, it is the custom to consider that these parts are not cut by the section plane. Nothing would be gained by showing the solid interior of such parts. A great deal of time is saved by eliminating large areas of cross-hatching. Figure 8.29 illustrates that the shaft, tapered pin, hexhead machine screws, key, gear teeth, and tapered rollers are not crosshatched.

8.18.2 Spokes of Wheels and Thin Webs. Spokes of wheels are not sectioned, even though the cutting plane passes through them. This not only saves time but also gives a method of distinguishing, in the sectioned view, between a wheel with spokes and a wheel with a web. Figure 8.30 illustrates the general practice in the arms of a sheave.

When the section plane passes through a thin web parallel to its larger dimensions, as in Fig. 8.31, the most commonly accepted practice is to omit the section lining on the web and show a visible outline at the intersection of the web with the body of the piece. This is the preferred method and is illustrated in Fig. 8.31a. Sometimes the line of intersection between the web and the body is drawn as an invisible outline and every other crosshatch line is extended through the web. This method is illustrated in Fig. 8.31b. When a section plane cuts through a web perpendicular to its larger dimensions, the web is crosshatched.

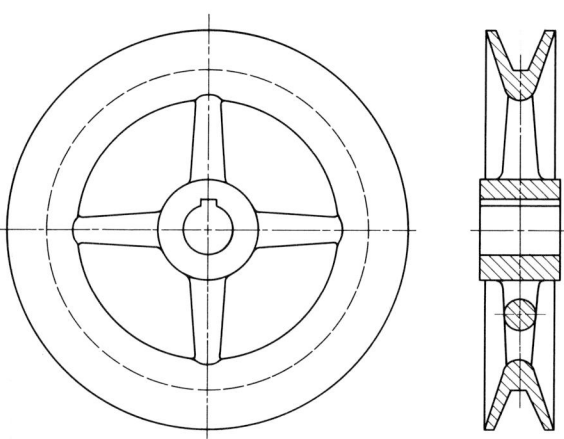

FIG. 8.30. Spokes of wheels in sectional views.

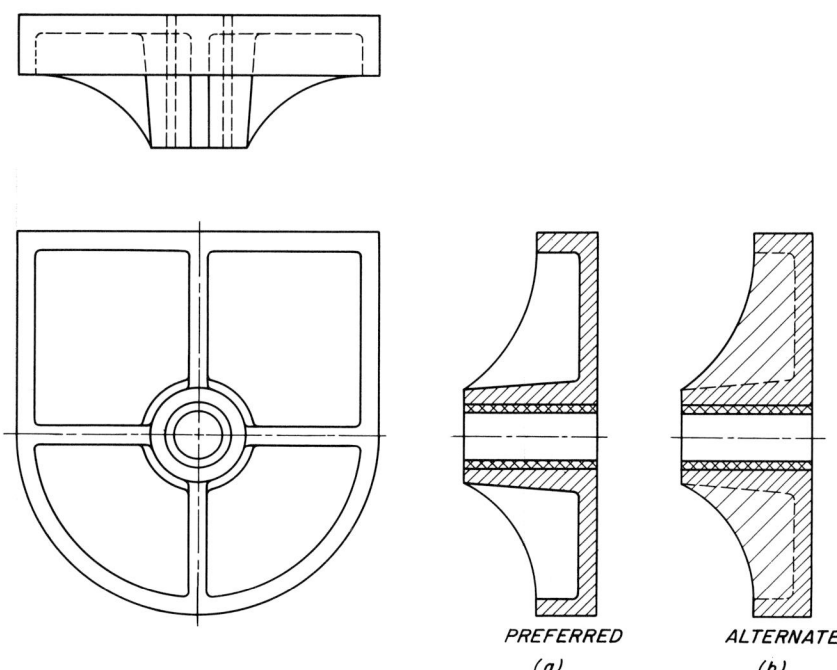

PREFERRED ALTERNATE
(a) (b)

FIG. 8.31. Thin webs in sectioned views.

8.19 OTHER CONVENTIONAL PRACTICES FOR ODD-NUMBERED AXES OF SYMMETRY

In passing a section plane through such objects as a cover plate with three webs, as in Fig. 8.32, or webs of couplings with an odd number of drilled holes, only one of the webs or holes will fall in the plane. To show the other webs or holes in their true positions by projecting them in the section view would usually require the use of a large number of hidden lines and difficult projections. This would make the object appear unsymmetrical. To avoid this confusion and to make the drawing easier to read, it is customary, in representing the areas not on the section plane, to consider them rotated into the section plane, as shown in Fig. 8.32b. Certain features contiguous to these areas must also be considered revolved at the same time.

This rotation of parts, though not in strict accord with the theory of orthographic projection, is so commonly used that it has come to be recognized and accepted as standard practice. It is used on both sectioned and unsectioned drawings. See Figs. 8.32c and 8.33. A part should not be revolved if its true relation to the rest of the object will be changed by the revolution; that is, only those areas that are on objects that fall in the class of cylindrical shapes can be treated in this manner.

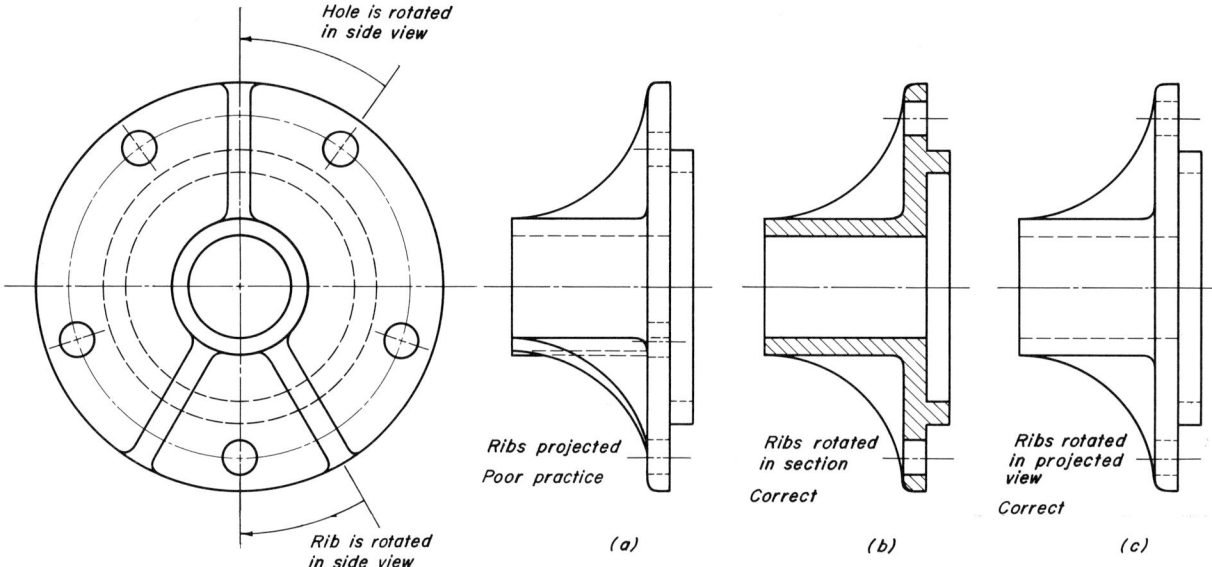

FIG. 8.32. Section of object with odd-numbered axes of symmetry.

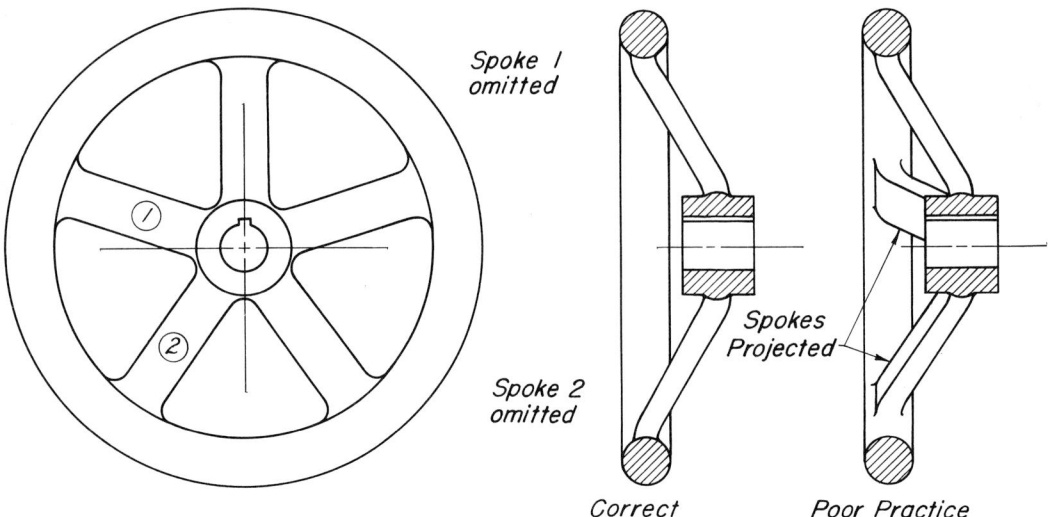

FIG. 8.33. Spokes of wheel in sectional view.

8.20 CONVENTIONAL BREAKS

Long members of uniform cross section are usually not drawn to scale lengthwise. This fact is brought out by showing a conventional break in the member, as illustrated in Fig. 8.34. The correct length of the member is shown by a dimension. A revolved section may or may not be interpolated in the break. However, the appearance of the break usually gives some indication of the shape of the cross section. Structural members when not to scale are not broken.

8.21 THREADS AND BOLTS IN SECTION

Threaded holes and bolts in section are represented by conventional symbols. A complete discussion of the methods of representing threads is given in Chapter 9. Typical sections involving threads, bolts, and screws are shown in Figs. 9.13 and 9.34 to aid the student in interpreting some of the drawings in this chapter. Notice that the bolts and screws are drawn as though the section plane did not pass through them.

8.22 MATERIAL SYMBOLS IN SECTIONS

Many industries indicate the kind of material by using a different type of section lining. The ANSI Drafting Standards Manual recommends the symbolic sections shown in Fig. 8.35. Their use is particularly valuable in assemblies where the drawing is complicated and several kinds of material are to be shown. For detail drawings it is usually preferable to use the standard symbol of parallel, uniformly spaced lines for all materials and specify the material with a note. An assembly is shown in Fig. 8.36, in which various materials are indicated by characteristic section lining.

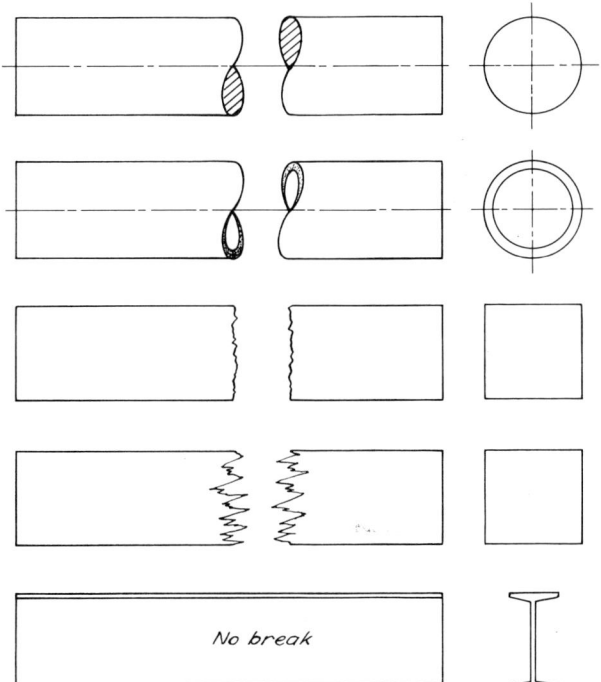

FIG. 8.34. Conventional breaks.

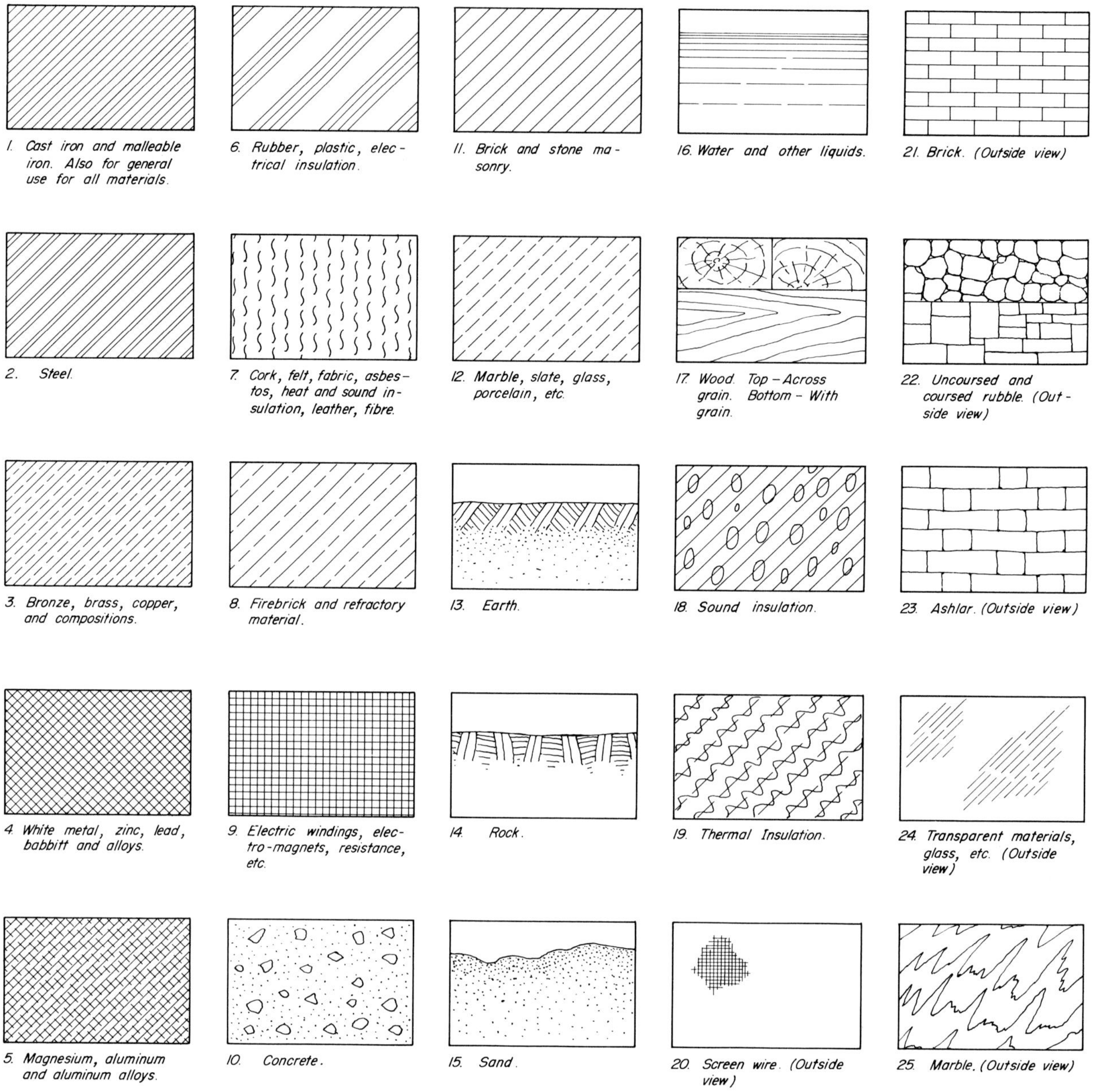

210 SECTIONAL VIEWS

1. Cast iron and malleable iron. Also for general use for all materials.	6. Rubber, plastic, electrical insulation.	11. Brick and stone masonry.	16. Water and other liquids.	21. Brick. (Outside view)
2. Steel.	7. Cork, felt, fabric, asbestos, heat and sound insulation, leather, fibre.	12. Marble, slate, glass, porcelain, etc.	17. Wood. Top – Across grain. Bottom – With grain.	22. Uncoursed and coursed rubble. (Outside view)
3. Bronze, brass, copper, and compositions.	8. Firebrick and refractory material.	13. Earth.	18. Sound insulation.	23. Ashlar. (Outside view)
4. White metal, zinc, lead, babbitt and alloys.	9. Electric windings, electro-magnets, resistance, etc.	14. Rock.	19. Thermal Insulation.	24. Transparent materials, glass, etc. (Outside view)
5. Magnesium, aluminum and aluminum alloys.	10. Concrete.	15. Sand.	20. Screen wire. (Outside view)	25. Marble. (Outside view)

FIG. 8.35. Standard symbols for sections and outside views (courtesy ANSI).

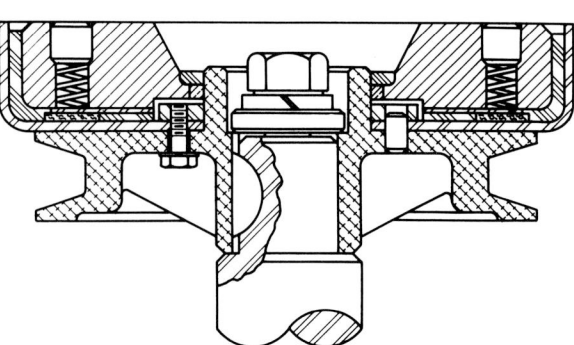

FIG. 8.36. Symbolic cross-hatching in an assembly.

SELF-STUDY QUESTIONS

Before trying to answer these questions, read the chapter carefully. Then, without reference to the text, answer as many questions as possible. For those that cannot be answered, the number in parentheses following the question number gives the article in which the answer can be found. Look it up and write down the answer. Check the answers that you did give to see that they are correct.

8.1 (**8.1**) The purpose of sectional views is to show the _____ shape of the part.

8.2 (**8.3**) When the cutting plane goes entirely across an object along an axis of symmetry, the _____ _____ of the cutting plane is not shown.

8.3 (**8.5**) The areas where material is cut by the section planes are _____.

8.4 (**8.5.1**) Crosshatch lines are usually drawn at an angle of _____ with the horizontal.

8.5 (**8.5.2**) Crosshatch lines are drawn about _____ to _____ apart.

8.6 (**8.5.4**) Crosshatch lines for adjacent parts should be approximately at _____ to each other.

8.7 (**8.7**) Visible lines behind a section plane (should/should not) be shown.

8.8 (**8.6**) Hidden lines on a sectioned view usually (are/are not) shown.

8.9 (**8.8**) A sectional view that cuts entirely across an object is called a _____ section.

8.10 (**8.8**) In making a full section, _____ of the object is imagined removed.

8.11 (**8.9**) When one fourth of a symmetrical part is imagined removed, the section is called a _____ section.

8.12 (**8.9d**) The line on the drawing separating the external half from the sectioned half may be either a _____ line or a _____ line.

8.13 (**8.10b**) The arrows at the ends of the cutting plane showing an offset section always point _____ from the sectioned view.

8.14 (**8.11**) When the cutting-plane line is turned at an angle one or more times, the section is known as an _____ section.

8.15 (**8.12**) A _____ section has the section rotated to show the true shape in the sectioned view.

8.16 (**8.13**) A removed section should be drawn in the _____ it would have been in the original view as a revolved section.

8.17 (**8.18.1**) When the cutting plane passes through the axis of solid shafts, bolts, pins, and screws, they are _____ _____.

8.18 (**8.18.2**) Spokes of wheels and thin webs are not crosshatched when the cutting plane passes through them in a direction _____ to their larger dimension.

8.19 (**8.19**) A part should not be revolved if its _____ to the rest of the object is changed by the revolution.

CHAPTER 9

FASTENERS

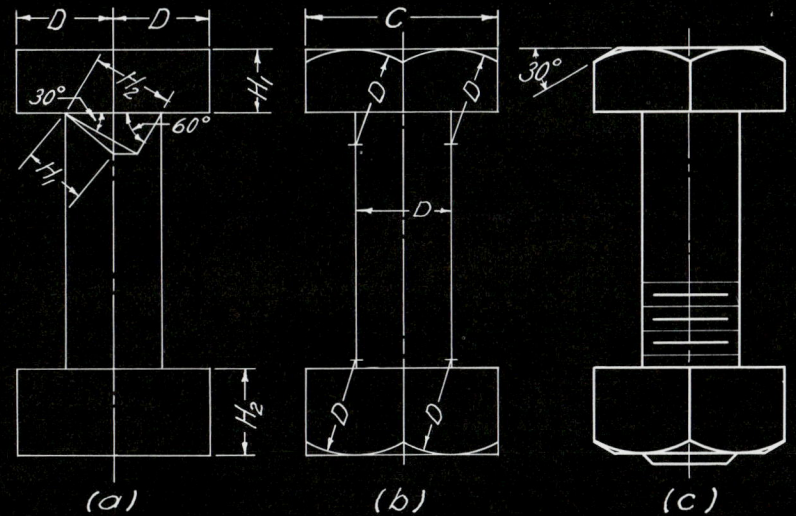

CHAPTER 9

9.1 INTRODUCTION
In all kinds of structures and machines designed by engineers, the various parts are held together by devices known as fasteners. These fasteners vary in kind and use, from ordinary nails and glue for wood structures to bolts, machine screws, keys, splines, and rivets for machines varying in size from locomotives to watches.

On working drawings the engineer may specify some fasteners by means of notes, without actually showing the fasteners themselves. Other fasteners may be shown in actual projected outline or in some common conventionalized form.

Both inch as well as metric fasteners will be discussed.

9.1.1 Fasteners Not Commonly Shown on Drawings. Such fasteners as brads, nails, spikes, cotterpins, dowel pins, and spring clips are specified as to size, kind, and use rather than being represented on a drawing. A few of the many kinds are illustrated in Fig. 9.1.

9.1.2 Fasteners Usually Shown by Symbols Only. On certain types of drawings some fasteners such as rivets, weldments, standard bolts, and screws are shown by symbols. These are conventionalized so that they are easily recognizable. See Fig. 9.2.

9.1.3 Fasteners Usually Detailed on Drawings. Special fasteners, which are designed for one particular usage, must be detailed. They may have standard parts such as a hexagon head or screw threads. A fastener of this type is shown in Fig. 9.3. Some stud bolts come under this class.

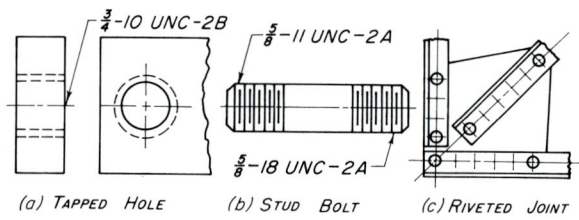

FIG. 9.2. Fasteners shown by symbols.

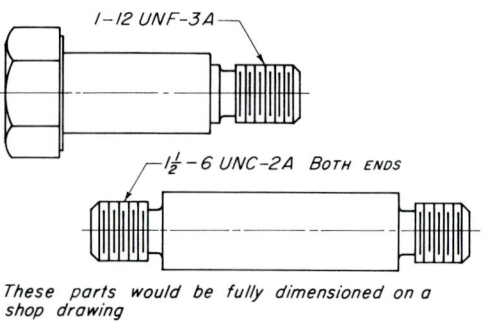

FIG. 9.3. Fasteners detailed on drawings.

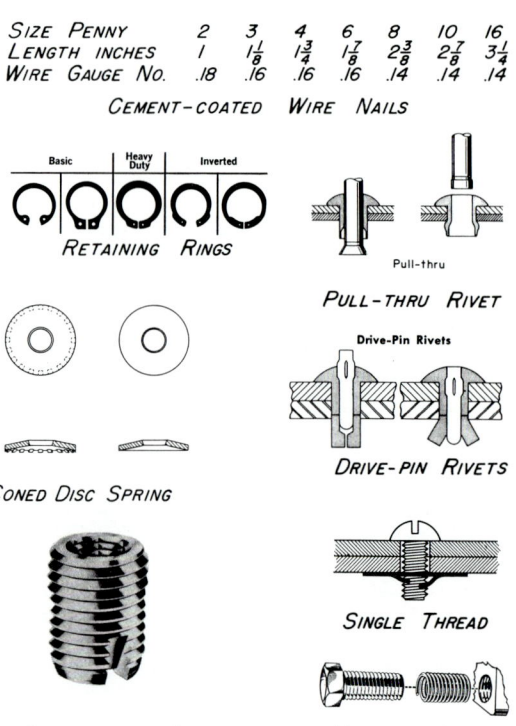

FIG. 9.1. Fasteners not commonly shown detailed on drawing (courtesy Machine Design Fastener Book).

9.2 STANDARD FASTENERS

It is estimated that there are approximately 500,000 standard fastener items that can be identified by name, type, size, and material. These can be found listed and shown in one or more published standards or specifications.

9.2.1 National Standards. A number of fastener standards are identified as bona fide national standards. Among them are those issued by the following.

a. American National Standards Institute.
b. National Bureau of Standards.
c. Index of Federal Specifications Standards and Handbooks.

Other fastener standards that have national recognition are issued by various technical and engineering societies such as the SAE (Society of Automotive Engineers), ASME (American Society of Mechanical Engineers, and ASTM (American Society of Testing Materials).

9.2.2 Industry Standards. Many large companies also have standards of their own. These usually conform to national standards where these exist but they also apply to company products and widely used components.

9.3 FASTENERS USING SCREW THREADS

Fasteners such as those described in Articles 9.1.2 and 9.1.3 need careful study from the standpoint of the drafting room methods and conventions employed in representing and specifying them on drawings. It is essential therefore to understand the generation and projections of the helix, which is the basic curve of all threaded fasteners.

9.3.1 The Helix. This curve is generated by a point moving on the surface of a cylinder or cone in a circumferential direction, at a constant angular speed, and with a simultaneous uniform rate of advance in an axial direction. A helix may be obtained by wrapping a string around a cylinder or cone in such a manner that the string advances parallel to the axis at a constant rate or, in other words, by the same amount for each revolution. The amount of this advance for one revolution is called the *pitch,* or *lead,* of the helix. See Fig. 9.4. If the cylinder were to be developed, the helix would become the hypotenuse of a right triangle whose base is the circumference of the cylinder and whose altitude is the lead of the helix. The angle between the hypotenuse and the base of the triangle is known as the helix angle. The geometrical construction of a helix is shown in Chapter 4.

9.4 THREAD TERMS

In a study of threads and the representation of them on a drawing, it is advisable to define a few terms. These definitions (abbreviated from ANSI Y14.6—1978 and Y14.6aM—1981) are illustrated in Fig. 9.5 and 9.6.

Screw thread. A ridge of uniform cross section in the form of helix on the surface of a cylinder (Fig. 9.5).

External thread. A thread on the outside of a member, such as a bolt (Fig. 9.5).

Internal thread. A thread on the inside of a member, such as a threaded hole or nut (Fig. 9.5).

Major diameter. The largest diameter of a screw thread (Fig. 9.5).

Minor diameter. The smallest diameter of a screw thread (Fig. 9.5).

Pitch diameter. On a cylindrical screw thread, the diameter of an imaginary cylinder, the surface of which would pass through the threads at such points as to make equal the width of the threads and the width of the spaces between the threads. See Figs. 9.5 and 9.6.

Pitch. The distance from a point on a screw thread to a corresponding point on the next thread measured parallel to the axis (Fig. 9.5).

Lead. The distance a screw thread advances axially in one turn (Figs. 9.5 and 9.7).

Angle of thread. The angle between the sides of a thread measured in an axial plane (Fig. 9.5).

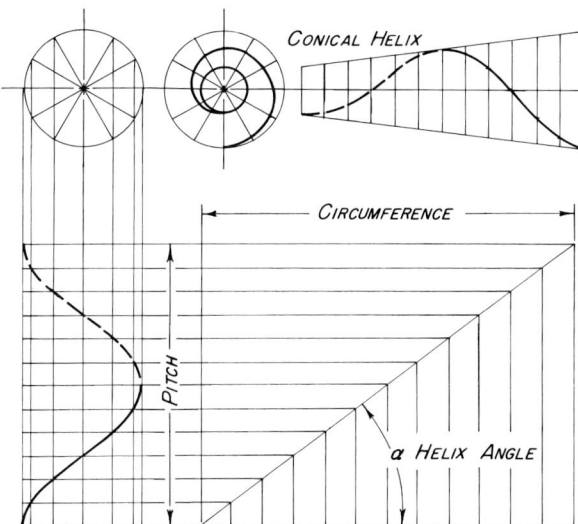

FIG. 9.4. Cylindrical and conical helix.

Crest. The top surface joining the two sides of a thread (Fig. 9.5).

Root. The bottom surface joining the sides of two adjacent threads (Fig. 9.5).

Base. Bottom section of thread. Greatest section between two adjacent roots (Fig. 9.6).

Depth of thread. The distance between the crest and root of a thread measured perpendicular to the axis (Fig. 9.5).

Single thread. All the threads on a member are built on a single helix. On a single thread the pitch is equal to the lead (Fig. 9.7).

Double thread. Two threads are wrapped around the cylinder on two parallel helices. The lead is twice the pitch (Fig. 9.7).

Multiple thread. Two or more separate threads on as many parallel helices are wrapped around the cylinder (Fig. 9.7).

Helix angle. Angle that the helix makes at any point with a plane perpendicular to the axis. The tangent of this angle is equal to the lead divided by the circumference of the helix cylinder. See Fig. 9.4.

Fit. The relative size of the pitch diameter of external and internal mating threads.

Clearance. An intentional difference between the major and minor diameters of the external and internal mating threads. Clearance is provided on the major and minor diameters of the nut as shown in Figs. 9.6 and 9.17.

Thread profile. The shape of the thread on a section plane containing the axis of the thread (Figs. 9.9 and 9.10).

Right-hand thread. A thread that advances into engagement in a direction away from the observer when turned in a clockwise direction (Fig. 9.8).

Left-hand thread. A thread that advances into

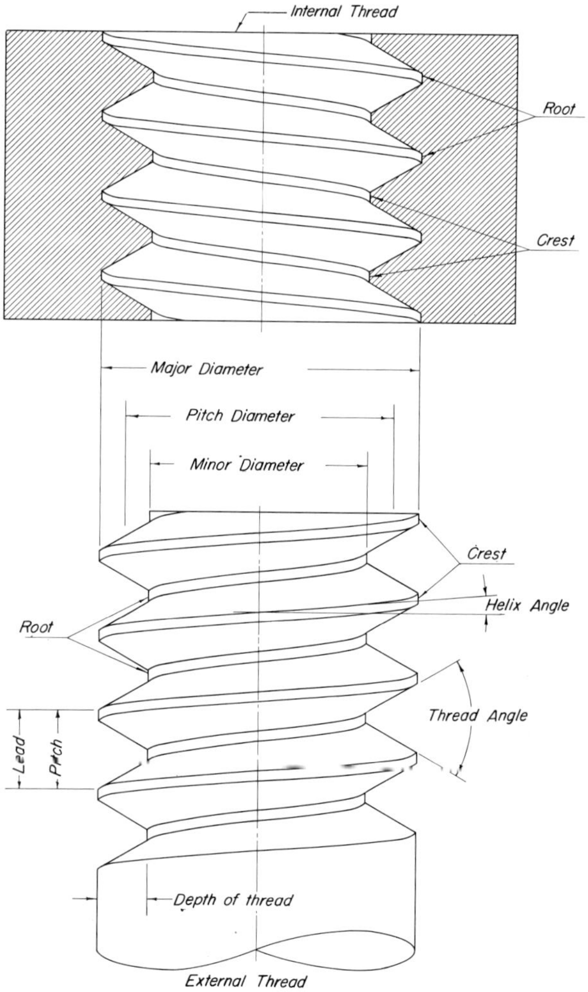

FIG. 9.5. Illustration of thread terms (courtesy ANSI).

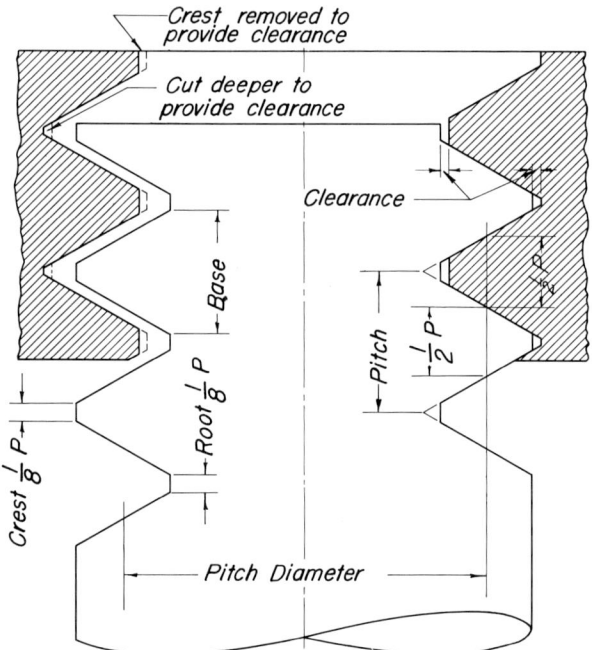

FIG. 9.6. Clearance on threads.

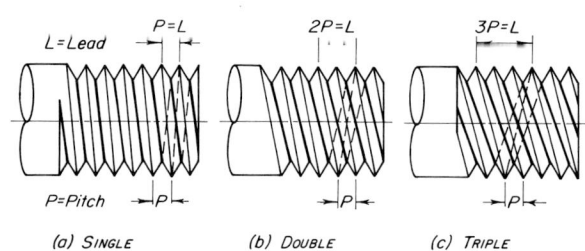

FIG. 9.7. Pitch and lead on single and multiple threads.

9.5 THREAD TYPES AND PROFILES

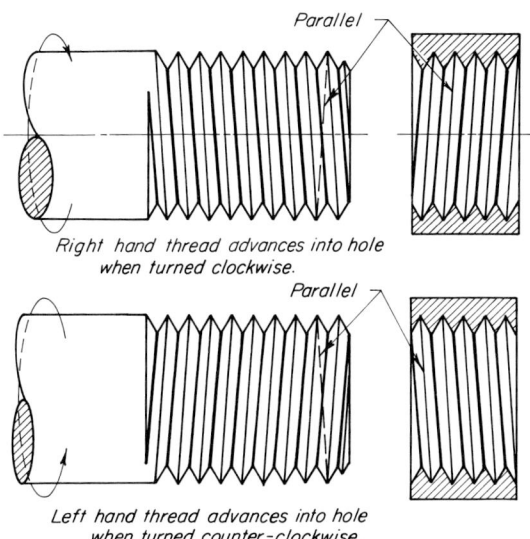

FIG. 9.8. Right-hand and left-hand threads.

engagement in a direction away from the observer when turned in a counterclockwise direction (Fig. 9.8).

9.5 THREAD TYPES AND PROFILES

The purpose for which the screw is to be used determines the profile of the thread. In the United States, general-purpose threads usually are of the Unified or American National thread form. The profile of the Unified thread is shown in Fig. 9.9. This thread form has been agreed upon as a standard for general-purpose threads by the United States, Great Britain, and Canada. It is therefore referred to as the *Unified thread*. Various other forms are shown in Fig. 9.10.

Among the thread profiles shown in Fig. 9.10, the Sharp-V is rarely used, but it is the basic form from which others were developed. The Unified National Thread has practically replaced the American National form. The Whitworth thread

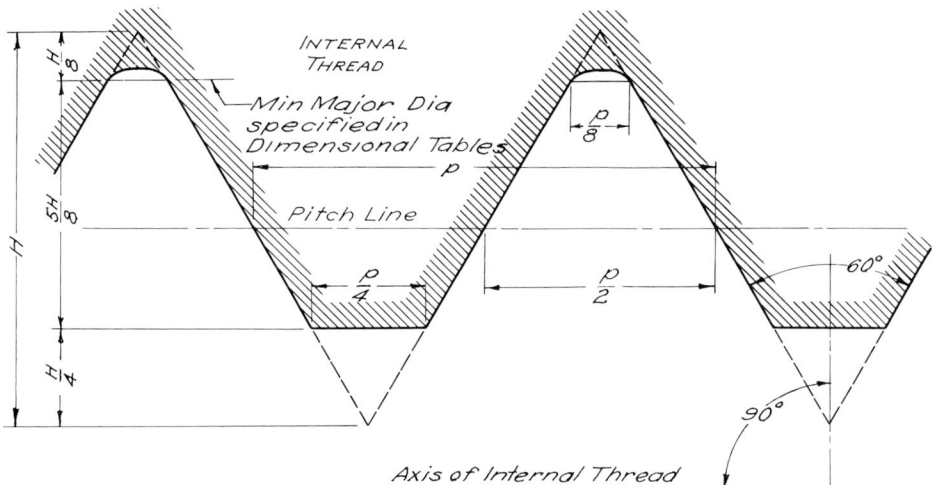

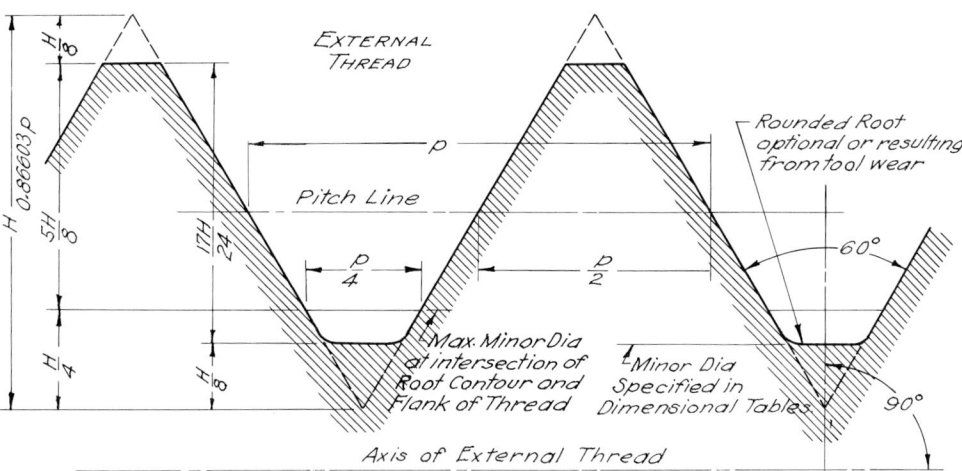

FIG. 9.9. Unified screw thread design forms (maximum material condition) (from ANSI B1.1-1960).

218 FASTENERS

is the general-purpose thread used in England. The knuckle thread has a rounded profile, which is convenient because it can be easily cast or rolled. The thread on the base of light bulbs is a form of the knuckle thread.

9.5.1 Power and Translating Threads. The Unified thread form is not suitable for transmitting power because it develops too great a radial force. The Square, Acme, Buttress, and Worm threads shown in Fig. 9.10 are used to transmit motion and power. The shape of these threads is shown in the figure. The Stub Acme and the Modified Square thread profiles are not shown. The Stub Acme has a depth of 0.3P. In other respects it is like the Acme. The Modified Square thread has flanks at a 10° included angle between them.

Translating threads, which must move freely for such things as steering gears, sometimes have ball bearings between the internal and external thread. Thus, the entire load is transmitted through the ball bearings that roll in the threads as one part moves in relation to the other. A tube is provided to carry the balls from one end to the other, thus keeping the same balls in continual motion.

9.6 DRAWING SHARP-V, UNIFIED AND AMERICAN NATIONAL THREADS, LARGE SIZE

Since the crest and root lines of a thread are always helices, it is a slow and difficult job to show a thread in its true projection. For this reason, the thread is seldom drawn in exact projection. Occasionally very large threads on display drawings may be drawn accurately, in which case one thread would be drawn carefully and the other threads reproduced by means of templates cut from the first thread.

On most working drawings the large threads would be drawn as shown in Fig. 9.11. This is a conventionalized form in which the profile for the Unified and American National thread is made the same as for the Sharp-V thread. The crest and root lines are drawn as straight lines instead of helices. The slope of these lines indicates the lead of the thread and shows the difference between right-hand and left-hand threads. For a right-hand single thread these lines on the bolt slant down to the right, in the position shown in Fig. 9.11, and advance parallel to the axis one half of the lead in going across the width of the bolt.

In a section view of a nut or internal thread the lines slope up to the right because these threads must fit the threads on the back of the bolt. The end view of the bolt or threaded hole is represented by two circles of the proper visibility, as shown in Fig. 9.11. The nominal size of the thread is always the size of the larger circle regardless of its visibility. This convention is preferred when the diameter of the bolt projects on the drawing as 1 in. or larger.

9.6.1 Conventional Thread Construction, Large Size. The steps to be followed in laying out this conventional thread are illustrated in Fig. 9.12.

a. Draw two light parallel lines so that the distance between them is equal to the diameter of the bolt.

b. On one of these lines mark off distances

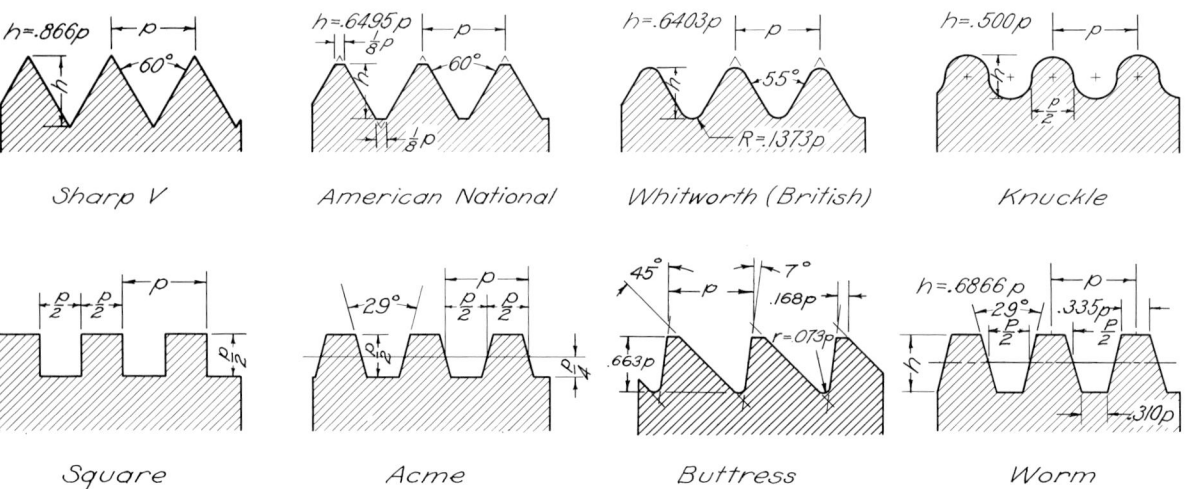

FIG. 9.10. Types of threads.

equal to the pitch of the thread. If convenient, the pitch should be laid out accurately, but if not convenient the nearest even number can be used, but the crest lines should not be closer than $\frac{1}{16}$ in.

c. At the first point erect a perpendicular to the lines, and on the other line mark off from this perpendicular a distance equal to one half the lead. See Fig. 9.12a. In a single thread this distance will be one half the pitch since pitch and lead are equal.

d. Draw a line joining the two end points m and n on the parallel lines. Draw lines parallel to this first line through the points previously marked off on the lower lines. These lines represent the crest lines of the thread (Fig. 9.12b).

e. With the 60° triangle, draw the V in the first space on each side of the bolt. Through the point of the V thus formed, draw light guidelines parallel to the sides of the bolt (Fig. 9.12c).

f. For the sake of speed, draw one side of each V from the ends of the crest lines to the guidelines on both sides as in Fig. 9.12d.

g. Complete the other side of the 60° V's and then draw the lines joining the bottoms of the V notches. These lines form the root lines of the thread (Fig. 9.12e).

h. Erase all construction lines; work over all lines to give good clean lines. Make the root lines and outlines heavier than the crest lines (Fig. 9.12f).

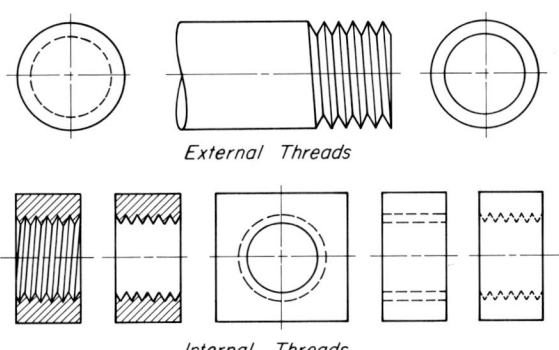

FIG. 9.11. V-Thread symbols.

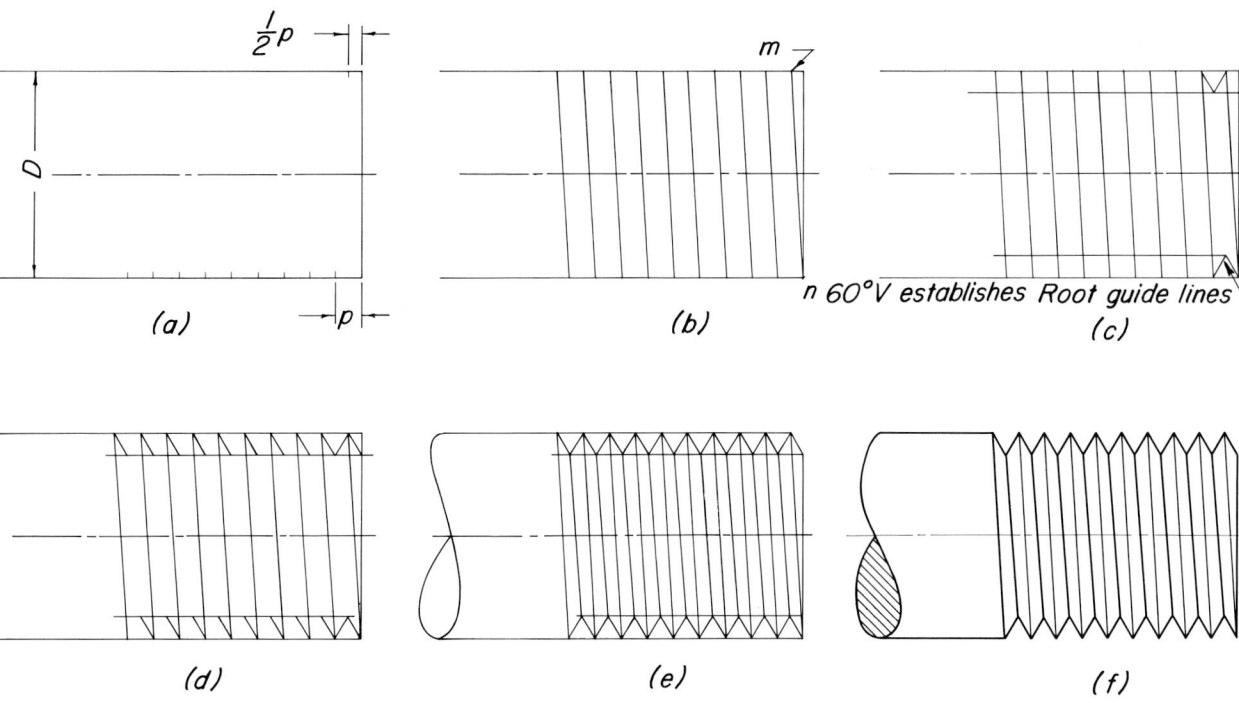

FIG. 9.12. Construction of V-thread symbols.

9.6.2 Representation and Construction of Unified and American National Threads, Small Size.
For threads that project smaller than 1-in. diameter the convention is still further simplified. The thread profile along the edge is replaced by a straight line at the edge of the bolt. The root and crest lines are drawn *perpendicular* to the edges instead of at a slant. The lines should be spaced at approximately the correct pitch. However, for convenience the pitch may be changed slightly. A minimum distance between crests should be set at approximately $\frac{1}{16}$ in., even though the threads may be actually much finer. An invisible threaded hole is represented by two sets of parallel lines, one representing the major diameter and the other the minor diameter. Figure 9.13 shows the conventional representation of threads under various conditions. When the thread does not go completely through a member, the drill point is always shown. The drilled hole should be deeper than the desired depth of thread because it takes an extra operation to thread a hole to the bottom. The internal angle for the point of the drill is 118°, but it is always drawn 120°. This conventional representation must always be accompanied by a thread specification.

The steps to be followed in laying out this conventional symbol are listed below and illustrated in Fig. 9.14.

a. Draw two parallel lines at a distance apart equal to the diameter of the thread. Using the pitch of the thread (actual or assumed) as a base, draw a 60° triangle to determine

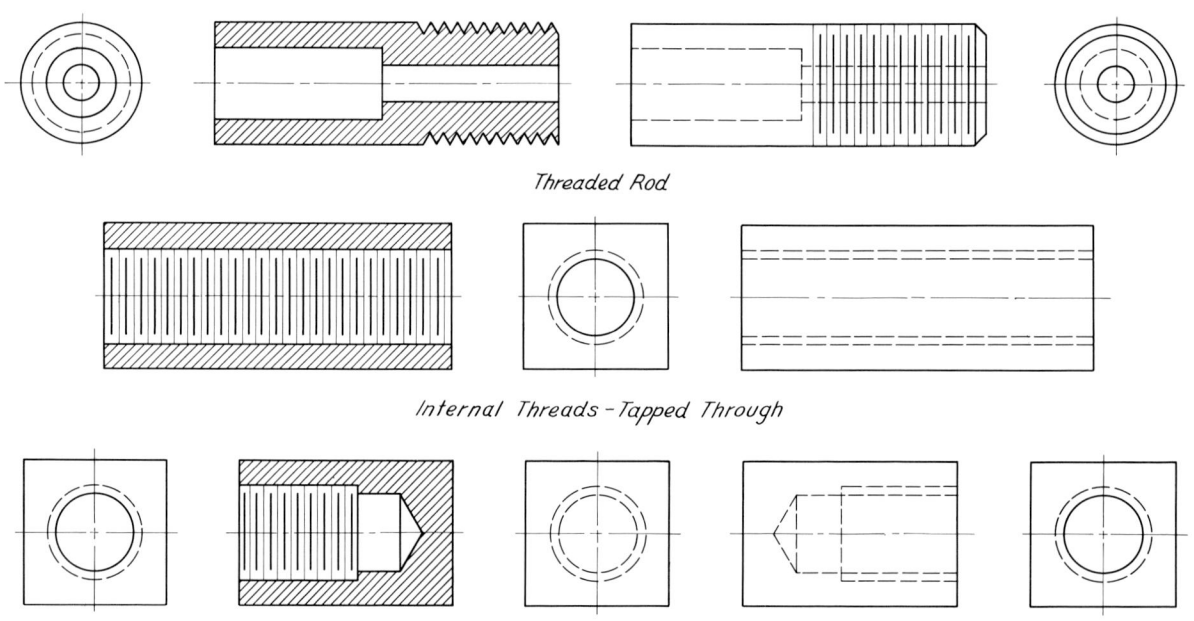

FIG. 9.13. Conventional thread symbols.

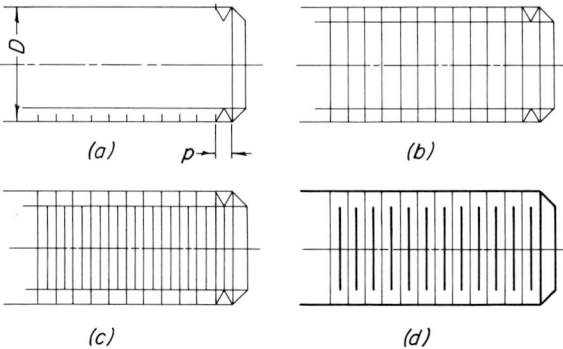

FIG. 9.14. Construction of conventional symbols.

the depth of thread on each side. Project the vertices of the triangles to the end of the bolt and from these points draw the 45° chamfer. This locates the first crest line. From this line mark off distances equal to the pitch of the thread. This spacing does not have to be exactly equal to the pitch, and the minimum spacing should be not less than $\frac{1}{16}$ in. (Fig. 9.14a).

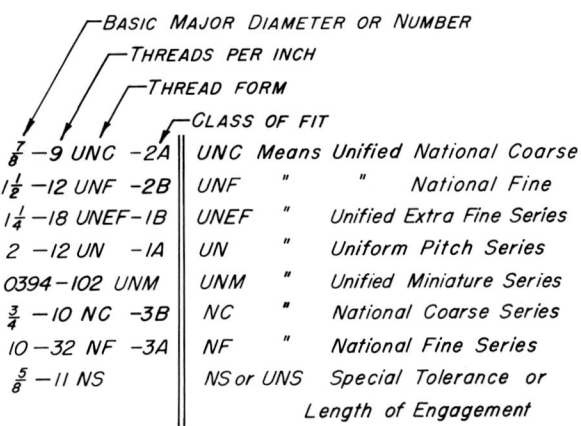

FIG. 9.15. Thread specification of threads and meaning (inch).

b. Through each of these points, draw lines perpendicular to the axis of the thread. These represent the crest lines of the thread. Through the points of the V's previously constructed, draw light guidelines parallel to the axis (Fig. 9.14b).

c. Draw the root lines midway between the crest lines and ending on the guidelines (Fig. 9.14c).

d. Clean up the drawing and go over all lines, making light lines for crest lines and heavy lines for outline and root lines (Fig. 9.14d).

9.6.3 Inch Thread Specifications. Since the thread symbol tells so little about the thread, it is necessary to give complete specifications in note form as shown in Fig. 9.15 and on other thread drawings. The meaning of each term is shown in the illustration. The terms should always appear in the order given. The last term indicating the class of fit is explained in detail in Articles 9.10.1 through 9.10.3.

9.6.4 Simplified Thread Symbols. In order to save time and money, another simplified set of thread symbols is coming into ever wider use. The symbols are shown in Fig. 9.16. They are

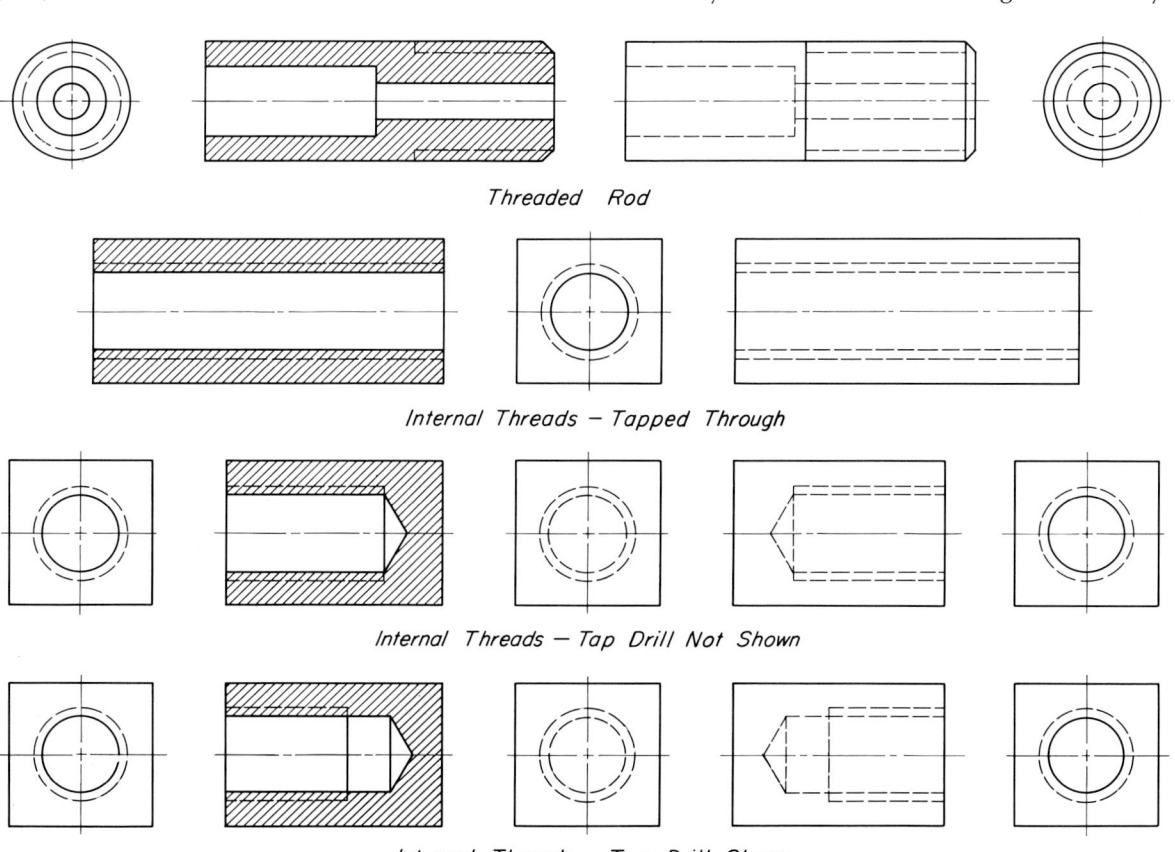

FIG. 9.16. Simplified thread symbols.

much faster to draw because they avoid the necessity of showing the root and crest lines. These symbols should be used with discretion on drawings involving a large number of hidden outlines.

9.7 MULTIPLE THREADS

Multiple Unified threads are shown in Fig. 9.7. They are used to secure faster movement in threads used for translation as on lathes and valves. Thus, with a double thread a valve will move twice as far in one revolution of the screw as it would for a single screw of the same pitch. An examination of Fig. 9.7 will show that on single- and odd-numbered multiple threads the crest is directly across from the root on opposite sides of the symbol. On double and other even-numbered threads crest is opposite crest. These facts are of assistance in drawing multiple threads. Square and Acme threads are used for translation of machine parts and power transmission. These threads are stronger than the Unified thread, of either single or multiple types. They are discussed in Article 9.12.

9.8 RIGHT- AND LEFT-HAND THREADS

All threads are assumed to be single and right-hand unless they are specifically indicated to be otherwise. See Article 9.4 and Fig. 9.8. A right-hand thread advances into a hole when turned clockwise. The left-hand thread advances when turned counterclockwise.

9.9 THREAD SERIES

Before the standardization of the American National thread, there were two main groups of threads in use in the United States. For most work, the United States Standard or Sellers thread was specified, but for automotive work a special thread was devised by the Society of Automotive Engineers. This thread had a smaller pitch than the United States Standard to give better resistance to vibration. These two groups were taken over almost entirely as the American National Coarse and American National Fine series. The SAE extra fine series and eight uniform series were added later. All of these are now included in the Unified National Standard. This means that this standard is used in the United States, Canada, and the United Kingdom. Since almost 99% of all threads manufactured in the United States conform to the Unified Standards, only these forms will be discussed. Dimensional specification for all thread series are given in the appendix.

9.9.1 Unified National Coarse Series.
These are used for bolts, screws, and nuts in general engineering application in materials such as malleable iron, soft metals, and plastics. This series is preferred where rapid assembly or disassembly is required. These threads are specified as 5/8-11 UNC-2A. See Fig. 9.15 for the meaning of this note and those that follow.

9.9.2 Unified National Fine Series.
This series is the same as the old SAE series with the addition of numbered sizes 0 to 12 below $\frac{1}{4}$ in. in diameter. These threads are recommended for bolts, screws, and nuts where the engagement of the thread is limited. This series is not used in soft metals and plastics. These threads are specified as 5/8-18 UNF-2A or 10-32 UNF-2A.

9.9.3 Unified National Extra Fine Series.
This series has more threads per inch than the other two series. It is used primarily in thin sections such as aircraft, missiles, and other aeronautical equipment where a maximum number of threads is desired. This thread is noted as 3/8-32 UNEF-2A.

9.9.4 Unified National Miniature Threads.
Specifications can be found in ANSI B1.10-1958. This thread is used for general-purpose screws in instruments and miniature devices. The diameters are usually specified in millimeters. The thread note is shown as 0.70-145 UNM.

9.9.5 Uniform Pitch Thread Series.
These threads have the same profile as other unified threads but the pitch is constant for all sizes. There are eight different pitches: 4, 6, 8, 12, 16, 20, 28, and 32 threads per inch. They are designed as $1\frac{1}{2}$-8 UN-2A.

a. *4 UN—4 threads per inch.* This series begins at $2\frac{1}{4}$ in. in diameter and is the same as the coarse series up to 4 in. It continues above that for all sizes up to 6 in. Basically it is an extension of the coarse thread series.

b. *6 UN—6 threads per inch.* Similar to others beginning at $1\frac{3}{8}$ in. diameter, but it is not a preferred series. Note that the beginning number after the diameter specification in these uniform pitch series means the number of threads per inch.

c. *8 UN—8 threads per inch* It is used primarily for bolts on high-pressure pipe flanges, cylinder head studs, and the like. This series begins at 1 in. diameter.

d. *12 UN—12 threads per inch.* The 12 UN and the 16 UN series are used for large-diameter fasteners that require a medium pitch such as thin nuts on shafts. Sizes from $\frac{9}{16}$ to $1\frac{3}{4}$ in. are used in boiler practice, which requires that more holes be retapped with a tap of the next larger size but having the same number of threads per inch.

  e. *20 UN, 28 UN, and 32 UN.* These uniform pitch threads are used where very fine pitch threads are needed. They are not recommended for regular fasteners. They are used in highly special situations.

9.10 CLASSES OF THREAD FITS

In a study of thread fit it is necessary to understand the meaning of a few terms. Fit is the relation between the size of two mating parts with reference to ease of assembly. In a thread the fit determines the pitch diameter of the screw. The quality of the fit is dependent on relative size, and quality of the finish, of mating parts.

Basic diameter is the theoretical or nominal standard size from which all variations are made. See Fig. 9.17.

Tolerance is the amount of variation allowed in manufacture.

Allowance is the intentional difference in the size between mating parts.

Crest clearance is the space between the crest of a thread and the root of its mating thread.

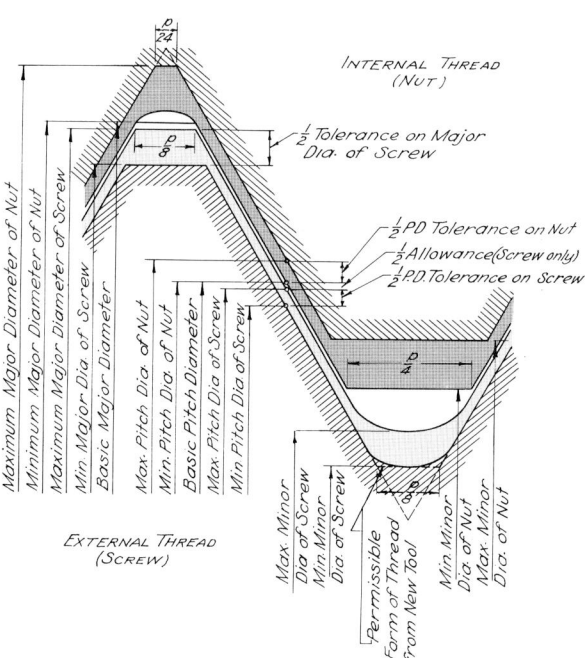

FIG. 9.17. Tolerance on Unified form screw threads (ANSI B1-1960).

The Unified National thread series provide for three principal classes of fits. These meet the requirements for most usage.

These classes are distinguished by numbers and letters. Classes 1A, 2A, and 3A apply to external threads only. Classes 1B, 2B, and 3B apply to internal threads only. The fits are obtained by the application of specific tolerances, or tolerances together with allowances to the basic pitch diameter of the thread. This basic pitch diameter is the same for both internal and external threads of like size and pitch. The tolerances throughout are applied plus to the hole and minus to the screw, as shown in Fig. 9.17. The allowances when used are applied to the bolt only. For this reason the external threads have been designated as A and the internal as B so that pitch diameters can be more easily specified.

An additional Class 5 fit, which is an interference fit, is used mostly for tap end stud bolts. In assembling a Class 5 fit the material of the thread is plastically deformed and makes a very tight fit.

9.10.1 Classes 1A and 1B, Loose Fit. The maximum pitch diameter of the screw is always smaller by a definite allowance than the minimum pitch diameter of the hole. Tolerances will tend to increase this difference so that there will always be a play or looseness between the mating parts when this fit is used. It is recommended only for work where clearance between mating threads is essential for rapid assembly and when shake or play is not objectionable.

9.10.2 Classes 2A and 2B, Free Fit. Maximum dimensions of Class 2A are reduced from the basic size by an allowance for clearance. This is 30% of the Class 2A pitch-diameter tolerance. This provides for plating or coating if desired. This class is the one most generally used for interchangeable works.

9.10.3 Classes 3A and 3B, Medium to Close Fit. There is no allowance in this class of fit. There is a variation from the tightest fit when both pitch diameters are at basic value to the loosest fit when Class 3A tolerance has been applied to both parts. The maximum play will be about 70% of that found in Class 2A fit.

This class of fit is used for an exceptionally high grade of commercial product and is recommended only where the high cost of precision tools and continual checking of tools and product are warranted.

9.10.4 Classes 2 and 3, Free and Medium Fits. Classes 2 and 3 are American National only. They are quite similar to 2A and 3A.

9.11 METRIC SCREW THREADS

With the large number of companies that have overseas plants and that sell finished products in overseas markets it is increasingly important to have a common standard in fasteners.

9.11.1 M Profile. The American National Standard Metric Screw Thread is designated the M profile. The M profile is a 60° V-type thread with basically the same dimensions as the Unified Screw Thread shown in Fig. 9.9. See Table 3 in the appendix for M profile dimensions.

Thread specifications and meanings are shown in Fig. 9.18. Note that lowercase letters are used for external threads and uppercase letters are used for internal threads for the tolerance position. See Article 9.11.3.

9.11.2 MJ Profile. The MJ profile screw thread is used in the aerospace industry in application under high stress and where a high fatigue strength is required. See Table 4 in the appendix for MJ profile dimensions.

9.11.3 Tolerance Position. The tolerance position is indicated by a letter and is an allowance (fundamental deviation). For internal threads G indicates a positive allowance and H indicates no allowance. For an external thread e indicates a large negative allowance, g a small negative allowance, and h indicates no allowance. If, as is often the case, pitch diameter and major diameter tolerances are the same, then it is not repeated.

9.12 REPRESENTATION OF SQUARE THREADS

The true Square thread profile shown in Fig. 9.10 is rarely used because of machining difficulties. The Modified Square thread is more generally used, but no attempt is made in representing

Internal Thread, Right Hand:

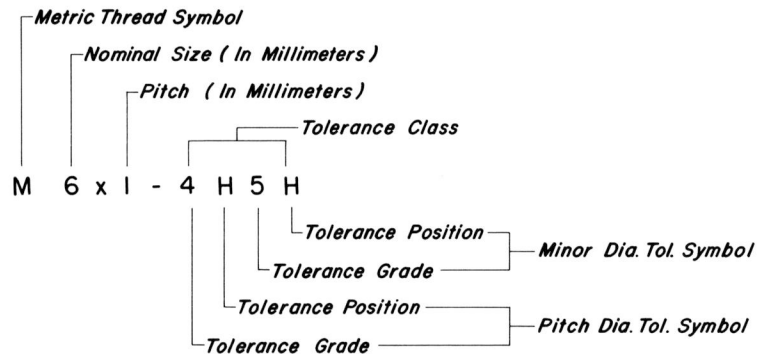

External Thread, Right Hand:

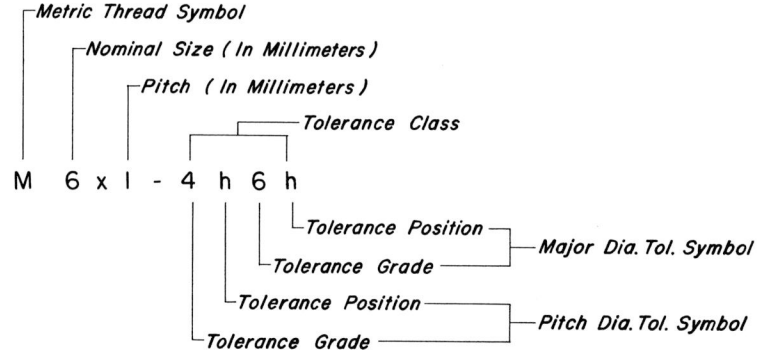

FIG. 9.18. Thread specification of threads and meaning (metric).

Square threads on drawings to show these refinements in detail. The thread is drawn as a square.

The representation of Square threads has been conventionalized as shown in Figs. 9.19e and 9.20d. When the diameter of the threads shows 1 in. or over, the form shown in Figs. 9.19 and 9.20 is preferred. When the diameter is less than 1 in. the simplified form shown in Fig. 9.21 may be used. On a long screw, ditto marks or phantom lines may be used to avoid the necessity of drawing all the threads. This is shown in Fig. 9.21d. The method of showing internal Square threads is shown in Figs. 9.20 and 9.21e–h.

9.12.1 Construction of the External Square Thread Symbol. The steps for making a right-hand, single, external thread are listed as follows and as shown in Fig. 9.19.

a. Lay out four parallel lines spaced as shown in Fig. 9.19a.

b. Mark off from a right section distances equal to one half the pitch on the two outside lines.

c. Draw the indicated inclined lines to follow one edge of a thread for one complete revolution.

d. Lay out the squares shown in Fig. 9.19b lightly, for the sides will be erased later.

e. For a single or triple thread, the opposite squares will be laid out in the same direction. See Figs. 9.19b and 9.22b.

f. For double or quadruple threads, the opposite squares will be laid out in opposite directions. See Fig. 9.22a.

g. From the external corners of the squares, draw lines parallel to the visible inclined line in Fig. 9.19a. These are the crest lines of the threads.

h. The root lines will be drawn to join the interior corners of the squares. Only the half of this line that lies outside the crest lines should be drawn, since the other half will be invisible, as in Fig. 9.19c.

i. From the external corners of the squares, draw lines as indicated in Fig. 9.19d, which are parallel to the dotted line shown in Fig. 9.19a.

j. Erase all unnecessary lines and go over the remaining lines to make them heavier. The drawing will now be complete, as shown in Fig. 9.19e.

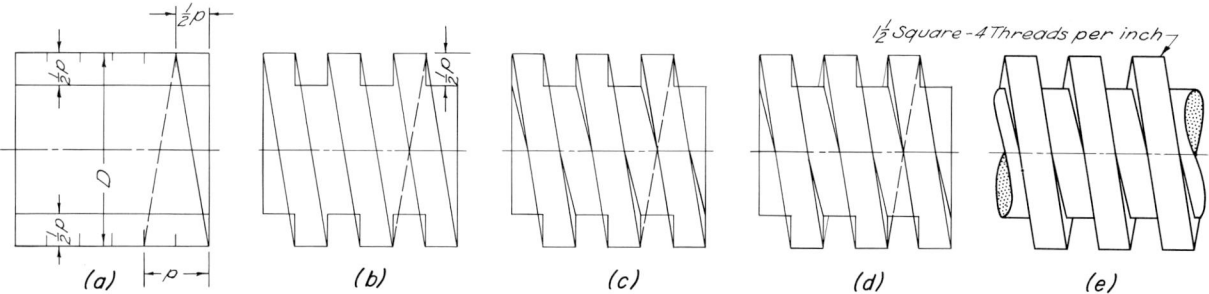

FIG. 9.19. Construction of external Square-thread symbol.

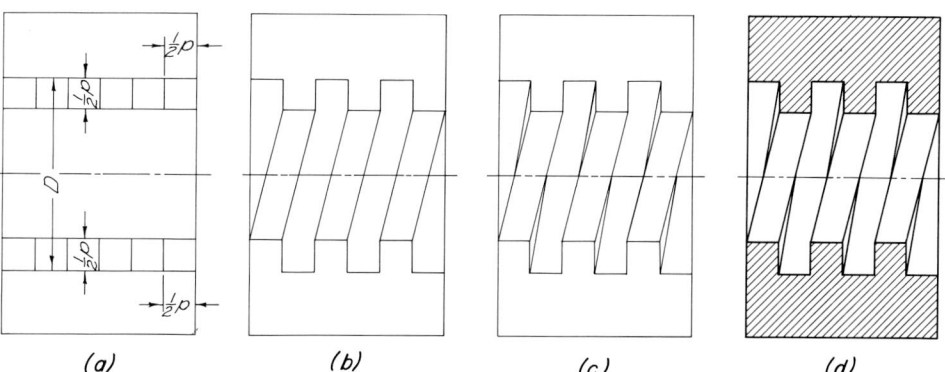

FIG. 9.20. Construction of internal Square-thread symbol.

9.12.2 Construction of Internal Square Thread Symbol. To make the internal Square thread symbol, the steps will be slightly different, as illustrated in Fig. 9.20.

a. Draw lightly the four lines representing the major and minor diameter of the thread as in Fig. 9.20a. Note that these lines are one half the pitch apart in pairs.
b. Measure off distances equal to one half the pitch on both outer lines and draw the squares as shown in Fig. 9.20b.
c. Connect the inside corners of the squares to represent the root lines of the threads.

Note that the slope is just the opposite of the external thread shown in Fig. 9.19.

d. Connect the outer corners of the squares but draw them only to the center line since the remaining part will be invisible and is not shown.
e. Make all visible outlines heavy and cross-hatch the proper areas.

9.12.3 Multiple Square Threads. Double and triple Square threads are shown in Fig. 9.22. The construction is similar to that in Fig. 9.19 except that for double threads crest is opposite crest. For triple threads crest is opposite root.

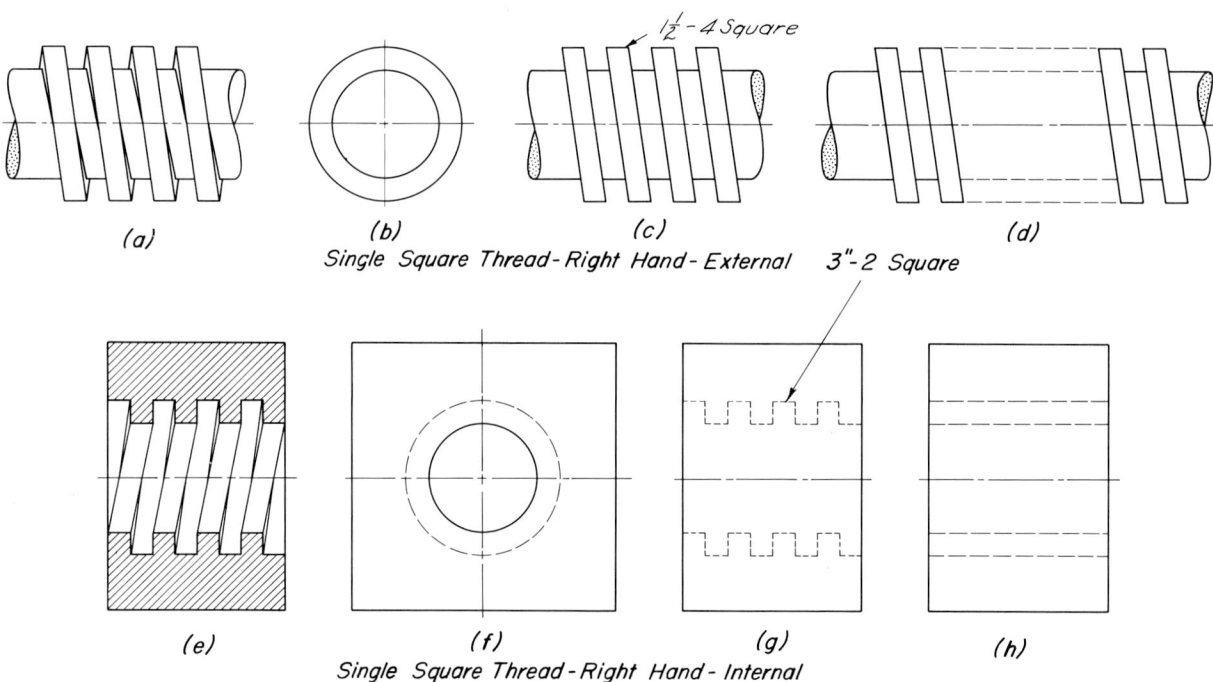

FIG. 9.21. Simplified Square-thread symbols.

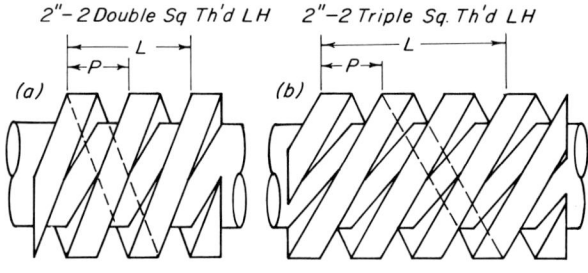

FIG. 9.22. Multiple Square-thread symbols. Double and triple left-hand.

9.13 AMERICAN PIPE THREADS

The pipe thread used in this country is known as the American, formerly the Briggs, standard. It differs from the British pipe thread in that the sides of the thread form an angle of 60°, whereas the British thread, which is built on the Whitworth system, shows an angle of 55°. The crest and root of the Standard American thread are slightly flattened, as shown in Fig. 9.23.

Pipe threads are cut on a taper of 1 in 16 measured on the diameter. Both internal and external threads are tapered. This taper allows the first few turns to be made by hand and ensures a tight joint when the threads are well engaged. A few of the threads on the pipe are slightly imperfect owing to the taper on the dies and taps used in making the threads. Figure 9.24 shows the American Standard pipe thread.

The pitch of pipe threads has been standardized for the various sizes of pipes, as shown in Table 35 in the appendix. They are specified on drawings by the letters NPT, National Pipe Tapered.

For certain purposes, for example, when a pipe must pass through the walls of a tank, for hose couplings, and free-fitting couplings, straight pipe threads are used. These threads have the same form as the standard pipe thread but do not have any taper. The straight pipe threads are designated by the letters NPS, National Pipe Straight. American Standard pipe threads are described in "ANSI B2.1 Pipe Threads." This is the standard used by the plumbing trade.

9.13.1 Representing Pipe Threads on Drawings — Regular, Tapered, and Straight. Pipe threads are shown and specified on drawings by either the conventional system shown in Fig. 9.25 or by the simplified system shown in Fig. 9.26.

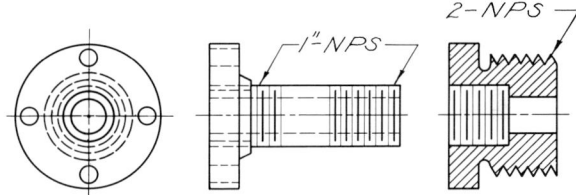

Regular Method - Straight Threads

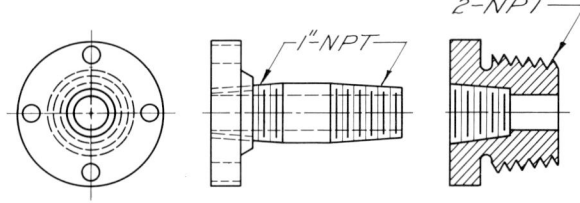

Regular Method - Taper Shown

FIG. 9.25. Pipe-thread symbols and notes.

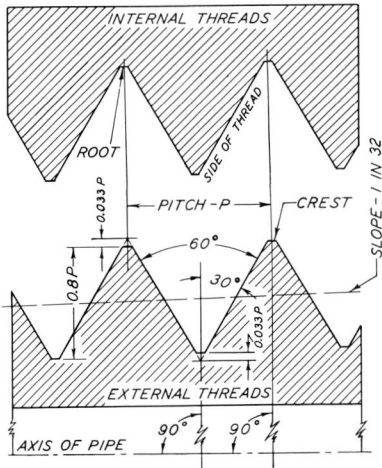

FIG. 9.23. American pipe-thread profile and dimensions.

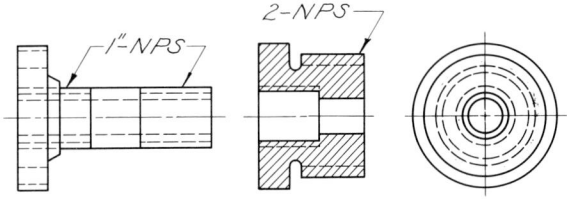

Simplified Method - Straight Threads

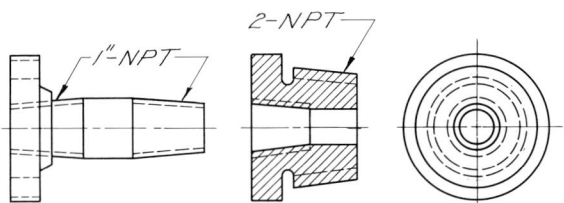

Simplified Method - Taper Shown

FIG. 9.26. Simplified pipe thread symbols.

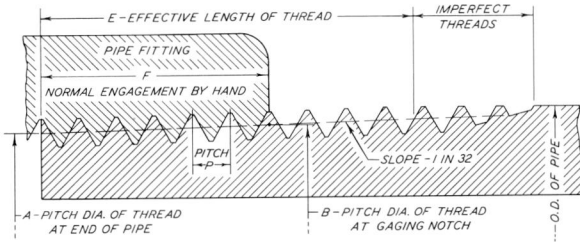

FIG. 9.24. American standard pipe thread.

228 FASTENERS

9.14 THREADED COMPONENTS
The foregoing paragraphs have shown methods of representing screw threads of various kinds. These threads are always a part of some component of a machine part or structure. These components include threaded holes as the simplest application, together with bolts and screws of many kinds and sizes. The methods of drawing and specifying these components are discussed in the following paragraphs.

9.15 THREADED HOLES
In many machines the parts are held together by machine screws, cap screws, or stud bolts. Where such fasteners are used the machine part usually has threaded holes to receive these bolts and screws. If a regular bolt is used, the nut, of standard form, has a threaded hole clear through.

9.15.1 Holes Clear Through. Unless the part through which a tapped hole is to go is very much thicker than needed by design requirements, the hole should be shown as threaded clear through. A complete thread note with pitch diameter tolerances is shown in Fig. 11.54. Regular thread notes are shown in Figs. 9.15, 9.18, and 9.22. Holes are usually countersunk on the side where the tap enters, unless the tap has a long taper of seven or eight threads.

9.15.2 Blind Holes. Blind holes should be avoided if possible. When possible blind holes should be drilled considerably deeper than the full thread engagement required. This allows for chips and makes tapping simpler.

Unless a bottoming tap is to be used, the drilled hole should be shown beyond the thread and dimensioned as shown in Fig. 11.54. The SAE Drawing Standard recommends drawing the run-out on threads as well as the countersink. Depth of drill and depth of thread may be dimensioned instead of being noted.

9.15.3 Internal Thread Relief. If internal thread relief is required to develop full engagement for the length of thread, this may be shown as in Fig. 11.59. The diameter of the relief must be greater than the major diameter of the internal thread. The depth of the relief must be not less than three times the thread pitch.

9.15.4 External Thread Relief. When external threads must be brought close to a shoulder as in Fig. 11.59, relief for the cutting tool must be provided as shown. Note that the diameter must be less than the minor diameter of the screw.

9.16 STANDARD BOLTS
Square-head and hexagon-head bolts with correspondingly shaped nuts are used to hold two or more parts of some assembly together.

The following types of bolts are produced by manufacturers in conformity with the American Standard ANSI B18.2.2 — 1972.

Regular square
Regular hexagon

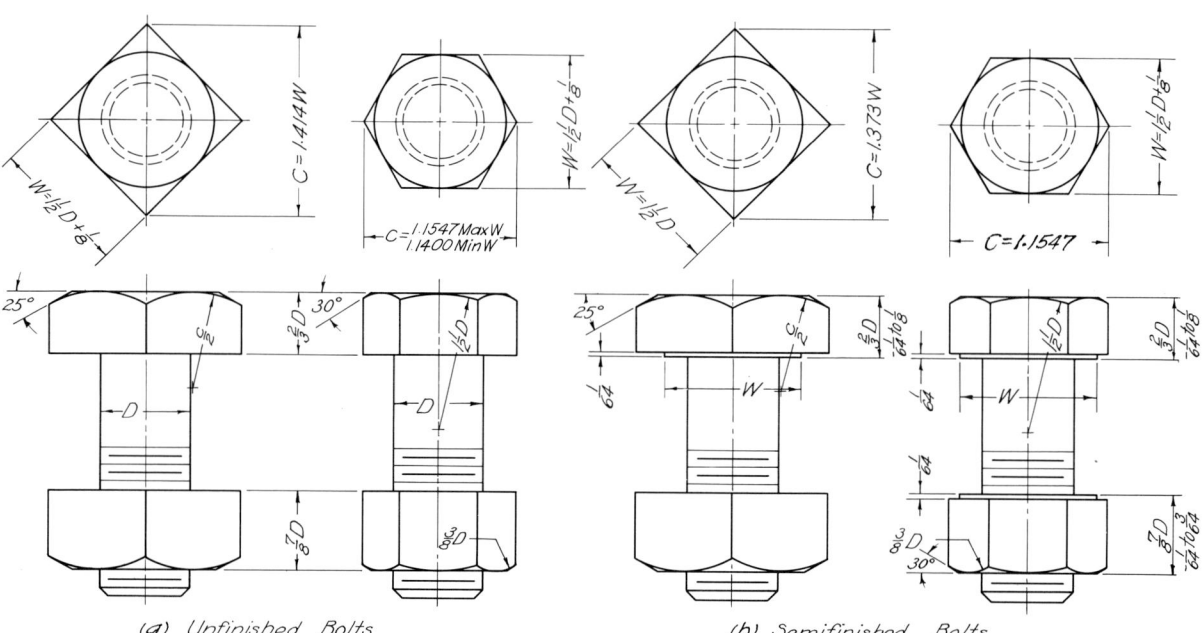

(a) Unfinished Bolts   (b) Semifinished Bolts

FIG. 9.27. Hexagon- and square-head bolts with dimensions for drawing.

Heavy hexagon
Regular semifinished hexagon
Heavy semifinished hexagon
Finished hexagon
Heavy finished hexagon

All dimensions for these bolts are given in the above standard. See Tables 8, 10, 12, and 14 in the appendix.

9.16.1 Regular Bolts. Regular bolts are not finished on any surface. The square bolts are supplied only in this form. The threads are coarse thread series Class 2A. See Fig. 9.27a.

9.16.2 Semifinished Bolts. Semifinished bolts are manufactured to produce a flat bearing surface under the head only. This is usually a circular washer-type face approximately $\frac{1}{64}$ in. thick with a diameter equal to the distance across flats. Coarse thread series Class 2A are used. See Fig. 9.27b.

9.16.3 Finished Bolts. Finished bolts are not necessarily completely machined. The term *finish* refers to the quality of manufacture and the closeness of the tolerance. The circular bearing face may be made as a washer-type face or by chamfering the corners to produce a circular bearing area. See Fig. 9.27b.

These bolts may be obtained with coarse, fine, or the eight uniform pitch threads Class 2A. Figures 9.27 through 9.29 show drafting methods for constructing standard bolt heads.

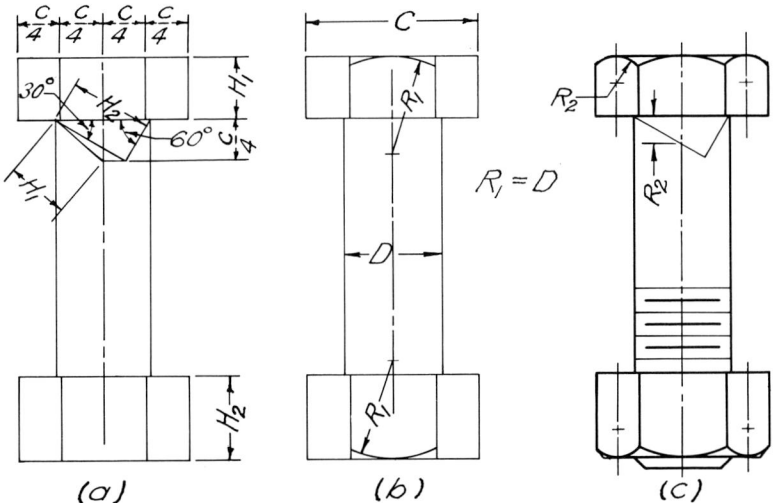

FIG. 9.28. Conventional construction of hexagon bolt head and nut.

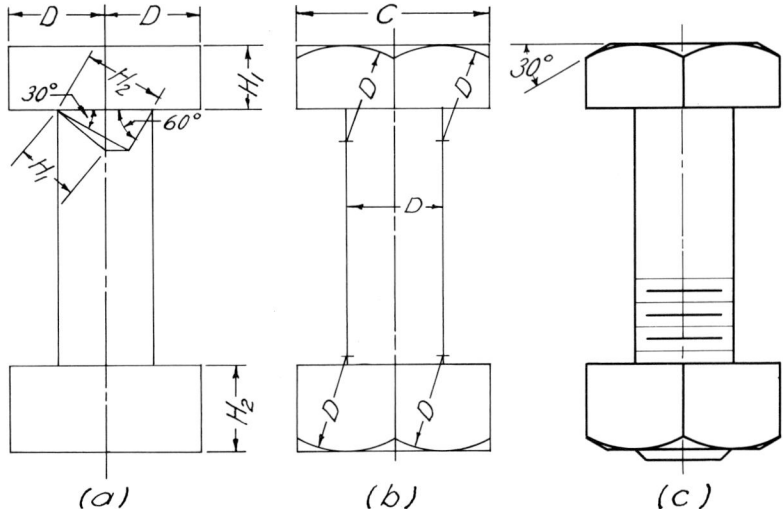

FIG. 9.29. Alternate construction of square-head bolt and nut.

9.17 STANDARD NUTS

The standard nut that is used on various types of bolts is a direct application of the threaded hole. The outside shape of the nut is standardized as well as the threads. Typical standard nuts are shown in Fig. 9.30. Dimensions of these nuts are specified in ANSI B18.2.2 — 1972.

Methods of drawing the regular and semifinished nuts are shown in Figs. 9.27 through 9.29. These are simplified methods, but they are sufficiently accurate for all drawing purposes.

9.18 LOCKING DEVICES

It is frequently necessary to provide some method of preventing the nut from unscrewing and thus leaving a loose connection. This is particularly true when the fastening is subject to vibration.

9.18.1 Jam Nuts. There are many different devices that have been designed for this purpose, the most common of which is the lock nut or jam nut. The American National jam nut has the same dimensions as the regular nut except for thickness, which is smaller because it is not designed to develop the full strength of the bolt but merely to hold the regular nut in place; Table 11 in the appendix gives dimensions of regular jam nuts such as that illustrated in Fig. 9.31a. Jam nuts may be obtained in the regular, heavy, and light series and also in unfinished, semifinished, and finished design.

9.18.2 Palnut. A patented lock nut known as Palnut, shown in Fig. 9.31i, is used principally in electrical work and holds by tightening sufficiently to deform the rounded web of the nut.

9.18.3 Lock Washers. Various kinds of lock washers are in use, some of which hold the nut by exerting a spring pressure on one side of the nut and others by a positive action that prevents the nut from turning. These are illustrated in Figs. 9.31b and c.

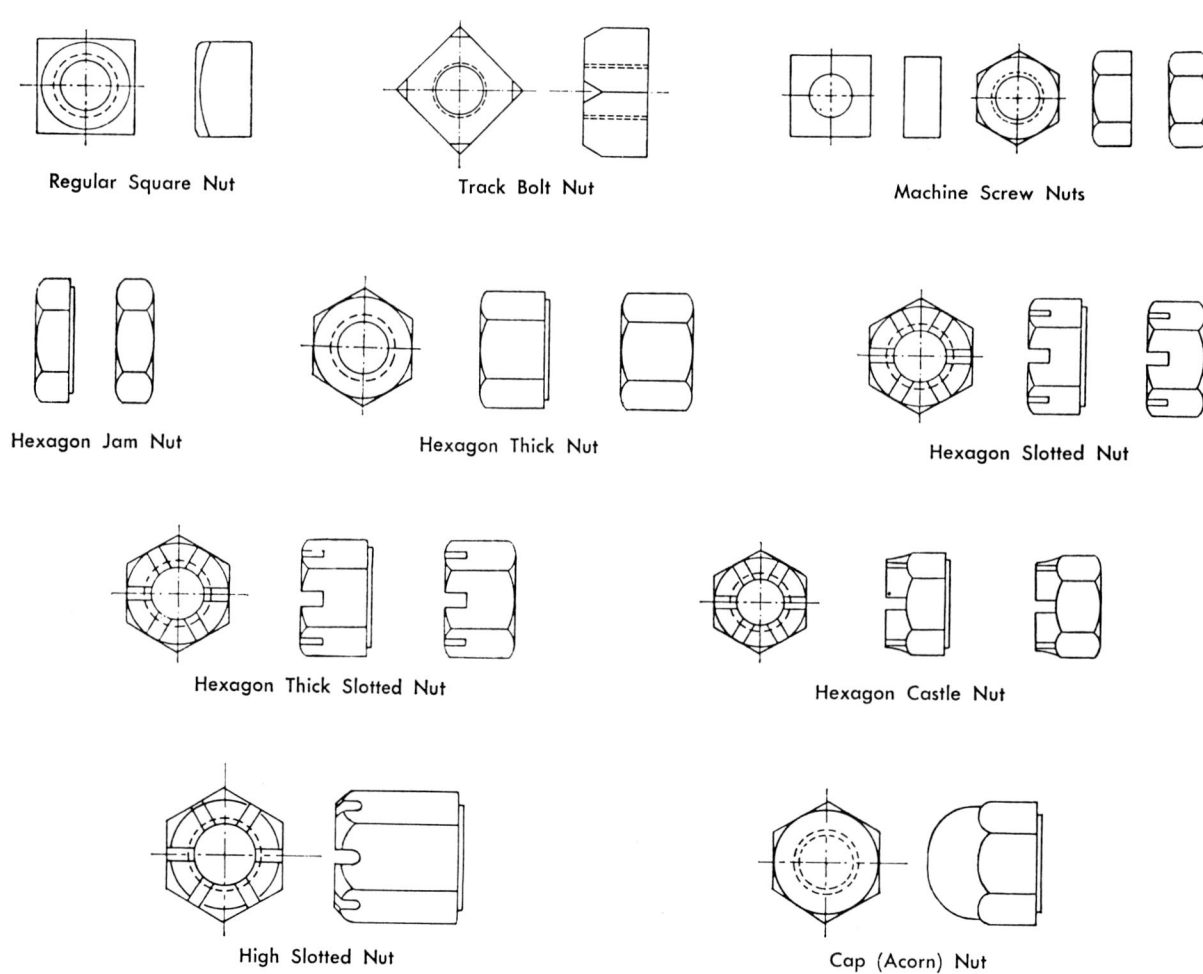

FIG. 9.30. Standard nut types (courtesy Machine Design Fastener Books).

9.18.4 Cotter Pins. The bolt may be drilled and a cotter pin or wire may be placed through the bolt and above the nut, as in Fig. 9.31d. This will prevent the nut from coming off but does not avoid a certain amount of loosening. A more definite lock is obtained by using a slotted nut so that the cotter pin may be placed in the slot and through the bolt, as shown in Fig. 9.31e.

9.18.5 Split Nuts. The nut may be split and deformed in various ways before it is in place. The deformation exerts a pressure on the threads that holds the nut in place by its friction. One method of doing this is shown in Fig. 9.31f.

9.18.6 Elastic Stop Nut. This is a patented fastener that has an elastic collar built into the nut, which grips the threads of the bolt when screwed on. This gripping action prevents the nut from backing off even when subjected to vibration. It has been used frequently in the airplane industry. Figure 9.31g shows a section of an elastic stop nut. Another form in which the elastic stop nut is used is shown in Fig. 9.31h. In aircraft work it is frequently used to fasten a bolt to a thin web, in which case a flanged nut is riveted to the web to provide a suitable length of penetration.

Various other methods of locking a nut are in use, some of which are as simple as the use of a set screw in a nut and others as complicated as the use of a specially designed thread that is cut on a taper so that the friction is increased when the load is applied.

9.19 STUD BOLTS

The stud bolt has threads on both ends so that one of the pieces being held together must be threaded to replace the head. See Fig. 9.32. The stud bolt is turned into the threaded hole by means of a pipe wrench or a special lock-nut device until the threads jam. The bolts then become an assembly guide by means of which the other part, which is drilled but not threaded, is

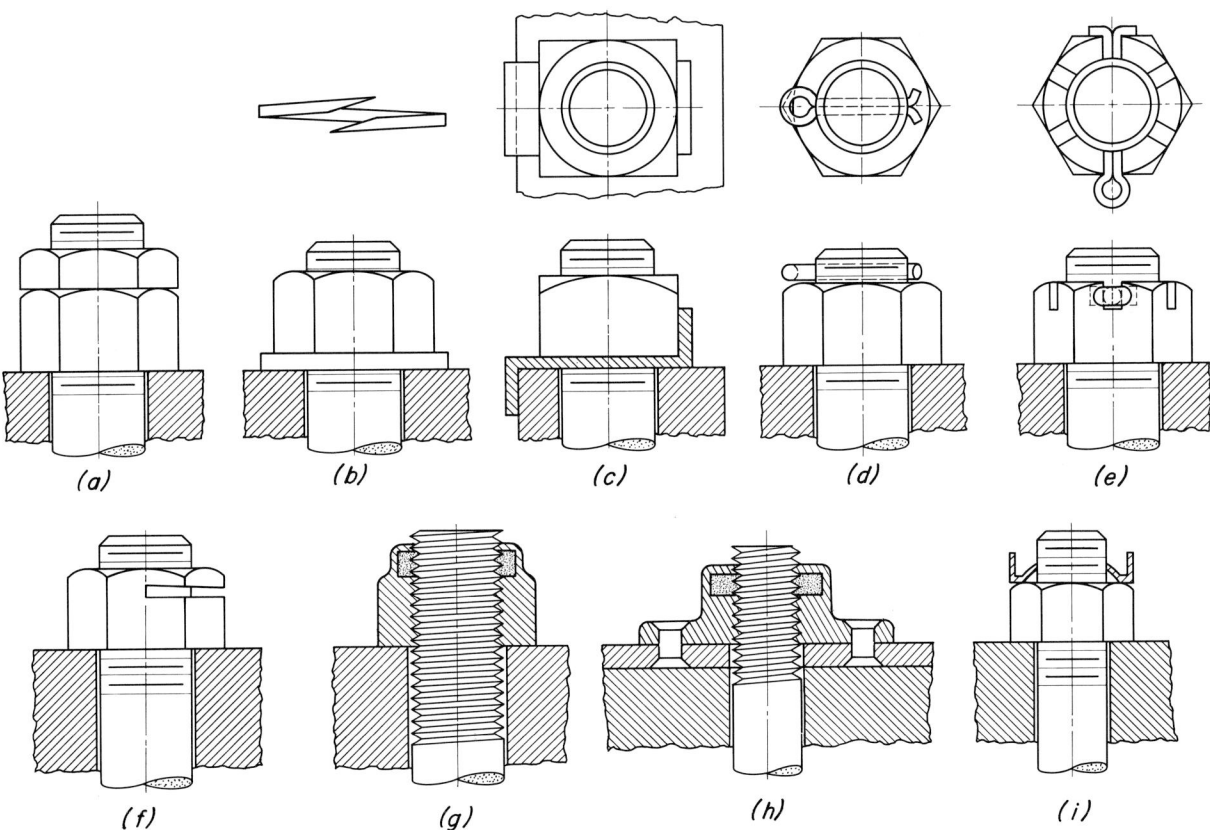

FIG. 9.31. Locking devices.

properly placed. A nut screwed on the other end of the bolt holds the two parts together. The Industrial Fastener Institute has established four basic classes of studs. These can be described only briefly.

9.19.1 Class 1 — Tap End Studs. This stud has a Class 5 interference fit on the tap end. The tap end is the part that screws into a threaded hole in a machine part. This hole has a Class 3B thread. During assembly the threads compress and flow elastically and plastically. This gives a very tight grip on the stud. The other end may have any type of relatively free-running thread.

9.19.2 Class 2 — Double-End Studs. This stud is similar to the Class 1A stud, except that it has free-running threads on both ends. These may be obtained in any one of four types. See Fig. 9.32 for a stud bolt of this type.

9.19.3 Class 3 — Bolt Studs for Pressure-Temperature Piping. These are threaded for their full length and are used under high-temperature, high-pressure conditions.

9.19.4 Class 4 — Continuous Threaded Studs. These are general-purpose studs ranging from large-diameter studs for heavy work to threaded wire.

9.20 OTHER TYPES OF BOLTS

Many other types of bolts for special purposes can be obtained from the trade. See Fig. 9.33. The Industrial Fastener Institute has standardized quite a number of these bolts.

9.20.1 Carriage Bolts. The carriage bolt is used principally for woodwork. It has a smooth oval head for a finished appearance and a square shank that prevents the bolt from turning as the nut is tightened. Four other types of heads, in addition to those shown in Fig. 9.33, are regularly manufactured.

9.20.2 Stove Bolts. The standard stove bolt (see Fig. 9.32) is peculiar in that it has a round flat head that is beveled on the underside to fit a countersunk hole. It is provided with a screwdriver slot for turning the threaded end into the nut. This bolt is used where it is desirable to have the head of the bolt flush with the surface of the metal held in place. A second kind of stove bolt has a button-shaped head with a slot for a screwdriver but is not beveled on the underside.

9.20.3 Plow Bolts. The plow bolt is similar to the stove bolt except that it does not have the screwdriver slot. The head has a projecting lug or square shank to keep the bolt from turning as the nut is being tightened. See Fig. 9.33. Three other types of plow bolts are regularly supplied by the trade.

9.21 DRAWING SQUARE- AND HEXAGONAL-HEAD BOLTS AND NUTS

Dimensions for regular square- and hexagonal-head unfinished and semifinished bolt heads and nuts are given in Tables 8 to 15 in the appendix.

On regular unfinished and semifinished bolt heads the distance across flats (W) as indicated in Fig. 9.27 is one and one half times the diameter, with adjustments to sixteenths to eliminate $\frac{1}{32}$-in. wrench openings. Regular nuts follow the same rule except that $\frac{1}{16}$ in. is added in sizes from $\frac{1}{4}$ to $\frac{5}{8}$ in.

The nominal height of the regular unfinished head is $\frac{2}{3}D$ and that of the semifinished head $\frac{2}{3}D$ minus $\frac{1}{64}$ to $\frac{1}{8}$, depending on the size. The height of the semifinished and finished bolt head includes the height of the washer face, which is usually $\frac{1}{64}$ in.

The nominal thickness of the regular unfinished nut is $\frac{7}{8}D$ and that of the unfinished nut $\frac{7}{8}D$ minus $\frac{1}{64}$ to $\frac{3}{64}$, depending on the size.

Bolt heads and nuts may be laid out on the drawing by means of dimensions taken from the tables, but for speed and convenience the method illustrated in Figs. 9.27 through 9.29, known as the Jorgensen method, is preferred. Although the dimensions obtained by this method are not exactly accurate, they are as close as would be required, except when clearances are involved. To lay out a hexagonal bolt head by this method, the following steps are necessary.

a. Draw the outline of the bolt and the base of the head and nut. With a 30-60 triangle, draw the sides of the triangle with the hypotenuse equal to the diameter of the bolt. See Fig. 9.28.

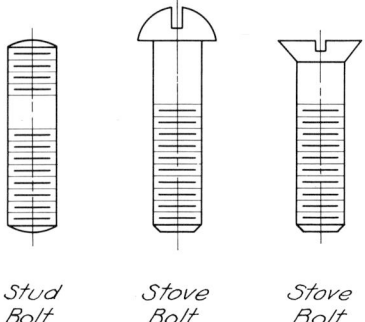

FIG. 9.32. Stud and stove bolts.

9.21 DRAWING SQUARE- AND HEXAGONAL-HEAD BOLTS AND NUTS   233

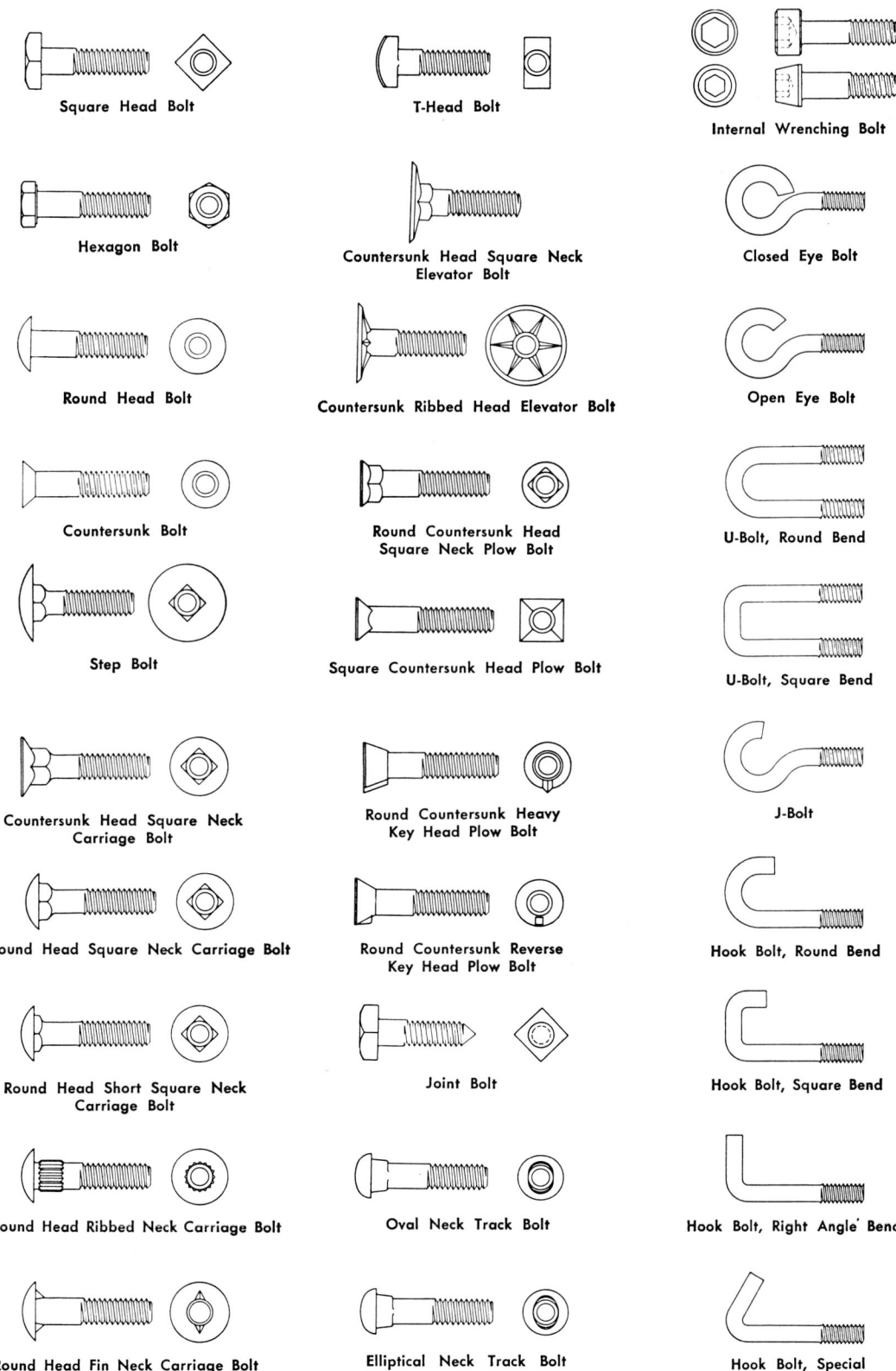

FIG. 9.33. Standard bolt forms (courtesy Machine Design Fastener Book).

b. The altitude of the triangle is equal to $C/4$ as indicated in Fig. 9.28a. This distance should be laid off twice on each side of the center line of the bolt to give the vertical lines of the bolt head and nut.

c. The long leg of the triangle marked H_2 gives the thickness of the nut, which may be laid off as shown in Fig. 9.28a.

d. By projecting the apex of the triangle horizontally to the center line, the distance H_1 is obtained, which is the height of the head, and is laid off as shown in Fig. 9.28a.

e. By using a radius $R_1 = D$, the curve on the front face of head and nut may be drawn as in Fig. 9.28b.

f. The distance R_2 is taken from the points where the long leg of the triangle crosses the center line of the bolt and is used as a radius for the curves on the side faces of head and nut. The center point is on the center line of the face as shown in Fig. 9.28c. The corners may be chamfered to finish the drawing if desired.

Similar steps for the square head and nut are illustrated in Fig. 9.29.

9.21.1 Standard Length of Threads on Bolts and Screws.
For all standard bolts the minimum length of thread for bolts up to and including 6-in. lengths shall be

$$L = 2D + \tfrac{1}{4} \text{ in.}$$

For lengths over 6 in. the threaded length shall be

$$L = 2D + \tfrac{1}{2} \text{ in.}$$

The tolerance on thread lengths shall be $+\tfrac{3}{16}$ in. or $2\tfrac{1}{2}$ threads, whichever is greater. For bolts too short to meet the above requirements, the distance from the bearing surface of the bolt to the first full thread shall not exceed $2\tfrac{1}{2}$ threads for diameters up to 1 in. and $3\tfrac{1}{2}$ threads for sizes greater than a 1-in. diameter.

9.21.2 Bolt Specifications.
In many instances the fasteners are not drawn in projection but are listed in the bill of material. All the information necessary for ordering the bolt must therefore be included. This will require specifying the following items:

Diameter is the diameter of the rod from which the bolt is made and is usually the same as the thread diameter. It is always in the specification.

Length is the distance from the underside of the head to the tip of the bolt. When a countersunk head is used, the length is to the top of the head. This dimension must be given in the specification.

Material is assumed to be steel unless otherwise specified.

Finish should be specified as semifinished or finished. If finished, the special kind of finish must be specified. If type of finish is not specified, the bolt is assumed to be unfinished.

Kind of head should be specified as hexagonal or square. If head and nut are different, this should be noted.

Series is assumed to be regular unless heavy or light is specified.

Length of thread is assumed to be normal unless otherwise noted.

Thread specification consists of the four items previously mentioned. A typical specification would be $\tfrac{3}{4}$-10 UNC-2A.

Typical bolt and screw specifications are given below. Abbreviations are usually used to save time and space.

$\tfrac{5}{8}$-11 UNC-2A × 3 Brass Fin Hex Hd Bolt.
$\tfrac{1}{2}$-13 UNC-2A × $2\tfrac{1}{2}$ Semifin Hex Nut.
$\tfrac{3}{8}$-16 UNC-2A × $1\tfrac{5}{8}$ Sq Hd Bolt.
$\tfrac{3}{8}$-16 UNC-2A × 2 Sq Hd Cup Point Set Scr.
$\tfrac{3}{4}$-10 UNC-2A × 3 Heavy Hex Hd Bolt.

9.21.3 Length of Engagement of Steel Bolts and Screws in Various Materials.
This is the item of first consideration in determining the full thread length of mating components. For steel bolts or screws the following lengths of engagement with other materials is recommended by the SAE Aerospace–Automotive Drawing Standard:

L = length of thread engagement.
D = nominal diameter of thread.
For steel $L = D$.
For cast iron, brass, bronze, or zinc, $L = 1.5D$.
For forged aluminum, $L = 2D$.
For cast aluminum or forged magnesium, $L = 2.5D$.
For cast magnesium or plastic, $L = 3D$.

9.22 CAP SCREWS

Cap screws are similar to bolts in that they have a head on one end and threads on the other. But, in the method of holding two pieces together, they differ widely. The bolt clamps two pieces between the head and the nut, and the cap screw is threaded into one of the pieces, thus clamping one piece between the head and the other piece. The bolt requires a smooth hole in both pieces slightly larger than the bolt, whereas the cap screw requires a smooth hole in one piece and a threaded hole in the other. Cap screws are manufactured with several styles of heads, as shown in Fig. 9.34, and in diameters varying from $\frac{1}{4}$ to $1\frac{1}{4}$ in. The point of all cap screws is flat and chamfered 35° (drawn 45°) to the flat surface and to a depth equal to the depth of the thread.

9.22.1 Length of Cap Screws. Cap screws may be obtained in lengths varying from $\frac{1}{2}$ to 6 in.

Lengths from $\frac{1}{2}$ to 1 in. change by $\frac{1}{8}$-in. increments.

Lengths from 1 to 4 in. change by $\frac{1}{4}$-in. increments.

Lengths from 4 to 6 in. change by $\frac{1}{2}$-in. increments.

The length of a cap screw is measured from the largest diameter of the bearing surface of the head to the end point of the screw and parallel to the axis of the screw. Dimensions of cap screws are given in Tables 6 and 7 in the appendix.

9.22.2 Cap Screw Threads. Slotted-head cap screws are regularly threaded with American National coarse threads with a usable length of thread equal to $2D + \frac{1}{4}$ in. Screws that are too short for this length of thread will be threaded as close to the head as practicable. Cap screws may also be obtained with threads of the fine-thread series. All dimensions for laying out cap screws are given in Tables 6 and 7 in the appendix. The standard specification for a cap screw would be

$\frac{5}{8}$-11 UNC-2A $\times$ $1\frac{5}{8}$ Brass Fillister Hd Cap Scr.

If made of steel, the material term is omitted.

9.23 MACHINE SCREWS

Machine screws are similar in function and operation to cap screws but are usually smaller in diameter. Machine screws are specified by numbers from 2 to 12 below the $\frac{1}{4}$-in. size and then by diameter up to $\frac{3}{8}$ in. The lengths of machine screws vary from $\frac{1}{8}$ to 3 in., changing by $\frac{1}{16}$ increments up to $\frac{1}{2}$ in. then by $\frac{1}{8}$ in. up to 1 in., and finally by $\frac{1}{4}$-in. up to 3 in. They are threaded not less than $1\frac{3}{4}$ in. for all screws over 2 in. in length, and within two threads of the bearing surface or closer if practicable for shorter screws. Machine screws may be obtained with either fine or coarse threads and with four types of heads, all slotted, as shown in Fig. 9.35. Square and hexagonal nuts may be obtained for all sizes of machine screws. The usual method of specifying machine screws is

No. 10-24 NC-3 $\times$ 1 Flat Hd Mach Scr.

The material should also be specified if other than steel. Dimensions for drawing machine screws are given in Table 5 in the appendix.

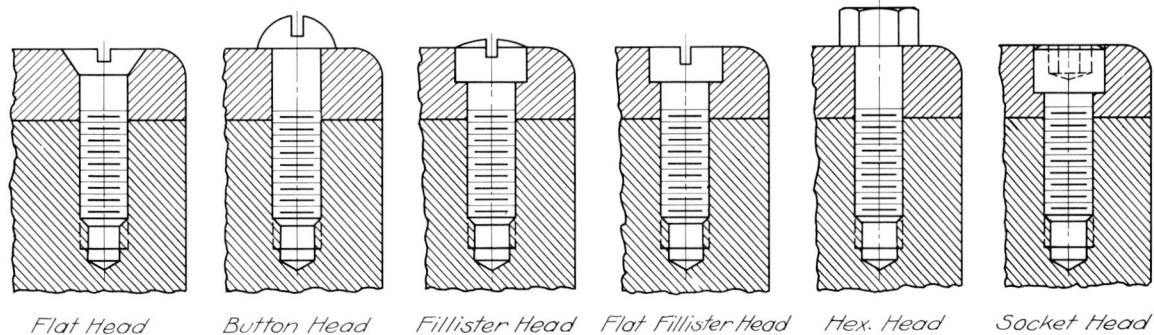

Flat Head   Button Head   Fillister Head   Flat Fillister Head   Hex. Head   Socket Head

FIG. 9.34. Cap screws.

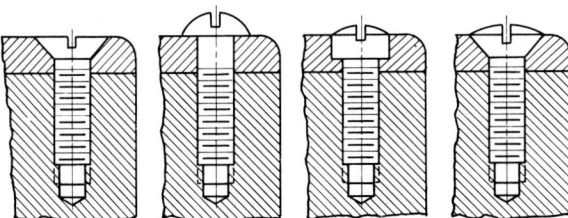

FIG. 9.35. Machine screws.

9.24 SET SCREWS

The purpose of a set screw is to prevent rotation or sliding between two parts. It is screwed into one part so that its point presses against the other, thus resisting relative motion between the two parts by means of the friction between the point of the set screw and one of the parts. The standard square-head set screw and several kinds of headless set screws are shown in Fig. 9.36. Information necessary for drawing set screws is given in Tables 16 and 17 in the appendix. They are specified on the drawing by giving the diameter, length, type of head, type of point, and thread specification. Thus, the specification should read

$\frac{1}{4}$-20 UNC-2B $\times \frac{3}{4}$ Socket Hd, Cone Pt, Set Scr.

9.25 OTHER TYPES OF STANDARD AND SPECIAL BOLTS AND SCREWS

Screw threads have been used on many kinds of screws and bolts that cannot be discussed in detail because of lack of space. Some of these are shown in Fig. 9.33 and discussed briefly in the following paragraphs.

9.25.1 Unslotted-Head Bolts.
Unslotted-head bolts such as carriage bolts, step bolts, tire bolts, plow bolts, and track bolts are used frequently in woodwork as well as in metalwork where the square shank keeps the bolt from turning. They may have round, oval, or countersunk heads, as shown in Fig. 9.33.

9.25.2 Slotted-Head Bolts.
Slotted-head bolts such as stove bolts are used chiefly for metalwork. They may have round or countersunk heads as shown in Fig. 9.32. The slot allows them to be tightened with a screwdriver.

9.25.3 Special-Purpose Bolts.
Special-purpose bolts having threads on one end only, such as eye bolts, U bolts, hook bolts, and thumb screws, are also illustrated in Fig. 9.33. They are used in many places where ordinary bolts would not be satisfactory.

9.25.4 Wood Screws.
Wood screws are usually designed with the thread forming a conical helix although they sometimes continue long threads on a cylindrical helix, as on the lag screw. They are designed in this manner so that they will cut their own threads in the wood as they are driven. The heads are usually slotted for a screwdriver and may be either round, elliptical, or countersunk. Wood screws are specified by number as shown in Table 9.1, which is abstracted from ANSI B18.6.1—1972.

Wood screws are threaded for approximately two thirds their length (Table 9.1). The length of screw is measured from the diameter of the widest bearing surface of the head to the point. Various types of screws are illustrated in Fig. 9.37. A special type head known as the Phillips recessed head is used frequently for production work.

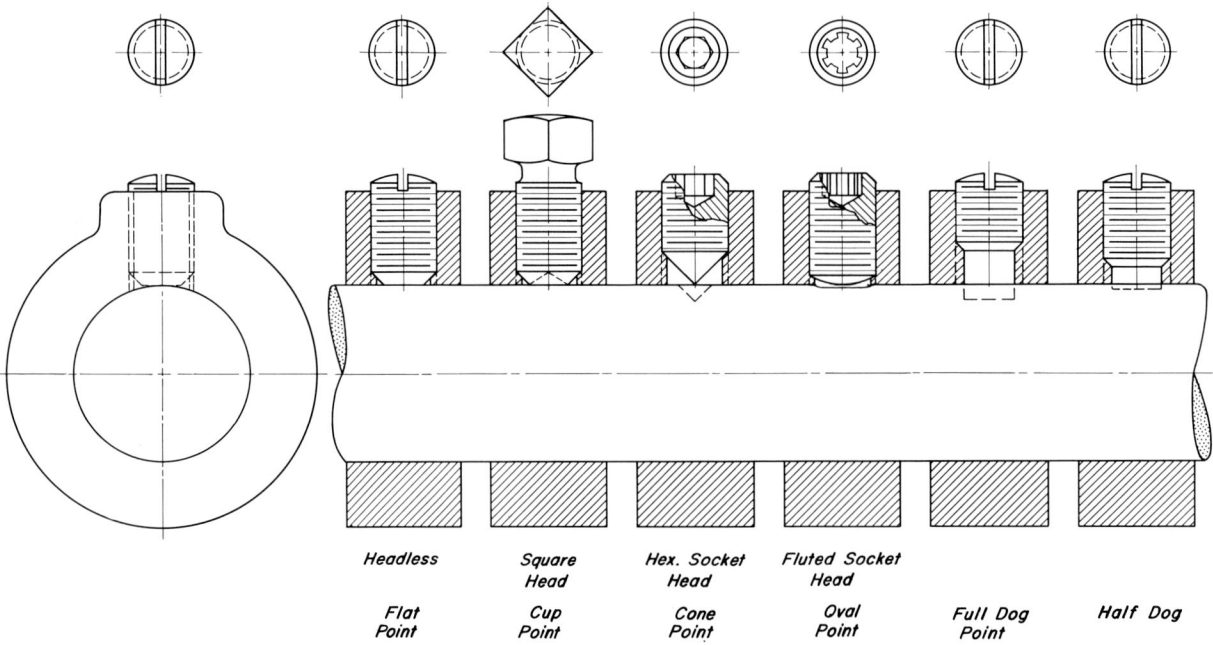

FIG. 9.36. Set screw heads and points.

TABLE 9.1 WOOD SCREW DIMENSIONS

Nominal size	0	1	2	3	4	5	6	7	8	9	10	12	14	16	18	20	24
Threads per inch	32	28	26	24	22	20	18	16	15	14	13	11	10	9	8	8	7
Basic diameter	0.060	0.073	0.086	0.099	0.112	0.125	0.138	0.151	0.164	0.177	0.190	0.216	0.242	0.268	0.294	0.320	0.372

9.25.5 Tapping Screws. These screws, sometimes called *self-tapping screws,* are made to either cut or form a mating thread, in metal, plastics, or other material, without having the holes previously tapped. This makes a very tight fit that prevents the screw from backing out even under vibrating conditions.

The dimensions of these standard screws are given in ANSI B18.6.4.—1958. Two general types are listed.

a. Thread-forming ANSI Types A, B, BP, C, and U.
b. Thread-cutting ANSI Types, D, F, G, T, BF, BG, and BT.

9.25.6 Rivnut. *Rivnut* is the patented name of a fastener that may be used to fasten two plates together and at the same time provide a nut plate to which other parts may be fastened. One of the important features of this fastener is that the entire operation can be completed from one side of the plate. Figure 9.38 shows the Rivnut being used as a rivet to hold two plates together and also as a nut plate by means of which the clip is fastened to the plate.

9.26 RIVETS

The chief purpose of rivets is to make a permanent fastening between plates or rolled sections. Rivets are made with various shapes of heads, but all function alike in their method of holding the parts together. They are manufactured with a head on one end only, the other head being formed in the driving. Holes for the rivets are punched slightly larger than the rivet so that they may be put in place easily. They are first heated and then inserted in the hole, after which the other head is formed by hammering with a pneumatic hammer or by pressure. The length of a rivet is figured from the area of greatest bearing on the head to the point. This length is specified so as to allow only enough material to form the head and fill the hole around the rivet.

The various shapes of rivet heads are shown in Fig. 9.39. Rivets are sometimes shown on the drawing and sometimes omitted entirely. When

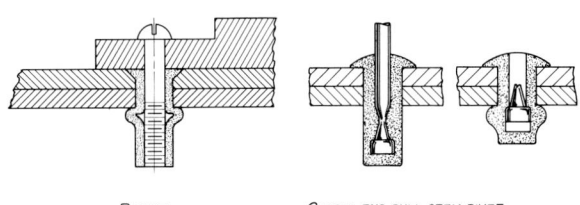

RIVNUT   CLOSED END PULL-STEM RIVET

FIG. 9.38. Self-locking screws.

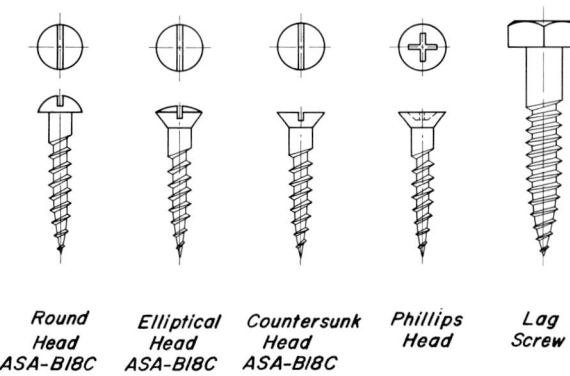

Round Head ASA-B18C   Elliptical Head ASA-B18C   Countersunk Head ASA-B18C   Phillips Head   Lag Screw

FIG. 9.37. Wood screws.

238 FASTENERS

shown they are represented by a circle as indicated in Fig. 9.40 or simply by center lines.

Since the pieces held together by rivets may be joined in the field or in the shop, it is necessary to have other symbols to indicate which way the riveting is to be done. Also rivet heads may be countersunk and/or chipped to provide clearances, and symbols have been devised to show these things. These symbols are shown in Fig. 9.40.

In boiler and tank work, considerable pains may be taken to show the actual shape of the rivet head. See Fig. 9.39. A drawing for several types of joints is shown in Fig. 9.41. In structural work the shape of the rivet head is seldom shown.

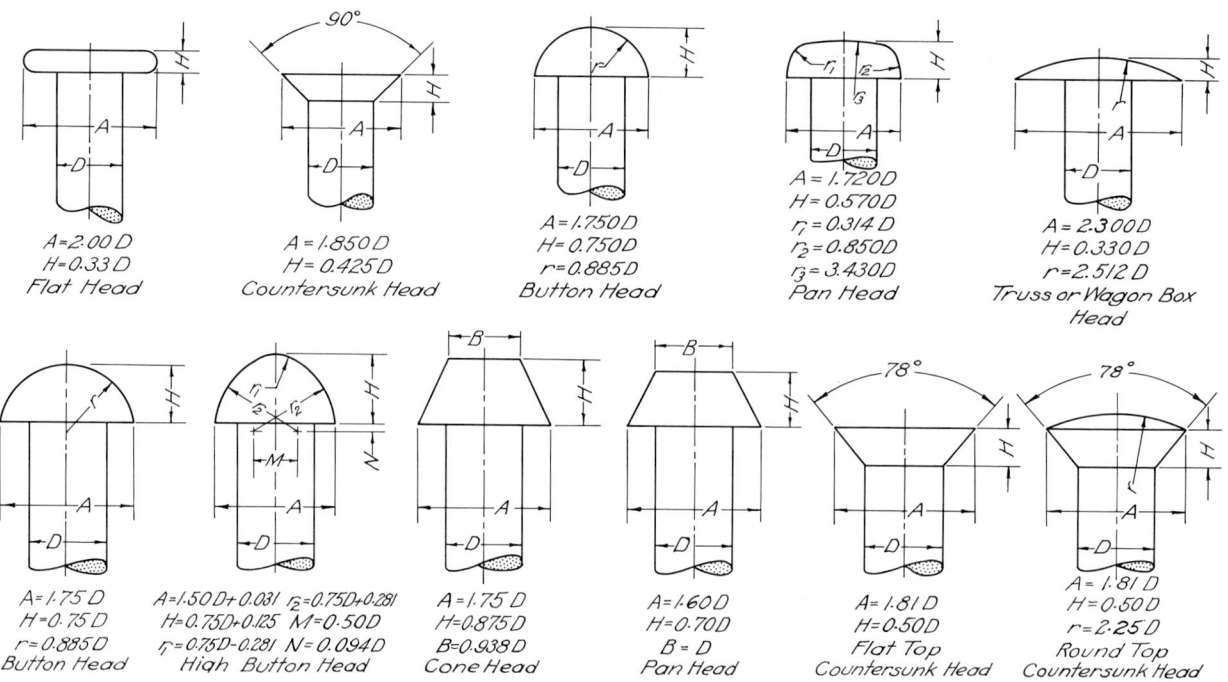

FIG. 9.39. Rivet heads (courtesy ANSI).

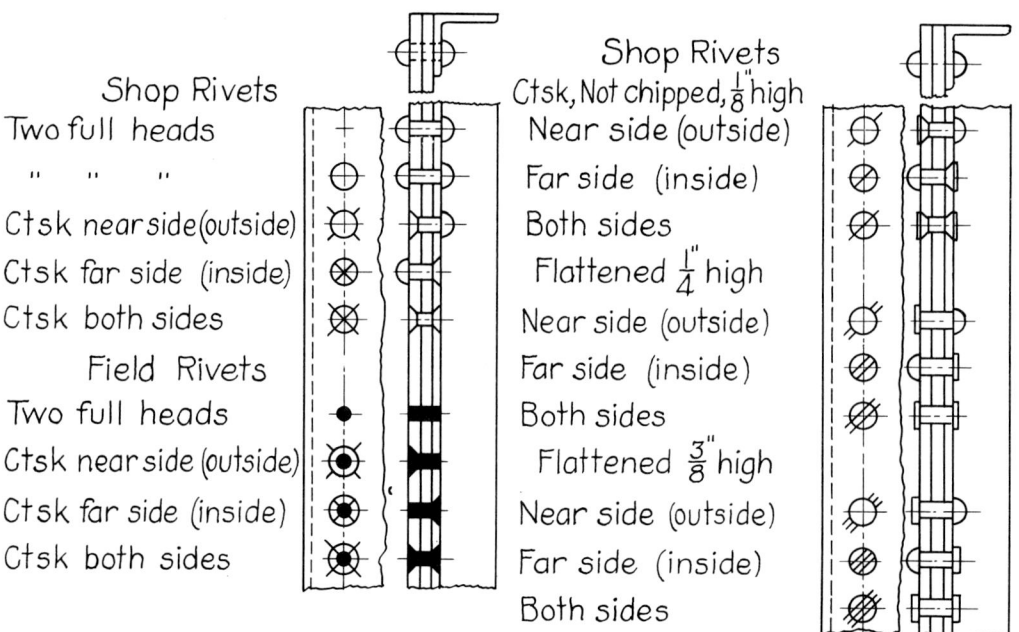

FIG. 9.40. Rivet symbols.

Explosive rivets are used in aircraft work where it is not convenient to work on both sides of the plate. These are aluminum rivets that have been hollowed out at the end to allow space for the explosive charge. When exploded the rivet is expanded to grip the plates as shown in Fig. 9.42.

9.27 KEYS

These fasteners called *keys* are chiefly used to hold pulleys, gears, and rocker arms on rotating shafts. There are several standard types.

Figure 9.43 shows five common types. Square-, flat-, taper-, and gib-head taper keys are dimensioned by notes giving the width, height, and length. Whenever possible, sizes should conform to standards. The keys need not be drawn except for special keys or when limits other than those of the standard are necessary. Patented keys such as the Woodruff and, Pratt and Whitney, shown in Fig. 9.43 are specified by number.

Keyways on shafts or internal members may be dimensioned as shown in Fig. 9.44a. On the hub or external member the keyway may be dimensioned as shown in Fig. 9.44b. The key seat for patented varieties may be specified by the key number.

Tables 18 through 20 in the appendix give dimensions for several standard keys. More complete information may be found in any mechanical engineer's handbook.

When heavy loads are to be transmitted from the shaft to the pulley or vice versa, two or more keys may be used or the Kennedy or Lewis systems may be practicable. See Fig. 9.45.

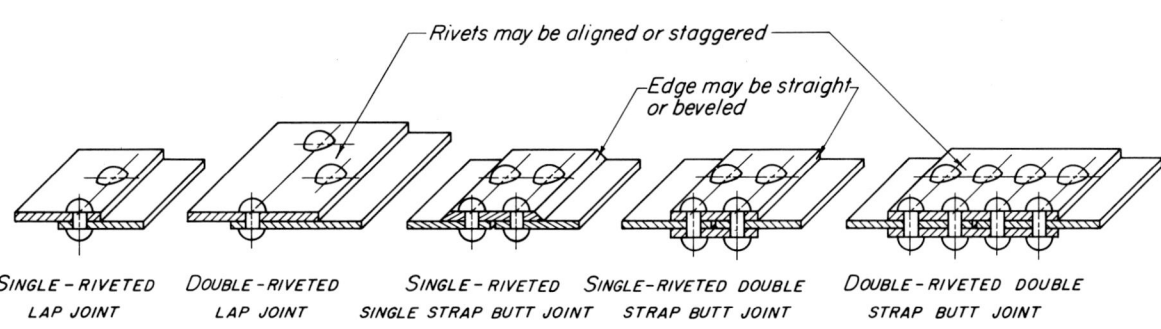

FIG. 9.41. Riveted joints.

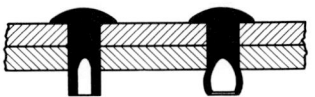

FIG. 9.42. Explosive rivet.

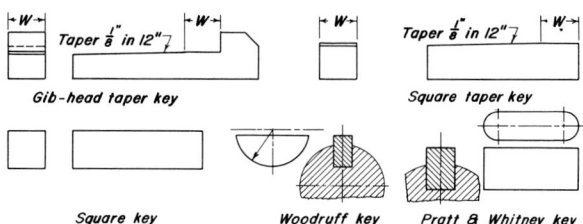

FIG. 9.43. Types of keys.

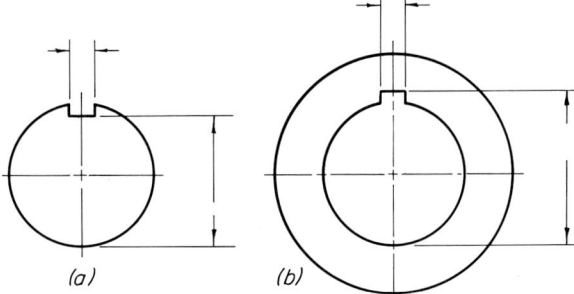

FIG. 9.44. Dimensioning keyways.

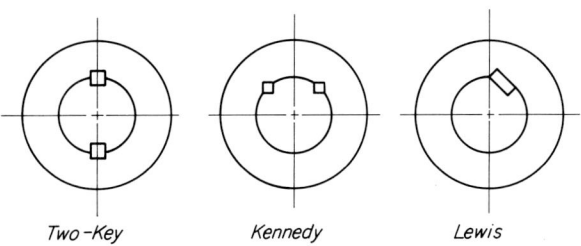

FIG. 9.45. Heavy-duty keys.

240 FASTENERS

9.28 INVOLUTE SPLINES
In many cases where heavy loads are to be transmitted, keys have been eliminated by the use of multiple spline shafts as illustrated in Figs. 9.46 and 9.47. Involute splines and serrations are commonly used.

9.29 TAPER PINS
In light work taper pins are sometimes used as fasteners in place of keys or as dowels. They taper ¼ in. in diameter per foot of length. Dimensions of standard taper pins may be obtained from Table 21 in the appendix.

9.30 SPRINGS
Although springs are not considered fasteners, the method of representing them is closely related to the drawing of screw threads. Springs are formed by winding the wire in the form of a cylindrical or conical helix. In actual practice, the projections of the spring are conventionalized in a manner similar to threads by using straight lines instead of helices. Figure 9.48 shows by steps the method of drawing a spring. Compression springs are usually ground on the ends to provide a flat bearing. Tension springs must have a loop of some kind on each end. In long springs a few turns may be shown on each end with ditto marks between, rather than drawing the entire spring. These things are illustrated in Figs. 9.48 through 9.51. For further information on springs see ANSI Y14.13M—1981.

In specifying a spring, the following information should be given: diameter and kind of wire, free length of spring, number of turns, and controlling diameter.

Springs are frequently represented by single lines as in Fig. 9.52 instead of the more complicated double line.

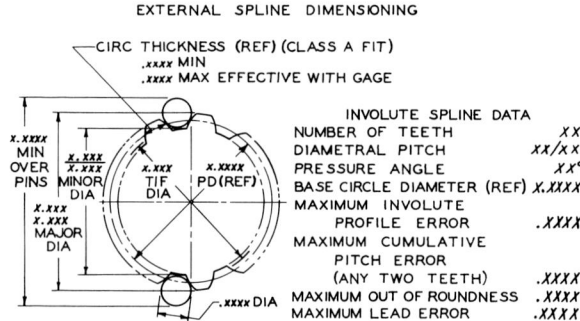

FIG. 9.46. External spline dimensioning (courtesy ANSI).

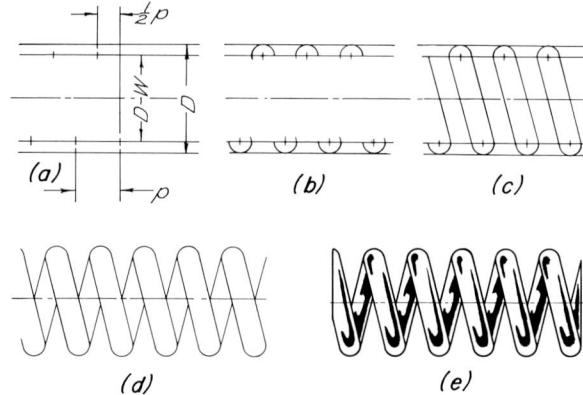

FIG. 9.48. Construction of a coiled spring symbol.

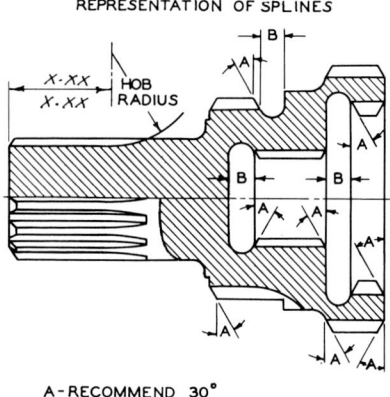

FIG. 9.47. Representation of splines (courtesy ANSI).

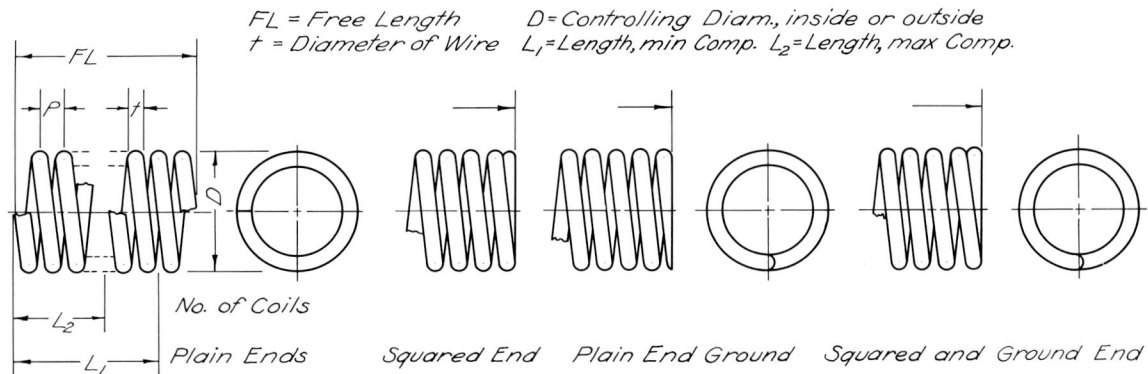

FIG. 9.49. Compression springs.

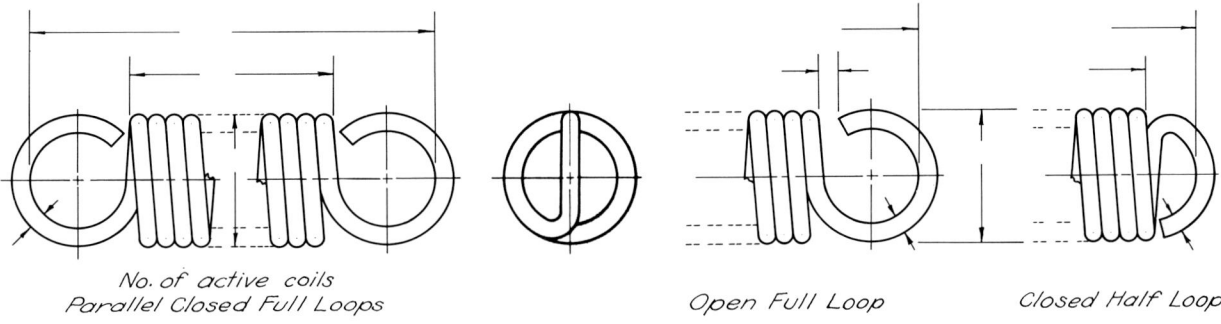

FIG. 9.50. Tension springs.

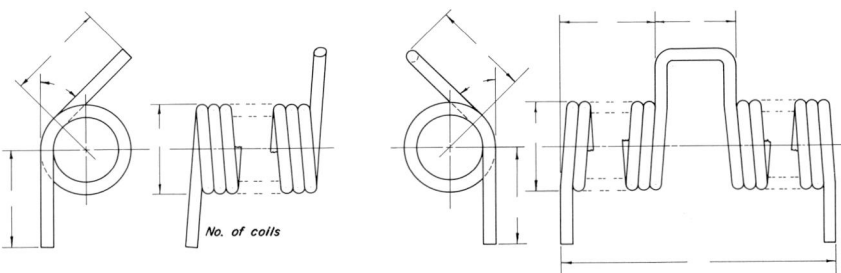

FIG. 9.51. Representing and dimensioning torsion springs.

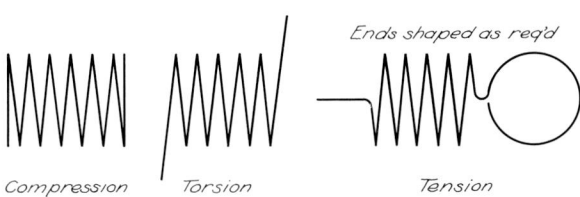

FIG. 9.52. Single line spring symbols.

9.30.1 Flat Springs. The coned disk or Belleville Spring is shown together with dimensions and other specifications in Fig. 9.53. The method of drawing and dimensioning an underslung leaf spring is shown in Fig. 9.54.

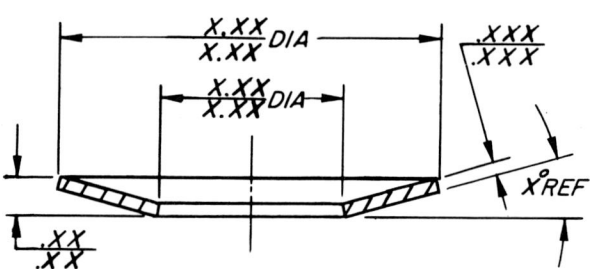

NOTE:-BREAK SHARP CORNERS .XX-.XX
(10-20% OF STOCK THICKNESS)
TEST LOAD XX LB FOR .XX-.XX DEFLECTION

FIG. 9.53. Coned disk or Belleville Spring. (Courtesy SAE.)

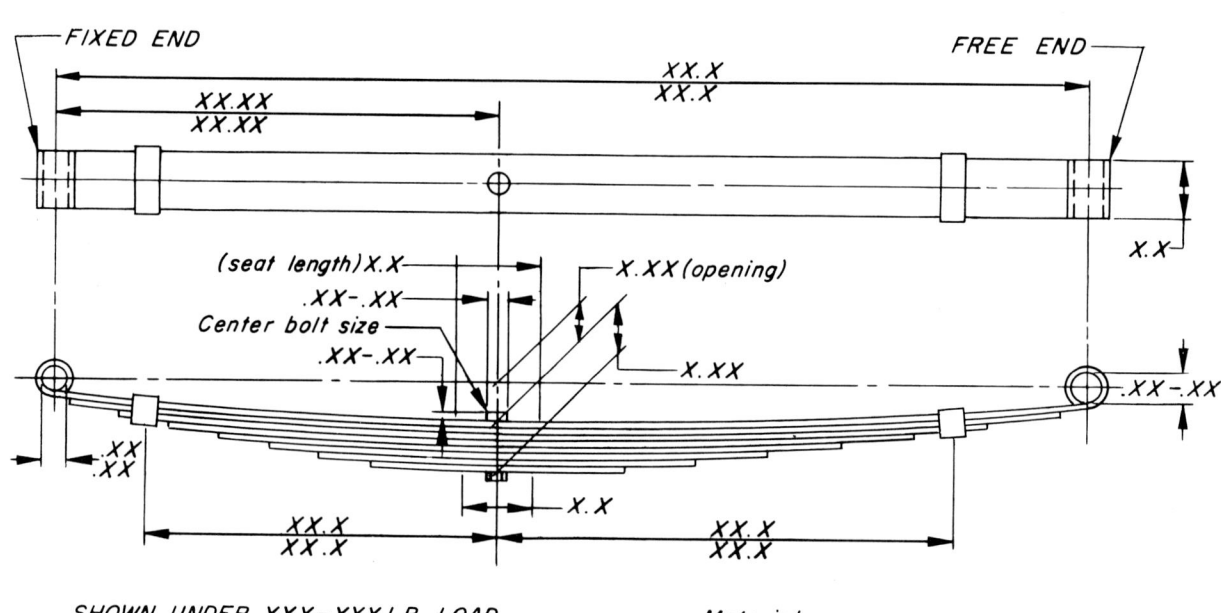

SHOWN UNDER XXX-XXX LB LOAD
CLEARANCE X.X IN. (metal to metal)
LEAVES NO. X-X-X-X SHOT PEENED
RATE XXX-XXX LB/IN

Material
Hardness specification
No. of leaves
Thickness of leaves

FIG. 9.54. Underslung leaf spring. (Courtesy SAE.)

SELF-STUDY QUESTIONS

Before trying to answer these questions, read the chapter carefully. Then, without reference to the text, answer as many questions as possible. For those that cannot be answered, the number in parentheses following the question number gives the article in which the answer can be found. Look it up and write down the answer. Check the answers that you did give to see that they are correct.

9.1 (**9.2.1**) Name one important source of national standards. _____

9.2 (**9.4**) The top edge or surface joining the two sides of a thread is called _____.

9.3 (**9.4**) The pitch of a Sharp-V-thread is the distance from crest to _____.

9.4 (**9.4**) The lead of a thread is the distance it advances in _____ turn.

9.5 (**9.4; 9.8**) A _____ hand thread advances into a hole when turned clockwise.

9.6 (**9.4**) The largest diameter of a thread is called the _____ diameter.

9.7 (**9.5.1**) Two thread types used for transmitting power are _____ and _____.

9.8 (**9.6.3; 9.9.1**) The letters UNC in a thread specification mean _____ _____ _____.

9.9 (**9.6.3; 9.9.2**) The letters UNF in a thread specification mean _____.

9.10 (**9.10**) The three classes of fits for an external thread are _____ _____ _____.

9.11 (**9.10**) The three classes of fit for an internal thread are _____ _____ _____.

9.12 (**9.10.1**) Classes 1A and 1B are described as a _____ fit.

9.13 (**9.10.2**) Classes 2A and 2B are commonly used for _____.

9.14 (**9.5**) Show by a small sketch the profile shape of a square thread.

9.15 (**9.11.1**) The _____ profile is a 60° V-type metric thread.

9.16 (**9.11.2**) The _____ profile screw thread is used in the aerospace industry.

9.17 (**9.15.2**) Holes that do not go clear through a part are called _____.

9.18 (**9.16.1**) Regular bolts are not _____ on any surface.

9.19 (**9.16.2**) Semifinished bolts have a circular _____ face under the head only.

9.20 (**9.18**) Two common locking devices for nuts are _____ and _____.

9.21 (**9.19**) Stud bolts are threaded on _____ _____.

9.22 (**9.20**) Name three other types of bolts besides the regular, semifinished, and finished hexagon-head bolts.

9.23 (**9.21.1**) The standard length of threads on bolts up to 6 in. in length is _____ $D+$ _____ in.

9.24 (**9.22**) The cap screw requires a _____ hole in one part and a threaded hole in the other of two parts to be held together.

9.25 (**9.23**) Machine screws differ from cap screws in that they are _____ than cap screws.

CHAPTER 10

SHOP TERMS AND PROCESSES

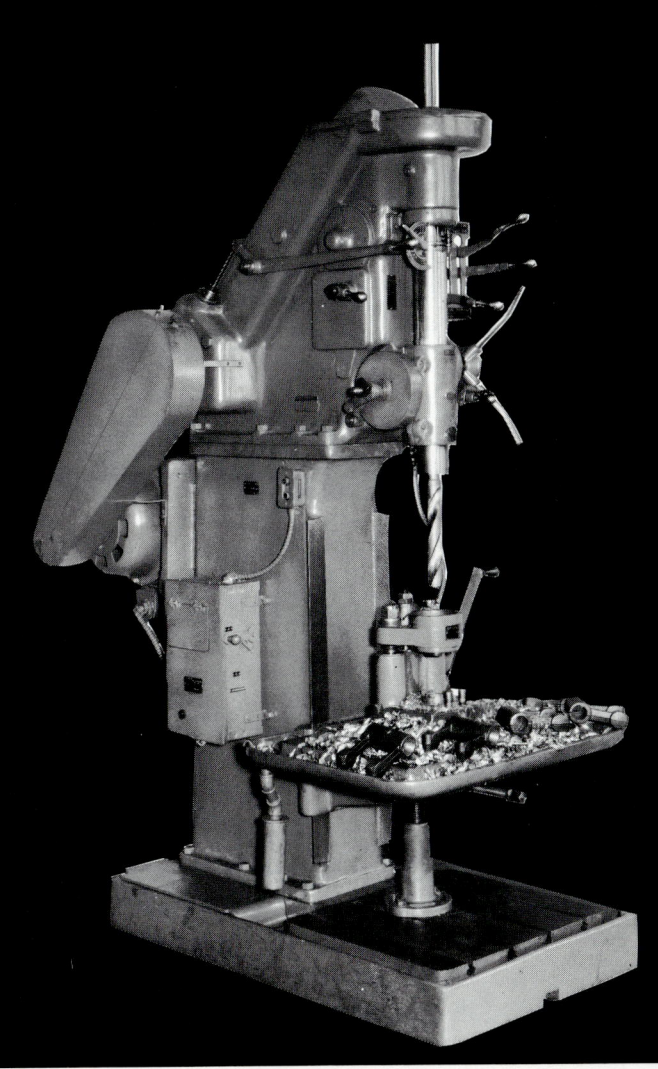

CHAPTER 10

10.1 INTRODUCTION

The relation of the drafting room to the various shops is often underestimated in teaching the drafter the techniques of the profession. No design has commercial value unless the object designed can be made in the shops and at a cost that will allow it to compete with similar products in the markets. Odd-size tools, impractical methods, and even impossible operations are often specified on drawings of the uninformed. The drafter must know the capabilities and limitations of the shops. The drawings must "talk" in the shop person's language.

10.2 CASTINGS

The genesis of all cast metal work is found in the pattern shop. All drawings specifying castings must *first go to the patternmaker who constructs a wood or plaster model of the object to be cast. This model, called a pattern,* is then sent to the foundry where the actual casting is made. Castings are made from various kinds of iron and steel and also from nonferrous metals, such as aluminum, magnesium, zinc, copper, bronze, and brass. To understand the processes that are carried on in the pattern shop and foundry, it is necessary to have a general knowledge of the terms employed there.

10.3 PATTERN DRAWING OR LAYOUT

The source from which the patternmaker obtains necessary information is the working drawing made in the drafting room. Since this working drawing contains information to be used in other

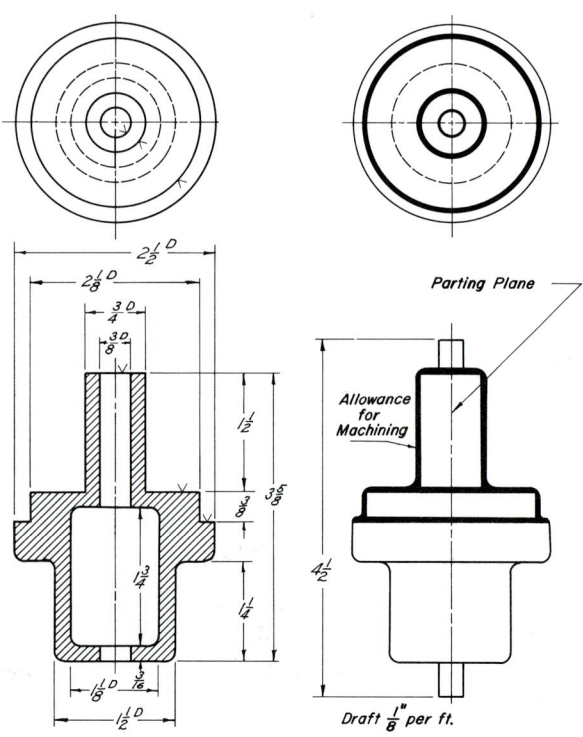

FIG. 10.1. Machine drawing and pattern drawing for same object.

246

shops but not needed by the patternmaker, sometimes a new drawing must be made that is called a *pattern drawing* or *layout*. This omits all unnecessary information and adds such items as parting plane, finish allowance, draft, and core prints. This drawing is made full-scale with the shrink rule. On the drawings many curves and intersections must be carefully constructed, since dimensions are taken directly from the pattern drawing. Sections may also be taken at different places on the pattern drawing for the purpose of cutting sheet metal templates with which to check the pattern. Figure 10.1 shows the working drawing of a simple object and also the pattern drawing for the same object.

10.4 PATTERNMAKER'S SHRINK RULE
When the metal in a casting is cooling, it continues to get smaller until room temperature is reached. The amount of this shrinkage varies with different metals, but in any case the patternmaker must allow for it by making the layout and the pattern oversize. This is done by using a shrink rule on which the divisions are all slightly larger than a normal rule. The amount of allowance made for the shrinkage of various metals is given below in inches per foot.

Cast iron	$\frac{1}{8}$
Cast steel	$\frac{1}{4}$
Aluminum alloys	$\frac{5}{32}$
Magnesium alloys	$\frac{11}{16}$

Thus a 12-in. shrink rule for cast iron would actually measure $12\frac{1}{8}$ in.

10.5 FINISH ALLOWANCES
Before the pattern drawing is complete, the patternmaker must add the "finish," which may be indicated by means of a heavy line. *The term finish as applied to pattern drawings means the amount of material added to the pattern to provide metal on the casting that is to be cut away in the finishing process.* The amount of this allowance varies from $\frac{1}{8}$ to $\frac{3}{4}$ in., depending on the size of the casting and the metal from which it is made. The finish allowance on the drawing in Fig. 10.2 has been indicated by dotted lines that show the outline of the finished piece. On the engineer's working drawing the finished surfaces must always be indicated by one of the standard methods explained in Chapter 12.

10.6 PARTING PLANE
Before the pattern drawing can be carried any further, the location of the parting line or plane must be determined. This is not indicated on the engineer's working drawing, but the designer must have considered the matter to avoid a design that is unusually hard to cast and therefore expensive. *The purpose of the parting plane is to enable the pattern to be removed from the mold without breaking the walls of the sand.* The parting plane should be at the largest part of the object and so arranged that there are no undercut faces or projections. In casting the object, the parting plane is made to coincide with the plane between the two parts of the mold or flask. In simple objects such as that shown in Fig. 10.3a, it is sometimes possible to use one face of the

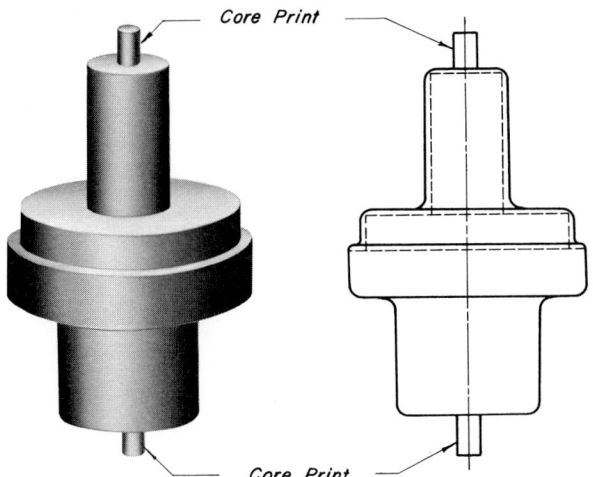

FIG. 10.2. Core prints on pattern.

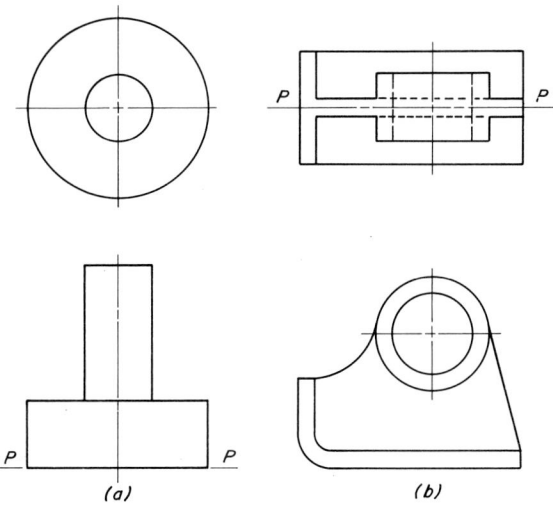

FIG. 10.3. Parting plane.

object as the parting plane, thus making the work of casting easier and less expensive. Usually one parting plane is necessary, as in Fig. 10.3b. Occasionally more than one plane is necessary, but this should be avoided if possible. The line on the drawing that shows the position of the parting plane is called the *parting line*. This line is marked on the patternmaker's layout.

10.7 CORE PRINTS

As soon as the position of the parting plane has been determined, the core prints should be added to the full-size drawing. *Core prints are projections from the pattern whose purpose is to make an impression in the sand mold into which the core will be placed.* Since the core will completely fill the impression made by the core print, the function of the core print is merely to hold the core in the proper position until the metal has cooled. Figure 10.2 shows the core prints added to the pattern layout and to the pattern. Core prints are not shown on the working drawing.

10.8 DRAFT

To make the removal of the pattern from the mold easier, the pattern is tapered away from the parting plane. This taper is called draft or draw. The draft can be added to the pattern by increasing the size at the parting plane, thus making the piece stronger and heavier, or by allowing the material to remain the same at the parting plane and decreasing it at the top or bottom. The latter method decreases the strength and weight. When a wood pattern is used, a draft of $\frac{1}{8}$ in. per foot is used, but with a metal pattern $\frac{1}{16}$ in. per foot is sufficient. Draft may be specified by degrees and will usually be from $\frac{1}{2}$ to 3°. The draft is shown on the full-size patternmaker's layout. It is not indicated in any way on the engineer's working drawing.

10.9 DESIGN DETAILS

In making a working drawing for a casting there are many design details with which the engineer should be familiar. These have been well standardized within the industry or by individual companies. Such things as fillets and rounded corners, bosses, pads, ribs, and rib intersections must be provided for by the drafter to facilitate production and ensure good quality.

10.9.1 Fillets and Rounds.

When the metal in a casting cools, the crystals tend to arrange themselves so that their lines of strength are perpendicular to the cooling surface, as depicted in Fig. 10.4. Therefore sharp angles tend to become planes of weakness where holes or cracks may occur during cooling. *For this reason the sharp internal angles on a pattern are filled with wood, leather, or wax, as illustrated in Fig. 10.5. This process is called filleting. Sharp external corners on a casting should also be rounded.* Careful attention to these details makes it easier to remove the pattern from the mold, allows the metal to flow more freely through the casting, and helps to avoid cracks and planes of weakness.

Each company usually has its own rules concerning size of fillets. Some require the fillets to have a radius equal to the thickness of the section, as shown in Fig. 10.6; others give the radii for fillets in tabular form. The following design data for minimum webs and fillet radii on aluminum allow castings are used by a large industrial company.

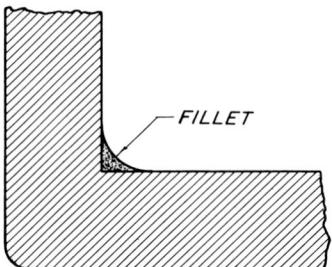

FIG. 10.5. Filleted corner.

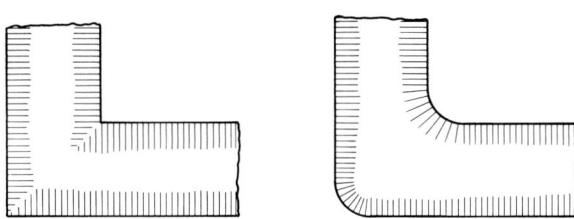

FIG. 10.4. Corners rounded and filleted to relieve stress.

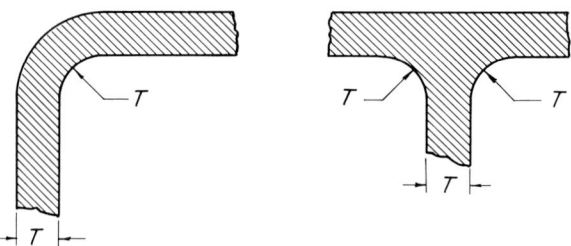

FIG. 10.6. Relation of fillet radii to thickness.

	Material No.				
	43	356	195	220	AM 265
t Min. web thickness	$\frac{5}{32}$	$\frac{5}{32}$	$\frac{5}{32}$	$\frac{5}{16}$	$\frac{5}{32}$
r Min. fillet radii	$\frac{5}{32}$	$\frac{3}{16}$	$\frac{3}{16}$	$\frac{3}{8}$	$\frac{3}{16}$

Although it is essential that all angles be filleted and corners rounded, it is also important to avoid using too large a radius for fillets with thin web sections. Too large fillets may cause cooling stresses in thin webs, owing to the heavy concentration of material at the intersections and consequent unequal cooling.

The engineer's working drawing should always show all fillets and carry a note such as "All fillets are (x) radius and rounds (x) radius unless otherwise specified."

10.9.2 Section Thickness in Castings. As the casting is being poured, the metal flows in various directions into the parts of the mold and gradually cools as it flows. *If sections are too thin, the metal may cool so much that it will not be hot enough to join properly when metal flowing in two directions comes together.* This forms a plane of weakness called a *cold shut*. The minimum thickness of webs varies with the kind of material and with company practice. For instance, one company recommends the following minimum thickness in inches: iron, $\frac{5}{32}$; brass and bronze, $\frac{3}{32}$; aluminum, $\frac{1}{8}$ to $\frac{3}{16}$.

A thin web intersecting a heavier member may

TABLE 10.1 LENGTH OF TAPER L-WALL THICKNESS W[a]

T	t-Aluminum and Magnesium Alloys										Max R
	$\frac{5}{32}$		$\frac{3}{16}$		$\frac{7}{32}$		$\frac{1}{4}$		$\frac{5}{16}$		
	L	W	L	W	L	W	L	W	L	W	
$\frac{5}{32}$											$\frac{3}{16}$
$\frac{3}{16}$					Values in this area						$\frac{3}{16}$
$\frac{7}{32}$					under critical						$\frac{7}{32}$
$\frac{1}{4}$					2:1 ratio					$\frac{1}{4}$	
$\frac{5}{16}$											$\frac{5}{16}$
$\frac{3}{8}$	$1\frac{1}{8}$	$\frac{3}{8}$									$\frac{3}{8}$
$\frac{7}{16}$	$1\frac{1}{4}$	$\frac{3}{8}$	$1\frac{1}{4}$	$\frac{13}{32}$							$\frac{3}{8}$
$\frac{1}{2}$	$1\frac{1}{4}$	$\frac{3}{8}$	$1\frac{1}{4}$	$\frac{13}{32}$	$1\frac{1}{4}$	$\frac{7}{16}$					$\frac{7}{16}$
$\frac{3}{4}$	$1\frac{1}{4}$	$\frac{3}{8}$	$1\frac{1}{4}$	$\frac{13}{32}$	$1\frac{1}{4}$	$\frac{7}{16}$	$1\frac{1}{4}$	$\frac{1}{2}$	$1\frac{1}{4}$	$\frac{17}{32}$	

[a] Courtesy Douglas Aircraft Co.

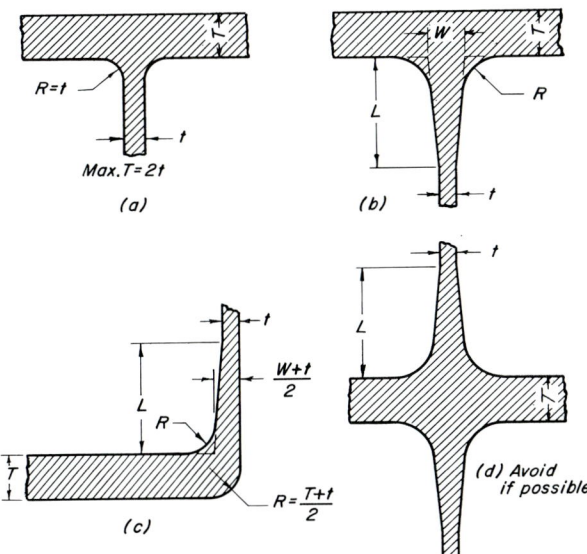

FIG. 10.7. Web and wall thickness.

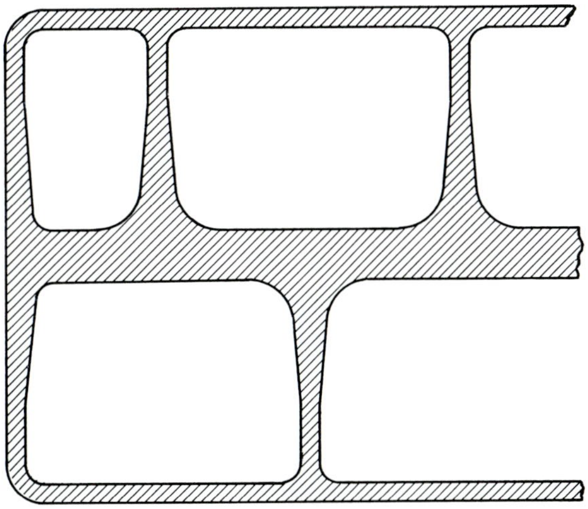

FIG. 10.8. Arrangement of interior webs.

develop cracks, owing to unequal cooling of the two parts. For this reason it is well to avoid too abrupt a change in cross section of the members. When such a change cannot be avoided, the thin member should be tapered to reduce the shrinkage stresses. It is recommended that the heavy section be not more than twice the thickness of the thin section, as shown in Fig. 10.7a. When the minimum 2:1 ratio cannot be maintained, the thin section shall be tapered as shown in Figs. 10.7b and c, according to the dimensions given in Table 10.1.

Intersecting webs may tend to cause cooling cracks because of the heavy concentration of material at the intersection. See Fig. 10.7d. This may be avoided or improved by alternating the webs as in Fig. 10.8, whenever possible.

10.9.3 Bosses. *Projections on a casting to allow for drilling holes or to provide bearing for a bolt head are called bosses.* When bosses occur they must be filleted to provide as gradual a change in cross section as possible, as shown in Fig. 10.9a. When bosses must be placed in webs, the web must be tapered, as in Fig. 10.9b, to provide the proper thickness.

10.9.4 Pads. Pads as illustrated in Fig. 10.10 will save in the cost of the part by eliminating some material and also eliminating large areas that must be machined. Pads increase the likelihood that the part will set flush on flat surfaces.

10.9.5 Ribs. On a casting, ribs perform two functions: to strengthen and stiffen the part and to prevent cooling cracks by acting as heat conductors, thus promoting the even cooling of a section.

10.10 DEFINITION OF PATTERN AND FOUNDRY TERMS

The following terms are used in the pattern shop, core shop, and foundry.

Boss. A projection on an object whose height is usually less than its diameter. It is placed there for the purpose of providing a bearing surface or to enable the shop to drill a hole to better advantage. It also elongates a hole in thin parts where additional bearing surface is needed. See Fig. 10.9.

Core. A sand model of the hollow interior of a casting. See Fig. 10.11b.

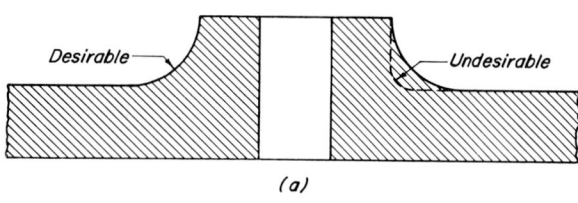
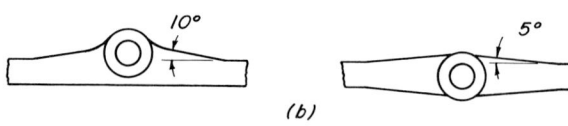

FIG. 10.9. Filleted bosses.

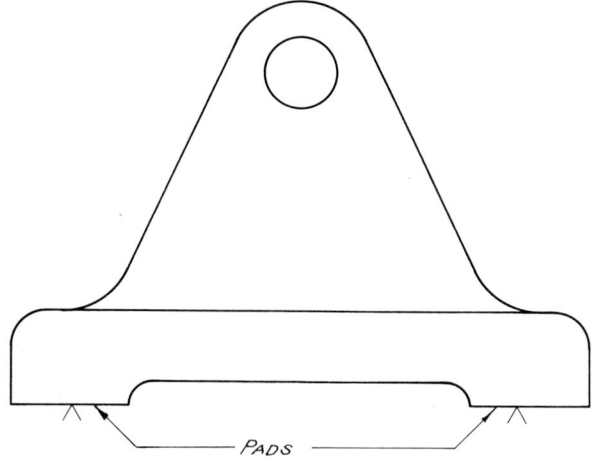

FIG. 10.10. Use of pads to reduce machining.

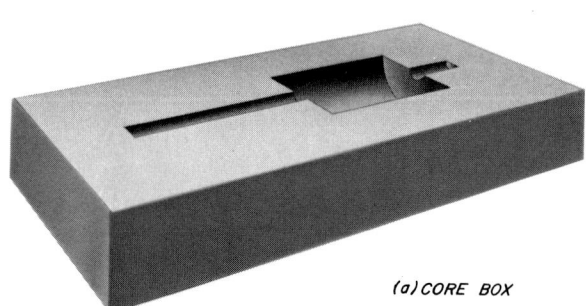
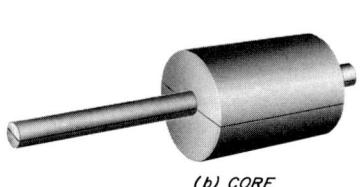

FIG. 10.11. Core box and core.

Core box. A wooden box whose internal shape is such that when it is packed with sand, the desired core is formed. See Fig. 10.11a.

Core print. A projecting part of the pattern that makes an impression in the sand mold into which the core is placed. See Fig. 10.2. This supports the core in the mold.

Cupola. A furnace in which the metal is melted in the foundry.

Draw or draft. The taper on a pattern that makes it easier to withdraw the pattern from the mold.

Fillet. The concave surface that fills in the sharp angles between two faces on a pattern. See Fig. 10.5.

Finish allowance. Extra material allowed on a pattern to provide additional metal for finishing a face on the casting. See Fig. 10.1.

Flask. Two or more boxlike parts having the same cross section into which the sand is packed to form the mold. See Fig. 10.12.

Cope. The upper part of the flask. See Fig. 10.12b.

Drag. The lower part of the flask. See Fig. 10.12a.

Gate. The opening in the sand through which the metal flows to the casting. See Fig. 10.13.

Parting plane. A plane on which the pattern can be divided so that both parts can be removed from the sand. See Fig. 10.3.

Pattern. A slightly oversize model of the object to be cast, usually made of wood. See Fig. 10.2.

Round. The rounded corner placed on an exterior corner of a part to remove the sharp corner. See Fig. 10.4.

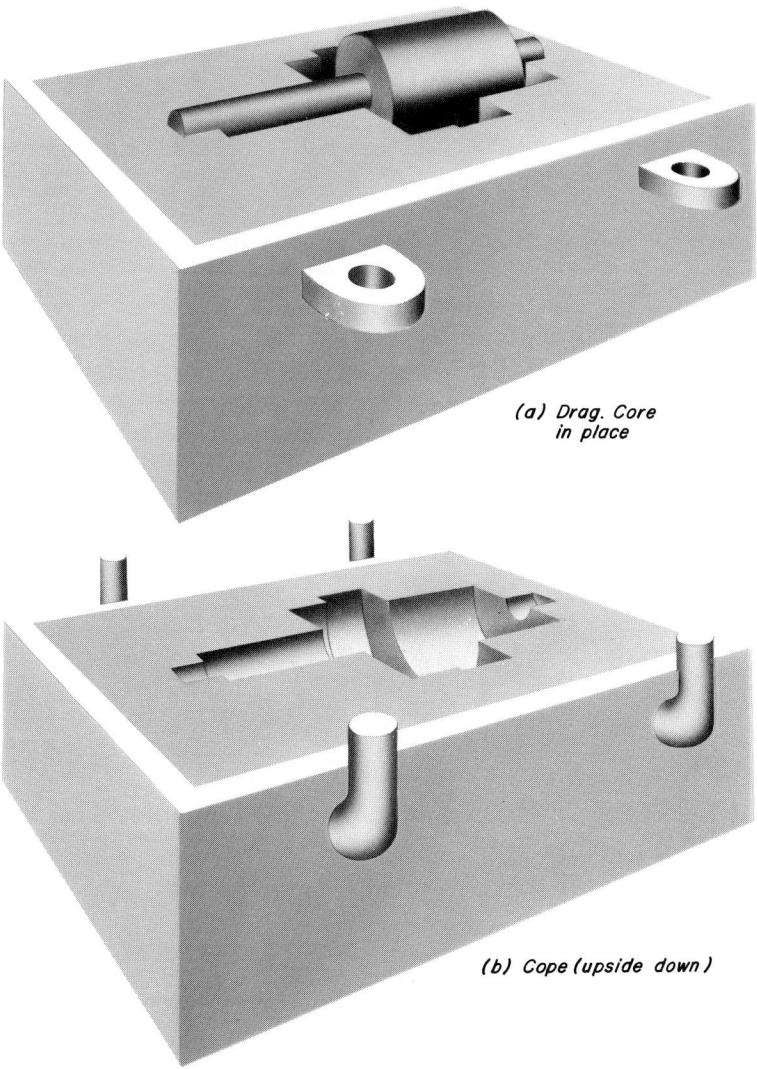

(a) Drag. Core in place

(b) Cope (upside down)

FIG. 10.12. Foundry flask.

Shrinkage allowance. The oversized measurement of the pattern to allow for the shrinkage of the metal when cooling.

Shrink rule. A rule used by the patternmaker that is made sufficiently oversize to allow for the shrinkage of the metal being used.

10.11 COLOR
For ease of interpretation, the complete pattern is stained in various colors. The parts that are to be unfinished are made black, those that are to be finished are red, and the core prints are yellow. Other color symbols are in use, but these are the most important.

10.12 CORE BOX
Since the pattern forms only the outer surface of a casting, it is necessary to have some method of forming the interior surfaces. The shape of these interior surfaces is determined by the shape of the core that is molded in the core box. It is part of the patternmaker's job to build up the core box, which is merely a hollow box whose interior shape conforms to the shape of the interior surfaces of the object to be cast. Since the core is usually made in two parts, which are later glued together, the construction of the core box will involve consideration of parting plane, shrinkage, draft, and finish, just as was done in the construction of the pattern itself. Figure 10.11a shows a core box.

10.13 CORES
After the patternmaker has completed the core box, it is sent to the core shop where the core itself is made. *The purpose of the core is to occupy the space in the mold where an opening is desired.* The engineer must design interior spaces so that it is possible to remove the core after the metal has been cast.

10.14 FLASK
The sand in which the imprint of the casting is made must have a strong boxlike container, which is called a flask. See Fig. 10.12. The flask is made in two parts that can be separated to remove the pattern and then be accurately put together again. See Fig. 10.13. The lower part of the flask is called the *drag*. The upper portion is referred to as the *cope*.

10.15 FOUNDRY
The drafter has little immediate connection with the foundry, since the patternmaker acts as an intermediary between him or her and the molder.

Shop drawings do not, as a rule, include any reference to foundry operation. However, there probably is no place where the item of cost is of more vital importance than in the foundry. Excessive metal and difficult shapes to cast make the manufactured product costly. The designer must always be on guard against these expensive items.

Examples of common parts that are poured in a foundry are engine blocks of cars and pistons.

10.16 PERMANENT MOLDS
When a large number of castings are to be made, a metal or permanent mold will decrease the cost and improve the quality of the casting. The metal mold must be thick and heavy enough to have a large heat-absorbing capacity and have enough cooling capacity so that the temperature of the metal mold does not get too high and lose its strength.

10.17 CENTRIFUGAL CASTINGS
In this type of casting the mold is rotated at a fairly high velocity while the metal is being poured. There are three types of centrifugal castings: (1) die molds; (2) semicentrifugal or center-pour; and (3) true centrifugal for cylindrical shapes where the inner diameter is controlled by the volume of metal poured.

Centrifugal casting is occasionally used as a substitute for forging, since it gives a product that has characteristics somewhat similar to a forging. It requires a rigid control of temperature and speed, but when it can be used it is sometimes cheaper than sand casting, because it is possible to hold the casting closer to finished specifications, thus saving in machining as well as metal.

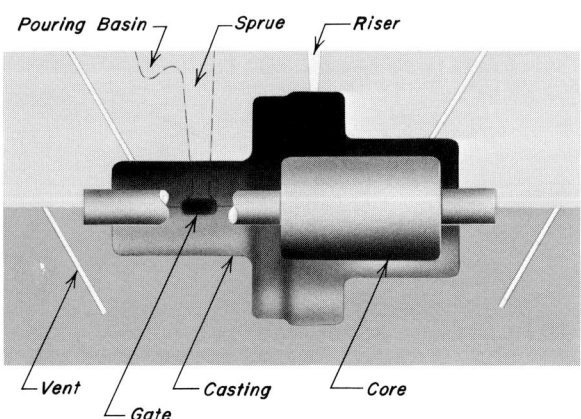

FIG. 10.13. Section through flask.

10.18 DIE CASTING

When a large number of castings of the softer metals are desired on which considerable machining would be necessary if sand castings were used, the least expensive and best method is die casting. This requires the construction of very accurate dies from high-grade steel. *The molten metal is forced into the dies under high pressure, forming a casting that is as accurate as would be obtained by ordinary machine work and harder than a sand casting made from the same metal.* Much machine work can thus be eliminated. Die castings are made from various alloys of zinc, aluminum, magnesium, and copper. This process is used for any small object that does not have undercut parts.

Examples of parts that are die cast are carburators, hood ornaments, and most plastic parts.

10.19 POWDER METALLURGY

A process that is being used extensively at the present time and that shows promise of still further usefulness is that of forming objects from metal powder. *The powder is placed in accurately cut dies and formed under heavy pressure varying from 5 to 100 tons per square inch. The object is then turned out of the die and heated or sintered at temperatures below the melting point until the grains of powder unite to form a solid piece.*

Iron cores for electrical components are often made by powder metallurgy.

10.20 FORGING

Many parts of a machine or structure must be designed to withstand shock or sudden stress, a characteristic for which castings are not recommended.

Hand forging is forming a piece to specified size by hammering or pressing with flat surfaces. *The hot metal is moved around on the anvil as desired and pounded into the desired shape either by hand or machine.* Considerable machining may be necessary to produce a finished piece from a hand forging, and the waste is apt to be large. This method is not used in mass production. Horseshoes are formed in this way.

Drop forging is a process in which the hot metal is forced into dies by means of drop hammers. The dies are similar in principle to those used in die casting or powder metallurgy; the metal, very hot but not in a molten condition, is forced into the die by pounding.

Because of the hammering that the metal receives in the forging process, it is much less porous than a casting and the consistency is more uniform. Another important characteristic of a forging that affects its physical properties is the fiber direction or grain of the metal.

Examples of parts that are forged are crankshafts and connecting rods although with many small engines today they are cast.

10.21 DESIGN DETAILS FOR FORGING

For this type of work a special forging drawing is made, which differs from the engineer's working drawing in that it shows the piece as it will be forged rather than the finished part after the machining has been completed. Figure 10.14 shows a forging drawing. Some of the practical considerations that must be kept in mind when making a forging drawing are given below.

10.21.1 Scale. All forging drawings should be made full-scale.

10.21.2 Draft. The draft angle is usually shown in degrees on the forging drawing. The slope begins at the parting line but must be measured from the direction of stroke of the forging press. For exterior surfaces the draft varies from 5 to 7°, depending on the shape of the piece, but for internal surfaces it should be about 10°, as shown in Figs. 10.15a and b. Bathtub-type fittings such as that in Fig. 10.15d should have approximately 5° draft for both interior and exterior surfaces.

10.21.3 Parting Plane. The parting plane should be indicated on the forging drawing, and

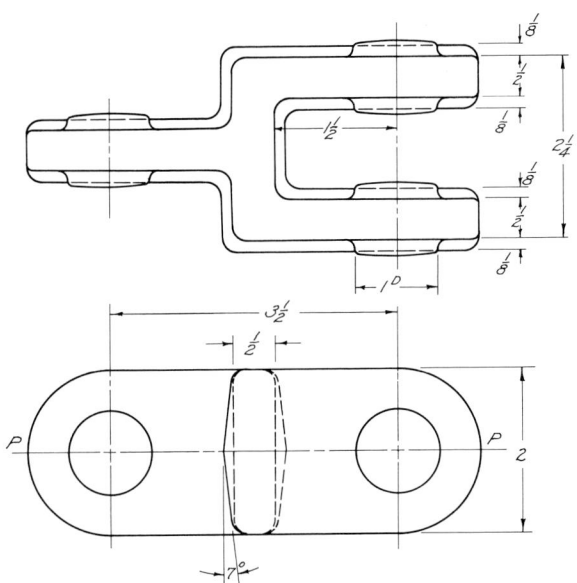

FIG. 10.14. Forging drawing.

when possible it should be located so that one die half contains all the impression. When both die halves contain part of the impression, the parting line will usually be placed on the center of a web, if there is a web in the object, but it need not be one continuous plane surface. See Fig. 10.16.

10.21.4 Forging Plane. The forging plane that is perpendicular to the direction of stroke, on which the dies come together, must be arranged to avoid unbalanced die load. Care should be exercised to avoid interference between die and piece and to take advantage of natural draft. Notice that the forging plane in Fig. 10.16g has been located to give a better balanced die load than would have occurred if the forging plane had been made to coincide with the main parting line. Figure 10.16d shows a forging plane located to take advantage of the natural draft angle of the piece.

10.21.5 Minimum Fillets. To facilitate the flow of metal through the die, the angles should be filleted. No fillet should be less than $\frac{1}{8}$ in. in radius. The chart shown in Fig. 10.17 gives recommended fillet radii.

10.21.6 Rounded Corners. To avoid excessive hammering and consequent breakage of

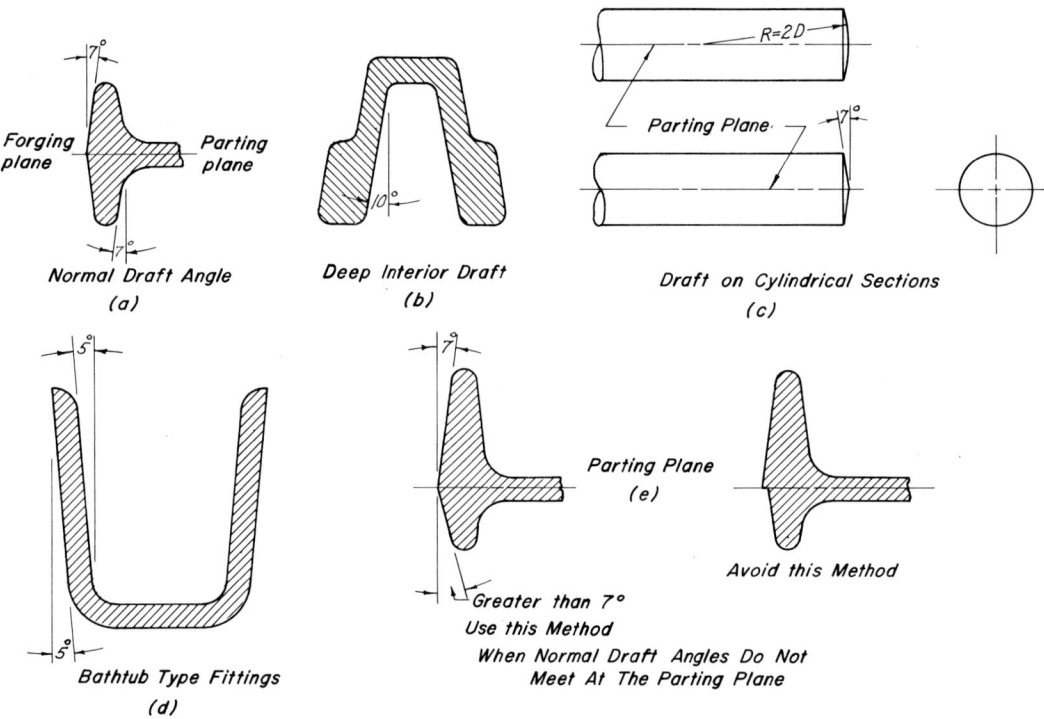

FIG. 10.15. Recommended draft angles (courtesy Product Engineering and O. A. Wheelon).

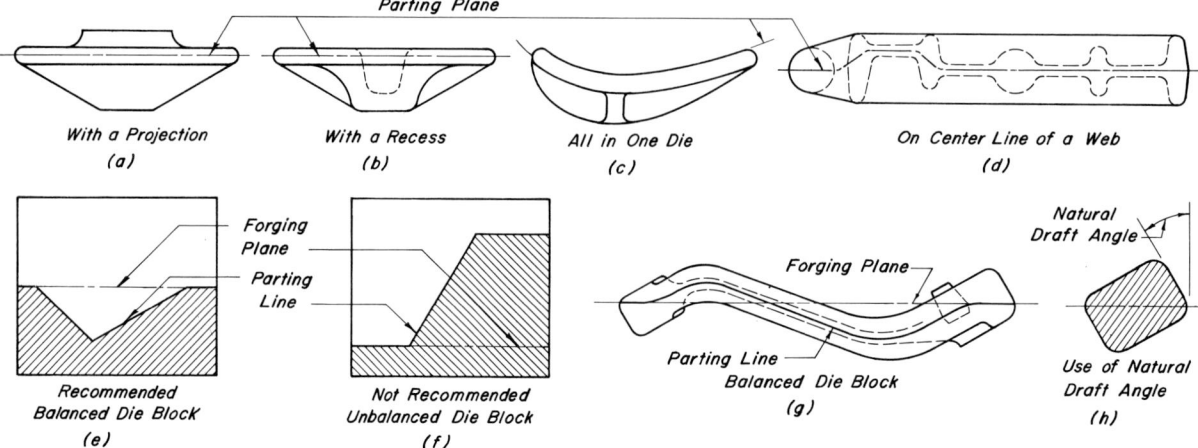

FIG. 10.16. Parting lines and forging planes (courtesy Product Engineering and O. A. Wheelon).

dies, the corners should be rounded to allow the metal to fill the die more easily. Figure 10.18 gives recommended edge radii.

10.21.7 Thin Webs on I-Beam Sections. In I-beam sections the web has a tendency to cool rapidly, owing to the large area of contact with the die, thus making it hard to forge. At the same time the flange tends to fill first, leaving no place for the metal to flow as the web is brought down to size. This sometimes causes the shearing of the flange, as indicated in Fig. 10.19a. The design suggestions given in Figs. 10.19b and c are recommended.

10.21.8 Tolerances. Dimensional tolerances should be as large as possible on all forgings to avoid excessive cost. The variations that must be allowed for are (1) shrinkage and warping as forging cools, (2) mismatching of dies, (3) failure to bring finish dies together, and (4) die wear. The following minimum dimensional tolerances, in inches, illustrated in Fig. 10.20, may be held without excessive cost.

a. Width for small forgings, $\pm \frac{1}{32}$.
b. Length for large forgings, $\pm \frac{1}{32}$ per foot of length.
c. Thickness across parting plane, $+\frac{1}{32}, -0$ for small forgings.
d. Thickness across parting plane, $+\frac{1}{16}, -\frac{1}{32}$ for large forgings.
e. Location of punched holes, at least $\pm \frac{1}{32}$, preferably $\pm \frac{1}{16}$.
f. When bosses that are to be bored out are located far apart, one should be made circular and the other elongated by an

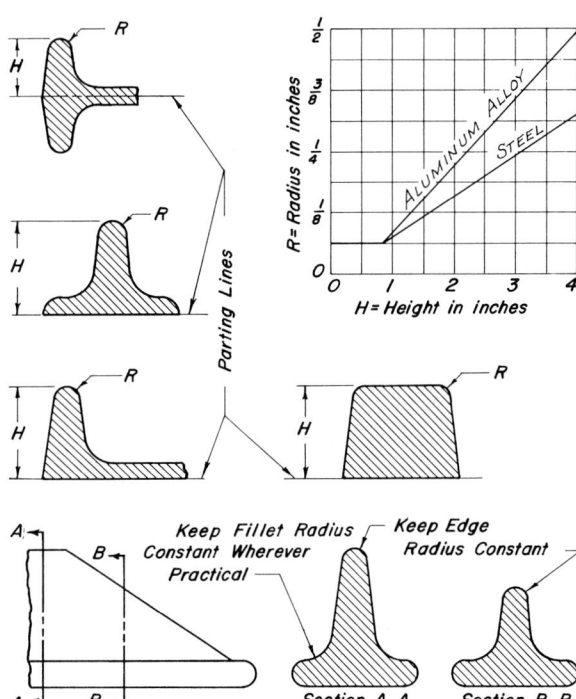

FIG. 10.18. Recommended edge radii for forgings (courtesy Product Engineering and O. A. Wheelon).

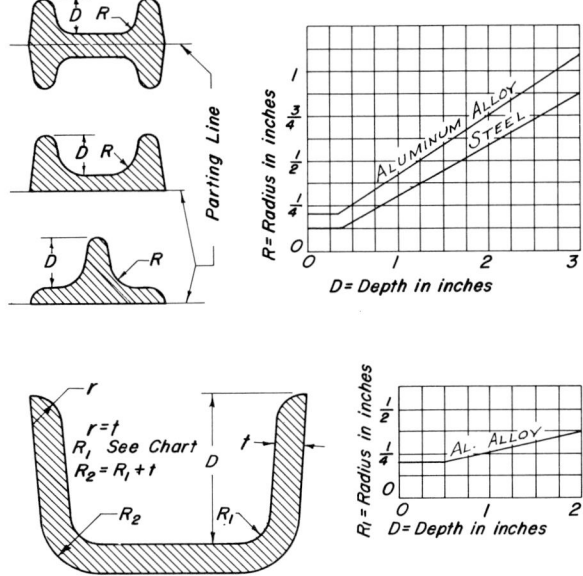

FIG. 10.17. Recommended fillet radii for forgings (courtesy Product Engineering and O. A. Wheelon).

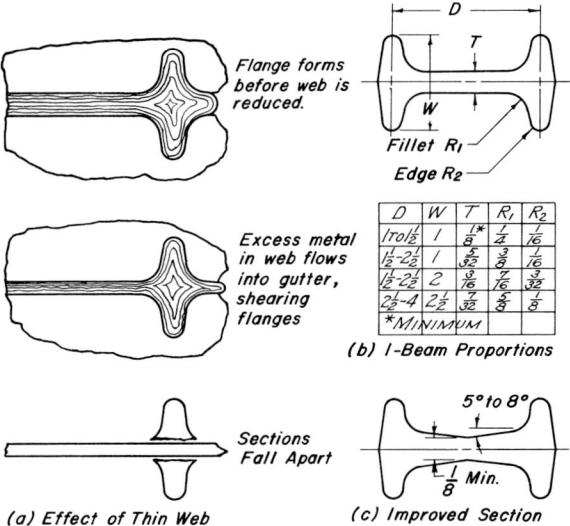

FIG. 10.19. Beam proportions (courtesy Product Engineering and O. A. Wheelon).

amount equal to twice the tolerance on the dimension between bosses, as shown in Fig. 10.21.

g. Tolerances for warping and mismatching of dies, shown in Figs. 10.20b and c, must be added to the above tolerance.

h. Allowance for Machining. The amount of material to be allowed for finishing operations should include the dimensional tolerances plus an allowance for warping and mismatching of dies, which may vary from 0 to $\frac{1}{8}$ in., depending on the size of the piece. If the corner radius is to be cut away, that radius or a portion of it must also be added to the allowance for machining. See Fig. 10.22.

10.22 STAMPING

The term stamping is applied to a variety of processes used in the forming of thin metal parts. It often includes cutting of the metal as well as shaping. A series of definitions with appropriate illustrations will give the best idea of some of the processes.

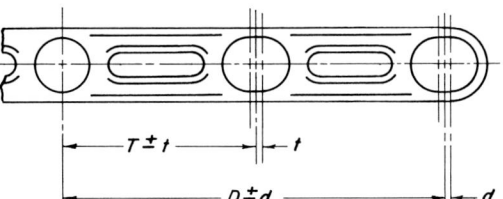

FIG. 10.21. Tolerance on location of bosses (courtesy Product Engineering and O. A. Wheelon).

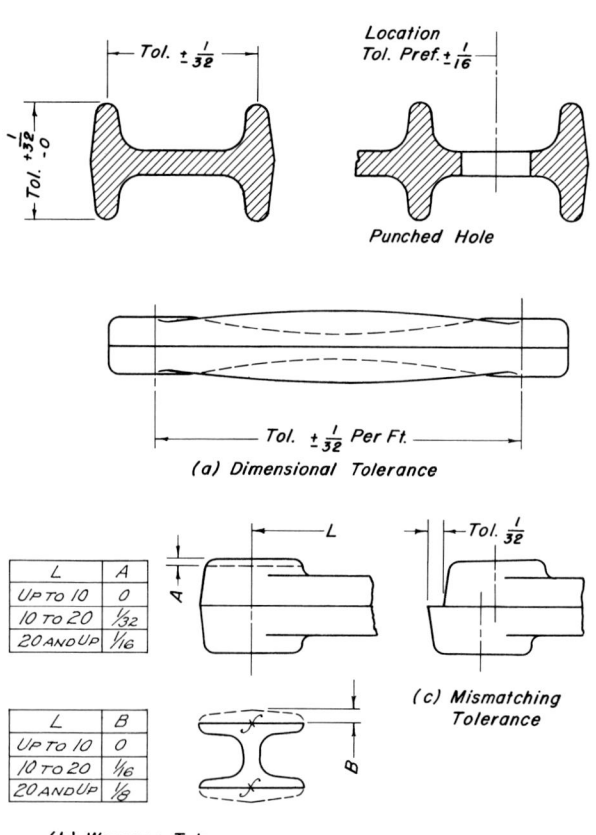

FIG. 10.20. Forging tolerances (courtesy Product Engineering and O. A. Wheelon).

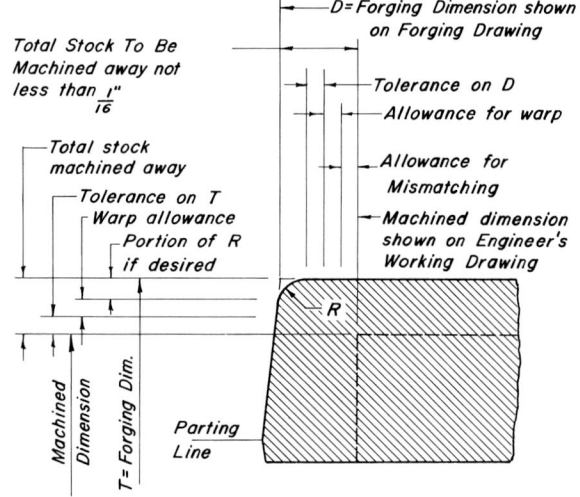

FIG. 10.22. Summary of tolerances and allowances needed on a forging (courtesy Product Engineering and O. A. Wheelon).

a. **Blanking** means cutting of the metal to the desired shape with one stroke of the press. See Fig. 10.23.
b. **Nesting** of blanks means that they should be designed so that they can be cut from the metal sheet with as little waste as possible. See Fig. 10.24.
c. **Punching** is a method of producing a hole in a part by one stroke of the press. A cylindrical punch produces a hole with practically smooth sides, as in Fig. 10.25a. A conical punch produces a flanged hole with ragged edges, as in Fig. 10.25b. A combination of the two punches having a cylindrical shape with a shoulder, as in Fig. 10.25c, produces a flanged hole with a fairly smooth edge.

d. **Trimming** is the process of removing excess metal from a stamping. Sometimes trimming is recommended rather than developing a blank.
e. **Shaving or burnishing** is a process that removes a very small amount of metal to produce a surface with a very close tolerance.
f. **Cutting off** is a process of cutting a blank to length from a strip of metal that has been slit or sheared to correct width.
g. **Notching** is done for the purpose of providing clearance, for attachment, or for locating elements, or to facilitate forming.
h. **Bending, forming, and embossing** are names applied to shaping a blank without materially changing the thickness of the metal.
i. **Drawing** is a process of forcing a metal blank to assume the shape of a die by stretching the metal. The depth of draw should be as shallow as practicable to keep down the number of operations.
j. **Coining** is a process by which great pressure forces the metal to flow in the die, thus making it thicker in some places and thinner in others.
k. **Swaging** is a cold forging operation in which the metal is squeezed to reduce the thickness in certain places. The metal flows outward and must be trimmed off.
l. **Extrusion** is a process whereby the metal is made to flow in a die either by pressure or impact.
m. **Necking, bulging, and curling** are processes for reducing, enlarging, or forming a rounded edge on drawn shells. Curling may be done on flat blanks for such purposes as hinge manufacture.

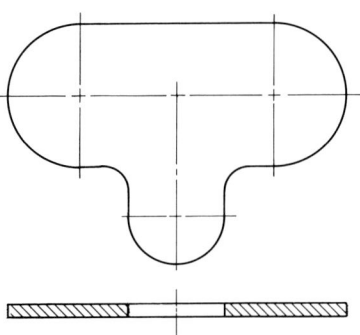

FIG. 10.23. Stamped blank.

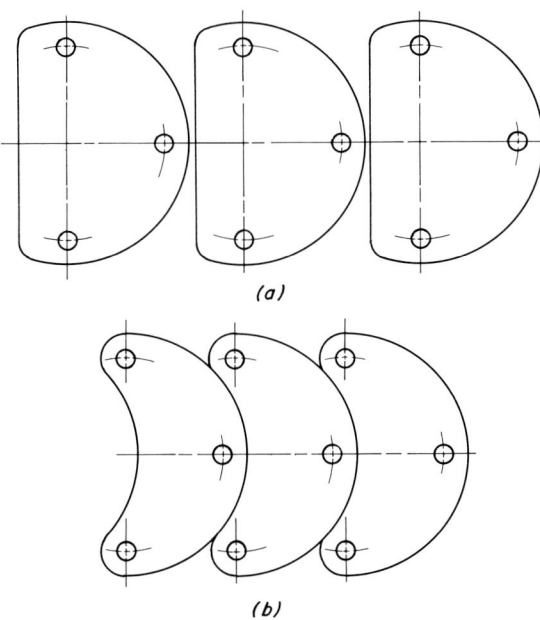

FIG. 10.24. Redesign for economy of material.

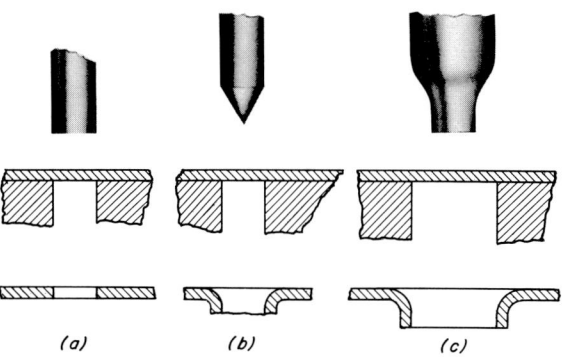

FIG. 10.25. Effect of type of punch.

10.22.1 Design Limitations for Stamping.
Certain limitations on the design of blanked parts must be considered for economical and efficient operation. A few of these will be mentioned.

a. *Grain.* Sometimes the functioning of a part requires that the grain of the metal run in a certain direction. This direction may be marked on the blank, but, as it can usually be allowed to vary by as much as 45° in either direction, it is usually specified by an arrow with a tolerance of ±45°.

b. *Radii of blanked parts.* To facilitate tool construction the radii should be as large as possible. The ends are usually made as semicircles, as shown in Fig. 10.23.

c. *Hole spacing.* To avoid distortion of the metal when holes are to be punched, the distance between centers of the holes should be a minimum of two times the thickness of the metal plus the sum of the two radii, but never less than $\frac{1}{32}$ in. plus the sum of the radii.

d. *Edge distance.* The clear distance between the edge of the metal and the center of a punched hole should be a minimum of two times the thickness of the metal plus the radius of the hole, but not less than $\frac{1}{32}$ in. plus the radius of the hole.

e. *Spacing for drawn holes.* To avoid distortion, the distance between drawn holes, or between a drawn hole and an edge, or between a drawn hole and another bend should not be less than that shown in Fig. 10.26.

f. *Clearance at bends.* Formed parts should be so designed that there is a clearance of at least half the thickness of the metal, in order that there is no metal interference and the tools can make a satisfactory bend. This is illustrated in Fig. 10.27.

g. *Radius for right-angle bends.* The radius varies with the kind and temper of the material, direction of the grain, and type of die used. In general it is well to allow a radius not less than the thickness of the material. As large a tolerance as possible should be allowed in the size of the angle of bend.

h. *Dimensions for drawing.* The depth of draw should be as small as possible. Figure 10.28 shows acceptable dimensions for single-operation draws.

i. *Size of punched holes.* Punched holes of a diameter less than the thickness of the material are not practical unless the punch is supported.

10.23 MACHINE SHOP

The drafter must be most familiar with the machine shop processes. Dimensions on the drawing must be arranged so that they can be conveniently used in the shop. Machine shop processes such as drill, ream, bore, and mill are sometimes indicated on the drawing. The degree and method of finish, and sometimes even the direction of the cutting strokes, are specified on the drawing. The drafter must therefore be familiar with the tools, the machines in which they are used, and the limitations and possibilities of each.

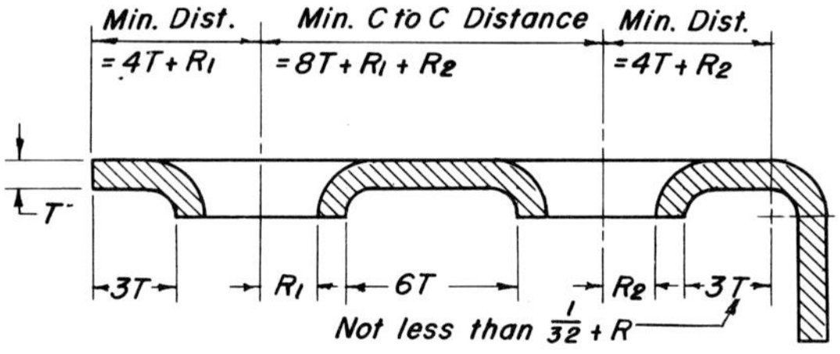

FIG. 10.26. Minimum material between drawn holes.

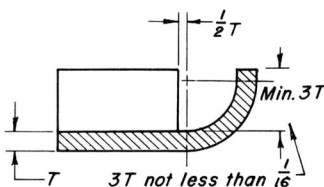

FIG. 10.27. Clearance for bending.

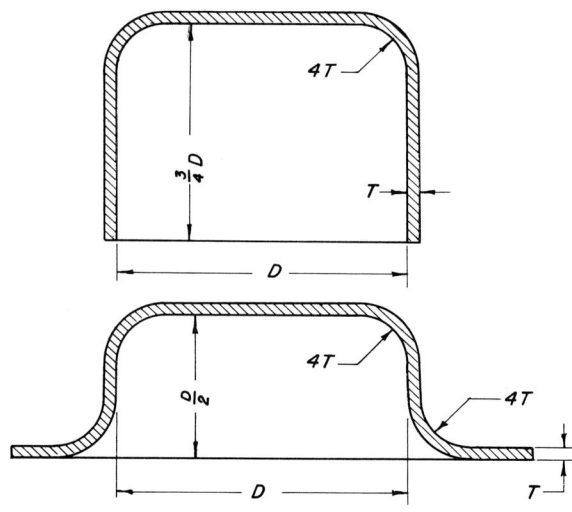

FIG. 10.28. Depth of draw.

Some companies do not specify shop operations on their drawings. In the following paragraphs a brief explanation of the important machines and tools are given, as well as the method of indicating on the drawing what processes are to be used.

10.23.1 Tooling Points. All location dimensions on a casting are measured from three datum planes that are usually mutually perpendicular. These planes are located by tooling points on the casting jig or fixture. These points form a contact between the jig or fixture and the object used to locate a casting for measuring or machining. The symbols suggested by the American Society for Metals to mark the tooling points are shown in Fig. 10.29.

a. The first datum plane is located by three points (actually minute areas) that are not

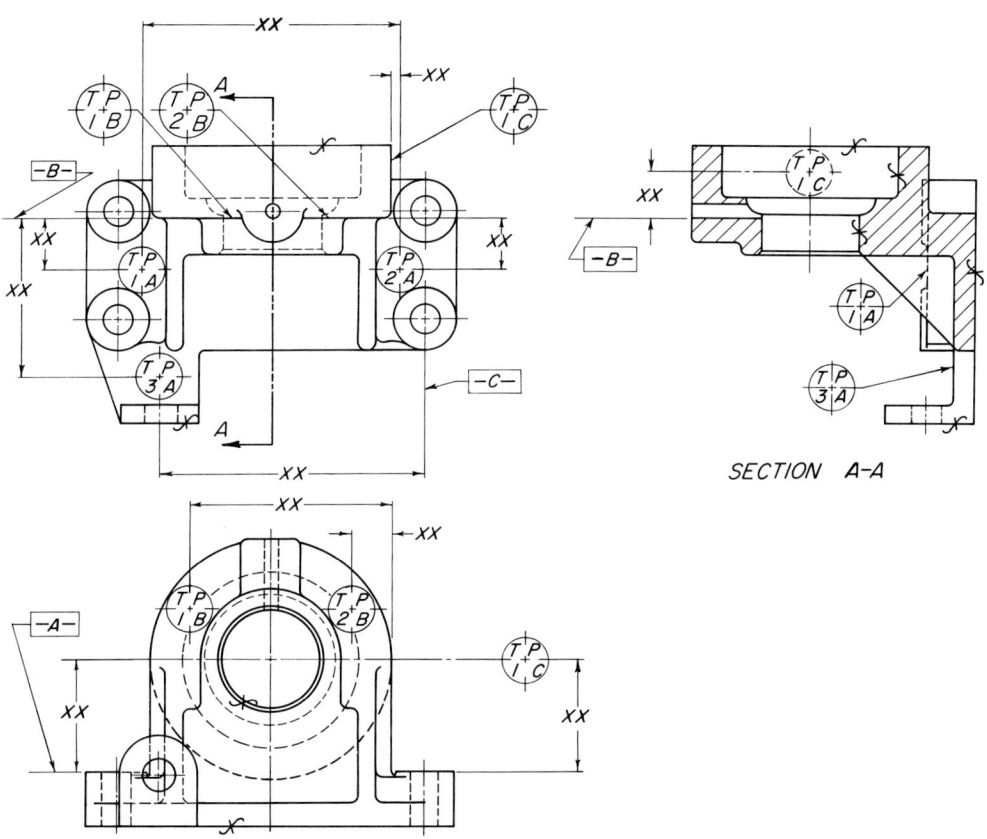

FIG. 10.29. Locating or tooling points (courtesy ANSI Y14.8).

260 SHOP TERMS AND PROCESSES

on a straight line. These three points are made to come in actual contact with a surface to determine the plane, as shown in Fig. 10.30a.

b. The second datum plane is defined by two other tooling points and its relation to the first plane. See Fig. 10.30b.

c. The third datum plane is located by one additional tooling point and its relation to the other two planes. See Fig. 10.30c.

10.24 MACHINE TOOLS

Many of the tools in common use in the machine shop may be used in more than one machine. The following tools may be used on either the drill press or the lathe.

10.24.1 Twist Drill. When a hole is to be drilled in a piece, it may be marked on the drawing by giving the diameter followed by the word "drill," thus $\frac{5}{8}$ Drill. Usually the diameter of the hole is specified with the required finish. This makes it possible for the machine shop to use any method desired as long as the result is satisfactory. Metal drills, such as that shown in Fig. 10.31, are obtainable in sizes varying from $\frac{1}{64}$ to 3 in. in diameter and in length up to 14 in. When used to precede a threading operation the tool is called a *tap drill*. See Fig. 10.38. Sizes of tap drills are given by numbers, letters, and fractional dimensions. Figure 10.32 gives some suggestions concerning design details for drilling operations.

10.24.2 Reamer. A hole that has been drilled is left with a rough and slightly scarred surface,

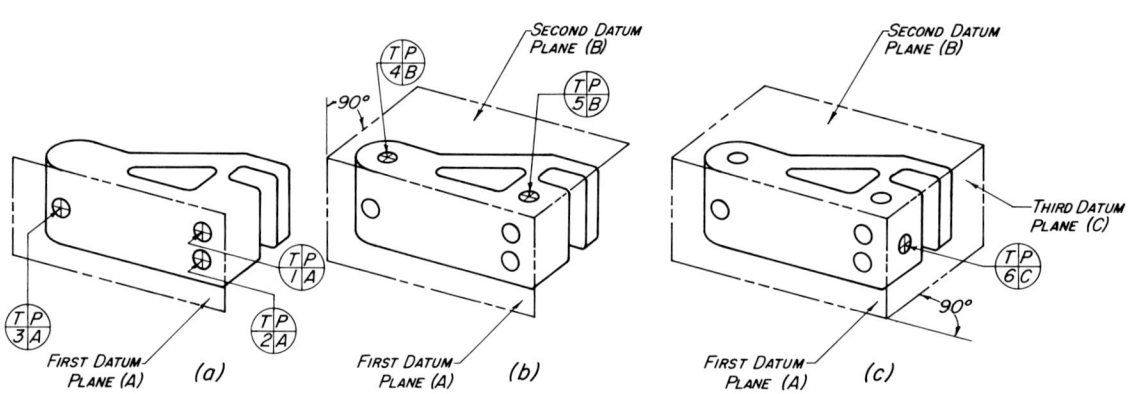

FIG. 10.30. Locating points on successive datum surfaces (courtesy ANSI Y14.8).

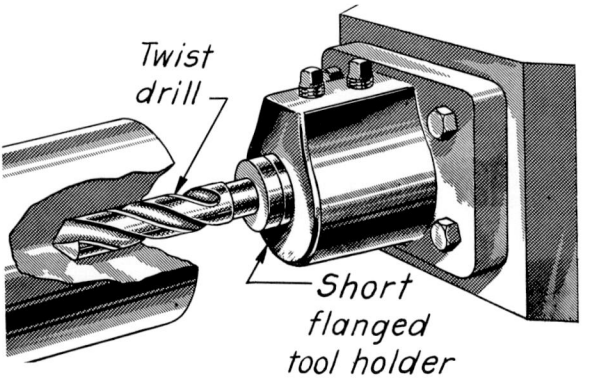

FIG. 10.31. Drilling on a lathe (courtesy Warner and Swazey Co.).

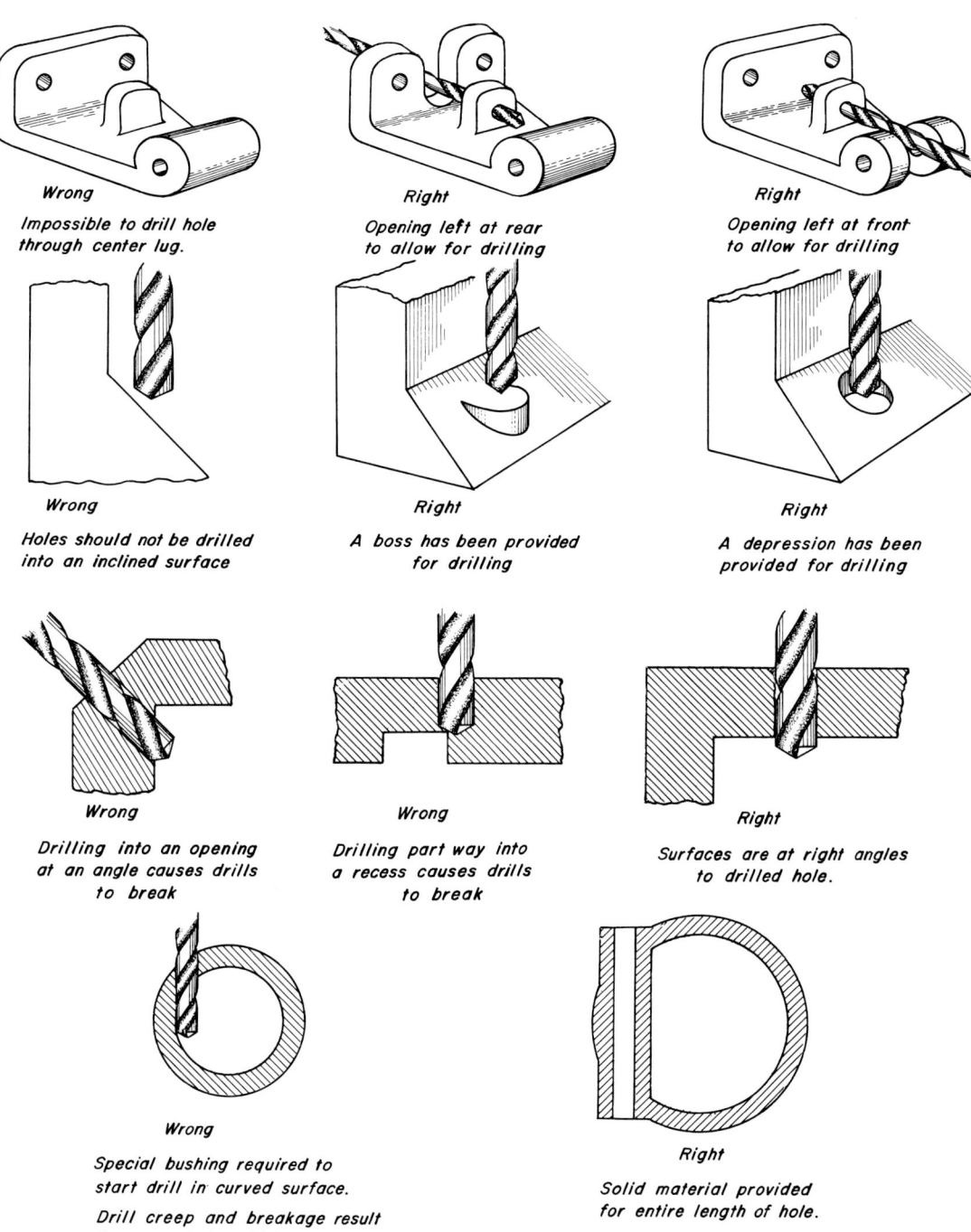

FIG. 10.32. Proper design for drilling (courtesy Curtiss–Wright Corp.).

262 SHOP TERMS AND PROCESSES

which is not suitable for close fits or for tapping if fine thread crests are desired. A reamer similar to the one illustrated in Fig. 10.33 is used to finish this rough surface. Reamers may be either straight or tapered for cylindrical or conical holes. See Fig. 10.34. The drawing usually indicates the diameter with the tolerance required. Occasionally it may be desirable to specify the operation in full as "$\frac{39}{64}$ Drill, $\frac{5}{8}$ Ream."

Reamers up to 3 in. in diameter and 17 in. in length can be secured. For holes up to $\frac{3}{4}$ in. in diameter, $\frac{1}{64}$ in. should be left for reaming; for holes over $\frac{3}{4}$ in. in diameter, $\frac{1}{32}$ in. may be left.

A few suggestions about the proper use of a reamer are given in Fig. 10.35.

10.24.3 Countersink. When a flat-headed screw is used, the hole must be enlarged in a

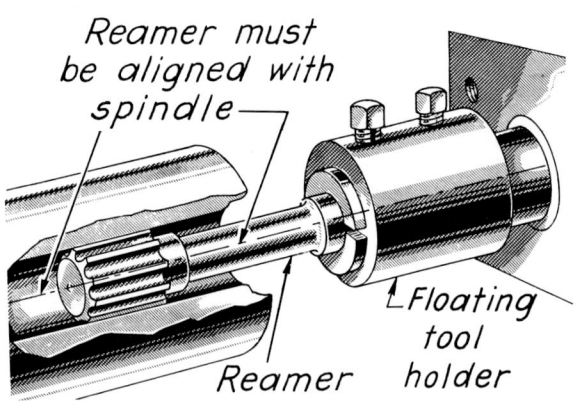

FIG. 10.33. Reaming (courtesy Warner and Swazey Co.).

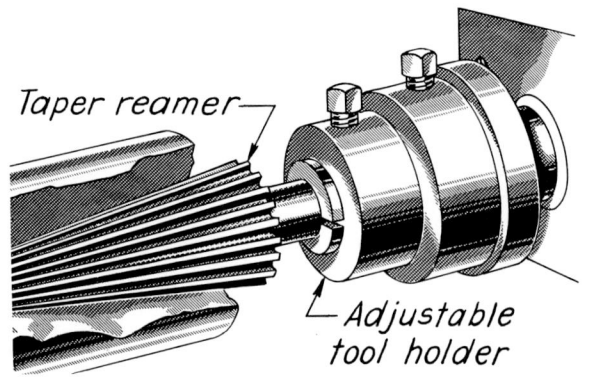

FIG. 10.34. Taper reaming (courtesy Warner and Swazey Co.).

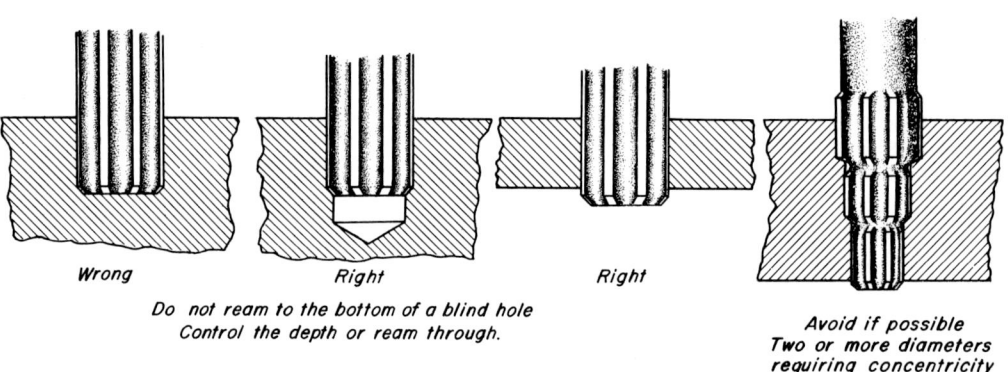

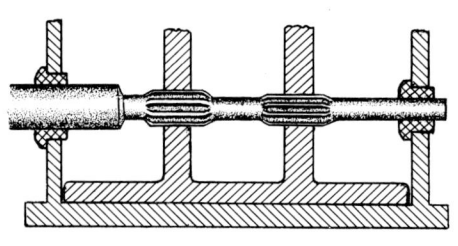

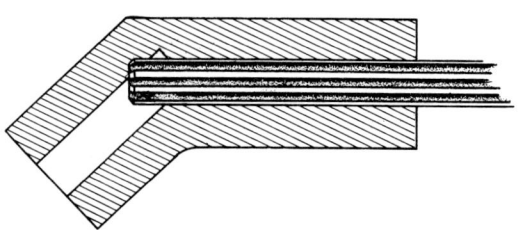

FIG. 10.35. Proper design for reaming (courtesy Curtiss–Wright Corp.).

conical manner to allow the top of the head to come flush with the surface of the piece. This enlarging is called *countersinking*. See Fig. 10.37. The note on the drawing should be similar to the following: "Countersink 82° to $\frac{7}{8}$" Diameter." For details of dimensioning a countersink, see Chapter 11.

10.24.4 Counterbore. Heads of bolts and screws may be brought level with the surface of the part by enlarging the hole to a depth equal to the height of the head. See Fig. 10.37. This operation is called *counterboring* and is done with a tool similar to the one illustrated in Fig. 10.36c. The pilot on the tool fits into a drilled hole and ensures concentricity. The specifications on the drawing are usually given by dimensions and tolerances, but occasionally the following note may be found: "$\frac{3}{8}$ Drill, $\frac{3}{4}$ Counterbore, $\frac{1}{2}$ Deep."

10.24.5 Spotface. Sometimes it is desired to make a smooth-bearing surface for the head of a bolt or a nut. This situation frequently occurs on a projection, called a *boss*. *Spotfacing* is accomplished by using a counterboring tool, as illustrated in Fig. 10.36. The note on the drawing

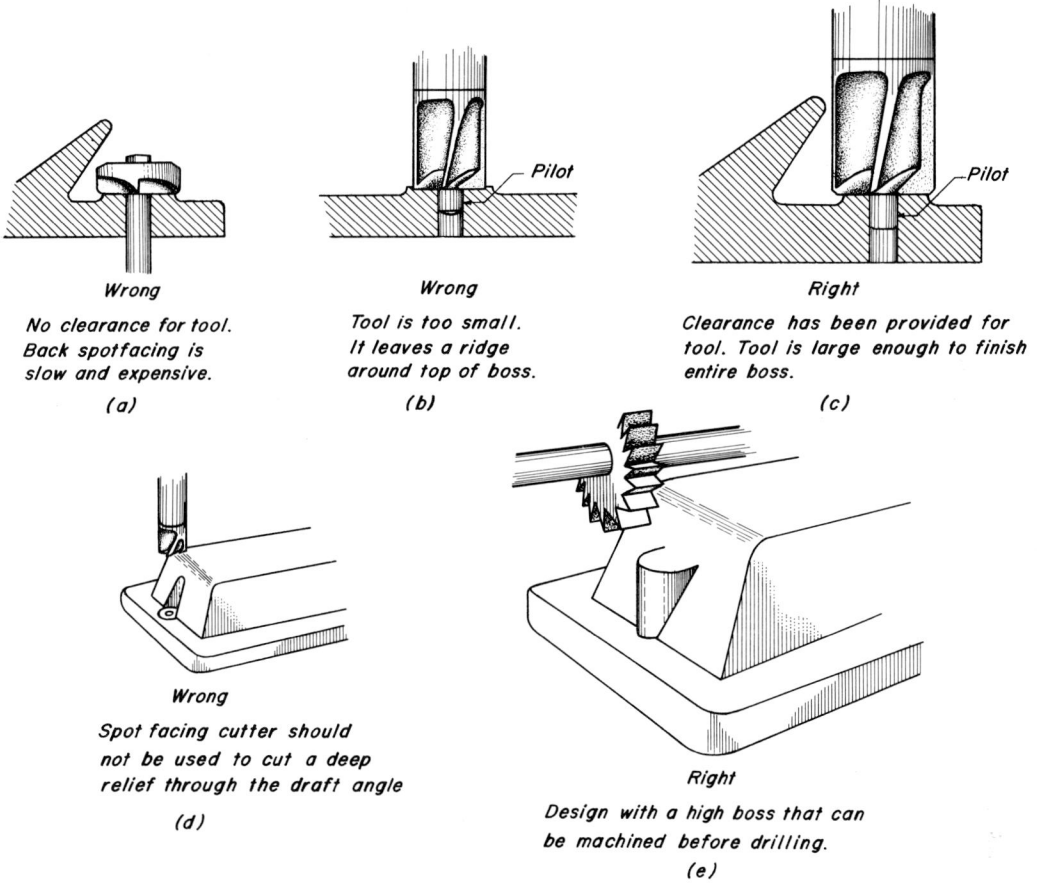

FIG. 10.36. Proper design for spot facing (courtesy Curtiss–Wright Corp.).

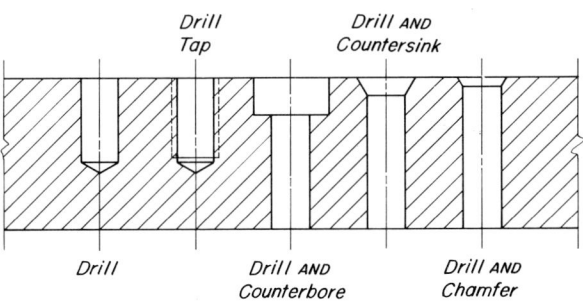

FIG. 10.37. Various types of holes.

should read "Spotface ¾" Diameter." Figure 10.36 gives a few design suggestions for pieces on which spotfacing is specified. A depth of spotface is not usually given but rather the operation is carried on to a depth sufficient to "clean up the surface."

10.24.6 Taps and Dies. These tools are used for cutting internal and external threads. For more detailed information, see Chapter 11. The use of a tap is shown in Fig. 10.38 and a die in Fig. 10.39.

10.25 THE DRILL PRESS

The single-spindle drill press shown in Fig. 10.40 is found in practically every machine shop. It may be used for drilling, reaming, countersinking, and counterboring. For special purposes where more than one hole is to be drilled at one time, multiple-spindle drill presses are used. Several design suggestions for efficient use of multiple-spindle drills are shown in Fig. 10.41. The radial drill press shown in Fig. 10.42 is a very useful type of machine because holes can be drilled in almost any part of a rather large piece without reclamping or moving it. This is accomplished by having the chuck mounted on an arm that can be revolved around and moved up and down on a vertical axis of the machine while at the same time the chuck may be given a horizontal motion along the arm by means of a screw and gear arrangement. The radial drill is a very versatile machine and can be used for such purposes as tapping and hollow milling. For practical purposes, the depth of the hole should not exceed five diameters.

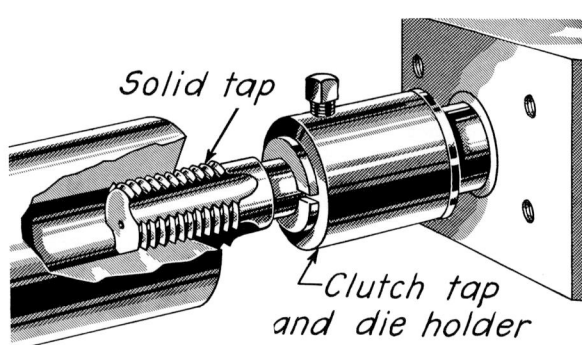

FIG. 10.38. Tapping internal threads (courtesy Warner and Swazey Co.).

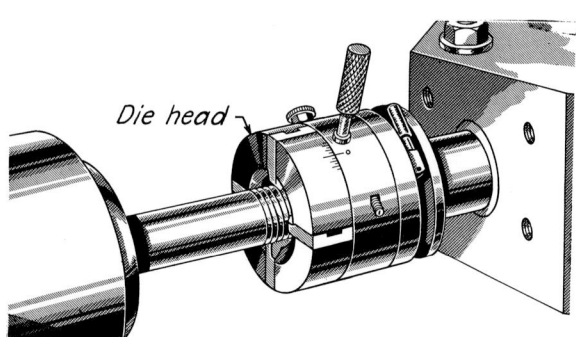

FIG. 10.39. Cutting external threads (courtesy Warner and Swazey Co.).

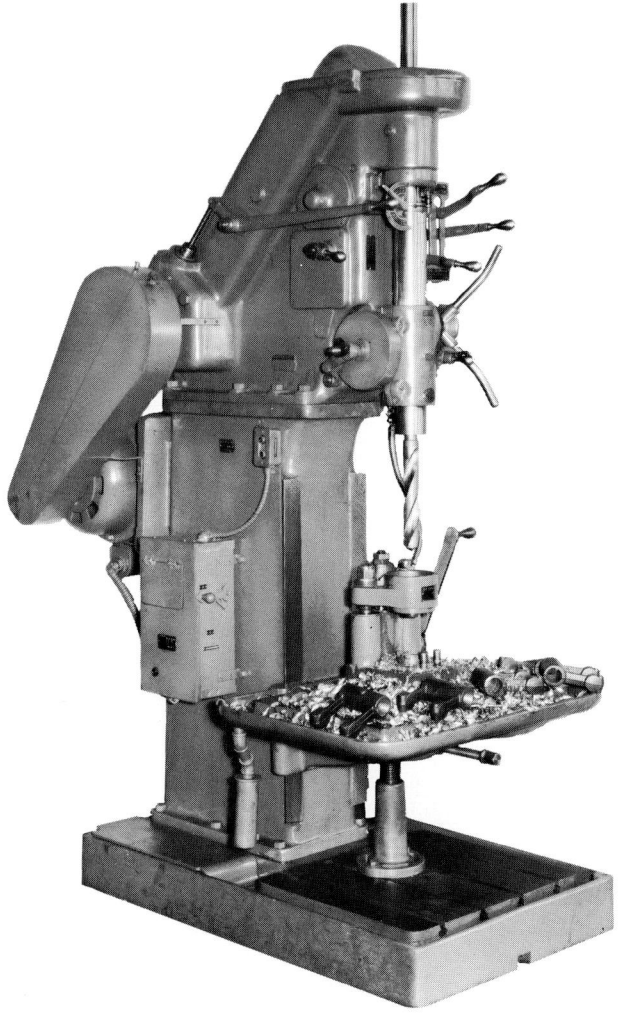

FIG. 10.40. Single-spindle drill press (courtesy Barnes Drill Co.).

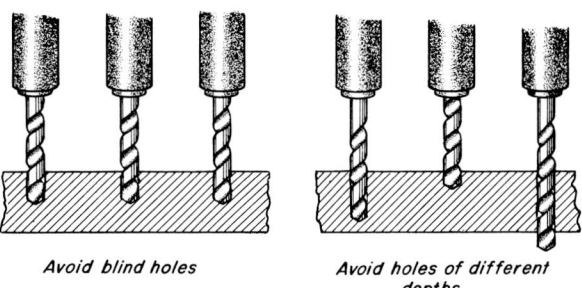
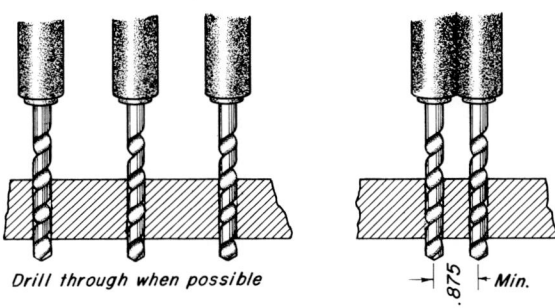

FIG. 10.41. Design suggested for multiple drilling (courtesy Curtiss–Wright Corp.).

FIG. 10.42. Radial drill (courtesy The American Tool Works Co.).

10.26 THE LATHE

The lathe is one of the most useful machines in the shop because of the many different operations that may be performed on it. The piece to be machined is supported between two centers, one in the tail stock and the other in the head stock, and then revolved by power supplied through the head stock. A tool post, which may be moved longitudinally along the lathe, carries the cutting tool. This tool removes a thin layer of metal each time it traverses the length of the surface being machined. This process is called *turning* and is used for machining practically all cylindrical surfaces. Figure 10.43 shows a lathe and Fig. 10.44 shows a close-up of an operation being performed on the lathe. In addition to turning, the lathe is used for drilling, reaming, boring, counterboring, facing, threading, knurling, and polishing. For rough turning, normal tolerances vary from 0.005 to 0.015 in., depending on the diameter. For the finish cut the tolerance may vary from 0.002 in. for a $\frac{1}{4}$-in. diameter to 0.007 in. for a 4-in. or larger diameter. Parts may also be clamped to the head stock alone by means of a chuck, and turning operations may be performed on the end face.

FIG. 10.43. A lathe (courtesy The American Tool Works Co.).

FIG. 10.44. Turning operation on a lathe (courtesy Warner and Swazey Co.).

10.27 THE BORING MILL

The boring machine does practically the same work as the lathe but is used for larger pieces. In the vertical boring machine, the table holding the work revolves while the cutting tool moves horizontally or vertically on the crossrail. A horizontal boring machine is also available on which an almost unlimited number of operations may be performed. Production tolerances of 0.003 in. for a 1-in. diameter to 0.01 in. for 54-in. diameter may be used.

10.28 THE MILLING MACHINE

A machine in which circular-type revolving cutters remove the metal as a worktable to which the piece is clamped moves under the cutter is called a *milling machine*. The machine is designed so that the worktable can be moved in three directions at right angles to each other, either manually or automatically. The rate of feed of the table and the speed of the cutter must be adjustable for various kinds of material and depth of cuts. Figure 10.45 shows a close-up of a milling machine cutting flutes on a drill, and Fig. 10.46 shows multiple operations being performed.

Many different kinds of cuts may be made on the milling machine, depending on the design of the cutting tool. These tools may have straight teeth or helical teeth, and the cutting edge may be ground on the tool itself or a cutting tooth may be inserted. Either the cylindrical surface of the tool or the end face may be used as the cutting surface. The milling machine may be used for cutting plane or irregular surfaces, slots, keyways, gears, and similar surfaces. Production tolerances of about 0.005 in. may be specified for work to be

FIG. 10.45. Milling flutes on a drill (courtesy Cincinnati Milling Machine Co.).

FIG. 10.46. Multiple operations on a milling machine (courtesy Kearney–Trecker Co.).

268 SHOP TERMS AND PROCESSES

done on a milling machine. Figure 10.47 gives a few suggestions for the design of parts that are to be finished on the milling machine.

10.29 THE GRINDING MACHINE

The wheel on a grinding machine may vary from the ordinary fine and coarse emery wheels to high-speed carborundum wheels. The purpose of grinding is to leave a finely finished surface on the metal and at the same time remove the small amount of stock left after the previous finishing operation has brought the piece almost to size. In grinding a cylindrical surface, the piece may be revolved between two centers during the grinding, as shown in Fig. 10.48, or it may be allowed to roll between the grinding wheel and a regulating wheel, as shown in Fig. 10.49. A work rest is used to help hold the work in place. The latter method is known as *centerless grinding*. Both external and internal grinders are used for cylindrical surfaces where fine finish and close tolerances are desired. For surface grinding, the piece is usually clamped to a movable table, which is traversed to bring the work under the grinding wheel. A surface grinder is shown in Fig. 10.50. The grinding machine is also used for grinding threads when close fits are desired and for making and sharpening tools for other machines. The drafter indicates the grinding operation on his/her drawing by means of a note when the limits and finish require its use. A tolerance of 0.0005 in. may be obtained by grinding.

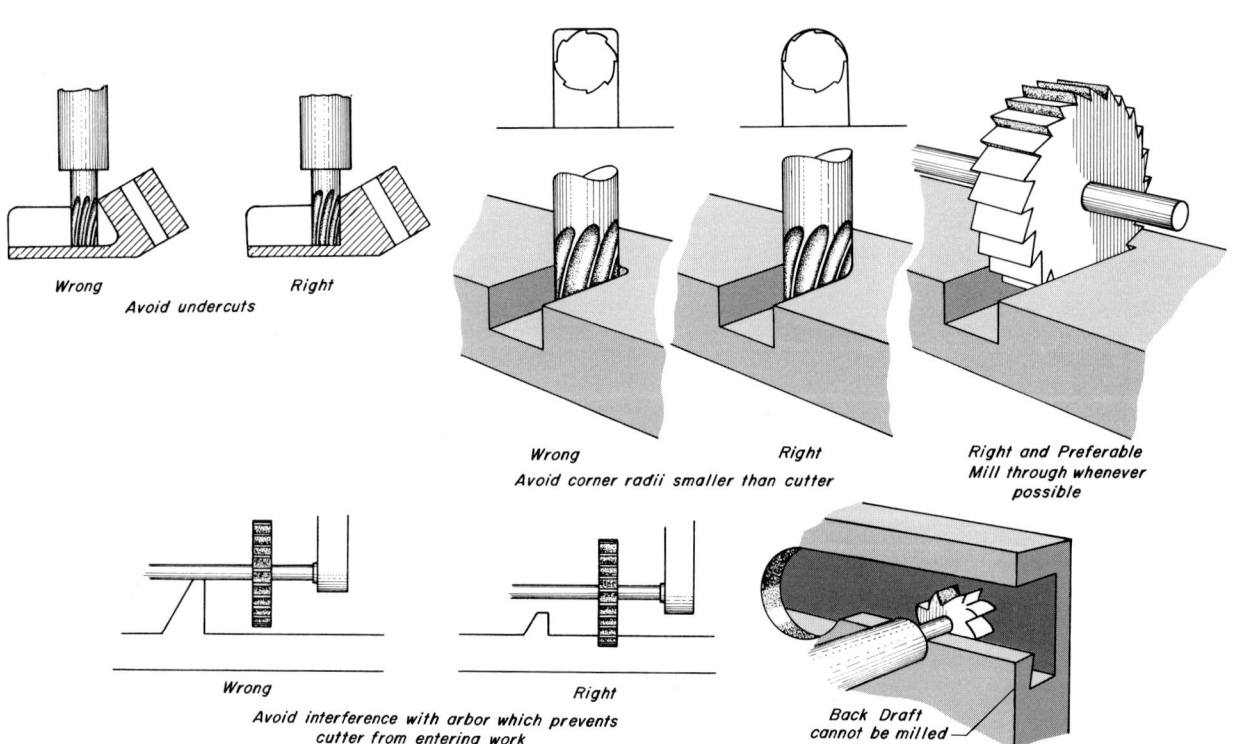

FIG. 10.47. Design for milling operations (courtesy Curtiss–Wright Corp.).

10.30 POLISHING

Polishing must not be confused with grinding. Although a ground surface is very smooth, it is not said to be a polished surface until it has been gone over carefully with a rapidly revolving disk of material like muslin or leather, containing a fine abrasive, which gives it a luster impossible to attain with the finest grinders. The drafter indicates such an operation by the note "grind and polish."

10.31 THE PLANER

When large flat surfaces are to be finished, a machine called a *planer* is used. The piece to be planed is mounted on a long horizontal bed that moves forward and backward under the cutting tool. The tool advances a small amount across the surface with each run of the bed, the width of each cut being determined by the distance that the tool advances. Numerous pieces of the same kind may be clamped to the bed and planed at the same time. The drafter makes no reference to this machine on the drawing but simply marks the surface to be finished by one of the standard symbols. Tolerances of 0.005 in. are obtained with this machine.

10.32 THE SHAPER

The shaper is used for finished surfaces on pieces that are smaller than those for which a planer is required or for surfaces that are curved. For pieces within the capacity, the shaper is preferred to the planer because it is less cumbersome and faster. In the shaper, the piece to be finished is fastened to a table while the cutting

FIG. 10.48. Cylindrical grinding: center type (courtesy The Carborundum Co.).

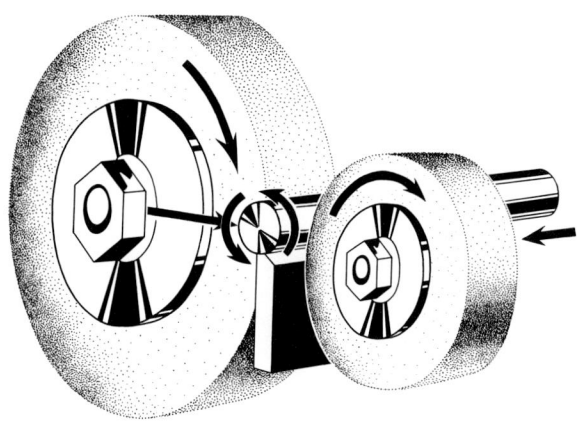

FIG. 10.49. Cylindrical grinding: centerless type (courtesy The Carborundum Co.).

FIG. 10.50. Surface grinding (courtesy The Carborundum Co.).

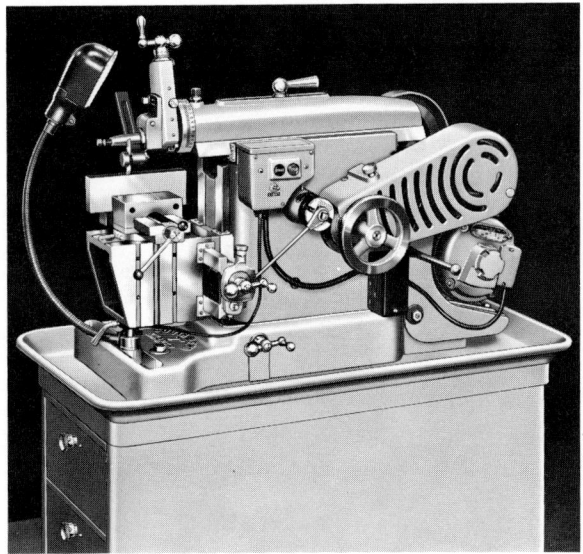

FIG. 10.51. A shaper (courtesy South Bend Lathe Works).

270 SHOP TERMS AND PROCESSES

tool moves backward and forward. The table advances the piece a small amount with each stroke of the cutting tool until the entire surface has been machined. Figure 10.51 shows a standard shaper. The shaper is particularly useful for cutting slots, keyways, and small flat or curved surfaces. Tolerances of 0.004 in. may be obtained with this machine. However, the drafter does not refer to this machine on the drawing.

10.33 THE BROACHING MACHINE

Originally the broach was used for cutting keyways and internal work such as forming square and hexagonal holes or holes of other shapes from a drilled hole, but now many external surfaces are machined by this method. The broach is a tool having a series of teeth or cutting edges that progressively increase in size so that each tooth removes a small amount of material, thus giving the desired surface quickly and accurately. Figure 10.52 illustrates the action of the broach. The broaching machine provides a method of holding the work and of supplying the power to force the broach through the work. This power is usually supplied hydraulically or by means of a screw. In cutting keyways, a guide bushing is inserted in the hole to hold the tool in the desired position. The broaching tool is rather expensive, but because it is especially useful for interchangeable work, it is extensively used in automotive work.

10.34 TOOLS—WORKPIECES

It is realized that it is at best difficult to understand the operations of these various machines

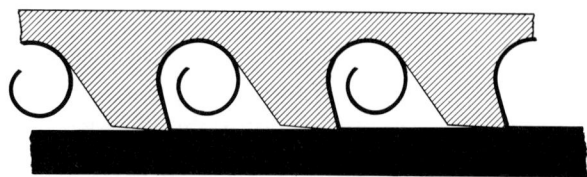

FIG. 10.52. Action of a broaching tool.

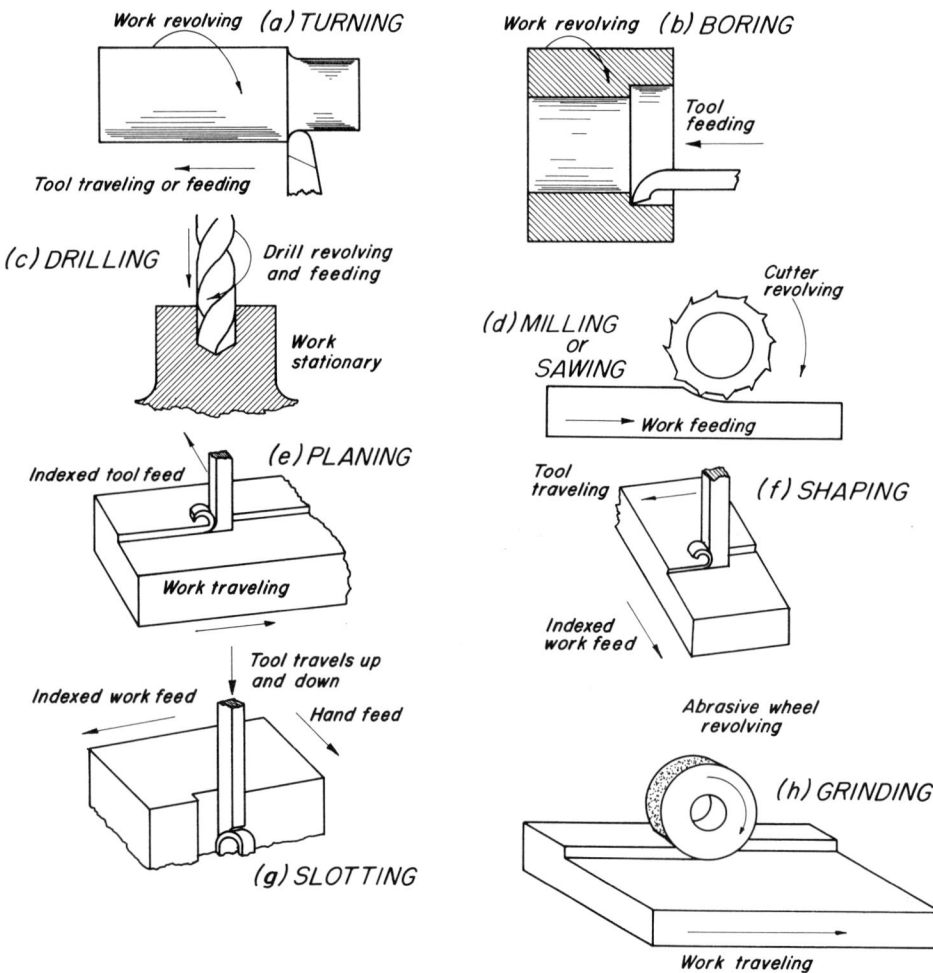

FIG. 10.53. Various tools and workpieces.

and tools without actually seeing them in action. If the student-engineer can arrange to tour a machine shop or foundry it is highly recommended. After seeing these various machines, the student will better understand what is being said in this chapter. Figure 10.53 shows the relationship and relative movements of various machine tools and their respective workpieces.

10.35 NUMERICALLY CONTROLLED MACHINES

With the proper selection of parts to be processed, numerically controlled or tape-controlled machines offer productivity increases of from 2 to 10 times those obtainable with conventional equipment. To justify their cost, the more expensive tape-controlled machines must have a productive capacity three times as high as regular equipment. For successful use of these machines, all cutting tools must be precision-ground and accurately preset in order that approximately 90% of the required setup can be accomplished beforehand. A clean, air-conditioned, temperature-controlled room for the machines is important. The work should be brought into this room and allowed to stabilize dimensionally for increased accuracy in machining. For cost economy it is important that the machine be kept operating as much as possible.

10.35.1 Kinds of Work Accomplished. A completely tape-controlled machine will position the various machine elements, select spindle speeds and feed rates, select and change tools, turn the coolant on and off, and start or stop the machine. The machines are controlled by a tape into which the necessary instructions have been punched.

10.35.2 Maintenance. With such a highly mechanized system, the necessary planned and preventive maintenance can be accurately scheduled. This is very important in keeping the machines in the best condition and preventing gradual deterioration of the product. Careful production scheduling and planned maintenance make it possible to keep the machines operating 75 to 90% of the time.

10.35.3 Benefits. The high cost of these machines is counteracted by many advantages.

a. *Increased flexibility.* Numerical control permits greater freedom in design with a minimum of delay and tool cost.
b. *Reduction in setup time.* Lead time is drastically reduced.
c. *Fewer fixtures are necessary.* Savings are made in the cost of making, handling, storing, and maintaining these items.
d. *Increased productivity.* Numerically controlled machines offer productivity increases of 2 to 10 times over manually controlled equipment.
e. *Less scrap loss.*
f. *Tapes are easily and inexpensively stored.*
g. *Higher quality parts.* Accuracy is increased by the repetitive precision assured by numerical controls.
h. *Inspection costs are reduced.*

10.35.4 Controls. There are various makes of numerically controlled machines, each with their own characteristics. Figure 10.54 shows one such machine that has five axes of motion. Machining can be done at almost any compound angle, and contouring can be done by combining any or all of the five axes.

The part shown being machined in Fig. 10.55 is completed in three operations in 14 hr. With

FIG. 10.54. Machine with five axes of motion (courtesy Machinery).

conventional equipment, 17 individual operations would be needed, each requiring a new setup. The time for the operation would be 56 hr.

The operations performed include face milling, contour milling, drilling, tapping, boring, and counterboring. Approximately 350 positionings, slide movements, tool changes, and other functions are completely tape controlled.

10.36 COMPUTER AIDED MANUFACTURE (CAM)

With CAD/CAM (computer aided design/computer aided manufacture) it is now possible to design an object and manufacture it without a set of drawings as we now know it. The design is stored in the computer's memory. The computer then directs the machine to perform various operations on the workpiece. Robots may move the workpiece to another station in the machine or to another machine entirely where the operations may be completed.

Parts that are manufactured in this manner are generally more accurate and have lower scrap rates and higher productivity, resulting in lower costs.

10.37 HEAT TREATMENT OF STEEL

The properties of steel, such as tensile strength, ductility, and hardness, depend on two items, namely, the chemical composition of the metal and the heat treatment to which it has been subjected. The principal heat treatments used in the production of steel products are annealing, normalizing, hardening, tempering, and case

FIG. 10.55. Turbojet engine frame (courtesy Machinery).

hardening. These processes involve the heating and cooling of the metal for the purpose of altering the grain structure and the amount of carbon or other substances dissolved in the steel.

10.38 ANNEALING
The steel is heated to a temperature close to the lower limit of the critical range, as shown in Fig. 10.56, and then cooled very slowly. Of the many purposes for annealing, the principal ones are probably to remove stresses and induce softness or increase ductility.

10.39 NORMALIZING
This involves heating the steel to the temperature indicated in Fig. 10.56 and cooling in air. This gives faster cooling than annealing. The treatment increases the tensile and yield strength over annealed steel while still retaining sufficient ductility for many purposes.

10.40 HARDENING
When steel is heated and cooled quickly, by quenching in water, oil, or some other cooling substance, the metal becomes hard and brittle. The temperature range for hardening is shown in Fig. 10.56.

10.41 TEMPERING
As hardened steels are too brittle for most uses, they must be tempered. This is done by heating the hardened steel to some point below the lower critical temperature, holding for a sufficient period of time, and cooling as desired. Quenching is sometimes used.

10.42 CASE HARDENING
For high-surface hardness and resistance to abrasion, case hardening is used. This is done by raising the temperature to 1700 or 1800°F and packing the steel in some carburizing compound. Carbon is absorbed to a certain depth, forming a hard case on the outside of the metal.

10.43 HEAT TREATMENT OF NONFERROUS ALLOYS
When the percentage of one metal dissolved in the other metal can be changed by varying the temperature, the properties of the alloy can usually be changed by heat treatment. For further information on this subject, see any good book on metallography.

10.44 MATERIAL SPECIFICATION
It is very important for the engineer to be able to specify the exact material that is to be used by the manufacturer in producing a certain product. Several societies have developed standards for this purpose. One of these which is frequently used is that of the Society of Automotive Engineers. They have a numbering system consisting of four digits. In its original conception the first digit represented the type to which the steel belonged. Thus 1 indicated carbon steel, 2 indicated nickel steel, and 3 was nickel chromium steel. In the case of simple alloy steels, the second digit generally indicated the approximate percentage of the predominant alloying element. Usually the last two digits indicated the approximate average carbon content in hundredths of 1%. Thus 2317 represented a nickel steel of approximately 3% nickel and 0.17% carbon.

Tables 10.2 and 10.3 give a few of the important steel numbers with their characteristics and uses. Table 10.4 gives the numbers and suggested uses for various kinds of grey iron. Table 10.5 lists some of the numbers, uses, and general properties of aluminum. (The tables are taken from the SAE Manual.)

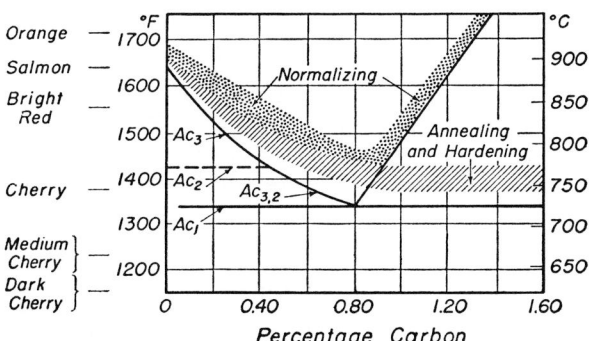

FIG. 10.56. Temperature ranges for heat treatment (courtesy ASM Handbook, 1948.)

TABLE 10.2[a] CHARACTERISTICS AND USES OF PLAIN CARBON STEELS

SAE	
1006 1008 1009 1010 1012 1015	These steels are the lowest carbon steels of the plain carbon type and are selected where cold formability is the primary requisite. They have relatively low tensile values, but may have excellent surface finish and good drawing qualities. These steels are nearly pure iron or ferritic in structure and do not machine freely; rimmed steel is used for cold-heading wire for tacks and rivets, for body and fender stock, hoods, lamps, oil pans, and other deep-drawn products.
1016 through 1027	Steels in this group, because of the carbon range covered, have increased strength and hardness and reduced cold formability. These steels are used for numerous forged parts. SAE 1020 is used for fan blades and some frame members. SAE 1024 may be used for such parts as transmission and rear axle gears.
1030 1033 1035 through 1043 1045 1046 1049 1050 1052	These steels, of the medium carbon type, are selected where their higher mechanical properties are needed. All steels in this class are used for forgings. As a class they are considered good for normal machining operations. SAE 1030 and 1035 are used for shifter forks and many small forgings. SAE 1038 is used for bolts and studs.
1055 1060 1064 1065 1070 1074 1078 1080 1084 1086 1090 1095	Steels in this group are of the high carbon type. They are used principally for applications where the higher carbon is needed to improve wear characteristics for cutting edges, to make springs of various types, and for special purposes, such as valve-spring wire and music wire. These steels find wide usage in the farm implement industry.

[a] Tables 10.2 through 10.5 are abstracted from the SAE Handbook of the latest years. This courtesy of the SAE is hereby acknowledged.

TABLE 10.3 CHARACTERISTICS AND USES OF FREE-CUTTING CARBON STEELS

SAE	
1111 1112 1113	These steels have excellent machining characteristics and are used for a wide variety of machined parts. They are not commonly used for vital parts owing to an unfavorable property of cold shortness.
1108 1109 1115 1117 through 1120 1126	Steels in this group are used where a combination of good machinability and uniform response to heat treatment is needed. These steels are used for small parts that are to be case-hardened.
1132 1137 1138 1140 1141 1144 1145 1146 1151	These steels are widely used for parts where a large amount of machining is necessary, or where threads, splines, or other operations offer special tooling problems. SAE 1137 is widely used for nuts, bolts, and studs with machined threads.

TABLE 10.4 SUGGESTED USES FOR AUTOMOTIVE GRAY IRON CASTINGS

SAE	
110	Miscellaneous soft iron castings in which strength is not of primary importance; exhaust manifolds
111	Small cylinder blocks, cylinder heads, air-cooled cylinders, pistons, clutch plates, oil-pump bodies, transmission cases, gear boxes, clutch housings, and lightweight brake drums
120	Automobile cylinder blocks, cylinder heads, flywheels, cylinder liners, and pistons
121	Truck and tractor cylinder blocks and heads, heavy flywheels, tractor transmission cases, differential carrier castings, and heavy gear boxes
122	Diesel-engine castings, liners, cylinders, pistons, and heavy parts in general

TABLE 10.5 PROPERTIES OF ALUMINUM

SAE	ASTM Designation	Usual Form	General Data
300	CS66A	Permanent mold castings	Pistons primarily
304	S5C	Die castings	Good to excellent casting characteristics and good to high corrosion resistance; suited for use in thin-walled or intricate castings
305	S12A		
306	SC84A		
308	SC84B		
309	SG100A		
310	ZG61A	Sand castings	General-purpose structural castings
320	G4A	Sand castings	Moderate strength, high corrosion resistance
321	SN122A	Permanent-mold castings	Pistons, low expansion
322	SC51A	Sand and permanent mold castings	High strength and pressure for general use, such as pump bodies and liquid-cooled cylinder heads
324	G10A	Sand castings	High strength and ductility, requires special foundry practice
326	SC64B	Sand and permanent mold castings	General-purpose alloy
328	SC122A	Permanent-mold castings	Pistons
330	SC64A	Permanent-mold castings	Moderate strength, general-purpose alloy
332		Permanent-mold castings	Automotive pistons
34	CG100A	Sand and permanent mold castings	Pistons, air-cooled cylinder heads, and valve tappet guides
35	S5B	Sand and permanent mold castings	Intricate castings having thin section
38	C4A	Sand castings	General structural castings; high strength and shock resistance
39	CN42A	Sand and permanent mold castings	Air-cooled cylinder heads, and high strength pistons

SELF-STUDY QUESTIONS

Before trying to answer these questions, read the chapter carefully. Then, without reference to the text, answer as many questions as possible. For those that cannot be answered, the number in parentheses following the question number gives the article in which the answer can be found. Look it up and write down the answer. Check the answers that you did give to see that they are correct.

10.1 (**10.2**) An object made by pouring molten metal into a mold is called a _____.

10.2 (**10.2**) A wooden mold of the object is called a _____.

10.3 (**10.10**) A sand model of the hollow interior of a casting is called a _____.

10.4 (**10.10**) Draft on a pattern is to make it easier to draw from the _____.

10.5 (**10.4**) A pattern is made slightly _____ to allow for _____ of the metal.

10.6 (**10.4**) A pattern is made oversize by using a _____ _____ in constructing it.

10.7 (**10.4**) The shrinkage allowed for cast iron is _____ _____ inch per _____.

10.8 (**10.7**) A _____ _____ is a projection on a pattern that makes an impression in the sand to support the core.

10.9 (**10.9.1**) Sharp angles on a casting are eliminated by using _____ and _____.

10.10 (**10.9.2**) A _____ "shut" is when metal flowing into a casting in opposite directions fails to _____.

10.11 (**10.13**) A hole is made in a casting by using a _____.

276 SHOP TERMS AND PROCESSES

10.12 (**10.16**) When a large number of castings are to be made a _____ mold may be used.

10.13 (**10.17**) A hollow pipe might be formed by _____ casting.

10.14 (**10.20**) When hot metal is forced into shape it is called a _____.

10.15 (**10.20**) When hot metal is forced into dies by drop hammer it is called a _____.

10.16 (**10.21.4**) The forging plane should be _____ to the direction of the stroke.

10.17 (**10.22**) Making a hole in a part by one stroke of the press is called _____.

10.18 (**10.22**) Forcing metal to flow through a die by pressure is called _____.

10.19 (**10.24.2**) To finish the surface of a hole, a _____ should be used.

10.20 (**10.24.3**) A hole should be _____ to fit the head of a flat-headed screw.

10.21 (**10.24.4**) A hole should be _____ to fit the head of a fillister-headed screw.

10.22 (**10.24.5**) A spot face makes a _____ _____ for the head of a bolt.

10.23 (**10.24.6**) A _____ is used to cut a thread in a hole.

10.24 (**10.24.6**) A _____ is used to cut a thread on a rod.

10.25 (**10.27**) The machine most commonly used to cut a large cylindrical hole in a piece of metal is called a _____ _____.

10.26 (**10.27**) The _____ _____ can be used for almost the same work as the lathe.

10.27 (**10.28**) The _____ _____ has circular-type revolving cutters.

10.28 (**10.29**) For very fine finish the _____ _____ may be used.

10.29 (**10.31**) The planar is used for milling large _____ _____.

10.30 (**10.31**) On the planer the work moves backward and forward under the _____.

10.31 (**10.32**) On the shaper the _____ _____ moves backward and forward.

10.32 (**10.33**) To cut a square hole, the _____ _____ is used.

10.33 (**10.38**) The process of relieving machining and temperature stresses is called _____.

10.34 (**10.41**) Reducing the brittleness of hardened steel is called _____.

10.35 (**10.35**) A numerically controlled machine must have a productivity of _____ to _____ times that of conventional equipment.

CHAPTER 11

BASIC DIMENSIONING

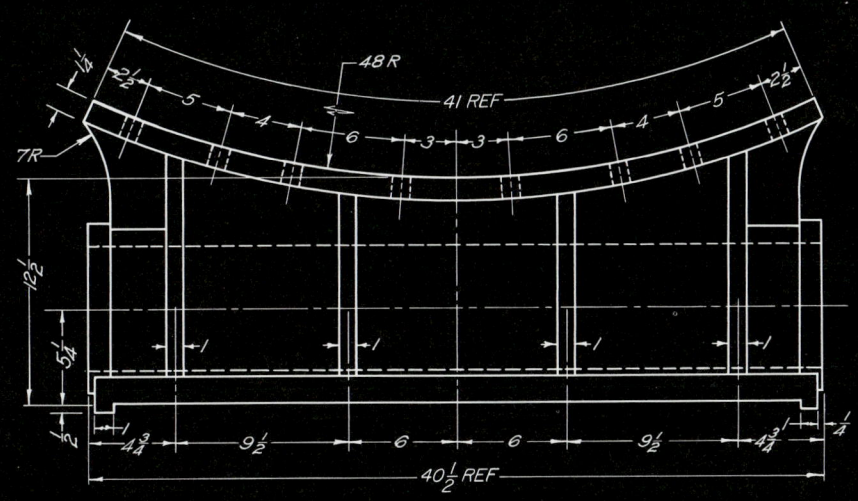

CHAPTER 11

11.1 INTRODUCTION

Engineering drawings may have the necessary views and details to describe completely the object to be manufactured, but if the dimensioning is not accurate and complete it will be impossible to fabricate the final product. Additional information that will be needed is the material from which the part is to be made, heat treatment, and finish.

In the manufacturing process it is impossible for workers to measure and reproduce dimensions exactly. Some variation in size, known as *tolerance*, must be permitted in the manufacture of mechanical parts. For accurate parts the tolerance is small and must be specified on the drawing. For architectural and civil engineering drawings, the tolerances are usually large and are covered by a good-workmanship clause within the specification.

Fractions have been the rule since the start of the industrial revolution. With the need for greater accuracy, for interchangeability of parts, the decimal inch has become widely accepted. Currently there is another change going on worldwide. Most countries have formally adopted the International System of Units (Systém International d'Unités) or SI. Although the United States has not completed the transition, more and more companies are making the change. This trend is being led by multinational corporations who have operations worldwide.

In this chapter and throughout the book all three systems are utilized since industry will continue to refer to "old" drawings that have not been updated to SI.

This chapter gives the fundamentals of dimensioning that apply to all branches of engineering.

11.2 GENERAL PRINCIPLES

Since mechanical parts usually require the most accurate dimensioning, this chapter will deal mostly with such parts. However, the following general principles may be applied to almost any kind of drawing.

a. There is an implied tolerance for every dimension.
b. Dimensions should be measured from some datum or reference plane that can be readily located on the part.
c. Dimensions should not be repeated.
d. On mechanical parts, tolerances should not be allowed to accumulate by giving a long, continuous series of dimensions.
e. Every feature on the object must be dimensioned for size and location.
f. Where practical, the finished part should be defined without specifying manufacturing methods.
g. No surface, line, or point should be located by more than one dimension in any given direction.
h. Dimensions for size, form, and location should be so complete that no scaling of a drawing is required.

11.3 DIMENSIONING A DRAWING

There are three factors that must be considered when placing dimensions on a drawing.

a. Technique of dimensioning.
b. The rules for placing the dimensions on a drawing.
c. The methods of deciding what dimensions are appropriate and necessary.

These techniques are discussed in the following paragraphs.

11.4 TECHNIQUE OF DIMENSIONING

The technique of dimensioning includes the mechanics of constructing dimension lines, leaders, arrowheads, and lettering. To facilitate the interpretation of a drawing, it is necessary for the drafter to follow certain standard procedures. These procedures are listed below.

11.4.1 Dimension Lines. *Dimension lines* are the lines with arrowheads on each end and that show the length of the dimension. See Fig. 11.1. They should be made according to the following instructions, which are illustrated in Fig. 11.1.

a. Draw them as thin, black lines, much thinner than the visible or invisible outlines, for contrast purposes.

b. Make them parallel to the line or length being dimensioned.

c. Construct an arrowhead at each end. See Fig. 11.8.

d. Do not allow them to coincide with a center line or an extension of a center line.

e. Do not place them on a line of the view or on an extension of such a line.

f. Break the dimension line to allow space for the numeral when dimensioning a mechanical part. See Fig. 11.1.

11.4.2 Extension Lines. *Extension lines,* sometimes called *witness lines,* are the lines that extend from the object to show the limits of a dimension. See Fig. 11.1.

The following rules concerning extension lines are illustrated in Fig. 11.2.

a. Make the extension lines the same weight as dimension lines.

b. Construct the extension lines perpendicular to the line being dimensioned. See Fig. 11.3.

c. Allow a gap of $\frac{1}{16}$ in. between the view and the extension line. See Fig. 11.2.

d. Carry the extension line $\frac{1}{8}$ in. beyond the dimension line, as shown in Fig. 11.1.

e. Do not cross extension lines unless necessary. If they must cross, they should do so without a break, as shown in Fig. 11.2.

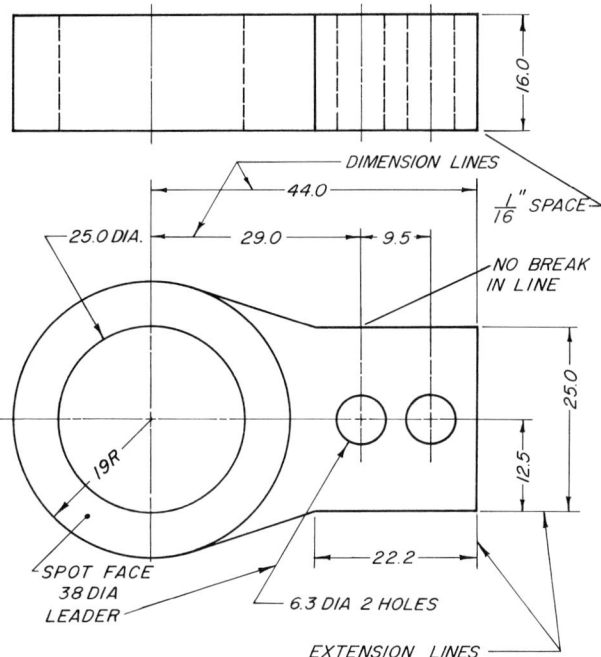

FIG. 11.1. Definition of dimensioning terms.

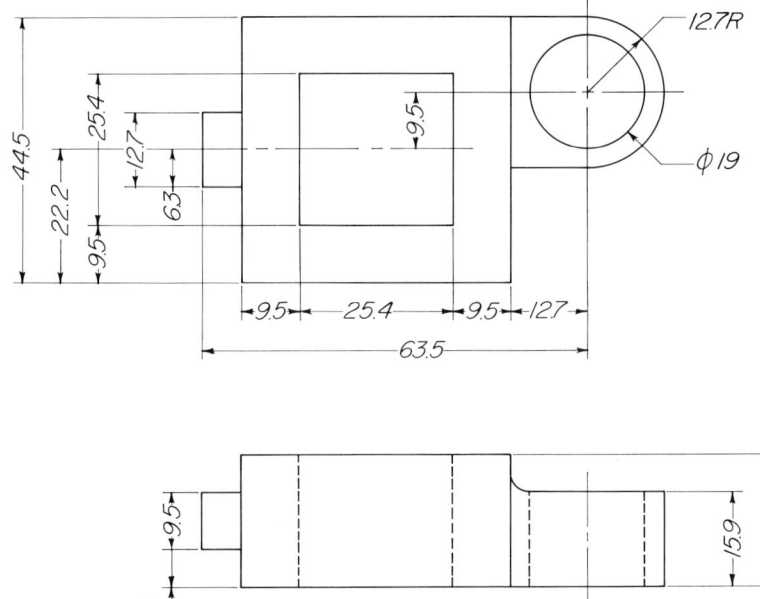

FIG. 11.2. Extension lines cross outlines and each other.

280 BASIC DIMENSIONING

f. Break the extension line when it crosses an arrowhead, as illustrated in Fig. 11.23.

g. Do not break the extension line when it crosses a line of the figure. See Fig. 11.2.

h. In special cases, such as flat curves, where there is not enough room for perpendicular extension lines, they are occasionally placed at an angle, as shown in Fig. 11.4. The dimension line is still parallel to the line being dimensioned.

11.4.3 Leaders. A leader is a line that runs from some part of a drawing, indicated by an arrow, to a note that concerns that part of the drawing. See Figs. 11.1 and 11.5. They are drawn in accordance with the rules stated below.

a. Make leaders the same weight as dimension lines.

b. Construct leaders with a straight edge, as indicated in Figs. 11.1 and 11.5.

c. When leaders point to a line, end them with an arrow. See Figs. 11.1 and 11.5.

d. When leaders point to an area, end them with a dot, as shown in Figs. 11.1 and 11.5.

e. Leaders are terminated on one end by an arrow or a dot and on the other end by a short horizontal line, to the middle of the letter as illustrated in Fig. 11.5.

f. Make the leaders parallel when two or more are close together. If possible, they should be 60° with the horizontal. See Fig. 11.5a.

g. Construct the leaders so that they do not make an angle of less than 30° with the line to which they point. See Fig. 11.5b.

h. When a leader points to a circle or an arc, construct the leader so that the arrow touches the circle or arc, not the center. Always make the leader point toward the center. This rule is illustrated in Fig. 11.6.

i. When there are two circles, as for a threaded hole, countersink, or counterbore, the leader points to the circle indicated in the first line of the note. See Fig. 11.6.

j. When the same note applies to a group of elements, the leader should point to one of the group, as shown in Fig. 11.7.

k. The following should be avoided:
Crossing leaders.
Excessively long leaders.

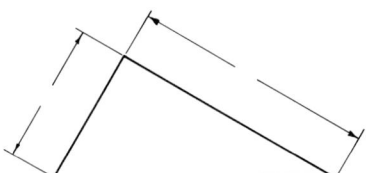

FIG. 11.3. Dimensioning inclined lines.

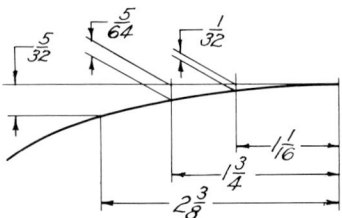

FIG. 11.4. Dimension lines.

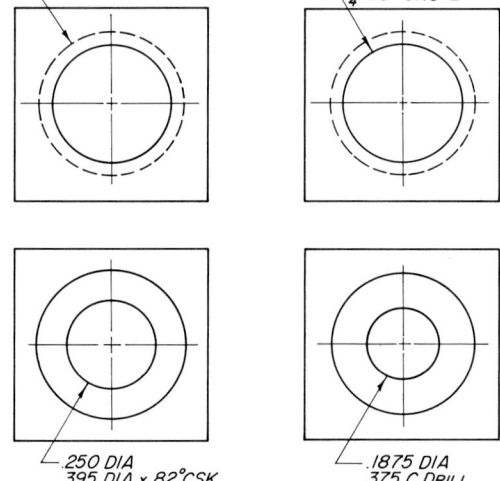

FIG. 11.6. Leaders to circular holes.

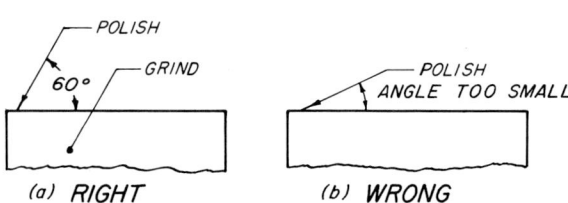

FIG. 11.5. Leaders steeply inclined to surface.

Leaders in horizontal or vertical position. Leaders parallel to adjacent dimension lines.

11.4.4 Arrowheads. An arrowhead is placed on both ends of a dimension line and on one end of a leader. They should be made according to the following rules.

a. The arrowhead must be at least three times as long as it is wide. See Fig. 11.8.
b. In most drawings, the arrowheads should be about $\frac{3}{16}$ in. long. The size may vary with the amount of space available, as in Fig. 11.8.
c. Arrowheads may be made open or solid according to the practice of the company. See Fig. 11.8.
d. Both barbs should be curved the same length and should be balanced about the dimension lines as in Fig. 11.8.

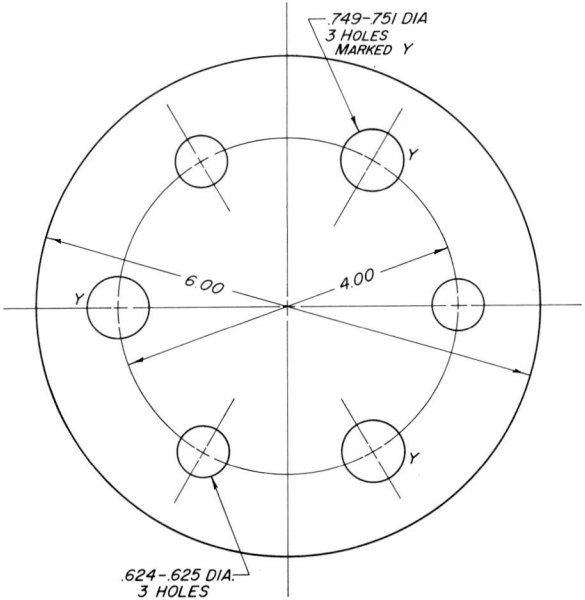

FIG. 11.7. Dimensioning diameters.

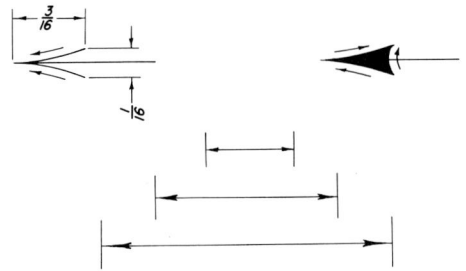

FIG. 11.8. Arrowheads.

e. The point of the arrow should just touch the extension line or the line to which the leader points. They should not overrun or stop short.

11.4.5 Lettering Notes. Letters must be large enough to be easily read and so carefully made that they cannot be misinterpreted. The following rules apply.

a. Notes always read horizontally. See Fig. 11.5. This means that they read in the same direction as the title block.
b. Capital letters are preferred, as indicated in Fig. 11.5a.
c. All lettering should be a minimum of $\frac{1}{8}$ in. high. It should be larger if the drawing is to be microfilmed with a large reduction.
d. Guidelines should be used for all lettering.
e. Numerals should be placed in a break in the dimension line usually in the center, as shown in Fig. 11.9.
f. In fractions, the bar is placed in line with the dimension line but is not a part of it. See Fig. 11.9.
g. The numerator and denominator of a fraction are made at least two thirds of the height of the whole number integers. They should not touch the bar. The total height of the fraction must be at least one and one half times the height of the integer. This rule is illustrated in Fig. 11.9.
h. Notes with leaders should begin with the midpoint of the first line of lettering at the level of the horizontal line that ends the leader. See Fig. 11.6.
i. When a dimension is not to scale, a wavy line is placed under the dimension or it is marked NTS (not to scale). See Fig. 11.10.
j. Abbreviations should be avoided. Only those recommended by the American National Standards Institute (ANSI Y1.1-1972). See Appendix Table 39.

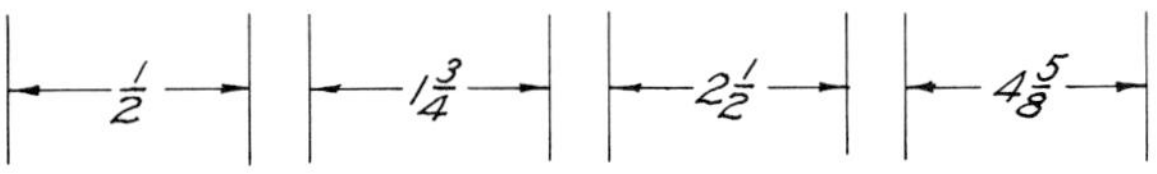

FIG. 11.9. Method of lettering fractions.

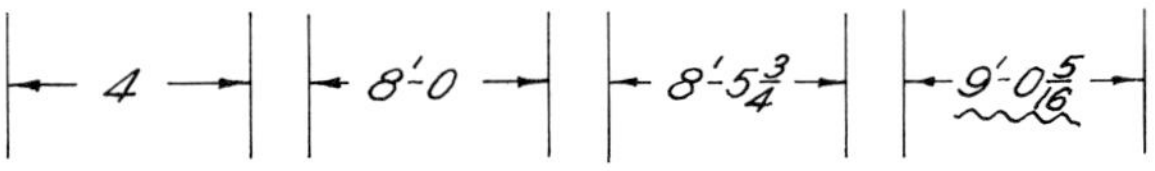

FIG. 11.10. Dimensioning in feet and inches.

282 BASIC DIMENSIONING

11.4.6 Units of Measurement in Dimensioning. The units of measurement specified by dimensions should be clearly stated. This may be accomplished as follows.

a. A general note in the title block may specify that "all dimensions are in inches" or "all dimensions are in millimeters." If this is done, inch marks or "mm" are not needed on the drawing.

b. When dimensions include both feet and inches, a single accent mark is used to designate feet and the inches are unmarked. The various situations that may arise are shown correctly in Fig. 11.10.

c. The double accent mark should be used to indicate inches if there is any danger of misinterpretation. Thus 1 Valve should read 1" Valve.

d. In dimensioning angles, the following symbols should be used: degrees, °; minutes, ';

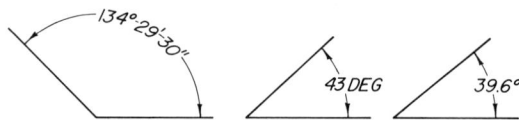

FIG. 11.11. Dimensioning angles.

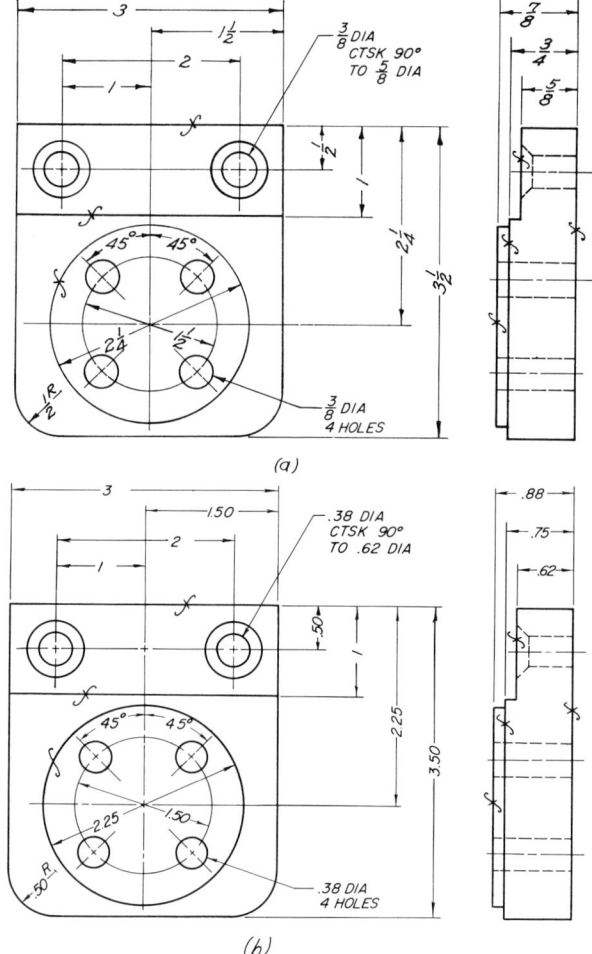

FIG. 11.12. Spacing dimensions.

seconds, ". Thus an angle could be dimensioned 134° 29' 30". If only degrees are used, the abbreviation DEG may be used, as 30 DEG. If minutes alone are to be specified, they should be preceded by 0°, thus 0° 17'. An angle also can be specified by degrees and decimals of a degree. See Fig. 11.11.

11.4.7 Fractions. It is very seldom that a dimension will be of a size that can be expressed completely in full inches. There must be some system for expressing parts of inches. At the present time there are two such systems.

Common fractions have been most frequently used in the past and are still used in many companies. When fractions are used they should be expressed in units of $\frac{1}{8}$, $\frac{1}{16}$, $\frac{1}{32}$, and $\frac{1}{64}$. Figure 11.12a shows an object dimensioned with fractions. Other fractions such as $\frac{1}{3}$, $\frac{1}{6}$, or $\frac{1}{7}$ are never used.

Fractions are difficult to add or subtract, and their use leads to many mistakes. When the distances become very small, as in modern industrial practice, fractions become very cumbersome and slow. For this reason decimals are more commonly used.

11.4.8 Decimals. In a complete decimal system the fundamental base is a two-place decimal. Whenever possible, the dimension should be an even two-place number, as in Fig. 11.12b. It can be divided easily by 2, for such purposes as converting diameters to radii, and still maintain a two-place decimal. In decimal dimensioning the decimal points must be sufficiently large so that they cannot be misread. The following rules apply to the use of decimals.

- a. *Decimal point.* The decimal point shall be in line with the bottom of the lettering and shall be given sufficient space.
- b. *Rounding off decimals.* When it becomes necessary to convert a fraction into a two-place decimal, the third place is dropped if it is less than 5. The next higher number in the second place is used if the third place is more than 5. Thus 1.5625 becomes 1.56, and 1.627 becomes 1.63. If the third place is exactly 5, the nearest even number in the second place is used. Thus 1.625 becomes 1.62, and 1.615 also becomes 1.62.

    The same rules for rounding off can be used for three-place decimals. Three- or four-place decimals may be needed when converting fractions to decimals or when requirements make greater accuracy necessary.
- c. *Advantages of the decimal system.* There are many advantages to the use of the decimal system for dimensioning. Some of them are listed below.
    1. It simplifies arithmetical computations.
    2. It greatly reduces mistakes.
    3. It decreases the time required for calculations.
    4. It is easier to read and understand.
    5. It simplifies conversion to the metric system.
    6. Decimals are used in data processing and in numerically controlled equipment.

11.4.9 Metric. The entire industrial world has or is converting to a common metric system of SI units. The United States, although not proceeding as fast as most countries, is slowly making the conversion.

The advantages of the SI system are the same as indicated above for the decimal system but in addition is necessary for world trade. Several countries will not allow imports of products that are not metric based.

11.5 WHERE TO PLACE DIMENSIONS ON A DRAWING

It is important to place dimensions on a drawing in such a manner that the drawing will be clear and easy to read. Because of long practice, workers have become accustomed to looking for dimensions in certain places. These customs should be followed unless there is a good reason for changing. However, almost all rules may be broken for the sake of clarity. The most important of these rules are given here.

- a. Dimension lines are spaced so that the first line is at least $\frac{3}{8}$ in. from the outline of the view. The second and successive lines are placed either $\frac{1}{4}$ or $\frac{3}{8}$ in. away from the preceding dimension lines. This is shown in Fig. 11.13.

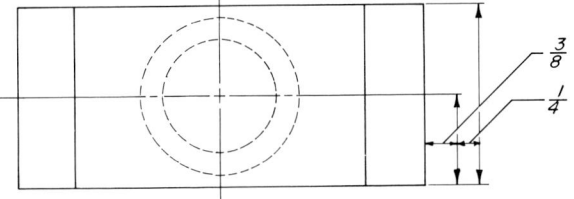

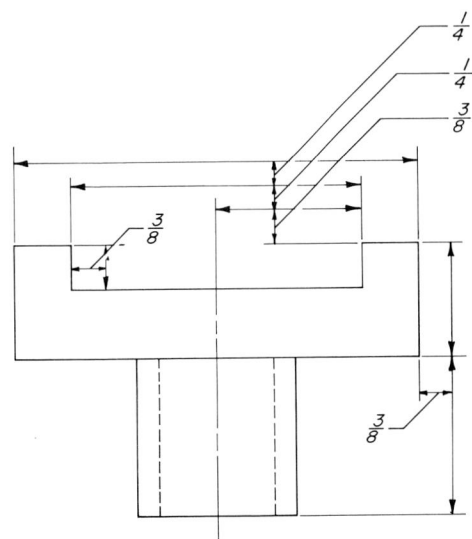

FIG. 11.13. Spacing dimension lines.

b. The numerals in adjacent dimensions should be staggered to avoid crowding, as in Fig. 11.14.
c. Dimensions should be placed outside the views whenever possible. To avoid long extension lines or crossing extension lines it is sometimes best to put a dimension line inside the view. See Fig. 11.15.
d. Dimensions should be placed between views but closest to the appropriate view. See Fig. 11.16.
e. Extension lines or center lines should not be carried through from one view to another.
f. When several dimensions are to be placed on the same side of a view, the smallest is placed closest to the view. This will avoid crossing of extension or dimension lines. See Fig. 11.16.
g. Related dimensions that are to be used in the same shop operation should be placed on a single line of dimensions. See Fig. 11.17. The series of dimensions placed below the front view in this drawing will be used in the machine shop.
h. Unrelated dimensions that refer to different operations or that are to be used by different workers should be placed on separate lines. See Fig. 11.17. The series of dimensions placed above the front view in this drawing will be used in the pattern and core shops.
i. Avoid duplicate dimensioning. No feature should be located in two different ways or places. This is illustrated in Fig. 11.17. When an overall dimension is given in conjunction with a series, one number in the series is always omitted or marked ref-

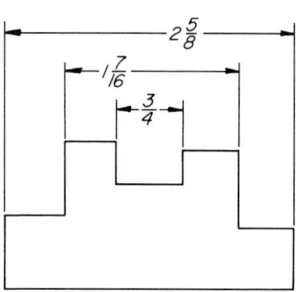 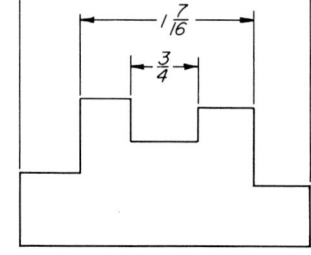

(a) GOOD PRACTICE   (b) POOR PRACTICE

FIG. 11.14. Staggering dimensions.

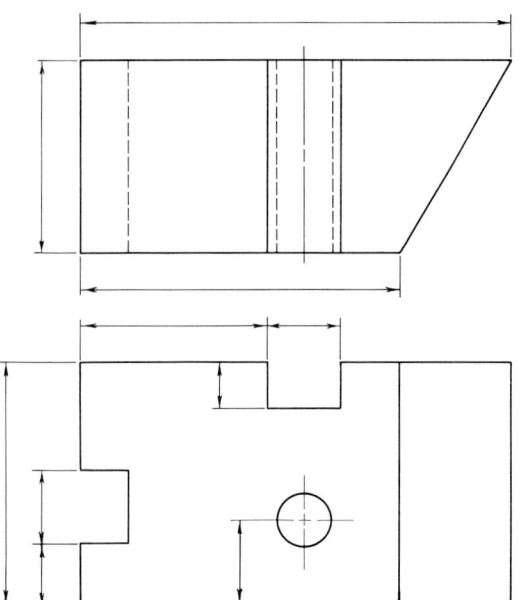

FIG. 11.15. Dimensioning inside a view

11.5 WHERE TO PLACE DIMENSIONS ON A DRAWING

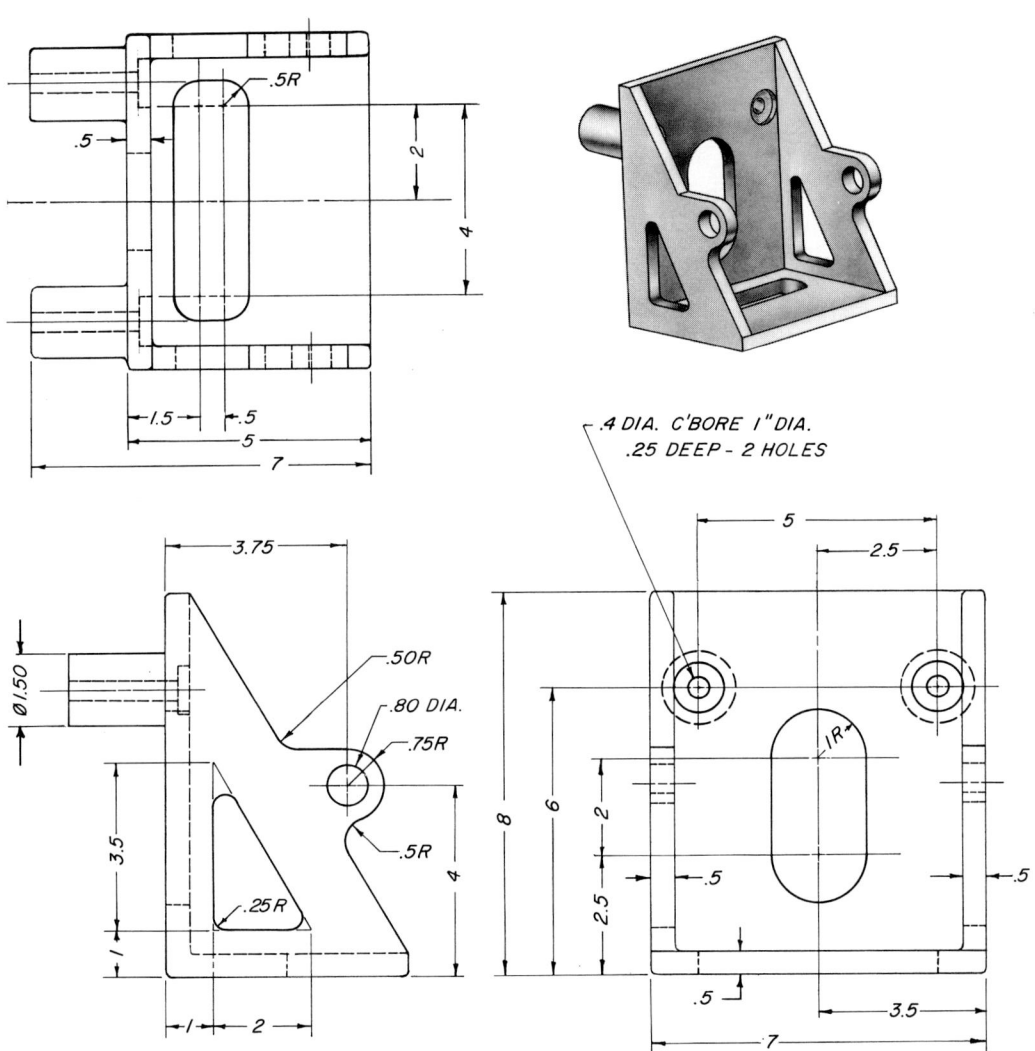

FIG. 11.16. Dimensioning between views.

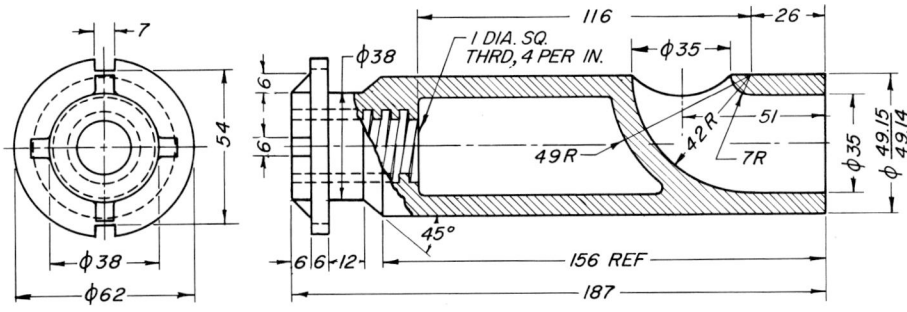

FIG. 11.17. Related dimensions in line.

erence (REF). In architectural and structural drawing, duplicate dimensioning is not only acceptable but also correct.

j. Avoid cumulative tolerances. Tolerance is the variation in size allowable in the manufacture of a part. A series of dimensions in one long line is known as *chain dimensioning*. Such an arrangement allows an accumulation of tolerances. See Fig. 11.18. This should be avoided.

k. Select three reference or datum planes before beginning the dimensioning. These should be finished surfaces or center planes. Center planes frequently make good datum planes, particularly if there are other features on them, such as holes, that enable the plane to be located in the shop. In dimensioning mating parts, the adjoining or functional surface should be used as a datum for both parts. Three possible reference planes are indicated in Figs. 11.19 and 11.42.

l. Individual features can be located from datum planes to save tolerance accumulation. This is illustrated in Fig. 11.20.

m. Dimensions should be placed on the view that is most descriptive or that shows the contour of the feature being dimensioned. Figure 11.20 shows the dimensions properly placed according to this rule.

n. Do not dimension to invisible lines. Hole patterns should be located on the view where the circles show. See Fig. 11.21, right-side view.

o. In a half section it is proper to dimension to an invisible line in the unsectioned part, if the dimension line is carried to a visible line in the sectioned half. This is also shown in Fig. 11.21, front view.

p. The diameter of a cylinder should be dimensioned on the view that shows as a rectangle, rather than the circular view. See Fig. 11.22. If only one view is given, it should be marked DIA or ϕ.

q. For dimensioning very narrow spaces, several methods are available, as shown in Fig. 11.23.

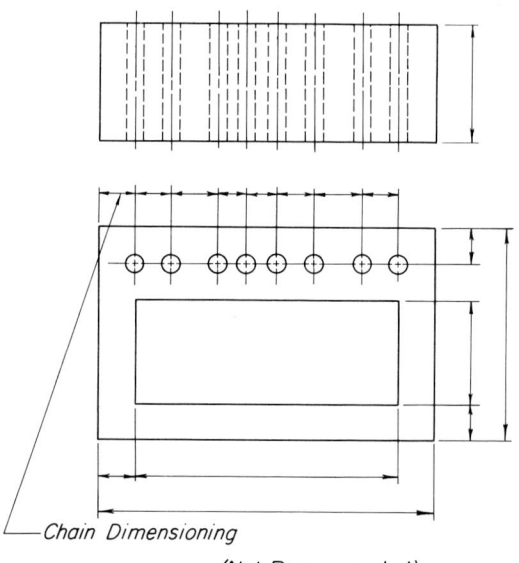

FIG. 11.18. Chain dimensioning.

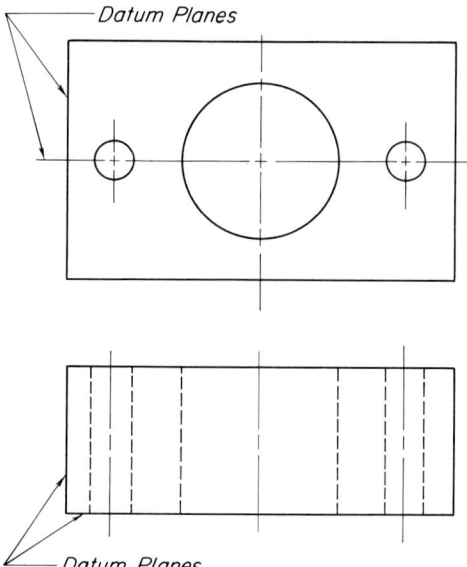

FIG. 11.19. Datum planes.

11.5 WHERE TO PLACE DIMENSIONS ON A DRAWING 287

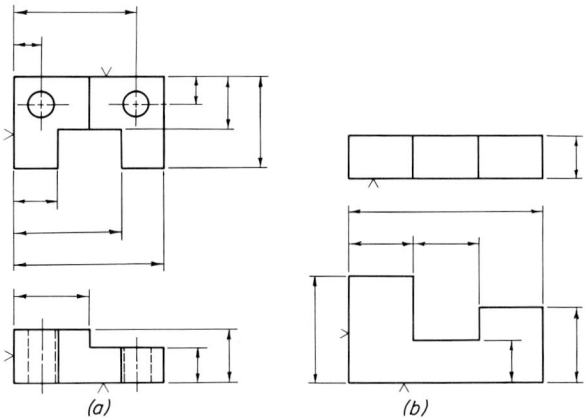

FIG. 11.20. Contour dimensioning. Features dimensioned from datum planes.

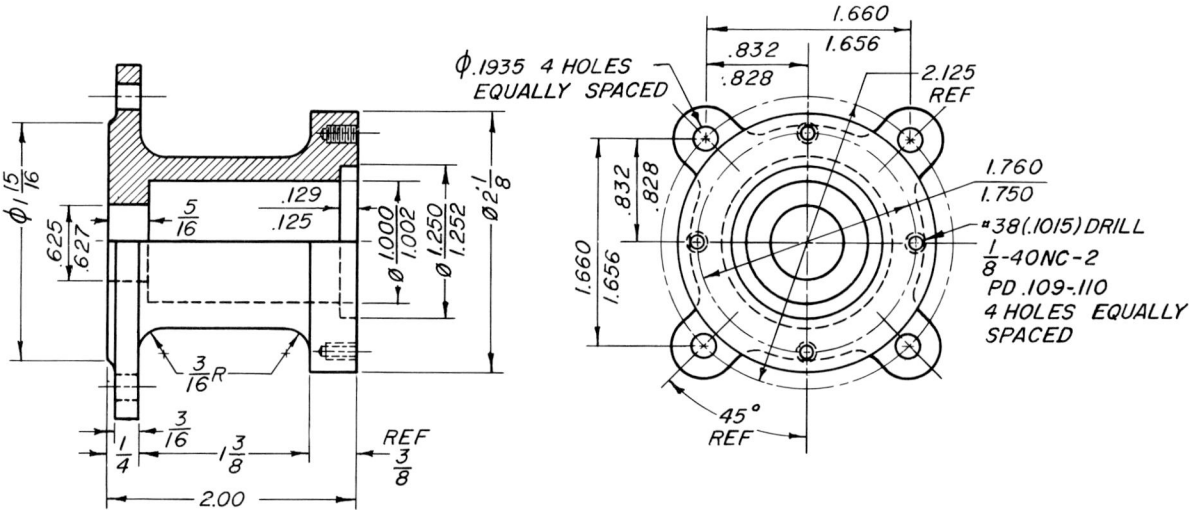

FIG. 11.21. Dimensioning a half-sectioned view.

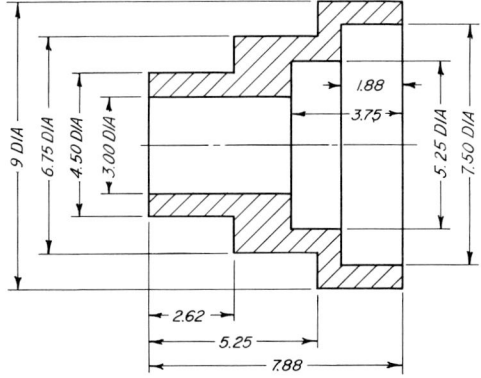

FIG. 11.22. Dimensioning a cylinder.

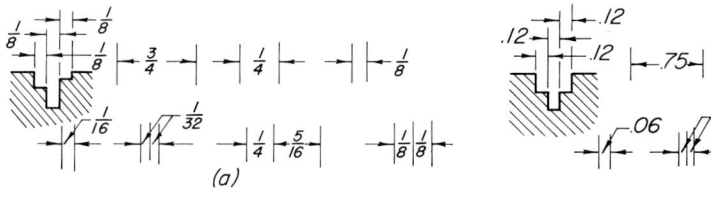

FIG. 11.23. Dimensioning narrow spaces.

11.6 DIMENSIONING SYSTEMS

There are two systems in use for placing the numerals in the dimension line. They are called the *aligned system* and the *unidirectional system*.

11.6.1 Aligned System. The aligned system has the dimensions reading from the bottom when the dimension line is horizontal and from the right when the dimension line is vertical. For inclined dimension lines, the numerals are parallel to the dimension line, but the preference is given to having them read from the bottom of the drawing. This method is illustrated in Fig. 11.24a.

11.6.2 Unidirectional System. When all of the numerals are placed so that they read from the bottom of the sheet regardless of the direction of the dimension line, the system is called *unidirectional*. This means that everything reads in the same direction as the title. Figure 11.24b illustrates the use of this method. The unidirectional system is becoming the most widely used system in industry.

11.7 SELECTION OF APPROPRIATE DIMENSIONS

A part must be dimensioned with four things in mind.

a. The part must operate properly.
b. The manufacturing process must be as economical as possible.
c. The inspection and gaging must be as simple as possible.
d. The number to be produced must be known.

When these things have been considered, the drafter must make sure that the size of the entire part and of every feature has been specified. Also the location of every feature must be completely specified relative to some datum.

11.7.1 Size Dimensioning. Giving the size dimensions of any object requires that these dimensions be placed on the proper views of the object. Since a view is a projection on a plane, it is first necessary to learn to dimension a plane figure, such as a triangle, rectangle, circle, or any combination of them. The proper method of dimensioning some basic plane figures is shown in Fig. 11.25. This involves careful following of the rules already given.

a. Dimension everything from a datum surface if possible.
b. Do not dimension anything twice.
c. Avoid cumulative tolerances.
d. The location of all features must be given.
e. The size of all features must be given.

11.7.2 Dimensioning Simple Objects. Every object must be dimensioned in three directions. This means that at least two views must be dimensioned unless a note can be used to replace one view. The dimensions must be so complete that it will not be necessary to add or subtract dimensions or scale the drawing.

a. *Prism and wedge.* For a prism, it is only necessary to dimension two rectangles. See Fig. 11.25. A wedge, a triangle, and a rectangle must be dimensioned, as in Figs. 11.25 and 11.26.

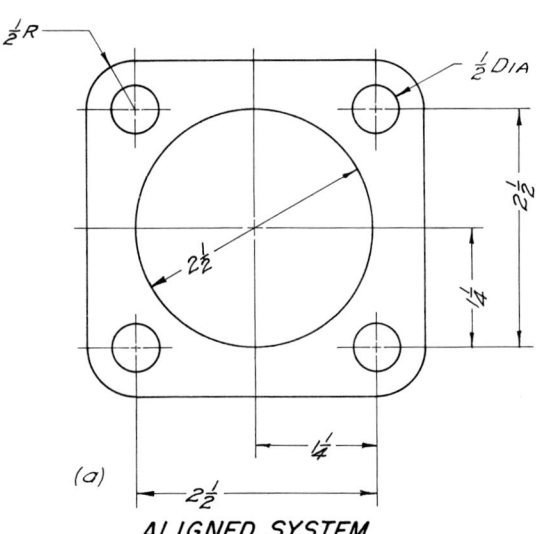
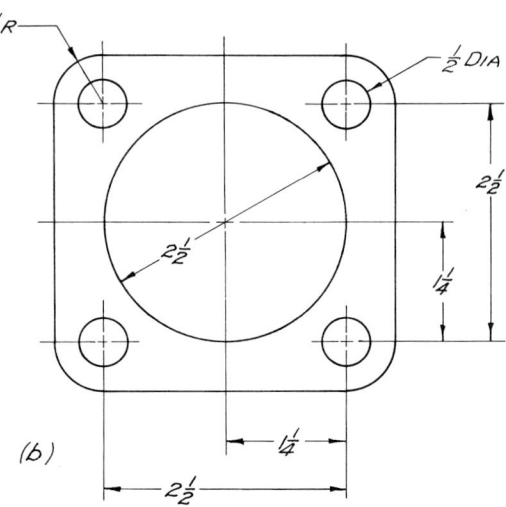

FIG. 11.24. Systems of placing dimension numerals.

11.7 SELECTION OF APPROPRIATE DIMENSIONS

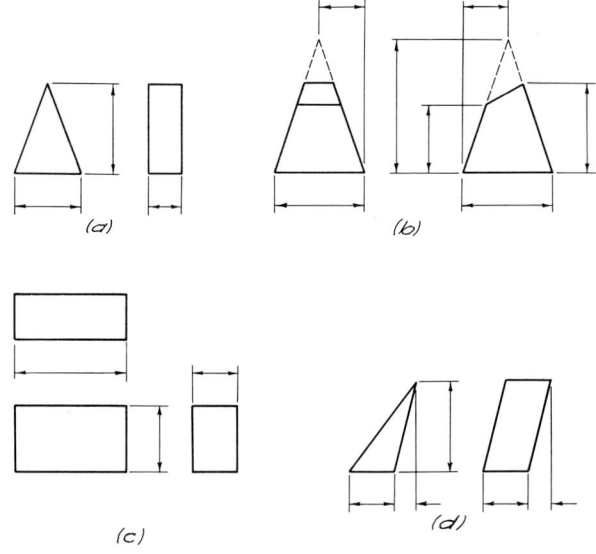

FIG. 11.25. Dimensioning solids.

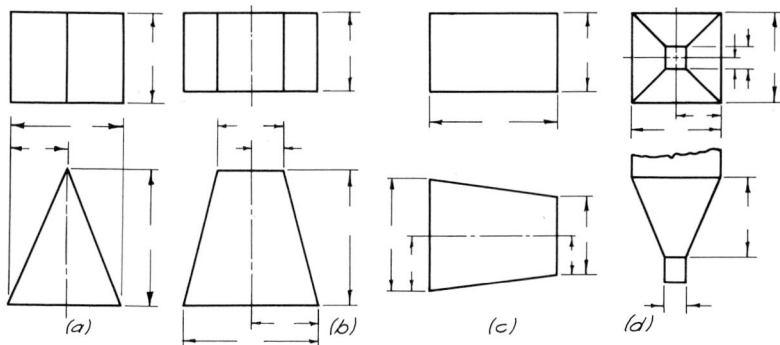

FIG. 11.26. Dimensioning wedges and pyramids.

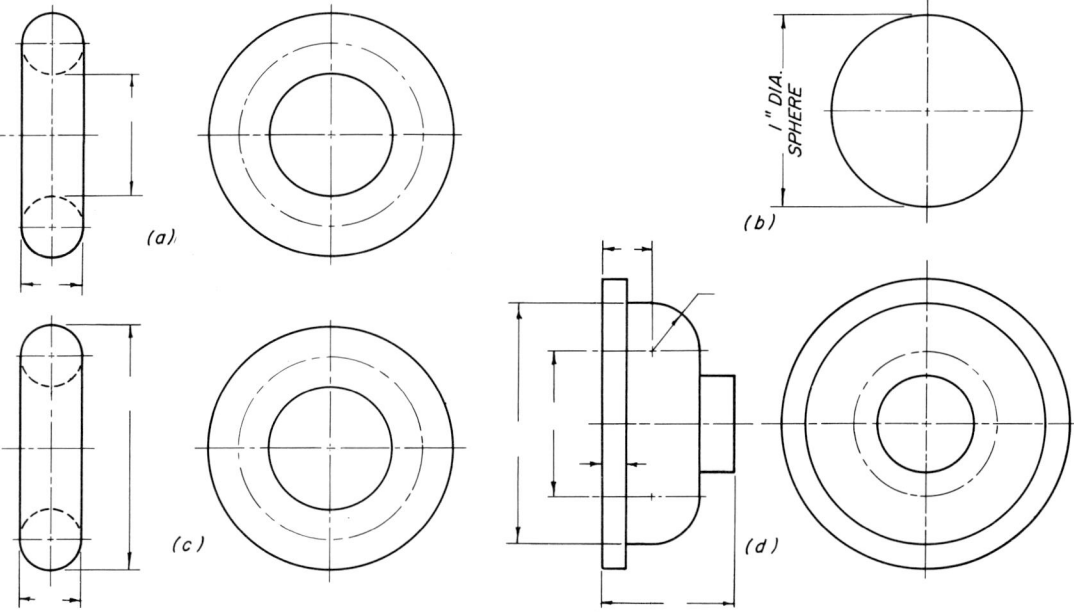

FIG. 11.27. Dimensioning the sphere and torus.

b. *Torus.* A torus can be dimensioned in three ways. The proper one to choose is determined by the method of manufacture. If the torus is to be formed by bending a rod around a mandrel, the method of Fig. 11.27a should be used. If the torus is to be formed by bending a rod to fit inside a cylinder, the method of Fig. 11.27c would be used. However, if the torus is to be turned on a lathe, the method of Fig. 11.27d might be preferable.

c. *Sphere.* Only one view of a sphere is necessary if the dimension is marked "spherical." See Fig. 11.27b.

d. *Cylinders and cones.* For a complete circle, the diameter should always be given. Methods for dimensioning cones and cylinders are shown in Fig. 11.28.

e. *Tubing details.* The minimum dimensions for describing a bent tube are shown in Fig. 11.29. Note that the center lines of straight portions of the tube, between curves, are carried to their points of intersection and that these points are dimensioned. In addition, this information may be supplemented by specifying the straight lengths, angles of bend, bend radii, and angles of twist or rotation for all portions of the tube. This is commonly done by means of auxiliary views.

11.7.3 Dimensioning Other Features. In addition to the simple solids dimensioned in the preceding article, there are many features that form only a part of a more complicated object that must be clearly dimensioned.

a. *Circular arcs.* For a circular arc, the radius rather than the diameter is given. The center may be marked with a small cross, if desired. The size of the radius is placed on a leader. The dimension must always be marked *R*. See Fig. 11.30. When the center is inaccessible or off the paper, Fig. 11.31 shows the method of dimensioning.

b. *Concentric circles.* Whenever possible, circular cylinders should be dimensioned as in Fig. 11.22. When it is necessary to dimension circles on the circular view, Fig. 11.32 shows the approved methods. When using aligned dimensioning, the shaded area of Fig. 11.32a should be avoided.

c. *Compound curves.* A series of tangent circular arcs may be dimensioned by giving the radii and locating the centers. The centers may be located by giving two rectangular coordinates as in Fig. 11.31a, by an angle and a distance as in Fig. 11.31b, or by two arcs as in Fig. 11.33.

d. *Diameters.* When it is obvious that the dimension refers to a diameter the abbreviation DIA or the diameter symbol ϕ may be omitted. If it is not obvious as in the noncircular view of a hole, then the dimension should be followed by DIA or preceded by the symbol ϕ

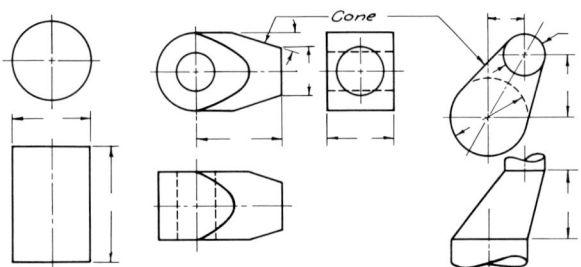

FIG. 11.28. Dimensioning cylinders and cores.

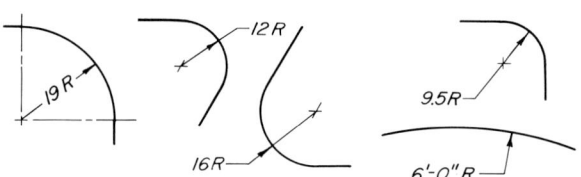

FIG. 11.30. Dimensioning arcs.

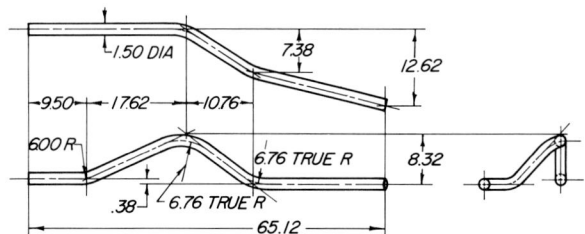

FIG. 11.29. Tubing layout (courtesy SAE).

11.7 SELECTION OF APPROPRIATE DIMENSIONS 291

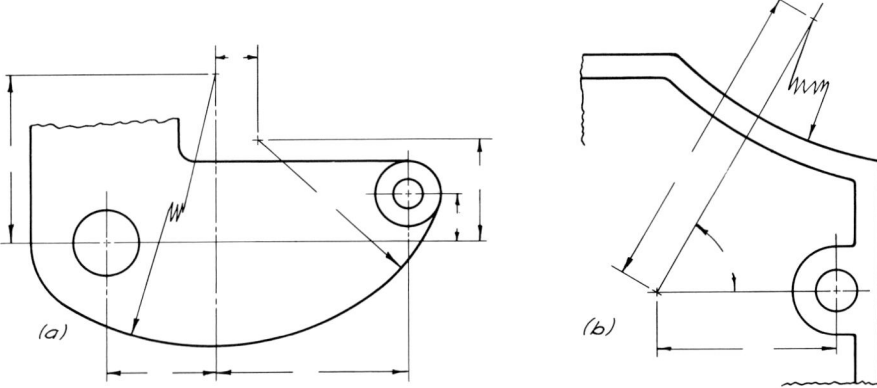

FIG. 11.31. Dimensioning long-radius arcs.

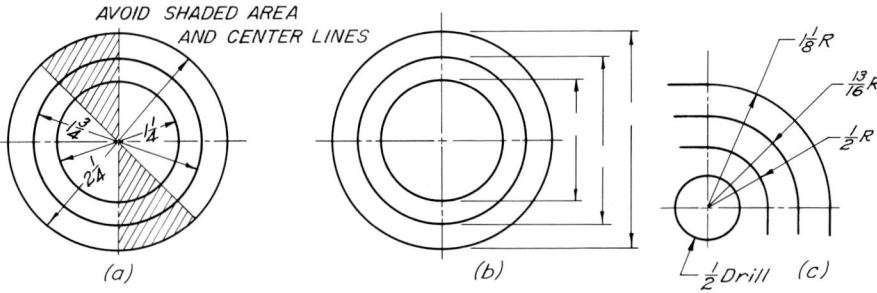

FIG. 11.32. Dimensioning concentric circles and arcs.

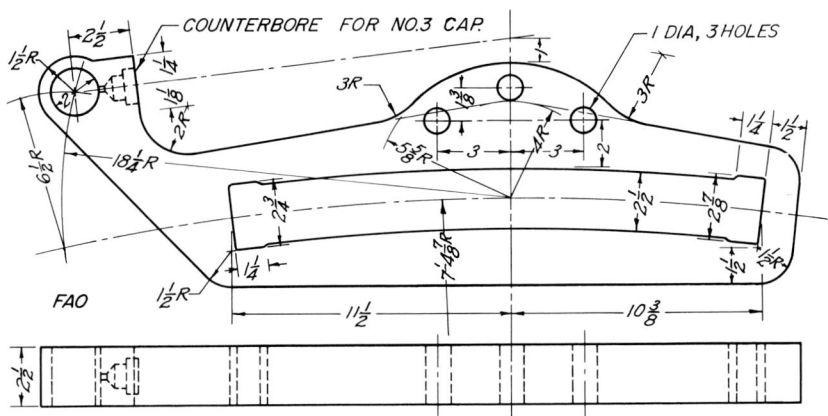

FIG. 11.33. Location dimensioning with arcs and radii.

e. *Noncircular curves.* Points on noncircular curves should be located by coordinates as in Figs. 11.34 and 11.35. This figure also shows a rather unusual use of dimension lines as extension lines.

f. *Progressive dimensioning.* When space is limited, a system known as *progressive dimensioning* may be used. This method places all of the numerals on one dimension line, but each one refers back to the datum plane. To indicate that this is the method used, only one arrowhead is drawn for each dimension. These arrows all point away from the datum. The only arrow pointing in the opposite direction is placed at the datum plane. See Fig. 11.35. This method is seldom used on machine drawings because of the danger of its being misread.

g. *Angles.* In dimensioning angles, as shown in Fig. 11.36, the number should read from the bottom of the drawing as in unidirec-

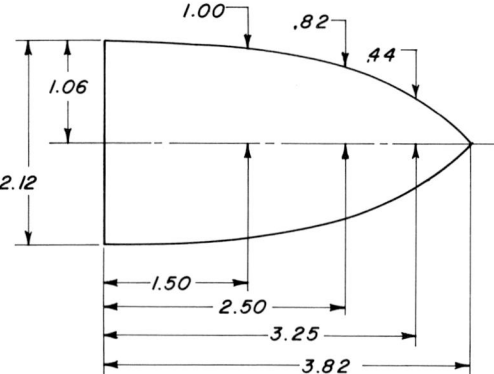

FIG. 11.34. Dimensioning noncircular curves.

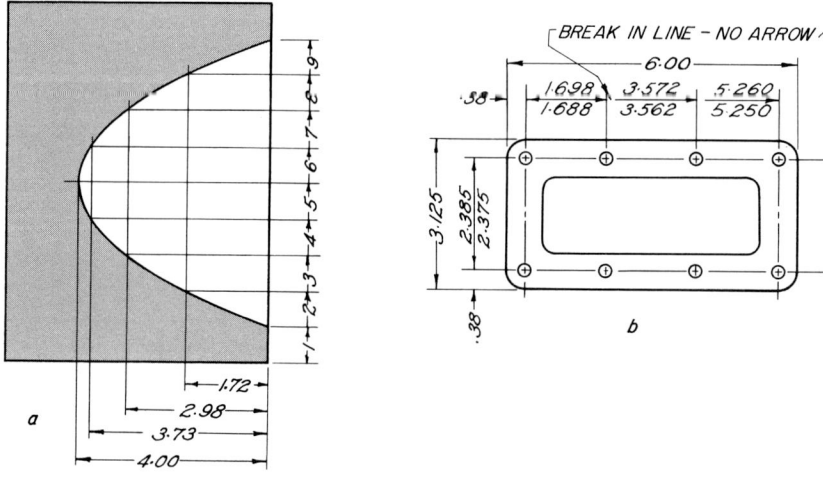

FIG. 11.35. Progressive dimensioning.

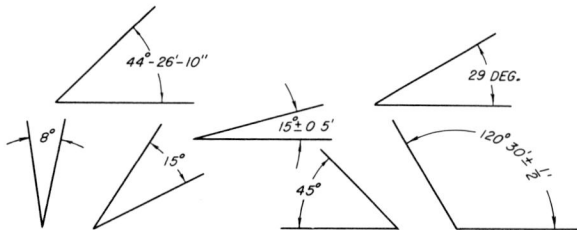

FIG. 11.36. Dimensioning angles.

tional dimensioning. For large angles, the number is sometimes placed in the dimension line as in aligned dimensioning.

An angle may also be expressed by giving the two legs of a right triangle, known as the *run* and *rise*. One of the legs is always made 12 in. Slopes may be expressed as the ratio of the horizontal distance to the vertical distance, as a batter of 1:4, or a slope of 1½:1. See Fig. 11.37. In architecture, the pitch or slope of a roof may be given as the ratio of the rise to the total span, such as ⅓ pitch. It may also be expressed as the number of inches of rise to 1 ft of horizontal distance.

h. *Tapers.* A taper is a conical surface on a machine part. It is specified as the change in diameter in inches per foot of length. In dimensioning a taper, there are four dimensions that can be given: the diameters at two different points, the taper in inches per foot, and the length. Any three of these will completely describe the part. All four should never be given, since that would be double dimensioning. The three that are given will be determined by the purpose of the part or by the method of manufacture. The various methods of dimensioning a taper are shown in Fig. 11.38. In this drawing the tolerances allowed in manufacture have been specified on the drawing. This is done by giving a top and bottom limit of size or by giving a plus or minus distance that will be allowed in the size.

i. *Machining centers.* Shaft centers may be required on shafts, spindles, and other cylindrical parts to receive machine centers on which the work pieces are supported during manufacture and inspection.

There are two types of center drilling: the regular type shown in Figs. 11.39a and b and the Bell type, which is countersunk

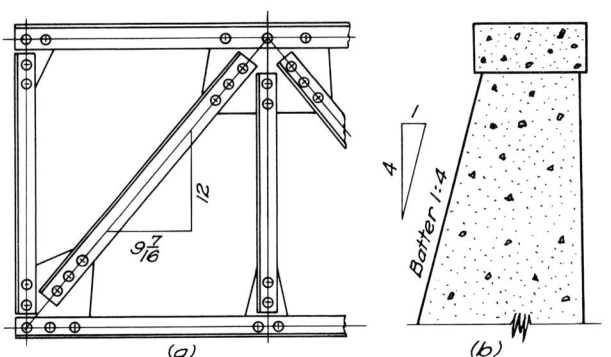

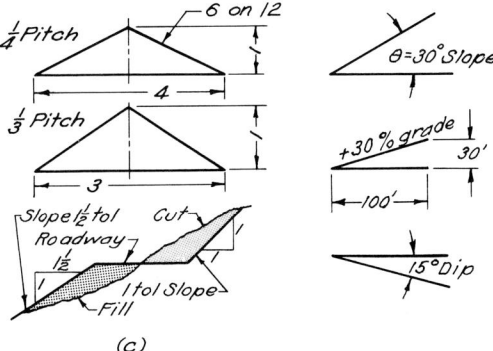

FIG. 11.37. Dimensioning slopes, pitch, and batter.

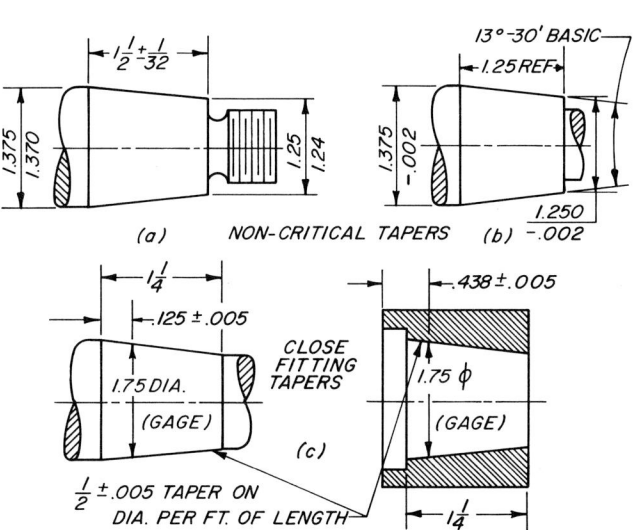

FIG. 11.38. Dimensioning tapers.

to provide protection to the working surface as shown in Fig. 11.39c. The method of dimensioning or specifying is shown in the figure.

11.7.4 Location Dimensioning. As previously stated, every feature of a part must be completely located by giving dimensions in three directions. These dimensions must be measured from properly selected datum planes and must be so chosen that the part will operate properly, may be made economically, and may be easily laid out and inspected.

- a. *Location of holes.* One very common problem in dimensioning is to locate a pattern of holes in a part. Locating dimensions to circles or holes should always be given to the centers on the view where they show as circles. If the holes are in a circular pattern, they may be dimensioned by giving the diameter of the center-line circle and distances or angles from a datum plane as shown in Fig. 11.40. They are also frequently located by giving the number of holes and specifying that they must be equally spaced. This method is illustrated in Fig. 11.41. When the holes are not in a circular pattern, they are usually located by dimensioning each hole from two datum planes. See Fig. 11.42.

If it should be necessary to keep the distance between the holes within specifications, it is possible to dimension one hole from the datum plane and the second from the first as shown in Figs. 11.43 and 11.44. This is the usual procedure for dimensioning parts that must fit together.

If still more accuracy is desired, it is best to use positional tolerancing dimensioning as explained in Chapter 12.

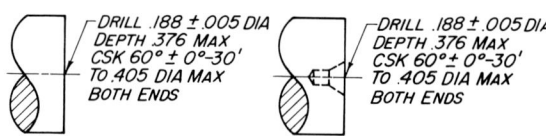
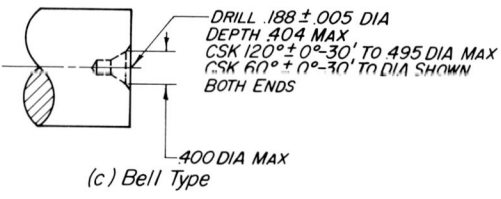

FIG. 11.39. Dimensioning machining centers.

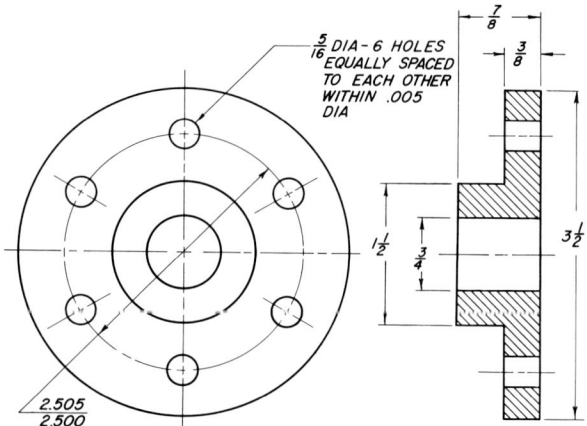

FIG. 11.41. Dimensioning holes in circular pattern-six holes, full section.

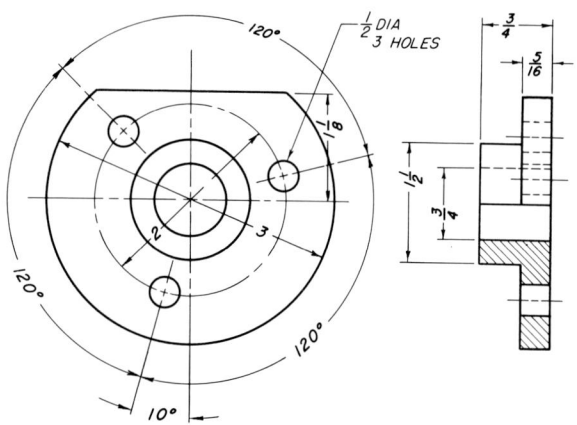

FIG. 11.40. Dimensioning holes in circular pattern-three holes, half section.

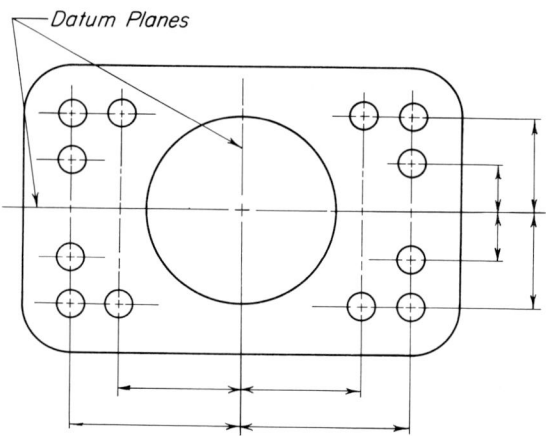

FIG. 11.42. Dimensioning pattern of holes that must be held within one tolerance from center lines.

b. *Symmetrical parts.* When the features of a part are symmetrical about a center line, the preferred method is to dimension one side from the center line and the other side from the first, as shown in Fig. 11.45a. The method of Fig. 11.45b is frequently used, but this requires the assumption that the center line indicates symmetry and that therefore the features would be made that way. The chain dimensioning shown in Fig. 11.45c should never be used.

c. *Partial circular parts.* Objects such as that shown in Fig. 11.46 may be dimensioned by giving the radius and the length or range of it by an angle. In other cases, such as that

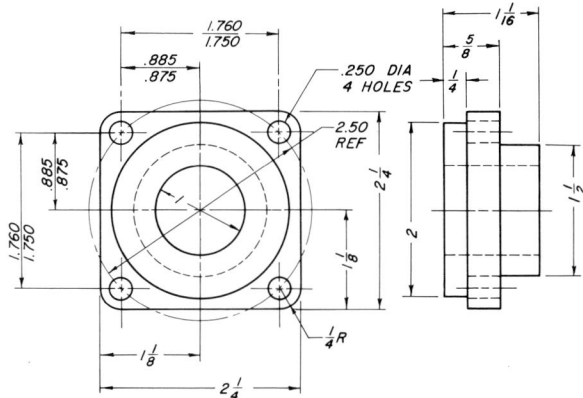

FIG. 11.43. Dimensioning a four-hole symmetrical flange.

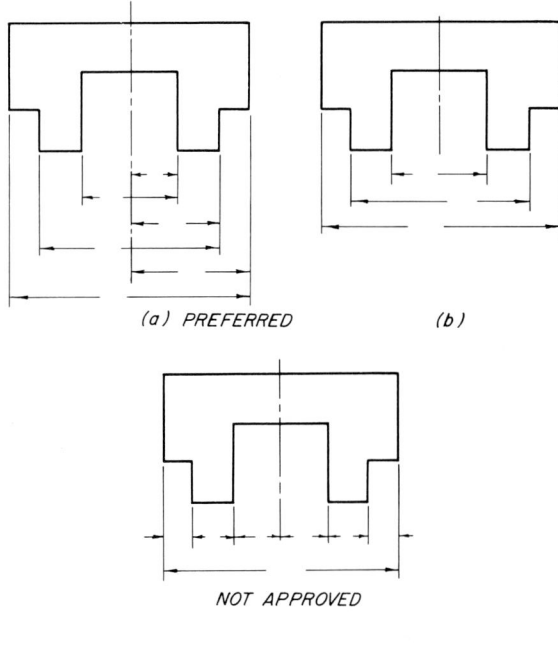

FIG. 11.45. Dimensioning symmetrical parts.

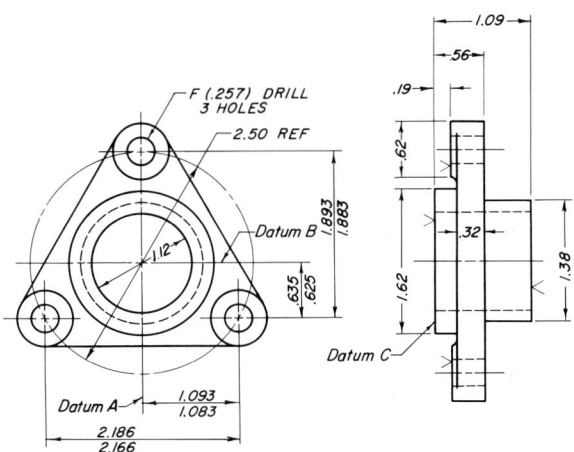

FIG. 11.44. Dimensioning a three-hole flange.

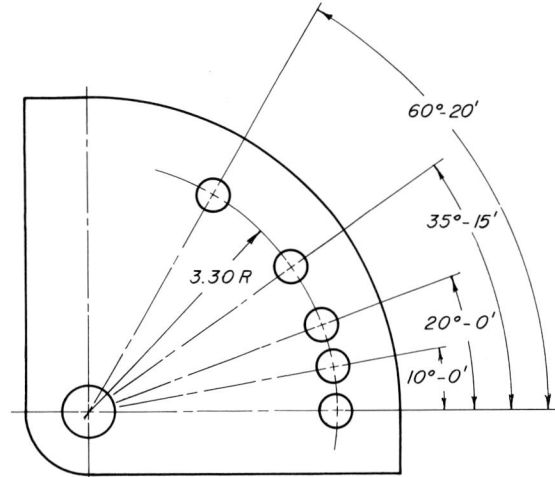

FIG. 11.46. Polar-coordinate dimensioning.

shown in Fig. 11.47, the radius may be given, and the distances along the surface may be dimensioned. In such a case it is understood that the distances are to be measured on the part, even though the dimension line is outside the given surface.

d. *Coordinate dimensioning.* When there are may holes or features to be located in a small area, this may be done by establishing two datum or reference lines and giving dimensions from them. They may be specified on the drawing or in a table.

When a table is used, each feature must be numbered or lettered as in Fig. 11.48. When the information is given on the drawing, the coordinate axes or datum lines are marked zero and distances are all shown from the zero points without putting in dimension lines. See Fig. 11.49.

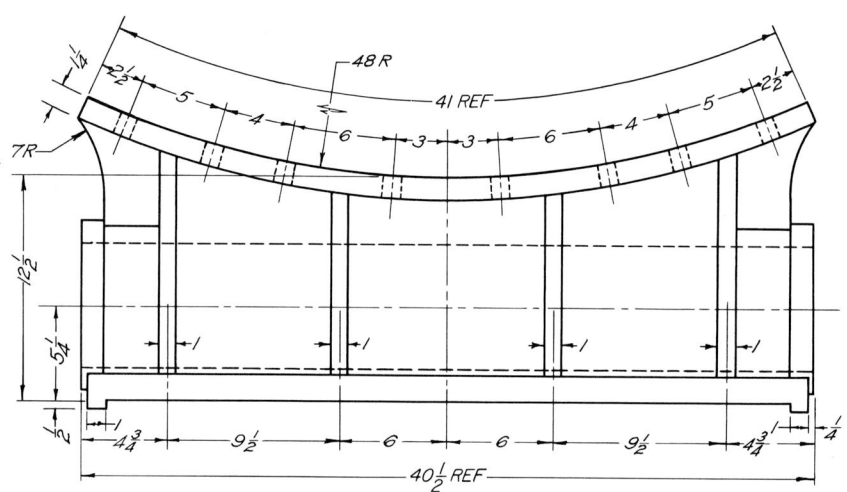

FIG. 11.47. Circumferential dimensioning.

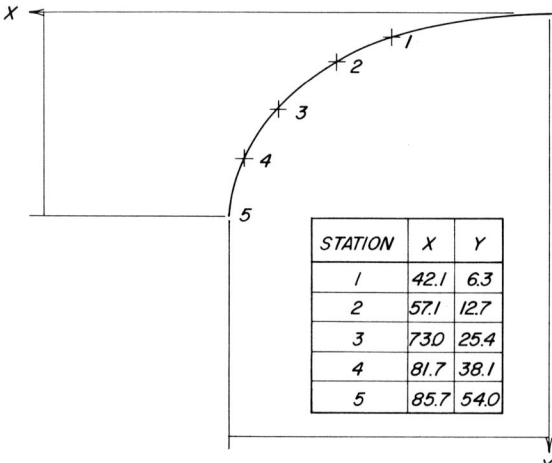

STATION	X	Y
1	42.1	6.3
2	57.1	12.7
3	73.0	25.4
4	81.7	38.1
5	85.7	54.0

FIG. 11.48. Location dimensions by coordinates.

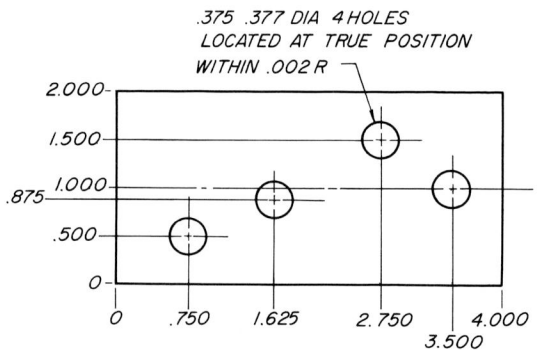

FIG. 11.49. Coordinate dimensioning without dimension lines.

e. *Slotted holes.* Slots of regular shape are dimensioned for size, by length and width, and shape of end by R for radius if this applies, as in Fig. 11.50.

For location, when there are several identical slots they are dimensioned along their longitudinal center lines to their near ends as shown in Fig. 11.50c.

f. *Parts with curved ends.* Parts of this type may be dimensioned as in Fig. 11.51. Three situations are shown.

1. When the holes near the ends have the same centers as the semicircular ends, the dimensioning should be as shown in Fig. 11.51a.
2. When the centers of the semicircular ends are not the same as the centers of the holes, the dimensioning of Fig. 11.51b should be used.
3. When the ends are short circular arcs not concentric with the end holes, the dimensioning should be as shown in Fig. 11.51c. In all cases duplicate dimensioning should be avoided.

g. *Toleranced dimensions.* When specific manufacturing tolerances are given on the dimensions, the method shown in Fig. 11.52 is used. This is discussed in Chapter 12.

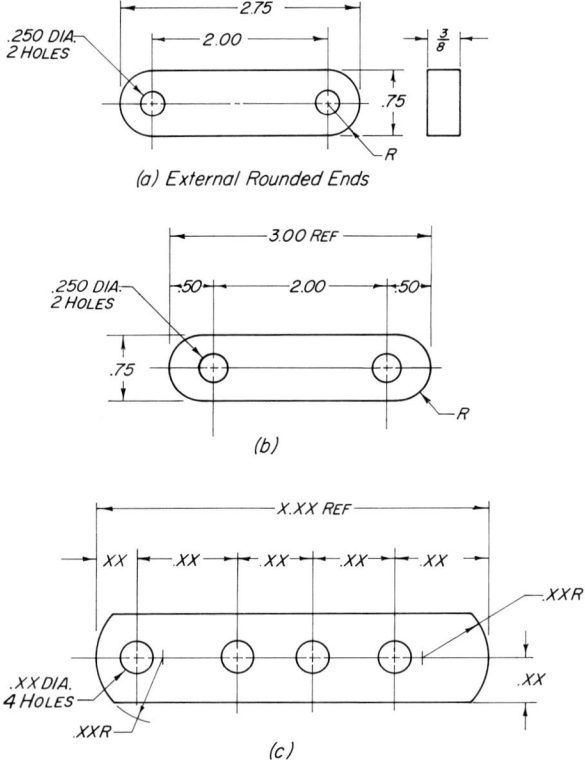

FIG. 11.51. Dimensioning parts with rounded ends.

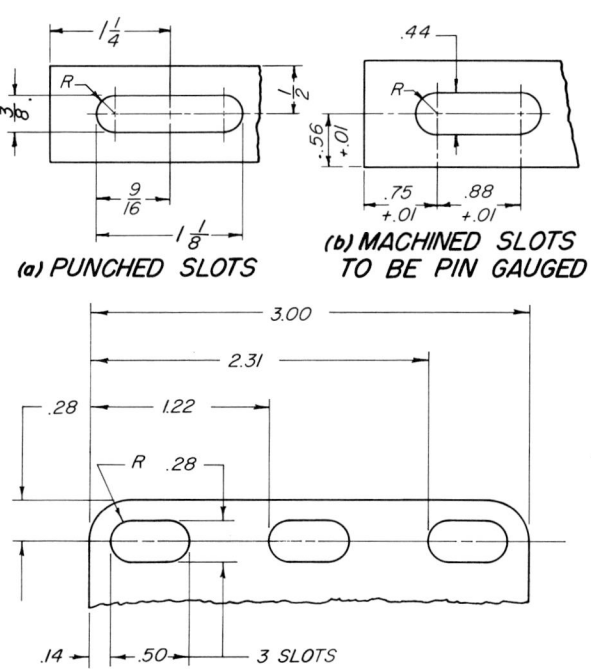

FIG. 11.50. Dimensioning slots with rounded ends.

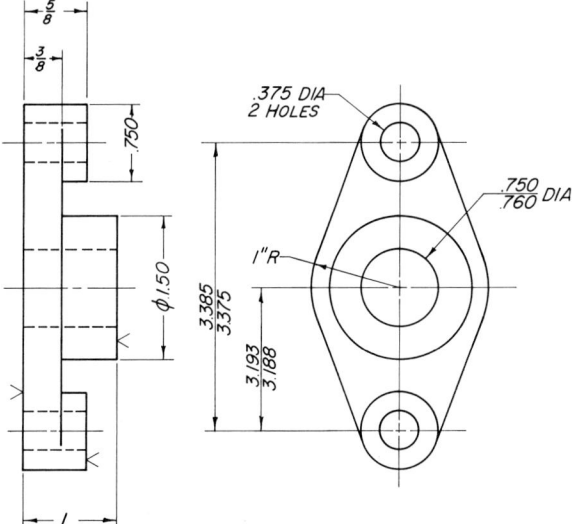

FIG. 11.52. Location of holes with limit dimensions.

11.8 DIMENSIONING FOR NUMERICALLY CONTROLLED MACHINES

Most of the basic dimensioning practices apply when the information is to be put into tapes used as manufacturing machine controls. However, to make the best use of this type of manufacturing, certain rules should be observed.

a. It is well to consult with manufacturing representatives before release of the final drawings.
b. Maximum use of standard tools, such as drills, reamers, and taps, should be part of the planning.
c. It is desirable to have all dimensions in decimals.
d. Angles may be specified but it is usually desirable to dimension by coordinates.
e. Tolerances should be based on design requirements and should not be made smaller than required just because the machine can handle them.
f. Coordinate dimensioning should be used from three main axes.
g. The reference planes may be common with the datum plane or they may be located outside of the object. They must be clearly marked on the drawing.
h. If possible, it is well to locate the reference planes so that all measurements will be positive.
i. Positional tolerancing may also apply to parts manufactured by tape-controlled machines.

11.9 DIMENSIONING STANDARD FEATURES

Many features of mechanical parts are produced by standard shop operations. The methods of specifying these procedures are given in the following paragraphs.

11.9.1 Blind Holes. When a hole does not go all the way through a part, it is referred to as a *blind hole*. It is specified by giving the diameter and the depth. The depth is given to the shoulder and not to the drill point. See Fig. 11.53.

11.9.2 Tapped or Threaded Holes. The method of specifying threads is given in Article 9.6.3. The accuracy and purpose of the work as well as the method of manufacture will determine the amount of information to be given about the thread. In case of doubt complete data should be given. Two methods are illustrated in Fig. 11.54.

For blind holes, the depth of the hole must be greater than the depth of the thread to allow for economical manufacture. See Fig. 11.54.

11.9.3 Countersinking. A conical hole cut so as to allow a beveled screw head to fit flush with the surface is called a *countersink*. Two methods of showing a countersink are given in Fig. 11.55. Although the actual angle is 82°, it is always drawn at 90° for simplicity.

11.9.4 Counterbore. A cylindrical hole made to allow a fillister-head screw to fit flush with the surface of the part is called a *counterbore*. Two methods of specifying a counterbore are shown in Figs. 11.56a and b. Clearance for screws or bolts must be allowed in both the drill and the counterbore. The amount of clearance varies with the size.

11.9.5 Spotface. To provide a smooth bearing for a bolt head or nut, an operation called *spotfacing* is done. The diameter of the spotface is given as shown in Fig. 11.57. The depth is seldom given since it is intended only to smooth a rough surface.

11.9.6 Chamfer. The ends of rods and bolts are usually beveled as shown in Fig. 11.58. When the bevel is at 45°, it may be shown as in Fig.

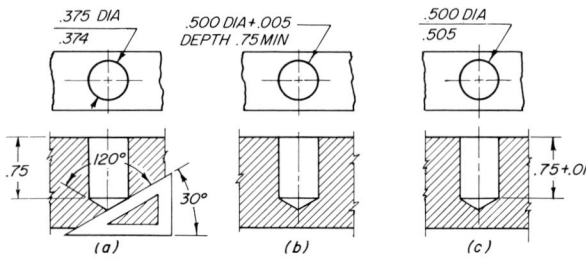

FIG. 11.53. Dimensioning blind holes.

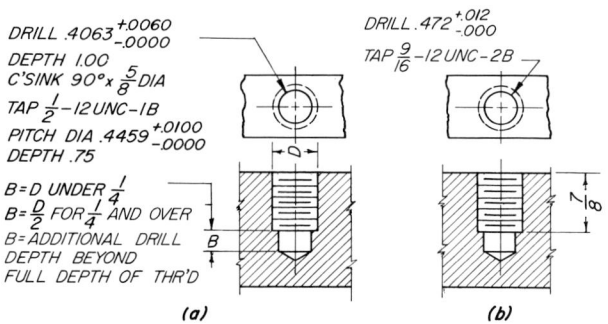

FIG. 11.54. Dimensioning threaded holes.

11.58a. *When the bevel is at any other angle, it must be shown as in Fig. 11.58b. Chamfering may be applied to a hole, as shown in Fig. 11.58c.*

11.9.7 Threading against a Shoulder. In this case it is necessary to cut a groove to provide clearance for the thread-cutting tool. This is known as *thread relief.* The method of dimensioning is shown in Fig. 11.59.

11.9.8 Keyways. The proper method for dimensioning keyways is shown in Fig. 11.60.

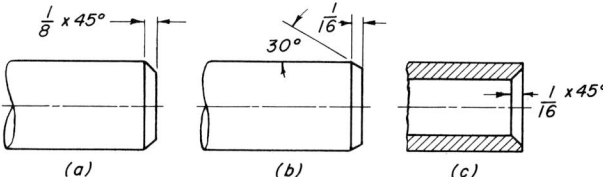

FIG. 11.58. Dimensioning a chamfer.

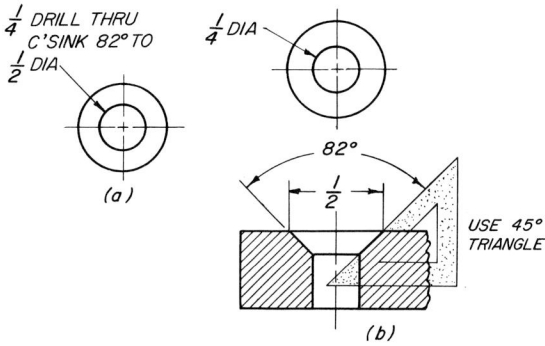

FIG. 11.55. Dimensioning a countersink.

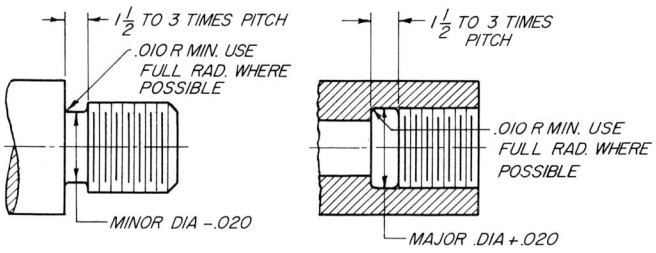

FIG. 11.59. Threading against a shoulder.

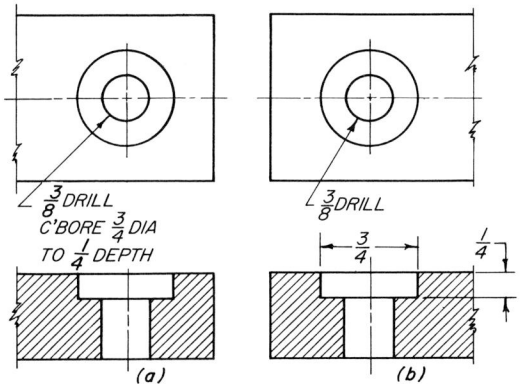

FIG. 11.56. Dimensioning a counterbore.

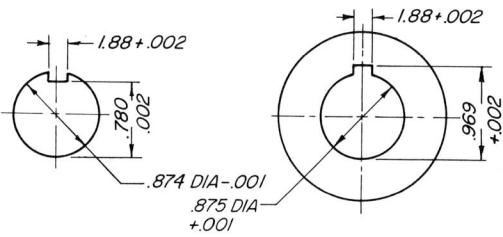

FIG. 11.60. Dimensioning a keyway.

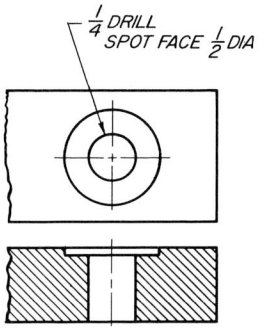

FIG. 11.57. Dimensioning a spotface.

11.9.9 Dovetail Tongue and Slot. The method of dimensioning a dovetail tongue and slot is shown in Fig. 11.61. The 60° angle marked "basic" means that the angle must fall within the tolerance zones established by the other dimensions with their tolerances. The datum surface must be carefully selected for proper functioning. In this case it was assumed that the bearing was to be maintained on the shoulder.

11.9.10 Knurls. Knurling is used to give a better grip on a part or for joining two parts together with a press fit. It is done by producing a series of ridges on the part. It may be dimensioned as shown in Fig. 11.62.

11.9.11 Parts Dimensioned with True Radius. Rods with spherical ends are dimensioned as shown in Fig. 11.63a, even though only one view of the spherical end is shown.

A bent bar that has a circular curve at the end that does not show in true shape in the end or side view may be dimensioned as shown in Fig. 11.63b. This eliminates an auxiliary view, which would be required to show the true shape of the curve.

11.10 NOTES

There are two kinds of notes that may be placed on a drawing: special notes and general notes.

11.10.1 Special Notes. Special notes refer to a specific part of the drawing. If the same note applies to several places, individual leaders may be carried to the same note. It is usually better to repeat the note if the leaders become long.

11.10.2 General Notes. General notes are placed in or near the title block. They are used to cover many items. Some of those most frequently used are as follows.

a. All dimensions in inches.
b. All dimensions in millimeters.
c. Finish all over or FAO.
d. Break sharp edges 0.01 or 0.03 R unless otherwise specified.
e. All small fillets $\frac{1}{8}$-in. radius.
f. All dimensions to be met after plating.
g. Remove burrs.
h. All draft angles 7° unless otherwise specified.

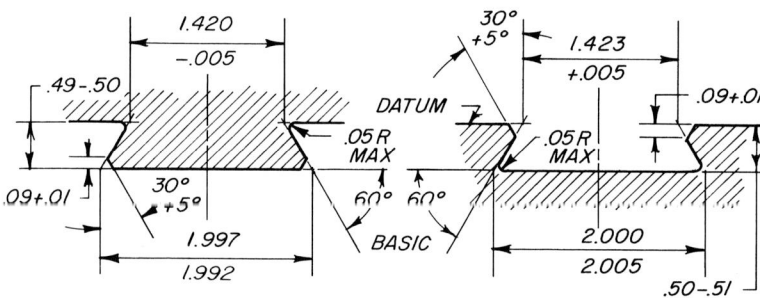

FIG. 11.61. Dimensioning a dovetail tongue and slot.

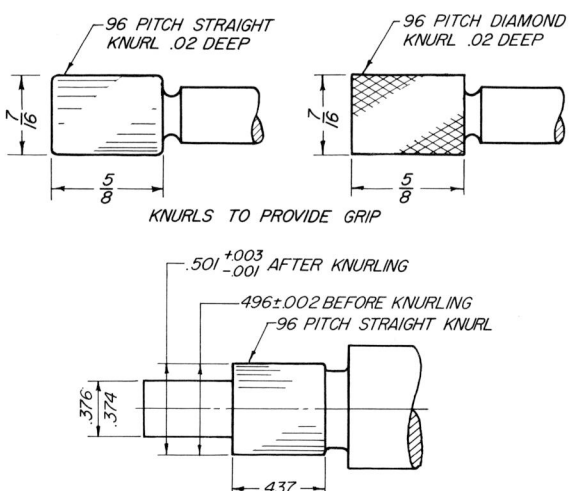

FIG. 11.62. Dimensioning a knurled part.

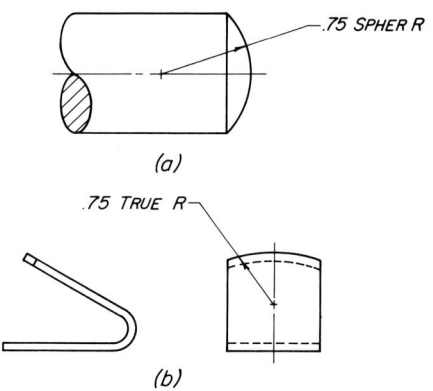

FIG. 11.63. Dimensioning special parts.

i. Paint one coat red lead.
j. Rivets ¾ DIA unless otherwise noted.

11.11 LIMIT DIMENSIONS
Dimensioning for interchangeable assembly is covered in Chapter 12. This includes such things as limits and fits, positional tolerancing, and geometric form or form tolerancing.

11.12 USE OF STANDARD PARTS
The drafter and engineer must always make use of standard parts, tools, and gages when designing a product. Such standard parts include.

a. Standard sizes of material such as bar stock, sheet metal, and wire.
b. Standard bolts, nuts, keys, washers, springs, ball bearings, and the like.
c. Part sizes that can be produced with standard tools and equipment.

11.13 FINISHED SURFACES
Certain surfaces of a part are made smooth or finished. This may be necessary for the proper functioning of the part, for appearance, or for ease of handling. The surfaces where mating parts come together must always be finished. In the machine shop this is done by cutting away a small amount of metal. It is very important that these finished surfaces be marked on the drawing so that the patternmaker will allow the extra material to be cut away and so that the machinist will know where to cut. The surfaces to be finished are marked with a 60°V, with the point resting on the edgewise view of the specified surface. See Fig. 11.64. For complete information about finish marks, see Article 12.39 in Chapter 12.

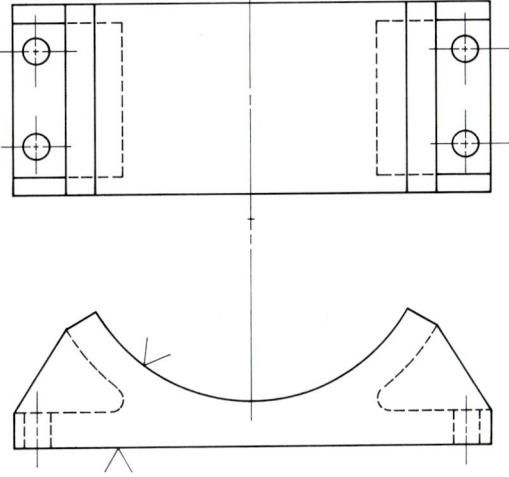

FIG. 11.64. Use of finish marks.

SELF-STUDY QUESTIONS

In the questions that follow fill in the blank spaces or underline the correct word where a choice is given. If you cannot answer, look up the paragraph number immediately following the question number.

11.1 (**11.2**) Dimensions should be measured from some _____ _____.

11.2 (**11.2**) Tolerances should not be allowed to _____.

11.3 (**11.2**) Every feature must be dimensioned for _____ and _____.

11.4 (**11.4.1**) The weight of a dimension line should be _____ than visible outlines.

11.5 (**11.4.1**) Dimension lines should not coincide or be an extension of a _____ _____ or a line on the object.

11.6 (**11.4.1**) An arrow should be placed on _____ end of a _____ line.

11.7 (**11.4.1**) The numeral should be placed in a _____ in the dimension line.

11.8 (**11.4.2**) A line that extends from the object to show the _____ of a dimension is called an _____.

11.9 (**11.4.3**) A line connecting a view to a note is called a _____.

11.10 (**11.4.3**) A line connecting a note to an area of a drawing should have a _____ on the end.

11.11 (**11.4.3**) A line connecting a note to a circle should end at the circle and point toward _____ _____.

302 BASIC DIMENSIONING

11.12 (**11.4.4**) An arrowhead should be _____ times as long as it is wide.

11.13 (**11.4.4**) On most drawings, the arrows should be about _____ _____ of an inch long.

11.14 (**11.4.4**) The point of the arrow ends exactly at the _____ _____.

11.15 (**11.4.5**) Notes should always read _____.

11.16 (**11.4.5**) No letter on a drawing should be smaller than _____ _____ inch.

11.17 (**11.4.5**) For all lettering, _____ lines should be used.

11.18 (**11.4.6**) Feet and inch marks should always be used on a drawing.   True   False

11.19 (**11.4.8**) The number 4.3175 should be rounded off to two places as _____.

11.20 (**11.4.8**) The number 2.9125 should be rounded off to two places as _____.

11.21 (**11.4.8**) The number 3.265 should be rounded off to two places as _____.

11.22 (**11.5**) The first dimension line should be _____ _____ of an inch from the object line of a view.

11.23 (**11.5**) Dimensions should be placed outside and _____ views when possible.

11.24 (**11.5**) Unrelated dimensions should be placed on _____ _____ of dimensioning.

11.25 (**11.5**) Dimensions should be measured from _____ _____ when possible.

11.26 (**11.5**) Dimensions should be placed on the view that shows the _____ of the object.

11.27 (**11.6.2**) When the numerals all face the bottom of the drawing in dimensioning, the system is called the _____ system.

11.28 (**11.7.4**) A pattern of holes should be dimensioned in the view that shows the holes as _____.

11.29 (**11.7.3**) In dimensioning a taper only _____ of the possible _____ dimensions should be given.

11.30 (**11.7.4**) When a graph of dimensions is given in a table, this is called _____ dimensioning.

11.31 (**11.8**) For numerically controlled machines:
   1. It is best to use _____ dimensions.
   2. _____ dimensioning should be used.
   3. _____ tools should be specified.
   4. Reference planes should be located so that all dimensions will be _____ if possible.

11.32 (**11.9.1**) The depth of a blind hole should be measured to the _____ rather than the _____.

11.33 (**11.9.2**) For blind threaded holes, the depth of the thread should be _____ than the depth of the hole.

11.34 (**11.9.3**) Cutting a conical hole to fit a beveled screwhead is called _____.

11.35 (**11.9.4**) A cylindrical hole to fit a fillister-head screw is called a _____.

11.36 (**11.9.5**) Smoothing a surface to allow for the bearing of a bolt head or nut is called _____.

11.37 (**11.9.10**) A rough surface provided to give a better grip is called a _____.

11.38 (**11.1**) The International System of Units is abbreviated _____.

CHAPTER 12
PRODUCTION DIMENSIONING

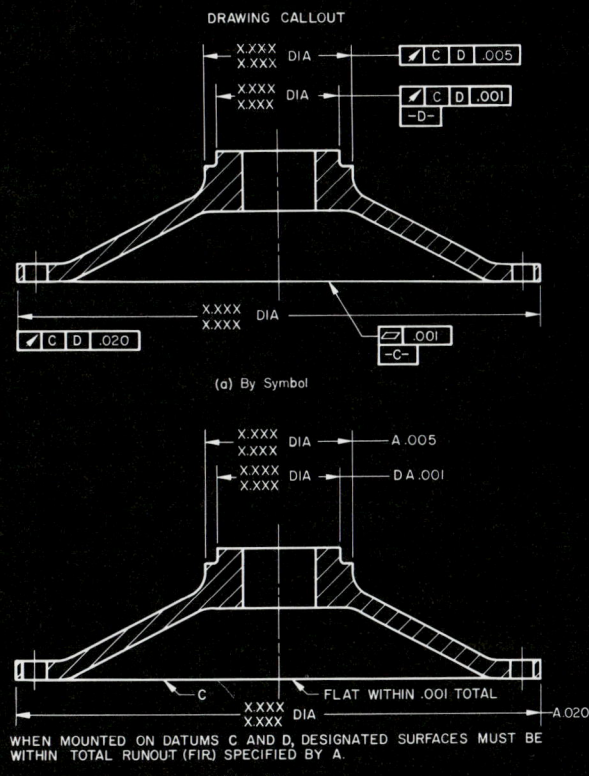

CHAPTER 12

12.1 INTRODUCTION

Production dimensioning is not really different from basic dimensioning, which is covered in Chapter 11, but rather an extension of basic dimensioning necessitated when parts are to be manufactured in large quantities. This requires first of all an understanding of what the finished product is to do and how its various parts fit and function together. In today's manufacturing operations, various parts may be produced at widely separated locations of a company or even by different companies. It is therefore absolutely necessary that the drawing be so clearly dimensioned that only one interpretation of it can be made. How this can be accomplished is the subject of this chapter. The entire process, however, begins with an assembly drawing or sketch.

12.2 ASSEMBLY DRAWINGS

Even the simplest machines have several parts. It is necessary, for purposes of both design and production, to know how the various parts fit together. Drawings made to show these relationships are called *assembly drawings*. As the name clearly implies, an assembly drawing shows the parts of a machine put together in their proper working position relative to each other. Drawings of this kind may serve a number of purposes as indicated in the following paragraphs. These purposes determine the character of the drawing and the dimensions placed upon it.

12.3 LAYOUT DRAWINGS

Most of the machines that an engineer is called upon to design have their beginning as an idea in

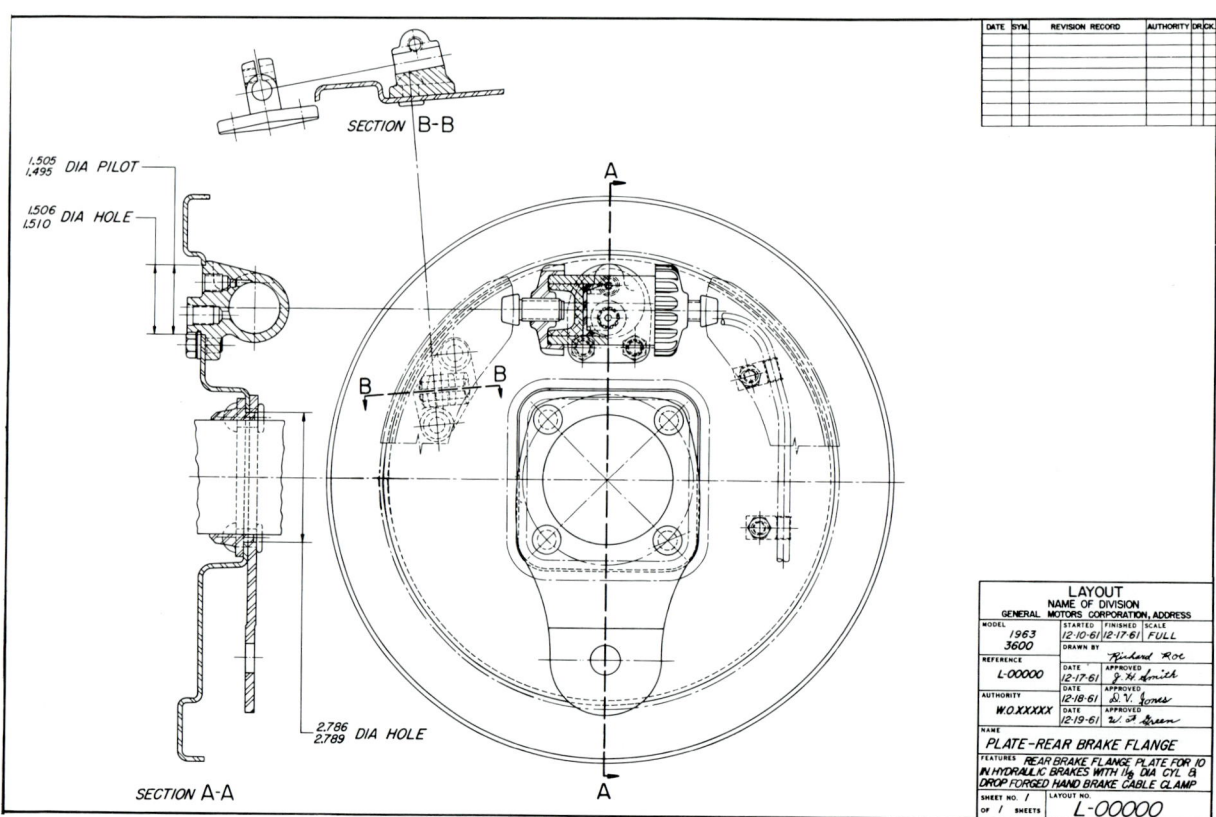

FIG. 12.1. Automotive part assembly layout.

someone's imagination. This idea is usually explained by a sketch. The drafter then makes a mechanical drawing showing the complete machine and how its parts are to function. See Fig. 12.1. In this first design assembly or layout, the size and shape of parts are determined by judgment based on past experience. Many times these drawings are schematic in character.

From these layout drawings three basic items must be considered in the design of the various parts: (1) the movement and velocity of the parts, which will ensure the proper performance of the machine; (2) design for strength and rigidity with minimum weight; and (3) dimensions and limits of size and location, which will permit economical production and satisfactory operation. The details of parts made from the layout of Fig. 12.1 are shown later in this chapter.

12.4 FINAL OR CHECK ASSEMBLIES

In some cases the original layout is sufficiently accurate so that no further work need be done. In other cases the actual design of parts will change their shape and dimensions from those shown in the original assembly. It then becomes necessary to redraw the assembly to make certain that the following things have been accomplished.

a. All the parts fit together as designed.
b. There is ample clearance for all moving parts.
c. Bolts and screws can actually be reached and tightened in the position indicated. An assembly of this type is shown in Fig. 12.2.

12.5 SHOP AND FIELD ASSEMBLIES

Assembly drawings may be used in the shop as a guide for the workmen who are putting the machine together. They are even more useful where a machine is shipped "knocked down" and assembled elsewhere. These drawings show each part numbered, and this number is placed on the part. The part can therefore be identified on the job, or the drawing of it can be located in the files. Such an assembly is shown in Fig. 12.2. In some cases exploded pictorial views are made,

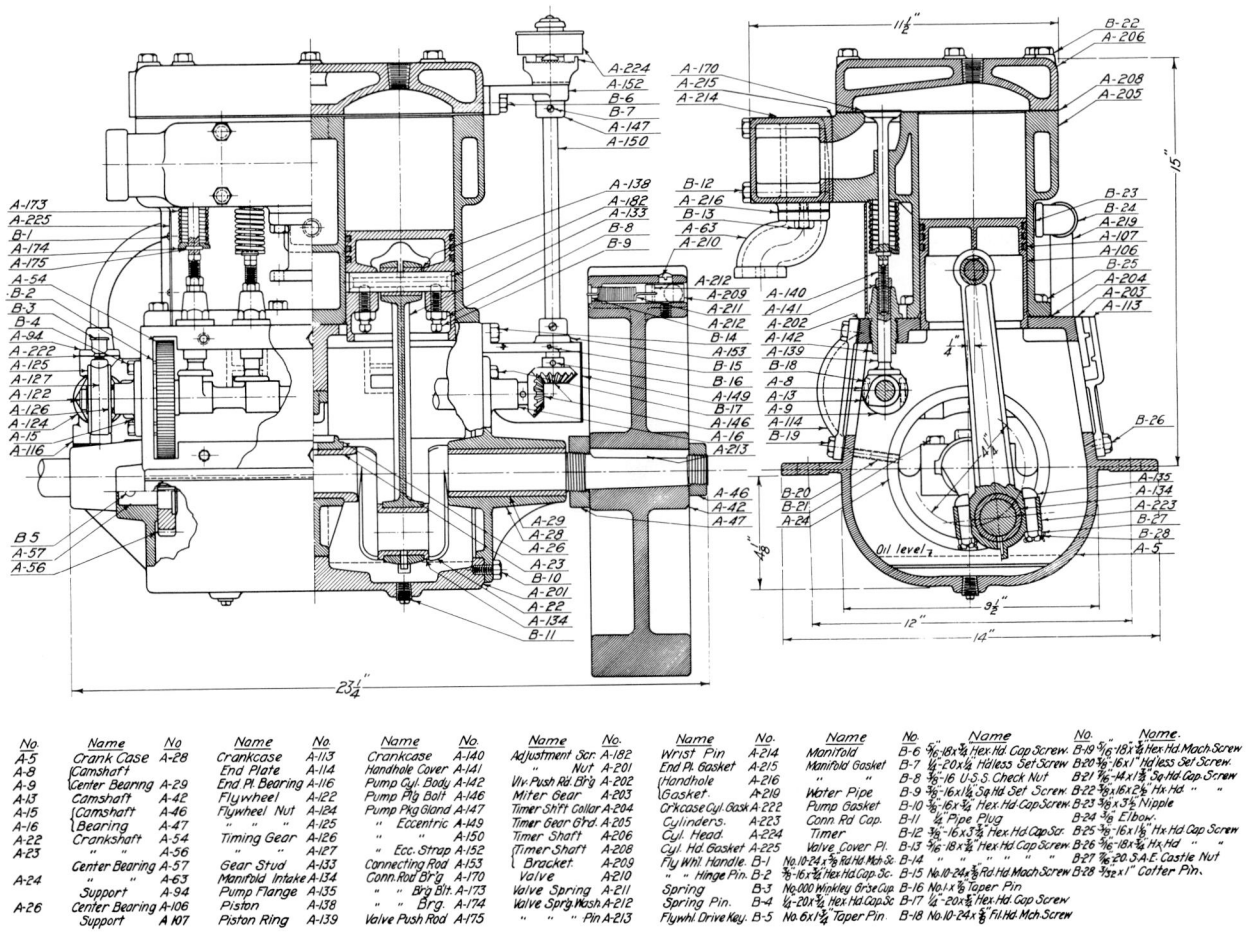

FIG. 12.2. Machine assembly drawings.

306 PRODUCTION DIMENSIONING

as in Fig. 12.3, which represents the same assembly as Fig. 12.7.

12.6 ERECTION DIAGRAMS
Large structures such as bridges and steel buildings also require a type of drawing that will enable the workers to assemble the parts properly. Such drawings are commonly referred to as *erection diagrams*. See Figs. 12.4.

12.7 INSTALLATION DRAWINGS
Frequently machine units are fitted into larger machines, and large machines require foundations or footings designed to support them. Assemblies are made showing principally the fastenings, anchor bolts, and clearances required and all necessary dimensions to show these things. Figures 12.5 and 12.6 are illustrations of this type of drawing.

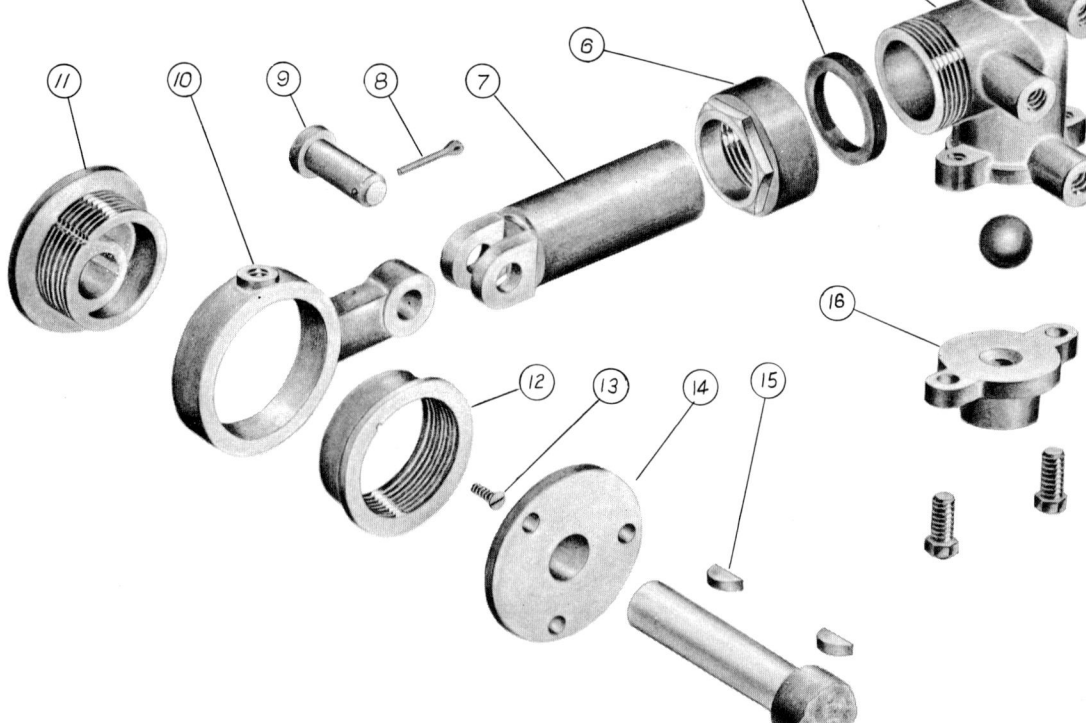

WATER PUMP

NO.	REQ.	DESCRIPTION
1	4	15/16" × 1" HEX-HEAD BOLTS
2	1	OUTLET FLANGE
3	2	STEEL BALLS
4	1	PUMP BODY
5	1	PACKING RING
6	1	PUMP PACKING GLAND
7	1	PUMP PLUNGER
8	1	COTTER PIN

NO.	REQ.	DESCRIPTION
9	1	PLUNGER BOLT
10	1	ECCENTRIC STRAP
11	1	ECCENTRIC (INSIDE)
12	1	ECCENTRIC (OUTSIDE)
13	1	FLAT HEAD SCREW
14	1	CAMSHAFT REAR BEARING
15	2	NO. 9 WOODRUFF KEYS
16	1	INTAKE FLANGE

FIG. 12.3. Pictorial exploded assembly drawing (courtesy E. R. Blackwell).

12.7 INSTALLATION DRAWINGS

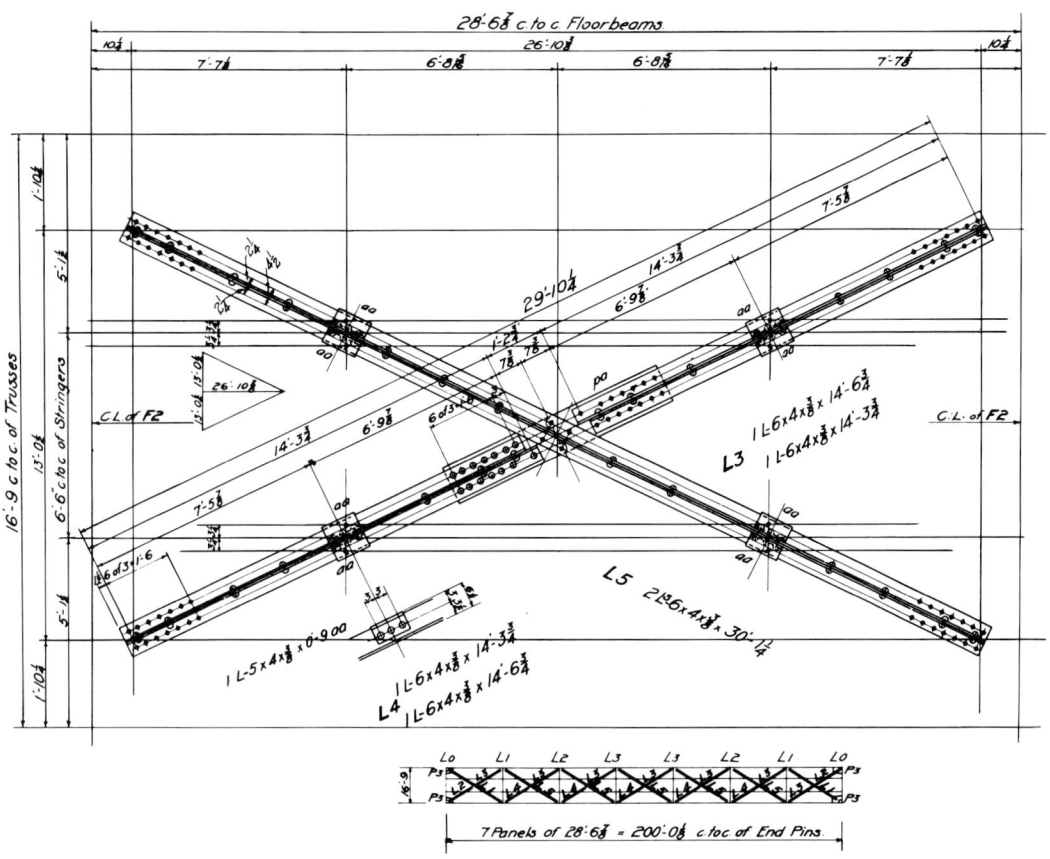

FIG. 12.4. Execution drawing of wind bracing.

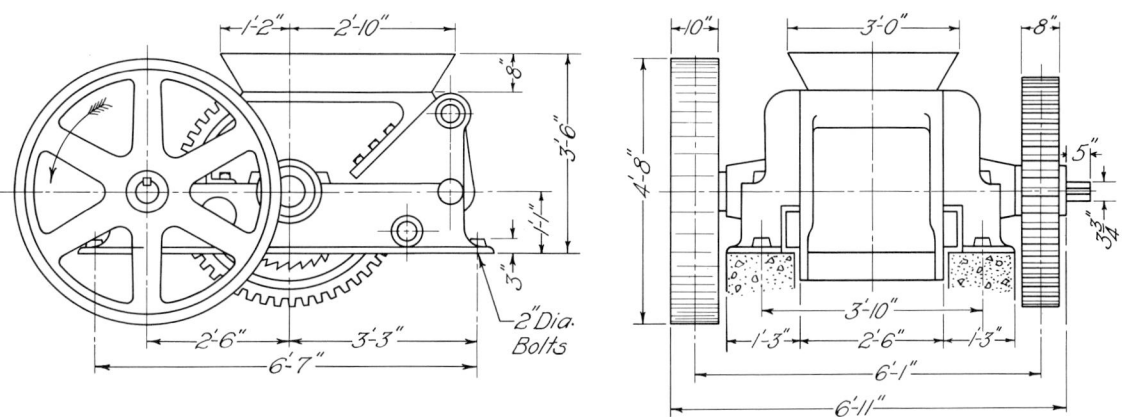

FIG. 12.5. Assembly drawing for installation.

12.8 DIMENSIONING ASSEMBLIES

The dimensioning of assembly drawings depends on the use for which they are intended. In general, it may be said that only the controlling dimensions, distance of travel of moving parts, and the like are shown on assemblies as illustrated in Figs. 12.5 and 12.7.

12.9 CROSS-HATCHING IN ASSEMBLIES

The system of usng only one type of cross-hatching is a common one on detail drawings. The use of different types of cross-hatching to represent different materials in the more common practice on assembly drawings. This calls attention in an unmistakable way to the difference in materials. Figure 12.7 illustrates the use of symbolic cross-hatching as approved by the ANSI and illustrated in Figs. 8.35 and 8.36 in Chapter 8.

12.10 HIDDEN LINES IN ASSEMBLIES

It is the usual practice to omit hidden lines entirely in assembly drawings unless it is necessary to show them for clarity. Then it is customary to show only the principal outlines of the object that may be needed.

12.11 REFERENCE OR PART NUMBERS

On assembly drawings for the shop, or for sales and service organizations, all parts are identified by numbers or letters. Each manufacturer has his/her own system of identification. In some shops the drawing number of the part is commonly used as a part number. See Figs. 12.2 and 12.7.

12.12 STANDARD DETAILS

Standard parts such as bolts, springs, and threaded parts are shown in conventional form. Hexagonal heads of bolts show three faces in both front and side views rather than as true projections. This identifies the bolt in either view. For economy in manufacturing, standard commercial parts should be used in design whenever possible.

Features produced by certain tools should also be specified and dimensioned in such size that standard drills, countersinks, counterbores, spot faces, milled slots, and the like can be produced by the tools carried in stock.

12.13 SUBASSEMBLIES

In larger and more complicated machines, it is impossible to show all parts in one assembly. The usual practice in such cases is to take groups of related parts that form a unit and make what is called a *subassembly*. See Fig. 12.7.

12.14 SELECTIVE ASSEMBLY

To make parts completely interchangeable there must be clearance between the parts at their maximum material condition (MMC). That is, when the allowance is positive, the largest shaft must be smaller than the smallest hole. When not at MMC, the hole is larger and the shaft smaller, so that there is more clearance, which is equal to the allowance plus both measured tolerances on the actual parts. For further discussion of the MMC principle, see Article 12.16.11. There are two methods of reducing this clearance to get a closer fit. One method is by reducing the tolerance. The other is by selecting assembling.

When selective assembly has been chosen, the parts are gaged and placed in various bins, depending on their actual size. In this manner the larger internal parts can be fitted to the external parts having the larger hole. The smaller internal parts can be fitted to the mating parts with the smaller holes. In this manner the excess clearance owing to parts not being at maximum mate-

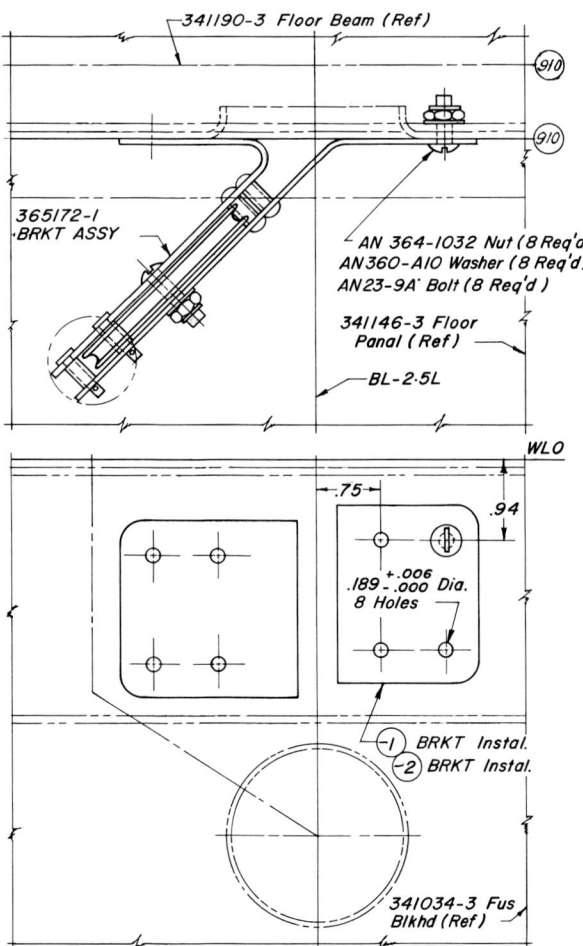

FIG. 12.6. Aircraft installation assembly.

rial condition can be made smaller because the actual measured allowance on the parts is reduced.

12.15 INTERCHANGEABLE ASSEMBLY
The assembly drawings as discussed in the preceding paragraphs show how the parts of an object fit together. Before any part can be manufactured, it must be detailed in a separate drawing. Usually each part is drawn on a sheet by itself, but occasionally the several parts of small assemblies may be detailed on one sheet.

Experience has show that parts cannot be manufactured economically to exact dimensions. Hence, when two or more parts must fit together, some latitude in the size of each part must be allowed the worker producing the parts. It is therefore necessary to dimension parts in such a manner that the worker knows exactly how much leeway is available in producing them. This leads to the use of some new terms in production dimensioning that must be thoroughly understood.

12.16 DEFINITION OF TERMS
In order to comprehend a discussion of methods of tolerancing dimensions, it is necessary to have a clear understanding of the following terms that are used in the discussion.

Feature. Features are specific parts of an object such as holes, bosses, lugs, and faces of a part.

Reference dimension. This is a dimension, without tolerance, used for information only. It is not used in manufacturing or inspection. On the drawing it is marked REF.

*Nominal size.** The nominal size is the designation used for the purpose of general identification. It is most often used to designate a commercial product and is not in any sense

---

* All definitions of terms, classes of fits, some sentences or phrases in this chapter, and drawings, when so indicated, have been taken from ANSI Y14.5-1973 with the permission of the publisher, The American Society of Mechanical Engineers.

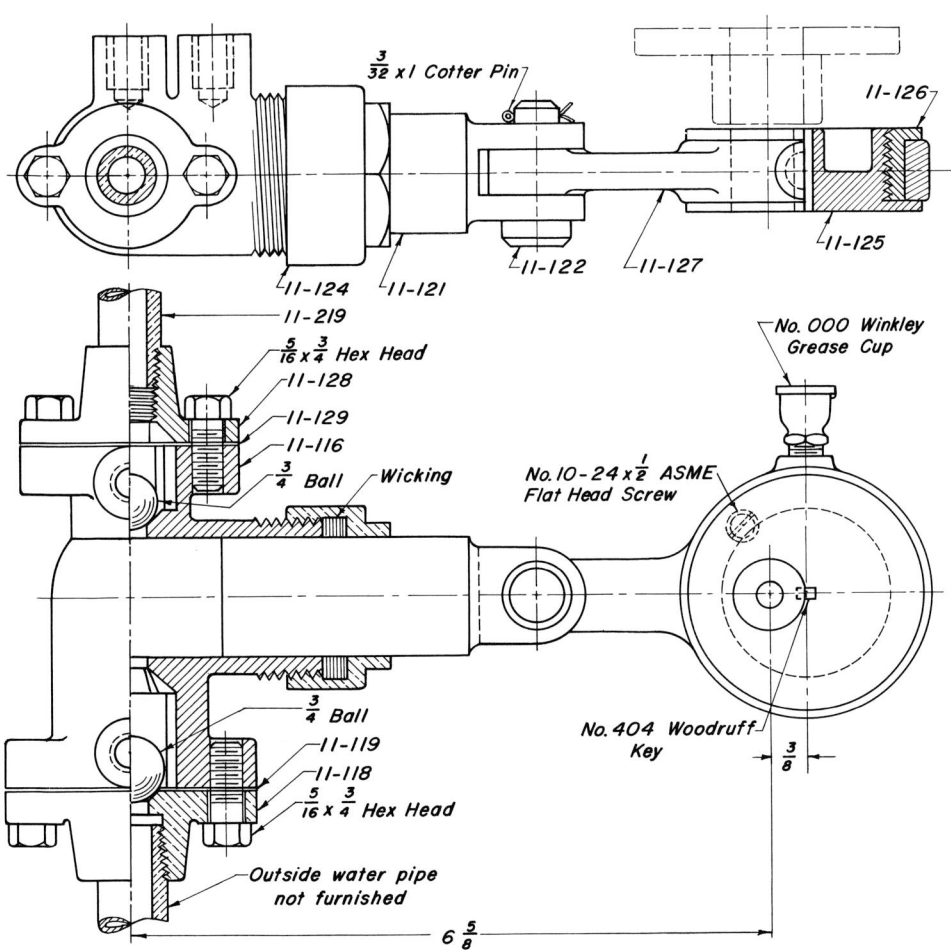

FIG. 12.7. Assembly of parts from Fig. 12.3.

the numerical size of the part. An example of a nominal size is ⅜-in. pipe, which actually has an outside diameter of 0.675 in. and an inside diameter of 0.493 in. Another is a piece of lumber called a 2 × 4, which is actually 1½ × 3½.

Basic size. This is the size of a part determined by design computations, from which the limits of size are determined by the application of allowances and tolerances. Thus the requirements of strength and stiffness may demand a 2-in. diameter shaft. This is the basic size. On the drawing, a dimension marked "Basic" is an untoleranced dimension giving the theoretically exact size or location. It may also be the basic size of the hole into which the shaft must fit, since the allowance is usually applied to the shaft. When it is necessary for special reasons, the allowance may be applied to the hole.

Design size. Design size is the size from which limits of size are derived by the application of tolerances. When there is no allowance, the design size is the same as the basic size. *Thus the application of an allowance to the basic size is considered a part of the design process.* In the previous illustration, if 2 in. is the basic size of the hole, then this is also the design size for the hole. If an allowance of 0.003 for clearance is applied to the shaft, then the design size of the shaft is 1.997 in. A tolerance is then applied to this dimension.

Actual size. The actual size is a measured size.

Fit. Fit is the general term used to specify the closeness (tightness or looseness) with which two mating parts are assembled. This results from a combination of allowances and tolerances. For types of fits see Section 12.18.

Limits. The term *limits may be defined as the extreme permissible dimensions of a part.* Two limit dimensions are always involved, a larger and a smaller or a maximum size and a minimum size.

Tolerance. *Tolerance is defined as the total amount of variation permitted in the size of a part.* It is the difference between the two limits of the same dimension. The variation permitted in the dimensions for locating holes or other parts is also called a tolerance. Again, it should be emphasized that *the tolerance on a dimension is the total variation permitted.* See Section 12.19 for method of tolerancing.

Allowance. Another important term used in production dimensioning is *allowance,* which is defined as the *intentional difference in the dimensions of mating parts* to provide the minimum clearance or the maximum interference that is intended between the parts. If allowance only exists between two parts, it represents the condition of the tightest permissible fit or the largest internal member mated with the smallest external member. This is sometimes referred to as the *maximum material condition* (MMC), since both parts contain the

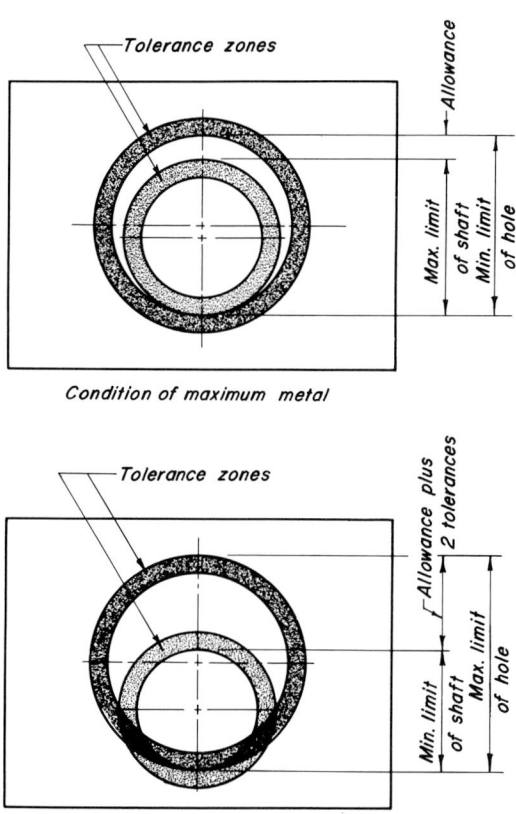

FIG. 12.8. Relation of tolerance and allowance to limits.

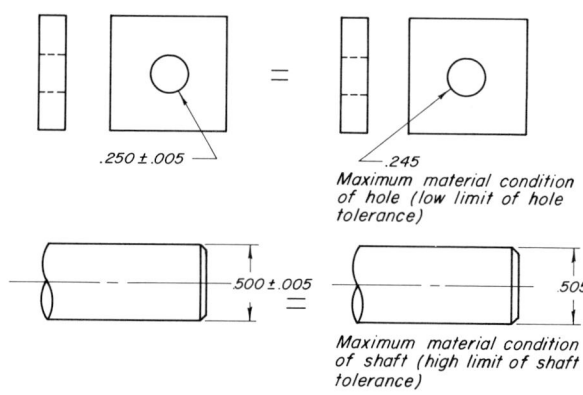

FIG. 12.9. Meaning of maximum material condition MMC.

maximum material. Allowance may be neutral or negative, thus providing interference fits for permanent assembly.

The student should note that the term "allowance" *refers to the difference in size between two different parts.* Thus the distinction between the three terms, *limits, tolerance,* and *allowance,* should be quite clear and unequivocal. They are illustrated in exaggerated form in Fig. 12.8. The purpose of an allowance is to provide for different classes of fits.

Maximum material condition. On the drawing this is designated MMC. This condition exists when external dimensions are at their maximum size and internal dimensions are at their minimum size, as in Fig. 12.9.

Full indicator movement (FIM). The total movement of the indicator when applied to a surface in an appropriate manner. The terms *full indicator reader* (FIR) and *total indicator reading* (TIR) were formerly used in previous editions of this standard and had identical meaning to FIM.

Datum. A datum is a surface or line from which dimensions are indicated on the drawing and measured in the shop. The surface may be a plane or cylinder. If it is a line, this is usually a center line. See Section 12.25 for further discussion.

12.17 SELECTING TOLERANCES

Great care and good judgment must be exercised in deciding upon tolerances that may be permitted on a part. The greater the demand for accuracy, the higher will be the cost of production. Since the specified tolerances will govern the method of manufacture, it is very important that tolerances he made as large as possible. When tolerances are reduced, the cost of manufacture rises very rapidly. The chart in Fig. 12.10 shows the accuracy that may be obtained economically in various common machine shop operations, assuming the machines to be in good condition.

In working out limit dimensions the drafter makes use of tables. Many companies have their own tables; others use the ANSI standards. In either case the method of computation is the same. Two systems are in use: the basic hole method and the basic shaft method. The choice of which to use depends upon the method of manufacture. In most cases the basic hole method is preferred, because standard tools can be used to produce the hole, whereas it is comparatively easy to turn or grind a shaft to any desired size. The methods of making computations for both systems are illustrated in Section 12.20.

12.18 CLASSES OF FITS: DESIGNATION OF STANDARD FITS

Standard fits are designated by means of the following symbols, which are used in the tables in the appendix. They are for educational purposes, to facilitate reference to various classes of fits. *They are not to be shown on manufacturing drawings.* Actual dimensions, as worked out from the tables, are used on drawings. The letter symbols have the following meanings.

RC   Running or sliding clearance fit
LC   Location clearance fit
LT   Transition clearance or interference fit
LN   Location interference fit
FN   Force or shrink fit

Range of Sizes From	To & Incl	Tolerances								
.000	.599	.00015	.0002	.0003	.0005	.0008	.0012	.002	.003	.005
.600	.999	.00015	.00025	.0004	.0006	.001	.0015	.0025	.004	.006
1.000	1.499	.0002	.0003	.0005	.0008	.0012	.002	.003	.005	.008
1.500	2.799	.00025	.0004	.0006	.001	.0015	.0025	.004	.006	.010
2.800	4.499	.0003	.0005	.0008	.0012	.002	.003	.005	.008	.012

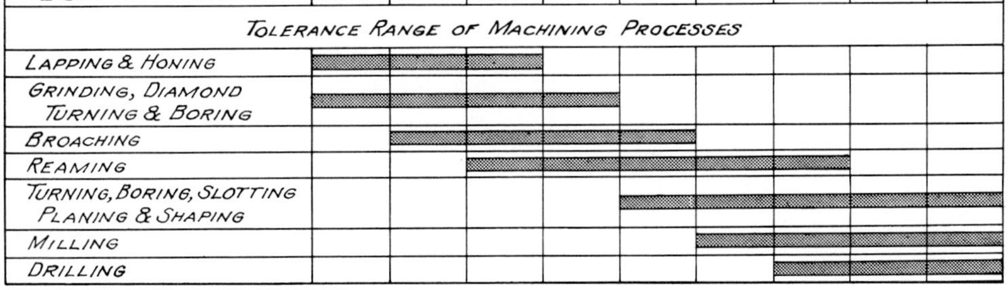

Tolerance Range of Machining Processes

Process										
Lapping & Honing										
Grinding, Diamond Turning & Boring										
Broaching										
Reaming										
Turning, Boring, Slotting Planing & Shaping										
Milling										
Drilling										

FIG. 12.10. Tolerance range for machining processes (courtesy Mil. Std. No. 8C-1962).

12.18.1 Running and Sliding Fits.
Running and sliding fits, for which limits of clearance are given in Table 22 in the appendix, are intended to provide an equivalent running performance, with suitable lubrication allowance, throughout the range of sizes.

RC1	*Close-sliding fits* are intended for the accurate location of parts that must assemble without perceptible play.
RC2	*Sliding fits* are intended for accurate location, but with greater maximum clearance than RC1. Parts made to this fit move and turn easily but are not intended to run freely and, in the larger sizes, may seize with small temperature changes.
RC3	*Precision-running fits* are the closest fits that can be expected to run freely and are intended for precision work at slow speeds and light journal pressures but are not suitable where appreciable temperature differences are likely to be encountered.
RC4	*Close-running fits* are intended chiefly for running fits on accurate machinery with moderate surface speeds and journal pressures, where accurate location and minimum play are desired.
RC5–RC6	*Medium-running fits* are intended for higher running speeds, heavy journal pressures, or both.
RC7	*Free-running fits* are intended for use where accuracy is not essential, where large temperature variations are likely to be encountered, or under both of these conditions.
RC8–RC9	*Loose-running fits* are intended for use where wide commercial tolerances may be necessary, together with an allowance, on the external member.

12.18.2 Locational Fits.
Locational fits are intended to determine only the location of the mating parts; they may provide rigid or accurate location, as with interference fits, or some freedom of location, as with clearance fits. Accordingly they are divided into three groups: clearance fits, transition fits, and interference fits. These are more fully described as follows.

LC	*Locational clearance fits* are intended for parts that are normally stationary but which can be freely assembled or disassembled. They run from snug fits for parts requiring accuracy of location, through the medium clearance fits for parts such as spigots, to the looser fastener fits where freedom of assembly is of prime importance.
LT	*Transition fits* are a compromise between clearance and interference fits, for application where accuracy of location is important, but a small amount of either clearance or interference is permissible.
LN	*Locational interference fits* are used where accuracy of location is of prime importance and for parts requiring rigidity and alignment with no special requirements for bore pressure. Such fits are not intended for parts designed to transmit frictional loads from one part to another by virtue of the tightness of fit. These conditions are covered by force fits.

12.18.3 Force Fits.
Force or shrink fits constitute a special type of interference fit, normally characterized by maintenance of constant bore pressures throughout the range of sizes. The interference therefore varies almost directly with diameter, and the difference between its minimum and maximum value is small in order to maintain the resulting pressures within reasonable limits. These fits may be described briefly as follows.

FN1	*Light-drive fits* are those requiring light assembly pressures, and they produce more or less permanent assemblies. The are suitable for thin sections or long fits or in cast-iron external members.
FN2	*Medium-drive fits* are suitable for ordinary steel parts or for shrink fits on light sections. They are about the tightest fits that can be used with high-grade, cast-iron external members.
FN3	*Heavy-drive fits* are suitable for heavier steel parts or for shrink fits in medium sections.
FN4–FN5	*Force fits* are suitable for parts that can be highly stressed or for shrink fits where the heavy pressing forces required are impractical.

12.19 TOLERANCING SYSTEMS FOR SIZE

There are three systems of expressing tolerances on drawings in addition to the specification of general tolerances by note. The note form usually applies to all dimensions not specifically toleranced and is placed in or near the title block. The three systems are known as the unilateral, bilateral, and limit systems.

12.19.1 Unilateral System. In the *unilateral system* the tolerance is shown in one direction only, either plus or minus as illustrated in Fig. 12.11a. It may also be expressed by giving the basic dimension and either a plus or a minus tolerance without indicating that the other is zero. This is shown in Fig. 12.11b.

12.19.2 Bilateral System. In the bilateral system the tolerance is divided into two parts, thus permitting a variation on either side of the basic dimension. The tolerance is usually divided equally, as shown in Fig. 12.12a, but it is not required that it should be. The tolerance of Fig. 12.12b is expressed in this manner. These deviations from the basic or design dimension are sometimes loosely called *bilateral tolerances*, but it should be noted that the tolerance is the total variation. Thus in Fig. 12.12a the tolerance is 0.002 and not 0.001. In no case may the two deviations be in the same direction, that is, both plus or both minus. The bilateral system is most useful for parts that are to be produced on machines with numerical control systems. Inspection or quality-control departments also find this method very convenient.

12.19.3 Limit System. In the limit system the extreme permissible dimensions are given on the drawing, as shown in Fig. 12.13. Note that in this case the tolerance is the difference between the limits. Two methods for placing limit dimensions on a drawing are approved by ANSI.

a. *Maximum material method.* In this method the number giving the maximum material size is placed above the line, that is, the largest dimension for a shaft and the smallest for a hole. This system lends itself well to situations where individual parts, perhaps only one, are produced and measured by the machinist. This method is shown in Fig. 12.14a.

b. *Maximum number method.* This second method is more commonly used on mass-production drawings, and in this scheme the largest number is always placed above

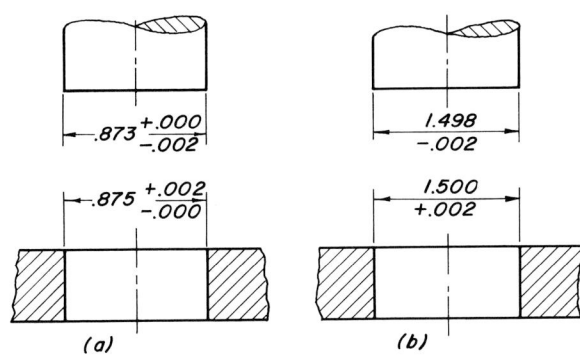

FIG. 12.11. Unilateral tolerance system.

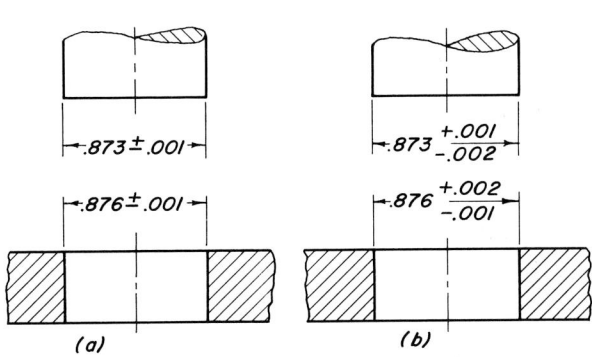

FIG. 12.12. Bilateral tolerance system.

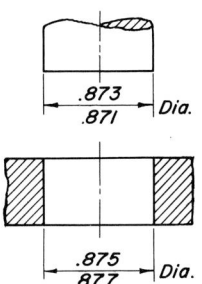

FIG. 12.13. Limit tolerancing system.

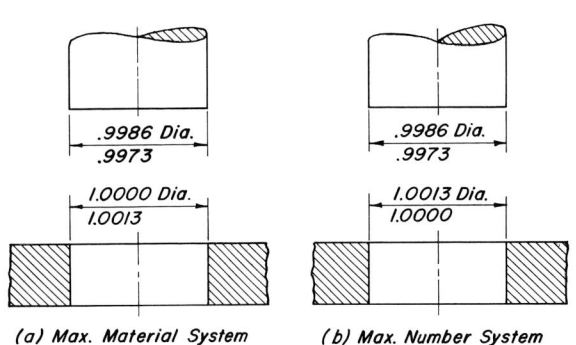

FIG. 12.14. Two methods of placing limit dimensions.

the line as shown in Fig. 12.14b. This is simpler for the drafter and is preferred by quality-control departments. Both methods should not be used on the same drawing.

12.19.4 Note Form. For dimensions that need not be held to close tolerances, the variation permitted is frequently specified by a general note in the form illustrated by the following examples.

Unless otherwise specified, tolerances are as follows.

Fractional dimensions $\pm \frac{1}{32}$.
Decimal dimensions ± 0.01.
Angular dimensions $\pm 0°30'$.
All diameters concentric within 0.001 FIM.

FIM means full indicator movement. The letters TIR and FIR were formerly used for the same purpose and mean total indicator reading and full indicator reading.

12.20 COMPUTING SIZE TOLERANCE DIMENSIONS FOR CYLINDRICAL MATING PARTS FROM STANDARD INCH TABLES

The computation of tolerances for mating parts may be determined in the following ways.

a. From ANSI B4.1-1967 (1974) Preferred Limits and Fits for Cylindrical Parts or from company standards.
b. By selecting standard parts such as bolts and then standard drills for holes of a somewhat larger size and computing tolerances from these dimensions.

See Tables 22 to 26 in the appendix. It should be noted that the top number in the limits of clearance column is the allowance (or tightest fit). This is the case for all clearance fits and transitional fits. Thus from Table 12.1 below 0.0040 (4.0) is the allowance (tightest fit) and 0.0088 (8.8) is the allowance and the shaft and hole tolerance. For interference and force fits the bottom number in the limits of interference column is the

TABLE 12.1[a]

Nominal Size Range (in.)	Limits of Clearance	Standard Limits	
		Hole	Shaft
1.97–3.15	4.0	+3.0	−4.0
	8.8	0.0	−5.8

[a] From Table 22 in the appendix.

TABLE 12.2[a]

Nominal Size Range (in.)	Limits of Interference	Standard Limits	
		Hole	Shaft
2.56–3.15	1.8	+1.2	+3.7
	3.7	−0	+3.0

[a] From Table 26 in the appendix.

negative allowance. Thus −0.0037 (3.7) is the allowance (tightest fit) (Table 12.2).

12.20.1 Basic Hole System. A basic hole system is used in calculations of limits in which the design size of the hole is the basic size and the allowance, if any, is applied to the shaft. In this method the computed size of the hole is considered as the basic size, and the size of the shaft is determined by subtracting the allowance from the hole size, thus giving the design size for the shaft. Tolerances are then applied to each part. For example, if the basic diameter of a hole is to be 3.0000 in. and a class RC7 fit is desired. Table 12.1 shows the following data from Table 22 in the appendix, ANSI B4.1-1967 (1974). Limits are in the thousandths of an inch.

For the hole the limits are

3.0000 + 0.0000 = 3.0000
3.0000 + 0.0030 = 3.0030

For the shaft the limits are

3.0000 − 0.0040 = 2.9960
3.0000 − 0.0058 = 2.9942

The tightest fit therefore is 3.0000 − 2.9960 = 0.0040, which is the allowance, and the loosest fit is 3.0030 − 2.9942 = 0.0088, which equals the allowance plus both tolerances. See second column in Table 12.1, also Fig. 12.15a.

As a second illustration, let us consider a force fit FN3 for a 3-in.-diameter hole. The data in Table 12.2 are from Table 26 in the appendix, ANSI B4.1-1967 (1974). Limits are in thousandths of an inch.

For the hole the limits are

3.0000 + 0.0012 = 3.0012
3.0000 + 0.0000 = 3.0000

For the shaft the limits are

3.0000 + 0.0037 = 3.0037
3.0000 + 0.0030 = 3.0030

The tightest fit occurs when the largest shaft, 3.0037, is placed in the smallest hole, 3.0000,

Table 12.3

Basic Hole System		Basic Shaft System	
Hole	Shaft	Hole	Shaft
3.0000	2.9960	3.0040	3.0000
3.0030	2.9942	3.0070	2.9982

TABLE 12.4

Basic Hole System		Basic Shaft System	
Hole	Shaft	Hole	Shaft
3.0000	3.0037	2.9963	3.0000
3.0012	3.0030	2.9975	2.9993

giving an interference of 0.0037, that is, an allowance of −0.0037. In a similar manner the loosest fit occurs with the smallest shaft and the largest hole, giving an interference of 0.0018, which equals the allowance plus both tolerances, that is, −0.0037 + 0.0012 + 0.0007 = −0.0018.

12.20.2 Basic Shaft System. A basic shaft system is used in calculations of limits in which the design size of the shaft is the basic size and the allowance, if any, is applied to the hole.

The tables in ANSI B4.1-1967 (1974) are designed specifically for tolerance by the basic hole system, which is usually preferred. The basic shaft system is used when purchasing finished parts, such as assembled ball bearings, which must be fitted into machined housings.

The simplest way for computing the limits for the basic shaft system in the case of clearance fits is to add the allowance to each of the limits obtained by the basic hole system. Using the illustration as given for the basic hole system, add

0.0040 to each dimension, thus finding the values shown in the column marked Basic Shaft System in Table 12.3. See Table 12.1.

The limits of clearance will be the same in both cases. See Fig. 12.15b.

For interference fits the allowance is added to the basic hole limits. Note the allowance is negative. Using the same illustration as before for an FN3 fit, add −0.0037 to each value listed under Basic Hole System in Table 12.4. See Table 12.2.

The loosest fit is −0.0018 and the tightest fit is −0.0037, which agrees with the data in the table under limits of interference.

12.21 COMPUTING SIZE TOLERANCE DIMENSIONS FOR CYLINDRICAL MATING PARTS FROM STANDARD METRIC TABLES

The computation of tolerances for mating parts may be determined in the following ways.

a. From ANSI B4.2-1978 "Preferred Metric Limits and Fits" or from company standards.
b. By selecting standard metric parts such as bolts and then standard drills for holes of a somewhat larger size and computing tolerances for those dimensions.

12.21.1 Metric Tables. The standard metric tables are set up much different from the inch tables and are much easier to use. There are two sets of tables, one for basic hole (preferred hole basis) and one for basic shaft (preferred shaft basis). Each table has 10 categories of fit from loose-running to force fit. The numbers given in the tables are the actual hole and shaft limits. The FIT column gives the maximum fit (allowance plus hole tolerance and shaft tolerance) and the minimum fit (allowance). Thus, one only needs to copy the numbers from the tables and does not need to add or subtract as with the inch tables.

12.21.2 Preferred Hole Method (Basic Hole). In this method the computed size of the hole is considered as the basic size. The basic diameter of the hole is 25 mm and a loose-run-

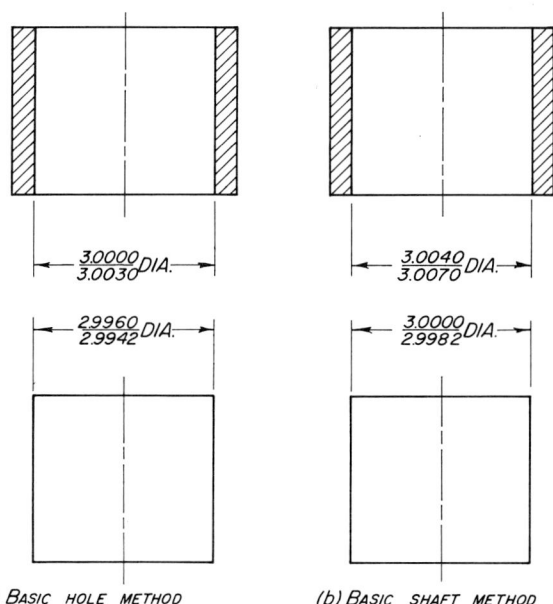

(a) BASIC HOLE METHOD   (b) BASIC SHAFT METHOD

FIG. 12.15. Basic hole and basic shaft methods of computing limits.

TABLE 12.5[a] Preferred Hole Method

Basic Size	Loose-Running		
	Hole	Shaft	Fit
25 Max	25.130	24.890	0.370
Min	25.000	24.760	0.110

[a] From Table 27 in the appendix.

TABLE 12.6[a] Preferred Hole Method

Basic Size	Medium Drive		
	Hole	Shaft	Fit
40 Max	40.025	40.059	−0.018
Min	40.000	40.043	−0.059

[a] From Table 28 in the appendix.

ning fit is assumed. The data in Table 12.5 are from Table 27 in the appendix, ANSI B4.2-1978 ("Preferred Metric Limits and Fits").

For the hole the limits are

25.130 and 25.000

For the shaft the limits are

24.890 and 24.760

The tightest fit is $25.000 - 24.890 = 0.110$, which is the allowance, and the loosest fit is $25.130 - 24.760 = 0.370$, which equals the allowance plus both tolerances. Note that these numbers are in the FIT column. See Fig. 12.16a.

As a second illustration let us consider a force fit (Medium Drive) with a 40-mm basic size. The data in Table 12.6 are from Table 28 in the appendix.

For the hole the limits are

40.025 and 40.000

For the shaft the limits are

40.059 and 40.043

The tightest fit therefore is $40.000 - 40.059 = -0.059$, which is the allowance, and the loosest fit is $40.025 - 40.043 = -0.018$, which equals the allowance plus both tolerances ($-0.059 + 0.025 + 0.016 = -0.018$). Note that the tightest fit -0.059 (allowance) and the loosest fit -0.018 are in the last column in the table. See Fig. 12.17a.

12.21.3 Preferred Shaft Method (Basic Shaft). In this method the shaft is usually purchased and the hole must be manufactured to fit the shaft. The basic diameter of the shaft is 25 mm and a loose-running fit is assumed.

The data in Table 12.7 are from Table 29 in the appendix, ANSI B4.2-1978 ("Preferred Metric Limits and Fits").

For the hole the limits are

25.240 and 25.110

For the shaft the limits are

25.000 and 24.870

The tightest fit is $25.110 - 25.000 = 0.110$, which is the allowance, and the loosest fit is

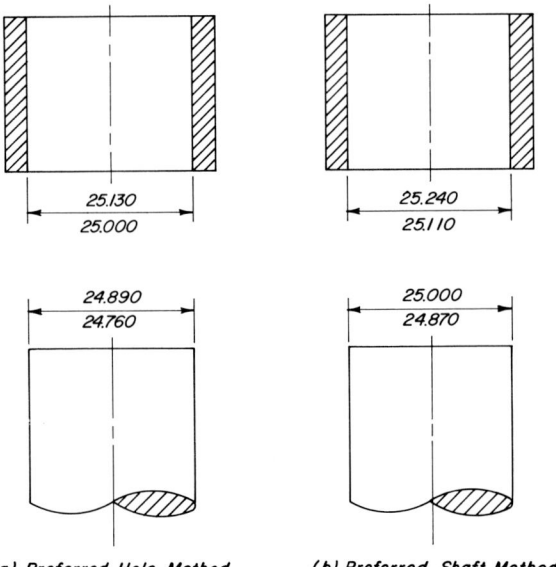

(a) Preferred Hole Method (b) Preferred Shaft Method

FIG. 12.16. Preferred hole and preferred shaft method of computing limits.

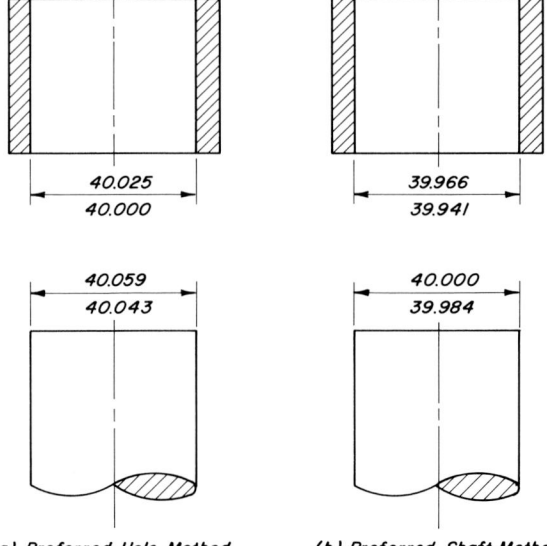

(a) Preferred Hole Method (b) Preferred Shaft Method

FIG. 12.17. Preferred hole and preferred shaft method of computing limits.

TABLE 12.7[a] Preferred Shaft Method

Basic Size	Loose-Running		
	Hole	Shaft	Fit
25 Max	25.240	25.000	0.370
Min	25.110	24.870	0.110

[a] From Table 29 in the appendix.

TABLE 12.8[a] Preferred Shaft Method

Basic Size	Medium-Drive		
	Hole	Shaft	Fit
40 Max	39.966	40.000	−0.018
Min	39.941	39.984	−0.059

[a] From Table 30 in the appendix.

$25.240 - 24.870 = 0.370$, which equals the allowance plus both tolerances. Note that these numbers are in the FIT column. See Fig. 12.16b.

As a second illustration let us consider a force fit (Medium Drive) with a 40-mm basic size. The data in Table 12.8 are from Table 30 in the appendix.

For the hole the limits are

39.966 and 39.941

For the shaft the limits are

40.000 and 39.984

The tightest fit is $39.941 - 40.000 = -0.059$, which is by definition the allowance. The loosest fit is $39.966 - 39.984 = -0.018$, which equals the allowance plus both tolerances ($-0.059 + 0.025 + 0.016 = -0.018$). See Fig. 12.17b.

12.21.4 Comparison of Calculations in Inch and Metric System. Calculations of limits in the metric system are much easier than calculations in the inch system. The reason for this is that the millimeter is much smaller than the inch, and the metric tables are set up using preferred sizes. In the inch tables, ranges are used and we have already seen the number taken from the tables must be added to or subtracted from the basic size. In the metric tables the actual limits are given. This does, however, necessitate two sets of metric tables, one for basic hole and one for basic shaft.

12.22 ANALYSIS OF DIMENSIONING

A technical drawing, as discussed previously, involves three elements.

  a. Shape description by means of two or more orthographic views, which completely define the geometric shape of an object and all of its related features or elements.
  b. Description of size of features and location of features relative to each other. This is done by means of dimensions, symbols, and notes, which clearly and unmistakably define the size and relationships of all features of an object.
  c. The specification of the kind and quality of materials to be used and of the finish, as to both surface quality and protective coatings such as plating or painting.

12.22.1 Methods of Inspection. In this chapter we are concerned primarily with dimensioning. If we consider only size dimensioning, this seems to be a simple matter, since the part is relatively easy to inspect. It should be noted, however, that the method of inspection may make a difference in the acceptance or rejection of parts.

In Fig. 12.18 a single hole is shown with limits (0.375 to 0.390) specified by note, but the part has been produced to the actual measurements shown. With a "Go/No-Go" gage, as illustrated in the sectioned front view, this part will be accepted, since the "Go" part enters the hole, but the "No-Go" part does not, even though it is oversize and out of round. But if it were inspected with a micrometer caliper, the reading would show 0.392, which indicates departure

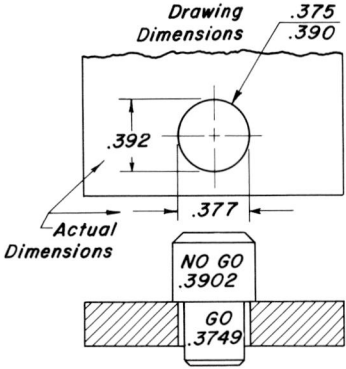

FIG. 12.18. Acceptance of part by plug gage not to specifications by reason of out of roundness.

from specified dimensions, and the part would be rejected.

To achieve the design intent, engineering, manufacturing, and inspection departments must be consulted if economical production is to result. Dimensioning with respect to size and location relationships must be so specified that manufacturing and inspection will use the same datum surfaces, points, or axes. It should be further emphasized that the method of verification must be agreed upon, whether it is to be "open set-up" inspections with standard gages, calipers, surface plates, and the like or with functional, fixed-pin acceptance gages. The method to be used will depend on the intent and function of the design and the number of parts to be made.

When location is involved, in addition to size, the problem becomes more complicated and the need for coordination of manufacturing and inspection becomes more important.

12.23 COORDINATE SYSTEMS OF DIMENSIONING

Two systems of coordinate dimensioning are in common use. Their application depends on the form and function of the part. These methods are discussed in the following paragraphs.

12.23.1 Rectangular Coordinates. Holes, for example, may be located by giving limit dimensions to their centers from each of two planes at right angles to each other as shown in Fig. 12.19a. This method results in a square tolerance zone for the location of centers, as shown in Fig. 12.19b. It will be noted in Fig. 12.20 that the tolerance along the diagonal of the square is 1.4 times the tolerance specified. This is not a desirable feature.

Methods of coordinate dimensioning are discussed in Article 11.7.4(d) and are illustrated in Figs. 11.48 and 11.49.

12.23.2 Polar Coordinates. Location tolerances may also be specified in polar-coordinate form by giving limits on the radial and angular dimensions as shown in Fig. 12.21a. This results in a sector tolerance zone, as shown in Fig. 12.21b. The tolerance along the diagonal is still nearly 1.4.

12.24 ACCUMULATION OF TOLERANCES IN THE COORDINATE SYSTEM

In coordinate dimensioning tolerances may accumulate and interfere with assembly or function of the parts or feature. This may occur in several ways.

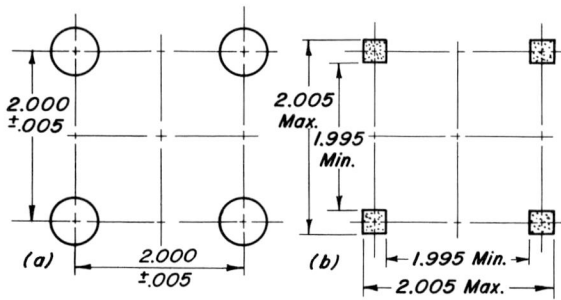

FIG. 12.19. Limited center distances give rectangular tolerance zones (courtesy ANSI).

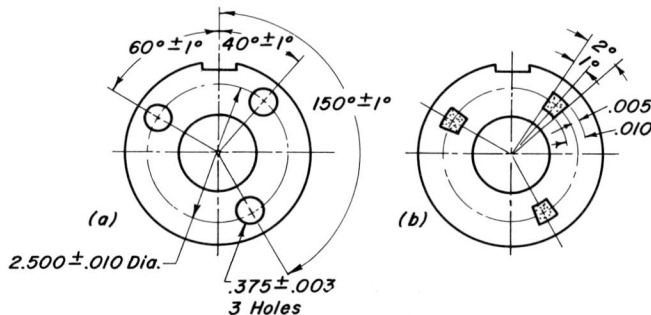

FIG. 12.21. Polar coordinates give sector tolerance zones (courtesy ANSI).

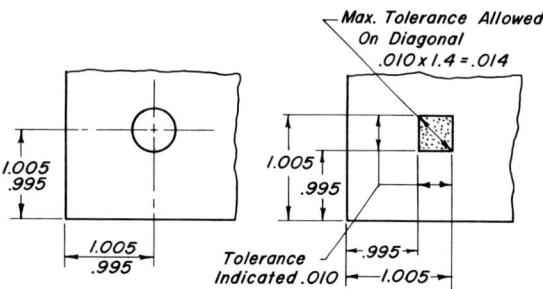

FIG. 12.20. Greater tolerance on diagonal then specified (courtesy ANSI).

12.24.1 Accumulation by Chain Dimensioning.
If chain dimensioning is used, an undesirable accumulation of tolerances results, as shown in Fig. 12.22. The tolerance between any two holes is not that specified on the drawing. Thus between A and B in Fig. 12.22 the variation in position could be from 1.95 to 2.05 in.

12.24.2 Accumulation by a Stack up of Toleranced Parts.
Tolerances may also accumulate simply as a result of the assembly of a group of toleranced pieces. Thus, in Fig. 12.23, several pieces, each with a tolerance of its own, are assembled with a bolt as shown. The toleranced dimensions are shown in Fig. 12.23a.

When all of the parts are at maximum size and the length of the bolt to the shoulder is at minimum size, as in Fig. 12.23b, the plates are firmly clamped together. When the parts are at minimum size and the length of the bolt to the shoulder is at maximum size, as shown in Fig. 12.23c, the plates are not clamped, and there will be some play between the parts. If it is intended that the parts be clamped together, the second situation will not be acceptable.

The stack up of parts in Fig. 12.23 could have been reduced by a choice of smaller tolerances. This method, of course, can be more expensive. In this case, however, a gasket having a thickness of 0.082 to 0.084 would have solved the problem. With the plates at minimum dimensions, the gasket between them also at minimum thickness, and the bolt at maximum length, the nut could be tightened enough to take up the 0.012-in. oversize and clamp the plates. With all parts at maximum size, there is an 0.024 take up for clamping and no problem.

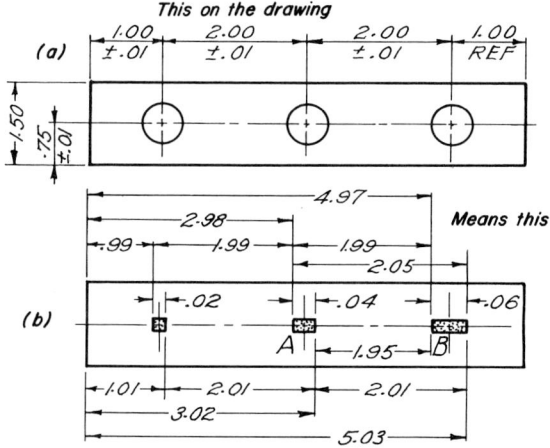

FIG. 12.22. Chain dimensioning produces cumulative tolerances.

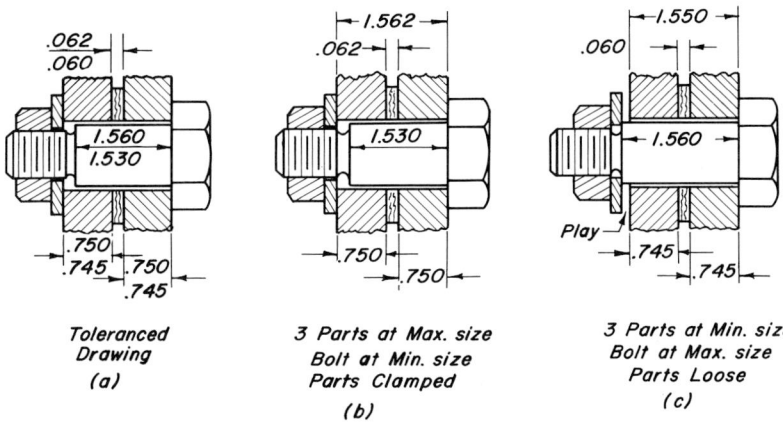

FIG. 12.23. Tolerance accumulation of three parts.

12.25 CONTROL OF CUMULATIVE TOLERANCES BY DATUM DIMENSONING

Some control of cumulative tolerances can be attained by referring all dimensions in a given direction to the same datum surface. Thus in Fig. 12.24 the left end of the part has been used as a datum and each hole has been located from this datum individually. Note that the distance between successive holes remains constant at 2.00 ± 0.02 in.

12.26 SELECTION OF DATUM SURFACES OR LINES

In dimensioning mating parts, the surfaces that will function or fit together must be used as datum surfaces for both parts. Thus in Fig. 12.25 if the dimension α in Fig. 12.25a is to be controlled and no bearing at surface B can be permitted, then surface C must be the datum for both parts as shown in Figs. 12.25b and c.

12.26.1 Identification of Datum by Dimensioning.

On some drawings the datum surface or surfaces can be readily implied by the position in which dimensions are placed. Thus in Fig. 12.24 the left end is clearly the datum surface for the holes.

In Fig. 12.26a, the left and bottom sides are indicated as the datum surfaces by the method of dimensioning that is used to locate the hole and the other two sides. While the roughness of the surface of the part is exaggerated in Fig. 12.26b, it does indicate that the real datum at inspection is on the gage and is identified on the part by the bearing points. This fact should not be overlooked, even though surfaces on the object must be used or marked as datums.

Features that are selected to serve as datum surfaces must be clearly identified and easily recognizable on the part. To be useful for measurement, a datum indicated on an actual piece must be accessible during manufacture so that measurement can be made readily. Sometimes a

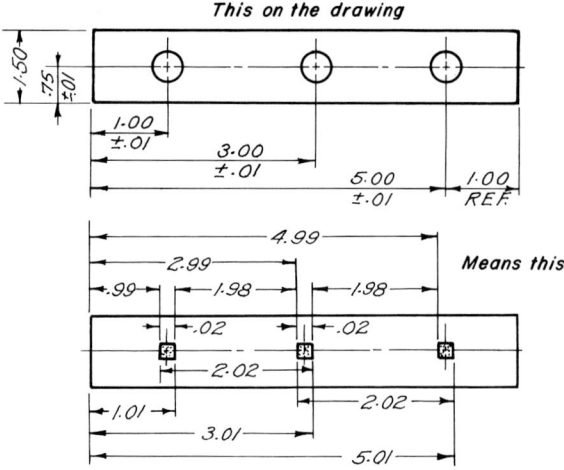

FIG. 12.24. Datum dimensioning reduces cumulative tolerances.

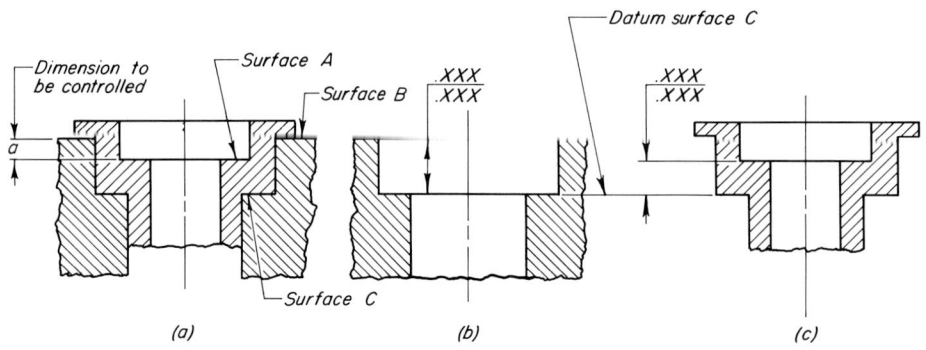

FIG. 12.25. Controlling part tolerances by selection of datum (courtesy SAE).

single pilot hole may not be adequate. Two aligned holes or a hole and a slot are preferable, since this provides both location and orientation. See Fig. 12.27.

12.26.2 Datum Indicated by Note. In more complicated parts the datum should be indicated by note, as in Fig. 12.25. The center line through the large pilot hole and the slot is thus specified for both location and orientation. Other methods of identifying datum surface and lines are discussed in Article 12.35.2.

12.26.3 Accuracy of Datum Surfaces. Even though the theoretical datum is on the gage, the datum surface indicated on an actual piece must nevertheless be more accurate than any location established from it. Thus if measurements are made from a datum surface to establish hole locations with a tolerance of 0.010, the total effect of surface inaccuracies on the measurements must be considerably less than 0.010, or the locations will not have the specified accuracy. It may be necessary to specify the accuracy of a datum surface by giving tolerances on features such as straightness, flatness, and roundness to assure that locations can be established with the specified accuracy.

12.26.4 Location of Surface Finished First on a Casting. Since most machined parts begin with a casting or a forging, it is necessary for the machinist to locate the first surface to be finished from some relatively rough surface on the original casting. This surface should therefore be located only once from one suitable, unfinished surface. The finished surface thus established may be the datum for the location of other finished features. See Fig. 12.28.

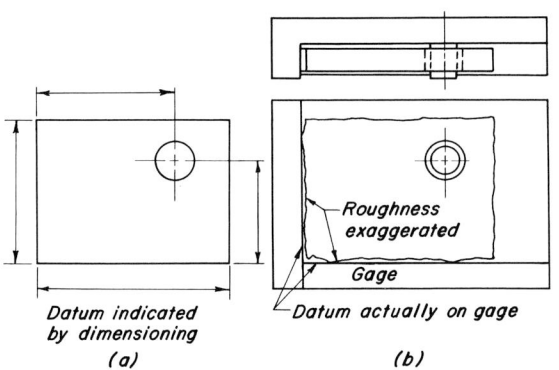

FIG. 12.26. Datum is actual on the gage.

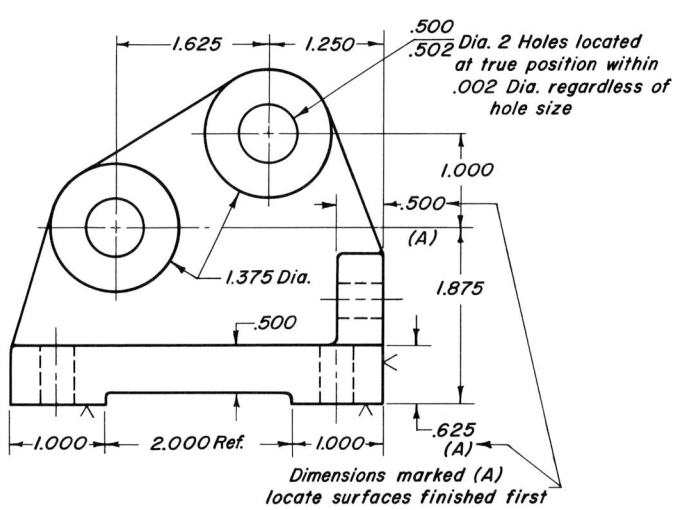

FIG. 12.28. Establishing first finished surface from one rough surface only.

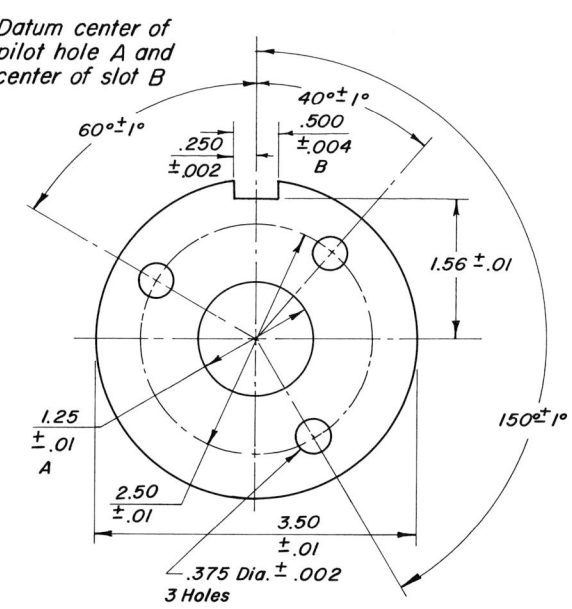

FIG. 12.27. Specification of datum by note.

12.27 GEOMETRIC FORM, TRUE-POSITION DIMENSIONING, AND TOLERANCING*

This form of dimensioning does not supersede or displace coordinate dimensioning discussed in preceding paragraphs. Rather, it supplements this system and produces greater efficiency in drafting and manufacturing since it frequently provides extra tolerances.

American, British, and Canadian industries, as well as the military services and governments, are essentially agreed upon this system. Subcontracting of work and the transfer of personnel between plants and industries make a uniform and clear drawing and dimensioning system essential.

12.28 WHEN TO USE GEOMETRIC FORM AND POSITIONAL TOLERANCING

In many cases where fit and function of parts are not critical the coordinate system of dimensioning and tolerancing is satisfactory and may be used.

Geometric form and true-position dimensioning and tolerancing are very useful and are recommended in the following situations.

a. Where the size, form, and location of part features are critical with respect to function or interchangeability.

b. When functional gaging or inspection techniques are desirable. See Fig. 12.29b.

c. When datum references are desirable to ensure compatability between manufacturing and inspection methods.

d. When interpretation of tolerances is not clearly evident or implied.

12.29 TRUE-POSITION DIMENSIONING

In this system the location of features is given by basic or untoleranced dimensions. These dimensions give the true position of the features, and the tolerance around these centers is given by a note along with the size of the feature, as shown in Fig. 12.29b. True-position dimensions, that is, those without tolerances, are enclosed in a rectangle in the symbolic system, as shown in Fig. 12.29b. When notes are used, the word *Basic* or the abbreviation *BSC* is placed under the dimension. This method results in a circular tolerance zone, as shown in Fig. 12.30b as distinguished from the rectangular zone shown in Fig. 12.30a. The true-position method has the following advantages over the coordinate method with limited center distances.

a. It corresponds to the distribution of errors that normally arise in production.

b. It corresponds to the control established by fixed-position gages.

c. It permits the use of chain dimensioning without the accumulation of tolerances.

d. It makes possible the specification of different tolerances for each of a number of features lying on a common center line.

e. It makes it simpler to determine the clearance between mating components because equal deviation is permitted in all directions.

---

*Some of the material in the following articles has been abstracted from *A Treatise on Geometric and Positional Dimensioning and Tolerancing*, by Lowell W. Foster, Senior Standardization Engineer, Honeywell, Inc., Minneapolis, Minn.

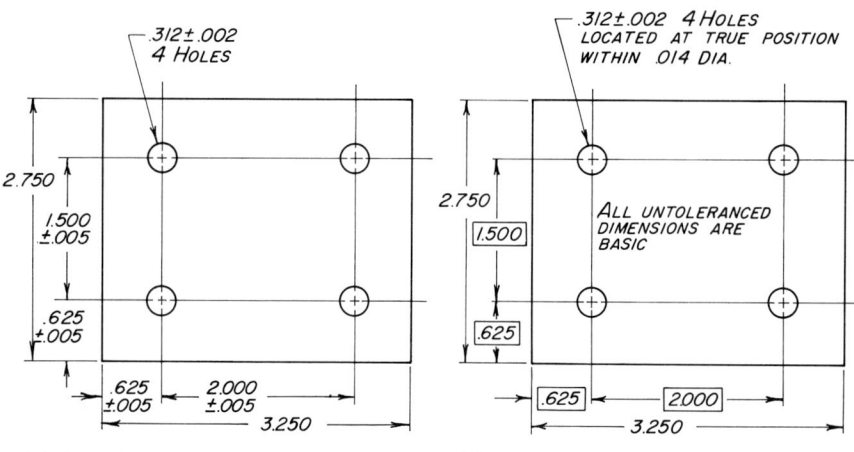

FIG. 12.29. Dimensioning methods compared.

12.29 TRUE-POSITION DIMENSIONING

12.29.1 Dimensioning Methods Compared. In Fig. 12.29 a simple plate with four holes has been dimensioned. In Fig. 12.29a the coordinate method has been used, whereas in Fig. 12.29b the true position has been used. Note the difference in the methods.

a. In the coordinate method the positional tolerance is given with the location dimension as 0.625 ± 0.005 and 2.000 ± 0.005.

b. In the true-position method the positional tolerance is given with the hole specification; see Fig. 12.29b. True-position dimensions are enclosed in a rectangle. The word Basic or abbreviation BSC may be used in the note form.

c. This results in a difference in the shape and size of the tolerance zones, as shown in Figs. 12.30a and b.

12.29.2 Tolerance Zones Compared. In Fig. 12.30a the tolerance zone is a square 0.01 in. on a side. The center of the holes could be anywhere within this square, even at the end of a diagonal. If it were at this point, it would be at 0.007 from the true position and would be acceptable, even though this is more than the 0.005 specified.

In the true-position method, the tolerance zone will be a circle, since the diameter is specified. To give the same extreme tolerance as the square, this zone diameter can be 0.014 in. This provides greater leeway in manufacturing, yet controls functioning as well as the square zone. With this larger tolerance more acceptable parts are produced and fewer parts must be scrapped.

12.29.3 Additional Tolerance Available When Feature Is Not at MMC. When true-position dimensioning is used, the tolerance is assumed to be with the features at MMC even though not specified. The maximum material condition may be specified, if desired, by adding "at MMC" to the tolerance note.

When the holes shown in two parts in Fig. 12.31 are away from MMC toward the largest permissible size, more positional tolerance is available, as shown in Fig. 12.31b.

If the design function of the part will prohibit this additional locational tolerance, the tolerance *must be marked* RFS, which means "regardless of feature size." This holds the tolerance to the specified size.

12.29.4 Size of Fixed-Position Functional Gage Pins. With the true-position dimensioning shown in Fig. 12.29, the size of the functional gage pin is shown in Fig. 12.30a. From this fixture it can be seen that the *size of gage pins for circular holes is equal to the minimum size of the hole minus the diameter of the positional tolerance zone.*

Thus if two circles representing the minimum size of the holes (MMC) are drawn with centers at the opposite ends of the tolerance zone diameter, it can be seen that the diameter of the gage pin must be

0.310	Smallest hole size
0.014	Diameter of tolerance zone
0.296	Diameter of gage pin

On the other hand, if the diameter of the holes at their largest size (minimum material) is used, as

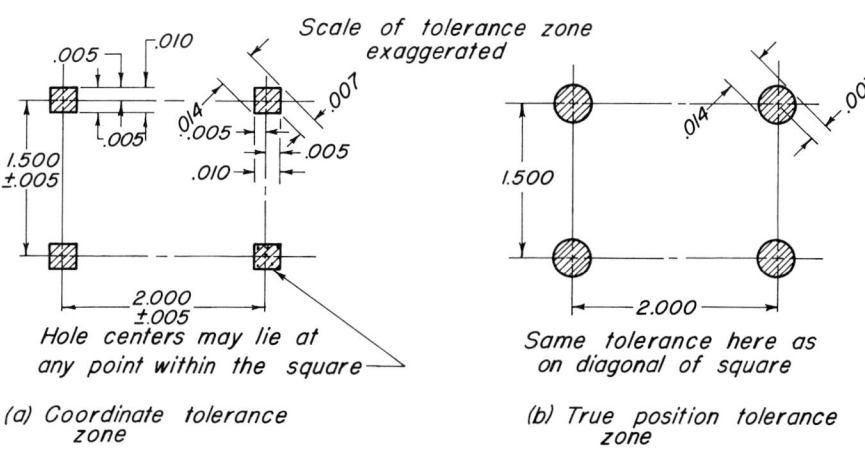

FIG. 12.30. Tolerance zones compared.

in Fig. 12.31b, and the same gage pin is used for checking, then the locational tolerance is

0.314   Hole size
0.296   Gage pin size
0.018   Diameter of tolerance zone
0.014   Specified tolerance
0.004   Increase of tolerance, which is equal to the hole size tolerance

12.29.5 Computation of Positional Tolerance from Bolt and Hole Sizes. Where positional tolerances do not need to be as close as those given in the ANSI Standard for cylindrical fits, as discussed in Sections 12.20 and 12.21, a satisfactory tolerance can be chosen by using standard parts such as bolts and standard tool sizes such as drills for making the holes.

a. The bolt size can be determined by computation for strength and rigidity. Thus computations might call for a bolt 0.367 in. in diameter. The next larger standard size would be chosen, namely 0.375 in. The actual body diameter is 0.388.

b. The hole could then be chosen to give approximately a $\frac{1}{32}$-in. clearance

$$0.388 + 0.031 = 0.419$$

A $\frac{27}{64}$-in. drill could be chosen. The diameter in decimals is 0.422. The actual clearance would then be $0.422 - 0.388 = 0.034$. Due to vibration and wear of the tools, the holes could be 0.422 ± 0.002. If the bolts also were permitted to vary between $0.388 + 0.002$ and $0.388 - 0.002$, the locational tolerance at MMC would then be $0.420 - 0.390 = 0.03$ or ± 0.015.

In many cases a much more generous clearance or allowance could be provided.

12.30 METHODS OF SPECIFYING TRUE-POSITION TOLERANCES

As already mentioned, a tolerance is the total variation permitted; hence, for location, the diameter of the tolerance zone is the total tolerance. The note, or other specification, should be unmistakable in this respect.

12.30.1 By Note Specifying the Diameter. One method of specifying the true-position tolerance is to give the diameter of the tolerance zone, as shown in Fig. 12.32.

12.30.2 By Note Specifying the Radius. Because of long practice in some industries, the specification of the radius of the tolerance zone is also approved, as shown in Fig. 12.33. The modifier MMC in Fig. 12.33 indicates the possi-

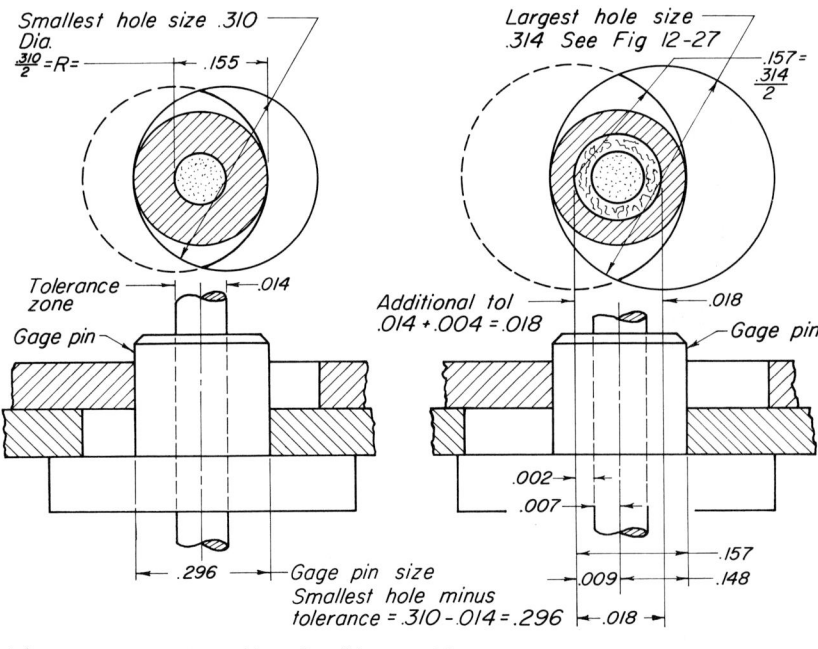

FIG. 12.31. Additional tolerance available when position tolerance is noted at MMC.

bility of a more liberal tolerance, as discussed in Article 12.29.3. The radius, however, must be recognized as a *deviation* from true position and the tolerance is twice the radius. The modifier, "regardless of feature or hole size," in Fig. 12.33, sometimes abbreviated to RFS, indicates restriction to the specified tolerance, no matter what the hole size.

12.31 APPLICATION OF TRUE-POSITION DIMENSIONING AND DATUM SELECTION

As a practical illustration of true-position dimensioning and of proper selection of datum surfaces, Figs. 12.34 through 12.36 are presented. These figures are detailed drawings needed to implement the design layout shown in Fig. 12.1.

In the subassembly layout shown in Fig. 12.1, the flange plate functions as the mounting base for the two other parts. It should therefore be detailed first, as shown in Fig. 12.34.

12.31.1 Primary Datum. The center of the large pilot hole A (2.786–2.789 Dia) and the square-plate mounting face are chosen as the primary datum for dimensioning, since they establish the position of the flange plate relative to the wheel hub and position the entire assembly.

The general shape of the plate is defined by dimensioning the square-shaped depression surrounding the pilot hole. Then the step to the next level (0.98) is given and finally the step to the outer rim (9.68 and 12.00), as shown in Fig. 12.34. Although the latter dimensions are not specifically tied into the center of the pilot hole by half-diameter dimensions, the symmetry of the piece indicates that this is the datum. It should be noted that all form dimensions are given to the same side of the metal. This is done to assist the die designer and to avoid unnecessary tolerance accumulation due to variation in stock thickness.

The four mounting holes (0.391–0.406 Dia) are located from the primary datum with basic dimensions, since their tolerance is given in relation to the pilot hole by the true-position tolerance (0.005R), given in the note for the diameters of the mounting holes. See Fig. 12.34.

12.31.2 Secondary Datum. The smaller the wheel-cylinder pilot hole (1.506–1.510 Dia B) and the two bolt holes are located as a group by the limit dimension (3.84–3.78 at the left) from the primary datum to the center of the wheel-cylinder pilot hole. See Fig. 12.35 for fitting spigot (1.495–1.505 Dia Y). The center of the wheel-cylinder pilot hole is the datum for the group, and the two bolt holes are dimensioned in relation to the group datum B with basic dimensions. The true-position tolerance for these holes is given with the note for these holes and with the note

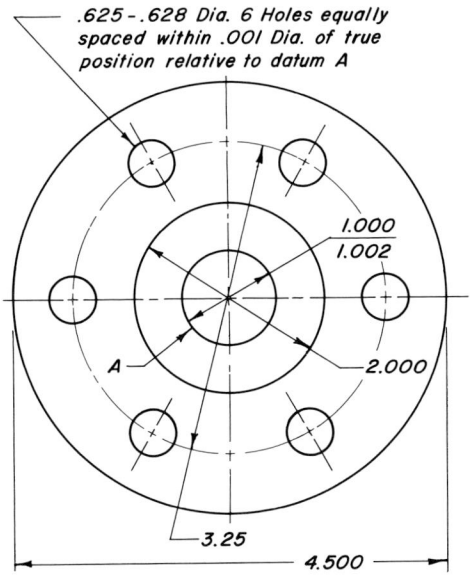

FIG. 12.32. Specifying tolerance by diameter of zone.

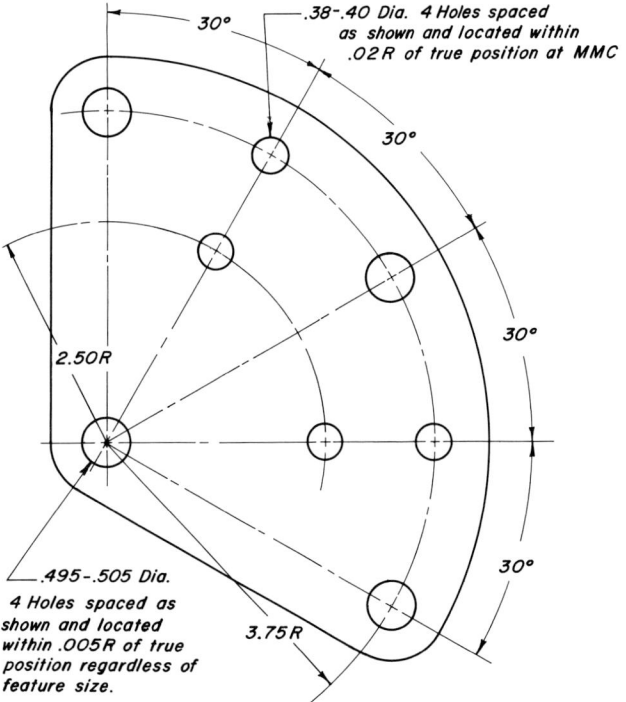

FIG. 12.33. Specifying tolerance by radius of zone (courtesy ANSI).

for the diameter (0.344–0.359) of the two bolt holes.

The two holes at the left on the flange-plate detail (Fig. 12.34), which are used for mounting the brake-cable clamp, are located as a group by limit dimensions from the primary datum. In this case the lower left hole becomes the group datum, and the other hole is located from the first by basic dimensions with a true-position tolerance.

The two holes at the right in Fig. 12.34 are clip holes used independently of each other and of other features. Therefore, they are located with liberally toleranced dimensions from the primary datum.

12.31.3 Form–Contour Tolerances.
Since the position and form–contour variations (flatness and parallelism) must be limited on both flange-plate mounting surface and the wheel-cylinder mounting surface, these tolerances are expressed with the contour-tolerance note, with the affected areas signaled ////// as shown in Fig. 12.34.

12.31.4 Wheel-Cylinder Body.
The second part of the subassembly shown in Fig. 12.35 mounts on the flange plate. The same fundamental method or pattern of dimensioning must be used. Note that the spigot (1.495–1.505 Dia Y) in Fig. 12.35 is toleranced to mate with pilot hole B (1.506–1.510) in Fig. 12.34. The two threaded holes are located by true-position dimensioning from the center of the spigot, ensuring assembly with the hole pattern on the flange plate. Note that the untoleranced basic dimensions from spigot and pilot hole are the same on both drawings and the true-position tolerance is also the same. The two threaded holes that enter the cylinder, in Section B-B, are located by basic or untoleranced dimensions so that the general tolerance of 0.01 given in the general notes applies.

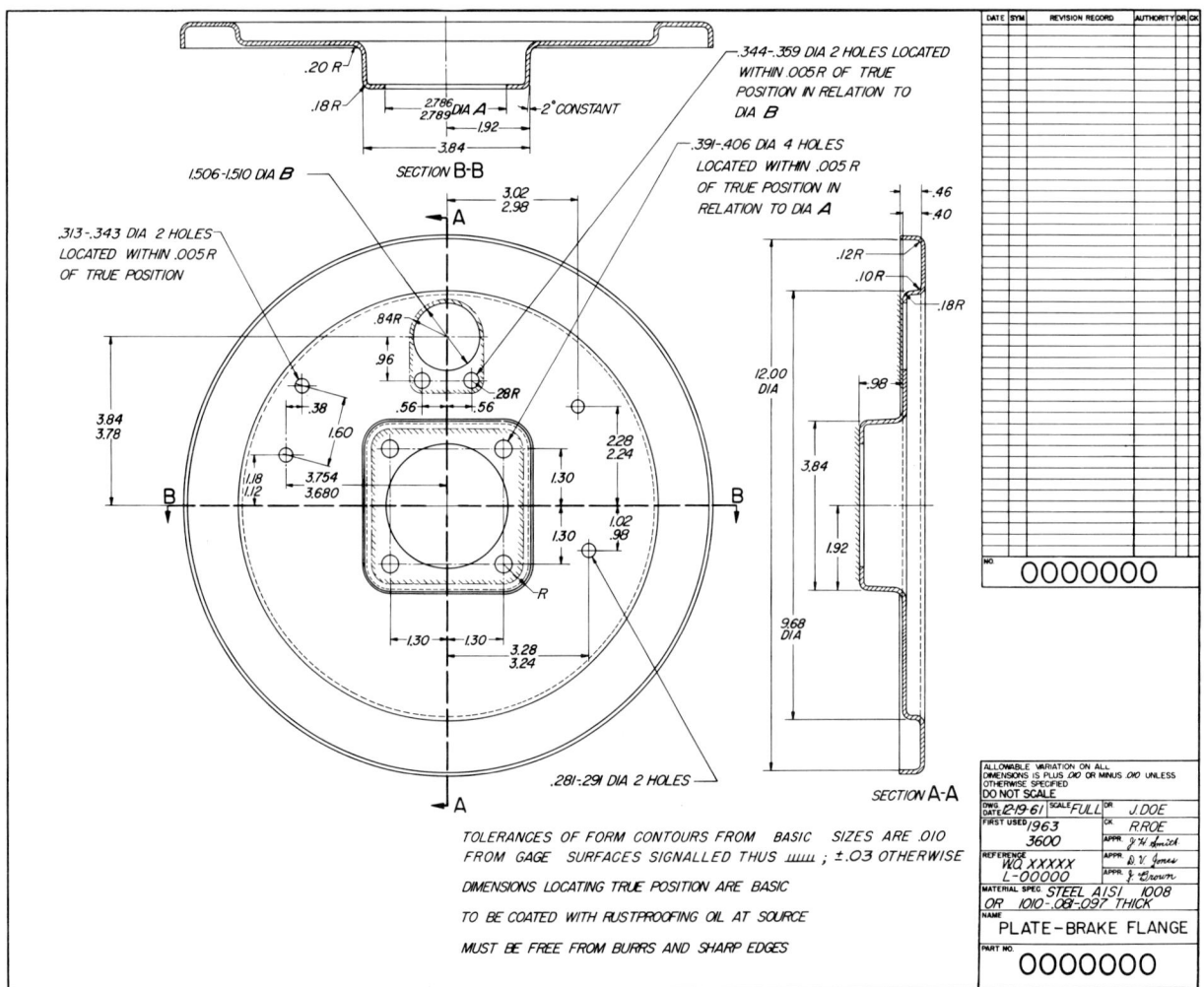

FIG. 12.34. Example of true-position tolerancing (courtesy General Motors Corp.).

12.31 APPLICATION OF TRUE-POSITION DIMENSIONING AND DATUM SELECTION

The surface of the mounting flange of the wheel-cylinder body and the center of the spigot Dia Y are chosen as primary datums for dimensions of this part, since it is located by these features at assembly to the flange plate. The cylinder bore (1.1250–1.1257 Dia) in turn becomes the datum for the location of the threaded holes that enter the bore.

12.31.5 Brake-Cable Clamp. Although this part is in an angular position on the layout (Fig. 12.1), it is drawn in relation to its own logical bases in Fig. 12.36. This is done to provide practical datums for dimensioning and to simplify manufacture and inspection of the detail part. The surface of the mounting flange and the center of one mounting hole are chosen as primary datum surfaces for dimensioning the part, since it is located by these features at assembly to the flange plate. The position of the plan view is determined by the center of the clamp hole. The flange thickness and the center of the clamp hole are located from the primary datum.

The clamp hole is located at one end by a height dimension (0.704–0.734), by the dimension 0.30 from the datum hole, and by an angle. See front view in Fig. 12.36. The use of an angle instead of a height dimension at the other end is justified in this case, since forging variations on the boss would not be compatible with the tolerances that would have to be applied to the second height dimension to control the finished hole properly.

Secondary Datum. The center of the clamp hole in turn becomes the datum for the boss, slot, and clamp-bolt hole.

12.31.6 Gaging Control. To make the drawing more effective for the control of inspection, the following note could be added to Fig. 12.34 with a leader to one of the four wheel-mounting holes.

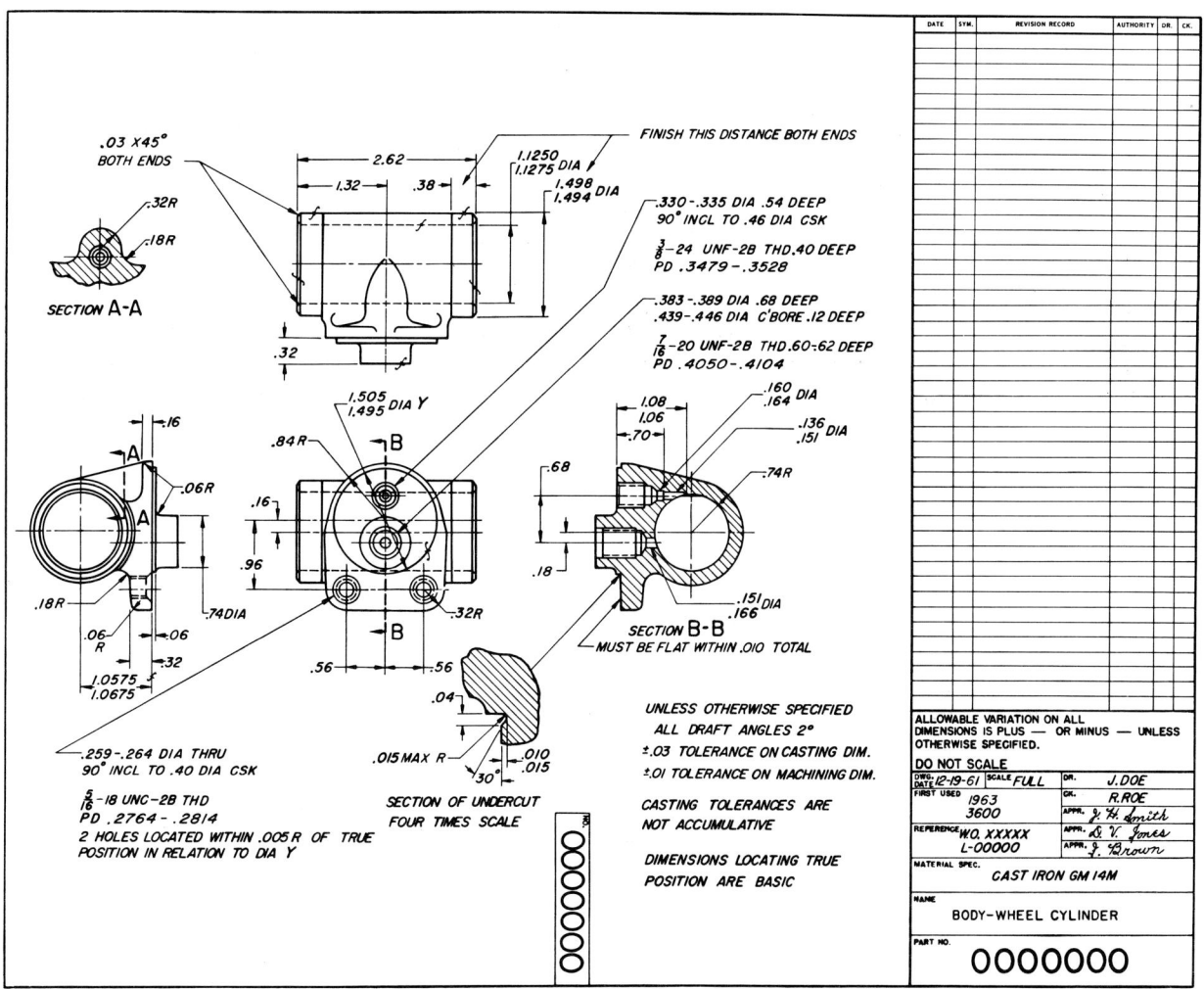

FIG. 12.35. Coordination of datum surfaces with Fig. 12.34 (courtesy General Motors Corp.).

328 PRODUCTION DIMENSIONING

$\frac{0.391}{0.406}$ Dia 4 holes must freely admit basically located gage pins 0.010 under minimum hole size when gage is piloted in Dia A with 2.776 Dia pilot.

Other holes, such as the wheel-cylinder mounting and the cable-clamp mounting, would be controlled in the same way.

12.32 TRUE-POSITION DIMENSIONING WITH SYMBOLS

In the foregoing articles true-position dimensioning has been accomplished by the use of notes. This can be simplified by the use of symbols, which are shown in Fig. 12.37. The application of the true-position symbol and the datum symbol are shown in Fig. 12.38.

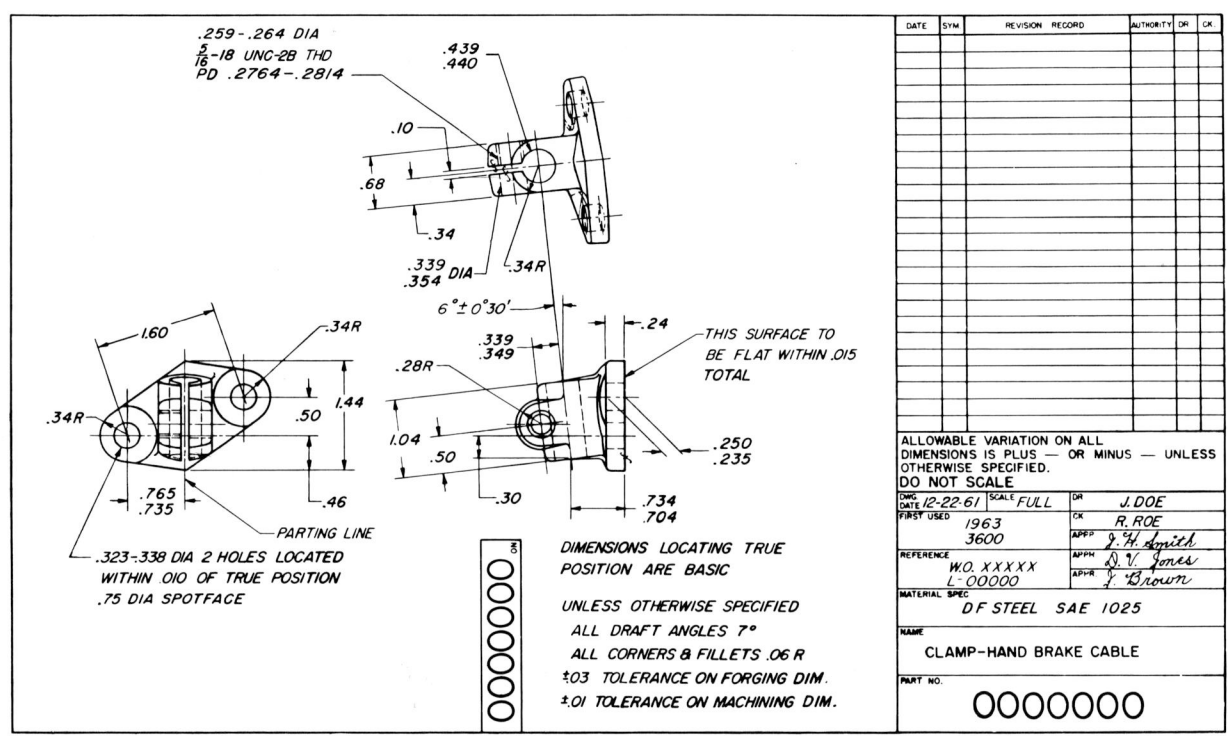

FIG. 12.36. True-position tolerancing (courtesy General Motors Corp.).

AMERICAN NATIONAL STANDARD
DIMENSIONING AND TOLERANCING

ANSI Y14.5–1973

		CHARACTERISTIC	SYMBOL	NOTES
INDIVIDUAL FEATURES	FORM TOLERANCES	STRAIGHTNESS	—	1
		FLATNESS	▱	1
		ROUNDNESS (CIRCULARITY)	○	
		CYLINDRICITY	⌭	
INDIVIDUAL OR RELATED FEATURES		PROFILE OF A LINE	⌒	2
		PROFILE OF A SURFACE	⌓	2
RELATED FEATURES		ANGULARITY	∠	
		PERPENDICULARITY (SQUARENESS)	⊥	
		PARALLELISM	//	3
	LOCATION TOLERANCES	POSITION	⊕	
		CONCENTRICITY	◎	3,7
		SYMMETRY	≡	5
	RUNOUT TOLERANCES	CIRCULAR	↗	4
		TOTAL	↗	4,6

Note: 1) The symbol ⌒ formerly denoted flatness.

The symbol ⌒ or — formerly denoted flatness and straightness.

2) Considered "related" features where datums are specified.

3) The symbol ‖ and ⦿ formerly denoted parallelism and concentricity, respectively.

4) The symbol ↗ without the qualifier "CIRCULAR" formerly denoted total runout.

5) Where symmetry applies, it is preferred that the position symbol be used.

6) "TOTAL" must be specified under the feature control symbol.

7) Consider the use of position or runout.

Where existing drawings using the above former symbols are continued in use, each former symbol denotes that geometric characteristic which is applicable to the specific type of feature shown.

GEOMETRIC CHARACTERISTIC SYMBOLS

FIG. 12.37. Form tolerancing symbols.

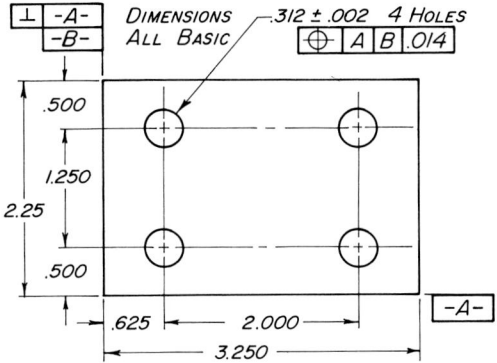

FIG. 12.38. True position with symbols.

12.33 FORM TOLERANCES CONTROLLED BY SIZE AND LOCATION TOLERANCE

In the past the geometric form of parts was assumed to be as shown by the drawing; that is, when two faces of a part are shown at right angles to each other on a drawing, these faces were assumed to be made at right angles to each other in the shop.

12.33.1 Control by Size Tolerance.
When there is doubt that ordinary shop practice will produce this form, it should be controlled in some way by a form tolerance. In many instances, the size tolerance given a part is sufficient to control the form. Thus in Fig. 12.39 the form implied by the drawing is a true hexagon, and the size tolerance given will control the shape within the limits shown in part b. In Fig. 12.40, straightness of a pin and hole is controlled by size tolerance only, and illustrations of permissible departure from straightness are shown in parts b, c, e, and f. Where greater accuracy of form is required, form tolerances must be specified. Such tolerances must naturally always be smaller than the size tolerances.

12.33.2 Form Control by Location Tolerance.
True-position tolerancing establishes a tolerance zone for the location of the axis of a feature such as a hole. The axis must lie within this tolerance zone throughout its length. Thus location tolerance automatically controls squareness of the axis of the hole with the surface to which the drawing shows it to be perpendicular, as shown in Fig. 12.41. It will also control the parallelism of one hole with another.

12.34 GEOMETRIC FORM TOLERANCING

When greater accuracy of form must be obtained than that given by size and location tolerances, specific form tolerancing must be used. Form tolerances are also necessary for some geometric shapes that cannot be controlled by size and location tolerancing. This may be done by the use of notes and the use of the symbols shown in Fig. 12.37.

In order to simplify the specification of form and also true-position tolerances, symbols have been designed, used, and standardized as shown in Figs. 12.37 and 12.42.

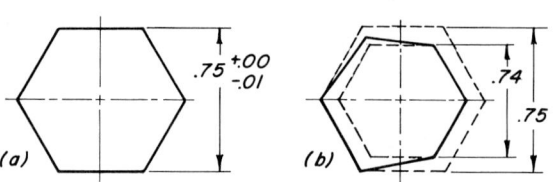

FIG. 12.39. Form controlled by size tolerance (courtesy ANSI).

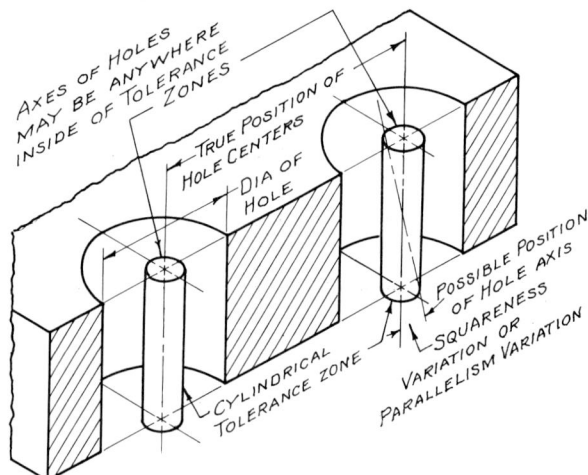

FIG. 12.41. Form tolerance controlled by location tolerance (courtesy ANSI).

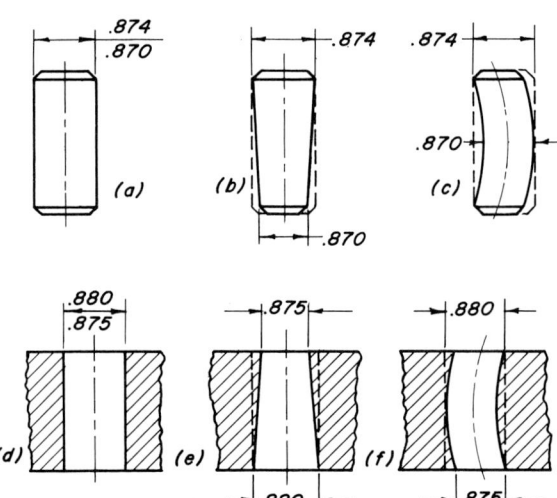

FIG. 12.40. Straightness controlled by size tolerance (courtesy ANSI).

12.34.1 Specifying Form Tolerance by Notes. In the following paragraphs and illustrations, form tolerance by the use of notes is shown. Their meaning or interpretation is also given.

12.34.2 Specifying Form Tolerance by Symbols. The specification of form tolerancing and true-positional tolerancing is more rapidly done by the use of symbols. The following paragraphs show how to use the symbols and where to place them.

12.35 GENERAL RULES FOR USE OF SYMBOLS*

When using symbols for geometric form and true-position tolerancing, the following rules should be observed.

a. When no modifier is specified, true-position tolerances apply at maximum material

---
* From *A Treatise on Geometric and Positional Dimensioning and Tolerancing,* by Lowell W. Foster, Senior Standards Engineer, Honeywell, Inc., Minneapolis, Minn.

condition. This condition also applies to datum references other than plane surfaces.

b. When no modifier is specified, all other positional or form tolerances apply regardless of feature size. See Fig. 12.42. "Regardless of feature size" also applies to datum surfaces.

c. All tolerances specified in connection with positional or form tolerancing are preferably totals, but radial deviations may be used if designated *R*. See ANSI Y14.5-1973.

12.35.1 Placing Symbols on Drawings. All symbols are placed in enclosing rectangular boxes in order to call attention to them and to separate them clearly from other notes. In Figs. 12.42 and 12.43 the minimum size of the enclosing rectangles has been shown. Other symbols and abbreviations and their meaning are also indicated in the figures.

As a further guide in the use of symbols and notes, the rules governing the use of the modifiers MMC and RFS and their symbols (M) and (S)

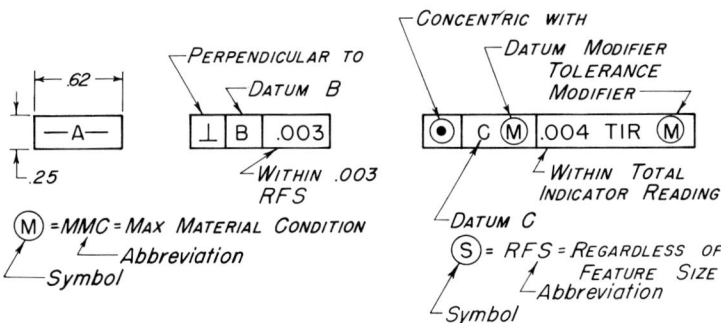

FIG. 12.42. Meaning and arrangement of symbols.

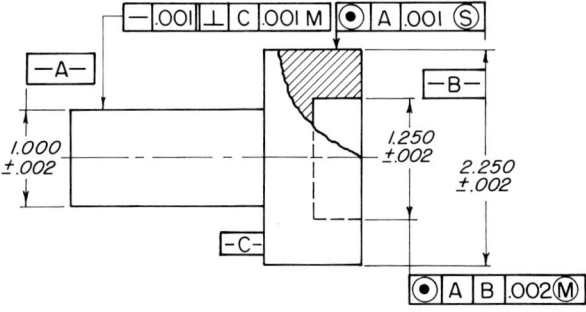

FIG. 12.43. Application of symbols.

should be noted. These rules are summarized in Fig. 12.44.

12.35.2 Datum Symbol (Symbol -A-). The datum used for dimensioning is always marked when using the symbolic method. A part may have one, two, or more datum surfaces. These are indicated by capital letters enclosed in a box as shown in Figs. 12.42 and 12.43. There may be primary, secondary, and other datum surfaces. The alphabetical sequence of letters has no significance in this respect. The letters I, O, and Q should not be used, since they may be confused with other letters. The letters R, M, and S should not be used, since they are already reserved for other uses. The placing of the datum box will be indicated in the illustrations that follow.

12.35.3 Straightness (Symbol ——). The straightness symbol is a single short ($\frac{1}{4}$-in.) line placed in a box as shown in Fig. 12.45. Any element of the feature must be between two parallel lines spaced at the specified distance apart, as in Fig. 12.45c. These lines lie in the same plane as the axis of the feature. If there is no modifier, straightness is implied to mean regardless of the size of the feature. The note form is shown in Fig. 12.45b.

In Fig. 12.45c the diametral dimensions could be 0.745 and 0.750 or any other pair differing by 0.005.

Straightness applies regardless of the length of the part. For long parts such as tubes or long shafts, the tolerance could be specified as "straight within 0.001 per fit." In this case the note form would be used.

Type of Tolerance	Applicability of MMC and RFS	
	For the Feature	For the Datum Reference(s)
Flatness Straightness Roundness Cylindricity	Not applicable	No datum reference
Profile of any line Profile of any surface	Not applicable	If a datum reference is necessary, RFS applies.
Perpendicularity Parallelism Angularity True Position	MMC or RFS applicable if tolerance applies to axis or center plane of considered feature; not applicable if considered feature is one plane surface.	MMC or RFS applicable if datum feature has an axis or center plane; not applicable if datum feature is one plane surface.
Runout	Not applicable	Not applicable
Concentricity Symmetry	Only RFS applicable	Only RFS applicable

FIG. 12.44. Application of tolerance modifiers (courtesy ANSI).

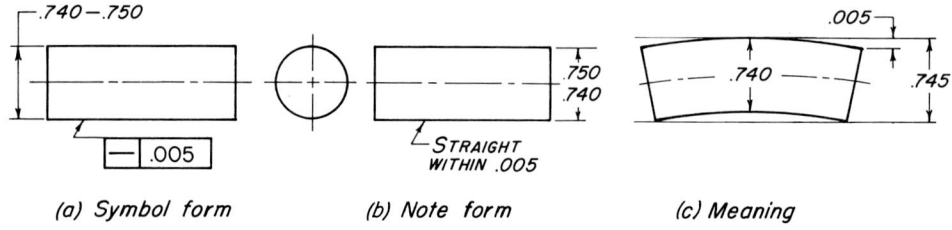

(a) Symbol form   (b) Note form   (c) Meaning

FIG. 12.45. Specifying straightness tolerance.

12.35.4 Flatness (Symbol ▱). The symbol form is shown in Fig. 12.46a, the note form in Fig. 12.46b, and the meaning in Fig. 12.46c. The flatness tolerance of a plane surface is the zone between two parallel planes the specified distance apart. If applicable, a note such as "must not be convex" or "must not be concave" may be added to the note form.

12.35.5 Parallelism (Symbol ∥). This symbol specifies a tolerance zone prescribed by two planes parallel to a datum between which the surface or axis must lie. Parallelism could be specified for one plane parallel to a datum plane, as in Fig. 12.47, or as an axis parallel to a datum plane. It could also be an axis parallel to a datum axis, as in Fig. 12.48. Either the feature or the datum could be specified at RFS or MMC if datum is other than a plane. A plane has only location, and measurements are made from it. In the case of an axis, the hole around it must be used for gaging, hence it can be specified RFS or MMC.

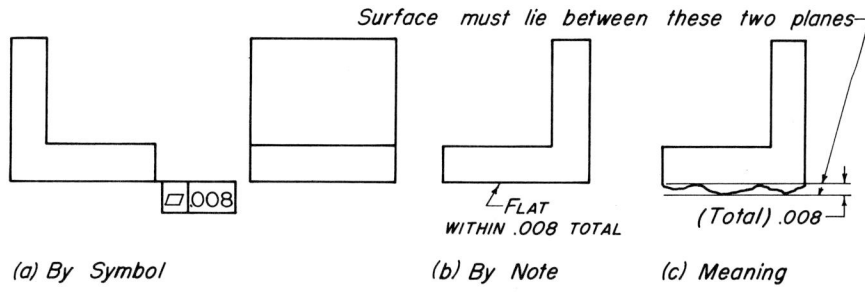

FIG. 12.46. Methods of specifying flatness.

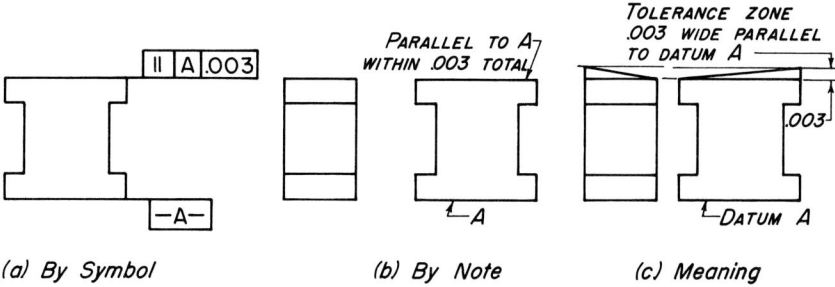

FIG. 12.47. Specifying parallelism between surfaces.

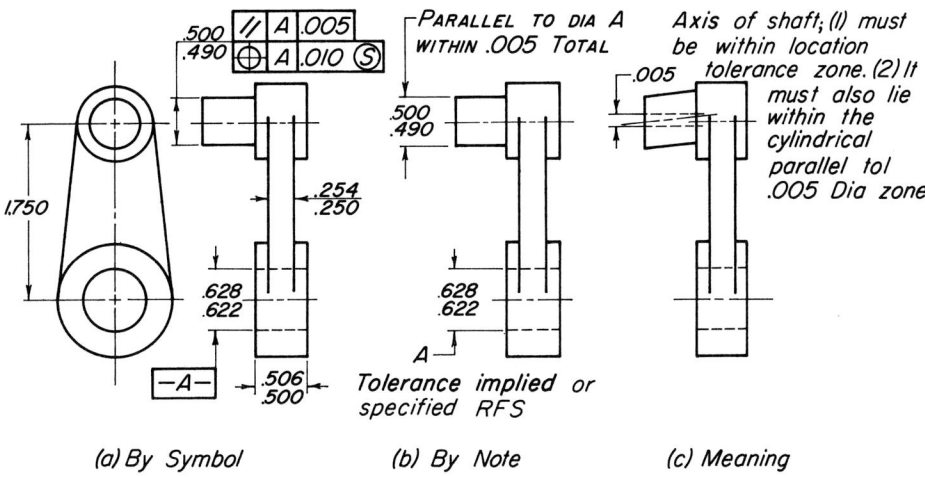

FIG. 12.48. Parallel form tolerance and location tolerance.

12.35.6 Squareness of Perpendicularity (Symbol ⊥).
The symbol form and note form for the squareness of one surface to a datum surface are shown in Fig. 12.49, as well as the interpretation.

For a slot, the specification is shown in Fig. 12.50 and for a cylinder with reference to a datum plane in Figs. 12.51 and 12.52. Note that when the feature is specified at MMC there is always a greater tolerance available when the part is not at MMC.

12.35.7 Specifying Dimensional Tolerance and Squareness.
In some situations, when size is involved as well as squareness, the squareness tolerance must lie within the size tolerance, otherwise it is meaningless. See Fig. 12.49.

12.35.8 Angularity (Symbol ∠).
Angularity may be specified by note giving an angular tolerance as in Fig. 12.53. It may also be specified by giving a tolerance zero between two parallel planes either by symbols as in Fig. 12.54a or by note as in Fig. 12.54b. Obviously the drafter must be governed by the practice of his/her company.

12.35.9 Tapers.
Applications of angularity tolerance are shown in Fig. 12.55. In Fig. 12.55 the gage diameter on the taper is not toleranced

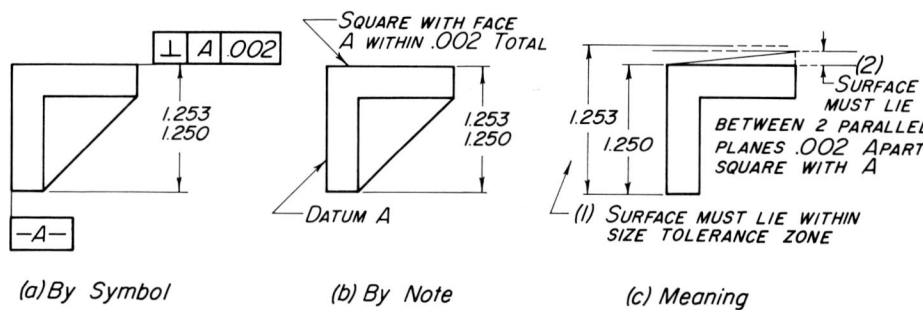

FIG. 12.49. Specification of squareness and size.

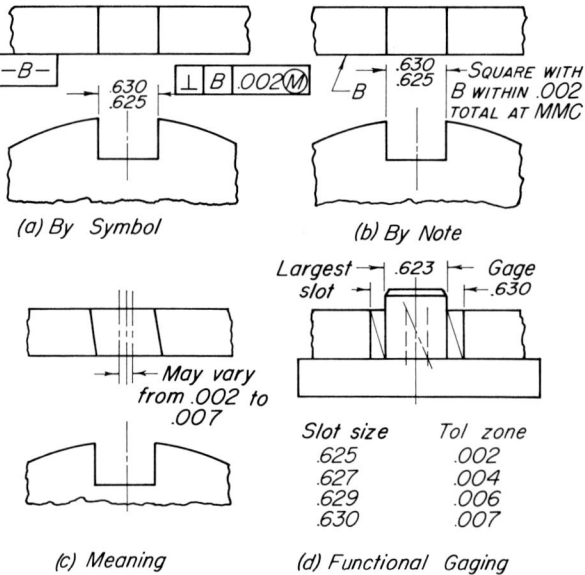

FIG. 12.50. Specifying squareness of MMC.

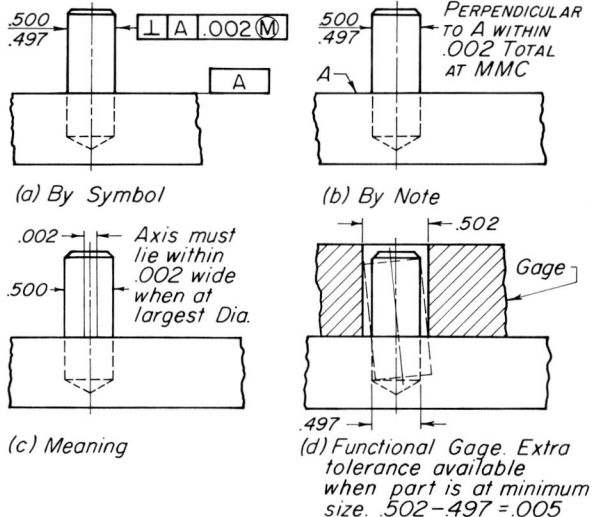

FIG. 12.51. Squareness tolerance for a pin at MMC.

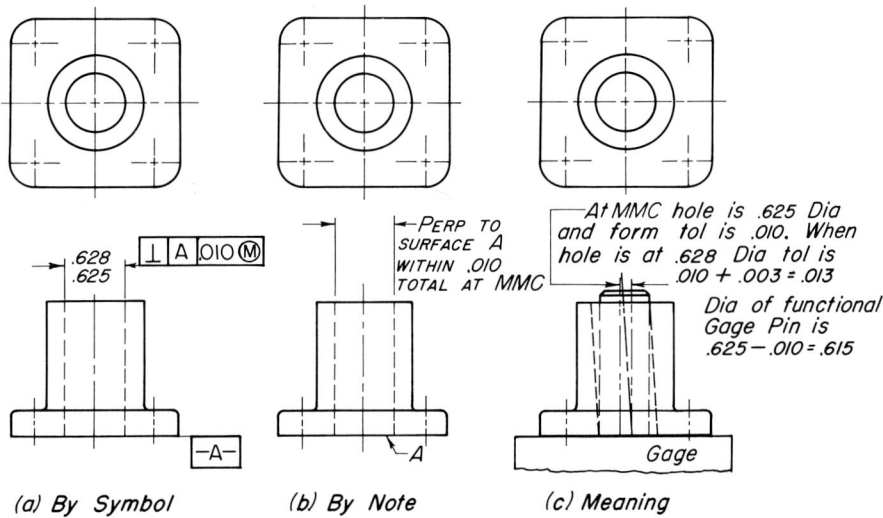

FIG. 12.52. Specifying perpendicularity of a cylindrical hole with plane datum.

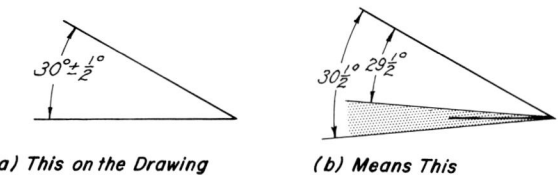

FIG. 12.53. Specifying angles.

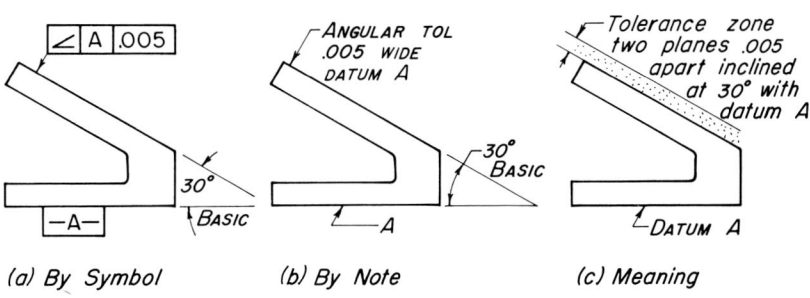

FIG. 12.54. Specifying angularity.

335

because doing so would give a cumulative tolerance. There are four possible dimensions that may be used to specify a taper. Any three may be employed, but the fourth may be given for reference only. Various combinations of the four dimensions are given in Figs. 12.55a–c. In Fig. 12.55d two dimensions are given for reference, and the location of the gaging point is toleranced.

12.35.10 Roundness (Symbol ○). Roundness may be specified as shown in Fig. 12.56, using either the radius or the diameter in describing the tolerance zone. The radius is preferred. Gaging, however, cannot be done with a micrometer caliper, since the lobed figures in Fig. 12.57 will measure the same as a true circle. Gaging should be done with three-point devices,

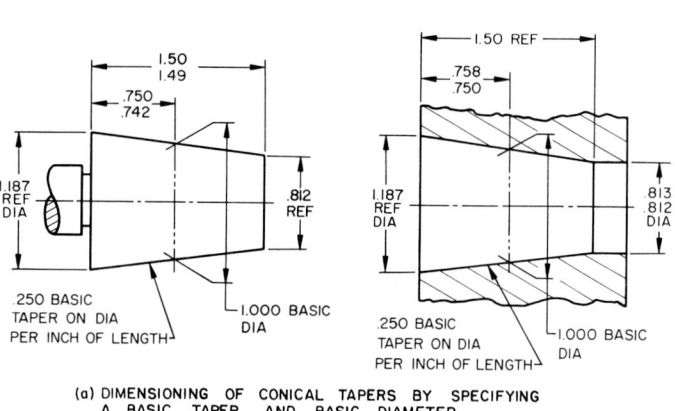
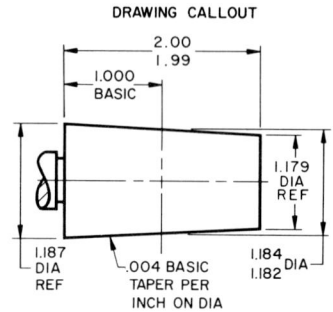

FIG. 12.55. Specification of taper tolerance (courtesy ANSI).

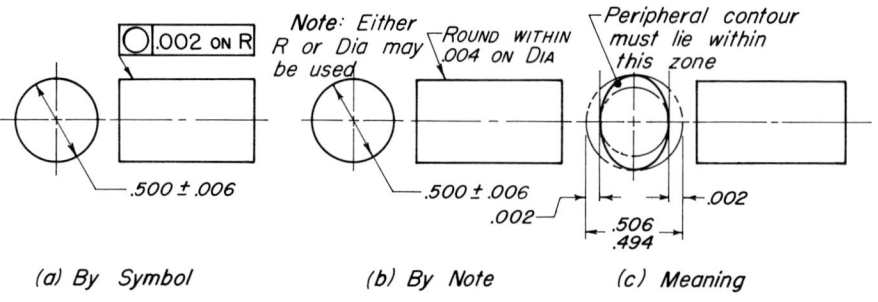

FIG. 12.56. Specifying roundness.

as shown in Fig. 12.58. Roundness on a cone may be specified as shown in Fig. 12.59.

12.35.11 Concentricity (Symbol ⌾). The tolerance governing the concentricity of axes is the diameter of the tolerance zone within which the axis must lie, as shown in Fig. 12.60. The zone for the outside surface where gaging must be done is shown in Fig. 12.61c. It should be noted that the actual eccentricity of the axis will be just one half the diametrical tolerance. The full indicator reading will be just twice the eccentricity.

Figure 12.62 gives an illustration of concentricity specifications when two datum surfaces are involved. The datum axis is then the mean axis of the two parts.

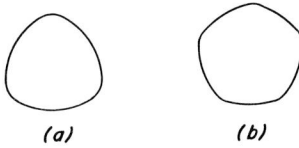

FIG. 12.57. Lobed shapes with constant diameters.

FIG. 12.58. Practical methods of checking roundness (courtesy British Standards Institution).

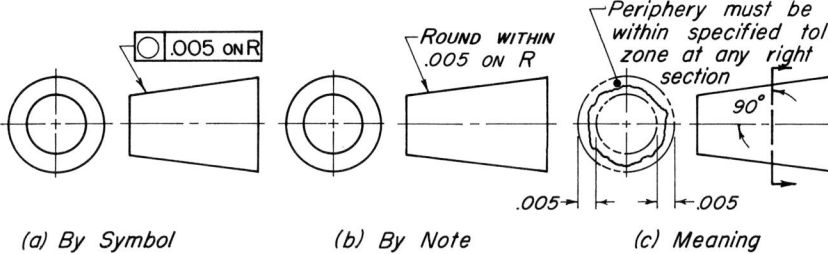

(a) By Symbol   (b) By Note   (c) Meaning

FIG. 12.59. Specifying roundness on a cone.

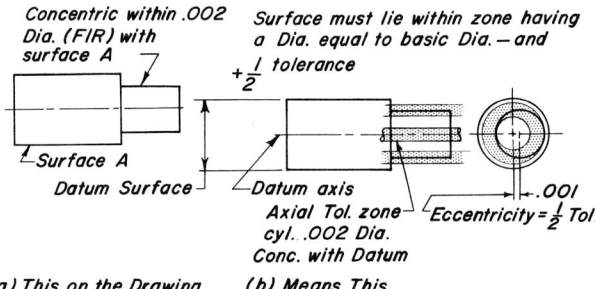

(a) This on the Drawing   (b) Means This

FIG. 12.60. Specifying concentricity.

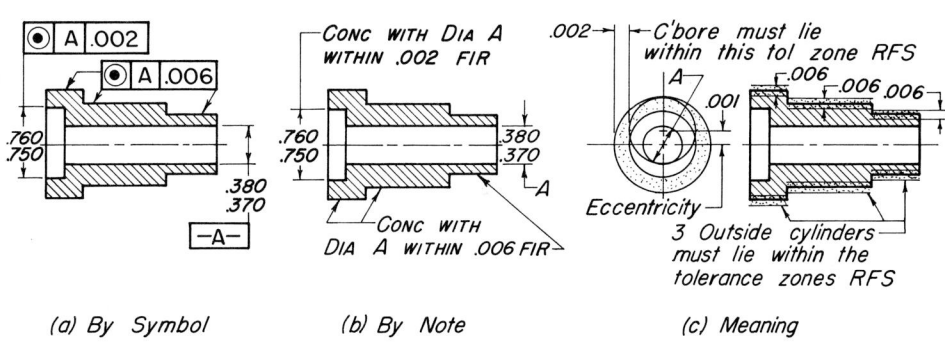

(a) By Symbol   (b) By Note   (c) Meaning

FIG. 12.61. Specifying concentricity.

When there are two datum surfaces, as in Fig. 12.62, the general procedure for checking is to mount the part on V blocks, one at X and one at Y. When the part is rotated on these V blocks, a gage on the surface under consideration will give a full indicator reading. However, this reading will include other items as well as concentricity, such as roundness and variation in angle between the center lines of surfaces X and Y. Because this fact is recognized, a composite tolerance is sometimes specified that will control all these variations. This is discussed in Articles 12.35.12 and 12.35.13 and illustrated in Fig. 12.64.

12.35.12 Symmetry (Symbol ⌯). The notes for symmetry, as well as all notes for other conditions, should state exactly what is meant. See Fig. 12.63. To say that the part is symmetrical within 0.001 would normally mean that it can vary by this amount on either side of the axis, making a tolerance zone 0.002 wide. The correct specification should read "symmetrical with A within 0.002 total or within 0.002-wide zone."

12.35.13 Cylindricity (Symbol ⌭). This symbol for cylindricity, if used for military contracts, must be defined on the drawing. Cylindricity tolerance specifies a tolerance zone confined to the annular space between two concentric cylinders within which the entire surface must lie. See Fig. 12.64. It should be noted that cylindricity tolerance simultaneously controls roundness, straightness, and parallelism of the elements of the surface.

12.35.14 Profile of a Surface (Symbol ⌒). This symbol for the control of the profile of a surface is taken from the ANSI Y14.5-1973. A contour tolerance is used on shapes other than cylindrical. It may be unilateral or bilateral. The zone is shown by phantom lines, as in Fig. 12.65. This also indicates whether the zone is on one side or both sides of the specified surface. Mea-

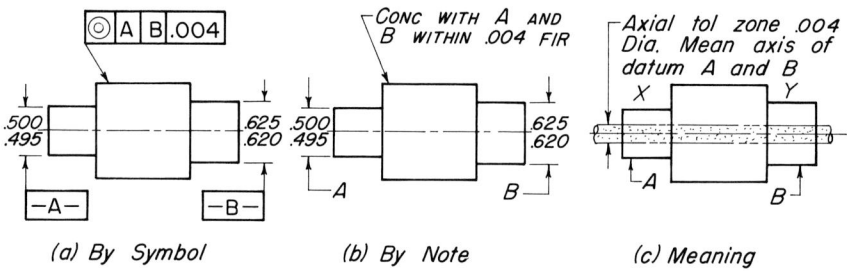

FIG. 12.62. Specifying concentricity with two datum references.

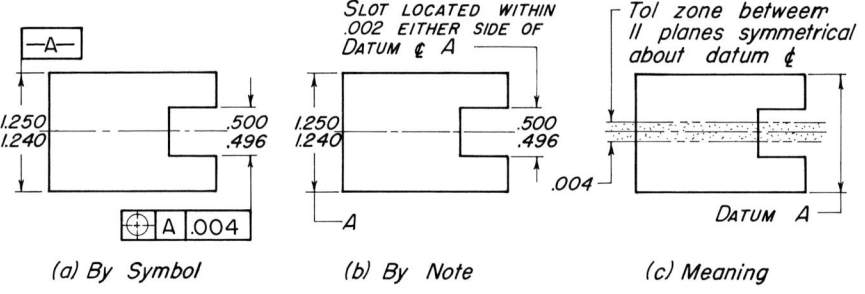

FIG. 12.63. Specifying symmetry (only RFS application).

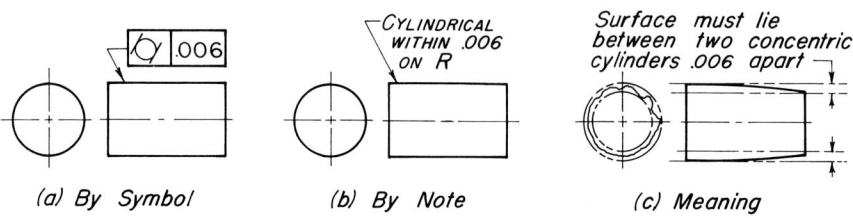

FIG. 12.64. Specifying cylindricity.

surements are made perpendicular to the contour at the point of measurement. An appropriate view or section showing the surface edgewise is drawn including the desired profile.

The profile is dimensioned by untoleranced dimensions referred to as basic. General tolerance notes do not apply and the drawing should be so noted.

12.35.15 Line Profile (Symbol ⌒). The profile of a line is specified as in Fig. 12.66. This can be used when the line profile of a surface requires closer control than in the other direction.

12.35.16 Runout (Symbol ↗). This symbol is from ANSI Y14.5-1973. The meaning of the runout tolerance is stated as in the following paragraphs.

a. *Runout tolerance.* A runout tolerance establishes a means of controlling the functional relationship of two or more features of a part within the allowable errors of concentricity, perpendicularity, and alignment of features. It also takes into account variations in roundness, straightness, flatness, and parallelism of individual surfaces. In essence, it establishes composite form control of those features of a part that have a common axis. Therefore, measurements should be taken under a single setup and normal to the true, desired geometrical shape. In applying a runout tolerance, the accessibility of datums for single, setup measurements should be given consideration.

b. *Basis of control.* As a basis for the control of the relationship of features, it is necessary to establish a datum axis about which the features are related. This axis may be established by a diameter of considerable length, two diameters having considerable axial separation, or a diameter and a face that is at right angles to it. Insofar as possi-

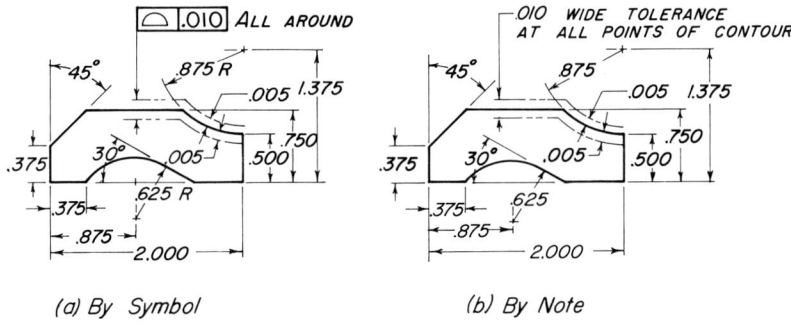

FIG. 12.65. Specifying surface profile tolerance.

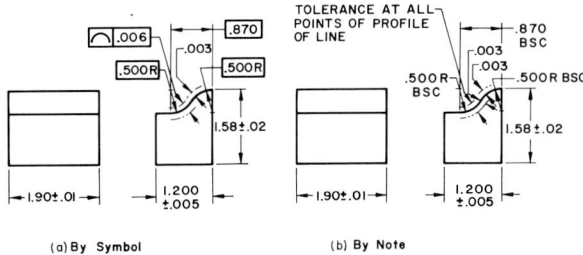

FIG. 12.66. Specifying line profile tolerance.

ble, surfaces used as datums for establishing axes should be functional. Various phases of mount control are illustrated in Figs. 12.67 and 12.68 as further explained in the text.

c. *Basic control notes.* Where the functional requirements of surfaces are in reference to a common axis, one of the basic form control notes stated below can be specified. The form control is assigned a suitable designation letter, such as A in Fig. 12.67, and the letter is applied to each related surface. The notes or symbols control concentricity, parallelism, and perpendicularity of specified surfaces to the mounting surface or surfaces and roundness, flatness, parallelism, and straightness of each specified surface.

Where the basic form control is specified by a feature control symbol, the runout symbol ↗ is used to designate the required form and notes are not used.

1. Where the related surfaces to be controlled have a single common axis, the general note is: *When mounted on datums C and D, designated surfaces must be within total mount specified by A. See Fig. 12.68.*

2. *Where datums C and D must be cylindrical within .XXXX on R.* This note is used where it is necessary to control roundness, parallelism, and straightness of an individual diameter more accurately than the diametrical tolerances. See Fig. 12.69.

d. *Applications.* Applications of the proper methods of using runout tolerances on drawings by notes or symbolically are shown in Figs. 12.68 through 12.70.

1. For a part mounted on machining centers, see Fig. 12.68.
2. For a part mounted on two bearing surfaces, see Fig. 12.69.
3. For a part mounted on a large flat surface with narrow finished diameters, see Fig. 12.70.

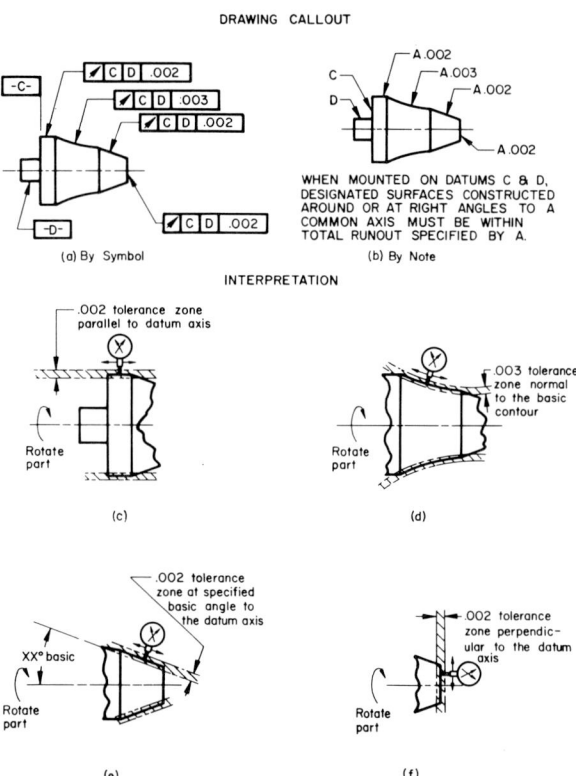

FIG. 12.67. Specifying runout tolerance zones (courtesy ANSI).

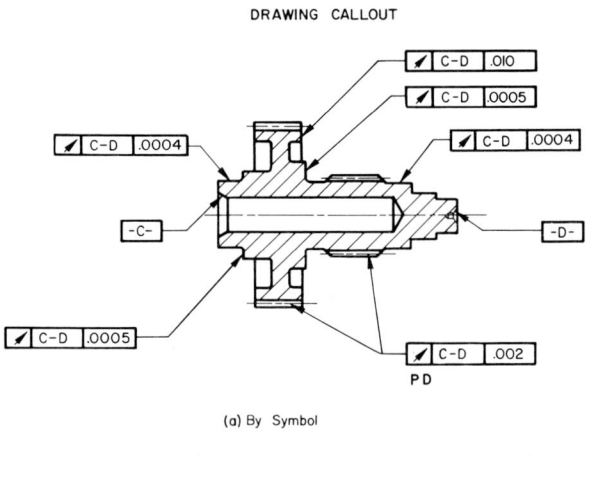

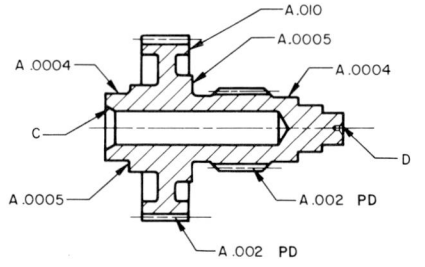

WHEN MOUNTED ON DATUMS C AND D, DESIGNATED SURFACES MUST BE WITHIN TOTAL RUNOUT (FIR) SPECIFIED BY A.

(b) By Note

FIG. 12.68. Runout tolerance for part with machine center datum (courtesy ANSI).

12.36 EFFECTIVE HOLE DIAMETER

Because of departure from true perpendicularity, there are situations in which a hole might meet specifications as to size when measured with a micrometer but still not accept a functional fixed position pin gage. When a hole is not square with a surface to which mating stud bolts or dowel pins are perpendicular, the effective hold diameter can be reduced as shown in Fig. 12.71.

Thus if the hole is in its smallest size, namely, 1.252, and the axis slopes across the full tolerance zone as shown in Fig. 12.71, the effective diameter of the hole for a perpendicular pin or stud has been reduced to 1.248. To ensure assembly, the minimum diameter of a clearance hole must equal the maximum diameter of the stud bolt plus twice the positional tolerance.

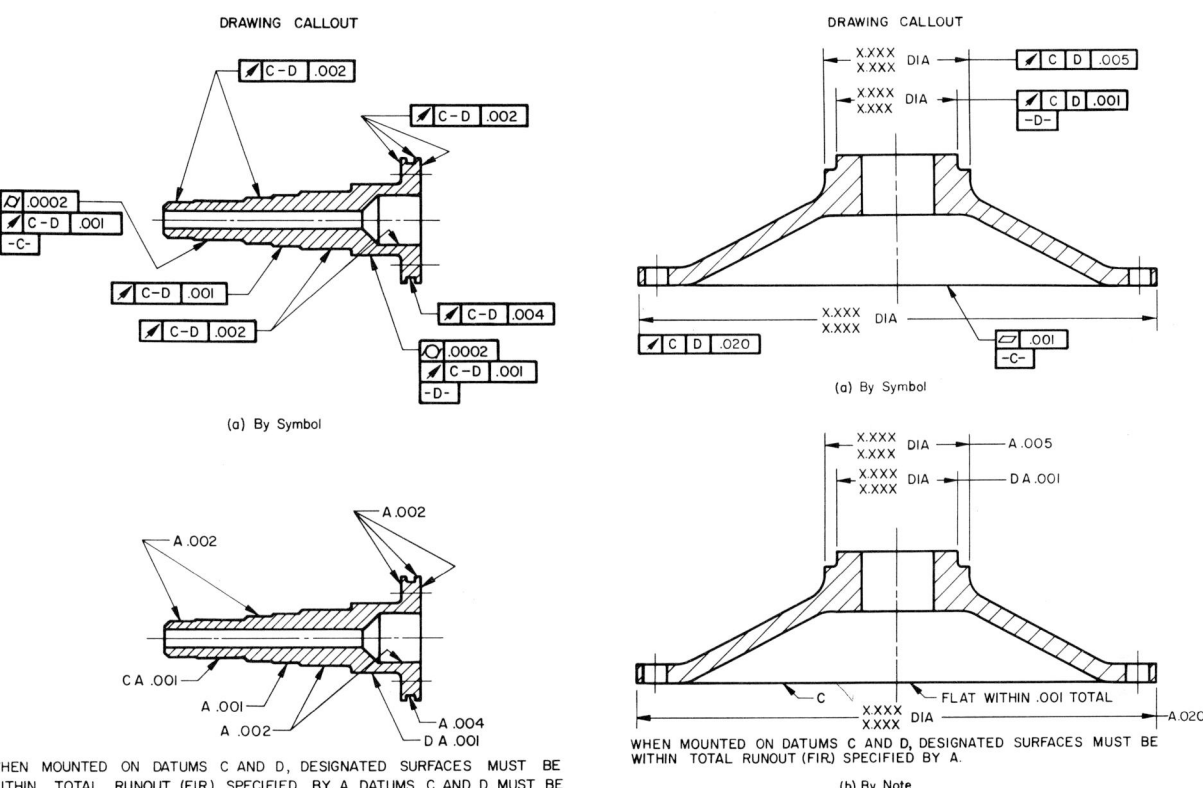

FIG. 12.69. Runout tolerance for part with two bearing surface datums (courtesy ANSI).

FIG. 12.70. Runout tolerance for part with large flat surface datum having narrow finished diameters (courtesy ANSI).

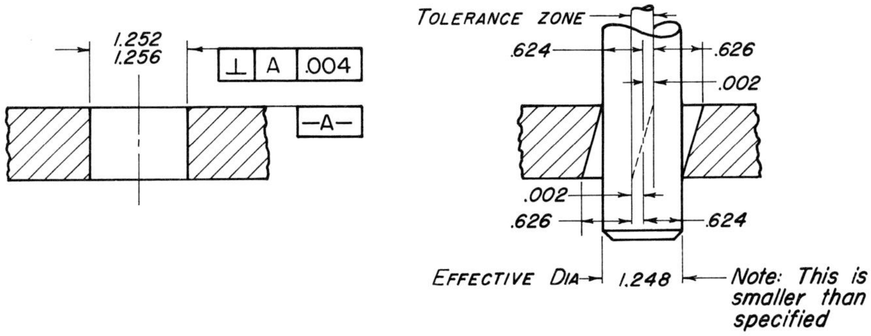

FIG. 12.71. Effective diameter of toleranced hole.

12.37 PROJECTED TOLERANCE ZONE FOR THREADED PARTS

In this discussion the following points are assumed.

a. That both parts 1 and 2 in Fig. 12.72 have the same locational tolerance.

b. That perpendicularity is controlled only by locational tolerance.

When a hole is threaded to take a cap screw or stud bolt, any departure from perpendicularity is magnified, as shown in Fig. 12.72c.

12.37.1 Minimum Size of Clearance Hole.
Figure 12.72a shows that with exact perpendicularity, the size of the clearance hole must be equal to the maximum diameter of the bolt plus twice the diameter of the tolerance zone. This is greater than the size required for two matching clearance holes.

12.37.2 Projected Tolerance Zone for Cap Screws.
Figure 12.73 shows that the tolerance zone for a cap screw or stud bolt is based on the full height of the screw. Figures 12.73b and c show that to permit engagement of the threaded screw through the upper mating part 2, the tolerance zone must be projected through the full thickness of the mating part.

12.37.3 Projected Tolerance on Stud Bolt.
Figure 12.72 shows that the tolerance zone must be projected or extended for the full length of the stud above the contact surface with part 1. Since the stud bolts must be assembled before part 2 can be put on, the bolt must pass through the effective diameter of the clearance hole.

Although the diameter of the clearance hole is as specified in Section 12.36, the effective diameter may not be less than the maximum diameter of the bolt plus the tolerance zone of part 2. See Fig. 12.72c.

12.37.4 Dimensioning and Gaging.
Figure 12.72d shows one method of dimensioning the part with threaded holes to take advantage of the projected tolerance zones. The gage for checking is shown in Fig. 12.72e.

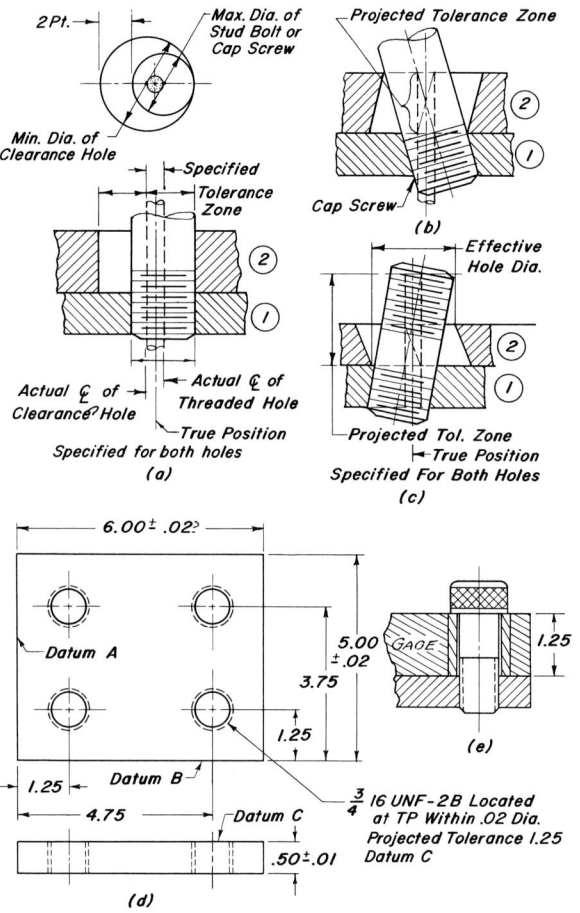

FIG. 12.72. Tolerance zones for threaded parts.

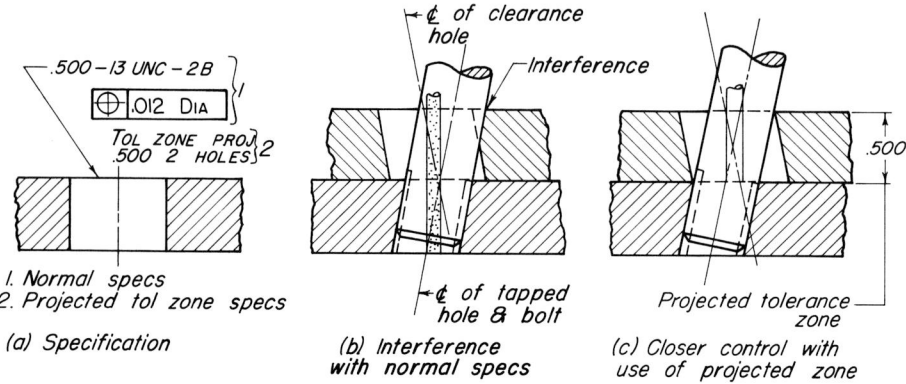

FIG. 12.73. Use of projected tolerance zone specifications.

From the various parts of Fig. 12.72 the following facts may be noted.

a. In b and c the most adverse assembly conditions have been shown.
b. The center of the threaded hole on the contact surface between parts 1 and 2 must be on or within the periphery of the tolerance zone.
c. The axis of the threaded hole may lie outside the tolerance zone. This gives greater manufacturing tolerance than the drawing would show.
d. To ensure assembly of either cap screws or stud bolts, the diameter of the gage-plate hole must be the effective diameter of the clearance hole rather than its true diameter. See Figs. 12.72 and 12.73.

This method of specifying the projected tolerance zone is shown in Fig. 12.73.

A complete drawing with the use of symbols for the specification of form tolerances and datum surfaces is shown in Fig. 12.74.

12.38 SURFACE FINISH

The engineer should understand clearly that the specification of surface finish is entirely distinct from specifying tolerances and limits. The finish of a surface determines its quality as to smoothness, surface marks, and the like, whereas tolerance refers to size and position only.

In certain applications the quality and degree of finish must be specified very clearly so that manufacturing processes may be determined and cost estimates prepared. Special operations must be employed to obtain very fine finishes, and consequently the cost increases rapidly as the finish is improved.

12.39 FINISH MARKS AND SPECIFICATIONS

One of the oldest methods of indicating a finished surface on a drawing is by placing the letter f across the edgewise view of the surface to be finished. The f is made as shown in Fig. 12.75a, and the correct method of placing it on the drawing, as well as several incorrect methods, is shown in Fig. 2.75b. This symbol calls for an ordinary machine finish and makes no attempt to indicate the quality of the surface finish.

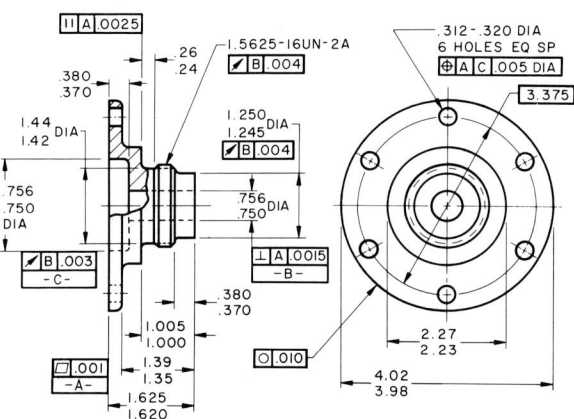

FIG. 12.74. Application of symbols for form- and position-tolerance specification (courtesy Mil. Std. 8C-1962).

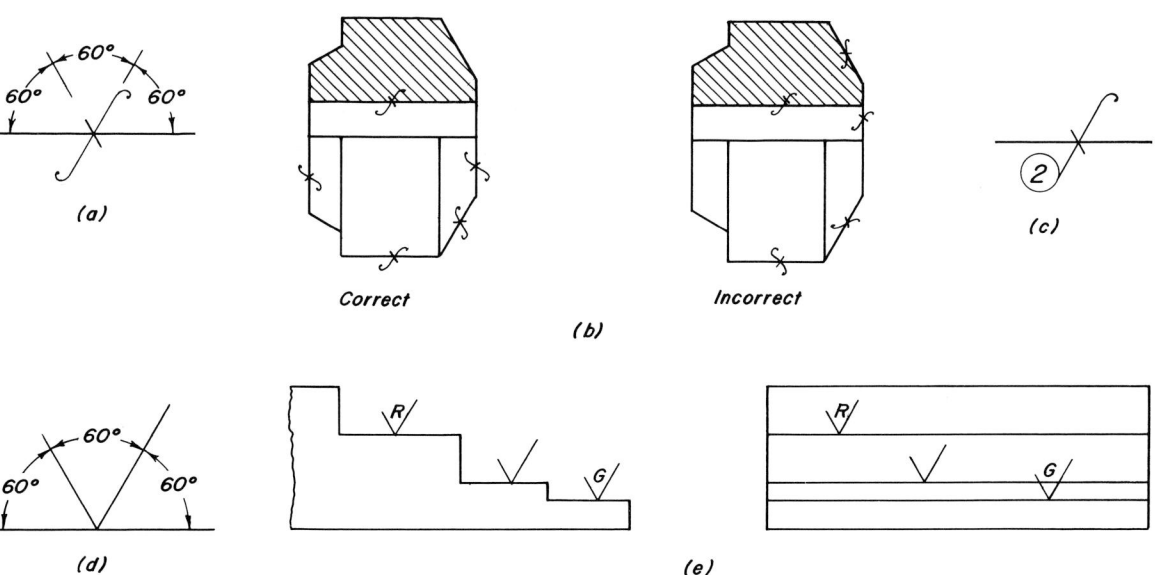

FIG. 12.75. General machining finish marks.

On some drawings this symbol is improved by adding a circle to the tail of the f, in which a number is placed, as shown in Fig. 12.75c. A note indicating the meaning of the number is placed on the drawing. By this means more specific information may be given concerning the character of the finish desired for that surface.

For some years American Standards have recommended the V symbol for indicating finish. The simplest form of this symbol is constructed as shown in Fig. 12.75d. Although it is not considered the best form, Fig. 12.75e shows one method of using the V symbol. The V is placed with its point touching the line that represents the edgewise view of the surface to be finished. The letters R and G mean "rough finish" and "grind." Other letters may be used to indicate certain operations or finishes.

The symbol for a finished surface should be placed wherever the surface shows edgewise as a visible or invisible line. This means that the symbol for finishing a single surface may be repeated in several views.

The more complete form of the V in which roughness, waviness, and lay are specified is shown in Figs. 12.76a through f. The first two forms have recently been adopted.

a. **Roughness** may be defined as the closely spaced surface irregularities produced by machining or grinding operations.
b. **Waviness** refers to the more widely spaced irregularities that may be produced by vibration, deflection of the part in machining, warping, or in the release of strains in the material.
c. **Lay** refers to the direction of the surface pattern of irregularities produced in the finishing of the surface.

The roughness height may be specified as the peak-to-valley height or as the average arithmetic deviation from the mean surface. Measurements are taken across the lay pattern. Roughness is specified in microinches (millionths of an inch). The values given in Tables 12.9 and 12.10 (from ANSI B46.1-1955) are commonly used in roughness and waviness specifications.

Waviness specifications are given in inches. Roughness width, as contrasted to depth, is also specified in inches, and this number appears after the lay symbol as shown in Fig. 12.76f.

In addition to roughness and waviness, it is sometimes necessary to specify the lay or the direction of the dominant lines of the surface. This is done by means of a set of symbols that may be placed to the right of the V as shown in Fig. 12.76e. The meaning of these symbols is explained as follows.

$\sqrt{=}$ Lines to be parallel to the boundary line representing the surface on the drawing to which the symbol is attached.

$\sqrt{\perp}$ Perpendicular to the boundary line.

$\sqrt{x}$ Angular in both directions to the boundary line.

$\sqrt{M}$ Multidirectional.

$\sqrt{C}$ Approximately circular relative to the center of the surface indicated.

$\sqrt{R}$ Approximately radial relative to the center of the surface indicated.

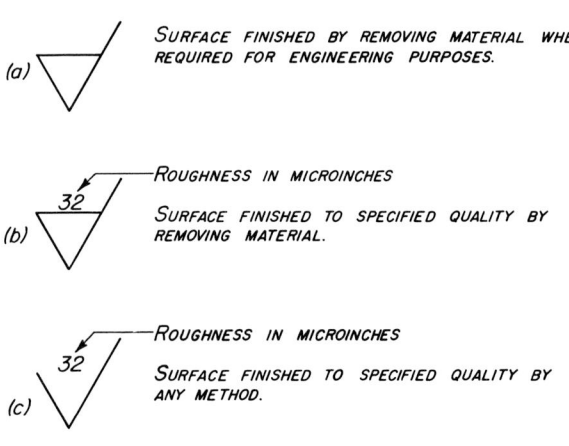
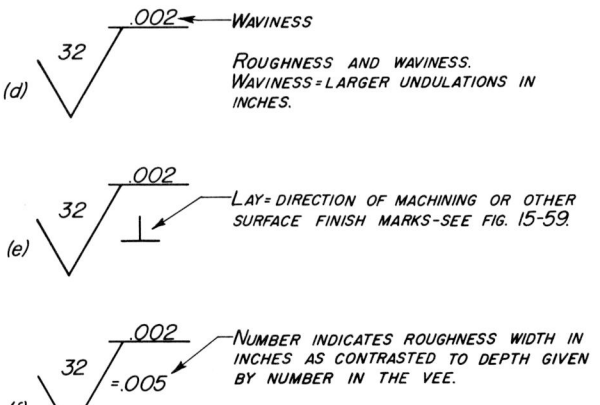

FIG. 12.76. Specific finish marks.

TABLE 12.9 WAVINESS HEIGHT VALUES (INCHES) TO BE USED WITH SYMBOL

0.00002	0.00008	0.0003	0.001	0.005
0.00003	0.0001	0.0005	0.002	0.008
0.00005	0.0002	0.0008	0.003	0.010

TABLE 12.10 PHYSICAL SPECIMENS OF SURFACE ROUGHNESS AND LAY[a]

Type of Surface	Roughness Height (μin)	Lay	Feed (in.)	Minimum Roughness-Width Cutoff (in.)
Honed, lapped, or polished	2	Parallel to long dimension of specimen	—	0.030
	4		—	0.030
	8		—	0.030
Ground with periphery of wheel	4	Parallel to long dimension of specimen	—	0.030
	8		—	0.030
	16		—	0.030
	32		—	0.030
	63		—	0.030
Ground with flat side of wheel	4	Angular in both directions	—	0.030
	8		—	0.030
	16		—	0.030
	32		—	0.030
	63		—	0.030
Shaped or turned	32	Parallel to long dimension of specimen	0.002	0.030
	63		0.005	0.030
	125		0.010	0.030
	250		0.020	0.100
	500		0.030	0.100
Side-milled, end-milled, or profiles	63	Circular	0.010	0.030
	125		0.020	0.100
	250	Angular in both directions	0.100	0.300
			0.100	0.300
Milled with periphery of cutter	63	Parallel to short dimension of specimen	0.050	0.300
	125		0.075	0.300
	250		0.125	1.000
	500		0.250	1.000

[a] Extracted from *American Standard Surface Roughness, Waviness and Lay*, ANSI B46.1—1955, with permission of the publisher, The American Society of Mechanical Engineers.

346 PRODUCTION DIMENSIONING

The final addition to this symbol is the roughness width, placed to the right of the lay symbol as shown in Fig. 12.77. The meanings of all these terms and a complete specification thereof are also shown in this figure.

This symbol may be used completely or any part may be used separately as occasion demands. It should be clearly understood that the specification of surface quality is not the same as specifying limits. In other words, a surface may be given both limit dimensions and surface quality specifications, as indicated in Fig. 12.78.

12.40 CHECKING A DRAWING
Before a drawing is released for production or construction, it must be very carefully checked. Personnel selected to do this work must be thoroughly familiar with construction methods or shop processes as well as being absolute masters of the theory and practice of drafting.

It is good practice for the checker to have an established routine to follow in order that he/she may not overlook some phase of the work. The following checking routine, slightly modified, is used by a large manufacturing concern. Corrections are usually noted on a print of the drawing, which is returned to the person who made the original drawing. This also makes it simpler for the checker to ascertain that all corrections have been made.

1. Does the general appearance of the drawing conform to the standard drafting practice?
2. Is the part sufficiently strong and suitable for the function it has to perform?

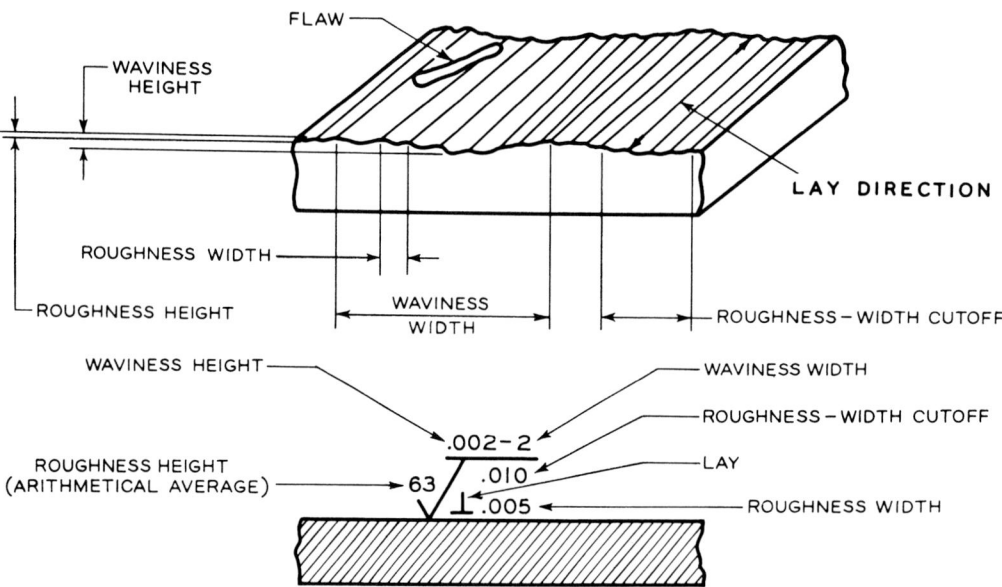

FIG. 12.77. Meaning of the terms roughness, waviness and lag (courtesy ANSI).

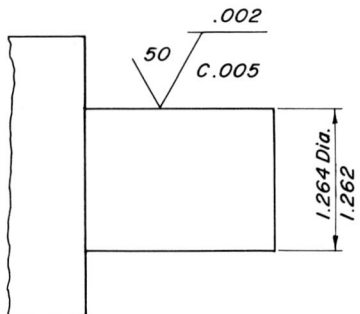

FIG. 12.78. Application of limit dimensions and finish marks.

3. Can the weight be reduced without sacrificing strength or function?
4. Does the drawing represent the most economical method of manufacture?
5. Are all necessary views and sections shown and are they in proper relation to one another?
6. Are all necessary dimensions shown?
7. Do the dimensions agree with the layout and related parts and are duplicate and unnecessary dimensions avoided?
8. Is the drawing to scale?
9. Is the drawing dimensioned to avoid unnecessary calculations in the shop?
10. Are stationary and operating clearances adequate?
11. Can the part or parts be assembled, disassembled, and serviced by the most economical methods?
12. Are proper limits or tolerances specified to produce the desired fits?
13. Have undesirable limit accumulations been avoided?
14. Are proper draft angles, fillets, and corner radii specified?
15. Are all necessary symbols for finishing, grinding, and so forth shown?
16. Are locating points and proper finish allowances provided?
17. Are sufficient notes, including concentricity, parallelism, squareness, flatness, and so forth shown?
18. Is the approximate developed length shown?
19. Is the stock size specified?
20. Are material and heat-treatment specifications given?
21. Are plating and painting specifications, either for protective or decorative purposes, given?
22. Are company trademark, part number, and manufacturer's identification shown according to divisional requirements?
23. Has the title block been filled in completely and is the information correct?
24. Are primary and secondary part numbers identical?
25. Are necessary part numbers of detail parts and subassemblies shown on assembly drawings?
26. Have original lines and drawing information damaged by erasures been properly restored?
27. Are revisions properly recorded?
28. Have all related drawings been revised to conform?

12.41 SIMPLIFIED DRAFTING

Because of high costs of drafting and their effect on total production costs, industrial drafting rooms are making efforts to reduce these costs by simplifying their drafting practice as much as possible without sacrificing clarity of meaning.

The first consideration in making any drawing should be the question, Who will use the drawing? With this question always in mind, the following list will suggest means of reducing drafting costs. Many of these are illustrated in this book.

1. Make freehand sketches of simple objects instead of instrumental drawings.
2. Take advantage of symmetry to reduce drawing time and drawing size.
3. Avoid the use of repetitive detail.
4. Omit unnecessary views and sections. Use partial views and sections where necessary for clarity.
5. Eliminate unnecessary use of letters and notes.
6. Use standard symbols wherever possible, for example, thread and welding symbols.
7. Eliminate inch marks; retain foot marks.
8. Omit drawing of standard bolts, nuts, and other hardware; list them.
9. Reduce hand lettering to a minimum.
10. Avoid ink drawings.
11. Use coordinate dimensioning where applicable.

SELF-STUDY QUESTIONS

Before trying to answer these questions, read the chapter carefully. Then, without reference to the text, answer as many questions as possible. For those that cannot be answered, the number in parentheses following the question number gives the article in which the answer can be found. Look it up and write down the answer. Check the answers that you did give to see that they are correct.

12.1 (**12.2**) Assembly drawings are used to show the parts of a machine in their _____ _____ position.

12.2 (**12.5**) Assembly drawings are used in the shop or field to show how parts are to be _____.

12.3 (**12.6**) Assembly drawings of such large structures as bridges are called _____ _____.

12.4 (**12.8**) Dimensioning of assembly drawings is based upon _____ for which they are intended.

12.5 (**12.10**) Hidden lines are _____ shown in assembly drawings.

12.6 (**12.12**) When selecting parts of a design such as bolts, screws, and springs, _____ sizes should be used.

12.7 (**12.12**) Holes, countersinks, counterbores, and other common features should be specified so that they can be produced by standard _____.

12.8 (**12.16.1**) The term *features* means parts of an object such as _____ and _____.

12.9 (**12.16.8**) The term *limits* is defined as the _____ permissible dimensions of a part.

12.10 (**12.16.9**) The term *tolerances* is defined as the _____ amount of variation permitted in the size of a part.

12.11 (**12.16.10**) The term *allowance* is the intentional _____ in the size of mating parts.

12.12 (**12.16.11**) A feature is said to be at maximum material condition when it contains the most _____.

12.13 (**12.16.13**) A line, plane, or surface from which other parts are dimensioned is called a _____.

12.14 (**12.19**) The three systems of tolerancing are known as _____, bilateral, or _____ systems.

12.15 (**12.20.1; 12.20.2**) The two systems for computing and tolerancing the fit between shafts and their associated bearing holes are known as the basic _____ method and the basic hole method.

12.16 (**12.23.1; 12.23.2**) Two systems of coordinate dimensioning are the _____ coordinate and the polar coordinate systems.

12.17 (**12.24.1**) If chain dimensioning is used in the coordinate systems, an _____ of tolerances may occur.

12.18 (**12.25**) An accumulation of tolerances may be prevented by using _____ surfaces.

12.19 (**12.28**) When the size, form, and location of a part are critical to function or interchangeable assembly, geometric form and _____ position dimensioning should be used.

12.20 (**12.29**) True-position dimensioning corresponds to the _____ of errors that normally arise in production.

12.21 (**12.29**) True-position dimensioning corresponds to the control established by _____ gages.

12.22 (**12.29.2**) For rectangular coordinate dimensioning the shape of the tolerance zone is a _____.

12.23 (**12.29.2**) For true-position dimensioning the shape of the tolerance zone is a _____.

12.24 (**12.34**) When greater _____ is required than that given by size and location tolerances, geometric form _____ should be used

In the following questions show your answer by a small sketch.

12.25 (**12.35.2**) The datum symbol is shown by

12.26 (**12.35.3; 12.35.4**) The straightness and flatness symbols are shown thus:

12.27 (**12.35.5**) Parallelism is shown by the symbol

12.28 (**12.35.6**) Squareness is shown by the symbol

CHAPTER 13

AXONOMETRIC PROJECTION

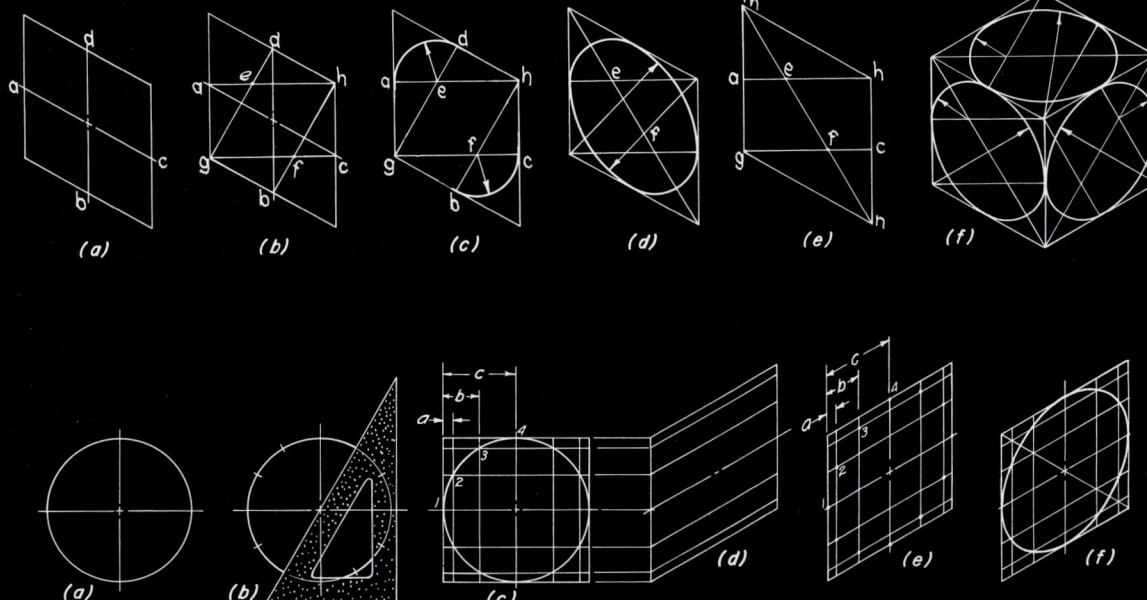

CHAPTER 13

13.1 INTRODUCTION

The term *axonometric projection* or drawing includes three forms of pictorial drawing. The most common, *isometric,* is widely used for the illustration of catalogues, assembly diagrams, piping diagrams and proposals for engineering projects. The other two, *dimetric* and *trimetric,* are seldom drawn by hand but are extensively used in computer generated drawings where the object is rotated until the most descriptive view is given. To determine this optimum position manually is nearly impossible.

13.2 ISOMETRIC PROJECTION BY ROTATION

In isometric projection, all three faces of an object shows in equal proportion, since the faces make equal angles with the plane of projection as illustrated for the cube in Fig. 13.1. The cube is in its normal position for orthographic projection in Fig. 13.1a. In Fig. 13.1b it has been rotated about the vertical edge OC (see top view) until the sides make 45° with the vertical plane. In Fig. 13.1c the cube has been tilted up until the body diagonal is perpendicular to the vertical plane, which is sometimes referred to as the picture plane. This places the three front edges of the cube at an angle of 35° 16′ with the picture plane. The angle marked in the side view of Fig. 13.1c is the only one that shows in true size. In the pictorial all edges project in equal length but not in the true length. Hence measurements can be made along these edges or axes at the same scale. In Fig. 13.1d the face diagonals AB, BC, and AC appear in true length.

13.3 ISOMETRIC PROJECTIONS BY AUXILIARY VIEWS

The same result can be more easily obtained by making an auxiliary view of a cube with the first auxiliary view parallel to the body diagonal as shown in Fig. 13.2. The second auxiliary plane is perpendicular to the body diagonal. The isometric views in both Figs. 13.1 and 13.2 are true projections. An examination of the two foregoing figures will show that for objects more complicated than a cube the process of making a true projection by either method would be quite tedious and time-consuming. Thus isometric projections are seldom made except when they are computer generated.

13.4 ISOMETRIC DRAWING COMPARED WITH ISOMETRIC PROJECTION

By observing certain facts from Figs. 13.1 and 13.2 it is possible to make a drawing that looks like an isometric projection but is not a true projection.

At this point the students should be cautioned not to permit themselves to quibble over the fact that any projection is a drawing. Every illustration in this book, regardless of kind or scale or object

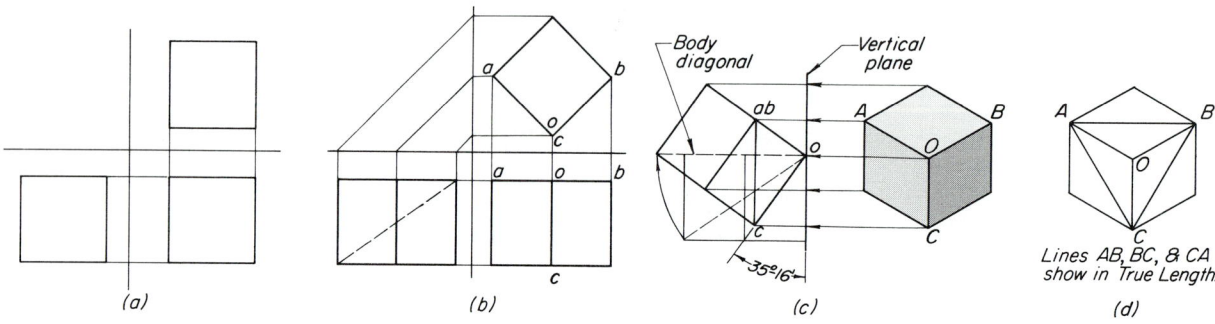

FIG. 13.1. Isometric view by turning cube.

represented, is a drawing but only certain ones are also true projections.

13.4.1 Isometric Projection. In an isometric projection the lines are foreshortened in a mathematical proportion, which can be obtained graphically by any one of four methods.

a. By rotation as in Fig. 13.1.
b. By auxiliary views as in Fig. 13.2.
c. By the use of an isometric scale as in Fig. 13.3.
d. By direct orthographic projection from properly placed orthographic views, as explained near the end of this chapter. See Article 13.21.2.

Direct projections can be made from the orthographic view of an object by methods a or b. Method c, although possible, is quite impractical since the isometric scales are not on the market and it is too time-consuming to make a scale like those of Fig. 13.3.

Method d is very accurate and useful for complicated objects. It is explained in Section 13.21.

13.4.2 Isometric Drawing. It is the common practice to make isometric, dimetric, and trimetric drawings rather than true projections. The axonometric drawings are made by "rule of thumb" based upon a careful study of axonometric projections. These methods are explained in the following paragraphs.

The only visual difference between isometric drawings and isometric projections is in their relative size. If an isometric drawing is constructed from a three-view orthographic drawing at the same scale as the three-view drawing, it will be 1.224 times as large as a direct projection made from the same three-view drawing. See Fig. 13.4.

In the case of dimetric and trimetric drawings there is an additional factor of a slight distortion because the scales used on the three axes are usually only convenient approximations of what they should be for true projection. In appearance, however, they may be equally satisfactory. In general, however, computer generated dimetric and trimetric drawings will not exhibit this distortion.

13.5 DEFINITION OF TERMS

At this point some of the more common terms used in this chapter are defined.

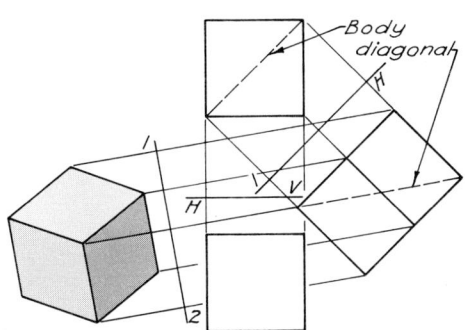

FIG. 13.2. Isometric view by auxiliary projection.

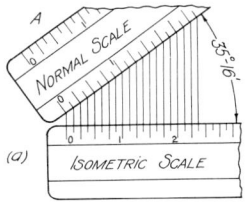
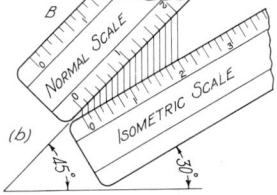

FIG. 13.3. Construction of isometric scales.

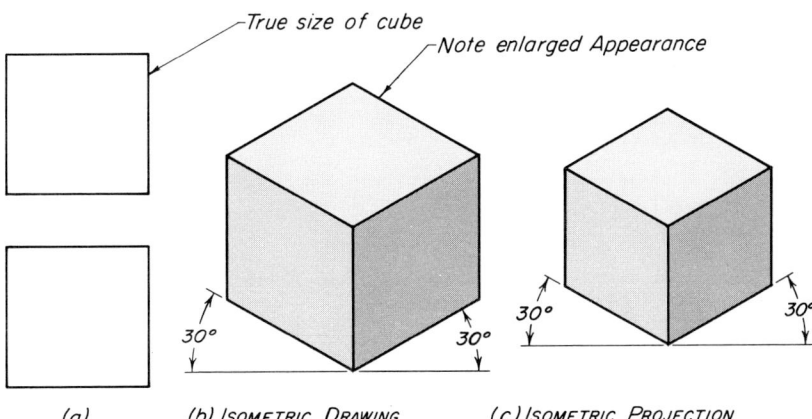

FIG. 13.4. Isometric projection and isometric drawing compared.

Axes. In Fig. 13.5a the three edges of the cube OX, OY, and OZ are referred to as axes. Much of the construction is built around these three lines.

Axonometric. This term includes isometric, dimetric, and trimetric. In all of these, the three edges of a box that are perpendicular to each other are called *axes*.

Isometric. This term literally means equal measurement. Measurements along the axes are made to the same scale.

Dimetric. The word implies two scales of measurement. The same scale is used along two axes but a different one is used on the third. The construction is the same as that for isometric except that two scales are used.

Trimetric. The term implies three scales of measurement. Different scales are used on each of the three axes. The construction is the same as that for isometric except that three scales are used.

Isometric lines. Lines parallel to any one of the isometric axes. See Figs. 13.5b and c.

Nonisometric lines. Any lines not parallel to one of the axes. See Figs. 13.5c and d.

(a) Inclined — in an isometric plane but not parallel to any axis. See Fig. 13.5d.

(b) Oblique — not in any isometric plane nor parallel to one. See Fig. 13.5c.

Isometric plane. Any plane containing two isometric axes or two lines each parallel to one of two axes. See Fig. 13.5a.

Nonisometric planes. It is useful to identify two types of nonisometric planes: those that are inclined to two isometric planes and perpendicular to the third as in Fig. 13.6a and those that are oblique to all isometric planes as in Fig. 13.6b.

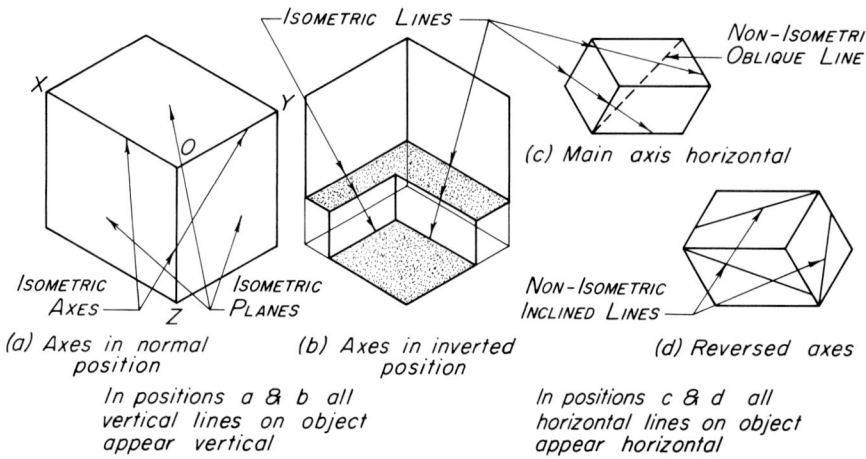

FIG. 13.5. Meaning of terms.

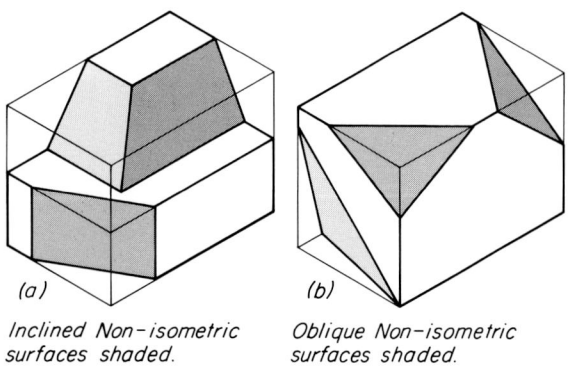

FIG. 13.6. Definition of terms.

13.6 ISOMETRIC OF PLANE FIGURES COMPOSED OF STRAIGHT LINES

The isometric drawing of a solid object consists mainly in representing three, more or less irregular plane faces, which are parallel or inclined to the faces of the isometric cube. As a prelude to the drawing of more complicated solids the construction of plane figures in isometric is presented first.

13.6.1 Details of Constructing a Plane Figure.
The construction proceeds as follows.

a. In Fig. 13.7a, an irregular seven-sided figure is shown in true shape.
b. In Fig. 13.7b, this figure has been enclosed in a rectangle and the coordinates of the corners of the figure relative to the box have been indicated.
c. In Figs. 13.7c and d, an isometric of the rectangle has been made in two different positions in order to show more than one arrangement.
d. Locate the points m, o, p, q, r, and s that lie on the edges of the rectangle. This is done by transferring the distances such as 1, 5, 4, 7, and 8 from Fig. 13.7b to both isometric rectangles on the proper sides.
e. Locate the corner n by transferring the distances 3 and 6 to the sides of the rectangle. Then draw isometric lines across the figure until they intersect at n.
f. Repeat the process for point r using distances 2 and 9 as shown in Figs. 13.7c and d.
g. Connect the plotted points.

13.6.2 Angles in Isometric.
It should be noted that angles cannot be shown in their true value in isometric drawing. Two points must be used. One may be assumed and the other laid out by coordinates from the first. For angles of 15, 30, 45, and 60° the values of the sine and tangent may be used to plot the points as shown in Fig. 13.8.

13.7 CIRCLES AND CURVES IN ISOMETRIC
A circle will appear as an ellipse in isometric. Several methods of constructing an ellipse in isometric are available. Some are exact and some approximate. Coordinate methods are exact.

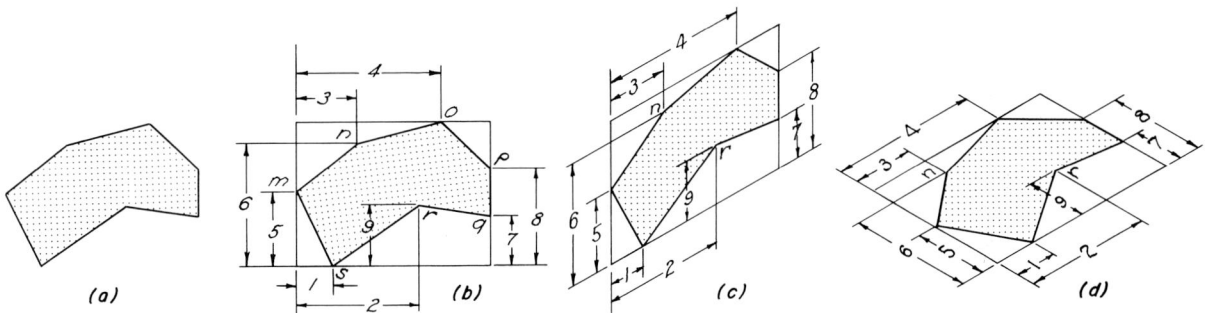

FIG. 13.7. Construction of plane figures in isometric.

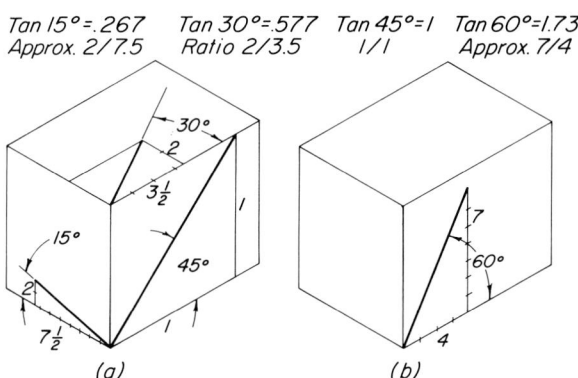

FIG. 13.8. Angles in isometric.

13.7.1 Isometric of Circles. The coordinate method 1. A circle may be constructed by the coordinate method as shown in Fig. 13.9. The procedure is as follows.

a. Divide the circle into 12 equal parts with a 30–60° triangle as shown in Fig. 13.9b.
b. Enclose the circle in a square as shown in Fig. 13.9c, and draw horizontal and vertical lines through the points on the circle to determine the coordinates of each point.
c. Draw the isometric square as shown in Fig. 13.9d. Note that the horizontal coordinate lines have been extended across from the circle in Fig. 13.9c.
d. Draw the vertical coordinates by transferring distances a, b, c, and so forth from the circle in Fig. 13.9c to the parallelogram as shown in Fig. 13.9e.
e. The intersection of the coordinates locates points 1, 2, 3, and so forth on the ellipse.
f. Draw a smooth curve through the points (Fig. 13.9f).

13.7.2 Isometric of Circles by the Four-Center Approximate Method. An approximate isometric of a circle may be drawn by the method shown in Fig. 13.10. This construction depends on the fact that the center of a circle that is tangent to a straight line lies on the perpendicular to the line at the point of tangency. Hence if we erect perpendiculars at the midpoints a, b, c, and d of the sides of the isometric square, as in Fig. 13.10b, these perpendiculars will intersect in pairs, thus locating the centers of the four arcs, e, f, g, and h, as in Fig. 13.10c. This will approximate the correct ellipse. It will be noted that in isometric two of these centers lie on the corners of the

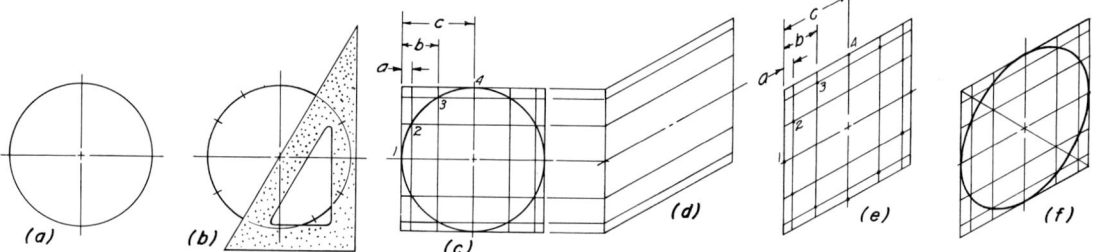

FIG. 13.9. Construction of a circle in isometric by the coordinate method.

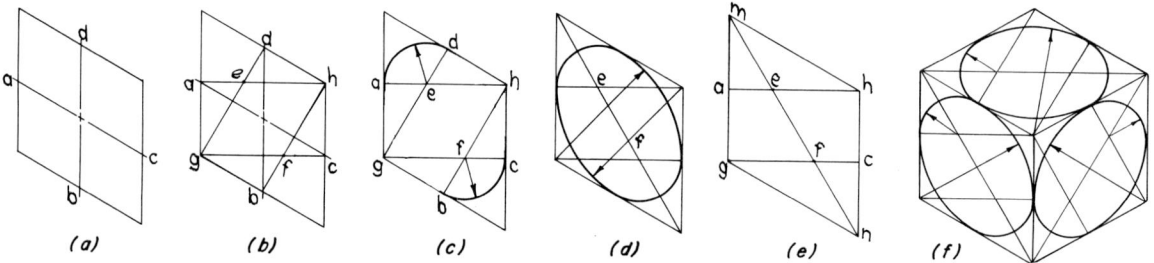

FIG. 13.10. Four-center approximate method of constructing a circle in isometric.

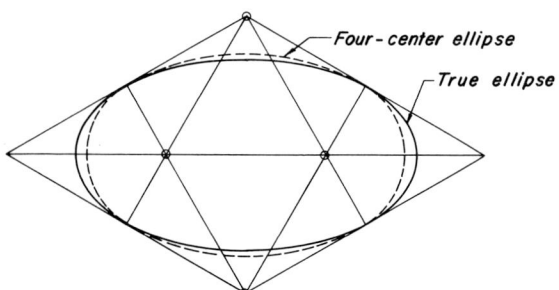

FIG. 13.11. True ellipse and four-center ellipse compared.

square and the other two lie on the long diagonal. Use of these facts enables the drafter to shorten the construction considerably by drawing only the lines *ah*, *gc*, and *mn*, as in Fig. 13.10e. This construction can be used in any isometric face of a cube, as illustrated in Fig. 13.10f. The method involves less labor than the coordinate method and is sufficiently accurate for most isometric work, except as noted in Article 13.7.4. This approximate ellipse has a shorter major axis and longer minor axis than the true ellipse, as shown in Fig. 13.11.

13.7.3 Isometric of Irregular Curves.
In Fig. 13.12, a section of an ogee molding is shown in orthographic and then in isometric. The coordinates to be used are shown in both the orthographic and the isometric. These may be transferred with dividers. The second curve is made parallel to the first by offsets as noted.

13.7.4 Limitation of the Four-Center Approximate Method.
The four-center method may be used for tangent circles or arcs only when they are tangent to each other at the midpoint of the sides of their enclosing rectangles, as shown in Fig. 13.13a. If the tangency points occur at other places, the circles will overlap or miss, as shown in Fig. 13.13b, because of their departure from the true ellipse. In such cases the coordinate method should be used or other approximations made.

The four-center method may be used in pictorial forms other than isometric but only when the scale on the two adjacent sides of the enclosing square is the same. Thus it can be used in the following cases.

a. Dimetric—in one face only.
b. Trimetric—not at all.
c. Only in the cavalier form of oblique projection.

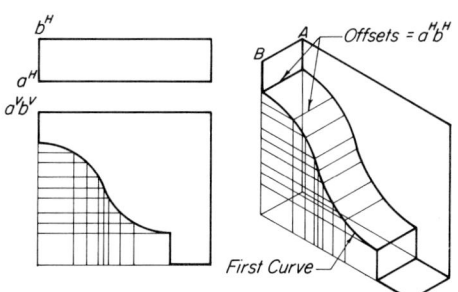

FIG. 13.12. Curves by the coordinate method.

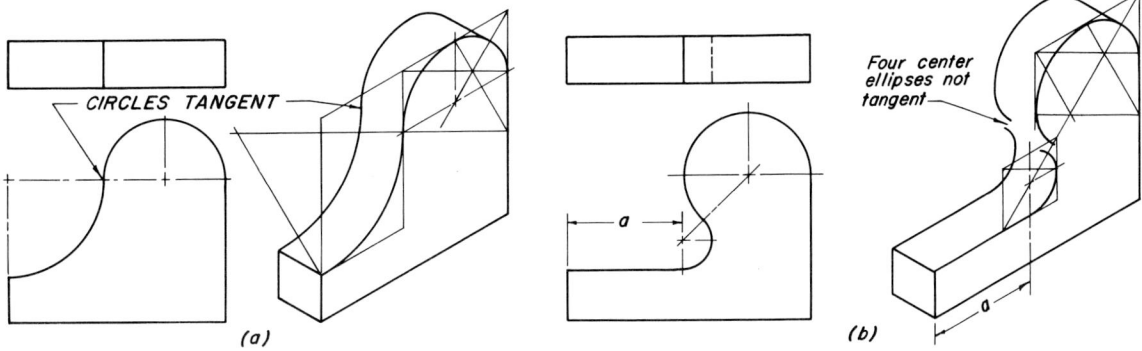

FIG. 13.13. Limitations of the four-center approximate method.

13.8 ISOMETRIC DRAWING OF SOLIDS: BOX METHOD

From the drawing of plane figures to the drawing of solid objects in isometric is but a simple step, involving only the use of a third coordinate distance. The steps in the procedure are as follows.

a. Draw the orthographic views of the object to the same scale as that to be used on the isometric.
b. Enclose the views in the smallest enclosing rectangular box.
c. Draw the enclosing box in isometric in the position that will best reveal the shape of the object, making the three edges at 120° with each other (Fig. 13.14b).
d. Draw the simple parts of the object that lie in or adjacent to the faces of the box.
e. Plot the curves and interior points, if any, by the coordinate method.
f. It is the usual practice to omit all invisible lines in pictorial drawing.

13.8.1 Isometric of a Block: Box Method of Construction.
An isometric drawing of the block, shown by three orthographic views in Fig. 13.14a, may be readily constructed in the following manner. The first step as outlined above consists of enclosing the orthographic views of the object in the smallest rectangular box that will just enclose it as shown by the light lines of Fig. 13.14a. This box serves as a reference frame from which dimensions can be measured in the orthographic views and plotted in the isometric.

The second step consists of drawing the isometric of the enclosing box in the position desired, as shown in Fig. 13.14b. Here the front orthographic view has been made the left face. The various parts that have been cut out of the block can now be cut from the isometric box in any order desired. Thus, in Fig. 13.14c, the distance (a^H-1) on the top view has been measured on the same line, AC, in the isometric and the distance (a^H-2) along the line AB. The diagonal line 1-2 can now be drawn in the isometric view. From the points 1 and 2 in the isometric, vertical lines can be dropped to the bottom of the box, and then the diagonal in the bottom face can be drawn.

In a similar manner the other cutouts can be transferred by direct measurement from the orthographic to the isometric, as illustrated in Figs. 13.14d and e. *It should be carefully noted that in all cases measurements are made on or parallel to the three isometric axes. They can be made in no other manner for no other lines are foreshortened in the same ratio as these lines.*

13.8.2 Solids Involving Nonisometric Lines.
As a second illustration, the construction of a truncated hexagonal pyramid is shown. This object has only two isometric lines. The remainder are all nonisometric. In Fig. 13.15a, the object is shown enclosed in a rectangular box; in Fig. 13.15b, the box has been drawn in isometric and the hexagonal base is shown in the bottom of the box. The measurements a and b for constructing

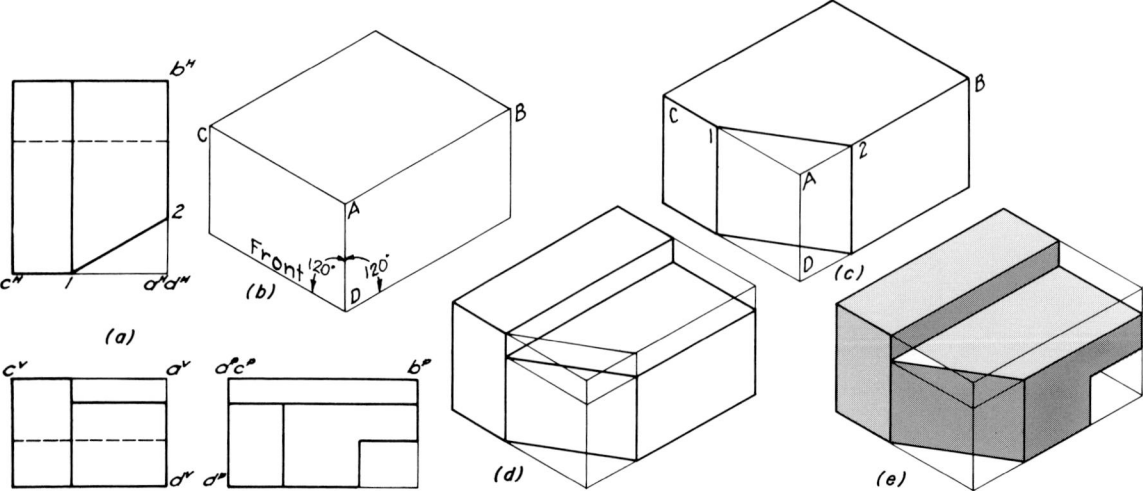

FIG. 13.14. Box method of drawing a solid in isometric.

this plane figure are obtained from the top view, as shown in the figure.

Whenever an object has a plane of symmetry, advantage should be taken of this fact to speed construction. Hence, in Fig. 13.15c, the central plane of symmetry has been established and the two points 2 and 9 located in it to give the center line of the truncated face (2-9). For example, point 2 is located by measuring the coordinate 1-2 in the central plane as indicated. Point 9 is located by going up along the center line from 0 to 9, using the distance 0-9 in the front view. Points 3, 4, and 5 are located by taking the measurements (1-3), (1-4), and (1-5) from the top or front views. By dropping perpendiculars from 3, 4, and 5 to the center line (2-9), points 6, 7, and 8 are located as shown in Fig. 13.15c. Isometric horizontal lines can then be drawn through points 6 and 7. Points 10 and 11 can be located by stepping off from point 6 on these lines the distances (6-10) and (6-11), which are equal. A similar procedure locates points 12 and 13 as shown in Fig. 13.15d. All these measurements are taken from the top view. The six points in the truncated face are then connected to form the sloping truncated face. The corners of this face are then joined to the corresponding corners of the base, thus completing the isometric as shown in Fig. 13.15e. It is customary in isometric drawing, as in all other pictorials, to omit hidden lines unless they are necessary to make clear the shape of the object. *Note again that all measurements were taken on or parallel to isometric lines.*

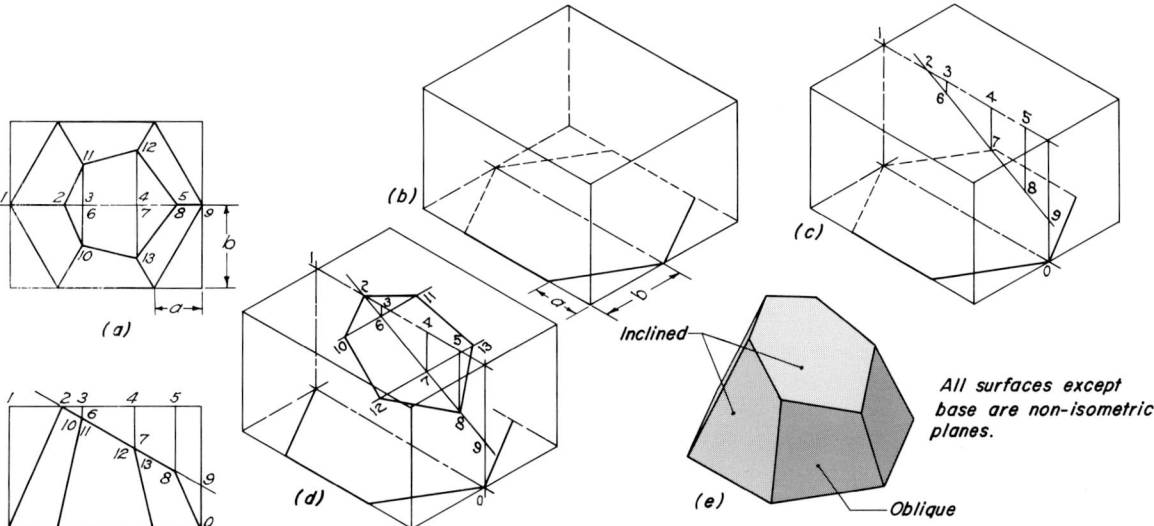

FIG. 13.15. Truncated pyramid in isometric by the box method.

13.9 SOLID OBJECTS INVOLVING CIRCLES

The objects illustrated thus far have been composed entirely of straight lines. Many objects, however, involve circles either singly or in groups. The following suggestions will assist in speeding construction and in avoiding the most common errors.

13.9.1 Parallel Circles or Other Curves.
In actual drawing, circles nearly always occur in

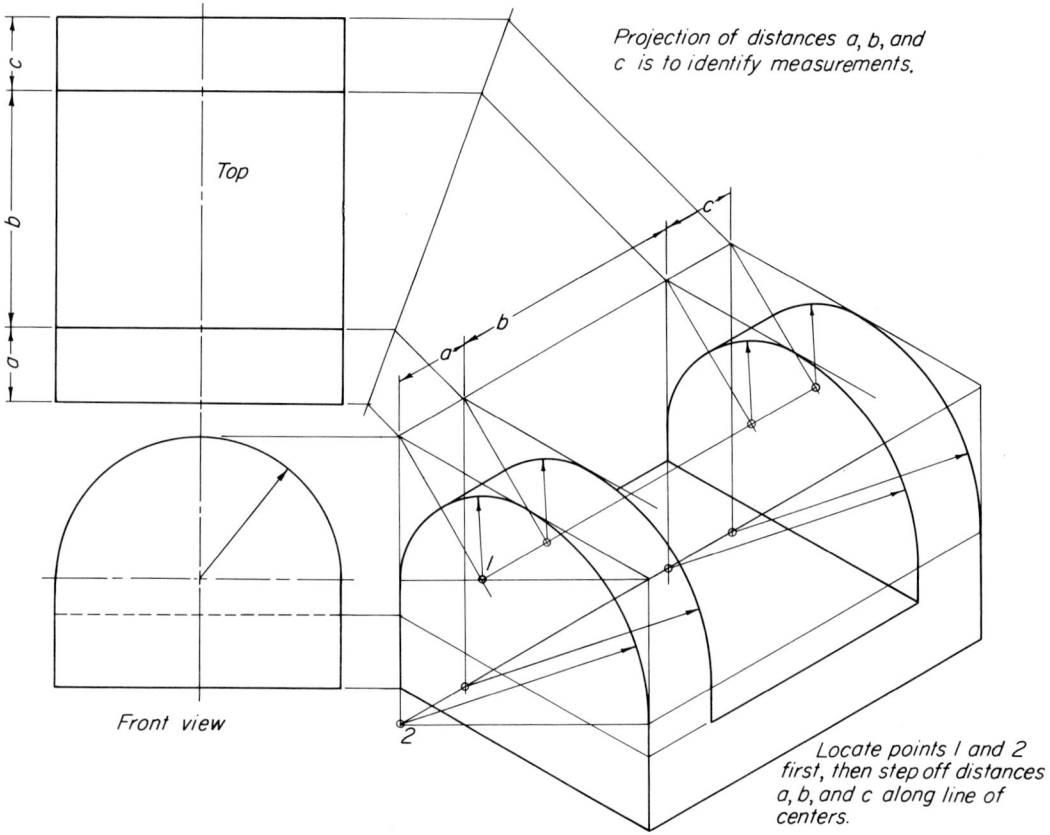

FIG. 13.16. Shortcut in drawing parallel circles in isometric.

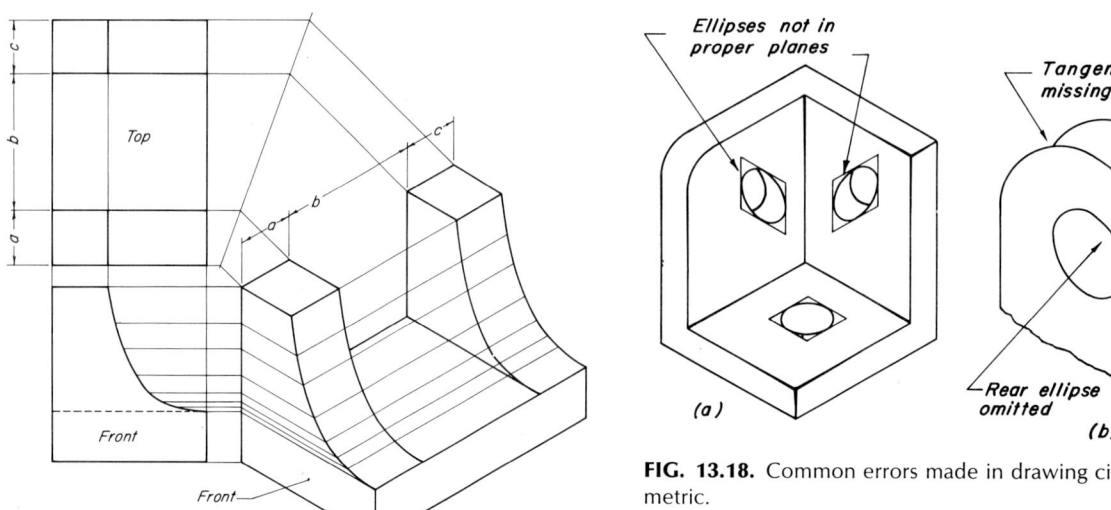

FIG. 13.17. Shortcut in drawing parallel curves in isometric.

FIG. 13.18. Common errors made in drawing circles in isometric.

pairs. Since rapidity in construction is always important, the suggestions for speeding the layout of circles parallel to each other, as shown in Fig. 13.16, are valuable.

In the four-center method the centers for the first circles are found in the usual way. Circles parallel to the first may be quickly found by drawing isometric lines from the original four centers and stepping off on them the distance between the circles, to locate the new centers, as shown in Fig. 13.16.

The same scheme may be used for a curve plotted by the coordinate methods, as shown in Fig. 13.17, for a noncircular curve. One curve is drawn in the usual way, and isometric lines are drawn from the plotted points. Each successive curve may be stepped off with one setting of the divider.

13.9.2 Common Errors in Drawing Circles.
Two common errors are frequently made by the student in drawing circles on various objects. One of these consists of drawing the circle out of the proper isometric plane, as shown in Fig. 13.18a. This can be avoided by making sure that the sides of the enclosing parallelogram are parallel to the isometric lines of the plane in which the circle lies, as shown for the circle in the lower face of the object in Fig. 13.18a.

A second error occurs in the drawing of short cylinders or cylindrical parts where the student fails to put in the isometric tangent line between the circles, as shown in Fig. 13.18b. The far side of small holes is omitted many times.

13.10 USE OF ISOMETRIC PLANES IN CONSTRUCTION

The simple bearing bracket shown in Fig. 13.19a will serve to illustrate further the method of construction, which has again been broken down into a series of successive steps, following the method previously suggested. Figure 13.19a shows the orthographic views enclosed in a box. Figure 13.19b shows the box in isometric with the base and cylindrical bearing partly completed. The circles are drawn in their proper planes by the four-center method based on the enclosing rectangles, which are shown.

In an object of this kind, considerable time can be saved by noting that much of the construction falls naturally into isometric planes. To plot the vertical and sloping webs, a series of horizontal planes d, e, f, and so forth are drawn in the

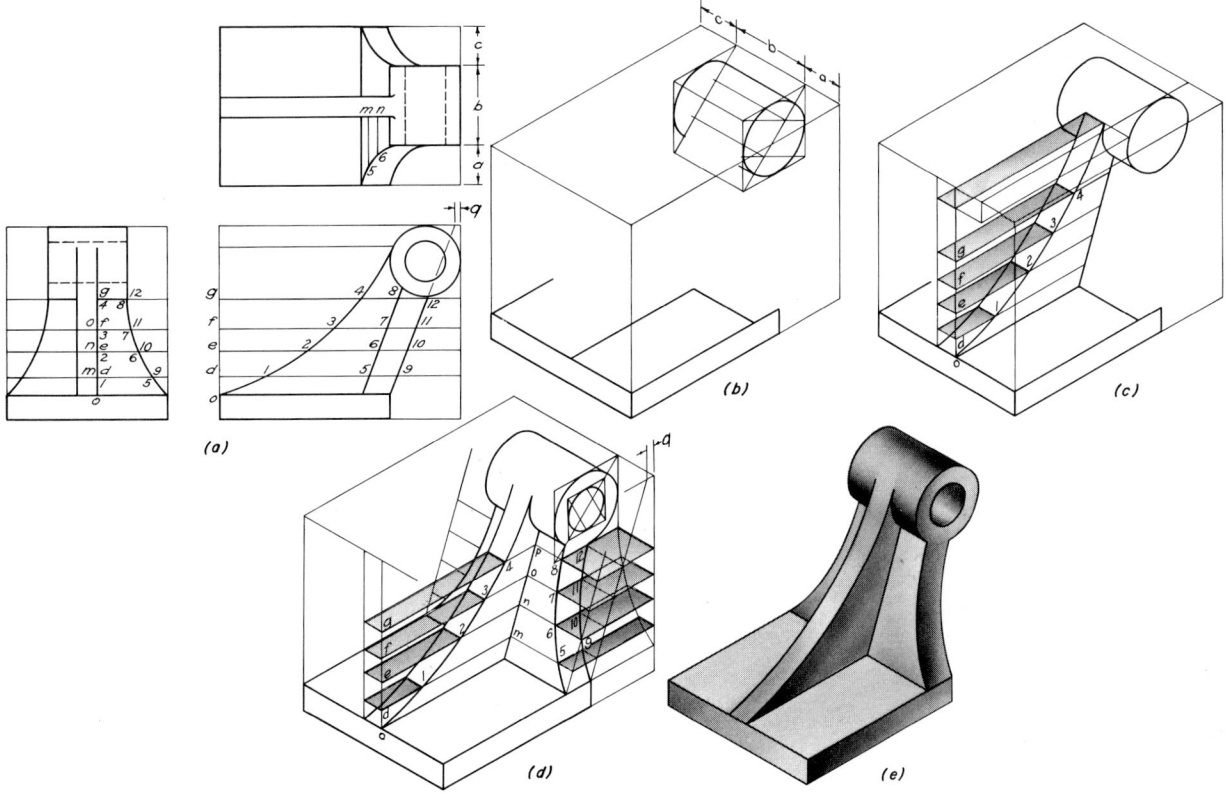

FIG. 13.19. Layout in parallel isometric planes that speeds construction.

orthographic views locating points 1 to 12 on the curves. In Fig. 13.19c, the end view of the vertical web is drawn in the end of the isometric box, and on it the points d, e, f, and so forth are located by measurements (o-d), (o-c), and so forth. From these points d, e, f, and so forth, isometric horizontal lines are drawn and measurements (d-1), (e-2), and so forth made on them, thus locating points 1, 2, 3, and so forth on the curve, which can then be drawn as in Fig. 13.19c.

The front view of the sloping web is next constructed in the right face of the isometric box, as shown in Fig. 13.19d. It is a simple matter to carry the m, n, o, p horizontal planes around the box from d, e, and so forth to locate the points (5, 6, 9, and 10), and so forth. From these points horizontal lines can be drawn in isometric, and the distances (m-5), (n-6), and so forth, obtained from the three-view drawing in Fig. 13.19a, can be measured on them, thus establishing points 5, 6, 7, and so forth. The other curves may be found in a similar manner. The completed drawing with all construction removed is shown in Fig. 13.19e.

13.11 CONSTRUCTION BY CENTER LINE LAYOUT

The box method of construction discussed in preceding paragraphs may be used for any type of object. However, when the object consists of a number of circular faces lying in the same or parallel planes, the center line layout shown in Fig. 13.20 is a convenient and rapid method of construction. In Fig. 13.20a the orthographic views are shown, and in Fig. 13.20b the isometric layout of the principal center lines are drawn. In Fig. 13.20b the parallelograms for some of the circles are drawn, and in Fig. 13.20c the drawing is completed.

In general, three-dimensional center lines for holes and cylinders will be shown so far as they do not make the drawing more difficult to read.

13.12 SECTIONAL VIEWS IN ISOMETRIC

The interior construction of complicated objects is best shown by sectional views. Half and full sections may be made by removing one fourth or one half of the object, respectively. The cutting

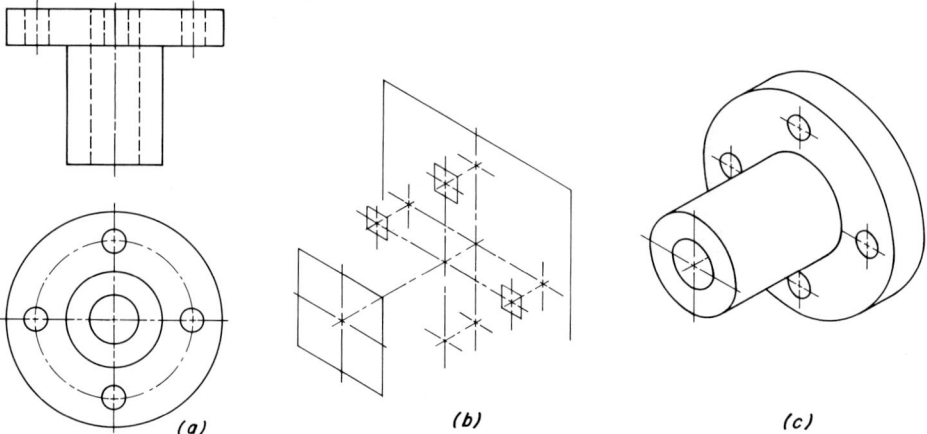

FIG. 13.20. Center-line method of constructing cylindrical objects in isometric.

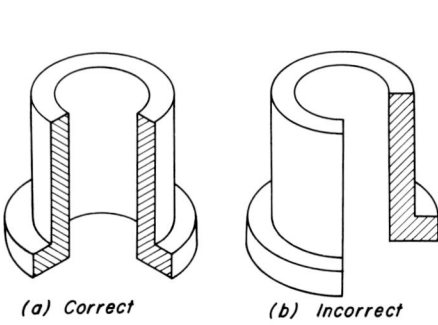

FIG. 13.21. Sectional views in isometric.

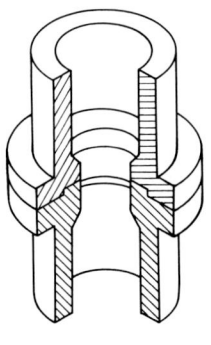

FIG. 13.22. Incorrect crosshatching.

13.14 ISOMETRIC OF DOUBLE-CURVED SURFACES

planes should always be isometric planes as shown in Fig. 13.21a. In a half section the cross-hatching lines should be drawn in a position to give the effect of coincidence if the two sectioned faces were revolved together. Correct and incorrect examples are given in Figs. 13.21 and 13.22 to illustrate this point. No new principles of construction are involved in making section views.

The step-by-step construction of a sectional view is shown in Fig. 13.23. By beginning with the sectioned parts, as shown in Figs. 13.23b and c, a minimum of construction lines need be used. Careful study of the figure will show the procedure. The finished sectional drawing is shown in Fig. 13.23f.

13.13 SPHERES AND OTHER CURVED PARTS IN ISOMETRIC

Spherical parts occur on pieces of machinery and can readily be drawn in isometric. The sphere appears as a true circle. A simple lever involving spheres and curved handle is shown in Fig. 13.24.

13.14 ISOMETRIC OF DOUBLE-CURVED SURFACES

On some objects, such as the pipe return bend shown in Fig. 13.25, an enveloping curve representing the outstanding contour of the object must be drawn. This curve does not lie in a single plane, hence it cannot be constructed by plotting points in the usual way.

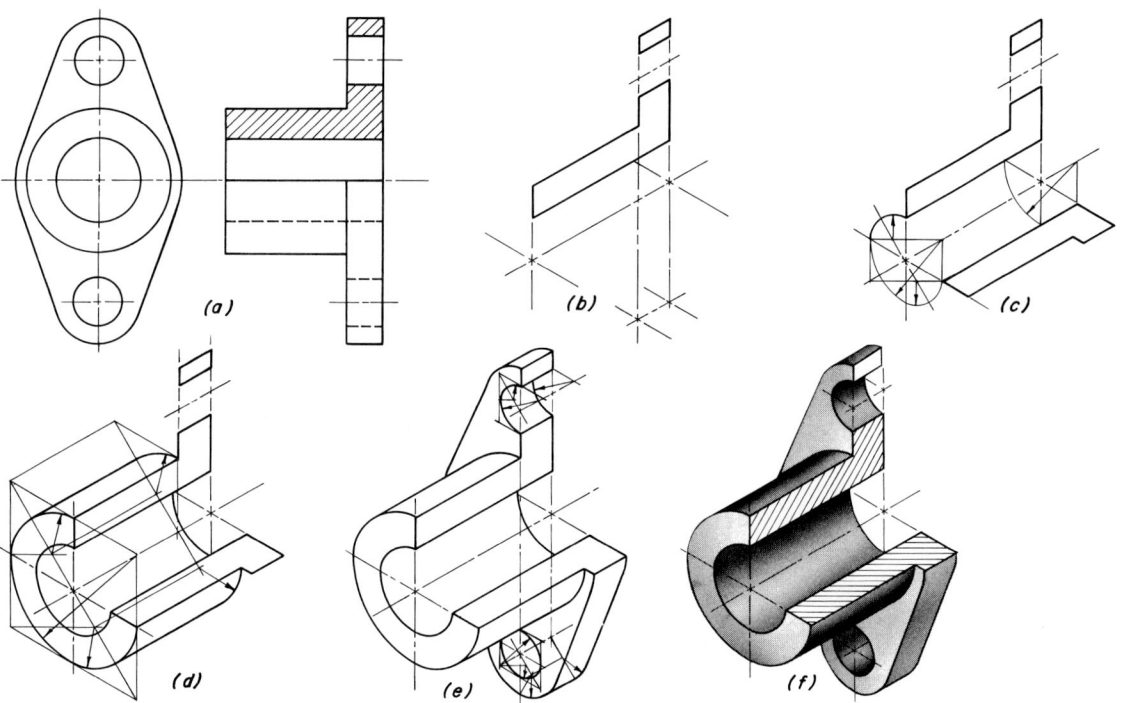

FIG. 13.23. Step-by-step construction of an isometric sectioned view.

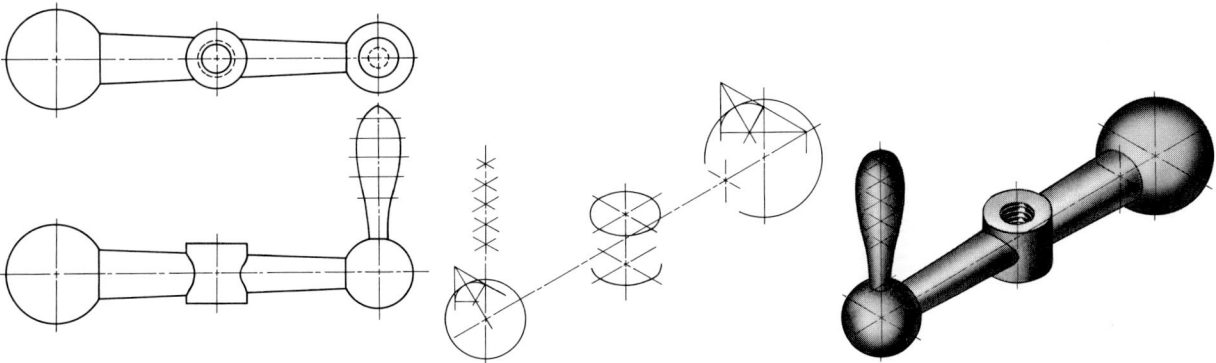

FIG. 13.24. Construction of object with curved surfaces.

Since a sphere projects as a circle in isometric, a simple method of making this, or any similar construction, is shown in Fig. 13.25a. A series of spheres may be imagined lying in the bend just tangent to it. The centers of five or six of these spheres may be located on the isometric of the center-line circle, as shown in the illustration by points a, b, and so forth. Next the size of the isometric sphere is obtained, as shown in Fig. 13.25b, by making the circle tangent to the isometric ellipse. Only the major axis of the ellipse needs to be drawn to determine the diameter of the spheres. With the radius thus determined, the arcs and a smooth curve tangent to them may be drawn.

13.15 SCREW THREADS IN ISOMETRIC

Screw threads could be accurately drawn in isometric, but the process is so laborious that a conventional scheme that is quite satisfactory has been adopted. Arcs of a series of parallel circles are used to represent the crest lines only because the root lines need not be shown. Any method of drawing the circles may be used, but the construction for the four-center approximate method is illustrated in Fig. 13.26.

Because of symmetry of construction, Square threads and Acme threads cannot be clearly shown in isometric. Dimetric or trimetric layouts are much more suitable for this purpose.

13.16 POSITION OF ISOMETRIC AXES

Thus far we have considered isometric drawing with the object always in one position. The three axes, however, may be drawn in an infinite number of positions so long as they always make equal angles (120°) with each other. Four easily drawn positions, as shown for the object and

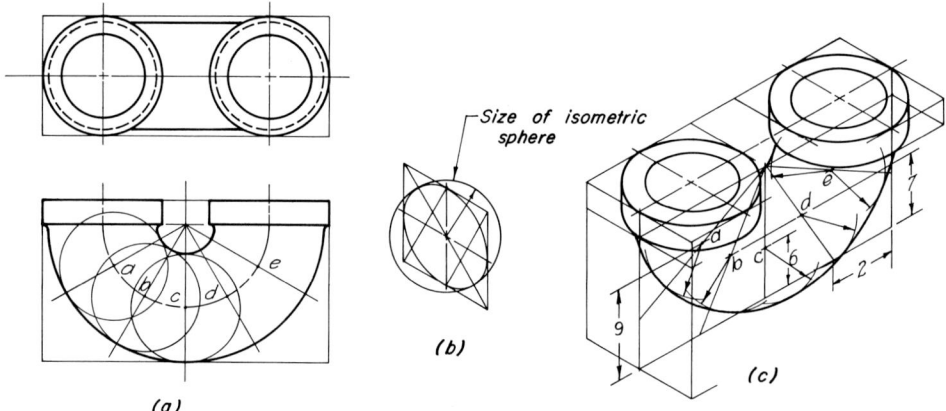

FIG. 13.25. Tangent sphere method of drawing a double-curved surface.

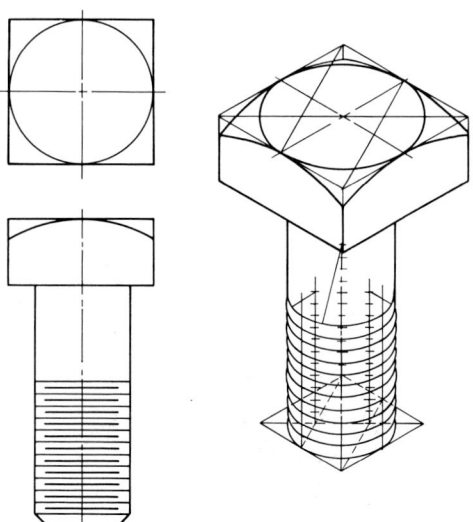

FIG. 13.26. Screw threads and square head bolt in isometric.

enclosing box in Fig. 13.27, are most commonly used.

The choice of the position of these axes will depend on the nature of the object. When the top and sides of the object contain most of the details, the position used thus far is best. If, on the other hand, the bottom contains the more important details, the position of Fig. 13.27b is by far the best. The object should, of course, be shown in the natural or normal position if it has one.

13.17 ADVANTAGES OF ISOMETRIC

As compared with two- and three-view orthographic projections, isometric has the advantage of showing three sides of the object in one view, thus giving a more realistic picture of it.

As compared with other forms of pictorial drawing, isometric has the advantage of being easily constructed since the same scale is used on all sides. Circles can be readily approximated by the four-center method. It can be scaled and dimensioned. It is flexible in the position in which an object may be shown but not as flexible as other types, particularly oblique projection described in a later chapter. Circles are not distorted as in oblique and sometimes in perspective.

Against these advantages may be placed definite disadvantages that limit its usefulness in certain situations. Long objects with parallel sides show a disagreeable distortion since the eye is accustomed to the perspective effect of long parallel lines that appear to approach each other. There is also an exactness of symmetry causing an overlaying of lines in some symmetrical objects, which makes the isometric difficult to read.

13.18 DIMENSIONING ISOMETRIC DRAWINGS

For shop purposes, other than assembly work, an isometric drawing must be dimensioned. The regular rules and suggestions for dimensioning two- or three-view working drawings hold for isometric drawing in a general way, but in addition the following rules must be observed.

13.18.1 Pictorial Plane Dimensioning. Dimensions on isometric drawings should be placed in such a way that they can be read from one point of view, which should be from the bottom of the sheet. This may be said to encompass all other rules in regard to the direction on which dimensions should read, and it is the only safe one to follow at all times. It is best to dimension the visible faces.

a. All dimension lines must be isometric lines and lie in isometric planes. This point must be carefully observed. Difficulty usually occurs in objects having nonisometric lines. Figure 13.28a illustrates a very com-

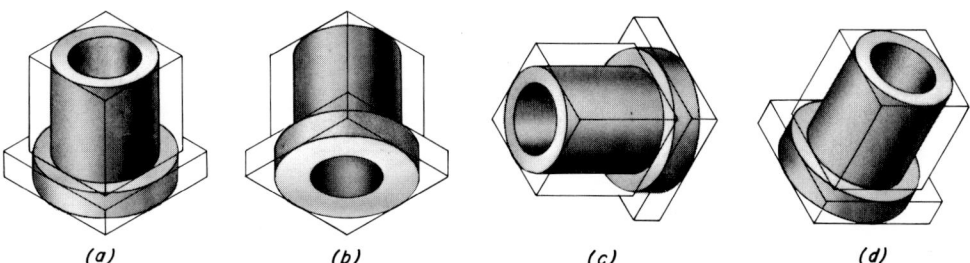

FIG. 13.27. Choice of position for isometric views.

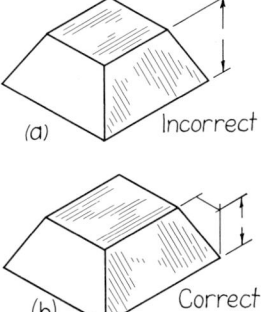

FIG. 13.28. Dimension lines in isometric planes.

366 AXONOMETRIC PROJECTION

mon error. The dimension line and the two witness lines do not lie in an isometric plane even though the dimension line is vertical. Figure 13.28*b* illustrates the correct method.

b. Figures and lettering of notes should be made to lie in isometric planes. Only vertical-style lettering should be used in isometric. Figure 13.29 shows how the parallelogram enclosing a letter or figure may be used as an aid in isometric lettering. The front views of the small cubes show the letters and their enclosing parallelograms orthographically; the two isometrics of the cubes show the six possible positions in which these parallelograms and figures may appear. Figure 13.30 illustrates the dimensioning of a rectangular object, placing the numerals in one or another of the positions shown in Fig. 13.29.

13.18.2 Unidirectional Dimensioning. ANSI has approved the placing of all dimensions and notes in one plane, as illustrated in Fig. 13.31. When using this system, only vertical letters or numerals should be employed. This method is simple and more rapid for production purposes.

13.19 DIMETRIC DRAWING

Somewhat the same distinction exists between dimetric projection and dimetric drawing as obtained between isometric projection and isometric drawing, namely, that scales approximating the projected scales are used in making the drawing. In Fig. 13.32*a* the conventional cube has been shown rotated from the position for isometric projection to a convenient dimetric position. The dimetric projection is shown in Fig. 13.32*b*.

In Fig. 13.33, four convenient positions for the dimetric axes are illustrated, with the approximate proportion of angles and scales for each axis indicated. The construction in conventional dimetric is carried on in the same manner as in isometric, except that on one axis the scale is changed. The simplest way of making a dimetric drawing is to proceed in the following manner.

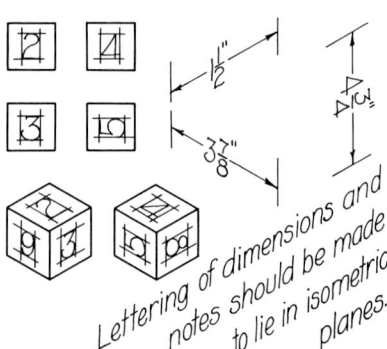

FIG. 13.29. Lettering and numerals in isometric.

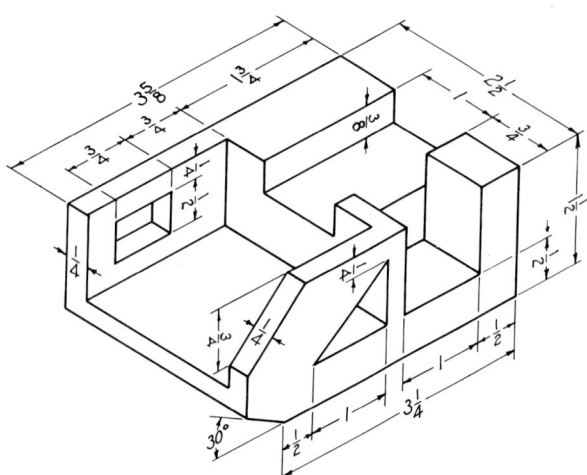

FIG. 13.30. Aligned system of dimensioning.

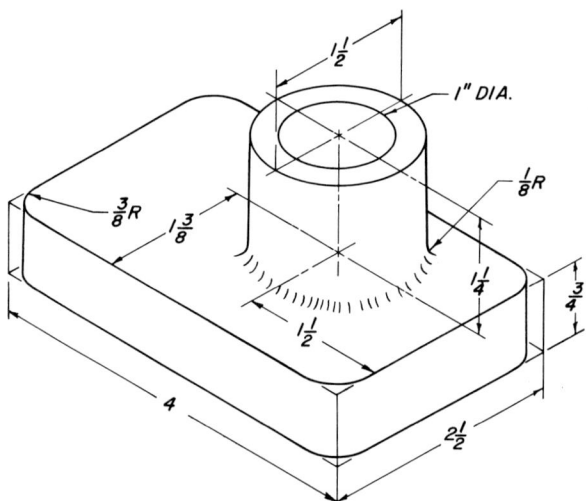

FIG. 13.31. Unidirection system of dimensioning.

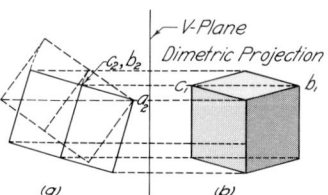

FIG. 13.32. Dimetric projection by rotation.

13.19.1 Construction of a Dimetric Drawing. For the construction of Fig. 13.34 the axes have been chosen in the position of Fig. 13.33c. The scale on the vertical edge and the left receding axis is full-size. The scale on the right receding axis is as shown, namely, five eighths of the full scale used on the front face. The construction is as follows.

a. Make the orthographic views to the scale desired for the two equal dimetric axes and then enclose the views in the smallest possible rectangular box, as shown in Fig. 13.34.

b. Draw the box in the desired dimetric position by transferring overall dimensions with dividers directly from the orthographic views for the equal axes and to the proper scale for the third axis. The scale to be used for this third axis is shown in Fig. 13.34 at the left side of the top view. In this case the scale on the short axis was made five eighths that of the other axes.

c. Plot points locating the corners and curves of the object just as in isometric, taking care to use the proper scale.

d. It should be noted that the four-center method of drawing circles can be used only in the face having equal scales on both sides. The coordinate method must be used in the other faces.

13.19.2 Advantages of Dimetric. As can be seen in Figs. 13.34 and 13.36, a face of an object may be emphasized by showing more or less of it than the other two sides. If all three faces are of unequal importance a trimetric may be called for.

Although the construction of dimetric and trimetric drawings is difficult, with the newer computer systems the necessary rotations of an object may be accomplished quite readily.

13.20 TRUE AXONOMETRIC PROJECTION

A simple method of making true axonometric projections was published by Theodor Schmid in 1922 and L. Eckhart in 1937. This method was presented in the United States by the authors in

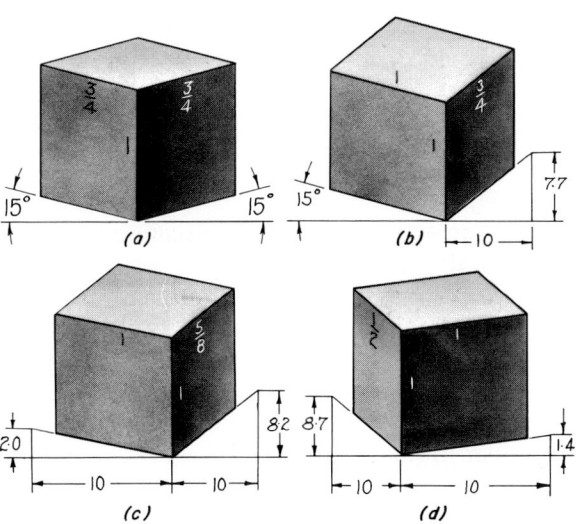

FIG. 13.33. Four convenient positions for dimetric drawing.

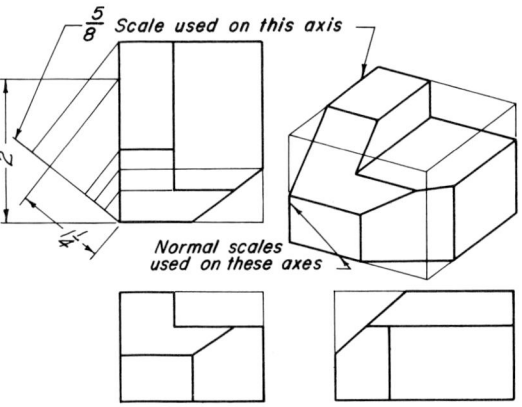

FIG. 13.34. Construction of a dimetric drawing by the box method.

1942 and has since been widely used. The method is applicable to isometric, dimetric, and trimetric projections and is particularly useful when orthographic views of an object are available. The question of scale on the various axes is automatically determined.

13.20.1 Definition of Terms. An axonometric projection may be defined as an orthographic projection upon a plane oblique to the three principal planes, as shown in Fig. 13.35. The object represented is usually assumed to have its principal faces parallel to the three principal planes of projection. Three types of views or projections may be obtained by varying the position of the axonometric plane. *When this plane makes equal angles with the principal planes, an isometric projection results.* The axes, or edges of the cube, are therefore equally foreshortened and make angles of 120° with each other in the projection.

When the axonometric plane is equally inclined to two of the principal planes, a dimetric projection is produced. Two of the axes project equally and the third is foreshortened by a different amount. In this case two of the angles between the axes are equal, whereas the third is different.

When the plane is unequally inclined to all three principal planes, a trimetric projection results. When this happens, the axes are all foreshortened by different amounts and the angles between the axes are all unequal. In no case can any of these angles be 90° or less.

13.20.2 Theory of Axonometric Projections. In Fig. 13.36, the position of the axonometric plane relative to the three principal planes is shown pictorially for each of the three types of projection. In each case the picture plane coincides with the axonometric plane. Therefore the axonometric triangle in each drawing is true size. The relative positions of the three principal views and the axonometric view are also shown. It should be noted at the beginning that axonometric projection as here discussed is orthographic projection, that is, the projection lines are perpendicular to the plane of projection. The basic principles of orthographic projection are involved in an understanding of axonometric projection. These principles are illustrated in the step-by-step solution in Section 13.21.

a. If a point is projected orthographically upon any two intersecting planes, as in Fig. 13.37, the two projections, when one of the planes is revolved into coincidence with the other, fall on a line that is perpendicular to the line of intersection of the two planes. Thus in Fig. 13.37, the two projections O^H and O_r^A lie on the same perpendicular to AB.

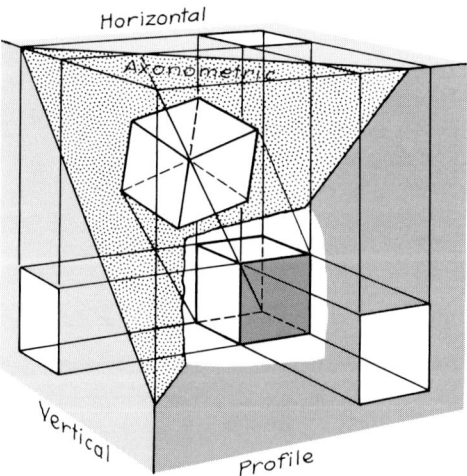

FIG. 13.35. Exact theory of isometric projection.

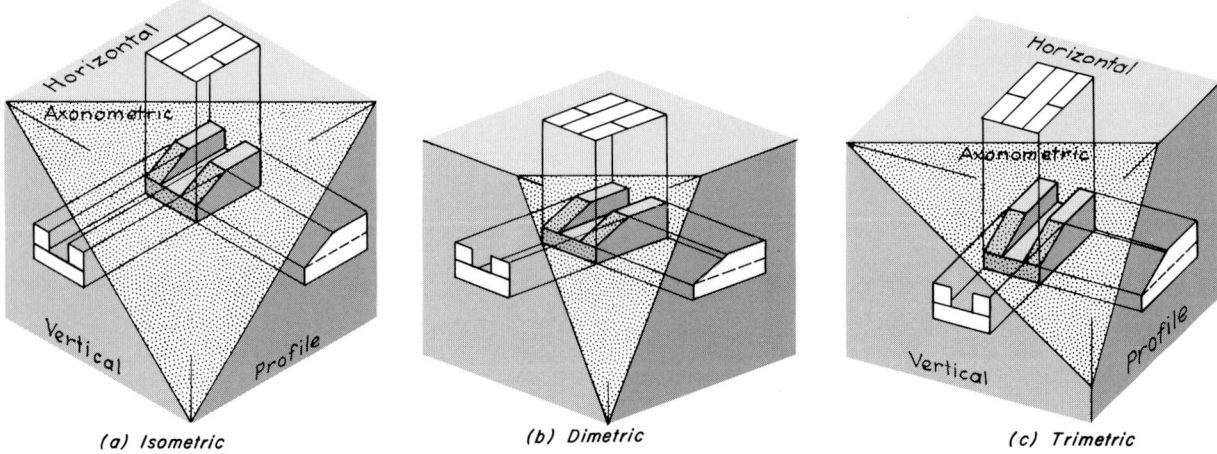

FIG. 13.36. Position of axonometric plane for isometric, dimetric, and trimetric projection.

In Figs. 13.38 and 13.39, the three principal planes, together with the projections of an object on them, have been revolved into the plane of the axonometric triangle. From Fig. 13.40 it can be seen that the axonometric projection can be obtained by direct projection from any two of the three revolved views.

b. The projecting lines are always perpendicular to the axis of rotation, which is the edge of the axonometric plane.

c. In each case, the revolved positions of the orthographic views have their principal edges parallel to the edges of the corresponding revolved plane.

d. One other geometric principle is involved. If three mutually perpendicular lines, for example, three intersecting edges of a cube, ox, oy, and oz, in Fig. 13.38a, are made to pass through the three corners of a triangle, xyz, the projections of these lines on the triangle will be perpendicular to the side opposite the corner through which the line passes. That is, ox will be perpendicular to yz, oy will be perpendicular to xz, and oz will be perpendicular to xz, and oz will be perpendicular to xy. This is shown in Fig. 13.38a.

e. Finally, in order to determine the position of right triangle XOY, of Fig. 13.38b, when revolved into the plane XYZ, it is necessary to use the geometric principle that any two lines drawn from the ends of a diameter of a circle to a point on the circumference make a right angle with each other. Then, since it is known that the angle XOY is a right angle, its true size must show where the plane has been revolved into the axonometric plane that has been set up to show in its true size.

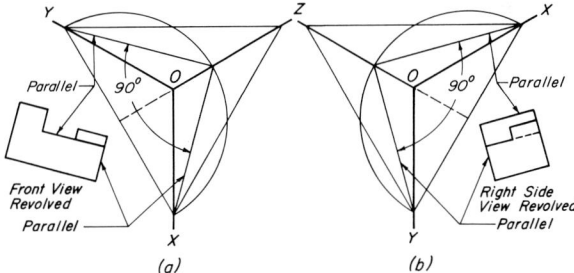

FIG. 13.39. Rotation of coordinate planes into the axonometric plane.

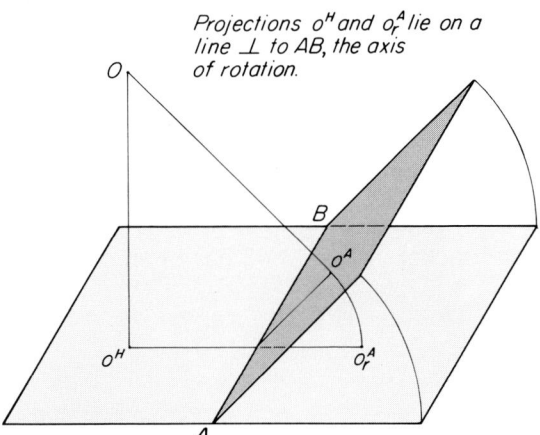

FIG. 13.37. Orthographic projection of a point on an inclined plane.

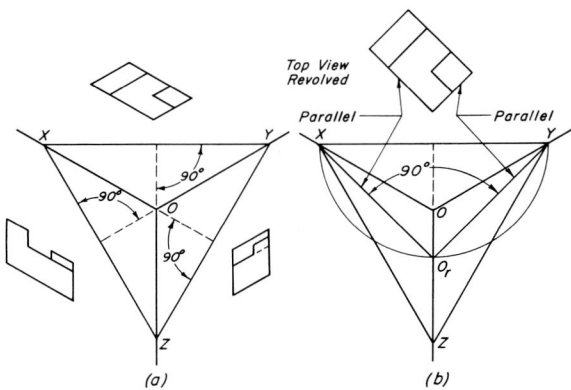

FIG. 13.38. Axonometric triangle for isometric projection.

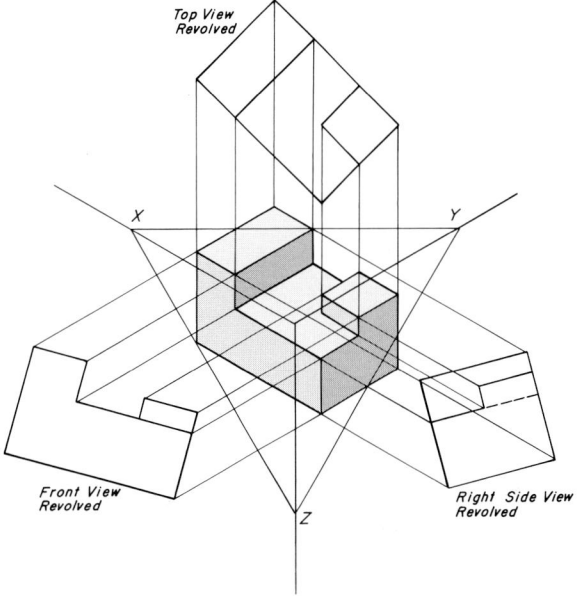

FIG. 13.40. Axonometric projection of a solid.

13.21 CONSTRUCTING AN ISOMETRIC PROJECTION

Making use of the principles outlined in the preceding article, an isometric projection may be made as shown in Figs. 13.38 through 13.40.

a. Draw the equilateral triangle XYZ that represents the isometric plane as in Figs. 13.38a and b.

b. With XY as a diameter draw the semicircle XO_rY. The point O has thus been revolved into the isometric plane at O_r, making a right angle XO_rY.

c. Draw the orthographic top view with the edges of the view respectively parallel to the lines O_rX and O_rY as shown in Fig. 13.38b.

d. Follow the same procedure as given in (b) and (c) to obtain the position of the orthographic front view as shown in Fig. 13.39a.

e. Follow the same procedure as in (b) and (c) to obtain the position of the orthographic side view as in Fig. 13.39b.

f. Having the three orthographic views located as in Fig. 13.40, draw projecting lines from each view perpendicular to the corresponding edge of the isometric plane.

g. Find the intersection of these projectors as shown in Fig. 15.40 to obtain the isometric projection of the object.

13.22 CONSTRUCTING A TRIMETRIC PROJECTION

Having these principles in mind, a step-by-step construction for making a trimetric projection may be made as follows. See Fig. 13.41.

a. Select the position of the three axes as in Fig. 13.41a. Note that the edges of the object in the finished drawing will be parallel to these lines. The angles between the lines may have any value greater than 90° except 180°. For trimetric the three angles must be unequal.

b. Draw lines at right angles to the axes across the opposite angles as shown in Fig. 13.41b. All three may be used but any two will be sufficient.

c. Determine the revolved position of the axes as shown in Figs. 13.41c and d. Note that this construction has been translated to a parallel position in order to leave the central area free.

d. Place the front and side views with the edges parallel to the revolved positions of the respective axes, as shown in Fig. 13.41e. Project from these views to form the axonometric projection.

As a further illustration of this method, the bearing bracket in Fig. 13.42 has been drawn in

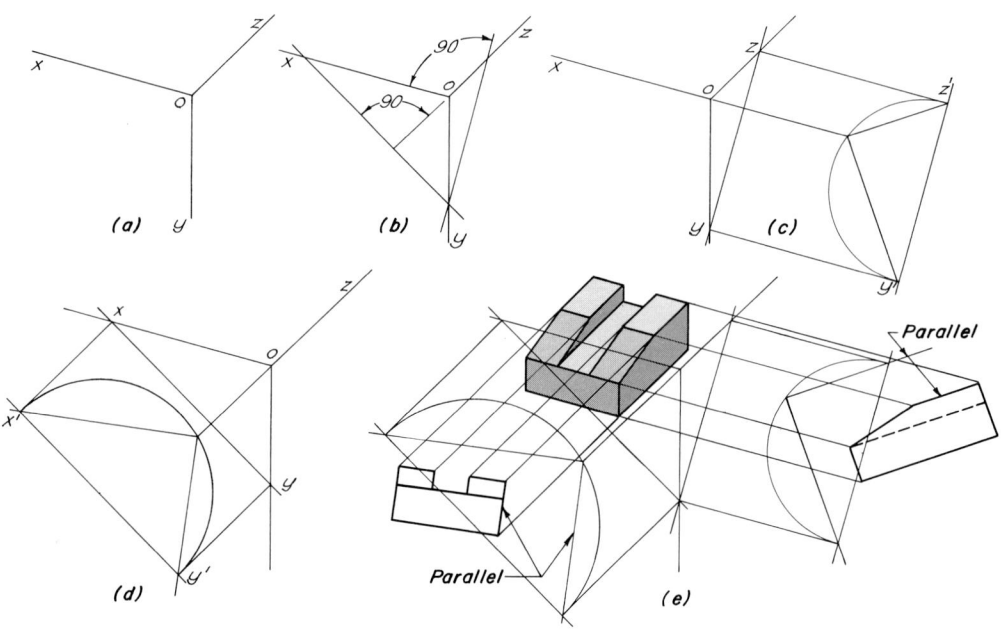

FIG. 13.41. Step-by-step construction of a trimetric projection.

dimetric. In this case two angles between the lines are equal.

The details of projection have been omitted. In order to orient the orthographic views properly, it is best to make a freehand thumbnail sketch of the axes and orthographic views in the position in which they are to be shown, as in Fig. 13.42a.

In Fig. 13.42b the top and front views have been used to make the construction. In drawing the circles with ellipse guides or by the trammel method, it is only necessary to find the location of the center and the major and minor axes in the pictorial. The length of the major axis is always equal to the diameter of the circle and its direction will be perpendicular to the axis of the right cylinder of which the circle is a base.

13.23 EXPLODED PICTORIAL ASSEMBLIES

The three forms of axonometric projection lend themselves very well to the preparation of so-called exploded assemblies. These drawings are

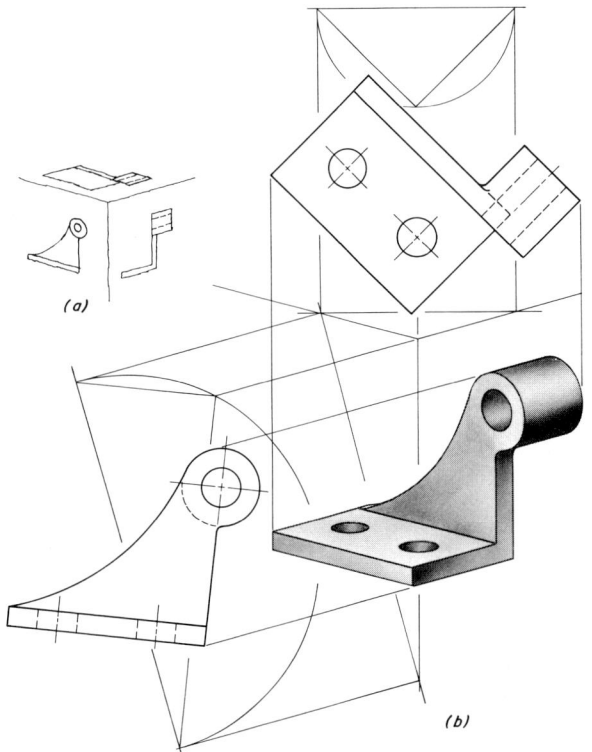

FIG. 13.42. Dimetric projection of a bearing bracket.

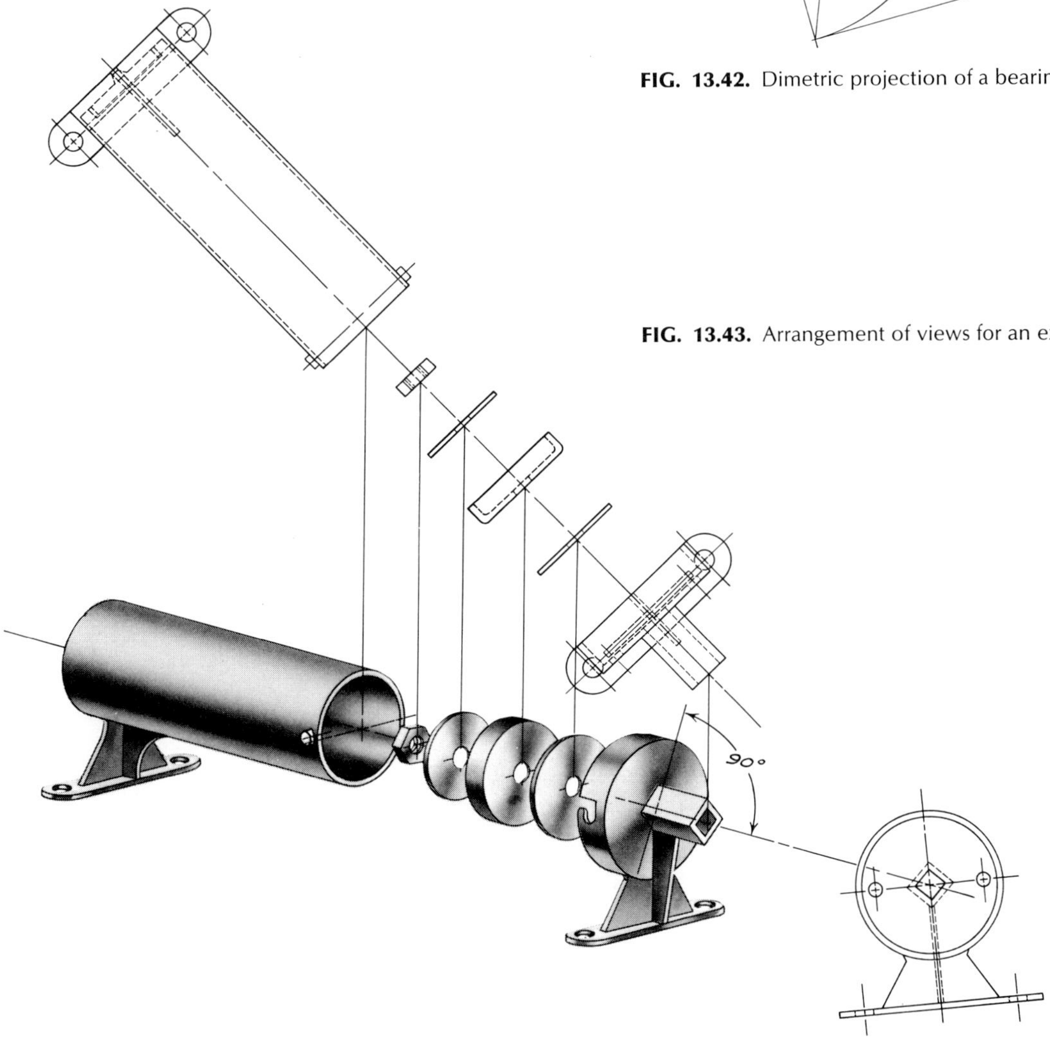

FIG. 13.43. Arrangement of views for an exploded assembly.

372 AXONOMETRIC PROJECTION

very useful in assisting persons who are not skilled in reading regular multiview drawings to perform work at the assembly bench. An illustration of this type of drawing is shown in Fig. 13.43. They can be made by direct projection from two orthographic views of each part as shown in Fig. 13.43. The parts must be arranged in the proper order for assembly.

13.24 PRACTICAL CONSIDERATION

In most pictorial drawings or exact projections the drawing of circles is one of the most time-consuming parts of the work. This can be speeded up by using mechanical aids such as the Lietz ellipse templates as shown in Fig. 13.44.

To use the ellipse guides properly, the axes of the projection must be carefully selected. For isometric projection there is only one choice, as shown in Fig. 13.45. The notes in the upper left part of the figure indicate the 35° Lietz ellipse guide and give the ratio of the major to minor axes. The position of the orthographic views is indicated and the scale comes directly as a result of projection from any two of the three orthographic views.

In Fig. 13.46 the same information is given for three different positions of the axes for dimetric projection. Note that in each case two faces have the same Lietz guide.

In Fig. 13.47 similar information is given for three positions of the axes for trimetric projection.

13.25 OTHER MECHANICAL AIDS

When numerous pictorial drawings are to be made, other mechanical aids in addition to the Lietz ellipse guide are very valuable as time-saving devices.

13.25.1 Special Boards. Several special drawing boards have been developed at the University of Illinois, Urbana, chiefly by Wayne L. Shick to make the use of axonometric projection simpler and more comprehensive. Boards similar to the one shown in Fig. 13.48 may be designed for any set of axes. The positions of the various views and the ellipse guides to be used on the various faces are marked clearly on the face of the board. The corners of the board are cut off perpendicular to the three axis so that all projection lines can be drawn with a T-square. If a blueprint of the object is available, the use of this board to make a trimetric projection may save up to 50% of the time required to make an isometric drawing of a complicated object.

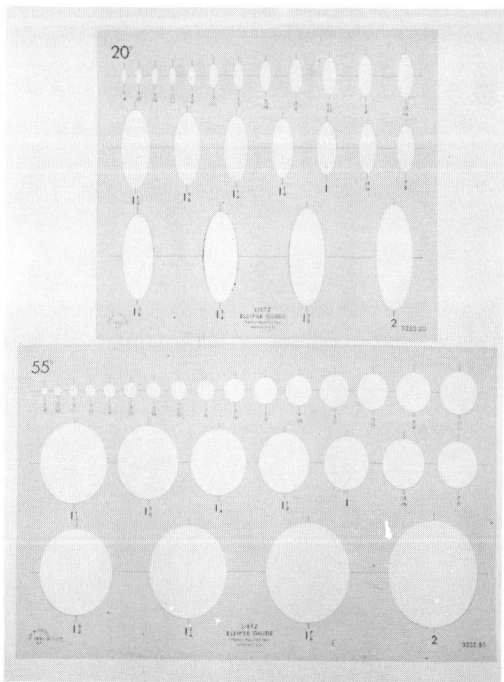

FIG. 13.44. Lietz ellipse template.

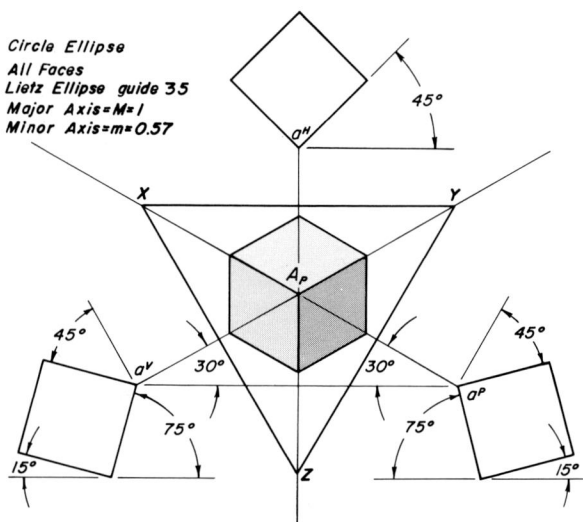

FIG. 13.45. Position of orthographic views for isometric projection.

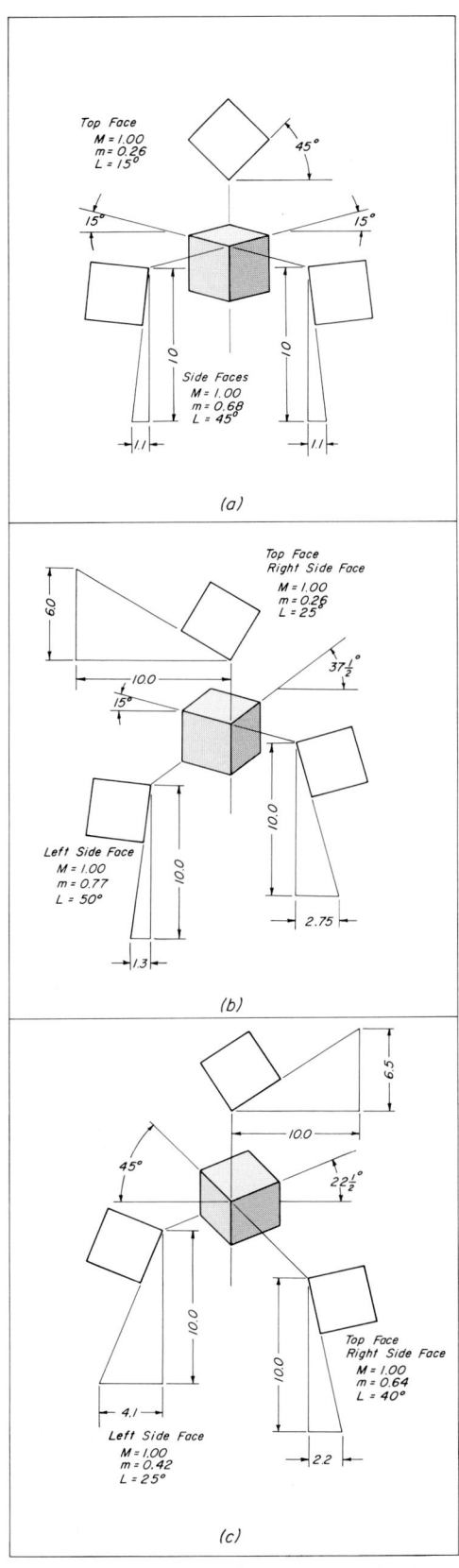

FIG. 13.46. Three positions for orthographic views for dimetric projection.

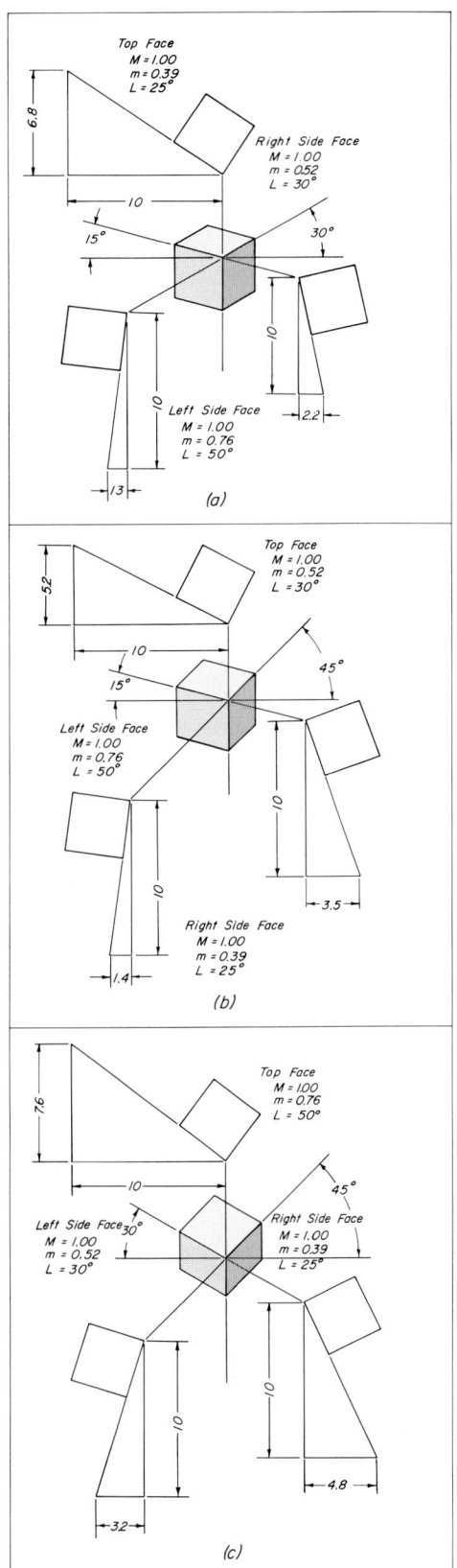

FIG. 13.47. Three positions for orthographic views for trimetric projection.

13.25.2 Special Board and Quadrangle for Isometric. By means of the specially designed board, shown in Fig. 13.49, it is possible to construct three orthographic views of an object and an isometric projection at the same time. This board was invented and patented by Wayne L. Shick of the University of Illinois, Urbana. It consists of a basic equilateral triangle on which the quadrangle slides. The quadrangle is a transparent plastic tool so designed that two sides may be

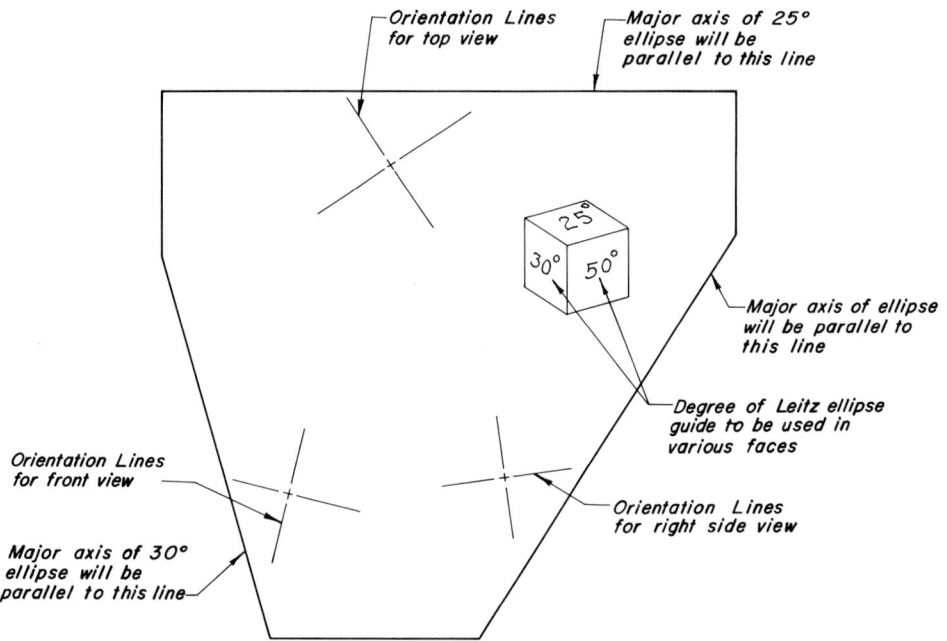

FIG. 13.48. Convenient trimetric board.

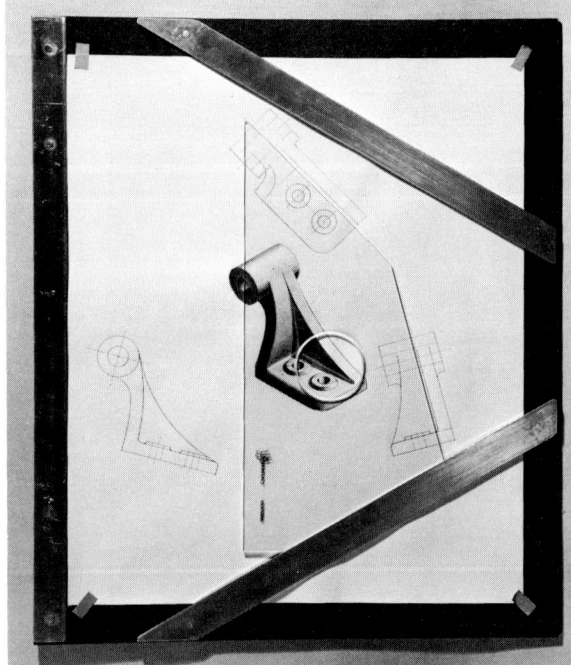

FIG. 13.49. Isometric board and drafting quadrangle.

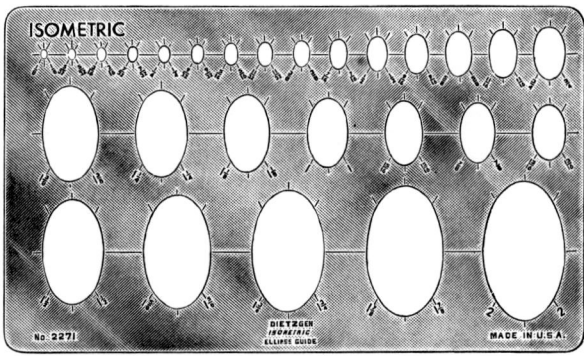

FIG. 13.50. Isometric stencil (courtesy Eugene Dietzgen Co.).

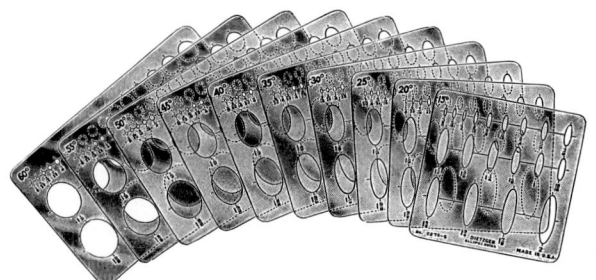

FIG. 13.51. Dimetric stencil (courtesy Eugene Dietzgen Co.).

used for projecting between two of the orthographic views, whereas the third side projects from one view to the isometric view. Since the angles are all equal in isometric, the quadrangle may be used on each of the three sides of the triangle for projection purposes. The advantages of the system are that direct projection is obtained between any two views and also that the isometric acts as a very positive check on the accuracy of all projections. The disadvantage is that the three-view drawings are not arranged in horizontal and vertical alignment as is customary in regular multiview projection.

13.25.3 Isometric Stencil. One ellipse stencil designed particularly for isometric drawings is shown in Fig. 13.50. Marks showing the major and minor axes and other marks showing the location of the isometric axes are marked along the edges of the ellipse. These marks enable the drafter to align the ellipse properly.

Figure 13.51 shows a set of 10 stencils that are very useful for making dimetric and trimetric drawings. The angles shown in the layouts of Figs. 13.46 and 13.47, as $L = 25$, $L = 30$, and so forth, are marked on these stencils.

SELF-STUDY QUESTIONS

Before trying to answer these questions, read the chapter carefully. Then, without reference to the text, answer as many questions as possible. For those that cannot be answered, the number in parentheses following the question number gives the article in which the answer can be found. Look it up and write down the answer. Check the answers that you did give to see that they are correct.

13.1 **(13.1)** The three forms of axonometric projection are _____, _____, and trimetric.

13.2 **(13.2)** A true isometric projection can be made by rotating the views of a cube until its body diagonal is _____ to the picture plane.

13.3 **(13.3)** Two _____ views of a cube can be made to produce an isometric projection.

13.4 **(13.5)** The three foremost edges of a cube drawn in isometric are called isometric _____.

13.5 **(13.5)** Lines parallel to the isometric axes are called _____.

13.6 **(13.5)** An isometric plane is one determined by two _____ lines.

13.7 **(13.5)** Any lines not parallel to the isometric axes are called non-_____ lines.

13.8 **(13.7.1)** Circles can be drawn in isometric by plotting points on the circle by the _____ method.

13.9 **(13.7.2)** A rapid way to draw circles in isometric by the use of a compass is the _____ _____ approximate method.

13.10 **(13.20.2)** All axonometric projections are true _____ projections.

13.11 **(13.7.4)** The four-center approximate method can be used in only one face of a _____ _____ and not at all in a trimetric projection.

13.12 **(13.8.1)** The _____ method of construction is useful for any type of axonometric drawing.

13.13 **(13.11)** For objects composed entirely of concentric cylindrical parts the _____ line method of construction is appropriate.

13.14 **(13.18)** Two methods of placing dimensions on an isometric drawing are the _____ method or the _____ method.

13.15 **(13.2)** In an isometric projection of a cube, face diagonals of the visible faces show in _____ length.

13.16 **(13.12)** In making sectional views in isometric, the cutting planes should be _____ to the isometric planes.

CHAPTER 14
OBLIQUE PROJECTION

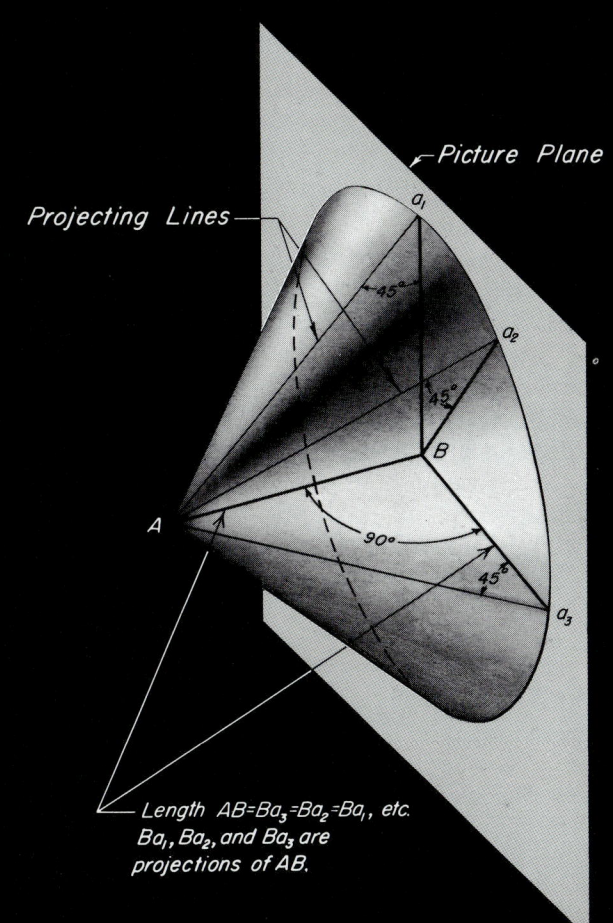

CHAPTER 14

14.1 FUNDAMENTAL PRINCIPLES

Multiple-view drawing and axonometric projection, which are both classified as orthographic projection, are similar to oblique projection, with one exception. While the point of sight (infinity) and the lines of sight (parallel) are similar, the difference is the angle between the lines of sight and the plane of projection. In orthographic the angle is 90°, whereas in oblique it can be any angle between 90° and 0°. One plane of projection is used in oblique as is the case for axonometric. The four most common types of oblique are shown in Table 14.1.

In oblique, the object may be placed in any position relative to the plane of projection. It is common practice to place the front face parallel to the picture plane in order to obtain the full advantage of this method of drawing.

Since the lines of sight are always parallel to each other, any line that is parallel to the picture plane will project in its true length. See Fig. 14.1. It will also project parallel to the original position of the line as shown in Fig. 14.2. Consequently, any face of an object that is parallel to the plane of projection will have exactly the same appear-

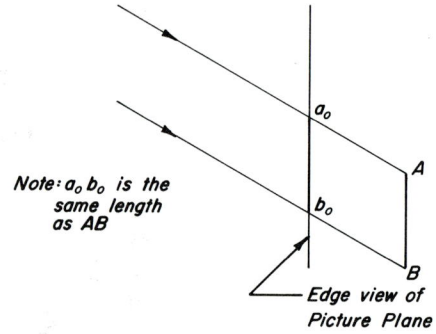

FIG. 14.1. Oblique projection.

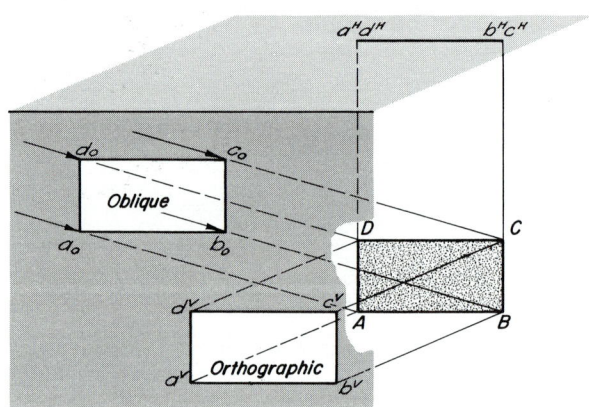

FIG. 14.2. Oblique projection of lines parallel to the plane of projection.

TABLE 14.1

Major Classifications	Subdivision	Number of Planes of Projection	Relation of Lines of Sight to Plane of Projection	Relation of Lines of Sight to Each other	Location of Point of Sight in Relation to Plane of Projection	Position of Enclosing Cube with Relation to Plane of Projection	Scale on Receding Axis	Angle of Receding Axis in Oblique View
Oblique	Cavalier	One	45°	Parallel	Infinite distance	Principle face parallel to plane	Full	Any angle
	Cabinet	One	63° 26′	Parallel	Infinite distance	Principal face parallel to plane	Half	Any angle
	Clinographic	One	80° 32′	Parallel	Infinite distance	Principal face at angle of 18° 26′	Varies	Varies
	General	One	Any angle except those above	Parallel	Infinite distance	Principal face parallel to plane	Varies but cannot be full or half	Any angle

ance in both oblique and multiview orthographic projection. This feature is the chief advantage of oblique projection over other forms of pictorial drawing.

14.2 PRACTICAL AND THEORETICAL METHODS OF CONSTRUCTION

Using the facts stated in Section 14.1, it is possible to make an oblique projection by drawing lines of sight from the object to the plane of projection. This theoretical method will be explained at the end of this chapter in Sections 14.14 and 14.15.

It is possible, however, to make an oblique projection by using a few simple rules of construction, which will be called the *practical method*. This method, which is very easy, is explained in the following articles and is used in solving all problems in the remainder of this chapter except Sections 14.14 and 14.15.

14.3 TYPES OF OBLIQUE PROJECTION

There are several types of oblique projection. Theoretically, they are distinguished from each other by the angle that the lines of sight make with the plane of projection. In the practical construction that we shall follow, they are distinguished by the ratio of the scale used on the receding axis as compared to the scale used on the front face.

14.3.1 Cavalier Projection.
In Cavalier projection the lines of sight make an angle of 45° with the plane of projection. Thus in Fig. 14.3 line AB, which is perpendicular to the plane of projection, is shown projected upon the vertical or picture plane in three places at Ba_1, Ba_2, and Ba_3. The lines of sight all make an angle of 45° with the picture plane, as indicated in the figure, but the projections of line AB thus found may have any position relative to the horizontal.

As noted in the figure, the projections of AB, namely, Ba_1, Ba_2, Ba_3, are all equal to the true length of the perpendicular line AB. This is true because the two sides of a 45° isosceles triangle, ABa_3, for example, are equal. Thus the same scale can be used on all three axes of the oblique projection. This is the chief advantage of Cavalier projection over other forms of oblique projection.

14.3.2 Cabinet Drawing.
A second type of oblique projection, which has been specifically named, is produced when the angle that the lines of sight make with the plane of projection is 63° 26′. See Fig. 14.4. The tangent of this angle is 2, which means that a line perpendicular to the picture plane is just twice as long as its projection on the plane. Stated in another way, the projection is just one half as long as the line. In making a projection of a perpendicular line, this result will be produced automatically if the scale used on the receding axis is just one half that on the front face. A drawing made in this way is called a *Cabinet drawing*. Note again that the angle that the lines of sight make with the plane of projection has nothing to do with the angle that the receding axis makes with the horizontal. Thus in Fig. 14.4 all three figures are Cabinet drawings, but the receding axis has a different angle in each case.

14.3.3 General Oblique Drawing.
Obviously, angles that the lines of sight make with the picture plane, other than the 45° and 63° 26′ angles, could be used to make oblique projections. From a practical point of view this simply means that different scales, other than full scale or one-half scale, can be used on the receding axis.

Any oblique drawing made on this basis is called a *general oblique projection* or *drawing*. A

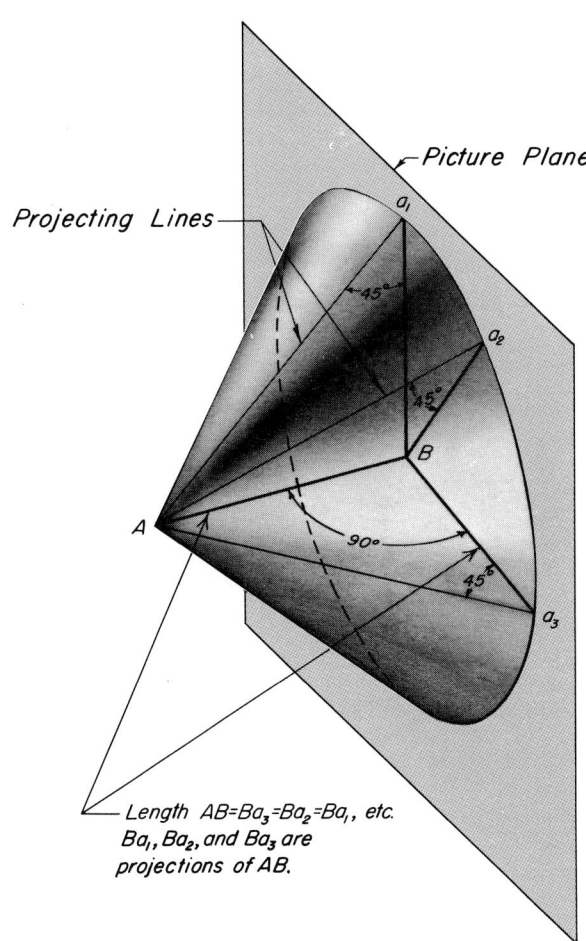

FIG. 14.3. Oblique projection of a line.

comparison of a drawing of a simple object representing each of the foregoing types of oblique projection is shown in Fig. 14.5. Note that in each case the receding axis has been made at the same angle with the horizontal.

14.3.4 Clinographic Projection. In this type of oblique there is not only a different angle for the lines of sight, but also the basic cube is turned at an angle to the picture plane. This type is used in crystallography. The drawing is made by exact projection. See Section 14.15 and Figs. 14.23 and 14.24.

14.4 CONVENTIONAL CONSTRUCTION OF CAVALIER PROJECTIONS

As discussed in Article 14.3.1, the advantage of using Cavalier projection lies in the fact that the same scale is used on the receding axis as on the front face. Cavalier projections, however, though easy to draw, have a distorted appearance. Suggestions for reducing distortion are given in Article 14.5.2.

14.4.1 Plane Figures. One of the advantages of oblique projection is that plane figures in the front face of an object, or parallel thereto, project

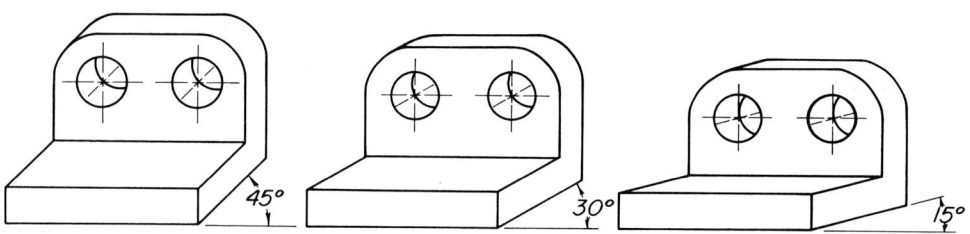

FIG. 14.4. Cabinet drawing with receding axis at various angles.

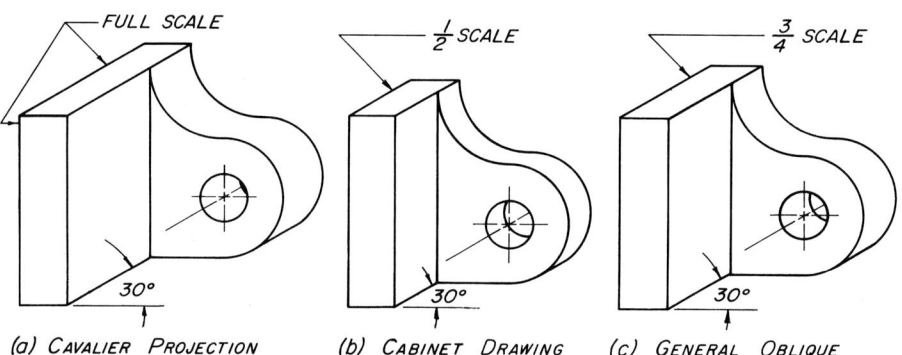

FIG. 14.5. Cavalier, Cabinet, and General oblique projections compared.

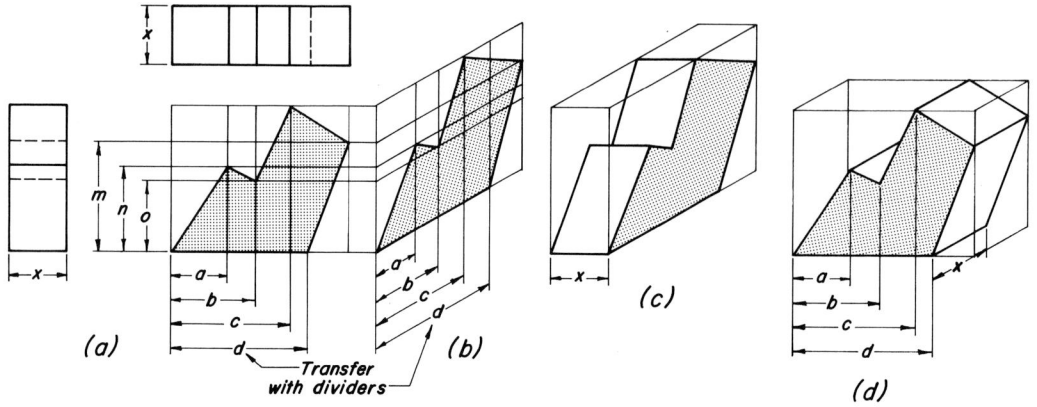

FIG. 14.6. Coordinate method of construction for plane figures.

in their true shape just as in the orthographic views, hence, they require little further explanation.

In Fig. 14.6 a plane figure composed of straight lines is shown. In Fig. 14.6d the true shape appears exactly as it would in the front view of the orthographic projection. Coordinates designated by letters have been indicated to locate the corners of the plane figure.

These coordinates can be used to draw the same figure in the side face of a box as shown in Figs. 14.6b and c. For a Cavalier projection these can be transferred from Fig. 14.6a to either of the other views with dividers, since the same scale is used on all axes.

14.4.2 Circles in the Front Face of an Object. In oblique projection, circles in the front face of an object can be drawn as true circles with a compass as shown in Fig. 14.7a. Figure 14.8 also shows circles and parts of circles in the front face of a link.

14.4.3 Circles in Faces Parallel to the Vertical Plane. In any face of an object parallel to the V-plane, circles will also show as true circles, like those of the rear face of the link in Fig. 14.8c.

14.4.4 Circles in the Side or Top Faces of Oblique Drawings. Such circles will always project as ellipses, as shown in Figs. 14.7a and b. They may be accurately drawn by the coordinate method.

Thus in Fig. 14.7a the ellipse representing the side face has been constructed as follows.

a. Divide the circle in the front face into a suitable number of parts. In this case 12 equal spaces were used.
b. Project these points to the right vertical edge.
c. From these points draw lines parallel to the receding axis across the side face.
d. Draw either diagonal of the side face. The long diagonal was used in this case.
e. Where the diagonal crosses the horizontal coordinates, draw vertical lines representing the other coordinates. Thus a'b' in the side face represents a coordinate similar to ab in the front face.
f. Mark the points and draw a smooth curve through them.

Figure 14.7b represents the cube with two cylindrical holes through it. The intersection of the cylinders was found by drawing the corre-

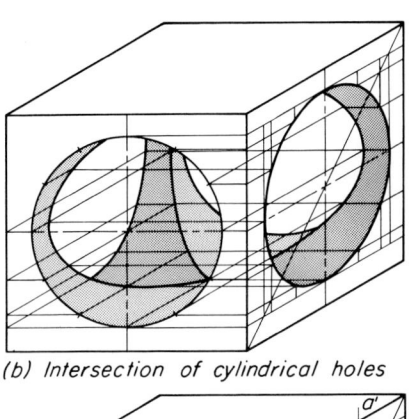

(b) Intersection of cylindrical holes

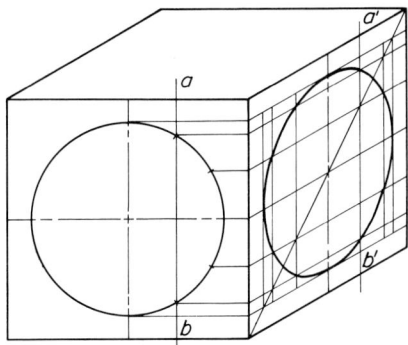

(a) Constructing circle in side face

FIG. 14.7. Construction of circles by coordinates.

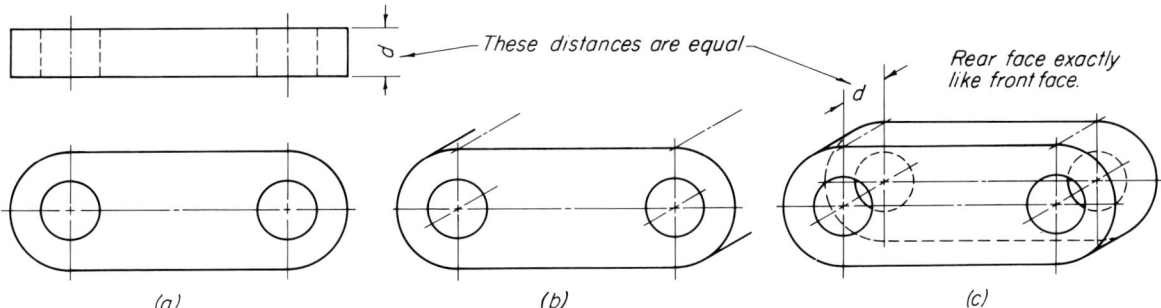

FIG. 14.8. Front face in true shape in oblique.

sponding elements of both cylinders to their crossing points as shown in the figure.

14.4.5 Circles by the Four-Center Approximate Method. A circle may be enclosed in a square, and since it is always tangent to the square at the midpoint of the sides, the Cavalier projection will show an ellipse that is tangent to the midpoint of the sides of the enclosing rhombus. Hence, to find the centers of the four arcs, erect perpendiculars to the midpoints of the four sides of the rhombus and find their intersections, as shown in Fig. 14.9. It should be noted that as the ratio of the major to the minor axis increases, the approximation becomes less accurate. Note, in oblique projection, the four-center method can be used only in Cavalier projection.

14.5 Drawing Solid Objects in Cavalier Projection. The drawing of solid objects involves more than adding a third dimension to the plane figures discussed heretofore. Some of the important considerations are

a. *The selection of the axes.* In oblique a wide choice is available. The proper selection is discussed in the following paragraphs.
b. *The position of the object.* This involves two things. One is to find the simplest construction and the other is to find the best appearance of the object. See Article 14.5.2.

14.5.1 Position of the Axes. As illustrated in Fig. 14.10, the receding axis may make any angle with the horizontal. This angle is not to be confused with the 45° angle that the lines of sight make with the plane of projection. The lines of sight that extend from the object to the plane of projection do not show in any illustration or construction except the theoretical. See Section 14.14.

As shown in Fig. 14.10, any face of the object

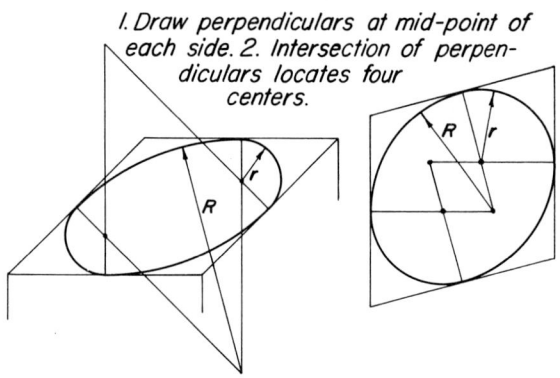

FIG. 14.9. Four-center approximate method of representing circles in Cavalier projection.

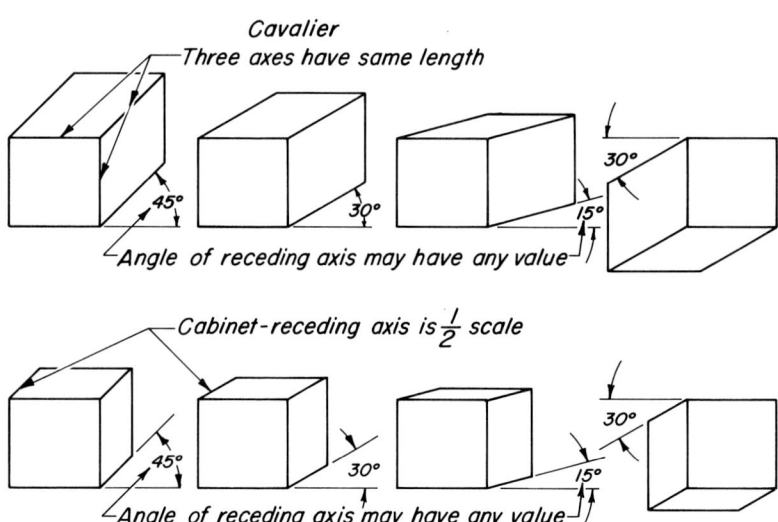

FIG. 14.10. A few positions of the receding axis.

may be emphasized by the proper selection of the receding axis. This must be considered first in beginning the layout of a drawing.

14.5.2 Position of Object.
a. *For simplicity of construction.* Since any face of an object that is parallel to the picture plane appears in its true shape in an oblique projection, the construction of many drawings can be kept quite simple by showing the face of the object that has the most circles, arcs, or other curves as the front face. This rule should be adhered to whenever possible, for it can be readily seen that the object in the position of Fig. 14.11a is not only easier to draw but also looks much better than the same object as represented in Fig. 14.11b.

b. *To reduce distortion.* Unpleasant distortion in Cavalier projection frequently can be reduced by placing an object that has one dimension much greater than the others with this long dimension parallel to the picture plane, as shown in Fig. 14.12. Some discrimination must be exercised, however, for this may not work so well with some objects as it does with others. A better method to reduce distortion is to reduce the scale on the receding axis, thus selecting a different type of oblique projection.

14.5.3 Drawing the Object by the Box Method.
With the foregoing paragraphs in mind, the steps in making a Cavalier projection of a solid or three-dimensioned object are illustrated in Fig. 14.13. They may be listed as follows.

a. Enclose the orthographic views of the object in the smallest possible rectangular box whose faces are parallel to the principal faces of the object. See Fig. 14.13a.

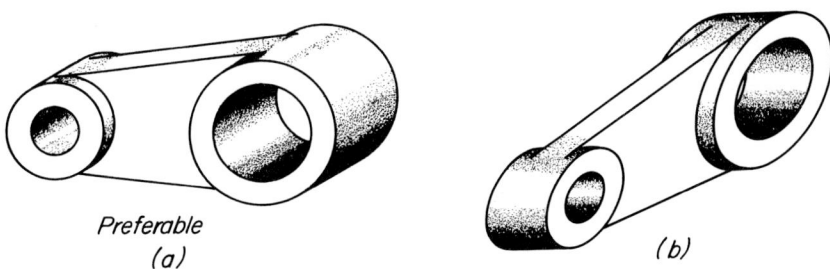

FIG. 14.11. Circles preferably parallel to the front face or plane of projection.

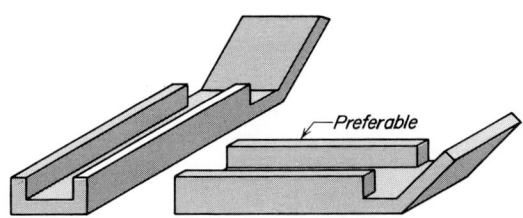

FIG. 14.12. Long dimension of objects preferably parallel to the plane of projection.

b. Draw this enclosing box in Cavalier projection, making the front of the box like the orthographic front view and the receding axis to the same scale as the front and at an angle that will show the side and top to the best advantage. See Fig. 14.13b.

c. Draw lightly all lines in the front face and establish position of planes parallel to the front face. See Fig. 14.13c.

d. Draw all lines in faces parallel to the front at the proper distances from the front face, as shown in Fig. 14.13d.

e. Construct coordinates for curves not in front face in the orthographic view as shown at 1-a, 2-b, 3-c, and so forth in Fig. 14.13a and then transfer these coordinates to the oblique as shown for one curve in Fig. 14.13e. Since the second curve is parallel to the first, points on it may be located by stepping off the thickness of the web (see front view) as bb' in Fig. 14.13e.

f. Complete all lines, and erase construction. After erasure make heavy the visible outlines of the drawing, producing the finished drawing of Fig. 14.13f.

14.6 CONSTRUCTION OF A CABINET DRAWING

A Cabinet drawing may be made in the same manner as a Cavalier projection, with the exception that the scale on the receding axis must be reduced one half. The steps in this procedure are shown in Fig. 14.14, which, though not a Cabinet drawing, illustrates the method of construction. Where curves are involved in the receding faces, a convenient scheme for obtaining the foreshortened or one-half-scale dimensions from the orthographic views is shown in Fig. 14.14a.

Thus, for a Cabinet drawing, if the total length ab on the receding axis is 3 in., line ac will be 1½

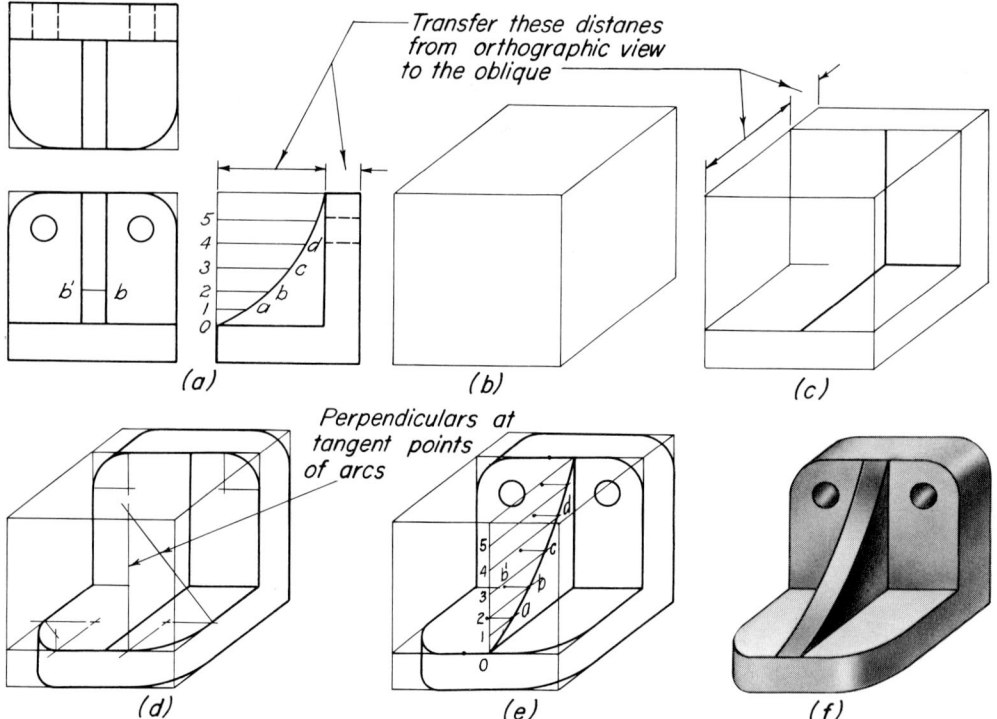

FIG. 14.13. Construction of curves in a receding plane using the box method.

in. long. This line may be laid out in any convenient direction from either end of line ab, as illustrated. Point c is then connected with b, and all other points to be plotted are transferred to line ac by lines parallel to bc. This device for making proportional divisions of any line should be thoroughly understood by the student.

The four-center approximate method of drawing circles cannot be used in Cabinet drawing since the sides of the enclosing parallelogram are not equal. A circle can be plotted, however, without referring to the original orthographic projections except to obtain the dimensions and position of the enclosing parallelogram. The method of plotting coordinates by means of a semicircle and a diagonal of the parallelogram is shown in Fig. 14.7.

14.7 GENERAL OBLIQUE
Obviously, scales other than those mentioned could be used on the receding axis. For example, if the scale on the front face were 1 in. = 1 in., a scale of $\frac{3}{4}$ in. = 1 in. could be used on the receding axis. A drawing made in this manner is neither a Cavalier projection nor a Cabinet drawing and is simply referred to as an *oblique projection*. The angle that the projecting lines make with the plane of projection can be determined by comparing the length of a perpendicular with the length of its projection. Thus, in Fig. 14.14 a perpendicular 1 in. long would project $\frac{3}{4}$ in. long. Hence the tangent of the angle that the projecting line makes with the plane of projection would be $1 \div \frac{3}{4}$ or 1.333. This angle is 53°8″. Its value, of course, is of theoretical interest only, since it does not enter into the actual construction of the drawing.

The construction of any oblique drawing follows the same pattern as the construction of a Cavalier projection, with the exception that the scale on the receding axis is changed to suit the conditions.

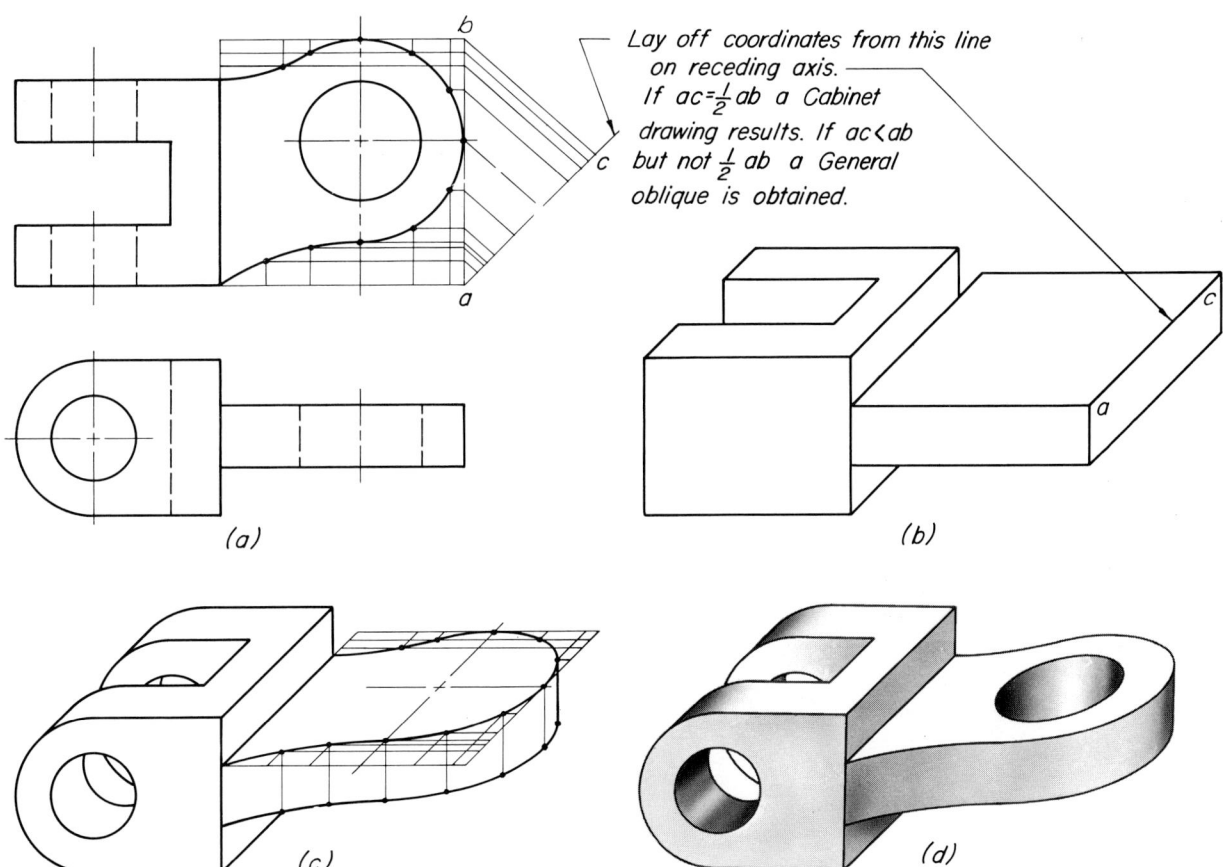

FIG. 14.14. Method of foreshortening the scale on the receding axis.

14.8 CENTER-LINE LAYOUT

When an object is composed of cylindrical parts, like the rocker arm in Fig. 14.15a, the construction can be based upon a center-line framework instead of the box construction previously explained. With the direction of the three axes chosen, the centers of all circular parts can be laid out as illustrated in Fig. 14.15b, using the proper scale on all axes. At each center, circles of the proper size can be drawn, as shown in Fig. 14.15c. The straight lines tangent to them can be drawn quickly, giving the final result shown in Fig. 14.15d.

14.9 SECTIONING OBLIQUE DRAWINGS

As in all pictorial drawing, hidden lines are not shown since it is very difficult to interpret them. Interior construction is best shown by making

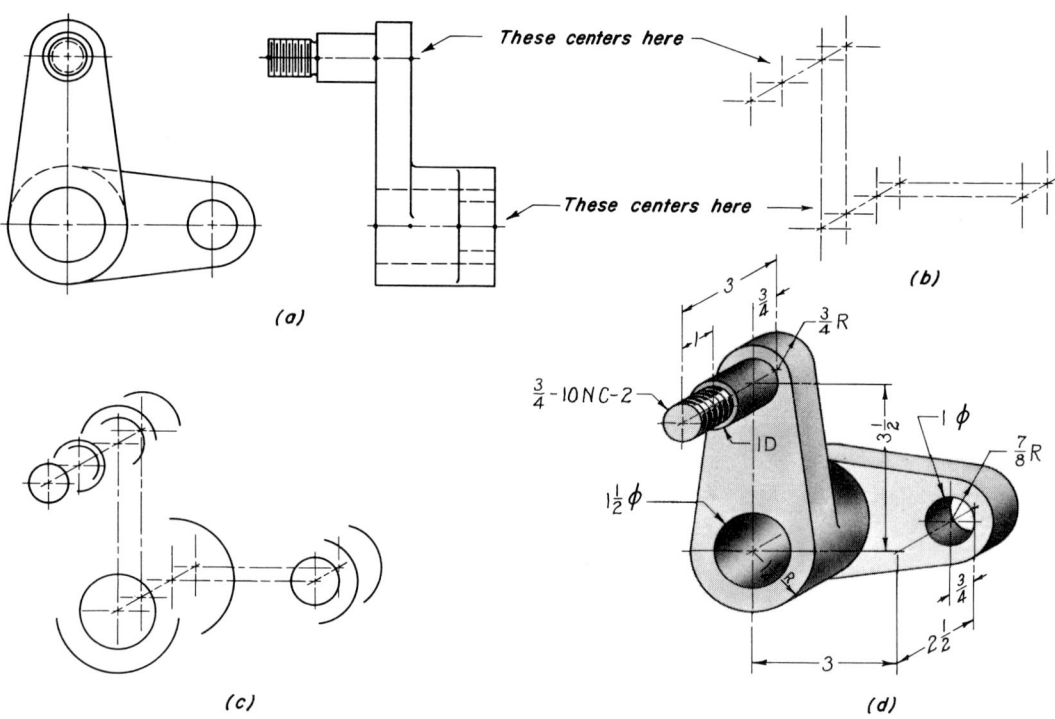

FIG. 14.15. Center-line method of construction: general oblique.

FIG. 14.16. Sectional view in oblique projection.

sectional views with either one fourth or one half the object removed. The cutting planes, in general, should be parallel to the oblique planes or faces of the enclosing box, as shown in Fig. 14.16. Other illustrations may be found in the problem sections of various chapters.

The cross-hatching lines lying in two planes that are at right angles to each other should be sloped in such a way that they would seem to coincide if the planes were rotated together. Correct section lining is shown in Fig. 14.16.

14.10 DIMENSIONING OBLIQUE DRAWINGS

The principles of dimensioning studied in connection with working drawings apply in general to oblique projections, but with the following additions.

a. Dimensions should be made to read from the bottom and right-hand side of the sheet so far as possible. See Figs. 14.15 and 14.17b.
b. Dimension lines and witness or extension lines must lie in the same oblique plane. See Fig. 14.18.
c. Dimensions must lie in the oblique plane determined by the dimension lines and extension lines.
d. Only vertical lettering and numerals should be used. Numerals and letters may be made to lie in oblique planes in the same manner as shown for isometric. The unidirectional system is also approved.
e. As far as possible, dimensions should be placed in the front face or parallel thereto since this makes the dimensioning similar to that in the orthographic views.
f. When notes are extensive and are not on the figure, they may be lettered neatly in slant style. In general, however, vertical lettering is preferred.

14.11 SCREW THREADS IN OBLIQUE PROJECTION

Since circles that are parallel to the plane of projection show as true circles in oblique projection, screw threads may be easily represented by a conventionalized scheme, if the axis of the thread is made parallel to the receding axis of the drawing. The axis of the bolt or nut, as in Fig. 14.19, becomes the line of centers for a series of circles representing the crests of the threads. The root line does not show. The spacing of the circles is made the same as the pitch of the thread until the pitch becomes too fine, in which case the smallest convenient spacing is used.

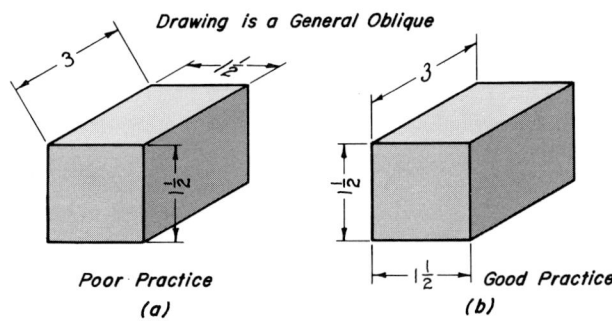

FIG. 14.17. Dimensioning in oblique projection.

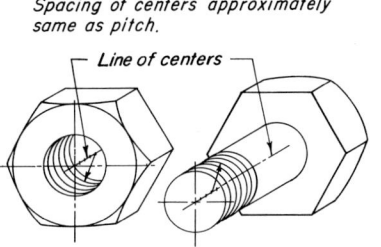

FIG. 14.19. Screw threads in oblique projection.

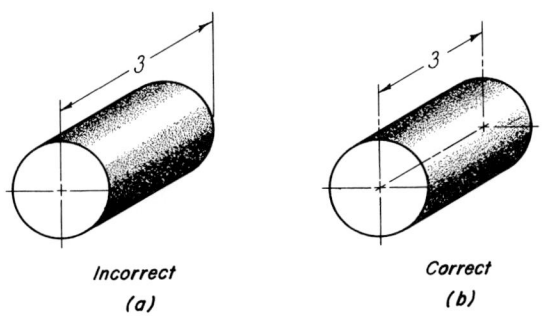

FIG. 14.18. Dimensions must lie in an oblique plane.

Whenever possible, bolts, screws, or nuts should be drawn in the position shown in Fig. 14.19. When placed with the axis parallel to the picture plane, the circles representing the thread crests must be plotted as ellipses and drawn with an irregular curve or ellipse guide, which is a slow process.

14.12 THREE-DIMENSIONAL CURVES

Three-dimensional curves, such as the conveyor shown in Fig. 14.20, can be drawn with comparative ease if the axis of the conveyor is placed on the receding axis of the drawing. The method is illustrated in the figure. The circles are first divided into 12 equal parts, and the pitch is likewise divided into the same number of parts. On the center line, step off intervals equal to one-twelfth the pitch. From these points radial lines can be drawn parallel to those used in dividing the circles. The construction is shown for the first four points on the curve. After the first turn has been completed, successive turns can be made by stepping off distances equal to the pitch from the previous curve. The thickness of the blade can also be stepped off at equal intervals from the first curve. In cases of this kind orthographic views are not required to make the construction, except to obtain the dimensions of the part.

If, on the other hand, the three-dimensional curve is irregular in shape, it may be necessary to resort to the box construction and plot three rectangular coordinates.

14.13 DOUBLE-CURVED SURFACES

The outline of a double-curved surface is represented by an enveloping curve tangent to imaginary curves in the surface. Thus a return pipe bend can have a series of semicircles drawn on its surface, as shown in Fig. 14.21a. These curves are then drawn in oblique and the enveloping curve is drawn tangent to them.

14.14 THEORETICAL CONSTRUCTION OF AN OBLIQUE PROJECTION

In order to give a more thorough understanding of oblique projection, the following discussion for making a projection from the orthographic views is presented. From this the rule-of-thumb methods used heretofore can be derived.

An oblique projection of an object may be constructed from its orthographic views by drawing the oblique lines of sight from these views and finding where they pierce the plane of projection, as shown in Fig. 14.22. Any two views could be used, but all three have been shown in the figure in order to illustrate the theory.

Since the vertical plane appears edgewise in both the side and the top views, the piercing points of the projecting lines can be seen in these views by inspection. Thus, in the pictorial top view of Fig. 14.22, the lines of sight from a^H pierce the plane at a_o^H, and the front view of this piercing point must lie in the perpendicular from a_o^H. Likewise, in the side view the projecting line from a^P pierces the picture plane at a_o^P, and the front view of this piercing point must lie horizontally across from a_o^P. The intersection of these two perpendiculars determines a_o^V, which is the oblique projection of point A on the object. The same procedure is used for all other points.

The orthographic construction of an oblique

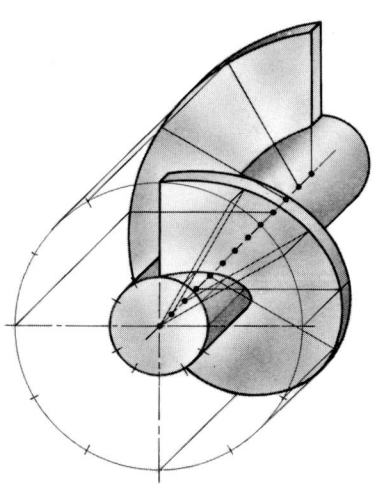

FIG. 14.20. Helicoid in oblique projection.

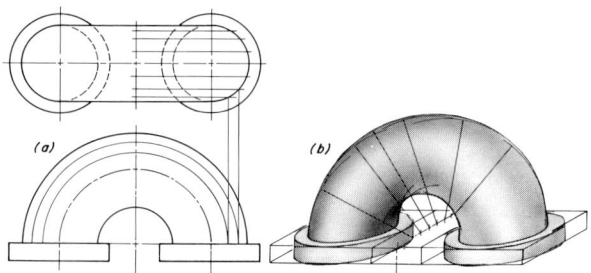

FIG. 14.21. A double-curved surface in oblique projection.

projection and the completed oblique projection are shown in heavy outline in Fig. 14.22b. Although this theoretical method can be used for so simple an object as a cube, it would be too cumbersome for more complicated objects. A conventional method of construction similar to that used in isometric has been used throughout all previous paragraphs.

It may be noted, in passing, that the true value of the angle that the lines of sight make with the V-plane does not show in any one of the three views of Fig. 14.22b. It may be found, however, by dropping a perpendicular to the V-plane from b^H in Fig. 14.22b and revolving the shaded triangle to $b^H c^H b_o^H$ around line BC until it is parallel to the H-plane. The true value of the angle then shows in the H-projection at b_{or}^H.

14.15 CLINOGRAPHIC PROJECTION

In the first part of this chapter, the object to be drawn was always placed with one face parallel to the plane of projection; this is an essential element in oblique projection, as there discussed. Although the advantages to be obtained by this means are of considerable practical importance in the way of convenience and speed, it must be borne in mind that oblique projections of an object can be made regardless of its position relative to the plane. The essential fundamental concept in oblique projection lies in the fact that the projecting lines are parallel to each other and oblique to the plane of projection. The position of the object relative to the plane may or may not facilitate the work of construction, but this does not change the method of projection.

For some purposes it may be desirable to turn the object at an angle to the plane of projection, particularly if there are no curved edges in its outlines. In fact, just such a system of oblique projection, called *clinographic projection,* has become firmly established in the field of mineralogy and crystallography in representing the various mineral crystals.

In clinographic projection, the angles that the three principal axes of the crystal make with the plane of projection, and the angle that the lines of sight make with the plane, have become fixed by usage, although there is no inherent reason why other angles might not have been used just as effectively.

These angles are shown in Fig. 14.23 in the clinographic projection of a cube. For construction purposes, it is well to remember that the tangent of 18° 26′ is one third and the tangent of 9° 28′ is one sixth. From an observation of the figure it is clear that the vertical line projects in true length, whereas the others are foreshortened, each by a different amount. In clinographic projection these foreshortened scales are determined accurately; approximations of scales or arbitrary assumptions are not permitted. In practice, the three coordinate axes shown inside of the cube are used as the skeleton upon which the crystal is drawn, as illustrated in Fig. 14.24.

Three of the six principal systems of crystals have their three axes of varying lengths and at

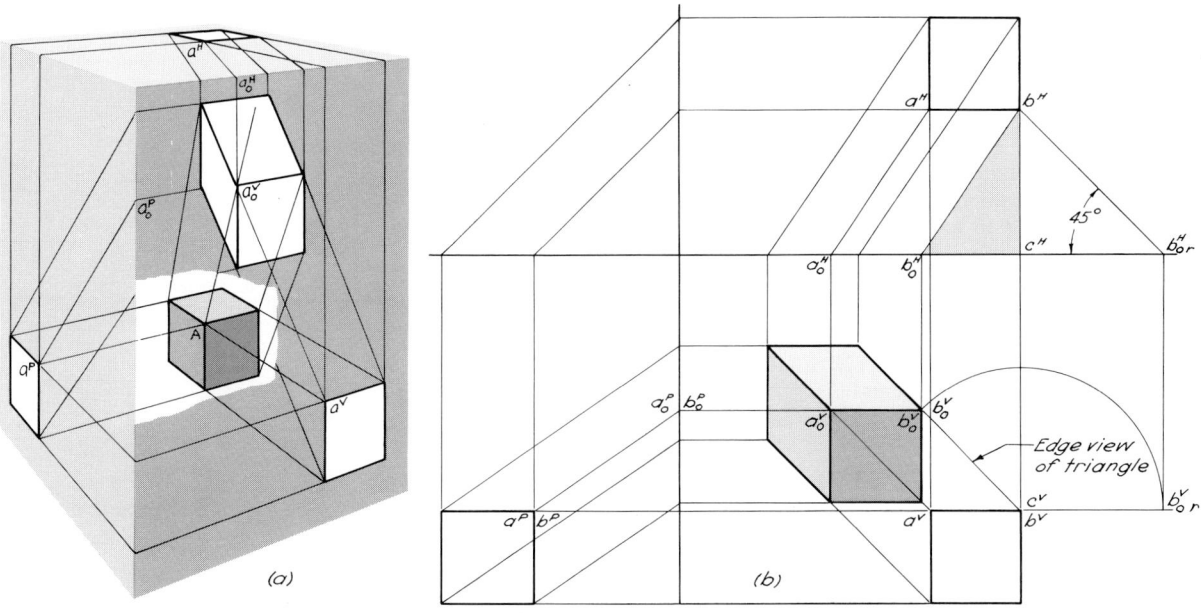

FIG. 14.22. Theory of oblique projection.

right angles to each other. The other three systems have axes that make angles with each other, differing from 90°, as shown in Fig. 14.25. All six sets of axes are drawn upon the cubic axes as a basis.

Clinographic projection is strictly an oblique projection, and although three different scales are used upon the axes, it is not for this reason to be called a trimetric projection, any more than we should call a Cavalier projection an isometric projection for the same reason. So far as appearance and actual construction are concerned, however, oblique projections can be made to appear quite similar to dimetric projection, as is evident from a comparison of the clinographic projection of the cube in Fig. 14.23 and the dimetric of the cube in Fig. 13.47.

14.16 ADVANTAGES OF OBLIQUE DRAWING

From the foregoing paragraphs it is clear that oblique drawings have several distinct advantages over other pictorial forms.

a. The front face of an object or any face parallel to it may be drawn like its true orthographic projection, hence circles may be drawn as true circles.

b. Distortion may be largely overcome by a careful foreshortening of the scale on the receding axis.

c. Dimensioning is simpler since only one set of dimensions need to be made in an oblique plane.

d. There is a greater range of choice of positions of the axes than in the other forms except trimetric. See Fig. 14.10.

To offset these advantages, it should be noted that oblique projections, even though foreshortened on the receding axis, have an unpleasing distortion. This is particularly true when circles must be shown in the receding faces.

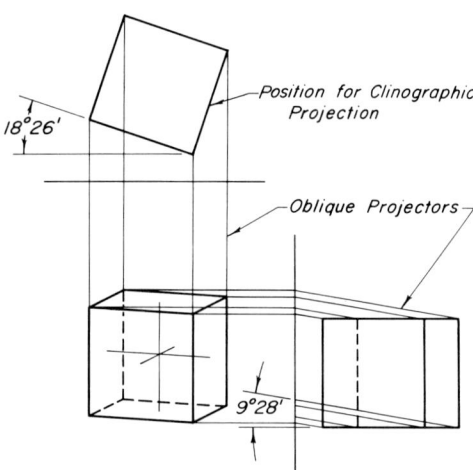

FIG. 14.23. Clinographic projection of a cube.

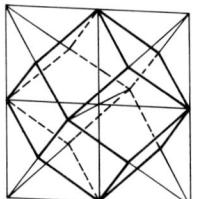

Rhombic Dodecahedron

FIG. 14.24. Crystal in clinographic projection.

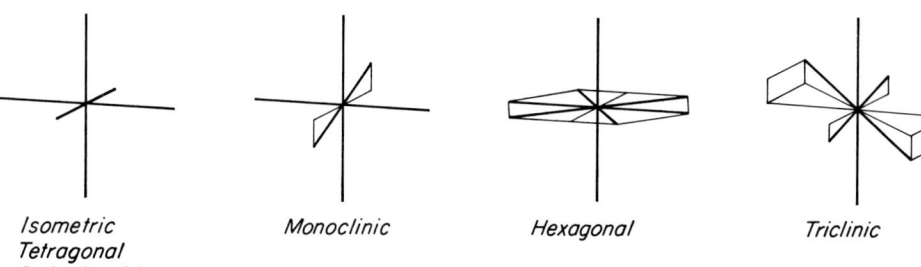

Isometric
Tetragonal
Orthorhombic

Monoclinic

Hexagonal

Triclinic

FIG. 14.25. Axes in six crystal systems.

SELF-STUDY QUESTIONS

Before trying to answer these questions, read the chapter carefully. Then, without reference to the text, answer as many questions as possible. For those that cannot be answered, the number in parentheses following the question number gives the article in which the answer can be found. Look it up and write down the answer. Check the answers that you did give to see that they are correct.

14.1 (**14.1**) In oblique projection the projecting lines are _____ to the plane of projection.

14.2 (**14.1**) The face of an object that is parallel to the vertical plane (picture plane) shows in _____ _____ in an oblique projection.

14.3 (**14.3.1–14.3.4**) Two of the four types of oblique projection are _____ and _____.

14.4 (**14.3.1**) In Cavalier projection the projecting lines make _____ _____ with the plane of projection.

14.5 (**14.3.1**) The _____ scale must be used on all axes of a Cavalier projection.

14.6 (**14.3.2**) In Cabinet drawing the scale on the receding axis is _____ _____ that on the front face.

14.7 (**14.4.2**) Circles in the front face of an object show as _____ in oblique projection.

14.8 (**14.4.5**) Circles in the side and top faces of an object can be drawn by the _____ _____ method in Cavalier projection.

14.9 (**14.6**) The four-center approximate method for drawing circles cannot be used in _____ drawing.

14.10 (**14.5.1**) The receding axes in a Cavalier projection can be made at _____ _____. It does not have to be at _____ degrees with the horizontal.

14.11 (**14.5.2**) For ease in construction, the front face of an object should be _____ to the plane of projection.

14.12 (**14.5.2**) To reduce distortion the _____ dimension of an object should be parallel to the plane of projection.

14.13 (**14.3.3**) In a general oblique projection, if the scale on the front face is 1 in. = 1 in., the scale of the receding axis could have _____ value except 1 in. = _____ or ½ in. = _____.

14.14 (**14.8**) If an object consists mostly of parallel circular parts, the _____ method of construction is very convenient.

14.15 (**14.9**) In making sectional views the cutting planes should preferably be _____ to the oblique faces of the enclosing box.

14.16 (**14.10**) Dimension lines and extension lines must be in the same _____ plane.

14.17 (**14.11**) If a bolt showing screw threads must be drawn in oblique, it is simplest to make the _____ of the bolt on the _____ axis of the drawing.

14.18 (**14.15**) In mineralogy and crystallography the shape of crystals is shown by a form of oblique drawing called _____ projection.

CHAPTER 15

PERSPECTIVE

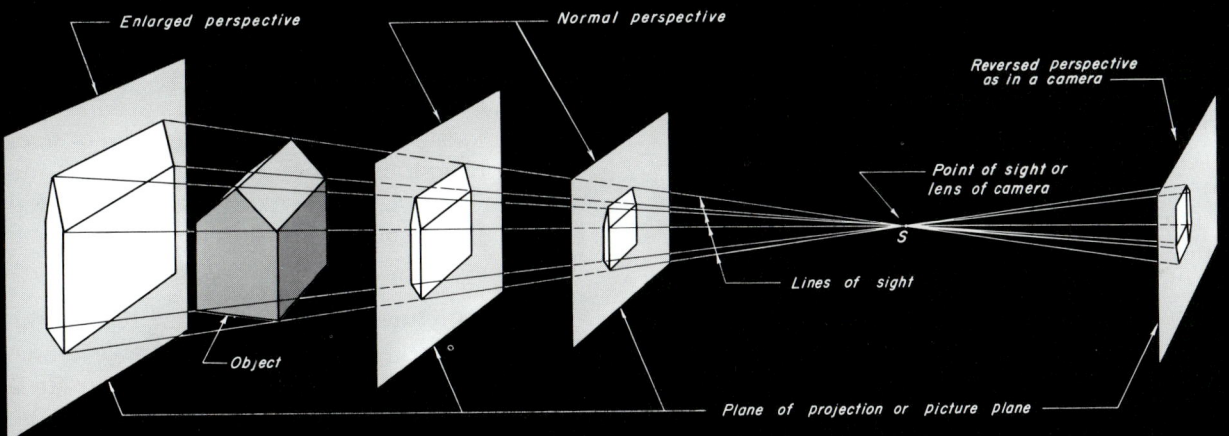

CHAPTER 15

15.1 DEFINITION OF PERSPECTIVE

When a person looks at an object, the light rays (visual rays) from the object are focused by the eye so that the picture is formed on the spherical rear surface of the eye known as the retina. Perspective is the form of pictorial drawing that most nearly approaches the picture as seen by the eye. Thus if we imagine a vertical plane between the eye and the cube, as in Fig. 15.1, the visual rays from the cube to the eye, if intercepted by the vertical or picture plane, will form an image that exactly coincides with the edges of the cube. This produces the same image as the cube itself if viewed from this one particular position. This image is called a *perspective*.

From Fig. 15.1 it will be observed that the major difference between perspective projection and the forms of projection studied heretofore lies in the fact that the point of sight is at a finite distance from the object. The visual rays or lines of sight from the object, therefore, converge to the point of sight instead of being parallel to each other as in other forms of projection.

The picture or perspective obtained will depend on the relative position of the object, picture plane, and point of sight.

15.2 BASIC TERMS AND RELATIONSHIPS OF PERSPECTIVE

Certain terms and the relationships between them are commonly used in a discussion of perspective. Some of these are presented in the following paragraphs.

15.2.1 The Point of Sight. The position of the eye of the observer in making a perspective is called the *point of sight*. It is sometimes referred to as the *station point*. In the illustrations of this book the point of sight will be designated by the letter S. Its projections will be s^H, s^V, and s^P, as is customary in orthographic projection. See Fig. 15.1.

15.2.2 Visual Rays. *Visual rays* are the straight lines from the object to the point of sight that represent the light rays that produce the image in the eye. They are the lines of sight used in making the actual construction of perspectives.

15.2.3 Picture Plane. The *picture plane* is the plane surface on which the perspective is drawn or projected. It could have any position. For a bird's-eye view a horizontal plane would be

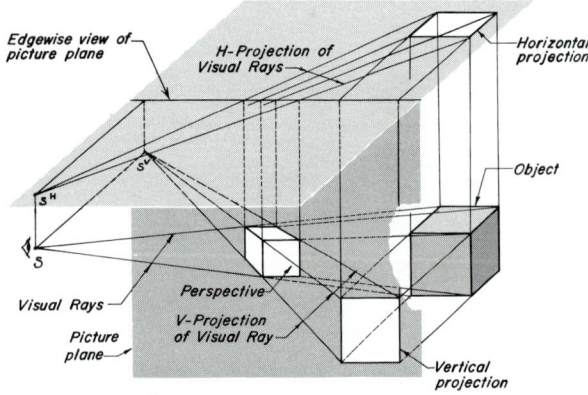

FIG. 15.1. Theory of perspective.

used. For most perspectives a vertical plane is used as the picture plane. Only this type will be discussed here.

15.2.4 Vanishing Points.
Vanishing points are the points in space where, by definition, parallel lines meet. On the drawing, the vanishing point, designated by the letters VP, is the perspective of the meeting point of the lines in space. Every set of parallel lines not parallel to the picture plane has a vanishing point.

15.2.5 Horizon Line.
The *horizon line* is a horizontal line that always passes through the vertical projection of the point of sight.

15.3 LOCATION OF THE PICTURE PLANE
If we assume the point of sight and the object in Fig. 15.2 to be fixed in position, the picture plane may have several positions relative to them.

15.3.1 Enlarged Perspective.
When the object is between the point of sight and the picture plane, an enlarged perspective is produced as in Fig. 15.2a. This arrangement is very rarely used.

15.3.2 Normal Perspective.
When the picture plane is between the object and the point of sight, the perspective is smaller than the object. The closer the plane is to the point of sight, the smaller the perspective becomes, as can be seen by comparing Figs. 15.2b and c. This is the arrangement usually used, and it is therefore called a *normal perspective*.

15.3.3 Reversed Perspective.
When the point of sight is between the object and the picture plane as in Fig. 15.2d, the perspective is inverted and reversed as in a camera. In the camera the lens represents the point of sight. In drawing, this arrangement is not used.

15.4 LOCATION OF THE POINT OF SIGHT
With the picture plane and the object in a definite relationship to each other the perspective can be greatly altered by a change in the position of the point of sight. The choice of the location of the point of sight is therefore very important in making an attractive perspective. A careful selection of the position of the point of sight makes it possible to emphasize the features in any visible face of the object. This is discussed in the following paragraphs.

15.4.1 Distance of the Point of Sight from the Object.
It has been observed from experience that when the viewer is at a certain point, the eye will see clearly all the picture contained within a right circular cone having its apex at the eye and an interior angle at the apex of approximately 30°. This condition is satisfied when the point of sight is placed at a distance from the object at least twice the longest dimension of the object, as shown in Fig. 15.3. For a good perspective, the point of sight should be located in front of the center of the object. Keeping these conditions in mind, the location of the point of sight may be selected to show prominently the more important faces of an object.

15.4.2 Position to the Right or Left of the Object.
In Fig. 15.4a the point of sight is to the right of the cube, thus revealing the right face. By moving the point of sight farther to the right, the right side of the cube could be made still more prominent.

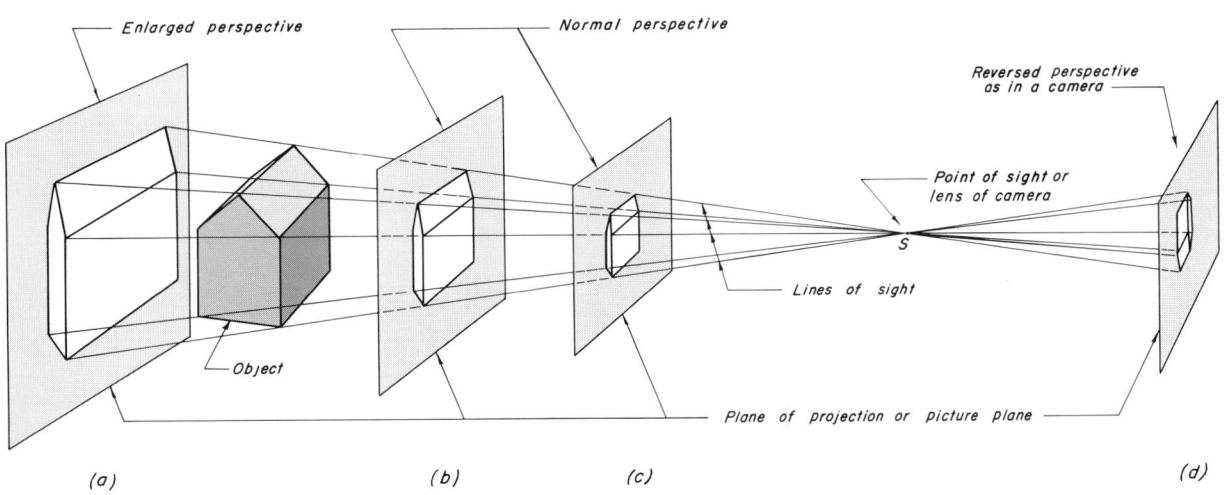

FIG. 15.2. Relationship of object, picture plane, and point of sight.

In Fig. 15.4b the left side of the cube is revealed by having the point of sight to the left of the cube.

15.4.3 Elevation of the Point of Sight.
In Fig. 15.5a the lower portion of the figure, the point of sight shown by projections s^V and s^P, is above the top of the cube. This can be seen in the side view at the left. The shaded perspective therefore shows the top of the cube.

In Fig. 15.5b the point of sight shown by the same projections s^V and s^P is below the side view of the cube. The perspective therefore shows the bottom of the cube.

15.5 POSITION OF THE OBJECT
One more preliminary consideration should be presented before proceeding with the actual construction of an object, namely, the effect of the position of the object.

By changing the position of the object relative to the picture plane, other changes in the appearance of the perspective can be made. These changes are so different from each other that they have been given special names.

15.5.1 Parallel or One-Point Perspective.
When the principal face of the object is parallel to the picture plane, as in Fig. 15.4, a parallel or

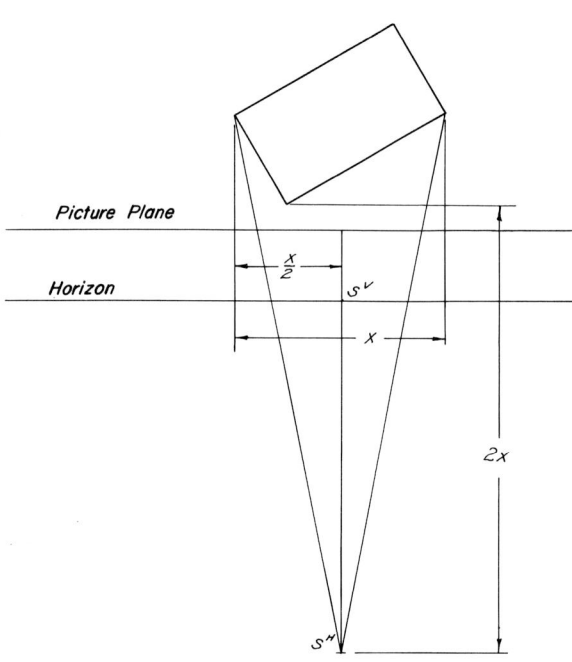

FIG. 15.3. Location of point of sight.

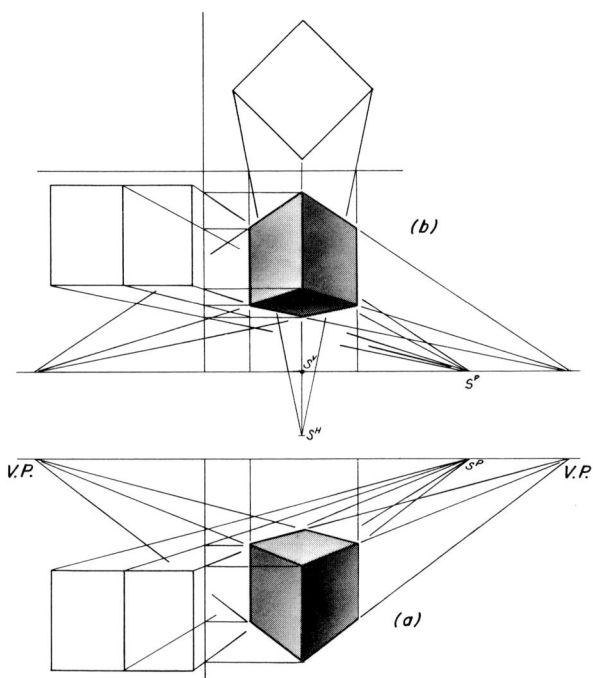

FIG. 15.5. Object above and below point of sight.

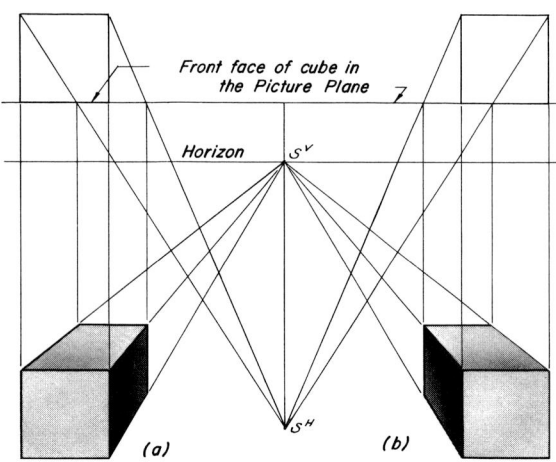

FIG. 15.4. Object right and left of point of sight.

one-point perspective is formed. The term *one-point* refers to the fact that such perspectives have only one principal vanishing point.

15.5.2 Angular or Two-Point Perspective.
When two faces of the object are inclined to the picture plane, as in Fig. 15.5, an angular or two-point perspective is formed. There are two principal vanishing points.

15.5.3 Oblique or Three-Point Perspective.
When all faces of the object are oblique to the picture plane, an oblique or three-point perspective results. There are three principal vanishing points, as shown in Fig. 15.6.

15.6 DEFINITION OF THE METHODS OF CONSTRUCTING A PERSPECTIVE
Four methods are commonly used for constructing perspective drawings. Each has its own advantages. A complete discussion of each method is given in later articles of the chapter.

15.6.1 Visual-Ray Method.
This method is based directly on the definition of a perspective. See Section 15.1 and Fig. 15.1. This method is very simple to understand and particularly useful for simple objects. It clarifies the meaning of perspective. See Fig. 15.7. It involves finding the piercing point of the visual ray with the picture plane. For the method of finding piercing points see Article 7.23.2.

15.6.2 Vanishing-Point Method.
The method consists of finding the perspective of two points in each line of an object. One of the points is the vanishing point for any series of parallel lines. This one point suffices for every line in the parallel group. See Fig. 15.8.

The other point is the piercing point of the lines on the object that are extended until they pierce the picture plane. The piercing point and the vanishing point determine the perspective of an infinite line. There is a separate piercing point for each line. See Fig. 15.14.

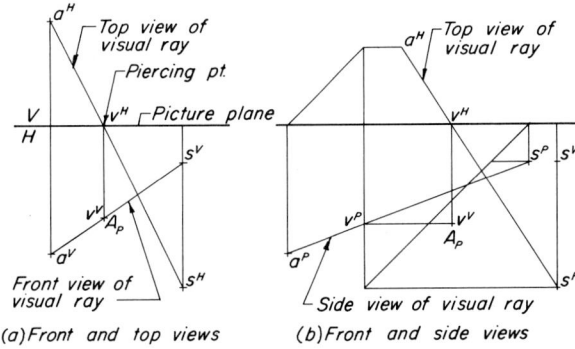

FIG. 15.7. Perspective of a point by the visual-ray method.

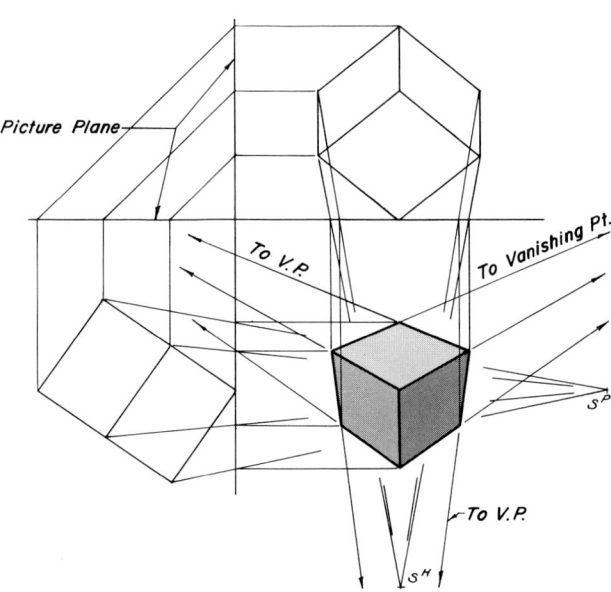

FIG. 15.6. Three-point perspective.

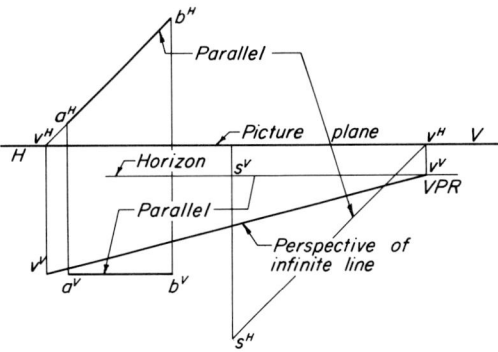

FIG. 15.8. Perspective of an infinite line by the vanishing point method.

15.6.3 Combination Method. This is the method most commonly used. In this method both the visual-ray method and the vanishing-point method are combined to obtain the greatest advantage from each. In Fig. 15.9 the perspective of the infinite line, of which AB is a segment, has been found by means of the vanishing-point method. The segment AB has been cut from the infinite line by visual rays. It is designated $A_p B_p$ in Fig. 15.9.

15.6.4 Measuring-Point and Measuring-Line Method. This is a more sophisticated method for finding perspectives. The orthographic views are used only to obtain dimensions and are otherwise not directly involved in the construction. See Article 15.16.3.

15.7 CONSTRUCTION BY THE VISUAL-RAY METHOD

Two methods of procedure may be employed, namely, by using the front and top views of both the object and the point of sight or by using the top and side views of both.

15.7.1 Front and Top Views. By this method it is only necessary to draw the visual rays from the object to the point of sight and find where they pierce the picture plane. This method has been used in Fig. 15.10 where visual rays from A to the point of sight S have been drawn in the top and front views. The vertical projection A_p of the point where this line pierces the vertical plane is the perspective of A. The remaining visible corners of the cube have been found in the same manner. They are then connected in the proper sequence to make the perspective. Only visible outlines are shown in perspective drawings.

15.7.2 Top and Side Views. In this method as in the preceding one it is necessary to lay out the top and side views of the object and of the point of sight all in proper relationship to the reference planes as shown in Fig. 15.11. As an example, note the following.

a. In the top view the visual ray goes from a^H to s^H and pierces the vertical plane at v^H.

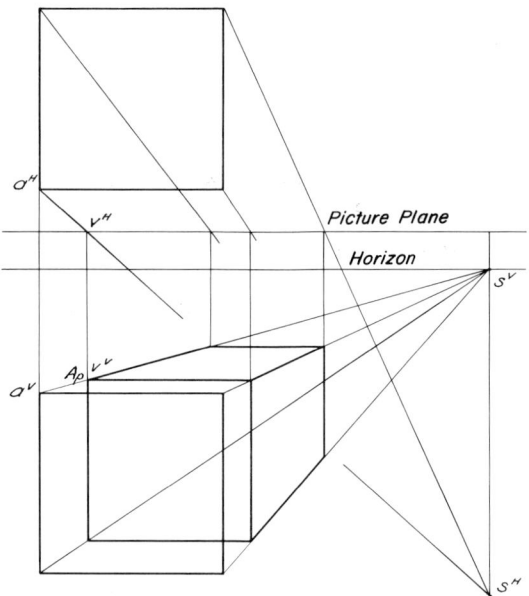

FIG. 15.10. Perspective of a cube. Visual-ray method using front and top views.

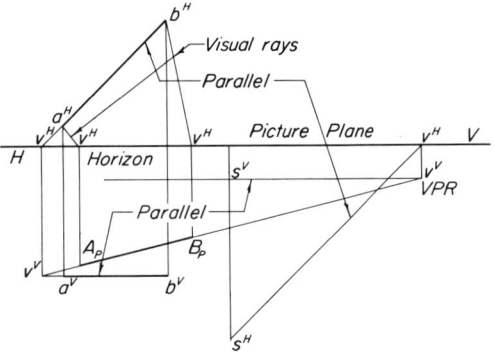

FIG. 15.9. Perspective of a line segment AB by the combination method.

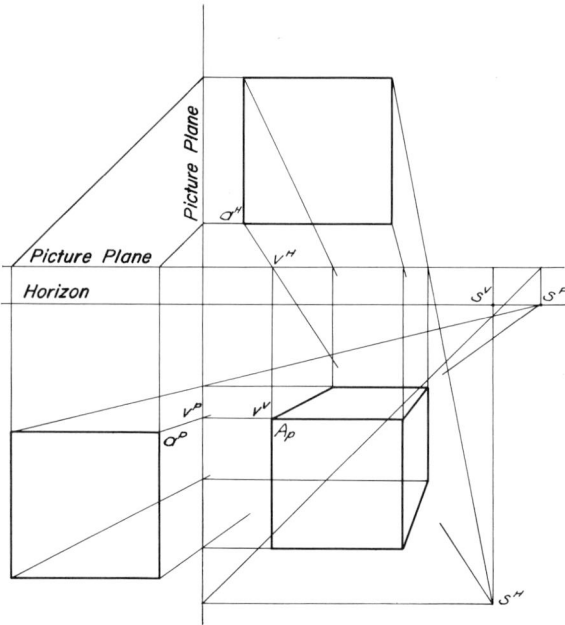

FIG. 15.11. Perspective of a cube. Visual-ray method using top and side views.

b. In the side view the visual ray goes from a^P to s^P and pierces the vertical or picture plane at v^P.

c. The vertical projection of the piercing point v^V is found by projecting down from v^H and across from v^P in the same manner as the front view of any point would be obtained from the top and side views.

It will be noted that the entire construction is simply orthographic projection throughout. The problem resolved itself into the simple one of finding the piercing point of visual rays with the V-plane.

15.8 CONSTRUCTION BY THE VANISHING-POINT METHOD

This method consists of the two operations listed in Article 15.6.2. Before proceeding with the perspective of an object it is necessary to be able to find the vanishing point for any kind of line.

15.8.1 General Principles for Finding Vanishing Points.

It can be observed in a number of the preceding figures that lines that are not parallel to the picture plane converge to a point in the perspective. By definition parallel lines meet at infinity. The perspective of this meeting place is called the *vanishing point of the lines*. To find the perspective of the vanishing point for any group of parallel lines, it is necessary to do the following.

a. To construct a line through the point of sight parallel to the group of parallel lines. This new line meets the others at the same point at infinity.

b. To find the point where this new line, through the point of sight, pierces the picture plane or vertical plane.

It should be noted that two projections of the given line or group of parallel lines must be available or must be drawn. There must also be two projections of the line through the point of sight.

15.8.2 Vanishing Points for Horizontal Lines.

The identifying feature of horizontal lines is that their vertical projections are always parallel to the H-V reference line. Given the projections of line AB in Fig. 15.8 and the projections of the point of sight, proceed as follows.

a. Through s^H draw a line parallel to $a^H b^H$ and extend it to the picture plane as seen edgewise in the top or plan view.

b. Through s^V draw a line parallel to $a^V b^V$.

c. The vertical projection v^V directly below v^H is the piercing point of the line with the picture plane. It is marked VPR since it is a vanishing point at the right. For all horizontal lines the vanishing point is always in the horizon line.

15.8.3 Vanishing Point of Oblique and Inclined Lines.

a. In Fig. 15.12a line CD is inclined upward to the rear and left. Follow the same three steps as in Article 15.8.2. Note that the vanishing point VPL is to the left of the point of sight and above the horizon line.

b. In Fig. 15.12b line EF is inclined downward to the right and rear. Follow the same three steps as in Article 15.8.2. Note that the vanishing point, marked VPR, is to the right of the point of sight and below the horizon line.

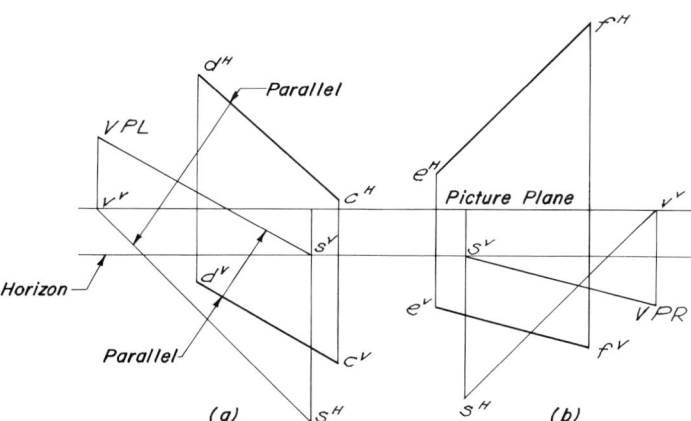

FIG. 15.12. Method of finding vanishing points.

c. In Fig. 15.13 line AB is horizontal and also perpendicular to the picture plane. Following the same procedure as in Article 15.8.2, it will be observed that *the vanishing point for lines perpendicular to the picture plane is always at the vertical projection of the point of sight s^V*.

15.9 CONSTRUCTION OF A PERSPECTIVE OF AN OBJECT BY THE VANISHING-POINT METHOD

This method is well adapted to the construction of angular perspective. With a cube shown in the position of Fig. 15.14 proceed as follows.

a. Find the vanishing points for the two sets of parallel lines like AB and AC. These are marked VPL and VPR in Fig. 15.14. See Article 15.8.2.

b. Extend all horizontal lines like AB and AC in Fig. 15.14 until they pierce the picture plane at v^H. There are eight lines, four in each direction, and eight piercing points all marked v^v.

c. Note that is is not necessary to have either a complete front or side view of the object. The elevation of the piercing points is obtained from the partial elevation of the cube at the left of Fig. 15.14.

d. We now have two points in every horizontal line on the cube, VPR being a point on all horizontal lines in the direction of AC and VPL being a point on every line in the direction of AB.

e. The crossing points of these lines in pairs locate the corners on the perspective of the cube such as A_p, B_p, and C_p.

f. Connecting the upper and lower corners of the cube by vertical lines completes the perspective. Note that vertical lines on the object are always vertical in the perspective since they are parallel to the picture plane.

15.10 COMBINATION METHOD

This method makes judicious use of both the visual-ray method and the vanishing-point method. The combination makes the most rapid and efficient use of both methods.

15.10.1 Perspective of an Inclined Line. To find the perspective of a finite segment of an inclined or oblique line proceed as follows.

a. From the two projections, s^H and s^V of the point of sight S, draw lines parallel to the corresponding projections of AB, as in Fig. 15.15a.

b. Find where this line pierces the picture plane. The vertical projection of this point v^v is the vanishing point VP for line AB. Since this line is oblique, the VP is not in the horizon line.

c. Extend line AB until it pierces the picture plane at v_1^H. A perpendicular from v_1^H to a^Vb^V extended locates v_1^V, the vertical projection of the piercing point of the line with picture plane.

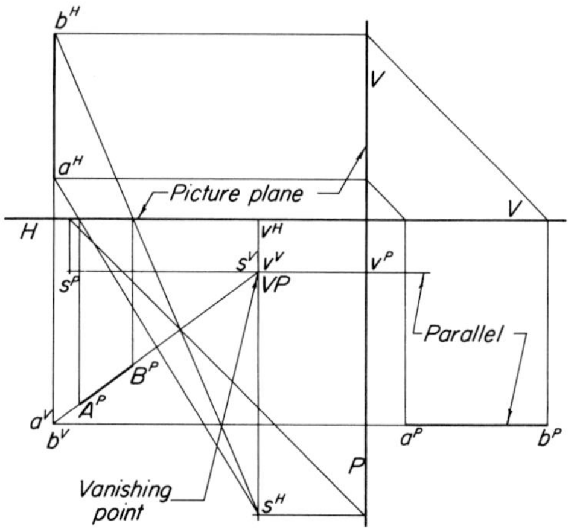

FIG. 15.13. Vanishing point for lines perpendicular to the picture plane.

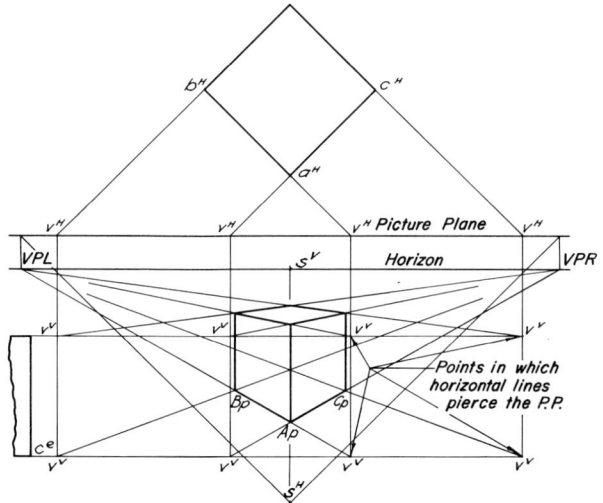

FIG. 15.14. Perspective by the vanishing-point method.

d. Connect VP and v_1^V. This is the perspective of the infinite line.
e. In Fig. 15.15b draw the visual rays from s^H to a^H and b^H.
f. Where these rays cross the reference line, drop perpendiculars to the infinite line cutting off the line segment A_pB_p, which is the required perspective.

15.10.2 Perspective of a Horizontal Line on an Object by the Combination Method. In the preceding problem it should be noted that two complete projections of the oblique line were required. For this problem and all problems involving horizontal lines it is not necessary to have the correctly located vertical projection of the horizontal lines on the object. It is only necessary to have the correctly located vertical projection of the horizontal lines on the object. It is only necessary to have a view that will give the correct elevation of the horizontal lines as shown in Fig. 15.16. Proceed as follows:

a. Find the vanishing point of line AB. This will be in the horizon line. See Fig. 15.14. It is therefore not necessary to have a true front view of the box.
b. Extend a^Hb^H until it pierces the picture plane at v^H. The actual piercing point v^V is at the level of AB in the view at the left and is simply projected across to the vertical line from v^H.

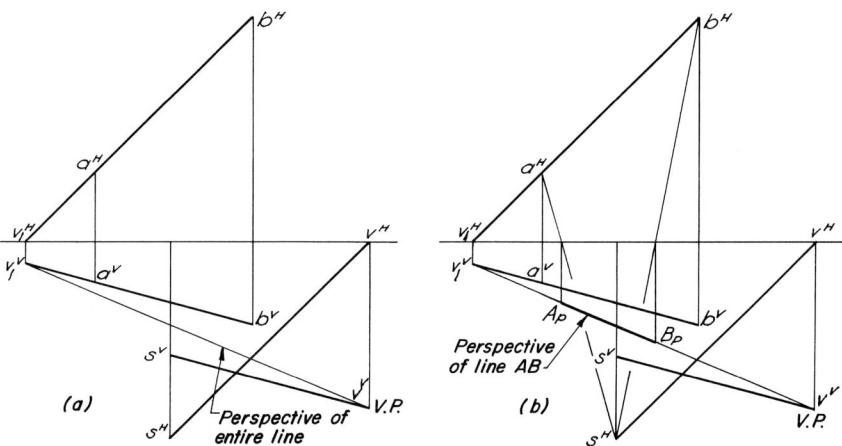

FIG. 15.15. Perspective of an inclined line by the combination vanishing-point and visual-ray method.

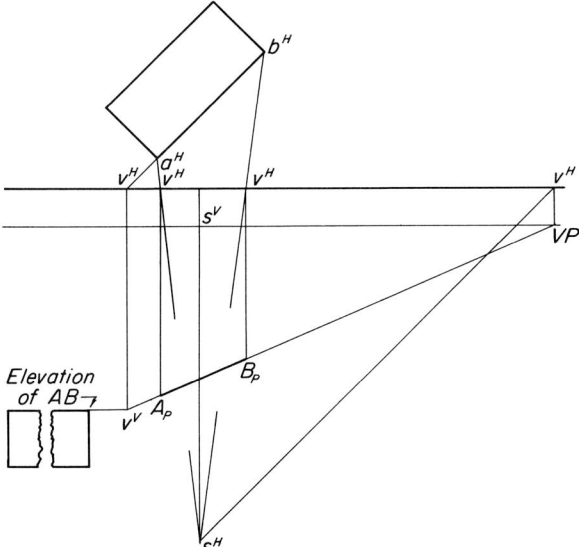

FIG. 15.16. Perspective of a horizontal line by the combination method.

c. Draw a line from v^V to VP.
d. Draw visual rays from s^H to a^H and b^H.
e. Drop perpendiculars from the points v^H, where the visual rays cross the picture plane, to the infinite line from v^V to VP.
f. This locates the perspective of the line segment A_pB_p.

15.11 KINDS OF PERSPECTIVE

The two common types of perspectives are the *parallel perspective* and *angular perspective*. The first is used mostly for interior views of buildings, whereas the latter gives more attractive results for the exterior views of buildings or larger civil engineering projects. A third type *three-point perspective* is discussed in Section 15.17.

15.12 PARALLEL OR ONE-POINT PERSPECTIVE

When the object is placed so that one face is parallel to the picture plane and the others perpendicular to it, the resulting picture is known as *parallel* or *one-point perspective*. Except for interior views this is not usually the most desirable position for the object. It tends to move the point of sight over to one side in order to show two exterior sides of an object as shown in Fig. 15.17 and thus distorts the picture.

For this construction the more practical combination method will be used.

It is necessary to have the horizontal projection drawn in the proper relation to the reference or picture plane and any elevation in its proper relation to the horizon. The horizontal and vertical projection of the point of sight must also be given. This information for a small house is shown in Fig. 15.17a.

The first step in the solution is always to find the vanishing points of the principal lines of the figure. In this case the lines parallel to the picture plane will have vanishing points at infinity, which means that any face parallel to the picture plane will show in the perspective in true shape but reduced in size, depending on the distance of the object behind the picture plane. The lines perpendicular to the picture plane will have a vanishing point that may be found by the method explained in Article 15.8.3c. In this case the lines drawn through S parallel to the given line must pierce the picture plane at a point whose vertical projection is at s^V, which therefore becomes the vanishing point for all lines perpendicular to the picture plane, as in Fig. 15.17a. In other words, the *vertical projection of the point of sight is always the vanishing point for lines perpendicular to the picture plane*.

The next step is to extend one of these lines

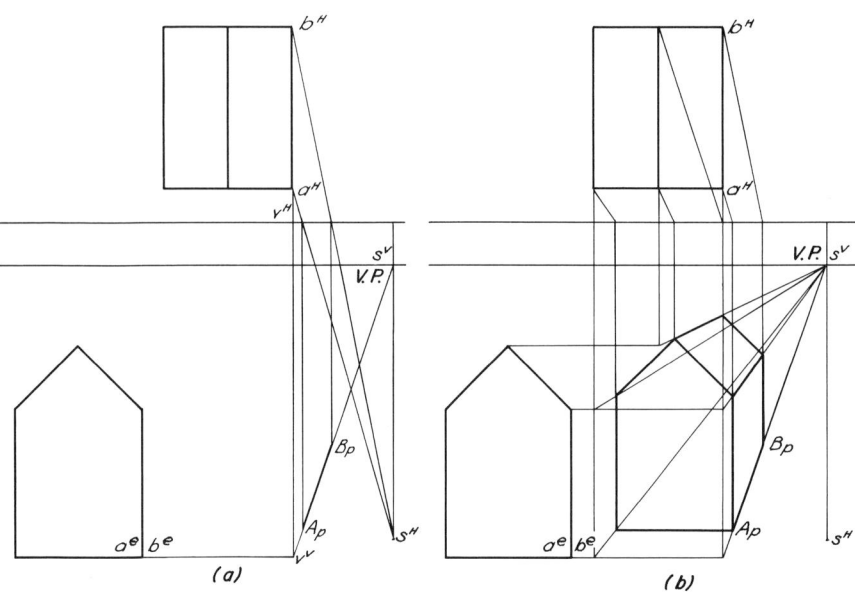

FIG. 15.17. Perspective of a building. Vanishing-point and visual-ray method.

until it pierces the picture plane. Since this is a horizontal line, the piercing point will be at the level of the line as shown in the given elevation. This piercing point is marked v^v in Fig. 15.17a. By joining the piercing point to the vanishing point the perspective of the complete line is formed. By visual rays the required points A_p and B_p on the line may be found to give one line on the building, as shown in Fig. 15.17a. Other lines on the building may be found in the same manner, as illustrated in Fig. 15.17b. It should be noticed that the front face of the building is true shape but not true size. The photograph in Fig. 15.18 is a parallel perspective.

FIG. 15.18. One-point photograph perspective of McGregor Lounge, Wayne State University (courtesy of Minoru, Yamasaki and Associates, architects, and Baltazor Korab, photographer).

15.13 ANGULAR OR TWO-POINT PERSPECTIVE BY THE COMBINATION METHOD

When one of the principal axes of the object is parallel and the other two are inclined to the picture plane, the resulting picture is called *two-point perspective*. In this case the required information, which is the same as that for parallel perspective, is shown in Fig. 15.19a. Note that a true profile projection is not required.

To find the perspective of any point, it is necessary to find the perspective of a line through the point and then locate the point on the line by a visual ray. Any two points on the line will determine the line, but the best ones are the vanishing point and the V-piercing point.

a. The first step is to find the vanishing points of both sets of inclined horizontal lines that are parallel respectively to AB and AC. These vanishing points have been located at VPR and VPL in Fig. 15.19b.

b. Then any line of the drawing such as AB in Fig. 15.19b may be extended to find its V-piercing point at v^V.

c. By connecting v^V with VPL, the vanishing point of AB, the perspective of the entire line is found.

d. Visual rays drawn to points A and B will serve to locate points A_p and B_p on the perspective of the line. These will be the perspectives of two corners of the building.

e. Another corner of the building, such as C in Fig. 15.19c, may be located by connecting A_p with VPR, which is the vanishing point of line AC, and drawing the visual ray through C, thereby locating C_p on line AC.

f. Corner D of the building must be located exactly as was done for point A. Any horizontal line through D may be used if its vanishing point has been located.

15.14 PERSPECTIVE OF CIRCLES OR CURVES

Thus far all objects shown have been made up of straight lines. In buildings, however, some circular parts such as arches are frequently involved. The circle will form the basis for the construction of many other curves.

15.14.1 Circle Parallel to the Picture Plane.
If a circle is parallel to the picture plane as in Fig. 15.20, the horizontal projection is the edge view that is a straight line parallel to the V-plane. If visual rays are drawn from s^H to the circle, they form an oblique cone. The intersection of the picture plane and the cone is a circle having a diameter as shown in Fig. 15.20. With the center line piercing the plane at C_p and a diameter as shown, the circle can be drawn with a compass.

15.14.2 Circle in a Vertical Position.
Figure 15.21 shows a circle in a vertical plane inclined to the picture plane. A series of points (12 in this case) are located and numbered in each view. Imaginary horizontal lines are drawn through these points and the perspective of the lines found in the usual way. The points are then located on these lines by visual rays. For this construction it is desirable to have all points located in each view.

15.14.3 Circle in a Horizontal Position.
In Fig. 15.22 the perspective has been found by the vanishing-point method only.

Twelve points have been located on the plan view of the circle. Only the elevation shown by the heavy line at the lower left is needed. Draw two sets of imaginary horizontal lines through the points as shown in Fig. 15.22. Find the perspectives of these two sets of lines by means of the

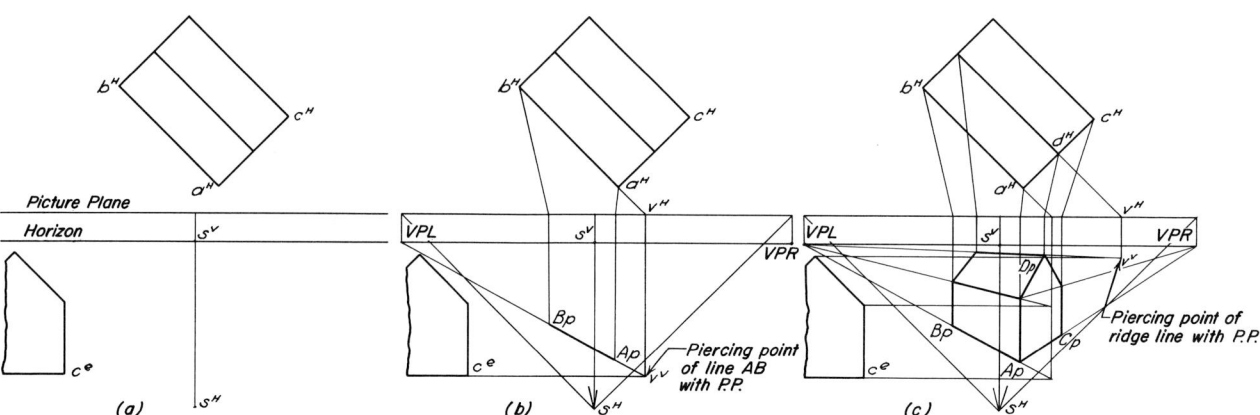

FIG. 15.19. Angular perspective: vanishing-point and visual-ray method.

piercing points and the vanishing points. Where the lines intersect in pairs locates the points on the perspective.

15.15 VANISHING-POINT TRACES

In perspective a vanishing point may be found for any line. Ordinarily they are useful only when it is necessary to find the perspective of a group of parallel lines. When two or more lines lie in a plane, the intersection of that plane with the plane of projection or picture plane forms a line. That line is the locus of vanishing points of all lines lying in that plane and is called a *vanishing-point trace*. Thus plane *ABCD* in Fig. 15.23 is a vertical plane that intersects the picture plane in the vertical line marked *VPL–VPCD*. This line then becomes a locus on which the piercing point or vanishing point of every line in plane *ABCD* will be found.

Any two points will determine a straight line; therefore, if it is possible to find any two vanishing points in a plane, the line joining them will be the vanishing-point trace for that plane. Thus since the vanishing point of *CD* is at *VPCD* and the vanishing point of *DN* is at *VPR*, the line joining *VPCD* with *VPR* will be the vanishing-point trace for place *CDNM*. In a similar manner it can be shown that *VPL–VPGK* is the vanishing-

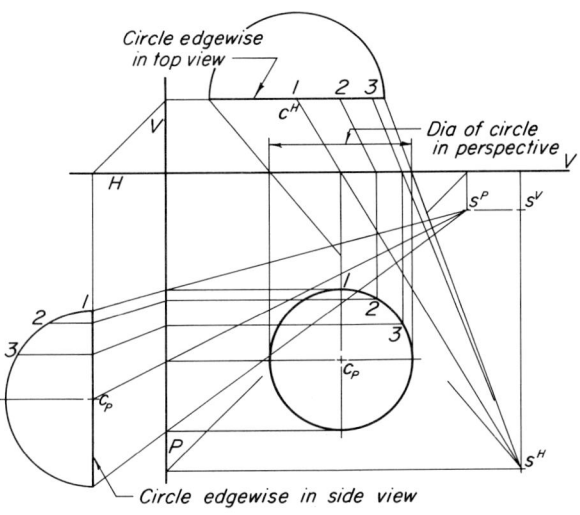

FIG. 15.20. Perspective of a circle parallel to the picture plane.

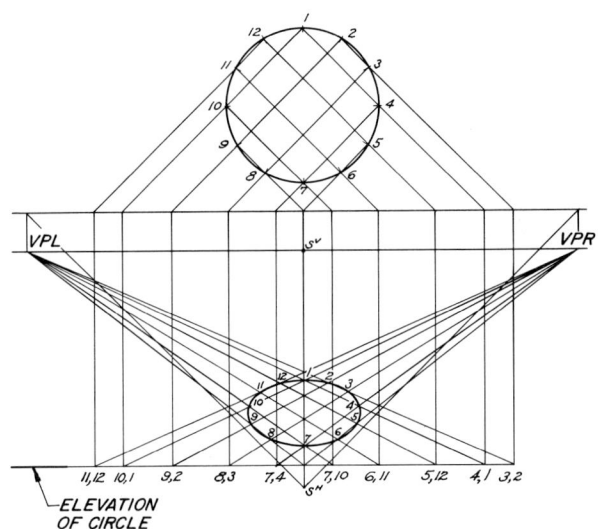

FIG. 15.22. Perspective of a circle in a horizontal position.

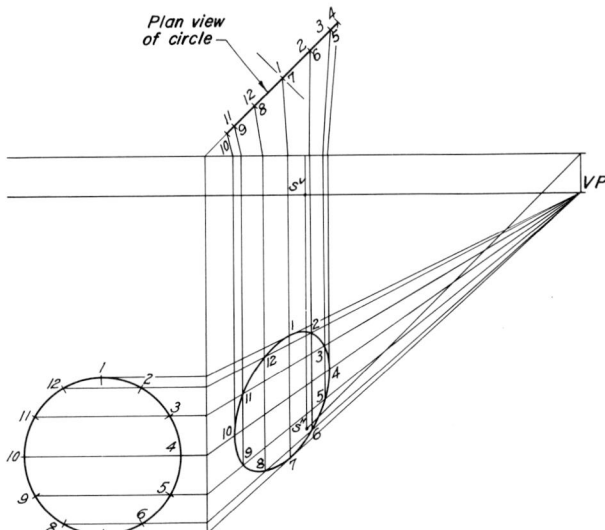

FIG. 15.21. Perspective of a circle in a vertical position inclined to the picture plane.

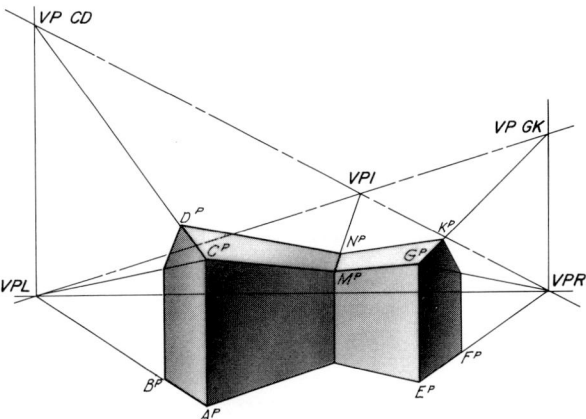

FIG. 15.23. Vanishing-point traces.

point trace for plane GKNM. It follows that the intersection of these two traces will be the vanishing point for line MN, which is the line of intersection of the two planes.

This principle is frequently useful in finding intersections of planes and in locating shadows in perspective.

15.16 MEASURING-POINT METHOD

The great advantage of this method over all other methods of perspective is that it is not necessary to set up the projections in any particular position and work from them by projection or visual rays.

15.16.1 Measuring Points and Lines.

When a vertical face that is inclined to the picture plane, such as face ABCD in Fig. 15.24, is rotated until it lies in the picture plane, the vertical projection of this face in the revolved position $a_r^V b_r^V c_r^V d_r^V$ is coincident with the perspective of the revolved position. If the vanishing point of the line joining b^H to b_r^H in Fig. 15.24 is found, the perspective of this line can be used to carry points from the revolved position to their correct position in the perspective. The vanishing point MPR of line $b^H b_r^H$ is therefore called a measuring point. Since the vertical projection of the revolved position $a_r^V b_r^V c_r^V d_r^V$ is in true size, the horizontal lines $a_r^V b_r^V$ and $c_r^V d_r^V$ are true length and any desired distances from A or D can be laid off along these projections. The projection $a_r^V b_r^V$ is therefore called a *measuring line*. The projection $c_r^V d_r^V$ could have been used equally well as a measuring line. In fact, a horizontal line can be drawn through any point that lies in the picture plane and used as a measuring line.

A simpler method for finding the measuring point MPR is illustrated in Fig. 15.24. If, through v^H, the H-projection of VPR, an arc having a radius $v^H s^H$ and center v^H, is drawn from s^H to the picture plane, this point on the picture plane may be projected straight down to MPR on the horizon.

Proof. Since $a^H b^H b_r^H$ was constructed as an isosceles triangle and since $v^H s^H v_1^H$ has its sides respectively parallel to $a^H b^H b_r^H$, the triangle $v^H s^H v_1^H$ is similar and is also an isosceles triangle. Therefore, $v^H s^H$ is equal to $v^H v_1^H$, and v_1^H may be located by constructing the arc shown on the figure.

15.16.2 Selection of Conditions for the Perspective by the Measuring Point.

a. The vanishing points right and left are usually taken as far apart as convenient.

b. The picture plane and horizon are made to coincide.

c. The angles that the sides of the object make with the picture plane are chosen.

d. A front corner of the object is selected at a certain place on the picture plane.

e. If the vanishing points are chosen first, s^H is located by drawing lines from the vanishing points parallel to the sides of the object, which are usually at right angles to each other.

15.16.3 Construction of a Perspective by the Measuring-Point Method.

In Fig. 15.25a the projections of an object are given. The problem is to draw the perspective so that the corner marked A will be in the picture plane 6 in. below the horizon and 1 in. right of the point of sight as

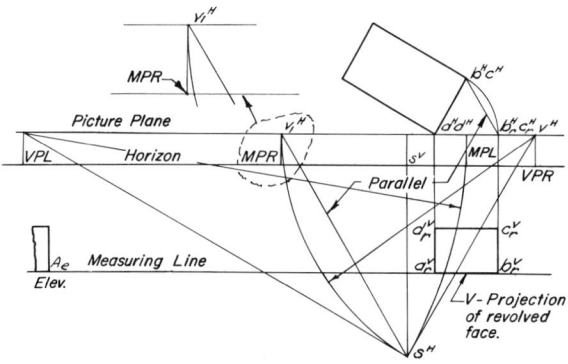

FIG. 15.24. Finding measuring points.

in Fig. 15.25d. Side AB is to make an angle of 60° with the picture plane, and the vanishing points are to be 18 in. apart. The construction will then proceed in the following steps.

  a. Figure 15.25b. Draw a horizontal line near the top of the sheet and mark off the vanishing points 18 in. apart.
  b. Figure 15.25b. Through VPR construct a line at 60° with the picture plane and through VPL a line making 30° with the picture plane. The intersection of these two lines will locate s^H. The vertical projection s^V will be on the horizon. This assures that the faces of the block will form the specified angles with the picture plane.
  c. Figure 15.25c. Using VPR as a center, swing an arc through s^H to the picture plane to locate MPR. Using VPL as a center, swing an arc through s^H to the picture plane to locate MPL.
  d. Figure 15.25d. Measure 6 in. below the horizon and 1 in. right of the point of sight to locate the perspective of point A. Through this point draw a horizontal line, which is the measuring line for horizontal distances.
  e. Figure 15.25e. On a vertical line through A_p, lay out the height of the object AD to locate D_p. Draw lines from these points to VPR and VPL.
  f. Figure 15.25f. Lay out the distance AB on the horizontal measuring line to the right of A_p. From this point draw a line to MPR to intersect line from A_p to VPR, thus locating B_p. Erect a vertical line through B_p to give the right edge of the object.
  g. Figure 15.25g. Lay out the distance AC on the horizontal measuring line to the left of A_p. From this point draw a line to MPL to intersect the line from A_p to VPL. This locates C_p. Erect a vertical line through C_p to give the left edge of the object. Complete the outline of the figure by drawing to the proper vanishing points.
  h. Figure 15.25h. From D_p lay out on the vertical line the distance FE, which is the depth of the slot. From this point draw a line to VPL. From A_p lay out the distance DE on the horizontal measuring line to the left of A_p. Connect this point to MPL to intersect the line from A_p to VPL. Erect a perpendicular to locate points E_p and F_p, which establishes the right side of the slot. In a similar manner locate the left side of the slot.
  i. Connect these points to the proper vanishing points to complete the picture.

15.16.4 Perspective of a Vertical Circle by the Measuring-Point Method. By taking measurements from the orthographic projections, points on a circle may be located in perspective by the use of coordinates, but this is a tedious process and should be avoided if possible.

The better method is to find the perspective of a square circumscribing the circle and then obtain the ellipse by the diagonal method. For a circle lying in a vertical face the procedure is illustrated in Fig. 15.26 and the steps are listed on the next page.

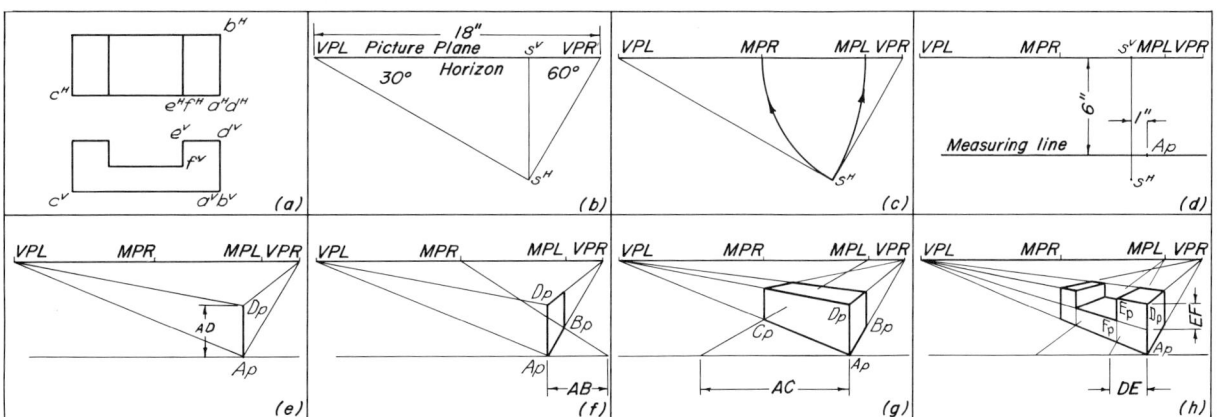

FIG. 15.25. Perspective by the measuring-point method.

a. Figure 15.26b. Find the perspective of the front face of the object shown in Fig. 15.26a by the method given in the preceding paragraph.
b. Figure 15.26c. On the measuring line through A_p lay out the horizontal distance A-2, from A to the center line of the circle, to the left of A_p. On either side of this point lay out the radius of the circle to locate points 1 and 3. Carry the three points back into the picture by drawing lines to MPL.
c. Figure 15.26d. From A_p lay out the vertical distance A-7, from A to the center line of the circle, on the front corner of the object. On either side of this point lay out the radius of the circle to locate points 6 and 8. From these three points draw to VPL to form the perspective of the square circumscribing the circle.
d. Figure 15.26e. On either of the vertical lines construct a semicircle as shown. Divide the semicircle into six equal parts and project these points horizontally to the vertical line that was used as the diameter of the construction circle. From these points on the vertical line draw lines to VPL.
e. Figure 15.26f. Construct the diagonal of the square and complete the grid by constructing vertical lines through the points where the horizontal lines cross the diagonal of the square and mark the points as shown in the figure.

15.16.5 Perspective of a Horizontal Circle by the Measuring-Point Method. When the circle lies in a horizontal plane, the procedure is similar except that the semicircle must be drawn on the measuring line, as illustrated in Fig. 15.27. When the points have been carried back into the perspective by means of the measuring points and vanishing points, the grid is constructed by using the diagonal as previously explained.

15.16.6 One-Point Measuring Point and Line. It is sometimes convenient to use the same measuring point for distances on both sides of an object in perspective. However, to be able to do this, it is necessary to locate a new measuring line.

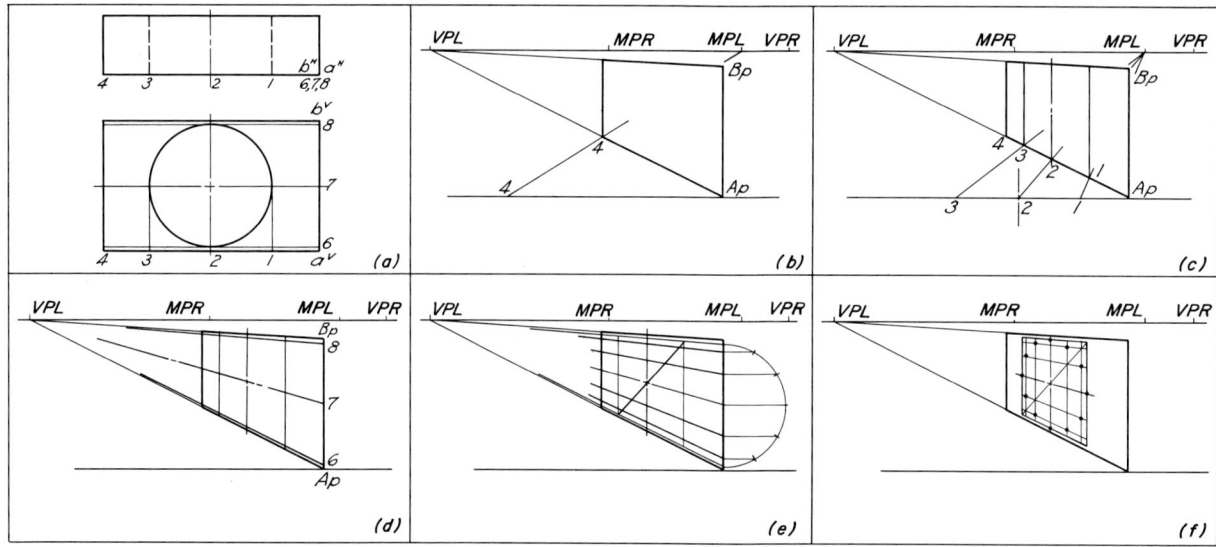

FIG. 15.26. Circle in a vertical plane by the measuring-point method.

The construction for this method is illustrated in Fig. 15.28. The steps are as follows.

a. Figure 15.28a. This figure shows the two-vanishing-point, VP 45°, for a line making 45° with the two principal lines has been added. In all of this construction the picture plane and the horizon are coincident.

b. Figure 15.28b. With VPL as a center, an arc is drawn through VP 45°. The intersection of this arc with the two-point measuring line locates point X.

c. Figure 15.28c. A line is constructed from VPL through X and is continued until it intersects an arc drawn through s^H with its center at VPL. Through the point of intersection marked y the one-point measuring line may be constructed parallel to the picture plane.

d. Figure 15.28d. If a line is drawn from VP 45° through O^P and continued until it intersects the one-point measuring line, point Z is located. Z is the point from which distances m and n may be laid out in opposite directions on the one-point measuring line. From the points thus obtained lines may be drawn to VP 45° to cut off the perspective lines through O^P to determine the size of the perspective. Visual rays have been added to show that both methods give the same results.

15.17 THREE-POINT PERSPECTIVE

When all three of the principal axes of an object are oblique to the picture plane, the resulting picture is called a *three-point perspective*. The theory of perspective as it has been developed for one- and two-point perspective can be extended to three-point. However, the actual construction is more complicated because the picture plane does not appear edgewise in the top view. It is possible to solve the problem by visual rays, as shown in Fig. 15.6, but the solution is tedious and will not be discussed here. If it is desired to specify the angle of tilt of the picture plane and the angle of rotation of the object, the reader should refer to *Industrial Production Illustration* by R. P. Hoelscher, C. H. Springer, and

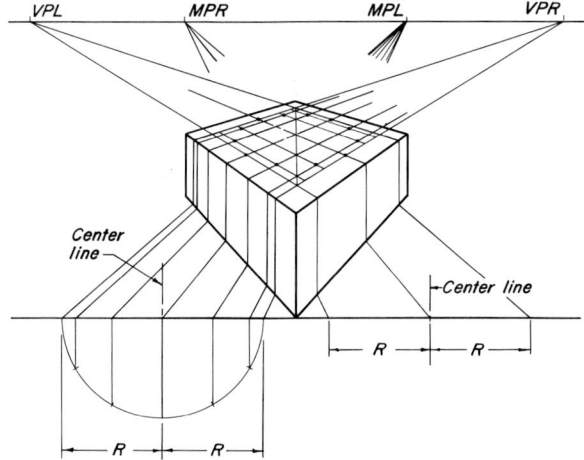

FIG. 15.27. Horizontal circle by the measuring-point method.

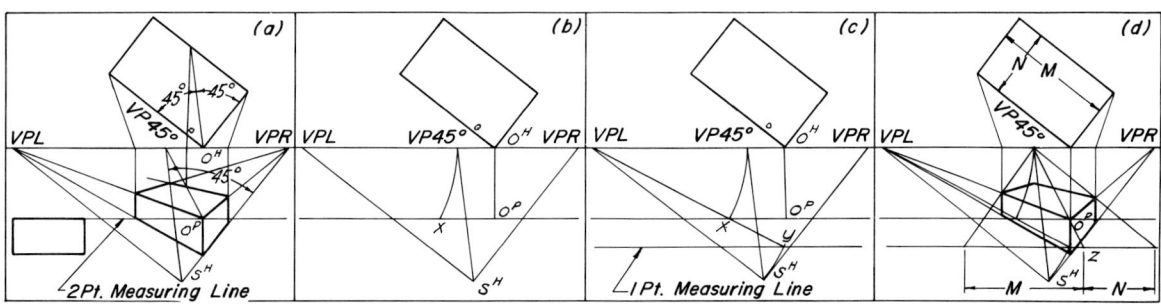

FIG. 15.28. One-point measuring point and line.

R. F. Pohle, McGraw–Hill, New York, 1943; and *Handbook of Perspective* by J. C. Morehead and J. C. Morehead, Jr., Pittsburgh, 1941.

The customary procedure is to select the three vanishing points as far apart as convenient and work from these points in the manner illustrated in Fig. 15.29 and explained below.

a. Figure 15.29a. Select the three vanishing points and draw the triangle connecting them. Through each corner construct a line perpendicular to the opposite side of the triangle. These three lines intersect at point S, which is the projection of the point of sight as determined by the selected vanishing points.

b. Figure 15.29b. Revolve the top plane into the picture plane as explained in Article 13.20.2 for axonometric. With VPR as a center and VPR-s_r as a radius, swing an arc to locate MPR. With VPL as a center and VPL-s_r as a radius, swing an arc to locate MPL.

c. Figure 15.29c. Revolve the right-side plane S, VPR, VPV into the picture plane as explained in Article 13.20.2. With VPV as a center and VPV-s_r as a radius, swing an arc to locate MPV. If desired, another MPR may be located, but this is not necessary since one has already been found.

d. Figure 15.29d. Select A_p as a point in the picture plane in such a position that the center of the picture will come approximately at S.

Through A_p, draw a line parallel to VPL–VPR. This is a measuring line, and distances may be laid out for right or left distances and front or back distances just as explained for two-point perspective in Article 15.16.3. The top of the box is obtained as shown in Fig. 15.29d.

e. Figure 15.29e. Construct a line through A_p parallel to VPR–VPV and on this line lay out the height of the box from A_p. From the point draw to MPV to intersect the line from A_p to VPV. This determines the height of the box in the perspective.

f. Figure 15.29f. From the points already located draw to the proper vanishing point to complete the perspective of the box.

15.17.1 Isometric Three-Point Perspective.* For the special case of three-point perspective in which the three principal vanishing points lie on the corners of an equilateral triangle, an easy construction has been developed that has the same relation to the theoretical three-point perspective that isometric drawing has to isometric projection.

The construction is as follows.

a. Figure 15.30a. Lay out an equilateral triangle of the size desired, usually the maximum a drawing board will accommodate, and let the corners be the three vanishing points.

b. Figure 15.30b. Select a point A_p in the picture plane and through this point construct two measuring lines, one parallel to VPL–VPR and the other parallel to VPR–VPV (or VPL–VPV).

c. Figure 15.3c. On the horizontal line lay out the right and left or length dimension of the

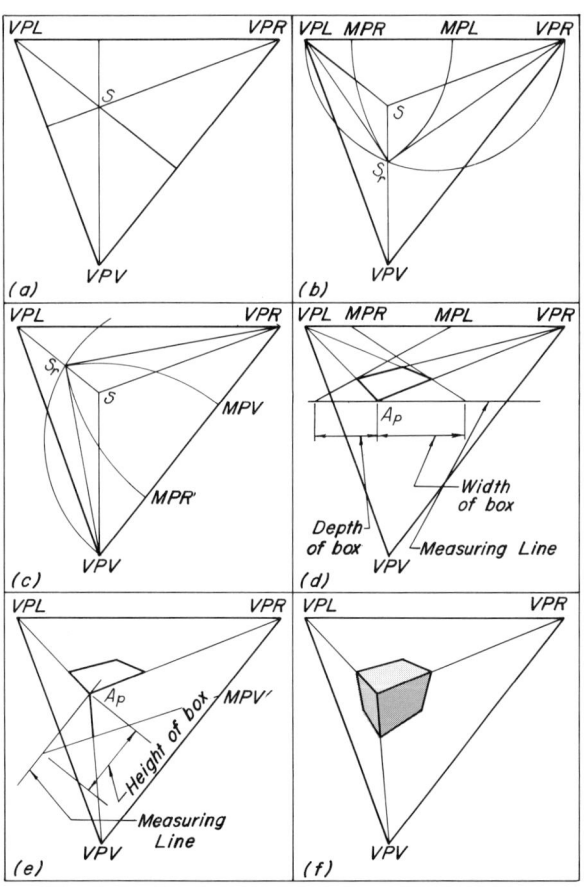

FIG. 15.29. Three-point perspective.

* This method was developed by Wayne Shick of the University of Illinois and is reproduced by his permission.

box to the left of A_p. Lay out the front and back dimensions or depth of the box to the right of A_p. Draw to the vanishing point as shown. This determines the top of the enclosing box for any object.

d. Figure 15.30d. On the measuring line parallel to VPR–VPV lay out the height of the box and draw to the vanishing point VPR as shown. The lower corner of the box is located at the point where this line crosses the line from A_p to VPV. Complete the box by drawing to the proper vanishing points. Other dimensions of an object may be laid out in the same manner.

15.17.2 Alternate Method for Three-Point Perspective. When the orthographic projections of the object are available at a scale that is convenient for use, the method illustrated in Fig. 15.31 is very expedient. In using this construction it is best to work on tracing paper placed directly over the appropriate views of the object. The procedure is as follows.

a. On the tracing paper lay out the three vanishing points as far apart as convenient and join them to form the triangle, VPL–VPR–VPV.
b. Revolve the top plane as shown in Fig. 15.31 to obtain s_r.
c. Assume point A^P, which is one corner of the object, in the picture plane.
d. Arrange the horizontal projection or plan view under the tracing paper with sides parallel to s_r-VPR and s_r-VPL, respectively.
e. Join A^P to VPR and VPL.
f. Join a^H, b^H, c^H, and d^H to s_r.
g. The perspective of B is located at B^P where $b^H s_r$ intersects A^P-VPR and the perspective of D at D^P where $d^H s_r$ intersects A^P-VPL.
h. This locates the top plane of the object at $A^P B^P C^P D^P$.
i. By revolving one of the side faces and proceeding in a similar manner, the vertical faces of the object can be located.

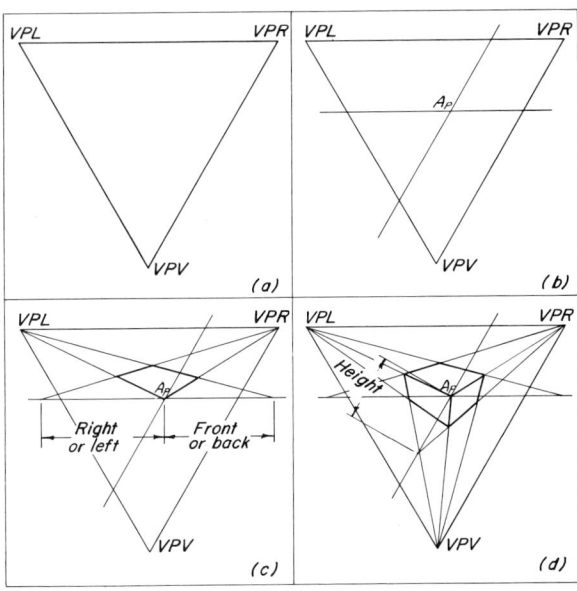

FIG. 15.30. Three-point perspective drawing.

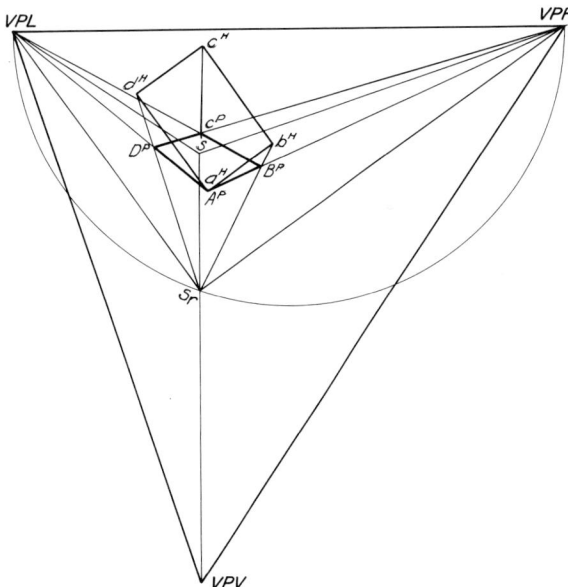

FIG. 15.31. Alternate method of three-point perspective.

15.17.3 Three-Point Perspective Board.

The big disadvantage of three-point perspective is that it is very difficult to get the vanishing points far enough apart to avoid excessive perspective effect. Even on a very large drawing board the triangle is so small that the lines vanish too rapidly for a good appearance. To avoid this, it is fairly easy to construct a special board by means of which the vanishing points can be widely separated. The board is shown in Fig. 15.32. The three arcs are constructed with their centers at the three vanishing points. Then by means of a special T-square so arranged that one edge is exactly centered between the two bearing points of the head, lines may be drawn that always point to the center of the circle. The use of this T-square is illustrated in the figure.

By means of the construction explained in Article 15.16.6, the 45° vanishing points are located for each face together with the corresponding one-point measuring lines. These lines are marked off in unit divisions as illustrated, and from these divisions the accurate perspective can be found directly. In this figure a rectangular solid $3 \times 4 \times 5$ has been drawn. With three measuring points and lines, it is possible to make each measurement in two different ways. Lines showing both sets appear in the figure.

15.18 SHADES AND SHADOWS IN PERSPECTIVE

When light shines on an object, part of the surface will be lighted and the remaining part will be dark. That part of the surface on which no light shines is said to be in shade. The lines on the object that separate the light areas from the shaded areas are called *shade lines*. When the object rests on or is adjacent to some other object, it casts a shadow, which may be outlined by finding the shadow of the shade lines. To find the shadow of an object, therefore, two things are necessary: first, to pick out the shade lines and, second, to find the shadow of these lines on the surfaces on which the shadow falls.

To recognize the shade line requires a knowledge of the direction of the light ray and the ability to visualize the object as it stands in space. In case of doubt it is possible to find the shadow of every line on the object, after which the largest area outlined by these shadows will be the shadow of the object.

When the light rays are tangent to any surface, that surface is said to be in shade.

5.18.1 Shadow of a Vertical Line.

A vertical line resting on a horizontal plane is used as the basic line in determining shadows. One reason for this is that since the horizontal projection of a vertical line is a point, the horizontal projection of the shadow of the line must coincide with the horizontal projection of a light ray. When the direction of the light ray is specified by two projections of one ray, as MN in Fig. 15.33, the shadow of line AB in the horizontal projection must be a line through $a^H b^H$ parallel to $m^H n^H$. Since that shadow is a horizontal line, it must have its vanishing point on the horizon. That vanishing point may be found at VPS in the usual

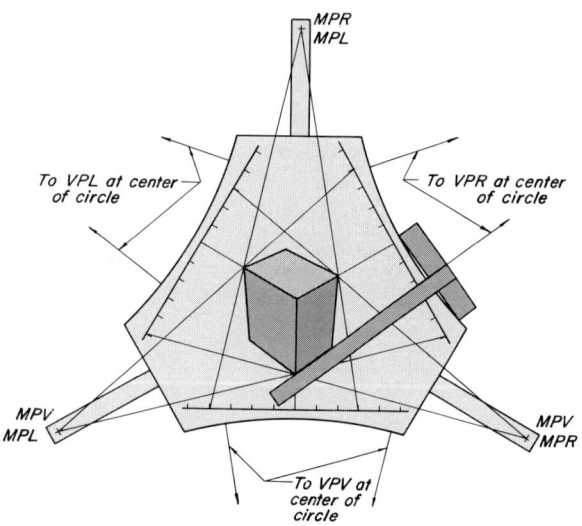

FIG. 15.32. Three-point perspective board.

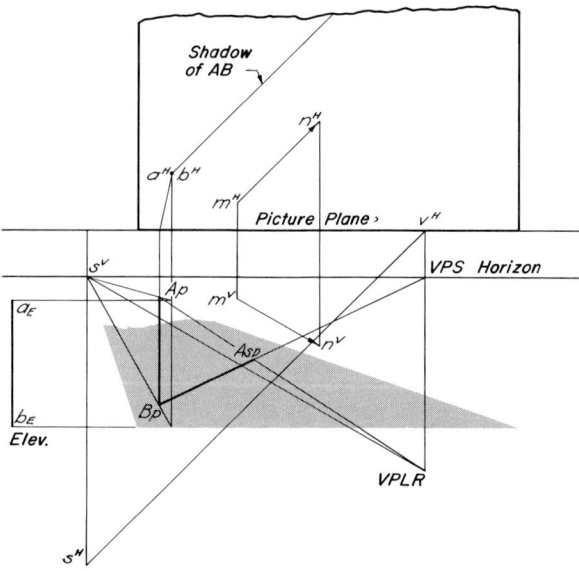

FIG. 15.33. Shadow of a vertical line on a horizontal plane.

manner, as shown in Fig. 15.33. Then the shadow of AB on the horizontal plane must vanish at VPS, and since B is on the horizontal plane, the shadow must start at B_p. By joining B_p to VPS, the shadow of a vertical line of infinite height is obtained. To find the shadow of A, it is then necessary to draw the perspective of the light ray through A and find the point where it intersects the shadow line. The vanishing point of the light ray MN, found as explained in Article 15.8.3, is located at the point marked VPLR. Then, by joining A_p to VPLR, point A_{sp} is located, and the actual shadow of the line lies between B_p and A_{sp}.

In finding shades and shadows in perspective, *it is always necessary to locate first the vanishing points VPS and VPLR. Always remember that VPS is the vanishing point of the shadow of the vertical line on a horizontal plane* and nothing more.

When the vertical line casts a shadow on a vertical plane, the shadow must be parallel to the line and in two-point perspective will show as a vertical line. The best way to find this shadow is to find the shadow of the line on the horizontal base plane until it crosses the base line of the vertical plane. From there the shadow will be vertical, as illustrated by the shadow or the flagpole in Fig. 15.34. The location of this vertical shadow can also be found by means of a visual ray from the plan or top view, as indicated in the figure.

When the shadow falls on an inclined surface such as the roof of the building in Fig. 15.34, it is necessary to locate two points on the shadow of the line or the line extended. One point on the eave line has already been located, and another can easily be located on the ridge line. One method is by visual ray, as shown in Fig. 15.34. The other method, which is better since it can also be used when working by measuring-point method, involves cutting a vertical plane through the center of the building, as indicated by the dotted line. Then by imagining the front of the building removed the shadow on the section plane can be found, which will locate the desired point on the ridge line. This construction is shown in dashed lines in Fig. 15.34. The shadow of the flagpole on the roof will be the line joining the point on the eave line to the point on the ridge line, and a light ray through the top of the flagpole to VPLR will locate the end of the shadow.

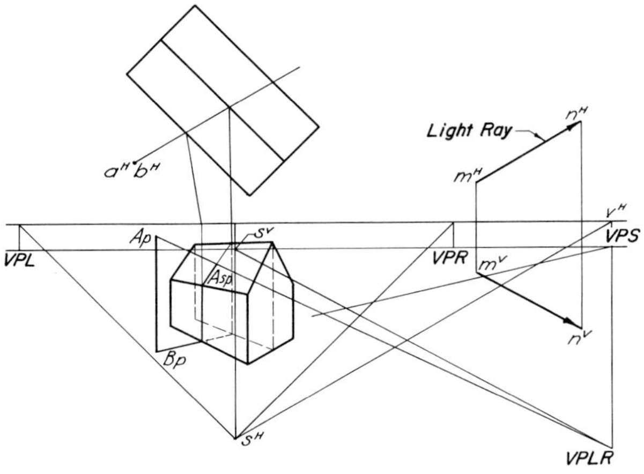

15.18.2 Shadow of a Horizontal Line. When a horizontal line casts a shadow on a horizontal plane, the shadow will be parallel to the line itself. In perspective this means that the two lines will have the same vanishing point. Thus, in Fig. 15.35, line AB vanishes at VPR, and consequently its shadow must also vanish at VPR. BC vanishes at VPL, and its shadow also vanishes at VPL.

When a horizontal line casts a shadow on a vertical or inclined plane, the shadow of two points on the line, or one point and the direction, must be found on the given plane. Thus, in Fig. 15.36, line AB casts a shadow on the horizontal, inclined, and vertical planes. The shadow on the horizontal plane vanishes at VPR, and if the inclined plane were removed, it would continue to

FIG. 15.34. Shadow of a vertical line on a vertical and an inclined surface.

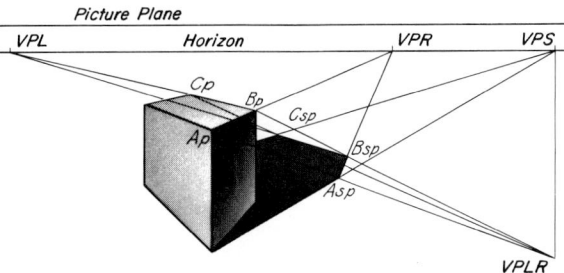

FIG. 15.35. Shadow of a horizontal line on a horizontal plane.

be the base of the vertical plane and from there would go to B_p because B is actually on the vertical face. The intersection of this shadow on the vertical face with the top line of the inclined face gives a second point on the inclined plane to determine the shadow of the horizontal line on the inclined plane. A vertical line through A_p can be used to locate A_{sp}, which is the shadow of A on the ground. The shadow of the inclined line AB is found by joining A_{sp} to VPR.

15.18.3 Shadow of an Inclined Line on a Horizontal Surface.
To find the shadow of an inclined line, it is usually best to assume as many vertical lines as necessary through points on the line and find the shadows of these vertical lines. For example, in Fig. 15.37 a vertical line was assumed through B_p. By means of this vertical line the shadow of B_p on the ground was found at B_{sp}. Note that the vertical line through B_p is an imaginary line.

15.18.4 Shadow of an Inclined Line on Horizontal, Vertical, and Inclined Planes.
The conical structure in Fig. 15.38 will cast a shadow on the house, but the actual lines that cast the shadow, called the shade lines, are elements of the cone and must be found as a part of the first step. As in the preceding illustration, it is necessary to establish a vertical line, which in this case is the altitude of the cone called AB. The shadow of AB on the ground is found in the usual manner, by drawing from B^P to VPS and from A^P to $VPLR$. The intersection of these two lines A_s^P gives the shadow of A on the ground. Since the base of the cone rests on the ground, the entire shadow of the cone on the ground can be found by drawing from A_s^P tangent to both sides of the base. The points of tangency C^P and D^P of those lines with the base locate the shade lines $A^P C^P$ and $A^P D^P$ on the cone. The problem then is to find the shadow of these inclined shade lines on the various planes.

The shadow of the shade lines on the ground has already been determined, and from these shadows points E^P and F^P at the base of the wall may be located. Next the imaginary shadow of A on the wall extended is found at A_{sw}^P by use of the vertical line AB. By joining this point with E^P and F^P the shadow of the cone on the vertical wall is found.

The shadow of AB on the inclined roof is found to be $K^P N^P$, as explained in Article 15.18.1. By extending $K^P N^P$ until it intersects the light ray through A^P, the shadow of point A on the inclined roof extended is located at A_{sr}^P. By connecting A_{sr}^P to R^P and S^P the shadow of the cone on the

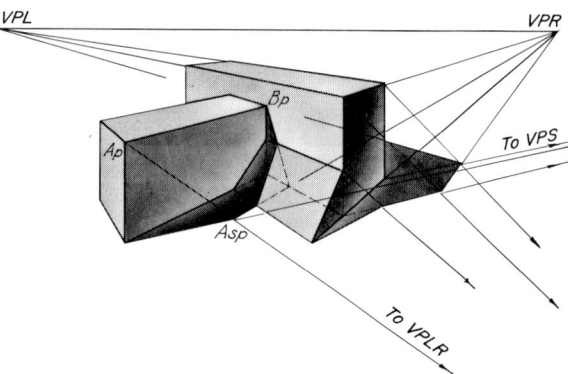

FIG. 15.36. Shadow of a horizontal line on various planes.

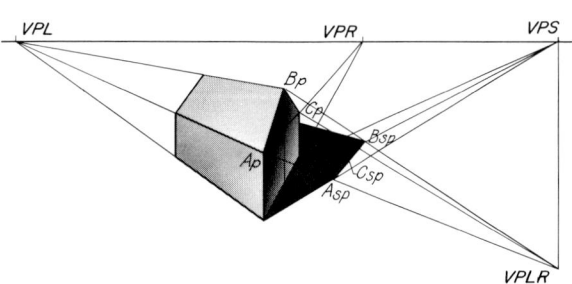

FIG. 15.37. Shadow of an inclined line on a horizontal plane.

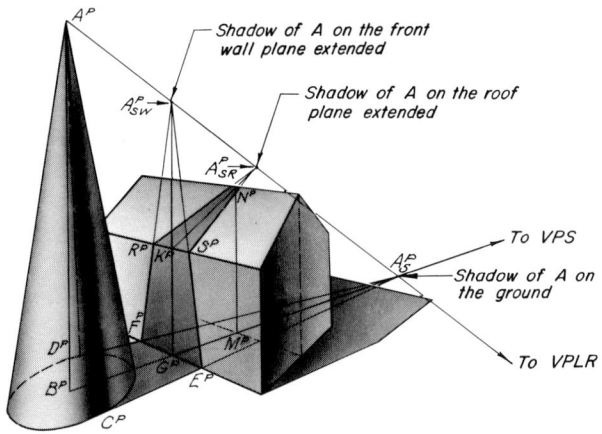

FIG. 15.38. Shadow of an inclined line on various planes.

inclined roof is completed. The various parts of the shadow may be shaded to give the complete picture as shown in Fig. 15.38.

15.18.5 Shadow of a Curved Line. The shadow of a curved line may be found by assuming a series of vertical lines through points on the curve. The shadow of these vertical lines may be found in the usual manner and the shadow of the point located on the shadow of the line. This is illustrated in Fig. 15.39.

15.18.6 Reflections. Whenever water appears in the foreground of a perspective, the building or structure shown in the picture will also show in the water as a reflection. To obtain the perspective of the reflection, the elevation of the structure being shown may be reversed about the waterline and the complete perspective of this inverted object may be constructed in the same manner as the original perspective. Figure 15.40 shows the elevation of the corner of a building standing on the bank of a body of water, with the picture plane and water surface shown edgewise. Point C is the corner of a building, C_w is the projection of C on the surface of the water, and C_r is the position of C in the reversed elevation.

Point S is the point of sight for the perspective and, therefore, the piercing points of the visual rays in the picture plane will locate the perspectives of C, C_w, and C_r at C_p, C_{wp}, and C_{rp}, respectively. By plane geometry it can be shown that since the two distances, y, were constructed equal, the three angles marked α must be equal. Then the perspective of C_r must coincide with the perspective of the reflection of C on the water surface, thus proving that the reflection of an object may be obtained in this manner.

By geometry it can also be shown that, since the two distances marked y were constructed equal, the two distances marked y' must also be equal. Since these distances are measured directly in the perspective, it becomes possible to locate the reflection of any point in a two-point perspective by measuring the distance, y', from the perspective of the point, C_p, to the perspective of its projection on the water surface, C_{wp}, and laying that distance below C_{wp} on the water surface to locate C_{rp}, which is the reflection of point C. This gives a rapid method of finding the reflection of any point, without drawing the reversed elevation. In a three-point perspective, the reversed elevation must be drawn and the reflection found in the same manner as the real perspective.

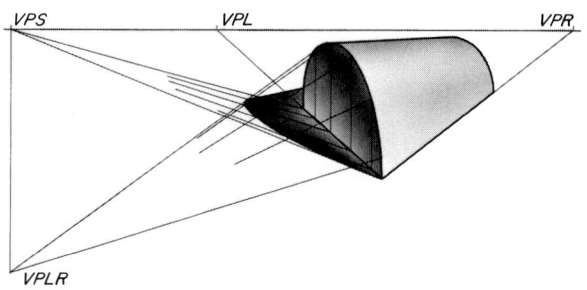

FIG. 15.39. Shadow of a curved line.

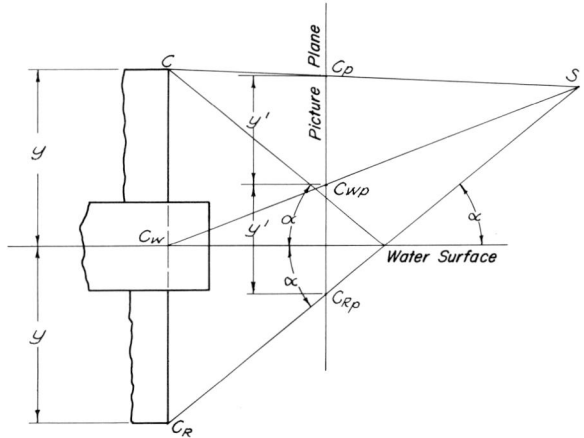

FIG. 15.40. Theory of reflections.

The lines in the reflection should be made irregular and broken to represent wave action. The surface of the water can be indicated by a series of irregular horizontal lines whose spacing increases toward the front of the drawing. Figure 15.41 shows a perspective with shades, shadows, and reflections. Figure 15.42 shows the method for finding reflections by reversing distances in the perspective.

15.19 SKETCHING IN PERSPECTIVE

Sketching in perspective is more difficult than sketching in any of the other pictorial forms. The student should be thoroughly familiar with the technique of sketching as discussed in Chapter 5.

15.19.1 Proportioning. As in all other sketching, proportioning of the sketch is the most important phase. Since the enclosing box method gives the best approach to the problem, the first step is to obtain the proper proportions for the box. Since dimensions are not the same along the three axes of the box, the construction is best made on the basis of a cube. The following suggestions give a choice of two positions of the basic cube and the proportions of the sides to the chosen height as shown in Fig. 15.43. The steps in Fig. 15.43a are as follows.

a. Select the two vanishing points as far apart as desired. Note that they must be on the same horizontal line.
b. Half-way between the vanishing points sketch a vertical line to represent the height of the basic cube you wish to draw.
c. Place the upper end of this line as far below the horizon line as you wish if the top of the cube is to be shown. The farther below the horizon line, the more the top shows.
d. Draw lines from the top and bottom of the vertical line to both vanishing points.
e. Sketch two other vertical lines on opposite

FIG. 15.41. Shades, shadows, and reflections in perspective.

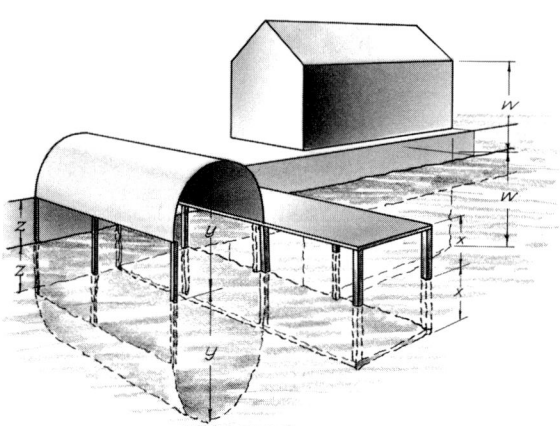

FIG. 15.42. Method of finding reflections.

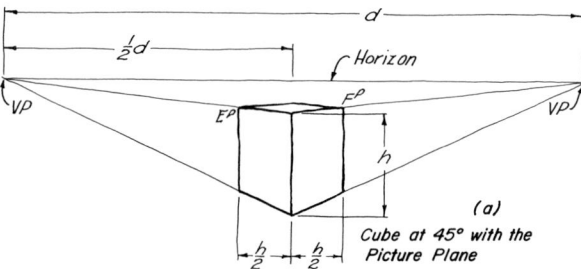

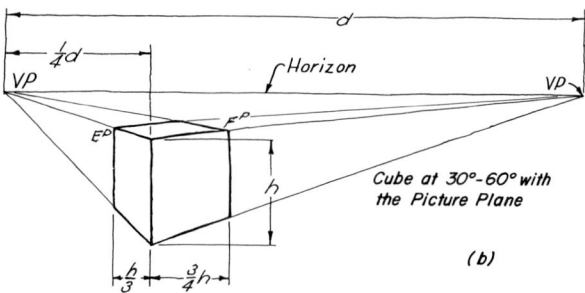

FIG. 15.43. Proportioning a cube in perspective sketching.

sides of the center, each at a distance of $\frac{h}{2}$ from it as shown in Fig. 15.43a.

f. From corners E_p and F_p draw lines to the vanishing points, thus completing the top of the cube.

This now represents a cube whose faces make 45° with the picture planes.

If it is desired to have a cube in the position shown in Fig. 15.43b, which makes the faces at 30° and 60° with the picture plane, use the same steps as given for the 45° cube but use the dimensions given in Fig. 15.43b.

15.19.2 Proportioning a Box by Repeating the Basic Cube. The basic cube must have been constructed on the smallest dimension of the box if the repetitive method is to be followed.

a. *Repeating the Cube.* When the sketch of the cube has been completed as in Fig. 15.44a, the general proportion of the object can be obtained by extending the lines and building other similar cubes. In Fig. 15.44b the diagonals and main center lines of one face of the original cube have been drawn. Next the diagonal line AB is drawn joining the top center point to the left center point and is continued to locate C. Through C the vertical line CDE is drawn to locate D and E. The procedure is continued until the desired number of cubes have been formed. In this case the proportion desired was $1 \times 1 \times 3$. In Fig. 15.44c the beginning of the same steps is shown in order to place another cube on top of the first one. When completed, this will give a box whose proportions are $1 \times 2 \times 3$. This method is discussed in Chapter 5.

b. *Proportioning a Box by Subdividing a Cube.* Figure 15.45 shows a more accurate method for obtaining a box in the proportions of $1 \times 2 \times 3$. In this case the original cube is made large enough to represent the largest dimension. By subdividing this cube, using the method explained in Article 4.2.11 and illustrated in Fig. 15.45, the proper proportion is obtained. This construction can be continued to locate the details of the object. The use of the construction is more important in perspective than in axonometric, because distances decrease in size as they get farther behind the picture plane.

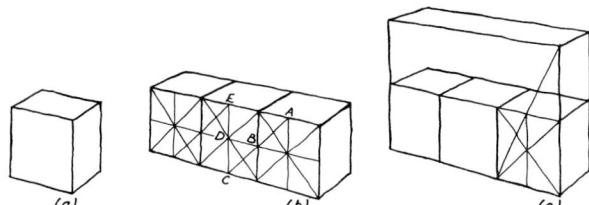

FIG. 15.44. Proportioning box in perspective sketching by adding cubes.

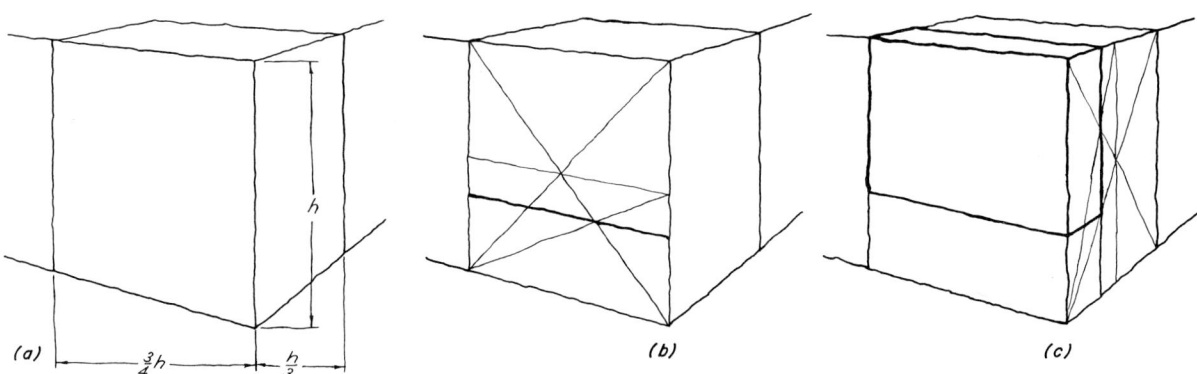

FIG. 15.45. Proportioning a box in perspective sketching by subdividing a cube.

418 PERSPECTIVE

15.19.3 Sketching Circles in Perspective.
The best method for sketching a circle in perspective is to sketch the square circumscribing the circle and then draw the circle inside the square. It is well to locate the center point of the sides by means of the diagonals and center lines. This is illustrated in Fig. 15.46a.

The proportions of circles in horizontal planes vary as the circle is placed above or below the horizon. When the circle is on the horizon it will be shown in perspective as a straight line. As it moves away from the horizon, the circle appears as an ellipse. The minor axes of the ellipse increase with respect to the major axes as the circle is moved farther from the horizon. This process is shown in Fig. 15.46b.

15.19.4 Steps in Sketching a Perspective.
When a perspective sketch of an object such as that shown in Fig. 15.47a is to be made, the drafter should follow the steps listed below.

a. Establish a basic proportion. In this case it would be approximately $1 \times 2 \times 3$.
b. Draw a horizontal line to represent the horizon and pick two points on it as far apart as possible to represent the vanishing points. See Fig. 15.47b.
c. Establish the front corner of the object for position and size.
d. Construct the basic cube and, from that, block in the outline of the object as discussed in Section 5.14. See Fig. 15.47c.
e. Locate the position of the slot by subdividing the left front face. See Fig. 15.47d.
f. Locate the center of the circle and construct a square circumscribing the circle by methods of proportions. This is illustrated in Fig. 15.47e.
g. Draw the circle. See Fig. 15.47e.
h. Clean up the drawing and make the lines heavy. Shade if desired. See Fig. 15.47f.

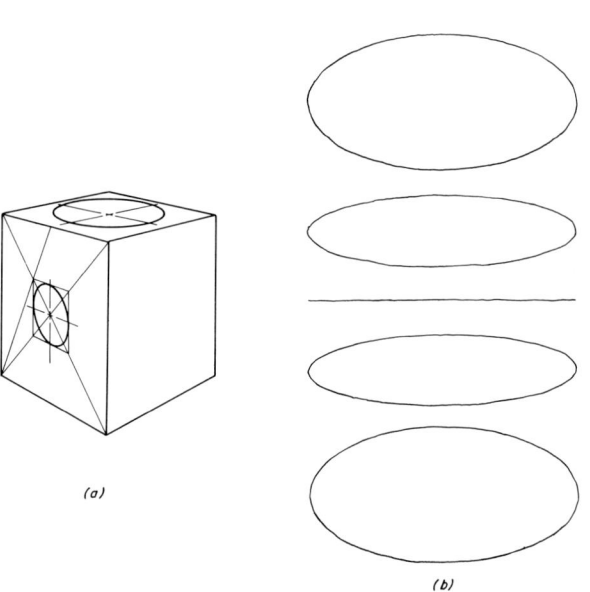

FIG. 15.46. Sketching circles in perspective.

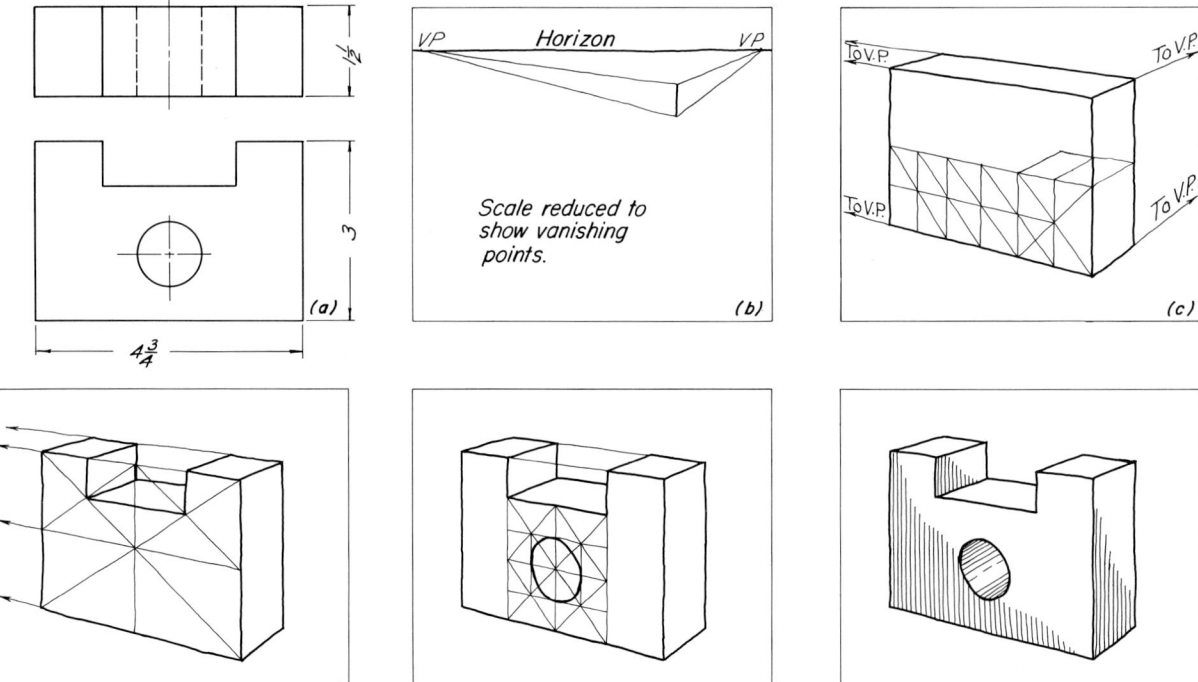

FIG. 15.47. Steps in making a perspective sketch.

15.20 PERSPECTIVE GRID

It is possible to purchase sets of ruled sheets similar to the one shown in Fig. 15.48 for the purpose of making perspective drawings. This grid is designed so that the faces make 30° and 60° with the picture plane.

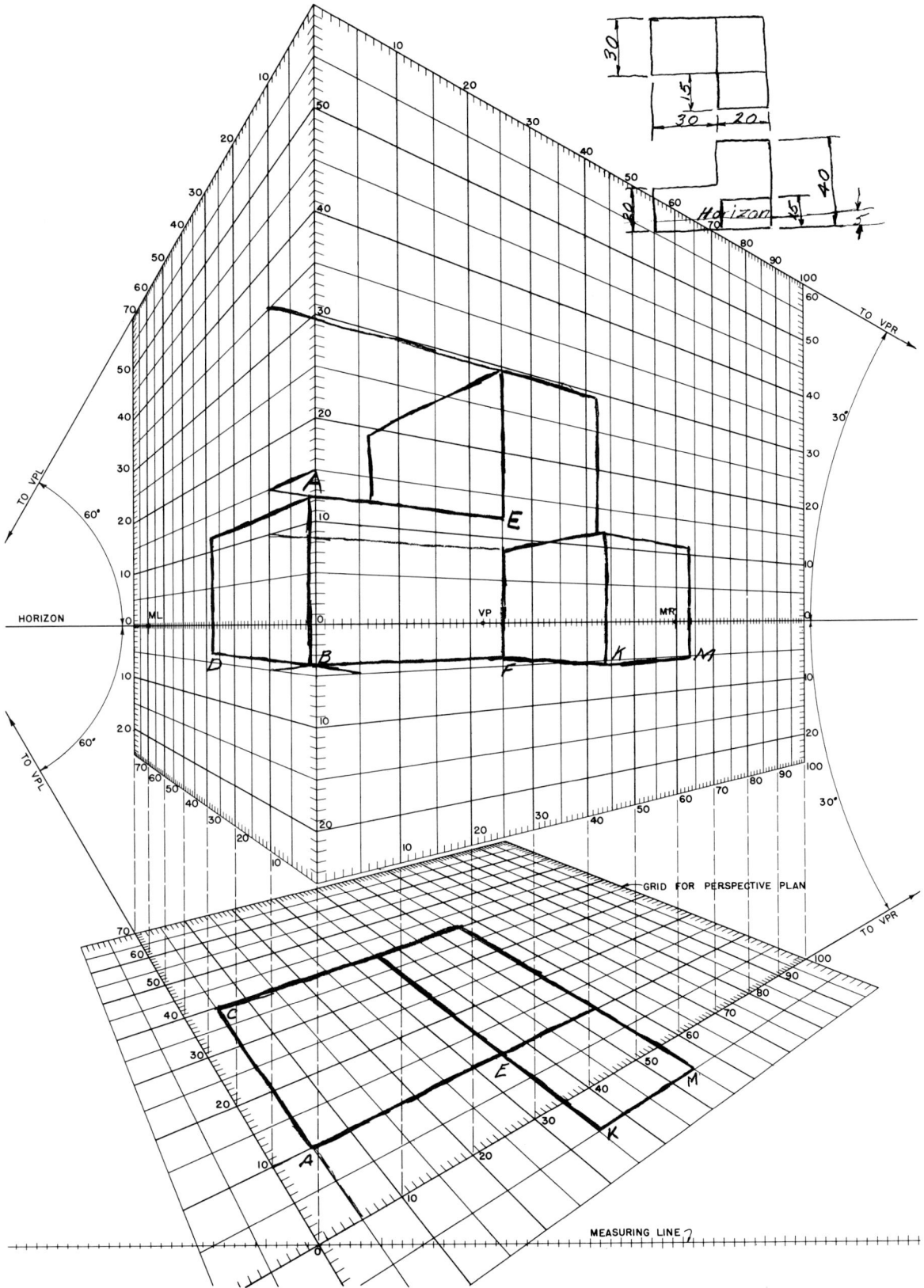

FIG. 15.48. Perspective sketch on grid (courtesy of Grace Wilson).

The best way to use the grid is to draw the perspective plan of the building on the grid at the bottom of the figure. These points may then be projected up to the perspective. It must be remembered that the vertical line marked zero is the only line on which vertical dimensions may be measured. Therefore, everything must be referred to that line. Thus to locate the perspective of point A it is necessary to start at a point on the zero vertical line at the proper elevation. From that point it is possible to move to the left on a line vanishing to the left, to the vertical line marked 10. From that point a line can be drawn to the right vanishing point and the perspective of A is directly above the plan of A in the lower grid.

The same procedure must be followed for every point.

SELF-STUDY QUESTIONS

Before trying to answer these questions, read the chapter carefully. Then, without reference to the text, answer as many questions as possible. For those that cannot be answered, the number in parentheses following the question number gives the article in which the answer can be found. Look it up and write down the answer. Check the answers that you did give to see that they are correct.

15.1 (**15.1**) The major difference between perspective and other forms of projections is that the point of _____ is at a _____ distance from the object.

15.2 (**15.2.2**) Visual rays are the lines connecting a point on the object with the _____ of sight.

15.3 (**15.2.3**) The _____ _____ is usually used as the _____ plane for most perspectives.

15.4 (**15.2.4**) Each set of parallel lines on an object, except those that are parallel to the picture plane, has a _____ point.

15.5 (**15.2.5**) The horizon line always passes through the _____ projection of the point of sight.

15.6 (**15.3.2**) For normal perspectives the picture plane is _____ the point of sight and the object.

15.7 (**15.4.1**) For good perspective the point of sight should be _____ the width of the object from it.

15.8 (**15.5.1**) If the object has one principal face _____ to the picture plane, a parallel or _____ _____ perspective will be produced.

15.9 (**15.5.2**) If two faces of an object are at _____ _____ with the picture plane, an angular or two-point perspective will be produced.

15.10 (**15.5.3**) An oblique perspective has _____ faces of the object inclined to the picture plane.

15.11 (**15.6.1**) The visual-ray method of finding a perspective involves finding the _____ point of the visual ray with the picture plane.

15.12 (**15.7.1; 15.7.2**) Either the front and top views or the _____ and side views may be used in making a perspective by the visual-ray method.

15.13 (**15.8.1**) Be definition _____ lines meet at infinity.

15.14 (**15.6.2**) To find the perspective of a line, the vanishing point, and the point where the line extended _____ the picture plane must be found.

15.15 (**15.8.2**) The vanishing points for horizontal lines are always in the _____ line.

15.16 (**15.8.3**) The vanishing points of oblique lines are either above or _____ the horizon line, never on it.

15.17 (**15.8.3c**) Lines that are perpendicular to the picture plane have their vanishing point at the _____ projection of the point of sight.

15.18 (**15.14.1**) The perspective of a circle that is parallel to the picture plane is a _____.

15.19 (**15.18.1**) The shadow of a vertical line on a horizontal plane will coincide with the _____ projection of the visual ray through the line.

15.20 (**15.18.2**) The shadow of a horizontal line on a horizontal plane will be _____ to the line. Therefore, the line and its shadow will have the same _____ point.

CHAPTER 16

CHARTS AND DIAGRAMS

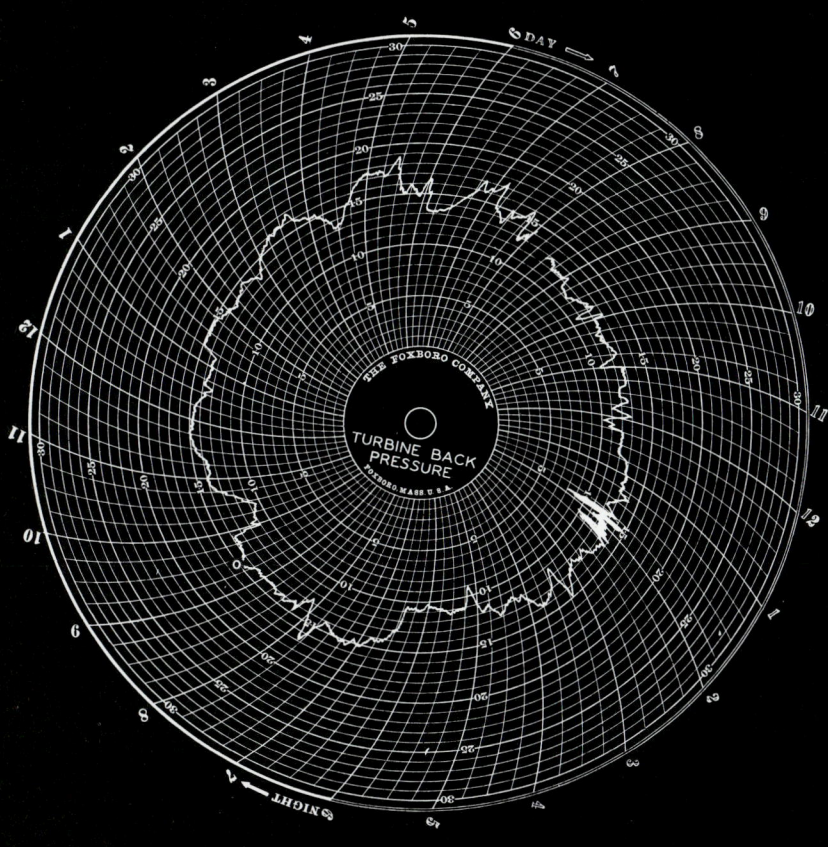

CHAPTER 16

16.1 USES OF CHARTS AND DIAGRAMS

The purpose of charts and diagrams is to present data and their significance in a more easily interpreted form than could be done with either words or tabular data. All technical and business publications as well as the popular press commonly present information by means of charts. Drawing inferences from tabular data are at best difficult. Figure 16.1 shows both data in tabular form and the data plotted on rectangular coordinate paper.

A few of the more common uses of charts are as follows.

a. To present results of test data obtained in experiments.
b. To correlate the observations of natural phenomena.
c. To present business statistics.
d. To determine trends in business.
e. To present equations graphically for computation uses.
f. To derive empirical equations.

16.2 CLASSIFICATION OF CHARTS

According to the method of presentation or drawing, with which this chapter is primarily concerned, charts may be readily classified in the following form:

a. Plane curves on rectangular coordinates, logarithmic, semilogarithmic, trilinear, polar coordinates, and others.
b. Bar charts of all kinds.
c. Pie or sector charts.
d. Computation charts, vector diagrams, and nomographs.
e. Flow charts and distribution diagrams.
f. Three-dimensional charts.
g. Map or distribution diagrams.

All of these are illustrated in the various figures of this chapter.

As with any form of communication, the audience to whom the information is intended must be considered. Data presented in bar, pie, or plane charts are most easily understood by non-

MARVIN'S FORMULA
$P = 0.004V^2$

Wind Velocity MPH	Pressure lb/in^2
0	0
10	0.4
20	1.6
30	3.6
40	6.4
50	10.0
60	14.4
70	19.6
80	25.6
90	32.4
100	40.0

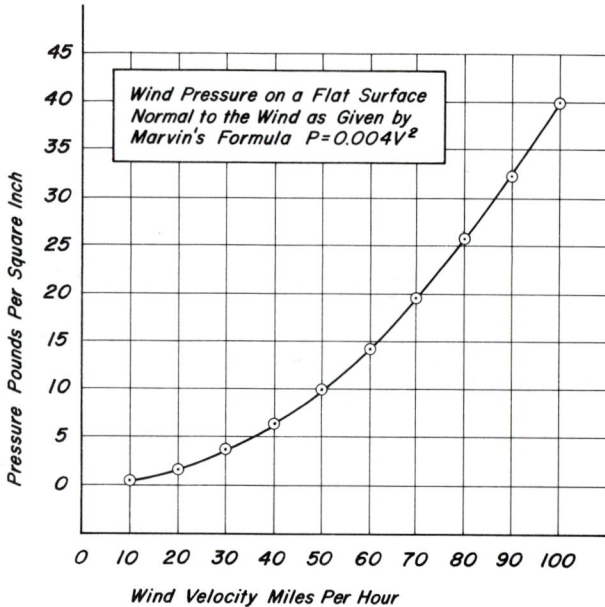

FIG. 16.1. Plotted data.

technically trained people. Trilinear charts, for instance, should not be used in presentation to the ordinary public as they will most likely not understand the information presented.

16.3 CHARTS ON RECTANGULAR COORDINATES

Charts are more commonly made on rectangular coordinates than on the other forms, as illustrated in Fig. 16.2. They are used to compare quantities. The impression given by such charts will depend on the scales selected for each of the coordinates.

16.4 HOW TO DRAW THE CHART

When one has the data for a chart given or collected in tabular form, the steps outlined in the following paragraphs must be taken to produce a chart that will give the desired effect. Two forms of presentation are possible, depending on the purpose of the chart. One form is the test or laboratory report; the other is designed for publication.

In this book we are concerned primarily with those prepared directly on commercially available printed coordinate papers.

16.4.1 Selection of Axes. Two variables are usually involved in a chart. It is therefore necessary to decide which variable will be placed along the vertical or y-axis and which on the horizontal or x-axis. It is standard practice to place the independent or controlled variable along the horizontal axis.

The location of the zero point or intersection of the axes must be so chosen that all values of the variables can be plotted. When only positive values are involved, this point is placed in the lower left corner of the chart about 1 in. in from the printed border, as shown in Fig. 16.3. This allows room for numerals and legends. Figure 16.4 is a similar graph to 16.3 but is not as detailed and is intended more to show trends than exact values.

16.4.2 Choice of Scales. The choice of scales materially affects the impression given by the chart. The scales on the two axes should be chosen to take maximum advantage of the space available. If the chart is to be made on printed coordinate paper, the scale units should be chosen to come upon the heavy printed lines. These units should be multiples of 1, 2, 4, or 5. Interpolation of the smaller divisions on the paper should be easy to make. If the chart is to be made for formal publication, coordinates are usually ruled upon blank paper. Figure 16.5 illustrates how the choice of the scale can distort the impression the curve gives. This is commonly used in the popular press in order to get across certain points of view, for example, the rate of inflation or rate of population growth (Fig. 16.13).

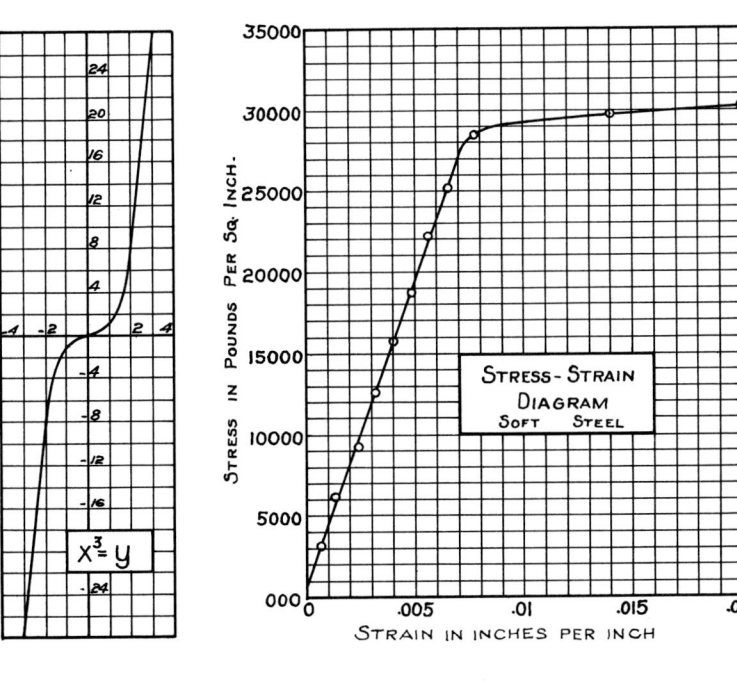
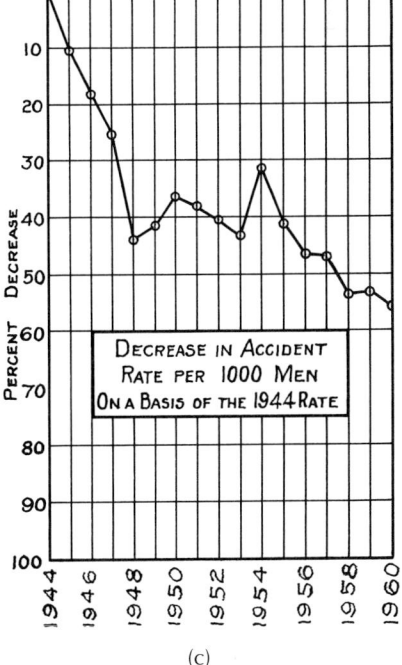

(a) (b) (c)

FIG. 16.2. Types of plane curves.

426 CHARTS AND DIAGRAMS

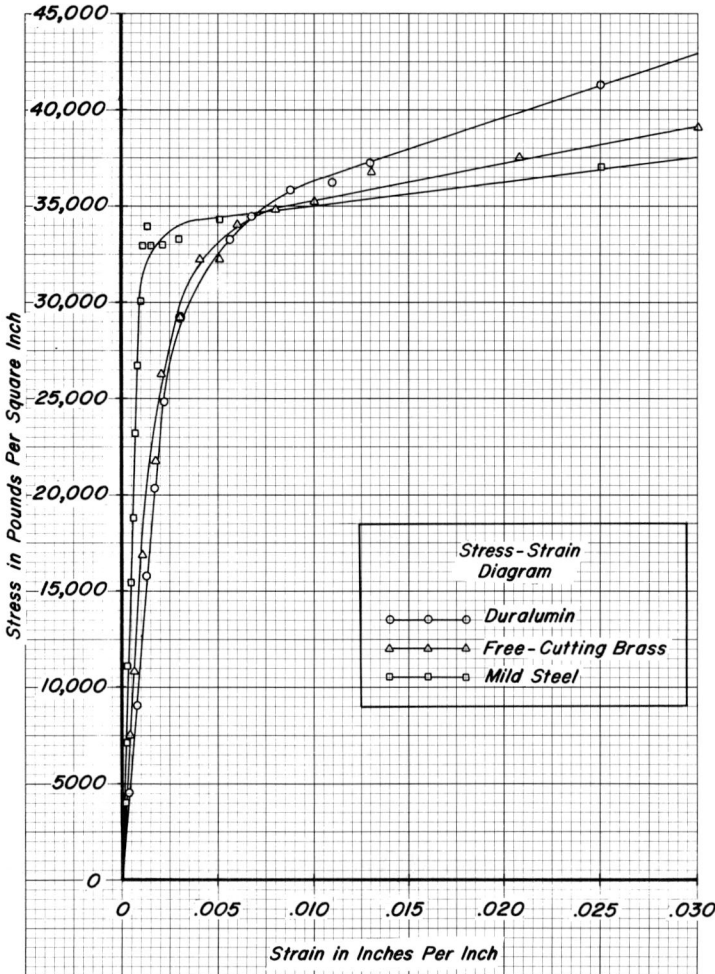

FIG. 16.3. Plane curve – rectangular coordinates.

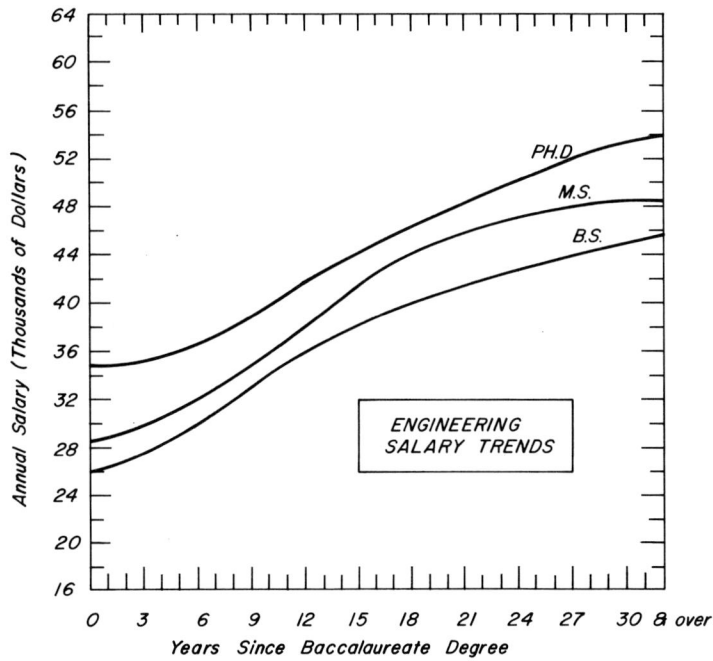

FIG. 16.4. Trend chart.

16.4.3 Marking Coordinates. Unit values of the coordinates should be marked on each axis, as shown in Fig. 16.3. A legend should indicate what the units are and the unit of measurement, for example, length in millimeters or meters, or time in days, hours, or minutes. Note in Fig. 16.3 that the smaller-ruled divisions have been used as guidelines for the lettering.

16.4.4 Showing Plotted Points. The plotted points are indicated by open circles 2 to 3 mm in diameter. If more than one curve is shown on a single chart, squares and triangles may be used for the points on the other curves, as shown in Fig. 16.3. If more curves are needed, solid circles, squares, and triangles may be used. They should be smaller than the open ones. If a curve is to be drawn to represent a mathematical equation, plotted points should not be shown on the finished chart (Fig. 16.2a).

16.4.5 Drawing the Curve. The nature of the curve to be drawn between plotted points will depend on the data involved. If there is no direct relationship between the variables, as for time and rainfall, straight lines will be drawn from point to point. Figure 16.2c, "decrease in accident rate," illustrates this principle as does Fig. 16.6. When there is a direct relationship, as in Fig. 16.3, a smooth curve will be drawn through the average of the points. The curve should not cross any data points, since this would obscure the accurately plotted point.

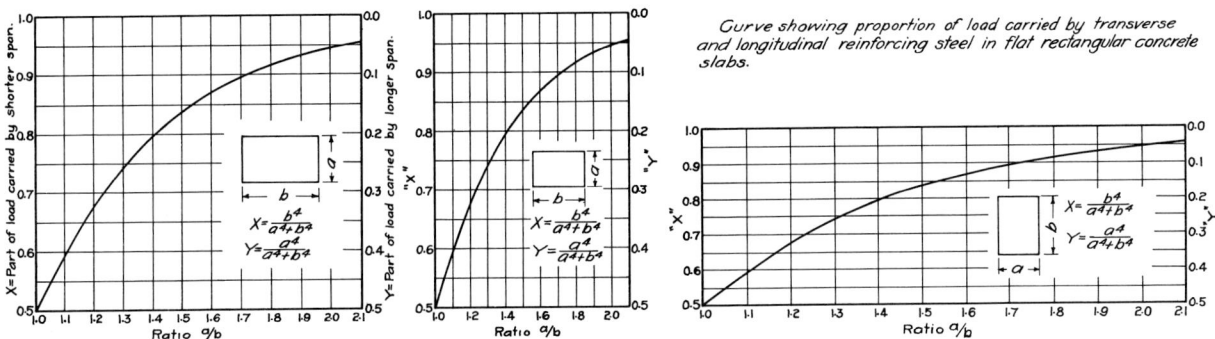

FIG. 16.5. Effect of scale on apparent slope of curve.

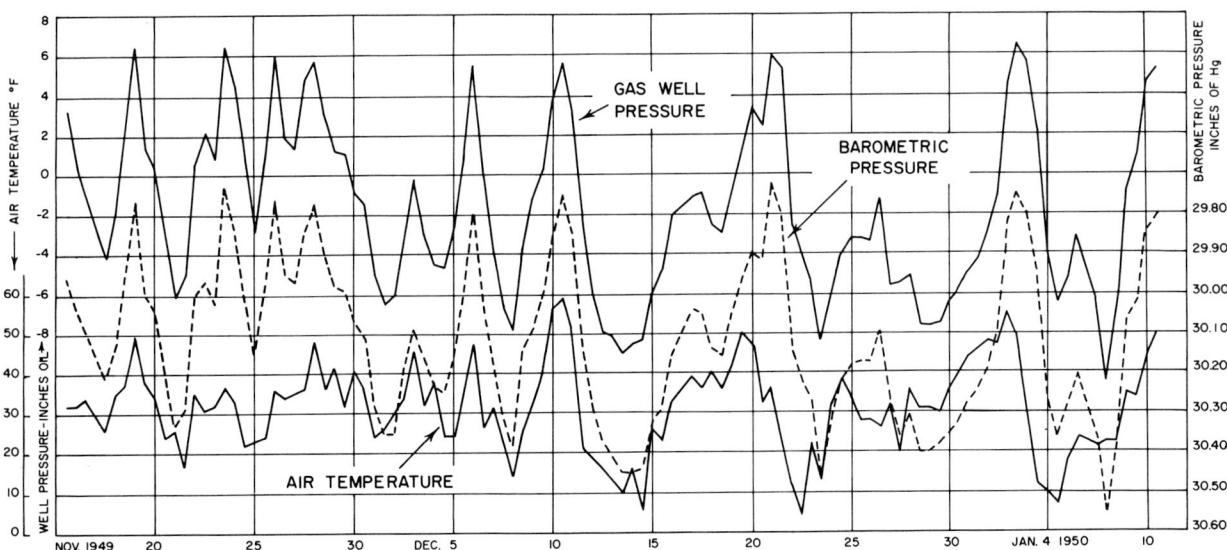

Record of gas-well pressures, barometer readings, and air temperatures during a two-month period, November 1949 to January 1950, for the Alvin Albrecht farm gas well, sec. 34, T. 15 N., R. 9 E., Bureau County.

FIG. 16.6. Gas well pressures (courtesy of Illnois State Geological Survey).

16.4.6 Titles. Every chart must have a well-thought-out title stating specifically what is represented. This should be placed in an open area to make the total effect one of a well-balanced sheet.

16.4.7 Sketches. If a small sketch will make the chart more intelligible, this may be placed on the sheet but tables of data and extensive explanatory matter should not be placed within the chart.

16.5 INTERPRETATION CHARTS

As indicated in Article 16.4.2, the choice and location of scales are very important. They can create an erroneous impression unless careful observations are made.

16.5.1 Location of the Origin. In Figs. 16.7 and 16.8 both figures have been correctly plotted, but in Fig. 16.8 the origin or zero point has been omitted. At a span of 20 ft, the deflection of a cedar beam would seem to be about three times that of oak. In Fig. 16.7 the deflection seems to be only twice as much for cedar as for oak.

16.5.2 Change of Scale. The effect of change of scale can also create an erroneous

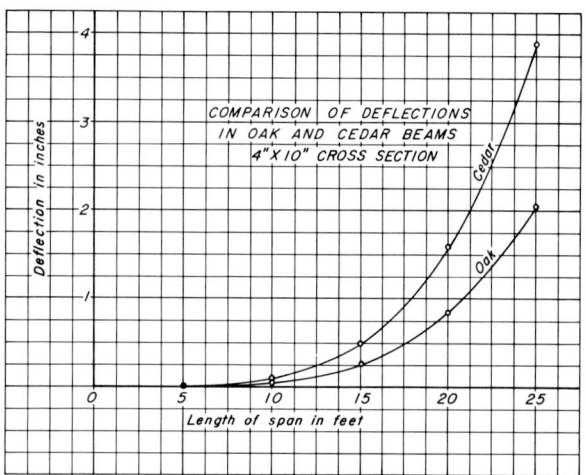

FIG. 16.7. Two plane curves in one chart.

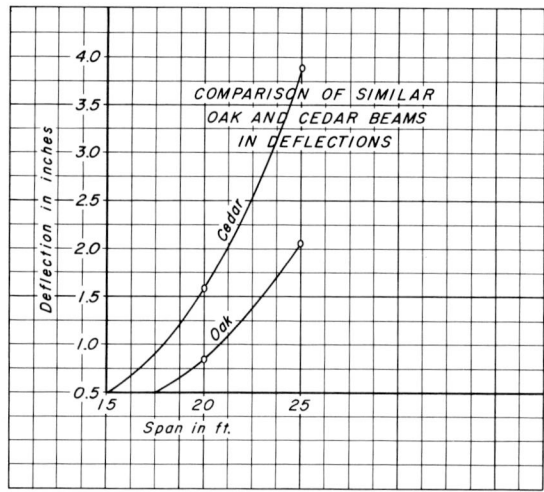

FIG. 16.8. Incorrect impression due to omission of origin.

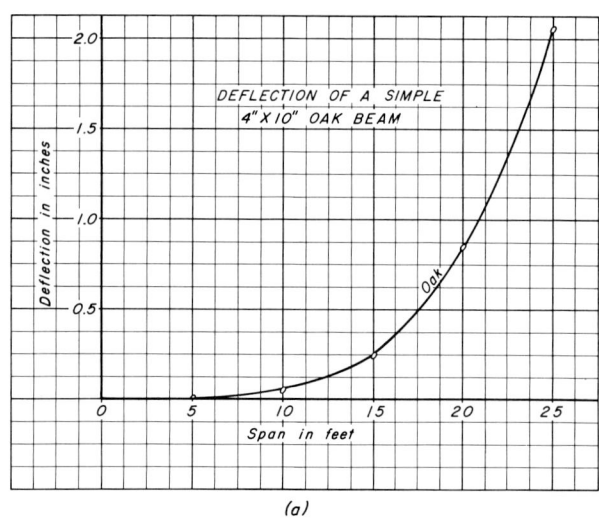

(a)

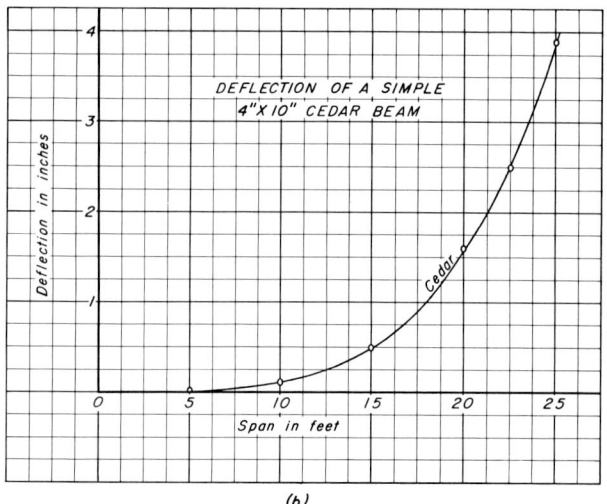

(b)

FIG. 16.9. Incorrect impression due to change of scale.

general impression. In Fig. 16.5 note the apparent flatness of the curve in the case of Fig. 16.5c compared to Fig. 16.5b. Again in Fig. 16.9, the two curves of Figs. 16.7 and 16.8 have been placed on separate grids with the vertical scale in Fig. 16.9b just twice that in Fig. 16.9a. The two curves seem to be almost alike even though Fig. 16.9a is for oak and Fig. 16.9b is for cedar. This distortion is commonly used in the popular press in order to reinforce or deprecate a particular point of view. In Fig. 16.10 one would think that the increase from 1970 to 1980 had been fivefold (length of the bar); this, however, is not the case as the increase has only been 24.3%.

16.6 LOGARITHMIC CHARTS

These charts are most useful in engineering work where the relationship of the variables is more complex, for example, a product, quotient, or exponential form of the variables. In such cases the chart becomes a straight line on logarithmic paper, as illustrated in Fig. 16.11. The ruled lines of this type of paper are spaced according to the logarithms of numbers. Such paper is commonly available with one, two, or three cycles in each direction.

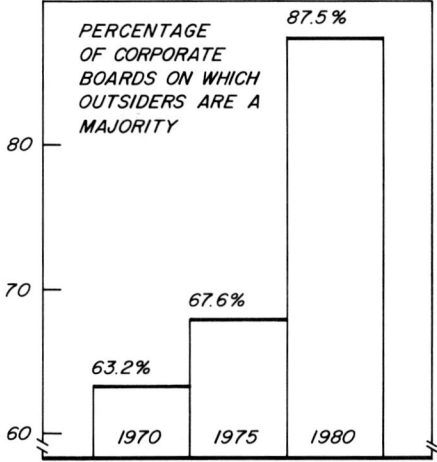

FIG. 16.10. Bar chart (origin not shown).

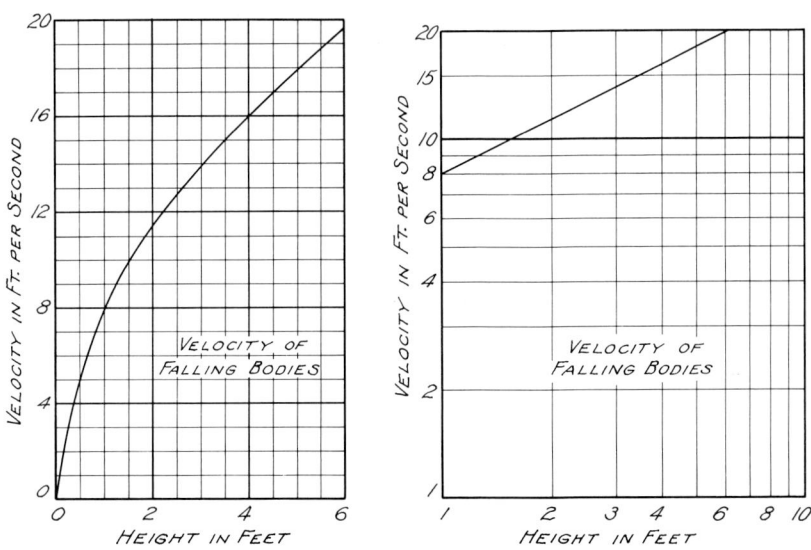

FIG. 16.11. Same curve on rectangular and logarithmic paper.

16.7 SEMILOGARITHMIC CHARTS

When the rate of change in two variables is more important than the quantitative change, semilogarithmic paper is used, since the slope of the tangent to the curve at any point gives the rate of change at that point and the whole chart indicates a trend in the rate more accurately than does the same data plotted on rectangular coordinates, as may be seen by comparing the two charts of Fig. 16.12. Semilogarithmic paper has a logarithmic scale in one direction and an arithmatic scale in the other. Equations of the general form $y = a10^{mx}$ will plot as straight lines on this paper. Empirical equations can frequently be determined by plotting data on this type of paper. A semilogarithmic scale is commonly used when the values to be plotted vary in order of magnitude. See Fig. 16.13. Note that all of the values

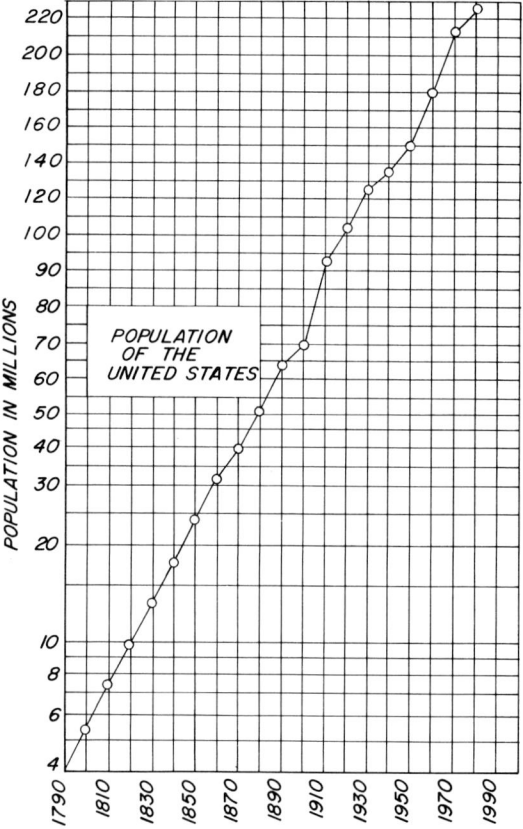
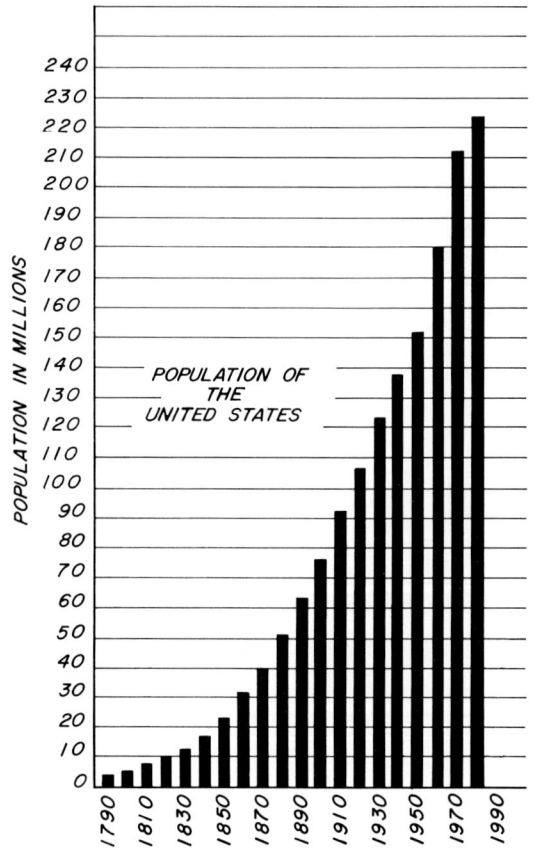

FIG. 16.12. Barograph and semilog chart of same data.

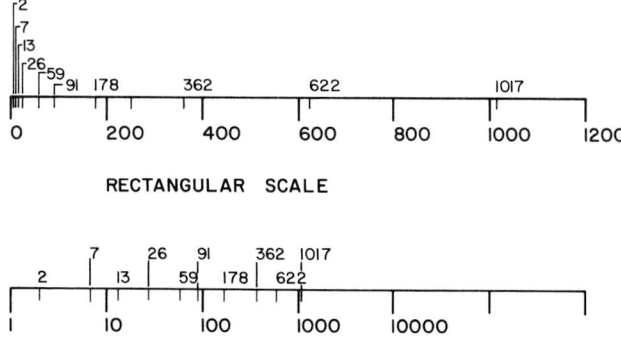

Data to Be Plotted	
2	91
7	178
13	362
26	622
59	1017

FIG. 16.13. Data on both rectangular and logarithmic scale.

may be plotted on the logarithmic scale, whereas on the rectangular scale the three smallest values are too close to zero to be accurately plotted.

16.8 TRILINEAR CHARTS

Trilinear charts are in the form of an equilateral triangle. The coordinates are rule parallel to the sides, and the altitude perpendicular to any side represents 100. These charts are useful in comparing properties of chemicals or alloys composed of three substances, as shown in Fig. 16.14. The equilateral triangle has the property that the sum of the perpendiculars to the three sides from any point inside is equal to the altitude of the triangle. It will be noted in Fig. 16.14 that only a portion of the complete triangle has been drawn since the curves lie in one corner and it would be a waste of space to show the remainder of the blank chart. In Fig. 16.14, read in the direction parallel to the arrows.

16.9 POLAR CHARTS

These charts are useful when equations are given in polar coordinate form or when quantities radiating from a center are involved, for example, in illumination or radiation charts, as in Fig. 16.15. They are also used in modified form on continuous recording devices, as illustrated in Fig. 16.16. The radiating lines are curved in this case because the recording stylus is pivoted at a fixed

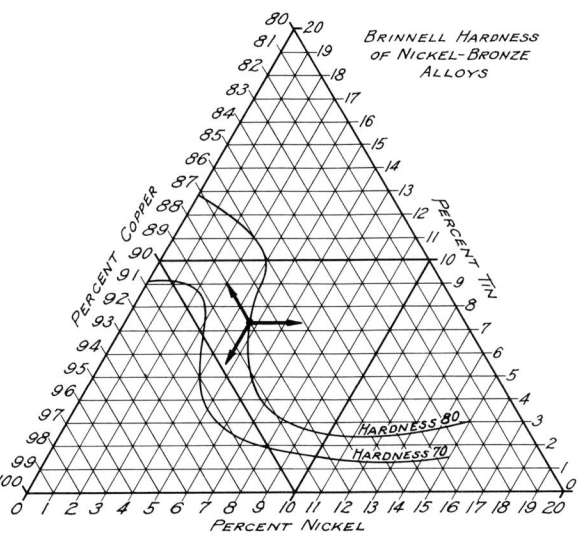

FIG. 16.14. Trilinear curve.

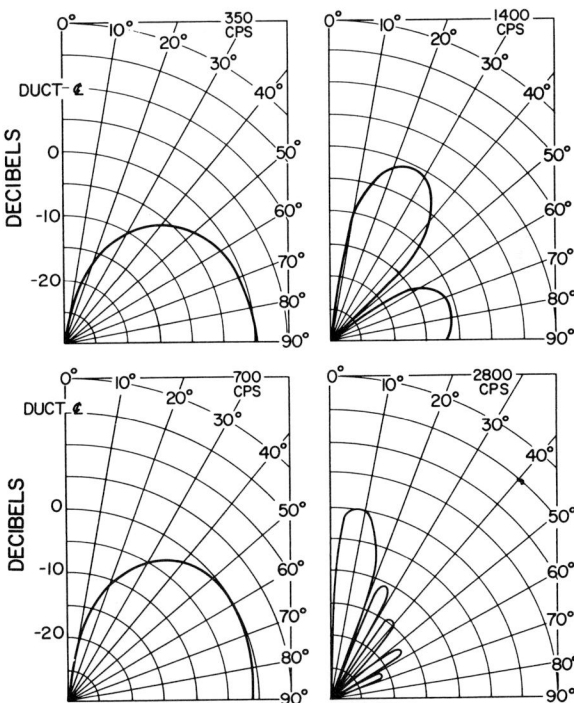

FIG. 16.15. Polar chart (courtesy SAE Journal, J. M. Tyler and T. G. Sofrin).

432 CHARTS AND DIAGRAMS

center. Typically, polar charts utilize one entire rotation as 24 hr or some other convenient period of time.

16.10 PIE DIAGRAMS OR SECTOR CHARTS
These charts, circular in form, are used to show the relative distribution of the parts of a whole, as illustrated in Fig. 16.17. Other common examples are the distribution of the tax dollar and the costs of production in an industry. They are quite effective and simple to make and are easily understood by the general public. The pictoral form of Fig. 16.18 is also very effective.

16.11 BAR CHARTS
Bar charts are commonly used for the popular presentation of facts since they are easy for the average layman to interpret. Bar charts may have

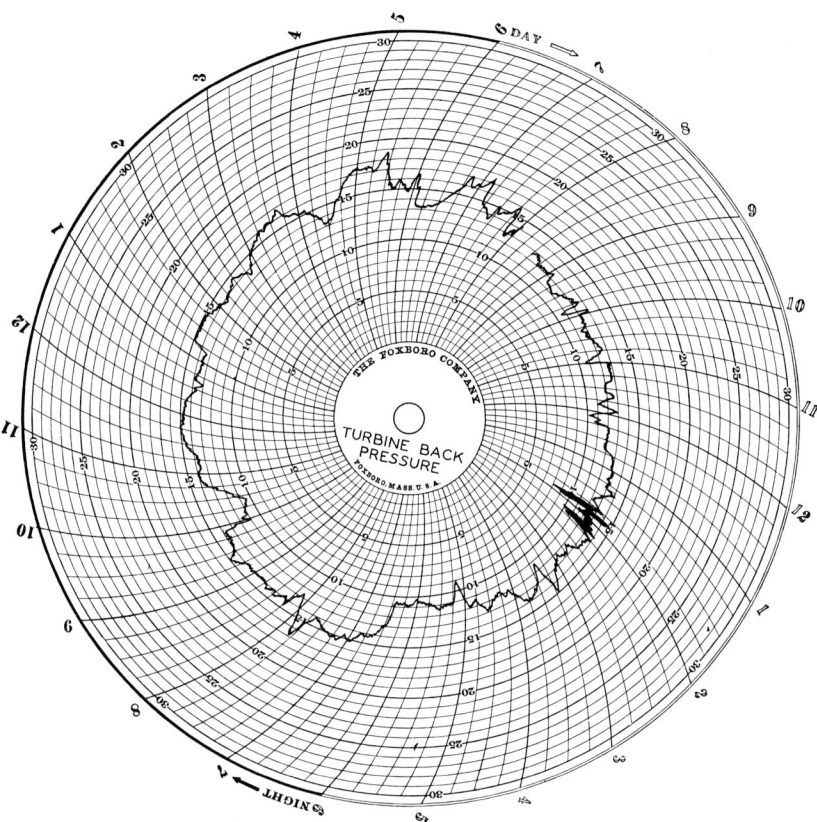

FIG. 16.16. Polar coordinate chart.

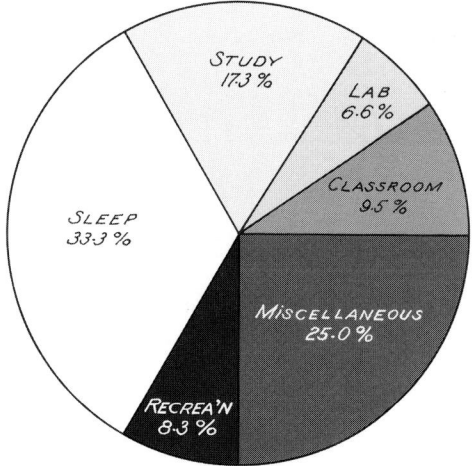

FIG. 16.17. Pie diagram or sector chart showing use of student's time.

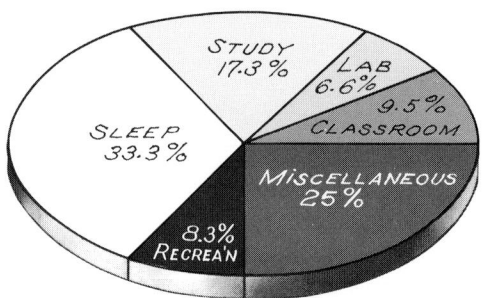

FIG. 16.18. Pictorial pie diagram (same data as for Fig. 16.17).

the bars either vertical or horizontal, as shown in Figs. 16.19 and 16.20. They may be additive, as in Fig. 16.21 (100% bar chart), or comparative, as in Fig. 16.22 (histogram). The same general rules used for rectangular coordinate charts apply here with only slight modification.

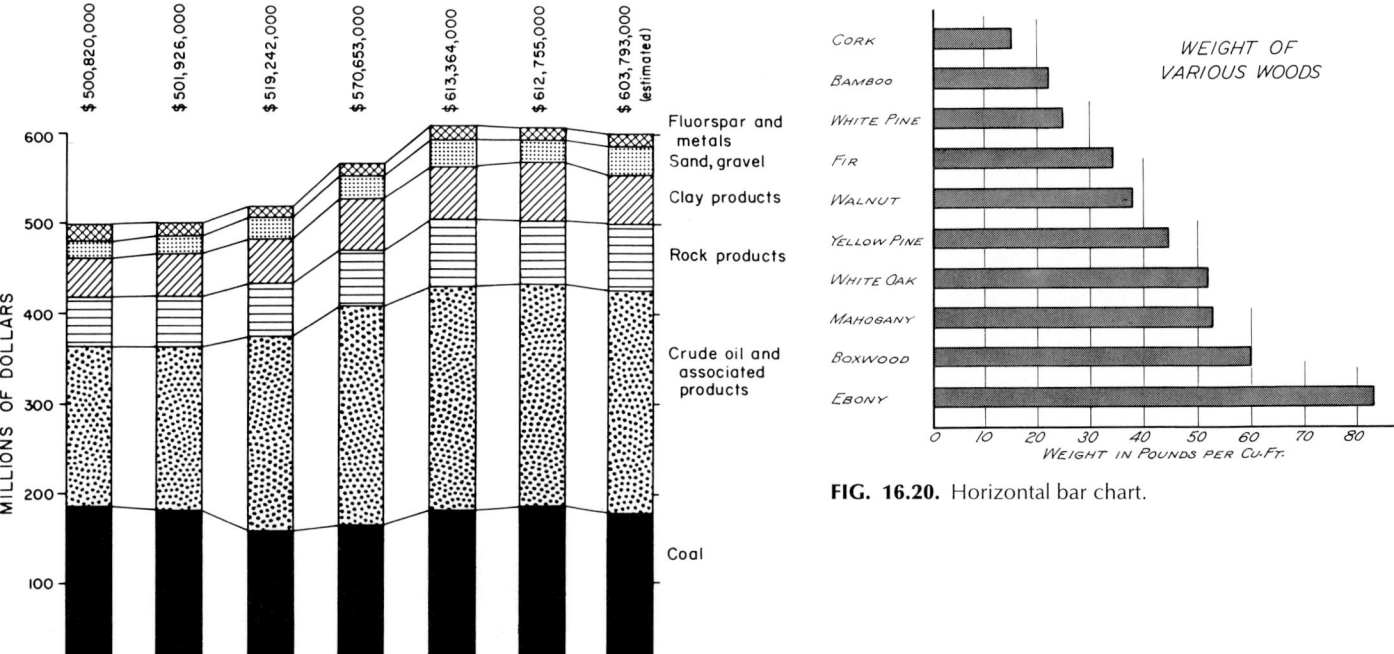

FIG. 16.19. Bar diagram (courtesy Illinois State Geological Survey).

FIG. 16.20. Horizontal bar chart.

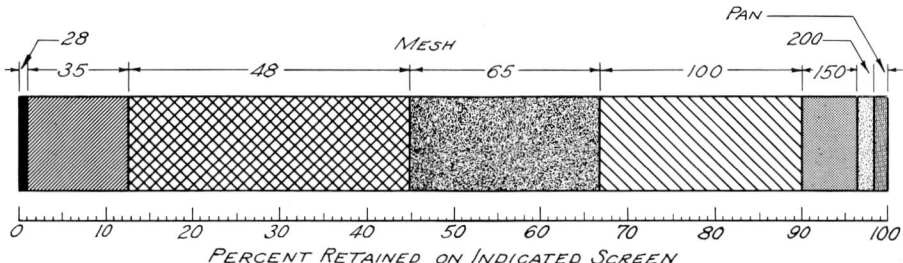

FIG. 16.21. A 100% bar chart.

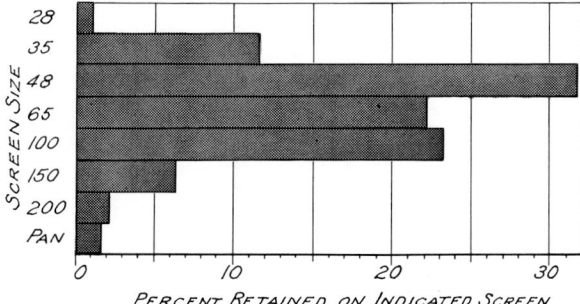

FIG. 16.22. Histogram.

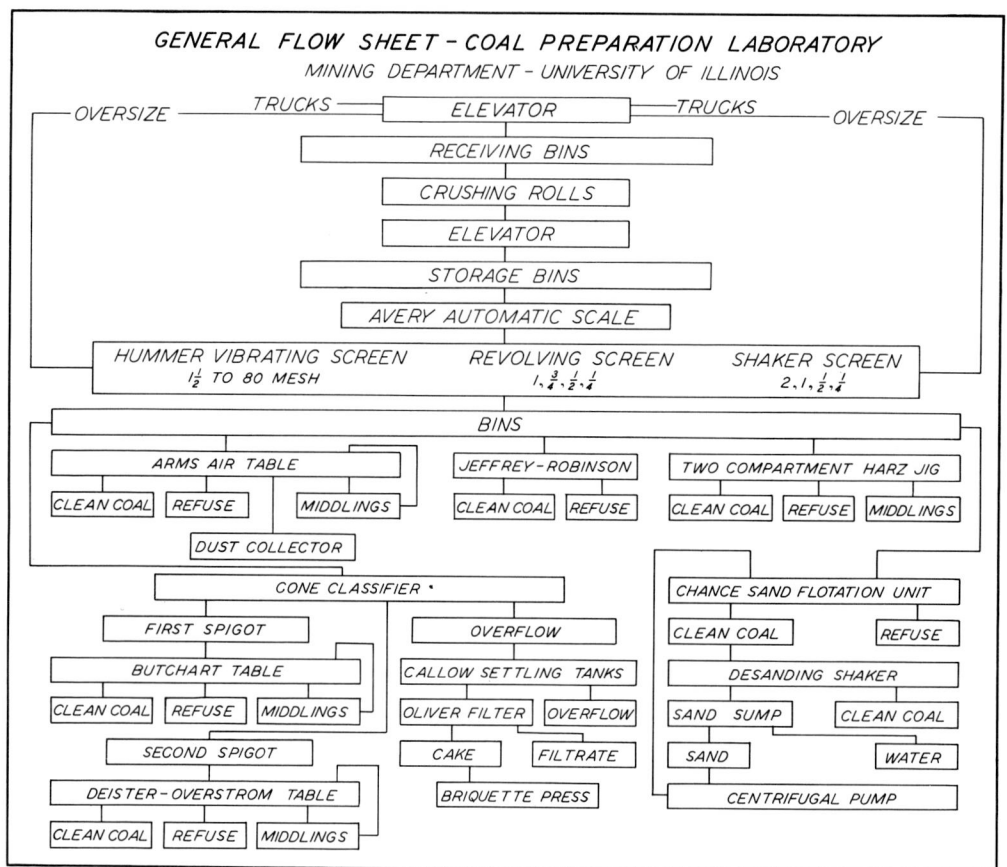

FIG. 16.23. Flow chart (material).

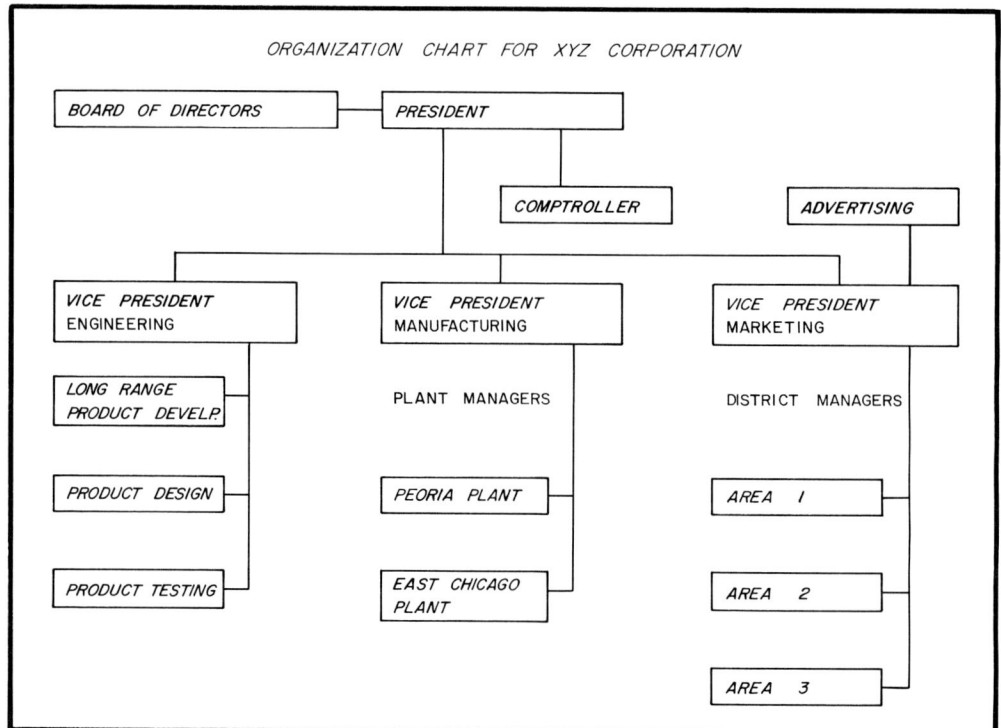

FIG. 16.24. Flow chart (authority).

16.12 FLOW AND ORGANIZATION CHARTS

In the process industries it is often desirable to trace the raw material through the various stages of handling to the finished product. This is readily accomplished by a flow chart, as illustrated in Fig. 16.23.

Charts showing the lines of authority or responsibility from chief executive to the minor departments can also be shown in this type of chart, which is then called an *organization chart,* as illustrated in Fig. 16.24. For formal presentation, the lettering on such charts is usually done mechanically with Wrico or Leroy lettering guides.

16.13 DISTRIBUTION DIAGRAMS

These diagrams usually take the form of maps and may show a wide variety of information useful in business operations, such as distribution of sales and density of population. A good map of the area under consideration is a basic requirement. These are usually traced from existing maps in atlases or the like. A typical example is shown in Fig. 16.25. In addition to location, quantitative values are usually involved. These values are shown by different types of shading.

16.14 THREE-DIMENSIONAL CHARTS

Three-dimensional charts are based on one of the pictorial forms of projection discussed in Chapters 13 and 14. They are useful for illustrating in a popular way the relationship between three variables. Figures 16.26 and 16.27 show charts of this type.

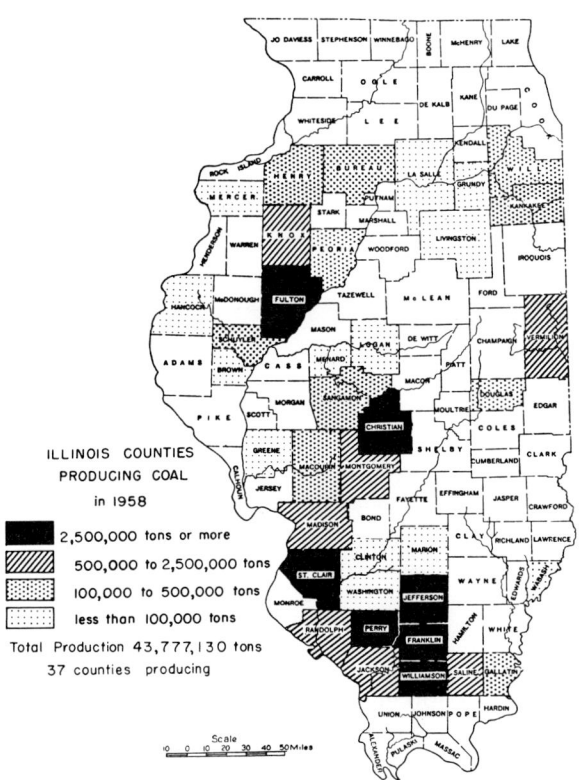

FIG. 16.25. Distribution chart (courtesy of Illinois State Geological Survey).

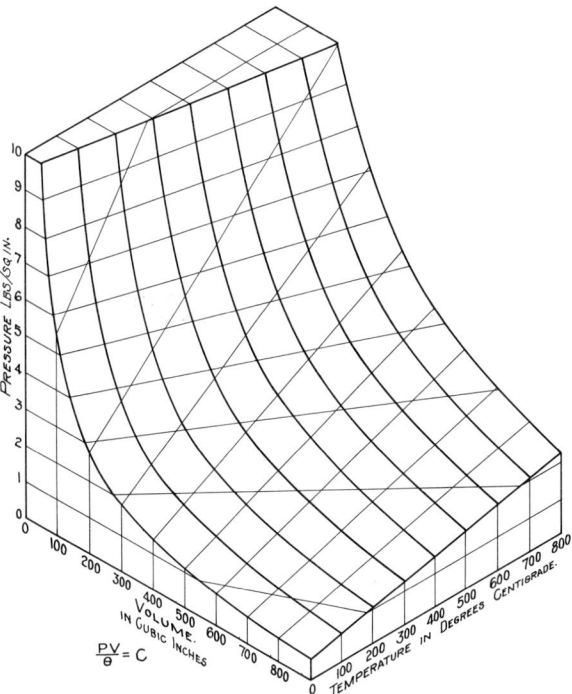

FIG. 16.26. Three-dimensional chart.

436 CHARTS AND DIAGRAMS

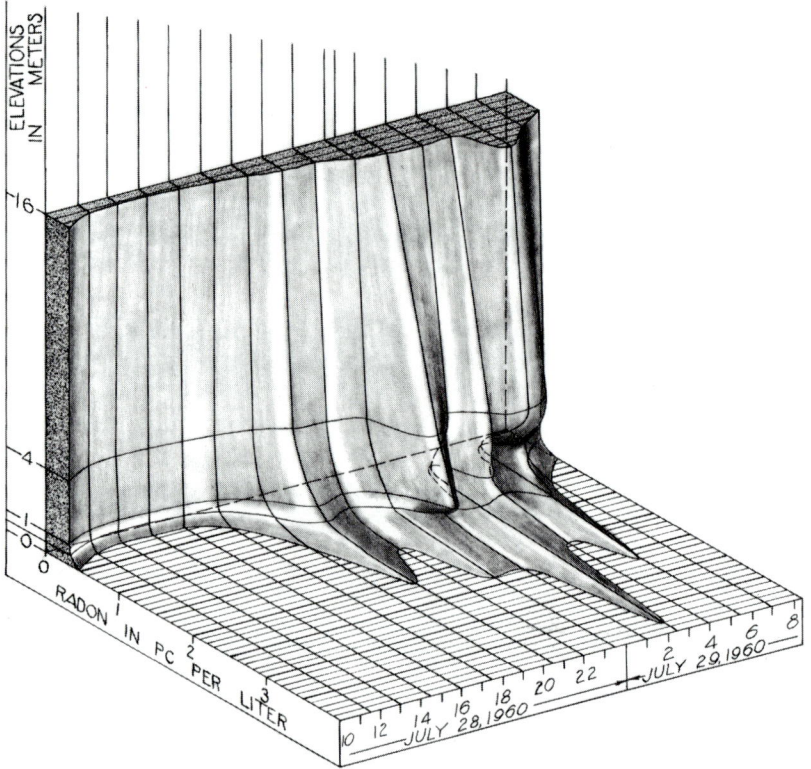

FIG. 16.27. Three-dimensional shaded chart.

SELF-STUDY QUESTIONS

Before trying to answer these questions, read the chapter carefully. Then, without reference to the text, answer as many questions as possible. For those that cannot be answered, the number in parentheses following the question number gives the article in which the answer can be found. Look it up and write down the answer. Check the answers that you did give to see that they are correct.

16.1 (**16.1**) A more understandable way of presenting tabular data is by the use of _____ and _____.

16.2 (**16.4**) One important use of a chart is in _____ reports, and the other in _____.

16.3 (**16.4.2**) The choice of _____ affects the impression given by the chart.

16.4 (**16.4.4**) Empirical data plotted on a chart have the points _____ on the curve.

16.5 (**16.4.5**) When there is a direct relationship between the variables being plotted a _____ _____ is used.

16.6 (**16.6**) Two- or three-cycle _____ paper is commercially available for plotting test data that may be complex in nature.

16.7 (**16.7**) To show data that indicate a rate of change in two variables _____ charts are used.

16.8 (**16.10**) Circular charts showing a distribution of a particular sample are called _____ charts.

16.9 (**16.11**) A popular manner of presenting data to the lay public is by the use of _____ charts.

16.10 (**16.12**) To show an administrative structure of a company an _____ chart is used.

16.11 (**16.12**) The process industries makes great use of _____ charts.

16.12 (**16.13**) A map showing geographic data is called a _____ diagram.

16.13 (**16.9**) Continuous recording devices often use _____ charts.

16.14 (**16.4.2**) In selecting _____ to use in plotting variables, care must be taken to show the data accurately.

16.15 (**16.7**) A change in the slope of a curve plotted on semilogarithmic paper indicates a variation in the _____ of change.

16.16 (**16.8**) A _____ chart is made from an equilateral triangle.

16.17 (**16.2**) An annual report of a corporation uses such charts as _____, _____, and _____.

CHAPTER 17

GRAPHIC VECTOR ANALYSIS

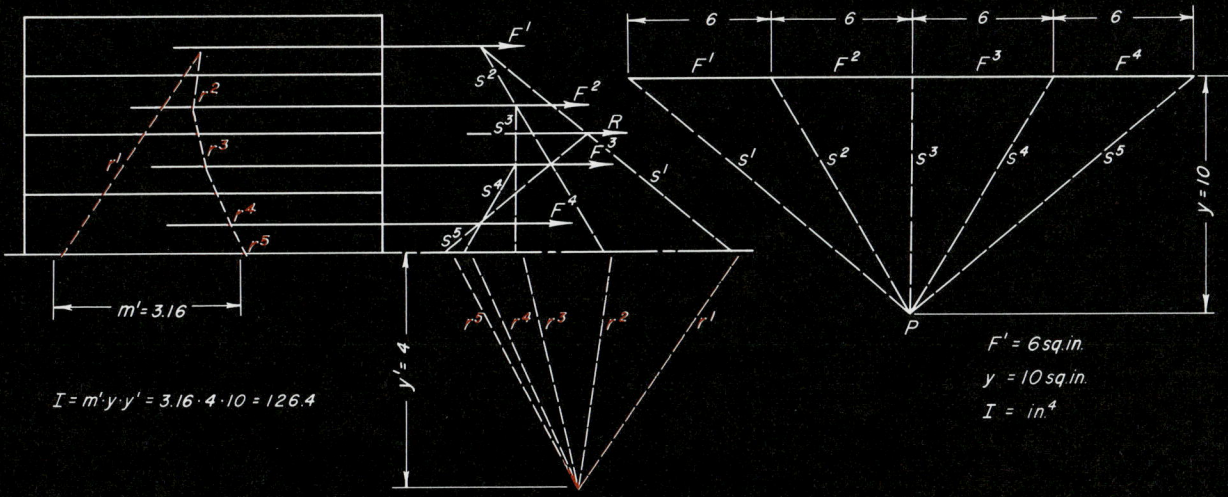

CHAPTER 17

17.1 INTRODUCTION
Many quantities that are important in engineering work are defined by two properties: direction and magnitude. In addition each has a point of application that locates the line of action and gives the quantity a definite position in space. These factors may be represented graphically by a line called a *vector*. The direction of the vector line represents the direction of the quantity, and the length of the line represents the magnitude. The vector as such has no relation to the line of action of the original quantity.

Such things as force, velocity, acceleration, displacement, and magnetic intensity fall into this class. Therefore, problems involving these factors may be solved graphically by means of a vector diagram. The graphical solution is usually comparatively easy and fast, and it can be made as accurate as necessary by increasing the scale of the drawing. Often the graphical solution is satisfactory, but at other times an analytical approach is preferred, in which case the graphical solution may be a valuable check. The graphical solution also frequently helps to visualize the action.

17.2 DEFINITIONS
In order to be able to understand the use and applications of vector diagrams, it is necessary to define a few terms.

  a. *Vector quantity.* Any quantity that may be completely described by direction, magnitude, and a definite position in space.
  b. *Line of action.* The line in space along which the vector quantity acts.
  c. *Point of application.* The point on the line of action where the action begins.
  d. *Vector.* A straight line that represents the vector quantity in direction and magnitude but not in line of action.
  e. *Direction.* The specification telling which way along the line of action the vector tends to produce results. It is usually specified by an arrow on the vector.
  f. *Concurrent vectors.* When all of the vectors meet in a single point, they are called *concurrent*.
  g. *Coplanar vectors.* When all of the vectors lie in a single plane, they are called *coplanar*.
  h. *Resultant.* A single vector that can be used to replace or *produce the same result* as a group or system of vectors.
  i. *Equilibrant.* A single vector that will just balance a group or system of vectors. It has the same magnitude as the resultant but the opposite direction.
  j. *Equilibrium.* A condition in a system of vectors where all resultants are zero.
  k. *Composition of vectors.* The process of combining a system of vectors into a smaller number, usually 1, which is the resultant.
  l. *Resolution of vectors.* The process of replacing a given system of forces by another system having a larger number of forces. The most common case of resolution is the breaking down of a single force into two or more components.
  m. *Vector diagram.* A continuous polygon of vectors. If the forces are concurrent and the polygon is a closed figure, the system is in equilibrium, provided that the vectors all point in a continuous directional pattern around the polygon.

17.3 COMPOSITION OF TWO FORCES
When two vectors act at a point, they form a concurrent coplanar system and the resultant may be obtained by means of a parallelogram. In Fig. 17.1a, a boat started at A and is being driven in the direction AB at 10 miles per hour, and the stream flows in the direction CG at 4 miles per hour. It is desired to determine the actual direction and speed of travel. The construction shown in Fig. 17.1b may be done in the following steps.

  a. Construct the vector ab parallel to AB.

b. Lay off 10 units on ab to any convenient scale (an engineer's scale is best).
c. Construct the vector ae parallel to CG through tail of vector ab.
d. Lay off 4 units on ae using the same scale as for ab.
e. Complete the parallelogram of forces.
f. Draw the diagonal ad, from the common tail toward the tip of the parallelogram.
g. Measure the length of ad using the same scale.
h. The vector ad gives the direction and velocity of the boat.

The usual method of solving this problem is by drawing only half of the parallelogram, as shown in Fig. 17.1c. The vector ab is constructed and then bd is drawn parallel to CG through the tip of ab and of the proper length to locate d and the resultant ad goes from the tail of ab to the tip of bd. This saves time and is especially convenient when more than two vectors are involved.

Line AD in Fig. 17.1a can be drawn parallel to ad to determine the point of landing on the opposite shore. The vector diagram could have been started at A to locate AD immediately, but this is not always possible or convenient. It is not necessary for the vector diagram to have any definite position with respect to the line of action which is AD.

In a coplanar system it is important that the vectors be laid out in the view that shows the true size of all vectors.

17.4 COMPOSITION OF CONCURRENT COPLANAR VECTORS

In Fig. 17.2 there are three ropes, AB, AC, and AD, pulling at point A, with the magnitude specified on each rope. Since the H- and V-projections do not show the ropes in true size, it will be necessary to project the system onto auxiliary plane 1, which is set up parallel to the three forces. In this situation it is customary to letter the spaces between the forces, as W, X, Y, and Z. Force AB will then be known as WX, force AC as XY, and AD as YZ. This makes the lettering of the vector diagram very simple. The forces may be taken in any order, but it is customary to take them either clockwise or counterclockwise. The construction is as follows.

a. Set up auxiliary plane 1 with the 1-V reference line parallel to $d^V a^V c^V e^V b^V$.
b. Project all forces onto plane 1.
c. Letter the spaces between the forces XYZW.
d. Construct vector WX parallel to $a'b'$.
e. Using any convenient engineer's scale, measure the 200 lb along WX.
f. Construct XY parallel to $a'c'$.
g. Measure from X on XY a distance of 150 lb, using the same scale.
h. Lay out YZ parallel to $a'd'$.
i. Measure from Y on YZ a distance of 100 lb, using the same scale.
j. Measure the length of WZ, using the same scale.

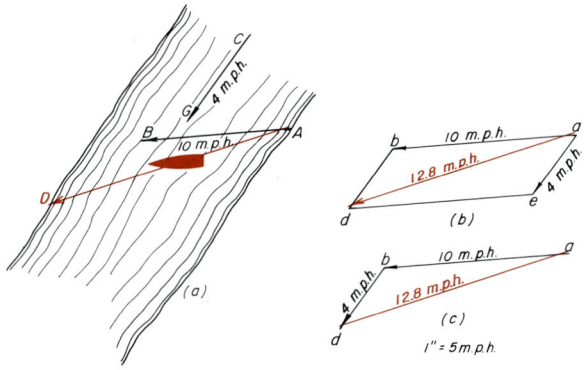

FIG. 17.1. Composition of two concurrent coplanar vectors.

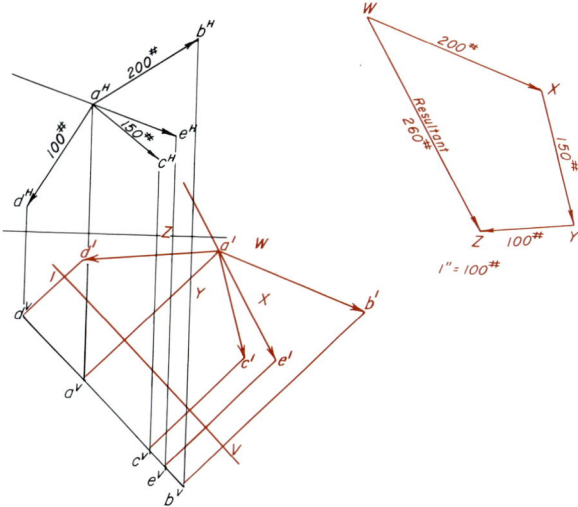

FIG. 17.2. Composition of three concurrent coplanar vectors.

442 GRAPHIC VECTOR ANALYSIS

The resultant or the force that will replace the three will be WZ and the magnitude, *260 lb, can be determined by measuring line WZ.* If desired, line AE may be drawn parallel to WZ and carried back to the H- and V-projections to locate the line of action of the resultant.

If the original forces are so located that the plane does not show edgewise in either the top or the front view, two auxiliary projections will be necessary to determine the true size of the plane and the true magnitude of the forces. This construction is shown in Fig. 17.3.

17.5 COMPOSITION OF CONCURRENT NONCOPLANAR VECTORS

When the vectors do not lie in one plane, it is impossible to find the true length of every vector in any one view. Therefore, the vector diagram must be drawn in two views at the same time. In Fig. 17.4a it is desired to find the direction and magnitude of the magnetic intensity at point P, due to AB, which is a long straight conductor that carries a current (i) and to CD, which is a bar magnet having point poles of strength m at each end. The values of H_i, H_c, and H_d represent the magnetic intensities at point P due to the conductor, pole C, and pole D, respectively. These values can be calculated by a well-known formula when the current i, the strength of the magnet, and all distances are known. Lines PE, PC, and PD give the respective lines of action of the forces. The values H_i, H_c, and H_d may be placed consecutively in a vector diagram whose resultant is the magnetic intensity acting on point P. The following steps, shown in Fig. 17.4, will give the value and direction of the resultant.

a. Letter the spaces between the forces.
b. Assume two projections of the starting point W.
c. Construct WX parallel to PE and lay out on it the value H_i, using any convenient engineer's scale. This can be measured in the V-projection because that shows in true length.
d. Construct XF parallel to PC.
e. Find the true length of XF at x'f'.
f. On x'f' measure the value H_e from X to locate Y.
g. From Y construct YG parallel to PD.
h. Find the true length of YG at y'g'.
i. On y'g' measure the value of H_d from Y to locate Z.
j. Construct the vector from W to Z.
k. Find the true length of WZ at w'z'.

The resultant then will be the vector from W to Z. Its true length must be determined to obtain

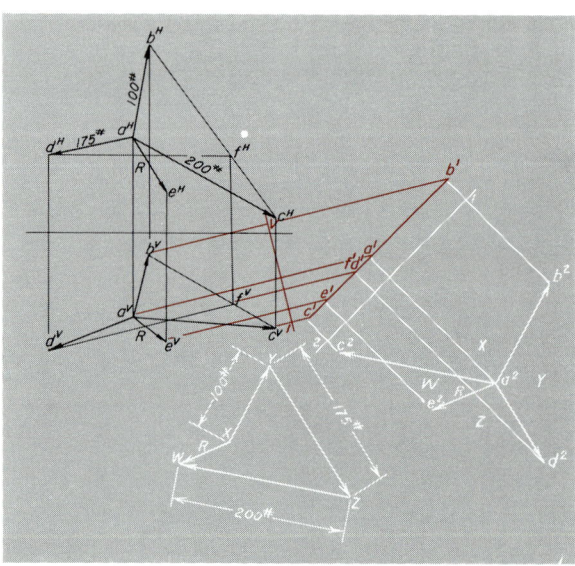

FIG. 17.3. Composition of concurrent coplanar vectors.

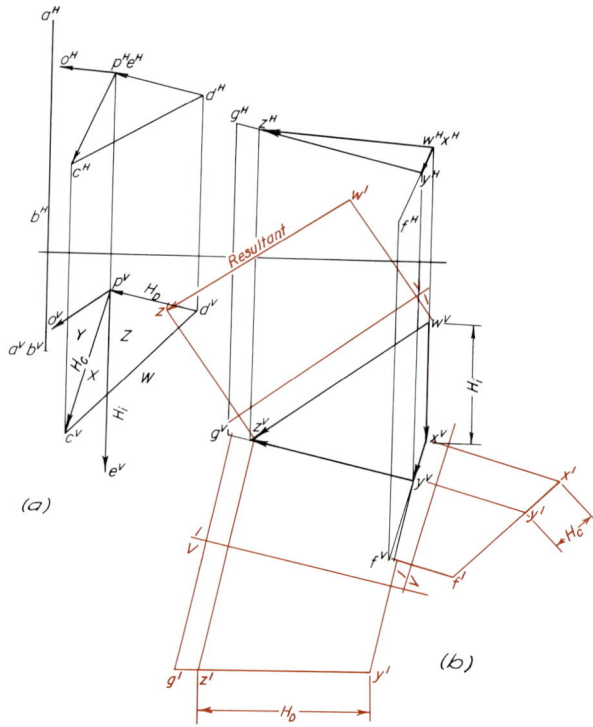

FIG. 17.4. Composition of concurrent noncoplanar vectors.

the magnitude of the magnetic intensity at point P. By drawing through P in Fig. 17.4a, a line parallel to WZ, line PO is established, which gives the direction of the magnetic intensity at P.

17.6 RESOLUTION OF FORCES

If it is desired to break a given force into a number of components, this may be done by considering the known force to be the resultant. From Fig. 17.5 it can be seen that there are any number of sets of components that may be drawn. The only conditions are that their lines of action all pass through the same point on the line of action of the original force and that in the vector diagram the original force turns out to be the resultant of all the components.

If the lines of action are given, the number of possible solutions is decreased. When there are only two components and the lines of action are given, there will be only one result. To resolve a known force into a set of components so that the system will be in equilibrium requires that the known force be the equilibrant of all the components instead of the resultant.

17.7 EQUILIBRIUM

When a condition exists in a system of vectors in which the resultant is zero, the system is said to be in *equilibrium*. The vector diagram of a system that is in equilibrium is a closed polygon with the arrows on the vectors all pointing around the polygon in the same direction. In each of the preceding problems an extra vector could have been added that would produce a state of equilibrium in the system. This vector is called the *equilibrant* and it is equal to the resultant but opposite in direction.

The system of forces in Fig. 17.6 in which a weight is suspended by two ropes is in a state of equilibrium. The vector diagram always considers the forces acting on a single joint, in this case on point A, and here they are concurrent coplanar forces. In considering the forces acting on this joint there are six factors that must be known. They are the magnitude and direction of each of the three forces. When it is known that the joint is in equilibrium, the problem can be solved if there are no more than two unknowns. If there are more unknowns at any one joint, the problem is said to be statically indeterminate. In Fig. 17.6 the magnitude and direction of the load are known, and the direction of each of the other forces is also known, leaving their two magnitudes unknown. Therefore the problem can be solved by the vector diagram shown in Fig. 17.6b. The steps are as follows.

a. Letter the spaces between the forces.
b. Lay out the vector for the known force by making XY parallel to a^Vb^V with a length of 100 lb to any convenient engineer's scale.
c. From Y construct a vector parallel to a^Vd^V of an indeterminate length.
d. From X construct a vector parallel to a^Vc^V of indeterminate length.
e. The intersection of these vectors is at Z. This completes the vector diagram for this joint.
f. Scale the lengths of XZ and YZ to determine the load on each rope. Use the same scale as for XY.

Since in equilibrium the arrows must all point around the diagram in the same direction, the direction of the forces is found to be from Y to Z and from Z to X. The arrows indicate the direction in which the forces act on the original joint in Fig. 17.6a. From this it can be seen that each force is pulling away from the joint and therefore they are all in tension. If the arrow is found to be pointing toward the joint, the member is in compression.

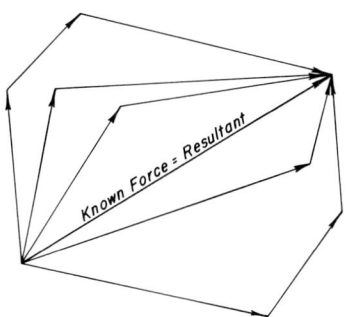

FIG. 17.5. Resolution of a vector.

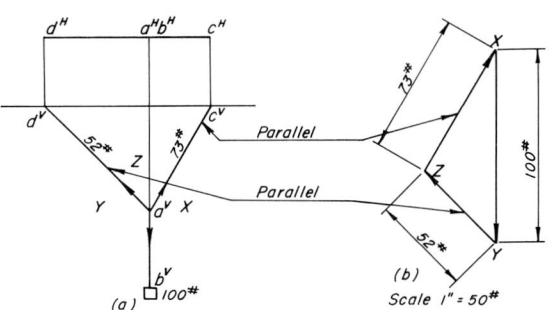

FIG. 17.6. Resolution of forces in equilibrium.

17.8 RESOLUTION OF VECTORS IN A THREE-DIMENSIONAL SYSTEM

When it becomes necessary to find the loads in a three-dimensional structure, the work of drawing the vector diagram becomes a little more complicated. It has been seen in a plane vector system that it is possible to construct the diagram if there are no more than two unknowns. *In a space or three-dimensional system it is possible to construct the diagram if there are no more than three unknowns.* In Fig. 17.7a a three-legged tripod is supporting two known forces of 100 lb each. There are 10 possible factors in this problem, of which 7 are known. They are the direction and magnitude of the known forces and the direction of the three components. The unknowns are the magnitudes of the components or, in other words, the load in each leg of the tripod. If the known forces are called AB and BC, the other three may be called CD, DE, and EA. The vector diagram may be constructed in the usual way. Assume point A in both projections of Fig. 17.7b. Draw AB in a vertical position and measure the length of 100 lb in the vertical projection. BX can be drawn parallel to the other known force and its true length determined at b^1x^1. On the true length lay out the value of 100 lb to locate c^1, and project it back to the horizontal and vertical views. Next a line AZ, of any length, must be drawn through point A parallel to leg EA of the tripod. Point E will be somewhere on that line. Then through C a line CY, of any length, must be drawn parallel to leg CD of the tripod. Point D will be somewhere on that line. The final step is to construct a line intersecting AZ and CY and at the same time parallel to leg DE of the tripod. The construction is shown in the figure, but the theory is discussed in detail in Article 7.24.5. In this case the point projection of line MN, which is parallel to leg DE, is found in the second auxiliary projection shown in Fig. 17.7c.

The vector diagram may be constructed in the usual way by the following steps.

a. Letter the spaces between the forces.
b. Assume two projections of the starting point A.
c. Draw AB parallel to the known vertical force and measure the known 100 lb on AB.
d. Construct BX parallel to the other known force.
e. Find the true length of BX at b^1x^1 and on b^1x^1 lay out the known value of 100 lb to locate c^1.
f. Project c^1 back to locate c^H and c^V.
g. From C construct a line CY parallel to leg CD of indeterminate length.
h. From A construct a line AZ parallel to EA of indeterminate length.
i. Construct MN parallel to DE of the tripod.

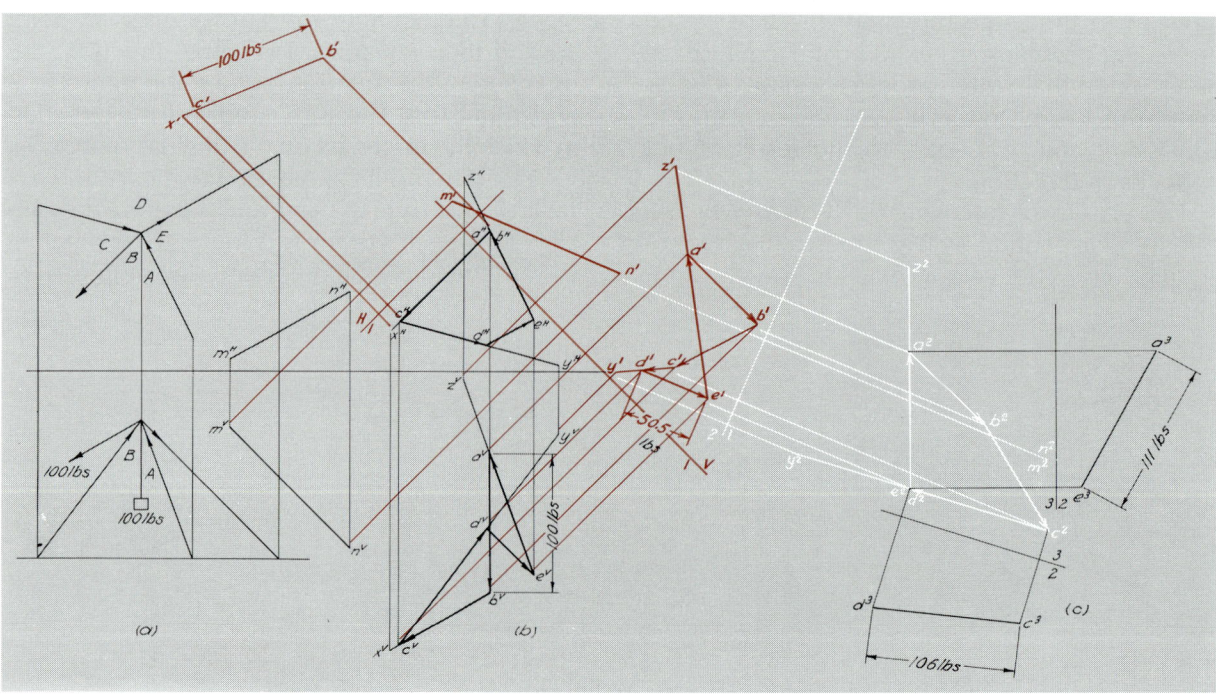

FIG. 17.7. Space vector diagram.

The next steps give the construction for drawing a line parallel to MN and intersecting AZ and CY.

j. Project MN, AZ, and CY on a plane parallel to MN.
k. Project MN, AZ, and CY on a plane perpendicular to MN.
l. Find the place where a^2z^2 and c^2y^2 intersect. This is the point projection of the line parallel to MN and intersecting AZ and CY.
m. Letter this projection e^2d^2.
n. Carry e^2d^2 back to the H- and V-projections.
o. ABCDEA is the completed vector diagram.
p. Find the true lengths of CD, DE, and EA. These will be the loads in the respective legs when scaled with the same scale as used for AB.
q. Transfer the arrows to the tripod in Fig. 17.7a to determine which is tension and which is compression. DE is pointing away from the joint and is in tension. The others are pointing toward the joint and are in compression.

17.8.1 Alternate Solution. A somewhat easier solution for this problem may be obtained by solving the following steps:

a. Find the resultant of all known forces. This is AB in Fig. 17.8a.
b. Construct a vector AX along resultant AB starting at the point where all forces meet. Make it of any length.
c. Find the true length of AX at a^1x^1.
d. On a^1x^1 lay out the true value of the resultant. This gives the vector AB.

Resolve the vector AB into two components. One component must be parallel to one of the legs of the tripod and the other component lies in the plane of the other two legs. The following steps will solve this problem.

e. Through point B construct BY parallel to leg 1.
f. Find the point, P, where BY pierces the plane of legs 2 and 3.
g. Through point P construct a line, PC, parallel to leg 3.
h. Find the point C where the line PC crosses leg 2 of the tripod.
i. The vector diagram is ABPCA (Fig. 17.8b).
j. Find the true length of each leg of the diagram to determine the load in each leg of the tripod.

By transferring the arrows to the tripod in Fig. 17.8a, it can be shown that leg 3 is in tension and the other two in compression.

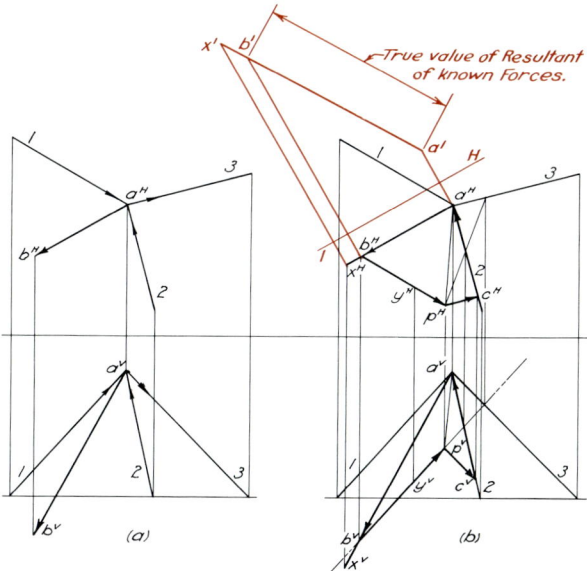

FIG. 17.8. Space vector diagram.

17.9 RESULTANT OF NONCONCURRENT COPLANAR VECTORS

When a series of nonconcurrent vectors lie in a single plane, they must intersect in pairs unless they are parallel. Therefore, the resultant can be found by combining two vectors as V^1 and V^2 in Fig. 17.9 to obtain the resultant R^1. Then V^3 may be combined with R^1 to obtain resultant R^2. The magnitude and direction of each resultant are found in the vector diagram in Fig. 17.9b and the line of action in the space diagram in Fig. 17.9a. If the resultant of the first two forces happens to be parallel and equal in magnitude to the third force but opposite in direction, the resultant will be a couple.

17.10 THE STRING POLYGON

The same result could be obtained by means of a string polygon or, as it is often called, a *funicular polygon*. This is a very important method of analysis and should be familiar to every engineer. The illustration in Fig. 17.10 is the same as Fig. 17.9, and although in this case the method of Fig. 17.9 may seem easier, the method of the string polygon can be used in many other problems that would be difficult or impossible to solve by the first method.

To solve a problem by the string polygon method, it is necessary to have some direct relationship between the original vectors. This is accomplished by resolving each force into two components in such a way that one of the components of the first force is also a component of the second force but opposite in direction. The vector AB in Fig. 17.10b represents the force V^1 of Fig. 17.10a, and by selecting a pole P at any convenient place the vector AB can be resolved into two components, AP, called string S^1, and PB, called string S^2. The vector BC, representing force V^2, can be resolved into two components BP, which is still called S^2, and PC, which is called S^3. From this it can be seen that PB and BP are equal and opposite and would therefore cancel out, leaving AP and PC as the components of the resultant R^1. In a similar manner the vector CD, representing force V^3, can be resolved into two

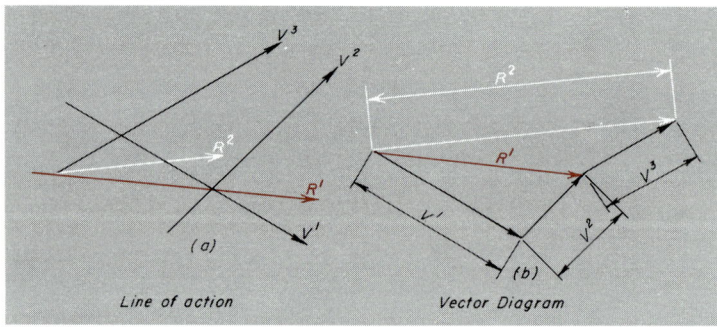

FIG. 17.9. Composition of nonconcurrent coplanar vectors.

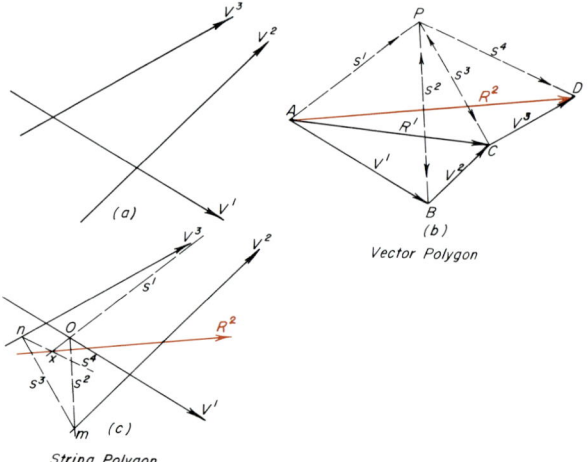

FIG. 17.10. Composition of nonconcurrent coplanar vectors: string polygon method.

components—CP, string S^3, and PD, string S^4. Again it can be seen that PC and CP cancel each other, leaving AP and PD as the components of the resultant R^2.

Lines of action of these various components can be determined as in Fig. 17.10c in the following steps.

a. Select a starting point, O, as any point on the line of action of force V^1.
b. Through point O construct lines parallel to strings S^1 and S^2 of Fig. 17.10b.
c. Locate the point, m, where the line parallel to string S^2 crosses the line of action of force V^2.
d. Through point m construct a line parallel to string S^3.
e. Locate the point, n, where the line parallel to string S^3 crosses the line of action of force V^3.
f. Through point n construct a line parallel to string S^4.
g. Locate the point where the line parallel to string S^4 crosses the line parallel to string S^1. This locates point X.
h. Through point X construct a line parallel to R^2 of Fig. 17.10b. This is the line of action of resultant R^2.

The first point O can be chosen any place on V^1 since any other locations would merely determine a different point on the line of action of R^2.

17.11 RESULTANTS OF VECTOR SYSTEMS

There are several possibilities that must be investigated when determining the resultant of a set of vectors. If, in each case, the vector polygon and the string polygon are drawn, it is possible to tell immediately what the form of the resultant will be. The three possibilities are as follows:

a. *If, as in Fig. 17.10b, the vector polygon does not close, the resultant must be a force.* A vector diagram is said to close when the end of the last vector coincides with the beginning of the first vector. In this case point D would have to fall on point A for closure.

b. *If the vector polygon closes with the vectors all pointing around the polygon in the same direction, but the string polygon does not close, the resultant will be a couple.* A couple is defined as two forces that are equal and opposite but whose lines of action are parallel and at a definite distance apart. In Fig. 17.11b the vector polygon closes, but the string polygon, shown in dashed line, in Fig. 17.11a does not close because there are two parallel positions of string S^1. In this case the resultant is a couple whose forces are both S^1 acting at a distance of a from each other.

c. *If the vector polygon closes with the vectors all pointing around the polygon in the same direction, and the string polygon also closes, the system is in equilibrium.* This condition is illustrated in Fig. 17.12.

17.12 PARALLEL FORCES

The problem of parallel forces is a special case of nonconcurrent forces and may be solved by the same method. Figure 17.13 shows a truss with vertical loads as specified. The appearance of the vector diagram, shown in Fig. 17.13b is different because it becomes a straight line with the panel loads all pointing down and the reactions pointing up. When in equilibrium, the sum of the reactions must equal the loads, so that the vector diagram will close at the top point of P^1. It will be necessary to determine the value of each reac-

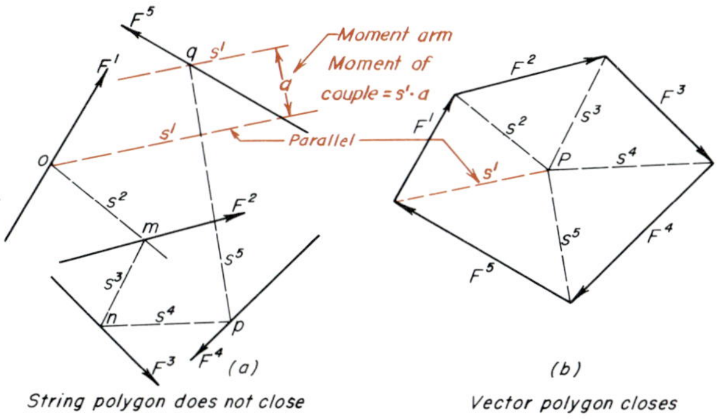

FIG. 17.11. Vector and string polygons of a force system when the resultant is a couple.

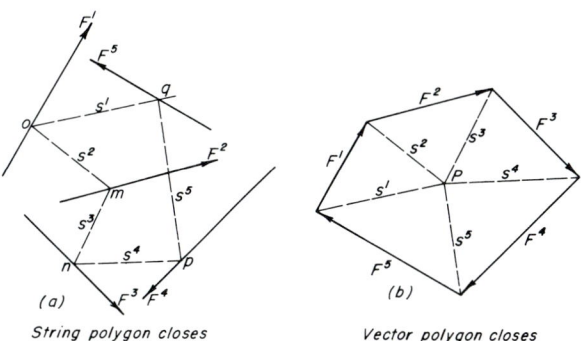

FIG. 17.12. Vector and string polygons of a force system in equilibrium. The resultant is zero.

448 GRAPHIC VECTOR ANALYSIS

tion. To do this the following steps are necessary, as illustrated in Fig. 17.13.

a. Construct the vector diagram of all forces. This falls on a straight line as shown in Fig. 17.13b. The loads are downward and the reactions upward.
b. Select a pole at some convenient place such as P.
c. Draw strings S^1 to S^5 so that each force is resolved into two forces represented by strings.
d. Construct the string polygon in the usual manner (Fig. 17.13c).
e. Extend strings S^1 and S^5 to their point of intersection. This locates a point on the line of action of the resultant. The value of the resultant can be found on the vector diagram from the beginning of P^1 to the end of P^4.
f. Continue S^1 to its intersection with the line of action of R^1 as in Fig. 17.13c.
g. Continue S^5 to its intersection with the line of action of R^2 as in Fig. 17.13c.
h. Join these two points to locate string S^6.
i. Draw string S^6 in the vector diagram (Fig. 17.13b) to locate a point in the vector diagram, which gives the magnitude of the two resultants. In the string polygon (Fig. 17.13c strings S^5 and S^6 intersect on R^2. Therefore these strings are components of R^2 and indicate on the vector diagram (Fig. 17.13b) the limits of R^2. R^1 whose components are S^1 and S^6 closes the diagram.
j. The vector diagram and the string polygon both close, so that the system is in equilibrium.

17.13 CENTROID OF AN AREA

The *centroid* or *center of gravity* of an area is defined as that point where the area would act if it were concentrated at a point. For simple figures it is usually quite easy to locate the centroid, but for irregular areas the string polygon forms a convenient method. The method given here is general and may be applied to any area, but for regular figures there are usually easier methods. For instance, if there is an axis of symmetry, the centroid must be on that axis.

The irregular figure shown in Fig. 17.14a will serve to show the general method. The steps are as follows.

a. Divide the area into fairly small parts by means of vertical lines.
b. Select the approximate center of each area by estimation and judgment. In a triangle the centroid will be one third of the altitude from the base. In a trapezoid the centroid will be near the center but closer to the larger base.
c. Draw a vertical line through the center of each area.
d. Calculate the area of each small part by scaling distances.

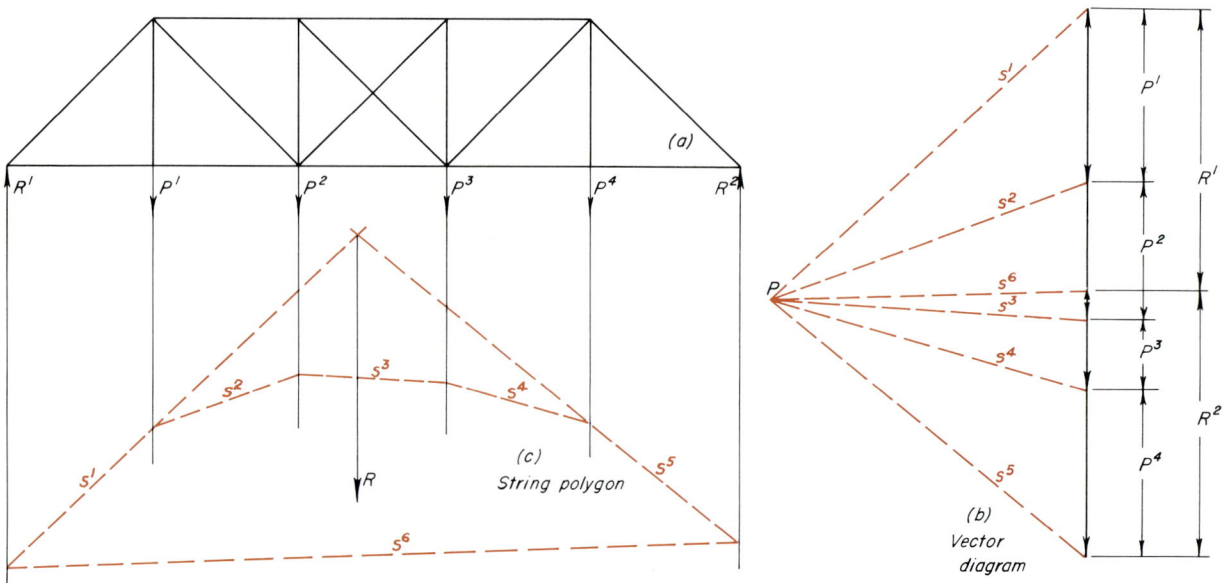

FIG. 17.13. Reactions of a truss.

e. Construct a vector diagram using the areas as forces. See Fig. 17.14b.
f. Select a pole and draw the strings S^1 to S^5. See Fig. 17.14b.
g. Construct the string polygon in the usual manner in Fig. 17.14c.
h. The intersection of strings S^1 and S^5 locates a point on the resultant. The centroid of the total area must lie on this resultant.
i. Divide the area into small sections by means of horizontal lines. See Fig. 17.14a.
j. Select the approximate center of each part.
k. Draw a horizontal line through each center.
l. Calculate the area of each small horizontal part.
m. Construct a vector diagram using the areas as forces. See Fig. 17.14d.
n. Assume a pole and draw strings S^1 to S^6.
o. Construct the string polygon as in Fig. 17.14e.
p. The intersection of strings S^1 and S^6 locates a point on the resultant of all the forces.
q. The centroid of the entire area must lie on that resultant.
r. The centroid C lies at the intersection of the two resultants.

17.14 MOMENTS OF A FORCE ABOUT A POINT

A *moment* is defined as the product of a force times the distance measured from the line of action of the force to some point about which the moments are to be taken. It is frequently necessary in engineering problems to find the moment of a force about some point. This may be done graphically as shown in Fig. 17.15. To find the moment of force F about point O, the following steps are necessary. See Fig. 17.15.

a. Construct a vector diagram as in Fig. 17.15b. If possible, the scale should be a convenient one such as 1 in. = 100 lb. Select the pole so that the pole distance y is

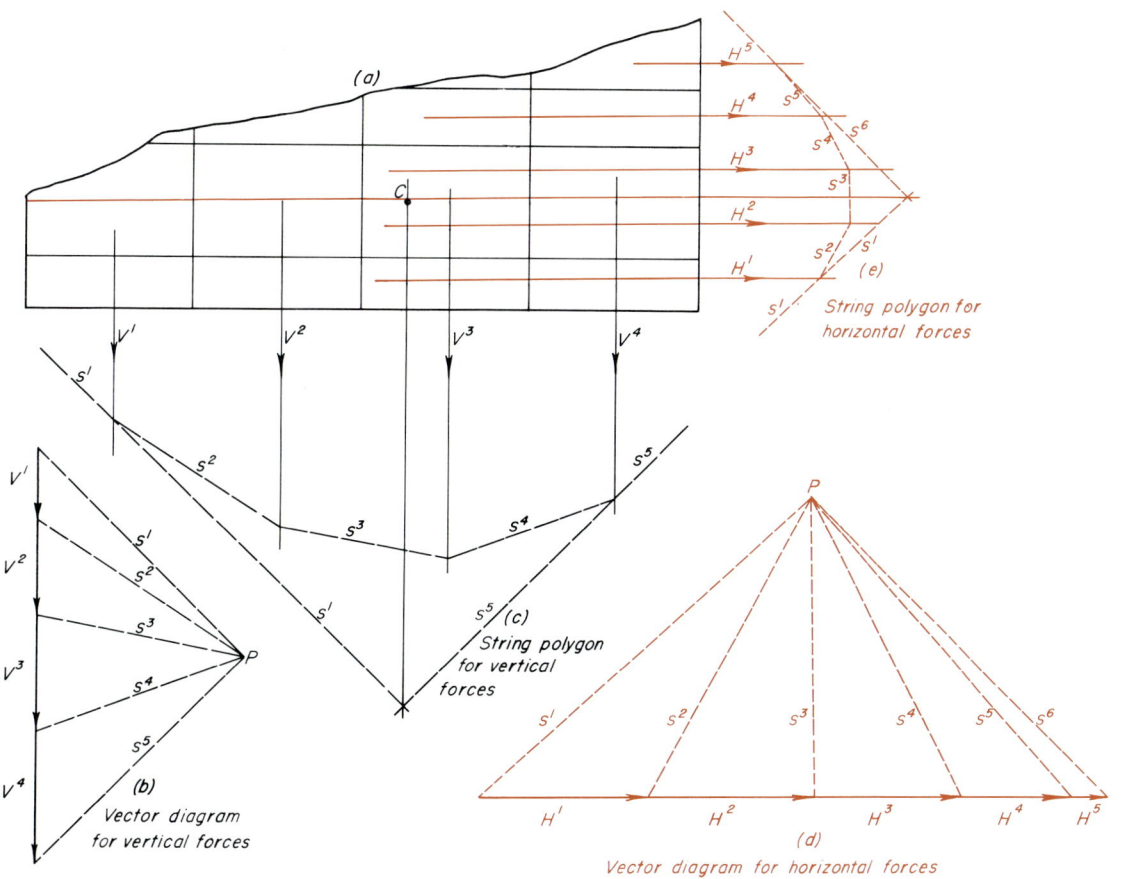

FIG. 17.14. Graphic method for locating a centroid.

an easily used number such as 10 or 100 lb. The components S^1 and S^2 complete the diagram.

b. Draw the string polygon ABC in Fig. 17.15a. Begin at any point A on the line of action of the force. Make the strings S^1 and S^2 parallel, respectively, to components S^1 and S^2.

c. Make the scale of the string polygon to some convenient scale such as 1 in. = 10 ft. Using this scale, lay out the distance X from the force to the point. The base of the polygon, BC, goes through the point O.

d. Scale the distance, m, in feet, using the scale to which the string polygon or force layout was made.

e. Scale the distance, y, in pounds, using the scale to which the vector diagram was made.

f. Multiply m by y to obtain the required moment in foot-pounds.

The theory involved is easily explained by geometry. The two triangles ABC and DEP are similar. Then

$$ED:y::BC:x$$

and

$$ED \cdot x = BC \cdot y$$

Since ED is the value of the force and x is the distance from O to the force, $ED \cdot x$ is the desired moment.

$\therefore BC \cdot y$ is the desired moment, which is also called $m \cdot y$.

It must be remembered that two scales are involved, one for m and a different one for y.

In Fig. 17.16 the moment of three forces about line XX is found to be $m \cdot y$ in foot-pounds. In this figure the vector diagram is drawn to a scale of 1 in. = 1000 lb and the pole distance, y, is scaled as 1400 lb. The string polygon is drawn to the scale of 1 in. = 10 ft and the distance between S^4 and S^1 on line XX, about which the moments are taken, is scaled as 21.2 ft. The moment is therefore $1400 \times 21.2 = 29{,}680$ ft-lb.

17.15 MOMENTS ON A BEAM

One of the most common problems of this kind occurs in finding the moments at various points in a beam. Figure 17.17 shows a beam loaded with a series of concentrated loads. It is desired to find the moment of these forces about the plane XX. First, the vector diagram (Fig. 17.17b) is drawn and the pole selected. The string polygon (Fig. 17.17c) is then drawn and string S^6 is located, from which the values of the reactions can be determined by means of the vector diagram. It was proved in Section 17.14 that the moment is the product of the pole distance and the intercept of the two strings of a force on the plane about which the moment is to be taken. Therefore, the moment of R^1 about XX is R_m^1 in Fig. 17.17c times y. In the same manner the moment of W^1 is W_m^1 times y, of W^2 it is W_m^2 times y, and

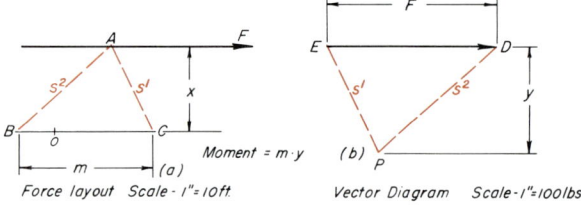
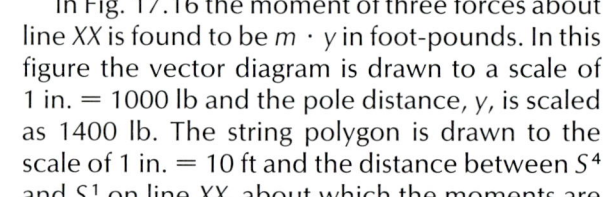

FIG. 17.15. Moment of a force about a point.

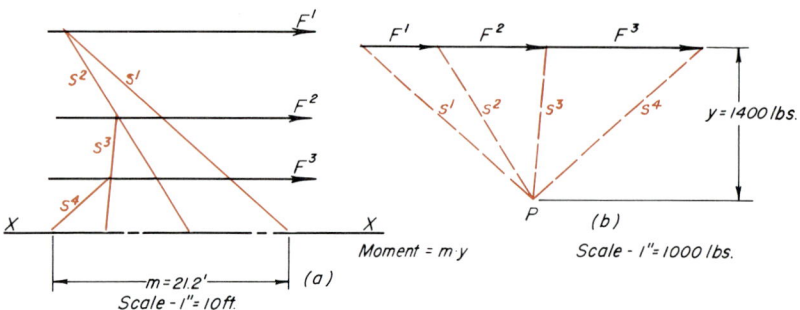

FIG. 17.16. Moment of forces about a line.

W^3 it is W_m^3 times y. However, since the moment of R^1 is clockwise and the other three are counterclockwise, they must be subtracted, leaving the actual moment as m^1 times y. It will be noted that m^1 is the vertical intercept on the string polygon, so that the moment at any point on the beam may be found by taking the vertical intercept on the string polygon at that point and multiplying it by the pole distance. It makes no difference whether S^6 is horizontal or not. Either end of the beam may be used and the result will be the same, since the end used may be considered as a free body with the beam broken on plane XX and acted on by the reaction and the specified loads. The free body is then held in equilibrium by means of the moment found on plane XX.

17.16 MOMENT OF INERTIA

Moment of inertia is defined as the product of the mass or area times the square of the distance from the center of gravity to the point about which moments are to be taken. The equation for the moment of inertia as given in textbooks is $I = \Sigma M(r)^2$, where M is the mass and r is the distance from the center of gravity to the point about which moments are to be taken. If the mass, assumed to be acting in a certain direction, is represented by a vector, F, as in Fig. 17.18a, and the distance r is marked x, it will be necessary to multiply the vector by $(x)^2$, if the moments are taken about point O. In Section 17.14 the method of multiplying the vector by x was explained. In Fig. 17.18 a method is shown whereby the vector can be multiplied by $(x)^2$. The necessary steps are shown in Fig. 17.18.

a. Lay out the vector F, which represents the mass to some even scale, such as 1 in. = 100 lb. See Fig. 17.18b.
b. Select pole distance y as some even distance such as 100 and draw the components S^1 and S^2, forming the triangle EDP.
c. Construct the string polygon ABC in Fig. 17.18a by starting at any point, A, on the line of action of the force F and drawing strings S^1 and S^2 parallel to the components S^1 and S^2 of Fig. 17.18b. Use some even scale such as 1 in. = 10 ft. Line BC goes through point O.
d. The moment of F about O will be $m \cdot y$. See Section 17.14.

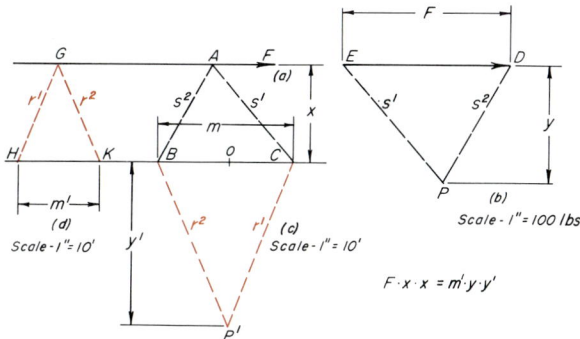

FIG. 17.18. Product of a force times the square of a distance.

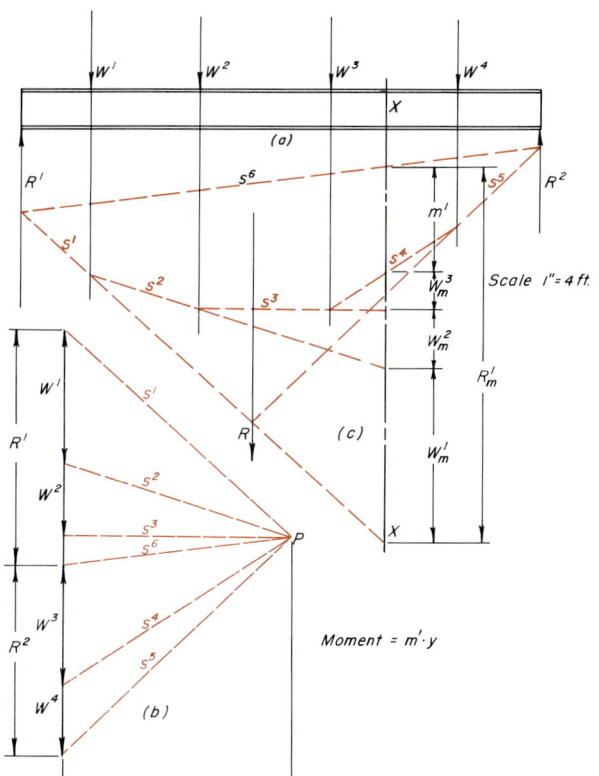

FIG. 17.17. Moments of a beam.

452 GRAPHIC VECTOR ANALYSIS

e. Assume another pole P' at some even distance from BC, using the same scale 1 in. = 10 ft. Figure 17.18c.

f. Draw the components r^1 and r^2 in Fig. 17.18c.

g. Using any point G on the line of action of the force F, construct another string polygon GHK by making the strings r^1 and r^2 in Fig. 17.18d parallel to the components r^1 and r^2 in Fig. 17.18c.

h. Scale HK (m'), using the same scale. 1 in. = 10 ft.

i. Scale y', using the scale to which it was drawn. 1 in. = 10 ft.

j. Scale y, using the scale to which it was drawn. 1 in. = 100 lb.

k. Multiply the terms $m' \cdot y' \cdot y$ to obtain the moment of inertia in terms of pounds (ft)².

The theory can be explained by geometry. GHK and BCP' are similar triangles. Therefore,

$BC : y' :: HK : x$

or

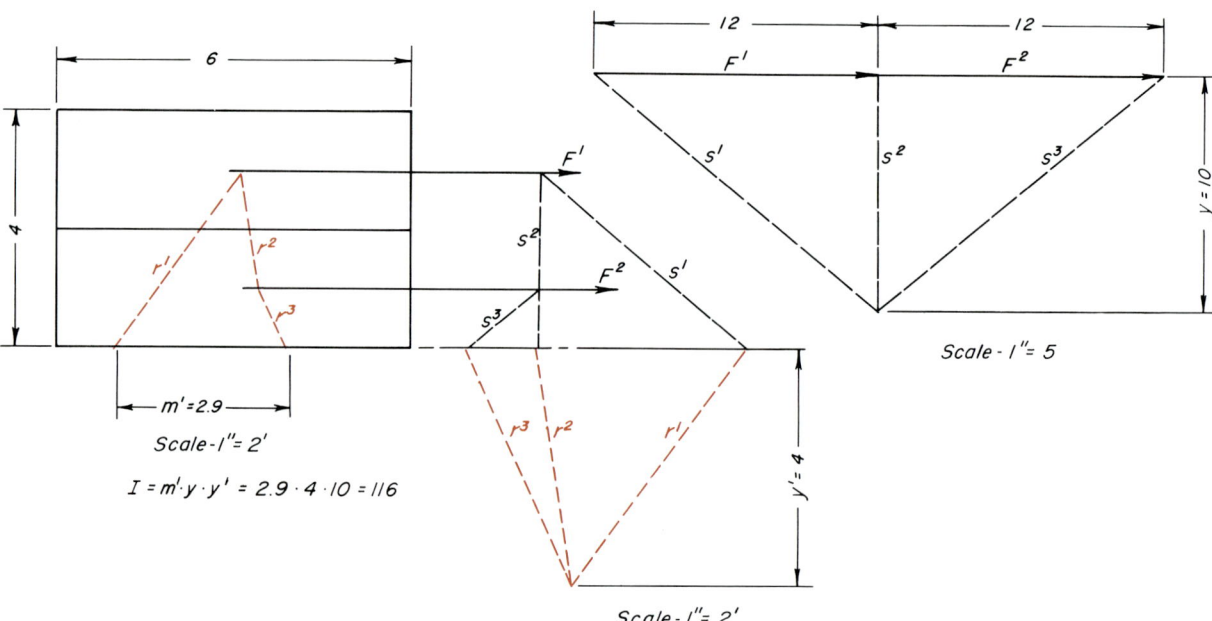

FIG. 17.19. Moments of inertia of a rectangle about one side – two divisions: Culman's method.

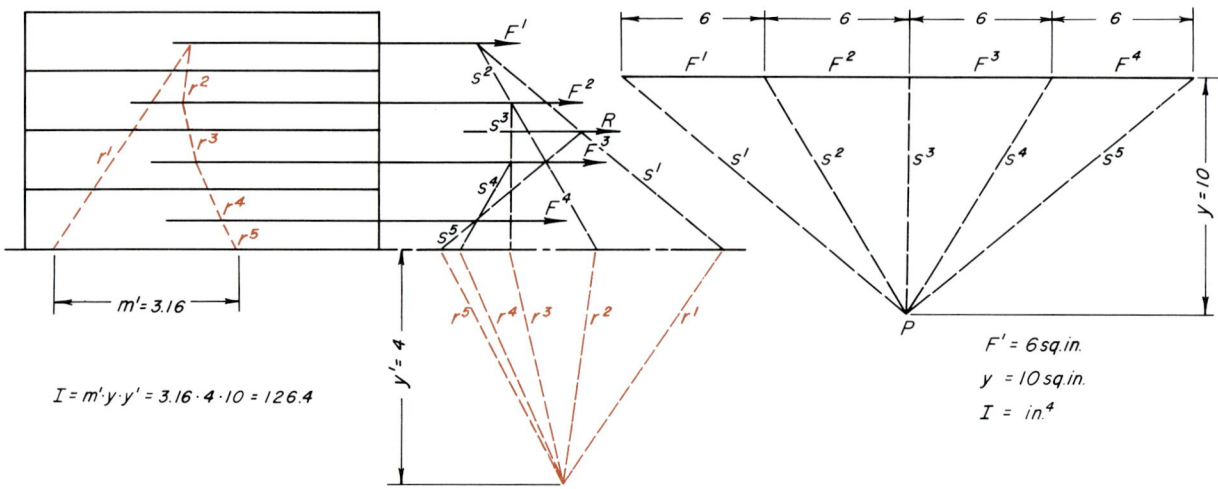

FIG. 17.20. Moment of inertia of a rectangle about one side – four division.

$$m : y' :: m' : x$$

multiplying

$$m \cdot x = m' \cdot y'$$

By definition,

$$I = F(x)^2 = F \cdot x \cdot x$$

By similar triangles EDP and ABC,

$$Fx = my$$

Then

$$I = m \cdot y \cdot x$$

By substitution,

$$I = m' \cdot y' \cdot y$$

For finding the moment of inertia of an area it is necessary to divide that area into rather small parts and to represent each part by means of a vector through the approximate centroid of the area. The method is approximate but becomes more accurate as the number of divisions is increased. To illustrate the effect of more divisions on the accuracy, Figs. 17.19 and 17.20 have been drawn. In each case the moment of inertia of a 4 × 6 rectangle has been found. In Fig. 17.19 the rectangle was divided into two parts and two vectors were used, whereas in Fig. 17.20 four divisions and four vectors were used. It can be seen that two different answers have been obtained, 116 and 126.4, respectively. If an infinite number of divisions could be used, the same result could be obtained as by analytical integration, and the result would be 128. Therefore, for an approximate answer it seems that, in this case, four divisions should be satisfactory. Actually, the graphical method would not be used for such simple areas. It is particularly valuable for irregular areas that are hard to solve analytically.

17.17 RADIUS OF GYRATION

The radius of gyration in Fig. 17.21a is the distance r from an axis about which moments are to be taken to a line R representing a summation of a series of parallel forces that is so located that the moment of inertia of the summation about the given axis is the same as the total moment of inertia of all the individual forces about the same

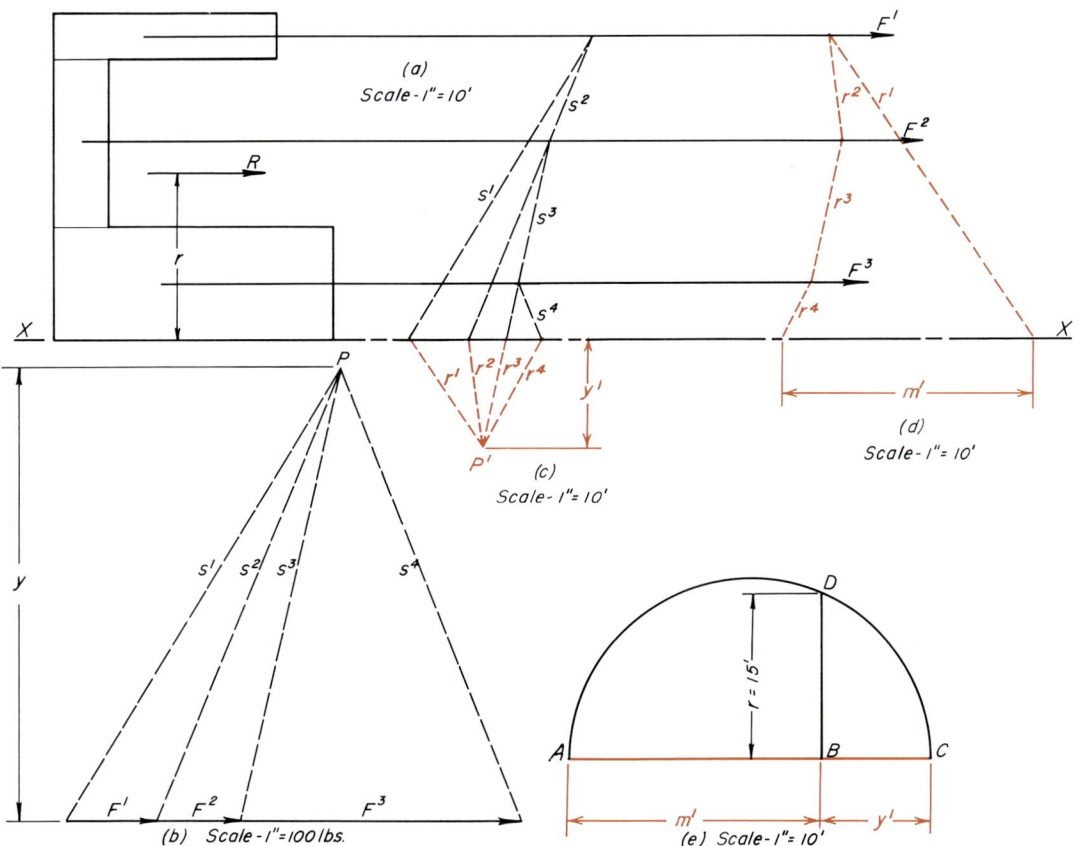

FIG. 17.21. Radius of gyration.

axis. In Fig. 17.21, R is a summation of all the forces ($F^1 + F^2 + F^3$), which acts at a distance r from the plane XX, about which moments are being taken. This distance r is by definition the radius of gyration if the moment of inertia of R about XX is the same as the sum of the moments of inertia of F^1, F^2, and F^3 about axis XX. By definition

$$I = R(r)^2$$

or

$$(r)^2 = \frac{I}{R}$$

By the previous paragraph

$$I = m^1 \cdot y^1 \cdot y$$

Then

$$(r)^2 = \frac{m^1 \cdot y^1 \cdot y}{R}$$

If the distance y is made equal to R, as has been done in Fig. 17.21b, the equation becomes

$$(r)^2 = m^1 \cdot y^1$$

The length of the radius of gyration may then be found graphically, as illustrated in Fig. 17.21e. To do this, the distance AB is made equal to m^1, and the distance BC is made equal to y^1. A semicircle is then drawn with AC as a diameter and with a perpendicular to that diameter erected at B. Then by geometry

$$AB : BD :: BD : BC$$

or

$$(BD)^2 = AB \cdot BC = m^1 \cdot y^1 = (r)^2$$

Therefore, the value of r may be scaled at BD on Fig. 17.21e. The steps needed in this construction are as follows.

a. Construct the vector diagram for the three forces F^1, F^2, and F^3. Use a convenient engineer's scale. See Fig. 17.21b.
b. Select the pole P at a distance y from the force. The distance y is made equal to R, which is the sum of F^1, F^2, and F^3. See Fig. 17.21b.
c. Construct the string polygon in Fig. 17.21c. Begin at any point on F^1.
d. Assume the pole P' at some even distance, y', from the plane XX. See Fig. 17.21c. Use any convenient engineer's scale.
e. Draw the components r^1 to r^4 from the points where the strings S^1 to S^4 cross plane XX. See Fig. 17.21c.
f. Construct the second string polygon shown in Fig. 17.21d.
g. Measure the distance m'. See Fig. 17.21d. Use the same scale as used in Fig. 17.21a.
h. On a horizontal line, lay out the distance m' at AB.
i. On the same line lay out the distance y' at BC.
j. Draw a semicircle with AC as a diameter.
k. Erect the line BD perpendicular to AC.
l. Measure BD. This is the radius of gyration r.

SELF-STUDY QUESTIONS

Before trying to answer these questions, read the chapter carefully. Then, without reference to the text, answer as many questions as possible. For those that cannot be answered, the number in parentheses following the question number gives the article in which the answer can be found. Look it up and write down the answer. Check the answers that you did give to see that they are correct.

17.1 (**17.2**) A vector can be used to represent any quantity that has _____ and _____.

17.2 (**17.2**) The line in space along which the vector acts is called the line of _____.

17.3 (**17.2**) When vector quantities meet in a single point they are called _____.

17.4 (**17.2**) When vector quantities act in a single plane they are said to be _____.

17.5 (**17.2**) A single vector that will produce the same result as a group of vectors is the _____.

17.6 (**17.2**) A single vector that will just balance a group of vectors is called the _____.

17.7 (**17.2**) The process of combining a group of vectors into a smaller number is called _____.

17.8 (**17.2**) The process of replacing a system of vectors by another system having a larger number of vectors is called _____.

17.9 (**17.2**) Any number of concurrent coplanar vectors can be combined into a single resultant: True   False

17.10 (**17.5**) Any number of concurrent noncoplanar vectors can be combined into a single resultant: True   False

17.11 (**17.7**) Equilibrium is the condition that exists in a system of vectors when the _____ is _____.

17.12 (**17.7**) In a system of concurrent coplanar vectors in equilibrium there can be only _____ unknown quantities.

17.13 (**17.7**) In the vector diagram of a system in equilibrium the arrows must point in the _____ _____.

17.14 (**17.7**) In the true-size view of the joint, the member is in _____ when the arrow points away from the joint.

17.15 (**17.8**) In a system of concurrent, noncoplanar forces that is known to be in equilibrium there can be _____ unknown quantities.

17.16 (**17.10**) The string polygon is used in solving a set of _____ _____ vectors.

17.17 (**17.11**) When a vector polygon does not close, the resultant is a _____.

17.18 (**17.11**) When the vector polygon closes but the string polygon does not close, the resultant is a _____.

17.19 (**17.11**) When the string polygon and the vector polygon both close, the system is in _____.

17.20 (**17.16**) Moment of inertia is the product of the _____ times the _____ of the _____ from the center of gravity to the point about which moments are taken.

CHAPTER 18

INTERSECTIONS AND DEVELOPMENTS

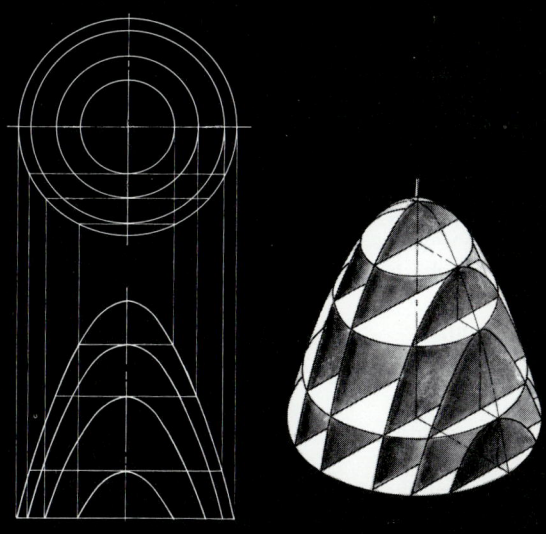

CHAPTER 18

18.1 INTRODUCTION

The problems involved in finding the intersections of surfaces and in developing surfaces of various kinds into flat patterns and templates have many applications in a wide range of industries. In some instances intersections are shown in a conventional way as discussed in Section 18.20, but in other cases it is necessary to find these intersections with accuracy. Boilers, smokestack breeching, ducts, and ventilators also involve problems of this kind. In ship, automotive, and aircraft drafting the problems are numerous and the intersections must be laid out full-size. In aircraft work the layouts on the loft floor are held to a very close tolerance.

In this chapter we discuss not only a variety of practical problems but also the general method of procedure that may be used to solve any problems of this type.

18.2 GEOMETRICAL SURFACES

Most structures involved in engineering practice are bounded by simple geometric surfaces or more complex combinations of them. The engineer should be familiar with these surfaces and the terminology connected with them. A classification of the more common surfaces shown in Fig. 18.1 is given in Table 18.1.

Many of the surfaces in the table are often found in practice to exist as the surface of solid objects. In dealing with them, however, the engineer is concerned only with their properties as surfaces. In this book the prisms and pyramids are treated as solids, and cylinders and cones are considered hollow surfaces without ends.

18.3 FINDING INTERSECTIONS

The intersection of two surfaces is the locus of all points common to both surfaces. The usual process of finding this line consists of passing planes that cut elements from both surfaces and finding the points where these elements intersect.

An element of a surface is any line lying wholly in that surface. It may be either curved or straight. In all ruled surfaces, the straight-line element is preferable since it is easiest to draw. On double-curved surfaces, the circle is the most practical element that may be used. The general method of finding intersections may be stated in the following steps.

a. Pass a cutting plane that cuts elements from both surfaces. These elements should project as straight lines or circles.
b. Find the several projections of each element.
c. Locate the points where these elements intersect. These are points on the line of intersection.
d. Repeat the process with as many cutting planes as necessary to obtain a good intersection.
e. Connect the points in the proper order and with the correct visibility.

TABLE 18.1 CLASSIFICATION OF SURFACES

Ruled surfaces (which can be generated by moving a straight line)	Plane surfaces	Five regular polyhedrons Prisms Pyramids
	Single curved	Cylinders Cones Convolutes
	Warped surfaces	Cylindroids Conoids Hyperbolic paraboloid Hyperboloid of revolution of one sheet Helicoid
Double-curved surfaces generated by revolving a curved line		Sphere Spheroids {oblate, prolate} Hyperboloid of two sheets Paraboloid Torus

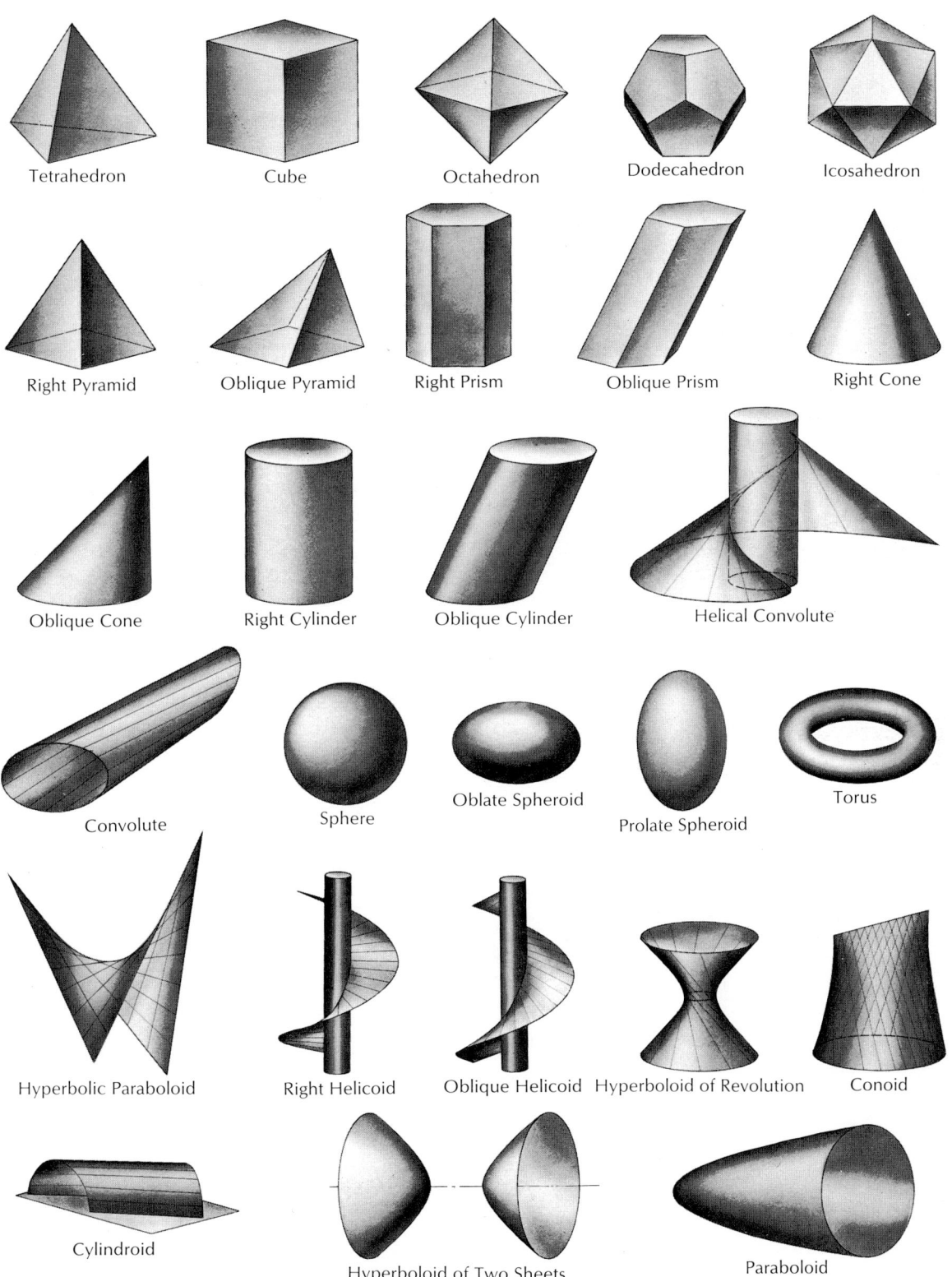

FIG. 18.1. Geometric surfaces.

In order that the construction be as easy as possible, the following rules of procedure should be observed.

f. The first cutting plane should be tangent to one or both surfaces, when possible.
g. The planes should be numbered in order from the first to the last. The last plane should also be tangent to one or both surfaces when possible.
h. The elements should be drawn showing the proper visibility in their own surface only.
i. The elements should be given the same number as the plane. There may be four elements having the same number.
j. The points where these elements cross should be given the same number as the plane and the elements.
k. The visibility of each projection of a point should be indicated. In this text a visible point is marked with a very small solid circle (●) and an invisible point is marked with an open circle (O).
l. When two visible elements cross, the point is visible. Any other combination gives an invisible point.
m. Connect the points in numerical order showing the proper visibility of the line of intersection.
n. The line of intersection can change visibility only where the curve of intersection touches a limiting element or crosses an open base.
o. The visibility of limiting elements must be determined and drawn properly.
p. The visible portion of the line of intersection must be a continuous line from one surface to the other unless it is interrupted by an open base.

18.4 INTERSECTION OF TWO PLANES

For a discussion of this topic, see Article 7.23.13 in Chapter 7.

18.5 INTERSECTION OF PLANE AND PRISM

There are two convenient methods for finding the line of intersection between a plane and a prism. The first is the edgewise view method and the second is the cutting plane method. The particular conditions of the problem will determine which method is the best.

18.5.1 The Edgewise View Method for Plane and Prism.
The edgewise view method is illustrated in Fig. 18.2. *The general analysis for*

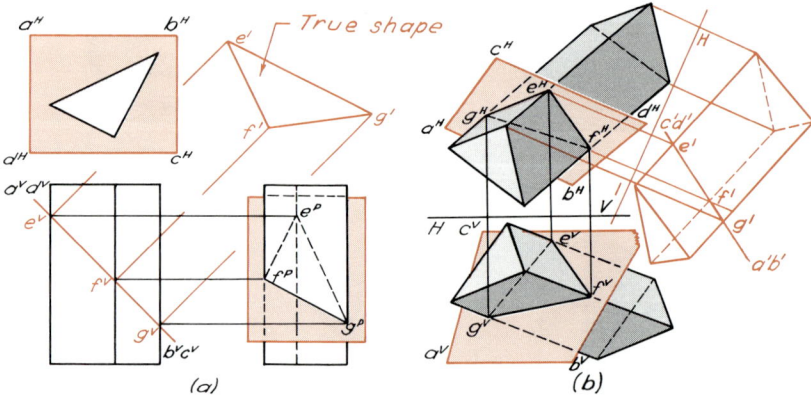

FIG. 18.2. Intersection of plane and prism: edgewise view method.

applying this method is to find the edge view of the given plane and in that view locate the piercing points of the edges of the prism and the plane. These are then projected to the various views to determine the line of intersection.

In Fig. 18.2a, the plane ABCD projects as an edge view in the vertical projection. The edges of the triangular prism pierce the plane at e^V, f^V, and g^V as can be determined by inspection. These points can be projected to the side view on the corresponding edges of the prism determining the profile projection of the line of intersection as $e^P f^P g^P$. The true shape of the intersection can be found by an auxiliary projection off the front view, as shown in $e^1 f^1 g^1$.

When the given views of the plane and prism do not show edgewise, as in Fig. 18.2b, an auxiliary projection is necessary to obtain an edgewise view of the plane. In this case the auxiliary plane is passed perpendicular to line AB since AB is parallel to the H-plane and therefore projects true length in the H-plane. In the auxiliary view the given plane ABCD appears as an edge view. The piercing points of the edges of the prism with plane ABCD are determined by inspection to be $e^1 f^1 g^1$. These points E, F, and G are projected back to the corresponding edges of the prism in the H- and V-views as shown in Fig. 18.2b. To determine the true shape of the intersection EFG, a second auxiliary plane parallel to $e^1 f^1 g^1$ would have to be used and EFG projected upon it.

18.5.2 The Cutting Plane Method for Plane and Prism. The cutting plane method is illustrated in Fig. 18.3 and follows the general procedure outlined in Section 18.3. The steps are as follows.

a. Pass a cutting plane through an element of the prism as in Fig. 18.3a. In this case a horizontal projecting plane is used and is called *cutting plane 1*.

b. Determine the line of intersection of cutting plane 1 with the given plane ABCD in the horizontal view as $e^H f^H$.

c. Project EF to the vertical view, $e^V f^V$, to determine point O, the piercing point of element 1 with plane ABCD.

d. Project point O to the horizontal view.

e. Repeat the above process by passing cutting planes 2 and 3 as shown in Fig. 18.3b to determine points M and N.

f. Points OMN determine the line of intersection between plane ABCD and the prism, which is shown with its proper visibility in Fig. 18.3c.

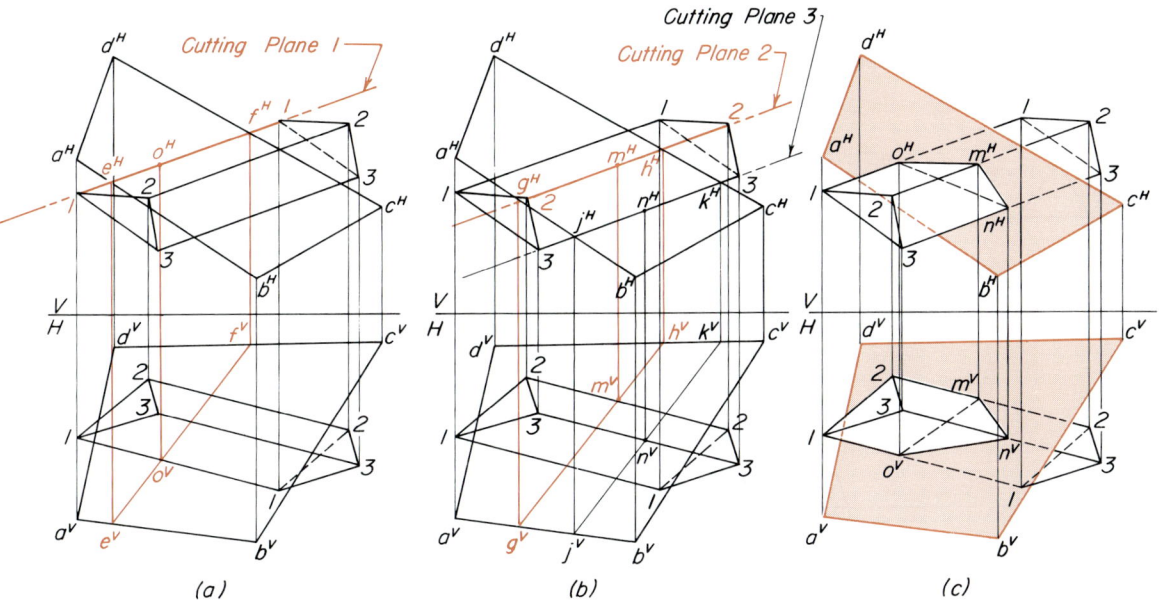

FIG. 18.3 Intersection of plane and prism: cutting plane method.

18.6 INTERSECTION OF A PLANE AND A PYRAMID

The same methods of procedure as discussed in the preceding article for prisms may be used for pyramids.

18.6.1 The Edgewise View Method for Plane and Pyramid.
The edgewise view method for finding the line of intersection between a plane and a pyramid is illustrated in Fig. 18.4. In Fig. 18.4a the plane shows edgewise in the front view. The piercing points between the edges of the pyramid and the plane are determined by inspection in front view. They are then projected to the respective views on the corresponding elements of the pyramid to determine the line of intersection in the other views.

When the plane does not appear edgewise in the given views, such a view must be obtained. In Fig. 18.4b, for example, a horizontal line can always be drawn lying in the plane by making the vertical projection $b^V d^V$ horizontal. From this the horizontal projection $b^H d^H$ is then located and the auxiliary plane set up perpendicular to BD by drawing the H-1 reference line perpendicular to $b^H d^H$. The plane then shows edgewise at $a^1 b^1 c^1 d^1$ and the piercing points may be found by inspection and projected back to the other view as shown in Fig. 18.4b.

18.6.2 The Cutting Plane Method for Plane and Pyramid.
This method follows the procedure outlined in Section 18.3 and is illustrated in Fig. 18.5. The steps are as follows.

a. Pass a cutting plane through an element of the pyramid as in Fig. 18.5a. In this case a vertical projecting plane is used and is called *cutting plane 1*.
b. Find the line of intersection of cutting plane 1 with the given plane PMNO in the vertical view as $e^V f^V$, as shown in Fig. 18.5a.
c. Project EF to the horizontal view, $e^H f^H$ to determine point 1, the piercing point of element 1 with plane PMNO.
d. Project point 1 to the vertical view.
e. Finding the piercing points of each of the other two limiting elements of the pyramid with the given plane LMNO as shown in Fig. 18.5b by using cutting planes 2 and 3.
f. Connect these piercing points to determine the line of intersection between the pyramid and plane and show proper visibility as in Fig. 18.5c.

18.7 INTERSECTION OF A PLANE AND CYLINDER

To find the line of intersection of a plane and cylinder requires the same procedure as followed in the example above. The main difference is that a cylinder has no edges and therefore a series of elements must be chosen lying in the surface of the cylinder. The line of intersection is determined by finding where these elements pierce the given plane. As a rule, 12 equally spaced elements will suffice as is shown in Fig. 18.6a. Note that if the upper part of the cylinder

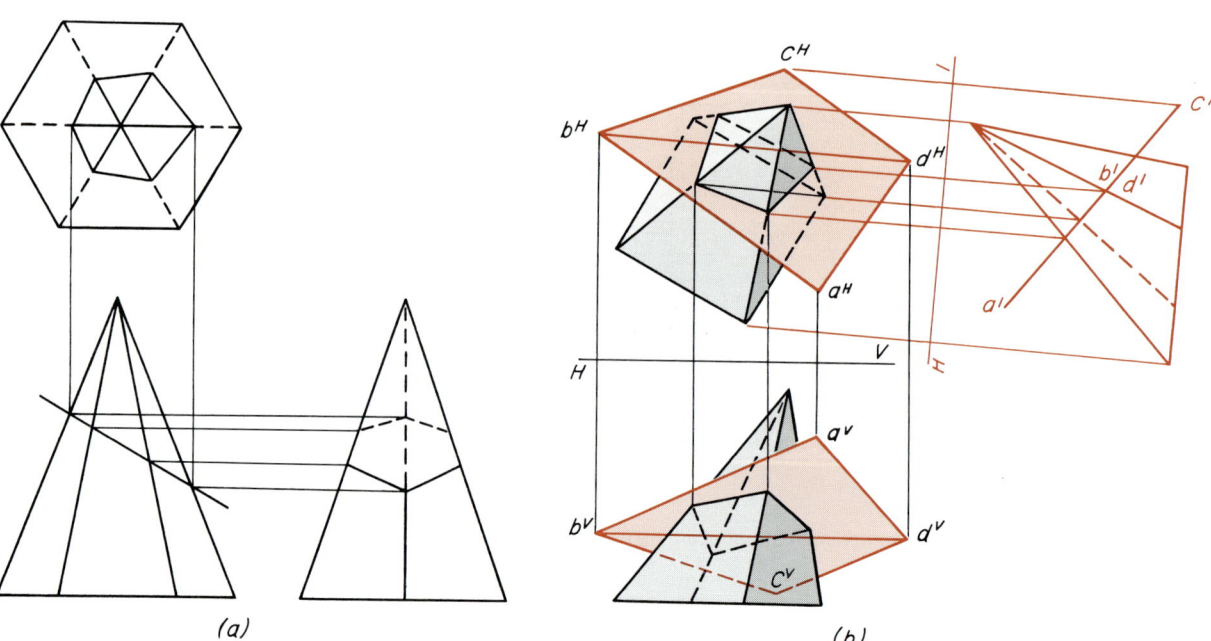

FIG. 18.4 Intersection of plane and pyramid: edgewise view method.

is detached and rotated 180°, a right-angled elbow is formed as shown.

18.7.1 Edgewise View Method for a Plane and Cylinder.
In Fig. 18.6a, the edge view of the plane appears in the front view. The 12 equally spaced elements are assumed in the top view and projected into the front and side views. The piercing points of these elements with the plane are determined by inspection in the front view. These are projected to the side view to determine the right-side view.

When the plane does not appear as an edge view, the first step is to pass an auxiliary plane to obtain the edge view. In Fig. 18.6b this has been done by passing the auxiliary view perpendicular to $a^H b^H$ since AB is parallel to the horizontal plane and is in true length in the top view. In the auxiliary view the piercing points of the elements with plane $ABCD$ can be determined by inspection. These are then projected back to the respective view on the corresponding elements to determine the line of intersection in the top and front views.

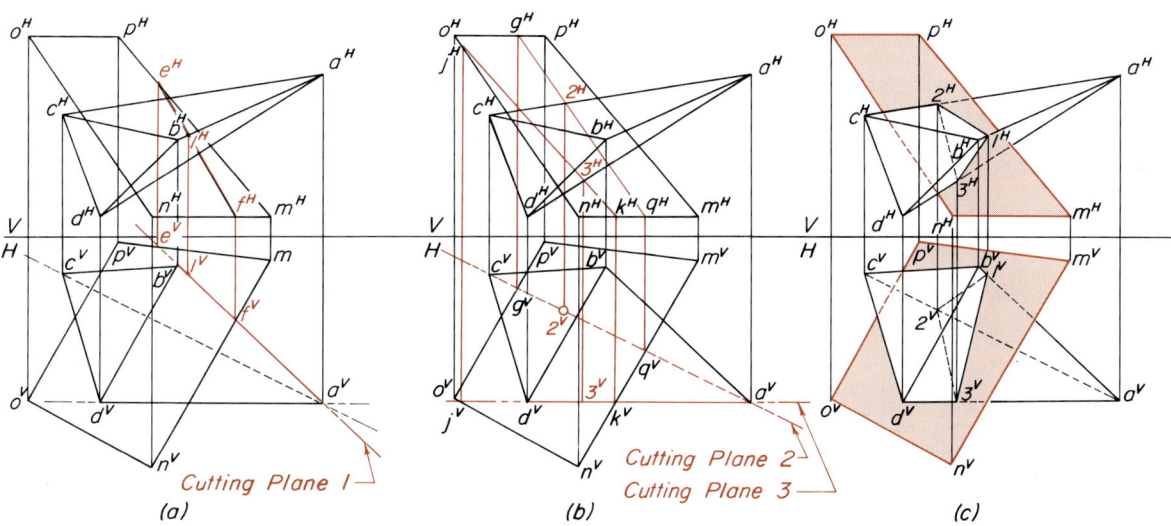

FIG. 18.5 Intersection of plane and pyramid: cutting plane method.

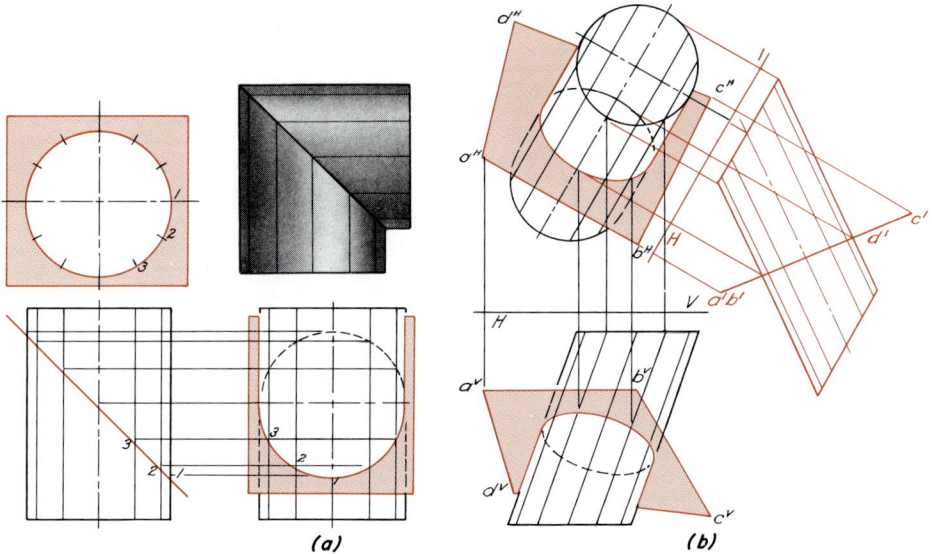

FIG. 18.6 Intersection of plane and cylinder: edgewise view method.

18.7.2 Cutting Plane Method for a Plane and Cylinder.

The method of obtaining the intersection of a cylinder and a plane by the cutting plane method is illustrated in Fig. 18.7. The cutting planes used in this case are horizontal projecting planes and are passed at regular or irregular intervals. They are passed closer together where the degree of curvature is changing rapidly and farther apart where the curve is flat. The procedure is as follows.

a. Pass the first projecting plane tangent to the cylinder. In Fig. 18.7a, plane 1 is a horizontal projecting plane that cuts one element from the cylinder.
b. Get both projections of this element as shown in Fig. 18.7a. Show the proper visibility of the element.
c. Find the line of intersection of the cutting plane with plane *ABCD*, as shown in Fig. 18.7b.
d. Find the point where the line of intersection crosses the element cut by plane 1. See Fig. 18.7b. The elements and point are both numbered 1 to agree with the number of the plane. Mark the projections of visible points with a solid circle and the invisible points with an open circle.
e. Pass successive planes such as planes 2 and 3 as shown in Fig. 18.7c and find points on the intersection in the same manner.
f. Pass other planes at the desired intervals to obtain the complete intersection. The last plane will be the one that is tangent to the other side of the cylinder. Only a part of the curve is shown.
g. Connect the points to form the complete line of intersection. The open circles will be connected with a dashed line and the solid circles with a solid line. The visibility of the curve can change only on the limiting element of the cylinder.

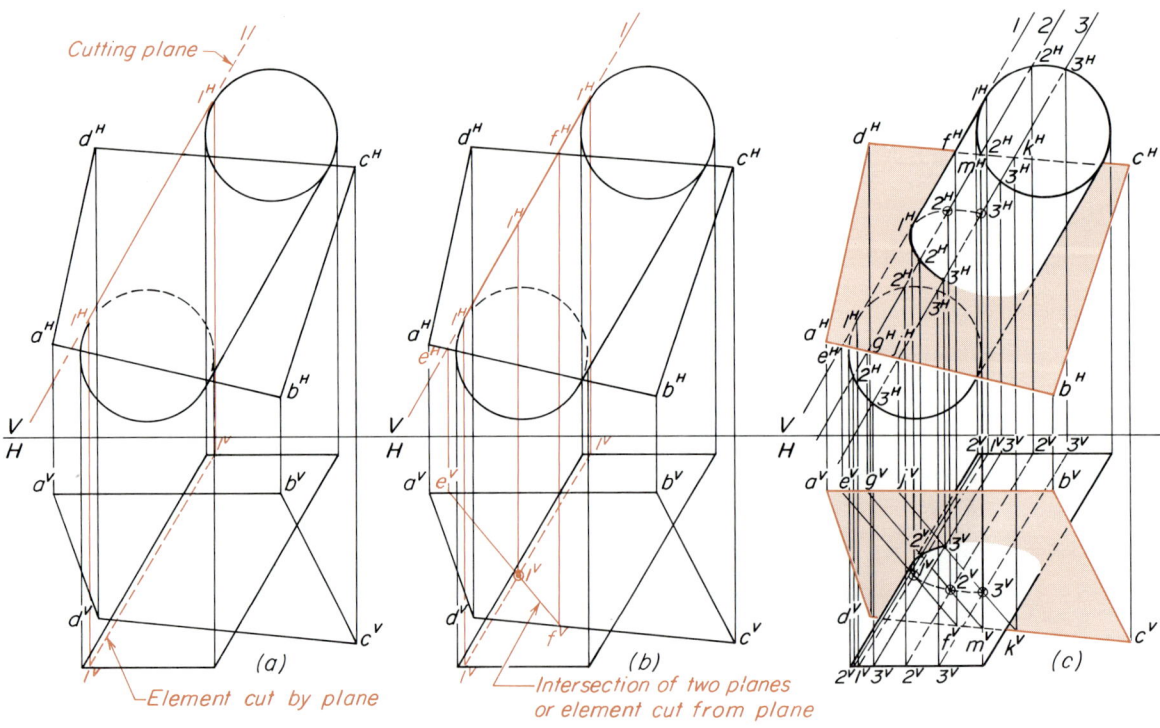

FIG. 18.7 Intersection of plane and cylinder: cutting plane method.

18.8 INTERSECTION OF PLANE AND CONE

The problem of finding the intersection of a plane and a cone is the same as for a plane and a pyramid. Elements are assumed in the cone. These are used to find where they pierce the given plane. These piercing points determine the line of intersection between the cone and plane.

18.8.1 Edge View Method for Plane and Cone.
The edge view method for determining the line of intersection of a plane and a cone is shown in Fig. 18.8. In making an auxiliary projection to obtain an edgewise view of the plane, it will facilitate the solution a great deal if the base of the cone (cylinder, prism, or pyramid) also appears edgewise in the auxiliary view. In Fig. 18.8, two solutions have been presented, the first of which, Fig. 18.8a, shows the auxiliary plane perpendicular to the V-plane, thus making the base of the cone an elliptical figure.

The position of the auxiliary plane was determined by drawing line AD in the plane parallel to the V-plane and making the auxiliary plane perpendicular to this line. Although the solution is correct, a less tedious solution is shown in Fig. 18.8b.

In the second solution, line AD was drawn in the plane parallel to the H-plane and the auxiliary plane placed perpendicular to this line. It can be observed that since the base of the cone is parallel to H, it will also appear as a straight line or edgewise in the auxiliary projection, thus making the solution much simpler.

18.8.2 Cutting Plane Method for Plane and Cone.
The cutting plane method follows the same pattern as given for the plane and cylinder. The method is as follows.

a. Pass the first cutting plane tangent to the cone as shown in Fig. 18.9a. Plane 1 is a vertical projecting plane that cuts one element from the cone.

b. Get both projections of this element as

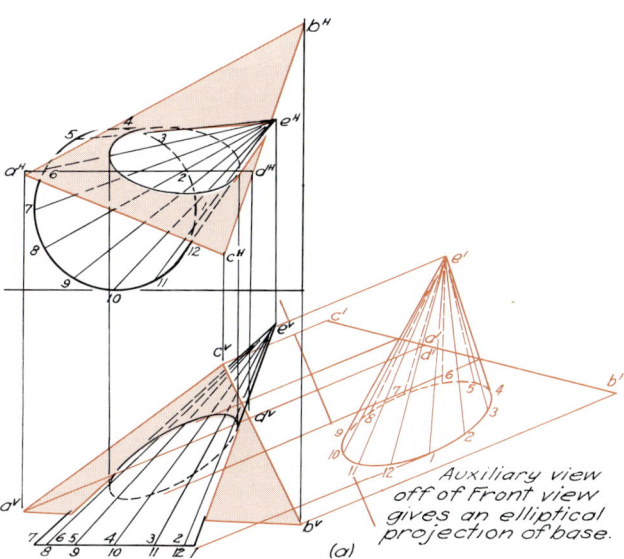
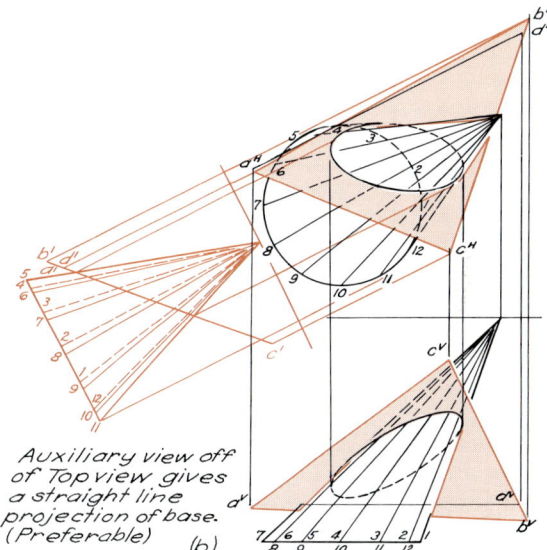

FIG. 18.8 Intersection of plane and cone: edgewise view method.

shown in Fig. 18.9a. Show the proper visibility of the element.

c. Find the line of intersection of the cutting plane with plane ABC, as shown in Fig. 18.9b.

d. Find the point where the line of intersection crosses the element cut by plane 1. See Fig. 18.9b. The element and point are both numbered 1 to agree with the number of the plane. Mark the projection of a visible point with a solid circle and an invisible point with an open circle.

e. Pass successive planes such as cutting planes 2 and 3 as shown in Fig. 18.8c and find points on the line of intersection in the same manner.

f. Pass other planes at the desired intervals to obtain the complete intersection. The last plane will be the one that is tangent to the other side of the cone.

g. Connect the points to form the complete line of intersection. The open circles will be connected with a dashed line and the solid circles with a solid line. The visibility of the curve of intersection can change only on a limiting element or where the curve crosses an open base.

18.9 INTERSECTION OF PLANE AND SURFACE OF REVOLUTION

To find the intersection of the plane ABCD and the sphere in Fig. 18.10a, a cutting plane MN has been passed through both surfaces. This cuts a circle from the sphere that appears as a straight line in the front view and a circle in the top view. The two planes intersect in the line GH. The intersection of the line and circle can be seen by inspection at $e^V f^V$ in the front view and e^H and f^H in the top view. A sufficient number of points to determine the curve of intersection may be obtained by repeating this process as shown in Fig. 18.10b. The curve is tangent to the great circle in

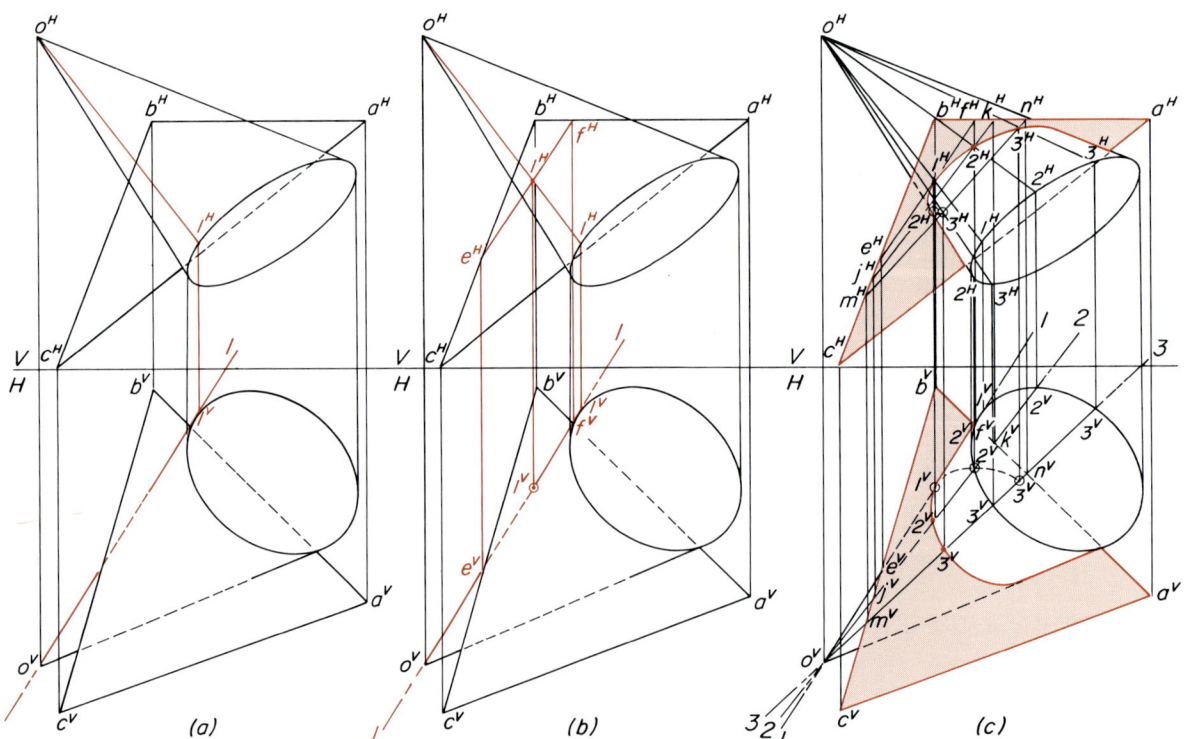

FIG. 18.9 Intersection of plane and cone: cutting plane method.

JOHN WILEY & SONS, INC.

A42860481

ONE WILEY DRIVE, SOMERSET, NEW JERSEY 08873 (201) 469-4400

10/12/84 SENT WITH COMPLIMENTS OF: S HOOPER   062

QUAN	AUTHOR AND SHORT TITLE	ISBN	LOCATION
1	DOBROVOLNY GRAPHICS 2 ED	0471-87124-9	1845A 0599
	POSTAGE	.83	

JOHN WILEY & SONS, INC.

10/12/84 S HOOPER 062

QUAN		ISBN	
1	DUBROVOLNY GRAPHICS 2 ED	0471-87124-5	1845A 0599
	POSTAGE	.83	

A42660481

the top view at points a^H and b^H where the cutting plane crosses the great circle at a^V and b^V in the front view.

The curve of intersection of a plane and sphere is always a circle but in our illustration it appears as an ellipse. In Figs. 18.10c through e, the plane forming the intersection has been shown edgewise in three different positions. Note the effect on the visibility of the ellipse and its tangency to the great circle in the different positions of the intersecting plane. In Fig. 18.10c, the intersection lies entirely above the great circle, which means that it is all in the upper half of the sphere. Therefore, it is entirely visible in the top view. In Fig. 18.10d, the intersecting plane passes through the center of the sphere and the curve is tangent to the great circle at the ends of the corresponding diameter in the top view. The upper half is visible but the lower half is invisible. The student should study these illustrations and clearly visualize each situation for himself.

A similar construction for the intersection of a plane and torus is shown in Fig. 18.11. In this case the true outline of the curve of intersection is not a circle. It may have a wide variety of shapes, depending on the position of the intersecting plane. The method of construction, however, is the same as for the sphere and is clearly shown in the figure. It should be noted that each construction plane, except the top and bottom ones, cuts two circles from the torus. Open circles indicate invisible points, and black circles indicate visible points.

18.10 INTERSECTION OF PLANE AND DOUBLE-CURVED SURFACE

Many practical problems involve finding the intersection of planes with curved surfaces that are considerably more complex, for example, the determination of station lines and waterlines in ship and aircraft construction. The principles involved, however, are exactly the same, the only difference being that in many surfaces straight-line elements cannot be drawn. For example, the

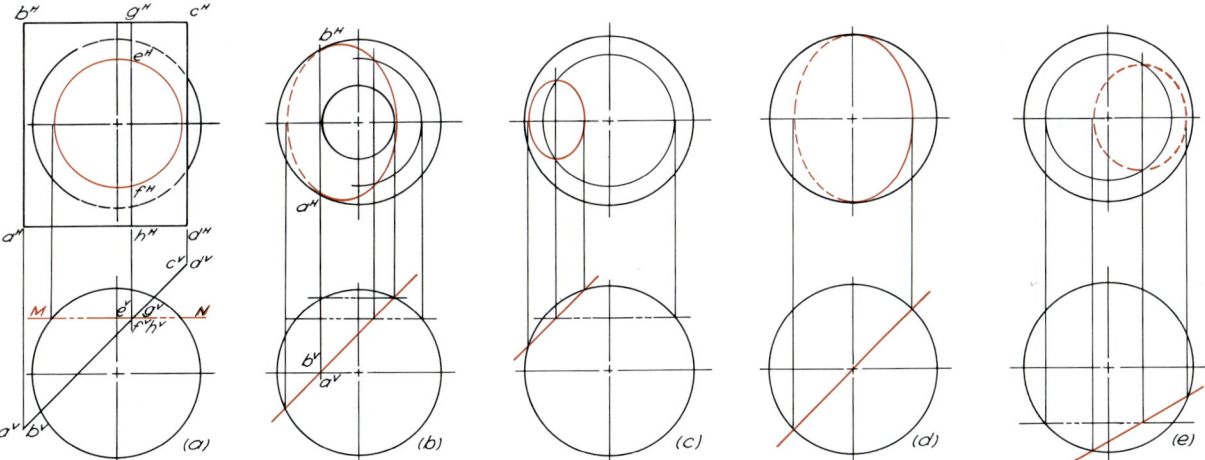

FIG. 18.10. Intersection of plane and sphere.

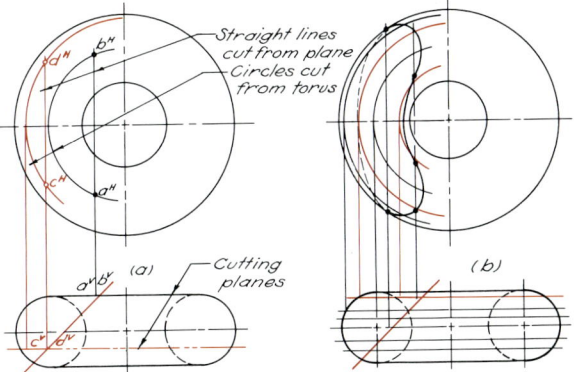

FIG. 18.11. Intersection of plane and torus.

simplest elements that may be drawn on a sphere or torus are circles. Figure 18.12 shows an aircraft surface that is determined by the body plan view and side view. To work with these surfaces, certain terms must be understood. In ship and aircraft drafting, a waterline represents the intersection of a horizontal plane with the hull or fuselage. A buttock line is the intersection of a vertical plane, running lengthwise, with the hull or fuselage, and a station line is the intersection of a transverse vertical plane with the hull or fuselage at right angles to the other two planes. All of them represent the intersection of a curved surface and a plane. These lines are used to define the shape of the surface. See Fig. 18.12.

The intersection of an inclined plane MN, shown edgewise in the side view with the surface, represented in Fig. 18.12, appears in the plan view and the body plan view. The central buttock line that pierces the plane at a^V is projected to a^P and a^H. The intersection of station line 4 with plane MN appears at b^V which is projected at b^P and b^H. Any number of points required to determine the curve can be found in the same manner, using waterlines, buttock lines, and station lines.

18.11 INTERSECTION OF TWO PRISMS

When it becomes necessary to find the intersection of two prisms, certain points can usually be found directly. This occurs when any of the planes appear edgewise in any view. Other points can be found by use of the analysis of Section 18.3. Figure 18.13 shows the intersection of two prisms. The top view shows the four faces of the square prism edgewise. Therefore, the projections of the points in which the three edges of the triangular prism pierce these faces are shown directly in the top view at points 1, 2, and 4 and 6, 8, and 9. The projections 1, 2, 4, 6, 8, and 9 in the front view are located on the corresponding projections of the same edge lines of the triangular prism. (Note that numbered points in this illustration and in succeeding ones are not given superscripts since the connection between the views is obvious.) It remains then to find where the two edges AE and CG of the square prism pierce the faces of the triangular prism. The construction is as follows.

a. Extend the plane of the face AEDH and use it as a cutting plane.
b. Find the intersection, 4UT, where this cutting plane intersects the triangular prism.
c. The points where element AE intersects triangle 4UT give points 3 and 5 on the intersection of the two prisms.
d. Using a cutting plane through BCGF, the same procedure can be used to locate points 7 and 10 on the right half of the intersection.

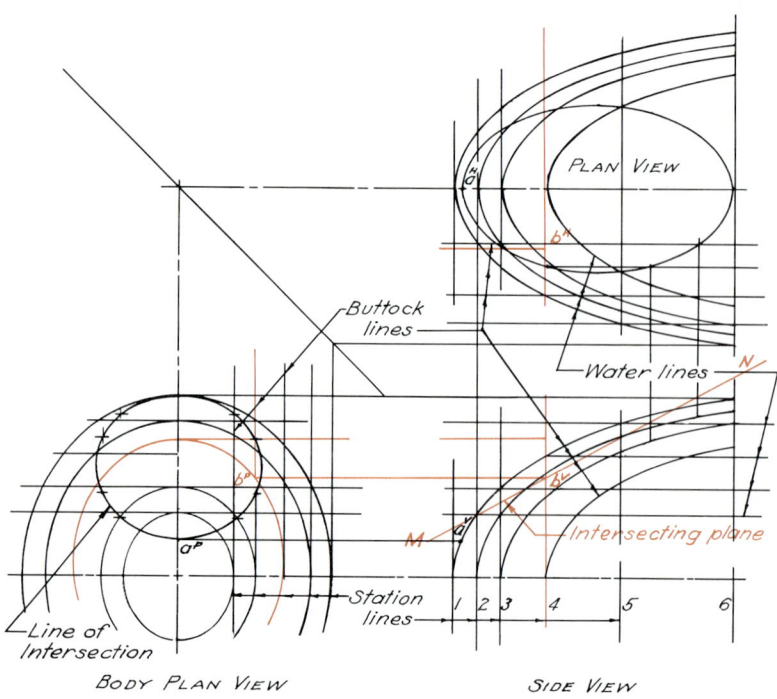

FIG. 18.12 Intersection of plane and double-curved surface.

When one of the prisms is shown endwise, as in Fig. 18.13, it is possible to determine the general shape of the intersection and also to number the points consecutively so that they may be connected in order. In this case it can be seen that the triangular prism goes completely through the square prism, which shows that there will be two completely separate parts to the intersection. Since the square prism shows endwise in the top view, all lines of the intersection must show on the outline of the square in the top view and cannot cut across the square. Thus planes ABEF and ADEH will form one part of the intersection and these planes must intersect the top part of the triangular prism, planes *MNOP* and *OPRS* first and then the bottom part, plane *MNRS*.

To number the points properly, it is best to start with point 1 in the top view and proceed around the square prism and on the top faces of the triangular prism first. Therefore, the numbering would go from 1 to 2, on plane *OPRS*, 2 to 3 and 3 to 4, on plane *MNOP*. From 4 the intersection will go to 5 and from 5 to 1 on plane *MNRS*. When the two points 3 and 5 have been obtained, as previously explained, it is essential that the higher point in the front view be marked 3 because this was taken as the point on the upper surface of the triangular prism.

In the same manner, points 6, 7, 8, 9, and 10 may be numbered on the right side of the intersection. When they have been numbered in this manner, they may be connected in order to give the correct intersection. If the end view of one of the prisms is not given, it is usually best to take enough auxiliary views to obtain an endwise view.

For visibility each line must be checked to see if it is on the front or back of both surfaces. If it is on the front of both surfaces, the line will be visible. All other combinations give an invisible line in the front view.

18.12 RULES FOR VISIBILITY

The visibility of points on each of two lines is determined simply by ascertaining which of the points is closest to the observer. Thus, in Fig. 18.14a, it can be observed from the top view that at the crossing point of the two rods in the front view the point 1 on rod *AB* is in front of point 2 on rod *CD*. Therefore, rod *AB* is visible in the front view and must be shown passing in front of *CD*. In Fig. 18.14b, an examination of the front view shows that rod *AB* is higher than *CD* at the crossing point shown in the top view. In the top view *AB* is entirely visible and must be shown passing over *CD*. The student should verify the visibility of the rods as shown in Figs. 18.14c and d.

It will be noted from the foregoing discussions that the visibility of a line in any view is always

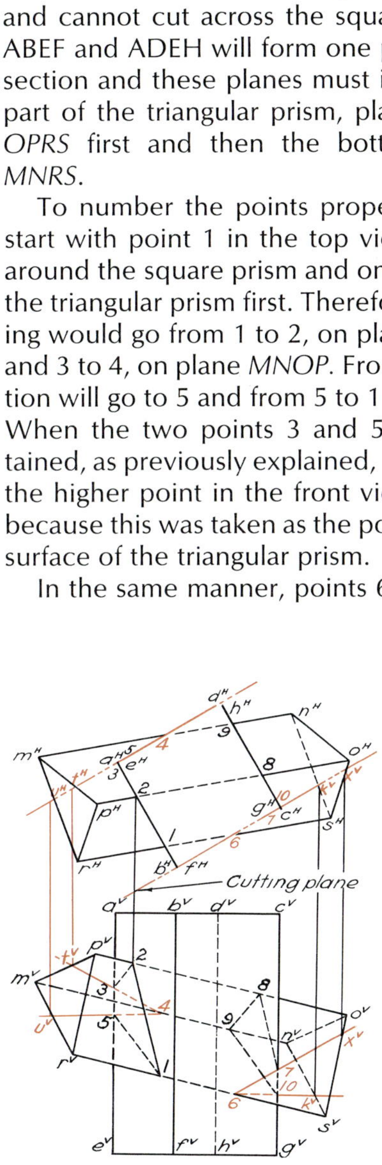

FIG. 18.13. Intersection of two prisms.

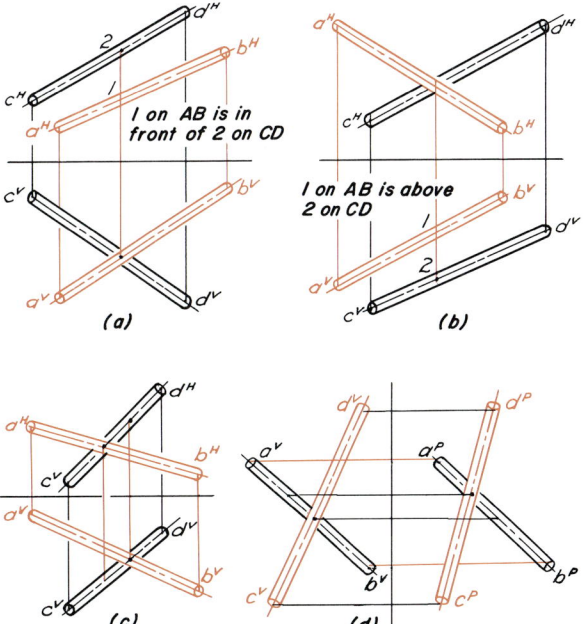

FIG. 18.14. Determining visibility.

determined by reference to the adjacent view. In Fig. 18.13, since line *BF* is in front of line *RS,* the latter enters the prism at point 1 and emerges again at point 6. Between these two points it is invisible. For the same reason line *PO* enters the prism at point 2 and emerges at point 8. Point 8, however, is invisible since it lies on the rear surface of the vertical prism, and line *PO* does not become visible again until it passes the edge *CG* of the vertical prism.

From Fig. 18.13 and subsequent figures, the following principles can be observed.

1. To be visible a point must lie on a visible edge or element of both intersecting surfaces.
2. For a line of intersection to be visible it must connect two visible points. If it connects one visible and one invisible point, it is entirely invisible.
3. A line of intersection can change from visible to invisible or vice versa only upon the outlines of one or the other of the two intersecting surfaces.

18.13 INTERSECTION OF PRISM AND PYRAMID

To find the intersection of a prism and a pyramid, certain points can be found directly if any plane shows edgewise. Other points can be found by using the analysis of Section 18.3. In Fig. 18.15, the cutting plane method is illustrated. The steps are as follows.

a. Pass a cutting plane *AB* through two edges of the pyramid. See Fig. 18.15a.
b. Find the triangle *ABC* that this plane cuts from the prism. See Fig. 18.15a.
c. Find the points 1, 2, 3, and 4 in the front view where this triangle crosses the edges of the pyramid. See Fig. 18.15a. Project these points to the top view. These are points on the line of intersection.
d. Pass another plane *CD* as in Fig. 18.15b.
e. Find the intersection *DEF* where this plane intersects the prism.
f. Find the points 5, 6, 7, and 8 where this triangle crosses the edges of the pyramid. These are points on the line of intersection.
g. Pass plane *EF* through the lower edge of the prism. See Fig. 18.15c.
h. Find the line of intersection, *MNOP,* of this plane with the pyramid. See Fig. 18.15c.
i. Find the points 9 and 10 where *MNOP* crosses the lower line of the prism. See Fig. 18.15c. These are points on the line of intersection.
j. Connect the points in the proper order and with the proper visibility to obtain the complete line of intersection.

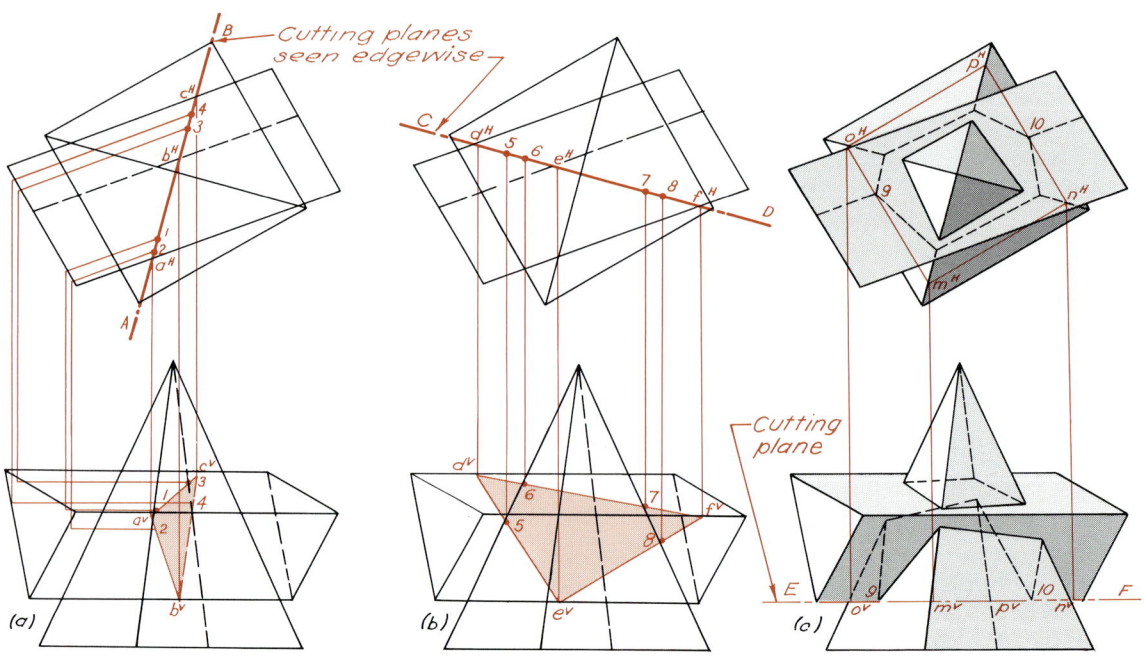

FIG. 18.15. Intersection of prism and pyramid.

The first points 1 to 8 in the problem could have been found quite conveniently by making an endwise view of the prism in an auxiliary projection. The simplest method for the remaining two points, 9 and 10, is by the cutting plane method used in Fig. 18.15c.

Had the prism in Fig. 18.15 been inclined to both H and V, the auxiliary plane method would have required two auxiliary views, whereas the cutting plane method could be applied without additional work.

18.14 INTERSECTION OF TWO PYRAMIDS

The method of cutting planes to obtain elements of both surfaces that intersect each other is further illustrated in determining the intersection of two pyramids as shown in Fig. 18.16. The cutting plane AB passing through the front edge of the horizontal pyramid in Fig. 18.16a cuts the shaded quadrilateral $a^V b^V c^V d^V$ from the other pyramid. The front element of the horizontal pyramid crosses this area at points 1 and 2 of the intersection.

In Fig. 18.16b, plane CD cuts a line from the upright pyramid and the shaded triangle RST from the other, thus locating points 3 and 4 of the intersection. Other points obtained in a similar manner give the final intersection shown in Fig. 18.16c.

18.15 POSSIBLE TYPES OF INTERSECTIONS

It is of considerable value to the drafter if he/she knows before beginning construction what general form the intersection will have. Only four forms are possible, and the type that any problem will give may be easily determined for cylinders, if an endwise view of one of the cylinders is obtained. The four types are shown in Fig. 18.17, with the cylinders placed in the most advantageous positions possible relative to the principal coordinate planes. With a complete penetration of one cylinder by the other, as in Fig. 18.17a, two closed curves are formed. With a partial penetration, as shown in Fig. 18.17b, one continuous closed curve is formed. With a partial penetration in which one cutting plane is tangent to both cylinders, as in Fig. 18.17c, a crossed curve with one point common to both parts like a figure 8 is formed. Finally, with a complete penetration of two cylinders of the same size with two cutting planes tangent to both cylinders, as in Fig. 18.17d, two closed curves are formed that cross each other at two points. In right circular cylinders and cones, these curves are ellipses. The above statements concerning intersecting cylinders apply equally to two cones, to a cone and a cylinder, or to prisms and pyramids, when under the same conditions as regards penetration and tangency of cutting planes. The determination of the form of the intersection can be readily made either from an auxiliary view or from the position

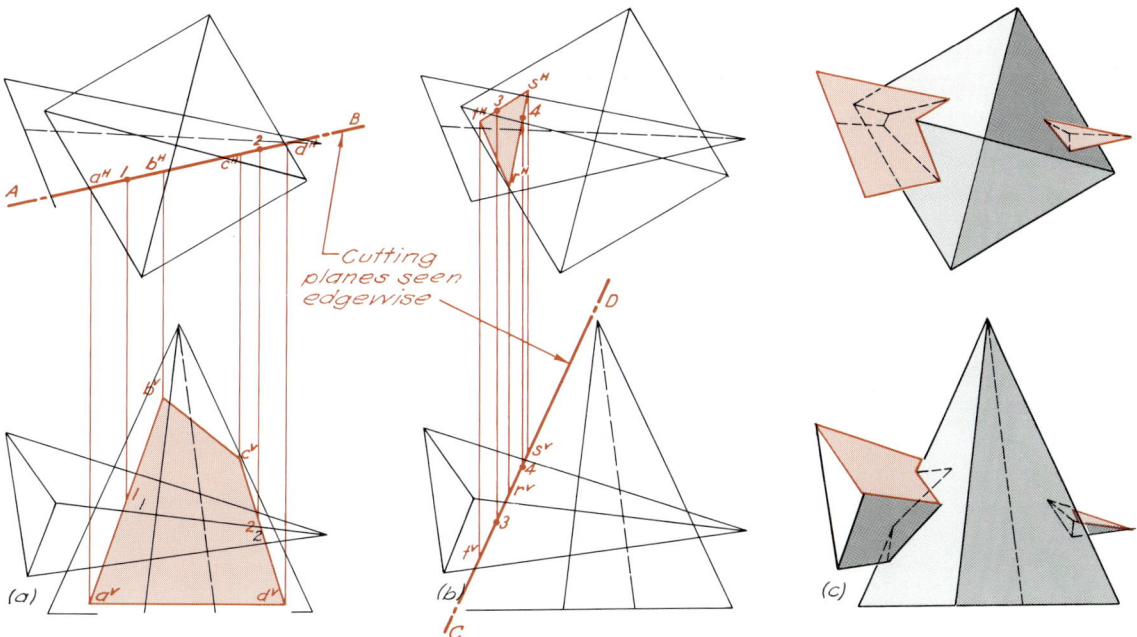

FIG. 18.16. Intersection of two pyramids.

472 INTERSECTIONS AND DEVELOPMENTS

of the limiting cutting planes, as shown in Fig. 18.33, where the first cutting plane, number 1, is tangent to both cylinders and the last is tangent to one and cuts the other, thus giving a crossed loop as in Fig. 18.17c.

18.16 THE INTERSECTION OF TWO CONES

The intersection of two cones is a common practical problem. In this case, if straight-line elements are to be cut from both surfaces, the cutting planes must pass through the vertices of both cones.

Thus in the pictorial drawing of Fig. 18.18, line AD passes through the vertices of B and C. Any plane such as ADE, which contains the line AD, will cut straight lines from both cones if it cuts the cones. Line ED in the plane of the base of cone B intersects the foot of elements 1 and 2 from this cone, and line EA in the plane of the base of cone C intersects the foot of elements 3 and 4 of cone C. The two pairs of elements, all lying in this one cutting plane, cross each other to give four points on the curve of intersection.

From Fig. 18.18 it can be observed that the crux of this problem lies in finding the foot of the elements cut by one plane. The method of procedure will depend on the situation in which the cones occur. In general the steps of the solution are as follows.

1. Draw a line through the vertices of both cones.
2. Find the piercing points of this line with the base of each cone.
3. Pass a cutting plane through this line and find the elements cut from each cone. These elements cross in points on the curve of intersection. A variety of procedures

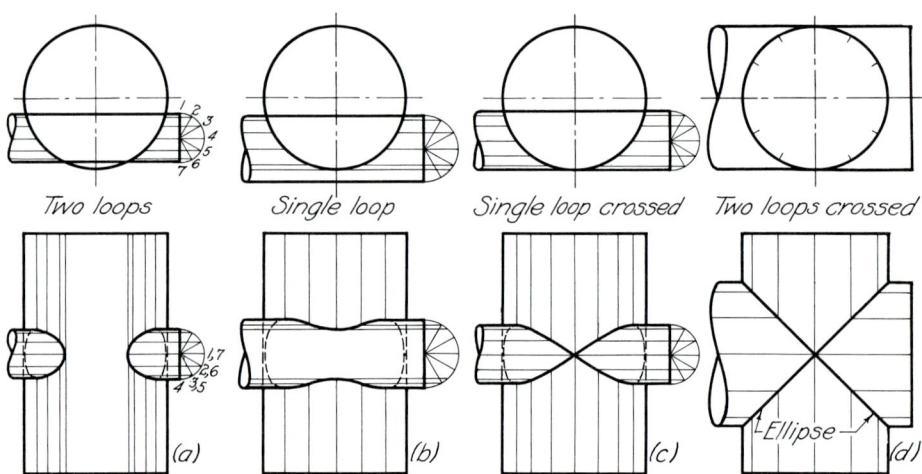

FIG. 18.17. Four types of intersections.

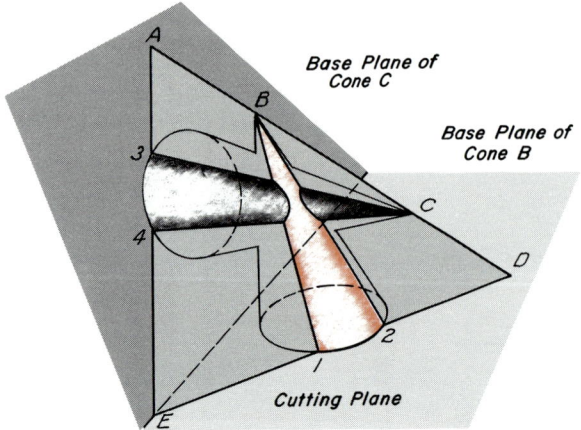

FIG. 18.18. Pictorial drawing of the intersection of two cones.

may be required to accomplish this second step. These are illustrated in the five following examples.

18.16.1 Endwise View of the Line Joining the Apexes. If the line joining the vertices of the cones is an H- or V-parallel, an endwise view of this line can be obtained in a first auxiliary view as shown in Fig. 18.19. The cutting planes then appear edgewise in the auxiliary view, and the foot of the elements cut from the cones can be obtained by inspection. See elements 1 and 4 in this figure.

18.16.2 Both Bases in the Same Plane. When the bases of both cones lie in the same plane, the cutting planes may be located by constructing a line joining the two apexes and finding the point where this line pierces the plane of the bases. In Fig. 18.20, line AB pierces the plane of the bases at point C. When a series of planes is passed through line AC, they will intersect the plane of the bases in a series of lines that meet at point C. From this point C, any number of lines may be drawn on the base plane so that they cut across both bases. Each of these lines when taken with the line AC determines an inclined plane containing both apexes. One such line is shown in Fig. 18.20, which locates points D, E, F, and G on the bases of the cones. From each of these points elements of the cones may be drawn to locate four points on the line of intersection of the two cones. The elements should be drawn showing their proper visibility in their own surface only, without regard to the other cone. Then when two visible elements cross, the point on the intersection will be visible. All other combinations will give an invisible point.

The problems occurring in engineering prac-

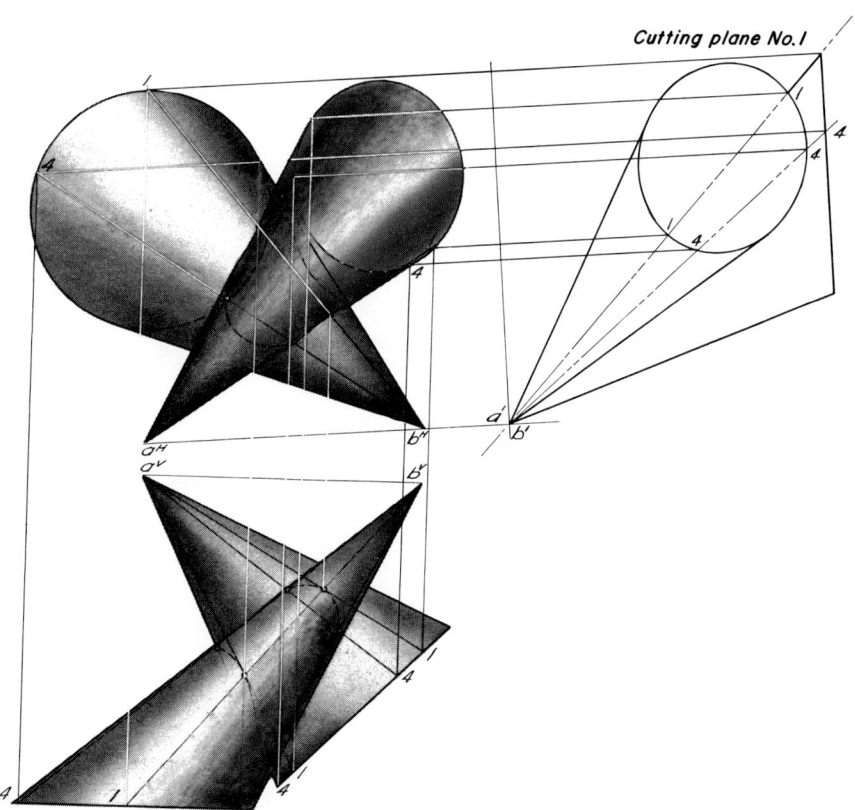

FIG. 18.19. Intersection of two cones: using auxiliary view.

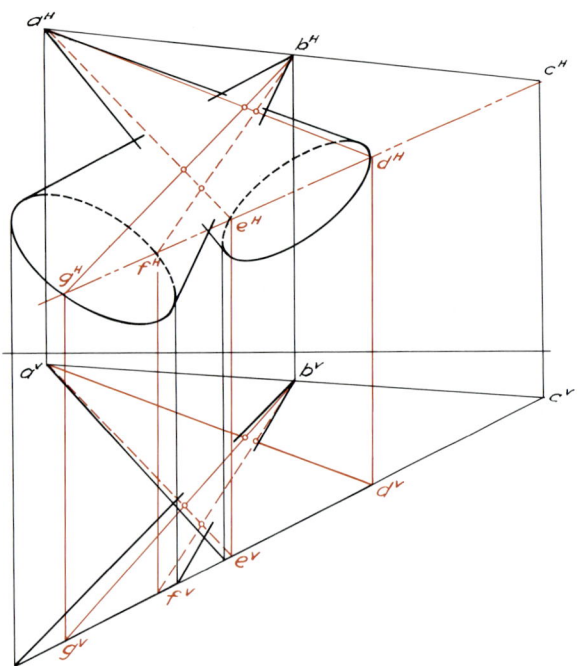

FIG. 18.20. Intersection of two cones: bases in same plane.

tice often involve frustums of cones. In cases of this kind the vertices must be found by extending two or more elements. An illustration of this type of problem is shown in Fig. 18.21. Here the two cones have a common base, which is, of course, one of the lines of intersection. When cones have a common base or common base plane, the solution is much simplified because elements cut out by any cutting plane are very easily found. The procedure is as follows. Extend the line joining the two vertices until it pierces the plane of the bases as at *P*. From this point draw a line across the bases. The intersections of this line with the curve of the bases locate the foot of all elements lying in the cutting plane determined by the two intersecting lines, that is, the one through the vertices and the one drawn across the bases. The elements cut by one such plane intersect in points on the curve of intersection. The construction for one cutting plane is shown in the figure. Line *XZ* pierces the plane of the bases at point *P*. *PB* is the line across the bases. Elements *XB* and *ZA* are cut from the cones. They intersect at point 3 on the surfaces of the cones. The other two elements that intersect at *A* and *B* on the common base have not been shown. Other points on

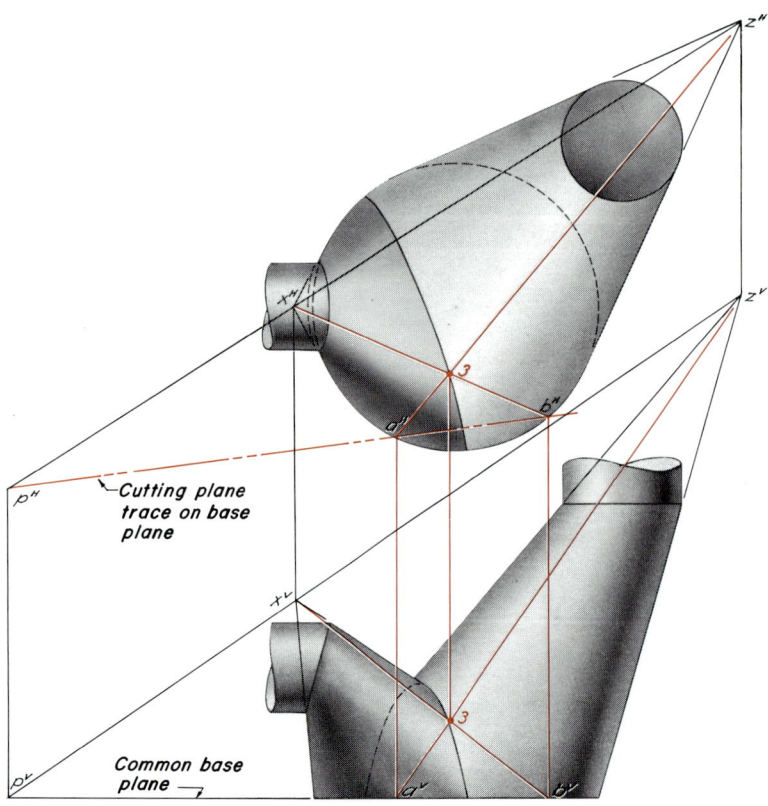

FIG. 18.21. Intersection of two cones with common bases.

the curve of intersection were obtained in a similar way.

18.16.3 Bases Edgewise in the Same View.
In Fig. 18.22, two cones are illustrated with bases that appear edgewise in the top view. To determine the elements that lie in a cutting plane, proceed as follows.

a. Draw a line AB connecting the vertices of the cones and extend it until it pierces the planes of both bases at C and D.

b. Find the line of intersection EF of the planes of the bases. It appears endwise at $e^H f^H$ in the top view and as a vertical line $e^V f^V$ in the front view.

c. Draw a line from d^V across the base of the cone up to the line of intersection at g^V and from this point across the other base to point c^V. These lines determine the elements 3 on both cones and the points on the curve of intersection, as shown in the figure. The entire curve has not been shown. As many planes as desired can be found in the same manner. A line is drawn through d^V intersecting the base of cone B to the point where it crosses $f^V g^V$. From this point another line is drawn to c^V crossing the base of the cone A. Through the points on the base, elements can be drawn and points on the line of intersection determined.

18.16.4 Bases Edgewise in Different Views.
In Fig. 18.23, one cone has its base edgewise in the top view and the other cone has its base edgewise in the front view. The procedure is exactly the same as in the preceding problem. It should be noted that the projections of the line of intersection of the base planes lie in the edgewise views of the bases since one is an H-projecting plane and the other a V-projecting plane. Line EF is the line of intersection of the base planes. Any line such as DE may be drawn in the plane of base 1 until it crosses the line of intersection at point E. Line EC can then be drawn in the plane of the other base, from e^H to c^H. If this plane is numbered 4, the elements 4 can be drawn in each cone through the points on the base of the cones where these lines DE and EC cross the respective bases. These elements locate four points on the line of intersection. Any other similar lines such as DF and FC can be drawn to locate other points which are marked 1.

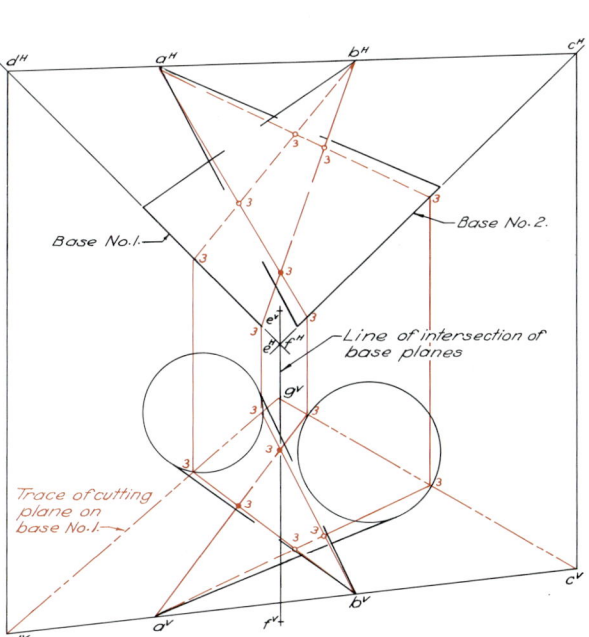

FIG. 18.22. Intersection of two cones: bases edgewise in same view.

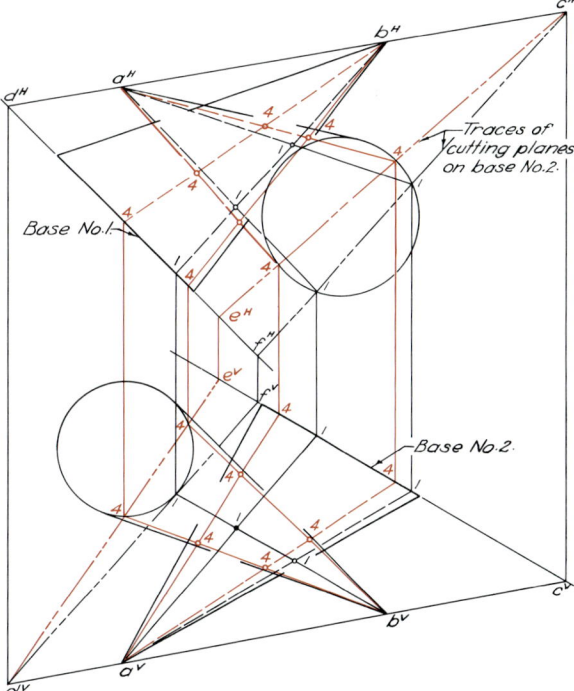

FIG. 18.23. Intersection of two cones: bases edgewise in different views.

18.16.5 Only One Base Shown Edgewise.
When the base of only one cone shows edgewise, the elements determined by a single cutting plane may be found as illustrated in Fig. 18.24. The procedure is as follows.

a. Extend the line *AB* joining the vertices of the two cones until it pierces the plane of the base of the cone that shows edgewise as at *D*.
b. Choose an element in the other cone as *AC* in Fig. 18.24 and extend it until it likewise pierces the plane of the edgewise base at *E*.
c. Draw the line *DE* that crosses the base of the cone at *F* and *G* in the front view. The cutting plane through the vertices is thus determined by lines *DE* and *AE*.
d. Draw the elements *BF* and *BG* that cross element *AC* at two points 1 on the curve of intersection. The curve is not shown in the figure.
e. Repeat the process with a succession of elements in cone *A*.

18.16.6 Neither Base Edgewise.
If neither base appears edgewise in any view, an auxiliary plane can be set up perpendicular to one of the bases and both cones projected to this auxiliary plane. When this is done, the procedure of Article 18.16.5 can be followed. Another method of solution would be to extend both cones to some common base plane and then proceed as in Fig. 18.20.

18.17 INTERSECTIONS OF CYLINDERS AND CONES

To cut straight lines from both a cone and a cylinder at the same time requires (1) *that the cutting plane pass through the vertex of the cone* and (2) *that it be parallel to the elements of the cylinder*.

18.17.1 Endwise View of Cylinder Method.
The line of intersection of a cone and cylinder can be obtained with relative ease if the end view of the cylinder can be obtained. Passing the required cutting planes can be accomplished by inspection if an endwise view of the cylinder can be obtained as in Fig. 18.25. Here the points obtained with one typical cutting plane have been shown.

18.17.2 Right Circular Cone with Cylinder Parallel to Base of the Cone.
When the cone is right circular and the cylinder is parallel to the base of the cone, as in Fig. 18.26, the cutting planes can be chosen so that they cut straight lines from the cylinder and circles from the cone that intersect in pairs to locate points on the curve of intersection.

18.17.3 Oblique Cutting Plane Method.
When the cone and cylinder are so situated that neither of the schemes used above can be readily applied, the cutting planes must be passed through a line drawn from the vertex of the cone and parallel to the elements of the cylinder. Any plane through this line will cut straight lines from both cone and cylinder if it cuts them at all. This is

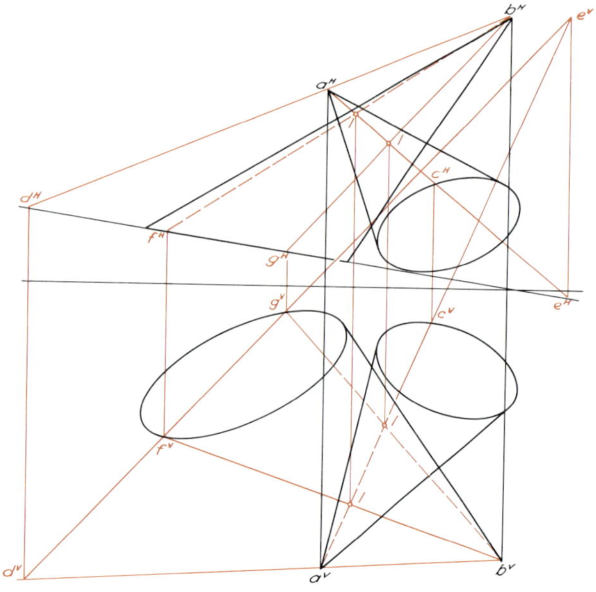

FIG. 18.24. Intersection of two cones: one base edgewise.

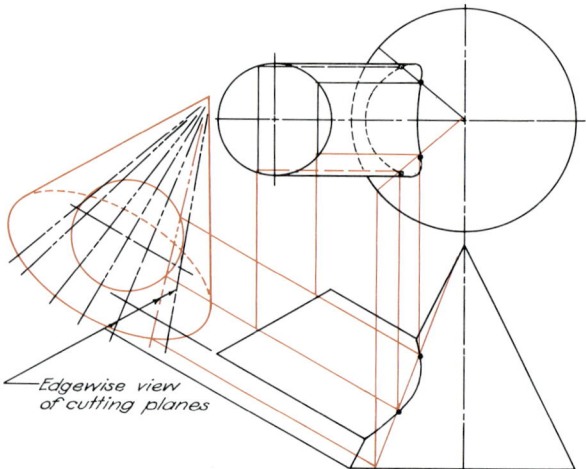

FIG. 18.25. Intersection of cone and cylinder: end view of cylinder.

illustrated pictorially in Fig. 18.27. To understand the theory and construction of this problem, it is necessary to follow both Figs. 18.27 and 18.28 at the same time. The steps required for the solution are as follows.

a. Construct a line through the apex of the cone parallel to the elements of the cylinder. In Fig. 18.27 this line is lettered *CAB*. In Fig. 18.28 the line is lettered *ACB*.

b. Find the point where this line pierces the plane of the base of the cylinder. In Fig. 18.27 this point is lettered *C*. In Fig. 18.28 the point is also labeled *C*, but in this figure it is located by its projections c^V and c^H.

c. Find the point where the line pierces the base of the cone. In Fig. 18.27 this point is labeled *B*. In Fig. 18.28 the point is also labeled *B*, but it is located by means of the two projections b^H and b^V.

d. Find the line of intersection of the plane of the base of the cone with the plane of the base of the cylinder. In Fig. 18.27 it can be seen that this line of intersection is the reference line. In Fig. 18.28 this line of intersection has its vertical projection in the reference line and also its horizontal projection in the reference line. Any two points on the reference line such as *E* and *F* will locate the line.

e. In Fig. 18.27, construct a line from *C* crossing the base of the cylinder and intersecting the line of intersection at point *D*. In Fig. 18.28 the horizontal projection of a line through *C*, intersecting line *EF* at *D*, is constructed at $c^H d^H$. This line crosses the base of the cylinder at two points marked 4. The vertical projection of this line is $c^V d^V$ in the reference line.

f. In Fig. 18.27 another line can be drawn through *D* to *B* crossing the base of the cone. In Fig. 18.28 a corresponding line can be drawn by constructing the vertical projection through b^V and d^V. The line in this

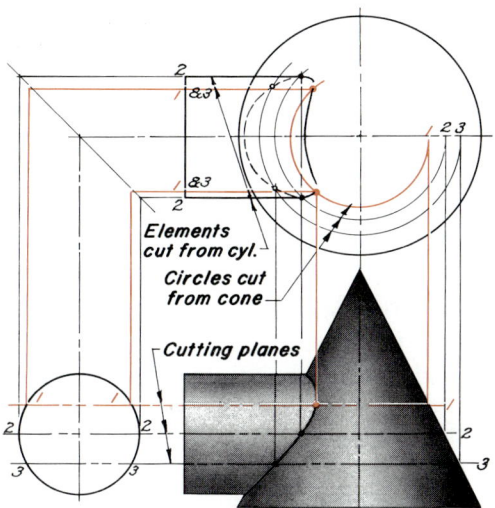

FIG. 18.26. Intersection of cone and cylinder: cylinder parallel to base of cone.

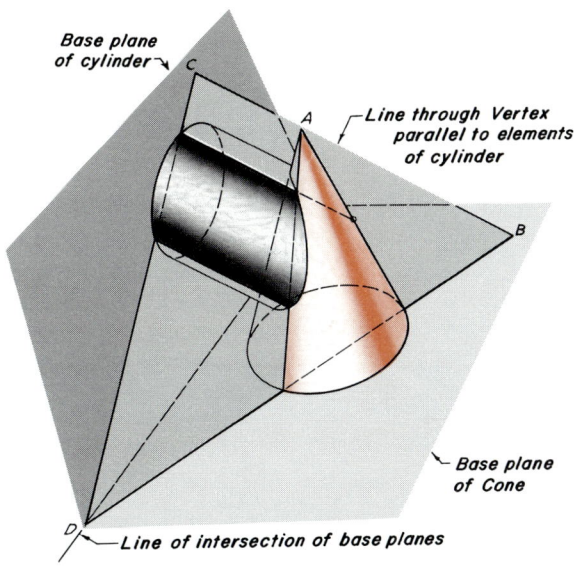

FIG. 18.27. Intersection of cone and cylinder: pictorial drawing.

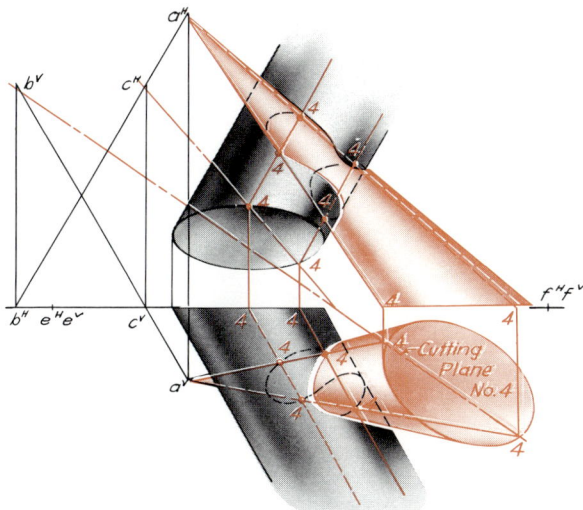

FIG. 18.28. Intersection of cone and cylinder: cutting plane method.

case must be extended to cross the base of the cone at the two points marked 4.

g. In Fig. 18.27, plane *BCD* must cut elements from both the cone and the cylinder because the plane passes through the apex of the cone parallel to the elements of the cylinder. These elements must pass through the points already found on the bases of the two surfaces. In Fig. 18.28 the plane is defined by lines *BD* and *DC* and cuts elements from both surfaces starting at the points marked 4. The elements should be drawn with the proper visibility in their own surface only.

h. Locate the points where the elements of the cone cross the elements of the cylinder. These points show plainly in the pictorial of Fig. 18.27. In Fig. 18.28 locate the points where the vertical projections of the elements of the cylinder cross the vertical projection of the elements of the cone. The same thing can be done in the horizontal projection. If the work is accurate the horizontal projection of each point must be directly above its vertical projection. If both elements are visible, the point will be visible. Any other combinations will give an invisible point.

i. Pass a series of planes in the same manner, each of which will go through a different point on the reference line.

j. Number the planes, elements, and points in order.

k. Connect the points in numerical order to form the line of intersection.

l. The final line of intersection is shown in Fig. 18.28 without all of the complicated construction.

18.18 INTERSECTION OF TWO CYLINDERS

Another illustration of the general method involving the intersection of two cylinders is shown pictorially and orthographically in Fig. 18.29. When obtaining the intersection of two cylinders, the cutting planes must be constructed parallel to the elements of both cylinders. This is illustrated pictorially in Fig. 18.30 where *CD* is parallel to one cylinder and *OA* is parallel to the other cylinder. These two lines determine the guide plane. All other cutting planes must be parallel to the guide plane.

18.18.1 Cutting Planes Edgewise. The cutting plane *AB* shown edgewise in the top view of Fig. 18.29 cuts elements numbered 3 and 7 from one cylinder and the element 3 only from the other. These elements cross each other at two points *c* and *d* on the curve of intersection. If the inclined cylinder passed entirely through the other, four points would have been determined by this one cutting plane. Other cutting planes are passed parallel to the first one until enough points have been obtained to determine a smooth curve. Two other common situations are shown in Figs. 18.31 and 18.32, with the method of solution indicated. When the two right circular cylinders are of the same size and their axes intersect, the curve of intersection is one half of each of two ellipses, unless the penetration is complete, in which case the two ellipses are also complete.

In the preceding illustrations the cylinders were so placed that planes that cut straight lines from both cylinders could be drawn by inspection. Although the drafter can, many times, draw intersecting cylinders in positions like these, sometimes this may not be convenient. To cover other situations, four illustrations are shown and discussed in the following paragraphs.

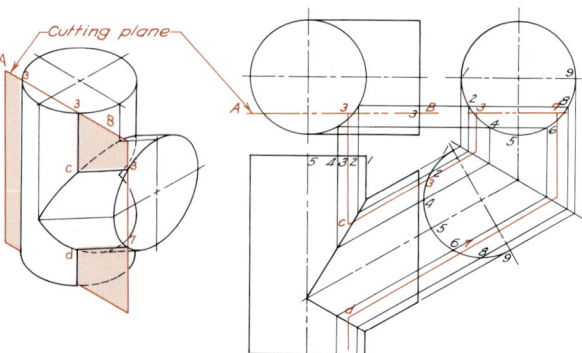

FIG. 18.29. Intersection of two cylinders.

18.18 INTERSECTION OF TWO CYLINDERS

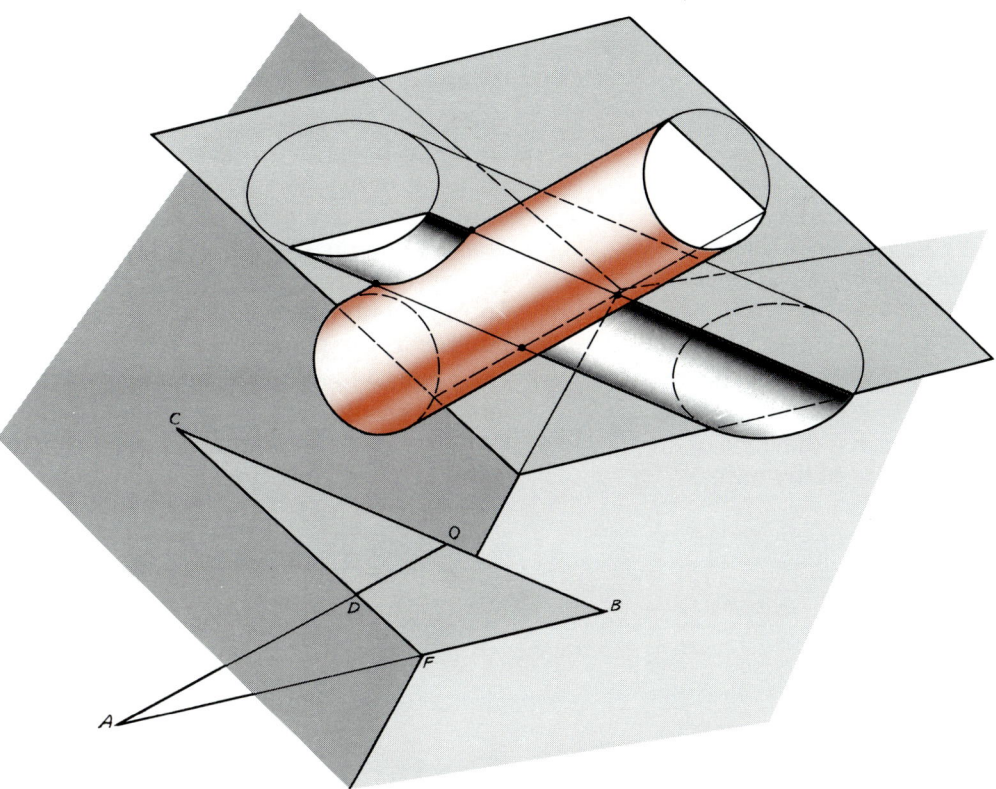

FIG. 18.30. Intersection of two cylinders: pictorial drawing.

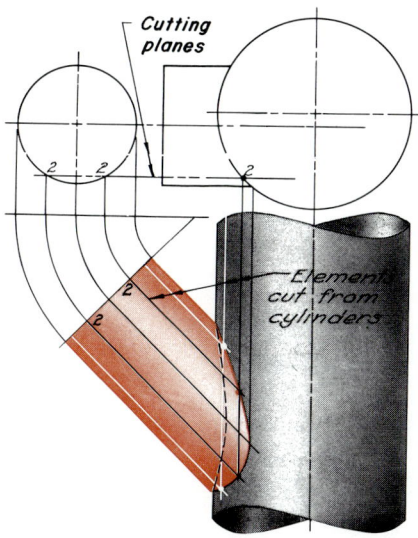

FIG. 18.31. Intersection of two cylinders: cutting plane method.

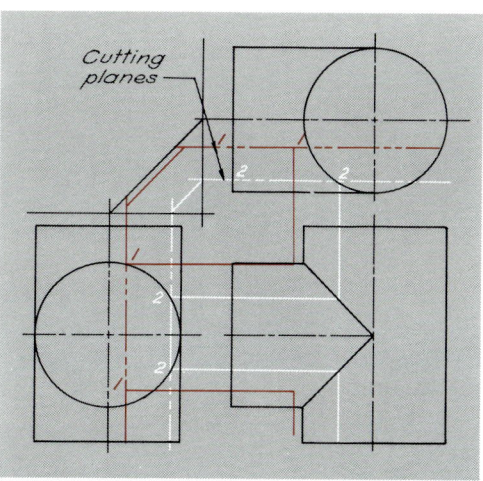

FIG. 18.32. Intersection of two cylinders of same size.

18.18.2 Bases in the Same Plane. In Fig. 18.33, two oblique cylinders are shown having a common vertical base plane. Up to this point it has been possible to draw the cutting plane by inspection since all of them have been *H-* or *V*-projecting planes. In this problem and those that follow, the cutting planes are oblique; hence they cannot be determined by inspection. It is necessary, therefore, to determine the position of the cutting planes that will cut straight lines from the cylinders or cones.

The following steps are necessary for this solution. See Fig. 18.33.

a. Assume point *A* and construct two lines *AB* and *AC* parallel, respectively, to the elements of the two cylinders. These lines determine a guide plane that is parallel to the elements of both cylinders.

b. Find points *B* and *C* where these two lines pierce the plane of the bases of the two cylinders.

c. Draw the line $b^V c^V$, which is the trace of the guide plane on the plane of the bases.

d. Construct other planes parallel to the guide plane by drawing traces parallel to *BC*, that is, parallel to $b^V c^V$. Cutting plane 3 in Fig. 18.33 is an example. It is understood that the other projection of these traces lies in the edgewise view of the bases. If the two cylinders intersect, it will be possible to draw these traces so that they cross both bases. The first plane should be tangent to one or both bases and the others numbered consecutively.

e. From the points where the guide traces cross the bases of the cylinders, draw the elements of the cylinders. For cutting plane 3 there are two elements in each cylinder. These elements should be drawn with the proper visibility in their own surface only. The elements should be given the same number as the plane.

f. The points where these elements cross determine points on the line of intersection. When two visible elements cross, the point will be visible. Any other combination will

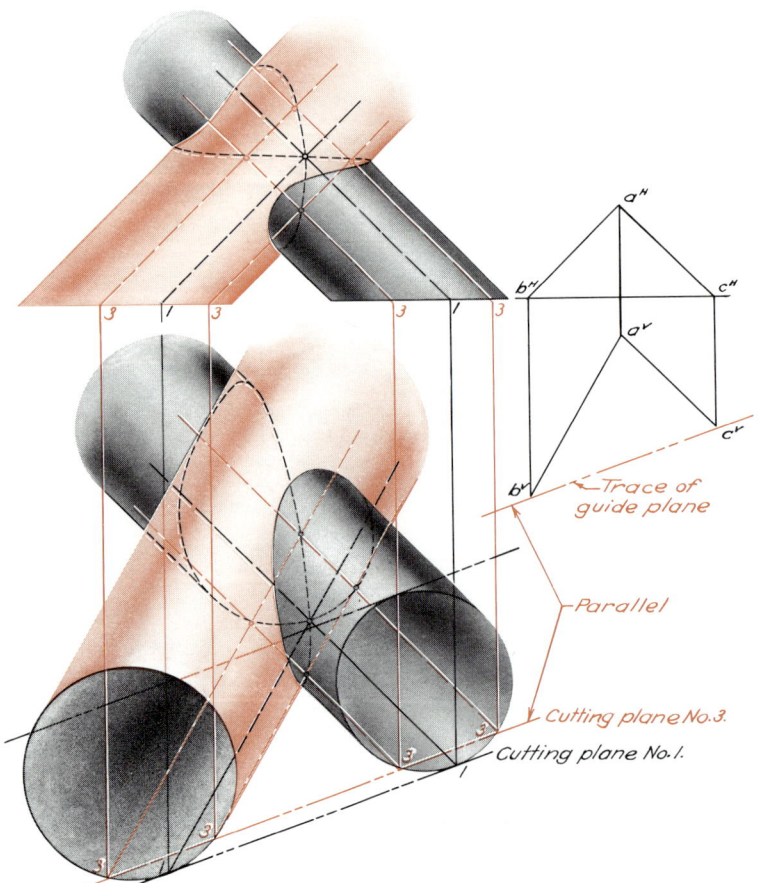

FIG. 18.33 Intersection of two cylinders with common base plane.

locate an invisible point. The points should be given the same number as the plane and elements.

g. Connect the points in numerical order to obtain the curve of intersection. Visibility of the curve can change only on the limiting element of one of the cylinders or where the curve crosses an open base.

18.18.3 Bases Edgewise on One Projection But Not in the Same Plane. In Fig. 18.34, the bases of the cylinders are in different planes but both appear edgewise in the top view. To find the intersection of these two surfaces it will be necessary to apply a procedure very similar to that used in Section 18.8.2.

a. Assume point O at any convenient place on the sheet, as shown in Fig. 18.34, on the left side of the figure. See also small illustration for guide plane at right side of figure.

b. Through point O at left construct a line BOC parallel to the elements of one of the cylinders. Then through point O construct another line AOD parallel to the elements of the other cylinder.

c. Find the points A and B where the two lines pierce the plane of the base of one of the cylinders.

d. Find the points C and D where the two lines pierce the plane of the base of the second cylinder.

e. Locate the line FG which is the line of intersection of the planes of the two bases.

f. Through a^V and b^V draw one trace of the desired guide plane on the plane of the base of cylinder 1.

g. Through c^V and d^V draw the trace of the desired guide plane on the plane of the base of cylinder 2.

h. These traces must intersect at f^V on line FG.

i. Traces of parallel planes may be drawn so that they intersect the bases of the cylinders. The two traces of all of these planes must intersect on the vertical projection of line FG. Plane 1 will be the first usable

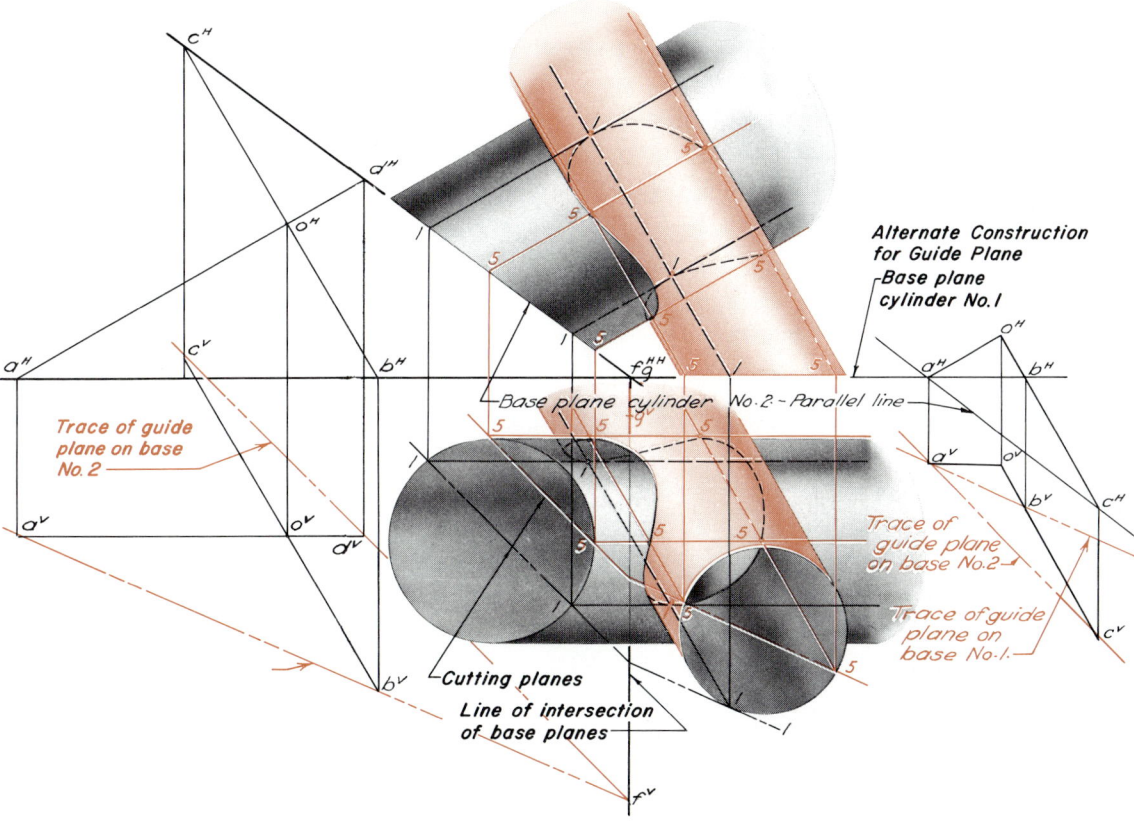

FIG. 18.34 Intersection of two cylinders: cutting plane method.

482 INTERSECTIONS AND DEVELOPMENTS

plane. Others should be numbered in order.

j. Through the points where the traces cross the bases, elements of the cylinders may be drawn. Show their visibility in their own cylinder only.

k. Find the point where these elements intersect. When two visible elements intersect, the point is visible. All other combinations give invisible points.

l. Join the points in numerical order. Visibility of the intersection can change only on a limiting edge of one of the cylinders or where the curve crosses an open base.

When it is not convenient to use the actual bases of the cylinders, parallel bases can be set up to locate the direction of the traces of the guide plane. This construction is shown at the right side of Fig. 18.34.

18.18.4 Bases Edgewise in Different Views. A third situation is shown in Fig. 18.35, where the base of one cylinder shows edgewise in the top view and the base of the other appears edgewise in the front view. The direction of the traces of the cutting plane on the base planes is shown in the construction at the right. In this construction it should be noted that one actual base plane has been used (base 1) and for convenience a second plane parallel to base plane 2 has been set up. The required steps are as follows.

a. Through point A construct two lines parallel, respectively, to the elements of the two cylinders.

b. Find the points where each line pierces the plane of base 1. These are points B and C in Fig. 18.35.

c. Find the point where each line pierces the plane that was set up parallel to base 2. These are points D and E in Fig. 18.35.

d. The projection $d^H e^H$ is the trace of the guide plane on base 2.

e. The projection $b^V c^V$ is the trace of the guide plane on base 1.

f. The line of intersection of the planes of base 1 and base 2 is line PN. The H-projection of this line must lie in the edgewise view of base 1. The V-projection of this line must lie in the edgewise view of base 2.

g. Construct a series of cutting planes similar to plane 6 in Fig. 18.35. Make a trace parallel to $b^V c^V$ cutting base 1 in the points marked 6.

h. Extend this trace until it crosses $p^V n^V$ at f^V.

i. Project f^V to f^H on $p^H n^H$.

j. From f^H draw a trace parallel to $d^H e^H$ until it crosses base 2 in the points marked 6.

k. Proceed from this point just as in j, k, and l in Section 18.8.3.

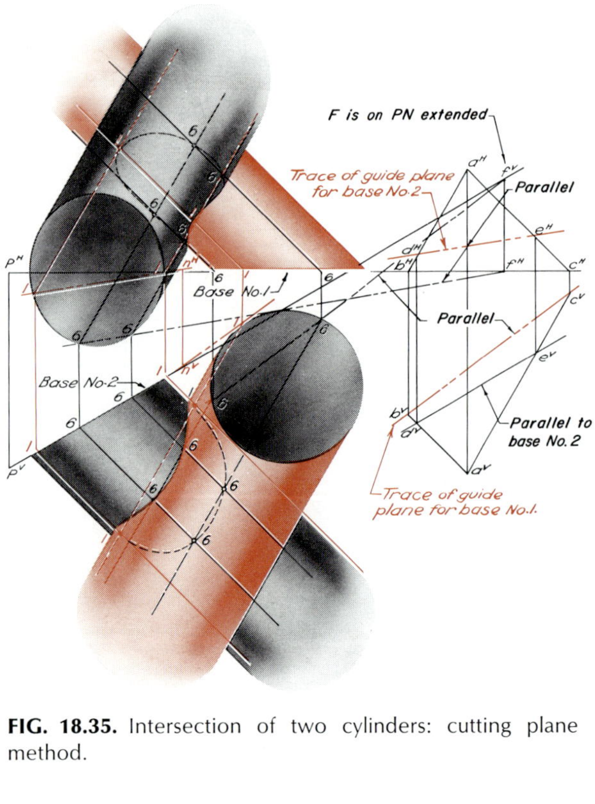

FIG. 18.35. Intersection of two cylinders: cutting plane method.

18.18.4 Neither Base Edgewise. In Fig. 18.36, two cylinders are shown in which neither base appears edgewise. As before, the problem is to determine cutting planes that will cut straight lines from both cylinders.

Several methods of solution are possible.

a. A plane could be chosen arbitrarily to cut across both cylinders, and the intersection of this plane with both cylinders could be found, thus making the new bases of both cylinders lie in the same plane, similar to the situation in Fig. 18.33. This having been done, the problem would be reduced to that of Fig. 18.33.

b. A second method consists of constructing the guide plane *ABC* in Fig. 18.36. An auxiliary plane perpendicular to this guide plane will show the edgewise view of all cutting planes. By making the views of both cylinders on this auxiliary plane the elements may be determined.

In Fig. 18.36, the direction of the horizontal trace of the guide plane has been determined at $b^H c^H$. The reference line *H-1* is drawn perpendicular to this line. Both cylinders are then projected in the auxiliary view. Here cutting planes like 4 and 7 are drawn edgewise to determine the foot of the elements in both cylinders. From an inspection of the auxiliary view it can be seen that the intersection will be a single continuous curve.

18.19 CYLINDER WITH A DOUBLE-CURVED SURFACE

A simple illustration of the intersection of a cylinder and sphere is shown in Fig. 18.37. Here the cylinder and sphere have been chosen in such position that the planes, which cut straight lines from the cylinder, cut circles from the sphere that project as circles in the top view. Two typical cutting planes are shown. In this position the problem becomes very simple. By the use of one or two auxiliary views the drafter can always reduce the problem to the situation shown in Fig. 18.37.

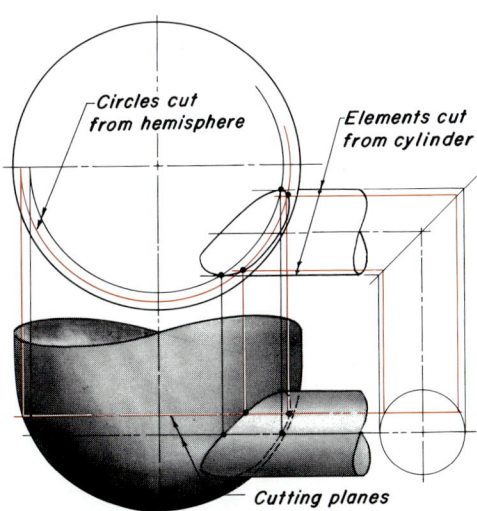

FIG. 18.37. Intersection of sphere and cylinder.

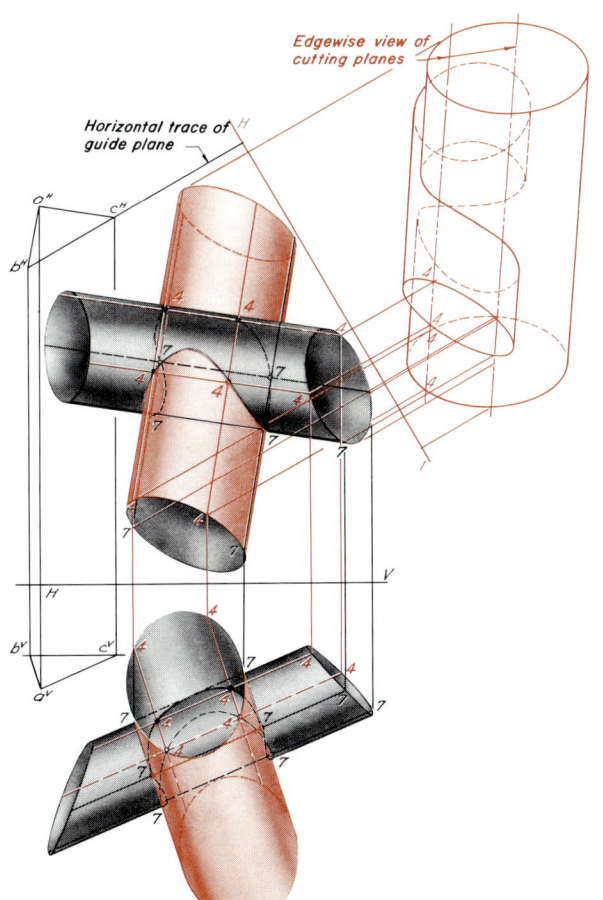

FIG. 18.36. Intersection of two cylinders: cutting plane method.

A more complex situation is shown in Fig. 18.38, where a curved pipe (geometrically a partial torus) passes through a double-curved surface. In this case, by a careful selection of the cutting plane and the auxiliary view, parts of circles are cut from the pipe and more complex curves from the other surfaces. The method for finding the curve on the larger surface is illustrated for two cutting planes in Fig. 18.38. The process must be repeated a sufficient number of times to determine the curve. In any event, it is a long and tedious process.

18.20 CONVENTIONAL INTERSECTION

On many machine parts, intersections occur that are automatically produced by machining operations. Thus a small hole drilled into a tube, for example, produces the intersection of two cylinders. Such intersections can be shown conventionally, as in Fig. 18.39. If the intersection is relatively large, an approximation may be made by locating three points and using an irregular curve, as shown in the lower right-hand illustration of Fig. 18.39. These conventional intersections are used to save time, but the actual intersections could have been found by the methods described in the previous article.

18.21 DEVELOPMENT OF SURFACES

Another practical problem that arises frequently in construction work and is commonly associated with the work in intersections is the development of surfaces. The term *development* means the laying out of flat patterns from which curved surfaces can be formed without stretching the material.

The method of making developments is best explained by concrete examples. One fundamental principle, however, may be noted, namely, *that every line used in making a development must represent the true length of that line on the actual surface.*

The two methods of finding the true length of a line will be found in Section 7.7 and 7.7.6, which should be thoroughly reviewed at this time.

18.22 DEVELOPMENT OF A PRISM

A right section of a prism develops as a straight line, known as a stretchout line. The elements of a prism develop as lines perpendicular to the stretchout line. The elements and the stretchout line both show in true length in the development.

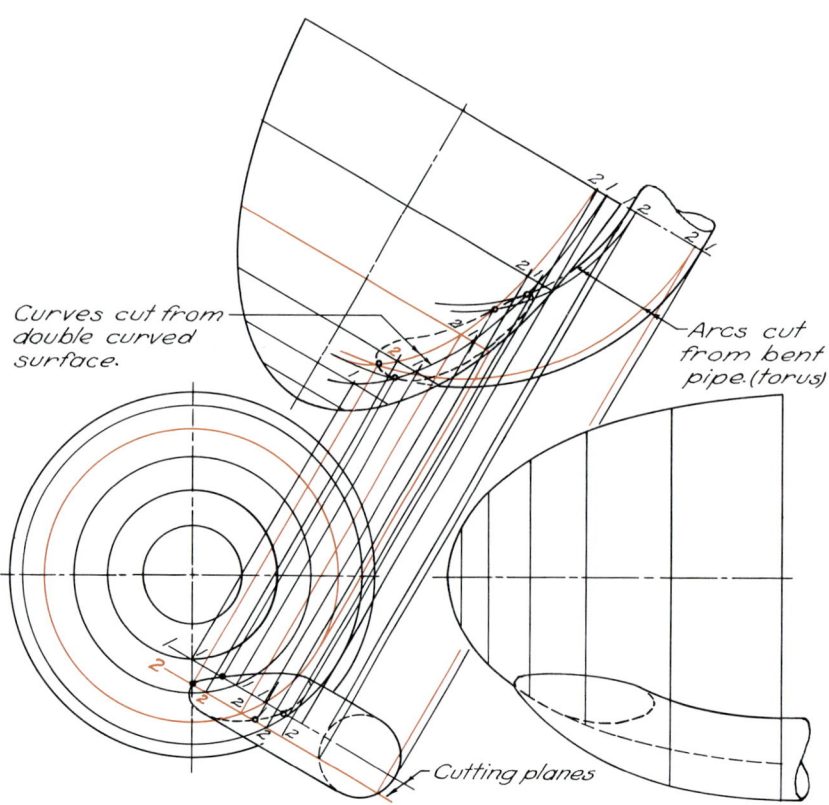

FIG. 18.38 Intersection of curved pipe and double-curve surface.

18.22.1 Right Prism. If the prism in Fig. 18.40 is cut along the edge *AE* and unfolded into a flat surface, the resulting pattern is called a development. In this case all the edges show in their true length in one or the other of the two original views. The plane of the base is perpendicular to the edges; hence this base line develops into a straight line perpendicular to the edges as shown in Fig. 18.40. The true size of the right section is laid out on the stretchout line at *ABCDA*. The true length of each element is then placed on perpendiculars to the stretchout line as shown for *AE* and *BF*.

18.22.2 Oblique Prism Parallel to a Coordinate Plane. When neither base of a prism is perpendicular to the elements, it is called an *oblique prism*. In this case it is necessary to find the true size of a right section. The best method of doing this is by setting up an auxiliary plane perpendicular to the elements as shown at $a^v b^v c^v d^v$ in Fig. 18.41a. In Fig. 18.41a, the elements show in their true length in the front view. The stretchout line *ABCDA* can be laid out on a perpendicular to the vertical projection of the elements. The length of the stretchout line is obtained from the true size of the right section $a^1 b^1 c^1 d^1$. The elements will then be laid out per-

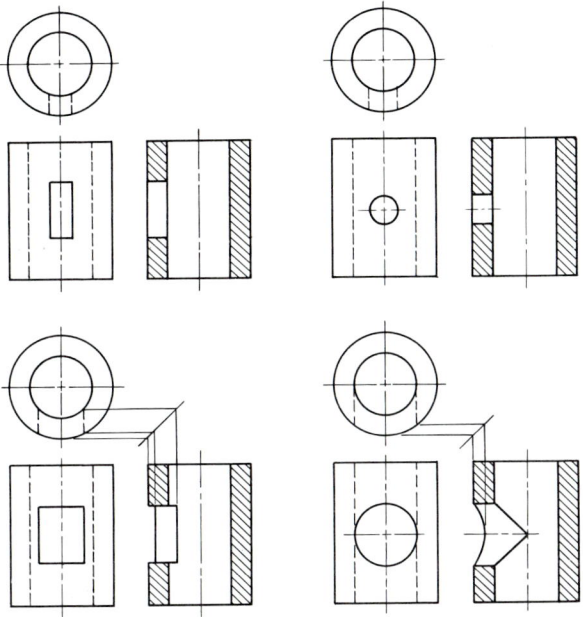

FIG. 18.39. Conventional intersections in sectional views.

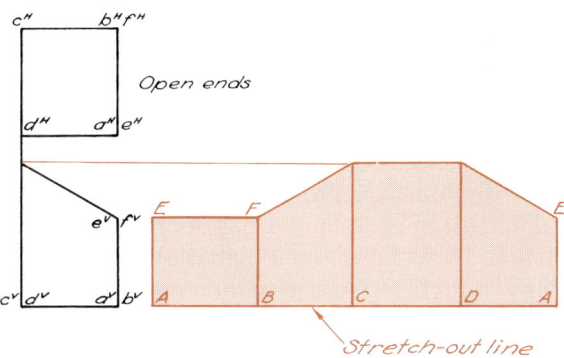

FIG. 18.40. Development of a right prism.

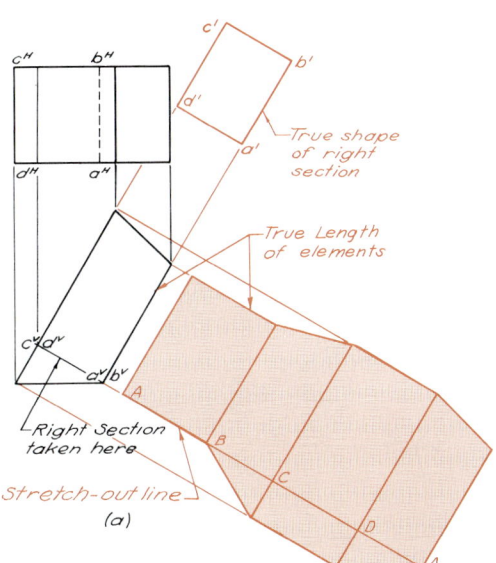

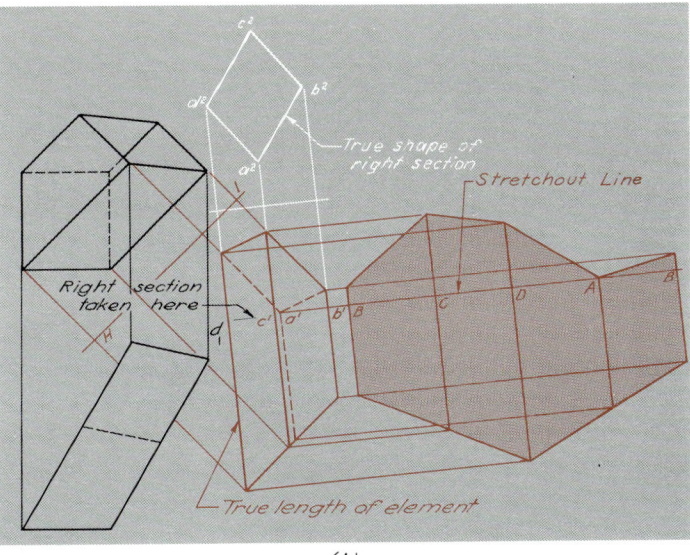

FIG. 18.41. Development of an oblique prism.

pendicular to the stretchout line and their length can be projected from the front view.

18.22.3 Oblique Prism Not Parallel to a Coordinate Plane.
In Fig. 18.41b, none of the edges of the prism show in true length in the front and top views. The true lengths of the four corner edges, however, are obtained in the first auxiliary view and the true shape of the right section in the second auxiliary view. These true lengths have then been used to obtain the development as shown. For convenience, the development has been projected from the first auxiliary, but it could have been laid out in any position by the use of dividers.

18.23 DEVELOPMENT OF A PYRAMID
A pyramid is developed by finding the true size of the base and the true length of all elements. These values are combined by building up a series of triangles in true size.

18.23.1 Development of a Right Pyramid.
Figure 18.42 shows two projections of a right square pyramid. The true size of the base shows in the top view. The faces of the pyramid are triangles. The true size of these triangles can be found by auxiliary projection or by finding the true length of each line as has been done in Fig. 18.42b. To find the true length of AE the projection a^H is revolved to a_r^H and then projected to a_r^V. The true length of AE shows at $a_r^V e^V$.

The development in Fig. 18.42a is made by laying out each complete triangle such as EDC, by using the true length of each line. Then the upper part of the pyramid that has been cut away is laid out in the same manner as for triangle EFG. The portion CDFG is the development of one face of the truncated pyramid.

18.23.2 Oblique Pyramid.
In the oblique pyramid in Fig. 18.42b, it was necessary to obtain the true length of the four edges by rotation as shown. The edges of the base appear in true length in the top view. Study carefully the method of obtaining the true lengths of the truncated portions. The true lengths having been obtained, the development is made as in the foregoing example. As in previous cases, the pyramid is cut on the shortest element FA of the truncated portion.

18.24 DEVELOPMENT OF A CYLINDER
The cylinder is a very common surface encountered in design. If its bases are parallel and perpendicular to the axis, it is easily seen that when split along any element and rolled out flat, as in Fig. 18.43, a rectangle is formed. The width of this rectangle is equal to the length of the cylinder, and its length is equal to the circumference of the cylinder. Both measurements are easily found from a working drawing. The development of a cylinder is the same as that of the prism. The cylinder is divided into a number of parts that actually form a prism that approximates the surface of the cylinder.

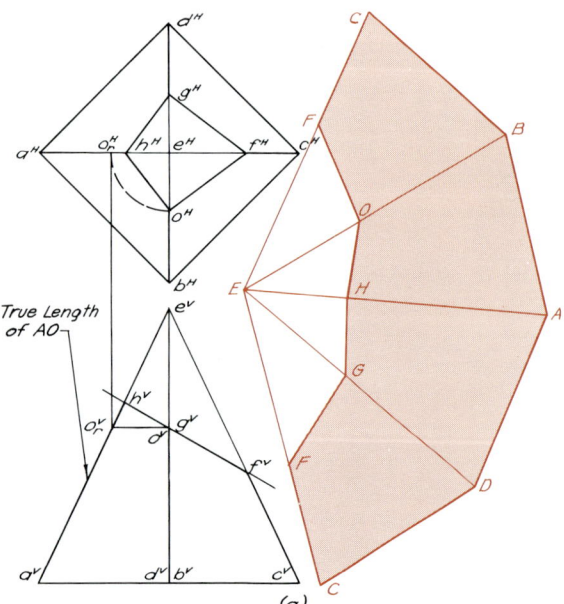
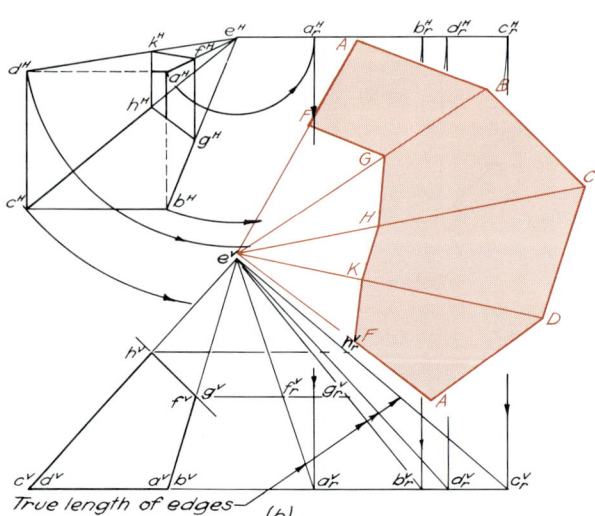

FIG. 18.42. Development of a pyramid.

18.24.1 Truncated Right Cylinder.

A truncated right circular cylinder has been developed in Fig. 18.44. Here the elements, 12 in number, show in their true length in the front view and may be projected directly to the development. The lower base shows the true size of the right section as a circle in the top view, and the true shape of the inclined face may be obtained by auxiliary projection if desired.

In stepping off the 12 spaces between elements with a divider, it is well to check the accuracy of the setting by stepping it off six times halfway around the true size of the base. If this does not check with the semicircle, the setting should be adjusted until a perfect check is obtained. The true size of the right section is laid out on the stretchout line and the length of each element is projected to a line perpendicular to the stretchout line.

18.24.2 Oblique Cylinder Parallel to a Coordinate Plane.

An oblique cylinder, that is, one that has its bases inclined to the elements, requires, first, the true length of all elements and, second, the true size of a right section similar to that explained in the discussion of the oblique prism. The right section may be taken at any convenient point and its true size may be obtained by auxiliary projection as in Fig. 18.45. In this case the elements show in their true length in the front view; they may be projected to the development in a direction at right angles to their length as illustrated for three of them or they may be transferred into another position with dividers as shown for the half development. The right section rolls out as a straight line, and the spaces between elements, obtained from the auxiliary view, may be laid out on this line. A practical application is shown in Fig. 18.46.

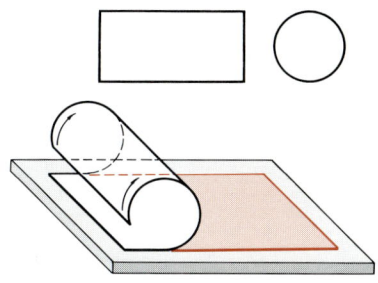

FIG. 18.43. Pictorial drawing of development of a right cylinder.

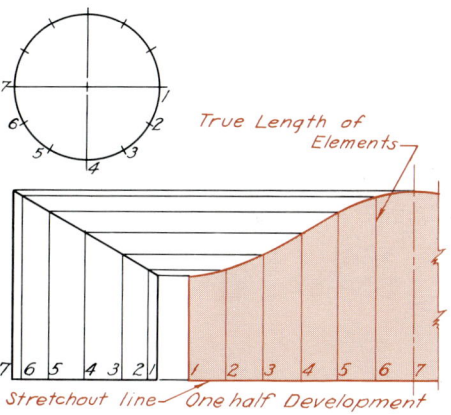

FIG. 18.44. Development of a right cylinder.

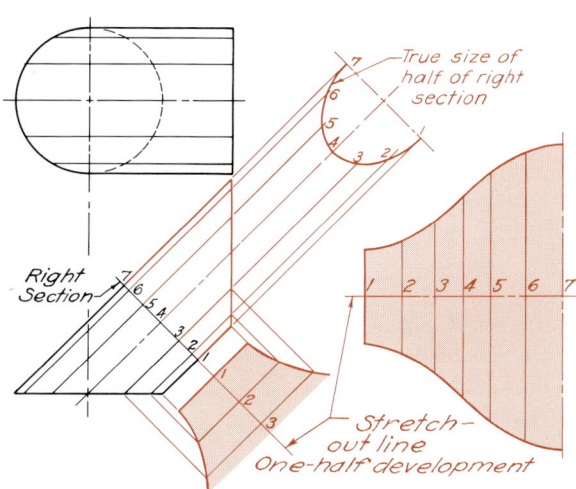

FIG. 18.45. Development of an oblique cylinder.

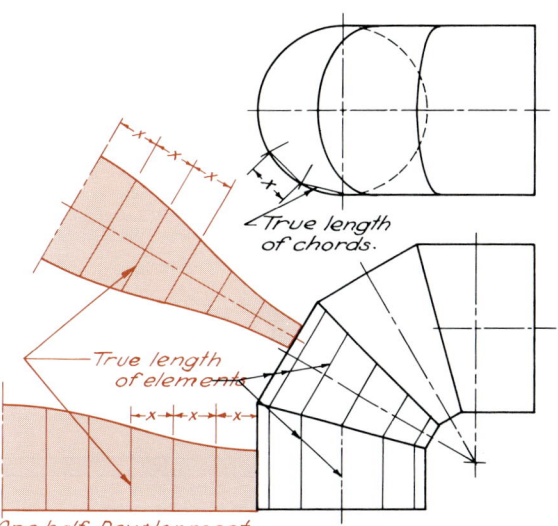

FIG. 18.46. Development of a pattern for a pipe elbow.

18.24.3 Oblique Cylinder Inclined to All Coordinate Planes.
A still more general case of the oblique cylinder is shown in Fig. 18.47. In this instance it is necessary to obtain the true length of the elements in the first auxiliary view and the true size of the right section by a second auxiliary view. These having been determined, the construction follows in the usual manner.

18.25 DEVELOPMENT OF A CONE
A cone may be regarded as a pyramid with an infinite number of sides. In practice it is divided into a practical number of parts that form a close approximation to the surface of the cone. Each section is developed as a triangle, as was explained for the pyramid.

18.25.1 Right Circular Cone.
The right circular cone develops into a sector of a circle as shown in Fig. 18.48. The front view shows the true length of the elements as y. The lengths of the chords of the base may be obtained from the top view. With y as a radius, an arc of indefinite length is drawn and on it a length equal to the chord x is stepped off 12 times. If these points are connected to the center, we have in reality 12 small triangles that approximate very closely the actual cone. This is the usual graphical method of development. A closer approximation can be obtained by dividing the base into a greater number of parts or the length of the base of the cone may be computed and laid out on an arc. The development of a cone intersected by a cylinder is shown in Fig. 18.49.

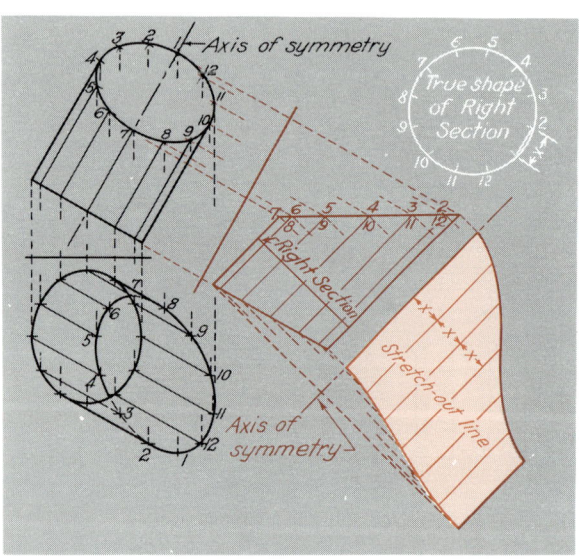

FIG. 18.47. Development of an oblique cylinder.

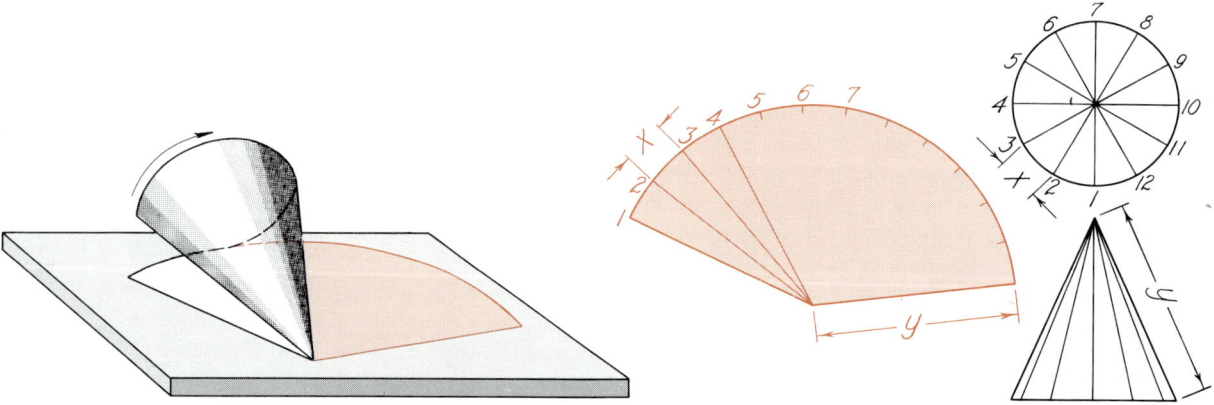

FIG. 18.48. Development of a right circular cone.

18.25.2 Oblique Cone. The oblique cone requires that the true length of all elements be found, and if the base is not parallel to one of the planes of projection, the true length around the base may also be found, preferably by auxiliary projection. Figure 18.50 illustrates the first type with the base parallel to *H*. Note again that the development consists of a series of triangles, the length of whose sides has been determined and that have been joined in consecutive order, beginning with the shortest element.

A second cone is shown in Fig. 18.51, in which the true shape of the base has been determined by auxiliary projection and the true length of the elements by rotation as usual. As soon as these things have been accomplished, the solution follows the customary procedure.

In several of the preceding figures the development has been begun on the shortest element. This is a practice feature. By splitting the part on the shortest element the least labor is required in welding or riveting the seam. Whenever possible, therefore, the shortest element should be chosen as the beginning of a development or template.

18.26 DEVELOPMENT OF TRANSITION PIECES

A large amount of sheet metal layout work consists of transition pieces or reducers as they are sometimes called. When it becomes necessary to join ducts of different shapes, the portion by which this change is accomplished is called a *transition piece*. If there is a change in size, the

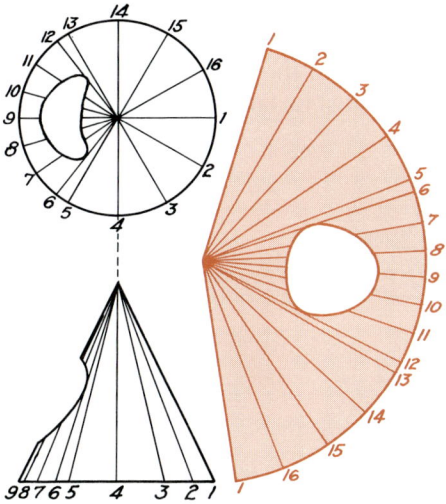

FIG. 18.49. Development of an intersected cone.

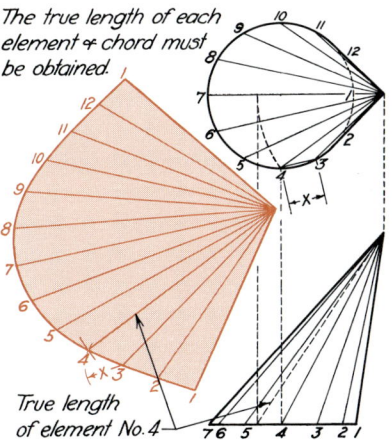

FIG. 18.50. Development of an oblique cone.

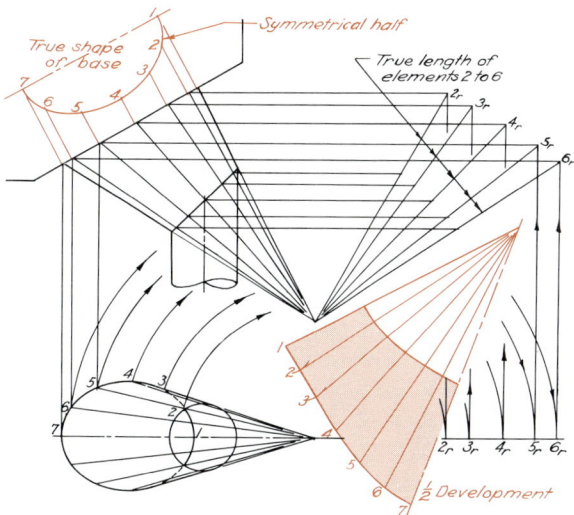

FIG. 18.51. Development of an oblique cone.

sections are commonly called *reducers*. Thus one may connect rectangular parts in different planes or change a section from square to round and so on through a wide range of combinations.

The problem in making these pieces lies, first, in recognizing and identifying the various component shapes. The transition pieces are composed of planes, cones, cylinders, and convolutes. The second step is the development of the several parts and connecting them in the proper order. It is also desirable to identify axes of symmetry since a pattern of one half will suffice if there is one axis of symmetry and one fourth may be enough if there are two axes of symmetry.

18.26.1 Rectangular Sections. A series of reducers connecting rectangular sections is shown in Fig. 18.52. In the first three of these the reducer is composed entirely of plane sections. In the last, Fig. 18.52d, two of the faces are warped quadrilaterals. Theoretically, warped surfaces cannot be formed without stretching the material but, practically, a development can be made if the warping is not too severe. In Fig. 18.52d, the warped surface can be made into two plane-triangular surfaces by bending slightly along the line between the triangles.

The axes of symmetry have also been shown in Fig. 18.52 by center lines. A little study of these illustrations and those that follow will be an aid in analyzing similar problems in practice.

The development of a reducer similar to that in Fig. 18.52b is shown in Fig. 18.53. Each part consists of a quadrilateral. To lay these out it is necessary to have the true length of the edges and a diagonal, unless some of the edges are at right angles to each other. This piece, having only one plane of symmetry, will be split on that plane and only one half need be developed. Parts 1 and 4 each have two edges perpendicular to the central plane of symmetry, hence a diagonal is not essential. Face 2–3, however, cannot be developed without one of the diagonals as shown.

The three parts must be laid out in their proper

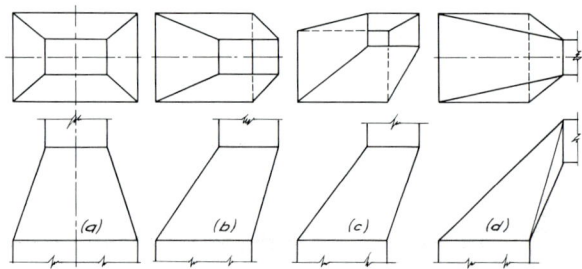

FIG. 18.52. Typical rectangular reducing sections.

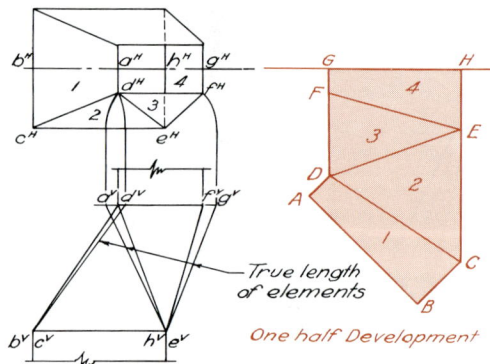

FIG. 18.53. Development of one half of a reducing section.

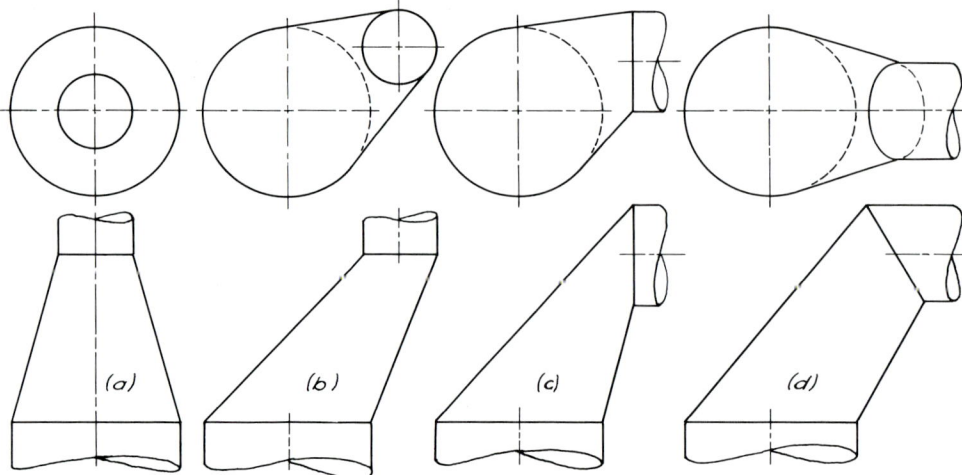

FIG. 18.54. Reducers between cylindrical surfaces.

relationship to each other. Although it would seem impossible for anyone to put them together in the wrong order, nevertheless this sometimes happens. If the drafter is not sure of him/herself, it is suggested that the corners be lettered on the two views and that these letters be carried through to the development as shown in Fig. 18.53.

18.26.2 Circular Sections. In Fig. 18.54, transition pieces connecting circular cylinders have been shown. In the first two cases the connector is a part of a cone because the circles at the top and bottom lie in parallel planes. In the last two cases, Figs. 18.54c and d, the surfaces are convolutes and may be developed by dividing the circular ends into an equal number of parts, beginning at any known element of the surface and then connecting the points on each end, making a series of quadrilaterals that can be developed by dividing each quadrilateral into two triangles as explained in Section 18.27.

18.26.3 Circular and Rectangular Sections. Reducers that change in shape from rectangular to circular are shown in Fig. 18.55. These are commonly found on the roofs of buildings as ventilators. In each case the reducing section consists of four partial cones and four triangular plane surfaces. The vertices of the cones lie at the corners of the rectangle. A typical development is shown in Fig. 18.56.

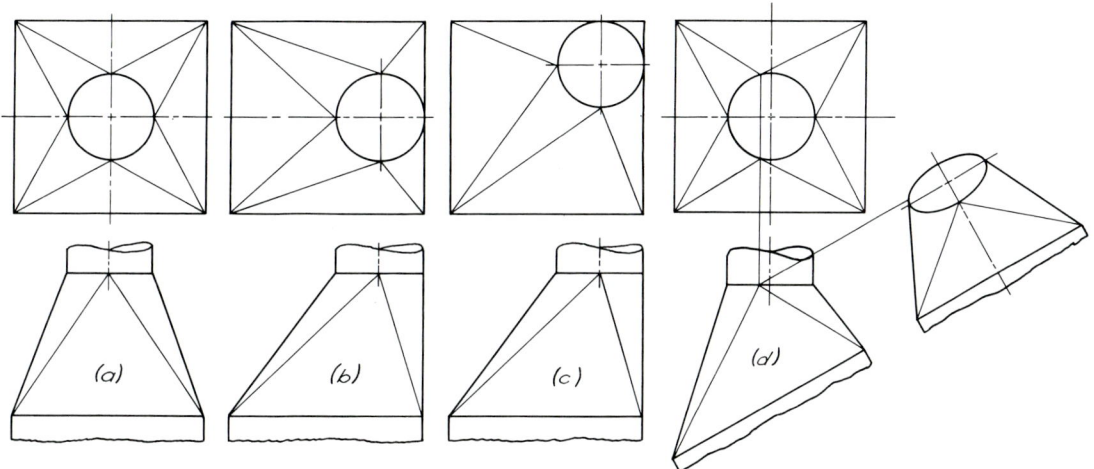

FIG. 18.55. Transition sections.

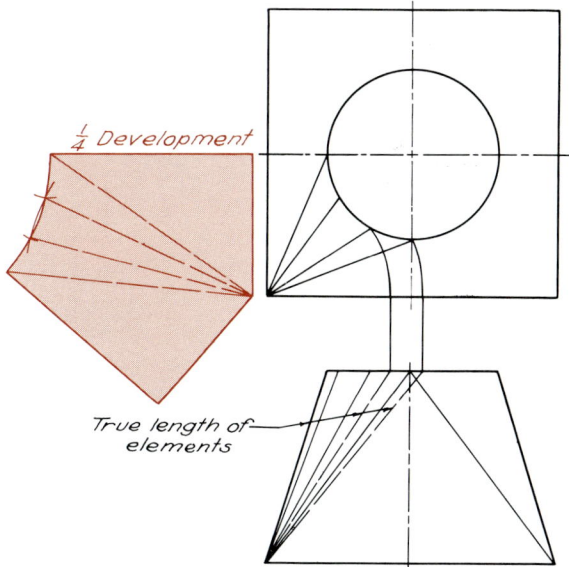

FIG. 18.56. Development of one fourth of a transition piece.

Another group of transition pieces involving parts of cylinders and planes and cones is shown in Fig. 18.57. The end part of the one in Fig. 18.57*d* is a portion of an oblique cone. With the vertex downward the chords of both bases show in true length in the top view; hence it is only necessary to obtain the true length of the segments of the elements.

In all cases developments should be made exactly to the planes of symmetry. If additional material is needed to make a joint, this should be added as a narrow strip in the developed pattern and not on the original two- or three-view drawing.

The reducer shown in Fig. 18.58 consists of parts of two oblique cylinders and two triangular plane sections. Note that the auxiliary view used to determine the length of the elements of the cylinder could be projected from either the front or top view. The top view was chosen because this makes the bases come out as straight lines in the auxiliary view, rather than curves. The right section is obtained in the second auxiliary view. Again, care must be exercised to connect the two triangular parts to the cylinder in proper order.

When it becomes desirable to design a transition surface to connect two cross sections composed of straight lines and curves, the surface must be divided into planes, cones, cylinders, or convolutes as necessary. In each case there will be only one solution that will give a perfectly smooth surface, although other subdivisions might be used to form a closed surface. The best procedure is to determine the plane surfaces

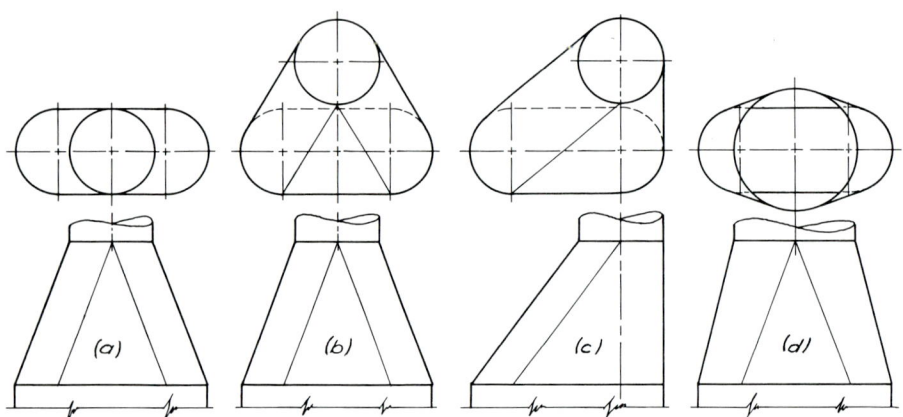

FIG. 18.57. Transition pieces between cylindrical surfaces.

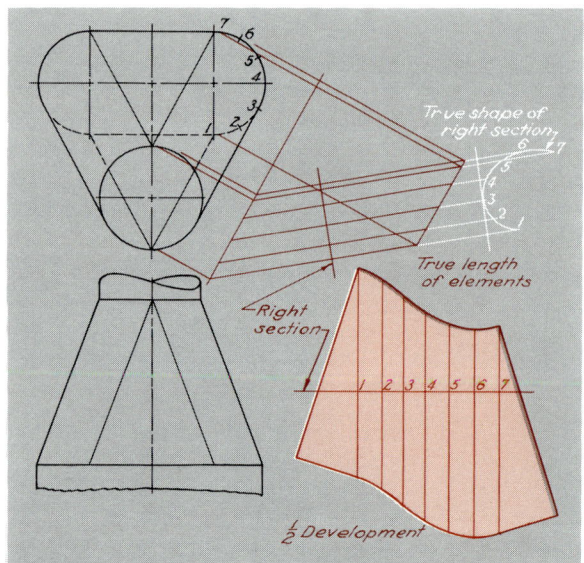

FIG. 18.58. Development of one half of a reducing section.

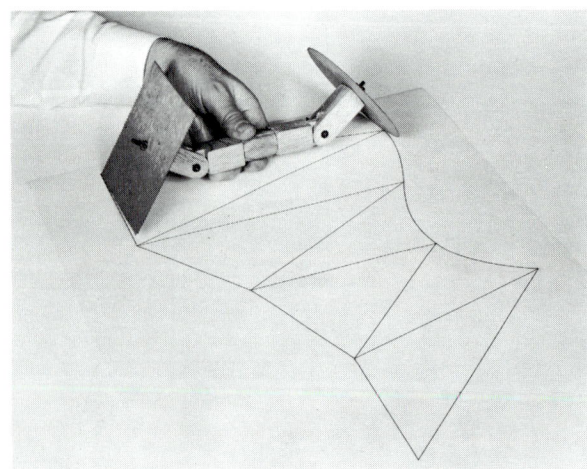

FIG. 18.59. Development by triangulation.

first, after which the remaining surfaces are usually quite evident. A few simple rules will help in making the divisions.

a. Every straight line in either base must lie in a plane surface. Determine the plane surfaces first.
b. When there are curves in one base but none in the other, there will be cones in the surface.
c. When there are circles in both bases of equal radius, there will probably be cylinders in the surface.
d. When there are circles of different radii or other curves in both bases or when the bases are not parallel, there may be frustums of cones of convolutes.

18.26.4 Mechanical Aid in Surface Development. Any surface made up of planes, cones, cylinders, or convolutes can be developed by means of the mechanical aid illustrated in Fig. 18.59. The bases are interchangeable and any plane figure can be used for either base. The bases can make any angle with each other or with the elements. Plane figures with reentrant angles and reverse curves are very difficult to use as bases and should be avoided. The mechanical aid illustrated in Fig. 18.59 must be rolled along with both bases in contact with the paper.

18.27 DEVELOPMENT BY TRIANGULATION

In many cases a transition piece cannot be divided into cylinders, cones, and planes. For example, Fig. 18.54 is composed entirely or in part of frustums of cones or convolutes. Such surfaces can be developed by a method commonly called triangulation. This method is frequently applied to cones if the vertex is too far removed for practical use. The method consists in dividing the surface into small triangles that will approximate the surface and then laying out these triangles in their true size and proper relative order.

This method is illustrated for a reducer connecting a circular and an elliptical section in Fig. 18.60. Beginning at any common element, usually on a plane of symmetry, the upper and lower bases are divided into the same number of equal parts and these points are connected in a manner similar to that for the elements of a cone. This divides the surface into a series of quadrilaterals. Next these quadrilaterals are divided into triangles by drawing diagonals. It is customary to keep all diagonals running in the same general direction. This method is simple but it cannot be regarded as a shortcut since it is still necessary to find the true length of every line used in the development. The method of finding the true length has again been illustrated for a few of the lines in Fig. 18.60.

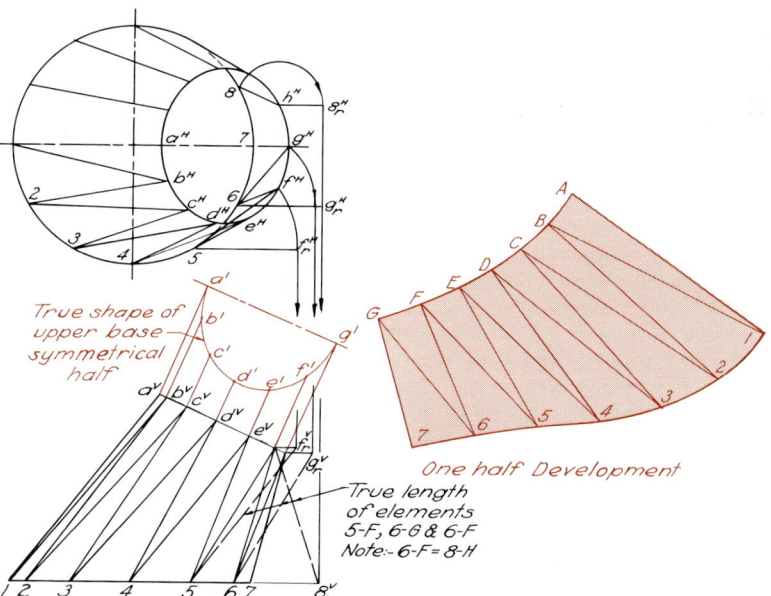

FIG. 18.60. Mechanical developing model.

18.28 DEVELOPMENT OF DOUBLE-CURVED SURFACES

Spheres and other double-curved surfaces can be only roughly approximated by development. If these surfaces are to be accurately reproduced, as many of them must be in aircraft work, the material must be stretched by forming in dies with a drop hammer, or hydropress, or by spinning.

The construction of these dies involves a knowledge of intersections as illustrated in Fig. 18.61, where the templates used in forming a plaster mold for a paraboloid of revolution are shown. The sections are taken at right angles to each other. The templates are cut from sheet metal and firmly joined together. Circular sections may be added to give accuracy to the shaping of the plaster cast. The spaces between the templates are filled with wire netting and excelsior over which the plaster can be placed. As the plaster dries, it can be scraped to the exact contour. It is then used to make a metal die, and from the die a metal punch similar to the plaster mold can be made. Allowance must be made for the thickness of metal to be formed.

Surfaces of revolution may be approximated in one or two ways, as illustrated in Figs. 18.62 and 18.63. When the sections are cut along meridian curves as in Fig. 18.62, the method is called the *gore method* and each section is referred to as a *gore*. When the sections are cut perpendicular to the axis of revolution as in Fig. 18.63, the scheme is referred to as the *zone method*. In either case the accuracy will obviously depend on the number of sections made. The greater the number, the closer will be the approach to the true surface. True lengths must again be used throughout.

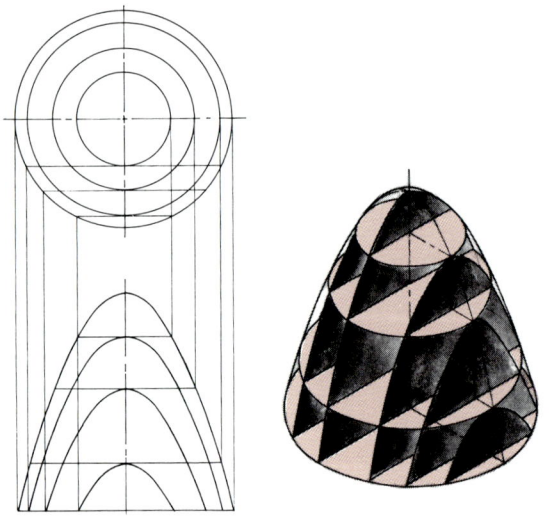

FIG. 18.61. Section of a paraboloid of revolution.

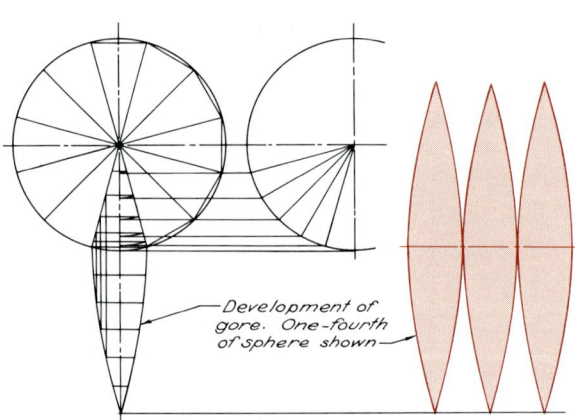

FIG. 18.62. Development of a sphere: gore method (approximate).

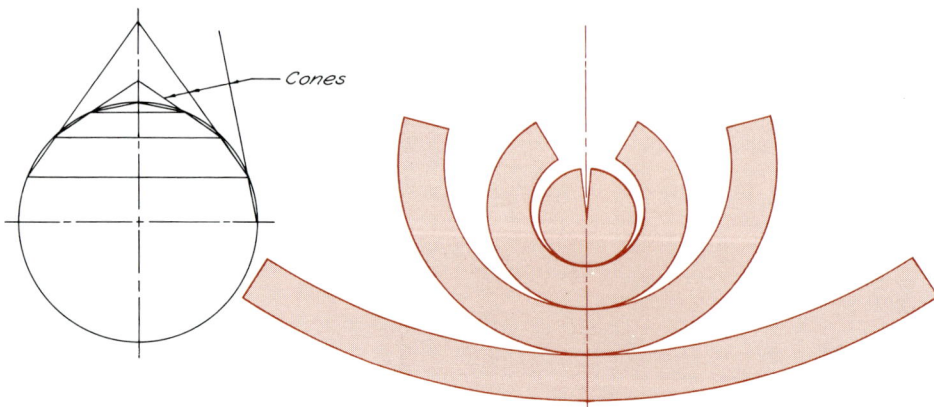

FIG. 18.63 Development of a sphere: zone method (approximate).

The zone method of development is in general similar to the polyconic system of map projection used for topographic maps of the United States and gives a good basis for further study of that system of map making.

18.29 SHEET METAL JOINTS

Sheet metal developments and templates would be of little value unless the ends could be fastened together. A wide variety of methods are used in sheet metal work. Seams may be made by bending, welding, or soldering and riveting, as shown in Fig. 18.64. Allowance must be made beyond the theoretical line of development to provide the necessary overlap to make these joints.

18.30 BEND RADII AND BEND ALLOWANCE

In bending or forming sheet metal parts the minimum radius to which the sheet can be bent is determined by the type of material, the thickness, and the equipment available. In any bend, the material on the outer portion of the bend is stretched and that on the inside is compressed. Somewhere between the two sides there is a line that has not been changed in length. This is referred to as the *neutral line* or *surface*. Experience has shown that this line is approximately 44% of the thickness from the inner or compressed side. In making a flat pattern, therefore, this line will develop in its true length.

Since the circumference of a circle is $2\pi r$, the length of a curve of 1° will be $(2\pi r)/360$ or $(\pi r)/180$. This reduces to $0.01745r$, where r is the radius to the neutral surface and is equal to $(R + 0.44t)$. If this quantity is multiplied by the angle of bend in degrees, the total length of material required to make the bend may be readily computed as $0.01745rA$, where A is the angle of bend. A simple illustration is shown in Fig. 18.65. To save the labor of computation, tables of bend allowances for 1° are usually used.

18.31 DEVELOPMENT BASED ON MOLD LINES

It will be noted in Fig. 18.65 that the distance to the bend line must be known to make use of the formula and table mentioned in the preceding article. Frequently, however, the distance to the mold line is known, rather than that to the bend line. Under these conditions it is simpler to use the distance between mold lines, called the *setback*, in determining the length of a flat pattern.

The mold line is the line of intersection of the two faces on each side of the bend, as shown in Fig. 18.66. Although there is a mold line for the

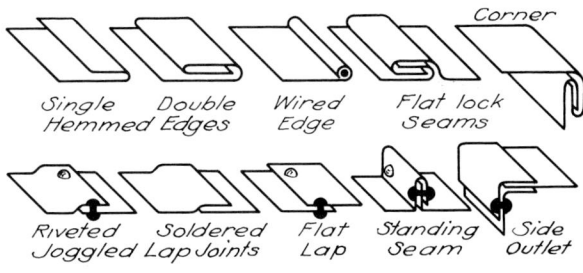

FIG. 18.64. Sheet metal joints.

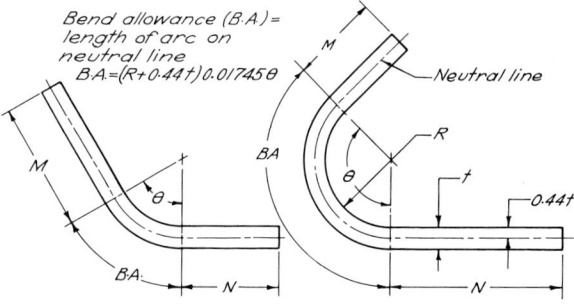

FIG. 18.65. Bend allowances.

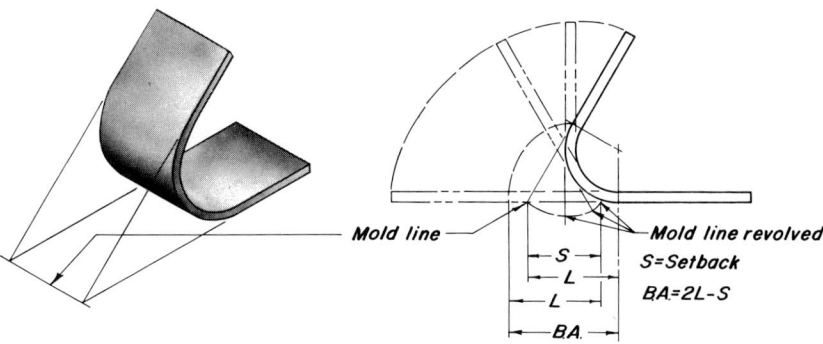

FIG. 18.66. Mold lines and setback.

inside surfaces as well as the outside surface, the outside mold line is generally used. It will be noted that although there is only one outside mold line for the part in its bent condition, there are two mold lines in the flat shape. In other words, there is a mold line for each face. The method of computing the setback and of using it in determining the developed length is shown in Fig. 18.66. The meaning of open and closed bevels is illustrated in Fig. 18.67.

18.32 BEND RELIEF

When two edges of a flat pattern are bent up, the corner is subjected to stresses that may tear the material. In order to avoid tearing, the corner is cut out on a circular arc to relieve this strain. This curve is frequently drawn tangent to the bend lines although it need not be. The chief purpose is to remove material that is subject to bending in two directions. A few simple layouts for the development of patterns are shown in Fig. 18.68.

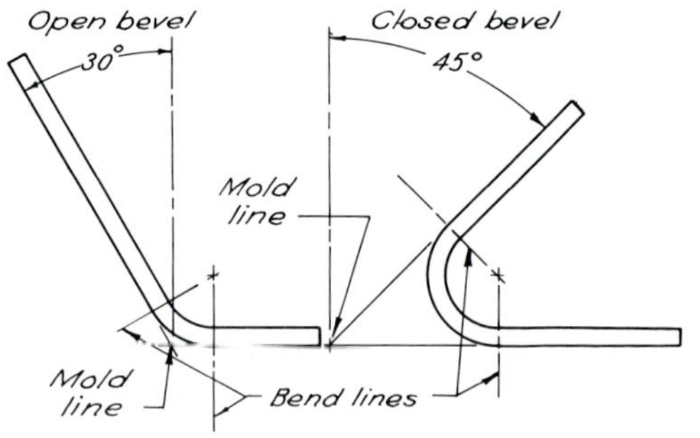

FIG. 18.67. Open and closed bevels.

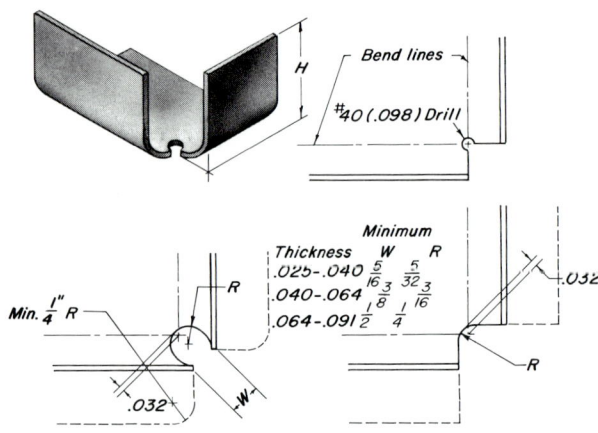

FIG. 18.68. Bend relief.

SELF-STUDY QUESTIONS

Before trying to answer these questions, read the chapter carefully. Then, without reference to the text, answer as many questions as possible. For those that cannot be answered, the number in parentheses following the question number gives the section in which the answer can be found. Look it up and write down the answer. Check the answers that you did give to see that they are correct.

18.1 (**18.2**) In this book prisms and pyramids are treated as _____.

18.2 (**18.2**) Cylinders and cones are considered to be _____ without ends.

18.3 (**18.3**) The usual process of finding intersections consists of passing _____ _____, which cut elements from both surfaces.

18.4 (**18.3**) The first cutting plane should be _____ to one or both surfaces.

18.5 (**18.3**) The cutting planes should be _____ in order.

18.6 (**18.3**) The elements should be drawn showing their proper _____ in their own surface only.

18.7 (**18.3**) The elements should be given the same _____ as the plane.

18.8 (**18.3**) The points should be given the same _____ as the elements.

18.9 (**18.3**) The points should be connected in _____ order.

18.10 (**18.3**) The line of intersection can change visibility only where the curve of intersection touches a _____ _____ or crosses an _____ _____.

18.11 (**18.5**) The intersection of a plane and prism can be solved by the _____ _____ method or by the _____ _____ method.

18.12 (**18.5.1**) When the intersection of a plane and object is found by getting the edgewise view of the plane, the _____ method is being used.

18.13 (**18.5.2**) The first step in the _____ _____ method of finding intersections is to pass a plane through an element.

18.14 (**18.9**) When working with a surface of revolution, the planes should be passed to cut _____ from the surface.

18.15 (**18.11**) When finding the intersection of a prism and some other surface, it is usually best to find the _____ _____ of the prism.

18.16 (**18.12**) In the top view the _____ points are visible.

18.17 (**18.12**) In the vertical projection the _____ points are visible.

18.18 (**18.16**) In getting the intersection of two cones, the _____ _____ method should be used.

18.19 (**18.17**) In finding the intersection of a cone and cylinder, the plane should be passed through the _____ of the cone and be _____ to the elements of the cylinder.

18.20 (**18.16**) In finding the intersection of two cones, the plane should be passed through _____ _____.

18.21 (**18.15**) If the first and last planes are tangent to both surfaces, the curve of intersection will be two _____ curves with _____ points in common.

18.22 (**18.21**) In developing a surface, every line in the development must show _____ _____.

18.23 (**18.22–18.22.3**) In developing a prism or a cylinder it is necessary to find the true size of a _____ and the _____ _____ of the elements.

18.24 (**18.25.2**) In developing a cone it is best to find the _____ _____ of the elements and the true size of the _____.

18.25 (**18.2**) The development of a double-curved surface can only be _____.

18.26 (**18.29**) When laying out a sheet metal development _____ _____ must be allowed for joints.

18.27 (**18.32**) When material is cut out to relieve stress while bending a flat pattern, this is known as _____ _____.

CHAPTER 19

MAP DRAWING

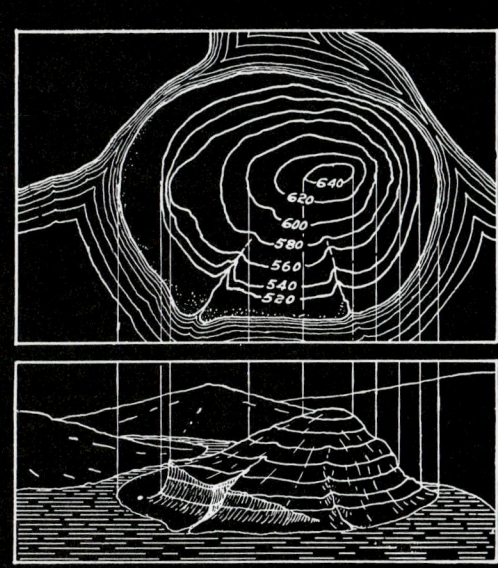

CHAPTER 19

19.1 INTRODUCTION

For the planning and construction of many engineering undertakings, it is necessary to have representations of the earth's surface. Such representations are called *maps*. When the area to be shown is small, a map is essentially a one-view orthographic projection, hence only two dimensions can be shown. This is frequently sufficient. Where the third dimension, namely, the difference in elevation of the earth's surface, is essential, symbols are used to give this information. If the area to be shown is large or if extreme accuracy is desired, other types of projection are used. Such projections are beyond the scope of this book.

19.2 CLASSIFICATION OF MAPS

Maps are conveniently classified on the basis of their purpose or intended use. On this basis maps may be divided into four classes: geographic, topographic, cadastral, and engineering. Although no hard and fast lines can be drawn between the various classes of maps, the distinctions are usually quite clear, as may be noted from the descriptions in the following paragraphs.

19.3 GEOGRAPHIC MAPS

Maps of this group show a comparatively large area and must, therefore, be drawn to very small scales, which means, of course, that only the more important features of the earth's surface can be shown, such as the larger rivers and lakes, mountains, cities, and railroads. On these maps, the cities are located by small circles, and only the larger curves in streams and the principal changes of direction of the railroads are shown. Examples of such maps are to be found in any atlas or geography textbook, hence they are familiar to everyone. The scales vary from a few miles to the inch to several hundred miles to the inch. Relief, or difference of elevation, is shown in a very general way, usually by hachures or shading.

19.4 TOPOGRAPHIC MAPS

The term *topography* means the configuration or shape of the land surface of any area. Because of the details that must be shown, the area covered by such maps is quite small as compared to geographic maps. The most widely known maps of this class are those prepared by the U.S. Geological Survey. A small portion of such a map is shown in Fig. 19.1, which is taken from the Urbana Quadrangle in Illinois in flat country.

U.S. geological maps are made to the following scales, determined by the needs in the economic development of the area shown. These maps are always bounded by meridians of longitude and parallels of latitude.

The larger scales shown below are used in the more highly developed areas and the smaller ones in more sparsely settled or desert regions. The arc of latitude and longitude covered is the same in both directions. A scale of 1:24,000, 1 in. = 2000 ft, covers $7\frac{1}{2}$ min of latitude and longitude.

$$1:62,500, 1 \text{ in.} = \text{nearly 1 mile}$$
$$1:125,000, 1 \text{ in.} = \text{nearly 2 mile}$$
$$1:250,000, 1 \text{ in.} = \text{nearly 4 mile}$$

Index maps and circulars of each state and Puerto Rico showing the areas covered by topographic and planimetric maps are available without charge from the U.S. Geological Survey, Washington, D.C. The charge for individual maps is given in the circulars.

The large-scale maps show all the natural features down to little streams that run dry in the summer. City streets, country roads and trails, tunnels, aqueducts, pipelines underground, bridges, houses, and all the works of humankind together with permanent vegetation such as forest areas are shown. A portion of a privately made topographic map is shown in Fig. 19.2.

19.5 CADASTRAL MAPS

Maps of this class are used primarily for showing political and civil boundaries, together with property lines, and are used for the purposes of

19.5 CADASTRAL MAPS **501**

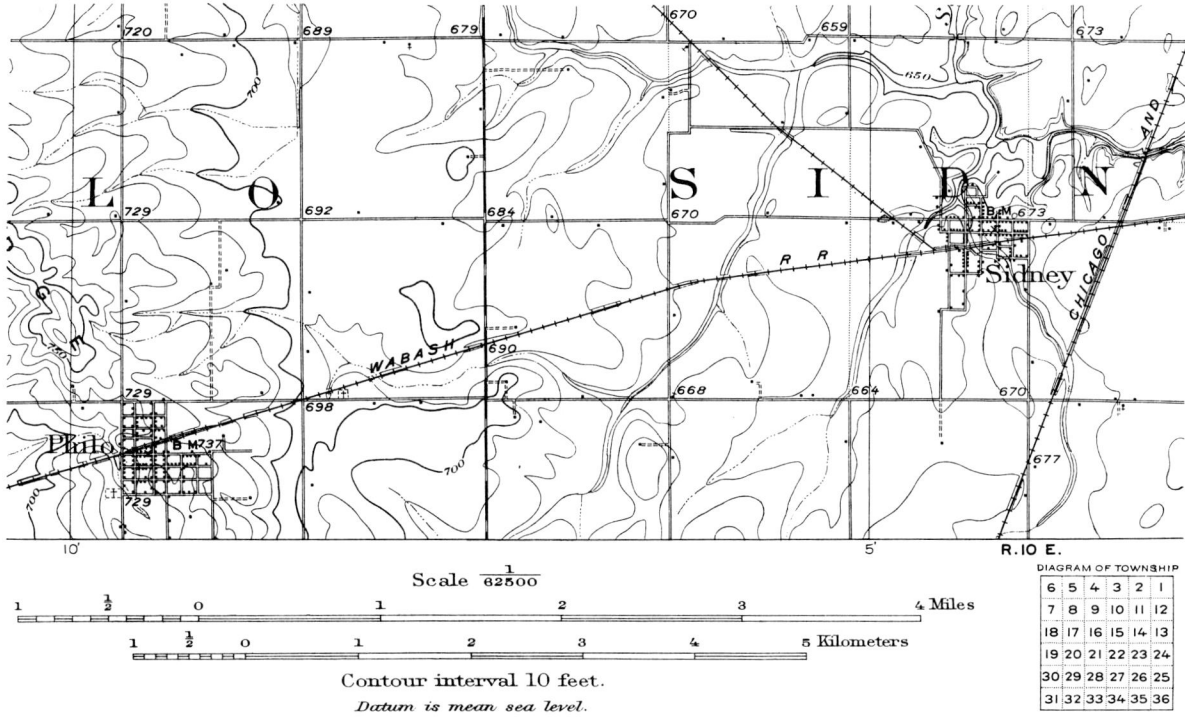

FIG. 19.1. Topographic map: part of Urbana, Illinois, Quadrangle.

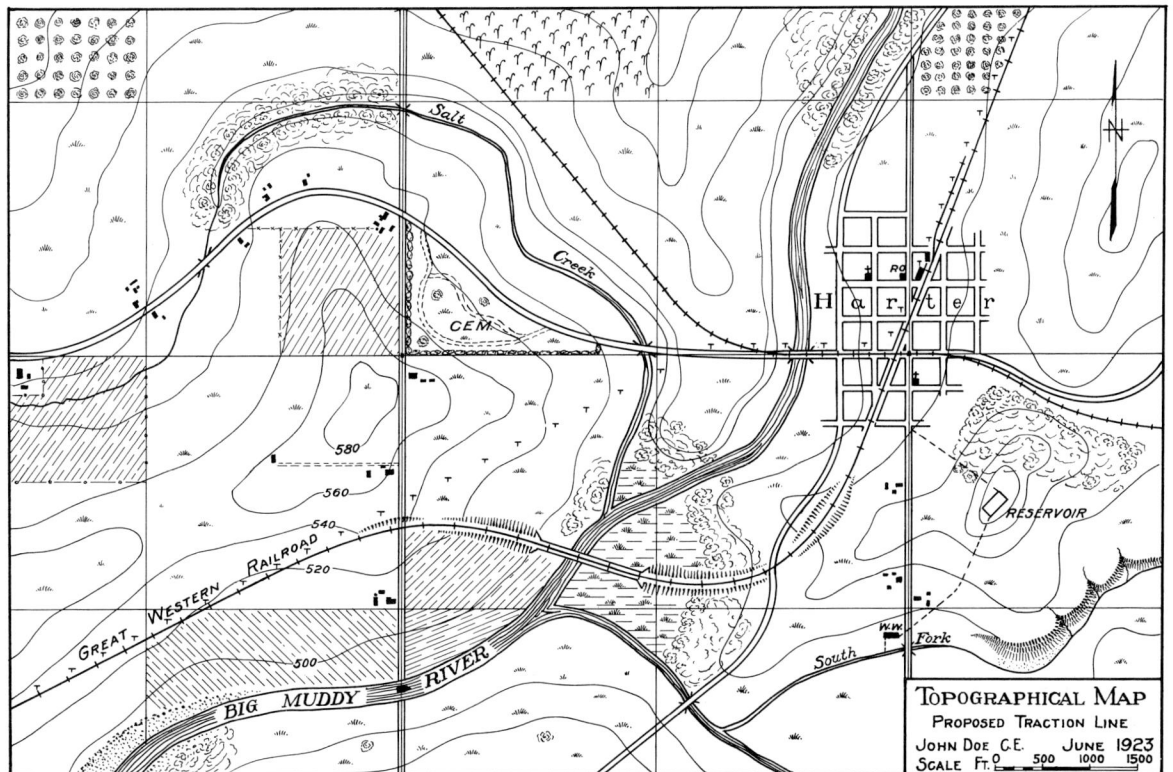

FIG. 19.2. Topographic map.

502 MAP DRAWING

taxation and the transfer of property. Hence, because of the accuracy required, such maps must be drawn on a still larger scale than either of the preceding classes. They contain, besides the property lines, only enough of the natural features, such as streams and roads, to enable one to locate the corresponding lines on the ground. Plats of city additions, mineral rights, farm surveys, and the like fall in this group. The scale for such maps is usually greater than 6 in. to 1 mi.

19.6 ENGINEERING MAPS

Maps drawn for reconnaissance, construction, or maintenance purposes are called *engineering maps*. The scale is seldom smaller than 1 in. equals 400 ft, and it may approach the architectural scales as the other limit, for example, $\frac{1}{8}$ in. equals 1 ft. Maps for railroad, highway, canal, or hydroelectric construction are excellent examples of this class of maps. Such maps frequently have the character of topographic maps in that they include the contour lines, the natural features, and works of man. Being on a larger scale, they are, of course, much more accurate in detail than the usual topographic map. A portion of a highway construction map is shown in Fig. 19.3.

Public utility companies use what are in effect large-scale cadastral maps to show the location of their lines and connections thereto. Figure 19.4 is a portion of a water company map showing the location of the water mains, fire hydrants, and connections to private property.

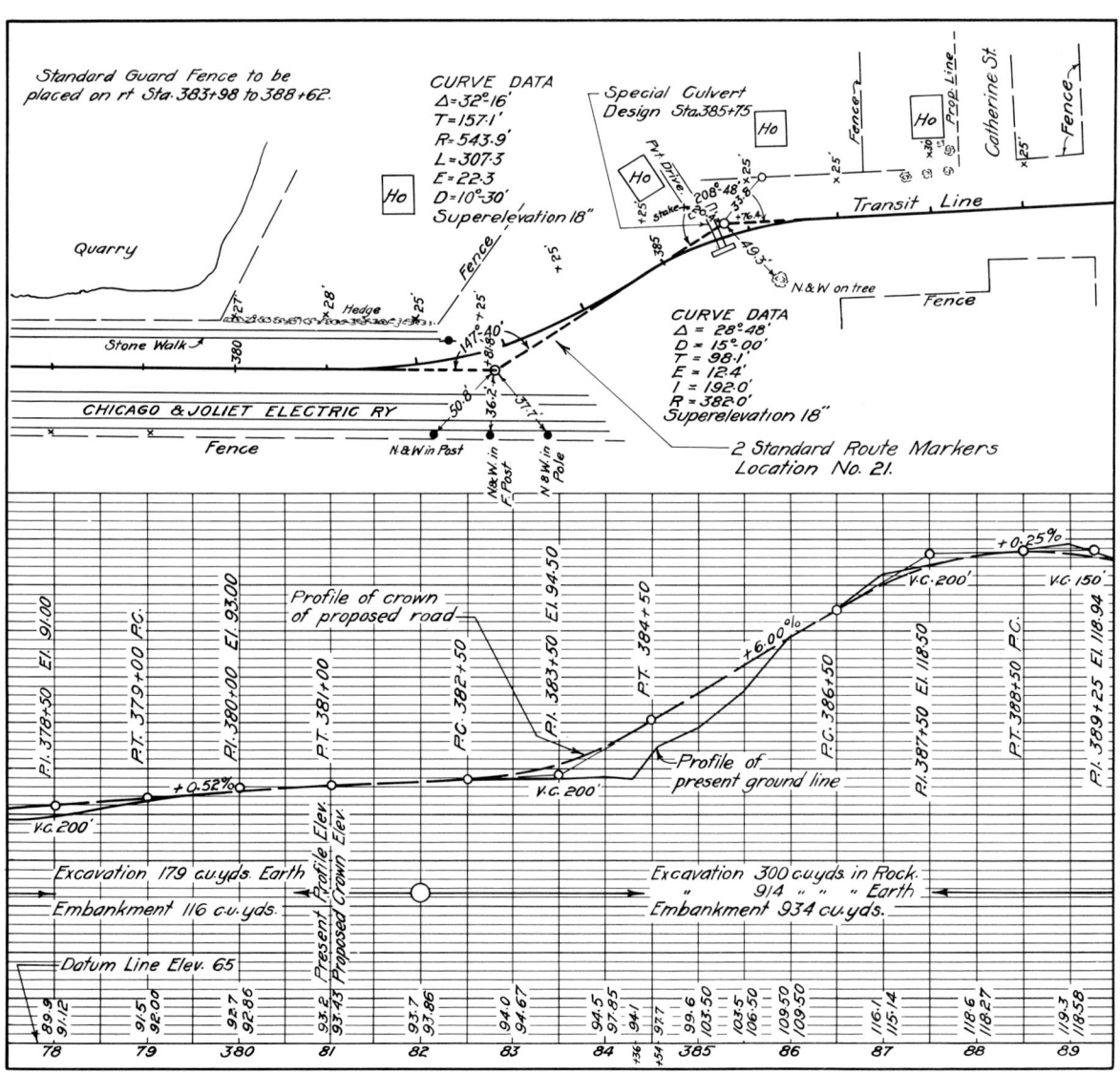

FIG. 19.3. Highway map and profile.

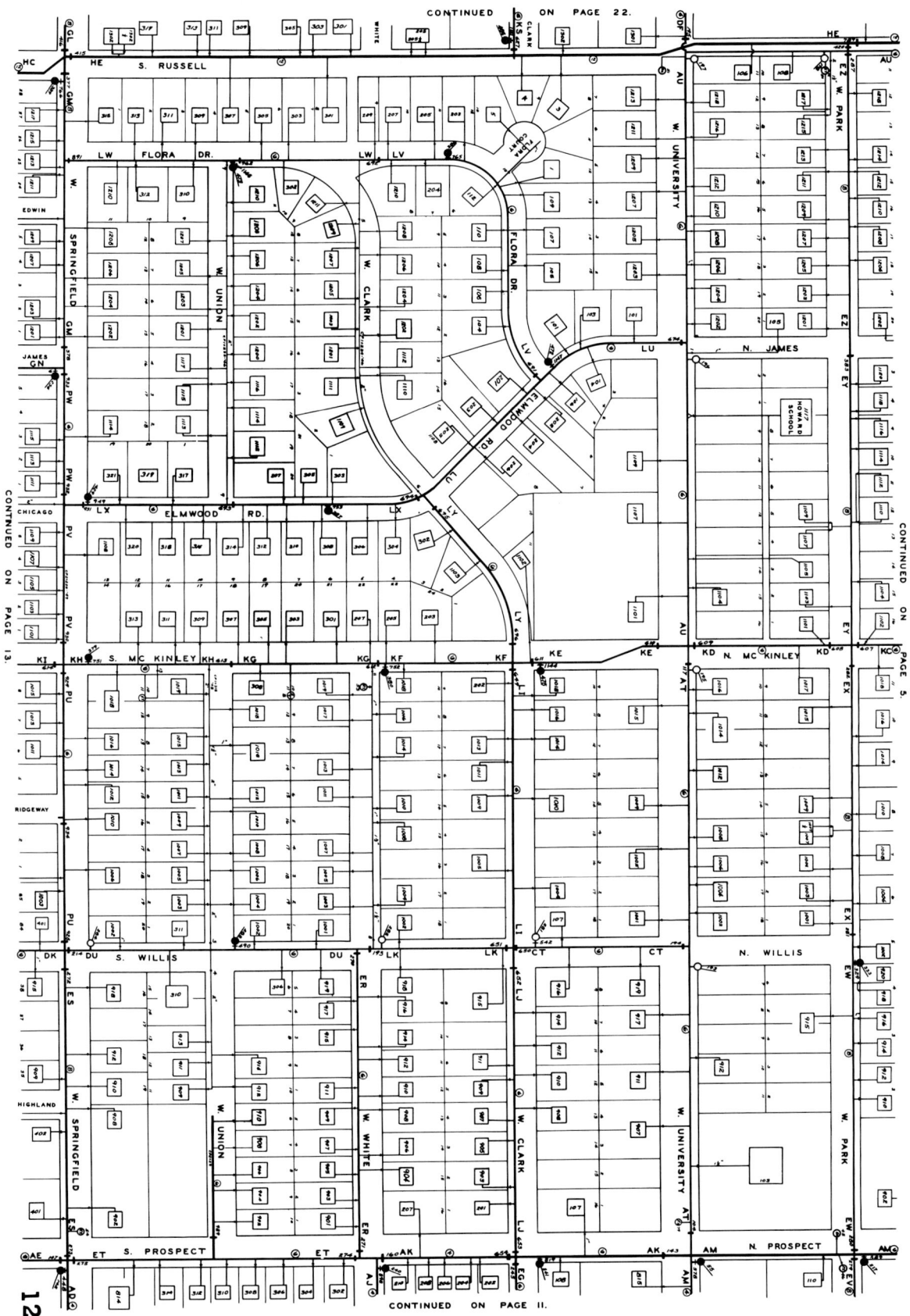

FIG. 19.4. Map of city water supply system (courtesy Northern Illinois Water Corp.).

504 MAP DRAWING

Contour maps, showing principally the elevation of the land, are used for location, estimating costs, and construction. A map of this type is shown in Fig. 19.5. Further details concerning the use of contour maps are discussed in Sections 19.25 to 19.27 inclusive.

19.7 MILITARY MAPS

This book is concerned primarily with engineering maps for civilian use, but there is a close connection between topographic maps for this use and military maps. Military maps may be classified according to the scale, which definitely determines the area that can be represented and the use of the map.

Small-scale maps with scales ranging from 1:1,000,000 to 1:7,000,000 are used for strategical planning by commanders of large forces.

Intermediate-scale maps with scales varying from 1:200,000 to 1:500,000 are used for planning operations, troop movements, concentrations, and supply.

Medium-scale maps with scales running from 1:50,000 to 1:125,000 are required for tactical and administrative studies for units of the size of regiments. U.S. Geological maps (1:62,500) are suitable for this purpose.

Large-scale maps of about 1:20,000 are used for tactical and technical battle needs.

Symbols for military maps are quite extensive and may be found in a booklet called *Military Symbols*.

19.8 MAP SCALES

From the foregoing classification of maps it will be noted that scales are used ranging from 0.1 in. equals 1 ft as the largest to 1 in. equals several hundred miles as the smallest. The scale of a map is shown both graphically and numerically at some place near the title or as a part of it, as shown in Fig. 19.1. Sometimes the numerical scale is stated as a ratio. Thus a scale of 1 in. to the mile is expressed as 1:63,360.

Survey measurements are made in feet and fractions of feet, which are expressed in feet and decimals of a foot instead of in feet and inches; hence the engineer's scales are in the decimal system and have 10, 20, 30, and so forth, divisions to the inch. These units may also represent 1, 2, or 3 ft to the inch, 10, 20, or 30 ft to the inch, or 100, 200, or 300 ft to the in. All these scales occur frequently in engineering work.

The selection of a scale for a map will be influenced by many factors, chief among which are the size and character of the area to be

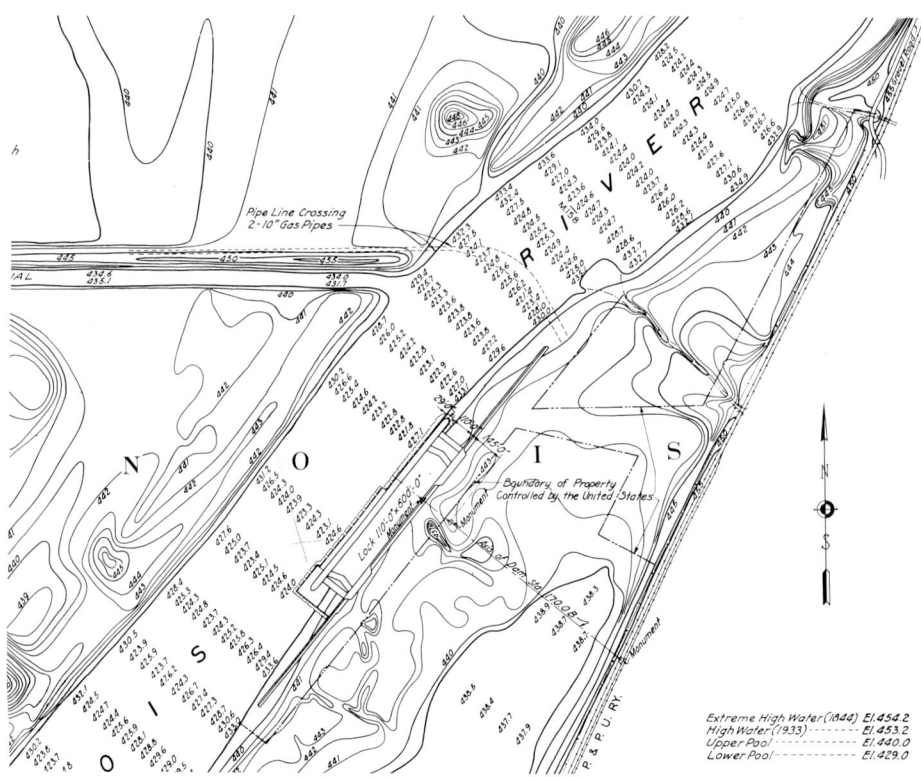

FIG. 19.5. Contour map of construction.

19.9 MAP SYMBOLS

Since the scale of any map is small, relatively speaking, the representation of objects upon it must be highly conventionalized. On all but the largest-scale engineering maps, even the largest objects must be shown by symbols rather than by plan views of them. The purpose of a map is, after all, not to show the exact appearance from above but, rather, to show the comparative size of objects and their position relative to one another. Hence, conventional symbols have been devised that bear some resemblance, where possible, to the objects themselves. The purpose of having this resemblance is for convenience in interpretation.

A well-standardized system of symbols, used by practically all map-making departments of the government, is published in a small booklet called *Military Symbols* published by the Department of the Army and the Air Force.

19.10 SIZE AND PROMINENCE OF SYMBOLS

The size of symbols should vary only slightly with the scale of the map, since, to almost any scale, most symbols are exaggerations, no matter how small they are made. The symbols shown in Figs. 19.6 to 19.17 are the proper size for the usual engineering maps.

Since the variation in size of symbols is quite limited, prominence may be secured by a variation in the weight of lines used. The purpose of the map will determine which symbols are to be made most prominent. On an oil property map, for example, flowing wells, dry wells, railroads, roads, and property lines are the important features. In practically all cases, the vegetable symbols are least important (military maps excepted) and therefore should be drawn lightly and not too closely together.

19.11 HOW TO DRAW SYMBOLS

There are two distinct steps in learning how to draw symbols, the first of which is a careful examination of a correct model or sample of the symbol, and the second, an endeavor to reproduce it. The more important of these steps is the examination of the sample, for upon the keenness and accuracy of the observation depend the effort at reproduction. For example, an examination of the water lining at the right in Fig. 19.6 will make clear that the first lines are drawn very near the shoreline and follow it around very closely, whereas the lines become farther apart and less irregular as we approach the center of the body of water. Examples of both correct and incorrect water lining are shown in Fig. 19.6.

Similarly, a careful examination of the symbol for grass in Fig. 19.7 will show that it consists of about seven short strokes, ranging in length from almost a dot at the ends to about $\frac{3}{32}$ in. at the center. It will also be observed that these strokes are slightly curved and seem to meet in a common center. The individual symbols are arranged at random and not in rows. The careless observer would fail to note many of these essential points, and consequently his attempts at imitation would lack the things that he overlooked. Common errors in making the symbols for grass are shown in Fig. 19.7, in contrast with a correct execution. Likewise, the result of poor observation of the tree symbol is shown in Fig. 19.8.

A large variety of map symbols, printed on cellophane for pasting on a drawing, may be obtained from the trade.

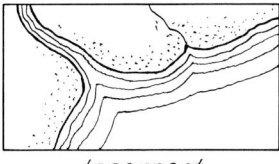
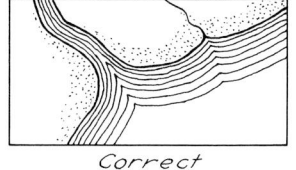
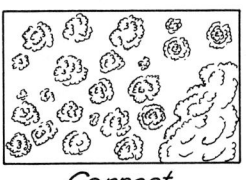
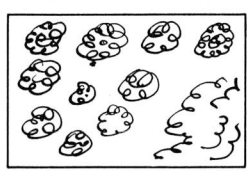

FIG. 19.6. Symbol for water lining.

FIG. 19.8. Deciduous trees.

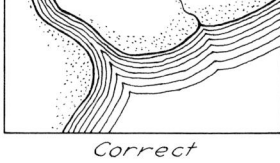

FIG. 19.7. Grass.

19.12 COLORS OF SYMBOLS

On a finished map the symbols should be shown in colors. The color for each symbol in Figs. 19.10 to 19.17 is indicated in the figure title. These colors may be readily remembered by four simple groupings: the artificial features, or works of man, are made in black; water features in blue; contours, sand, washes, and so forth in brown; and vegetation in green. In printing maps, each color requires a separate printing; therefore, a reduction in the number of colors used reduces the cost. Since vegetation, except for very large forests, is not permanent, the green is usually omitted.

19.13 SPACING OF SYMBOLS

One of the greatest difficulties in drawing symbols is to learn how to space those that do not have plotted locations. The general tendency is to cover the sheet too thickly. The drafter must constantly be on guard against this practice for two reasons. First, the more symbols are drawn the longer it takes and the more it costs to produce the map. Second, it is more difficult to produce uniformity of texture when the symbols are crowded. The heavy and light areas on the map are disagreeably noticeable when symbols are placed too closely together. When there are large areas to be covered with symbols involving the use of parallel lines, as in the case of marshlands, a section liner should be used.

19.14 POSITION OF SYMBOLS

Another very important point is the position of the symbols on the sheet. All symbols that have a definite base, for example, grass, marsh, palm trees, and corn, should be drawn with the base parallel to the bottom of the sheet so that the symbols appear in a natural upright position. They should never be placed with their bases parallel to roadways or property lines that run diagonally across the sheet. An illustration of this point is shown in Fig. 19.9. Symbols for vegetation that occur in rows, however, may have the rows running in any direction.

19.15 SPECIAL SYMBOLS

For some purposes, special symbols must be devised. Thus for purposes of aerial navigation a map must show clearly the landing fields of various kinds, beacon, and other aids to navigation, as well as those objects that project up into the air

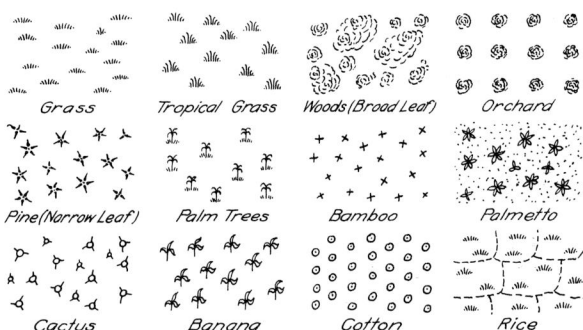

FIG. 19.10. Vegetation symbols (green).

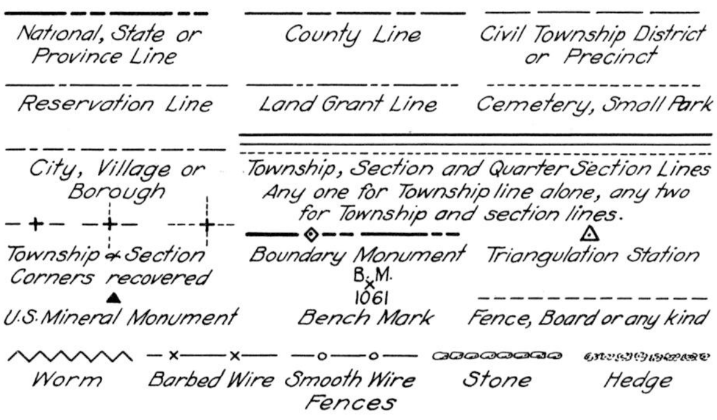

FIG. 19.11. Civil boundaries (black).

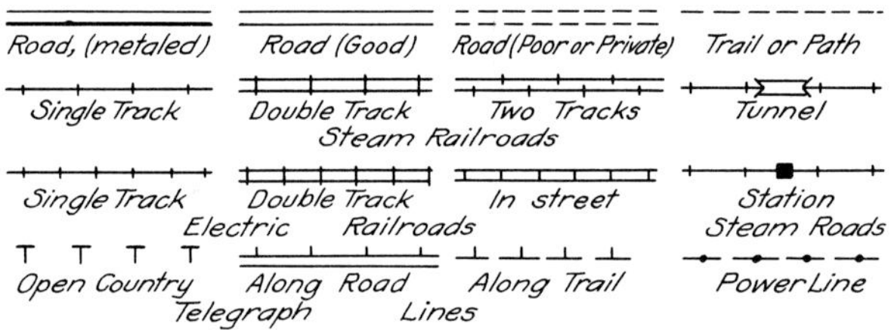

FIG. 19.12. Road and communication symbols (black).

FIG. 19.9. Palm trees and tropical grass.

and may be obstructions to flight. Some of these symbols are shown in Fig. 19.16. Such objects as roads, railroads, railway stations, rivers, lakes, woods, and telegraph lines, which will not interfere with flight, are shown by the usual symbols.

Property maps of various industries may also require the engineer to devise symbols to show certain features that are not included among the standard symbols. It should be the engineer's purpose always to make such symbols unmistakable as to meaning and easy to interpret.

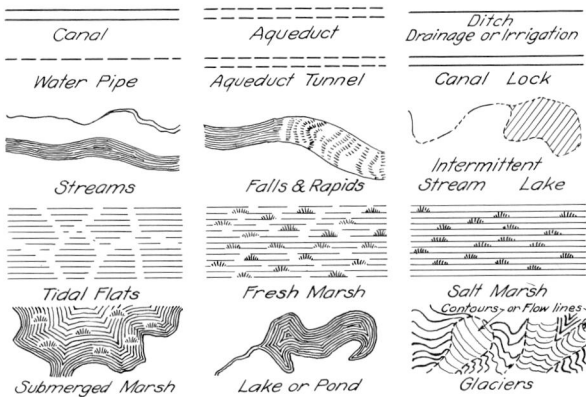

FIG. 19.13. Hydrographic symbols (blue).

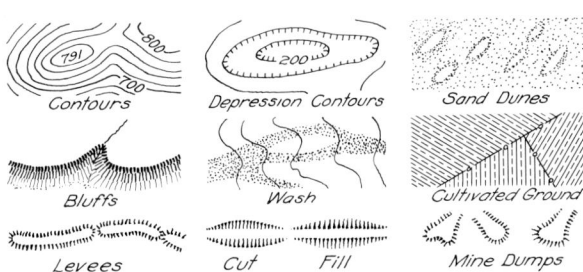

FIG. 19.14. Relief symbols (brown).

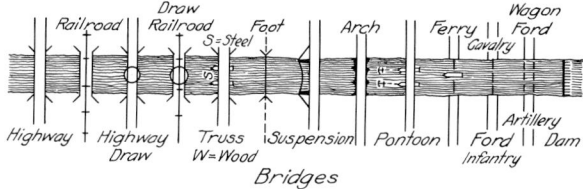

FIG. 19.15. Bridges and fords (black).

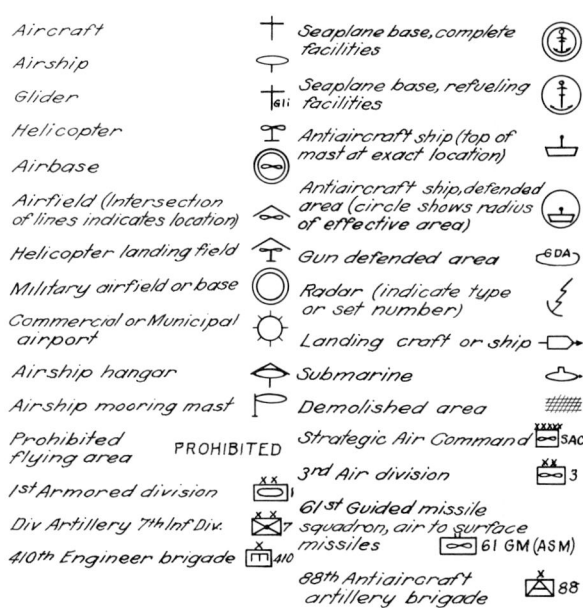

FIG. 19.16. Aeronautical symbols, Departments of the Army and Air Force (black: friendly, single line; enemy, double line); (color: friendly, blue; enemy, red).

19.16 DEFINITION OF TERMS

The following terms are commonly used in surveying and map drafting.

a. *Azimuth.* The azimuth of a line is the angle the line makes with a north and south line measured clockwise from the north. See Fig. 19.18. In older survey notes the azimuth angle is frequently given from the south.

b. *Back azimuth.* The angle measured to the line running in the opposite direction from the azimuth measurement. The back azimuth is therefore equal to the azimuth plus or minus 180°. See Fig. 19.18.

c. *Bearing.* The angle that a line makes with a north and south line measured either from the north or south. It is always less than 90°. The bearing of a line making 57° to the east of north would be specified as North 57° East or N 57° E. See Fig. 19.18.

d. *Backsight.* A sight looking back to the last point or station previously occupied. It is 180° from the foresight. See Fig. 19.19.

e. *Foresight.* In occupying a new point the surveyor orients his transit by sighting back on the point previously occupied. This is the backsight. The telescope of the transit is then plunged 180° to give the foresight. See Fig. 19.19.

f. *Deflection angle.* The angle of a line in a survey measured to the right or left of the foresight. See Fig. 19.19.

g. *Magnetic north.* The north point as indicated by the needle of a magnetic compass that points toward the magnetic North Pole. This varies from place to place.

h. *True north.* A north line established by observations on Polaris, the North Star.

i. *Traverse.* A broken line measured by observation of angles and distances in the field. In property surveys it is usually the boundary line. A traverse may be open or closed. See Fig. 19.19. It is said to be closed when it ends upon the point of beginning or upon a point whose location has been previously determined.

j. *Stations.* The turning points of a traverse. In railroad and highway surveys points on the center line at 100-ft intervals are also called *stations* and are identified during the survey and construction by stakes with the station numbers on them. Points between stations are given the last station number with the distance from that station as a plus quantity.

The following abbreviations are commonly used on maps and map notes.

P.C.   Point of curvature: the point at which the tangent ends and the

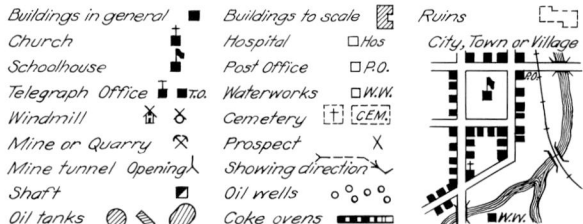

FIG. 19.17. Building symbols (black).

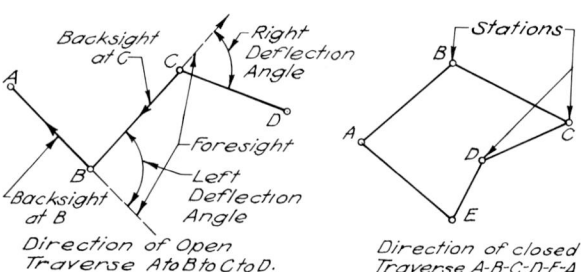

FIG. 19.19. Traverse stations.

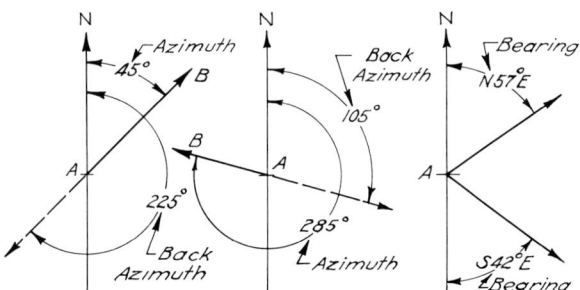

FIG. 19.18. Azimuth and bearing of a line.

P.T. Point of tangency: the point where the curve ends and the tangent begins

P.I. Point of intersection: the point where the tangents to a curve intersect

P.S. Point of switch: the point at which a switch diverges from the main line of a railroad

P.R.C. Point of reverse curve: the point at which one curve ends and another of opposite curvature begins

19.17 PLOTTING A MAP TRAVERSE

The method of plotting map notes depends upon that of making the survey and upon the accuracy required. In general, the plotting on a map indicates or duplicates in miniature the work carried out in the field. Thus an angle measured in the field with a transit may be measured on the map with a protractor, and a distance measured by tape or stadia in the field is measured on the map with a scale.

The three most common methods of plotting transit surveys, in an ascending order of accuracy, are (1) protractor and scale method, (2) tangent method, and (3) rectangular coordinate method.

19.18 PROTRACTOR AND SCALE METHOD

Where great accuracy is not required, the survey notes may be plotted by means of the protractor and scale. The degree of accuracy depends both on the kind of instruments used and on the skill of the drafter. Any errors made are, of course, carried forward but are not necessarily cumulative, since the possibility of error in either direction is the same, and in a large number of measurements these errors will to some extent balance each other, unless the errors are due to a personal and constant tendency of the drafter to overestimate or to underestimate in plotting angles or distances.

For ordinary work, nothing less than a 6-in. celluloid protractor should be used, and this should be tested to see that the 180° and 90° angles, at least, are correct. Steel protractors, with straight edge and vernier attached, are, of course, much more desirable.

A portion of a page of survey notes is shown in Fig. 19.20. Let it be required to lay out the angle at Station 2 to locate Station 3 from these notes. Assume the point B in Fig. 19.21 to be Station 2 and the line BC the "backsight." The first step, then, in laying out the angle is to extend the line CB so far past B that both ends of the protractor may be on the line when the center is at B. Deflection angles are measured to the right and left of the sight line produced; that is, a deflection angle marked right should be laid off to the right of the extended line (Foresight) BA when looking forward along that line in the direction A. Hence, the protractor must be laid on the right or left side of the line from which the angle is measured, according as the notes describe the angle, as measured in the field, to the right or left of the line. Having the protractor set as described above, mark off the angle of 59°40' as accurately as possible, with a very sharp pencil, at the point D. With a straight edge and pencil, draw the line through the point B and the new point just located, and on this line scale off the proper length, namely, 652 ft to locate Station 3.

℄ of R.R. bridge		79°40'L	603'
℄ of Park Road bridge		138°50'L	231'
Shore of island at fork		171°20'L	220'
⊙ 3		59°40'R	652'
℄ of Park Road No.1. on curve		151°45'R	119'
North corner of boat house		121°25'R	575'
⊙ 2			

FIG. 19.20. Survey notes.

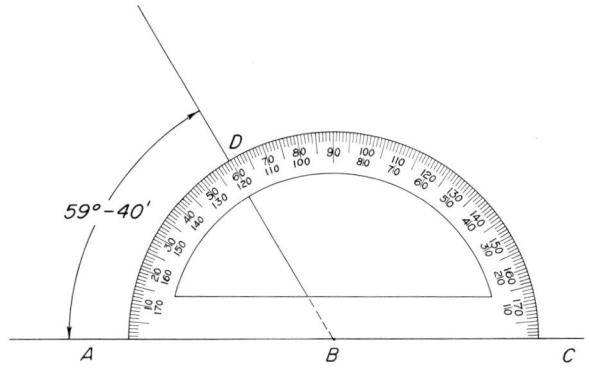

FIG. 19.21. Plotting angles by the protractor method.

19.19 TANGENT METHOD

This method is more accurate for the plotting of angles but requires more time; hence, it is generally used for plotting traverses. The protractor method may be used for plotting in the details. The tangent method requires a table of natural tangents and is, in brief, simply the plotting of an angle on the basis of the definition of the tangent. Again assuming the same angle as in the previous case, draw the line AB (see Fig. 19.22), extend it beyond B, and lay off on it 10 units with any one of the engineer's scales. The larger the scale the greater the accuracy. At the end of these 10 units, erect a perpendicular to the right of the line, since the deflection is marked right, and on it scale off the natural tangent of the angle multiplied by 10, which is 17.0901. Then through the point thus located and the point B, draw the line required, and on it scale off the distance 652 ft.

If the deflection angle is much greater than 45°, greater accuracy may be obtained by first erecting a perpendicular at B and laying off on it 10 units, and then at the end of this ten-unit line erecting another perpendicular on which may be laid off a distance equal to 10 times the tangent of the complement of the angle, as shown in Fig. 19.23.

If the deflection angle is greater than 135°, the 10-unit line should be laid off between B and E, as shown in Fig. 19.24, and the tangent of the supplementary angle must be used.

19.20 RECTANGULAR COORDINATE METHOD

Inasmuch as this method requires considerable trigonometric calculation and is used only when great accuracy is required, it will not be discussed in this book. Complete information concerning this method may be found in surveying texts.

19.21 REPRESENTATION OF ELEVATION ON MAPS

A one-plane projection can show only two dimensions, but for many purposes it is highly desirable that a map shall show three dimensions, namely, length, breadth, and difference of elevation. This object may be attained by two conventional schemes, that is, by the use of hachures or contours.

19.22 HACHURES

If only a general idea of the elevation of the country is desired, the method of hachures is satisfactory, since it gives the effect of relief and is readily understood by the average person. Differences in elevation between any two points, however, can be shown only in a very relative manner. An example of this method of representation is shown in Fig. 19.25, from an examination of which it will be noted that the strokes are

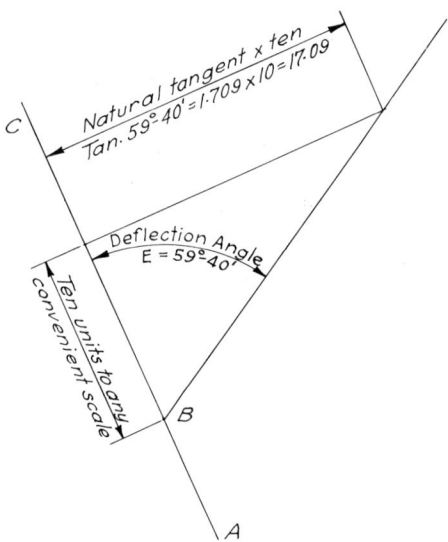

FIG. 19.22. Plotting angles by the tangent method.

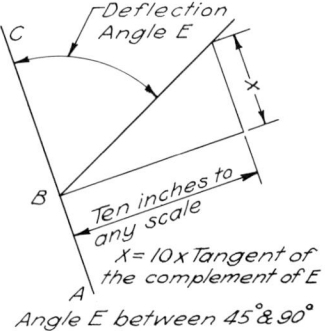

FIG. 19.23. Plotting angles from 45° to 90° by the tangent method.

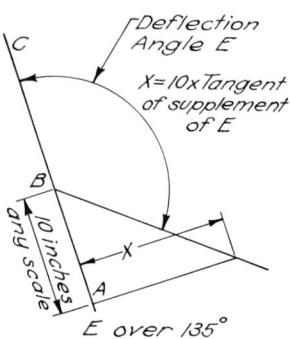

FIG. 19.24. Plotting large angles by the tangent method.

short, heavy, and close together where the slope is steep, becoming gradually longer, lighter, and farther apart as the slope becomes more gentle and approaches the horizontal. The direction of the stroke should be the same as that in which water would flow on the slope. Care should be exercised not to have a continuous white line between the several rows of short strokes.

19.23 CONTOUR LINES

A contour line on a map is the projection of an imaginary line on the earth's surface that passes through all points of the same elevation. The meaning of a contour line may perhaps become a little clearer from an examination of Fig. 19.26, the lower part of which shows a landscape in perspective, and the upper part of which shows the same landscape in map form with elevations indicated by contour lines at intervals of 20 ft. The shoreline is in reality a contour line. The first contour line above the shore represents what the shoreline would be if the water rose vertically 20 ft. To put it in other words, if a man could walk along such a line on the ground, he would go neither up nor down but proceed always on a level and eventually he would return to the place from which he started.

With a little reflection the following rules will be observed to be true, both of imaginary contour lines on the ground and of their projection on a map. The rules are stated as applied to a map.

a. Every contour line must either close upon itself or extend to the edge of the map.

b. When a contour line closes, it usually indicates a summit but it may indicate a depression. When it indicates a depression, this is made clear by the symbol shown in Figs. 19.14 and 19.27.

c. Contour lines never cross, except for an overhanging ledge or a cave.

d. Where contour lines are close together, the surface is steep, and where they are far apart, the surface is gently sloping.

e. When contour lines are close together, they are in a sense parallel to each other (not parallel in a strictly mathematical sense). When they are far apart, they need not necessarily be parallel.

f. Contour lines approaching a stream go upstream before reaching the water's edge,

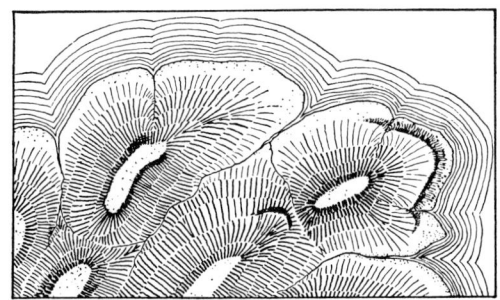

FIG. 19.25. Hachures (brown).

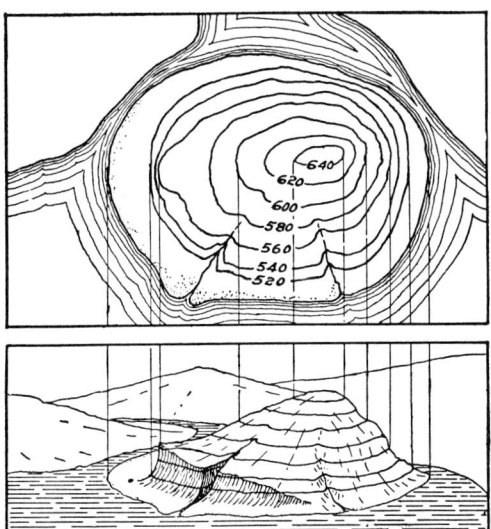

FIG. 19.26. Meaning of contour lines.

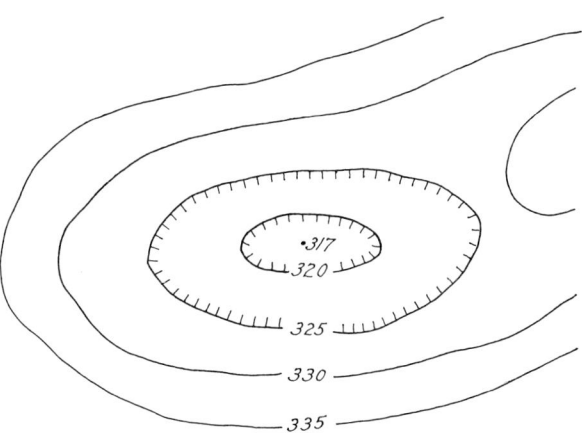

FIG. 19.27. Depression contours.

where they stop at points directly opposite each other at right angles to the stream. If the stream is shown by a single line, they cross it at right angles; if shown by two lines, they do not cross.

g. Contour lines cannot run into the shore of a lake or other still body of water, since the water surface is at the same level at all points.

h. The first contour lines from the water's edge, on opposite sides of a still body of water, must be of the same number or elevation.

i. It is customary to make every fifth contour line heavier than the rest. This line is broken at some convenient place, and the number representing its elevation is inserted in the break in ink of the same color as the line. Where the contour lines are far apart, each one may be numbered.

The numbers indicating the elevation of the contour lines are lettered parallel to the contour line, and, where possible, the numbers for consecutive lines are placed in rows, as shown in Fig. 19.26. If it is possible, these numbers should read from the bottom of the sheet. Contour lines may be drawn with a lettering pen or with a pen designed especially for the purpose and called a *contour pen*. The point of this pen is on a swivel that allows it to turn freely in any direction.

The contour interval may be any desired value, from 1 ft in very flat country up to 200 ft in rough mountainous country. Contour interval means the vertical distance between the planes of consecutive contour lines; 5, 10, 20, 100, and 200 ft are the intervals most frequently used.

On a summit or depression, the last contour line is numbered, or the elevation of the high or low point within the contour is given. See Figs. 19.26 and 19.27.

19.24 PLOTTING CONTOUR LINES

The data for making a contour map may be obtained in the field by obtaining the horizontal location and the elevation of a series of points sufficiently close together so that reasonable interpolations may be made between them. These points may be taken in a rectangular pattern at intervals suitable to the terrain. They may also be taken in radial patterns from traverse points. In the latter case it is usual to choose points on the surface where there is a change in slope. Thus, in Fig. 19.28 elevations were obtained at four points as indicated at the top of the figure. A sectional view shows the slope of the ground surface, and the light horizontal lines are contour planes at 1-ft intervals. A sectional view, of course, is not necessary since the horizontal distance between points can be divided proportionately. Thus in Fig. 19.28 the 605 contour is two thirds the distance from 603 to 606.

With this explanation, the actual work of plotting contour lines from survey data is illustrated in Fig. 19.29. Here the traverse Stations 1, 2, and 3 have been shown with the observations of elevations on the earth's surface plotted from each station in the usual radial pattern. Assuming the ground slope to be uniform between plotted points, the space between points can be divided into a number of spaces equal to the difference in elevation between points as indicated for a number of spaces in Fig. 19.29.

In the upper portion of the map, completed contours have been shown, and in the other portions the lines are sketched in only between plotted elevations. In general, it will be noted that between the radial lines of plotted points the contour lines follow the stream pattern.

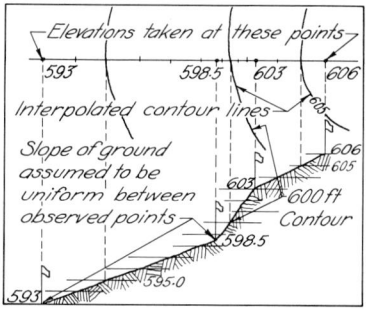

FIG. 19.28. Plotting contour lines by interpolation.

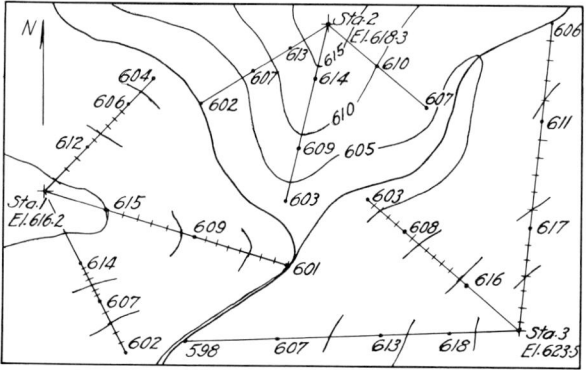

FIG. 19.29. Plotting contours.

19.25 USE OF CONTOUR MAPS

Contour maps are used in engineering work to make preliminary estimates of excavations for structures, in locating dams and computing the volumes of water stored behind them, in computing the area of watersheds, and in many other kinds of work. Figure 19.1 shows a portion of a topographic map taken from a U.S. Geological map of the Urbana Quadrangle in Illinois.

19.26 OUTCROP FROM CONTOUR MAPS

Contour maps also are used extensively in geology and mining. Figure 19.30 illustrates the use of a contour map to determine data concerning a stratum of limestone. The upper surface of the layer was observed to outcrop on the 500-ft contour indicated at a^H. At other places the stratum was covered by overlaying material. Borings were made at B and C at elevations shown on the map. At B the top surface was encountered at elevation 580 and the bottom at 465. At C these values were, respectively, 420 and 305.

Using the map, an elevation view was made of the three points a^V, b^V, and c^V at the known elevations of the top surface of the stratum. A horizontal line drawn across this triangle from a^V to d^V determines the strike line $a^H d^H$, which upon measurement from the north line shows the strike to be S 52°E. By making an endwise view of the strike line and again plotting the known points as at (b) the stratum is shown edgewise and its thickness and slope or dip can be determined as shown in Fig. 19.30b. The dip is shown on the map by drawing an arrow perpendicular to the strike line pointing in the downward direction with the value of the dip lettered on it.

Having the edgewise view of the stratum, the outcrop can be determined by finding where the top and bottom surfaces of the limestone bed cross the 400- to 700-ft contour planes in the edge view and projecting these back to the corresponding contour lines. These are shown by the small circles at the edge of the shaded area for the top surface and the black dots for the bottom.

If, on the other hand, an outcrop of a bed is shown on a contour map, the strike can be determined at once by connecting the points where any one contour line crosses the upper line of the outcrop. Having the strike line, an edgeview of the bed or layer may be obtained and from this the dip and thickness.

19.27 CUT AND FILL FROM CONTOUR MAPS

Contour maps are also used to determine the cut and fill required in the construction of a railroad or highway, as illustrated in Figs. 19.31 and 19.32. Since the sides of a cut or fill are plane surfaces, contour lines on them will be parallel and equally spaced. Finding the outline of a cut or fill therefore is simply a problem in the intersection of surfaces. Having the contours on the map, it is only necessary to draw the contours in the cut or fill at the same levels as the map contours and find where these lines intersect.

If the roadbed is level, the contour lines in cut or fill are parallel to the edge of the roadbed. The spacing of these contour lines depends on the slope of the cut or fill. If the slope is 45° or 1:1, the spacing will be the same as the contour interval of the map, whereas if it is $1\frac{1}{2}$:1, the spacing will be $1\frac{1}{2}$ times the contour interval. The solution of a problem of this type is shown in Fig. 19.31a.

In hilly country, however, level roadbeds seldom occur. There is usually a definite slope, and this slope or grade is expressed in per cent. Thus

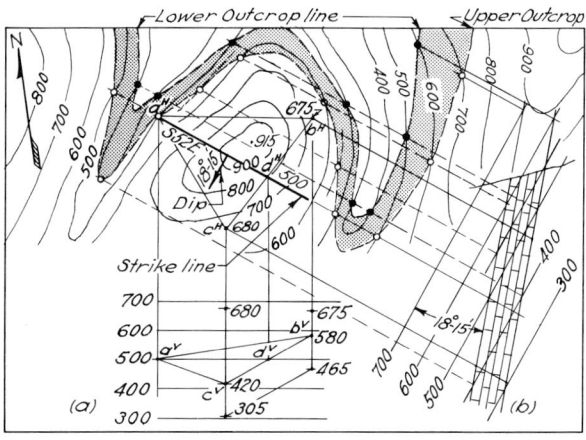

FIG. 19.30. Finding strike and dip from an outcrop.

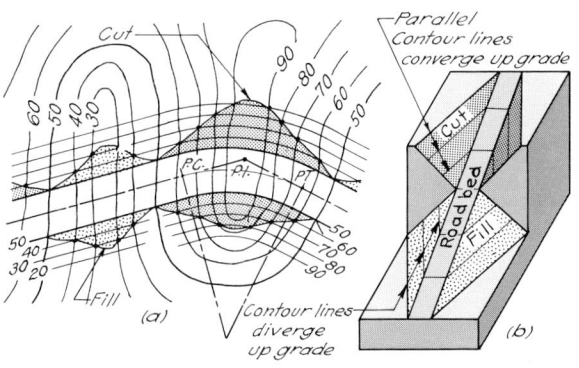

FIG. 19.31. Cut and fill on (a) level grade, (b) inclined grade.

a slope or rise of 1 ft in 100 ft of horizontal distance is called a 1% grade. A rise of 2 ft in 100 ft is a 2% grade, and so on. An ascending grade is marked plus and a descending grade minus.

When the roadbed is on a grade, the contour lines of the cut and fill are not parallel to the edge of the roadbed, as can be seen in Fig. 19.31b. On an upgrade, the contour lines in a cut converge toward the edge of the roadbed and those on a fill diverge. The rate of divergence depends on the grade and the slope of the cut. On a 1 : 1 slope the divergence will be equal to the grade. Thus in a 1% grade and a 1 : 1 slope the contour line will diverge 10 ft from the edge of the roadbed in 1000 ft. With the same grade and a $1\frac{1}{2}$: 1 slope the divergence in 1000 ft would be 15 ft. On curves, the divergence must be plotted at each station and a smooth curve drawn. This has been done on one side of the line shown in Fig. 19.32. Theoretically, cut and fill will meet at the edge of the roadbed on each side. In order not to complicate the illustration, the culvert necessary for drainage was omitted.

Where necessary, additional contours may be interpolated on the map as well as on cut and fill.

A more customary method for determining the outline of the cut and fill is by taking cross sections in the field. This is done by establishing a line perpendicular to the center line at each station and obtaining elevations at selected points on this perpendicular line. These cross sections are then plotted on paper and the roadway placed in its proper position, as shown for the two sections in Fig. 19.32. These sections can be used to determine the edge of the cut or fill and to calculate the quantities of earth to be moved.

19.28 PROFILES

A profile is a line showing the elevations of the ground along some one particular line on the earth's surface. Although a profile represents something entirely different from a contour, the two are nevertheless related in such a manner that one may be obtained from the other. A profile usually accompanies a map showing a road, railroad, sewer, water-supply line, or canal location. If the profile is to be made very accurately, the elevation of points on the line should be obtained by means of a level in the field. However, a profile for preliminary purposes may be obtained from a contour map as shown in Fig. 19.33.

When elevations are obtained with an instrument in the field, readings are taken every 100 ft in flat country and at closer intervals of 50, 25, or 10 ft in rough country, depending on the ruggedness of the slope. These readings are then plotted on a special coordinate paper called *profile paper*, in which the spacing of coordinates is different in the two directions.

When the elevations are obtained from a contour map, the proposed line is drawn on the map and the intersections of this line with the contour lines give the elevations of points whose dis-

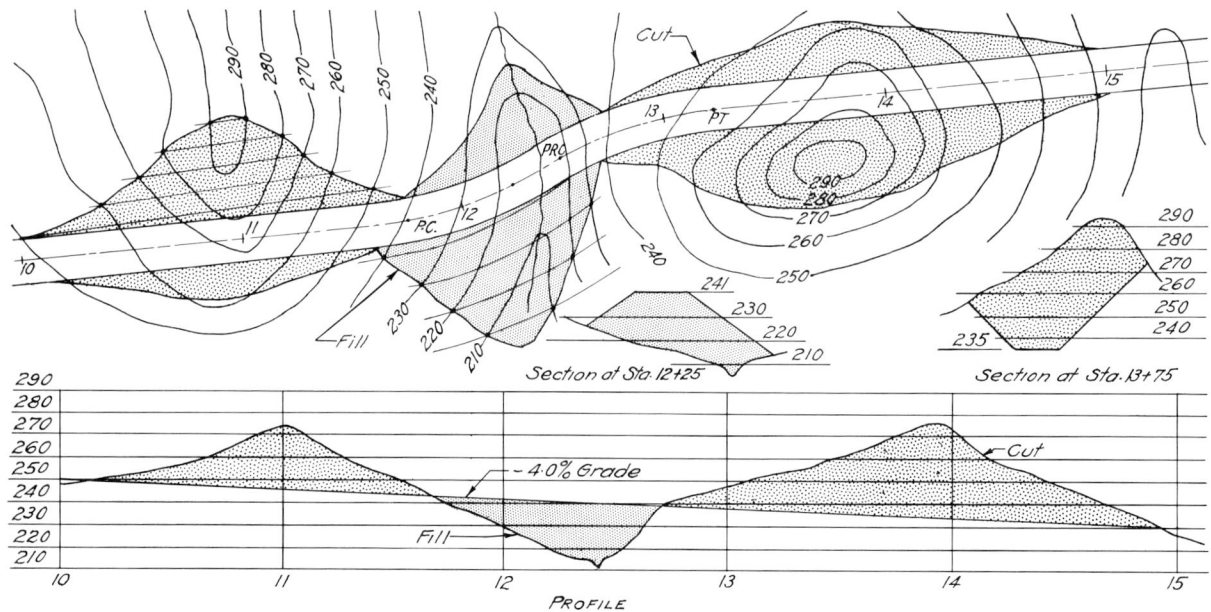

FIG. 19.32. Plotting cut and fill on an inclined grade line.

tances apart are obtained by scaling the map. These points are then plotted on the profile paper.

Profiles, as indicated in a preceding paragraph, are usually plotted to two different scales, the larger of which is used on the vertical axis. The purpose of the two scales is to show the variations of elevation more clearly. Since a profile is usually thousands of feet or several miles in length, whereas the difference of elevation varies only over a few hundred feet, the scale that would bring the horizontal length within workable limits would make the vertical distances so small as to be insignificant.

19.29 PROFILE ON CURVES

When a profile is made of a line, a portion of which is curved, like a railroad line, for example, the developed length of the curve is shown in profile and not the projected length. In other words, the length of the profile is the same as the true length of the line. The beginning and ending of the curve are shown, and the degree of curvature is indicated, as in Fig. 19.34.

19.30 GRADE LINES

In engineering work, where maps are used, a profile is seldom drawn except for the purpose of establishing the grade line of some such structure

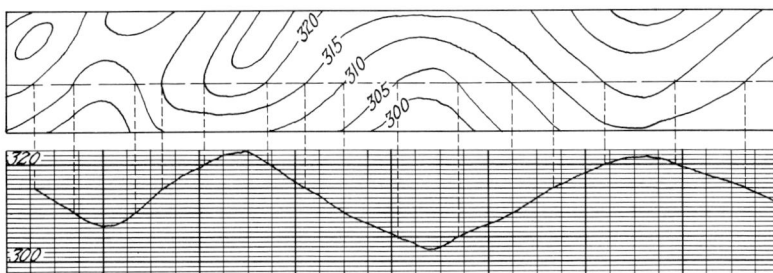

FIG. 19.33. Profile determined from a contour map.

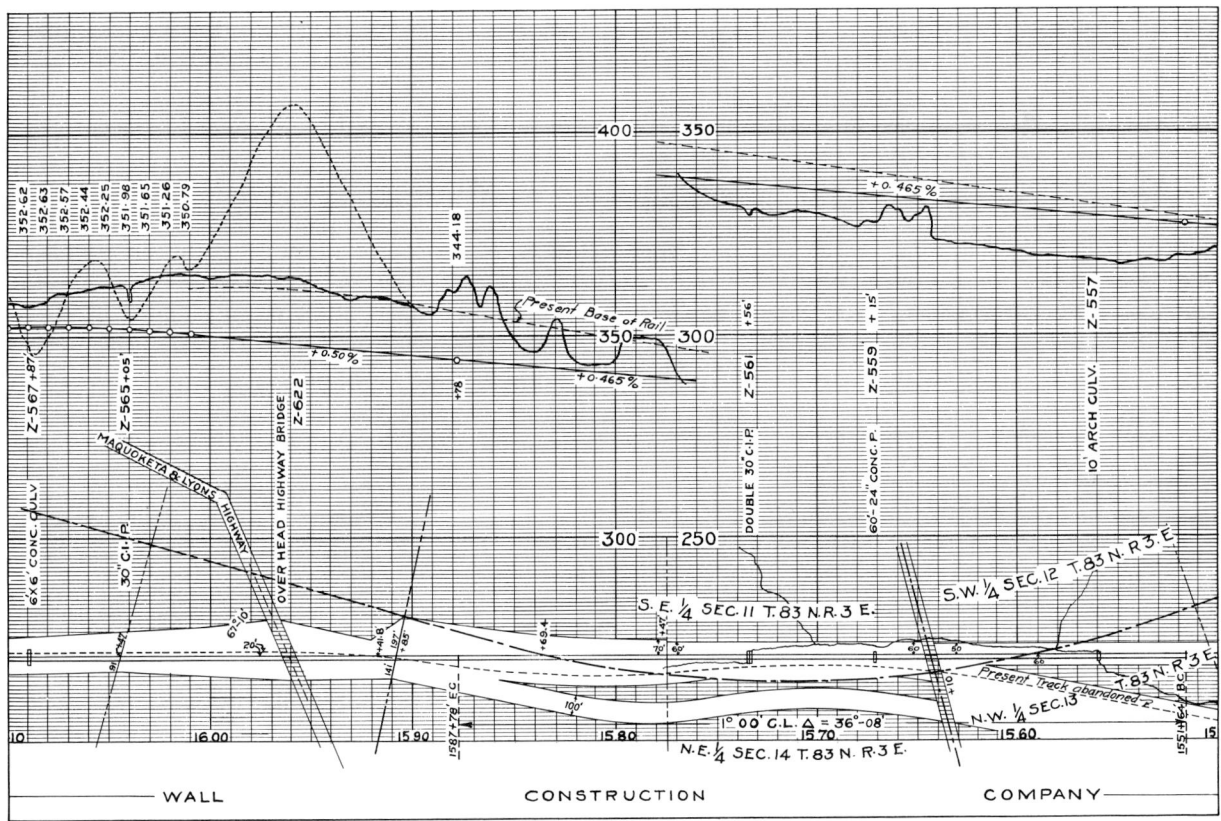

FIG. 19.34. Railway profile.

as a railroad, highway, sewer, or other engineering project. The grade line is the controlling line in construction of the types of structures mentioned. It establishes the slope or deviation from the horizontal. The grades of lines are specified in percentages and represent the number of feet of vertical rise or fall in 100 ft horizontal distance. Grades are specified as plus when the slope is upward, and minus when the slope is downward, in the direction in which the line is laid out. See Fig. 19.32. Thus a −4% grade means a fall of 4 ft in 100 ft horizontal distance.

19.31 VERTICAL CURVES

In lines of any considerable length a uniform grade cannot be maintained from end to end. Although two grade lines of different slope will intersect in a point, in actual construction they must be joined by a vertical curve in order to smooth out the otherwise abrupt change of direction which would be disastrous on highways and railroads. These vertical curves are usually laid out as parabolas in the following manner and as indicated in Fig. 19.35. Lay out on opposite sides of the point of intersection of the grade lines the same number of 10-, 25-, or 50-ft spaces. In practice both the length and number of these spaces are arbitrarily selected to suit the length of the curve and the nature of the work. The elevation of the end points E and D of the curve in Fig. 19.35 may be determined from the grade lines, and the elevation of the midpoint C of the line ED computed. The parabola passes halfway between A and C at B. With this point established, other points on the parabola may be determined by the fact that the offset from the tangent to a parabola varies as the square of the distance along the tangent. The value of the offset at B being known, offsets at the other points may be computed as shown in Fig. 19.35.

19.32 HORIZONTAL CURVES

When railroads and highways change direction, the change is accomplished by means of circular curves that join the straight parts of the line that are called *tangents*. The curvature is specified in degrees, as for example, a 3° curve. A 3° curve is one on which a chord of 100 ft subtends an angle of 3° at the center. Circular curves are joined to the tangent by an easement or spiral curve, but this spiral portion is not shown in the usual map. The radii of curves for various degrees is given in the appendix.

19.33 LETTERING

Engineering maps, particularly those drawn to a large scale for the purpose of construction, are usually lettered in single-stroke Reinhardt letters, either slant or vertical, except the titles, which may be made in a more ornamental style. On geographic and U.S. government maps, the lettering is in modern roman with certain variations designed for special purposes. A competent map drafter must be a master of this style of lettering.

Although the lettering is about the last thing to be inked on a map, the placing of it must receive attention during the preliminary pencil work; otherwise, there will often be no place to put some very essential information when the work is nearing completion. As in all other types of drawing, lettering should be placed, as far as conditions will permit, so that it may be readable from the bottom and right-hand side of a drawing.

19.34 TITLES

The title of a map is usually placed in the lower right-hand corner, if possible. It should contain a statement of what the map is, that is, Plat of Jones Subdivision, the location of the ground, the name of the person or company for whom the map was made, the date of the survey, the scale, the north

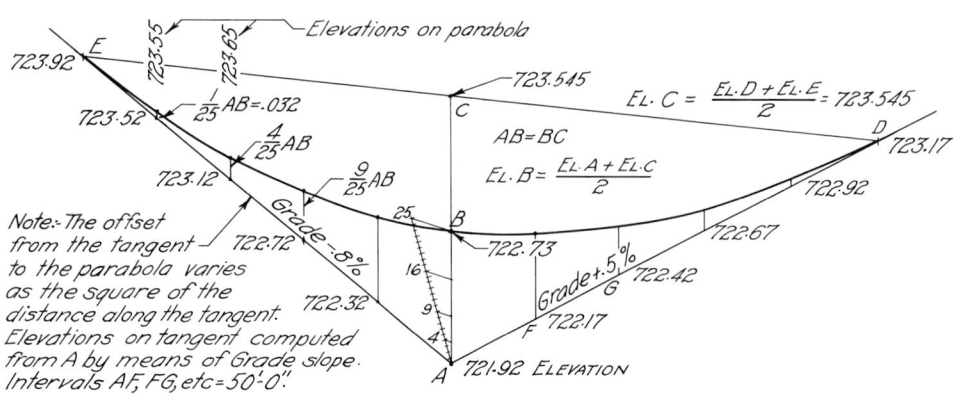

FIG. 19.35. Plotting vertical curves.

point, and the name or initials of the drafter. The name of the surveyor may be in the title or it may occur only in a statement that certifies that the survey and map are correct. This statement and signature must be written. They are usually placed near the title but are not a part of it. The scale is frequently represented graphically below the title. In no case should the title be enclosed in a box.

On engineering maps Gothic letters are used; on the more highly finished maps the roman style is preferred.

19.35 NORTH POINTS

Every map should have the direction of the meridian indicated by means of a suitable arrow. Unless otherwise specified, this arrow points true north. The south portion of the arrow should be somewhat longer than the north portion to give it a balanced appearance. The barb and tail should be narrow and graceful, thus avoiding an arrow that is too bold or conspicuous. Sample arrows are shown in Fig. 19.36.

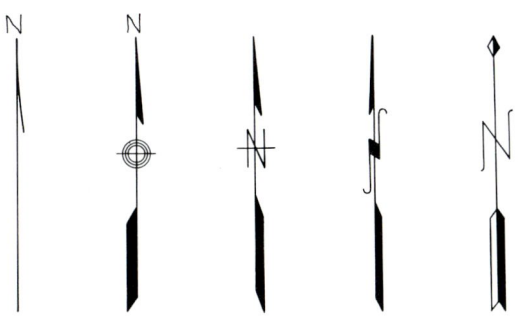

FIG. 19.36. North points.

SELF-STUDY QUESTIONS

Before trying to answer these questions, read the chapter carefully. Then, without reference to the text, answer as many questions as possible. For those that cannot be answered, the number in parentheses following the question number gives the section in which the answer can be found. Look it up and write down the answer. Check the answers that you did give to see that they are correct.

19.1 **(19.1)** When the area to be shown on a map is small, the map is essentially a one-view _____ projection.

19.2 **(19.3)** Maps that show large areas of the earth's surface are called _____ maps.

19.3 **(19.4)** Maps prepared by the U.S. Geological Survey are classed as _____ maps.

19.4 **(19.6)** Engineering maps are drawn for the purpose of _____.

19.5 **(19.6)** The scale of an engineering map is seldom smaller than 1 in. equals _____.

19.6 **(19.7)** Many U.S. geological maps have a scale expressed as 1 : _____.

19.7 **(19.7)** Large-scale military maps have a scale of 1 : _____.

19.8 **(19.9)** On even large-scale engineering maps objects are shown by _____ rather than as true projections.

19.9 **(19.10)** The size of symbols should vary only _____ with the scale of the map.

19.10 **(19.12)** When topographic maps are drawn in color the following colors are used:
   (a) Water features _____
   (b) Contour and sand _____
   (c) Vegetation _____
   (d) Works of man _____

19.11 **(19.14)** All symbols that have a definite base line should have this line _____ to the bottom of the sheet.

19.12 **(19.16)** The azimuth of a line is the angle that a line makes with a north and south line measured _____ from the north.

19.13 (**19.16**) The bearing of a line is the acute angle that a line makes with a north and south line measured _____ from the north or south.

19.14 (**19.16**) The bearing angle is always _____ than 90°.

19.15 (**19.16**) A deflection angle in a survey is measured to the right or _____ of the _____.

19.16 (**19.18**) Angles may be plotted by the _____ method when great accuracy is not required.

19.17 (**19.19**) The tangent method of plotting angles requires the use of a table of _____.

19.18 (**19.23**) Differences in the elevation of the land can be shown on a map by _____ lines.

19.19 (**19.23**) When contour lines are close together the slope of the land surface is relatively _____.

19.20 (**19.23**) Contour lines _____ cross.

19.21 (**19.23**) Contour lines cannot _____ upon the edge of a still body of water.

19.22 (**19.28**) A profile is a line showing the _____ of the earth's surface along some specific line such as a railroad center line.

CHAPTER 20

COMPUTER GRAPHICS AND CAD/CAM

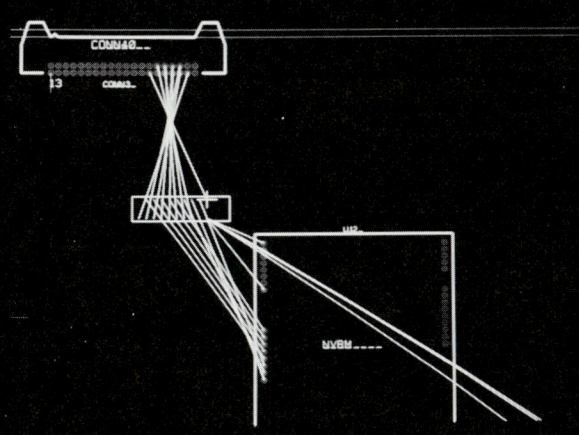

CHAPTER 20*

20.1 INTRODUCTION

This chapter addresses the subject of computer graphics and its use in the expanding field of computer aided design and computer aided manufacturing (CAD/CAM). It will attempt to give some notions of what computer graphics is and how it works. In addition, it will explore some of its representative uses in CAD/CAM. Finally, it will offer a perception of the important ties between traditional engineering graphics and computer graphics in CAD/CAM.

It is an inescapable fact that computer graphics is changing the way in which the engineering profession and related activities are being conducted. Computer graphics is now being widely used and will become an even more extensively used tool in the years to come. Thus, it is quite important to gain some insight into what computer graphics really is, how it works, what roles it can play, and how it relates to the subjects of traditional engineering graphics covered in the preceding chapters of this text.

To be sure, the subjects of computer graphics and CAD/CAM are vast. Moreover, they are ones of continuing change brought about by technological and computer programming advances. As a result, this chapter, nestled within a broader text, must necessarily be restricted in scope and depth. The detailed study of the subjects rightfully belongs to current texts and classes devoted to the topics. Some pertinent textbooks available up to publication time are cited in the Selected Bibliography at the end of the chapter. Some useful reading may also be found in popular journals and programming texts such as those available for home computers.

No doubt some readers will be involved in classes that utilize computer graphics equipment. In those situations, this chapter should provide a helpful framework as you become familiar with some of the principles, methods, and uses of the systems employed.

---

* This chapter was written by Dr. Michael Pleck, Associate Professor, University of Illinois at Urbana-Champaign.

20.2 HISTORICAL PERSPECTIVE

Simply stated, computer graphics is the marriage of visual or graphical information and computer equipment and methods. Computer graphics as a discipline is over 30 years old, its present modern form over 20. Its impetus came from the need of computer users to comprehend more easily the results of vast amounts of tabularized computer computations. It was, indeed, a classic manifestation of the old Chinese proverb that states, "a picture is worth a thousand words." Gradually, computer graphics began to expand its role, fostering graphical interactions and the concomitant storage and retrieval of information. But overall, in its early years computer graphics was stymied by costly or inappropriate equipment. Only recently has it become successful in practical and commercial terms in a number of fields. This success has been due in part to technological advances and cost reductions in computer equipment and a maturing of computer program strategies to take full advantage of the emerging hardware. It has also been due to a closer working relationship between the developers and the users of computer graphics systems to define and satisfy needs of the computing community.

Computer graphics is a far-reaching subject that in its own right falls under the heading of computer science. But it is also a powerful tool applied to many disciplines, engineering for one. In this regard, computer graphics is like other subjects, such as mathematics or chemistry. But unlike them, *computer graphics,* as a term, tends to be more broadly used and to be synonymous with its applications.

Within the field of engineering, in its broadest sense, computer graphics is being applied to numerous tasks. These include initial layout, analysis, manufacturing process control, and assembly, just to name a few. For many of these, the use of computer graphics has or is becoming the accepted and standard method for accomplishing the tasks at hand. In many other cases computer graphics is augmenting the use of traditional engineering graphic methods. The forecast for the future is that the use of computer graphics will become even more widespread.

For many engineering functions, special terms and acronyms have arisen when computer graphics is used. There are too many to list completely. Among them are CAD/CAM, computer aided engineering (CAE), computer integrated manufacturing (CIM), and computer aided design drafting (CADD). Unfortunately, the definitions of some of these are not clear-cut and overlap others. Some acronyms may fall into disuse or be replaced by new ones. With the passage of time, these problems may be resolved. In this chapter, the most widely used acronym, CAD/CAM, will be applied in a generic sense to cover all of these functions.

Many of us, whether fully aware of it or not, have already encountered computer graphics in our daily lives. One of its most widespread uses, experienced first-hand by many of us, is in so-called video games. This popular form of entertainment is found in microprocessor-based arcade consoles, personal computer systems, and so forth, as depicted in Plate 1. Its major attractions are fast-paced action, eye-catching visual effects (though "rough" in appearance in early systems), and a high level of user involvement or "interaction" to control the course of events.

A more sophisticated use of computer graphics seen by many readers resides in the filmmaking and videotechnology industries. This use is what is regarded as "special effects," the portrayals of science-fiction movie or video scenes and TV network logo and advertising segments, and more, by computer graphics methods—all of which have become quite commonplace, as suggested by Plate 2. Prior to this implementation of computer graphics, similar animations (though seldom as complex) were usually created by the painstaking technique of "cartooning"—rendering each film or video frame in the sequence. But in contrast to some animations created this way, animations produced by computer graphics can be very realistic. This is because the content of each frame of the animation can be made very sophisticated by applying more computer resources. In addition, the dynamic continuity between frames can be closely controlled by interaction with the computer graphics system.

These two examples of commonplace uses of computer graphics are by no means the only ones. But they should serve to make a point. Computer graphics is a part of the fabric of our daily lives and probably will become more so in the future. But, more importantly, this exposure has instilled a familiarity with some of the capabilities of computer graphics systems that are also found in professional applications such as CAD/CAM. This will be helpful in learning more about such applications.

20.3 BACKGROUND AND OVERVIEW

This section sets forth a brief description of a computer system that will be helpful in discussing computer graphics. Further, it presents a capsule look into the issues of what constitutes a computer graphics system and its functions. This preview gives a road map of what is treated in more detail in following sections. Some other relevant issues and terminology are also introduced.

20.3.1 Computing Concepts. No doubt some readers have had an exposure to computers or computer programming. For those who have not, or for those seeking a refresher course, the following highlights of some of the more relevant concepts will be of help in understanding some of the workings of computer graphics systems.

By the use of the word *computer*, *digital computer* is meant. It is one that stores and retrieves user programs and data that are encoded internally by groups of binary digits or *bits*. A grouping of eight bits is called a *byte*, which is an often-used unit of information used to represent characters and numbers. A *word* is another, often associated with the design of a computer. Word lengths of 8, 16, and 32 bits have been the most commonly used. The more bits in a word, the more accurately numbers can be represented. But the tradeoff is a higher cost to store more bits. However, technology has provided a miracle of decreasing costs, year after year, of such storage.

A computer program, or *software*, is a list of instructions that directs the operations that the computer performs. So-called "user" programs are translated into internal binary form by special programs called *compilers*. Some common ones are BASIC, FORTRAN, and PASCAL. In more sophisticated computer systems, including groups of them termed *networks*, a special computer program called an *operating system* is used. Much like a traffic cop, it may direct the flow of information between various devices, the simultaneous running of different computer programs, and the sharing of computer resources among different users, just to name a few of its typical functions.

In contrast to software, the tangible elements of a computer system are known as *hardware*. From a physical standpoint, the computer itself may be likened to an electronic racetrack that manipulates information and shuttles it between different locations. The pathways along which the information flows are miniaturized electrical circuits of a very special kind. The storage locations are termed *memory*. Memory is analogous to the array of rooms in a hotel, each with an *address* and contents of one computer word rather than furnishings. Memory is accessed under strict control to produce the desired *write* and *read* modes associated with the placement and subsequent retrieval of the information, respectively.

Memory is constructed in one of several ways to incorporate magnetic, optical, or other principles to represent bits by "on/off" or "yes/no" encodings corresponding to binary ones and zeros. Many thousands and perhaps millions of memory locations may comprise a typical installation, giving designers and manufacturers real challenges! Some impressive-sounding names, such as RAM (random access memory) and ROM (read only memory), are used to describe types of memory.

At the heart of every computer is the *central processing unit* (CPU) or *processor* for short. This is a hardware device in which the operating system and various other programs perform their functions. The processor performs operations on numbers, such as addition and multiplication, by binary arithmetic. It also can perform comparisons of numbers and characters, and many other tasks. The processor employs a region of memory to store currently used program instructions and data.

Some computers use auxiliary storage media as ways of keeping programs and data not in current use. Examples of such devices storing binary information are magnetic tapes and disks. Their analogies in the audio realm are cassette tapes (which in fact are used in some computer systems) and phonograph records. The former require *sequential access,* that is, movement along the tape, which is relatively slow but inexpensive. The latter may use *random access,* that is, skipping to a particular track, which is fast but relatively expensive.

Input and output devices, often referred to as *peripherals,* are also necessary hardware components of computer systems. Some typical input devices are keyboards, magnetic tape or disk readers, and a variety of contrivances that move a *cursor,* a visible symbol or mark, on a viewing screen or *terminal*. Some common output devices are printers, plotters, magnetic tape or disk writing units, and one of several types of terminals. Some devices can serve both functions.

Some work in conjunction with others, especially terminals. Those associated with the user aspects of graphical input and output will be discussed in Article 20.6.2. The bridges between the peripheral devices and the processor (which controls them) are called *interfaces*. They are physical connections that may employ logic devices.

Aside from an input/output distinction, peripherals may be associated with *passive* or *interactive* systems. The difference lies in whether or not the user can "interact" with the computer system. The early days of computing were largely passive in nature. Total programs were submitted (often via punch cards), executed, and summarized in static form (e.g., on paper)—if they worked—all without any intervention by the user. In contrast, interactive computing permits the submittal, execution, and results of a computer program to be in a hands-on partnership with the wishes of the user to control and interpret the course of events. Obviously, advances in hardware and software made this possible and easier as time passed. One significant consequence of this has been the use of computer graphics to convey ideas in more meaningful ways than mere text input and output, giving rise to *interactive computer graphics*—or interactive graphics for short.

20.3.2 Entities, Models, and Views. In using a computer graphics system, we should be aware of the three levels at which information is considered. The terms *entity, model,* and *view* are used to characterize these levels. Two situations illustrating them are given in Figs. 20.1 and 20.2.

The *entity level* is associated with observed or imagined objects, phenomena, or ideas. It really has nothing to do with computers at all. In fact, the entity to be portrayed need not even be visual, as indicated in Fig. 20.1a. There the cloud and rain streaks are only a cartoon suggestive of rainfall accumulation, something not normally seen except in puddles! The solid object of Fig. 20.2a might have variations in density, temperature, or stress distributions, all which are not intrinsically visual but might be desirable to portray by means of visual presentations.

The *model* level is associated with a representation of the entity that will result in a visual presentation. It may be symbolic, such as the graph of Fig. 20.1b, suggestive and approximative, such as the "stick" model of Fig. 20.2b, or realistic. The illustrations of Figs. 20.1b and 20.2b, however, tell only a part of the model story. The model must be stored in the computer as information, usually by means of a *data structure,* not as a tangible, visual thing as the illustrations suggest! What the illustrations do suggest, though, is how we may interpret the data structure. It is as if we could physically construct the model from tangible ingredients such as wire, paper sheets, blocks, and so forth from directions prescribed by the data structure.

Models for a given entity are not unique. Consider the graph of Fig. 20.1b. It could be replaced by a monochrome bar chart, or a multicolored variation of it in which color, rather than height, represents the magnitude of rainfall. In the realm of three-dimensional objects, there are a number of different ways to create models, including "stick," surface, and solid representations. The methods to do this, together with a deeper look at models are found in Section 20.4.

The method of modeling chosen may depend on factors beyond alternate visual presentations alone. This is because the model may be used for other engineering purposes including design analysis (e.g., determining stresses, temperatures, volumes, motion clearances, and aerodynamic lift) or manufacturing process planning (e.g., metal cutting and assembling). The data structures of such models may also contain nongeometric information necessary for those tasks. If so, they are more complex than models intended for views only. For these tasks, however, the more complex engineering models are often processed by a computer program to create additional view-only models.

The *view* level is associated with the actual visual presentation of the model. The presentation itself may be done on different types of devices (e.g., *plotters* or *display terminals*) and in different modes (e.g., line drawings or continuous tone color images). Associated with each device and mode is something akin to the data structure for models. It is called *display file* or *display buffer* depending upon the situation. It contains the data that, when fed to the device, result in a picture or view. As in the data structure

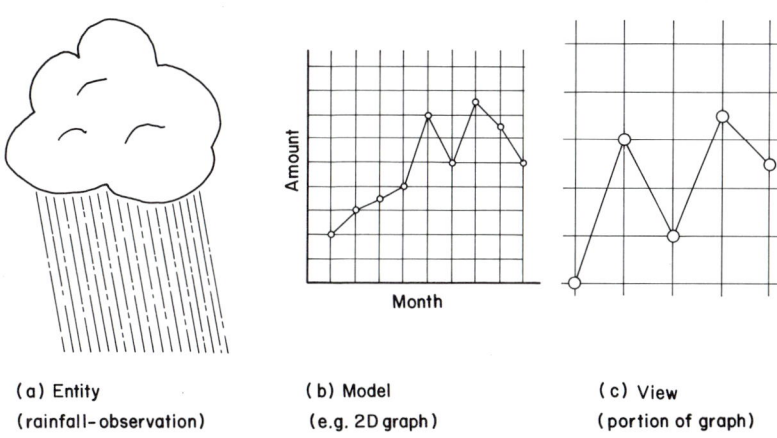

(a) Entity
(rainfall-observation)

(b) Model
(e.g. 2D graph)

(c) View
(portion of graph)

FIG. 20.1. Entity, two-dimensional model, and view.

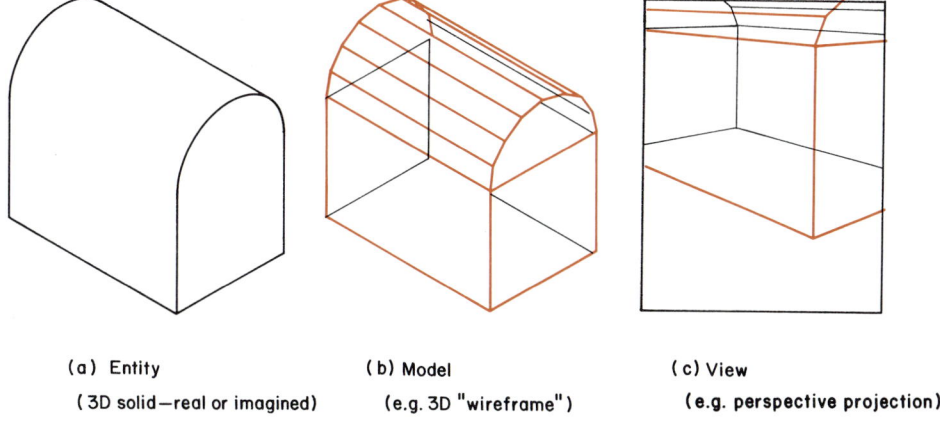

(a) Entity
(3D solid—real or imagined)

(b) Model
(e.g. 3D "wireframe")

(c) View
(e.g. perspective projection)

FIG. 20.2. Entity, three-dimensional model, and view.

case, the data in display files or buffers are not visual per se. These issues will be treated in more detail in Section 20.5.

As Figs. 20.1c and 20.2c indicate, the view is two dimensional in nature, appearing on a sheet of paper, and so forth or on a flat (or nearly so) display screen. (Holography and experimental three-dimensional illusion-producing devices are not considered here.) Moreover, the view is a "picture" of the visualized model from a particular vantage point and with a variable scale. The computer graphics system user must often indicate parameters that specify the desired view.

The concepts of models and views can be explained by a useful analogy involving a camera. Imagine the model to be physical as the illustrations suggest. Then imagine taking snapshots of the model with a camera. With different vantage points and a zoom lens allowing narrow or wide-angle effects, the snapshots obtained could be wide-ranging in visual effect. But we must allow the camera to have a somewhat magical power in this situation, in order to produce different types of projections, such as axonometric, perspective, and oblique, as we desire. Finally, the snapshots themselves could be made to any size, in effect matching the actual size of the device view area (or a desired portion of it).

20.3.3 Transformations. Inherent but not necessarily obvious in the modeling and viewing stages of computer graphics are a variety of *transformations,* or *mappings* as they are sometimes called. These include changing coordinates of points from one set of coordinate axes to another in two or three dimensions, in either the modeling or the viewing stages. They also include reducing, in a "projective" sense, three- to two-dimensional coordinates in order to obtain a view. Elementary explanations of these transformations will be found in Sections 20.4 and 20.5 that give a more detailed look at models and views.

20.3.4 User Interaction. An important aspect of the growing acceptance of computer graphics is the degree of "friendliness" exhibited by the systems that produce it. Although this is an inexact term and sometimes quantified in different ways, it purports to convey the ease and quickness with which users can learn and operate the system. Often considered as an important factor in this so-called friendliness are the methods and devices for interaction on a graphical level with the models and their subsequent views. A brief survey of these will be found in Section 20.6. Another important factor of the friendliness is the way in which the "dialog" between user and the system is structured. One that guides the user in clear, unambiguous fashion is sought but not often 100% achieved. Issues in this realm will not be delved into in this chapter.

20.4 MODELS

The model, as introduced in Article 20.3.2, is a representation of an entity inside a computer in the form of a data structure. This section deals with some interesting aspects of model creation and manipulation. For the most part, these considerations pertain to the internal workings of a computer graphics system. An awareness of them, however, can promote more effective system use. A variety of user-oriented schemes are usually programmed into the system to cause the creation and manipulation of a data structure comprising a model. In some instances they parallel the internal notions to be discussed here.

20.4.1 Representations. There are several ways to represent models by data in a computer. The most straightforward is by a "map," which is not a model in a true sense. Modeling usually refers to one of three general ways to represent entities. These include *wireframe, surface,* and *solid* models.

20.4.1.1 Direct Map. In a *direct map,* the notions of a model and view are regarded as synonymous. The model or view may be likened to a painting. Examined closely, the painting is imagined to consist of mosaic elements or tiles. For each tile in the mosaic, necessary "painting" information is recorded as data. Because this representation is more often considered at the view level, its details and examples will be deferred until Article 20.5.1.2.

20.4.1.2 Wireframe. A *wireframe model* is a collection of points or vertices and the straight-line (or "vector") connections between them. Figure 20.3a shows the visual interpretation of a simple two-dimensional wireframe model. Figure 20.3b indicates the ingredients of a typical data structure corresponding to the model. It is because any subsequent view of the model (or our visual perception of it) looks stick- or wire-like in appearance that the name "wireframe" has stuck. A very good physical analogy to a wireframe model is a Tinkertoy or similar construction kit using "knobs" and "connectors" in two or three dimensions. The set of connectors defines a set of vectors that give "substance" to the model, corresponding to two-dimensional boundaries or three-dimensional edges. But despite this, the notions of surface and solid interiors must be inferred by the creator or observer. Indeed, some object wireframes can be interpreted differently, as shown in Figure 20.4.

For convenience in a wireframe model, curved lines may be approximated by a sequence of short straight lines or vectors. We have been introduced to this trick in elementary geometry: thinking of a circle as a regular polygon as its number of sides becomes large. Practically speaking, the number of vectors may be limited as a means of keeping the amount of data manageable. But if the number of vectors is too few, the discontinuity of the approximation may be highly visible in a subsequent computer generated view. Some data structures allow curved lines to be represented in terms of geometric data (e.g., for a circle, the center, and radius) from which the curved line is constructed when needed by equations. The subsequent construction may be a collection of vectors.

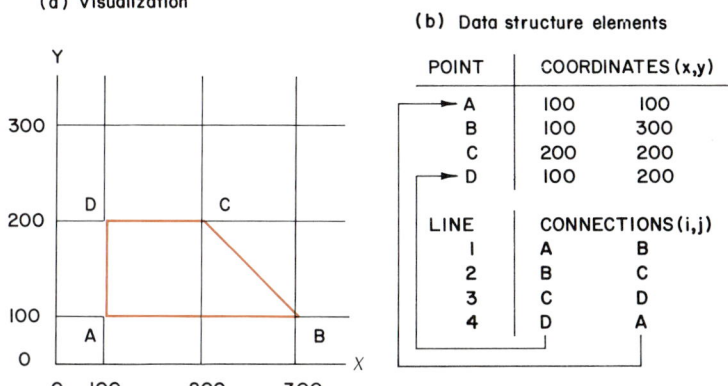

FIG. 20.3. Wireframe model.

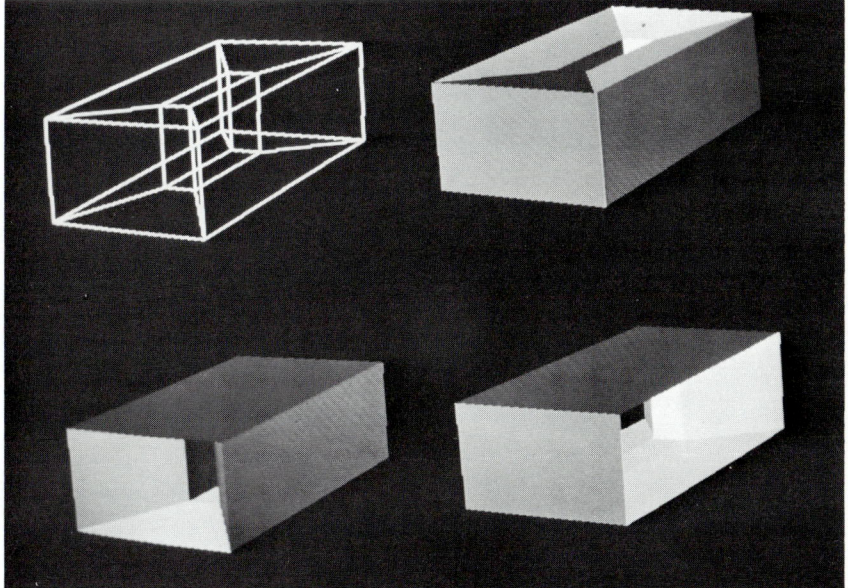

FIG. 20.4. A wireframe model may have multiple interpretations (courtesy Production Automation Project of the University of Rochester).

20.4.1.3 Surface.
A *surface model* involves levels of description beyond what are contained in a wireframe model. Figure 20.5 gives the visual and suggestive data structure interpretations for surface model of a simple two-dimensional case that could just as well be three dimensional by changing the coordinate data. In the model, one or more sets of edges, referred to as *loops,* define "faces" in either two- or three-dimensional contexts. Flat polygons with straight edges are the simplest type of surface model. Increasing the complexity, straight edges may be replaced with curved ones as in wireframe models. Even more complex curved surfaces (such as in a cone or sphere) may be represented by a collection of flat polygons. Alternatively, they may be described in precise form by geometric data and have constructions computed from them when needed.

The notions of the "inside" or "outside" of a face can be built into the data structure of a surface model. This can be done by assigning "inside" to one side of a loop-traversing arrow defined by the connection order as suggested in Fig. 20.5a. "Holes" can be introduced in this fashion. However, the determination of what constitutes the solid portion of a model composed of joined faces is generally left up to the model's user.

20.4.1.4 Solid.
A *solid model* is the highest form of modeling description. It may be visualized as a solid form such as in Fig. 20.6a, but its underlying data structure can be extremely complex. Solid models currently fall into two classifications: boundary and constructive geometry representations. A *boundary* representation consists of faces and their associated edges and vertices as indicated by Fig. 20.6b. A *constructive geometry* representation consists of combinations of imagined solid components (sometimes called *elements* or *primitives* as indicated by Fig. 20.6c. The components can be taken from a "library" of possible shapes or may be created in other ways. Some examples of this will be found in Section 20.7. In constructive geometry models, the components are imagined to be combined into creative forms by the Boolean operations of union, difference, and intersection (analogous to the operators of Boolean algebra). Figure 20.6c shows a union. A hole would be formed by a difference, in effect subtracting one shape from another. An intersection is more abstract, namely, the volume common to overlapping shapes.

The data structures for solid models must keep track of the elements of either the boundary or the constructive geometry representations, as well as the associations between the elements. With this information, together with rules and procedures incorporated into computer programs, it is possible to ascertain if a point in space is inside or outside the solid model, or on its boundary surface. This is very important information for creating certain kinds of subsequent views, as well as performing engineering-related tasks in CAD/CAM.

20.4.2 Transformations.
In the creation of model representations, coordinate data changes occur or are set up frequently, even though a computer graphics system user may not be fully aware of them. The only case in which they are not used is when point coordinate data for a

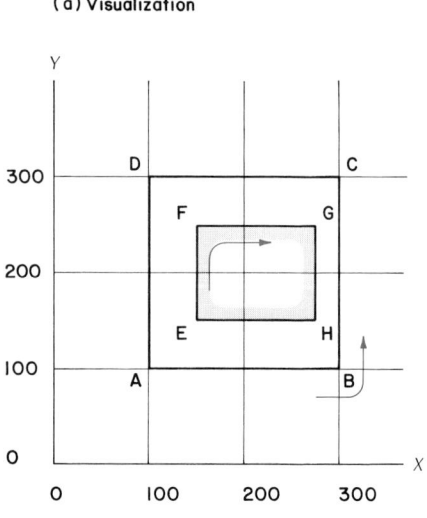

FIG. 20.5. Surface model.

model (or subportions of it) are specified in absolute coordinates, already compatible with the model or world space of the representation. In other situations, it is necessary to obtain the true absolute coordinates from other coordinate data. This procedure is called a *transformation* or *mapping* of coordinates. The most frequently used transformations are scaling, rotation, and translation.

One elementary instance of transformation use involves so-called relative or incremental coordinate usage. In Fig. 20.7 the triangle model is first described in absolute coordinates, and then in relative coordinates. The coordinate origin remains unchanged at point O for the absolute case. The coordinate origin takes on a new position at point A after its coordinates are specified in the relative case. Variations are possible in the latter case; the user often has control over the selection.

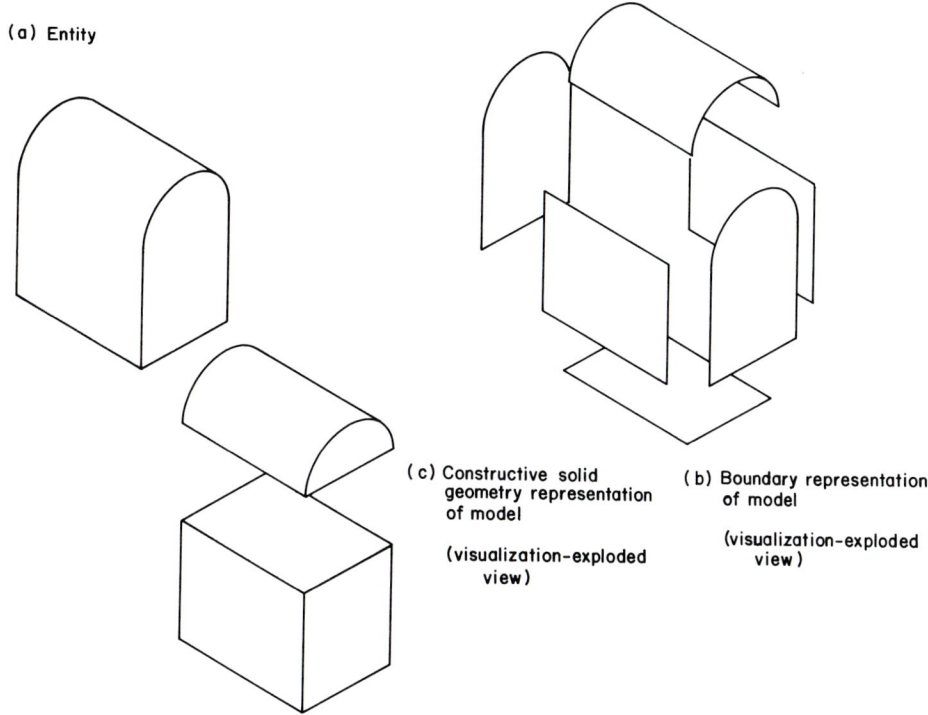

FIG. 20.6. Solid model.

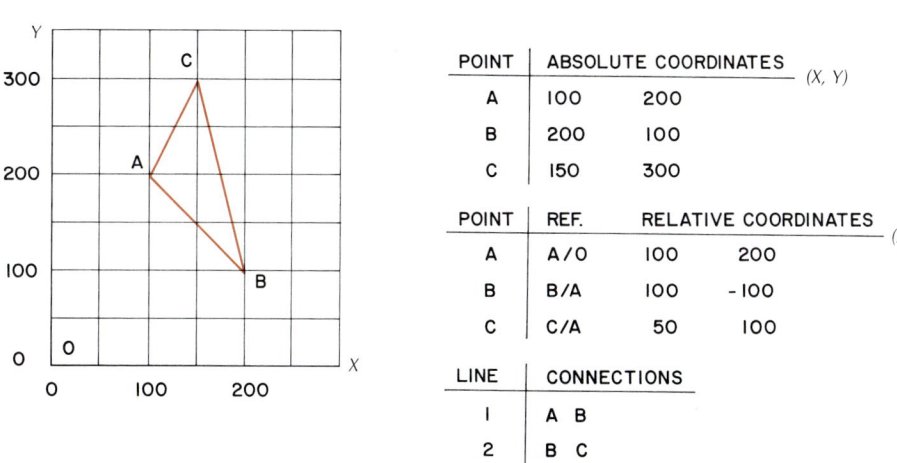

FIG. 20.7. Absolute and relative coordinates.

528 COMPUTER GRAPHICS AND CAD/CAM

To be usable, the model or world space must be in absolute coordinates. This means the points located by relative coordinates must be mapped into absolute coordinates. For the example of Fig. 20.7, the transformation to do this is a simple translation. For points B and C, it is expressed by

$$x_i \text{ (abs)} = x_i \text{ (rel)} + 100 \qquad i = B,C$$
$$y_i \text{ (abs)} = y_i \text{ (rel)} + 200 \qquad (1)$$

The translation of Eq. 1 can be applied in one of two basic ways. The data can be altered before they are put into the data structure of the model. Alternately, the transformation parameters (100 and 200 in this case) can be included in the data structure and applied via the equations later when the absolute data is needed, for example, when a view is desired.

In creating models, groups of points belonging to a part of the entity are often manipulated together. Such elements of an entity may currently be part of the model, or merely exist as a predefined "menu" item. Either way, a replica is being added to the model, although perhaps altered in shape and position. In these situations, suggested by the simple two-dimensional example of Fig. 20.8, it may be necessary to invoke scaling, rotation, and translation.

The total transformation of the two-dimensional wireframe component from its initial position to its replicated position is shown in Fig. 20.8a. The step-by-step scaling, rotation, and translation—in that order—achieving the total transformation are depicted in Fig. 20.8b. The equations to effect the transformation steps of Fig. 20.8b are

$$x' = s_x \times x \qquad \text{(scaling)}$$
$$y' = s_y \times y \qquad (2)$$
$$x'' = \cos(\theta) \times x' - \sin(\theta) \times y' \qquad \text{(rotation)}$$
$$y'' = \sin(\theta) \times x' + \cos(\theta) \times y' \qquad (3)$$
$$X = x'' + t_x \qquad \text{(translation)}$$
$$Y = y'' + t_y \qquad (4)$$

where the notation is clarified in the figures. Equations 2 to 4 may be combined by successive substitution to yield

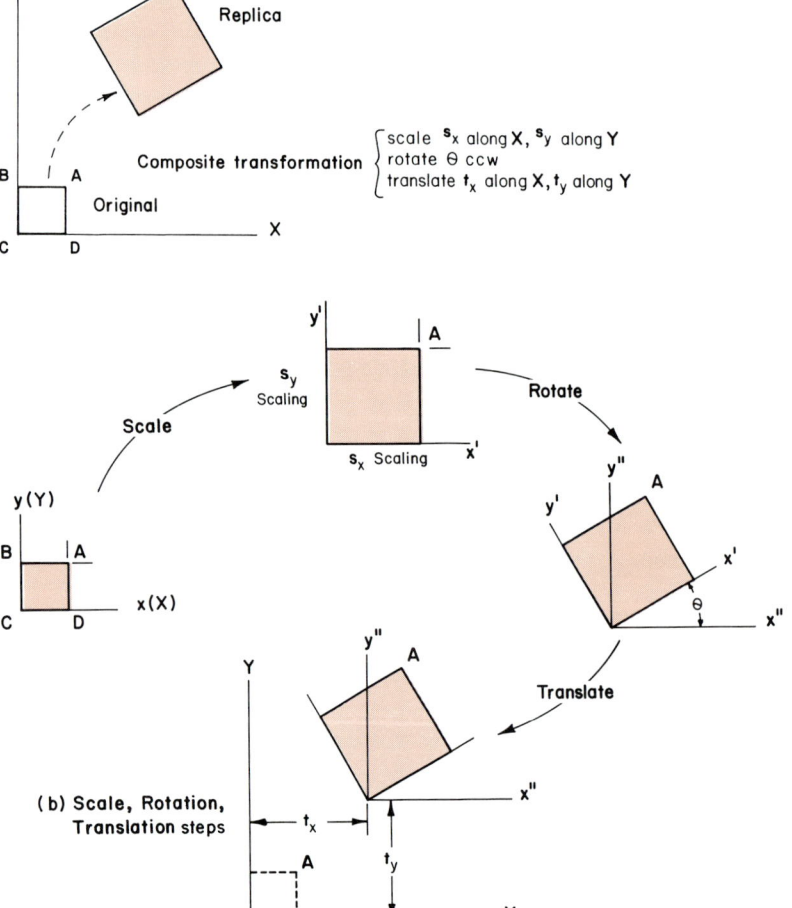

FIG. 20.8. Transformation of a two-dimensional replica.

$$X = \cos(\theta) \times s_x \times x - \sin(\theta) \times s_y \times y + t_x$$
$$Y = \sin(\theta) \times s_x \times x + \cos(\theta) \times s_y \times y + t_y \quad (5)$$

The extension of these transformation ideas to the third dimension is not difficult. Scaling and translation are simply expanded to include the third dimension. However, the rotational aspect for three dimensions must include three independent axis rotations. This means two more sets of equations similar to Eq. 3 must be added.

An interesting aspect of transformation theory, though not explained here, is that *matrices* can be used to represent transformations in a uniform way. As a result, it is possible to combine many successive transformations, even of different types, as a series of matrix multiplications. This procedure, presented in appropriate texts, is analogous to the successive substitution used to get Eq. 5.

The basic examples of transformations considered are often used, but are only at the tip of the iceberg insofar as possible complexity is concerned. The level to which complexity can rise can be suggested by the modeling of a robot arm, consisting of many components—forearm, hand, gripper, and so forth—all which move with respect to each other. To model the component placement successfully, it is necessary to introduce intermediate coordinate systems at each connection. The position of the outermost part is referenced to the part adjacent to its connection, and so on down the line. Thus, a vast sequence of successive transformations must be applied to find the position of the outermost robot arm part in the absolute coordinate system of the model. These transformations can be handled in matrix form as mentioned earlier. They are all done internally and, as in previous cases, with minimal direct knowledge of the user.

20.5 VIEWS

The topic of views encompasses the conversion of model data into a visible picture. This conversion involves representation methods and transformations necessary, as well as the devices on which the views appear.

20.5.1 Representations. It may be recalled that the model of an entity is stored within the memory of a computer in the form of a data structure. The information in the data structure may be interpreted in a visual way to foster an understanding of the model. However, as we know, the data structure is not a real picture in any sense of the word. In order to get a picture, or view as it is often called, the data structure of the model must be converted into a form suitable for graphic presentation on a device. The form it is converted to, a display file or buffer, depends on the technique used to render the view physically on the device. Two techniques prevail to do this: vector and raster. Both involve what takes place on a plane surface corresponding to the view.

20.5.1.1 Vector. In many respects, the *vector* technique is analogous to conventional drafting procedure. It relies on the placement of successive "line" strokes as indicated in Fig. 20.9a. The order of the strokes is not unique. These facts give rise to *stroke* and *random scan* as alternate names for the technique. The repertoire of

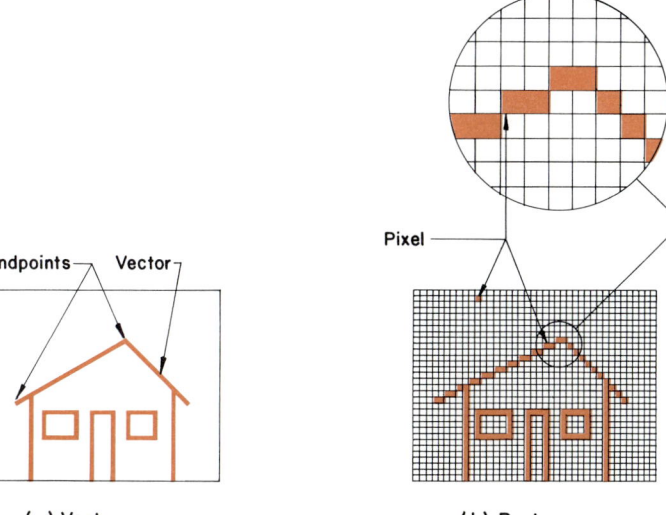

FIG. 20.9. View representation techniques.

strokes available generally includes straight, curved (usually circles), or stroke combinations (e.g., characters and symbols). Straight strokes can vary in length and direction. Special strokes may vary in size and inclination.

The data underlying the strokes are kept in what is termed a *display file*. For straight strokes, this file consists of planar coordinate values and the indication of connections constituting the strokes. For special strokes, it contains coordinates and geometric values, such as size and inclination. By and large, the display file is similar to the data structure of a planar wireframe model.

20.5.1.2 Raster. The *raster* technique for view-rendering may be likened to the creation of a mosaic picture. In Section 20.4 a painted tile surface was introduced as an example of a mosaic. Some additional examples can be mentioned. In the acromatic realm (sometimes regarded as black and white), an example is the electronic scoreboard or marquee using a grid of lights. In the color realm an example is the product of needlecraft or a grid-coloring arts-and-crafts kit.

The basis for the raster technique has already been introduced in the context of a direct map model (Article 20.4.1.1). It utilizes a finite grid of uniform planar elements, as shown in Fig. 20.9b. The elements are often referred to as *pixels* or *pels* (contractions of "picture elements"). Each pixel is "painted" in such a way so as to have — or give the illusion of having — a uniform brightness (conversely, darkness) or color. Acromatic and color examples are given in Fig. 20.9b and Plate 3, respectively. Both should seem coarse in appearance. But if they are imagined to be blow-ups of small regions of bigger views in which there may be a 500×500 grid of pixels, the realistic nature of raster-produced views should be appreciated. Look at the examples from a distance of a meter or more to get the effect.

There are several ways of creating uniform pixel brightness/darkness or color with certain view devices. These will be dealt with in the next article. But there are some view devices that rely on the strategy of giving the *illusion* of uniform brightness (or darkness) levels or colors. Some of these devices will be presented in the next article too. Long-standing in this category are line printers, producing mosaics such as in Fig. 20.10. In this application, a pixel is represented by a character position. Each may have a different character or several characters overstruck to produce the illusion of varying gray levels. In several view devices, a more sophisticated pixel grouping method is used. Each pixel can be assigned a uniform brightness or primary color. But clusters of the pixels then give the illusion of varying brightness or blended color for the pixel group. This effect is seen everyday in newsprint graphics. A variation of the principle involves subdividing each pixel. This is done in color television monitors with color triads.

The data supporting a raster rendition are stored in a *display buffer*. The buffer typically consists of one or more memory bits for each pixel, representing the brightness or color. Because of the correspondence of the pixels and buffer storage locations, the buffer may be imagined to be a grid also, giving rise to its functional name, a *bit map*. In the simplest application of "on/off" pixel activation, such as in a black/white (no gray) rendition, a nominal 500×500 bit map with one brightness bit per pixel would require a quarter-million-bit display buffer! If color or gray scale choices are permitted, the buffer becomes, say, $500 \times 500 \times N$ in size, where N is the number of bits describing the choice for each pixel. Alternate schemes may be used, but at the expense of increased logic and computer processing time.

There is a fundamental problem in creating raster renditions that has not been mentioned so far. It involves the conversion of model data into the raster format. This is a complex process that is referred to as *raster scan conversion* or just scan conversion. Essentially, it involves determining what features of a model (other than a direct map one) "fall" into each pixel (or a pixel group if blending is used to achieve gray levels or colors). One way of approaching this task is to "scan" each row of pixels for "intersections." This is compatible with many devices that produce a raster pictured by "painting" it in successive rows of pixels.

There are several additional aspects of raster-produced views that are worthy of mention. These focus on the potential drawbacks of raster renditions and how they can be overcome. Some of us who have seen early versions of video games or have worked with raster art are surely aware of the potential drawbacks. These include an insufficient number of pixels (a raster grid that is too coarse), limited monochrome shade or color choices, and a general discontinuous line or edge effect (likened to a jagged sawtooth or staircase appearance). All can affect the quality of the rendering adversely as Fig. 20.9b and Plate 3 show.

The same problems loom when the raster technique is used in computer graphics view devices. But the use of more memory, made

PLATE 1. Typical video game scenes.

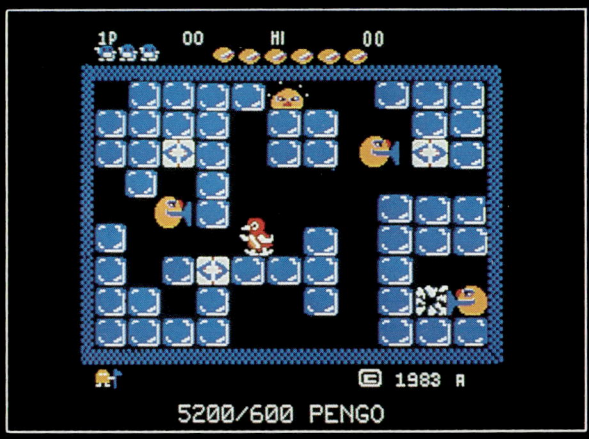

a. A two-dimensional representation. (Pengo, a trademark of Sega Enterprises Inc. and used by Atari Inc. under license. Courtesy Atari Inc.)

b. A three-dimensional illusion. (POLE POSITION, engineered and designed by Namco Ltd., manufactured under license by Atari Inc. Trademark and copyright Namco 1982. Courtesy Atari Inc.)

PLATE 2. Typical media uses.

a. Print media illustration. ("East/West", copyright 1982 Digital Effects Inc.; C. Robert Hoffman III, Alan Green, Gene Miller illlustrators.)

b. Video animation scene. ("Subway", copyright 1981 Digital Effects Inc./Mark Lindquist.)

PLATE 3. Color raster picture art form with each color square composed of numerous pixels. Shown at three scales to portray the effect of different raster size. (Copyright 1982 Digital Effects Inc.)

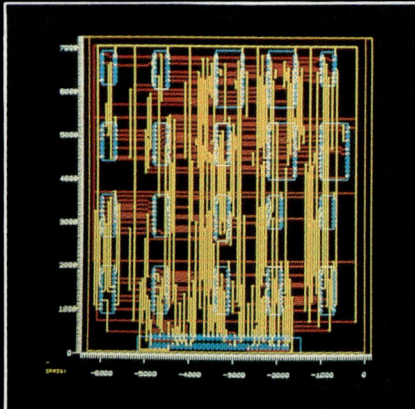
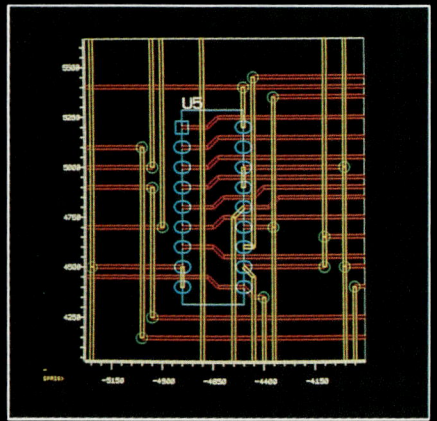

PLATE 4. View interactions for a two-dimensional model of a printed circuit board layout. Note the use of color to highlight different wiring layers, components, holes and text. Each such group might be separately displayable if an overlay feature exists. (Courtesy IBM.)

a. Overall view.

b. Close-up view of a selected region (note coordinates along the axes) via the plan and zoom technique.

PLATE 5. View interactions for a three-dimensional model of a vehicle. (Sequence courtesy Structural Dynamics Research Corporation (SDRC).)

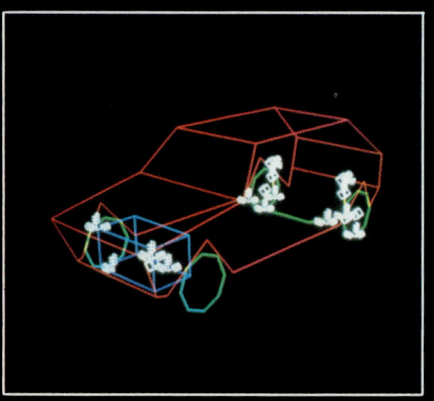
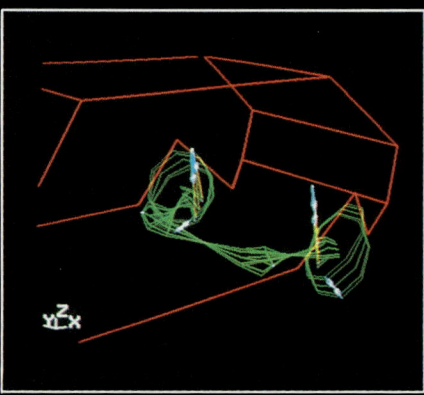

a. Colored-coded component systems superimposed in overlay fashion.

b. "Time lapse" animation of rear axle dynamic behavior overlaid on the vehicle outline, with a close-up effect achieved by panning and zooming.

PLATE 6. Data representation is an integral part of CAD/CAM activities. (Examples using DISSPLA, a registered trademark of Integrated Software Systems Corporation. Copyright 1984 Integrated Software Systems Corporation.)

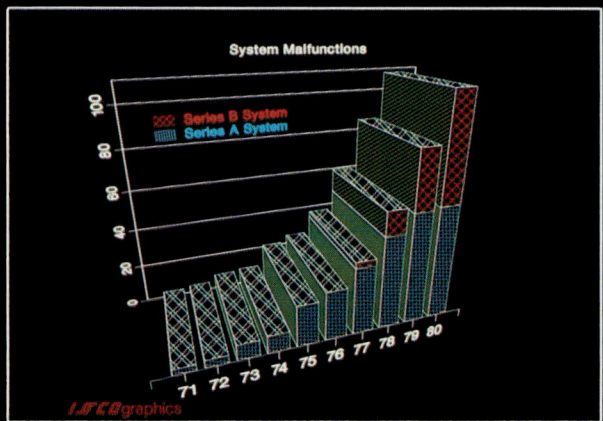
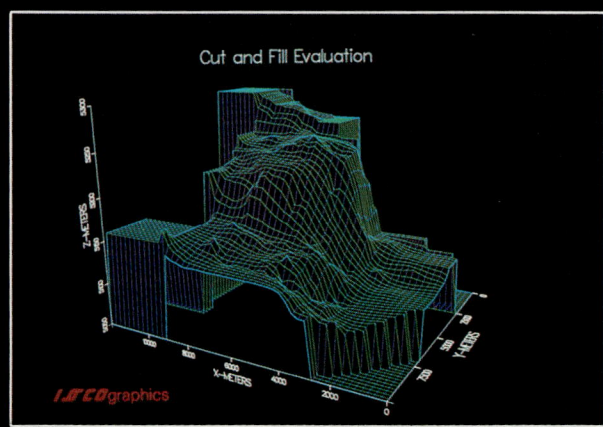

a. Data of many forms may be modeled as charts.

b. Entities of observed or imagined objects, phenomena, or ideas may be modeled as surfaces.

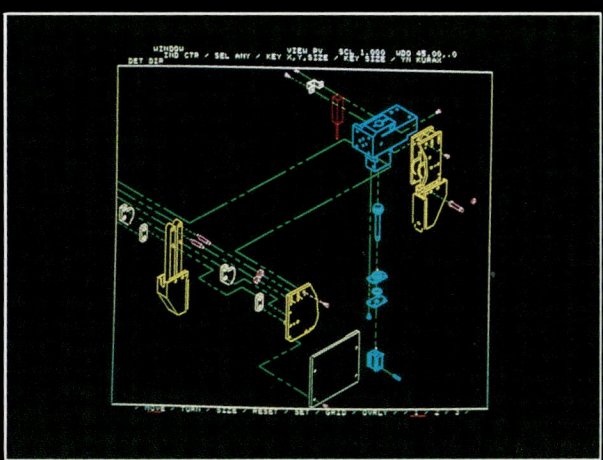

PLATE 7. Mechanical assembly/layout drafting. (Courtesy IBM.)

a. Cutaway or sectioned views aid understanding. (Courtesy McDonnell Douglas Automation Company (MCAUTO).)

PLATE 8. Mechanical part assemblies may be studied before any manufacturing takes place.

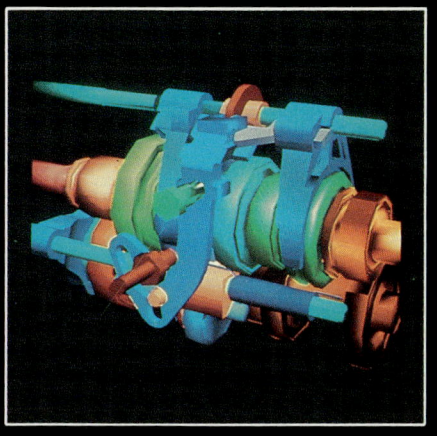

b. Solid modeling can be used to check for interferences between components and subassemblies. (Courtesy IBM.)

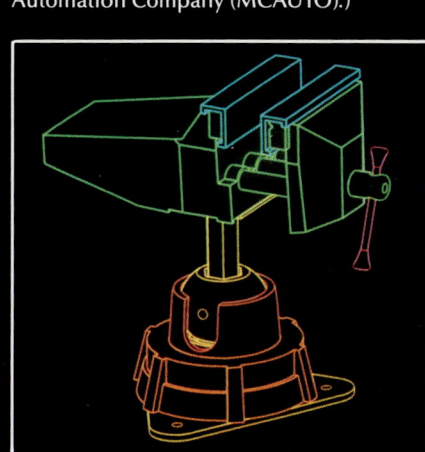

c. A boundary representation is an alternate form of solid model (compare to the wireframe of Fig. 20.13). (Courtesy Evans and Sutherland.)

PLATE 9. Material removal analysis is one facet of CAM. Here, the results of tool path sequencing are shown. (Sequence courtesy Intergraph Corporation.)

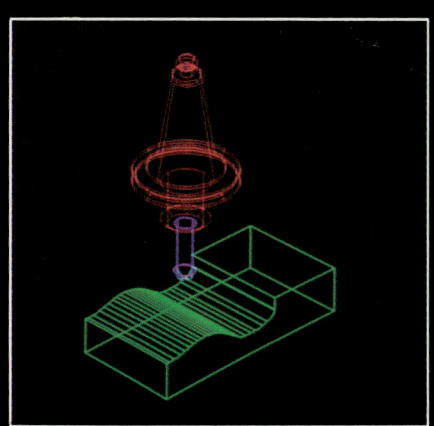

a. Wireframe model.

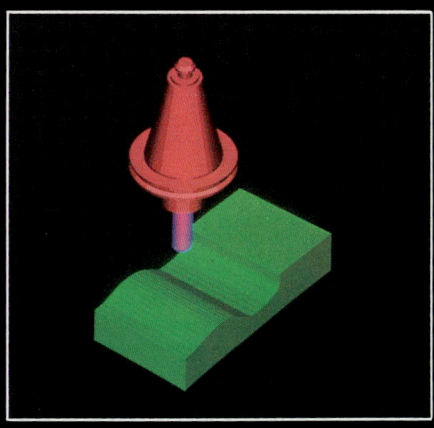

b. Solid model.

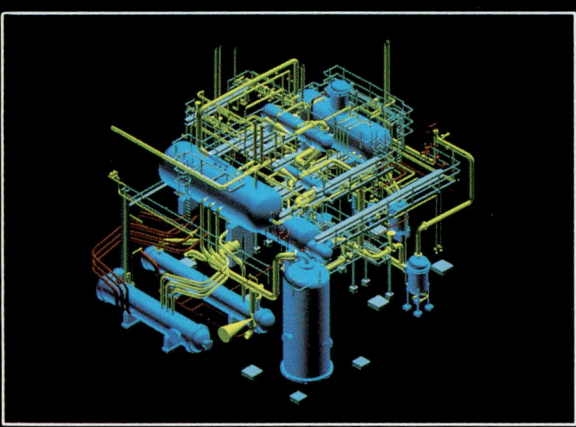

PLATE 10. Piping and conduit layouts may be conceptualized and analyzed for performance. (Courtesy Computervision Corporation.)

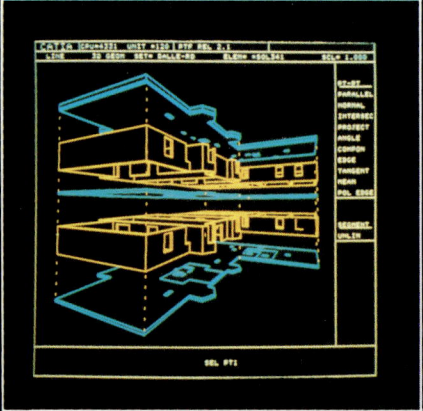

PLATE 11. In architectural designs and studies, buildings and their contents may be created and simulations of walking through them may be generated. (Courtesy IBM.)

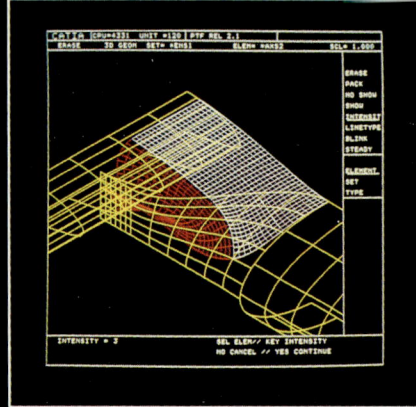

PLATE 12. Complex aerodynamic and other surface shapes, including intersections, can be created and analyzed. (Courtesy IBM.)

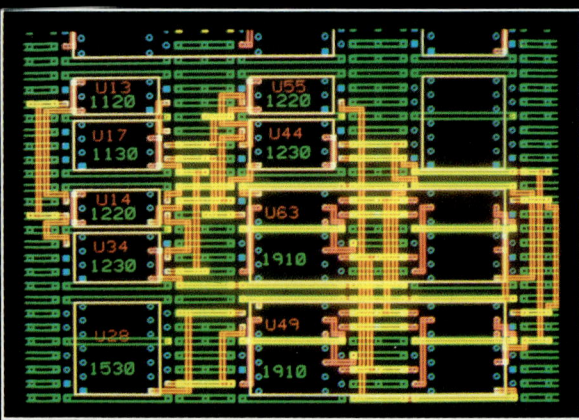

PLATE 13. Electronic MOS gate array design, shown here, and other integrated circuit design procedures greatly enhance the reliability of IC chips and productivity in their manufacture. (Courtesy Intergraph Corporation.)

PLATE 14. The power of solid modeling may be applied to the design of complex items such as satellite. (Sequence courtesy Structural Dynamics Research Corporation (SDRC).)

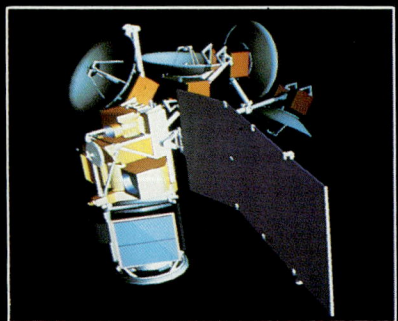

a. A model of the total assembly, showing a step in the unfolding of the antenna.

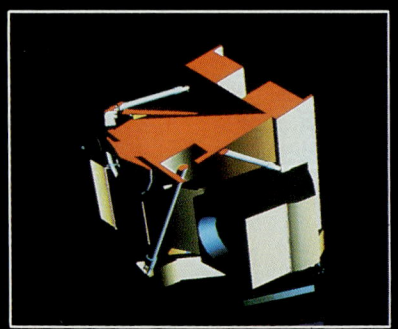

b. A sectional view of a subassembly.

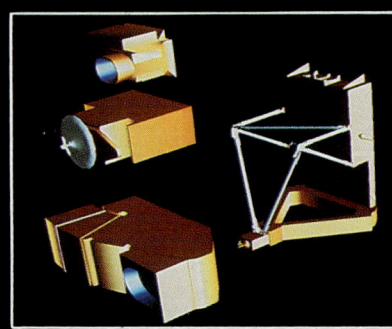

c. An exploded view of a subassembly's components.

cheaper each year by technological advances, has ushered in the use of finer raster grids, virtually eliminating the coarseness problem. The same advances have permitted more bits of pixel color or gray scale information. In dealing with the problem of the staircase effect, the technique of illusion introduced in the discussion of pixel blending can be used. It is based on the idea of adjusting the intensities of pixels along the lines or edges of areas they represent. This process has the effect of smoothing out the step-like appearance of lines and colored area boundaries.

20.5.2 View Devices. Insofar as possible, the discussion of the previous article sought to remain aloof from specific view devices. This article focuses attention on the devices themselves. It gives an overview of some of the more prominent types, including what can be expected from the views they produce.

There are several ways to classify view devices. Vector versus raster representation might be one. Color versus noncolor rendition might be another. But perhaps the most meaningful one to a computer graphics system user is based on view device purpose. This classification separates devices into two groups: *hardcopy* and *display*. Hardcopy devices are intended to produce a permanent record of a view. On the other hand, display devices are intended to produce views that may be seen, rapidly changed, or interacted with in a transient fashion.

In many instances, both device types may be present in a computer graphics installation. In some cases, the view seen on one device may also be seen on or transferred to another. This is advantageous if initial viewing is done in an inexpensive mode for checking purposes. If conditions then warrant it, the viewing can be reproduced in a higher quality or permanent fashion.

FIG. 20.10. Line printer raster picture. Character overstrikes may be used to produce more pixel contrast. (courtesy S. P. Harbison)

And although not discussed specifically here, the views on displays can be permanently recorded by photography or videotechnology, as the examples of Section 20.2 indicate.

20.5.2.1 Hardcopy. *Line printers* are sometimes useful as hardcopy view devices. They use a raster approach as described in the previous article. The views they produce do not have high quality and resolution, but they are handy for graphs, charts, and pictures.

Vector plotters such as in Figs. 20.11 and 20.12 are vector devices found in a variety of forms, sizes, and speeds. The two widely used forms are *flatbed* and *drum*. The flatbed form has a stylus that is moved across the view surface by electromechanical or magnetic means. Stylus motions may occur in any direction under the simultaneous control of conventional *x* and *y* coordinate positions on the view surface. In the drum form, the motion of the stylus is confined to a track parallel to one coordinate axis. But the drum-guided drawing medium moves simultaneously in a perpendicular direction to provide movement along the other coordinate axis.

In some plotters of either form the lines drawn may have the staircase effect familiar to raster displays. This occurs if the driving mechanisms for the component motions produce movements with discrete increments or steps. Although small, the steps may be visible in inclined and curved lines. For the most part, however, the resolution and quality of plotter outputs can be very high.

The majority of vector plotters use pens as styli. They can be of different types, point widths, and colors. They may be changed manually or mounted and controlled in groups. Drawing media used in pen plotters can be any of the conventional drafting types. In an important application, a light-emitting stylus is used to expose photographic film for printed circuit board masks.

Vector plotters come in desk top to room-filling sizes, producing every size of drawing imaginable. Speeds and accuracies are cost related, and there are varieties to suit a wide range of needs.

Dot matrix plotters such as in Fig. 20.13 are raster devices. They come in a range of types, paper sizes, and raster resolutions. In the *electrostatic plotter,* paper passes unidirectionally beneath a *writing head* composed of numerous

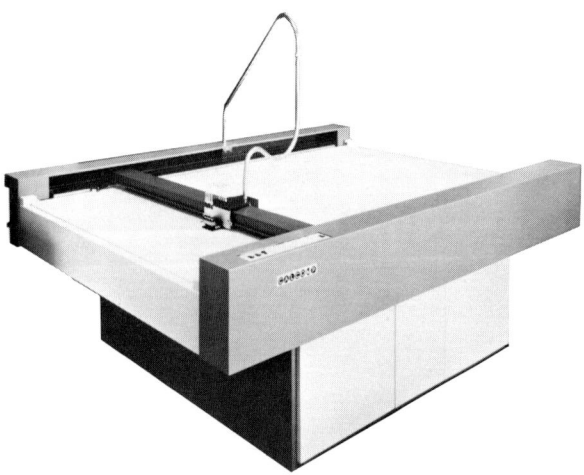

FIG. 20.11. Vector flatbed plotter: The pen holder moves along the dark guide, which in turn moves laterally [courtesy CALCOMP (California Computer Products Inc.)].

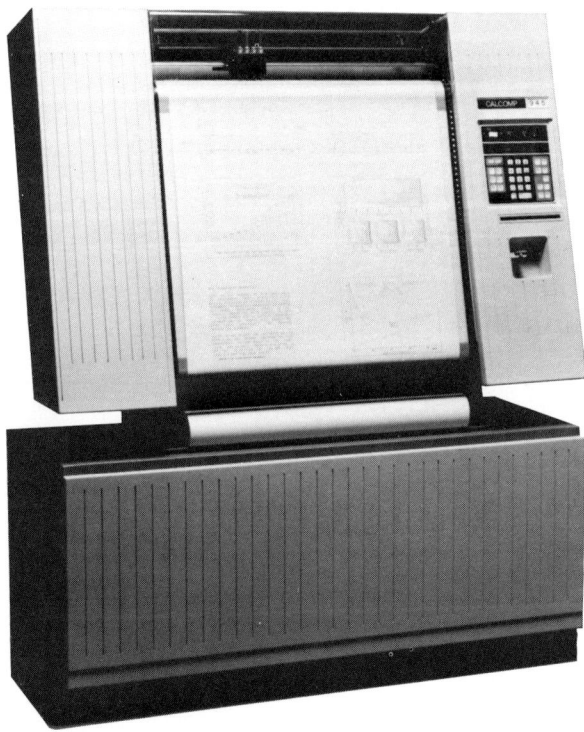

FIG. 20.12. "Belt-bed" variation of a vector drum plotter: The pen holder (at the top) moves horizontally above the drum, over which the drawing medium moves vertically (courtesy CALCOMP).

electric contacts. Electrical impulses are timed to each contact as the paper passes to cause a negative charge to be deposited as a "dot" on the paper. A liquid carbon toner then contacts the paper. The toner, positively charged, sticks to the negatively charged spots. There are nominally 100 to 200 contacts per inch in the writing head, and the pulses are timed to produce a grid of dots. Each dot, or group of dots, may be treated as a raster pixel.

The *impact* plotter employs a fixed or traversing head of mechanical pins that strike an ink ribbon. Each pin corresponds to a dot. As each row or a group of dot rows is completed, the paper advances. The *thermal* plotter is similar to the impact plotter but relies on heat-sensitive paper to produce the dots.

The single-pass electrostatic plotter makes plots many times faster than vector plotters. But it, like impact and thermal plotters, is limited to black-and-white images, generally of lower contrast than vector plotters. The other devices are relatively slow and may have larger dots than the electrostatic plotter.

Emerging technologies for hardcopy devices include three raster-based systems that permit color. One involves the use of color xerography. In essence it is an electrostatic process repeated for each of the three primary colors. Dry-pigmented toners are used in this process. The second is the *ink-jet plotter*. It squirts pulses of three colored inks at paper fastened to a rapidly rotating drum. The ink jets move steadily along a track as in a drum plotter from one side to the other, depositing the three colors simultaneously. Quality and resolution in both devices are similar to those of the electrostatic plotter. The third is a simple variation of the impact plotter. It uses multiple passes of the mechanical pin strikers with a multicolored ink ribbon to "mix" colors at each dot.

20.5.2.2 Display. *Cathode ray tubes* or CRTs are the most common form of display device where images can be changed quickly. The CRTs found in home black-and-white and color TVs are in most respects identical to those used for computer graphic displays, such as in Fig. 20.14. The CRT shoots a stream of electrons, varying in density, toward a phosphor-coated screen. The stream is focused and aimed by an $x-y$ deflection system that is electrostatically or magnetically controlled. The density of the electron stream

FIG. 20.13. Dot matrix electrostatic plotter: The paper moves unidirectionally past an electrically charged writing head (courtesy Versatec, a Xerox Co.).

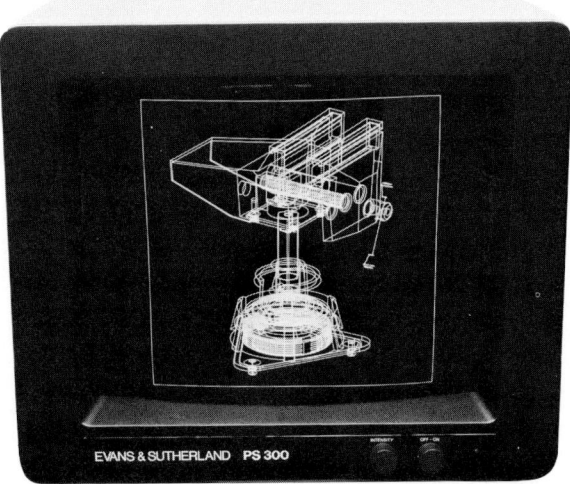

FIG. 20.14. Cathode ray tube (CRT) (courtesy Evans and Sutherland).

controls the intensity of the light emitted by the phosphor spot struck by the stream. The light emission decays quickly and must be rekindled many times per second as the electron beam is directed across the screen.

In vector-based CRTs, the electron beam is aimed in random fashion to produce vectors, that is, line strokes, on the screen. From a practical point of view, the resolution of the vector-based screen as well as the number and total length of vectors drawn on it are limited. If either or both is too high, the image that is refreshed will "flicker."

In raster-based CRTs, the electron beam is swept across the screen from left to right, in line-by-line fashion from top to bottom. Each line corresponds to a raster row. As the beam sweeps across, it is modulated to produce different light intensities at spots corresponding to the individual pixels in the raster row.

In one form of color CRTs, three different light-emitting phosphors (usually red, green, and blue) are used in color triads corresponding to each pixel. Three electron guns are synchronized to strike the desired component of each triad to produce varying intensities of the three colors that are in turn perceived as a uniform color mixture by a viewer. In another form, used for random vector images, multiple layers of different phosphors are used. They have different excitation levels and the beam intensity thus controls the color mixes produced.

The *direct view storage tube* or DVST is similar to the black-and-white CRT. But unlike it, the DVST does not need to have its image refreshed many times each second. This is because the electron beam "charges" a secondary surface parallel to the phosphor screen. This charged surface continuously excites the phosphor, generally in an on/off mode and in monochrome only. A special "write-through" feature allows the stored image to be unaffected as low-energy-level electrons strike the phosphor screen to produce a dynamic effect, such as a moving cursor. But by its nature, the DVST must completely erase the stored image and redraw it to effect any change in it. But this disadvantage is offset by the fact that the DVST does not need a buffer from which data are grabbed to refresh the image as is the case for normal CRTs. In addition, the resolution of the DVST can be increased to high levels, such as 4096 × 4096 addressable points, and many vectors can be displayed without concern for flickering images.

The *plasma display panel* has a viewing screen that consists of a grid of tiny neon-filled chambers. The chambers can be turned on and off individually and remain in the desired state until changed. Like the DVST, the plasma panel needs no refresh buffer, but it has limited resolution, seldom more than 64 chambers, that is, pixels, per inch.

Two emerging solid-state display technologies should be mentioned. They are the light-emitting diode (LED) and liquid crystal display (LCD), commonly found in calculator and watch read-out panels. Both rely on the arrangement of cells in closely spaced grids, ideal for raster display. Their attractive features include high resolution, low-power needs, and portability. But they are currently limited to monochrome images.

20.5.3 Transformations. The second major context in which transformations are used is in mapping model data into a view. This may be regarded to take place in two stages for either two- or three-dimensional models, though the first stage is different for each. We will briefly analyze the two stages for the simpler two-dimensional case first and then extend the concepts to the three-dimensional case.

20.5.3.1 Two-Dimensional Model Case. For the two-dimensional model shown in Fig. 20.15, the first stage is a mapping of some or all of the model into a planar *window* defined by the user. In this case, the window, which may be thought of as a piece of window glass sliding over a drawing, delineates what portion of the total world space is of interest. Though not shown, it might be tilted. Unknown to the user, the coordinates of the world system may be mapped via translation as in Eq. 4 to a reference corner of the window. If the window is tilted, rotations analogous to Eq. 3 would also be applied. But more importantly, the contents of the data structure are modified to exclude the portions of the model outside the window, since they will not be needed in the view to be produced. The relevant data structure itself is not changed, since it will no doubt be needed for other views or purposes. Rather, a copy of it is changed and sent to the display file or buffer of the viewing device. Using a wireframe model as an example, the net effect of the modification may be the loss of some points and lines entirely, or the change of one or both endpoint positions of some line connections as the "windowing" of the graph in Fig. 20.15 suggests. This process is often called *clipping* or *scissoring*.

The second stage of obtaining a view of a two-dimensional model is to map the coordinates of the windowed portion of the model into a suitable *viewport*, that is, region of the viewing device surface that may be all of the available surface, or just a part of it. This mapping usually involves a scaling transformation of the form given in Eq. 2. This may be because the "size" of the viewport on the surface of a viewing device is different from its counterpart contained within the windowed portion of the model. But more often it is due to a difference in coordinate units. Users may be well aware of the physical units associated with a model but not aware of those for the devices. An internal transformation takes care of this automatically. Likewise, translations may be in order to accommodate the device screen origin or to permit arbitrary positioning of the viewport.

20.5.3.2 Three-Dimensional Model Case. The view-mapping process for three-dimensional models is illustrated in Figs. 20.16 and 20.17. In contrast to two-dimensional models, three-dimensional models require a change in coordinates to put them into a viewer's or observer's coordinate reference frame. In turn, the model in these new world coordinates must be reduced onto a plane prior to obtaining the view. The reduction is a "projection" in the same sense as used in conventional engineering graphics. These two steps are condensed in Fig. 20.16a, with the result being the projection plane of Fig. 20.16b. The projection plane, only outlined in Fig. 20.16a, is parallel to the observer's x–y axes. To achieve the transformation from world to viewer coordinates as indicated in Fig. 20.16a, translations and rotations similar to those of Eqs. 4 and 3 — but extended to three dimensions — would be applied automatically by the computer graphics system. The computer graphics system user, that is, the "observer" or "viewer," would typically specify the vantage point and direction of his or her observation to create input parameters for the transformations.

In addition to observation parameters, the

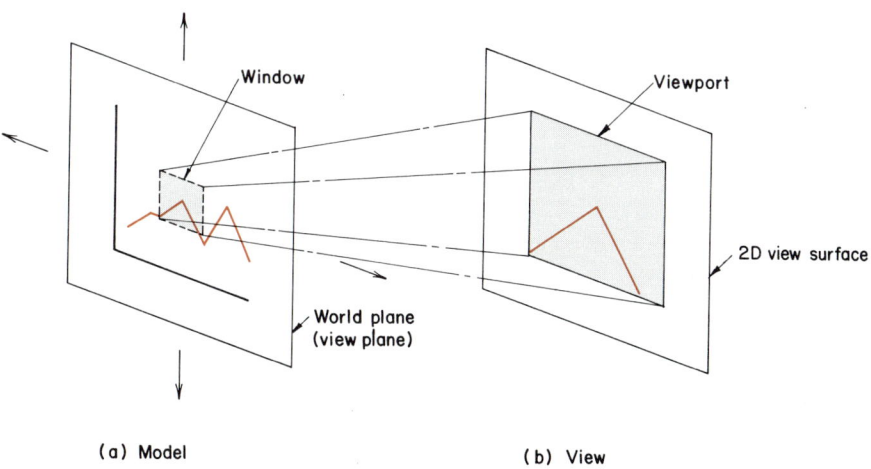

(a) Model (b) View

FIG. 20.15. Two-dimensional view transformation.

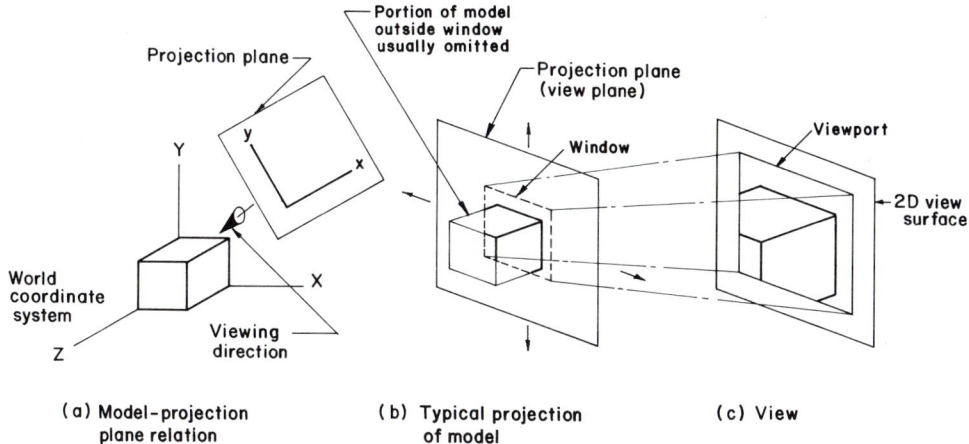

(a) Model-projection plane relation (b) Typical projection of model (c) View

FIG. 20.16. Three-dimensional view transformation.

viewer usually specifies the type of projection to be made. Figure 20.17 indicates three popular types. The nature of the projective transformations for these three common projections should be familiar. They were treated in spatial terms in earlier chapters to explain how they could be used to construct views on a viewplane legitimately. The situation is really no different here, except that mathematical manipulation of coordinate data, rather than graphical projection via sight lines, is used to accomplish the task. The axonometric transformation is the simplest, a mere dropping of the z coordinate as Fig. 20.17a shows. The perspective case, indicated by Fig. 20.17b, involves the scaling of x and y coordinates in three dimensions dictated by similar triangles in the z–x and z–y planes. For oblique, depicted in Fig. 20.17c, it is a translation of x and y coordinates in three dimensions as a function of the z coordinate and the angle between the view direction and the z axis.

The explicit equations embodying these mathematical projections can be found in various texts. They are not as important as the realization that they are easily incorporated into computer graphics programs. Noting that the common projective transformations discussed involve specialized types of scaling and translation, it will be recalled that they can be included in a sequence of transformations and handled as matrix multiplications as mentioned in an earlier section.

Associated with each projective transformation is a *view volume*. The view volume defines the extent to which the world space is to be

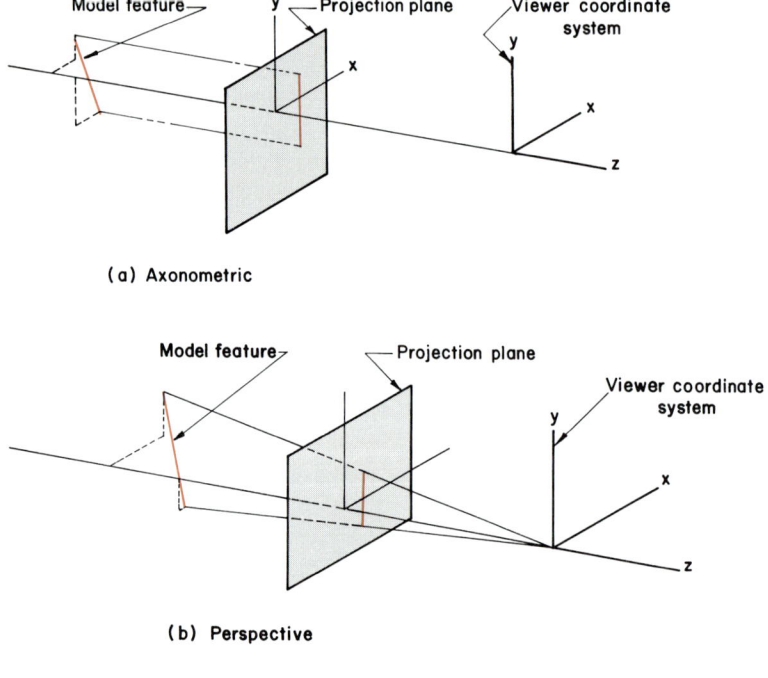

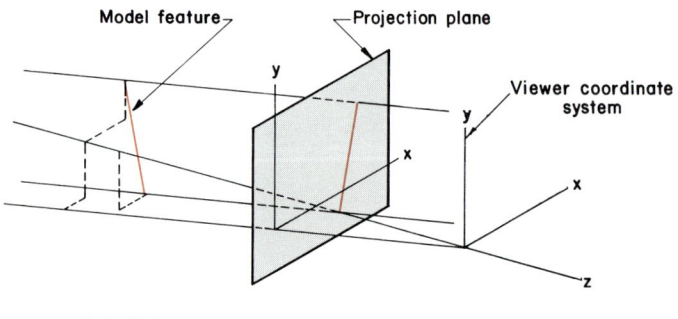

FIG. 20.17. Three-dimensional projective transformations.

considered for viewing. One may think of the view volume as a kind of tunnel. The viewer peering from one end would only see items (or parts of them) within the tunnel. For the axonometric case, Fig. 20.18a, the view volume is thought of as a rectangular prism, with projection (or sight) lines parallel to each other and perpendicular to the projection plane. For the perspective case, Fig. 20.18b, the view volume is imagined to be a pyramid, symmetric to the central viewing direction along the z axis. The projection plane is normal to the central viewing direction. The view volume is assumed pyramidal so the resulting view will be rectangular. Admittedly, this is a concession to view devices. In reality, human vision has a cone-like field of vision. But either way, the projection lines emanate from a single point imagined to be coincident with the viewer's eye point. For oblique projection, Fig. 20.18c, a view volume similar to that of axonometric is used, but with the view direction, that is, its axis, inclined to the projection plane.

The intersection of the view volume with the projection plane defines a "window" on the plane. In the perspective case, Fig. 20.18b, it is exactly like a window frame separating an observer inside a building with the world seen on the other side. In fact, the "world seen on the other side" could be traced onto the window glass with a marker by an observer with one eye closed, if it were within the observer's reach, to produce a correct perspective projection. Similar "window frame" interpretations could be made for axonometric and oblique projections as Figs. 20.18a and c suggest. However, projection or sight lines would, like we think of the sun's rays, have to be imagined to be parallel.

An interesting fact emerges after all the effort in obtaining the mapping of a three-dimensional model into the projection plane window. The subsequent mapping onto the viewport of the view surface is identical to that for two-dimensional model view creation. That is, the second stage of view mapping is identical for the two- and three-dimensional cases. In both cases, the image in the window is two-dimensional. A comparison of Figs. 20.15a and b with Figs. 20.16b and c shows this. Only in one case, the image captured in the window represents a replica of a two-dimensional model; in the other, a projection of a three-dimensional one. Thus for both cases the data transmitted to the view device are planar coordinate data and elements derived from the data structure to complete the view of the model.

20.5.4 Visibility. An integral aspect of view creation involves the determination of the proper visibility of three-dimensional model features. The task in solving for proper visibility may be stated in simple terms. But achieving it is quite a different matter!

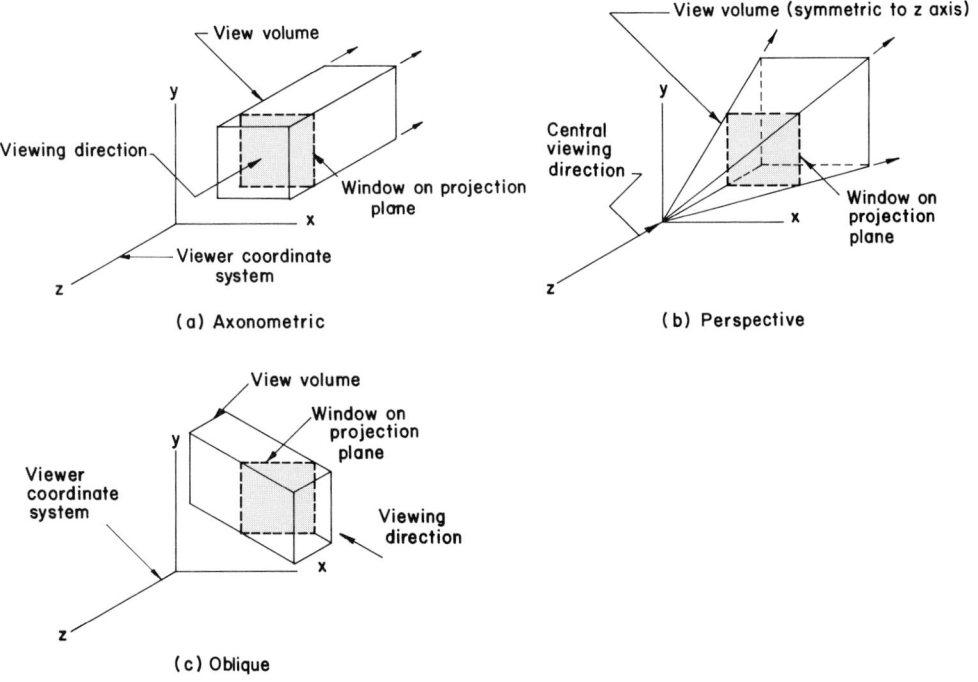

FIG. 20.18. View volumes and windows.

Taskwise, in wireframe models that represent solids, edges that are partially or wholly hidden from sight as the model is viewed must not be included in the display file or buffer for the view. In surface and solid models, the task becomes more challenging. Whole or partial surfaces must be removed. In addition, if the intersections of model elements are not explicitly contained in the data structure of the model, they must be made to appear in the view.

A commonly used technique to achieve these goals can be based upon the concept of comparing the directions of outward-pointing surface normals (thought of as nails partially driven perpendicularly into each surface) with each other and with the vector parallel with the sight direction. This works well for simple polyhedra but becomes involved if the model is composed of multiple ones. In the case of curved surfaces, the success of the method hinges upon the approximate representation of the surface by real or imagined flat elements. Another method compares the intersections of sightlines with surfaces of the model, much like probing a physical object or rock strata with a drill. Implicitly, the surface associated with the intersection nearest to the observer is the one hiding others along each sightline. Many sightlines may be required to resolve total visibility, especially in the vicinity of model element intersections.

Visibility determination is indeed a complex subject. In most instances, its implementation takes relatively large amounts of computational resources — time and memory. Invariably, these implementations are attuned to the type of model, view device, and features of the view rendition, for example, continuous color shading.

20.6 INTERACTION MODES, DEVICES, AND TECHNIQUES

One of the intriguing aspects of interactive computer graphics is embodied in the word *interactive*. Technology and software developments have resulted in the introduction and widespread use of devices and methods to manipulate models and views dynamically while working at a display terminal. By its very nature, this interaction is graphical in action or result. Its purpose is really no different than dealing with a drawing prepared manually. But that is where the similarity ends. For the computer allows things to be done that are far more sophisticated or powerful and faster than in the manual realm. The popular modes, devices, and techniques for this interaction are examined briefly in this section.

The reader should take note. Despite the words and figures that follow, very little justice can be done in conveying the exciting and appealing nature of the interactions found in interactive computer graphics. Hands-on experience is the best way to achieve that goal. To be sure, some of us already have — in the form of video games. Indeed, some of the important ingredients of interactive computer graphics are found in them, as we shall now see.

20.6.1 Modes. Interaction modes may be thought of as tasks, as opposed to techniques in which they are used. Generally speaking, there are five modes.

a. Choosing a course of action from a group of alternatives, often via what is termed *menu selection*.
b. Indicating position, which is termed *locating*.
c. Selecting an element displayed, referred to as *picking*.
d. Inputting numerical values.
e. Inputting text.

These mode descriptions do not fully convey the diversity of interactions, but it is helpful to try to classify them in this way. As we shall see, many forms of devices can be used to accomplish them, often in alternative ways. In some instances, the choice of particular devices and the ease with which they can be used and accomplish the intended goal may affect the overall success of a computer graphics system in the marketplace.

Inherent to interaction is a means of seeing or pointing to a position on a display terminal screen so that the tasks listed above can be done. In many systems, a visible mark or *cursor* appears on the screen. It is driven by actions taken on the interaction device. Video games often use an entity in the game as the cursor. Cursors are often used when the interaction device is not physically an integral part of the display screen. Thus, while your hands are operating the device, your eyes watch the cursor on the screen. In some systems, there exists a direct screen pointing ability. Here, the device and screen are both in the field of vision, so a cursor is not essential. In other situations, the device and screen do not have to be watched simultaneously, and, again, a cursor is not needed.

20.6.2 Devices. There are a number of commonly used interaction devices. The following ones generally do not use a screen cursor.

a. The so-called *light pen,* such as shown in Fig. 20.19, detects a spot on the CRT screen when the phosphor is illuminated by the electron beam striking it. It may have a depressable point, switch, or button to permit interactions to be activated (cf. item c below). It is useful for picking, since coordinate positions associated with the beam are known.

b. The *touch panel* is a device coincident with the display terminal screen. It often goes unnoticed. A finger or stylus is pointed at the screen, and its presence is detected by the device as it nears or touches the screen. High resolution is not generally found in such devices.

c. The *function keyboard* or "button box," such as shown in Fig. 20.20, has keys or buttons that can be depressed or released. Each corresponds to a menu item. These items may include commands to be entered, alternatives to be selected, or activity to be started, stopped, reset, or altered. The menu may be changed to correspond to a different program. Often function buttons are incorporated into or used in conjunction with other devices, such as the light pen.

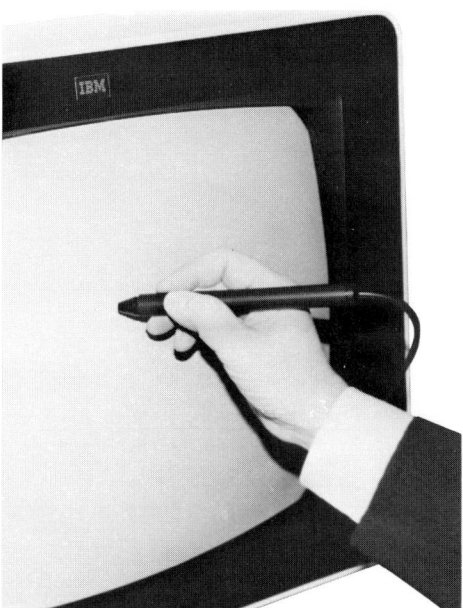

FIG. 20.19. Light pen used against the surface of a CRT.

FIG. 20.20. Function keyboard: Interchangeable labeled menu templates may be placed onto the keyboard and active keys may be backlighted (courtesy Evans and Sutherland).

540 COMPUTER GRAPHICS AND CAD/CAM

d. The *dial,* often found in clusters such as in Fig. 20.21, is used to input numerical values. Analog motion of the dial is converted into digital values by a potentiometer. Typical values are position, angle, and rate that are often associated with movement displayed on the screen.

The following devices usually drive a cursor on the screen.

a. The *graphics tablet,* such as shown in Fig. 20.22, has a stylus or other locating device used in conjunction with the tablet's flat surface. Absolute coordinate position is

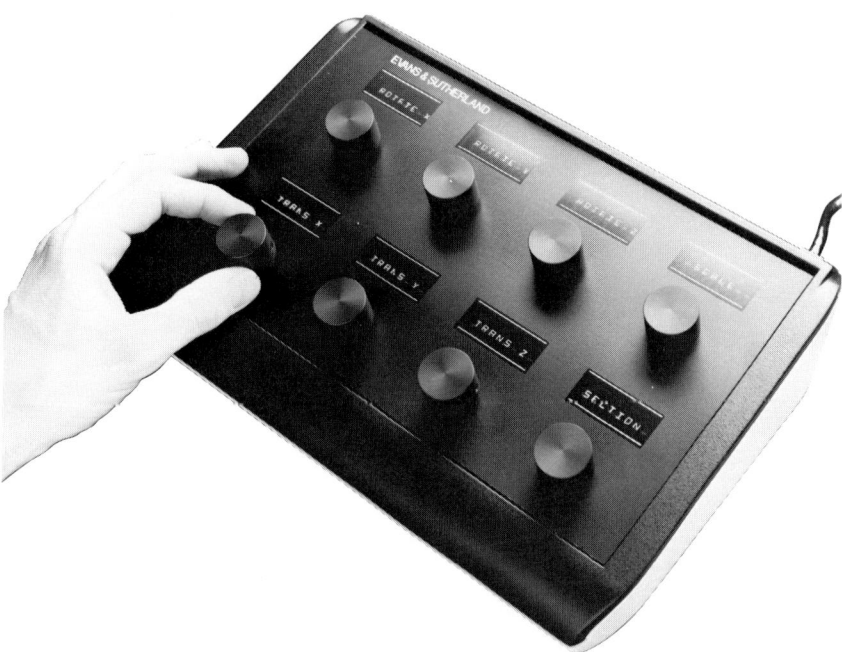

FIG. 20.21. Dials: Each may be labeled interchangeably (courtesy Evans and Sutherland).

FIG. 20.22. Graphics tablet for graphic input: Optional menu items may appear on the terminal screen or on a template placed over the tablet surface (courtesy Evans and Sutherland).

detected by an electrical or other sensing mechanism often embedded in the tablet's surface. The stylus may have a depressable point, serving the role of a function button. Alternate locators used with tablets may be outfitted with function buttons or their equivalent.

b. The so-called *mouse,* such as in Fig. 20.23, is a hand-sized device moved over a flat surface. Relative coordinates may be determined by the motion of one or more rolling balls or disks on its underside. It is normally rigged with a few function buttons.

c. The *joystick,* such as in Fig. 20.24, has a hand-guided shaft protruding from a housing. It is tilted radially outward as if from the center of a clock face toward its numerals. Vertical position or direction on the screen is coincident with 12 o'clock in the joystick reference. The amount of tilting may either have no effect on the rate of motion or be proportional to it. Because cursor movements are usually magnifications of joystick motion, overshoot can easily occur. A joystick is usually used in conjunction with a menu-selecting device.

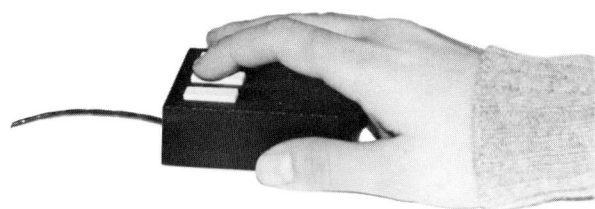

FIG. 20.23. Mouse with function keys under fingertip control.

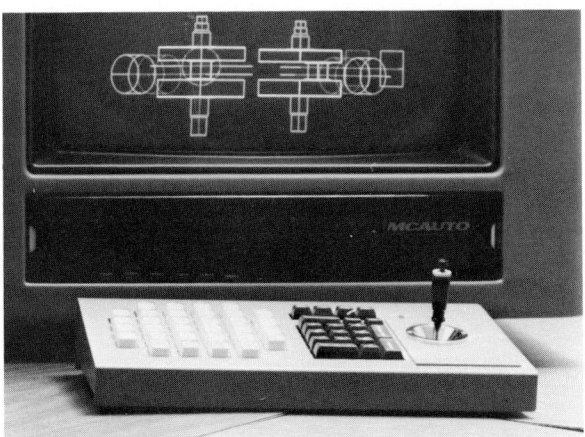

FIG. 20.24. Joystick, lower right, accompanied by a function keyboard and numeric keypad [courtesy McDonnell Douglas Automation Company (MCAUTO)].

d. The *trackball,* such as in Fig. 20.25, has a freely moving ball in a housing. The direction and velocity with which the ball is moved by hand action are transformed into corresponding cursor movement. Large movements are awkward. Like the joystick, the trackball is usually accompanied by menu selection interaction.

e. The *alphanumeric keyboard* is for text and value input, often in response to programmed queries. The text may be interpreted as displayable character strings or menu items such as those associated with a function keyboard.

In a number of instances, these devices are adapted to simulate other interaction modes. For example, consider menu selection. In one case, a graphic representation of a menu table is displayed on a terminal screen, appearing much like a transparent template. It can be picked by a light pen or pointed to by a cursor-driven device and selected by a button hit. In another case, a region of a graphics tablet can be associated with a menu selection template superimposed upon it. Positioning the tablet's stylus or locator over a menu item results in its selection. Other simulations abound.

Voice recognition systems are being tested as a means of interactive input at the time of publication. They may become prominent for some interaction modes in the years to come.

20.6.3 Techniques. The modes and devices just described can be employed in a number of interesting interactive techniques. They allow a view to be manipulated directly or its underlying

FIG. 20.25. Trackball, center, and auxiliary numeric keypads (courtesy Atari Inc.).

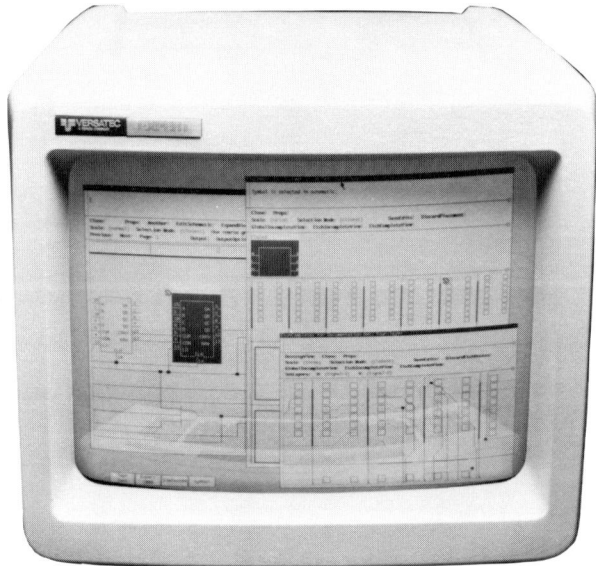

FIG. 20.26. View interaction via "window management" of superimposed views (courtesy of Versatec, a Xerox Company).

model to be manipulated indirectly by actions upon its view, which changes as a result.

View interaction techniques may employ the view transformation concepts of Article 20.5.3 to provide a desired viewplane window and viewport on the terminal screen. A few common ones will be mentioned. Figures 20.26 and 20.27 and Plates 4 and 5 illustrate some of these.

a. *Zooming* is enlarging or reducing the window on the viewplane, resulting in a scaling effect for a given viewport so that more or less of the model is seen. A zoom lens on a camera produces this kind of effect.

b. *Panning* is sweeping a window across a viewplane so that what is in the window at any instant is seen in the viewport. Panning a scene with a film or video camera is a direct analogy.

c. *Windowing* is defining one or more suitable viewports on a terminal screen for showing related or unrelated views simultaneously. It is becoming popular to allow the windows to overlap if desired like worksheets on a desk so that the top ones hide portions of those below.

d. *Overlaying* is the superposition of views to produce a combined view. Thought of another way, a view may be composed of layers, each of which has its own purpose but which is often related to the others.

Model interaction techniques frequently require the application of the model transformations introduced in Article 20.4.2. The techniques are used most often in the construction or modification of models, whether two or three-dimensional. Several commonly used techniques will be indicated. Figures 20.27 to 20.29 and

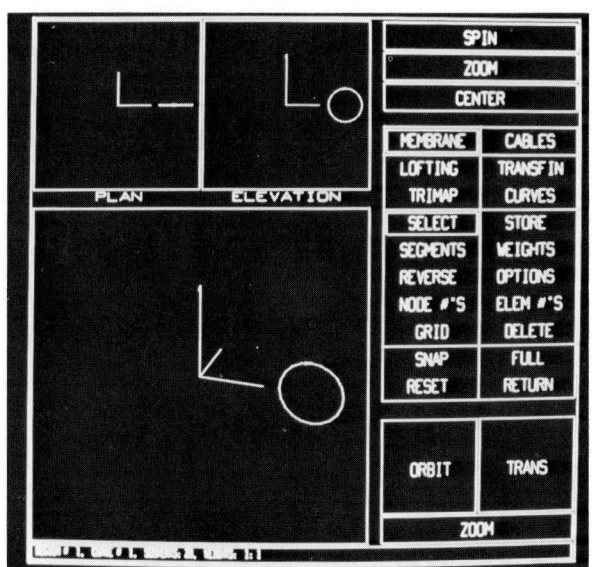
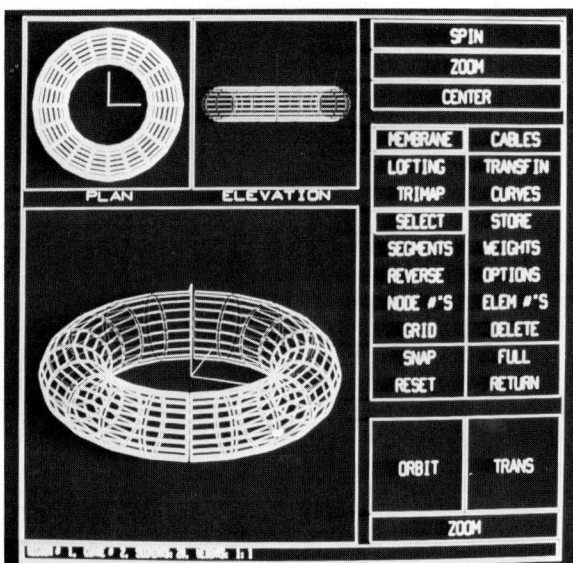

FIG. 20.27. View and model interaction via multiple windows and menu display on a terminal screen (sequence courtesy Robert Haber, University of Illinois — Urbana/Cornell University Program for Computer Graphics). (a) Initial geometry layout for a torus accomplished by defining. (b) Torus resulting from the automatic generation of surface geometry via a revolution of the cross section.

544 COMPUTER GRAPHICS AND CAD/CAM

Plate 8 illustrate some of them. Often, they may be achieved by different interactive devices or require more than one to be used simultaneously. See Figs. 20.30 and 20.31.

  a. *Defining* is the process specifying the elements of a model, ranging from elementary points involving coordinates to surfaces and solids involving geometric parameters.
  b. *Digitizing* is the process of inputting coordinate data by a locating device.
  c. *Dragging* is the process of moving an image with a locating device. A dynamic effect is achieved by automatically sampling the locator's position frequently and redisplaying the image at successive new positions.

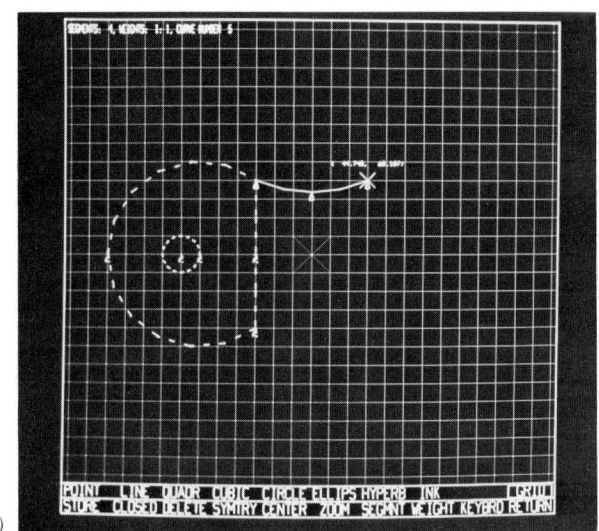
(a)

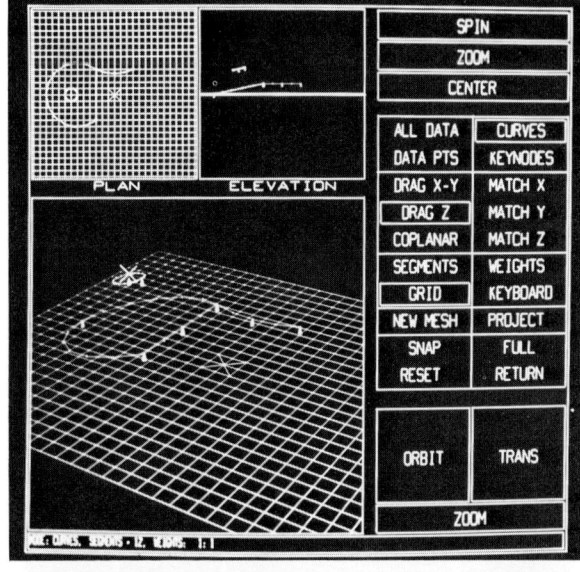
(b)

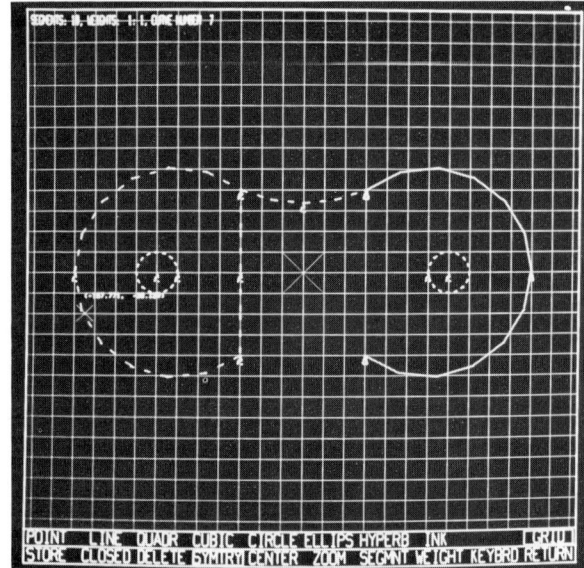
(c)

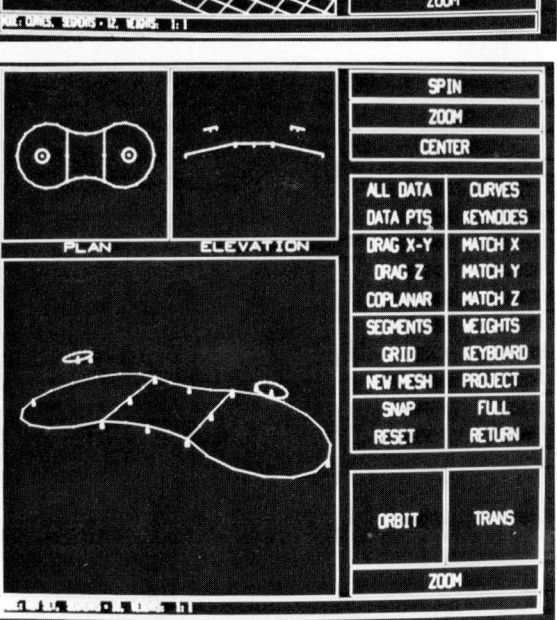
(d)

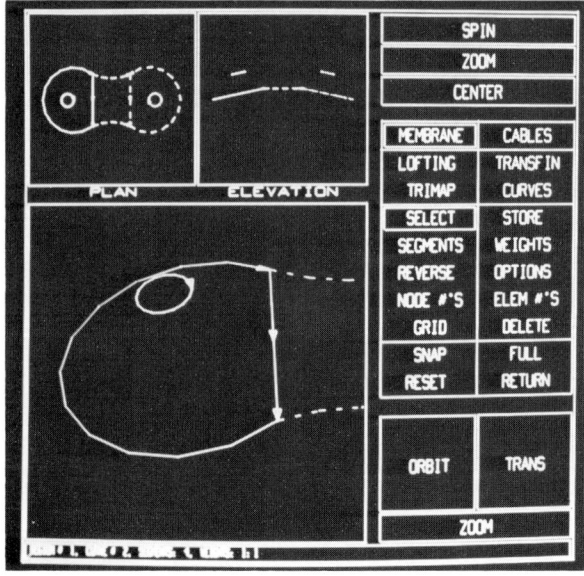
(e)

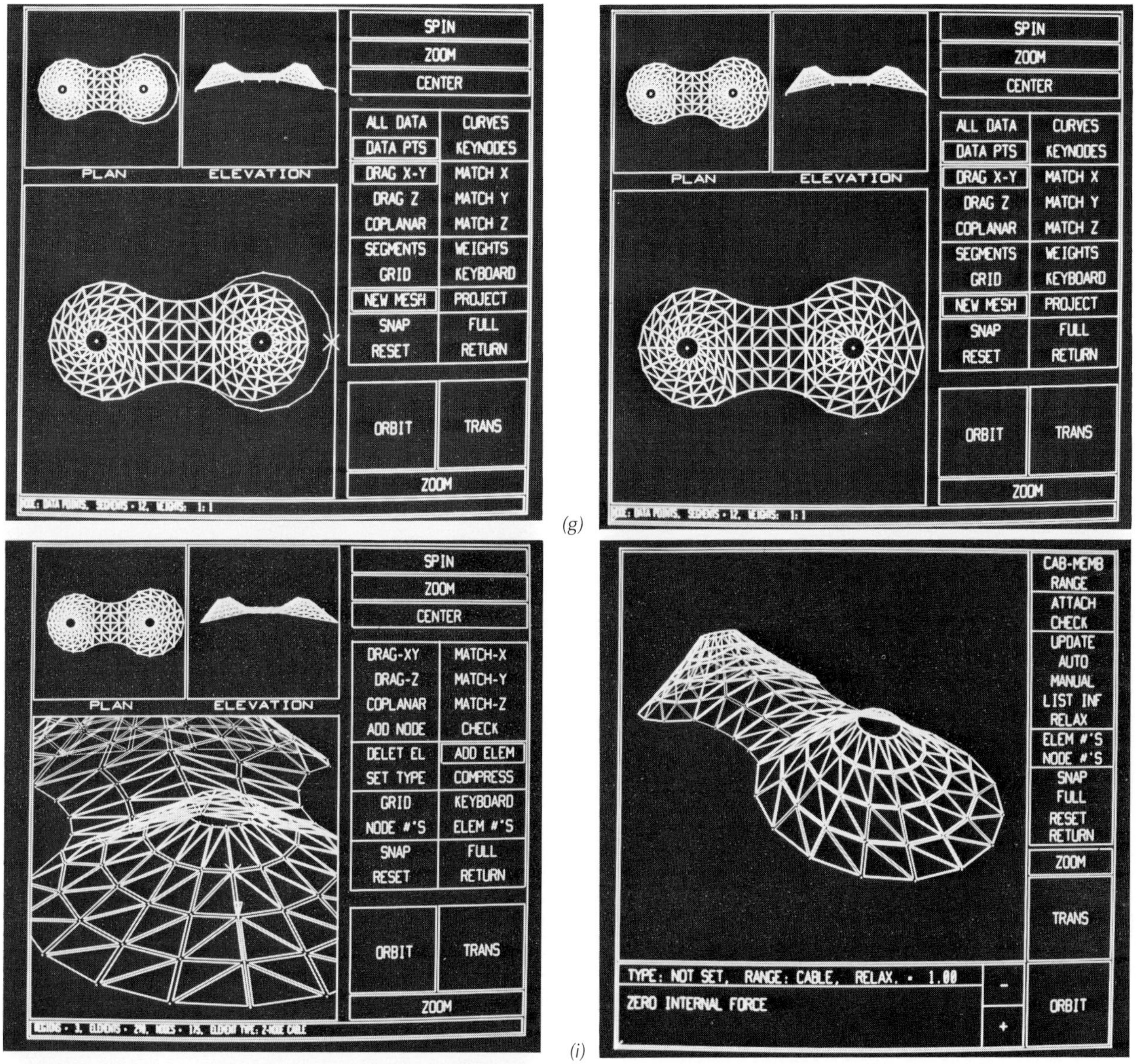

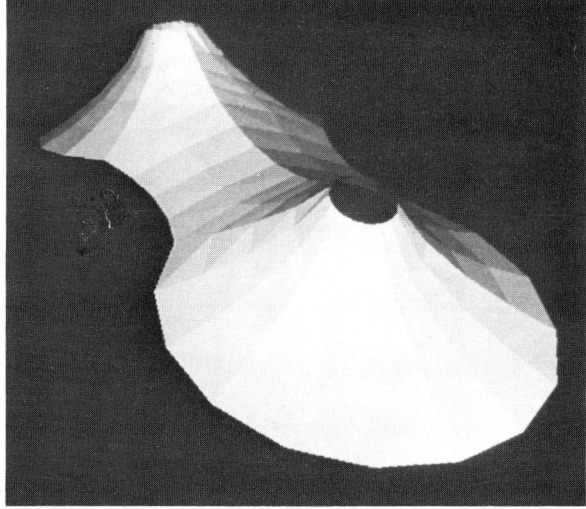

FIG. 20.28. Model interaction techniques applied to the development of a cable-reinforced, tension-membrane roof. Note displayed menus on the terminal screen change (sequence courtesy Robert Haber, University of Illinois—Urbana/Cornell University Program for Computer Graphics). (a) Defining points and curves; note the gridding feature. (b) Observing from a different vantage point. (c) Defining of points and curves to form loops, continued; note the straight line segments comprising the curves. (d) Completed loops of the surface model. (e) Observing the result of the last step from another vantage point; note the directional arrows on the loops. (f) Dragging one loop with a bounding constraint of maintaining a circular arc. (g) Modification of the finite element mesh for the surface to conform with the changed loop. (h) Closeup observation from another vantage point. (i) Overall observation from yet another vantage point. (j) A shaded surface representation of the final model.

d. *Rubberbanding* is the process of allowing one endpoint of a line (or a vertex for several lines) to be moved by a locating device. The net effect is that each line is dynamically relocated, stretching or shrinking in the process.

e. *Sketching* is the process of producing discrete or continuous-looking lines with the freehand motion of a locating device. In the discrete mode, the user exercises control over the position of each new endpoint. Rubberbanding or after-the-fact line appearance may take place. In the continuous mode, small straight-line segments are automatically created as the locator is moved.

f. *Gridding* is the process of constraining endpoints to fall on the nearest point of a defined grid that may or may not be displayed. Gridding is one method to ensure that endpoints are made coincident when using a locating device (with a cursor or not).

g. *Bounding* is the process of using limits, often defined as straight lines or planes that may or may not be shown. Bounding may be used in several ways. One is to constrain lines constructed by sketching or other

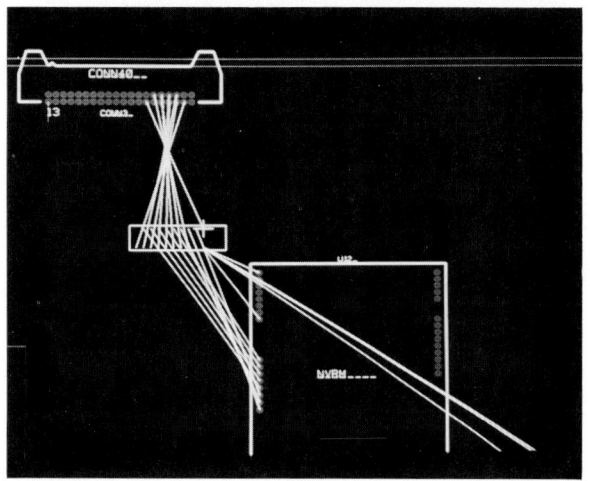
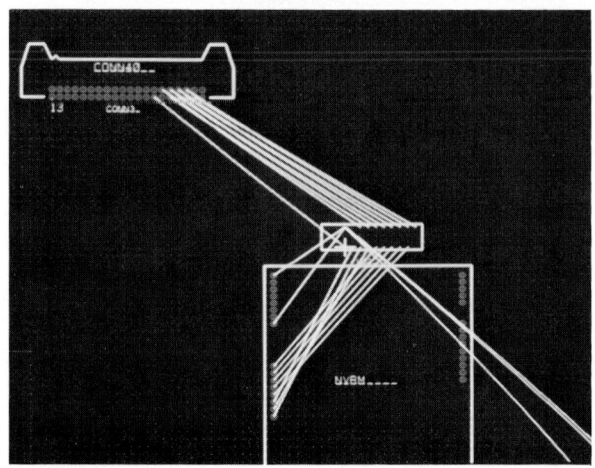

(a) (b)

FIG. 20.29. Dragging and rubberbanding model interactions in a component design system (sequence courtesy Intergraph Corp.). (a) Initial orientation; note cursor location relative to the small rectangle. (b) Small rectangle component translated and rotated with line connections automatically maintained.

20.6 INTERACTION MODES, DEVICES, AND TECHNIQUES 547

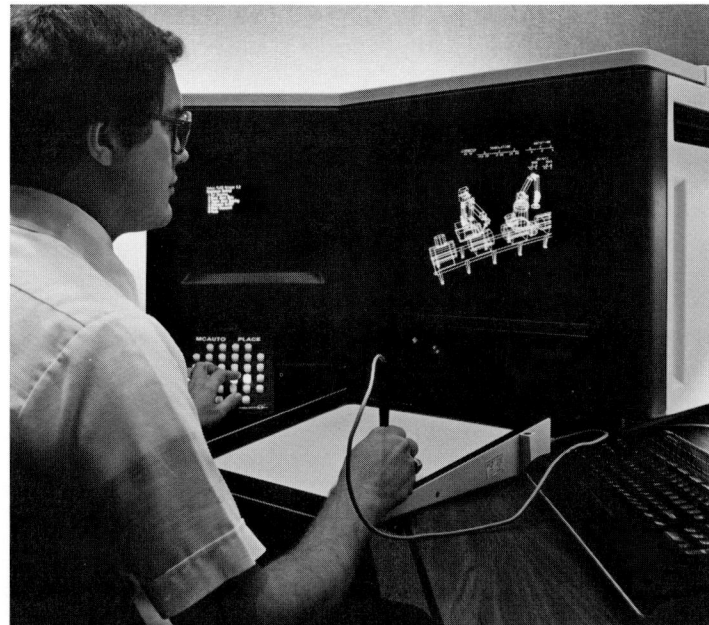

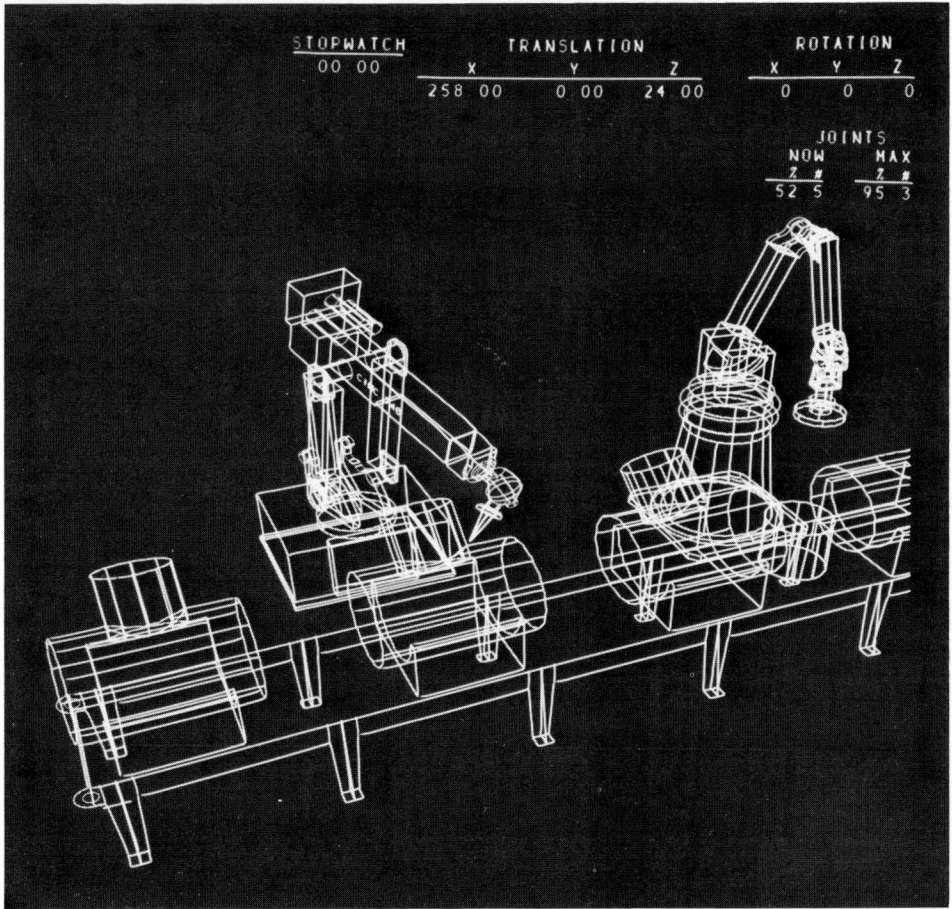

FIG. 20.30. Interactive analysis of robot-manned assembly process (sequence courtesy MCAUTO). (a) Workstation interaction. (b) Closeup of terminal screen.

techniques to be parallel to a certain limit. Another is to clip the endpoints of edges to their intersections with limiting lines or planes, no matter how far past the limits they were initially constructed.

h. *Highlighting* is the process of invoking attention-getting or distinguishing visual cues in static or dynamic views. The simplest example is that of a blinking cursor. Other examples include the use of increased intensity, color variations, blinking, or alternately flashing colors for selected elements of a model. Highlighting is also effectively used with nongraphic aspects of screen dialogs accompanying the graphics, such as menus.

i. *Observing* is the process of manipulating models in order to facilitate understanding. This is especially useful in the three-dimensional realm. It is achieved by the selection of multiple static views, simultaneously or in sequence, from different vantage points. Extended to dynamically changing views, observing is done by the simulation of observer or model movement, or both. This is different from panning and zooming of a static viewplane image. The simple cases of "walking" into a computer-modeled building or "watching" a moving mechanism are examples.

20.7 SOME CAD/CAM APPLICATIONS

Interactive computer graphics, whose fundamentals have been introduced in the previous sections, stands at the heart of CAD and CAM. As previously stated, the acronym CAD/CAM is used here to cover a broad range of activities in engineering and other disciplines.

Two of the most prominent areas of early CAD/CAM activity were electronic and mechanical applications. Both of these developed initially from the idea of adapting two-dimensional drafting techniques to the interactive graphics environment. In the early days of CAD/CAM, it was natural to match the two-dimensional nature of drafting activities to the inherent two-dimensional characteristic of computer display devices. This was and still is the case in the electronics environment where there is a strong correlation between the componentry being worked with and the "drawing" surface it is portrayed upon.

In the three-dimensional mechanical realm, computer graphics focused on dealing with planar views, usually orthographic, manipulating elements such as coordinates in the viewplane. The dimension perpendicular to the viewplane was usually not interacted with until seen in another view. The model used was generally the wireframe (Article 20.4.1.2), and it was often projected pictorially during the interactions or as an after effect.

In many respects, interactive graphics afforded early CAD/CAM users a sophisticated set of drafting templates and instruments. But in the late 1970s, mechanical CAD began to take on a distinctive nature of its own that involved thinking and interacting in three-dimensional terms. That shift has also profoundly influenced the development of CAD/CAM techniques in other application areas.

Mechanical CAD/CAM took on its new character with the advent of surface and solid models (Article 20.4.1) and faster and more powerful computers to accommodate them. These models are more complex but can be interacted with in more sophisticated ways. Also, particularly for solid models, construction and alteration activities can take place without losing the integrity of the model. This is especially important when the model may be used as the basis for activities in addition to graphic display.

FIG. 20.31. Molds and dies for many material forming processes are an important design activity product. Here, a mold is designed for a plastic handle (courtesy MCAUTO).

CAD activities include the design of components and systems that are structural, mechanical, electrical, electronic, or otherwise physical in nature. Representative areas of application include automobiles, aircraft, space vehicles, ships, buildings, bridges, industrial and agricultural machinery, heavy construction equipment, robots, home appliances and products, computers, microprocessors, products for the military, scientific and business markets, and many more.

In many instances, the products designed by CAD interactive graphics require several types of activity. For example, a commercial aircraft has a structural frame, body, engines, mechanisms, electrical, electronic and hydraulic systems, control systems, radar, on-board computers, and so forth. CAD has made it possible not only to design the subsystems of aircraft and other complex products well but also to integrate them effectively into an overall design.

The realm of CAM covers a diverse range of activities. Traditional machining processes such as those described in Chapter 10 constitute one group. Their automation is based on the principles of *numerical control*, an instruction-based mechanical control scheme. Production steps for printed circuit boards found in televisions and other electronic gear are in another. Highly specialized manufacturing steps for integrated circuit chips at the heart of microprocessors and computer memory are in yet another. Automatic equipment to bend conduit, hydraulic tubing, piping, and so forth is in another. Additional groups exist and more are being added to the list each year.

CAM may also involve the sequencing of operations as the basis for a *production process* that can be automated. A *machining center* in which parts may be loaded, repositioned, or moved from one machine tool to another, and tools changed, all under computer control, is a significant example. There are others.

Also in the broad interpretation of CAM is *assembly*. A few examples will be cited. At the task level, printed circuit boards can have transistors and other components automatically inserted. Fastening machines can be used to place and insert screws in complex assemblies such as aircraft bodies. Robot welders can execute difficult weld patterns around automotive vehicle frames. In a broader vein, vehicles, devices, and products are partially or fully assembled in modern production plants with little or no human intervention.

Robots are being used increasingly in the au-

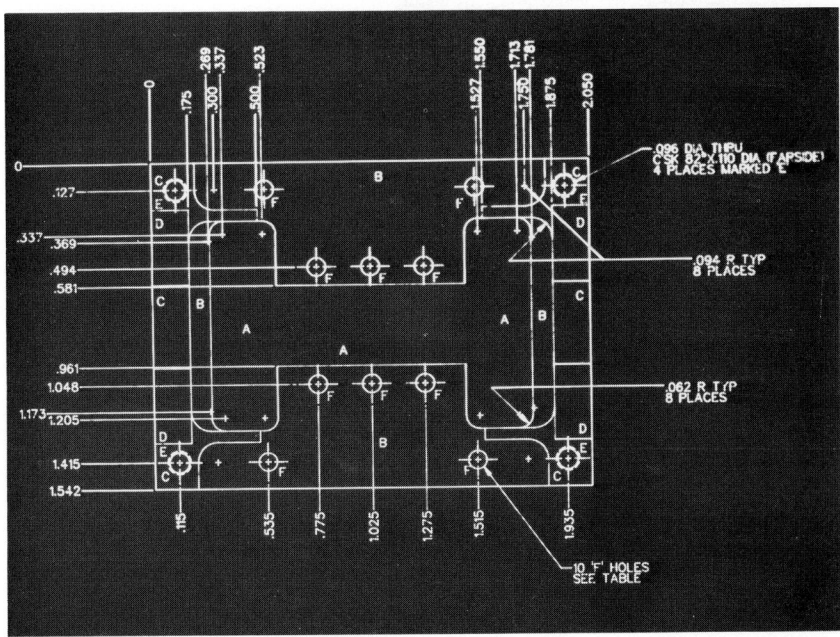

FIG. 20.32. Mechanical part drafting [courtesy Structural Dynamics Research Corporation (SDRC)].

tomation of production processes and assembly lines. Though they may not seem to be, current production robots are blind. This means they must be preprogrammed, sometimes for a variety of tasks that may be alternated. Unanticipated situations, as well as errors in a robot's driving program, may have serious consequences.

In the future, *flexible manufacturing systems* may become viable. They are envisioned to coordinate the spectrum of machining operations, manufacturing processes, assembly, inspection, and so forth for the production of diverse parts and products under computer control.

The basis for CAM, as its is for CAD, is computerization. The fabrication, process, and assembly equipment for CAM are under the control of computers that follow instructions precisely. These instructions may be referred to as a *CAM database*. In a growing number of cases, such instructions are being generated directly by computer programs that process the *CAD database* created during design activity. The CAD database includes a model data structure, discussed in Article 20.4.1, and other information. In other cases, humans intervene by interactive graphics to assist in the creation of the CAM database from the CAD database. This process may be, for example, merely specifying tooling parameters or other conditions. On the other hand, it may require the delineation of a step-by-step "walk-through" of a sequence of events. Such "teaching" is commonly used with current robots.

Whether the development of a CAM database is automatic or not, the database provides a

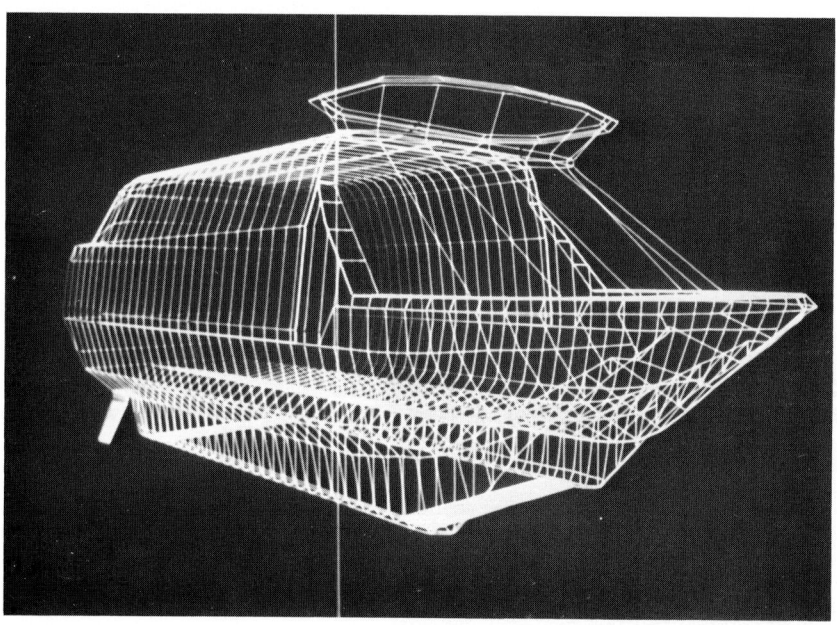

FIG. 20.33. Ship hull designing (courtesy Evans and Sutherland/University of Washington-Seattle).

means of previewing the CAM operation. Computer graphics is used to watch the operation via a sequence of "time-lapse" views or a dynamic simulation. Such previews are extremely valuable. They can show mistakes instantly, preventing errors that may be costly in terms of time, bad products, and damage to manufacturing equipment—in sum, money! This is but one ingredient in the total assessment of the impact of CAD/CAM. Overall, CAD/CAM offers the advantages of time savings, productivity increases, and better designs and finished products. In most cases, cost-effectiveness can be included in the list of advantages, especially as computer graphics system sophistication and power increase while costs continue a long-standing trend of coming down.

As an interesting aside, color is finding an increasing acceptance and use in CAD/CAM. In engineering in particular, this change has been somewhat revolutionary as color seldom found its way into traditional drafting rooms. It is very useful in the technique of highlighting (Article 20.6.3) to foster quicker and better understanding of what is seen. Consider just one situation: a complex electrical or integrated circuit schematic or mechanical part drawing. If it is zoomed in on interactively, a sense of what lines seen represent what is easily lost because of the lack of a total picture. But if they are color coded, the confusion may be greatly diminished. Color codings are also useful in visualizing physical quantities and effects such as temperature, stress, different machined surfaces, and so on.

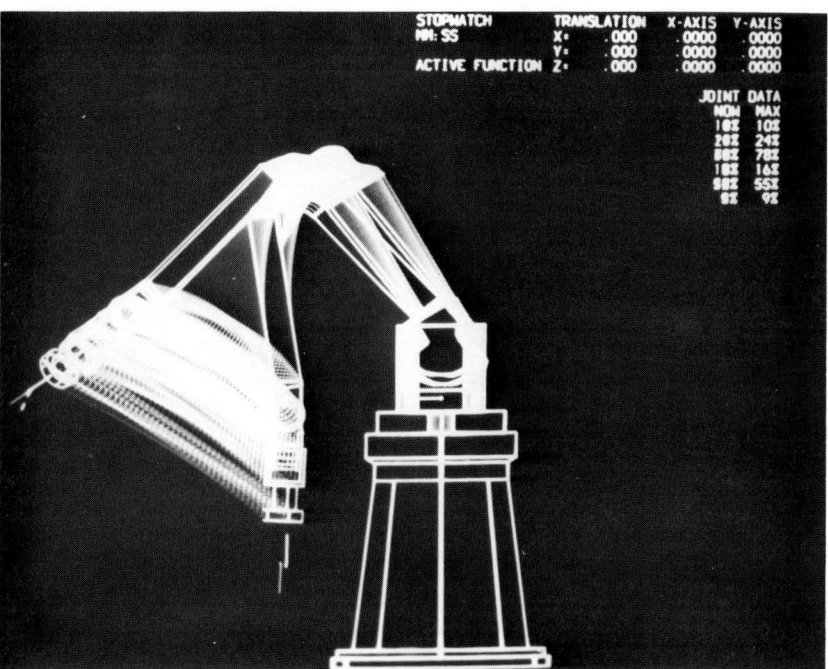

FIG. 20.34. Motion analysis of a robot arm (time exposure) (courtesy Evans and Sutherland/MCAUTO).

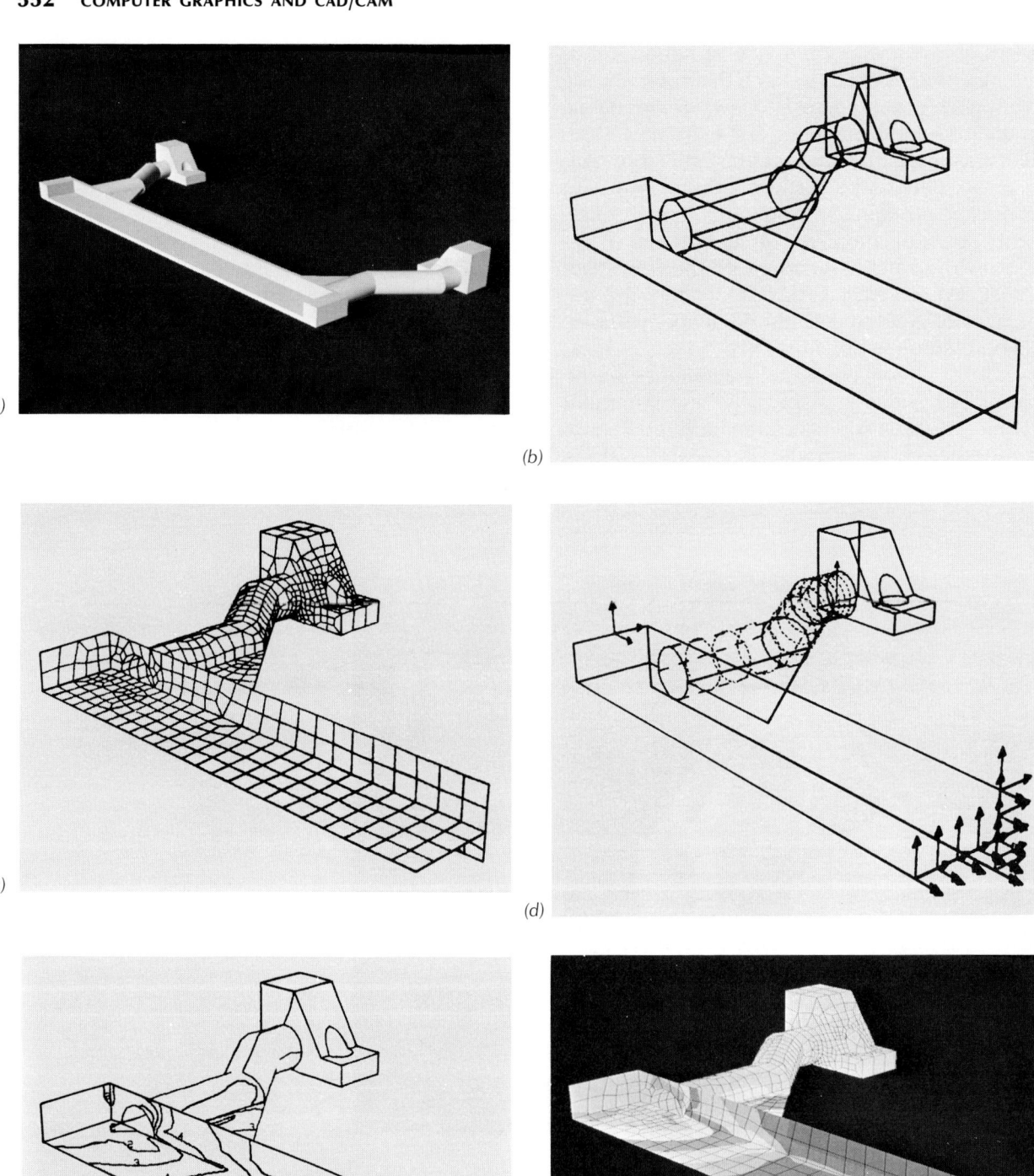

FIG. 20.35. Vehicle rear axle modeling and analysis (sequence courtesy SDRC). (a) Complete axle solid model. (b) Simplified model of a symmetric half of the axle. (c) Finite element mesh created for the simplified model. (d) Loads and restraints added to the model. (e) Stress contours determined from analysis of the finite element model. (f) Alternate representation of stress levels (detailed graphs and tables of the stresses may also be generated).

20.7 SOME CAD/CAM APPLICATIONS 553

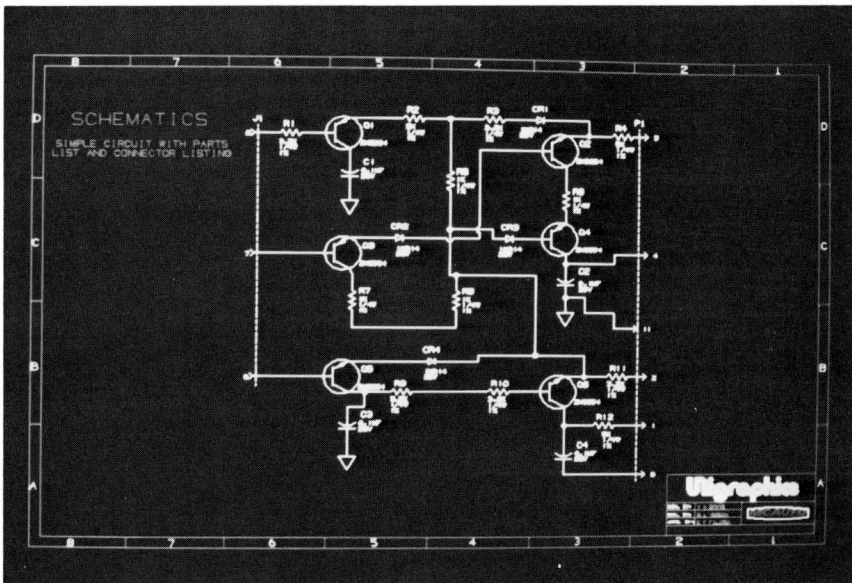

FIG. 20.36. Electric circuit design schematic. Analysis programs may be used to check the circuit's performance (courtesy MCAUTO).

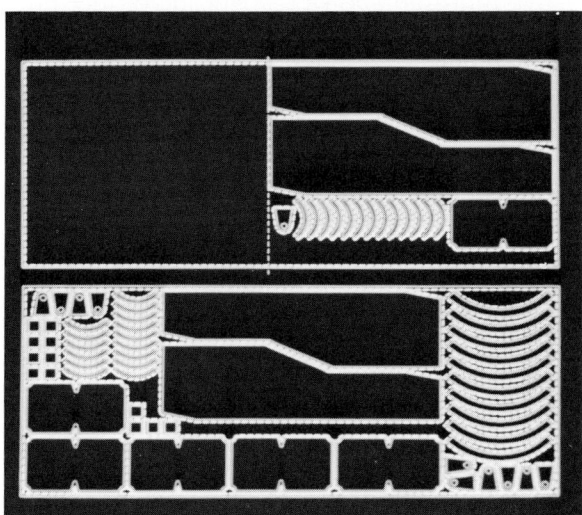

FIG. 20.37. The optimum arrangement of parts to be obtained from a piece of stock material—parts nesting—is easily obtained to yield maximum material use (courtesy Integraph corp.).

A selection of photographs depicting a representative sampling of CAD/CAM activities will be found in Figs. 20.27 to 20.37 and Plates 4 to 14.

20.8 COMPUTER GRAPHICS AND TRADITIONAL METHODS

The introduction of computer graphics in CAD/CAM has created a controversy among engineers, particularly those dealing with traditional engineering graphics as embodied in the previous chapters of this text. That controversy involves the respective roles of traditional engineering graphics and computer graphics. A feeling has arisen among some individuals that computer graphics will outmode the need to study and practice the traditional discipline. Such a concern is unjustified. The basis for this assertion comes from vast experience by none other than CAD/CAM users themselves.

It can be argued that the use of computer graphics in CAD/CAM will reduce or eliminate time spent manually creating drawings in the traditional sense. In fact, this has been witnessed in a number of settings where there are no longer vast rooms of drafters working over drawing tables. But this does not mean that the principles and use of engineering graphics have fallen by the wayside.

It will be observed in CAD/CAM environments that graphics is the key issue insofar as the engineer or designer-drafter is concerned. At all stages of interaction with a CAD/CAM system, graphics are displayed. In fact, it is certain that engineers and designer-drafters encounter a lot more graphics in CAD/CAM environments than without them. With the power of the computer put to the task, graphics are easy and quick to produce, coordinated in useful user–computer interactions or dialogs.

As a result, the need to understand engineering graphics has become even more important under CAD/CAM situations. At each stage of interaction with a CAD/CAM system, evaluative judgments are necessary. The graphics is created to afford that opportunity. These judgments may apply to two- or three-dimensional applications. They may apply to constructive layout procedures similar to manual drafting or mere requests for steps to be performed. The types of judgments may encompass geometric correctness or appropriateness. The former refers to whether what is displayed is actually what was intended; the latter refers to whether what is displayed is reasonable in terms of satisfying design or other objectives.

Other types of judgments may be necessary as well. When wireframe displays are present, it may be necessary to provide mentally the proper visibility. This may include removing hidden edges if the computer system has not been requested or is not able to do so. It may also include adjusting mentally the wireframe representation to account for incorrect, extraneous, or missing edges caused by the computer system (contrary to belief, "bugs" do appear in even well-tested and expensive systems). In *raster* or "painted" displays, analogous situations may occur with surfaces.

The keen and discerning eye of the computer graphics system user is essential in these and other situations. Realizing that the equipment being used is expensive and highly productive, it is important that judgments exercised be informed and rapid. The basis for this ability is nothing other than a firm grasp of many fundamentals of traditional engineering graphics, such as understanding orthographic and pictorial projective representations, recognizing proper proportions and constructs such as tangencies, and realizing the applicability of documentation, design, manufacturing, and production standards.

The arguments above should clearly indicate that a successful entry into the CAD/CAM field will be influenced in part by experience gained with traditional engineering graphics. But for all the excitement and promise of computer graphics in CAD/CAM, one important point is often overlooked. It is not likely that computer graphics will totally replace conventional drafting. Traditional graphics has been around for a long time. Practices have become entrenched in businesses and the like. Total changeovers to computer graphic methods in overnight fashion are impractical. Many factors must enter into the decision to make such changeovers, such as costs, retraining, and so forth.

Moreover, there are situations in which access to computer graphics tools is not practical nor warranted. This is especially true in impromptu group discussions, whether in offices, hallways, or any variety of out-of-the-way places. Some of us have even experienced situations of substituting something other than the usual paper and pencil when an unexpected need to communicate something graphical arose. The same sort of situation exists when alone in thought, too. When exploring initial concepts or making preliminary assessments, the old standbys—pencil, paper, and eraser—still retain great utility in quickly creating incisive graphics, possibly leading to the formulation of an idea that might be explored more formally, perhaps with a CAD/CAM system. For these reasons, the study of traditional engineering graphics in its own right retains great importance.

20.9 CONCLUSION

This chapter has introduced some fundamental principles of and notions about computer graphics. It has described the use of computer graphics in the field of CAD/CAM as practiced by engineers, architects, and others. It has also demonstrated that an understanding of the principles underlying the presentation of ideas by traditional drawing methods is fundamental to the successful use of computer graphics. Finally, it has pointed out that the increasingly widespread use of computer graphics does not diminish the need for manually produced drawings and sketches in many situations. In summary, computer graphics has shifted the emphasis of graphical activity and provided a powerful means of increasing design and manufacturing productivity.

Although the material in this chapter has tried to provide some meaningful ideas and insights, it has only touched lightly upon the vast and complex fields of computer graphics and CAD/CAM. Readers interested in learning more are encouraged to browse through the books suggested below. In addition, excellent videotapes and films demonstrating computer graphics and CAD/CAM can often be obtained for educational showings. Articles and advertisements in trade publications and popular journals will indicate potential vendor and consumer sources of these.

SELECTED BIBLIOGRAPHY

Prince, D., *Interactive Graphics for Computer Aided Design,* Addison–Wesley, Reading, Mass., 1971.

Rogers, D. F., and Adams, J. A., *Mathematical Elements for Computer Graphics,* McGraw-Hill, New York, 1976.

Newman, W. M., and Sproull, R. F., *Principles of Interactive Computer Graphics,* 2nd ed., McGraw-Hill, New York, 1979.

Machover, C., and Blauth, R. E., eds., *The CAD/CAM Handbook,* Computervision Corp., Bedford, Mass., 1980.

Foley, J. D., and van Dam, A., *Fundamentals of Interactive Computer Graphics,* Addison-Wesley, Reading, Mass., 1982.

Demel, J. T., and Miller, M. J., *Introduction to Computer Graphics,* Brooks/Cole Engineering Division, Monterey, CA., 1984.

APPENDIX

APPENDIX

1. General Dimensions of Straight Shank Twist Drills Taper Length — Through $\frac{1}{2}$ in. (12.7-mm)-Diameter Fractional, Number, and Metric Sizes
2. Unified Standard Screw Thread Series
3. American National Standard Metric Screw Threads — M Profile
4. American National Standard Metric Screw Threads — MJ Profile
5. Slotted Head Machine Screws (American Standard)
6. Slotted and Hexagonal-Head Cap Screws (American Standard)
7. Dimensions of Hexagon and Spline Socket Head Cap Screws (1960 Series)
8. Regular Square Bolts
9. Regular Square Nuts
10. Regular Hexagon Bolts
11. Regular Hexagon and Hexagon-Jam Nuts
12. Regular Semifinished Hexagon Bolts
13. Regular Semifinished Hexagon and Hexagon-Jam Nuts
14. Finished Hexagon Bolts
15. Finished Hexagon and Hexagon-Jam Nuts
16. Square Head Set Screws
17. Square Head Set Screw Points
18. Proportions of Keys in the Pratt and Whitney System
19. Woodruff Keys
20. Plain Parallel Stock Keys
21. Dimensions of Taper Pins
22. Running and Sliding Fits (Inch)
23. Locational Clearance Fits (Inch)
24. Locational Transition Fits (Inch)
25. Locational Interference Fits (Inch)
26. Force and Shrink Fits (Inch)
27. Preferred Hole Basis Clearance Fits (Metric)
28. Preferred Hole Basis Transition and Interference Fits (Metric)
29. Preferred Shaft Basis Clearance Fits (Metric)
30. Preferred Shaft Basis Transition and Interference Fits (Metric)
31. Dimensions of Plane Washers
32. American Standard Steel Pipe Data
33. ASTM Standard Brass and Copper Pipe Data
34. ASTM Standard Copper Water Tube Data
35. American Standard Taper Pipe Thread Data
36. Wire Gages
37. Areas and Volumes
38. Flat Pattern Development
39. List of Common Abbreviations on Drawings
40. ANSI Standards of Interest to Designers and Drafters
41. International, American, British, and Canadian Symbols for Positional and Form Tolerances

APPENDIX 559

TABLE 1 GENERAL DIMENSIONS OF STRAIGHT SHANK TWIST DRILLS TAPER LENGTH—THRU ½-in. (12.7-mm) Dia. FRACTIONAL, NUMBERS, AND METRIC SIZES

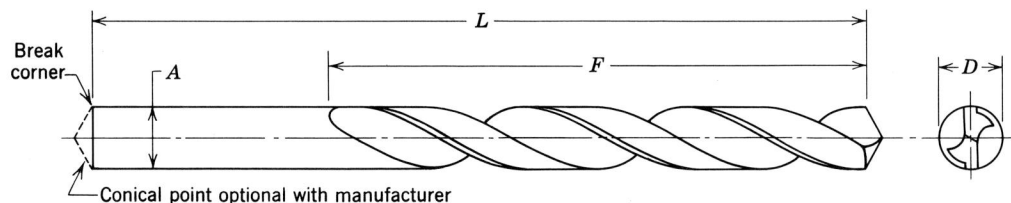

Conical point optional with manufacturer

Diameter of Drill					Flute Length		Overall Length	
	D		Decimal Inch Equivalent	Millimeter Equivalent	F		L	
Fraction	No.	mm			Inch	mm	Inch	mm
		1.85	0.0728	1.850	2	51	3¾	95
	49		0.0730	1.854	2	51	3¾	95
		1.90	0.0748	1.900	2	51	3¾	95
	48		0.0760	1.930	2	51	3¾	95
		1.95	0.0768	1.950	2	51	3¾	95
⁵⁄₆₄			0.0781	1.984	2	51	3¾	95
	47		0.0785	1.994	2¼	57	4¼	108
		2.00	0.0787	2.000	2¼	57	4¼	108
		2.05	0.0807	2.050	2¼	57	4¼	108
	46		0.0810	2.057	2¼	57	4¼	108
	45		0.0820	2.083	2¼	57	4¼	108
		2.10	0.0827	2.100	2¼	57	4¼	108
		2.15	0.0846	2.150	2¼	57	4¼	108
	44		0.0860	2.184	2¼	57	4¼	108
		2.20	0.0866	2.200	2¼	57	4¼	108
		2.25	0.0886	2.250	2¼	57	4¼	108
	43		0.0890	2.261	2¼	57	4¼	108
		2.30	0.0906	2.300	2¼	57	4¼	108
		2.35	0.0925	2.350	2¼	57	4¼	108
	42		0.0935	2.375	2¼	57	4¼	108
³⁄₃₂			0.0938	2.383	2¼	57	4¼	108
		2.40	0.0945	2.400	2½	64	4⅝	117
	41		0.0960	2.438	2½	64	4⅝	117
		2.45	0.0965	2.450	2½	64	4⅝	117
	40		0.0980	2.489	2½	64	4⅝	117
		2.50	0.0984	2.500	2½	64	4⅝	117
	39		0.0995	2.527	2½	64	4⅝	117
	38		0.1015	2.578	2½	64	4⅝	117
		2.60	0.1024	2.600	2½	64	4⅝	117
	37		0.1040	2.642	2½	64	4⅝	117
		2.70	0.1063	2.700	2½	64	4⅝	117
	36		0.1065	2.705	2½	64	4⅝	117
⁷⁄₆₄			0.1094	2.779	2½	64	4⅝	117
	35		0.1100	2.794	2¾	70	5⅛	130
		2.80	0.1102	2.800	2¾	70	5⅛	130
	34		0.1110	3.819	2¾	70	5⅛	130
	33		0.1130	2.870	2¾	70	5⅛	130
		2.90	0.1142	2.900	2¾	70	5⅛	130
	32		0.1160	2.946	2¾	70	5⅛	130
		3.00	0.1181	3.000	2¾	70	5⅛	130

ANSI B94.11M-1979

TABLE 1 (cont'd)

Diameter of Drill					Flute Length		Overall Length	
D			Decimal Inch Equivalent	Millimeter Equivalent	F		L	
Fraction	No.	mm			Inch	mm	Inch	mm
	31		0.1200	3.048	$2^3/_4$	70	$5^1/_8$	130
		3.10	0.1220	3.100	$2^3/_4$	70	$5^1/_8$	130
$1/_8$			0.1250	3.175	$2^3/_4$	70	$5^1/_8$	130
		3.20	0.1260	3.200	3	76	$5^3/_8$	137
	30		0.1285	3.264	3	76	$5^3/_8$	137
		3.30	0.1299	3.300	3	76	$5^3/_8$	137
		3.40	0.1339	3.400	3	76	$5^3/_8$	137
	29		0.1360	3.464	3	76	$5^3/_8$	137
		3.50	0.1378	3.500	3	76	$5^3/_8$	137
	28		0.1405	3.569	3	76	$5^3/_8$	137
$9/_{64}$			0.1406	3.571	3	76	$5^3/_8$	137
		3.60	0.1417	3.600	3	76	$5^3/_8$	137
	27		0.1440	3.658	3	76	$5^3/_8$	137
		3.70	0.1457	3.700	3	76	$5^3/_8$	137
	26		0.1470	3.734	3	76	$5^3/_8$	137
	25		0.1495	3.797	3	76	$5^3/_8$	137
		3.80	0.1496	3.800	3	76	$5^3/_8$	137
	24		0.1520	3.861	3	76	$5^3/_8$	137
		3.90	0.1535	3.900	3	76	$5^3/_8$	137
	23		0.1540	3.912	3	76	$5^3/_8$	137
$5/_{32}$			0.1562	3.967	3	76	$5^3/_8$	137
	22		0.1570	3.988	$3^3/_8$	86	$5^3/_4$	146
		4.00	0.1575	4.000	$3^3/_8$	86	$5^3/_4$	146
	21		0.1590	4.039	$3^3/_8$	86	$5^3/_4$	146
	20		0.1610	4.089	$3^3/_8$	86	$5^3/_4$	146
		4.10	0.1614	4.100	$3^3/_8$	86	$5^3/_4$	146
		4.20	0.1654	4.200	$3^3/_8$	86	$5^3/_4$	146
	19		0.1660	4.216	$3^3/_8$	86	$5^3/_4$	146
		4.30	0.1693	4.300	$3^3/_8$	86	$5^3/_4$	146
	18		0.1695	4.305	$3^3/_8$	86	$5^3/_4$	146
$11/_{64}$			0.1719	4.366	$3^3/_8$	86	$5^3/_4$	146
	17		0.1730	4.394	$3^3/_8$	86	$5^3/_4$	146
		4.40	0.1732	4.400	$3^3/_8$	86	$5^3/_4$	146
	16		0.1770	4.496	$3^3/_8$	86	$5^3/_4$	146
		4.50	0.1772	4.500	$3^3/_8$	86	$5^3/_4$	146
	15		0.1800	4.572	$3^3/_8$	86	$5^3/_4$	146
		4.60	0.1811	4.600	$3^3/_8$	86	$5^3/_4$	146
	14		0.1820	4.623	$3^3/_8$	86	$5^3/_4$	146
	13	4.70	0.1850	4.700	$3^3/_8$	86	$5^3/_4$	146
$3/_{16}$			0.1875	4.762	$3^3/_8$	86	$5^3/_4$	146

ANSI B94.11M-1979

TABLE 1 (cont'd)

Diameter of Drill					Flute Length		Overall Length	
	D		Decimal Inch Equivalent	Millimeter Equivalent	F		L	
Fraction	No.	mm			Inch	mm	Inch	mm
	12	4.80	0.1890	4.800	$3^5/_8$	92	6	152
	11		0.1910	4.851	$3^5/_8$	92	6	152
		4.90	0.1929	4.900	$3^5/_8$	92	6	152
	10		0.1935	4.915	$3^5/_8$	92	6	152
	9		0.1960	4.978	$3^5/_8$	92	6	152
		5.00	0.1969	5.000	$3^5/_8$	92	6	152
	8		0.1990	5.054	$3^5/_8$	92	6	152
		5.10	0.2008	5.100	$3^5/_8$	92	6	152
	7		0.2010	5.105	$3^5/_8$	92	6	152
$13/_{64}$			0.2031	5.159	$3^5/_8$	92	6	152
	6		0.2040	5.182	$3^5/_8$	92	6	152
		5.20	0.2047	5.200	$3^5/_8$	92	6	152
	5		0.2055	5.220	$3^5/_8$	92	6	152
		5.30	0.2087	5.300	$3^5/_8$	92	6	152
	4		0.2090	5.309	$3^5/_8$	92	6	152
		5.40	0.2126	5.400	$3^5/_8$	92	6	152
	3		0.2130	5.410	$3^5/_8$	92	6	152
		5.50	0.2165	5.500	$3^5/_8$	92	6	152
$7/_{32}$			0.2188	5.558	$3^5/_8$	92	6	152
		5.60	0.2205	5.600	$3^3/_4$	95	$6^1/_8$	156
	2		0.2210	5.613	$3^3/_4$	95	$6^1/_8$	156
		5.70	0.2244	5.700	$3^3/_4$	95	$6^1/_8$	156
	1		0.2280	5.791	$3^3/_4$	95	$6^1/_8$	156
		5.80	0.2283	5.800	$3^3/_4$	95	$6^1/_8$	156
		5.90	0.2323	5.900	$3^3/_4$	95	$6^1/_8$	156
$15/_{64}$			0.2344	5.954	$3^3/_4$	95	$6^1/_8$	156
		6.00	0.2362	6.000	$3^3/_4$	95	$6^1/_8$	156
		6.10	0.2402	6.100	$3^3/_4$	95	$6^1/_8$	156
		6.20	0.2441	6.200	$3^3/_4$	95	$6^1/_8$	156
		6.30	0.2480	6.300	$3^3/_4$	95	$6^1/_8$	156
$1/_4$			0.2500	6.350	$3^3/_4$	95	$6^1/_8$	156
		6.40	0.2520	6.400	$3^7/_8$	98	$6^1/_4$	159
		6.50	0.2559	6.500	$3^7/_8$	98	$6^1/_4$	159
$17/_{64}$			0.2656	6.746	$3^7/_8$	98	$6^1/_4$	159
		6.80	6.2677	6.800	$3^7/_8$	98	$6^1/_4$	159
		7.00	0.2756	7.000	$3^7/_8$	98	$6^1/_4$	159
$9/_{32}$			0.2812	7.142	$3^7/_8$	98	$6^1/_4$	159
		7.20	0.2835	7.200	4	102	$6^3/_8$	162
		7.50	0.2953	7.500	4	102	$6^3/_8$	162
$19/_{64}$			0.2969	7.541	4	102	$6^3/_8$	162

ANSI B94.11M-1979

TABLE 1 (cont'd)

Diameter of Drill			Decimal Inch Equivalent	Millimeter Equivalent	Flute Length F		Overall Length L	
Fraction	No.	mm			Inch	mm	Inch	mm
		7.80	0.3071	7.800	4	102	$6^3/_8$	162
$5/_{16}$			0.3125	7.938	4	102	$6^3/_8$	162
		8.00	0.3150	8.000	$4^1/_8$	105	$6^1/_2$	165
		8.20	0.3228	8.200	$4^1/_8$	105	$6^1/_2$	165
$21/_{64}$			0.3281	8.334	$4^1/_8$	105	$6^1/_2$	165
		8.50	0.3346	8.500	$4^1/_8$	105	$6^1/_2$	165
$11/_{32}$			0.3438	8.733	$4^1/_8$	105	$6^1/_2$	165
		8.80	0.3465	8.800	$4^1/_4$	108	$6^3/_4$	171
		9.00	0.3543	9.000	$4^1/_4$	108	$6^3/_4$	171
$23/_{64}$			0.3594	9.129	$4^1/_4$	108	$6^3/_4$	171
		9.20	0.3622	9.200	$4^1/_4$	108	$6^3/_4$	171
		9.50	0.3740	9.500	$4^1/_4$	108	$6^3/_4$	171
$3/_8$			0.3750	9.525	$4^1/_4$	108	$6^3/_4$	171
		9.80	0.3858	9.800	$4^3/_8$	111	7	178
$25/_{64}$			0.3906	9.921	$4^3/_8$	111	7	178
		10.00	0.3937	10.000	$4^3/_8$	111	7	178
		10.20	0.4016	10.200	$4^3/_8$	111	7	178
$13/_{32}$			0.4062	10.317	$4^3/_8$	111	7	178
		10.50	0.4134	10.500	$4^5/_8$	117	$7^1/_4$	184
$27/_{64}$			0.4219	10.716	$4^5/_8$	117	$7^1/_4$	184
		10.80	0.4252	10.800	$4^5/_8$	117	$7^1/_4$	184
		11.00	0.4331	11.000	$4^5/_8$	117	$7^1/_4$	184
$7/_{16}$			0.4375	11.112	$4^5/_8$	117	$7^1/_4$	184
		11.20	0.4409	11.200	$4^3/_4$	121	$7^1/_2$	190
		11.50	0.4528	11.500	$4^3/_4$	121	$7^1/_2$	190
$29/_{64}$			0.4531	11.509	$4^3/_4$	121	$7^1/_2$	190
		11.80	0.4646	11.800	$4^3/_4$	121	$7^1/_2$	190
$15/_{32}$			0.4688	11.908	$4^3/_4$	121	$7^1/_2$	190
		12.00	0.4724	12.000	$4^3/_4$	121	$7^3/_4$	197
		12.20	0.4803	12.200	$4^3/_4$	121	$7^3/_4$	197
$31/_{64}$			0.4844	12.304	$4^3/_4$	121	$7^3/_4$	197
		12.50	0.4921	12.500	$4^3/_4$	121	$7^3/_4$	197
$1/_2$			0.5000	12.700	$4^3/_4$	121	$7^3/_4$	197

ANSI B94.11M-1979

TABLE 2 UNIFIED STANDARD SCREW THREAD SERIES

| SIZES | | BASIC | THREADS PER INCH | | | | | | | | | | | SIZES |
| Primary | Secondary | MAJOR DIAMETER | Series with graded pitches | | | Series with constant pitches | | | | | | | | |
			Coarse UNC	Fine UNF	Extra fine UNEF	4UN	6UN	8UN	12UN	16UN	20UN	28UN	32UN	
0		0.060	—	80	—	—	—	—	—	—	—	—	—	0
	1	0.073	64	72	—	—	—	—	—	—	—	—	—	1
2		0.086	56	64	—	—	—	—	—	—	—	—	—	2
	3	0.099	48	56	—	—	—	—	—	—	—	—	—	3
4		0.112	40	48	—	—	—	—	—	—	—	—	—	4
5		0.125	40	44	—	—	—	—	—	—	—	—	—	5
6		0.138	32	40	—	—	—	—	—	—	—	—	UNC	6
8		0.164	32	36	—	—	—	—	—	—	—	—	UNC	8
10		0.190	24	32	—	—	—	—	—	—	—	—	UNF	10
	12	0.216	24	28	32	—	—	—	—	—	—	UNF	UNEF	12
1/4		0.250	20	28	32	—	—	—	—	—	UNC	UNF	UNEF	1/4
5/16		0.3125	18	24	32	—	—	—	—	—	20	28	UNEF	5/16
3/8		0.375	16	24	32	—	—	—	—	UNC	20	28	UNEF	3/8
7/16		0.4375	14	20	28	—	—	—	—	16	UNF	UNEF	32	7/16
1/2		0.500	13	20	28	—	—	—	—	16	UNF	UNEF	32	1/2
9/16		0.5265	12	18	24	—	—	—	UNC	16	20	28	32	9/16
5/8		0.625	11	18	24	—	—	—	12	16	20	28	32	5/8
	11/16	0.6875	—	—	24	—	—	—	12	16	20	28	32	11/16
3/4		0.750	10	16	20	—	—	—	12	UNF	UNEF	28	32	3/4
	13/16	0.8125	—	—	20	—	—	—	12	16	UNEF	28	32	13/16
7/8		0.875	9	14	20	—	—	—	12	16	UNEF	28	32	7/8
	15/16	0.9375	—	—	20	—	—	—	12	16	UNEF	28	32	15/16
1		1.000	8	12	20	—	—	UNC	UNF	16	UNEF	28	32	1
	1 1/16	1.0625	—	—	18	—	—	8	12	16	20	28	—	1 1/16
1 1/8		1.125	7	12	18	—	—	8	UNF	16	20	28	—	1 1/8
	1 3/16	1.1875	—	—	18	—	—	8	12	16	20	28	—	1 3/16
1 1/4		1.250	7	12	18	—	—	8	UNF	16	20	28	—	1 1/4
	1 5/16	1.3125	—	—	18	—	—	8	12	16	20	28	—	1 5/16
1 3/8		1.375	6	12	18	—	UNC	8	UNF	16	20	28	—	1 3/8
	1 7/16	1.4375	—	—	18	—	6	8	12	16	20	28	—	1 7/16
1 1/2		1.500	6	12	18	—	UNC	8	UNF	16	20	28	—	1 1/2
	1 9/16	1.5625	—	—	18	—	6	8	12	16	20	—	—	1 9/16
1 5/8		1.625	—	—	18	—	6	8	12	16	20	—	—	1 5/8
	1 11/16	1.6875	—	—	18	—	6	8	12	16	20	—	—	1 11/16
1 3/4		1.750	5	—	—	—	6	8	12	16	20	—	—	1 3/4
	1 13/16	1.8125	—	—	—	—	6	8	12	16	20	—	—	1 13/16
1 7/8		1.875	—	—	—	—	6	8	12	16	20	—	—	1 7/8
	1 15/16	1.9375	—	—	—	—	6	8	12	16	20	—	—	1 15/16
2		2.000	4 1/2	—	—	—	6	8	12	16	20	—	—	2
	2 1/8	2.125	—	—	—	—	6	8	12	16	20	—	—	2 1/8
2 1/4		2.250	4 1/2	—	—	—	6	8	12	16	20	—	—	2 1/4
	2 3/8	2.375	—	—	—	—	6	8	12	16	20	—	—	2 3/8
2 1/2		2.500	—	—	—	UNC	6	8	12	16	20	—	—	2 1/2
	2 5/8	2.625	—	—	—	4	6	8	12	16	20	—	—	2 5/8
2 3/4		2.750	—	—	—	UNC	6	8	12	16	20	—	—	2 3/4
	2 7/8	2.875	—	—	—	4	6	8	12	16	20	—	—	2 7/8
3		3.000	—	—	—	UNC	6	8	12	16	20	—	—	3
	3 1/8	3.125	—	—	—	4	6	8	12	16	—	—	—	3 1/8
3 1/4		3.250	—	—	—	UNC	6	8	12	16	—	—	—	3 1/4
	3 3/8	3.375	—	—	—	4	6	8	12	16	—	—	—	3 3/8
3 1/2		3.500	—	—	—	UNC	6	8	12	16	—	—	—	3 1/2
	3 5/8	3.625	—	—	—	4	6	8	12	16	—	—	—	3 5/8
3 3/4		3.750	—	—	—	UNC	6	8	12	16	—	—	—	3 3/4
	3 7/8	3.875	—	—	—	4	6	8	12	16	—	—	—	3 7/8
4		4.000	—	—	—	UNC	6	8	12	16	—	—	—	4
	4 1/8	4.125	—	—	—	4	6	8	12	16	—	—	—	4 1/8
4 1/4		4.250	—	—	—	4	6	8	12	16	—	—	—	4 1/4
	4 3/8	4.375	—	—	—	4	6	8	12	16	—	—	—	4 3/8
4 1/2		4.500	—	—	—	4	6	8	12	16	—	—	—	4 1/2
	4 5/8	4.625	—	—	—	4	6	8	12	16	—	—	—	4 5/8
4 3/4		4.750	—	—	—	4	6	8	12	16	—	—	—	4 3/4
	4 7/8	4.875	—	—	—	4	6	8	12	16	—	—	—	4 7/8
5		5.000	—	—	—	4	6	8	12	16	—	—	—	5
	5 1/8	5.125	—	—	—	4	6	8	12	16	—	—	—	5 1/8
5 1/4		5.250	—	—	—	4	6	8	12	16	—	—	—	5 1/4
	5 3/8	5.375	—	—	—	4	6	8	12	16	—	—	—	5 3/8
5 1/2		5.500	—	—	—	4	6	8	12	16	—	—	—	5 1/2
	5 5/8	5.625	—	—	—	4	6	8	12	16	—	—	—	5 5/8
5 3/4		5.750	—	—	—	4	6	8	12	16	—	—	—	5 3/4
	5 7/8	5.875	—	—	—	4	6	8	12	16	—	—	—	5 7/8
6		6.000	—	—	—	4	6	8	12	16	—	—	—	6

Excerpt from ANSI B1.1-1974. Entries in column 'BASIC MAJOR DIAMETER' have been modified from the original by deletion of the fourth place decimal zero to emphasize the practice (described in para. 6.2.1.3 of this standard) of omitting any fourth place decimal zero on drawings. It is not intended to require conformance with this practice in the presentation of tabulated data.

ANSI Y14.6-1978

TABLE 3 AMERICAN NATIONAL STANDARD METRIC SCREW THREADS — M PROFILE
Internal Thread—Limiting Dimensions M Profile

Basic Thread Designation	Tol Class	Minor Diameter D_1 Min	Max	Pitch Diameter D_2 Min	Max	Tol	Major Diameter D Min	Max[b] For Reference
M1.6 x 0.35	6H	1.221	1.321	1.373	1.458	0.085	1.600	1.736
M2 x 0.4	6H	1.567	1.679	1.740	1.830	0.090	2.000	2.148
M2.5 x 0.45	6H	2.013	2.138	2.208	2.303	0.095	2.500	2.660
M3 x 0.5	6H	2.459	2.599	2.675	2.775	0.100	3.000	3.172
M3.5 x 0.6	6H	2.850	3.010	3.110	3.222	0.112	3.500	3.699
M4 x 0.7	6H	3.242	3.422	3.545	3.663	0.118	4.000	4.219
M5 x 0.8	6H	4.134	4.334	4.480	4.605	0.125	5.000	5.240
M6 x 1	6H	4.917	5.153	5.350	5.500	0.150	6.000	6.294
M8 x 1.25	6H	6.647	6.912	7.188	7.348	0.160	8.000	8.340
M8 x 1	6H	6.917	7.153	7.350	7.500	0.150	8.000	8.294
M10 x 1.5	6H	8.376	8.676	9.026	9.206	0.180	10.000	10.396
M10 x 1.25	6H	8.647	8.912	9.188	9.348	0.160	10.000	10.340
M10 x 0.75	6H	9.188	9.378	9.513	9.645	0.132	10.000	10.240
M12 x 1.75	6H	10.106	10.441	10.863	11.063	0.200	12.000	12.453
M12 x 1.5	6H	10.376	10.676	11.026	11.216	0.190	12.000	12.406
M12 x 1.25	6H	10.647	10.912	11.188	11.368	0.180	12.000	12.360
M12 x 1	6H	10.917	11.153	11.350	11.510	0.160	12.000	12.304
M14 x 2	6H	11.835	12.210	12.701	12.913	0.212	14.000	14.501
M14 x 1.5 M14 x 1.25 (a)	6H	12.376	12.676	13.026	13.216	0.190	14.000	14.406
M15 x 1	6H	13.917	14.153	14.350	14.510	0.160	15.000	15.304
M16 x 2	6H	13.835	14.210	14.701	14.913	0.212	16.000	16.501
M16 x 1.5	6H	14.376	14.676	15.026	15.216	0.190	16.000	16.406
M17 x 1	6H	15.917	16.153	16.350	16.510	0.160	17.000	17.304
M18 x 1.5	6H	16.376	16.676	17.026	17.216	0.190	18.000	18.406
M20 x 2.5	6H	17.294	17.744	18.376	18.600	0.224	20.000	20.585
M20 x 1.5	6H	18.376	18.676	19.026	19.216	0.190	20.000	20.406
M20 x 1	6H	18.917	19.153	19.350	19.510	0.160	20.000	20.304
M22 x 2.5	6H	19.294	19.744	20.376	20.600	0.224	22.000	22.585
M22 x 1.5	6H	20.376	20.676	21.026	21.216	0.190	22.000	22.406
M24 x 3	6H	20.752	21.252	22.051	22.316	0.265	24.000	24.698
M24 x 2	6H	21.835	22.210	22.701	22.925	0.224	24.000	24.513
M25 x 1.5	6H	23.376	23.676	24.026	24.226	0.200	25.000	25.416
M27 x 3	6H	23.752	24.252	25.051	25.316	0.265	27.000	27.698
M27 x 2	6H	24.835	25.210	25.701	25.925	0.224	27.000	27.513
M30 x 3.5	6H	26.211	26.771	27.727	28.007	0.280	30.000	30.785
M30 x 2	6H	27.835	28.210	28.701	28.925	0.224	30.000	30.513
M30 x 1.5	6H	28.376	28.676	29.026	29.226	0.200	30.000	30.416
M33 x 2	6H	30.835	31.210	31.701	31.925	0.224	33.000	33.513
M35 x 1.5	6H	33.376	33.676	34.026	34.226	0.200	35.000	35.416
M36 x 4	6H	31.670	32.270	33.402	33.702	0.300	36.000	36.877
M36 x 2	6H	33.835	34.210	34.701	34.925	0.224	36.000	36.513
M39 x 2	6H	36.835	37.210	37.701	37.925	0.224	39.000	39.513
M40 x 1.5	6H	38.376	38.676	39.026	39.226	0.200	40.000	40.416
M42 x 4.5	6H	37.129	37.799	39.077	39.392	0.315	42.000	42.965

(cont'd.)

ANSI B1.13M-1979

TABLE 3 (cont'd)

External Thread—Limiting Dimensions M Profile

Basic Thread Designation	Tolerance Class	Allowance es(b)	Major Diameter(a) d		Pitch Diameter(a) d_2			Minor Diameter(a) d_1 (Flat Root) Max	Minor Diameter(c) d_3 (Rounded Root) Min —For Reference—
			Max	Min	Max	Min	Tol		
M1.6 x 0.35	6g	0.019	1.581	1.496	1.354	1.291	0.063	1.202	1.075
M1.6 x 0.35	4g6g	0.019	1.581	1.496	1.354	1.314	0.040	1.202	1.098
M2 x 0.4	6g	0.019	1.981	1.886	1.721	1.654	0.067	1.548	1.408
M2 x 0.4	4g6g	0.019	1.981	1.886	1.721	1.679	0.042	1.548	1.433
M2.5 x 0.45	6g	0.020	2.480	2.380	2.188	2.117	0.071	1.993	1.840
M2.5 x 0.45	4g6g	0.020	2.480	2.380	2.188	2.143	0.045	1.993	1.866
M3 x 0.5	6g	0.020	2.980	2.874	2.655	2.580	0.075	2.439	2.272
M3 x 0.5	4g6g	0.020	2.980	2.874	2.655	2.607	0.048	2.439	2.299
M3.5 x 0.6	6g	0.021	3.479	3.354	3.089	3.004	0.085	2.829	2.635
M3.5 x 0.6	4g6g	0.021	3.479	3.354	3.089	3.036	0.053	2.829	2.667
M4 x 0.7	6g	0.022	3.978	3.838	3.523	3.433	0.090	3.220	3.002
M4 x 0.7	4g6g	0.022	3.978	3.838	3.523	3.467	0.056	3.220	3.036
M5 x 0.8	6g	0.024	4.976	4.826	4.456	4.361	0.095	4.110	3.869
M5 x 0.8	4g6g	0.024	4.976	4.826	4.456	4.396	0.060	4.110	3.904
M6 x 1	6g	0.026	5.974	5.794	5.324	5.212	0.112	4.891	4.596
M6x 1	4g6g	0.026	5.974	5.794	5.324	5.253	0.071	4.891	4.637
M8 x 1.25	6g	0.028	7.972	7.760	7.160	7.042	0.118	6.619	6.272
M8 x 1.25	4g6g	0.028	7.972	7.760	7.160	7.085	0.075	6.619	6.315
M8 x 1	6g	0.026	7.974	7.794	7.324	7.212	0.112	6.891	6.596
M8 x 1	4g6g	0.026	7.974	7.794	7.324	7.253	0.071	6.891	6.637
M10 x 1.5	6g	0.032	9.968	9.732	8.994	8.862	0.132	8.344	7.938
M10 x 1.5	4g6g	0.032	9.968	9.732	8.994	8.909	0.085	8.344	7.985
M10 x 1.25	6g	0.028	9.972	9.760	9.160	9.042	0.118	8.619	8.272
M10 x 1.25	4g6g	0.028	9.972	9.760	9.160	9.085	0.075	8.619	8.315
M10 x 0.75	6g	0.022	9.978	9.838	9.491	9.391	0.100	9.166	8.929
M10 x 0.75	4g6g	0.022	9.978	9.838	9.491	9.428	0.063	9.166	8.966
M12 x 1.75	6g	0.034	11.966	11.701	10.829	10.679	0.150	10.072	9.601

ANSI B1.13M-1979

(cont'd.)

TABLE 3 (cont'd)

External Threads—Limiting Dimensions M Profile (Continued)

Basic Thread Designation	Tolerance Class	Allowance es (b)	Major Diameter (a) d		Pitch Diameter (a) d_2			Minor Diameter (a) d_1 (Flat Root) Max	Minor Diameter (c) d_3 (Rounded Root) Min —For Reference—
			Max	Min	Max	Min	Tol		
M12 × 1.75	4g6g	0.034	11.966	11.701	10.829	10.734	0.095	10.072	9.656
M12 × 1.5	6g	0.032	11.968	11.732	10.994	10.854	0.140	10.344	9.930
M12 × 1.25	6g	0.028	11.972	11.760	11.160	11.028	0.132	10.619	10.258
M12 × 1.25	4g6g	0.028	11.972	11.760	11.160	11.075	0.085	10.619	10.305
M12 × 1	6g	0.026	11.974	11.794	11.324	11.206	0.118	10.891	10.590
M12 × 1	4g6g	0.026	11.974	11.794	11.324	11.249	0.075	10.891	10.633
M14 × 2	6g	0.038	13.962	13.682	12.663	12.503	0.160	11.797	11.271
M14 × 2	4g6g	0.038	13.962	13.682	12.663	12.563	0.100	11.797	11.331
M14 × 1.5	6g	0.032	13.968	13.732	12.994	12.854	0.140	12.344	11.930
M14 × 1.5	4g6g	0.032	13.968	13.732	12.994	12.904	0.090	12.344	11.980
M14 × 1.25 (d)									
M15 × 1	6g	0.026	14.974	14.794	14.324	14.206	0.118	13.891	13.590
M15 × 1	4g6g	0.026	14.974	14.794	14.324	14.249	0.075	13.891	13.633
M16 × 2	6g	0.038	15.962	15.682	14.663	14.503	0.160	13.797	13.271
M16 × 2	4g6g	0.038	15.962	15.682	14.663	14.563	0.100	13.797	13.331
M16 × 1.5	6g	0.032	15.968	15.732	14.994	14.854	0.140	14.344	13.930
M16 × 1.5	4g6g	0.032	15.968	15.732	14.994	14.904	0.090	14.344	13.980
M17 × 1	6g	0.026	16.974	16.794	16.324	16.206	0.118	15.891	15.590
M17 × 1	4g6g	0.026	16.974	16.794	16.324	16.249	0.075	15.891	15.633
M18 × 1.5	6g	0.032	17.968	17.732	16.994	16.854	0.140	16.344	15.930
M18 × 1.5	4g6g	0.032	17.968	17.732	16.994	16.904	0.090	16.344	15.980
M20 × 2.5	6g	0.042	19.958	19.623	18.334	18.164	0.170	17.252	16.624
M20 × 2.5	4g6g	0.042	19.958	19.623	18.334	18.228	0.106	17.252	16.688
M20 × 1.5	6g	0.032	19.968	19.732	18.994	18.854	0.140	18.344	17.930
M20 × 1.5	4g6g	0.032	19.968	19.732	18.994	18.904	0.090	18.344	17.980
M20 × 1	6g	0.026	19.974	19.794	19.324	19.206	0.118	18.891	18.590
M20 × 1	4g6g	0.026	19.974	19.794	19.324	19.249	0.075	18.891	18.633

ANSI B1.13M-1979

(cont'd.)

TABLE 3 *(cont'd)*

External Threads—Limiting Dimensions M Profile (Continued)

Basic Thread Designation	Tolerance Class	Allowance es (b)	Major Diameter (a) d		Pitch Diameter (a) d_2			Minor Diameter (a) d_1 (Flat Root) Max	Minor Diameter (c) d_3 (Rounded Root) Min —For Reference—
			Max	Min	Max	Min	Tol		
M22 × 2.5	6g	0.042	21.958	21.623	20.334	20.164	0.170	19.252	18.624
M22 × 1.5	6g	0.032	21.968	21.732	20.994	20.854	0.140	20.344	19.930
M22 × 1.5	4g6g	0.032	21.968	21.732	20.994	20.904	0.090	20.344	19.980
M24 × 3	6g	0.048	23.952	23.577	22.003	21.803	0.200	20.704	19.955
M24 × 3	4g6g	0.048	23.952	23.577	22.003	21.878	0.125	20.704	20.030
M24 × 2	6g	0.038	23.962	23.682	22.663	22.493	0.170	21.797	21.261
M24 × 2	4g6g	0.038	23.962	23.682	22.663	22.557	0.106	21.797	21.325
M25 × 1.5	6g	0.032	24.968	24.732	23.994	23.844	0.150	23.344	22.920
M25 × 1.5	4g6g	0.032	24.968	24.732	23.994	23.899	0.095	23.344	22.975
M27 × 3	6g	0.048	26.952	26.577	25.003	24.803	0.200	23.744	22.955
M27 × 2	6g	0.038	26.962	26.682	25.663	25.493	0.170	24.797	24.261
M27 × 2	4g6g	0.038	26.962	26.682	25.663	25.557	0.106	24.797	24.325
M30 × 3.5	6g	0.053	29.947	29.522	27.674	27.462	0.212	26.158	25.306
M30 × 3.5	4g6g	0.053	29.947	29.522	27.674	27.542	0.132	26.158	25.386
M30 × 2	6g	0.038	29.962	29.682	28.663	28.493	0.170	27.797	27.261
M30 × 2	4g6g	0.038	29.962	29.682	28.663	28.557	0.106	27.797	27.325
M30 × 1.5	6g	0.032	29.968	29.732	28.994	28.844	0.150	28.344	27.920
M30 × 1.5	4g6g	0.032	29.968	29.732	28.994	28.899	0.095	28.344	27.975
M33 × 2	6g	0.038	32.962	32.682	31.663	31.493	0.170	30.797	30.261
M33 × 2	4g6g	0.038	32.962	32.682	31.663	31.557	0.106	30.797	30.325
M35 × 1.5	6g	0.032	34.968	34.732	33.994	33.844	0.150	33.344	33.920
M36 × 4	6g	0.060	35.940	35.465	33.342	33.118	0.224	31.610	30.654
M36 × 4	4g6g	0.060	35.940	35.465	33.342	33.202	0.140	31.610	30.738
M36 × 2	6g	0.038	35.962	35.682	34.663	34.493	0.170	33.797	33.261
M39 × 2	4g6g	0.038	35.962	35.682	34.663	34.557	0.106	33.797	33.325
M39 × 2	6g	0.038	38.962	38.682	37.663	37.493	0.170	36.797	36.261
M39 × 2	4g6g	0.038	38.962	38.682	37.663	37.557	0.106	36.797	36.325

ANSI B1.13M-1979

(cont'd.)

TABLE 4 AMERICAN NATIONAL STANDARD METRIC SCREW THREADS — MJ PROFILE
LIMITING DIMENSIONS OF AEROSPACE METRIC SCREW THREAD STANDARD SERIES
(millimeters)

Basic Thread Designation	Tol Class	INTERNAL THREAD					
		Minor Dia, D_1		Pitch Dia, D_2			Major Dia, D	
		Min	Max	Min	Max	Tol	Min	Max
*MJ1.6 x 0.35	4H6H	1.259	1.359	1.373	1.426	0.053	1.600	1.704
MJ1.8 x 0.35	4H6H	1.459	1.559	1.573	1.626	0.053	1.800	1.904
*MJ2 x 0.4	4H6H	1.610	1.722	1.740	1.796	0.056	2.000	2.114
MJ2.2 x 0.45	4H6H	1.762	1.887	1.908	1.968	0.060	2.200	2.325
*MJ2.5 x 0.45	4H6H	2.062	2.187	2.208	2.268	0.060	2.500	2.625
*MJ3 x 0.5	4H6H	2.513	2.653	2.675	2.738	0.063	3.000	3.135
*MJ3.5 x 0.6	4H6H	2.915	3.075	3.110	3.181	0.071	3.500	3.658
*MJ4 x 0.7	4H6H	3.318	3.498	3.545	3.620	0.075	4.000	4.176
MJ4.5 x 0.75	4H6H	3.769	3.959	4.013	4.088	0.075	4.500	4.683
*MJ5 x 0.8	4H6H	4.221	4.421	4.480	4.560	0.080	5.000	5.195
MJ6 x 0.75	4H5H	5.269	5.419	5.513	5.598	0.085	6.000	6.193
*MJ6 x 1	4H5H	5.026	5.216	5.350	5.445	0.095	6.000	6.239
MJ7 x 0.75	4H5H	6.269	6.419	6.513	6.598	0.085	7.000	7.193
*MJ7 x 1	4H5H	6.026	6.216	6.350	6.445	0.095	7.000	7.239
MJ8 x 0.75	4H5H	7.269	7.419	7.513	7.598	0.085	8.000	8.193
*MJ8 x 1	4H5H	7.026	7.216	7.350	7.445	0.095	8.000	8.239
MJ8 x 1.25	4H5H	6.782	6.994	7.188	7.288	0.100	8.000	8.280
MJ9 x 0.75	4H5H	8.269	8.419	8.513	8.598	0.085	9.000	9.193
MJ9 x 1	4H5H	8.026	8.216	8.350	8.445	0.095	9.000	9.239
MJ9 x 1.25	4H5H	7.782	7.994	8.188	8.288	0.100	9.000	9.280
MJ10 x 0.75	4H5H	9.269	9.419	9.513	9.598	0.085	10.000	10.193
MJ10 x 1	4H5H	9.026	9.216	9.350	9.445	0.095	10.000	10.239
*MJ10 x 1.25	4H5H	8.782	8.994	9.188	9.288	0.100	10.000	10.280
MJ10 x 1.5	4H5H	8.539	8.775	9.026	9.138	0.112	10.000	10.328
MJ11 x 0.75	4H5H	10.269	10.419	10.513	10.598	0.085	11.000	11.193
MJ11 x 1	4H5H	10.026	10.216	10.350	10.445	0.095	11.000	11.239
MJ11 x 1.25	4H5H	9.782	9.994	10.188	10.288	0.100	11.000	11.280
MJ11 x 1.5	4H5H	9.539	9.775	10.026	10.138	0.112	11.000	11.329
MJ12 x 1	4H5H	11.026	11.216	11.350	11.450	0.100	12.000	12.244
*MJ12 x 1.25	4H5H	10.782	10.994	11.188	11.300	0.112	12.000	12.292
MJ12 x 1.5	4H5H	10.539	10.775	11.026	11.144	0.118	12.000	12.335
MJ12 x 1.75	4H5H	10.295	10.560	10.863	10.988	0.125	12.000	12.378
MJ14 x 1	4H5H	13.026	13.216	13.350	13.450	0.100	14.000	14.244
MJ14 x 1.25	4H5H	12.782	12.994	13.188	13.300	0.112	14.000	14.292
*MJ14 x 1.5	4H5H	12.539	12.775	13.026	13.144	0.118	14.000	14.334
MJ14 x 2	4H5H	12.051	12.351	12.701	12.833	0.132	14.000	14.421
MJ15 x 1	4H5H	14.026	14.216	14.350	14.450	0.100	15.000	15.244
MJ15 x 1.5	4H5H	13.539	13.775	14.026	14.144	0.118	15.000	15.335
MJ16 x 1	4H5H	15.026	15.216	15.350	15.450	0.100	16.000	16.244
*MJ16 x 1.5	4H5H	14.539	14.775	15.026	15.144	0.118	16.000	16.334
MJ16 x 2	4H5H	14.051	14.351	14.701	14.833	0.132	16.000	16.421
MJ17 x 1	4H5H	16.026	16.216	16.350	16.450	0.100	17.000	17.244
MJ17 x 1.5	4H5H	15.539	15.775	16.026	16.144	0.118	17.000	17.335
MJ18 x 1	4H5H	17.026	17.216	17.350	17.450	0.100	18.000	18.244
*MJ18 x 1.5	4H5H	16.539	16.775	17.026	17.144	0.118	18.000	18.334
MJ18 x 2	4H5H	16.051	16.351	16.701	16.833	0.132	18.000	18.421
MJ18 x 2.5	4H5H	15.564	15.919	16.376	16.516	0.140	18.000	18.501
MJ20 x 1	4H5H	19.026	19.216	19.350	19.450	0.100	20.000	20.244
*MJ20 x 1.5	4H5H	18.539	18.775	19.026	19.144	0.118	20.000	20.334
MJ20 x 2	4H5H	18.051	18.351	18.701	18.833	0.132	20.000	20.421
MJ20 x 2.5	4H5H	17.564	17.919	18.376	18.516	0.140	20.000	20.501

*Standard series for aerospace nuts.

ANSI B1.21M-1978

TABLE 4 *(cont'd)*
LIMITING DIMENSIONS OF AEROSPACE METRIC SCREW THREAD STANDARD SERIES
(millimeters)

Basic Thread Designation	Tol Class	EXTERNAL THREAD					
		Major Dia, d		Pitch dia, d_2			Minor Dia, d_3	
		Max	Min	Max	Min	Tol	Max	Min
*MJ1.6 x 0.35	4h6h	1.600	1.515	1.373	1.333	0.040	1.196	1.135
MJ1.8 x 0.35	4h6h	1.800	1.715	1.573	1.533	0.040	1.396	1.335
*MJ2 x 0.4	4h6h	2.000	1.905	1.740	1.698	0.042	1.538	1.472
MJ2.2 x 0.45	4h6h	2.200	2.100	1.908	1.863	0.045	1.681	1.608
*MJ2.5 x 0.45	4h6h	2.500	2.400	2.208	2.163	0.045	1.980	1.908
*MJ3 x 0.5	4h6h	3.000	2.894	2.675	2.627	0.048	2.423	2.344
*MJ3.5 x 0.6	4h6h	3.500	3.375	3.110	3.057	0.053	2.807	2.718
*MJ4 x 0.7	4h6h	4.000	3.860	3.545	3.489	0.056	3.192	3.093
MJ4.5 x 0.75	4h6h	4.500	4.360	4.013	3.957	0.056	3.634	3.533
*MJ5 x 0.8	4h6h	5.000	4.850	4.480	4.420	0.060	4.076	3.968
MJ6 x 0.75	4h6h	6.000	5.860	5.513	5.450	0.063	5.134	5.026
*MJ6 x 1	4h6h	6.000	5.820	5.350	5.279	0.071	4.845	4.714
MJ7 x 0.75	4h6h	7.000	6.860	6.513	6.450	0.063	6.134	6.026
*MJ7 x 1	4h6h	7.000	6.820	6.350	6.279	0.071	5.845	5.714
MJ8 x 0.75	4h6h	8.000	7.860	7.513	7.450	0.063	7.134	7.026
*MJ8 x 1	4h6h	8.000	7.820	7.350	7.279	0.071	6.845	6.714
MJ8 x 1.25	4h6h	8.000	7.788	7.188	7.113	0.075	6.557	6.406
MJ9 x 0.75	4h6h	9.000	8.860	8.513	8.450	0.063	8.134	8.026
MJ9 x 1	4h6h	9.000	8.820	8.350	8.279	0.071	7.845	7.713
MJ9 x 1.25	4h6h	9.000	8.788	8.188	8.113	0.075	7.557	7.406
MJ10 x 0.75	4h6h	10.000	9.860	9.513	9.450	0.063	9.134	9.026
MJ10 x 1	4h6h	10.000	9.820	9.350	9.279	0.071	8.845	8.713
*MJ10 x 1.25	4h6h	10.000	9.788	9.188	9.113	0.075	8.557	8.406
MJ10 x 1.5	4h6h	10.000	9.764	9.026	8.941	0.085	8.268	8.092
MJ11 x 0.75	4h6h	11.000	10.860	10.513	10.450	0.063	10.134	10.026
MJ11 x 1	4h6h	11.000	10.820	10.350	10.279	0.071	9.845	9.713
MJ11 x 1.25	4h6h	11.000	10.788	10.188	10.113	0.075	9.557	9.406
MJ11 x 1.5	4h6h	11.000	10.764	10.026	9.941	0.085	9.268	9.092
MJ12 x 1	4h6h	12.000	11.820	11.350	11.275	0.075	10.845	10.710
*MJ12 x 1.25	4h6h	12.000	11.788	11.188	11.103	0.085	10.557	10.396
MJ12 x 1.5	4h6h	12.000	11.764	11.026	10.936	0.090	10.268	10.087
MJ12 x 1.75	4h6h	12.000	11.735	10.863	10.768	0.095	9.979	9.778
MJ14 x 1	4h6h	14.000	13.820	13.350	13.275	0.075	12.845	12.709
MJ14 x 1.25	4h6h	14.000	13.788	13.188	13.103	0.085	12.557	12.396
*MJ14 x 1.5	4h6h	14.000	13.764	13.026	12.936	0.090	12.268	12.087
MJ14 x 2	4h6h	14.000	13.720	12.701	12.601	0.100	11.691	11.469
MJ15 x 1	4h6h	15.000	14.820	14.350	14.275	0.075	13.845	13.710
MJ15 x 1.5	4h6h	15.000	14.764	14.026	13.936	0.090	13.268	13.087
MJ16 x 1	4h6h	16.000	15.820	15.350	15.275	0.075	14.845	14.709
*MJ16 x 1.5	4h6h	16.000	15.764	15.026	14.936	0.090	14.268	14.087
MJ16 x 2	4h6h	16.000	15.720	14.701	14.601	0.100	13.691	13.469
MJ17 x 1	4h6h	17.000	16.820	16.350	16.275	0.075	15.845	15.710
MJ17 x 1.5	4h6h	17.000	16.764	16.026	15.936	0.090	15.268	15.087
MJ18 x 1	4h6h	18.000	17.820	17.350	17.275	0.075	16.845	16.709
*MJ18 x 1.5	4h6h	18.000	17.764	17.026	16.936	0.090	16.268	16.087
MJ18 x 2	4h6h	18.000	17.720	16.701	16.601	0.100	15.691	15.469
MJ18 x 2.5	4h6h	18.000	17.665	16.376	16.270	0.106	15.113	14.856
MJ20 x 1	4h6h	20.000	19.820	19.350	19.275	0.075	18.845	18.710
*MJ20 x 1.5	4h6h	20.000	19.764	19.026	18.936	0.090	18.268	18.087
MJ20 x 2	4h6h	20.000	19.720	18.701	18.601	0.100	17.691	17.469
MJ20 x 2.5	4h6h	20.000	19.665	18.376	18.270	0.106	17.113	16.856

*Standard series for aerospace screws and bolts.

ANSI B1.21M-1978

TABLE 5 SLOTTED HEAD MACHINE SCREWS (AMERICAN STANDARD)

Nominal size	Maximum Diameter	Threads per inch (coarse Series)	Maximum diameter of head			Maximum height of head			Maximum width of slot	Maximum depth of slot				Maximum height of the head	
			Flat and oval	Round	Fillister	Flat and oval	Round	Fillister		Flat	Round	Fillister	Oval	Oval	Fillister
	(D)		(A)	(A)	(A)	(H)	(H)	(H)	(J)	(T)	(T)	(T)	(T)	(F)	(F)
2	0.086	56	0.172	0.162	0.140	0.051	0.070	0.055	0.036	0.023	0.048	0.037	0.045	0.036	0.028
3	.099	48	.199	.187	.161	.059	.078	.063	.038	.027	.053	.043	.052	.038	.032
4	.112	40	.225	.211	.183	.067	.086	.072	.040	.030	.058	.048	.059	.040	0.35
5	.125	40	.252	.236	.205	.075	.095	.081	.043	.034	.062	.054	.067	.043	.039
6	.138	32	.279	.260	.226	.083	.103	.089	.045	.038	.067	.060	.074	.045	.043
8	.164	32	.352	.309	.270	.100	.119	.106	.050	.045	.076	.071	.088	.050	.050
10	.190	24	.385	.359	.313	.116	.136	.123	.055	.053	.086	.083	.103	.055	.057
12	.216	24	.438	.408	.357	.132	.152	.141	.059	.060	.095	.094	.117	.059	.064
1/4	.250	20	.507	.472	.414	.153	.174	.163	.066	.070	.108	.109	.136	.066	.074
5/16	.3125	18	.636	.591	.519	.192	.214	.205	.077	.088	.130	.137	.171	.077	.092
3/8	.375	16	.762	.708	.622	.230	.254	.246	.088	.106	.153	.164	.206	.088	.109

TABLE 6 SLOTTED AND HEXAGONAL-HEAD CAP SCREWS (AMERICAN STANDARD)

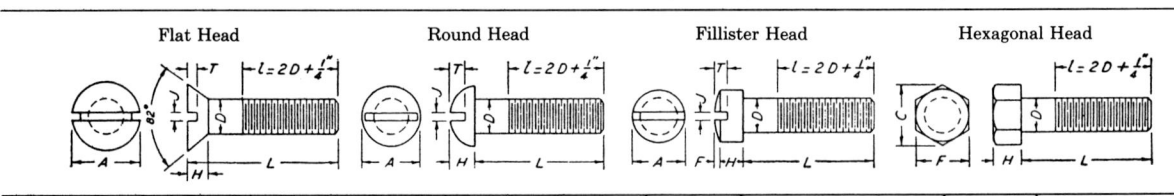

Nominal size	Maximum diameter	Threads per inch	Maximum diameter of head			Maximum height of head			Maximum width of slot	Maximum depth of slot			Maximum height of fillister head oval	Finished hexagonal-head cap screw		
			Flat	Round	Fillister	Flat (Nominal)	Round	Fillister		Flat	Round	Fillister		Maximum width across flats	Minimum width across corners	Maximum height
	(D)		(A)	(A)	(A)	(H)	(H)	(H)	(J)	(T)	(T)	(T)	(F)	(F)	(C)	(H)
1/4	0.2500	20	0.500	0.437	0.375	0.140	0.191	0.172	0.075	0.068	0.117	0.097	0.044	0.4375	0.488	0.163
5/16	0.3125	18	.625	.562	.437	.177	.245	.203	.084	.086	.151	.115	.050	0.5000	0.577	.211
3/8	0.3750	16	.750	.625	.562	.210	.273	.250	.094	.103	.168	.142	.064	0.5625	0.628	.243
7/16	0.4375	14	.813	.750	.625	.210	.328	.297	.094	.103	.202	.168	.071	0.6250	0.698	.291
1/2	0.5000	13	0.875	0.812	0.750	0.210	0.354	0.328	0.106	0.103	0.218	0.193	.084	0.7500	0.840	0.323
9/16	0.5625	12	1.000	.937	.812	.244	.409	.375	.118	.120	.252	.213	.091	0.8125	0.910	.371
5/8	0.6250	11	1.125	1.000	.875	.281	.437	.422	.133	.137	.270	.239	.099	0.8750	0.980	.403
3/4	0.7500	10	1.325	1.250	1.000	.352	.546	.500	.149	.171	.338	.283	.112	1.0000	1.121	.483
7/8	0.8750	9	1.625		1.125	.423		.594	0.167	0.206		.334	.126	1.1250	1.261	0.563
1	1.0000	8	1.875		1.312	.494		.656	.188	.240		.371	.146	1.3125	1.473	.627
1⅛	1.1250	7	2.062			.529			.257					1.5000	1.684	.718
1¼	1.2500	7	2.312			.600			.291					1.6875	1.896	.813

TABLE 7 DIMENSIONS OF HEXAGONAL- AND SPLINE SOCKET-HEAD CAP SCREWS (1960 SERIES)

AMERICAN STANDARD

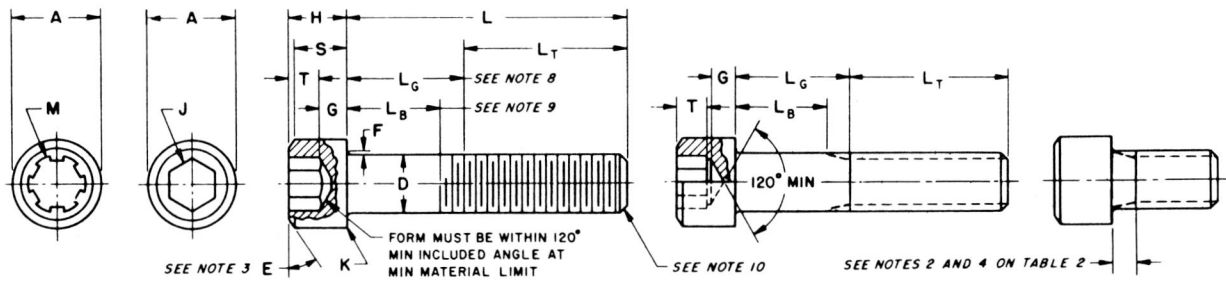

Nominal size	D Body diameter		A Head diameter		H Head height		S Head side height	M Spline socket size	J Hexagon socket size	T Key engagement	G Wall thickness	F Fillet		K Chamfer or radius	Basic thread length
	Max	Min	Max	Min	Max	Min	Min	Nom	Nom	Min	Min	Max	Min	Max	
0	0.0600	0.0568	0.096	0.091	0.060	0.057	0.054	0.062	0.050	0.025	0.019	0.007	0.003	0.003	0.500
1	0.0730	0.0695	0.118	0.112	0.073	0.070	0.066	0.074	1/16	0.031	0.023	0.007	0.003	0.003	0.625
2	0.0860	0.0822	0.140	0.134	0.086	0.083	0.077	0.098	5/64	0.038	0.028	0.008	0.004	0.003	0.625
3	0.0990	0.0949	0.161	0.154	0.099	0.095	0.089	0.098	5/64	0.044	0.032	0.008	0.004	0.003	0.625
4	0.1120	0.1075	0.183	0.176	0.112	0.108	0.101	0.115	3/32	0.051	0.036	0.009	0.005	0.005	0.750
5	0.1250	0.1202	0.205	0.198	0.125	0.121	0.112	0.115	3/32	0.057	0.040	0.010	0.006	0.005	0.750
6	0.1380	0.1329	0.226	0.218	0.138	0.134	0.124	0.137	7/64	0.064	0.044	0.010	0.006	0.005	0.750
8	0.1640	0.1585	0.270	0.262	0.164	0.159	0.148	0.173	9/64	0.077	0.052	0.012	0.007	0.005	0.875
10	0.1900	0.1840	5/16	0.303	0.190	0.185	0.171	0.188	5/32	0.090	0.061	0.014	0.009	0.005	0.875
1/4	0.2500	0.2435	3/8	0.365	1/4	0.244	0.225	0.221	3/16	0.120	0.080	0.014	0.009	0.008	1.000
5/16	0.3125	0.3053	15/32	0.457	5/16	0.306	0.281	0.298	1/4	0.151	0.100	0.017	0.012	0.008	1.125
3/8	0.3750	0.3678	9/16	0.550	3/8	0.368	0.337	0.380	5/16	0.182	0.120	0.020	0.015	0.008	1.250
7/16	0.4375	0.4294	21/32	0.642	7/16	0.430	0.394	0.463	3/8	0.213	0.140	0.023	0.018	0.010	1.375
1/2	0.5000	0.4919	3/4	0.735	1/2	0.492	0.450	0.463	3/8	0.245	0.160	0.026	0.020	0.010	1.500
5/8	0.6250	0.6163	15/16	0.921	5/8	0.616	0.562	0.604	1/2	0.307	0.200	0.032	0.024	0.010	1.750
3/4	0.7500	0.7406	1 1/8	1.107	3/4	0.740	0.675	0.631	5/8	0.370	0.240	0.039	0.030	0.010	2.000
7/8	0.8750	0.8647	1 5/16	1.293	7/8	0.864	0.787	0.709	3/4	0.432	0.280	0.044	0.034	0.015	2.250
1	1.0000	0.9886	1 1/2	1.479	1	0.988	0.900	0.801	3/4	0.495	0.320	0.050	0.040	0.015	2.500
1 1/8	1.1250	1.1086	1 11/16	1.665	1 1/8	1.111	1.012	.....	7/8	0.557	0.360	0.055	0.045	0.015	2.812
1 1/4	1.2500	1.2336	1 7/8	1.852	1 1/4	1.236	1.125	.....	7/8	0.620	0.400	0.060	0.050	0.015	3.125
1 3/8	1.3750	1.3568	2 1/16	2.038	1 3/8	1.360	1.237	.....	1	0.682	0.440	0.065	0.055	0.015	3.437
1 1/2	1.5000	1.4818	2 1/4	2.224	1 1/2	1.485	1.350	.....	1	0.745	0.480	0.070	0.060	0.015	3.750
1 3/4	1.7500	1.7295	2 5/8	2.597	1 3/4	1.734	1.575	.....	1 1/4	0.870	0.560	0.080	0.070	0.015	4.375
2	2.0000	1.9780	3	2.970	2	1.983	1.800	.....	1 1/2	0.995	0.640	0.090	0.075	0.015	5.000
2 1/4	2.2500	2.2280	3 3/8	3.344	2 1/4	2.232	2.025	.....	1 3/4	1.120	0.720	0.100	0.085	0.031	5.625
2 1/2	2.5000	2.4762	3 3/4	3.717	2 1/2	2.481	2.250	.....	1 3/4	1.245	0.800	0.110	0.095	0.031	6.250
2 3/4	2.7500	2.7262	4 1/8	4.090	2 3/4	2.730	2.475	.....	2	1.370	0.880	0.120	0.105	0.031	6.875
3	3.0000	2.9762	4 1/2	4.464	3	2.979	2.700	.....	2 1/4	1.495	0.960	0.130	0.115	0.031	7.500
3 1/4	3.2500	3.2262	4 7/8	4.837	3 1/4	3.228	2.925	.....	2 1/4	1.620	1.040	0.140	0.125	0.031	8.125
3 1/2	3.5000	3.4762	5 1/4	5.211	3 1/2	3.478	3.150	.....	2 3/4	1.745	1.120	0.150	0.135	0.031	8.750
3 3/4	3.7500	3.7262	5 5/8	5.584	3 3/4	3.727	3.375	.....	2 3/4	1.870	1.200	0.160	0.145	0.031	9.375
4	4.0000	3.9762	6	5.958	4	3.976	3.600	.....	3	1.995	1.280	0.170	0.155	0.031	10.000

Courtesy USASI

TABLE 8 REGULAR SQUARE BOLTS

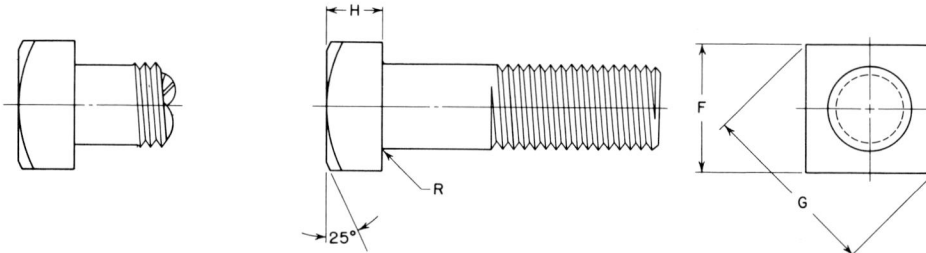

Nominal size or basic major diameter of thread	Body diam.	Width across flats F		Width across corners G		Height H			Radius of fillet R
	Max.	Max (basic)	Min	Max	Min	Nom	Max	Min	Max
1/4	0.2500	3/8	0.3750	0.362	0.530	0.498	11/64	0.188	0.156	0.031
5/16	0.3125	1/2	0.5000	0.484	0.707	0.665	13/64	0.220	0.186	0.031
3/8	0.3750	9/16	0.5625	0.544	0.795	0.747	1/4	0.268	0.232	0.031
7/16	0.4375	5/8	0.6250	0.603	0.884	0.828	19/64	0.316	0.278	0.031
1/2	0.5000	3/4	0.7500	0.725	1.061	0.995	21/64	0.348	0.308	0.031
5/8	0.6250	15/16	0.9375	0.906	1.326	1.244	27/64	0.444	0.400	0.062
3/4	0.7500	1⅛	1.1250	1.088	1.591	1.494	1/2	0.524	0.476	0.062
7/8	0.8750	1 5/16	1.3125	1.269	1.856	1.742	19/32	0.620	0.568	0.062
1	1.0000	1½	1.5000	1.450	2.121	1.991	21/32	0.684	0.628	0.093
1⅛	1.1250	1 11/16| 1.6875	1.631	2.386	2.239	3/4	0.780	0.720	0.093
1¼	1.2500	1⅞	1.8750	1.812	2.652	2.489	27/32	0.876	0.812	0.093
1⅜	1.3750	2 1/16	2.0625	1.994	2.917	2.738	29/32	0.940	0.872	0.093
1½	1.5000	2¼	2.2500	2.175	3.182	2.986	1	1.036	0.964	0.093
1⅝	1.6250	2 7/16	2.4375	2.356	3.447	3.235	1 3/32| 1.132	1.056	0.125

Courtesy USASI

TABLE 9 REGULAR SQUARE NUTS

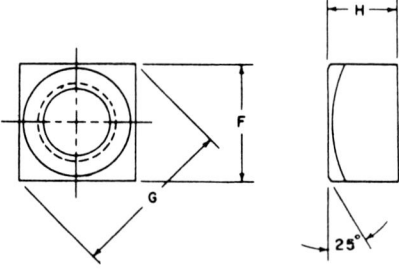

Nominal size or basic major diameter of thread		Width across flats F		Width across corners G		Thickness H		
		Max (basic)	Min	Max	Min	Nom	Max	Min
1/4	0.2500	7/16	0.4375	0.425	0.619	0.584	7/32	0.235	0.203
5/16	0.3125	9/16	0.5625	0.547	0.795	0.751	17/64	0.283	0.249
3/8	0.3750	5/8	0.6250	0.606	0.884	0.832	21/64	0.346	0.310
7/16	0.4375	3/4	0.7500	0.728	1.061	1.000	3/8	0.394	0.356
1/2	0.5000	13/16	0.8125	0.788	1.149	1.082	7/16	0.458	0.418
5/8	0.6250	1	1.0000	0.969	1.414	1.330	35/64	0.569	0.525
3/4	0.7500	1⅛	1.1250	1.088	1.591	1.494	21/32	0.680	0.632
7/8	0.8750	1 5/16	1.3125	1.269	1.856	1.742	49/64	0.792	0.740
1	1.0000	1½	1.5000	1.450	2.121	1.991	7/8	0.903	0.847
1⅛	1.1250	1 11/16| 1.6875	1.631	2.386	2.239	1	1.030	0.970
1¼	1.2500	1⅞	1.8750	1.812	2.652	2.489	1 3/32	1.126	1.062
1⅜	1.3750	2 1/16	2.0625	1.994	2.917	2.738	1 13/64| 1.237	1.169
1½	1.5000	2¼	2.2500	2.175	3.182	2.986	1 5/16	1.348	1.276

Courtesy USASI

TABLE 10 REGULAR HEXAGON BOLTS

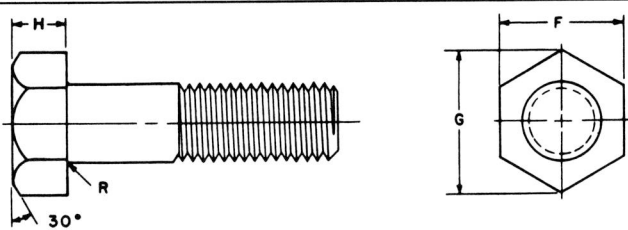

Nominal size or basic major diameter of thread		Body diam.	Width across flats F			Width across corners G		Height H			Radius of fillet R
		Max.		Max (basic)	Min	Max	Min	Nom	Max	Min	Max
1/4	0.2500	0.260	7/16	0.4375	0.425	0.505	0.484	11/64	0.188	0.150	0.031
5/16	0.3125	0.324	1/2	0.5000	0.484	0.577	0.552	7/32	0.235	0.195	0.031
3/8	0.3750	0.388	9/16	0.5625	0.544	0.650	0.620	1/4	0.268	0.226	0.031
7/16	0.4375	0.452	5/8	0.6250	0.603	0.722	0.687	19/64	0.316	0.272	0.031
1/2	0.5000	0.515	3/4	0.7500	0.725	0.866	0.826	11/32	0.364	0.302	0.031
5/8	0.6250	0.642	15/16	0.9375	0.906	1.083	1.033	27/64	0.444	0.378	0.062
3/4	0.7500	0.768	1⅛	1.1250	1.088	1.299	1.240	1/2	0.524	0.455	0.062
7/8	0.8750	0.895	1 5/16	1.3125	1.269	1.516	1.447	37/64	0.604	0.531	0.062
1	1.0000	1.022	1½	1.5000	1.450	1.732	1.653	43/64	0.700	0.591	0.093
1⅛	1.1250	1.149	1 11/16	1.6875	1.631	1.949	1.859	3/4	0.780	0.658	0.093
1¼	1.2500	1.277	1⅞	1.8750	1.812	2.165	2.066	27/32	0.876	0.749	0.093
1⅜	1.3750	1.404	2 1/16	2.0625	1.994	2.382	2.273	29/32	0.940	0.810	0.093
1½	1.5000	1.531	2¼	2.2500	2.175	2.598	2.480	1	1.036	0.902	0.093
1¾	1.7500	1.785	2⅝	2.6250	2.538	3.031	2.893	1 5/32	1.196	1.054	0.125

Courtesy USASI

TABLE 11 REGULAR HEXAGON AND HEXAGON-JAM NUTS

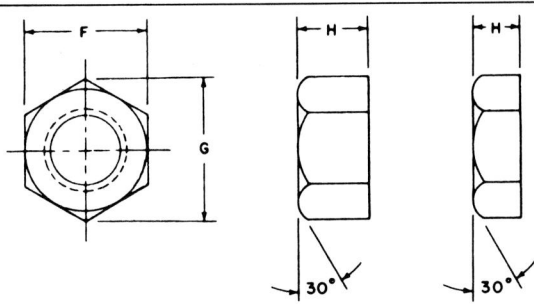

Nominal size or basic major diameter of thread		Width across flats F			Width across corners G		Thickness regular nuts H			Thickness regular jam nuts H		
			Max (basic)	Min	Max	Min	Nom	Max	Min	Nom	Max	Min
1/4	0.2500	7/16	0.4375	0.425	0.505	0.484	7/32	0.235	0.203	5/32	0.172	0.140
5/16	0.3125	9/16	0.5625	0.547	0.650	0.624	17/64	0.283	0.249	3/16	0.204	0.170
3/8	0.3750	5/8	0.6250	0.606	0.722	0.691	21/64	0.346	0.310	7/32	0.237	0.201
7/16	0.4375	3/4	0.7500	0.728	0.866	0.830	3/8	0.394	0.356	1/4	0.269	0.231
1/2	0.5000	13/16	0.8125	0.788	0.938	0.898	7/16	0.458	0.418	5/16	0.332	0.292
9/16	0.5625	7/8	0.8750	0.847	1.010	0.966	1/2	0.521	0.479	11/32	0.365	0.323
5/8	0.6250	1	1.0000	0.969	1.155	1.104	35/64	0.569	0.525	3/8	0.397	0.353
3/4	0.7500	1⅛	1.1250	1.088	1.299	1.240	21/32	0.680	0.632	7/16	0.462	0.414
7/8	0.8750	1 5/16	1.3125	1.269	1.516	1.447	49/64	0.792	0.740	1/2	0.526	0.474
1	1.0000	1½	1.5000	1.450	1.732	1.653	7/8	0.903	0.847	9/16	0.590	0.534
1⅛	1.1250	1 11/16	1.6875	1.631	1.949	1.859	1	1.030	0.970	5/8	0.655	0.595
1¼	1.2500	1⅞	1.8750	1.812	2.165	2.066	1 3/32	1.126	1.062	3/4	0.782	0.718
1⅜	1.3750	2 1/16	2.0625	1.994	2.382	2.273	1 13/64	1.237	1.169	13/16	0.846	0.778
1½	1.5000	2¼	2.2500	2.175	2.598	2.480	1 5/16	1.348	1.276	7/8	0.911	0.839

Courtesy USASI

TABLE 12 REGULAR SEMIFINISHED HEXAGON BOLTS

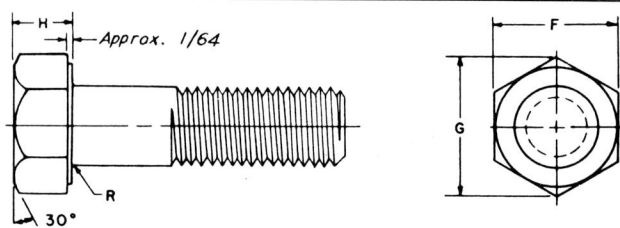

Nominal size or basic major diameter of thread		Body diam.	Width across flats F			Width across corners G		Height H			Radius of fillet R	
		Max.	Max (basic)		Min	Max	Min	Nom	Max	Min	Max	Min
1/4	0.2500	0.260	7/16	0.4375	0.425	0.505	0.484	5/32	0.163	0.150	0.009	0.031
5/16	0.3125	0.324	1/2	0.5000	0.484	0.577	0.552	13/64	0.211	0.195	0.009	0.031
3/8	0.3750	0.388	9/16	0.5625	0.544	0.650	0.620	15/64	0.243	0.226	0.009	0.031
7/16	0.4375	0.452	5/8	0.6250	0.603	0.722	0.687	9/32	0.291	0.272	0.009	0.031
1/2	0.5000	0.515	3/4	0.7500	0.725	0.866	0.826	5/16	0.323	0.302	0.009	0.031
5/8	0.6250	0.642	15/16	0.9375	0.906	1.083	1.033	25/64	0.403	0.378	0.021	0.062
3/4	0.7500	0.768	1 1/8	1.1250	1.088	1.299	1.240	15/32	0.483	0.455	0.021	0.062
7/8	0.8750	0.895	1 5/16	1.3125	1.269	1.516	1.447	35/64	0.563	0.531	0.031	0.062
1	1.0000	1.022	1 1/2	1.5000	1.450	1.732	1.653	39/64	0.627	0.591	0.062	0.093
1 1/8	1.1250	1.149	1 11/16	1.6875	1.631	1.949	1.859	11/16	0.718	0.658	0.062	0.093
1 1/4	1.2500	1.277	1 7/8	1.8750	1.812	2.165	2.066	25/32	0.813	0.749	0.062	0.093
1 3/8	1.3750	1.404	2 1/16	2.0625	1.994	2.382	2.273	27/32	0.878	0.810	0.062	0.093
1 1/2	1.5000	1.531	2 1/4	2.2500	2.175	2.598	2.480	15/16	0.974	0.902	0.062	0.093
1 3/4	1.7500	1.785	2 5/8	2.6250	2.538	3.031	2.893	1 3/32	1.134	1.054	0.078	0.125

Courtesy USASI

TABLE 13 REGULAR SEMIFINISHED HEXAGON AND HEXAGON-JAM NUTS

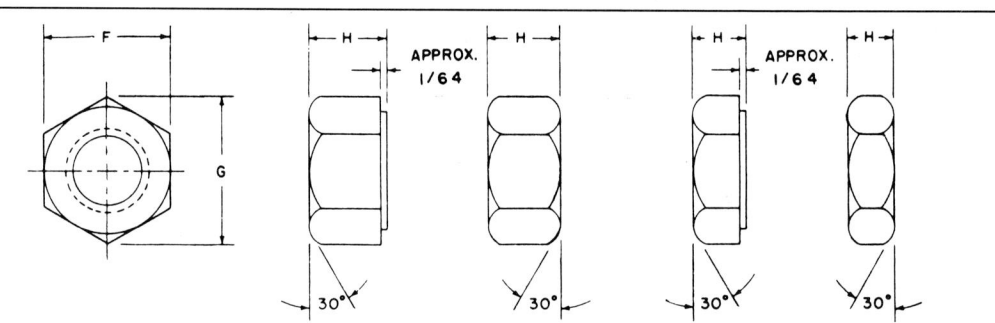

Nominal size or basic major diameter of thread		Width across flats F			Width across corners G		Thickness regular nuts H			Thickness regular jam nuts H		
		Max (basic)		Min	Max	Min	Nom	Max	Min	Nom	Max	Min
1/4	0.2500	7/16	0.4375	0.425	0.505	0.485	13/64	0.219	0.187	9/64	0.157	0.125
5/16	0.3125	9/16	0.5625	0.547	0.650	0.624	1/4	0.267	0.233	11/64	0.189	0.155
3/8	0.3750	5/8	0.6250	0.606	0.722	0.691	5/16	0.330	0.294	13/64	0.221	0.185
7/16	0.4375	3/4	0.7500	0.728	0.866	0.830	23/64	0.378	0.340	15/64	0.253	0.215
1/2	0.5000	13/16	0.8125	0.788	0.938	0.898	27/64	0.442	0.402	19/64	0.317	0.277
9/16	0.5625	7/8	0.8750	0.847	1.010	0.966	31/64	0.505	0.463	21/64	0.349	0.307
5/8	0.6250	1	1.0000	0.969	1.155	1.104	17/32	0.553	0.509	23/64	0.381	0.337
3/4	0.7500	1 1/8	1.1250	1.088	1.299	1.240	41/64	0.665	0.617	27/64	0.446	0.398
7/8	0.8750	1 5/16	1.3125	1.269	1.516	1.447	3/4	0.776	0.724	31/64	0.510	0.458
1	1.0000	1 1/2	1.5000	1.450	1.732	1.653	55/64	0.887	0.831	35/64	0.575	0.519
1 1/8	1.1250	1 11/16	1.6875	1.631	1.949	1.859	31/32	0.999	0.939	39/64	0.639	0.579
1 1/4	1.2500	1 7/8	1.8750	1.812	2.165	2.066	1 1/16	1.094	1.030	23/32	0.751	0.687
1 3/8	1.3750	2 1/16	2.0625	1.994	2.382	2.273	1 11/64	1.206	1.138	25/32	0.815	0.747
1 1/2	1.5000	2 1/4	2.2500	2.175	2.598	2.480	1 9/32	1.317	1.245	27/32	0.880	0.808
1 5/8	1.6250	2 7/16	2.4375	2.356	2.815	2.686	1 25/64	1.429	1.353	29/32	0.944	0.868
1 3/4	1.7500	2 5/8	2.6250	2.538	3.031	2.893	1 1/2	1.540	1.460	31/32	1.009	0.929

Courtesy USASI

TABLE 14 FINISHED HEXAGON BOLTS

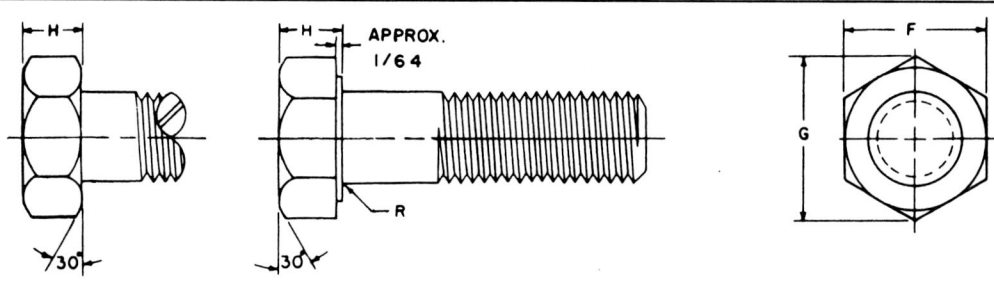

Nominal size or basic major diameter of thread		Body diameter min (maximum equal to nominal size)	Width across flats F			Width across corners G		Height H			Radius of fillet R	
			Max (basic)		Min	Max	Min	Nom	Max	Min	Min	Max
1/4	0.2500	0.2450	7/16	0.4375	0.428	0.505	0.488	5/32	0.163	0.150	0.009	0.023
5/16	0.3125	0.3065	1/2	0.5000	0.489	0.577	0.557	13/64	0.211	0.195	0.009	0.023
3/8	0.3750	0.3690	9/16	0.5625	0.551	0.650	0.628	15/64	0.243	0.226	0.009	0.023
7/16	0.4375	0.4305	5/8	0.6250	0.612	0.722	0.698	9/32	0.291	0.272	0.009	0.023
1/2	0.5000	0.4930	3/4	0.7500	0.736	0.866	0.840	5/16	0.323	0.302	0.009	0.023
9/16	0.5625	0.5545	13/16	0.8125	0.798	0.938	0.910	23/64	0.371	0.348	0.021	0.041
5/8	0.6250	0.6170	15/16	0.9375	0.922	1.083	1.051	25/64	0.403	0.378	0.021	0.041
3/4	0.7500	0.7410	1⅛	1.1250	1.100	1.299	1.254	15/32	0.483	0.455	0.021	0.041
7/8	0.8750	0.8660	1 5/16	1.3125	1.285	1.516	1.465	35/64	0.563	0.531	0.041	0.062
1	1.0000	0.9900	1½	1.5000	1.469	1.732	1.675	39/64	0.627	0.591	0.062	0.093
1⅛	1.1250	1.1140	1 11/16	1.6875	1.631	1.949	1.859	11/16	0.718	0.658	0.062	0.093
1¼	1.2500	1.2390	1⅞	1.8750	1.812	2.165	2.066	25/32	0.813	0.749	0.062	0.093
1⅜	1.3750	1.3630	2 1/16	2.0625	1.994	2.382	2.273	27/32	0.878	0.810	0.062	0.093
1½	1.5000	1.4880	2¼	2.2500	2.175	2.598	2.480	15/16	0.974	0.902	0.062	0.093
1¾	1.7500	1.7380	2⅝	2.6250	2.538	3.031	2.893	1 3/32	1.134	1.054	0.062	0.093

Note: Boldface indicates products unified dimensionally with British and Canadian Standards.

Courtesy USASI

TABLE 15 FINISHED HEXAGON AND HEXAGON-JAM NUTS

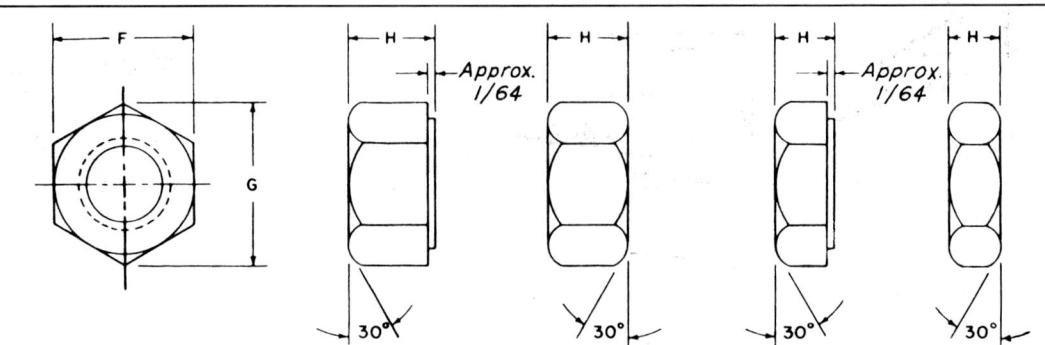

Nominal size or basic major diameter of thread		Width across flats F			Width across corners G		Thickness nuts H			Thickness jam nuts H		
		Max (basic)		Min	Max	Min	Nom	Max	Min	Nom	Max	Min
1/4	0.2500	7/16	0.4375	0.428	0.505	0.488	7/32	0.226	0.212	5/32	0.163	0.150
5/16	0.3125	1/2	0.5000	0.489	0.577	0.557	17/64	0.273	0.258	3/16	0.195	0.180
3/8	0.3750	9/16	0.5625	0.551	0.650	0.628	21/64	0.337	0.320	7/32	0.227	0.210
7/16	0.4375	11/16	0.6875	0.675	0.794	0.768	3/8	0.385	0.365	1/4	0.260	0.240
1/2	0.5000	3/4	0.7500	0.736	0.866	0.840	7/16	0.448	0.427	5/16	0.323	0.302
9/16	0.5625	7/8	0.8750	0.861	1.010	0.982	31/64	0.496	0.473	6/16	0.324	0.301
5/8	9.6250	15/16	0.9375	0.922	1.083	1.051	35/64	0.559	0.535	3/8	0.387	0.363
3/4	0.7500	1⅛	1.1250	1.088	1.299	1.240	41/64	0.665	0.617	27/64	0.446	0.398
7/8	0.8750	1⁵⁄₁₆	1.3125	1.269	1.516	1.447	3/4	0.776	0.724	31/64	0.510	0.458
1	1.0000	1½	1.5000	1.450	1.732	1.653	55/64	0.887	0.831	35/64	0.575	0.519
1⅛	1.1250	1¹¹⁄₁₆	1.6875	1.631	1.949	1.859	31/32	0.999	0.939	39/64	0.639	0.579
1¼	1.2500	1⅞	1.8750	1.812	2.165	2.066	1 1/16	1.094	1.030	23/32	0.751	0.687
1⅜	1.3750	2¹⁄₁₆	2.0625	1.994	2.382	2.273	1¹¹⁄₆₄	1.206	1.138	25/32	0.815	0.747
1½	1.5000	2¼	2.2500	2.175	2.598	2.480	1⁹⁄₃₂	1.317	1.245	27/32	0.880	0.808
1¾	1.7500	2⅝	2.6250	2.538	3.031	2.893	1½	1.540	1.460	31/32	1.009	0.929

Note: Boldface indicates products unified dimensionally with British and Canadian Standards.

Courtesy USASI

TABLE 16 SQUARE HEAD SET SCREWS

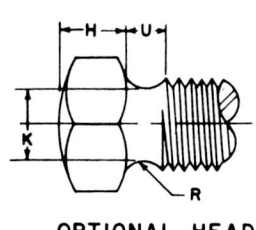

OPTIONAL HEAD

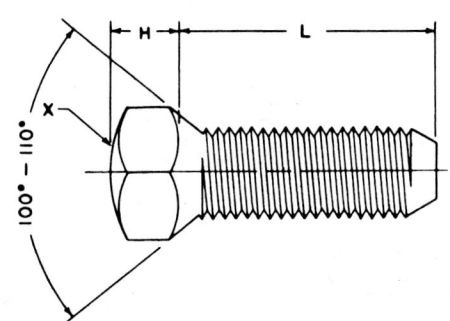

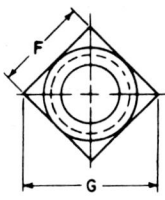

Nominal size		F Width across flats		G Width across corners	H Height of head			K Diameter of neck relief		X Radius of head	R Rad. of neck relief	U Width of neck relief
		Max	Min	Min	Nom	Max	Min	Max	Min	Nom	Max	Min
#10	0.190	0.1875	0.180	0.247	9/64	0.148	0.134	0.145	0.140	15/32	0.027	0.083
#12	0.216	0.216	0.208	0.292	5/32	0.163	0.147	0.162	0.156	35/64	0.029	0.091
1/4	0.250	0.250	0.241	0.331	3/16	0.196	0.178	0.185	0.170	5/8	0.032	0.100
5/16	0.3125	0.3125	0.302	0.415	15/64	0.245	0.224	0.240	0.225	25/32	0.036	0.111
3/8	0.3750	0.375	0.362	0.497	9/32	0.293	0.270	0.294	0.279	15/16	0.041	0.125
7/16	0.4375	0.4375	0.423	0.581	21/64	0.341	0.315	0.345	0.330	1 3/32	0.046	0.143
1/2	0.500	0.500	0.484	0.665	3/8	0.389	0.361	0.400	0.385	1 1/4	0.050	0.154
9/16	0.5625	0.5625	0.545	0.748	27/64	0.437	0.407	0.454	0.439	1 13/32	0.054	0.167
5/8	0.6250	0.625	0.606	0.833	15/32	0.485	0.452	0.507	0.492	1 9/16	0.059	0.182
3/4	0.750	0.750	0.729	1.001	9/16	0.582	0.544	0.620	0.605	1 7/8	0.065	0.200
7/8	0.875	0.875	0.852	1.170	21/32	0.678	0.635	0.731	0.716	2 3/16	0.072	0.222
1	1.000	1.000	0.974	1.337	3/4	0.774	0.726	0.838	0.823	2 1/2	0.081	0.250
1 1/8	1.125	1.125	1.096	1.505	27/32	0.870	0.817	0.939	0.914	2 13/16	0.092	0.283
1 1/4	1.250	1.250	1.219	1.674	15/16	0.966	0.908	1.064	1.039	3 1/8	0.092	0.283
1 3/8	1.375	1.375	1.342	1.843	1 1/32	1.063	1.000	1.159	1.134	3 7/16	0.109	0.333
1 1/2	1.500	1.500	1.464	2.010	1 1/8	1.159	1.091	1.284	1.259	3 3/4	0.109	0.333

All dimensions given in inches.

Threads shall be coarse-, fine-, or 8-thread series, class 2A; unless otherwise specified, coarse-thread series will be furnished. Square head set screws 1/4 in. size and larger are normally stocked in coarse thread series only.

Tolerance on screw length for sizes up to and including 5/8 in. shall be; minus 1/32 in. for lengths up to and including 1 in.; minus 1/16 in. for lengths over 1 in. to and including 2 in.; and minus 3/32 in. for lengths over 2 in. The tolerance shall be doubled for larger size screws of comparable length.

Square head set screws shall be made from alloy or carbon steel suitably hardened. Screws made from nonferrous material or corrosion-resisting steel shall be made from a material mutually agreed upon by manufacturer and user.

Courtesy USASI

TABLE 17 SQUARE HEAD SET SCREW POINTS

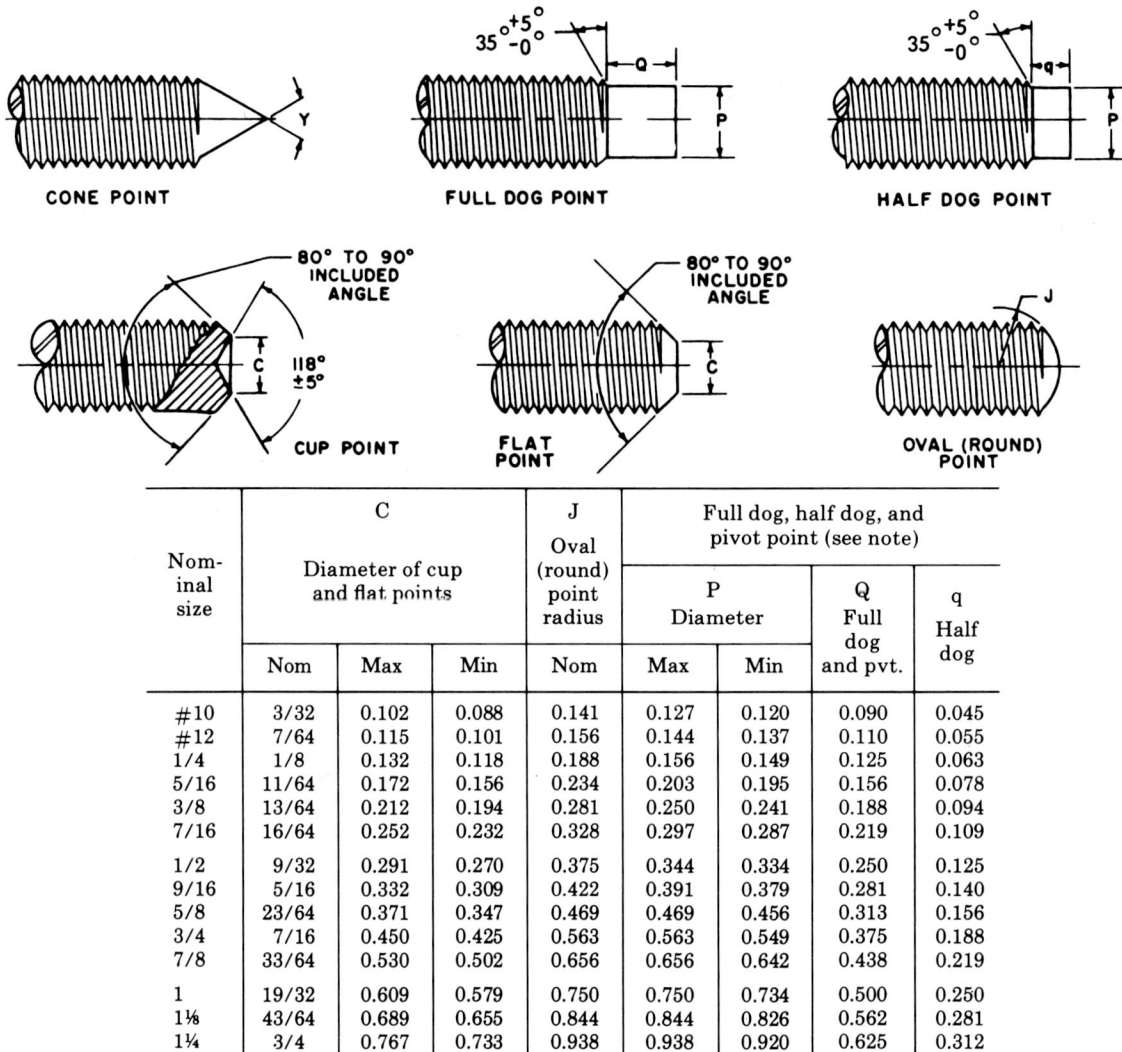

Nominal size	C Diameter of cup and flat points			J Oval (round) point radius	Full dog, half dog, and pivot point (see note)			
					P Diameter		Q Full dog and pvt.	q Half dog
	Nom	Max	Min	Nom	Max	Min		
#10	3/32	0.102	0.088	0.141	0.127	0.120	0.090	0.045
#12	7/64	0.115	0.101	0.156	0.144	0.137	0.110	0.055
1/4	1/8	0.132	0.118	0.188	0.156	0.149	0.125	0.063
5/16	11/64	0.172	0.156	0.234	0.203	0.195	0.156	0.078
3/8	13/64	0.212	0.194	0.281	0.250	0.241	0.188	0.094
7/16	16/64	0.252	0.232	0.328	0.297	0.287	0.219	0.109
1/2	9/32	0.291	0.270	0.375	0.344	0.334	0.250	0.125
9/16	5/16	0.332	0.309	0.422	0.391	0.379	0.281	0.140
5/8	23/64	0.371	0.347	0.469	0.469	0.456	0.313	0.156
3/4	7/16	0.450	0.425	0.563	0.563	0.549	0.375	0.188
7/8	33/64	0.530	0.502	0.656	0.656	0.642	0.438	0.219
1	19/32	0.609	0.579	0.750	0.750	0.734	0.500	0.250
1⅛	43/64	0.689	0.655	0.844	0.844	0.826	0.562	0.281
1¼	3/4	0.767	0.733	0.938	0.938	0.920	0.625	0.312
1⅜	53/64	0.848	0.808	1.031	1.031	1.011	0.688	0.344
1½	29/32	0.926	0.886	1.125	1.125	1.105	0.750	0.375

All dimensions are given in inches.

Pivot points are similar to full dog point except that the point is rounded by a radius equal to J.

Where usable length of thread is less than the nominal diameter, half-dog point shall be used.

When length equals nominal diameter or less, $Y = 118$ deg $\pm$ 2 deg; when length exceeds nominal diameter, $Y = 90$ deg $\pm$ 2 deg.

Courtesy USASI

TABLE 18 PROPORTIONS OF KEYS IN THE PRATT AND WHITNEY SYSTEM

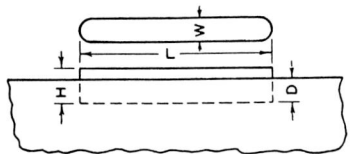

No. of key	L	W	H	D	No. of key	L	W	H	D
1	1/2	1/16	3/32	1/16	22	1⅜	1/4	3/8	1/4
2	1/2	3/32	9/64	3/32	23	1⅜	5/16	15/32	5/16
3	1/2	1/8	3/16	1/8	F	1⅜	3/8	9/16	3/8
4	5/8	3/32	9/64	3/32	24	1½	1/4	3/8	1/4
5	5/8	1/8	3/16	1/8	25	1½	5/16	15/32	5/16
6	5/8	5/32	15/64	5/32	G	1½	3/8	9/16	3/8
7	3/4	1/8	3/16	1/8	51	1¾	1/4	3/8	1/4
8	3/4	5/32	15/64	5/32	52	1¾	5/16	15/32	5/16
9	3/4	3/16	9/32	3/16	53	1¾	3/8	9/16	3/8
10	7/8	5/32	15/64	5/32	26	2	3/16	9/32	3/16
11	7/8	3/16	9/32	3/16	27	2	1/4	3/8	1/4
12	7/8	7/32	21/64	7/32	28	2	5/16	15/32	5/16
A	7/8	1/4	3/8	1/4	29	2	3/8	9/16	3/8
13	1	3/16	9/32	3/16	54	2¼	1/4	3/8	1/4
14	1	7/32	21/64	7/32	55	2¼	5/16	15/32	5/16
15	1	1/4	3/8	1/4	56	2¼	3/8	9/16	3/8
B	1	5/16	15/32	5/16	57	2¼	7/16	21/32	7/16
16	1⅛	3/16	9/32	3/16	58	2½	5/16	15/32	5/16
17	1⅛	7/32	21/64	7/32	59	2½	3/8	9/16	3/8
18	1⅛	1/4	3/8	1/4	60	2½	7/16	21/32	7/16
C	1⅛	5/16	15/32	5/16	61	2½	1/2	3/4	1/2
19	1¼	3/16	9/32	3/16	30	3	3/8	9/16	3/8
20	1¼	7/32	21/64	7/32	31	3	7/16	21/32	7/16
21	1¼	1/4	3/8	1/4	32	3	1/2	3/4	1/2
D	1¼	5/16	15/32	5/16	33	3	9/16	27/32	9/16
E	1¼	3/8	9/16	3/8	34	3	5/8	15/16	5/8

TABLE 19 WOODRUFF KEYS

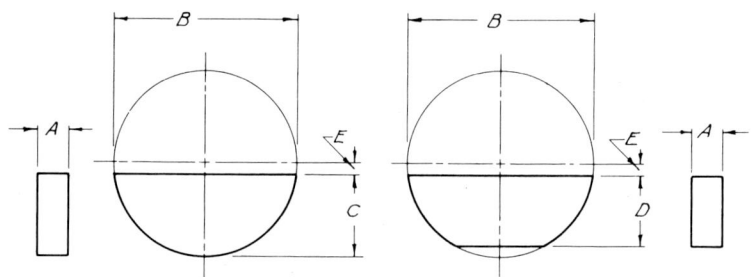

Woodruff Key Dimensions

Key* number	Nominal key size A × B	Width of key A		Diam. of key B		Height of key				Distance below center E
		Max	Min	Max	Min	C		D		
						Max	Min	Max	Min	
204	1/16 × 1/2	0.0635	0.0625	0.500	0.490	0.203	0.198	0.194	0.188	3/64
304	3/32 × 1/2	.0948	.0938	0.500	0.490	.203	.198	.194	.188	3/64
305	3/32 × 5/8	.0948	.0938	0.625	0.615	.250	.245	.240	.234	1/16
404	1/8 × 1/2	.1260	.1250	0.500	0.490	.203	.198	.194	.188	3/64
405	1/8 × 5/8	.1260	.1250	0.625	0.615	.250	.245	.240	.234	1/16
406	1/8 × 3/4	.1260	.1250	0.750	0.740	.313	.308	.303	.297	1/16
505	5/32 × 5/8	.1573	.1563	0.625	0.615	.250	.245	.240	.234	1/16
506	5/32 × 3/4	.1573	.1563	0.750	0.740	.313	.308	.303	.297	1/16
507	5/32 × 7/8	.1573	.1563	0.875	0.865	.375	.370	.365	.359	1/16
606	3/16 × 3/4	.1885	.1875	0.750	0.740	.313	.308	.303	.297	1/16
607	3/16 × 7/8	.1885	.1875	0.875	0.865	.375	.370	.365	.359	1/16
608	3/16 × 1	.1885	.1875	1.000	0.990	.438	.433	.428	.422	1/16
609	3/16 × 1⅛	.1885	.1875	1.125	1.115	.484	.479	.475	.469	5/64
807	1/4 × 7/8	.2510	.2500	0.875	0.865	.375	.370	.365	.359	1/16
808	1/4 × 1	.2510	.2500	1.000	0.990	.438	.433	.428	.422	1/16
809	1/4 × 1⅛	.2510	.2500	1.125	1.115	.484	.479	.475	.469	5/64
810	1/4 × 1¼	.2510	.2500	1.250	1.240	.547	.542	.537	.531	5/64
811	1/4 × 1⅜	.2510	.2500	1.375	1.365	.594	.589	.584	.578	3/32
812	1/4 × 1½	.2510	.2500	1.500	1.490	.641	.636	.631	.625	7/64

All dimensions given in inches.

* Note: Key numbers indicate the nominal key dimensions. The last two digits give the nominal diameter (B) in eighths of an inch and the digits preceding the last two give the nominal width (A) in thirty-seconds of an inch. Thus, 204 indicates a key 2/32 × 4/8 or 1/16 × 1/2 inches; 1210 indicates a key 12/32 × 10/8 or 3/8 × 1¼ inches.

Courtesy USASI

TABLE 20 PLAIN PARALLEL STOCK KEYS

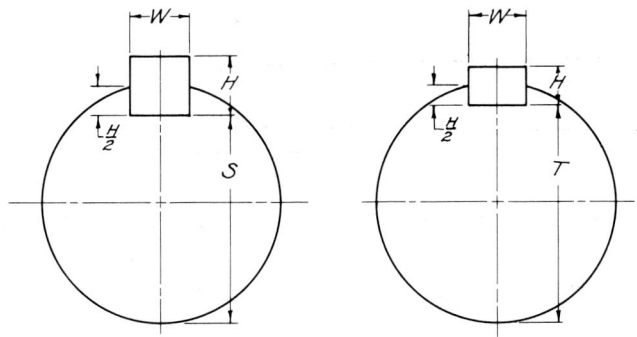

Dimensions of Square and Flat Plain Parallel Stock Keys

Shaft diameter	Square key $W \times H$	Flat key $W \times H$	Tolerance on W and H $(-)$	Bottom of keyseat to opposite side of shaft	
				Square key S	Flat key T
1/2	1/8 × 1/8	1/8 × 3/32	0.0020	0.430	0.445
9/16	1/8 × 1/8	1/8 × 3/32	.0020	0.493	0.509
5/8	3/16 × 3/16	3/16 × 1/8	.0020	0.517	0.548
11/16	3/16 × 3/16	3/16 × 1/8	.0020	0.581	0.612
3/4	3/16 × 3/16	3/16 × 1/8	.0020	0.644	0.676
13/16	3/16 × 3/16	3/16 × 1/8	.0020	0.708	0.739
7/8	3/16 × 3/16	3/16 × 1/8	.0020	0.771	0.802
15/16	1/4 × 1/4	1/4 × 3/16	.0020	0.796	0.827
1	1/4 × 1/4	1/4 × 3/16	.0020	0.859	0.890
1 1/16	1/4 × 1/4	1/4 × 3/16	.0020	0.923	0.954
1 1/8	1/4 × 1/4	1/4 × 3/16	.0020	0.986	1.017
1 3/16	1/4 × 1/4	1/4 × 3/16	.0020	1.049	1.081
1 1/4	1/4 × 1/4	1/4 × 3/16	.0020	1.112	1.144
1 5/16	5/16 × 5/16	5/16 × 1/4	.0020	1.137	1.169
1 3/8	5/16 × 5/16	5/16 × 1/4	.0020	1.201	1.232
1 7/16	3/8 × 3/8	3/8 × 1/4	.0020	1.225	1.288
1 1/2	3/8 × 3/8	3/8 × 1/4	.0020	1.289	1.351
1 9/16	3/8 × 3/8	3/8 × 1/4	.0020	1.352	1.415
1 5/8	3/8 × 3/8	3/8 × 1/4	.0020	1.416	1.478
1 11/16	3/8 × 3/8	3/8 × 1/4	.0020	1.479	1.542
1 3/4	3/8 × 3/8	3/8 × 1/4	.0020	1.542	1.605

Courtesy USASI

TABLE 21 DIMENSIONS OF TAPER PINS

Number	7/0	6/0	5/0	4/0	3/0	2/0	0	1	2	3	4	5	6	7	8	9	10
Size (large end)	0.0625	0.0780	0.0940	0.1090	0.1250	0.1410	0.1560	0.1720	0.1930	0.2190	0.2500	0.2890	0.3410	0.4090	0.4920	0.5910	0.7060
Length, L																	
0.375	X	X															
0.500	X	X	X	X	X	X	X										
0.625	X	X	X	X	X	X	X										
0.750		X	X	X	X	X	X	X	X	X							
0.875					X	X	X	X	X	X							
1.000			X	X	X	X	X	X	X	X	X	X					
1.250						X	X	X	X	X	X	X	X				
1.500							X	X	X	X	X	X	X				
1.750								X	X	X	X	X	X				
2.000								X	X	X	X	X	X	X	X		
2.250									X	X	X	X	X	X	X		
2.500									X	X	X	X	X	X	X		
2.750										X	X	X	X	X	X	X	
3.000										X	X	X	X	X	X	X	
3.250													X	X	X	X	
3.500													X	X	X	X	X
3.750													X	X	X	X	X
4.000													X	X	X	X	X
4.250														X	X	X	X
4.500														X	X	X	X
4.750															X	X	X
5.000															X	X	X
5.250																X	X
5.500																X	X
5.750																X	X
6.000																X	X

All dimensions are given in inches.
Standard reamers are available for pins given above the line.
Pins Nos. 11 (size 0.8600), 12 (size 1.032), 13 (size 1.241), and 14 (1.523) are special sizes—hence their lengths are special.
To find small diameter of pin, multiply the length by 0.02083 and subtract the result from the large diameter.

TYPES	COMMERCIAL TYPE	PRECISION TYPE
Sizes	7/0 to 14	7/0 to 10
Tolerance on diameter	(+0.0013, −0.0007)	(+0.0013, −0.0007)
Taper	¼ in. per ft	¼ in. per ft
Length tolerance	(±0.030)	(±0.030)
Concavity tolerance	None	0.0005 up to 1 in. long
		0.001 1 1/16 to 2 in. long
		0.002 2 1/16 and longer

APPENDIX 583

TABLE 22 RUNNING AND SLIDING FITS (INCH)

Limits are in thousandths of an inch.
Limits for hole and shaft are applied algebraically to the basic size to obtain the limits of size for the parts.
Data in bold face are in accordance with ABC agreements.
Symbols H5, g5, etc., are Hole and Shaft designations used in ABC System

Nominal Size Range Inches		Class RC 1			Class RC 2			Class RC 3			Class RC 4		
		Limits of Clearance	Standard Limits		Limits of Clearance	Standard Limits		Limits of Clearance	Standard Limits		Limits of Clearance	Standard Limits	
Over	To		Hole H5	Shaft g4		Hole H6	Shaft g5		Hole H7	Shaft f6		Hole H8	Shaft f7
0	−0.12	0.1 / 0.45	+0.2 / 0	−0.1 / −0.25	0.1 / 0.55	+0.25 / 0	−0.1 / −0.3	0.3 / 0.95	+0.4 / 0	−0.3 / −0.55	0.3 / 1.3	+0.6 / 0	−0.3 / −0.7
0.12	−0.24	0.15 / 0.5	+0.2 / 0	−0.15 / −0.3	0.15 / 0.65	+0.3 / 0	−0.15 / −0.35	0.4 / 1.12	+0.5 / 0	−0.4 / −0.7	0.4 / 1.6	+0.7 / 0	−0.4 / −0.9
0.24	−0.40	0.2 / 0.6	+0.25 / 0	−0.2 / −0.35	0.2 / 0.85	+0.4 / 0	−0.2 / −0.45	0.5 / 1.5	+0.6 / 0	−0.5 / −0.9	0.5 / 2.0	+0.9 / 0	−0.5 / −1.1
0.40	−0.71	0.25 / 0.75	+0.3 / 0	−0.25 / −0.45	0.25 / 0.95	+0.4 / 0	−0.25 / −0.55	0.6 / 1.7	+0.7 / 0	−0.6 / −1.0	0.6 / 2.3	+1.0 / 0	−0.6 / −1.3
0.71	−1.19	0.3 / 0.95	+0.4 / 0	−0.3 / −0.55	0.3 / 1.2	+0.5 / 0	−0.3 / −0.7	0.8 / 2.1	+0.8 / 0	−0.8 / −1.3	0.8 / 2.8	+1.2 / 0	−0.8 / −1.6
1.19	−1.97	0.4 / 1.1	+0.4 / 0	−0.4 / −0.7	0.4 / 1.4	+0.6 / 0	−0.4 / −0.8	1.0 / 2.6	+1.0 / 0	−1.0 / −1.6	1.0 / 3.6	+1.6 / 0	−1.0 / −2.0
1.97	−3.15	0.4 / 1.2	+0.5 / 0	−0.4 / −0.7	0.4 / 1.6	+0.7 / 0	−0.4 / −0.9	1.2 / 3.1	+1.2 / 0	−1.2 / −1.9	1.2 / 4.2	+1.8 / 0	−1.2 / −2.4
3.15	−4.73	0.5 / 1.5	+0.6 / 0	−0.5 / −0.9	0.5 / 2.0	+0.9 / 0	−0.5 / −1.1	1.4 / 3.7	+1.4 / 0	−1.4 / −2.3	1.4 / 5.0	+2.2 / 0	−1.4 / −2.8
4.73	−7.09	0.6 / 1.8	+0.7 / 0	−0.6 / −1.1	0.6 / 2.3	+1.0 / 0	−0.6 / −1.3	1.6 / 4.2	+1.6 / 0	−1.6 / −2.6	1.6 / 5.7	+2.5 / 0	−1.6 / −3.2
7.09	−9.85	0.6 / 2.0	+0.8 / 0	−0.6 / −1.2	0.6 / 2.6	+1.2 / 0	−0.6 / −1.4	2.0 / 5.0	+1.8 / 0	−2.0 / −3.2	2.0 / 6.6	+2.8 / 0	−2.0 / −3.8
9.85	−12.41	0.8 / 2.3	+0.9 / 0	−0.8 / −1.4	0.8 / 2.9	+1.2 / 0	−0.8 / −1.7	2.5 / 5.7	+2.0 / 0	−2.5 / −3.7	2.5 / 7.5	+3.0 / 0	−2.5 / −4.5
12.41	−15.75	1.0 / 2.7	+1.0 / 0	−1.0 / −1.7	1.0 / 3.4	+1.4 / 0	−1.0 / −2.0	3.0 / 6.6	/ 0	−3.0 / −4.4	3.0 / 8.7	+3.5 / 0	−3.0 / −5.2
15.75	−19.69	1.2 / 3.0	+1.0 / 0	−1.2 / −2.0	1.2 / 3.8	+1.6 / 0	−1.2 / −2.2	4.0 / 8.1	+1.6 / 0	−4.0 / −5.6	4.0 / 10.5	+4.0 / 0	−4.0 / −6.5
19.69	−30.09	1.6 / 3.7	+1.2 / 0	−1.6 / −2.5	1.6 / 4.8	+2.0 / 0	−1.6 / −2.8	5.0 / 10.0	+3.0 / 0	−5.0 / −7.0	5.0 / 13.0	+5.0 / 0	−5.0 / −8.0
30.09	−41.49	2.0 / 4.6	+1.6 / 0	−2.0 / −3.0	2.0 / 6.1	+2.5 / 0	−2.0 / −3.6	6.0 / 12.5	+4.0 / 0	−6.0 / −8.5	6.0 / 16.0	+6.0 / 0	−6.0 / −10.0
41.49	−56.19	2.5 / 5.7	+2.0 / 0	−2.5 / −3.7	2.5 / 7.5	+3.0 / 0	−2.5 / −4.5	8.0 / 16.0	+5.0 / 0	−8.0 / −11.0	8.0 / 21.0	+8.0 / 0	−8.0 / −13.0
56.19	−76.39	3.0 / 7.1	+2.5 / 0	−3.0 / −4.6	3.0 / 9.5	+4.0 / 0	−3.0 / −5.5	10.0 / 20.0	+6.0 / 0	−10.0 / −14.0	10.0 / 26.0	+10.0 / 0	−10.0 / −16.0
76.39	−100.9	4.0 / 9.0	+3.0 / 0	−4.0 / −6.0	4.0 / 12.0	+5.0 / 0	−4.0 / −7.0	12.0 / 25.0	+8.0 / 0	−12.0 / −17.0	12.0 / 32.0	+12.0 / 0	−12.0 / −20.0
100.9	−131.9	5.0 / 11.5	+4.0 / 0	−5.0 / −7.5	5.0 / 15.0	+6.0 / 0	−5.0 / −9.0	16.0 / 32.0	+10.0 / 0	−16.0 / −22.0	16.0 / 36.0	+16.0 / 0	−16.0 / −26.0
131.9	−171.9	6.0 / 14.0	+5.0 / 0	−6.0 / −9.0	6.0 / 19.0	+8.0 / 0	−6.0 / −11.0	18.0 / 38.0	+8.0 / 0	−18.0 / −26.0	18.0 / 50.0	+20.0 / 0	−18.0 / −30.0
171.9	−200	8.0 / 18.0	+6.0 / 0	−8.0 / −12.0	8.0 / 22.0	+10.0 / 0	−8.0 / −12.0	22.0 / 48.0	+16.0 / 0	−22.0 / −32.0	22.0 / 63.0	+25.0 / 0	−22.0 / −38.0

TABLE 22 (cont'd)

RUNNING AND SLIDING FITS

Limits are in thousandths of an inch.

Limits for hole and shaft are applied algebraically to the basic size to obtain the limits of size for the parts

Data in bold face are in accordance with ABC agreements

Symbols H8, e7, etc., are Hole and Shaft designations used in ABC System

	Class RC 5			Class RC 6			Class RC 7			Class RC 8			Class RC 9		Nominal Size Range Inches	
Limits of Clearance	Standard Limits		Limits of Clearance	Standard Limits		Limits of Clearance	Standard Limits		Limits of Clearance	Standard Limits		Limits of Clearance	Standard Limits			
	Hole H8	Shaft e7		Hole H9	Shaft e8		Hole H9	Shaft d8		Hole H10	Shaft c9		Hole H11	Shaft	Over	To
0.6 1.6	+0.6 −0	−0.6 −1.0	0.6 2.2	+1.0 −0	−0.6 −1.2	1.0 2.6	+1.0 0	−1.0 −1.6	2.5 5.1	+1.6 0	−2.5 −3.5	4.0 8.1	+2.5 0	−4.0 −5.6	0	−0.12
0.8 2.0	+0.7 −0	−0.8 −1.3	0.8 2.7	+1.2 −0	−0.8 −1.5	1.2 3.1	+1.2 0	−1.2 −1.9	2.8 5.8	+1.8 0	−2.8 −4.0	4.5 9.0	+3.0 0	−4.5 −6.0	0.12−	0.24
1.0 2.5	+0.9 −0	−1.0 −1.6	1.0 3.3	+1.4 −0	−1.0 −1.9	1.6 3.9	+1.4 0	−1.6 −2.5	3.0 6.6	+2.2 0	−3.0 −4.4	5.0 10.7	+3.5 0	−5.0 −7.2	0.24−	0.40
1.2 2.9	+1.0 −0	−1.2 −1.9	1.2 3.8	+1.6 −0	−1.2 −2.2	2.0 4.6	+1.6 0	−2.0 −3.0	3.5 7.9	+2.8 0	−3.5 −5.1	6.0 12.8	+4.0 −0	−6.0 −8.8	0.40−	0.71
1.6 3.6	+1.2 −0	−1.6 −2.4	1.6 4.8	+2.0 −0	−1.6 −2.8	2.5 5.7	+2.0 0	−2.5 −3.7	4.5 10.0	+3.5 0	−4.5 −6.5	7.0 15.5	+5.0 0	−7.0 −10.5	0.71−	1.19
2.0 4.6	+1.6 −0	−2.0 −3.0	2.0 6.1	+2.5 −0	−2.0 −3.6	3.0 7.1	+2.5 0	−3.0 −4.6	5.0 11.5	+4.0 0	−5.0 −7.5	8.0 18.0	+6.0 0	−8.0 −12.0	1.19−	1.97
2.5 5.5	+1.8 −0	−2.5 −3.7	2.5 7.3	+3.0 −0	−2.5 −4.3	4.0 8.8	+3.0 0	−4.0 −5.8	6.0 13.5	+4.5 0	−6.0 −9.0	9.0 20.5	+7.0 0	−9.0 −13.5	1.97−	3.15
3.0 6.6	+2.2 −0	−3.0 −4.4	3.0 8.7	+3.5 −0	−3.0 −5.2	5.0 10.7	+3.5 0	−5.0 −7.2	7.0 15.5	+5.0 0	−7.0 −10.5	10.0 24.0	+9.0 0	−10.0 −15.0	3.15−	4.73
3.5 7.6	+2.5 −0	−3.5 −5.1	3.5 10.0	+4.0 −0	−3.5 −6.0	6.0 12.5	+4.0 0	−6.0 −8.5	8.0 18.0	+6.0 0	−8.0 −12.0	12.0 28.0	+10.0 0	−12.0 −18.0	4.73−	7.09
4.0 8.6	+2.8 −0	−4.0 −5.8	4.0 11.3	+4.5 −0	−4.0 −6.8	7.0 14.3	+4.5 0	−7.0 −9.8	10.0 21.5	+7.0 0	−10.0 −14.5	15.0 34.0	+12.0 0	−15.0 −22.0	7.09−	9.85
5.0 10.0	+3.0 0	−5.0 −7.0	5.0 13.0	+5.0 0	−5.0 −8.0	8.0 16.0	+5.0 0	−8.0 −11.0	12.0 25.0	+8.0 0	−12.0 −17.0	18.0 38.0	+12.0 0	−18.0 −26.0	9.85−	12.41
6.0 11.7	+3.5 0	−6.0 −8.2	6.0 15.5	+6.0 0	−6.0 −9.5	10.0 19.5	+6.0 0	−10.0 −13.5	14.0 29.0	+9.0 0	−14.0 −20.0	22.0 45.0	+14.0 0	−22.0 −31.0	12.41−	15.75
8.0 14.5	+4.0 0	−8.0 −10.5	8.0 18.0	+6.0 0	−8.0 −12.0	12.0 22.0	+6.0 0	−12.0 −16.0	16.0 32.0	+10.0 0	−16.0 −22.0	25.0 51.0	+16.0 0	−25.0 −35.0	15.75−	19.69
10.0 18.0	+5.0 0	−10.0 −13.0	10.0 23.0	+8.0 0	−10.0 −15.0	16.0 29.0	+8.0 0	−16.0 −21.0	20.0 40.0	+12.0 0	−20.0 −28.0	30.0 62.0	+20.0 0	−30.0 −42.0	19.69−	30.09
12.0 22.0	+6.0 0	−12.0 −16.0	12.0 28.0	+10.0 0	−12.0 −18.0	20.0 36.0	+10.0 0	−20.0 −26.0	25.0 51.0	+16.0 0	−25.0 −35.0	40.0 81.0	+25.0 0	−40.0 −56.0	30.09−	41.49
16.0 29.0	+8.0 0	−16.0 −21.0	16.0 36.0	+12.0 0	−16.0 −24.0	25.0 45.0	+12.0 0	−25.0 −33.0	30.0 62.0	+20.0 0	−30.0 −42.0	50.0 100	+30.0 0	−50.0 −70.0	41.49−	56.19
20.0 36.0	+10.0 0	−20.0 −26.0	20.0 46.0	+16.0 0	−20.0 −30.0	30.0 56.0	+16.0 0	−30.0 −40.0	40.0 81.0	+25.0 0	−40.0 −56.0	60.0 125	+40.0 0	−60.0 −85.0	56.19−	76.39
25.0 45.0	+12.0 0	−25.0 −33.0	25.0 57.0	+20.0 0	−25.0 −37.0	40.0 72.0	+20.0 0	−40.0 −52.0	50.0 100	+30.0 0	−50.0 −70.0	80.0 160	+50.0 0	−80.0 −110	76.39−	100.9
30.0 56.0	+16.0 0	−30.0 −40.0	30.0 71.0	+25.0 0	−30.0 −46.0	50.0 91.0	+25.0 0	−50.0 −66.0	60.0 125	+40.0 0	−60.0 −85.0	100 200	+60.0 0	−100 −140	100.9	−131.9
35.0 67.0	+20.0 0	−35.0 −47.0	35.0 85.0	+30.0 0	−35.0 −55.0	60.0 110	+30.0 0	−60.0 −80.0	80.0 160	+50.0 0	−80.0 −110	130 260	+80.0 0	−130 −180	131.9	−171.9
45.0 86.0	+25.0 0	−45.0 −61.0	45.0 110.0	+40.0 0	−45.0 −70.0	80.0 145.0	+40.0 0	−80.0 −105.0	100 200	+60.0 0	−100 −140	150 310	+100 0	−150 −210	171.9	−200

TABLE 23 LOCATIONAL CLEARANCE FITS (INCH)

Limits are in thousandths of an inch.
Limits for hole and shaft are applied algebraically to the basic size to obtain the limits of size for the parts.
Data in bold face are in accordance with ABC agreements.
Symbols H6, h5, etc., are Hole and Shaft designations used in ABC System

Nominal Size Range Inches Over / To	Class LC 1 Limits of Clearance	Class LC 1 Hole H6	Class LC 1 Shaft h5	Class LC 2 Limits of Clearance	Class LC 2 Hole H7	Class LC 2 Shaft h6	Class LC 3 Limits of Clearance	Class LC 3 Hole H8	Class LC 3 Shaft h7	Class LC 4 Limits of Clearance	Class LC 4 Hole H10	Class LC 4 Shaft h9	Class LC 5 Limits of Clearance	Class LC 5 Hole H7	Class LC 5 Shaft g6
0 – 0.12	0 / 0.45	+0.25 / −0	+0 / −0.2	0 / 0.65	+0.4 / −0	+0 / −0.25	0 / 1	+0.6 / −0	+0 / −0.4	0 / 2.6	+1.6 / −0	+0 / −1.0	0.1 / 0.75	+0.4 / −0	−0.1 / −0.35
0.12– 0.24	0 / 0.5	+0.3 / −0	+0 / −0.2	0 / 0.8	+0.5 / −0	+0 / −0.3	0 / 1.2	+0.7 / −0	+0 / −0.5	0 / 3.0	+1.8 / −0	+0 / −1.2	0.15 / 0.95	+0.5 / −0	−0.15 / −0.45
0.24– 0.40	0 / 0.65	+0.4 / −0	+0 / −0.25	0 / 1.0	+0.6 / −0	+0 / −0.4	0 / 1.5	+0.9 / −0	+0 / −0.6	0 / 3.6	+2.2 / −0	+0 / −1.4	0.2 / 1.2	+0.6 / −0	−0.2 / −0.6
0.40– 0.71	0 / 0.7	+0.4 / −0	+0 / −0.3	0 / 1.1	+0.7 / −0	+0 / −0.4	0 / 1.7	+1.0 / −0	+0 / −0.7	0 / 4.4	+2.8 / −0	+0 / −1.6	0.25 / 1.35	+0.7 / −0	−0.25 / −0.65
0.71– 1.19	0 / 0.9	+0.5 / −0	+0 / −0.4	0 / 1.3	+0.8 / −0	+0 / −0.5	0 / 2	+1.2 / −0	+0 / −0.8	0 / 5.5	+3.5 / −0	+0 / −2.0	0.3 / 1.6	+0.8 / −0	−0.3 / −0.8
1.19– 1.97	0 / 1.0	+0.6 / −0	+0 / −0.4	0 / 1.6	+1.0 / −0	+0 / −0.6	0 / 2.6	+1.6 / −0	+0 / −1	0 / 6.5	+4.0 / −0	+0 / −2.5	0.4 / 2.0	+1.0 / −0	−0.4 / −1.0
1.97– 3.15	0 / 1.2	+0.7 / −0	+0 / −0.5	0 / 1.9	+1.2 / −0	+0 / −0.7	0 / 3	+1.8 / −0	+0 / −1.2	0 / 7.5	+4.5 / −0	+0 / −3	0.4 / 2.3	+1.2 / −0	−0.4 / −1.1
3.15– 4.73	0 / 1.5	+0.9 / −0	+0 / −0.6	0 / 2.3	+1.4 / −0	+0 / −0.9	0 / 3.6	+2.2 / −0	+0 / −1.4	0 / 8.5	+5.0 / −0	+0 / −3.5	0.5 / 2.8	+1.4 / −0	−0.5 / −1.4
4.73– 7.09	0 / 1.7	+1.0 / −0	+0 / −0.7	0 / 2.6	+1.6 / −0	+0 / −1.0	0 / 4.1	+2.5 / −0	+0 / −1.6	0 / 10	+6.0 / −0	+0 / −4	0.6 / 3.2	+1.6 / −0	−0.6 / −1.6
7.09– 9.85	0 / 2.0	+1.2 / −0	+0 / −0.8	0 / 3.0	+1.8 / −0	+0 / −1.2	0 / 4.6	+2.8 / −0	+0 / −1.8	0 / 11.5	+7.0 / −0	+0 / −4.5	0.6 / 3.6	+1.8 / −0	−0.6 / −1.8
9.85– 12.41	0 / 2.1	+1.2 / −0	+0 / −0.9	0 / 3.2	+2.0 / −0	+0 / −1.2	0 / 5	+3.0 / −0	+0 / −2.0	0 / 13	+8.0 / −0	+0 / −5	0.7 / 3.9	+2.0 / −0	−0.7 / −1.9
12.41– 15.75	0 / 2.4	+1.4 / −0	+0 / −1.0	0 / 3.6	+2.2 / −0	+0 / −1.4	0 / 5.7	+3.5 / −0	+0 / −2.2	0 / 15	+9.0 / −0	+0 / −6	0.7 / 4.3	+2.2 / −0	−0.7 / −2.1
15.75– 19.69	0 / 2.6	+1.6 / −0	+0 / −1.0	0 / 4.1	+2.5 / −0	+0 / −1.6	0 / 6.5	+4 / −0	+0 / −2.5	0 / 16	+10.0 / −0	+0 / −6	0.8 / 4.9	+2.5 / −0	−0.8 / −2.4
19.69– 30.09	0 / 3.2	+2.0 / −0	+0 / −1.2	0 / 5.0	+3 / −0	+0 / −2	0 / 8	+5 / −0	+0 / −3	0 / 20	+12.0 / −0	+0 / −8	0.9 / 5.9	+3.0 / −0	−0.9 / −2.9
30.09– 41.49	0 / 4.1	+2.5 / −0	+0 / −1.6	0 / 6.5	+4 / −0	+0 / −2.5	0 / 10	+6 / −0	+0 / −4	0 / 26	+16.0 / −0	+0 / −10	1.0 / 7.5	+4.0 / −0	−1.0 / −3.5
41.49– 56.19	0 / 5.0	+3.0 / −0	+0 / −2.0	0 / 8.0	+5 / −0	+0 / −3	0 / 13	+8 / −0	+0 / −5	0 / 32	+20.0 / −0	+0 / −12	1.2 / 9.2	+5.0 / −0	−1.2 / −4.2
56.19– 76.39	0 / 6.5	+4.0 / −0	+0 / −2.5	0 / 10	+6 / −0	+0 / −4	0 / 16	+10 / −0	+0 / −6	0 / 41	+25.0 / −0	+0 / −16	1.2 / 11.2	+6.0 / −0	−1.2 / −5.2
76.39– 100.9	0 / 8.0	+5.0 / −0	+0 / −3.0	0 / 13	+8 / −0	+0 / −5	0 / 20	+12 / −0	+0 / −8	0 / 50	+30.0 / −0	+0 / −20	1.4 / 14.4	+8.0 / −0	−1.4 / −6.4
100.9 – 131.9	0 / 10.0	+6.0 / −0	+0 / −4.0	0 / 16	+10 / −0	+0 / −6	0 / 26	+16 / −0	+0 / −10	0 / 65	+40.0 / −0	+0 / −25	1.6 / 17.6	+10.0 / −0	−1.6 / −7.6
131.9 – 171.9	0 / 13.0	+8.0 / −0	+0 / −5.0	0 / 20	+12 / −0	+0 / −8	0 / 32	+20 / −0	+0 / −12	0 / 8	+50.0 / −0	+0 / −30	1.8 / 21.8	+12.0 / −0	−1.8 / −9.8
171.9 – 200	0 / 16.0	+10.0 / −0	+0 / −6.0	0 / 26	+16 / −0	+0 / −10	0 / 41	+25 / −0	+0 / −16	0 / 100	+60.0 / −0	+0 / −40	1.8 / 27.8	+16.0 / −0	−1.8 / −11.8

TABLE 23 (cont'd)

LOCATIONAL CLEARANCE FITS

Limits are in thousandths of an inch.
Limits for hole and shaft are applied algebraically to the basic size to obtain the limits of size for the parts.
Data in bold face are in accordance with ABC agreements.
Symbols H9, f8, etc., are Hole and Shaft designations used in ABC System

Class LC 6			Class LC 7			Class LC 8			Class LC 9			Class LC 10			Class LC 11			Nominal Size Range Inches	
Limits of Clearance	Standard Limits		Limits of Clearance	Standard Limits		Limits of Clearance	Standard Limits		Limits of Clearance	Standard Limits		Limits of Clearance	Standard Limits		Limits of Clearance	Standard Limits			
	Hole H9	Shaft f8		Hole H10	Shaft e9		Hole H10	Shaft d9		Hole H11	Shaft c10		Hole H12	Shaft		Hole H13	Shaft	Over	To
0.3 / 1.9	+1.0 / 0	−0.3 / −0.9	0.6 / 3.2	+1.6 / 0	−0.6 / −1.6	1.0 / 3.6	+1.6 / −0	−1.0 / −2.0	2.5 / 6.6	+2.5 / −0	−2.5 / −4.1	4 / 12	+4 / −0	−4 / −8	5 / 17	+6 / −0	−5 / −11	0	0.12
0.4 / 2.3	+1.2 / 0	−0.4 / −1.1	0.8 / 3.8	+1.8 / 0	−0.8 / −2.0	1.2 / 4.2	+1.8 / −0	−1.2 / −2.4	2.8 / 7.6	+3.0 / −0	−2.8 / −4.6	4.5 / 14.5	+5 / −0	−4.5 / −9.5	6 / 20	+7 / −0	−6 / −13	0.12	0.24
0.5 / 2.8	+1.4 / 0	−0.5 / −1.4	1.0 / 4.6	+2.2 / 0	−1.0 / −2.4	1.6 / 5.2	+2.2 / −0	−1.6 / −3.0	3.0 / 8.7	+3.5 / −0	−3.0 / −5.2	5 / 17	+6 / −0	−5 / −11	7 / 25	+9 / −0	−7 / −16	0.24	0.40
0.6 / 3.2	+1.6 / 0	−0.6 / −1.6	1.2 / 5.6	+2.8 / 0	−1.2 / −2.8	2.0 / 6.4	+2.8 / −0	−2.0 / −3.6	3.5 / 10.3	+4.0 / −0	−3.5 / −6.3	6 / 20	+7 / −0	−6 / −13	8 / 28	+10 / −0	−8 / −18	0.40	0.71
0.8 / 4.0	+2.0 / 0	−0.8 / −2.0	1.6 / 7.1	+3.5 / 0	−1.6 / −3.6	2.5 / 8.0	+3.5 / −0	−2.5 / −4.5	4.5 / 13.0	+5.0 / −0	−4.5 / −8.0	7 / 23	+8 / −0	−7 / −15	10 / 34	+12 / −0	−10 / −22	0.71	1.19
1.0 / 5.1	+2.5 / 0	−1.0 / −2.6	2.0 / 8.5	+4.0 / 0	−2.0 / −4.5	3.0 / 9.5	+4.0 / −0	−3.0 / −5.5	5 / 15	+6 / −0	−5 / −9	8 / 28	+10 / −0	−8 / −18	12 / 44	+16 / −0	−12 / −28	1.19	1.97
1.2 / 6.0	+3.0 / 0	−1.2 / −3.0	2.5 / 10.0	+4.5 / 0	−2.5 / −5.5	4.0 / 11.5	+4.5 / −0	−4.0 / −7.0	6 / 17.5	+7 / −0	−6 / −10.5	10 / 34	+12 / −0	−10 / −22	14 / 50	+18 / −0	−14 / −32	1.97	3.15
1.4 / 7.1	+3.5 / 0	−1.4 / −3.6	3.0 / 11.5	+5.0 / 0	−3.0 / −6.5	5.0 / 13.5	+5.0 / −0	−5.0 / −8.5	7 / 21	+9 / −0	−7 / −12	11 / 39	+14 / −0	−11 / −25	16 / 60	+22 / −0	−16 / −38	3.15	4.73
1.6 / 8.1	+4.0 / 0	−1.6 / −4.1	3.5 / 13.5	+6.0 / 0	−3.5 / −7.5	6 / 16	+6 / −0	−6 / −10	8 / 24	+10 / −0	−8 / −14	12 / 44	+16 / −0	−12 / −28	18 / 68	+25 / −0	−18 / −43	4.73	7.09
2.0 / 9.3	+4.5 / 0	−2.0 / −4.8	4.0 / 15.5	+7.0 / 0	−4.0 / −8.5	7 / 18.5	+7 / −0	−7 / −11.5	10 / 29	+12 / −0	−10 / −17	16 / 52	+18 / −0	−16 / −34	22 / 78	+28 / −0	−22 / −50	7.09	9.85
2.2 / 10.2	+5.0 / 0	−2.2 / −5.2	4.5 / 17.5	+8.0 / 0	−4.5 / −9.5	7 / 20	+8 / −0	−7 / −12	12 / 32	+12 / −0	−12 / −20	20 / 60	+20 / −0	−20 / −40	28 / 88	+30 / −0	−28 / −58	9.85	12.41
2.5 / 12.0	+6.0 / 0	−2.5 / −6.0	5.0 / 20.0	+9.0 / 0	−5 / −11	8 / 23	+9 / −0	−8 / −14	14 / 37	+14 / −0	−14 / −23	22 / 66	+22 / −0	−22 / −44	30 / 100	+35 / −0	−30 / −65	12.41	15.75
2.8 / 12.8	+6.0 / 0	−2.8 / −6.8	5.0 / 21.0	+10.0 / 0	−5 / −11	9 / 25	+10 / −0	−9 / −15	16 / 42	+16 / −0	−16 / −26	25 / 75	+25 / −0	−25 / −50	35 / 115	+40 / −0	−35 / −75	15.75	19.69
3.0 / 16.0	+8.0 / 0	−3.0 / −8.0	6.0 / 26.0	+12.0 / −0	−6 / −14	10 / 30	+12 / −0	−10 / −18	18 / 50	+20 / −0	−18 / −30	28 / 88	+30 / −0	−28 / −58	40 / 140	+50 / −0	−40 / −90	19.69	30.09
3.5 / 19.5	+10.0 / 0	−3.5 / −9.5	7.0 / 33.0	+16.0 / −0	−7 / −17	12 / 38	+16 / −0	−12 / −22	20 / 61	+25 / −0	−20 / −36	30 / 110	+40 / −0	−30 / −70	45 / 165	+60 / −0	−45 / −105	30.09	41.49
4.0 / 24.0	+12.0 / 0	−4.0 / −12.0	8.0 / 40.0	+20.0 / −0	−8 / −20	14 / 46	+20 / −0	−14 / −26	25 / 75	+30 / −0	−25 / −45	40 / 140	+50 / −0	−40 / −90	60 / 220	+80 / −0	−60 / −140	41.49	56.19
4.5 / 30.5	+16.0 / 0	−4.5 / −14.5	9.0 / 50.0	+25.0 / −0	−9 / −25	16 / 57	+25 / −0	−16 / −32	30 / 95	+40 / −0	−30 / −55	50 / 170	+60 / −0	−50 / −110	70 / 270	+100 / −0	−70 / −170	56.19	76.39
5.0 / 37.0	+20.0 / 0	−5 / −17	10.0 / 60.0	+30.0 / −0	−10 / −30	18 / 68	+30 / −0	−18 / −38	35 / 115	+50 / −0	−35 / −65	50 / 210	+80 / −0	−50 / −130	80 / 330	+125 / −0	−80 / −205	76.39	100.9
6.0 / 47.0	+25.0 / 0	−6 / −22	12.0 / 67.0	+40.0 / −0	−12 / −27	20 / 85	+40 / −0	−20 / −45	40 / 140	+60 / −0	−40 / −80	60 / 260	+100 / −0	−60 / −160	90 / 410	+160 / −0	−90 / −250	100.9	131.9
7.0 / 57.0	+30.0 / 0	−7 / −27	14.0 / 94.0	+50.0 / −0	−14 / −44	25 / 105	+50 / −0	−25 / −55	50 / 180	+80 / −0	−50 / −100	80 / 330	+125 / −0	−80 / −205	100 / 500	+200 / −0	−100 / −300	131.9	171.9
7.0 / 72.0	+40.0 / 0	−7 / −32	14.0 / 114.0	+60.0 / −0	−14 / −54	25 / 125	+60 / −0	−25 / −65	50 / 210	+100 / −0	−50 / −110	90 / 410	+160 / −0	−90 / −250	125 / 625	+250 / −0	−125 / −375	171.9	200

TABLE 24 LOCATIONAL TRANSITION FITS (INCH)

Limits are in thousandths of an inch.
Limits for hole and shaft are applied algebraically to the basic size to obtain the limits of size for the mating parts.
Data in bold face are in accordance with ABC agreements.
"Fit" represents the maximum interference (minus values) and the maximum clearance (plus values).
Symbols H7, js6, etc., are Hole and Shaft designations used in ABC System

Nominal Size Range Inches		Class LT 1			Class LT 2			Class LT 3			Class LT 4			Class LT 5			Class LT 6		
			Standard Limits			Standard Limits			Standard Limits			Standard Limits			Standard Limits			Standard Limits	
Over	To	Fit	Hole H7	Shaft js6	Fit	Hole H8	Shaft js7	Fit	Hole H7	Shaft k6	Fit	Hole H8	Shaft k7	Fit	Hole H7	Shaft n6	Fit	Hole H7	Shaft n7
0	0.12	-0.10 +0.50	+0.4 -0	+0.10 -0.10	-0.2 +0.8	+0.6 -0	+0.2 -0.2							-0.5 +0.15	+0.4 -0	+0.5 +0.25	-0.65 +0.15	+0.4 -0	+0.65 +0.25
0.12	0.24	-0.15 +0.65	+0.5 -0	+0.15 -0.15	-0.25 +0.95	+0.7 -0	+0.25 -0.25							-0.6 +0.2	+0.5 -0	+0.6 +0.3	-0.8 +0.2	+0.5 -0	+0.8 +0.3
0.24	0.40	-0.2 +0.8	+0.6 -0	+0.2 -0.2	-0.3 +1.2	+0.9 -0	+0.3 -0.3	-0.5 +0.5	+0.6 -0	+0.5 +0.1	-0.7 +0.8	+0.9 -0	+0.7 +0.1	-0.8 +0.2	+0.6 -0	+0.8 +0.4	-1.0 +0.2	+0.6 -0	+1.0 +0.4
0.40	0.71	-0.2 +0.9	+0.7 -0	+0.2 -0.2	-0.35 +1.35	+1.0 -0	+0.35 -0.35	-0.5 +0.6	+0.7 -0	+0.5 +0.1	-0.8 +0.9	+1.0 -0	+0.8 +0.1	-0.9 +0.2	+0.7 -0	+0.9 +0.5	-1.2 +0.2	+0.7 -0	+1.2 +0.5
0.71	1.19	-0.25 +1.05	+0.8 -0	+0.25 -0.25	-0.4 +1.6	+1.2 -0	+0.4 -0.4	-0.6 +0.7	+0.8 -0	+0.6 +0.1	-0.9 +1.1	+1.2 -0	+0.9 +0.1	-1.1 +0.2	+0.8 -0	+1.1 +0.6	-1.4 +0.2	+0.8 -0	+1.4 +0.6
1.19	1.97	-0.3 +1.3	+1.0 -0	+0.3 -0.3	-0.5 +2.1	+1.6 -0	+0.5 -0.5	-0.7 +0.9	+1.0 -0	+0.7 +0.1	-1.1 +1.5	+1.6 -0	+1.1 +0.1	-1.3 +0.3	+1.0 -0	+1.3 +0.7	-1.7 +0.3	+1.0 -0	+1.7 +0.7
1.97	3.15	-0.3 +1.5	+1.2 -0	+0.3 -0.3	-0.6 +2.4	+1.8 -0	+0.6 -0.6	-0.8 +1.1	+1.2 -0	+0.8 +0.1	-1.3 +1.7	+1.8 -0	+1.3 +0.1	-1.5 +0.4	+1.2 -0	+1.5 +0.8	-2.0 +0.4	+1.2 -0	+2.0 +0.8
3.15	4.73	-0.4 +1.8	+1.4 -0	+0.4 -0.4	-0.7 +2.9	+2.2 -0	+0.7 -0.7	-1.0 +1.3	+1.4 -0	+1.0 +0.1	-1.5 +2.1	+2.2 -0	+1.5 +0.1	-1.9 +0.4	+1.4 -0	+1.9 +1.0	-2.4 +0.4	+1.4 -0	+2.4 +1.0
4.73	7.09	-0.5 +2.1	+1.6 -0	+0.5 -0.5	-0.8 +3.3	+2.5 -0	+0.8 -0.8	-1.1 +1.5	+1.6 -0	+1.1 +0.1	-1.7 +2.4	+2.5 -0	+1.7 +0.1	-2.2 +0.4	+1.6 -0	+2.2 +1.2	-2.8 +0.4	+1.6 -0	+2.8 +1.2
7.09	9.85	-0.6 +2.4	+1.8 -0	+0.6 -0.6	-0.9 +3.7	+2.8 -0	+0.9 -0.9	-1.4 +1.6	+1.8 -0	+1.4 +0.2	-2.0 +2.6	+2.8 -0	+2.0 +0.2	-2.6 +0.4	+1.8 -0	+2.6 +1.4	-3.2 +0.4	+1.8 -0	+3.2 +1.4
9.85	12.41	-0.6 +2.6	+2.0 -0	+0.6 -0.6	-1.0 +4.0	+3.0 -0	+1.0 -1.0	-1.4 +1.8	+2.0 -0	+1.4 +0.2	-2.2 +2.8	+3.0 -0	+2.2 +0.2	-2.6 +0.6	+2.0 -0	+2.6 +1.4	-3.4 +0.6	+2.0 -0	+3.4 +1.4
12.41	15.75	-0.7 +2.9	+2.2 -0	+0.7 -0.7	-1.0 +4.5	+3.5 -0	+1.0 -1.0	-1.6 +2.0	+2.2 -0	+1.6 +0.2	-2.4 +3.3	+3.5 -0	+2.4 +0.2	-3.0 +0.6	+2.2 -0	+3.0 +1.6	-3.8 +0.6	+2.2 -0	+3.8 +1.6
15.75	19.69	-0.8 +3.3	+2.5 -0	+0.8 -0.8	-1.2 +5.2	+4.0 -0	+1.2 -1.2	-1.8 +2.3	+2.5 -0	+1.8 +0.2	-2.7 +3.8	+4.0 -0	+2.7 +0.2	-3.4 +0.7	+2.5 -0	+3.4 +1.8	-4.3 +0.7	+2.5 -0	+4.3 +1.8

TABLE 25 LOCATIONAL INTERFERENCE FITS (INCH)

Limits are in thousandths of an inch.
Limits for hole and shaft are applied algebraically to the basic size to obtain the limits of size for the parts.
Data in bold face are in accordance with ABC agreements,
Symbols H7, p6, etc., are Hole and Shaft designations used in ABC System

Nominal Size Range Inches Over — To	Class LN 1			Class LN 2			Class LN 3		
	Limits of Interference	Standard Limits		Limits of Interference	Standard Limits		Limits of Interference	Standard Limits	
		Hole H6	Shaft n5		Hole H7	Shaft p6		Hole H7	Shaft r6
0 — 0.12	0 / 0.45	+0.25 / −0	+0.45 / +0.25	0 / 0.65	+0.4 / −0	+0.65 / +0.4	0.1 / 0.75	+0.4 / −0	+0.75 / +0.5
0.12 — 0.24	0 / 0.5	+0.3 / −0	+0.5 / +0.3	0 / 0.8	+0.5 / −0	+0.8 / +0.5	0.1 / 0.9	+0.5 / −0	+0.9 / +0.6
0.24 — 0.40	0 / 0.65	+0.4 / −0	+0.65 / +0.4	0 / 1.0	+0.6 / −0	+1.0 / +0.6	0.2 / 1.2	+0.6 / −0	+1.2 / +0.8
0.40 — 0.71	0 / 0.8	+0.4 / −0	+0.8 / +0.4	0 / 1.1	+0.7 / −0	+1.1 / +0.7	0.3 / 1.4	+0.7 / −0	+1.4 / +1.0
0.71 — 1.19	0 / 1.0	+0.5 / −0	+1.0 / +0.5	0 / 1.3	+0.8 / −0	+1.3 / +0.8	0.4 / 1.7	+0.8 / −0	+1.7 / +1.2
1.19 — 1.97	0 / 1.1	+0.6 / −0	+1.1 / +0.6	0 / 1.6	+1.0 / −0	+1.6 / +1.0	0.4 / 2.0	+1.0 / −0	+2.0 / +1.4
1.97 — 3.15	0.1 / 1.3	+0.7 / −0	+1.3 / +0.7	0.2 / 2.1	+1.2 / −0	+2.1 / +1.4	0.4 / 2.3	+1.2 / −0	+2.3 / +1.6
3.15 — 4.73	0.1 / 1.6	+0.9 / −0	+1.6 / +1.0	0.2 / 2.5	+1.4 / −0	+2.5 / +1.6	0.6 / 2.9	+1.4 / −0	+2.9 / +2.0
4.73 — 7.09	0.2 / 1.9	+1.0 / −0	+1.9 / +1.2	0.2 / 2.8	+1.6 / −0	+2.8 / +1.8	0.9 / 3.5	+1.6 / −0	+3.5 / +2.5
7.09 — 9.85	0.2 / 2.2	+1.2 / −0	+2.2 / +1.4	0.2 / 3.2	+1.8 / −0	+3.2 / +2.0	1.2 / 4.2	+1.8 / −0	+4.2 / +3.0
9.85 — 12.41	0.2 / 2.3	+1.2 / −0	+2.3 / +1.4	0.2 / 3.4	+2.0 / −0	+3.4 / +2.2	1.5 / 4.7	+2.0 / −0	+4.7 / +3.5
12.41 — 15.75	0.2 / 2.6	+1.4 / −0	+2.6 / +1.6	0.3 / 3.9	+2.2 / −0	+3.9 / +2.5	2.3 / 5.9	+2.2 / −0	+5.9 / +4.5
15.75 — 19.69	0.2 / 2.8	+1.6 / −0	+2.8 / +1.8	0.3 / 4.4	+2.5 / −0	+4.4 / +2.8	2.5 / 6.6	+2.5 / −0	+6.6 / +5.0
19.69 — 30.09		+2.0 / −0		0.5 / 5.5	+3 / −0	+5.5 / +3.5	4 / 9	+3 / −0	+9 / +7
30.09 — 41.49		+2.5 / −0		0.5 / 7.0	+4 / −0	+7.0 / +4.5	5 / 11.5	+4 / −0	+11.5 / +9
41.49 — 56.19		+3.0 / −0		1 / 9	+5 / −0	+9 / +6	7 / 15	+5 / −0	+15 / +12
56.19 — 76.39		+4.0 / −0		1 / 11	+6 / −0	+11 / +7	10 / 20	+6 / −0	+20 / +16
76.39 — 100.9		+5.0 / −0		1 / 14	+8 / −0	+14 / +9	12 / 25	+8 / −0	+25 / +20
100.9 — 131.9		+6.0 / −0		2 / 18	+10 / −0	+18 / +12	15 / 31	+10 / −0	+31 / +25
131.9 — 171.9		+8.0 / −0		4 / 24	+12 / −0	+24 / +16	18 / 38	+12 / −0	+38 / +30
171.9 — 200		+10.0 / −0		4 / 30	+16 / −0	+30 / +20	24 / 50	+16 / −0	+50 / +40

TABLE 26 FORCE AND SHRINK FITS (INCH)

Limits are in thousandths of an inch.
Limits for hole and shaft are applied algebraically to the basic size to obtain the limits of size for the parts.
Data in bold face are in accordance with ABC agreements.
Symbols H7, s6, etc., are Hole and Shaft designations used in ABC System

Nominal Size Range Inches Over To	Class FN 1			Class FN 2			Class FN 3			Class FN 4			Class FN 5		
	Limits of Interference	Standard Limits		Limits of Interference	Standard Limits		Limits of Interference	Standard Limits		Limits of Interference	Standard Limits		Limits of Interference	Standard Limits	
		Hole H6	Shaft		Hole H7	Shaft s6		Hole H7	Shaft t6		Hole H7	Shaft u6		Hole H8	Shaft x7
0 – 0.12	0.05 / 0.5	+0.25 / −0	+0.5 / +0.3	0.2 / 0.85	+0.4 / −0	+0.85 / +0.6				0.3 / 0.95	+0.4 / −0	+0.95 / +0.7	0.3 / 1.3	+0.6 / −0	+1.3 / +0.9
0.12 – 0.24	0.1 / 0.6	+0.3 / −0	+0.6 / +0.4	0.2 / 1.0	+0.5 / −0	+1.0 / +0.7				0.4 / 1.2	+0.5 / −0	+1.2 / +0.9	0.5 / 1.7	+0.7 / −0	+1.7 / +1.2
0.24 – 0.40	0.1 / 0.75	+0.4 / −0	+0.75 / +0.5	0.4 / 1.4	+0.6 / −0	+1.4 / +1.0				0.6 / 1.6	+0.6 / −0	+1.6 / +1.2	0.5 / 2.0	+0.9 / −0	+2.0 / +1.4
0.40 – 0.56	0.1 / 0.8	+0.4 / −0	+0.8 / +0.5	0.5 / 1.6	+0.7 / −0	+1.6 / +1.2				0.7 / 1.8	+0.7 / −0	+1.8 / +1.4	0.6 / 2.3	+1.0 / −0	+2.3 / +1.6
0.56 – 0.71	0.2 / 0.9	+0.4 / −0	+0.9 / +0.6	0.5 / 1.6	+0.7 / −0	+1.6 / +1.2				0.7 / 1.8	+0.7 / −0	+1.8 / +1.4	0.8 / 2.5	+1.0 / −0	+2.5 / +1.8
0.71 – 0.95	0.2 / 1.1	+0.5 / −0	+1.1 / +0.7	0.6 / 1.9	+0.8 / −0	+1.9 / +1.4				0.8 / 2.1	+0.8 / −0	+2.1 / +1.6	1.0 / 3.0	+1.2 / −0	+3.0 / +2.2
0.95 – 1.19	0.3 / 1.2	+0.5 / −0	+1.2 / +0.8	0.6 / 1.9	+0.8 / −0	+1.9 / +1.4	0.8 / 2.1	+0.8 / −0	+2.1 / +1.6	1.0 / 2.3	+0.8 / −0	+2.3 / +1.8	1.3 / 3.3	+1.2 / −0	+3.3 / +2.5
1.19 – 1.58	0.3 / 1.3	+0.6 / −0	+1.3 / +0.9	0.8 / 2.4	+1.0 / −0	+2.4 / +1.8	1.0 / 2.6	+1.0 / −0	+2.6 / +2.0	1.5 / 3.1	+1.0 / −0	+3.1 / +2.5	1.4 / 4.0	+1.6 / −0	+4.0 / +3.0
1.58 – 1.97	0.4 / 1.4	+0.6 / −0	+1.4 / +1.0	0.8 / 2.4	+1.0 / −0	+2.4 / +1.8	1.2 / 2.8	+1.0 / −0	+2.8 / +2.2	1.8 / 3.4	+1.0 / −0	+3.4 / +2.8	2.4 / 5.0	+1.6 / −0	+5.0 / +4.0
1.97 – 2.56	0.6 / 1.8	+0.7 / −0	+1.8 / +1.3	0.8 / 2.7	+1.2 / −0	+2.7 / +2.0	1.3 / 3.2	+1.2 / −0	+3.2 / +2.5	2.3 / 4.2	+1.2 / −0	+4.2 / +3.5	3.2 / 6.2	+1.8 / −0	+6.2 / +5.0
2.56 – 3.15	0.7 / 1.9	+0.7 / −0	+1.9 / +1.4	1.0 / 2.9	+1.2 / −0	+2.9 / +2.2	1.8 / 3.7	+1.2 / −0	+3.7 / +3.0	2.8 / 4.7	+1.2 / −0	+4.7 / +4.0	4.2 / 7.2	+1.8 / −0	+7.2 / +6.0
3.15 – 3.94	0.9 / 2.4	+0.9 / −0	+2.4 / +1.8	1.4 / 3.7	+1.4 / −0	+3.7 / +2.8	2.1 / 4.4	+1.4 / −0	+4.4 / +3.5	3.6 / 5.9	+1.4 / −0	+5.9 / +5.0	4.8 / 8.4	+2.2 / −0	+8.4 / +7.0
3.94 – 4.73	1.1 / 2.6	+0.9 / −0	+2.6 / +2.0	1.6 / 3.9	+1.4 / −0	+3.9 / +3.0	2.6 / 4.9	+1.4 / −0	+4.9 / +4.0	4.6 / 6.9	+1.4 / −0	+6.9 / +6.0	5.8 / 9.4	+2.2 / −0	+9.4 / +8.0
4.73 – 5.52	1.2 / 2.9	+1.0 / −0	+2.9 / +2.2	1.9 / 4.5	+1.6 / −0	+4.5 / +3.5	3.4 / 6.0	+1.6 / −0	+6.0 / +5.0	5.4 / 8.0	+1.6 / −0	+8.0 / +7.0	7.5 / 11.6	+2.5 / −0	+11.6 / +10.0
5.52 – 6.30	1.5 / 3.2	+1.0 / −0	+3.2 / +2.5	2.4 / 5.0	+1.6 / −0	+5.0 / +4.0	3.4 / 6.0	+1.6 / −0	+6.0 / +5.0	5.4 / 8.0	+1.6 / −0	+8.0 / +7.0	9.5 / 13.6	+2.5 / −0	+13.6 / +12.0
6.30 – 7.09	1.8 / 3.5	+1.0 / −0	+3.5 / +2.8	2.9 / 5.5	+1.6 / −0	+5.5 / +4.5	4.4 / 7.0	+1.6 / −0	+7.0 / +6.0	6.4 / 9.0	+1.6 / −0	+9.0 / +8.0	9.5 / 13.6	+2.5 / −0	+13.6 / +12.0
7.09 – 7.88	1.8 / 3.8	+1.2 / −0	+3.8 / +3.0	3.2 / 6.2	+1.8 / −0	+6.2 / +5.0	5.2 / 8.2	+1.8 / −0	+8.2 / +7.0	7.2 / 10.2	+1.8 / −0	+10.2 / +9.0	11.2 / 15.8	+2.8 / −0	+15.8 / +14.0
7.88 – 8.86	2.3 / 4.3	+1.2 / −0	+4.3 / +3.5	3.2 / 6.2	+1.8 / −0	+6.2 / +5.0	5.2 / 8.2	+1.8 / −0	+8.2 / +7.0	8.2 / 11.2	+1.8 / −0	+11.2 / +10.0	13.2 / 17.8	+2.8 / −0	+17.8 / +16.0
8.86 – 9.85	2.3 / 4.3	+1.2 / −0	+4.3 / +3.5	4.2 / 7.2	+1.8 / −0	+7.2 / +6.0	6.2 / 9.2	+1.8 / −0	+9.2 / +8.0	10.2 / 13.2	+1.8 / −0	+13.2 / +12.0	13.2 / 17.8	+2.8 / −0	+17.8 / +16.0
9.85 – 11.03	2.8 / 4.9	+1.2 / −0	+4.9 / +4.0	4.0 / 7.2	+2.0 / −0	+7.2 / +6.0	7.0 / 10.2	+2.0 / −0	+10.2 / +9.0	10.0 / 13.2	+2.0 / −0	+13.2 / +12.0	15.0 / 20.0	+3.0 / −0	+20.0 / +18.0
11.03 – 12.41	2.8 / 4.9	+1.2 / −0	+4.9 / +4.0	5.0 / 8.2	+2.0 / −0	+8.2 / +7.0	7.0 / 10.2	+2.0 / −0	+10.2 / +9.0	12.0 / 15.2	+2.0 / −0	+15.2 / +14.0	17.0 / 22.0	+3.0 / −0	+22.0 / +20.0
12.41 – 13.98	3.1 / 5.5	+1.4 / −0	+5.5 / +4.5	5.8 / 9.4	+2.2 / −0	+9.4 / +8.0	7.8 / 11.4	+2.2 / −0	+11.4 / +10.0	13.8 / 17.4	+2.2 / −0	+17.4 / +16.0	18.5 / 24.2	+3.5 / +0	+24.2 / +22.0
13.98 – 15.75	3.6 / 6.1	+1.4 / −0	+6.1 / +5.0	5.8 / 9.4	+2.2 / −0	+9.4 / +8.0	9.8 / 13.4	+2.2 / −0	+13.4 / +12.0	15.8 / 19.4	+2.2 / −0	+19.4 / +18.0	21.5 / 27.2	+3.5 / −0	+27.2 / +25.0
15.75 – 17.72	4.4 / 7.0	+1.6 / −0	+7.0 / +6.0	6.5 / 10.6	+2.5 / −0	+10.6 / +9.0	9.5 / 13.6	+2.5 / −0	+13.6 / +12.0	17.5 / 21.6	+2.5 / −0	+21.6 / +20.0	24.0 / 30.5	+4.0 / −0	+30.5 / +28.0
17.72 – 19.69	4.4 / 7.0	+1.6 / −0	+7.0 / +6.0	7.5 / 11.6	+2.5 / −0	+11.6 / +10.0	11.5 / 15.6	+2.5 / −0	+15.6 / +14.0	19.5 / 23.6	+2.5 / −0	+23.6 / +22.0	26.0 / 32.5	+4.0 / −0	+32.5 / +30.0

TABLE 27 PREFERRED HOLE BASIS CLEARANCE FITS (METRIC)

Dimensions in mm.

BASIC SIZE		LOOSE RUNNING Hole H11	Shaft c11	Fit	FREE RUNNING Hole H9	Shaft d9	Fit	CLOSE RUNNING Hole H8	Shaft f7	Fit	SLIDING Hole H7	Shaft g6	Fit	LOCATIONAL CLEARANCE Hole H7	Shaft h6	Fit
1	MAX	1.060	0.940	0.180	1.025	0.980	0.070	1.014	0.994	0.030	1.010	0.998	0.018	1.010	1.000	0.016
	MIN	1.000	0.880	0.060	1.000	0.955	0.020	1.000	0.984	0.006	1.000	0.992	0.002	1.000	0.994	0.000
1.2	MAX	1.260	1.140	0.180	1.225	1.180	0.070	1.214	1.194	0.030	1.210	1.198	0.018	1.210	1.200	0.016
	MIN	1.200	1.080	0.060	1.200	1.155	0.020	1.200	1.184	0.006	1.200	1.192	0.002	1.200	1.194	0.000
1.6	MAX	1.660	1.540	0.180	1.625	1.580	0.070	1.614	1.594	0.030	1.610	1.598	0.018	1.610	1.600	0.016
	MIN	1.600	1.480	0.060	1.600	1.555	0.020	1.600	1.584	0.006	1.600	1.592	0.002	1.600	1.594	0.000
2	MAX	2.060	1.940	0.180	2.025	1.980	0.070	2.014	1.994	0.030	2.010	1.998	0.018	2.010	2.000	0.016
	MIN	2.000	1.880	0.060	2.000	1.955	0.020	2.000	1.984	0.006	2.000	1.992	0.002	2.000	1.994	0.000
2.5	MAX	2.560	2.440	0.180	2.525	2.480	0.070	2.514	2.494	0.030	2.510	2.498	0.018	2.510	2.500	0.016
	MIN	2.500	2.380	0.060	2.500	2.455	0.020	2.500	2.484	0.006	2.500	2.492	0.002	2.500	2.494	0.000
3	MAX	3.060	2.940	0.180	3.025	2.980	0.070	3.014	2.994	0.030	3.010	2.998	0.018	3.010	3.000	0.016
	MIN	3.000	2.880	0.060	3.000	2.955	0.020	3.000	2.984	0.006	3.000	2.992	0.002	3.000	2.994	0.000
4	MAX	4.075	3.930	0.220	4.030	3.970	0.090	4.018	3.990	0.040	4.012	3.996	0.024	4.012	4.000	0.020
	MIN	4.000	3.855	0.070	4.000	3.940	0.030	4.000	3.978	0.010	4.000	3.988	0.004	4.000	3.992	0.000
5	MAX	5.075	4.930	0.220	5.030	4.970	0.090	5.018	4.990	0.040	5.012	4.996	0.024	5.012	5.000	0.020
	MIN	5.000	4.855	0.070	5.000	4.940	0.030	5.000	4.978	0.010	5.000	4.988	0.004	5.000	4.992	0.000
6	MAX	6.075	5.930	0.220	6.030	5.970	0.090	6.018	5.990	0.040	6.012	5.996	0.024	6.012	6.000	0.020
	MIN	6.000	5.855	0.070	6.000	5.940	0.030	6.000	5.978	0.010	6.000	5.988	0.004	6.000	5.992	0.000
8	MAX	8.090	7.920	0.260	8.036	7.960	0.112	8.022	7.987	0.050	8.015	7.995	0.029	8.015	8.000	0.024
	MIN	8.000	7.830	0.080	8.000	7.924	0.040	8.000	7.972	0.013	8.000	7.986	0.005	8.000	7.991	0.000
10	MAX	10.090	9.920	0.260	10.036	9.960	0.112	10.022	9.987	0.050	10.015	9.995	0.029	10.015	10.000	0.024
	MIN	10.000	9.830	0.080	10.000	9.924	0.040	10.000	9.972	0.013	10.000	9.986	0.005	10.000	9.991	0.000
12	MAX	12.110	11.905	0.315	12.043	11.950	0.136	12.027	11.984	0.061	12.018	11.994	0.035	12.018	12.000	0.029
	MIN	12.000	11.795	0.095	12.000	11.907	0.050	12.000	11.966	0.016	12.000	11.983	0.006	12.000	11.989	0.000
16	MAX	16.110	15.905	0.315	16.043	15.950	0.136	16.027	15.984	0.061	16.018	15.994	0.035	16.018	16.000	0.029
	MIN	16.000	15.795	0.095	16.000	15.907	0.050	16.000	15.966	0.016	16.000	15.983	0.006	16.000	15.989	0.000
20	MAX	20.130	19.890	0.370	20.052	19.935	0.169	20.033	19.980	0.074	20.021	19.993	0.041	20.021	20.000	0.034
	MIN	20.000	19.760	0.110	20.000	19.883	0.065	20.000	19.959	0.020	20.000	19.980	0.007	20.000	19.987	0.000
25	MAX	25.130	24.890	0.370	25.052	24.935	0.169	25.033	24.980	0.074	25.021	24.993	0.041	25.021	25.000	0.034
	MIN	25.000	24.760	0.110	25.000	24.883	0.065	25.000	24.959	0.020	25.000	24.980	0.007	25.000	24.987	0.000
30	MAX	30.130	29.890	0.370	30.052	29.935	0.169	30.033	29.980	0.074	30.021	29.993	0.041	30.021	30.000	0.034
	MIN	30.000	29.760	0.110	30.000	29.883	0.065	30.000	29.959	0.020	30.000	29.980	0.007	30.000	29.987	0.000

ANSI B4.2-1978

TABLE 27 (cont'd)

APPENDIX 591

ANSI B4.2-1978

Dimensions in mm.

BASIC SIZE		LOOSE RUNNING			FREE RUNNING			CLOSE RUNNING			SLIDING			LOCATIONAL CLEARANCE		
		Hole H11	Shaft c11	Fit	Hole H9	Shaft d9	Fit	Hole H8	Shaft f7	Fit	Hole H7	Shaft g6	Fit	Hole H7	Shaft h6	Fit
40	MAX	40.160	39.880	0.440	40.062	39.920	0.204	40.039	39.975	0.089	40.025	39.991	0.050	40.025	40.000	0.041
	MIN	40.000	39.720	0.120	40.000	39.858	0.080	40.000	39.950	0.025	40.000	39.975	0.009	40.000	39.984	0.000
50	MAX	50.160	49.870	0.450	50.062	49.920	0.204	50.039	49.975	0.089	50.025	49.991	0.050	50.025	50.000	0.041
	MIN	50.000	49.710	0.130	50.000	49.858	0.080	50.000	49.950	0.025	50.000	49.975	0.009	50.000	49.984	0.000
60	MAX	60.190	59.860	0.520	60.074	59.900	0.248	60.046	59.970	0.106	60.030	59.990	0.059	60.030	60.000	0.049
	MIN	60.000	59.670	0.140	60.000	59.826	0.100	60.000	59.940	0.030	60.000	59.971	0.010	60.000	59.981	0.000
80	MAX	80.190	79.850	0.530	80.074	79.900	0.248	80.046	79.970	0.106	80.030	79.990	0.059	80.030	80.000	0.049
	MIN	80.000	79.660	0.150	80.000	79.826	0.100	80.000	79.940	0.030	80.000	79.971	0.010	80.000	79.981	0.000
100	MAX	100.220	99.830	0.610	100.087	99.880	0.294	100.054	99.964	0.125	100.035	99.988	0.069	100.035	100.000	0.057
	MIN	100.000	99.610	0.170	100.000	99.793	0.120	100.000	99.929	0.036	100.000	99.966	0.012	100.000	99.978	0.000
120	MAX	120.220	119.820	0.620	120.087	119.880	0.294	120.054	119.964	0.125	120.035	119.988	0.069	120.035	120.000	0.057
	MIN	120.000	119.600	0.180	120.000	119.793	0.120	120.000	119.929	0.036	120.000	119.966	0.012	120.000	119.978	0.000
160	MAX	160.250	159.790	0.710	160.100	159.855	0.345	160.063	159.957	0.146	160.040	159.986	0.079	160.040	160.000	0.065
	MIN	160.000	159.540	0.210	160.000	159.755	0.145	160.000	159.917	0.043	160.000	159.961	0.014	160.000	159.975	0.000
200	MAX	200.290	199.760	0.820	200.115	199.830	0.400	200.072	199.950	0.168	200.046	199.985	0.090	200.046	200.000	0.075
	MIN	200.000	199.470	0.240	200.000	199.715	0.170	200.000	199.904	0.050	200.000	199.956	0.015	200.000	199.971	0.000
250	MAX	250.290	249.720	0.860	250.115	249.830	0.400	250.072	249.950	0.168	250.046	249.985	0.090	250.046	250.000	0.075
	MIN	250.000	249.430	0.280	250.000	249.715	0.170	250.000	249.904	0.050	250.000	249.956	0.015	250.000	249.971	0.000
300	MAX	300.320	299.670	0.970	300.130	299.810	0.450	300.081	299.944	0.189	300.052	299.983	0.101	300.052	300.000	0.084
	MIN	300.000	299.350	0.330	300.000	299.680	0.190	300.000	299.892	0.056	300.000	299.951	0.017	300.000	299.968	0.000
400	MAX	400.360	399.600	1.120	400.140	399.790	0.490	400.089	399.938	0.208	400.057	399.982	0.111	400.057	400.000	0.093
	MIN	400.000	399.240	0.400	400.000	399.650	0.210	400.000	399.881	0.062	400.000	399.946	0.018	400.000	399.964	0.000
500	MAX	500.400	499.520	1.280	500.155	499.770	0.540	500.097	499.932	0.228	500.063	499.980	0.123	500.063	500.000	0.103
	MIN	500.000	499.120	0.480	500.000	499.615	0.230	500.000	499.869	0.068	500.000	499.940	0.020	500.000	499.960	0.000

TABLE 28 PREFERRED HOLE BASIS TRANSITION AND INTERFERENCE FITS (METRIC)

ANSI B4.2-1978

Dimensions in mm.

BASIC SIZE		LOCATIONAL TRANSN. Hole H7	Shaft k6	Fit	LOCATIONAL TRANSN. Hole H7	Shaft n6	Fit	LOCATIONAL INTERF. Hole H7	Shaft p6	Fit	MEDIUM DRIVE Hole H7	Shaft s6	Fit	FORCE Hole H7	Shaft u6	Fit
1	MAX	1.010	1.006	0.010	1.010	1.010	0.006	1.010	1.012	0.004	1.010	1.020	-0.004	1.010	1.024	-0.008
	MIN	1.000	1.000	-0.006	1.000	1.004	-0.010	1.000	1.006	-0.012	1.000	1.014	-0.020	1.000	1.018	-0.024
1.2	MAX	1.210	1.206	0.010	1.210	1.210	0.006	1.210	1.212	0.004	1.210	1.220	-0.004	1.210	1.224	-0.008
	MIN	1.200	1.200	-0.006	1.200	1.204	-0.010	1.200	1.206	-0.012	1.200	1.214	-0.020	1.200	1.218	-0.024
1.6	MAX	1.610	1.606	0.010	1.610	1.610	0.006	1.610	1.612	0.004	1.610	1.620	-0.004	1.610	1.624	-0.008
	MIN	1.600	1.600	-0.006	1.600	1.604	-0.010	1.600	1.606	-0.012	1.600	1.614	-0.020	1.600	1.618	-0.024
2	MAX	2.010	2.006	0.010	2.010	2.010	0.006	2.010	2.012	0.004	2.010	2.020	-0.004	2.010	2.024	-0.008
	MIN	2.000	2.000	-0.006	2.000	2.004	-0.010	2.000	2.006	-0.012	2.000	2.014	-0.020	2.000	2.018	-0.024
2.5	MAX	2.510	2.506	0.010	2.510	2.510	0.006	2.510	2.512	0.004	2.510	2.520	-0.004	2.510	2.524	-0.008
	MIN	2.500	2.500	-0.006	2.500	2.504	-0.010	2.500	2.506	-0.012	2.500	2.514	-0.020	2.500	2.518	-0.024
3	MAX	3.010	3.006	0.010	3.010	3.010	0.006	3.010	3.012	0.004	3.010	3.020	-0.004	3.010	3.024	-0.008
	MIN	3.000	3.000	-0.006	3.000	3.004	-0.010	3.000	3.006	-0.012	3.000	3.014	-0.020	3.000	3.018	-0.024
4	MAX	4.012	4.009	0.011	4.012	4.016	0.004	4.012	4.020	0.000	4.012	4.027	-0.007	4.012	4.031	-0.011
	MIN	4.000	4.001	-0.009	4.000	4.008	-0.016	4.000	4.012	-0.020	4.000	4.019	-0.027	4.000	4.023	-0.031
5	MAX	5.012	5.009	0.011	5.012	5.016	0.004	5.012	5.020	0.000	5.012	5.027	-0.007	5.012	5.031	-0.011
	MIN	5.000	5.001	-0.009	5.000	5.008	-0.016	5.000	5.012	-0.020	5.000	5.019	-0.027	5.000	5.023	-0.031
6	MAX	6.012	6.009	0.011	6.012	6.016	0.004	6.012	6.020	0.000	6.012	6.027	-0.007	6.012	6.031	-0.011
	MIN	6.000	6.001	-0.009	6.000	6.008	-0.016	6.000	6.012	-0.020	6.000	6.019	-0.027	6.000	6.023	-0.031
8	MAX	8.015	8.010	0.014	8.015	8.019	0.005	8.015	8.024	0.000	8.015	8.032	-0.008	8.015	8.037	-0.013
	MIN	8.000	8.001	-0.010	8.000	8.010	-0.019	8.000	8.015	-0.024	8.000	8.023	-0.032	8.000	8.028	-0.037
10	MAX	10.015	10.010	0.014	10.015	10.019	0.005	10.015	10.024	0.000	10.015	10.032	-0.008	10.015	10.037	-0.013
	MIN	10.000	10.001	-0.010	10.000	10.010	-0.019	10.000	10.015	-0.024	10.000	10.023	-0.032	10.000	10.028	-0.037
12	MAX	12.018	12.012	0.017	12.018	12.023	0.006	12.018	12.029	0.000	12.018	12.039	-0.010	12.018	12.044	-0.015
	MIN	12.000	12.001	-0.012	12.000	12.012	-0.023	12.000	12.018	-0.029	12.000	12.028	-0.039	12.000	12.033	-0.044
16	MAX	16.018	16.012	0.017	16.018	16.023	0.006	16.018	16.029	0.000	16.018	16.039	-0.010	16.018	16.044	-0.015
	MIN	16.000	16.001	-0.012	16.000	16.012	-0.023	16.000	16.018	-0.029	16.000	16.028	-0.039	16.000	16.033	-0.044
20	MAX	20.021	20.015	0.019	20.021	20.028	0.006	20.021	20.035	-0.001	20.021	20.048	-0.014	20.021	20.054	-0.020
	MIN	20.000	20.002	-0.015	20.000	20.015	-0.028	20.000	20.022	-0.035	20.000	20.035	-0.048	20.000	20.041	-0.054
25	MAX	25.021	25.015	0.019	25.021	25.028	0.006	25.021	25.035	-0.001	25.021	25.048	-0.014	25.021	25.061	-0.027
	MIN	25.000	25.002	-0.015	25.000	25.015	-0.028	25.000	25.022	-0.035	25.000	25.035	-0.048	25.000	25.048	-0.061
30	MAX	30.021	30.015	0.019	30.021	30.028	0.006	30.021	30.035	-0.001	30.021	30.048	-0.014	30.021	30.061	-0.027
	MIN	30.000	30.002	-0.015	30.000	30.015	-0.028	30.000	30.022	-0.035	30.000	30.035	-0.048	30.000	30.048	-0.061

TABLE 28 (cont'd)

Dimensions in mm.

BASIC SIZE		LOCATIONAL TRANSN. Hole H7	Shaft k6	Fit	LOCATIONAL TRANSN. Hole H7	Shaft n6	Fit	LOCATIONAL INTERF. Hole H7	Shaft p6	Fit	MEDIUM DRIVE Hole H7	Shaft s6	Fit	FORCE Hole H7	Shaft u6	Fit
40	MAX	40.025	40.018	0.023	40.025	40.033	0.008	40.025	40.042	-0.001	40.025	40.059	-0.018	40.025	40.076	-0.035
	MIN	40.000	40.002	-0.018	40.000	40.017	-0.033	40.000	40.026	-0.042	40.000	40.043	-0.059	40.000	40.060	-0.076
50	MAX	50.025	50.018	0.023	50.025	50.033	0.008	50.025	50.042	-0.001	50.025	50.059	-0.018	50.025	50.086	-0.045
	MIN	50.000	50.002	-0.018	50.000	50.017	-0.033	50.000	50.026	-0.042	50.000	50.043	-0.059	50.000	50.070	-0.086
60	MAX	60.030	60.021	0.028	60.030	60.039	0.010	60.030	60.051	-0.002	60.030	60.072	-0.023	60.030	60.106	-0.057
	MIN	60.000	60.002	-0.021	60.000	60.020	-0.039	60.000	60.032	-0.051	60.000	60.053	-0.072	60.000	60.087	-0.106
80	MAX	80.030	80.021	0.028	80.030	80.039	0.010	80.030	80.051	-0.002	80.030	80.078	-0.029	80.030	80.121	-0.072
	MIN	80.000	80.002	-0.021	80.000	80.020	-0.039	80.000	80.032	-0.051	80.000	80.059	-0.078	80.000	80.102	-0.121
100	MAX	100.035	100.025	0.032	100.035	100.045	0.012	100.035	100.059	-0.002	100.035	100.093	-0.036	100.035	100.146	-0.089
	MIN	100.000	100.003	-0.025	100.000	100.023	-0.045	100.000	100.037	-0.059	100.000	100.071	-0.093	100.000	100.124	-0.146
120	MAX	120.035	120.025	0.032	120.035	120.045	0.012	120.035	120.059	-0.002	120.035	120.101	-0.044	120.035	120.166	-0.109
	MIN	120.000	120.003	-0.025	120.000	120.023	-0.045	120.000	120.037	-0.059	120.000	120.079	-0.101	120.000	120.144	-0.166
160	MAX	160.040	160.028	0.037	160.040	160.052	0.013	160.040	160.068	-0.003	160.040	160.125	-0.060	160.040	160.215	-0.150
	MIN	160.000	160.003	-0.028	160.000	160.027	-0.052	160.000	160.043	-0.068	160.000	160.100	-0.125	160.000	160.190	-0.215
200	MAX	200.046	200.033	0.042	200.046	200.060	0.015	200.046	200.079	-0.004	200.046	200.151	-0.076	200.046	200.265	-0.190
	MIN	200.000	200.004	-0.033	200.000	200.031	-0.060	200.000	200.050	-0.079	200.000	200.122	-0.151	200.000	200.236	-0.265
250	MAX	250.046	250.033	0.042	250.046	250.060	0.015	250.046	250.079	-0.004	250.046	250.169	-0.094	250.046	250.313	-0.238
	MIN	250.000	250.004	-0.033	250.000	250.031	-0.060	250.000	250.050	-0.079	250.000	250.140	-0.169	250.000	250.284	-0.313
300	MAX	300.052	300.036	0.048	300.052	300.066	0.018	300.052	300.088	-0.004	300.052	300.202	-0.118	300.052	300.382	-0.298
	MIN	300.000	300.004	-0.036	300.000	300.034	-0.066	300.000	300.056	-0.088	300.000	300.170	-0.202	300.000	300.350	-0.382
400	MAX	400.057	400.040	0.053	400.057	400.073	0.020	400.057	400.098	-0.005	400.057	400.244	-0.151	400.057	400.471	-0.378
	MIN	400.000	400.004	-0.040	400.000	400.037	-0.073	400.000	400.062	-0.098	400.000	400.208	-0.244	400.000	400.435	-0.471
500	MAX	500.063	500.045	0.058	500.063	500.080	0.023	500.063	500.108	-0.005	500.063	500.292	-0.189	500.063	500.580	-0.477
	MIN	500.000	500.005	-0.045	500.000	500.040	-0.080	500.000	500.068	-0.108	500.000	500.252	-0.292	500.000	500.540	-0.580

TABLE 29 PREFERRED SHAFT BASIS CLEARANCE FITS (METRIC) — ANSI B4.2-1978

Dimensions in mm.

BASIC SIZE		LOOSE RUNNING Hole C11	LOOSE RUNNING Shaft h11	Fit	FREE RUNNING Hole D9	FREE RUNNING Shaft h9	Fit	CLOSE RUNNING Hole F8	CLOSE RUNNING Shaft h7	Fit	SLIDING Hole G7	SLIDING Shaft h6	Fit	LOCATIONAL CLEARANCE Hole H7	LOCATIONAL CLEARANCE Shaft h6	Fit
1	MAX	1.120	1.000	0.180	1.045	1.000	0.070	1.020	1.000	0.030	1.012	1.000	0.018	1.010	1.000	0.016
	MIN	1.060	0.940	0.060	1.020	0.975	0.020	1.006	0.990	0.006	1.002	0.994	0.002	1.000	0.994	0.000
1.2	MAX	1.320	1.200	0.180	1.245	1.200	0.070	1.220	1.200	0.030	1.212	1.200	0.018	1.210	1.200	0.016
	MIN	1.260	1.140	0.060	1.220	1.175	0.020	1.206	1.190	0.006	1.202	1.194	0.002	1.200	1.194	0.000
1.6	MAX	1.720	1.600	0.180	1.645	1.600	0.070	1.620	1.600	0.030	1.612	1.600	0.018	1.610	1.600	0.016
	MIN	1.660	1.540	0.060	1.620	1.575	0.020	1.606	1.590	0.006	1.602	1.594	0.002	1.600	1.594	0.000
2	MAX	2.120	2.000	0.180	2.045	2.000	0.070	2.020	2.000	0.030	2.012	2.000	0.018	2.010	2.000	0.016
	MIN	2.060	1.940	0.060	2.020	1.975	0.020	2.006	1.990	0.006	2.002	1.994	0.002	2.000	1.994	0.000
2.5	MAX	2.620	2.500	0.180	2.545	2.500	0.070	2.520	2.500	0.030	2.512	2.500	0.018	2.510	2.500	0.016
	MIN	2.560	2.440	0.060	2.520	2.475	0.020	2.506	2.490	0.006	2.502	2.494	0.002	2.500	2.494	0.000
3	MAX	3.120	3.000	0.180	3.045	3.000	0.070	3.020	3.000	0.030	3.012	3.000	0.018	3.010	3.000	0.016
	MIN	3.060	2.940	0.060	3.020	2.975	0.020	3.006	2.990	0.006	3.002	2.994	0.002	3.000	2.994	0.000
4	MAX	4.145	4.000	0.220	4.060	4.000	0.090	4.028	4.000	0.040	4.016	4.000	0.024	4.012	4.000	0.020
	MIN	4.070	3.925	0.070	4.030	3.970	0.030	4.010	3.988	0.010	4.004	3.992	0.004	4.000	3.992	0.000
5	MAX	5.145	5.000	0.220	5.060	5.000	0.090	5.028	5.000	0.040	5.016	5.000	0.024	5.012	5.000	0.020
	MIN	5.070	4.925	0.070	5.030	4.970	0.030	5.010	4.988	0.010	5.004	4.992	0.004	5.000	4.992	0.000
6	MAX	6.145	6.000	0.220	6.060	6.000	0.090	6.028	6.000	0.040	6.016	6.000	0.024	6.012	6.000	0.020
	MIN	6.070	5.925	0.070	6.030	5.970	0.030	6.010	5.988	0.010	6.004	5.992	0.004	6.000	5.992	0.000
8	MAX	8.170	8.000	0.260	8.076	8.000	0.112	8.035	8.000	0.050	8.020	8.000	0.029	8.015	8.000	0.024
	MIN	8.080	7.910	0.080	8.040	7.964	0.040	8.013	7.985	0.013	8.005	7.991	0.005	8.000	7.991	0.000
10	MAX	10.170	10.000	0.260	10.076	10.000	0.112	10.035	10.000	0.050	10.020	10.000	0.029	10.015	10.000	0.024
	MIN	10.080	9.910	0.080	10.040	9.964	0.040	10.013	9.985	0.013	10.005	9.991	0.005	10.000	9.991	0.000
12	MAX	12.205	12.000	0.315	12.093	12.000	0.136	12.043	12.000	0.061	12.024	12.000	0.035	12.018	12.000	0.029
	MIN	12.095	11.890	0.095	12.050	11.957	0.050	12.016	11.982	0.016	12.006	11.989	0.006	12.000	11.989	0.000
16	MAX	16.205	16.000	0.315	16.093	16.000	0.136	16.043	16.000	0.061	16.024	16.000	0.035	16.018	16.000	0.029
	MIN	16.095	15.890	0.095	16.050	15.957	0.050	16.016	15.982	0.016	16.006	15.989	0.006	16.000	15.989	0.000
20	MAX	20.240	20.000	0.370	20.117	20.000	0.169	20.053	20.000	0.074	20.028	20.000	0.041	20.021	20.000	0.034
	MIN	20.110	19.870	0.110	20.065	19.948	0.065	20.020	19.979	0.020	20.007	19.987	0.007	20.000	19.987	0.000
25	MAX	25.240	25.000	0.370	25.117	25.000	0.169	25.053	25.000	0.074	25.028	25.000	0.041	25.021	25.000	0.034
	MIN	25.110	24.870	0.110	25.065	24.948	0.065	25.020	24.979	0.020	25.007	24.987	0.007	25.000	24.987	0.000
30	MAX	30.240	30.000	0.370	30.117	30.000	0.169	30.053	30.000	0.074	30.028	30.000	0.041	30.021	30.000	0.034
	MIN	30.110	29.870	0.110	30.065	29.948	0.065	30.020	29.979	0.020	30.007	29.987	0.007	30.000	29.987	0.000

TABLE 29 (cont'd)

Dimensions in mm.

BASIC SIZE		LOOSE RUNNING			FREE RUNNING			CLOSE RUNNING			SLIDING			LOCATIONAL CLEARANCE		
		Hole C11	Shaft h11	Fit	Hole D9	Shaft h9	Fit	Hole F8	Shaft h7	Fit	Hole G7	Shaft h6	Fit	Hole H7	Shaft h6	Fit
40	MAX	40.280	40.000	0.440	40.142	40.000	0.204	40.064	40.000	0.089	40.034	40.000	0.050	40.025	40.000	0.041
	MIN	40.120	39.840	0.120	40.080	39.938	0.080	40.025	39.975	0.025	40.009	39.984	0.009	40.000	39.984	0.000
50	MAX	50.290	50.000	0.450	50.142	50.000	0.204	50.064	50.000	0.089	50.034	50.000	0.050	50.025	50.000	0.041
	MIN	50.130	49.840	0.130	50.080	49.938	0.080	50.025	49.975	0.025	50.009	49.984	0.009	50.000	49.984	0.000
60	MAX	60.330	60.000	0.520	60.174	60.000	0.248	60.076	60.000	0.106	60.040	60.000	0.059	60.030	60.000	0.049
	MIN	60.140	59.810	0.140	60.100	59.926	0.100	60.030	59.970	0.030	60.010	59.981	0.010	60.000	59.981	0.000
80	MAX	80.340	80.000	0.530	80.174	80.000	0.248	80.076	80.000	0.106	80.040	80.000	0.059	80.030	80.000	0.049
	MIN	80.150	79.810	0.150	80.100	79.926	0.100	80.030	79.970	0.030	80.010	79.981	0.010	80.000	79.981	0.000
100	MAX	100.390	100.000	0.610	100.207	100.000	0.294	100.090	100.000	0.125	100.047	100.000	0.069	100.035	100.000	0.057
	MIN	100.170	99.780	0.170	100.120	99.913	0.120	100.036	99.965	0.036	100.012	99.978	0.012	100.000	99.978	0.000
120	MAX	120.400	120.000	0.620	120.207	120.000	0.294	120.090	120.000	0.125	120.047	120.000	0.069	120.035	120.000	0.057
	MIN	120.180	119.780	0.180	120.120	119.913	0.120	120.036	119.965	0.036	120.012	119.978	0.012	120.000	119.978	0.000
160	MAX	160.460	160.000	0.710	160.245	160.000	0.345	160.106	160.000	0.146	160.054	160.000	0.079	160.040	160.000	0.065
	MIN	160.210	159.750	0.210	160.145	159.900	0.145	160.043	159.960	0.043	160.014	159.975	0.014	160.000	159.975	0.000
200	MAX	200.530	200.000	0.820	200.285	200.000	0.400	200.122	200.000	0.168	200.061	200.000	0.090	200.046	200.000	0.075
	MIN	200.240	199.710	0.240	200.170	199.885	0.170	200.050	199.954	0.050	200.015	199.971	0.015	200.000	199.971	0.000
250	MAX	250.570	250.000	0.860	250.285	250.000	0.400	250.122	250.000	0.168	250.061	250.000	0.090	250.046	250.000	0.075
	MIN	250.280	249.710	0.280	250.170	249.885	0.170	250.050	249.954	0.050	250.015	249.971	0.015	250.000	249.971	0.000
300	MAX	300.650	300.000	0.970	300.320	300.000	0.450	300.137	300.000	0.189	300.069	300.000	0.101	300.052	300.000	0.084
	MIN	300.330	299.680	0.330	300.190	299.870	0.190	300.056	299.948	0.056	300.017	299.968	0.017	300.000	299.968	0.000
400	MAX	400.760	400.000	1.120	400.350	400.000	0.490	400.151	400.000	0.208	400.075	400.000	0.111	400.057	400.000	0.093
	MIN	400.400	399.640	0.400	400.210	399.860	0.210	400.062	399.943	0.062	400.018	399.964	0.018	400.000	399.964	0.000
500	MAX	500.880	500.000	1.280	500.385	500.000	0.540	500.165	500.000	0.228	500.083	500.000	0.123	500.063	500.000	0.103
	MIN	500.480	499.600	0.480	500.230	499.845	0.230	500.068	499.937	0.068	500.020	499.960	0.020	500.000	499.960	0.000

TABLE 30 PREFERRED SHAFT BASIS TRANSITION AND INTERFERENCE FITS (METRIC)

Dimensions in mm.

BASIC SIZE		LOCATIONAL TRANSN. Hole K7	LOCATIONAL TRANSN. Shaft h6	LOCATIONAL TRANSN. Fit	LOCATIONAL TRANSN. Hole N7	LOCATIONAL TRANSN. Shaft h6	LOCATIONAL TRANSN. Fit	LOCATIONAL INTERF. Hole P7	LOCATIONAL INTERF. Shaft h6	LOCATIONAL INTERF. Fit	MEDIUM DRIVE Hole S7	MEDIUM DRIVE Shaft h6	MEDIUM DRIVE Fit	FORCE Hole U7	FORCE Shaft h6	FORCE Fit
1	MAX	1.000	1.000	0.006	0.996	1.000	0.002	0.994	1.000	0.000	0.986	1.000	-0.008	0.982	1.000	-0.012
	MIN	0.990	0.994	-0.010	0.986	0.994	-0.014	0.984	0.994	-0.016	0.976	0.994	-0.024	0.972	0.994	-0.028
1.2	MAX	1.200	1.200	0.006	1.196	1.200	0.002	1.194	1.200	0.000	1.186	1.200	-0.008	1.182	1.200	-0.012
	MIN	1.190	1.194	-0.010	1.186	1.194	-0.014	1.184	1.194	-0.016	1.176	1.194	-0.024	1.172	1.194	-0.028
1.6	MAX	1.600	1.600	0.006	1.596	1.600	0.002	1.594	1.600	0.000	1.586	1.600	-0.008	1.582	1.600	-0.012
	MIN	1.590	1.594	-0.010	1.586	1.594	-0.014	1.584	1.594	-0.016	1.576	1.594	-0.024	1.572	1.594	-0.028
2	MAX	2.000	2.000	0.006	1.996	2.000	0.002	1.994	2.000	0.000	1.986	2.000	-0.008	1.982	2.000	-0.012
	MIN	1.990	1.994	-0.010	1.986	1.994	-0.014	1.984	1.994	-0.016	1.976	1.994	-0.024	1.972	1.994	-0.028
2.5	MAX	2.500	2.500	0.006	2.496	2.500	0.002	2.494	2.500	0.000	2.486	2.500	-0.008	2.482	2.500	-0.012
	MIN	2.490	2.494	-0.010	2.486	2.494	-0.014	2.484	2.494	-0.016	2.476	2.494	-0.024	2.472	2.494	-0.028
3	MAX	3.000	3.000	0.006	2.996	3.000	0.002	2.994	3.000	0.000	2.986	3.000	-0.008	2.982	3.000	-0.012
	MIN	2.990	2.994	-0.010	2.986	2.994	-0.014	2.984	2.994	-0.016	2.976	2.994	-0.024	2.972	2.994	-0.028
4	MAX	4.003	4.000	0.011	3.996	4.000	0.004	3.992	4.000	0.000	3.985	4.000	-0.007	3.981	4.000	-0.011
	MIN	3.991	3.992	-0.009	3.984	3.992	-0.016	3.980	3.992	-0.020	3.973	3.992	-0.027	3.969	3.992	-0.031
5	MAX	5.003	5.000	0.011	4.996	5.000	0.004	4.992	5.000	0.000	4.985	5.000	-0.007	4.981	5.000	-0.011
	MIN	4.991	4.992	-0.009	4.984	4.992	-0.016	4.980	4.992	-0.020	4.973	4.992	-0.027	4.969	4.992	-0.031
6	MAX	6.003	6.000	0.011	5.996	6.000	0.004	5.992	6.000	0.000	5.985	6.000	-0.007	5.981	6.000	-0.011
	MIN	5.991	5.992	-0.009	5.984	5.992	-0.016	5.980	5.992	-0.020	5.973	5.992	-0.027	5.969	5.992	-0.031
8	MAX	8.005	8.000	0.014	7.996	8.000	0.005	7.991	8.000	0.000	7.983	8.000	-0.008	7.978	8.000	-0.013
	MIN	7.990	7.991	-0.010	7.981	7.991	-0.019	7.976	7.991	-0.024	7.968	7.991	-0.032	7.963	7.991	-0.037
10	MAX	10.005	10.000	0.014	9.996	10.000	0.005	9.991	10.000	0.000	9.983	10.000	-0.008	9.978	10.000	-0.013
	MIN	9.990	9.991	-0.010	9.981	9.991	-0.019	9.976	9.991	-0.024	9.968	9.991	-0.032	9.963	9.991	-0.037
12	MAX	12.006	12.000	0.017	11.995	12.000	0.006	11.989	12.000	0.000	11.979	12.000	-0.010	11.974	12.000	-0.015
	MIN	11.988	11.989	-0.012	11.977	11.989	-0.023	11.971	11.989	-0.029	11.961	11.989	-0.039	11.956	11.989	-0.044
16	MAX	16.006	16.000	0.017	15.995	16.000	0.006	15.989	16.000	0.000	15.979	16.000	-0.010	15.974	16.000	-0.015
	MIN	15.988	15.989	-0.012	15.977	15.989	-0.023	15.971	15.989	-0.029	15.961	15.989	-0.039	15.956	15.989	-0.044
20	MAX	20.006	20.000	0.019	19.993	20.000	0.006	19.986	20.000	-0.001	19.973	20.000	-0.014	19.967	20.000	-0.020
	MIN	19.985	19.987	-0.015	19.972	19.987	-0.028	19.965	19.987	-0.035	19.952	19.987	-0.048	19.946	19.987	-0.054
25	MAX	25.006	25.000	0.019	24.993	25.000	0.006	24.986	25.000	-0.001	24.973	25.000	-0.014	24.960	25.000	-0.027
	MIN	24.985	24.987	-0.015	24.972	24.987	-0.028	24.965	24.987	-0.035	24.952	24.987	-0.048	24.939	24.987	-0.061
30	MAX	30.006	30.000	0.019	29.993	30.000	0.006	29.986	30.000	-0.001	29.973	30.000	-0.014	29.960	30.000	-0.027
	MIN	29.985	29.987	-0.015	29.972	29.987	-0.028	29.965	29.987	-0.035	29.952	29.987	-0.048	29.939	29.987	-0.061

TABLE 30 *(cont'd)*

Dimensions in mm.

BASIC SIZE		LOCATIONAL TRANSN. Hole K7	Shaft h6	Fit	LOCATIONAL TRANSN. Hole N7	Shaft h6	Fit	LOCATIONAL INTERF. Hole P7	Shaft h6	Fit	MEDIUM DRIVE Hole S7	Shaft h6	Fit	FORCE Hole U7	Shaft h6	Fit
40	MAX	40.007	40.000	0.023	39.992	40.000	0.008	39.983	40.000	-0.001	39.966	40.000	-0.018	39.949	40.000	-0.035
	MIN	39.982	39.984	-0.018	39.967	39.984	-0.033	39.958	39.984	-0.042	39.941	39.984	-0.059	39.924	39.984	-0.076
50	MAX	50.007	50.000	0.023	49.992	50.000	0.008	49.983	50.000	-0.001	49.966	50.000	-0.018	49.939	50.000	-0.045
	MIN	49.982	49.984	-0.018	49.967	49.984	-0.033	49.958	49.984	-0.042	49.941	49.984	-0.059	49.914	49.984	-0.086
60	MAX	60.009	60.000	0.028	59.991	60.000	0.010	59.979	60.000	-0.002	59.958	60.000	-0.023	59.924	60.000	-0.057
	MIN	59.979	59.981	-0.021	59.961	59.981	-0.039	59.949	59.981	-0.051	59.928	59.981	-0.072	59.894	59.981	-0.106
80	MAX	80.009	80.000	0.028	79.991	80.000	0.010	79.979	80.000	-0.002	79.952	80.000	-0.029	79.909	80.000	-0.072
	MIN	79.979	79.981	-0.021	79.961	79.981	-0.039	79.949	79.981	-0.051	79.922	79.981	-0.078	79.879	79.981	-0.121
100	MAX	100.010	100.000	0.032	99.990	100.000	0.012	99.976	100.000	-0.002	99.942	100.000	-0.036	99.889	100.000	-0.089
	MIN	99.975	99.978	-0.025	99.955	99.978	-0.045	99.941	99.978	-0.059	99.907	99.978	-0.093	99.854	99.978	-0.146
120	MAX	120.010	120.000	0.032	119.990	120.000	0.012	119.976	120.000	-0.002	119.934	120.000	-0.044	119.869	120.000	-0.109
	MIN	119.975	119.978	-0.025	119.955	119.978	-0.045	119.941	119.978	-0.059	119.899	119.978	-0.101	119.834	119.978	-0.166
160	MAX	160.012	160.000	0.037	159.988	160.000	0.013	159.972	160.000	-0.003	159.915	160.000	-0.060	159.825	160.000	-0.150
	MIN	159.972	159.975	-0.028	159.948	159.975	-0.052	159.932	159.975	-0.068	159.875	159.975	-0.125	159.785	159.975	-0.215
200	MAX	200.013	200.000	0.042	199.986	200.000	0.015	199.967	200.000	-0.004	199.895	200.000	-0.076	199.781	200.000	-0.190
	MIN	199.967	199.971	-0.033	199.940	199.971	-0.060	199.921	199.971	-0.079	199.849	199.971	-0.151	199.735	199.971	-0.265
250	MAX	250.013	250.000	0.042	249.986	250.000	0.015	249.967	250.000	-0.004	249.877	250.000	-0.094	249.733	250.000	-0.238
	MIN	249.967	249.971	-0.033	249.940	249.971	-0.060	249.921	249.971	-0.079	249.831	249.971	-0.169	249.687	249.971	-0.313
300	MAX	300.016	300.000	0.048	299.986	300.000	0.018	299.964	300.000	-0.004	299.850	300.000	-0.118	299.670	300.000	-0.298
	MIN	299.964	299.968	-0.036	299.934	299.968	-0.066	299.912	299.968	-0.088	299.798	299.968	-0.202	299.618	299.968	-0.382
400	MAX	400.017	400.000	0.053	399.984	400.000	0.020	399.959	400.000	-0.005	399.813	400.000	-0.151	399.586	400.000	-0.378
	MIN	399.960	399.964	-0.040	399.927	399.964	-0.073	399.902	399.964	-0.098	399.756	399.964	-0.244	399.529	399.964	-0.471
500	MAX	500.018	500.000	0.058	499.983	500.000	0.023	499.955	500.000	-0.005	499.771	500.000	-0.189	499.483	500.000	-0.477
	MIN	499.955	499.960	-0.045	499.920	499.960	-0.080	499.892	499.960	-0.108	499.708	499.960	-0.292	499.420	499.960	-0.580

TABLE 31 DIMENSIONS OF PLANE WASHERS

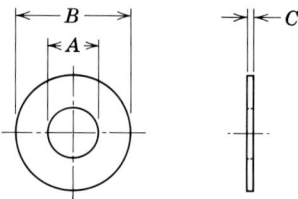

Screw or bolt size	Series	Inside diameter A Nom	Outside diameter B Nom	Thickness C Nom	Screw or bolt size	Series	Inside diameter A Nom	Outside diameter B Nom	Thickness C Nom
5/16	Narrow Regular Wide	0.344	5/8 7/8 1 1/8	0.063 0.063 0.063	1 3/8	Narrow Regular Wide	1 7/16	2 1/2 3 1/4 3 3/4	0.160 0.160 0.250
3/8	Narrow Regular Wide	0.406	47/64 1 1 1/4	0.063 0.063 0.100	1 1/2	Narrow Regular Wide	1 9/16	2 3/4 3 1/2 4	0.160 0.250 0.250
7/16	Narrow Regular Wide	0.480	7/8 1 1/8 1 15/32	0.063 0.063 0.100	1 5/8	Narrow Regular Wide	1 3/4	3 3 3/4 4 1/4	0.160 0.250 0.250
1/2	Narrow Regular Wide	0.540	1 1 1/4 1 3/4	0.063 0.100 0.100	1 3/4	Narrow Regular Wide	1 7/8	3 1/4 4 4 1/2	0.160 0.250 0.250
9/16	Narrow Regular Wide	0.604	1 1/8 1 15/32 2	0.063 0.100 0.100	1 7/8	Narrow Regular Wide	2	3 1/2 4 1/4 4 3/4	0.250 0.250 0.250
5/8	Narrow Regular Wide	0.666	1 1/4 1 3/4 2 1/4	0.100 0.100 0.160	2	Narrow Regular Wide	2 1/8	3 3/4 4 1/2 5	0.250 0.250 0.250
3/4	Narrow Regular Wide	0.812	1 3/8 2 2 1/2	0.100 0.100 0.160	2 1/4	Narrow Regular Wide	2 3/8	4 5 5 1/2	0.250 0.250 0.375
7/8	Narrow Regular Wide	0.938	1 15/32 2 1/4 2 3/4	0.100 0.160 0.160	2 1/2	Narrow Regular Wide	2 5/8	4 1/2 5 1/2 6	0.250 0.375 0.375
1	Narrow Regular Wide	1 1/16	1 3/4 2 1/2 3	0.100 0.160 0.160	2 3/4	Narrow Regular Wide	2 7/8	5 6 6 1/2	0.250 0.375 0.375
1 1/8	Narrow Regular Wide	1 3/16	2 2 3/4 3 1/4	0.100 0.160 0.160	3	Narrow Regular Wide	3 1/8	5 1/2 6 1/2 7	0.375 0.375 0.375
1 1/4	Narrow Regular Wide	1 5/16	2 1/4 3 3 1/2	0.160 0.160 0.250					

All dimensions are given in inches.
Inside and outside diameters shall be concentric within at least the inside diameter tolerance.
Washers shall be flat within 0.005 for outside diameters up to and including 7/8 inch, and 0.010 for outside diameters greater than 7/8 inch.
Tolerance on inside diameter 5/16 to 7/8 is −0.010.
Tolerance on inside diameter 1 to 3 is ±0.010.
Tolerance on outside diameter ±0.010.

Courtesy USASI

TABLE 32 AMERICAN STANDARD STEEL PIPE DATA
(All dimensions in inches. Weights in pounds.)

Nominal size	Actual outside diameter	Standard weight (40)*			Extra-strong (80)†			Double-extra strong‡		
		Inside diameter	Wall thickness	Weight per foot§	Inside diameter	Wall thickness	Weight per foot	Inside diameter	Wall thickness	Weight per foot
1/8	0.405	0.269	0.068	0.244	0.215	0.095	0.314	.....	.....	.....
1/4	0.540	0.364	.088	0.424	0.302	.119	0.535	.....	.....	.....
3/8	0.675	0.493	.091	0.567	0.423	.126	0.738	.....	.....	.....
1/2	0.840	0.622	.109	0.850	0.546	.147	1.087	0.252	0.294	1.714
3/4	1.050	0.824	.113	1.130	0.742	.154	1.473	0.434	.308	2.440
1	1.315	1.049	.133	1.678	0.957	.179	2.171	0.599	.358	3.659
1¼	1.660	1.380	.140	2.272	1.278	.191	2.996	0.896	.382	5.214
1½	1.900	1.610	.145	2.717	1.500	.200	3.631	1.100	.400	6.408
2	2.375	2.067	.154	3.652	1.939	.218	5.022	1.503	.436	9.029
2½	2.875	2.469	.203	5.79	2.323	.276	7.66	1.771	.552	13.70
3	3.500	3.068	.216	7.58	2.900	.300	10.25	2.300	.600	18.58
3½	4.000	3.548	.226	9.11	3.364	.318	12.51	.....	.....	.....
4	4.500	4.026	.237	10.79	3.826	.337	14.98	3.152	.674	27.54
5	5.563	5.047	.258	14.62	4.813	.375	20.78	4.063	.750	38.55
6	6.625	6.065	.280	18.97	5.761	.432	28.57	4.897	.864	53.16
8	8.625	7.981	.322	28.55	7.625	.500	43.39	6.875	.875	72.42
10	10.750	10.020	.365	40.48	9.750	.500	54.74	.....	.....	.....
12	12.750	12.000	.375	49.56	11.750	.500	65.42	.....	.....	.....

* Same as **USASI** B36.10—"Schedule 40" except 12-inch diameter.
† Same as **USASI** B36.10—"Schedule 80" except 10- and 12-inch diameter.
‡ Not identified with **USASI** Schedule number, but available as indicated.
§ Plain ends.

TABLE 33 ASTM STANDARD BRASS AND COPPER PIPE DATA
(All dimensions in inches. Weights in pounds.)

Nominal size	Outside diameter	Regular			Extra-strong		
		Wall thickness	Weight per foot		Wall thickness	Weight per foot	
			Red brass	Copper		Red brass	Copper
1/8	0.405	0.062	0.253	0.259	0.100	0.363	0.371
1/4	0.540	0.082	0.447	0.457	0.123	0.611	0.625
3/8	0.675	0.090	0.627	0.641	0.127	0.829	0.847
1/2	0.840	0.107	0.934	0.955	0.149	1.23	1.25
3/4	1.050	0.114	1.27	1.30	0.157	1.67	1.71
1	1.315	0.126	1.78	1.82	0.182	2.46	2.51
1¼	1.660	0.146	2.63	2.69	0.194	3.39	3.46
1½	1.900	0.150	3.13	3.20	0.203	4.10	4.19
2	2.375	0.156	4.12	4.22	0.221	5.67	5.80
2½	2.875	0.187	5.99	6.12	0.280	8.66	8.85
3	3.500	0.219	8.56	8.75	0.304	11.6	11.8
3½	4.000	0.250	11.2	11.4	0.321	14.1	14.4
4	4.500	0.250	12.7	12.9	0.341	16.9	17.3
5	5.562	0.250	15.8	16.2	0.375	23.2	23.7
6	6.625	0.250	19.0	19.4	0.437	32.2	32.9
8	8.625	0.312	30.9	31.6	0.500	48.4	49.5
10	10.750	9.365	45.2	46.2	0.500	61.1	62.4
12	12.750	0.375	55.3	56.5			

TABLE 34 ASTM STANDARD COPPER WATER TUBE DATA
(All dimensions in inches. Weights in pounds.)

Nominal size	Outside diameter	Type K		Type L		Type M	
		Wall thickness	Weight per foot	Wall thickness	Weight per foot	Wall thickness	Weight per foot
1/8	0.250	0.032	0.085	0.025	0.068	0.025	0.068
1/4	0.375	0.032	0.134	0.030	0.126	0.025	0.107
3/8	0.500	0.049	0.269	0.035	0.198	0.025	0.145
1/2	0.625	0.049	0.344	0.040	0.285	0.028	0.204
5/8	0.750	0.049	0.418	0.042	0.362	0.030	0.263
3/4	0.875	0.065	0.641	0.045	0.455	0.032	0.328
1	1.125	0.065	0.839	0.050	0.655	0.035	0.465
1¼	1.375	0.065	1.04	0.055	0.884	0.042	0.682
1½	1.625	0.072	1.36	0.060	1.14	0.049	0.940
2	2.125	0.083	2.06	0.070	1.75	0.058	1.46
2½	2.625	0.095	2.93	0.080	2.48	0.065	2.03
3	3.125	0.109	4.00	0.090	3.33	0.072	2.68
3½	3.625	0.120	5.12	0.100	4.29	0.083	3.58
4	4.125	0.134	6.51	0.110	5.38	0.095	4.66
5	5.125	0.160	9.67	0.125	7.61	0.109	6.66
6	6.125	0.192	13.9	0.140	10.2	0.122	8.92
8	8.125	0.271	25.9	0.200	19.3	0.170	16.5
10	10.125	0.338	40.3	0.250	30.1	0.212	25.6
12	12.125	0.405	57.8	0.280	40.4	0.254	36.7

TABLE 35 AMERICAN STANDARD TAPER PIPE THREAD DATA
(All dimensions in inches. Weights in pounds.)

Nominal size	Outside diameter	Inside diameter	Threads per inch	Tap drill	Weight per foot thds. and couplings	Normal engagement by hand
1/8	0.405	0.269	27	11/32	0.245	0.180
1/4	0.540	0.364	18	7/16	0.425	0.200
3/8	0.675	0.493	18	37/64	0.568	0.240
1/2	0.840	0.622	14	23/32	0.852	0.320
3/4	1.050	0.824	14	59/64	1.134	0.339
1	1.315	1.049	11½	1 5/32	1.684	0.400
1¼	1.660	1.380	11½	1½	2.281	0.420
1½	1.900	1.610	11½	1 47/64	2.731	0.420
2	2.375	2.067	11½	2 7/32	3.678	0.436
2½	2.875	2.469	8	2 5/8	5.82	0.682
3	3.500	3.068	8	3¼	7.62	0.766
3½	4.000	3.548	8	3¾	9.20	0.821
4	4.500	4.026	8	4¼	10.89	0.844
5	5.563	5.047	8	5 5/16	14.81	0.937
6	6.625	6.065	8	6 5/16	19.18	0.958
8	8.625	7.981	8		29.35	1.063
10	10.750	10.020	8		41.85	1.210
12	12.750	12.000	8		51.15	1.360

TABLE 36 WIRE GAGES

There has come about, through lack of standardization, a great deal of confusion concerning wire gages to be specified on the engineer's drawings. Until wire manufacturers have agreed to some national standard it would be well to specify on the drawing the exact diameter of the wire wanted. In the case of steel wires, the Bureau of Standards at Washington has recommended that the American Steel and Wire Co.'s gage be adopted as the Steel Wire Gage. This gage is given in the table below in decimals of an inch, and is the same as the Washburn & Moen gage. When there is danger of confusion with the British gage, it should be called the United States Steel Wire Gage.

In the case of copper wire, the American Wire Gage is standard throughout the United States and is the same as the Brown & Sharpe gage. It is also given in the table below in decimals of an inch.

Sheet and Plate Metal Gage

Congress legalized the United States Standard Gage for sheet and plate iron and steel, March 3, 1893. The various gage sizes are given in decimals of an inch in the table below.

WIRE AND SHEET METAL GAGES

No. of gage	Steel wire gage	American copper or B. & S. wire gage	British imperial wire gage	U. S. St'd. gage for plate	No. of gage	Steel wire gage	American wire gage	British imperial wire gage	U. S. St'd. gage for plate
0000000	0.4900	......	0.5000	0.5000	23	0.0258	0.0226	0.0240	0.0281
000000	0.4615	0.5800	0.4640	0.4688	24	0.0230	0.0201	0.0220	0.0250
00000	0.4305	0.5165	0.4320	0.4375	25	0.0204	0.0179	0.0200	0.0219
0000	0.3938	0.4600	0.4000	0.4063	26	0.0181	0.0159	0.0180	0.0188
000	0.3625	0.4096	0.3720	0.3750	27	0.0173	0.0142	0.0164	0.0172
00	0.3310	0.3648	0.3480	0.3438	28	0.0162	0.0126	0.0148	0.0156
0	0.3065	0.3249	0.3240	0.3125	29	0.0150	0.0113	0.0136	0.0141
1	0.2830	0.2893	0.3000	0.2813	30	0.0140	0.0100	0.0124	0.0125
2	0.2625	0.2576	0.2760	0.2656	31	0.0132	0.0089	0.0116	0.0109
3	0.2437	0.2294	0.2520	0.2500	32	0.0128	0.0080	0.0108	0.0102
4	0.2253	0.2043	0.2320	0.2344	33	0.0118	0.0071	0.0100	0.0094
5	0.2070	0.1819	0.2120	0.2188	34	0.0104	0.0063	0.0092	0.0086
6	0.1920	0.1620	0.1920	0.2031	35	0.0095	0.0056	0.0084	0.0078
7	0.1770	0.1443	0.1760	0.1875	36	0.0090	0.0050	0.0076	0.0070
8	0.1620	0.1285	0.1600	0.1719	37	0.0085	0.0045	0.0068	0.0066
9	0.1483	0.1144	0.1440	0.1563	38	0.0080	0.0040	0.0060	0.0063
10	0.1350	0.1019	0.1280	0.1406	39	0.0075	0.0035	0.0052	......
11	0.1205	0.0907	0.1160	0.1250	40	0.0070	0.0031	0.0048	......
12	0.1055	0.0808	0.1040	0.1094	41	0.0066	0.0028	0.0044	......
13	0.0915	0.0720	0.0920	0.0938	42	0.0062	0.0025	0.0040	......
14	0.0800	0.0641	0.0800	0.0781	43	0.0060	0.0022	0.0036	......
15	0.0720	0.0571	0.0720	0.0703	44	0.0058	0.0020	0.0032	......
16	0.0625	0.0508	0.0640	0.0625	45	0.0055	0.00176	0.0028	......
17	0.0540	0.0453	0.0560	0.0563	46	0.0052	0.00157	0.0024	......
18	0.0475	0.0403	0.0480	0.0500	47	0.0050	0.00140	0.0020	......
19	0.0410	0.0359	0.0400	0.0438	48	0.0048	0.00124	0.0016	......
20	0.0348	0.0320	0.0360	0.0375	49	0.0046	0.00099	0.0012	......
21	0.0317	0.0285	0.0320	0.0344	50	0.0044	0.00088	0.0010	......
22	0.0286	0.0253	0.0280	0.0313					

TABLE 37 AREAS AND VOLUMES

TRIANGLE	Area $= \tfrac{1}{2}bh$ $= \tfrac{1}{2}ab\sin C$	REGULAR RIGHT PRISMS	Surface Area $= nah$ $n=$ number of sides	REGULAR RIGHT PRISMS	Volume $= Bh$ $B=$ area of base
RECTANGLE	Area $= bh$	REGULAR RIGHT PYRAMIDS	Surface Area $= \tfrac{1}{2}san$ $n=$ number of sides $s = h/\sin\alpha$	REGULAR RIGHT PYRAMID	Volume $= \tfrac{1}{3}Bh$
PARALLELOGRAM	Area $= bh$ $= ab\sin C$	RIGHT CIRCULAR CYLINDER	Surface Area $= 2\pi rh$	RIGHT CIRCULAR CYLINDER	Volume $= \pi r^2 h$
TRAPEZOID	Area $= \tfrac{1}{2}(a+b)h$ $A = \tfrac{1}{2}d_1 d_2 \sin C$	SEGMENT OF CYLINDER	For Area of end segment see 1st Col.		Volume $= Bl$ $B=$ area of segment see bottom of Col. 1.
ANY REGULAR POLYGON	Area $= \tfrac{1}{2}nbh$ $n=$ number of sides $h = (b/2)(\tan\alpha)$ $\alpha = ((n-2)/2n)180°$	HOLLOW CYLINDER			Volume $= \pi l(R^2 - r^2)$
CIRCLE	Circum. $= \pi d = 2\pi r$ Area $= \pi r^2 = \tfrac{\pi}{4}d^2$	RIGHT CIRCULAR CONE	Surface Area $= \pi rs$	RIGHT CIRCULAR CONE	Volume $= \tfrac{1}{3}\pi r^2 h$
ANNULUS	Area $= \pi(R^2 - r^2)$ $= \tfrac{\pi}{4}(D^2 - d^2)$	FRUSTUM OF CONE	Surface Area $= \pi s(R+r)$	FRUSTUM OF CONE	Volume $= \tfrac{1}{3}\pi h(R^2 + Rr + r^2)$
SECTOR OF CIRCLE	Area $= \tfrac{1}{2}rs$ $= \pi r^2 (\alpha/360°)$ $s = 2\pi r(\alpha/360°)$	SPHERE	Surface Area $= 4\pi r^2$	SPHERE	Volume $= \tfrac{4}{3}\pi r^3$
SEGMENT OF CIRCLE	Area $= \tfrac{1}{2}[r(s-c)+ch]$ $c = 2r\sin\alpha$ $s = 4\pi r(\alpha/360°)$ $\cos\alpha = (r-h)/r$	SPHERICAL SEGMENT	Surface Area $= 2\pi rh$	SPHERICAL SEGMENT	Volume $= \tfrac{1}{6}\pi h(3a^2 + h^2)$ $a^2 = (h(2r-h))$

TABLE 38 FLAT PATTERN DEVELOPMENT

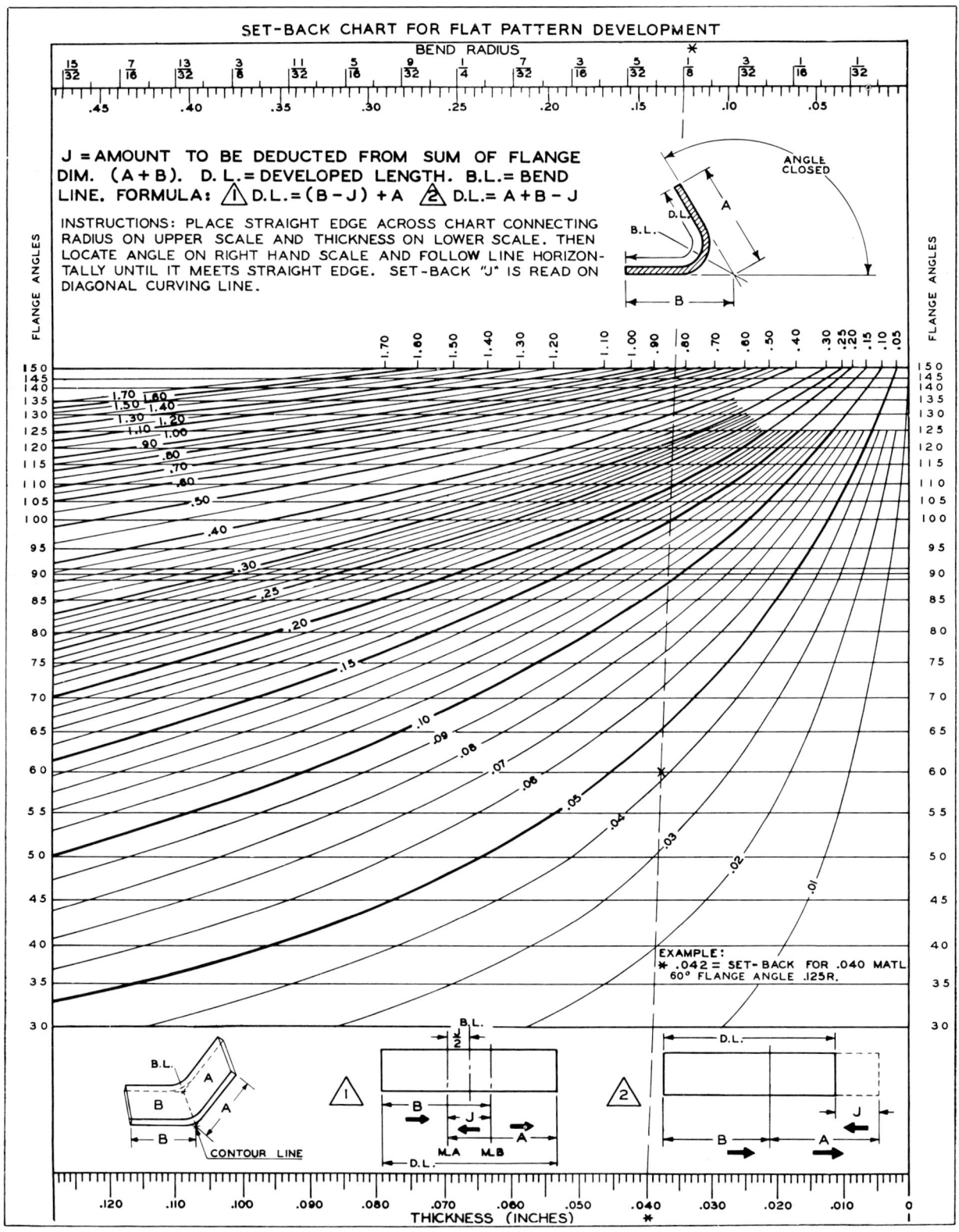

TABLE 39 LIST OF COMMON ABBREVIATIONS ON DRAWINGS

Adjust	ADJ	Harden	HDN
Allowance	ALLOW	Head	HD
Alteration	ALTRN	Heat Treat	HT TR
Aluminum	AL	High-Speed Steel	HSS
American Standard	AMER STD	Hot Rolled Steel	HRS
American Wire Gage	AWG	Inside Diameter	ID
Approximate	APPROX	Left-Hand Thread	LH THD
Assemble	ASSEM	Locate	LCT
Assembly	ASSY	Lubricate	LUBT
Auxiliary	AUX	Malleable Iron	MI
Balance	BAL	Material	MATL
Base Line	BL	Maximum	MAX
Base Plate	BP	Micrometer	MIC
Between Centers	BC	Minimum	MIN
Bill of Materials	B/M	National Coarse	NC
Blueprint	BP	National Extra Fine	NEF
Bureau of Standards	BU STD	National Fine	NF
Bushing	BSHG	National Special	NS
Carburize	CARB	On Center	OC
Cast Iron	CI	Outside Diameter	OD
Cast Steel	CS	Overall	OA
Center	CTR	Pattern	PATT
Center Line	CL	Piece	PC
Center to Center	C TO C	Place	PLC
Chamfer	CHAM	Plate	PL
Cold Rolled Steel	CRS	Punch	PCH
Counterbore	C'BORE	Radius	R
Countersink	CSK	Ream	RM
Detail	DET	Reference	REF
Diameter	DIA	Required	REQD
Dimension	DIM	Right-Hand Thread	RH THD
Dowel	DWL	Rivet	RVT
Drawing	DWG	Screw	SCR
Drill	DR	Section	SECT
Drop Forge	DF	Set Screw	SSCR
Eccentric	ECC	Shaft	SFT
Finish All Over	FAO	Socket Head	SCH
Fixture	FXTR	Spotface	SF
Flat Head	FLH	Stainless Steel	SST
Forged Steel	FST	Stock	STK
Forging	FORG	Taper	TPR
Foundry	FDRY	Temper	TMP
Full Indicator Reading	FIR	Tolerance	TOL
Gage	GA	Tool Steel	TS
Grind	GRD	Wrought Iron	WI

Abbreviations should be used only where their meaning is unquestionably clear. WHEN IN DOUBT, SPELL OUT.
ANSI Y1.1-1972

TABLE 40 ANSI STANDARDS OF INTEREST TO DESIGNERS AND DRAFTERS

USA Standard Drafting Practices
- Section 1   Size and Format   Y14.1-1980
- Section 2   Line Conventions and Lettering   Y14.2M-1979
- Section 3   Multi- and Sectional View Drawings   Y14.3-1975
- Section 4   Pictorial Drawing   Y14.4-1957
- Section 5   Dimensioning and Notes   Y14.5-1966
- Section 6   Screw Threads Representation   Y14.6-1978
- Section 7   Part 1 Spur, Helical, Rack   Y14.7.1-1971
-           Part 2 Gear, Spline Bevel Hypoid   Y14.7.2-1978
- Section 8   Castings   In Preparation
- Section 9   Forgings   Y14.9-1958
- Section 10 Metal Stampings   Y14.10-1959
- Section 11 Plastics   Y14.11-1958
- Section 12 Die Castings   In Preparation
- Section 13 Springs, Helical and Flat   Y14.13M-1981
- Section 14 Mechanical Assemblies   Y14.14-1961
- Section 15 Electrical and Electronic Diagrams   Y14.15-1966
- Section 16 Tools, Dies, and Gages   In Preparation
- Section 17 Fluid Power Diagrams   Y14.17-1966
- Section 18 Drawings for Optical Parts   In Preparation
- Section 19 Engineering Drawings for Photographic Reproduction   In Preparation
-             Digital Representation for Communication of Product Definition Data   Y14.26M-1981
-             Dictionary of Terms, Computer-Aided Preparation of Product Definition Data   Y14.26.3-1975

Graphical Symbols for:
- Metallizing Symbols   Y32.12-1960
- Electrical and Electronics Diagrams   ANSI Y32.2-1975
- Welding   Y32.3-1959
- Plumbing   Y32.4-1955
- Pipe Fittings, Valves and Piping   Z32.2.3-1949 (Reaffirmed 1953)
- Heating, Ventilating and Air Conditioning   Z32.2.4-1949
- Use on Railroad Maps and Profiles   Y32.7-1957
- Heat-Power Apparatus   Z32.2.6-1950 (Reaffirmed 1956)
- Fluid Power Diagrams   Y32.10-1958
- Process Flow Diagrams in Petroleum and Chemical Industries   Y32.11-1961
- Nondestructive Testing Symbols   Y32.17-1962
- Abbreviations for Use on Drawings   ANSI Y1.1-1972

Letter Symbols for:
- Hydraulics   Y10.2-1958
- Rocket Propulsion   Y10.14-1959
- Mechanics for Solid Bodies   Z10.3-1948
- Structural Analysis   Z10.8-1949
- Heat and Thermodynamics   Y10.4-1957
- Physics   Z10.6-1948
- Aeronautical Sciences   Y10.7-1954
- Radio   Y10.9-1953
- Meteorology   Y10.10-1953
- Acoustics   Y10.11-1953 (Reaffirmed 1959)
- Chemical Engineering   Y10.12-1955 (Reaffirmed 1961)
- Petroleum Reservoir Engineering and Electric Logging   Y10.15-1958
- Abbreviations for Scientific and Engineering Terms   Z10.1-1941
- Guide for Selecting Greek Letters Used as Letter Symbols for Engineering Mathematics   Y10.17-1961
- Shell Theory   Y10.16-1964
- Surface Texture Symbols   ANSI Y14.36-1978

Preferred Limits and Fits for Cylindrical Parts   USAS B4.1-1967(1974)
Preferred Metric Limits and Fits   ANSI B4.2-1978
Metric Screw Threads M Profile   ANSI B1.13M-1979
Metric Screw Threads MJ Profile   ANSI B1.21M-1978
Nomenclature, Definitions, Letter Symbols for Screw Threads   ANSI B1.7-1977
General Tolerances for Metric Dimensioned Products   ANSI B4.3-1978
Twist Drills   ANSI B94.11M-1979

TABLE 41 INTERNATIONAL, AMERICAN, BRITISH AND CANADIAN SYMBOLS FOR POSITIONAL AND FORM TOLERANCES

	DIFFERENCES IN SYMBOLIZATION FOR POSITIONAL AND FORM TOLERANCES			
Item	International ISO/TC 10 (Symbols mandatory)	American USASI Y14.5 (Symbols optional)	British BS 308	Canadian B 78.1
			(No symbols-equivalent terms)	
Straightness	—	—	STR TOL	STRAIGHT WITHIN
Roundness (circularity)	○	○	RD TOL	ROUND TOL
Profile of any line	⌒	⌒	Not specified	Not specified
Flatness	▱	▱	FLAT TOL	FLAT WITHIN
Cylindricity	⌭	⌭	CYL TOL	Not specified
Profile of any surface	⌒	⌒	TOL ZONE	TOL ZONE
Parallelism	//	‖	PAR TOL	PARALLEL TO
Perpendicularity (squareness)	⊥	⊥	SQ TOL	SQUARE WITH
Angularity	∠	∠	ANG TOL	ANG TOL
Runout	↗	↗	Not specified	Not specified
True position	⊕	⊕	POSN TOL	LOCATE WITHIN
Concentricity	◎	◎	CONC TOL	CONCENTRIC TO
Symmetry	≡	≡	SYM TOL	SYMMETRICAL WITHIN
Datum feature	[A]	-A-	⟋ A	⟋ A
True position dimension	[127]	[5.000]	5.000 TP	5.000 (TP)
Maximum material condition	Ⓜ	Ⓜ	MMC	MMC
Regardless of feature size	Not specified	Ⓢ	Not specified	Not specified
Tolerance	Total value specified	Total value specified except where radial (half) value is an option.	Total value specified	Total value specified
Diameter	⌀	DIA	DIA or ⌀	DIA or ⌀
Shape of tolerance zone	Zone is a width in direction of leader arrow. ⌀ specified where zone is circular or cylindrical.	DIA, TOTAL, or R specified as applicable for positional tolerances. Otherwise, not specified and diameter or width implied as applicable.	DIA or WIDE specified as applicable.	FIM or DIA specified when considered necessary.
Sequence within geometric tolerance notation	1st-Geometric characteristic symbol 2nd-Tolerance value & modifier 3rd-Datum reference & modifier	1st-Geometric characteristic symbol 2nd-Datum reference & modifier 3rd-Tolerance value & modifier	1st-Geometric characteristic term 2nd-Tolerance value & modifier 3rd-Datum reference & modifier	1st-Geometric characteristic term 2nd-Datum reference & modifier 3rd-Tolerance value & modifier

INDEX

Abbreviations, 604
Accumulation of tolerances, 318
Actual size, 310
Adjacent parts, crosshatching of, 197
Adjacent projections, 137
Adjacent views, 107
Aids, mechanical, axonometric
    projection, 372
Aligned sections, 201
    rules for, 201
Aligned system of dimensioning, 288
Allowance, 310
    finish, 247
    shrinkage, 251
    thread, 223
Alternate positions, 132
Aluminum, general properties of, 275
Ames lettering guide, 16
Angle:
    bisecting an, 56
    deflection, 508
    equal to a given angle, 56
    between a line and plane, 109
    between lines, 152
    between a plane and coordinate
        planes, 160
    between planes, 113
    between two planes, 172
    by protractor, 57
    by sine method, 57
    by tangent method, 57
Angles, 103
    definition of terms, 56
    dimensioning of, 392
    in isometric drawing, 355
Angularity, 334
Annealing, 273
ANSI standards, list of, 605
Appendix, index of, 558
Arc:
    equal to a straight line, 67
    rectifying an, 66
Archimedes, spiral of, 82
Arrangement of views, 106
    alternate, 122

Arrowheads, 281
Assemblies:
    check, 305
    crosshatching in, 308
    dimensioning, 308
    pictorial, exploded, 371
    shop, 305
Assembly, drawings, 304
    interchangeable, 309
    selective, 308
Auxiliary plane, relation to object,
    136
Auxiliary projection, elimination of
    measurements in, 147
Auxiliary projections, theory of,
    138
Auxiliary sections, 205
Auxiliary view:
    layout on one auxiliary plane,
        148
    point projection of line, 141
    true length of line, 138
Auxiliary views:
    for constructing principal views, 146
    notation, 137
    placing of, 136
    on two auxiliary planes, 149
    uses of, 138
Axes, axonometric, 354
    odd number of, 130
    selection of on charts, 425
Axonometric, axes, 354
    exploded assemblies, 371
    projection, 8, 352, 367
    self-study questions, 375
Axonometric projection:
    boards, 372
    mechanical aids, 372
    rotation of coordinate planes, 369
    theory, 368
Azimuth, 508
    of line, 139

Back azimuth, 508
Backsight, 508

Ball bearings, 779
Bar charts, 432
Basic hole method, size tolerances, 314
Basic shaft method, size tolerances, 315
Basic size, 310
Bearing of a line, 139, 508
Bend allowance, 495
Bending, 257
Bend radii, 495
Bend relief, 496
Bilateral system for size tolerancing,
    313
Blanking, 257
Boards:
    axonometric projection, 372
    drawing, 36
    three-point perspective, 412
Bolts:
    carriage, 232
    finished, 229
    hexagon head, 573
    length of engagement, 234
        thread, 234
    plow, 232
    regular, 229
    in section, 207, 209
    semi-finished, 229, 574
    slotted head, 236
    specifications, 233
    square head, 572
    standard, 228
    stud, 231
    types of, 232
Boring mill, 267
Bosses, 250
Bottom view, 106
Braddock lettering triangle, 15
Brass and copper pipe data, 599
Breaks, conventional, 209
Broaching machine, 270
Broken out sections, 205

Cabinet drawing, 379
    construction of, 384
Cabinet projection, 6

608 INDEX

Cadastral maps, 500
CAD/CAM, 520, 521, 548, 554
   applications of, 548
   assembly, 549
   color use importance, 551
   Computer-Aided Design (CAD), 520, 549
   Computer-Aided Design Drafting (CADD), 521
   Computer-Aided Engineering (CAE), 521
   Computer-Aided Manufacturing (CAM), 520, 549
   Computer-Integrated Manufacturing (CIM), 521
   flexible manufacturing system, 550
   importance of traditional graphics in, 554
   numerical control, 549
   relation to computer graphics, 520
   robots, 549
   see also Computer Graphics
Cap screws, 234, 853
   length of, 235
   threads, 235
Case hardening, 273
Castings, 246
   centrifugal, 252
   design details, 248
      die, 253
         gray iron, automotive, suggested uses, 274
         permanent mold, 252
         section thickness, 249
Cavalier projection, 5
   conventional construction, 380
   drawing solid objects, 382
Celluloid tools, 48
Center line, 46
   reference plane, 146
Center line layout, oblique, 386
Centers, machining, dimensioning, 293
Centrifugal castings, 252
Chain dimensioning, 319
Chamfer, dimensioning of, 298
Charts, bar, 432
   change of scale, effect of, 428
   choice of scales, 425
   classification of, 424
   computation, 435
   drawing the curve, 427
   flow, 435
   how to draw, 425
   interpretation of, 428
   location of origin, 428
   logarithmic, 429
   marking coordinates, 427
   organization, 435
   polar, 431
   on rectangular coordinates, 425
   sector, 432
   selection of axes, 425
   semilogarithmic, 430
   sketches on, 428
   three dimensional, 435
   titles on, 428
   trilinear, 431
   uses of, 424
Charts and diagrams, 424
   self-study questions, 436, 437
Checking, engineering, method of, 10
Checking a drawing, 346
Circle:
   horizontal, perspective by measuring point method, 408
   in a horizontal position, perspective of, 404
   in isometric, 356
   isometric drawing of, 360
   lines tangent to, 77
   parallel to picture plane, perspective of, 404
   sketching in perspective, 418
   within a square, 72
   tangent to a line and circle, 73
      to two circles, 73–74
      to two lines, 73
   through a point tangent to circle, 72
   through a point tangent to line, 72
   through three points, 72
   vertical, perspective by measuring point method, 545
   in vertical position, perspective of, 404
Circles or curves, perspective of, 404
Clearance, crest, 223
   locational fits, 585
Clinographic projection, 6, 380, 389
Coining, 257
Combination method of perspective, 398, 400
   horizontal line, 401
Compass, 42
Composition:
   of concurrent coplanar vectors, 441
   of concurrent noncoplanar vectors, 442
   of vectors, 440
Computer aided, 272
   design, 11
   manufacture, 272
Computer graphics:
   computing concepts for, 521
      address, 522
      auxiliary storage, 522
      bit, 521
      byte, 521
      Central Processing Unit (CPU), 522
      compiler, 521
      computer, 521
      computer program, 521
      digital computer, 521
      hardware, 521, 522
      input, 522
      interactive system, 522
      interface, 522
      memory, 522
      network, 521
      operating system, 521
      output, 522
      passive system, 522
      peripherals, 522
      program, 521
      random access, 522
      sequential access, 522
      software, 521
      terminal, 522
      user program, 521
      word, 521
   coordinates:
      absolute, 527, 528
      incremental, 527
      relative, 527
      world, 527
   cursor, 538, 540, 542, 546
   data structure, 522, 523, 524, 525, 526
   devices, see Interaction, View devices
   display buffer, 523, 529, 530, 534
   display file, 523, 529, 530, 534
   display terminal, 523. See also Computer graphics, view devices
   entity, 522
   historical perspective of, 520
   interaction, 524
      Alphanumeric Keyboard, 542
      bounding, 546
      button box, 539
      defining, 544
      devices, 538
      dial, 540
      digitizing, 544
      dragging, 544
      friendliness, 524
      function keyboard, 539
      gridding, 546
      highlighting, 546
      inputting, 538
      joystick, 541
      light pen, 539
      locating, 538
      menu selecting, 538
      modes, 538
      mouse, 541
      observing, 546
      overlaying, 543
      panning, 543
      picking, 538
      rubberbanding, 546
      sketching, 546
      tablet, 540
      techniques for models, 543
      techniques for views, 543
      touch panel, 539
      trackball, 542
      voice recognition, 542
      windowing, 543
      zooming, 543
   interactive, 522
   mapping, 524, 527, see Transformations
   menu, 528, 538
   model, 522, 524
      boundary, 526
      constructive geometry, 526
      data structure, 524, 525, 526

INDEX

direct map, 524
interaction techniques, 543. *See also* Computer graphics, interaction
representation, 524
solid, 526
stick, 522, 523
surface, 526
wireframe, 524, 525
pel, 530
pixel, 530
plotter, 523, 532. *See also* Computer graphics, view devices
raster, 530, 532
raster scan conversion, 530
relation to CAD/CAM, 520
relation to traditional graphics, 554
transformations, 524, 527, 534
   absolute coordinates, 527, 528
   axonometric, 536
   incremental coordinates, 527
   matrix use in, 529
   oblique, 536
   perspective, 536
   projections (3D), 535
   relative coordinates, 527
   rotation, 527, 528
   scaling, 527, 528
   translation, 527, 528
vector, 524, 525, 529
view, 523, 524, 529
view devices, 531
   Cathode Ray Tube (CRT), 533
   Direct-View Storage Tube (DVST), 534
   display, 531, 533
   dot matrix plotter, 532
   drum plotter, 532
   electrostatic plotter, 532
   flatbed plotter, 532
   hardcopy, 531, 532
   impact plotter, 533
   ink-jet plotter, 533
   interaction techniques, 543. *See also* Computer graphics, interaction
   LCD, 534
   LED, 534
   line printer, 531
   plasma display panel, 534
   thermal plotter, 533
   vector plotter, 532
viewport, 535
view representation, 529
   random scan, 529
   raster, 530
   stroke, 529
   vector, 529
view volume, 536, 537
visibility, 537
window, 534, 537
Concentricity, 337
Concurrent coplanar force systems, 157
Concurrent noncoplanar forces, 158
Cones, construction, 176
Conic, through five points, 76

Conic sections, definition of, 67
Conoid, 187
Construction cones, 176
Contour lines, 511
   plotting of, 512
Contour maps, 503
   use of, 513
Conventional practices, 130
Conventional practice in sectioning, 206
Copper water tube data, 600
Coordinate dimensioning, 296, 318
Coordinates:
   marking on charts, 427
   polar, 318
   rectangular, 318
Cope, 251, 266
Core, 250
Core box, 252
Core print, 248, 251, 265
Cores, 252
Cotter pins, 214, 231
Counterbore, 263
   dimensioning of, 298
Countersink, 262
   dimensioning of, 298
Crosshatching adjacent parts, 197
Crosshatching assemblies, 309
Crosshatch lines, 197
Cupula, 249
Curved lines, projection of, 126
Curved surfaces, projection of, 128
Cut and fill, from contour maps, 513
Cutting planes:
   indication of, 195
   location of, 195
Cutting plane symbols, 195
Cycloid, 81
Cylindricity, 338
Cylindroid, 188

Datum, 311
   planes, 285
   primary, 325
   secondary, 325
   selection, 325
Datum surfaces:
   accuracy of, 321
   identification of by dimensioning, 320
   indicated by note, 321
   selection of, 320
Datum symbol, 332
Decimals, 283
Descriptive geometry, 6
Design, engineering, 10
Design details, casting, 248
Design size, 310
Details, design, casting, 248
   sketches, 96
   standard, 308
Development, based on mold lines, 495
   of a cone, 488
   of a cylinder, 486
   of double curved surfaces, 494
   of an oblique cone, 489

   of an oblique cylinder, 487
   of an oblique prism, 485
   of an oblique pyramid, 486
   of a right circular cone, 489
   of a right prism, 485
   of a right pyramid, 486
   of a truncated right cylinder, 487
Development of surfaces, 494
   mechanical aid in, 493
   by triangulation, 493
Development of transition surfaces, 489, 491
Devices, locking, 230
Diagrams:
   distribution, 435
   erection, 306
   pie, 432
   vector, 156, 440
Die, 264
Die casting, 253
Dimensioning:
   aligned system, 287
   analysis of, 317
   angles, 292
   arrowheads for, 281
   assemblies, 308
   blind holes, 298
   centers machining, 293
   chain, 319
   chamfers, 298
   circular arcs, 290
   compound curves, 290
   concentric circles, 290
   cones, 290
   on contour view, 285
   coordinate, 296, 318
   counterbore, 298
   countersink, 298
   cumulative tolerances in, 285
   cylinders, 286, 290
   from datum planes, 285
   in decimals, 283
   dovetails, 300
   drawings, 278
   extension lines for, 279
   finished surfaces, 301
   in fractions, 282
   half sections, 285
   invisible lines, 285
   isometric drawings, 365
   keyways, 299
   knurls, 300
   leaders for, 280
   location, 294
   methods compared, 323
   narrow spaces, 286
   noncircular curves, 292
   notes, general, 300
      lettering in, 281
      special, 300
   for numerically controlled machines, 298
   oblique drawings, 387
   parts with curved ends, 295, 297
   part with true radius, 300
   placing numerals, 284

Dimensioning: (Continued)
  prisms, 290
  production, 304
  progressive, 292
  projected tolerance zones, 343
  reference planes in, 285
  self-study questions, 301
  simple objects, 290
  size, 290
  slotted holes, 297
  spheres, 290
  spotfaces, 298
  symmetrical parts, 295
  tapers, 293
  technique of, 278
  threaded holes, 298
  thread relief, 299
  torus, 290
  true position, 322
  tubing, 290
  unidirectional system, 288
  units of measurement in, 282
  use of standard parts in, 301
Dimension lines, 278
  crossing of, 284
  spacing of, 283
Dimensions, between views, 284
  duplicate, 284
  limit, 301
  reference, 309
  related, 284
  selection of, 288
  toleranced, 297
  unrelated, 284
  where to place, 283
Dimetric, 354
Dimetric drawing, 366
Dimetric projection, 5
  special positions, 373
Dip of a plane, 161
Distance:
  between parallel lines, 153
  between parallel planes, 162
  from a point to a line, 153
Distribution diagrams, 435
Dividers, 44
Double-curved surfaces:
  in isometric, 363
  in oblique, 388
Dovetail, dimensioning of, 300
Dowel pins, 214
Draft, 251
  forging, 253
  of a pattern, 248
Drafting, simplified, 347
Drafting machine, 37
Drafting technique, 47
Drag, 251
Draw, 251
  of a pattern, 248
Drawing, 257
  checking a, 346
  first-quadrant, 132
  maps, 499
  multiview, selection of views, 118
  reading a, 116, 118

Drawing, zoned, 204
Drawing media, paper, 38
Drawings, assembly, 304
  installation, 306
  layout, 304
  one-view, 120
  projection, 2
  three-view, 122
  two-view, 120
Drill, tap, 260
  twist, 260
Drill press, 264

Edgewise view of a plane, 138, 143
Elevations:
  on maps by contour lines, 511
  on maps by hachures, 250
  of points in a plane, 162
Elimination of invisible lines, 124
Ellipse:
  four center in a rhombus, 72
  line tangent to, 76
Ellipse construction, as section of a
    cone, 68
  by diagonal method, 71
  by four center method, 70
  having conjugate axes, 70
  by intersection method, 70
  by trammel method, 70
  by two circles, 69
  by use of foci, 69
Ellipse guides, Lietz, 372
Engagement, length of for bolts, 234
Engineering design, 10
Engineering maps, 502
Epicycloid, 82
Equilibrant of vectors, 440
Equilibrium, 440, 443
Equipment, regular, 32
Erasers, 35
Erasing tools, 36
Erection diagrams, 306
Evaluation, engineering, method of,
    10
Exploded pictorial assemblies, 371
Extension lines, 279
  between views, 284
Extrusion, 257

Face of object used as reference plane,
    146
Fasteners, 214
  detailed on drawing, 214
  self-study questions, 243
  shown by symbol, 214
  standards:
    industry, 215
    national, 215
  using screw threads, 215
Feature, 309
Fillets, 248, 251
  minimum, forging, 254
Finish, surface, 301, 343
  allowance, 247, 251
  marks, 343
  specifications, 343

First quadrant, 103
  drawing, 132
Fits, 311
  inch:
    classes of, 311
    clearance locational, 312, 585
    force, 311, 312
    force and shrink, 312, 589
    interference locational, 312, 588
    locational, 311, 312
    running and sliding, 311, 312, 583
    thread, classes of, 223
    transition locational, 587
  metric:
    clearance locational, 590
    force, 590
    interference locational, 592
    medium drive, 592
    running, 590
    sliding, 590
    transitional locational, 592
Flask, 251, 252
Flatness, 333
Flow charts, 435
Forces:
  concurrent coplanar, 157
  concurrent noncoplanar, 158
  in oblique positions, 158
  parallel to V plane, 158
  resolution of, 443
Force and shrink fits, 589
Forging, 253
  design details, 253
  draft, 253
  drop, 253
  fillets, minimum, 254
  hand, 253
  parting plane, 253
  plane, 253
  rounded corners, 254
  thin webs, 255
  tolerances, 255
Form-contour tolerances, 326
Form tolerance:
  controlled by location tolerance, 330
  controlled by size tolerance, 330
  geometric, 322
  specified by notes, 322
  specified by symbols, 322
Foundry, 252
  terms defined, 250
Four center ellipses in oblique, 382
Fourth quadrant, 103
Fractions, 282
Freehand drawing, 4
Freehand sketching, uses of, 86
French curves, 44
Front view, 105
Front view sketches, 96
Full section, 199
  rules for, 199
Functions of engineering drawing, 3
  aerospace, 4
  Department of Defense, 4
  design of buildings, 4
  design of machines, 3

detail drawings, 4
fabrication details, 3
highway and railroad plans, 4
sales, 4
Funicular polygon, 446

Gage pins, size of, 323
Gages, for inspection, 317
  wire, 601
Gaging, projected tolerance zones, 342
Gaging control, 327
Gate, 251
General oblique, construction of, 385
General oblique drawing, 379
Geographic maps, 500
Geometrical constructions, 51
  self-study questions, 83
Geometrical form tolerancing, 322, 330
Geometrical surfaces, 139
Geometry, engineering, fundamental operations in, 139
Geometry of engineering drawing, 136
  self-study questions, 189
Glass cloth, 39
Gothic style, 14
Grade lines, 515
Graphics, engineering, 4
Grinding machine, 268
Guide lines, lettering, 15

Half sections, 200
Hardening, 273
Heat treatment, nonferrous alloys, 273
Heat treatment of steel, 272
Helicoid, 187
Helix, 82
  angle, 215
Hexagon:
  construction with distance across corners, 60
  construction with distance across flats, 61
  construction with length of side, 60
Hexagon bolts:
  finished, 575
  semifinished, 574
Hexagon head bolts, 573
Hexagon nuts:
  finished, 576
  regular and jam, 573
  semifinished, 574
Hidden lines in sectioned views, 198
Hole clearance, minimum size of, 342
Hole diameter, effective, 341
Holes, blind:
  threaded, 228
  threaded through, 228
Horizontal curves, railway, 516
Horizontal plane, 103
Hyperbola:
  construction as section of a cone, 75
  construction with asymptotes and one point, 76
  construction with foci and vertices, 75
  find directrix and asymptotes, 80
  line tangent to, 80

Hyperbolic paraboloid, 184
Hyperboloid of revolution, 185
Hypocycloid, 82

Inertia, moment of, 593
Ink, drawing, 48
Inking, 48
  order of, 49
Ink line, quality of, 48
Inspection, methods of, 336
Installation drawings, 306
Instrumental drawing, 4
Instruments, bow, 44
  case of, 42
  self-study questions, 50
Interchangeable assembly, 309
Interference locational fits, 312, 588
Interpretation of lines, 116
Interpretation of planes, 116
Interpretation of solids, 118
Intersecting lines, 152
Intersection:
  cylinder and cone:
    cylinder parallel to base of cone, 476
    endwise view of cylinder method, 476
    oblique cutting plane method, 476
  cylinder with a double curved surface, 483
  plane and cone, 465
    cutting plane method, 465
    edgewise view method, 465
  plane and cylinder, 462
    cutting plane method, 464
    edgewise view method, 463
  plane and double curved surface, 467
  plane and prism:
    cutting plane method, 460
    edgewise view method, 460
  plane and pyramid:
    cutting plane method, 462
    edgewise view method, 462
  plane and surface of revolution, 466
  prism and pyramid, 470
  two cones:
    bases edgewise:
      in different views, 475
      in same view, 475
    bases in same plane, 473
    cutting plane method, 471
    end view of line joining apexes, 473
    neither base edgewise, 476
    only one base edgewise, 476
  two cylinders:
    bases edgewise:
      in different views, 475
      in same plane, 482
      not in same plane, 480
    cutting plane edgewise, 481
    cutting plane method, 476
    neither base edgewise, 483
  two planes, 460
  two pyramids, 471
Intersections, conventional, 484

method of finding, 458
possible types of, 471
Intersections and developments, 458
  self-study questions, 496, 497
Invisible lines, 103
  elimination of, 124
  in sectioned views, 198
  technique of, 124
Involute, 81
Involute splines, 240
Irregular curves, 44
Isometric, 354
  comparison of drawing and projection, 352
Isometric axes, position of, 364
Isometric drawing, 353
  advantages of, 365
  angles in, 355
  box construction of a block, 358
  center line layout, 362
  circles:
    coordinate method, 356
    four center method, 356
    four center methods, limitations, 357
    true diameter, four center method, 357
  circles and curves, 355
  common errors involving circles, 359
  dimensioning in, 365
  double curved surfaces, 363
  irregular curves, 357
  parallel circles, 360
  plane figure, 355
  pictorial plane, dimensioning in, 365
  screw threads in, 364
  sectional views, 362
  solids, box method, 358
    with circles, 360
    with non-isometric lines, 358
  spheres, 363
  unidirectional dimensioning in, 366
Isometric lines, 354
  sketching, 97
Isometric plane, 354
  use in construction, 361
Isometric projection, 5, 353, 368
  by auxiliary views, 352
  construction of, 370
  quadrangle, 374
  by rotation, 352
  special board, 374
Isometric sketching, 96
Isometric stencil, 375
Isometric three-point perspective, 410

Keys, 239
  dimensioning of, 239
  plain parallel stock, 581
  Pratt and Whitney, 239
  sectional views, 207
  Woodruff, 239
Keyways, dimensioning of, 299
Knurl, dimensioning a, 300

Lathe, 266

Lay, 345
Layout drawings, 304
Layout of three-view drawing, 132
Lead, 215
Leaders, 280
Left side view, 105
Legal aspects of drawing, 9
Leroy lettering device, 25
Lettering:
    art of, 18
    ascenders, 15
    capitals with straight and curved lines, 23
    combination of stem and oral, 17
    compressed, 19
    descenders, 15
    elements, 16
    expanded, 19
    fractions, 15, 23
    Gothic, double stroke, 27
    guidelines, 15
    large and small capitals, 24
    Leroy, 25
    lower case, 23
    maps, 499
    mechanical, 25
    mechanics of, 20
    for microfilming, 15
    modern roman, 26
    notes, dimensioning, 281
    numerals, 15
    old English, 28
    ovals, 17
    parts list, 25
    pencils:
        selection of, 17
        sharpening of, 17
    pens, selection of, 17
    point projections, 107
    position of hand, 18
    revision block, 25
    sample alphabet, 20–23
    self-study questions, 29
    size, 15
    slope, 16
    stability, rule of, 23
    stems, 17
    style, 14
    technique, 17
    titles, 24
    uniformity of shape, 18
    uniformity of size, 18
    uniformity of slope, 18
    uniformity of spacing, 18
    uniformity of style, 18
    uniformity of weight, 18
    Wrico, 25
Lietz ellipse guides, 372
Limits, 310
Limit system for size tolerancing, 313
Line, angle with a plane, 109, 167
    of action of a vector, 440
    azimuth of, 139, 508
    bearing of, 139, 508
    bisecting a, 52
    center, used as reference feature, 146
    contour, 511
    curved, shadow in perspective, 413
    definition of, 108
    distance from a point, 153
    divided into fractional parts, 55
    divided into proportional parts, 54
    grade, 515
    horizontal, shadow in perspective, 413
    inclined:
        perspective of, 400
        shadow in perspective, 414
    intersecting skew lines at specified angles, 179
    intersecting two other lines, 55
    invisible, 103
    isometric, 354
    making angles with two intersecting lines, 178
    making specified angle with a line and plane, 181
    oblique, point projection of, 141
    oblique to principal planes, 111
    parallel:
        to a curved line, 53
        to H-plane, 109
        to a line, 52
        to a plane, 166
        to P-plane, 110
        to V-plane, 110
    perpendicular to a principal plane, 108
        to H-plane, 108
        to a line, 53, 155
        to another line, 56
        to a plane, 168
        to P-plane, 108
        to V-plane, 108
    piercing principal planes, 164
    piercing point in a plane, 163
    in a plane, 143
    point projection of, 138, 141
    slope of, 138
    through a point making a given angle with a coordinate plane, 176
    through a point making a given angle with a line, 152
    through a point making given angles with any two planes, 177
    through a point making given angles with two coordinate planes, 176
    through a point, parallel to a line, 152
    true length of, 138, 141
    vertical, shadow in perspective, 412
Lines, angle between, 152
    crosshatch, 197
    dimension, 278
    distance between parallel, 153
    extension, 279
    hidden:
        in assemblies, 308
        in sectioned views, 198
    interpretation of, 116
    intersecting, 152
    non-isometric, 354
    parallel, 151
    position of, 108
    projection, 106
    relationship between, 151
    relationship with planes, 163
    runout, 129
    skew, 174
        grade line with given bearing between, 175
        line parallel to a given line and intersection, 176
        shortest distance between, 174
        shortest horizontal line between, 174
        shortest line of specified slope between, 174
    visible, 103
    visible behind section plane, 198
    of sight, 102
Lining, section, 197
Locking devices, 230
Locus problems, 183
Logarithmic scales, 430
    construction of, 54

Machines, numerically controlled, 271
    benefits, 271
Machine screws, 235
Machine shop, 258
Machining centers, dimensioning of, 293
Magnetic north, 508
Maps, cadastral, 500
    drawing of, 500
    engineering, 502
    geographic, 500
    lettering, 516
    military, 503
    topographic, 500
Map scales, 500
Map symbols, 505
    color of, 506
    how to draw, 505
    size and prominence, 505
    spacing, 506
Map traverse, 509
Material specification, 273
Material symbols in sections, 209
Mating parts, cylindrical, size dimensions, 314
Maximum material conditions (MMC), 311, 323
Meaning of terms, axonometric projection, 5
    cabinet, 5
    cavalier, 5
    clinographic, 6
    descriptive geometry, 6
    dimetric, 5
    engineering graphics, 4
    freehand drawing, 4
    instrumental drawing, 4
    isometric, 5
    multiview drawings, 5
    nonprojection drawings, 4
    oblique, 5
    perspective, 6

pictorial projections, 5
projection drawing, 5
trimetric, 5
Measurement, units of, 282
Measurements, elimination of in auxiliary projection, 147
Measuring-line in perspective, 398
Measuring lines, 406
Measuring point and line, one point, 408
Measuring point method of perspective, 406
  of horizontal circle, 408
  of vertical circle, 407
Measuring-point in perspective, 398
Measuring points, 406
Metallurgy, powder, 253
Method, engineering:
  checking, 10
  evaluation, 10
  proposed solution, 10
  report, 10
  statement of problem, 10
Metric screw threads, 224
  M profile, 224
  MJ profile, 224
Military maps, 503
Milling machine, 267
Molds, permanent, 252
Moment of a force about a point, 449
Moment of inertia, 451
Moments on a beam, 450
Motion, problems of, 7
Multiview drawing, 5
  selection of views, 118
Multiview sketches, 94

Nails, 214
Necking, 257
Nesting, 257
Nomenclature, 107
Nominal size, 309
Non-isometric lines, 354
Non-isometric plane, 354
Normalizing, 273
North points on maps, 517
Notation, auxiliary views, 137
Notching, 257
Note form, for tolerancing, 314
Notes, lettering, 281
Notes for dimensioning, 300
Numbers, part, 308
  reference, 308
Nuts:
  drawing of, 232
  elastic stop, 231
  jam, 230
  split, 231
  standard, 230

Object, 102
  face as reference plane, 146
  placing, 103
Oblique, 378
  fundamental principles, 378
  practical methods, 379

Oblique drawing, general, 379
Oblique line, true length of, 138
Oblique plane, true size of, 145
Oblique projection, 5
  advantages of, 390
  box method of construction, 383
  cabinet, 384
  cavalier, 379
  center line layout, 386
  circles by the four-center method, 382
  circles in front face, 381
  circles in top or side faces, 381
  clinographic, 380, 389
  dimensioning in, 387
  double curved surfaces in, 388
  general, construction of, 385
  plane figures, 380
  position of axes, 382
  position of object, 383
  reducing distortion in, 383
  screw threads in, 387
  sectioning in, 386
  self-study questions, 391
  theoretical construction, 388
  three-dimensional curves, 388
  types of, 379
Oblique sketching, 98
Octagon:
  construction with distance across corners, 62
  construction with distance across flats, 62
Odd number of axes, 130
Offset sections, 201
  rules for, 201
One-view drawings, 120
Organization charts, 435
Origin, location of, on charts, 428
Orthographic projection, 5
  self-study questions, 133, 134
Outcrop, from contour maps, 513
Outcrop of a plane, 171
Outline section lining, 197

Pads, 250
Palnut, 230
Paper, drawing, 39
Paper fasteners, 40
Parabola:
  construction as section of cone, 73
  construction by offsets, 74
    with apex and two symmetrical points, 74
    with focus and directrix, 74
    with hyperbolic paraboloid, 74
  to find the axis, 79
  to find the directrix, axis, focus and vertex given, 80
  to find the focus with the axis given, 80
  line tangent to, 79
Parallel:
  forces, 447
  lines, 151
  rules, 37
  symbol for, 333

Parallel lines, distance between, 153
Partial section lining, 197
Partial views, 128, 147
Parting plane, 247, 251
  forging, 253
Part numbers, 308
Parts, right and left-hand, 132
Parts list, 25
Pascal line, 77
Pattern, 246
  color, 252
  drawing, 246
  makers shrink rule, 247
  terms defined, 250
Pencils, 33
Pencil technique, 47
Pens, lettering, 36
Pentagon, construction in a circle, 60
Permanent molds, 252
Perpendicular, symbol and notes, 334
Perpendicular lines, 155
Perspective:
  angular or two-point, 6, 397
    construction by, 404
  combination method, 398, 400
  definition, 394
  enlarged, 395
  grid, 419
  horizon line, 395
  isometric three point, 410
  kinds of, 402
  location of picture plane, 395
  location of point of sight, 395
  measuring-point and measuring-line method, 398
  measuring-point method, 406
  methods of constructing, 397
  normal, 395
  oblique or three-point, 397
  one-point measuring point and line, 408
  parallel or one-point, 396, 402
  picture plane, 394
  point of sight, 394
  position of object, 396
  proportioning a cube, 416
  proportioning by repeating cube, 417
  proportioning by subdividing a cube, 417
  reversed, 395
  self-study questions, 420, 421
  shades and shadows in, 411
  sketching, 100, 416
    of circles, 418
  three-point, 409
    alternate method, 411
  three-point board, 412
  vanishing point construction, 397, 399
  vanishing point method, 400
  vanishing points, 395
  visual ray method, 394, 397
Perspective of:
  an inclined line, 400
  circle in horizontal position, 404
  circle parallel to picture plane, 404
  circles or curves, 404

Perspective of: (Continued)
  circle in vertical position, 404
  horizontal circle by measuring point method, 408
  horizontal line by combination method, 401
  reflections, 415
  shadow of curved line, 415
    of horizontal line, 413
    of inclined line, 414
    of a vertical line, 414
  vertical circle by measuring-point method, 407
Phantom sections, 206
Phi, 108
Pi, 108
Pictorial, exploded assemblies, 371
  projections, 5
  sketches, 96
Picture plane:
  location of, 395
  perspective, 394
Pie diagrams, 432
Pins, cotter, 231
  in section, 207
  taper, 240
Pipe, threads, American, 227
  representation of, 227
Pitch, 215
Placing dimensions on a drawing, 283
Placing the object, 103
Plane, 269
  angle with coordinate planes, 160
  definition of, 111
  dip of, 161
  edgewise view of, 143
  horizontal, 103
  isometric, 354
  line in a, 143
  making specified angles with two given lines, 181
  nonisometric, 354
  oblique, true size of, 145
  oblique to principal planes, 114
  outcrop of, 171
  parallel to a line, 166
    to H, 111
    to P, 112
    to two lines, 167
    to V, 112
  parting, 247
  perpendicular to a line, 168
    to H, 113
    to P, 114
    to a plane, 170
    to V, 114
  piercing point of a line in, 164
  profile, 103
  of projection, 102
  strike of, 161
  through a line making specified angle with given plane, 181
  through a point parallel to a plane, 163
  true shape of, 138, 144
  true size by rotation, 159

  true size of, with edge-view given, 144
  vertical, 103
  with specified strike and dip, 162
Plane figure:
  construction of, 159
  isometric drawing of, 353
  oblique, 380
  reproduction of, 62
Plane and line, angle between, 109, 167
Planes:
  angle between, 113, 142
  cutting, in isometric, 361
  datum, 283
  inclined, 160
  interpretation of, 116
  intersection of, 170
  oblique, 160
  parallel, distance between, 162
  principal, 103
  reference, 285
  relationship of, 159
  relationship with lines, 163
  revolution of, 105
Plotting points on charts, 427
Plotting a traverse, 509
Point, distance from a line, 153
  piercing:
    of a line in coordinate plane, 154
    of a line in a plane, 163
  projection of, 106
  a line, 138, 141
  projection on a plane, 168
  revolved around a line, 155
Points, in a plane:
  elevation of, 162
  lettering of, 107
  location of, 107
Point of sight, 102
  above or below object, 396
  distance from object, 395
  location of, 395
  perspective, 394
  right or left of object, 395
Polar charts, 431
Polishing, 269
Polygon:
  construction of, 57, 61
  construction with length of size, 61
  inscribe in a circle, 61
  string, 446
Position of a line, 108
Positions, alternate, 132
Powder metallurgy, 253
Powder threads, 218
Practices, conventional, 130
Pratt and Whitney keys, 239
Precedence of lines in visibility, 123, 124
Principal planes, 103
Principles, engineering scientific method, 10
Problem, engineering statement of, 10
Production dimensioning, 304
  self-study questions, 348
Professional aspects of drawing, 8

Profile:
  line, 339
  of a surface, 338
  plane, 103
  railway, 515
Profiles:
  from contour maps, 514
  on curves, 515
Progressive dimensioning, 292
Projected tolerance zone, 342
  gaging, 342
  for threaded parts, 342
Projections, 103
  adjacent, 137
  curved lines, 126
  curved surfaces, 126
  drawing, origin, 2
  lettering of, 107
  lines, 106
  orthographic, 102
  plane, 102
  of points, 106
  revolution of plane of, 105
  types of, 4
  views in, 106
Proportioning:
  in perspective:
    repeating a cube, 417
    subdividing a cube, 417
  in sketching, 93
Proportioning a cube in perspective, 416
Protractors, 48
Punching, 257

Quadrants, 103
Questions, self-study for auxiliary, 189

Reading a drawing, 116
Reamer, 261
Rear view, 105
Rectifying, an arc, 66
Reference, dimension, 309
  lines, elimination of, 145
  numbers, 308
  planes, 285
Reference plane:
  center line, 146
  face of object, 146
Reflections in perspective, 415
Regardless of feature size (RFS), 323
Relief, thread, 228
Removed section:
  on separate sheet, 204
  on zoned drawing, 204
Removed sections, 203
  drawing of, 204
Report, engineering, method, 10
Resolution:
  of forces, 443
  of vectors, 440
  of vectors in three-dimensional system, 444
Resultant of:
  nonconcurrent coplanar vectors, 446
  vectors, 440
  vector systems, 445

Reverse curve:
　connecting nonparallel lines, 66
　connecting parallel lines, equal radii, 66
　connecting parallel lines, unequal radii, 66
Revision block, 25
Revolve a point around a line, 155
Revolved sections, 202
　advantages of, 202
Ribs, cast iron, 250
Right and left-hand parts, 132
Right side view, 105
Rivets, 237
　explosive, 239
Rivnut, 237
Roughness, 344
Roundness, 336
Rounds, 248
Rule:
　patternmaker's shrink, 247
　shrink, 252
Running and sliding fits, 312, 583
Runout, 339
Runout lines, 129

Scales, 40
　architect's, 40
　choice of, on charts, 425
　civil engineer's, 41
　decimal, 41
　logarithmic, 54
　mechanical engineer's, 41
　use of, 42
Screws:
　cap, 234, 570, 571
　　length of, 235
　machine, 235, 570
　in section, 207
　set, 236
　tapping, 237
　wood, 236
Screw threads, 234
　classes of fit, 223
　in isometric drawing, 364
　left-hand, 222
　in oblique, 387
　representation of, 218–220
　right-hand, 222
　series, 222
　specifications, 221
　symbols, 221
　terminology, 215, 216
　types, 217
Second quadrant, 104
Sectional views, 194
　definition, 194
　in isometric, 362
　purpose, 194
Sectioned, in oblique, 386
Sectioned view, conventional practice, 206
Sectioned views:
　drawing of, 197
　hidden lines in, 198
　removed, on zoned drawing, 204

Section liners, 48
Section lining, 197
　direction of, 197
　partial, 197
　spacing of, 197
　weight of, 197
Section planes:
　indication of, 196
　location of, 195
　visible lines behind, 198
Sections:
　aligned, 201
　auxiliary, 205
　broken out, 205
　with conventional breaks, 209
　full, 199
　　rules for, 200
　half, 200
　material symbols in, 209
　with odd numbered axes of symmetry, 208
　offset, 201
　　rules for, 201
　phantom, 206
　removed, 203
　revolved, 202
　　advantages of, 202
　self-study questions, 211
　solid bars, shafts, etc., 207
　spokes, 207
　thin webs, 206, 207
　threads and bolts, 209
Sector charts, 432
Selective assembly, 308
Self-study questions:
　auxiliary, 189, 190
　axonometric, 375
　charts and diagrams, 436, 437
　dimensioning, 301
　engineering geometry, 189, 190
　fasteners, 243
　geometrical constructions, 83
　instruments, use of, 50
　intersections and developments, 496, 497
　lettering, 29
　oblique, 391
　orthographic projection, 133, 134
　perspective, 420, 421
　production dimensioning, 348
　sectioning, 211
　shop terms and processes, 275
　sketching, 100
　vector analysis, 454, 455
Semifinished:
　hexagon bolts, 574
　hexagon nuts, 574
Semilogarithmic charts, 430
Set screw:
　points, 578
　square heads, 577
Set screws, 236
Shades and shadows in perspective, 412
Shadows:
　curved line in perspective, 415
　horizontal line in perspective, 413
　inclined line in perspective, 414
　vertical line in perspective, 412
Shafts in section, 207
Shaper, 269
Sharp V threads, 218
Shaving, 257
Sheet metal joints, 495
Sheet sizes, 38
Shop methods, 8
Shop terms and processes, 246
　self-study questions, 275
Shrinkage allowance, 251
Shrink rule, 252
　patternmaker's, 252
Side view sketches, 96
Simplified drafting, 347
Size:
　actual, 310
　basic, 310
　design, 310
　nominal, 309
Size dimensioning, 290
Size dimensions, cylindrical mating parts, 314
Size tolerancing, 313
Sketches, on charts, 428
Sketching, 86
　border lines, 91
　circles, 91
　　in perspective, 418
　　in pictorial, 92
　cylindrical objects, 98
　details, 96
　ellipses, 92
　free arm movement, 92
　front view, 96
　horizontal lines, 90
　inclined lines, 90
　isometric, 96
　materials for, 87
　multiview, 94
　nonisometric lines, 97
　oblique, 98
　paper, 89
　parallel lines, 90
　perspective, 100, 417
　　steps in, 418
　pictorial, 96
　proportioning, 93
　self-study questions, 100
　side view, 96
　straight lines, 89
　tools for measurement, 89
　top view, 94
　types of, 94
　uses of, 86
　vertical lines, 90
Skew lines, 174
　grade line with given bearing between, 175
　line parallel to a given line and intersecting, 176
　shortest distance between, 174
　shortest horizontal line between, 174
　shortest line of given slope between, 175

Slope, methods of expressing, 139
Slope of a line, 138
Slotted head machine screws, 570
Socket head cap screws, 570
Solids, 114
    bounded by double curved surfaces, 114
        by plane and single curved surfaces, 115
            plane surfaces, 115
    box method of isometric drawing, 358
    construction in fixed position, 169
    interpretation of, 118
    involving warped surfaces, 116
Solution engineering, proposed, method, 10
Space relationships, 7
Specification, of threads, 221
Specifications, bolt, 232
    finish, 343
    material, 273
Spheres in isometric drawing, 363
Spiral of Archimedes, 82
Splines, involute, 240
Spokes in section, 207
Spotface, 263
    dimensioning a, 298
Springs, 214, 240
    compression, 240
    flat, 242
    representation of, 241
    specification of, 240
    tension, 241
Square:
    construction with length of diagonal, 59
    construction with length of side, 59
Square head bolts, 572
Squareness symbol, 334
Square nuts, regular, 572
Square thread, construction of, 224, 225
Stack-up, accumulation of tolerances, 319
Stamping, 256
    design limitations, 258
Standardization of drawing, 8
Standards, fasteners, 215
Steel:
    free cutting carbon characteristics of, 274
    pipe data, 599
    plain carbon, characteristics of, 274
Stencil, isometric, 375
Straightness symbol, 332
Straight shank twist drills, 559
Strike of a plane, 161
String polygon, 446
Stud bolts, 231
Style, Gothic, 14
Styles, in lettering, uses of, 14
Subassemblies, 308
Surface finish, 343
Surface finished first, location of, 321
Surfaces:
    classification of, 460
    development of, 484

double curved in oblique, 388
finished, 301
geometrical, 458
warped, 184
Swaging, 257
Symbol:
    angularity, 334
    concentricity, 337
    cylindricity, 338
    datum, 332
    flatness, 333
    line profile, 339
    parallelism, 333
    perpendicularity, 334
    profile of a surface, 338
    roundness, 336
    runout, 339
    squareness, 334
    straightness, 332
    symmetry, 338
Symbols:
    cutting plane, 195
    form tolerancing, 606
    general rules for use of, 331
    map, 505
    material, in section, 209
    placing on drawings, 331
    thread, simplified, 221
    true position, 328
Symmetrical parts, dimensioning of, 295
Symmetry, 338
    odd numbered axes of, 208

Tangent line:
    to a circle through a point, 78
    to an ellipse through a point, 79
    to a hyperbola through a point, 80
    to a parabola through a point, 79
    to two circles, 78
Tap, 264
Taper pins, 240
Taper pipe thread data, 599
Tapers, 334
    dimensioning of, 293
Technique, drafting, 47
    of dimensioning, 278
    of drawing invisible lines, 124
    microfilm, 47
    pencil, 47
Tempering, 273
Theta, 108
Thin sections, 206
Third quadrant, 103
Thread:
    acme, 218
    allowance, 273
    American national, 217
    American pipe, 227
    angle of, 215
    base, 216
    basic diameter, 223
    buttress, 218
    cap screw, 234
    classes of fit, 223
    clearance, 216

    crest, 216
    crest clearance, 223
    depth of, 216
    double, 216
    external, 215
    fit, 216
    helix angle of, 216
    internal, 215
    lead, 215
    left-hand, 216, 222
    length on bolts and screws, 233
    major diameter, 215
    minor diameter, 215
    multiple, 216, 222
    pipe, representation of, 227
    pitch, 215
    pitch diameter, 215
    power, 218
    profile, 217
    representation of unified, small size, 221
    right-hand, 216, 222
    root, 216
    screw, 215
    sellers, 222
    series, 222
    sharp V, 217
    simplified symbols, 221
    single, 216
    specifications:
        inch, 222
        metric, 224
    square, 218
        construction of, 225
        representation of, 224
    styles, 217
    tolerance, 223
    translating, 218
    unified national:
        coarse, 222
        extra fine, 22
        fine, 222
        miniature, 222
    uniform pitch series, 222
    United States standard, 222
    Whitworth, 217
    worm, 218
Threaded components, 228
Threaded holes, 228
    blind, 228
Thread fit:
    free, 223
    loose, 223
    medium, 223
    1A, 1B; 2A, 2B; 3A, 3B, 223
Thread relief, 228
    dimensioning of, 299
Threads:
    drawing of, 218
    in section, 209
Three-dimensional charts, 435
Three-dimensional curves, in oblique, 388
Three point perspective, 409
    alternate method of, 411
Three-view drawing, layout, 132

Three-view drawings, 122
Titles, 24
  on charts, 428
Tolerance, 310
  additional, not at MMC, 323
  form:
    by notes, 331
    by symbols, 331
    position, 224
    positional, computation of, 324
Tolerances, accumulation of, 318
  cumulative, 285
    control of, 320
    forging, 255
  form:
    control by size tolerance, 330
    controlled by location tolerance, 330
  form-contour, 326
  selecting, 311
  size:
    basic hold method, 314
    basic shaft method, 315
  size dimensions, mating parts, 314
  by stack-up, 319
Tolerance zones, compared, 323
Tolerancing, geometric form, 322, 330
  systems for size, 313
Tooling, points, 259
Topographic maps, 500
Top view, 105
Top view sketches, 96
Traces, vanishing point, 405
Tracing cloth, 39
Transition locational fits, 312, 587
Translating threads, 218
Traverse, map, 509
  plotting:
    protractor and scale method, 509
    rectangular coordinate method, 510
    tangent method, 510
Triangle:
  construction of, 58
  construction with three sides, 59
  construction with three sides and included angle, 58
  construction with two angles and included side, 59
Triangles, 37
Trilinear charts, 431
Trimetric, 354
Trimetric projection, 5
  construction of, 370
  special positions, 373
Trimming, 257

True length of a line, 138
True North, 508
True position dimensioning, 322
  application of, 325
  when to use, 322
True position symbols, 328
True position tolerances:
  specified by diameter, 324
  specified by radius, 325
True shape of a plane, 138
True size of oblique plane, 144
True size of plane, with edgewise view given, 143
T-square, 37
Twist drills, 559
Two-view drawings, 120
Types of intersections, 471

Unidirectional system of dimensioning, 288
Unified and American threads, 563
Unified national thread sizes, 222
Unilateral system for size tolerancing, 313

Vanishing point:
  for horizontal lines, 399
  method, construction by, 399
  method of perspective, 397
  for oblique lines, 399
  perspective, 395
  principles for finding, 398
  traces, 405
Vector, 440
  direction, 440
  line of action, 440
  point of application, 440
  quantity, 440
  systems, resultants, 447
Vector analysis, graphic, 440
  self-study questions, 454, 455
Vector diagram, 10, 156, 440
  for centroid of an area, 448
  moment of a force about a point, 449
  moment of inertia, 451
  moments on a beam, 450
  parallel forces, 447
  radius of gyration, 453
Vectors:
  composition of, 440, 441
  concurrent noncoplanar, composition of, 442
  coplanar, 440
  equilibrant, 440

  nonconcurrent coplanar, resultant of, 446
  resolution of, 440
  resultant of, 440
  in a three-dimensional system, resolution of, 444, 446
Vertical plane, 103
Views:
  adjacent, 107
  alternate arrangement of, 122
  arrangement of, 105, 106
  auxiliary layout, 148, 149
  bottom, 106
  front, 105
  left side, 105
  number to be drawn, 120
  partial, 128, 147
  principal, 105
  use of auxiliary for constructing, 146
  in projection, 106
  rear, 105
  right side, 105
  sectional, 194
  selection of, 118
  top, 105
Visibility, 166
  determination of, 122
  rules for, 469
Visible lines, 103
  behind section plane, 198
Visual ray method, construction by, 397
  of perspective, 397
  using top and front views, 398
  using top and side views, 398
Visual rays, perspective, 394

Warped surfaces, 184
  general procedure in drawing, 188
Washers:
  lock, 230
  plain, 598
Waviness, 344
Webs, thin:
  forging, 255
  in section, 207
Wire gages, 601
Witness lines, 279
Woodruff keys, 239
Wood screws, 236
Wrico lettering device, 25

Zone, tolerance, projected, 342
Zoned drawing, 204
Zones, tolerance compared, 323